(*continued on back*)

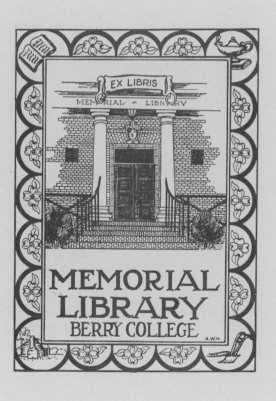

ORGANIC SYNTHESES

ORGANIC SYNTHESES

Collective Volume 5

A REVISED EDITION OF
ANNUAL VOLUMES 40–49

JOHN WILEY AND SONS
NEW YORK · LONDON · SYDNEY · TORONTO · SINGAPORE

THIS BOOK IS DEDICATED TO THE MEMORY OF
PROFESSOR ROGER ADAMS (1889–1971),
ONE OF THE FOUNDERS OF ORGANIC SYNTHESES, INC.,
AND ITS LEADER DURING THE PERIOD, 1920–1971.

PREFACE

Collective Volume 5 of *Organic Syntheses* consists of the procedures previously published in annual Volumes 40–49, revised and updated where necessary. Errors found in the original printings have been corrected, calculations and references have been checked, and modifications and improvements brought to the attention of the Editorial Board have been incorporated in the procedures. This general plan is similar to that followed for the preceding four collective volumes; however, because of the change in emphasis on the types of procedures chosen for recent annual volumes, the specific plan followed in the revision of procedures in this volume was somewhat different than that for previous volumes.

Beginning in Volume 41 the Editorial Board made a deliberate change of emphasis away from procedures describing the preparation of specific organic compounds and toward model procedures that illustrate important types of reactions. This change was the result of the reduced need for directions for the preparation of many specific organic chemicals because of the increasingly effective efforts of the manufacturers of fine organic chemicals in the U.S.A. and abroad to provide chemists with the chemicals needed for their research at reasonable prices. Thus, this volume contains both procedures directed toward the preparation of a specific chemical or reagent and those intended to illustrate a reaction or method of synthesis. Revision of the procedures based largely on information gathered by the Editor-in-Chief from *Chemical Abstracts* (as was done in previous collective volumes) appeared inadequate for the expanded scope of the procedures in this collective volume. Therefore, the Editor-in-Chief attempted to obtain from the original submitters any new information on their procedures of value to the users of *Organic Syntheses*. Although not all submitters could be contacted, replies were received from about 80% of the submitters. Suggestions received from the submitters ranged from no change to complete replacement of the original procedure. The Editor-in-Chief is indebted to the submitters for their generous assistance in bringing these procedures up to date. Additional changes in the Methods of Preparation, Merits of the Preparation, and Discussion Sections have been made by the Editor-in-Chief based largely on information taken from *Chemical Abstracts*, *Organic Reactions*, and the Fieser's invaluable work, *Reagents for Organic Synthesis* (Volumes 1–3). Apart from numerous minor revisions and a number of suggestions for possibly superior alternative methods, new, checked procedures are given to replace the preparations published in the annual volumes for 2,5-dihydro-2,5-dimethoxyfuran, methanesulfinyl chloride, and *p*-toluenesulfonylhydrazide. Checked procedures for the preparation of 2,4,6-trimethylpyrilium tetrafluoroborate and 2,4,6-trimethylpyrilium trifluoromethanesulfonate, which may be used as substitutes for the more hazardous 2,4,6-trimethylpyrilium perchlorate, are given here for the first time. The procedure for the preparation of α-chloroanisole which appeared in Volume 47 is not included in this volume because the procedure has been found to give largely *p*-chloroanisole rather than the title compound.

The practice has been continued of noting which of the compounds, whose preparations have been described, have become commercially available. This information was obtained from *Chemical Sources* (1972 edition) and the catalogs issued through November 1972 by the principal suppliers of fine organic chemicals in the U.S.A. Available compounds are indicated by asterisks after the titles of the preparations.

A considerable number of warning notices have been added to the procedures in this volume on the basis of information supplied by users of the annual volumes. During the period covered by this volume *Organic Syntheses* began the practice of publishing in the annual volumes warning and correction notices pertaining to procedures published in any previous volume, annual or collective. Previously the notices had been distributed as separates for incorporation in the volume referred to. This practice has been continued along with the publication in the current annual volumes. Those notices published within the time span of this volume, 1959–1972, and referring to any procedure published through annual Volume 49 are republished in this volume. The continuing appearance of hazard notices suggests that many of the procedures and chemicals described in this series may be hazardous and should be treated with due caution, as dictated by one's experience and knowledge and by good organic chemical practice.

All of the services provided to users in previous collective volumes are continued in this volume. Thus, the first reference number in each procedure is used to indicate the laboratory in which the directions were developed. Indices included in this volume are Type of Reaction Index, Type of Compound Index, Formula Index, Preparation or Purification of Solvents and Reagents Index, Apparatus Index, Author Index, and General Index. A collective index covering collective Volumes 1–5 is in the process of preparation and is expected to be published in the near future.

Readers interested in the history and evolution of *Organic Syntheses* may find a brief description written by one of the founders, Roger Adams, in annual Volume 50 (p. vii). Further information may be obtained from the prefaces in the 52 annual volumes published to date.

The editors of *Organic Syntheses* are indebted to the contributors and users of these volumes and appreciate the assistance that they have provided during the half century that the series has been published. Additional suggestions, comments, and corrections will be welcomed and should be sent to the Secretary to the Editorial Board.

To reduce the probability of introducing new errors and to keep the cost to the user down, this volume was planned to use as much of the previously published copy as was feasible. For this reason the procedures appear in several formats and the journal abbreviations (which have been made internally consistent) in the references are those in vogue toward the end of the period 1959–1969. Potential contributors to this series should refer to the most recent annual volume and to the *Organic Syntheses* style sheet (available from the Secretary) for the current style and format. Submissions of directions for the annual volumes should also be sent to the Secretary. Details about submissions may also be obtained from the preface to the most recent annual volume.

The Editor-in-Chief takes this opportunity to acknowledge the invaluable service provided to the publication of this series during the past thirteen years by Mrs. June A. Splichal, who handled all of the typing and reproduction of submitted procedures as well as all correspondence with submitters and users.

February, 1973   HENRY E. BAUMGARTEN

CONTENTS

CONTENTS

CONTENTS

2-ACETAMIDO-3,4,6-TRI-*O*-ACETYL-2-DEOXY-α-D-GLUCOPYRANOSYL CHLORIDE

(Glucopyranosyl chloride, 2-acetamido-2-deoxy-, triacetate, α-D-)

Submitted by Derek Horton [1]
Checked by A. L. Johnson and B. C. McKusick

1. Procedure

In a 500-ml. round-bottomed flask equipped with a magnetic stirrer bar and a reflux condenser protected by a tube of calcium chloride is placed 100 ml. of acetyl chloride; this operation and the subsequent reaction are conducted in a hood. The condenser is temporarily removed, and 50 g. (0.226 mole) of dried 2-acetamido-2-deoxy-D-glucose (*N*-acetylglucosamine) (Note 1) is added in the course of 2 or 3 minutes with good stirring. The mixture is stirred for 16 hours without external heating at a room temperature of approximately 25°. The mixture boils spontaneously during the first hour of reaction. It is a clear, viscous, amber liquid at the end of the reaction (Note 2).

Through the condenser there is added 400 ml. of chloroform (U.S.P. grade), and the solution is poured with vigorous stirring onto 400 g. of ice and 100 ml. of water in a 3-l. beaker. The mixture is transferred to a 1-l. separatory funnel and shaken. The organic solution is drawn off without delay into a 3-l. beaker containing ice and 400 ml. of saturated sodium bicarbonate solution. The mixture in the beaker is stirred, and the neutralization is completed by shaking the mixture in the separatory funnel. The organic layer is run directly into a flask containing

1

about 25 g. of anhydrous magnesium sulfate. The entire washing procedure should be completed within 15 minutes (Note 3). The solution is shaken or stirred with the drying agent for 10 minutes (Note 4). The drying agent is separated on a 7.5-cm. Büchner funnel and is well washed with *dry, alcohol-free* chloroform or methylene chloride (Note 5). The filtrate passes through an adaptor directly into a 1-l. round-bottomed flask. The filtrate is concentrated to 75 ml. on a rotary evaporator at 50°, and dry ether (500 ml.) is rapidly added with swirling to the warm solution (Note 6). Crystallization usually begins after about 30 seconds. The flask is stoppered and set aside for 12 hours at room temperature.

The product is scraped from the walls of the flask and broken up by means of a curved spatula. The solid is collected on a 12.5-cm. Büchner funnel, washed with two 150-ml. portions of dry ether, dried by suction on the filter for 5 minutes, and stored in a desiccator over sodium hydroxide and phosphorus pentoxide. Analytically pure 2-acetamido-3,4,6-tri-O-acetyl-2-deoxy-α-D-glucopyranosyl chloride is obtained; weight 55–65 g. (67–79%); m.p. 127–128° (Fisher-Johns apparatus) (Note 7); typical —NHCOCH$_3$ absorptions at 6.09 μ and 6.49 μ in the infrared; nmr (CDCl$_3$) δ 6.25 and J$_{1,2}$ 3.5 Hz for the H-1 doublet. Evaporation of the mother liquors and addition of ether to the concentrated solution gives an additional 4–6 g. (5–7%) of crystalline product, m.p. 125–127°, that is sufficiently pure for most purposes. The pure product may be stored in an open dish in a desiccator at room temperature for at least 3 years without decomposition (Note 8).

2. Notes

1. Suitable material is available from Pfanstiehl Laboratories, Waukegan, Illinois. It may also be prepared from the hydrochloride of 2-amino-2-deoxy-D-glucose (D-glucosamine) in 95% yield by a facile procedure.[2] The 2-acetamido-2-deoxy-D-glucose should be dried at 25° (1 mm.) for at least 12 hours before use. If this material is in the form of a powder rather than compact crystals, more acetyl chloride may

have to be added in order to get a stirrable mixture; the checkers found that an extra 50 ml. of acetyl chloride did not lower the yield.

2. The reaction mixture may be left for longer periods, as over a weekend, without adverse effect. If the ambient temperature is too low, undissolved material may be present after 16 hours, in which case a longer period of stirring is indicated, or the reaction mixture may be gently heated (not above 30°).

3. It is essential that isolation of the product be conducted rapidly and at 0° throughout, especially while the solution is acidic. All apparatus and solutions should be at hand before the reaction mixture is poured on ice. The product reacts fairly rapidly with water in the presence of an acid catalyst, undergoing acetyl migration to give the water-soluble 1,3,4,6-tetra-O-acetyl-2-amino-2-deoxy-α-D-glucopyranose hydrochloride.

4. An extended period of drying is unnecessary and should be avoided.

5. Commercial methylene chloride is usually sufficiently dry to use without pretreatment in place of dry, alcohol-free chloroform. The checkers used a pressure funnel under dry nitrogen for the filtration; filtration was rapid and exposure to atmospheric moisture was slight.

6. The solution must not be evaporated to a volume that permits crystallization to begin before the ether is added. The addition of ether should be sufficiently rapid that the heavy syrup is diluted to a clear, homogeneous solution before crystallization begins.

7. The checkers observed m.p. 118–119° when an open capillary tube containing a sample of analytical purity was placed in a stirred oil bath at 100° with the temperature rising several degrees a minute. The melting point of benzoic acid, determined simultaneously, was 122–123°. The checkers found $[\alpha]^{24}_{D}|+110°$ (c. 1, CHCl$_3$); literature values range from +109.7° to +118°.

8. Material of lesser purity may decompose within a much shorter time. If the product is exposed to moist air, it is converted into 1,3,4,6-tetra-O-acetyl-2-amino-2-deoxy-α-D-glucopyranose hydrochloride, which is insoluble in chloroform.

3. Methods of Preparation

The direct one-step preparation of 2-acetamido-3,4,6-tri-*O*-acetyl-2-deoxy-α-D-glucopyranosyl chloride was reported by Micheel and co-workers,[3] and the described procedure is essentially the method of Horton and Wolfrom.[4] The product was first prepared through a two-step route from 2-amino-2-deoxy-D-glucose hydrochloride by Baker and co-workers,[5] and a number of adaptations of this method have been described.[6-8]

4. Merits of the Preparation

The procedure permits acetylation of the sugar and replacement of the 1-acetoxy group by chlorine in one operation in only 2–3 hours of working time, gives good yields of pure product, and does not require gaseous hydrogen chloride. The two-step procedure from 2-amino-2-deoxy-D-glucose hydrochloride [5-7] is time-consuming, and yields are very low if the acetylated intermediate is isolated.[8] The yield is better when the second stage is performed without isolation of the intermediate,[9] but gaseous hydrogen chloride is required, and the preparation takes considerably more working time than the method described.

The product is used in the preparation of glycoside, thioglycoside, oligosaccharide, and glycosylamine derivatives of 2-acetamido-2-deoxy-D-glucose.[10] A number of these compounds are of current interest; several seem to be involved in viral penetration of cells, and others are of interest in the synthesis of model substrates for enzymes. The product has the α-D configuration and normally reacts to give glycosides with the β-D configuration, presumably through participation of the acetamido group in a bicyclic, closed-ion intermediate. Under controlled conditions it reacts with water to give 1,3,4,6-tetra-*O*-acetyl-2-amino-2-deoxy-α-D-glucopyranose hydrochloride.[11]

1. Department of Chemistry, The Ohio State University, Columbus, Ohio.
2. Y. Inouye, K. Onodera, S. Kitaoka, and S. Hirano, *J. Am. Chem. Soc.*, 78, 4722 (1956); D. Horton, *Biochem. Prep.*, 11, 1 (1966).
3. F. Micheel, F.-P. van de Kamp, and H. Petersen, *Ber.*, 90, 521 (1957).
4. D. Horton and M. L. Wolfrom, *J. Org. Chem.*, 27, 1794 (1962).

5. B. R. Baker, J. P. Joseph, R. E. Schaub, and J. H. Williams, *J. Org. Chem.*, **19**, 1786 (1954).
6. D. H. Leaback and P. G. Walker, *J. Chem. Soc.*, 4754 (1957).
7. Y. Inouye, K. Onodera, S. Kitaoka, and H. Ochiai, *J. Am. Chem. Soc.*, **79**, 4218 (1957).
8. J. Conchie and G. A. Levvy, *Methods Carbohyd. Chem.*, **2**, 332 (1963).
9. M. Akagi, S. Tejima, and M. Haga, *Chem. Pharm. Bull. (Tokyo)*, **9**, 360 (1961); D. H. Leaback, *Biochem. Prep.*, **10**, 118 (1963).
10. D. Horton, R. W. Jeanloz, in "The Amino Sugars," Vol. IA, Academic Press, New York, 1969, Chapter 1.
11. D. Horton, W. E. Mast, and K. D. Philips, *J. Org. Chem.*, **32**, 1471 (1967).

ACETONE DIBUTYL ACETAL*

(Propane, 2,2-dibutoxy-)

$$(CH_3)_2C(OCH_3)_2 + 2C_4H_9OH \xrightarrow{H^+}$$

$$(CH_3)_2C(OC_4H_9)_2 + 2CH_3OH$$

Submitted by N. B. LORETTE and W. L. HOWARD.[1]
Checked by MAX TISHLER and STANLEY NUSIM.

1. Procedure

A mixture of 312 g. (3 moles) of acetone dimethyl acetal (Note 1), 489 g. (6.6 moles) of butanol, 1.0 l. of benzene, and 0.2 g. of *p*-toluenesulfonic acid is placed in a 3-l. flask. The flask is connected to a packed fractionating column and the solution distilled until the azeotrope of benzene and methanol, boiling at 58°, is completely removed (Note 2). The contents of the boiler are then cooled below the boiling point and a solution of 0.5 g. of sodium methoxide in 20 ml. of methanol (Note 3) is added all at once with stirring. The flask is replaced for further distillation, and most of the remaining benzene is distilled at atmospheric pressure. The pressure is then reduced, and the remaining benzene and unreacted butanol are removed (Note 4). Finally, the pressure is reduced to 20 mm., the last traces of low-boiling materials are taken to the cold trap, and the product is distilled. After a small fore-run, acetone dibutyl acetal is collected at 88–90°/20 mm. The yield is 421–453 g. (74.6–80.3%), n_D^{25} 1.4105, d_4^{25} 0.8315.

2. Notes

1. Commercial acetone dimethyl acetal (2,2-dimethoxypro-pane) from the Dow Chemical Company was used without further treatment.

2. About 570 ml. of this azeotrope is obtained. The methanol produced may be estimated by washing an aliquot with about two volumes of water in a graduated cylinder. The methanol content is approximately the difference between the initial volume and that of the residual benzene phase, and about 230 ml. is obtained, depending on the efficiency of fractionation. Other hydrocarbons, e.g., hexane or cyclohexane, can be used for the removal of methanol.

The submitters' distillation was carried out in a 19 x 1200-mm. vacuum-jacketed silvered column fitted with a magnetically oper-ated vapor-takeoff head controlled by a timed relay. The checkers found that a 19 x 340-mm. vacuum-jacketed column fitted with a magnetically operated liquid takeoff and packed with $\frac{1}{4}$-in. glass Raschig rings was sufficient for carrying out the distil-lation. The checkers, using a reflux ratio of 2.7 to 1 throughout the distillation, found the total time required to be 19 hours.

Since the required separations are not difficult, any reasonably efficient fractionating column may be used.

3. Other soluble non-volatile bases may be used.

4. It is best to keep the temperature of the distilland below 125–150°, because pyrolysis of the product becomes progressively more serious at higher temperatures. The pressure is reduced to a convenient value when the distilland temperature reaches 125°. For example, a pressure of 200 mm. will allow the condensation of the benzene without resort to special cooling.

3. Methods of Preparation

Acetone dibutyl acetal has been prepared from isopropenyl acetate and butanol,[2] from butanol and isopropenyl butyl ether obtained from the reaction of butanol with propyne,[3] and by orthoformic ester synthesis.[4,5]

4. Merits of Preparation

The preparation described here is a modification of previously used alkoxyl interchange reactions, but it is more convenient because the use of the azeotrope-forming solvent permits the virtually complete removal of the by-product alcohol under mild conditions. The method is general for most primary and secondary alcohols, including those with functional groups which are stable under the mild conditions used.

1. The Dow Chemical Company, Texas Division, Freeport, Tex.
2. W. J. Croxall, F. J. Glavis, and H. T. Neher, *J. Am. Chem. Soc.*, **70**, 2805 (1948).
3. M. F. Shostakovskii and E. P. Gracheva, *Zhur. Obshch. Khim.*, **23**, 1320 (1953) [C.A., 48, 9899 (1954)].
4. O. Grummitt and J. A. Stearns, Jr., *J. Am. Chem. Soc.*, **77**, 3136 (1955).
5. C. A. MacKenzie and J. H. Stocker, *J. Org. Chem.*, **20**, 1695 (1955).

3β-ACETOXYETIENIC ACID

(3β-Acetoxy-5-androstene-17β-carboxylic acid)

$$2NaOH + Br_2 \longrightarrow NaOBr + NaBr + H_2O$$

Submitted by J. STAUNTON and E. J. EISENBRAUN.[1]
Checked by W. G. DAUBEN and J. H. E. FENYES.

1. Procedure

A solution of 42 g. (1.05 moles) of sodium hydroxide in 360 ml. of water is placed in a 1-l. three-necked, round-bottomed flask fitted with a mechanical stirrer and a thermometer and is cooled to −5° in an ice-salt bath. The stirrer is started, and 43 g. (0.263 mole) of bromine is added from a separatory funnel at such a rate that the temperature remains below 0° (addition time about 5 minutes). The ice-cold solution is diluted with 240 ml. of dioxane (Note 1) that has previously been cooled to 13–14° (Note 2). This solution is kept at 0° until required.

A solution of 28.8 g. (0.08 mole) of 3β-acetoxy-5-pregnen-20-one (pregnenolone acetate) (Note 3) in 1.1 l. of dioxane (Note 1) is diluted with 320 ml. of water and placed in a 5-l. three-necked, round-bottomed flask fitted with a mechanical stirrer and a thermometer (Note 4). The stirrer is started and the mixture is cooled in ice. When the internal temperature has fallen to 8°, the cold hypobromite solution is added in a steady stream. The temperature of the reaction mixture is maintained below 10° throughout the reaction. A white precipitate begins to form after 10 minutes, and the solution becomes colorless during 1 hour. The mixture is stirred for an additional 2 hours, and then the excess sodium hypobromite is destroyed by the addition of a solution of 10 g. of anhydrous sodium sulfite in 100 ml. of water (Note 5).

The stirrer and thermometer are removed and the flask is fitted with a condenser for reflux. The mixture is heated under reflux for 15 minutes, and the solution, while still hot (90°), is acidified by the cautious addition of 50 ml. of concentrated hydrochloric acid (Note 6). The clear yellow solution is kept at 5° for 24 hours. The crystalline precipitate is collected by suction filtration, washed with water, and dried at 100° at atmospheric pressure. The yield of 3β-hydroxyetienic acid, m.p. 274–276°, is 18–20 g. An additional 3–5 g. of product can be obtained by subjecting the filtrate to steam distillation until a white precipitate is formed. The etienic acid collected from the cooled solution melts at 268–272°. The total yield is 23–24 g. (91–95%).

The 3β-hydroxyetienic acid is placed in a 500-ml. round-bottomed flask fitted with a condenser protected with a drying tube and is dissolved with warming in 150 ml. of dry pyridine. After the solution has cooled to room temperature, 20 ml. of acetic anhydride is added; a white crystalline precipitate starts to form immediately. After the mixture has stood for 18–24 hours, it is treated with 20 ml. of water and boiled until the precipitate has dissolved (Note 7). The clear solution is diluted with 70 ml. of water and allowed to cool. The crystalline product is collected by suction filtration, washed with water, and dried in a vacuum oven at 105°/20 mm. The yield of 3β-acetoxyetienic acid, m.p. 235–238°, is 23–24 g. Recrystallization from glacial acetic acid gives a purer product, m.p. 238–240°. The yield is 16–18 g. (55–63% based on the amount of pregnenolone acetate used).

2. Notes

1. Dioxane as supplied by Matheson-Coleman Bell Co. was used without purification.

2. The temperature of the hypobromite solution is kept below 10° to avoid the formation of sodium bromate.

3. Pregnenolone acetate (3β-acetoxy-5-pregnen-20-one) supplied by Syntex S. A., Apartado Postal 2679, Mexico, D. F., was used.

4. It is advisable to carry out any operation involving dioxane in a fume hood.

5. Although this amount of sodium sulfite is sufficient to destroy the excess sodium hypobromite, the solution may still give a positive test with starch-iodide paper because of the presence of peroxides in the dioxane used. It is not necessary to destroy these peroxides before proceeding.

6. The solution should be swirled gently during the addition of the hydrochloric acid. Since this operation causes the dioxane to boil, it must be carried out in a fume hood.

7. The anhydride of etienic acid is hydrolyzed in this process to give the soluble acid. Prolonged boiling should be avoided to prevent extensive attack on the less readily hydrolyzed acetate group.

3. Methods of Preparation

3β-Hydroxy-Δ^5-etiocholenic acid has been prepared from pregnenolone acetate by the action of sodium hypoiodite;[2] by oxidation of the furfurylidene derivative;[3] and by oxidation of the benzylidene derivative of the 5,6-dibromide followed by debromination.[4] The side chain of 3β-hydroxy-Δ^5-bisnorcholenic acid has been systematically degraded to give the etienic acid.[5] Two synthetic approaches have involved, respectively, the replacement of the halogen in 17-chloro-3-acetoxy-Δ^5-androstene by an alkali metal followed by treatment with carbon dioxide [6] and the conversion of dehydroandrosterone acetate to its cyanohydrin, which then was successively dehydrated, hydrolyzed, and selectively hydrogenated to furnish 3β-hydroxyetienic acid.[7, 8]

4. Merits of Preparation

3β-Acetoxyetienic acid has been found to be particularly suitable for the resolution of alcohols. Thus it was employed by Woodward and Katz for the resolution of 1α-hydroxydicyclopentadiene; [9] by Djerassi, Warawa, Wolff, and Eisenbraun for the resolution of *trans-3-tert*-butylcyclohexanol; [10] and by Djerassi and Staunton for the resolution of *cis,cis*-1-decalol.[11]

1. Department of Chemistry, Stanford University, Stanford, California.
2. R. E. Marker and R. B. Wagner, *J. Am. Chem. Soc.*, 64, 1842 (1942).
3. W. C. J. Ross, U. S. pat. 2,470,903 [C.A., 43, 7519 (1949)].
4. R. E. Marker, E. L. Wittle, E. M. Jones, and H. M. Crooks, *J. Am. Chem. Soc.*, 64, 1282 (1942).
5. M. Steiger and T. Reichstein, *Helv. Chim. Acta*, 20, 1040 (1937).
6. Organon, French pat. 834, 940 [C.A., 34, 4744 (1939)]; Organon, Dutch Pat. 47,317 [C.A., 34, 2538 (1940)]; Schering, Brit. pat. 516,030 [C.A., 35, 6058 (1941)].
7. A. Butenandt and J. Schmidt-Thome, *Ber.*, 71, 1487 (1938).
8. L. Ruzicka, E. Hardegger, and C. Kauter, *Helv. Chim. Acta*, 27, 1164 (1944).
9. R. B. Woodward and T. J. Katz, *Tetrahedron*, 5, 70 (1959).
10. C. Djerassi, E. J. Warawa, R. E. Wolff, and E. J. Eisenbraun, *J. Org. Chem.*, 25, 917 (1960).
11. C. Djerassi and J. Staunton, *J. Am. Chem. Soc.*, 83, 736 (1961).

3-ACETYLOXINDOLE

(Oxindole, 3-acetyl-)

$$2K + 2NH_3 \rightarrow 2KNH_2 + H_2$$

$$+ 2KNH_2 \rightarrow$$

$$+ 2NH_3 + KCl$$

$$+ HCl \rightarrow$$

$$+ KCl$$

Submitted by J. F. BUNNETT, B. F. HRUTFIORD and S. M. WILLIAMSON.[1]
Checked by B. C. McKUSICK and D. C. BLOMSTROM.

1. Procedure

An apparatus resembling that pictured by Schlatter [2] is assembled in a good hood. Two 5-l. three-necked flasks are mounted side by side about 10 cm. apart and about 10 cm. above the bench top or stand base. These are referred to as the "left" and "right" flasks. Each flask is provided with a dry ice condenser in the outermost neck, and each condenser is protected from the

air by a soda-lime drying tube. Each flask is provided through
the center neck with a motor-driven stirrer. The left stirrer
should have a large sweep blade, and the right stirrer should have
a small propeller-type blade. The bearing on each stirrer should
be capable of holding a small positive pressure (the submitters
used ball-joint bearings). The innermost neck of each flask is
fitted with a two-holed rubber stopper. One hole in each stopper
is for nitrogen supply; a short piece of glass tubing is inserted
through each stopper, and these pieces of glass tubing are connected by rubber tubes to a glass "Y" tube which in turn is
connected by rubber tubing to a tank of dry nitrogen. The
rubber tubes between the "Y" tube and the flasks are provided
with pinch clamps so that the flow of nitrogen can be directed
into either flask or into both at once. The other hole in each
stopper is for transfer of liquid ammonia from the right flask to
the left. A glass tube reaching to the very bottom of the right
flask is inserted through the right stopper. A glass tube is inserted through the left stopper so that it projects only a few
centimeters into the left flask. These glass tubes are bent so
that they point toward each other, and they are connected by
a piece of rubber tubing provided with a pinch clamp.

With nitrogen flowing and all pinch clamps open, the apparatus is flamed to drive away traces of moisture. The condensers
are then provided with dry ice covered by isopropyl alcohol, and
the lower part of the right flask is embedded in crushed dry ice.
Liquid ammonia (4 l.) is introduced into the right flask through
the nitrogen inlet from which the rubber tubing is temporarily
disconnected, and 105.8 g. (0.5 mole) of o-acetoacetochloroanilide
(Note 1) is placed in the left flask. In order to destroy any
water in the ammonia, the right stirrer is started and small pieces
of potassium metal are dropped into the ammonia, by briefly
lifting the right two-holed stopper, until the blue color persists
for 3 minutes. The nitrogen connection to the left flask is
clamped shut and, by partially blocking the escape of nitrogen
from the right drying tube, about 1 l. of ammonia is forced into
the left flask. The connection between the two flasks is now
clamped shut. Brief operation of the left stirrer facilitates solution of the o-acetoacetochloroanilide in the ammonia.

Potassium metal (78 g.; 2 moles) is cut into chunks just small enough to pass through the neck of the right flask; these are stored in a beaker under xylene until needed. About 5 g. of potassium is introduced into the right flask by briefly lifting the two-holed stopper. The right stirrer is started and the potassium is allowed to dissolve. To the resulting deep blue solution is added 0.1 g. of finely crushed ferric nitrate hydrate, a catalyst for the reaction of potassium with ammonia. The solution should begin to boil with evolution of hydrogen. (*Caution: No flames or sparks should be nearby.*) The rest of the potassium is added at such a rate as to maintain active gas evolution (Note 2). Stirring is continued in the right flask until all the potassium is consumed, i.e., until the blue color disappears. The right flask now contains a solution of potassium amide in liquid ammonia; 30–60 minutes is required for its preparation.

The tube between the two flasks is opened by releasing the pinch clamp, and the left stirrer is started. With nitrogen flow to the left flask still blocked, the potassium amide solution is caused to flow into the left flask by partially blocking the right nitrogen exit. The solution in the left flask slowly assumes a chartreuse color. As soon as the right flask is as nearly empty as the apparatus will permit, nitrogen flow is opened to the left flask and closed to the right flask and the connection between the two flasks is clamped shut. The right flask is then disconnected and immediately cleaned by rinsing it carefully with ethyl or isopropyl alcohol to destroy potassium amide and then washing it with water. (*Caution: Potassium amide is inflammable and will ignite on contact with moisture.*)

The solution in the left flask is stirred for 30 minutes after all the potassium amide has been added. The nitrogen inlet is briefly removed and 120 g. (1.5 moles) of ammonium nitrate is added; this discharges the chartreuse color. (*Caution: Vigorous foaming occurs.*) Ethyl ether (500 ml.) is added and the dry ice condenser is replaced by a standard water-cooled condenser. The ammonia is evaporated by allowing the stirred reaction mixture to warm to room temperature; this takes several hours and it is convenient to have it occur overnight.

Water (1.5 l.) is added and the mixture is transferred to a

separatory funnel. The lower aqueous layer, which contains the potassium salt of 3-acetyloxindole, is separated and is then extracted with ethyl ether three times to remove a purple impurity. The aqueous layer is then made acidic to litmus by addition of hydrochloric acid; this causes precipitation of crude, tan-colored 3-acetyloxindole. The mixture is chilled, and the product is collected by suction filtration and washed well on the filter with water. The yield of crude 3-acetyloxindole, m.p. 204–206°, is 65–68 g. (74–78%). It may be purified by recrystallizing it from 1.7 l. of chloroform in the presence of 2 g. of decolorizing carbon. A heated filter funnel must be used in separating the carbon because the product starts to crystallize only slightly below the boiling point of chloroform. The recrystallized 3-acetyloxindole weighs 53–59 g. (61–67%) and is in the form of white needles, m.p. 204–205.5°.

2. Notes

1. The o-acetoacetochloroanilide used was the technical product of Union Carbide Chemicals Co.; m.p. 107–109°.

2. If the reaction of potassium with liquid ammonia slows down before all the potassium is consumed, an additional pinch of ferric nitrate hydrate is added.

3. Methods of Preparation

3-Acetyloxindole has been made by condensing ethyl acetate with oxindole in the presence of sodium ethoxide [3] and by heating N-acetyloxindole with sodium amide in xylene.[4] The present method was developed by Hrutfiord and Bunnett.[5] It illustrates a general principle for the synthesis of heterocyclic and homocyclic compounds. This principle involves the creation of an intermediate species that is of the benzyne type and has a nucleophilic center located so that it can add, intramolecularly, to the "triple bond" of the benzyne structure. Other applications of the principle using essentially the present procedure are the conversion of thiobenz-o-bromoanilide or thiobenz-m-bromoanilide to 2-phenylbenzothiazole (90% and 68% respectively), of benz-*o*-chloroanilide to 2-phenylbenzoxazole (69%),[5] of *o*-chlorohydrocinnamonitrile to 1-cyanobenzocyclobutene (61%),[6] and of methanesulfone(N-methyl-*o*-chloro)anilide to 1-methyl-2,1-benzisothiazoline 2,2-dioxide (66%).[7]

1. University of North Caroline, Chapel Hill, North Carolina.
2. M. Schlatter, *Org. Syntheses,* Coll. Vol. 3, 223 (1955).
3. L. Horner, *Ann.,* 548, 131 (1941).
4. H. Behringer and H. Weissauer, *Ber.,* 85, 774 (1952).
5. J. F. Bunnett and B. F. Hrutfiord, *J. Am. Chem. Soc.,* 83, 1691 (1961).
6. J. F. Bunnett and J. A. Skorcz, *J. Org. Chem.,* 27, 3836 (1962).
7. J. F. Bunnett, T. Kato, R. R. Flynn, and J. A. Skorcz, *J. Org. Chem.,* 28, 1 (1963).

ADAMANTANE*

(Tricyclo[3.3.1.1 ³,⁷] decane)

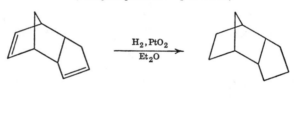

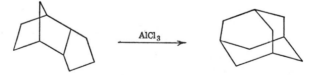

Submitted by PAUL VON R. SCHLEYER, M. M. DONALDSON, R. D. NICHOLAS, and C. CUPAS.[1]
Checked by WILLIAM G. DAUBEN and FRED G. WILLEY.

1. Procedure

A. *endo-Tetrahydrodicyclopentadiene.* A solution of 200 g. (1.51 moles) of purified dicyclopentadiene (Note 1) in 100 ml. of dry ether containing 1.0 g. of platinum oxide is hydrogenated at 50 p.s.i. hydrogen pressure using a Parr apparatus. The reaction mixture becomes quite warm during the initial stage of the hydrogenation,[2] and the uptake of 2 mole equivalents of hydrogen requires 4–6 hours. The catalyst is removed by suction filtration, and the filtrate is distilled at atmospheric pressure through a 30-cm. Vigreux column.

When the removal of the ether is complete, the condenser at the top of the column is replaced by a wide-diameter adapter the bottom of which is placed in a receiver flask immersed in an ice bath. The adapter is heated (Note 2) to prevent premature solidification of the distillate. The distillation is continued and the *endo*-tetrahydrodicyclopentadiene, b.p. 191–193°, is collected. The yield is 196–200 g. (96.5–98.4%). The melting point depends on the purity of the starting material but generally is above 65°.

B. *Adamantane*. In a 500-ml. Erlenmeyer flask having a 24/40 standard taper joint are placed 200 g. (1.47 moles) of molten *endo*-tetrahydrodicyclopentadiene and a magnetic stirring bar. A well-greased inner joint (2.2 x 15 cm., 24/40) is fitted into the top of the flask to serve as an air condenser, and 40 g. of anhydrous aluminum chloride is added through the opening (Note 3). The reaction mixture is simultaneously stirred and heated at 150–180° (Notes 4, 5) by means of a combination magnetic stirrer-hot plate. Aluminum chloride sublimes to the top of the flask, especially at the beginning of the reaction, and the accumulated sublimate is, from time to time, pushed down into the reaction liquid. After the mixture has been heated for 8–12 hours, the flask is removed from the hot plate-stirrer and the black contents upon cooling separate into two layers. The upper layer, a brown mush of adamantane and other products, is decanted carefully from the lower black tarry layer into a 600-ml. beaker. The Erlenmeyer flask is rinsed five times with a total of 250 ml. of petroleum ether (b.p. 30–60°) with decantation into the same beaker (Notes 6, 7). The petroleum ether suspension is warmed until all the adamantane is in solution; there should be an appreciable excess of solvent. The solution is decolorized by careful addition of 10 g. of chromatography-grade alumina, the hot solution filtered, and the alumina and the beaker washed thoroughly with solvent. The nearly colorless filtrate (Note 8) is concentrated to a volume of about 200 ml. by distillation and then cooled in a Dry Ice-acetone bath. The solid adamantane is removed by suction filtration and there results 27–30 g. (13.5–15.0%) of crystals, melting point about 255–260° (Notes 9, 10). One recrystallization from petroleum ether raises the melting point to 268–270° (Notes 11, 12).

2. Notes

1. Technical grade dicyclopentadiene is purified by distillation at water pump pressure through a 30-cm. Vigreux column, and the fraction boiling at 64–65°/14 mm. (72–73°/22 mm.) is used in the reaction. The best material is solid or semisolid at room temperature.

2. The adapter can readily be heated by placing an infrared lamp above it.

3. The evolution of heat initially observed is due to the exothermic rearrangement of *endo*-tetrahydrodicyclopentadiene to its *exo*-isomer.[3]

4. The temperature of the reaction is followed by inserting a thermometer into the reaction flask through the joint.

5. Other methods of heating and stirring may be used but are more troublesome because of the tendency of aluminum chloride to sublime and clog top-mounted stirrers.

6. If appreciable amounts of tar have been transferred to the beaker, repeat the decantation and washing process into a clean beaker.

7. The tarry flasks and beakers can be cleaned easily with acetone. *Do not use water* until all the tar has been removed.

8. If necessary, the filtrate should be warmed to dissolve all the adamantane.

9. The melting point must be taken in a sealed capillary, and the sealed portion must be completely immersed in the liquid of the melting-point bath.

10. Additional adamantane, 2–6 g., can be obtained by distilling the mother liquors through a 10-cm. Vigreux column and chilling the fraction boiling between 180° and 200°. The filtrate from the collection of the second portion of adamantane consists mostly of *exo*-tetrahydrodicyclopentadiene.[3] The amount of this frac-

tion, 30–100 g., depends on the severity of the rearrangement conditions. Much non-distillable residue is obtained. Conversion of the 180–200° fractions to adamantane is brought about by treating them with aluminum chloride as before, and the yields are comparable.

11. The recrystallization is not necessary unless material of the highest purity is desired.

12. 1-Methyladamantane and 1,3-dimethyladamantane can be prepared by analogous isomerizations.[4]

3. Methods of Preparation

Adamantane can be isolated from petroleum, where it is found in minute yield.[5] Two multistep syntheses starting with tetra-ethyl bicyclo[3.3.1]nonane-2,6-dione-1,3,5,7-tetracarboxylate have been reported.[6] Also, starting with endo-tetrahydrodicyclopentadiene, it has been found that a catalyst composed of aluminum chloride and hydrogen chloride will bring about the rearrangement to adamantane in 30–40% yield, but the reaction must be performed in a hydrogen atmosphere at high pressure.[7] A recent patent describes the conversion in up to 30% yield using a boron trifluoride-hydrogen fluoride catalyst under pressure.[8] The present method is based on the published procedure of the submitters and is the preferred method by virtue of the greater convenience.[4]

1. Department of Chemistry, Princeton University, Princeton, New Jersey.
2. K. Alder and G. Stein, Ber., 67, 613 (1934); G. Becker and W. A. Roth, Ber., 67, 627 (1934).
3. P. von R. Schleyer and M. M. Donaldson, J. Am. Chem. Soc., 82, 4645 (1960); P. von R. Schleyer, J. Am. Chem. Soc., 79, 3292 (1957).
4. P. von R. Schleyer, Tetrahedron Lett., 305 (1961).
5. S. Landa and V. Macháček, Collection Czechoslov. Chem. Communs., 5, 1 (1960); S. Landa and S. Hála, Collection Czechoslov. Chem. Communs., 24, 93 (1959).
6. V. Prelog and R. Seiwerth, Ber., 74, 1644, 1769 (1941); H. Stetter, O. E. Bander, and W. Neumann, Ber., 89, 1922 (1956).
7. Belgian pat. 583, 519, Oct. 12, 1959, to E. I. du Pont and Co.; H. Koch and J. Franken, Brennstoff-Chem., 42, 90 (1961).
8. U.S. pat. 2,937,211 (May 17, 1960), R. E. Ludwig to E. I. du Pont de Nemours and Co. [C.A., 54, 19540c (1960)].

1-ADAMANTANECARBOXYLIC ACID*

$$\xrightarrow[-(CH_3)_3CH, -H_2O]{(CH_3)_3COH + HCO_2H + H_2SO_4}$$ —CO_2H

Submitted by H. Koch and W. Haaf [1]
Checked by W. W. Prichard and B. C. McKusick

1. Procedure

Caution! Because carbon monoxide is evolved, the reaction should be carried out in a good hood.

A 1-l. three-necked flask equipped with stirrer, thermometer, dropping funnel, and gas-outlet tube is charged with 470 g. (255 ml., 4.8 moles) of 96% sulfuric acid (Note 1), 100 ml. of carbon tetrachloride (Note 2), and 13.6 g. (0.100 mole) of adamantane.[2] The well-stirred mixture is cooled to 17–19° in an ice bath, and 1 ml. of 98% formic acid is added. Then a solution of 29.6 g. (38 ml., 0.40 mole) of *t*-butyl alcohol in 55 g. (1.2 moles) of 98–100% formic acid is added dropwise; the rate of addition and the cooling are regulated so that the addition requires 1–2 hours, and the temperature of the reaction mixture is kept at 17–25°. The reaction mixture is stirred for an additional 30 minutes and poured onto 700 g. of crushed ice. The layers are separated, and the upper, acid layer is extracted with three 100-ml. portions of carbon tetrachloride.

The combined carbon tetrachloride layers are shaken with 110 ml. of 15N ammonium hydroxide (Note 3), and the crystalline ammonium 1-adamantanecarboxylate that separates is collected on a Büchner funnel having a coarse fritted disk. The salt is washed with 20 ml. of cold acetone and suspended in 250 ml. of water. The suspension is made strongly acidic with 25 ml. of 12N hydrochloric acid and extracted with 100 ml. of chloroform. The chloroform layer is dried over anhydrous sodium sulfate and evaporated to dryness on a steam bath (Note 4). The residue is crude 1-adamantanecarboxylic acid; weight 12–13 g. (67–72%) (Note 5); m.p. 173–174°. Recrystallization of this product from a mixture of 30 ml. of methanol and about 10 ml. of water gives 10–11 g. (56–61%) of pure acid, m.p. 175–176.5° (Note 6).

2. Notes

1. Acid concentrations of 95–98% are satisfactory. The yield falls with concentrations lower than 95%.

2. Cyclohexane or *n*-hexane can be used in place of carbon tetrachloride. Technical "normal hexane" may contain substantial amounts of methylcyclopentane and isohexane that lower the yield through formation of C_7-acids that are hard to remove.

3. A large amount of trimethylacetic acid and a small amount of at least one C_9-acid and one C_{13}-acid are formed from the *t*-butyl alcohol. The treatment with ammonia separates 1-adamantanecarboxylic acid from these acids, the ammonium salts of which remain in solution.

4. Acid that is satisfactory for most purposes may be obtained by interrupting the evaporation of the chloroform solution when crystals start to appear, cooling the concentrated chloroform solution to 0–5°, and collecting the acid on a Büchner funnel. The acid melts at 173–174°.

5. The checkers obtained similar yields when the quantity of reactants was increased fivefold.

6. As an alternative purification procedure, the checkers have esterified the crude acid by refluxing it for 2 hours with three times its weight of methanol and 2 ml. of 98% sulfuric acid. The solution is poured into 10 volumes of water and extracted with the minimum amount of chloroform required to give a clean separation of layers. The chloroform solution is washed with water, dried over calcium chloride, and distilled from a Claisen flask with an indented neck. Methyl 1-adamantanecarboxylate is collected at 77–79° (1 mm.); m.p. 38–39°. Hydrolysis of the ester with the calculated amount of $1N$ potassium hydroxide followed by acidification yields 1-adamantanecarboxylic acid; m.p. 175–176.5°; 90% overall recovery.

3. Methods of Preparation

1-Adamantanecarboxylic acid can be prepared by carboxylation of 1-adamantanol [8] or 1-bromoadamantane [3,4] by formic acid and 96% sulfuric acid; by carboxylation of adamantane by formic acid, t-butyl alcohol, and 96% sulfuric acid; [5] and by carboxylation of adamantane by formic acid and 130% sulfuric acid. [6]

4. Merits of the Preparation

This procedure illustrates a general method of carboxylating saturated hydrocarbons that have a tertiary hydrogen. [7] It has been used to convert isopentane to 2,2-dimethylbutanoic acid, 2,3-dimethylbutane to 2,2,3-trimethylbutanoic acid, and methylcyclohexane to 1-methylcyclohexanecarboxylic acid. The preparation of 1-methylcyclohexanecarboxylic acid by a variation of this procedure is described on p. 739 of this volume.

1. Max-Planck Institute für Kohlenforschung, Mülheim-Ruhr, Germany.
2. P. R. Schleyer, M. M. Donaldson, R. D. Nicholas, and C. Cupas, this volume, p. 16.
3. H. Stetter, M. Schwarz, and A. Hirschhorn, *Ber.,* 92, 1629 (1959).
4. H. Stetter and E. Rauscher, *Ber.,* 93, 1161 (1960).
5. H. Koch and W. Haaf, *Angew. Chem.,* 72, 628 (1960).
6. C. Wulff, Doctoral Thesis, Technische Hochschule, Aachen, Germany, "Uber Substitution-reaktionen des Adamantans," September, 1961, p. 65.
7. W. Haaf and H. Koch, *Ann.,* 638, 122 (1960).
8. H. Stetter, *Angew. Chem.,* 74, 361 (1962).

ALLENE*

$$CH_2{=}C{-}CH_2{-}Cl + Zn \xrightarrow[H_2O]{EtOH} CH_2{=}C{=}CH_2$$
$$\underset{Cl}{|}$$

Submitted by H. N. Cripps [1] and E. F. Kiefer. [2]
Checked by W. E. Russey, R. D. Birkenmeyer, and F. Kagan. [3]

1. Procedure

A 1-l. three-necked flask is equipped with a Hershberg stirrer operating in a ground-glass bearing (Note 1), a 250-ml. pressure-equalizing dropping funnel, and a coil condenser. The exit from the condenser is connected to a train consisting of a trap (of at least 50-ml. capacity below the bottom of the inlet tube) cooled in ice, a drying tube (about 6 in. long by 1 in. I.D.) filled with indicating Drierite and calcium chloride, an efficient trap of at least 150-ml. capacity cooled in Dry Ice-acetone to $-70°$ or below, and a drying tube containing Drierite. A mixture of 95% ethanol (400 ml.), water (80 ml.), and 300 g. (4.6 g. atoms) of zinc dust is placed in the reaction flask. The addition funnel is charged with 260 g. (2.34 moles) of 2,3-dichloropropene (Note 2), the reaction mixture is stirred and heated to reflux, and the 2,3-dichloropropene is added dropwise at such a rate that reflux is maintained without external heating (2–3 hours). After the addition is complete, heating is resumed for 1 hour. The ice-cooled trap is warmed to about 25°, and the residual allene is purged from the reaction flask with a very slow stream of nitrogen.

The trap cooled in Dry Ice-acetone contains about 105 g. of crude product which, when distilled through a column packed with glass helices (Note 3), yields about 75 g. (80%) of allene (Note 4). No external heat is needed during the distillation. The distillation flask is allowed to warm to room temperature, the distillation beginning at a liquid temperature of $-34°$ and virtually stopping at about 10°. The distilled product contains no detectable ethanol, water, 2,3-dichloropropene, or methylacetylene as determined by gas-liquid chromatography (Note 5).

2. Notes

1. The stirrer should be smooth running and gas-tight. The stirring motor (air-driven) should have a high torque because the reaction mixture tends to agglomerate as the reaction proceeds.

2. 2,3-Dichloropropene from Distillation Products or Columbia Chemicals was employed.

3. A vacuum-jacketed column 1 ft. long by 1 in. I.D. packed with glass helices (4 mm. O.D.) is satisfactory for this distillation. It is fitted with a cold finger in the top of the column cooled by means of acetone that has been cooled in a Dry Ice bath. The fraction cutter is jacketed and similarly cooled. A small circulat-

ing pump is used to circulate acetone successively through copper coils in a Dry Ice bath, the fraction cutter, and the cold finger. When the fraction cutter is full, the bottom may be attached to a cooled, evacuated gas cylinder and the allene sucked into the cylinder.

4. The allene contains up to 3% of 2-chloropropene, determined by its vapor-phase infrared spectrum and by vapor-phase chromatography (cf. Note 5).

5. The checkers used an F and M Model 500 gas chromatographic apparatus (F and M Scientific Corporation, P. O. Box 245, Avondale, Penn.) equipped with a polyester column (pentaerythritol adipate, 20% W/W on Chromasorb P, LAC-2-R446) 1 ft. by $\frac{1}{4}$ in. O.D., helium flow 45 cc. per minute, column temperature 50°, block temperature 215°, injector temperature 225°. This system was able to separate allene from methylacetylene and 2-chloropropene.

3. Methods and Merits of Preparation

Although many routes to allene are described in the literature, most preparations give a mixture of allene and methylacetylene. The virtue of the present preparation, which is essentially that described by Gustavson and Demjanoff,[4] is that it gives allene in a reproducible manner with 2-chloropropene as its only impurity.

Allene is an extremely useful reagent for cycloaddition reactions giving cyclobutane derivatives.[5] Allene dimer is also a useful and versatile starting material.[6]

1. Contribution No. 566 from the Central Research Department, Experimental Station, E. I. du Pont de Nemours and Co., Wilmington, Delaware.
2. Gates and Crellin Laboratories of Chemistry, California Institute of Technology, Pasadena, California. Present Address: Department of Chemistry, University of Hawaii, Honolulu, Hawaii.
3. The Upjohn Company, Kalamazoo, Michigan.
4. G. Gustavson and N. Demjanoff, *J. Prakt. Chem.*, [2] 38, 202 (1888).
5. J. D. Roberts and C. M. Sharts, *Org. Reactions*, 12, 23 (1962).
6. J. K. Williams and W. H. Sharkey, *J. Am. Chem. Soc.*, 81, 4269 (1959); S. Lebedev and B. K. Merezhkovskii, *J. Russ. Phys. Chem. Soc.*, 45, 1249 (1913) [C.A., 8, 320 (1914)].

2-ALLYLCYCLOHEXANONE

(Cyclohexanone, 2-allyl-)

$$\text{(Cyclohexane ring)}(OCH_2CH{=}CH_2)_2 \xrightarrow{H^+}$$

$$\text{(Cyclohexanone ring)}{-}CH_2CH{=}CH_2 + CH_2{=}CHCH_2OH$$

Submitted by W. L. HOWARD and N. B. LORETTE.[1]
Checked by MELVIN S. NEWMAN and W. S. GAUGH.

1. Procedure

A solution of 196 g. (1 mole) of cyclohexanone diallyl acetal (Note 1), 150 g. of toluene, and 0.10 g. of p-toluenesulfonic acid is distilled through a good fractionating column (Note 2). In about 3 hours, 110 g. of distillate boiling at 91–92° (Note 3) is obtained and the temperature in the head then rises abruptly. The residue in the distilling flask is cooled and washed with 5 ml. of aqueous potassium carbonate to remove the acid. The remaining solution is passed through a filter containing anhydrous powdered magnesium sulfate and returned to the still. Most of the remaining toluene is removed by distillation at 100 mm. pressure (b.p. 52°). The receiver is changed, the pressure is reduced to 15 mm., and the last of the toluene is collected in a cold trap. The residual oil is rapidly vacuum-distilled to separate the product from a higher-boiling residue. Redistillation yields 117–126 g. (85–91%) of 2-allylcyclohexanone, b.p. 86–88°/15 mm., n_D^{25} 1.4670.

2. Notes

1. The preparation of cyclohexanone diallyl acetal is described on p. 292.

2. A 14-in. helices-packed column is sufficient.

3. This distillate is the azeotrope of toluene and allyl alcohol whose composition is about 50% allyl alcohol by weight.[2]

3. Methods of Preparation

2-Allylcyclohexanone has been prepared from the sodium derivative of cyclohexanone by alkylation with allyl bromide [3] or with allyl iodide,[4] and by ketonic hydrolysis of ethyl 1-allyl-2-ketocyclohexanecarboxylate.[5,6]

4. Merits of Preparation

This procedure, when combined with the preparation of allyl ketals (p. 292), provides a general method for obtaining allyl substitution alpha to a carbonyl group. A discussion of some of these applications, as well as the vinyl allyl ether rearrangement which is involved, has been given by Hurd and Pollack.[7] Also, the procedure can be repeated to allow the introduction of more than one allyl group.

1. The Dow Chemical Company, Texas Division, Freeport, Texas.
2. L. H. Horsley and co-workers, *Advances in Chem. Ser.*, No. 6, Azeotropic Data, American Chemical Society, Washington, 1955, p. 83.
3. C. A. VanderWerf and L. V. Lemmerman, *Org. Syntheses* Coll. Vol. 3, 44 (1955).
4. R. Cornubert, *Ann. Chim.*, [9] 16, 145 (1921).
5. A. C. Cope, K. E. Hoyle, and D. Heyl, *J. Am. Chem. Soc.*, 63, 1848 (1941).
6. R. Grewe, *Ber.*, 76, 1075 (1943).
7. C. D. Hurd and M. A. Pollack, *J. Am. Chem. Soc.*, 60, 1905 (1938).

AMINOACETONE SEMICARBAZONE HYDROCHLORIDE

(Amino-2-propanone, semicarbazone hydrochloride)

$$H_2NCH_2CO_2H + 2(CH_3CO)_2O \xrightarrow[\substack{-CO_2 \\ -2CH_3CO_2H}]{Pyridine} \begin{array}{l} CH_2COCH_3 \\ | \\ NHCOCH_3 \end{array}$$

$$\downarrow \text{HCl, H}_2\text{O}$$

$$\begin{array}{l} CH_3 \\ | \\ CH_2-C=NNHCONH_2 \\ | \\ NH_3{}^+ \, Cl^- \end{array} \xleftarrow{NH_2CONHNH_2} \begin{array}{l} CH_2COCH_3 \\ | \\ NH_3{}^+ \, Cl^- \end{array}$$

Submitted by John D. Hepworth [1]
Checked by W. T. Nolan and V. Boekelheide

1. Procedure

A. *Acetamidoacetone.* A mixture of 75.0 g. (1.0 mole) of glycine (Note 1), 475 g. (485 ml., 6 moles) of pyridine (Note 1), and 1190 g. (1.1 l., 11.67 moles) of acetic anhydride (Notes 1 and 2) is heated under reflux with stirring for 6 hours (Note 3) in a 3-l., three-necked, round-bottomed flask. The reflux condenser is replaced by one set for downward distillation, and the excess pyridine, acetic anhydride, and acetic acid are removed by distillation under reduced pressure. The residue is transferred to a simple distillation apparatus such as a Claisen flask and is distilled to give 80–90 g. (70–78%) of a pale yellow oil, b.p. 120–125° (1 mm.). This product is of satisfactory purity for use in step B.

B. *Aminoacetone hydrochloride.* A mixture of 175 ml. of concentrated hydrochloric acid and 175 ml. of water is added to 52 g. (0.45 mole) of the acetamidoacetone from step A contained in a 1-l. round-bottomed flask. The mixture is boiled under reflux under a nitrogen atmosphere (Note 4) for 6 hours. The resulting solution is concentrated using a flash evaporator held below 60° and with the condensation trap for solvent being cooled by a dry ice-acetone bath. The dark red oily residue (40–45 g.) is satisfactory for use in step C (Note 5).

C. *Aminoacetone semicarbazone hydrochloride.* The product from step B is dissolved in 250 ml. of absolute alcohol in a 1-l. Erlenmeyer flask, and to this solution is added a solution of 48 g. of semicarbazide hydrochloride (Note 1) in 100 ml. of water. The mixture is allowed to stand at room temperature for 2 hours, the crystalline precipitate is collected by suction filtration, and the off-white product is washed on the filter with absolute alcohol. The crystals, after air-drying, amount to 54–58 g. (72–78%) and melt at 208–210°. The product is essentially pure and can be used for most purposes without further purification (Note 6).

2. Notes

1. The glycine, pyridine, acetic anhydride, and semicarbazide hydrochloride employed were of reagent grade and were used directly as supplied.

2. This ratio of pyridine to acetic anhydride has been found to be the most satisfactory.

3. It is necessary that the mixture actually boil under reflux or the yield may drop to 25–30%.

4. The checkers used high-purity nitrogen. If ordinary commercial nitrogen is employed, the oxygen should be removed by passing the gas through Fieser's solution.

5. Aminoacetone hydrochloride is very hygroscopic and is best stored as the semicarbazone. If the compound itself is desired, however, the dark red oil is dried under reduced pressure over phosphorus pentoxide. The resulting crystalline aminoacetone hydrochloride can be purified by dissolving it in absolute ethanol and precipitating it by the addition of dry ether.

6. For further purification, the semicarbazone hydrochloride may be recrystallized from aqueous ethanol to give colorless crystals, m.p. 212°.

3. Methods of Preparation

This preparation is based on the procedure used to synthesize 3-acetamido-2-butanone.[2] Aminoacetone hydrochloride has been prepared from isopropylamine via the N,N-dichloroisopropyl-amine,[3] from hexamethylenetetramine and chloroacetone,[4] by reduction of nitroacetone [5] or isontirosoacetone,[6] and from phthal-imidoacetone by acid hydrolysis,[6] cited as the most convenient method of preparation.[7] The semicarbazone has been prepared previously in the same manner.[8]

4. Merits of the Preparation

Aminoacetone is a versatile starting material for many syntheses, particularly for the preparation of heterocycles. The present procedure describes a convenient method for its preparation in a form suitable for storage. The aminoacetone can be generated from aminoacetone semicarbazone hydrochloride *in situ* as needed.

1. Department of Chemistry, College of Technology, Huddersfield, England; present address, North Lindsey Technical College, Scunthorpe, England.
2. R. H. Wiley and O. H. Borum, *Org. Syntheses*, Coll. Vol. **4**, 5 (1963).
3. H. E. Baumgarten and F. A. Bower, *J. Am. Chem. Soc.*, **76**, 4561 (1954).
4. C. Mannich and F. L. Hahn, *Ber.*, **44**, 1542 (1911).
5. Ad. Lucas, *Ber.*, **32**, 3181 (1899).
6. S. Gabriel and G. Pinkus, *Ber.*, **26**, 2197 (1893).
7. A. W. Johnson, C. E. Dalgliesh, W. E. Harvey, and C. Buchanan, Amino-aldehydes and aminoketones, in E. H. Rodd, ed., "Chemistry of Carbon Compounds," Vol. 1, Elsevier Publishing Company, 1951, Part A, p. 714.
8. W. R. Boon and T. Leigh, *J. Chem. Soc.*, 1497 (1951).

2-AMINOFLUORENE *

(2-Fluorenylamine)

$$2 \text{ [structure with NO}_2\text{]} + 3 N_2H_4 \rightarrow$$

$$2 \text{ [structure with NH}_2\text{]} + 3 N_2 + 4 H_2O$$

Submitted by P. M. G. BAVIN.[1]
Checked by JOHN C. SHEEHAN and
ROGER E. CHANDLER.

1. Procedure

In a 2-l. three-necked round-bottomed flask, equipped with a mechanical stirrer (Note 1), reflux condenser, and dropping funnel, are placed 30 g. of pure 2-nitrofluorene, m.p. 157° [*Org. Syntheses*, Coll. Vol. **2**, 447 (1943)], and 250 ml. of 95% ethanol. After warming to 50° on a steam bath, 0.1 g. of palladized charcoal catalyst (previously moistened with alcohol) is added (Note 2) and the stirrer is started. About 15 ml. of hydrazine hydrate is added from the dropping funnel during 30 minutes (Note 3). At this point an additional 0.1 g. of catalyst (previously moistened with alcohol) is added and the mixture is heated until the alcohol refluxes gently. After 1 hour the nitrofluorene has dissolved completely and the supernatant liquor is almost colorless.

The catalyst is removed by filtration with gentle suction through a thin layer of Celite (Note 4). The flask is rinsed with 30 ml. of hot alcohol which is then used to wash the catalyst and Celite. The combined filtrates are concentrated under reduced pressure to about 50 ml. (Note 5) and then heated to boiling at atmospheric pressure. When 250 ml. of hot water is added slowly, 2-aminofluorene is precipitated as a colorless, crystalline powder. After cooling in an ice bath, the 2-aminofluorene is collected, washed with water, and dried in the dark in a vacuum desiccator. The product melts at 127.8–128.8° (Note 6) and amounts to 24–25 g. (93–96%).

2. Notes

1. If the stirring is omitted, the nitrofluorene takes longer to dissolve.

2. A suitable catalyst is 10% palladium-on-charcoal, such as is supplied by Baker and Company, Inc., 113 Astor Street, Newark 5, New Jersey.

3. The reaction is exothermic, and too rapid addition of the hydrazine may cause the mixture to foam out of the condenser.

4. *Caution! The catalyst is often pyrophoric and should be kept moistened with alcohol.* Celite is a diatomaceous earth filter aid.

5. A rotary evaporator is very convenient for the concentration since some of the amine invariably crystallizes toward the end.

6. The melting point is that reported in *Organic Syntheses*, Coll. Vol. 2, 448 (1943), for a recrystallized sample.

3. Discussion

The preparation of 2-aminofluorene reported previously in *Organic Syntheses* [Coll. Vol. 2, 448 (1943)] we based on the method of Diels.[2]

The present procedure illustrates a general method for the reduction of aromatic nitro compounds to aromatic amines using hydrazine and a hydrogenation catalyst such as palladium, platinum, nickel, iron, or ruthenium. The literature on this procedure up to 1963 has been reviewed.[3] In many instances the catalytic hydrazine reductions give yields of amine equal to or better than those obtained by direct catalytic hydrogenation or other reduction methods. Both the apparatus and the procedure are simple. Under appropriate conditions the method may be used for the dehalogenation of aliphatic and aromatic halides,[3] a reaction for which palladium appears to be a specific catalyst. The method has also been used for the reduction of azobenzene and azoxybenzene to hydrazobenzene (80-90%),[4] as well as for the synthesis of steroid aziridines by reduction of mesylate esters by vicinal azido alcohols (using Raney nickel).[5]

1. National Research Council of Canada Post-doctorate Fellow, 1954-56, at the University of Ottawa, Ottawa, Ontario.
2. O. Diels, *Ber.*, 34, 1758 (1901).
3. A. Furst, R. C. Berlo, and S. Hooton, *Chem. Rev.*, 65, 51 (1965).
4. P. M. G. Bavin, *Can. J. Chem.*, 36, 238 (1958).
5. K. Ponsold, *Ber.*, 97, 3524 (1964).

AMINOMALONONITRILE p-TOLUENESULFONATE

(Malononitrile, amino-, p-toluenesulfonate)

$$CH_2(CN)_2 + NaNO_2 + CH_3COOH \rightarrow$$
$$HON{=}C(CN)_2 + CH_3CO_2Na + H_2O$$

$$3HON{=}C(CN)_2 + 4Al + 9H_2O \rightarrow$$
$$3NH_2CH(CN)_2 + 4Al(OH)_3$$

$$NH_2CH(CN)_2 + p\text{-}CH_3C_6H_4SO_3H \rightarrow$$
$$p\text{-}CH_3C_6H_4SO_3^- \overset{+}{N}H_3CH(CN)_2$$

Submitted by J. P. Ferris, R. A. Sanchez, and R. W. Mancuso [1]
Checked by O. W. Webster and R. E. Benson

1. Procedure

A. *Oximinomalononitrile.* Malononitrile (Note 1) (25 g., 0.38 mole) is dissolved in a mixture of 20 ml. of water and 100 ml. of acetic acid in a 1-l. round-bottomed flask equipped with a stirrer, a thermometer, and a powder funnel. The solution is cooled to $-10°$ with a dry ice-acetone bath, and 50 g. (0.72 mole) of granulated sodium nitrite is added in approximately 2-g. portions over a 30-minute period while the temperature is maintained at $0°$ to $-10°$. After the addition is complete a wet ice bath is used to maintain the temperature below $5°$ while the mixture is stirred for 4 hours. Four hundred milliliters of tetrahydrofuran (Note 2) and 400 ml. of ether are added in separate portions, and the mixture is stored at $-40°$ overnight. The mixture is filtered rapidly, and the solid is washed with a mixture of 200 ml. of tetrahydrofuran (Note 2) and 200 ml. of ether. The filtrate and washings are combined and concentrated by distillation to a volume of 250 ml. by the use of a water aspirator and a bath at 40° (Note 3). This solution of oximinomalononitrile is used directly in the next step.

B. *Aminomalononitrile p-toluenesulfonate.* Aluminum foil (13.7 g., 0.51 g. atom) is cut into half-inch squares and is covered with a 5% aqueous solution of mercuric chloride until a mercury coating is visible on the aluminum (*ca.* 30 seconds). The mercuric chloride solution is decanted, and the amalgamated aluminum is washed twice with water, once with ethanol, and twice with tetrahydrofuran (Note 2). The amalgamated aluminum is transferred to a 2-l. round-bottomed flask fitted with a condenser, a stirrer, and a 250-ml. addition funnel and is covered immediately with 300 ml. of tetrahydrofuran (Note 2). The mixture is cooled in a dry ice-acetone bath, and the solution of oximinomalononitrile from procedure A is added with stirring over a 15-minute period while the temperature is maintained at −15° to −30°. Stirring is continued for an additional 5 minutes. The dry-ice acetone bath is then removed, and the mixture is allowed to warm to room temperature. (*Caution! Cooling with a dry ice-acetone bath is usually needed to control the reaction.*) After the spontaneous reaction subsides, the mixture is warmed to reflux until most of the aluminum is consumed (45 minutes). The reaction mixture is cooled to room temperature, 200 ml. of ether is added with stirring, and the aluminum salts are removed by vacuum filtration through Celite filter aid. The solid is washed with 250 ml. of tetrahydrofuran (Note 2) followed by 500 ml. of ether (Notes 3 and 4). The original filtrate and washings are combined and concentrated to about 250 ml. by the use of a water aspirator and a bath at 40°. To the resulting brown solution is slowly added with stirring a mixture of 60 g. (0.32 mole) of *p*-toluenesulfonic acid monohydrate as a slurry in 250 ml. of ether (Note 5). The total volume is brought to 1 l. with ether, the mixture is cooled to 0°, and the crystalline solid is collected by vacuum filtration. The product is washed successively with 200 ml. of ether, 200 ml. of cold (0°) acetonitrile, and 200 ml. of ether and dried at 25° (1 mm.) to give light tan crystals, m.p. 169–171° (dec.); yield, 75–79 g. (78–82%).

This product is suitable for most synthetic purposes. An almost colorless product may be obtained by recrystallization from boiling acetonitrile (100 ml. dissolves 1.8 g. of product) with treatment with activated carbon. The recovery of aminomalononitrile p-toluenesulfonate, m.p. 172° (dec.), is *ca.* 80%.

2. Notes

1. Commercial malononitrile is purified by dissolving 260 g. in 1 l. of ether, refluxing the solution with 5 g. of activated carbon for 10 minutes, and filtering through Celite under vacuum. The malononitrile crystallizes from the filtrate as a result of the cooling and concentration during the filtration. It is collected by filtration and washed with 350 ml. of cold (−20°) ether to give 214 g. of white crystals.

2. Tetrahydrofuran from Fisher Scientific Co. was used by the checkers. [*Caution! See page 976 for a warning regarding the purification of tetrahydrofuran.*]

3. Occasionally a precipitate may form in the filtrate. It is removed by filtration before proceeding to the next step.

4. Additional washing is necessary if the washings are not colorless at this point.

5. One can check for complete precipitation of the aminomalononitrile by adding p-toluenesulfonic acid to the clear supernatant liquid.

3. Methods of Preparation

The present procedure is a modification of the original synthesis.[2] Previous reports of the synthesis of aminomalononitrile are in error.[2] Oximinomalononitrile was prepared by a modification of the procedure of Ponzio.[3]

4. Merits of the Preparation

This procedure provides a convenient synthesis of aminomalononitrile, which has been demonstrated to be a useful intermediate for the preparation of substituted imidazoles, thiazoles, oxazoles, purines, and purine-related heterocycles.[2] It is also a convenient starting material for the preparation of diamino-maleonitrile.[2, 4]

1. The Salk Institute for Biological Studies, San Diego, California [Present address (J.P.F.): Department of Chemistry, Rensselaer Polytechnic Institute, Troy, New York 12181].
2. J. P. Ferris and L. E. Orgel, *J. Am. Chem. Soc.*, 88, 3829 (1966); 87, 4976 (1965).
3. G. Ponzio, *Gazz. Chim. Ital.*, 61, 561 (1931).
4. J. P. Ferris and R. A. Sanchez, this volume, p. 344.

1-AMINO-1-METHYLCYCLOHEXANE

(Cyclohexylamine, 1-methyl-)

Submitted by PETER KOVACIC and SOHAN S. CHAUDHARY[1]
Checked by R. A. HAGGARD and WILLIAM D. EMMONS

1. Procedure

Caution! *The reactions should be carried out in a hood behind a protective screen since trichloramine is noxious and potentially explosive; however, no difficulties from decomposition have been encountered under the conditions described.*

A. *Trichloramine.* A mixture of 600 ml. of water (Note 2), 900 ml. of methylene chloride (Note 3), and 270 g. (1.32 moles) of calcium hypochlorite (Note 4) is cooled to 0–10° in a 3-l., three-necked, vented flask equipped with a stirrer, a thermometer, and a dropping funnel. A solution of 66.0 g. (1.23 moles) of ammonium chloride in 150 ml. of concentrated hydrochloric acid and 450 ml. of water is added dropwise with stirring over a 1-hour period at 0–10°. After an additional 20 minutes of stirring, the organic layer is separated, washed with three 200-ml. portions of cold water, and dried over anhydrous sodium sulfate. The yellow solution is filtered, and the trichloramine concentration is determined by iodometric titration (Note 5).

B. *1-Amino-1-methylcyclohexane.* A 3-l. three-necked flask is fitted with a paddle stirrer, a condenser, a thermometer, and a dropping funnel with an extension for below-surface addition. Provision is made for introduction of nitrogen by use of a side-arm adapter. The vessel is charged with 196 g. (2.0 moles) of methylcyclohexane (Note 6) and 106 g. (0.80 mole) of anhydrous aluminum chloride. A solution (*ca.* 600 ml.) of trichloramine (0.40 mole) in methylene chloride is added with efficient stirring over a period of 2 hours at −5° to 5° (Note 7). Throughout the reaction a stream of nitrogen is passed through the flask (Note 8). The brown mixture is stirred for an additional 20–30 minutes at the same temperature.

The reaction mixture is then added with good stirring to a slurry of 800–900 g. of ice and 50 ml. of concentrated hydrochloric acid (Note 9). The layers are separated, and the dark organic layer is washed with three 100-ml. portions of 5% hydrochloric acid and discarded. Traces of non-basic organic material are removed from the combined aqueous layer and washings by extraction with pure ether (Note 10) until the extract is colorless. The aqueous solution is treated with 600 ml. of 50% aqueous sodium hydroxide (Note 11) with cooling, and the basic organic product is extracted with three 125-ml. portions of pure ether (Note 10). The ethereal solution is dried over sodium sulfate, and the solvent is distilled on the steam bath to give 42–46 g. of a clear, amber product (Note 12). To this crude product is added 10 g. of triethylenetetramine (Note 13). Distillation through a small Vigreux column yields 21.5–30 g. (48–67%, based on trichloramine) of 1-amino-1-methylcyclohexane, b.p. 44–49° (20–25 mm.), n^{22}D 1.4516 (Note 14).

2. Notes

1. The stoichiometry of the reaction is not known.
2. Deionized water is used throughout.
3. Commercial methylene chloride was distilled before use by the submitters. The checkers used reagent grade methylene chloride without distillation.

4. Calcium hypochlorite is obtained as "HTH" (Olin Mathieson Chemical Co., 70% purity).

5. Iodometric determination of positive chlorine is carried out as follows: 2.0 g. of potassium iodide or sodium iodide is dissolved in 10 ml. of water, and 40 ml. of glacial acetic acid is added. Into this solution is pipetted 1.0 ml. of the methylene chloride solution of trichloramine. The liberated iodine is titrated with $0.100N$ sodium thiosulfate. The solution is found to be $0.6–0.7M$ in trichloramine. Storage for several days at 0–5° results in negligible decomposition, although it is not recommended unless adequate safety precautions are observed. Excess methylene chloride-trichloramine solution can be conveniently disposed of by its slow addition to a cold, stirred, dilute aqueous solution of sodium metabisulfite.

6. A pure grade of methylcyclohexane (Eastman Organic Chemicals) is used. Subsequent to the checking of this preparation, the submitters reported 69–72% yields with 78.4 g. (0.80 mole) of methylcyclohexane.[2] In this case a 1-l. three-necked flask is employed for the reaction; the remainder of the procedure is unchanged.

7. Cooling is accomplished with either an ice-salt bath or preferably a dry ice-acetone bath. The time of addition can be reduced to 1 hour by use of the latter. However, if the temperature is much below that designated, unchanged trichloramine accumulates, resulting eventually in an uncontrollable reaction.

8. Purging with nitrogen results in some increase in yield. If the flow is too vigorous, trichloramine is lost by volatilization.

9. The mixture can be stored overnight at this stage.

10. High-purity ether (*e.g.*, Baker Analyzed Reagent) is used since a grade of lower quality gives a product that is more difficult to purify because of contamination with alcohol.

11. Excess sodium hydroxide is needed to dissolve the aluminum-containing precipitate.

12. The last portion of solvent is carefully removed at the water aspirator.

13. Triethylenetetramine (redistilled, Eastman Organic Chemicals) prevents bumping and foaming and acts as a chaser for the distillation.

14. The product contains less than 10% of lower-boiling impurities determined (by the checkers) by vapor-phase chromatography with a column packed with 15% XF-1150 on Chromosorb W. Further purification can be effected readily with good recovery by drying over sodium hydroxide pellets and fractionating at atmospheric pressure through an efficient spinning band column, with collection of the fraction, b.p. 142–146°, n^{22}D 1.4522.

3. Methods of Preparation

In addition to the present method,[2] 1-amino-1-methylcyclohexane has been synthesized by the following procedures: Ritter reaction, e.g., with 1-methylcyclohexanol (76%, 67%)[3, 4] or 1-methylcyclohexene (35%);[4] Hofmann reaction with 1-methylcyclohexanecarboxamide (80% as hydrochloride);[5] reduction of 1-methyl-1-nitrocyclohexane (63%);[5] Schmidt reaction with 1-methylcyclohexanecarboxylic acid (42%).[6]

4. Merits of the Preparation

This procedure constitutes the first example of one-step conversion of a t-alkane to the corresponding t-alkylamine. Other hydrocarbons in this class, such as isobutane, have also been aminated with good results.[7] Only a very limited number of convenient routes, e.g., the Ritter reaction, are available for the preparation of t-carbinamines. The present preparation illustrates a simple method that utilizes a novel substrate.

1. Department of Chemistry, Case Western Reserve University, Cleveland, Ohio 44106.
2. P. Kovacic and S. S. Chaudhary, *Tetrahedron*, **23**, 3563 (1967).
3. H. J. Barber and E. Lunt, *J. Chem. Soc.*, 1187 (1960).
4. W. Haaf, *Ber.*, **96**, 3359 (1963).
5. K. E. Hamlin and M. Freifelder, *J. Am. Chem. Soc.*, **75**, 369 (1953).
6. C. Schuerch, Jr., and E. H. Huntress, *J. Am. Chem. Soc.*, **71**, 2233 (1949).
7. P. Kovacic and S. S. Chaudhary, unpublished work.

3(5)-AMINOPYRAZOLE*

[Pyrazole, 3(or 5)-amino-]

$$CH_2{=}CHCN \ + \ NH_2NH_2 \longrightarrow NH_2NHCH_2CH_2CN \xrightarrow{H_2SO_4}$$

$$\xrightarrow[\text{NaHCO}_3]{p-CH_3C_6H_4SO_2Cl} \qquad \xrightarrow{(CH_3)_2CHONa}$$

(diagram: 3-amino-3-pyrazoline with H$_2$SO$_4$ converting to tosyl derivative)

$$+ \quad p\text{-}CH_3C_6H_4SO_2Na \quad + \quad (CH_3)_2CHOH$$

(diagram: H$_2$N-substituted pyrazole)

Submitted by H. Dorn and A. Zubek [1]
Checked by L. G. Vaughan and R. E. Benson

1. Procedure

A. *β-Cyanoethylhydrazine.* To a 2-l. two-necked flask fitted with a thermometer and a pressure-equalizing funnel are added a large magnetic stirring bar and 417 g. (6.00 moles of $N_2H_4 \cdot H_2O$) of 72% aqueous hydrazine hydrate. Acrylonitrile (318 g., 6.00 moles) is gradually added with stirring during 2 hours. The internal temperature is kept at 30–35° by occasional cooling of the flask. The funnel is replaced by a distillation condenser. Removal of water by distillation at 40 mm. at a bath temperature of 45–50° gives 490–511 g. (96–100%) of β-cyanoethylhydrazine as a yellow oil that is suitable for use in the next step. This product can be purified by distillation; b.p. 76–79° (0.5 mm.).

B. *3-Amino-3-pyrazoline sulfate.* In a 2-l. four-necked flask equipped with a reflux condenser, a dropping funnel, a thermometer, and a mechanical stirrer with four blades (Note 1) is placed 308 g. (169 ml., 3.0 moles) of 95% sulfuric acid (sp. gr. 1.834). Absolute ethanol (450 ml.) is added dropwise over 20–30 minutes. The internal temperature is maintained at 35° by cooling. A solution of 85.1 g. (1.00 mole) of β-cyanoethylhydrazine in 50 ml. of absolute ethanol is added with vigorous stirring over 1–2 minutes without further cooling (Note 1). The mixture warms spontaneously to 88–90° and is kept at this temperature for 3 minutes until the product begins to crystallize. The temperature of the stirred mixture is gradually lowered during the next hour to 25° by cooling with water, and the mixture is then allowed to stand at room temperature for 15–20 hours. The crystals are collected by filtration and washed three times with 80 ml. of absolute ethanol and finally with 80 ml. of ether. After being dried at 80° the product weighs 177–183 g. (97–100%), m.p. 143–144° (Note 2). The product is sufficiently pure for use in the following step; it may be recrystallized from methanol to give white needles, m.p. 144–145° (Note 2).

C. *3-Imino-1-(p-tolylsulfonyl)pyrazolidine.* To a 3-l. four-necked flask fitted with a condenser, a thermometer, a wide-mouthed funnel, and a high-speed mechanical stirrer having five pairs of blades are added 183 g. (1.00 mole) of 3-amino-3-pyrazoline sulfate and 1 l. of water. Sodium bicarbonate (210 g., 2.5 moles) is gradually added during 10 minutes with stirring. The rate of stirring is increased to 5000–6000 r.p.m., and a solution of 229 g. (1.20 moles) of *p*-toluenesulfonyl chloride in 400 ml. of benzene containing 0.5 g. of sodium dodecylbenzenesulfonate (Note 3) is added at one time. Three further portions of sodium bicarbonate are added sequentially: 25.2 g. (0.30 mole) after 15 minutes; 16.8 g. (0.20 mole) after 30 minutes; 16.8 g. (0.20 mole) after 55 minutes. The mixture is stirred for 5 hours at 18 25°, occasional cooling being required. Sodium bicarbonate (8.4 g., 0.10 mole) is added, then 200 ml. of ether, and stirring is continued for another hour. The colorless product is collected by filtration on a sintered-glass funnel, washed with three 50-ml. portions of ether followed by 50 ml. of water, and dried at 90°. The yield is 139–180 g. (58–75%); m.p. 183–185° (Note 4). The product is used directly in the next step.

D. *3(5)-Aminopyrazole* (Note 5). (*Caution! Because hydrogen gas is evolved, this reaction should be conducted in an efficient hood in the absence of an ignition source.*) A solution of sodium isopropoxide is prepared from 18.4 g. (0.80 g. atom) of sodium and 500 ml. of isopropyl alcohol in a 2-l. four-necked flask fitted with a mechanical stirrer, a thermometer, a reflux condenser, and a stopper. The reflux condenser is fitted with a nitrogen-inlet line attached to a bubbler device to maintain an anhydrous atmosphere. After all the sodium has dissolved, the temperature is adjusted to 60–70°, the stopper is replaced by a wide-mouthed funnel, and 191 g. (0.80 mole) of 3-imino-1-(p-tolylsulfonyl)-pyrazolidine is added gradually over 10 minutes to the hot solution under a blanket of nitrogen. The funnel is replaced by the stopper, and the mixture is stirred vigorously and then refluxed briefly. Stirring is continued, and the mixture is allowed to cool to room temperature during 2 hours. The precipitated sodium p-toluenesulfinate (140–142 g.) is removed by filtration and washed with a total of 100 ml. of isopropyl alcohol in several portions. The filtrate is treated twice with 4-g. portions of Norit activated carbon. The solvent is removed by distillation, the final trace being removed at a bath temperature of 50° (20 mm.) to give 62–66 g. (93–99%) of 3(5)-aminopyrazole as a light yellow oil. This is purified by distillation to give the product as a yellow oil, b.p. 100–102° (0.01 mm.), in 74–84% recovery (Note 6). The product crystallizes on cooling; m.p. 37–39° (Note 7). Its n.m.r. spectrum (60 MHz, dimethyl sulfoxide-d_6) shows two one-proton doublets at δ 7.33 and 5.52 p.p.m. ($J = 2$ Hz) and a broad three-proton singlet at δ 7.05 p.p.m. that is absent after addition of D_2O.

2. Notes

1. A stirrer with large blades operating at high speed is essential. Inadequate stirring results in solidification of the reaction mixture and makes proper washing of the product very difficult.

2. The checkers found melting points of 138–141° and 140–142°. After three recrystallizations from methanol the product has a melting point of 139.7–140°. The product appeared to be unstable to prolonged heating in methanol.

3. This salt serves as an emulsifying agent.

4. A sample, m.p. 184–185°, prepared by recrystallization of the product from nitromethane, gives satisfactory elemental analytical data. Its n.m.r. spectrum (60 MHz, dimethyl sulfoxide-d_6) reveals that the compound exists in the iminopyra-

zolidine form under these conditions; signals at δ 7.72 p.p.m. (doublet, $J = 8.4$ Hz), 7.40 p.p.m. (doublet, $J = 8.4$ Hz), 6.1 p.p.m. (broad singlet; absent after addition of D_2O), 3.4 p.p.m. (triplet, $J = 9.0$ Hz), and 2.4 p.p.m. (sharp singlet superimposed on triplet) with relative intensities of 2:2:2:2:5. The signals at 7.72 and 7.40 p.p.m. are assigned to the four aromatic protons, that at 6.1 p.p.m. to the two N—H protons, that at 3.4 p.p.m. to one pair of methylene protons, and that at 2.4 p.p.m. to the second pair of methylene protons plus the protons of the methyl group.

5. 3(5)-Aminopyrazole may also be obtained by hydrolysis of 3-imino-1-(p-tolylsulfonyl)pyrazolidine with aqueous alkali. In this case the pyrazolidine (239 g., 1.00 mole) is added to a solution of 40 g. (1.0 mole) of sodium hydroxide in 250 ml. of water at 75°, the resulting solution is stirred briefly, and the water is removed at reduced pressure. 3(5)-Aminopyrazole is separated from the sodium p-toluenesulfinate by several extractions with isopropyl alcohol.

6. In order to obtain maximum recovery the submitters conducted the distillation of 120 g. of crude product for 7–10 hours.

7. The checkers observed b.p. 119–121° (1.0 mm.) and m.p. 34–37°.

3. Methods of Preparation

3(5)-Aminopyrazole has been prepared by a Curtius degradation of pyrazole-3(5)-carboxylic acid hydrazide,[2, 3] by saponification and decarboxylation of ethyl 3-aminopyrazole-4-carboxylate[4] obtained from ethyl ethoxymethylenecyanoacetate and hydrazine, and by the present procedure.[5, 6]

4. Merits of the Preparation

This procedure represents the most convenient synthesis of 3(5)-aminopyrazole. It employs readily available starting materials and gives excellent yields in all steps.[5, 6] p-Toluenesulfonyl chloride can be replaced by other arenesulfonyl chlorides. 3-Imino-1-arylsulfonylpyrazolidines can be alkylated with dimethyl sulfate or with alkyl p-toluenesulfonates in dimethylformamide to give salts of 1-alkyl-2-arylsulfonyl-5-amino-4-

pyrazolines from which arenesulfinate can be eliminated as described in procedure D.   In this fashion 1-alkyl-5-aminopyrazoles can be easily prepared.[6]

1. Institut für Organische Chemie der Deutschen Akademie der Wissenschaften, Berlin-Adlershof, Germany.
2. L. Knorr, *Ber.*, **37**, 3520 (1904).
3. H. Reimlinger, A. van Overstraeten, and H. G. Viehe, *Ber.*, **94**, 1036 (1961).
4. P. Schmidt and J. Druey, *Helv. Chim. Acta*, **39**, 986 (1956).
5. H. Dorn, G. Hilgetag, and A. Zubek, *Angew. Chem.*, **76**, 920 (1964); *Angew. Chem. Intern. Ed. Engl.*, **3**, 748 (1964).
6. H. Dorn, G. Hilgetag, and A. Zubek, *Ber.*, **98**, 3368 (1965).

1-AMINOPYRIDINIUM IODIDE

(Pyridinium, 1-amino-, iodide)

Submitted by R. Gösl and A. Meuwsen.[1]
Checked by N. A. Fedoruk and V. Boekelheide.

1. Procedure

To a freshly prepared solution of 11.3 g. (0.10 mole) of hydroxylamine-O-sulfonic acid (Note 1) in 64 ml. of cold water there is added 24 ml. (24 g., 0.30 mole) of pyridine (Note 2).   The mixture is heated at about 90° on a steam bath for 20 minutes.   It is then cooled to room temperature with stirring, and 13.8 g. (0.10 mole) of potassium carbonate is added.   The water and excess pyridine are removed from the mixture by heating it at 30–40° in a rotatory evaporator in conjunction with a water aspirator. The residue is treated with 120 ml. of ethanol, and the insoluble precipitate of potassium sulfate is removed by filtration.

Fourteen milliliters (22 g., 0.10 mole) of 57% hydriodic acid is added to the filtrate, and the resulting solution is stored at −20° for 1 hour (Note 3).   The solid that separates is collected; weight 15.5–17.5 g.   Recrystallization of this solid from about 100 ml. of absolute ethanol gives 14–16 g. (63–72%) of 1-aminopyridinium iodide as almost-white crystals, m.p. 160–162° (Note 4).

2. Notes

1. Hydroxylamine-O-sulfonic acid may be purchased from Ventron Corporation or prepared according to the directions in *Inorganic Syntheses.*[2]

Because aqueous solutions of hydroxylamine-O-sulfonic acid are not very stable, it is very important to use freshly prepared solutions. The purity of hydroxylamine-O-sulfonic acid should be checked by iodometric titration. If it is less than 85–90% pure, the yield of 1-aminopyridinium iodide will suffer. The acid can be purified by dissolving it in an equal weight of water and then precipitating it by stirring 7 volumes of acetic acid into the solution.

2. The pyridine was distilled before use. When the conversion is carried out in the presence of potassium carbonate using an equimolar amount of pyridine instead of an excess, the yields obtained are 20–30% lower.[3]

3. The temperature is kept at −20° or lower by a bath of dry ice and methanol. If the temperature rises above −20°, an appreciable quantity of 1-aminopyridinium iodide may redissolve and be lost.

4. The melting point recorded for 1-aminopyridinium iodide is 161–162°.[3]

3. Methods of Preparation

The formation of 1-aminopyridinium chloride has been accomplished by the acid hydrolysis of N-(*p*-acetaminobenzene-sulfonimido)pyridine.[4] Also, the rearrangement of a substituted diazepine has been observed to give a 1-aminopyridine derivative.[5] The present procedure is an adaptation of that described by Gösl and Meuwsen.[3]

4. Merits of the Preparation

This procedure is a convenient and general method for preparing asymmetrically substituted hydrazines.[3] This is illustrated by the following examples reported by the submitters [3] (% yields in parentheses): methylamine to methylhydrazinium hydrogen sulfate (49–53%); ethylamine to ethylhydrazinium hydrogen oxalate (51%); butylamine to butylhydrazinium hydrogen sulfate (49–56%); piperidine to 1-aminopiperidinium hydrogen oxalate (32%); dibutylamine to 1,1-dibutylhydrazinium hydrogen oxalate (34%); trimethylamine to 1,1,1-trimethylhydrazinium hydrogen oxalate (79–85%); 2-picoline to 1-amino-2-methylpyridinium iodide (57%); 2,4-lutidine to 1-amino-2,4-dimethylpyridinium iodide (40%); 2,6-lutidine to 1-amino-2,6-dimethylpyridinium iodide (34%); 2,4,6-collidine to 1-amino-2,4,6-trimethylpyridinium iodide (30%); and quinoline to 1-aminoquinolinium iodide (32%).

Primary, secondary, and tertiary amines can be aminated by chloramine also, but pyridine nitrogens have been aminated only by hydroxylamine-O-sulfonic acid.

It has been shown that, on treatment with base, 1-aminopyridinium iodide undergoes 1,3-dipolar addition with ethyl propiolate or dimethyl acetylenedicarboxylate; thus the N-aminoheterocycles may serve as convenient starting materials for the synthesis of a variety of unusual fused heterocycles.[6]

1. Institut für Anorganische Chemie der Universität Erlangen, Erlangen, Germany.
2. H. J. Matsuguma and L. Audrieth, *Inorg. Syntheses*, 5, 122 (1957).
3. A. Meuwsen and R. Gösl, *Angew. Chem.*, 69, 754 (1957); R. Gösl and A. Meuwsen, *Chem. Ber.*, 92, 2521 (1959).
4. J. N. Ashley, G. L. Buchanan, and A. P. T. Easson, *J. Chem. Soc.*, 60 (1947).
5. J. A. Moore, *J. Am. Chem. Soc.*, 77, 3417 (1955); J. A. Moore and J. Binkert, *J. Am. Chem. Soc.*, 81, 6045 (1959).
6. R. Huisgen, R. Grashey, and R. Krischke, *Tetrahedron Lett.*, 387 (1962).

o-ANISALDEHYDE*

Submitted by A. J. SISTI [1]
Checked by G. L. WALFORD and PETER YATES

1. Procedure

A. *4-Dimethylamino-2'-methoxybenzhydrol.* An ethereal solution of o-methoxyphenylmagnesium bromide is prepared in the
usual manner [2] with 250 ml. of anhydrous ether, 14.5 g. (0.60 g.
atom) of magnesium, and 100 g. (0.53 mole) of o-bromoanisole
(Note 1). A solution of 60 g. (0.40 mole) of p-dimethylaminobenzaldehyde (Note 2) in 200 ml. of anhydrous benzene is added
dropwise to the Grignard reagent (Note 3). After the addition
is completed, the reaction mixture is stirred for 10 hours at room

temperature. The magnesium complex, which forms a very thick suspension, is decomposed with a solution of 75 g. of ammonium chloride in 450 ml. of water. The ether-benzene layer is separated, washed with 100 ml. of water, and dried over calcium sulfate (Note 4). The solvent is removed under reduced pressure, and the residue is induced to crystallize by trituration with a little petroleum ether (30–60°). Recrystallization of the solid from benzene-petroleum ether (30–60°) gives 4-dimethyl-amino-2'-methoxybenzhydrol (59–60 g., 57–58%), m.p. 75–80°.

B. *o-Anisaldehyde.* In a 3-l. three-necked flask fitted with a mechanical stirrer and a nitrogen inlet tube are placed 60 g. (0.35 mole) of sulfanilic acid (Note 5), 18 g. (0.17 mole) of anhydrous sodium carbonate, and 400 ml. of water. Stirring is started, and the resulting solution is cooled to 0–5° in an ice bath. Nitrogen is passed into the reaction flask, and a nitrogen atmosphere is maintained throughout the reaction. To the cooled solution is added three-quarters of a solution of 24.2 g. (0.35 mole) of sodium nitrite in 75 ml. of water, followed by 32 ml. of concentrated hydrochloric acid. During the diazotization the temperature of the solution is maintained below 5° by the addition of ice in small pieces. After a few minutes another 32 ml. of acid is added. Further additions of the sodium nitrite solution are made slowly until a positive test for excess nitrous acid is observed (Note 6). The diazonium solution is buffered to pH ∼6 by the addition of a cooled solution of 50 g. of sodium acetate in 125 ml. of water. A solution of 52 g. (0.20 mole) of 4-dimethylamino-2'-methoxybenzhydrol in 500 ml. of acetone is added rapidly, followed by an additional 500 ml. of acetone. The reaction mixture becomes red almost immediately, and stirring is continued for 30 minutes at 0–5°. The cooling bath is replaced by a warm water bath (50–60°), and stirring is continued for an additional 30 minutes. The reaction mixture is diluted with an equal volume of water and extracted with three 750-ml. portions of ether. The combined ethereal extracts are washed with water until all the dissolved methyl orange is removed, then dried over calcium sulfate. The ether is removed under reduced pressure, and the residue is distilled to yield 19–20.5 g. (69–75%) of colorless liquid, b.p. 79–80° (1.5 mm.), n^{25}D 1.5586 (Note 7).

2. Notes

1. o-Bromoanisole obtained from Eastman Kodak Company was used without further purification.

2. A good commercial grade (Matheson, Coleman and Bell) of p-dimethylaminobenzaldehyde was used without further purification.

3. In one run the checkers cooled the reaction mixture in an ice bath throughout the addition. In another run only initial cooling was used. There was no difference in yield.

4. The checkers found that separation of the aqueous and organic phases is very difficult if the mixture is shaken. In one run shaking and washing were omitted without affecting the yield or purity of the product.

5. Eastman white label sulfanilic acid was used without purification.

6. Excess nitrous acid causes an *immediate* blue color at the point of contact with starch-iodide test paper. At all times there must be an excess of mineral acid (Congo red test paper).

7. The submitters found for the 2,4-dinitrophenylhydrazone m.p. 252–254° (lit.[2] m.p. 249–250°). The checkers found m.p. 34–36° for o-anisaldehyde (lit.[3] m.p. 37°) and m.p. 249–251° for the 2,4-dinitrophenylhydrazone.

3. Methods and Merits of Preparation

o-Anisaldehyde is commercially available. However, this procedure illustrates a method of general applicability[4-6] for the preparation of aromatic, aliphatic, and unsaturated aldehydes and ketones.

1. Department of Chemistry, Adelphi University, Garden City, Long Island, New York.
2. E. K. Harvill and R. M. Herbst, *J. Org. Chem.*, 9, 21 (1944).
3. F. B. Garner and S. Sugden, *J. Chem. Soc.*, 2877 (1927).
4. M. Stiles and A. J. Sisti, *J. Org. Chem.*, 25, 1691 (1960).
5. A. Sisti, J. Burgmaster, and M. Fudim, *J. Org. Chem.*, 27, 279 (1962).
6. A. J. Sisti, J. Sawinski, and R. Stout, *J. Chem. Eng. Data*, 9, 108 (1964).

AROMATIC ALDEHYDES.   MESITALDEHYDE*

(Benzaldehyde, 2,4,6-trimethyl-)

$$+ \ Cl_2CHOCH_3 \ \xrightarrow{\ TiCl_4\ }$$

$$+ \ HCl \ \xrightarrow{\ H_2O\ }$$

$$+ \ CH_3OH + HCl$$

Submitted by A. Rieche, H. Gross, and E. Höft[1]
Checked by G. N. Taylor and K. B. Wiberg

1. Procedure

A solution of 72 g. (0.60 mole) of mesitylene in 375 ml. of dry methylene chloride is placed in a 1-l. three-necked flask equipped with a reflux condenser, a stirrer, and a dropping funnel.   The solution is cooled in an ice bath, and 190 g. (110 ml., 1.0 mole) of titanium tetrachloride is added over a period of 3 minutes. While the solution is stirred and cooled, 57.5 g. (0.5 mole) of dichloromethyl methyl ether [2] is added dropwise over a 25-

minute period. The reaction begins (as indicated by evolution of hydrogen chloride) when the first drop of chloro ether is added. After the addition is complete, the mixture is stirred for 5 minutes in the ice bath, for 30 minutes without cooling, and for 15 minutes at 35°.

The reaction mixture is poured into a separatory funnel containing about 0.5 kg. of crushed ice and is shaken thoroughly. The organic layer is separated, and the aqueous solution is extracted with two 50-ml. portions of methylene chloride. The combined organic solution is washed three times with 75-ml. portions of water. A crystal of hydroquinone is added to the methylene chloride solution (Note 1) which is then dried over anhydrous sodium sulfate. After evaporation of the solvent, the residue is distilled to give the crude product, b.p. 68–74° (0.9 mm.).   After redistillation there is obtained 60–66 g. (81–89%) of mesitaldehyde; b.p. 113–115° (11 mm.), n^{20}D 1.5538.

2. Note

1. Hydroquinone retards the autoxidation of the aldehyde.

3. Methods of Preparation

Mesitaldehyde may be prepared from mesitylmagnesium bromide by the reaction with orthoformate esters[3] or ethoxymethyleneaniline;[3] from acetylmesitylene by oxidation with potassium permanganate;[4] from mesitoyl chloride by reduction;[5] from mesityllithium by the reaction with iron pentacarbonyl;[6] and from mesitylene by treatment with formyl fluoride and boron trifluoride,[7] by treatment with carbon monoxide, hydrogen chloride, and aluminum chloride,[8] or by various applications of the Gatterman synthesis.[9–11]

4. Merits of the Preparation

The preparation of mesitaldehyde is an example of a generally applicable method for the preparation of aromatic aldehydes by treatment of aromatic compounds with dichloromethyl methyl ether.[12]   Aldehydes derived from polynuclear aromatic compounds,[12,13] phenols,[14] phenol ethers,[12] and hetero-aromatic compounds [12] are also obtained using this procedure.   In addition, colchicine derivatives have been formylated [15] by means of dichloromethyl methyl ether.

1. Institute for Organic Chemistry of the German Academy of Sciences at Berlin, Berlin-Adlershof, Germany.
2. H. Gross, A. Rieche, E. Höft, and E. Beyer, this volume, p. 365.
3. L. I. Smith and J. Nichols, *J. Org. Chem.*, **6**, 489 (1941).
4. R. P. Barnes, C. I. Pierce, and C. C. Cochrane, *J. Am. Chem. Soc.*, **62**, 1084 (1940).
5. R. P. Barnes, *Org. Syntheses*, Coll. Vol. **3**, 551 (1955).
6. M. Ryang, I. Rhee, and S. Tsutsumi, *Bull. Chem. Soc. Japan*, **37**, 341 (1964).
7. G. A. Olah and S. J. Kuhn, *J. Am. Chem. Soc.*, **82**, 2380 (1960).
8. L. Gattermann, *Ann.*, **347**, 347 (1906).
9. R. C. Fuson, E. C. Horning, S. P. Rowland, and M. L. Ward, *Org. Syntheses*, Coll. Vol. **3**, 549 (1955).
10. L. E. Hinkel, E. E. Ayling, and W. H. Morgan, *J. Chem. Soc.*, 2793 (1932).
11. L. E. Hinkel, E. E. Ayling, and J. H. Beynon, *J. Chem. Soc.*, 339 (1936).
12. A. Rieche, H. Gross, and E. Höft, *Ber.*, **93**, 88 (1960).
13. H. Reimlinger, J. P. Golstein, J. Jadot, and P. Jung, *Ber.*, **97**, 349 (1964).
14. H. Gross, A. Rieche, and G. Matthey, *Ber.*, **96**, 308 (1963).
15. G. Muller, A. B. Font, and R. Bardoneschi, *Ann.*, **662**, 105 (1963).

ARYLBENZENES: 3,4-DICHLOROBIPHENYL

(Biphenyl, 3,4-dichloro-)

Submitted by D. H. HEY and M. J. PERKINS [1]
Checked by K. K.-W. SHEN and K. B. WIBERG

1. Procedure

Thirty-eight grams (0.1 mole) of bis-3,4-dichlorobenzoyl peroxide (Note 1) is added to a boiling solution of 3 g. of *m*-dinitro-

benzene in 800 ml. of dry reagent grade benzene contained in a
1-l. round-bottomed flask, and the resulting solution is boiled
under reflux for 40 hours. The solvent is then distilled from the
red solution until the residual volume is about 200 ml. (Note 2),
and the mixture is allowed to cool. The 3,4-dichlorobenzoic acid
which separates is removed by suction filtration, washed with a
little cold benzene, and then with 100 ml. of petroleum ether
(b.p. 80–100°). The combined filtrate and washings are further
concentrated by distillation (Note 2) to about 60 ml., cooled,
and a small second crop of 3,4-dichlorobenzoic acid is removed
and washed with a little benzene followed by a little petroleum
ether. The total yield of acid, m.p. 208–210° (Lit.[2] m.p. 208–
209°), is 18.2 g. (95%) (Note 3). The filtrate and washings are
combined (Note 4) and chromatographed on a column of basic
alumina (30 cm. x 3.5 cm.) which is eluted with petroleum ether
(b.p. 40–60°). Solvent is distilled from the eluate (Note 5), and
the residual crude 3,4-dichlorobiphenyl is distilled under reduced
pressure using a short air condenser and a receiver chilled in ice.
There is obtained 17.3–18.0 g. (78–81%) of almost pure 3,4-
dichlorobiphenyl (b.p. 146–150° at 2 mm.) which sets to a colorless
solid, m.p. 44–47° (Note 6).

2. Notes

1. The bis-3,4-dichlorobenzoyl peroxide may be prepared as
follows.[3, 4] To a 2-l. beaker containing 400 ml. of water which
is cooled to 0–5° in an ice bath is added slowly 40 g. (0.51 mole)
of sodium peroxide. A dropping funnel containing 167.6 g.
(0.8 mole) of 3,4-dichlorobenzoyl chloride in 400 ml. of dry toluene
is supported over the beaker. The peroxide solution is cooled
and stirred vigorously while the toluene solution is added dropwise over a 1-hour period. The solution is stirred for an additional 2 hours. The precipitate is filtered using a suction funnel,
washed with 600 ml. of cold water, and dried in air overnight.
There is obtained 114 g. (75%) of bis-3,4-dichlorobenzoyl peroxide, m.p. 135° dec. The crude product is purified by dissolving
it in chloroform and precipitating by the addition of methanol,
m.p. 139° dec.

2. Solvent removal may conveniently be carried out with a rotary evaporator to obviate bumping caused by separation of the dichlorobenzoic acid from the boiling solution.

3. In syntheses of other arylbenzenes, in which the acid by-product is more soluble, it may be extracted from the reaction mixture with aqueous sodium bicarbonate or removed in the subsequent chromatography.

4. At this point the solvent is largely petroleum ether. Appreciable quantities of benzene in the mixture to be chromatographed tend to carry some of the dinitrobenzene through the alumina.

5. After removal of the solvent from the eluate, almost pure white 3,4-dichlorobiphenyl crystallizes, m.p. 44–48°. It may be purified by recrystallization from methanol as an alternative to vacuum distillation.

6. The product may be freed from a trace (∽1%) of biphenyl present as an impurity by recrystallization from methanol which raises the melting point to 47–48°. The literature values range from 46° [5] to 49–50°. [6]

3. Discussion

3,4-Dichlorobiphenyl has been prepared by the arylation of benzene using the Gomberg procedure starting with 3,4-dichloroaniline,[5] and using the acid-catalyzed decomposition of 1-(3,4-dichlorophenyl)-3,3-dimethyltriazene in benzene.[6] The arylation procedure given above, which utilizes a diaroyl peroxide as the aryl radical source, provides a route to arylbenzenes which involves simple operations and gives a good yield of pure product. In the absence of the nitro compound, the mode of action of which has been discussed in terms of two somewhat different mechanisms,[7, 8] the yields of aroic acids and arylbenzenes are commonly below 50%,[9] and the arylbenzene may be contaminated with aryldihydrobenzenes.[10] The present procedure has been used to prepare a variety of simple arylbenzenes in isolated yields ranging from 70 to 90%. If a nitro substituent is present in the peroxide, high yields are obtained without the addition of further nitro compound.[9]

1. Department of Chemistry, King's College, University of London.
2. G. M. Kraay, *Rec. Trav. Chim.*, **49**, 1083 (1930).
3. A. V. Tobolsky and R. B. Mesrobian, "Organic Peroxides," Interscience Publishers, New York, 1954, p. 38.
4. A. I. Vogel, "Practical Organic Chemistry," 3rd. ed., Longmans, London, 1956, p. 808.
5. H. A. Scarborough and W. A. Waters, *J. Chem. Soc.*, 557 (1926).
6. U. S. Patent 2,280,504 (1942) [*Chem. Abstr.*, **36**, 5658 (1942)]; Ger. Patent 870,106 (1953) [*Chem. Abstr.*, **50**, 16863 (1956)].
7. G. B. Gill and G. H. Williams, *J. Chem. Soc.*, B, 880 (1966).
8. G. R. Chalfont, D. H. Hey, K. S. Y. Liang, and M. J. Perkins, *J. Chem. Soc.*, (B), 233 (1971).
9. K. H. Pausacker, *Australian J. Chem.*, **10**, 49 (1957).
10. D. F. DeTar and R. A. J. Long, *J. Am. Chem. Soc.*, **80**, 4742 (1958).

BENZENEDIAZONIUM-2-CARBOXYLATE

(Benzenediazonium, *o*-carboxy-, hydroxide, inner salt)

AND BIPHENYLENE

Submitted by Francis M. Logullo, Arnold H. Seitz, and Lester Friedman [1]
Checked by G. D. Abrams, Hermann Ertl, and Peter Yates

1. Procedure

Caution! *Benzenediazonium-2-carboxylate when dry detonates violently on being scraped or heated, and it is strongly recommended that it be kept wet with solvent at all times. It should be prepared and used in a hood behind a safety screen. A wet towel or sponge should be kept within easy reach with which to deactivate any spilled material, which should then be disposed of by flooding with water.*

A. *Benzenediazonium-2-carboxylate.* A solution of 34.2 g. (0.25 mole) of anthranilic acid (Note 1) and 0.3 g. of trichloroacetic acid (Note 2) in 250 ml. of tetrahydrofuran (Note 3) is prepared in a 600-ml. beaker equipped with a thermometer and cooled in an ice-water bath. The solution is stirred magnetically, and 55 ml. (48 g., 0.41 mole) of isoamyl nitrite (Note 4) is added over a period of 1–2 minutes. A mildly exothermic reaction occurs, and the reaction mixture is maintained at 18–25° and stirred for a further 1–1.5 hours. A transient orange to brick-red precipitate may appear (Note 5) which is slowly converted to the tan product. When the reaction is completed, the mixture is cooled to 10°, and the product is collected by suction filtration on a plastic Buchner funnel and washed on the funnel with cold tetrahydrofuran until the washings are colorless. (*Caution! The filter cake should not be allowed to become dry.*) The benzenediazonium-2-carboxylate is then washed with two 50-ml. portions of 1,2-dichloroethane to displace the tetrahydrofuran, and the solvent-wet material is used in the next step (Notes 6, 7, and 8).

B. *Biphenylene.* The solvent-wet benzenediazonium-2-carboxylate is washed from the funnel into a 400-ml. beaker with *ca.* 150 ml. of 1,2-dichloroethane, dispensed from a plastic wash bottle, with the aid of a plastic spatula (Note 9). The resultant slurry is added during 3–5 minutes to 1250 ml. of gently boiling, stirred 1,2-dichloroethane in a 2-l. beaker on a magnetic stirrer-hot plate in the hood (Note 10). Frothing ceases a few minutes after completion of the addition, and the mixture assumes a clear red-brown color, signaling the end of the reaction.

A 1-l., two-necked, round-bottomed flask is equipped with a 1-l. addition funnel and a Claisen distillation head and water-cooled condenser. The cooled reaction mixture is transferred to the funnel, and enough of it is admitted to the flask to half-fill the latter. The 1,2-dichloroethane, b.p. 83–84°, is distilled with the use of magnetic stirring to maintain even ebullition; the remainder of the reaction mixture is added from the funnel at a rate such that the flask remains about half-full. When *ca.* 75 ml. of dark residue remains in the flask, 300 ml. of ethylene glycol is added. An air condenser is substituted for the water-cooled condenser, and distillation is recommenced. A forerun, b.p. <150°, is discarded, and the fraction, b.p. 150–197°, is collected (Note 11). The distillate is cooled to 10°, and the product is collected by suction filtration, washed with 10–15 ml. of cold ethylene glycol and several times with water, and dried at atmospheric pressure over phosphorus pentoxide. The yield of biphenylene, m.p. 109–112°, is 4.0–5.6 g. (21–30%, based on anthranilic acid). Additional biphenylene (0.2–0.5 g.) can be obtained from the mother liquor and ethylene glycol washings by redistillation or dilution with water (Note 12).

2. Notes

1. The submitters used practical grade anthranilic acid from Mallinckrodt Chemical Works.

2. Perfluorobutyric or trifluoroacetic acid may be used in place of trichloroacetic acid. Strong mineral acids and acetic acid are wholly unsatisfactory. If a catalyst is not used, the product is of poor quality and the yield only 30%. Trichloroacetic acid is conveniently added as a solution in tetrahydrofuran (0.01 g./ml.).

3. The submitters used commercial tetrahydrofuran. The checkers found that the product yield was the same when either practical grade or Fisher Certified reagent grade tetrahydrofuran was used.

4. The submitters used "amyl nitrite" U.S.P. from Mallinckrodt Chemical Works; isoamyl nitrite supplied by Matheson, Coleman and Bell is apparently the same material. They found that other alkyl nitrites (ethyl, *n*-butyl, *t*-butyl, *n*-amyl) may be used with equal success. Subsequent to the checking of this procedure, they reported that the amount of nitrite can be reduced to a 20% molar excess.

5. This precipitate is apparently 2,2'-dicarboxydiazoaminobenzene.

6. The product should be used immediately because it deteriorates slowly at room temperature. It is freed of tetrahydrofuran by washing with the solvent to be used in subsequent reactions and transferred as a slurry in that solvent (cf. procedure B). Traces of water, if present, do not appear to interfere with subsequent reactions of diazonium carboxylates, as observed in the submitters' laboratory.

7. The submitters have found that this procedure works equally well with many substituted anthranilic acids; however, it does not work with 3-chloro-, 5-chloro-, 4-nitro-, 5-nitro-, and 4,5-benzoanthranilic acids.

8. Although it strongly recommended that the product not be dried, particularly when prepared on the scale described here, the following slightly modified procedure can be used for the preparation of solvent-free benzenediazonium-2-carboxylate. A solution of 2.74 g. (0.020 mole) of anthranilic acid and 0.030 g. of trichloroacetic acid in 30 ml. of tetrahydrofuran is prepared in a 100-ml. beaker equipped with a thermometer and cooled in a bath of *ca.* 25 g. of crushed ice. The solution is stirred magnetically, and 5 ml. (4.4 g., 0.038 mole) of isoamyl nitrite (Note 4) is added during *ca.* 0.5 minute. The mixture is stirred and allowed to warm to room temperature over a period of 1 hour. It is cooled to 10°, and the product is collected by suction filtration with the use of a plastic Buchner funnel and plastic spatula and washed with ice-cold tetrahydrofuran until the washings are colorless. The yield of air-dried (30 minutes) benzenediazonium-2-carboxylate is 2.55–2.88 g. (86–97%). (*Caution! Danger of detonation! See above.*) (Note 13).

9. The checkers transferred the benzenediazonium-2-carboxylate to the beaker with the aid of gentle air pressure (cf. Note 13) and then slurried it with *ca.* 150 ml. of 1,2-dichloroethane.

10. The checkers added the slurry via a large, medium-bore, glass funnel with fire-polished edges. In one of three runs a small, sharp report was heard, apparently from a source above the liquid in the beaker; the yield of biphenylene in this run did not differ significantly from that obtained in the other runs.

11. The checkers found it necessary to heat the condenser with a microburner from time to time to prevent clogging with

biphenylene. The submitters have reported that this can be avoided by connecting the Claisen head via an adapter to a two-necked receiving flask fitted with an upright water-cooled condenser and cooled by immersion in ice-water.

12. For convenient preparation and workup of larger amounts of biphenylene, several runs can be combined after the decomposition of the benzenediazonium-2-carboxylate. Thus the submitters obtained 19.1 g. (25%) of air-dried biphenylene by combining four batches. They found that the use of larger amounts of 1,2-dichloroethane resulted in a moderate increase in yield; by combining four batches, each prepared in 2.75 l. of 1,2-dichloroethane in a 4-l. beaker, they obtained 22.8 g. (30%) of product.

13. The checkers transferred the solvent-moist product to a tared Petri dish by means of a gentle puff of compressed air through the stem of the funnel; solid adhering to the filter paper and funnel was transferred to the dish with the aid of a soft rubber policeman, which was also used to spread the product over the surface of the dish. The product was then air-dried for 30 minutes in the hood.

3. Methods of Preparation

Benzenediazonium-2-carboxylate [2] and its substituted derivatives [3] have been prepared by diazotization of anthranilic acids in the presence of hydrochloric acid followed by dehydrochlorination of the resultant diazonium carboxylate hydrochlorides with silver oxide.

Biphenylene has been prepared in low yield by the reaction of 2,2'-dibromobiphenyl or 2,2'-biphenyliodonium iodide with cuprous oxide,[4] by the action of cupric chloride on 2,2'-biphenyldimagnesium dibromide,[5] from 2,2'-diiodobiphenyl via dibenzomercurole (o,o'-biphenylenemercury) (49%),[6] by pyrolysis or photolysis of phthaloyl peroxide (27%),[7] by reaction of o-fluorobromobenzene with lithium amalgam (24%),[8] by reaction of o-bromoiodobenzene with magnesium (12%),[9] and by the decomposition of diphenyliodonium-2-carboxylate,[10] 1,2,3-benzothiadiazole-1,1-dioxide,[11] benzenediazonium-2-carboxylate,[12,13] and by the oxidation of 1-aminobenzotriazole with lead tetraacetate (83%).[14]

4. Merits of the Preparation

These procedures illustrate facile methods for the preparation of benzenediazonium-2-carboxylate and its derivatives[15] and of biphenylene and certain biphenylene derivatives.[13] The latter preparation is far more convenient and proceeds in much better yield than do previous syntheses, which involve more steps, less accessible intermediates, and more complicated techniques.

1. Department of Chemistry, Case Western Reserve University, Cleveland, Ohio 44106.
2. A. Hantzsch and W. B. Davidson, *Ber.*, **29**, 1522 (1896).
3. M. Stiles, R. G. Miller, and U. Burckhardt, *J. Am. Chem. Soc.*, **85**, 1792 (1963).
4. W. C. Lothrup, *J. Am. Chem. Soc.*, **63**, 1187 (1941); W. Baker, M. P. V. Boarland, and J. F. M. McOmie, *J. Chem. Soc.*, 1476 (1954).
5. W. S. Rapson, R. G. Shuttleworth, and J. N. van Niekerk, *J. Chem. Soc.*, 326 (1943).
6. G. Wittig and W. Herwig, *Ber.*, **87**, 1511 (1954).
7. G. Wittig and H. F. Ebel, *Ann.*, **650**, 20 (1961).
8. G. Wittig and L. Pohmer, *Ber.*, **89**, 1334 (1956).
9. H. Heany, F. G. Mann, and I. T. Millar, *J. Chem. Soc.*, 3930 (1957).
10. E. Le Goff, *J. Am. Chem. Soc.*, **84**, 3786 (1962).
11. G. Wittig and R. W. Hoffmann, *Ber.*, **95**, 2718 (1962).
12. R. S. Berry, G. N. Spokes, and M. Stiles, *J. Am. Chem. Soc.*, **84**, 3570 (1962).
13. L. Friedman and A. H. Seitz, to be published.
14. C. D. Campbell and C. W. Rees, *J. Chem. Soc.* (C), 742 (1969).
15. L. Friedman, D. F. Lindow, and F. M. Logullo, to be published.

1,2,3-BENZOTHIADIAZOLE 1,1-DIOXIDE

Submitted by G. Wittig and R. W. Hoffmann [1]
Checked by C. D. Smith, R. A. Clement, and B. C. McKusick

1. Procedure

A. *2-Nitrobenzenesulfinic acid* (Note 1). *Caution! This re-action should be done in a good hood because noxious fumes are released.*

2-Nitroaniline (13.8 g., 0.10 mole) (Note 2) is dissolved in a hot solution of 75 ml. of 96% sulfuric acid, 100 ml. of phosphoric acid (density 1.7), and 50 ml. of water in a 1-l. beaker. A stirrer and a thermometer are introduced into the mixture, and the beaker is immersed in an ice bath. A solution of 8.3 g. (0.12 mole) of sodium nitrite in 25 ml. of water is added dropwise to the well-stirred solution at such a rate that the temperature is maintained at 10–15°. Excess nitrite is destroyed by adding sulfamic acid in small portions (Note 3). The mixture is cooled to −10° in an ice-salt bath, and about 50 ml. of liquid sulfur dioxide

(Note 4) is poured into the well-stirred reaction. The product is immediately poured onto a mixture of 55.6 g. (0.20 mole) of $FeSO_4 \cdot 7 H_2O$ and 1 g. of defatted copper powder in a wide 2-l. beaker. Nitrogen and excess sulfur dioxide bubble off with much foaming.

After 30 minutes the solid sulfinic acid is separated on a fritted-glass filter. The sulfinic acid is dissolved from the filter by a mixture of 750 ml. of ether and 750 ml. of methylene chloride. The solution is dried over calcium chloride and evaporated to dryness under reduced pressure (bath temperature 25°) (Note 5). The residue is suspended in 50 ml. of water, and small portions of dilute ammonia are added to the well-stirred suspension until it has a pH of 9 (Note 6). Insoluble impurities are separated by filtration, and 2-nitrobenzenesulfinic acid is precipitated from the filtrate by adding 5-ml. portions of $6N$ hydrochloric acid with cooling; the sulfinic acid precipitated by each portion of acid is separately collected on a Buchner funnel (Note 7). The acid, a pale yellow solid, is dried on a clay plate in a vacuum desiccator over potassium hydroxide pellets, m.p. 120–125° (dec.), weight 9.4–14.9 g. (50–80%). If the 2-nitrobenzenesulfinic acid is to be used for the hydrogenation of the next step high purity is required, and it is generally advisable to reprecipitate the acid once more in the same way (Note 8).

B. *Sodium 2-aminobenzenesulfinate.* 2-Nitrobenzenesulfinic acid (3.74 g., 0.020 mole) is suspended in 10 ml. of water, and sufficient $1N$ NaOH (about 20 ml.) is added to the well-stirred mixture to dissolve the acid and bring the pH to 9. Palladium oxide (0.2–1.0 g., Note 9) is suspended in 20 ml. of water in a 200-ml. glass hydrogenation bottle. The bottle is attached to a hydrogenation apparatus such as that of Adams and Voorhees,[2] and the suspension is shaken with hydrogen under a pressure of 1–3 atm. until the palladium oxide is reduced. The solution of 2-nitrobenzenesulfinic acid is added, and the mixture is shaken under a hydrogen pressure of 1–3 atm. The solution becomes completely decolorized in 2–6 hours, during which time about 95% of the calculated amount of hydrogen is absorbed. The catalyst is separated by filtration and washed with two 20-ml. portions of water, which are added to the filtrate. The filtrate, which may have a yellowish color, is evaporated to dryness under reduced pressure (bath temperature 45°). The residue, a white or light yellow solid, is sodium 2-aminobenzenesulfinate. After being dried in a desiccator over calcium chloride, it weighs 3.05–3.20 g. (85–89%).

C. *1,2,3-Benzothiadiazole 1,1-dioxide. Caution! 1,2,3-Benzo-thiadiazole 1,1-dioxide in the solid state can explode spontaneously, particularly on being warmed, jolted, or scratched. For most purposes it need not be isolated, but can be used in solutions, which are relatively safe. Any operations involving the solid material should be done very carefully, using good shielding.*

A solution of 1.43 g. (0.0080 mole) of sodium 2-aminobenzene-sulfinate in the least possible amount of water is combined with a solution of 0.55 g. (0.0080 mole) of sodium nitrite in the least amount of water. A mixture of 16 ml. of $2N$ sulfuric acid and 22 ml. of glycerol is placed in a 250-ml. three-necked flask equipped with a dropping funnel, a low-temperature thermometer, and a stirrer, and the flask is immersed in a bath of acetone and dry ice. The stirred mixture is cooled to $-15°$, and the solution of sodium 2-aminobenzenesulfinate and sodium nitrite is added dropwise over a period of about 5 minutes; the cooling and rate of addition are such as to maintain the temperature at $-15° \pm 3°$. The mixture is stirred for an additional 2 hours at this temperature, and 30 ml. of ether is added. The product is stirred vigorously for a few minutes and then allowed to warm to $-6°$ with gentle stirring. The ether layer is decanted or transferred by means of a chilled pipet into a vessel cooled in a dry ice bath, and the reaction mixture is again cooled to $-15°$. In this way the reaction mixture is extracted with five 20-ml. portions of ether. After the last extraction the aqueous layer is frozen solid and the ether layer is poured off. The combined extracts are dried at $-20°$, first over calcium chloride and then over phosphorus pentoxide; a cold room at $-20°$ is particularly convenient for this operation. The solution is transferred to a tared distillation flask immersed in an ice bath (Note 10), and the ether is removed by evaporation under reduced pressure. The flask is weighed rapidly and dried in a desiccator over phosphorus pentoxide at $-20°$ (*Caution!* Notes 11, 12). The residue is 1,2,3-benzothiadiazole 1,1-dioxide in the form of yellow-brown needles; weight 0.77–1.04 g. (57–77%). It explodes between 45° and 60° (Note 13).

2. Notes

1. This is essentially the method of J. Lange.[3]

2. Technical material of Badische Anilin & Soda-Fabrik is satisfactory.

3. To detect nitrous acid, a drop of the mixture is diluted with water and tested with starch iodide paper.

4. It is convenient to condense sulfur dioxide from a cylinder in a calibrated trap cooled in a dry ice bath.

5. After the procedure had been checked, the submitters recommended the following time-saving modification. The methylene chloride-ether solution of the sulfinic acid is extracted with one 80-ml. portion and two 35-ml. portions of $2N$ sodium hydroxide solution. The extracts are combined and the sulfinic acid is precipitated with 5-ml. portions of $6N$ hydrochloric acid as described in the text.

6. An excess of ammonia leads to products that are contaminated with ammonium chloride.

7. In this way one avoids an excess of hydrochloric acid which, if it adheres to the product, causes its gradual decomposition.

8. The checkers dissolved the crude acid in the minimum amount of $2N$ sodium hydroxide (about 3 ml./g.) and reprecipitated it in 5 portions with $2N$ hydrochloric acid; recovery 75–85%. Alternatively, they added the acid to boiling ethyl acetate (9 ml./g.), added decolorizing carbon to the solution, boiled the mixture for 5 minutes, separated the carbon by filtration, and cooled the hot filtrate; recovery 45–55%. The checkers found no difference in the infrared spectra of material purified in the two ways, but recrystallized material was reduced more quickly by hydrogen.

9. The submitters used 0.2 g. of palladium oxide prepared by the method of Shriner and Adams [4] and required 2 hours for complete hydrogenation under a hydrogen pressure of 1 atm. The checkers used 1.0 g. of palladium oxide (75.7%) from

Engelhard Industries, 113 Astor Street, Newark, New Jersey, and required 4 hours for complete hydrogenation under a hydrogen pressure of 2–3 atm. Conditions should be chosen to give complete hydrogenation within 6 hours or colored by-products may be formed.

10. Not quite half of the flask should dip into the ice water or the layer of ice forming on the flask may be hard to remove.

11. 1,2,3-Benzothiadiazole 1,1-dioxide slowly decomposes even at 0°; hence it should always be used on the day on which it is made. For most purposes it is not necessary to isolate the dioxide; the ether solution can be used, or solutions in other solvents can be prepared by adding the other solvent and distilling off the ether under reduced pressure (bath temperature 0°). In this way larger amounts of the dioxide than are described in this procedure can be handled without danger.

12. 1,2,3-Benzothiadiazole 1,1-dioxide can be conveniently assayed and characterized without isolation by forming its adduct with cyclopentadiene.[5] The following procedure illustrates characterization; for assay the same procedure can be applied to an aliquot, with all amounts scaled down in proportion. The dried ether extract of 1,2,3-benzothiadiazole 1,1-dioxide prepared from 1.43 g. (0.0080 mole) of sodium 2-aminobenzenesulfinate is concentrated to about 20 ml. at 0°, and 20 ml. of acetonitrile at −20° is added. Twenty milliliters of cold, freshly prepared cyclopentadiene [6] is added. The mixture is kept overnight at −10° to 0°. Solvent and excess cyclopentadiene are removed by evaporation at 0° under reduced pressure to leave 1.20–1.28 g. (64–68% based on sodium 2-aminobenzenesulfinate) of crude 1:1 adduct, m.p. 87° (dec.). For purification it is dissolved in 20 ml. of methylene chloride, 70 ml. of ether is added, and the solution is kept at −70°. Adduct decomposing at 90° crystallizes; recovery is about 75%. From pure, crystalline 1, 2, 3-benzothiadiazole 1,1-dioxide the yield of adduct is 92–98%.

13. Purer product can be obtained by reducing 1,2,3-benzothiadiazole 1,1-dioxide with zinc and acetic acid to 1,2,3-benzothiadiazoline 1,1-dioxide, which is oxidized back with lead tetraacetate.[5]

3. Methods of Preparation

1,2,3-Benzothiadiazole 1,1-dioxide has been prepared only by the present method.[5]

4. Merits of the Preparation

1,2,3-Benzothiadiazole 1,1-dioxide decomposes smoothly in solution at 10° to give dehydrobenzene ("benzyne"), nitrogen, and sulfur dioxide.[5, 7] In this way, as well as by the thermal

decomposition of benzenediazonium-2-carboxylate,[8, 9] it is possible to obtain dehydrobenzene in the absence of organometallic or strongly alkaline reagents; for this reason the choice of the reaction partner for dehydrobenzene is hardly limited at all. Compared to dry benzenediazonium-2-carboxylate, 1,2,3-benzothiadiazole 1,1-dioxide possesses the following advantages as a source of dehydrobenzene: the explosive compound does not need to be isolated and the decomposition temperature is lower. Solvent wet benzenediazonium 2 carboxylate, being insoluble in most organic media, is less generally useful than 1,2,3-benzothiadiazole 1,1-dioxide but is more convenient to prepare.[9] Because of their special reaction conditions, other methods of obtaining dehydrobenzene without using an organometallic compound [10] are not so generally applicable. Earlier volumes of *Organic Syntheses* illustrate the preparation of dehydrobenzene by the action of magnesium on *o*-fluorobromobenzene [11] and a type of ring closure in which a dehydrobenzene is an intermediate.[12] Methods of generating dehydrobenzenes and the reactions of these reactive substances were recently reviewed.[13]

1. Institut für Organische Chemie, Universität Heidelberg, Heidelberg, Germany.
2. R. Adams and V. Voorhees, *Org. Syntheses*, Coll. Vol. **1**, 61 (1941).
3. J. Lange, Dissertation, Universität Marburg, 1951; Houben-Weyl, *Methoden der Organischen Chemie*, **9**, 323 (1955), Georg Thieme Verlag, Stuttgart, Germany.

4. R. L. Shriner and R. Adams, *J. Am. Chem. Soc.*, **46**, 1683 (1924); D. Starr and R. M. Hixon, *Org. Syntheses*, Coll. Vol. **2**, 566 (1943).
5. G. Wittig and R. W. Hoffmann, *Ber.*, **95**, 2718 (1962).
6. M. Korach, D. R. Nielsen, and W. H. Rideout, *Org. Syntheses*, **42**, 50 (1962); R. B. Moffett, *Org. Syntheses*, Coll. Vol. **4**, 238 (1963); G. Wilkinson, *Org. Syntheses*, Coll. Vol. **4**, 473 (1963).
7. G. Wittig and R. W. Hoffmann, *Ber.*, **95**, 2729 (1962).
8. M. Stiles and R. G. Miller, *J. Am. Chem. Soc.*, **82**, 3802 (1960); R. S. Berry, G. N. Spokes, and M. Stiles, *J. Am. Chem. Soc.*, **84**, 3570 (1962).
9. L. Friedman and F. M. Logullo, *J. Org. Chem.*, **34**, 3089 (1969).
10. G. Köbrich, *Ber.*, **92**, 2985 (1959); H. E. Simmons, *J. Org. Chem.*, **25**, 691 (1960); G. Wittig and H. F. Ebel, *Ann.*, **650**, 20 (1961).
11. G. Wittig, *Org. Syntheses*, Coll. Vol. **4**, 964 (1963).
12. J. F. Bunnett, B. F. Hrutfiord, and S. M. Williamson, this volume, p. 12.
13. R. W. Hoffmann, "Dehydrobenzene and Cycloalkynes," Academic Press, New York, 1967.

BENZOYL FLUORIDE*

$$C_6H_5COCl + HF \rightarrow C_6H_5COF + HCl$$

Submitted by GEORGE A. OLAH and STEPHEN J. KUHN [1]
Checked by JOHN A. DUPONT and WILLIAM D. EMMONS

1. Procedure

Caution! Anhydrous hydrogen fluoride is toxic and in contact with skin can cause serious burns. This preparation should be carried out in a well-ventilated hood. Rubber gloves and safety goggles should be worn by the operator. In case of contact with hydrogen fluoride wash the affected skin area immediately with copious amounts of water, and apply a calcium gluconate paste (Note 1).

Hydrogen fluoride (50 g., 2.5 moles) is distilled from the cylinder through a polyethylene tube into a 250-ml. polyethylene transfer bottle which has been previously weighed and calibrated. A vent is provided during this process by inserting a large-gauge hypodermic needle through the bottle cap. No provision against atmospheric moisture is necessary. The bottle is cooled in a dry ice-acetone bath, and 45–50 ml. of liquid hydrogen fluoride is collected. The amount of liquid obtained can be determined by weight difference; however, since an excess of hydrogen fluoride is employed, the exact weight need not be determined. The time required for collection of the hydrogen fluoride can be appreciably shortened by placing the cylinder in a pan of warm water.

The reaction itself is carried out in a 1-l. polyolefin bottle (Note 2) or fused silica flask (Note 3) fitted with an inlet tube (Note 4) leading to the bottom of the reaction vessel and a reflux condenser which is connected to a hydrogen chloride absorber or which leads directly to the hood. A condenser suitable for work with anhydrous hydrogen fluoride can easily be prepared from a glass-jacketed polyolefin, Teflon®, silica, or copper tube (Note 5).

Benzoyl chloride (281 g., 2.0 moles) is placed in the reaction vessel, and the hydrogen fluoride gas is then introduced by its distillation from the transfer bottle through the inlet tube. Prior to this distillation the hypodermic needle is closed off by a metal cap. The hydrogen fluoride is added over a period of approximately 1 hour. Generally, external cooling is not needed, as the evaporating hydrogen chloride cools the reaction mixture. When the addition is completed, the reaction mixture is warmed to 30–40° and kept at this temperature for 1 hour. The mixture is then washed in an ordinary glass separatory funnel (Note 6) with 500 ml. of ice water in which 12.5 g. (0.2 mole) of boric acid is dissolved (Note 7). The organic layer is quickly separated, and to it are added 10 g. of anhydrous sodium fluoride and 10 g. of anhydrous sodium sulfate (Note 7). The mixture is allowed to stand for 30 minutes and is then filtered and distilled through a short Vigreux column. The yield of benzoyl fluoride, b.p. 159–161°, n^{15}D 1.4988 (Note 8), is 187–200 g. (75–80%).

2. Notes

1. An alternative treatment which has been used with good results at Rohm and Haas Company is, after thoroughly washing the exposed area with tap water, to soak the burned area in an ice-cold 0.2% solution of Hyamine 1622 (a product of Rohm

and Haas Company) in 70% aqueous ethanol for 1 hour. It has also been stated that soaking the affected area with ice and water for 1 hour is almost as effective.[2]

2. Polyolefin bottles of suitable size are commercially available. One inconvenience occasionally observed with bottles which have not previously been in contact with hydrogen fluoride is the formation of a slight pink color in the reaction mixture, possibly due to the plasticizers. This coloration does not affect either the yields or the purity of the product, however, because the color is generally eliminated after the product is washed and treated with sodium fluoride.

3. No color problem exists when fused silica equipment, preferably with normal joints lubricated with a fluorinated grease, is used.

4. The inlet tube can be either polyolefin, Teflon®, fused silica, or copper.

5. Silica or copper gives much better heat transfer than do plastic tubes. The checkers found, however, that the use of a condenser was superfluous and that substitution of a simple polyethylene tube long enough to vent the off-gas away from the operator and apparatus was quite satisfactory.

6. Although some slight etching can take place, at this stage glass equipment is entirely safe, and no contamination of the product occurs.

7. The crude product contains hydrogen fluoride which is removed by the addition of boric acid to the wash water ($H_3BO_3 + 4HF \rightarrow HBF_4 + 3H_2O$). The sodium fluoride disposes of any hydrogen fluoride remaining in the benzoyl fluoride ($NaF + HF \rightarrow NaHF_2$).

8. Benzoyl fluoride is a potent lachrymator and is undoubtedly toxic. It is advisable to rinse all glassware with acetone followed by 10% aqueous ammonia before removing the glassware from the hood.

3. Methods of Preparation

Benzoyl fluoride can also be prepared by the reaction of anhydrous hydrogen fluoride [3-5] or potassium fluoride [6] with benzoic anhydride and by the halogen exchange of benzoyl chloride with alkali fluorides, such as NaF,[7] KF,[6] KHF_2,[8] Na_2SiF_6,[9] or various other metal fluorides.[10]

4. Merits of the Preparation

The described procedure, first applied by Colson and Fredenhagen,[3,4] is useful for the preparation of a wide variety of acyl fluorides.[5] The yields are normally 80–90%. Some examples of acyl fluorides prepared are listed in Table I. Benzoyl fluoride can also be employed as a convenient source of acetyl fluoride by reaction with acetic acid.[11]

TABLE I

Product	B.P., °C.
Propionyl fluoride	43
n-Butyryl fluoride	69
Isobutyryl fluoride	61
Valeryl fluoride	90
Isovaleryl fluoride	81
Caproyl fluoride	122
Heptanoyl fluoride	40 (15 mm.)
Octanoyl fluoride	62 (15 mm.)
Pelargonyl fluoride	81 (15 mm.)
Decanoyl fluoride	92 (15 mm.)
Fluoroacetyl fluoride	54
Chloroacetyl fluoride	77
Dichloroacetyl fluoride	85
Trichloroacetyl fluoride	67
Bromoacetyl fluoride	104
Phthaloyl fluoride	84 (15 mm.)
Phenylacetyl fluoride	85 (15 mm.)

1. Contribution No. 78 from the Exploratory Research Laboratory, Dow Chemical of Canada, Limited, Sarnia, Ontario.
2. B. C. McKusick, E. I. du Pont de Nemours and Co., Wilmington 98, Delaware, private communication.
3. A. Colson, *Bull. Soc. Chim. France*, [3] **17**, 55 (1897); *Ann. Chim. (Paris)*, [7] **12**, 255 (1897).
4. K. Fredenhagen, *Z. Physik. Chem. (Leipzig)*, **A164**, 189 (1933); K. Fredenhagen and G. Cadenbach, *Z. Physik. Chem. (Leipzig)*, **A164**, 201 (1933).
5. G. A. Olah and S. J. Kuhn, *J. Org. Chem.*, **26**, 237 (1961).
6. A. I. Mashentsev, *J. Gen. Chem. USSR (Engl. Transl.)*, **11**, 1135 (1941) [*C.A.*, **37**, 2716 (1943)]; *J. Gen. Chem. USSR (Engl. Transl.)*, **15**, 915 (1945) [*C.A.*, **40**, 6443 (1946)].

7. G. W. Tullock and D. D. Coffman, *J. Org. Chem.*, **25**, 2016 (1960).
8. G. A. Olah, S. Kuhn, and S. Beke, *Ber.*, **89**, 862 (1956).
9. J. Dahmlos, *Angew. Chem.*, **71**, 274 (1959).
10. A. M. Lovelace, D. A. Rausch, and W. Postelnek, "Aliphatic Fluorine Compounds," Reinhold Publishing Co., New York, 1958.
11. G. A. Olah, S. J. Kuhn, W. S. Tolgyesi, and E. B. Baker, *J. Am. Chem. Soc.*, **84**, 2733 (1962).

3-BENZOYLOXYCYCLOHEXENE

(2-Cyclohexen-1-ol, benzoate)

$$\text{(cyclohexene)} + C_6H_5\overset{O}{\overset{\|}{C}}OOC(CH_3)_3 \xrightarrow{Cu^+/Cu^{2+}} \text{(3-benzoyloxycyclohexene, } OCOC_6H_5\text{)} + (CH_3)_3COH$$

Submitted by Knud Pedersen, Preben Jakobsen, and Sven-Olov Lawesson [1]
Checked by R. Schöllhorn and R. Breslow

1. Procedure

Caution! This reaction should be carried out behind a safety screen. The solvent removal and product distillation steps should also be carried out behind a screen to minimize danger due to contamination of the product with undetected peroxides.

A 250-ml., three-necked, round-bottomed flask equipped with a sealed mechanical stirrer, a reflux condenser, and a pressure-equalizing dropping funnel is set up for conducting a reaction in an atmosphere of nitrogen by fitting into the top of the condenser a T-tube attached to a low-pressure supply of nitrogen and to a mercury bubbler. In the flask are placed 41 g. (0.50 mole) of cyclohexene and 0.05 g. (0.00035 mole) of cuprous bromide, and the mixture is heated in an oil bath at 80–82°. When the temperature of the mixture reaches that of the oil bath, 40 g. (0.21 mole) of *t*-butyl perbenzoate (Note 1) is added dropwise with stirring over a 1-hour period, during which the color of the now homogeneous solution becomes blue. Stirring and heating are continued for an additional 3.5 hours (Note 2). The cooled reaction mixture is washed with two 50-ml. portions of dilute aqueous sodium carbonate to remove benzoic acid (Note 3). The remaining organic phase is washed with water until neutral and dried over anhydrous sodium sulfate. The excess of cyclohexene is removed by distillation under reduced pressure, and the residue (Note 4) is distilled through a short Vigreux column to give 29–33 g. (71–80%) of 3-benzoyloxycyclohexene, b.p. 97–99° (0.15 mm.), n^{20}D 1.5376–1.5387 (Note 5).

2. Notes

1. t-Butyl perbenzoate is supplied by Lucidol Division, Wallace and Tiernan, Inc., Buffalo 5, New York, or L. Light and Co., Ltd., Colorbrook, Bucks, England. The Lucidol product contains 98% t-butyl perbenzoate.

2. The progress of the reaction can most conveniently be followed by periodic examination of the infrared spectrum of the mixture ($\nu_{C=O}$ for peroxybenzoate: 1775 cm.$^{-1}$). After all of the perester has been added, $ca.$ 3 hours is required for its consumption.

3. After acidification of the aqueous phase 1.5–2 g. of benzoic acid can be isolated.

4. It is recommended that an infrared spectrum be run on the residue before the distillation to check for the absence of perester (see Note 2).

5. The same yield is obtained when the scale is increased threefold.

3. Methods of Preparation

The procedure is that of Kharasch, Sosnovsky, and Yang.[2]

4. Merits of the Preparation

The reaction described is of considerable general utility for the preparation of benzoyloxy derivatives of unsaturated hydrocarbons.[2-8] Reactions of t-butyl perbenzoate with various other classes of compounds in the presence of catalytic amounts of copper ions produce benzoyloxy derivatives. Thus this reaction can also be used to effect one-step oxidation of saturated hydrocarbons,[9, 10] esters,[5, 11] dialkyl and aryl alkyl ethers,[12-14] benzylic ethers,[11, 15] cyclic ethers,[13, 16] straight-chain and benzylic sulfides,[12, 17-19] cyclic sulfides,[11, 19] amides,[11] and certain organosilicon compounds.[20] Reviews[20,21] of these reactions are available.

1. Department of Chemistry, University of Aarhus. 8000 Aarhus C., Denmark.
2. M. S. Kharasch, G. Sosnovsky, and N. C. Yang, *J. Am. Chem. Soc.*, 81, 5819 (1959).
3. D. B. Denney, D. Z. Denney, and G. Feig, *Tetrahedron Lett.*, No. 15, 19 (1959).
4. A. L. J. Beckwith and G. W. Evans, *Proc. Chem. Soc.*, 63 (1962).
5. G. Sosnovsky and N. C. Yang, *J. Org. Chem.*, 25, 899 (1960).
6. B. Cross and G. H. Whitham, *J. Chem. Soc.*, 1650 (1961).
7. J. K. Kochi, *J. Am. Chem. Soc.*, 84, 774 (1962).
8. D. Z. Denney, A. Appelbaum, and D. B. Denney, *J. Am. Chem. Soc.*, 84, 4969 (1962).
9. M. S. Kharasch and A. Fono, *J. Org. Chem.*, 23, 324 (1958).
10. T. I. Wang, Ph.D. Thesis, University of Chicago, 1962.
11. C. Berglund and S.-O. Lawesson, *Arkiv Kemi*, 20, 225 (1963).
12. S.-O. Lawesson, C. Berglund, and S. Grönwall, *Acta Chem. Scand.*, 15, 249 (1961).
13. G. Sosnovsky, *Tetrahedron*, 13, 241 (1961).
14. S.-O. Lawesson and C. Berglund, *Arkiv Kemi*, 17, 465 (1961).
15. S.-O. Lawesson and C. Berglund, *Arkiv Kemi*, 16, 287 (1960).
16. S.-O. Lawesson and C. Berglund, *Arkiv Kemi*, 17, 475 (1961).
17. S.-O. Lawesson and C. Berglund, *Acta Chem. Scand.*, 15, 36 (1961).
18. G. Sosnovsky, *Tetrahedron*, 18, 15 (1962).
19. G. Sosnovsky, *Tetrahedron*, 18, 903 (1962).
20. G. Sosnovsky and S.-O. Lawesson, *Angew. Chem.*, 76, 218 (1964).
21. S.-O. Lawesson and G. Schroll, in S. Patai, "The Chemistry of Carboxylic Acids and Esters," Wiley-Interscience, London, 1969, p. 669.

N-BENZYLACRYLAMIDE*

(Acrylamide, N-benzyl-)

$$CH_2=CHCN + \langle\ \rangle CH_2OH \xrightarrow{H_2SO_4}$$

$$\langle\ \rangle CH_2NHCOCH=CH_2$$

Submitted by CHESTER L. PARRIS.[1]
Checked by WILLIAM E. PARHAM, WAYLAND E. NOLAND, and
JOAN M. WEINMANN.

1. Procedure

In a 1-l., three-necked, round-bottomed flask equipped with a sealed Hershberg stirrer,[2] a 125-ml. dropping funnel, and a thermometer is placed 200 g. (250 ml., 3.78 moles) of acrylonitrile (Note 1). The flask is immersed in an ice-water bath, and then 75 ml. of concentrated sulfuric acid is added dropwise over a period of about 1 hour while the temperature is maintained at 0–5°. From a clean dropping funnel (Note 2), 108.1 g. (105 ml., 1.0 mole) of benzyl alcohol (Note 3) is added dropwise over about 1 hour at the same temperature. The clear, yellow mixture is held below 5° for about 3 hours longer and is then allowed to warm slowly to room temperature. After 2 days of stirring at room temperature the mixture is poured into a 2-l. separatory funnel containing about 1 l. of water and chopped ice. The mixture is shaken thoroughly and the resulting oil is taken up with 200 ml. of ethyl acetate. The aqueous phase is separated and extracted twice more with 200-ml. portions of solvent. The organic extracts are combined and washed successively with four 250-ml. portions of saturated sodium chloride solution, four 250-ml. portions of saturated sodium bicarbonate solution, and again with four portions of the salt solution. The neutral ethyl acetate extract is dried over 20 g. of anhydrous magnesium sulfate and filtered. The filtrate is concentrated and the residue is distilled under reduced pressure. A fore-run of 1–3 g. of semisolid is obtained up to 120°/0.02 mm. The product is then collected as a light–yellow oil, b.p. 120–130°/0.01–0.02 mm., which solidifies in the chilled receiver. The distillate (97–101 g.) is melted on a steam bath and dissolved in a mixture of 50 ml. of benzene and 50 ml. of hexane. The solution is transferred quantitatively to a 500-ml. Erlenmeyer flask and the solvent evaporated on a steam bath. The oily residue is placed in a refrigerator for at least 1 day to ensure complete crystallization. The white solid is transferred to a Büchner funnel with the aid of a little ice-cold hexane. After drying in air, the yield is 95–100 g. (59–62%) of N-benzylacryl-amide, m.p. 65–68° (Note 4).

2. Notes

1. Commercial acrylonitrile from American Cyanamid Company was redistilled, b.p. 77–78°.
2. The dropping funnel which has been wetted with sulfuric acid should be washed and thoroughly dried, or a fresh funnel employed.
3. Benzyl alcohol from Fisher Scientific Company was used without further purification.
4. The product is of suitable purity for further reactions. It may be obtained analytically pure, m.p. 70–72°, by recrystallization from benzene. The reported melting point is 69°.[3]

3. Methods of Preparation

N-Benzylacrylamide has been prepared by dehydrohalogenation of N-benzyl-β-chloropropionamide with aqueous potassium hydroxide,[3] and by the reaction of acetylene with carbon monoxide and benzylamine.[4] The procedure described is the method of Parris and Christenson.[5]

4. Merits of Preparation

The alternative methods of preparation of N-benzylacrylamide are reported in patents, and no yields are given. One of them requires two steps and costlier intermediates; the other appears to be more suitable for plant than for laboratory preparation. The procedure presented involves a simple, one-step reaction taking place under mild conditions, employing inexpensive reactants and affording satisfactory yields.

The N-alkylation of nitriles with aralkyl alcohols, a special case of the Ritter reaction,[6] is a novel general reaction. The following compounds were prepared by this procedure in the corresponding yields: N-benzylacetamide (48%), N-(2,4-dimethylbenzyl)-acetamide (40%), N-(4-methoxybenzyl)-acetamide (60%), N,N'-diacrylyl-p-xylene-α-α'-diamine (64%), N,N'-diacetyl-4,6-dimethyl-m-xylene-α,α'-diamine (62%). Another example of the Ritter reaction is given elsewhere in this volume.[7]

The title compound is of special interest and utility as a polymerizable monomer.

1. Pittsburgh Plate Glass Company, Springdale, Pa.
2. E. B. Hershberg, *Ind. Eng. Chem. Anal. Ed.,* 8, 313 (1936).
3. G. Kranzlein and M. Corell (to I. G. Farbenindustrie, Akt.-Ges.), Ger. pat. 752,481 (Nov. 10, 1952).
4. E. H. Specht, A. Neuman, H. T. Neher (to Rohm and Haas Co.), U. S. pat. 2,773,063 (Dec. 4, 1956).
5. C. L. Parris and R. M. Christenson, *J. Org. Chem.,* 25, 331, 1888 (1960).
6. L. I. Krinsen and D. J. Cota, *Org. Reactions,* 17, 213 (1969).
7. J. J. Ritter and J. Kalish, this volume, p. 471.

2-BENZYLCYCLOPENTANONE

Submitted by FRITZ ELSINGER [1]
Checked by WILLIAM G. DAUBEN and W. TODD WIPKE

1. Procedure

A. *2-Benzyl-2-carbomethoxycyclopentanone.* A dry 2-l. three-necked flask is fitted with a Vibromischer stirrer (Note 1), a reflux condenser, and a 250-ml. dropping funnel with a pressure-equalizing side tube.[2] A nitrogen-inlet tube is connected to the top of the dropping funnel, and an outlet tube is placed on the top of the condenser and connected to a mercury valve. The latter consists of a U-tube the bend of which is just filled with mercury.

To the flask are added 13.4 g. (0.58 mole) of clean sodium and 200 ml. of absolute toluene. The Vibromischer stirrer is activated, the toluene heated to reflux, and the agitation continued at this temperature until all the sodium is pulverized into a very fine sand. The agitation is ceased, and the solution is allowed to cool to room temperature. The nitrogen flow rate is increased, the Vibromischer stirrer is replaced with a conventional sealed mechanical stirrer with a Teflon® blade, and a solution of 85 g. (0.6 mole) of 2-carbomethoxycyclopentanone (Note 2) in 450 ml. of absolute benzene is placed in the addition funnel.

The stirrer is started, and the solution in the addition funnel is added over a 2-hour period without external heating (Note 3).

After the addition is complete, the mixture is heated under reflux for 2.5 hours, at the end of which time the mixture has a pasty consistency. A solution of 106 g. (0.84 mole) of benzyl chloride in 100 ml. of dry benzene is added in one portion, the mixture heated under reflux for 14 hours, and the solution (Note 4) poured into 600 ml. of water. The benzene layer is separated, the aqueous layer extracted twice with ether, and the combined benzene-ether extract washed with 100 ml. of water and dried over anhydrous sodium sulfate. The solvent is removed under reduced pressure using a rotary evaporator, and the residual liquid distilled to yield 108–116 g. (81–86%) of colorless 2-benzyl-2-carbomethoxycyclopentanone, b.p. 126–128° (0.5 mm.) (Note 5).

B. *2-Benzylcyclopentanone.* A mixture of 30 g. (0.177 mole) of lithium iodide dihydrate (Notes 6 and 7) and 140 ml. of dry 2,4,6-collidine (Note 8) in a 300-ml. three-necked flask fitted with a dropping funnel, a reflux condenser, and a nitrogen-inlet system (as in step A) is heated to reflux. As soon as all the lithium iodide has dissolved (Note 9), 30 g. (0.129 mole) of 2-benzyl-2-carbomethoxycyclopentanone dissolved in 30 ml. of 2,4,6-collidine (Note 10) is added to the boiling, faintly yellow solution; and during this process the solution turns darker in color and a precipitate forms (Note 11). Evolution of carbon dioxide begins immediately, and its formation can be followed by passing the nitrogen flush through a saturated barium hydroxide solution. The mixture is heated under reflux and a nitrogen atmosphere for 19 hours, at the end of which time the evolution of carbon dioxide is very slow (Note 12).

The mixture is cooled and poured onto a mixture of 200 ml. of 6N hydrochloric acid, 200 ml. of ether, and 100 g. of ice. The residue in the flask is dissolved in a mixture of 6N hydrochloric acid and methylene chloride, and this mixture is added to the main reaction. The aqueous layer is separated and extracted with two 100-ml. portions of ether. The combined ethereal solution is washed once with 70 ml. of 6N hydrochloric acid, once with 2N sodium carbonate solution, twice with saturated sodium chloride solution, and dried over anhydrous sodium sulfate. The solvent is removed under reduced pressure, and the residue is distilled to yield 16–17 g. (72–76%) of colorless 2-benzylcyclo-pentanone, b.p. 83–85° (0.3 mm.), 108–110° (0.75 mm.) (Note 13).

2. Notes

1. This stirring apparatus is available from Ag. für Chemie Apparatebau, Mannedorf, Zurich, Switzerland.

2. The submitter prepared the material from dimethyl adipate following the procedure published by Pickney [3] for the diethyl ester. The checkers obtained their material by fractional distillation of mixed carbomethoxy- and carbethoxycyclo-pentanone available from Arapahoe Chemical Co., Boulder, Colorado.

3. If the 2-carbomethoxycyclopentanone is added in one portion, the yield of the product drops to 67%.

4. At the end of the reflux period, the reaction mixture is a nonviscous solution containing a white precipitate.

5. The semicarbazone melts at 168–170°.

6. Lithium iodide dihydrate is available from Fluka A.G., Buchs, S.G., Switzerland. The checkers used the trihydrate and, by means of a Dean Stark trap [4] attached between the flask and the condenser, 1 mole of water was removed via azeotropic distillation with collidine.

7. In cases where a carbomethoxy group is desired to be selectively cleaved in the presence of a readily hydrolyzed ester group, such as an acetate of a secondary alcohol, anhydrous lithium iodide must be employed.[5] In order to avoid partial decomposition of the salt to iodine, it is best dried by slowly heating it to 150° in a high vacuum. The solubility of anhydrous lithium iodide in boiling collidine or lutidine is slightly less than that of the dihydrate, but it still is adequate for the reaction. In the present case, the use of the anhydrous salt lowers the yield of the 2-benzylcyclopentanone to 67%, and a large amount of a product, believed to be a dimer, boiling around 200° (0.5 mm.) is obtained.

8. For the cleavage of less hindered esters, the lower-boiling 2,6-lutidine (b.p. 143°) can be used as the solvent.

9. The development of a small amount of iodine is difficult to

avoid. The nitrogen atmosphere is essential to keep this salt decomposition to a minimum.

10. Methyl esters react more rapidly with lithium iodide than do ethyl esters, which in turn react more rapidly than esters of secondary alcohols. On the other hand, t-butyl esters are cleaved very readily with a catalytic amount of lithium iodide.

11. A precipitate remains throughout the reaction.

12. By using three mole equivalents of lithium iodide dihydrate, at the end of 6.5 hours of reflux a 77% yield of 2-benzylcyclopentanone is obtained.

13. The semicarbazone melts at 204–205°.

3. Methods of Preparation

This preparation of 2-benzyl-2-carbomethoxycyclopentanone is based on a procedure described by Baker and Leeds [6] for the ethyl ester, and the methyl ester has not been previously prepared. The ethyl ester, also, has been prepared by the alkylation of 2-carbethoxycyclopentanone with benzyl chloride in the presence of potassium hydroxide in acetaldehyde dipropylacetal.[7] The preparation and isolation of the potassium salt of 2-carbethoxycyclopentanone can be readily achieved in a very simple way using aqueous alcoholic potassium hydroxide; by reaction of this salt with a variety of different halides in anhydrous media many 2-alkyl-2-carbethoxycyclopentanones have been prepared.[8, 9]

The preparation of 2-benzylcyclopentanone from 2-benzyl-2-carbomethoxycyclopentanone has not been previously reported. Starting with the ethyl ester, however, the compound has been prepared by heating the ester for many hours with concentrated hydrochloric acid.[6, 10] The direct alkylation of cyclopentanone with benzyl chloride in the presence of sodium amide in liquid ammonia goes only in a poor yield.[11]

4. Merits of the Preparation

This procedure illustrates a general method for the selective splitting of a carbomethoxy group in the presence of easily hydrolyzed esters of other alcohols, such as the easily hydrolyzed equatorial acetoxy group. The specificity of the reaction is not affected by steric hindrance, and a highly hindered methyl ester can be split in the presence of other less hindered esters of second-ary alcohols. Normal alkaline saponification goes in exactly the opposite way.

The present case simply illustrates another utility of the ester cleavage reaction, *i.e.*, the cleavage of a β-keto ester with con-comitant decarboxylation under only slightly basic conditions. The method should be particularly applicable to systems which are prone to undergo reverse Claisen reactions.

1. Oesterr. Institut für Hämoderivate, Industriestrasse 72, Wien XXII, Austria.
2. K. B. Wiberg, "Laboratory Technique in Organic Chemistry," McGraw-Hill Book Company, Inc., New York, 1960, p. 207.
3. P. S. Pickney, *Org. Syntheses*, Coll. Vol. 2, 119 (1943).
4. See ref. 2, p. 215.
5. F. Elsinger, J. Schreiber, and A. Eschenmoser, *Helv. Chim. Acta*, **43**, 113 (1960).
6. W. Baker and W. G. Leeds, *J. Chem. Soc.*, 974 (1948).
7. Ch. Weizmann, E. Bergmann, and M. Sulzbacher, *J. Org. Chem.*, **15**, 918 (1950).
8. R. Mayer and E. Alder, *Ber.*, **88**, 1866 (1955).
9. R. Mayer, G. Wenschuh, and W. Töpelmann, *Ber.*, **91**, 1616 (1958).
10. W. Treibs, R. Mayer, and M. Madejski, *Ber.*, **87**, 356 (1954).
11. T. M. Harris and C. R. Hauser, *J. Am. Chem. Soc.*, **81**, 1160 (1959).

α-BENZYLIDENE-γ-PHENYL-$\Delta^{\beta,\gamma}$-BUTENOLIDE

(Cinnamic acid, α-(β-hydroxystyryl)-, γ-lactone)

$$C_6H_5CHO \ + \ \underset{\underset{CH_2COC_6H_5}{|}}{CH_2COOH} \ \xrightarrow[CH_3CO_2Na]{(CH_3CO)_2O} \ C_6H_5CH{=}C{-}C{=}O \ + \ 2H_2O$$

Submitted by Robert Filler, Edmund J. Piasek, and Hans A. Leipold.[1]
Checked by S. Trofimenko and B. C. McKusick.

1. Procedure

The apparatus consists of a 200-ml., three-necked, round-bot-tomed flask fitted with thermometer, reflux condenser, and gas-inlet tube. The flask is charged with 17.8 g. (0.10 mole) of 3-benzoylpropionic acid (Note 1), 10.6 g. (10.6 ml., 0.10 mole) of benzaldehyde, 61.3 g. (57 ml., 0.60 mole) of acetic anhydride, and 8.2 g. (0.10 mole) of powdered anhydrous sodium acetate (freshly fused). The flask is placed in an oil bath maintained at a temperature of 95–100° and is kept there for 2 hours while dry oxygen-free nitrogen is passed through the reaction mixture (Note 2). At the end of this time the flask is removed from the oil bath, and the hot solution is decanted from the sodium acetate into a 250-ml. Erlenmeyer flask. The solution is kept at 0–5° in a refrigerator for 1 hour, during which time α-benzylidene-γ-phenyl-Δ^{β,γ}-butenolide separates as an orange solid.

About 40 ml. of 95% ethanol is added to the contents of the flask, and the butenolide is brought into suspension by thoroughly breaking up all lumps with a spatula. The suspension is filtered with suction, and the filter cake is washed with 30 ml. of cold 95% ethanol and then with 100 ml. of boiling water to remove any sodium acetate present. The butenolide is obtained as a yellow solid, m.p. 149–154°, weight 11.1–12.4 g. (45–50%), after being dried overnight in a vacuum desiccator. This product, which is pure enough for most purposes, may be further purified by crystallization from 95% ethanol (Note 3).

2. Notes

1. 3-Benzoylpropionic acid [2] is available from Aldrich Chemical Co., Milwaukee, Wisconsin.

2. Oxygen is removed from the nitrogen gas by passing the latter through Brady solution, which consists of zinc amalgam, sodium hydroxide, and sodium anthraquinone-β-sulfonate.[3] It has been shown that oxidizing agents induce formation of a Pechmann dye, a deep red substance which is difficult to remove from the butenolide.[4]

3. About 75 ml. of ethanol is used for every gram of butenolide to be dissolved. Clarification of the solution with charcoal should be avoided because the butenolide tends to separate from solution during filtration and clogs the steam-jacketed funnel. The crystallized butenolide melts at 150–152°.

3. Discussion

α-Benzylidene-γ-phenyl-$\Delta^{\beta,\gamma}$-butenolide has been prepared by the condensation of benzaldehyde with 3-benzoylpropionic acid in the presence of acetic anhydride and sodium acetate[5,6] or in the presence of a mixture of dimethylformamide and sulfur trioxide.[7]   The butenolide has also been obtained by reaction of a-chloromethylene-γ-phenyl-$\Delta^{\beta,\gamma}$-butenolide with benzene in the presence of anhydrous aluminum chloride.[8]

The method described above may be used for the preparation of a wide variety of butenolides substituted in the arylidene ring with either electron-withdrawing or electron-releasing substituents.   γ-Lactones such as α-benzylidene-γ-phenyl-$\Delta^{\beta,\gamma}$-butenolide are isoelectronic with azlactones, but have received much less attention.   Like the azlactone ring, the butenolide ring may be opened readily by water, alcohols, or amines to form keto acids, keto esters, or keto amides.[9] a-Benzylidene-γ-phenyl-$\Delta^{\beta,\gamma}$-butenolide is smoothly isomerized by aluminum chloride to 4-phenyl-2-naphthoic acid[10] in 70% yield via intramolecular alkylation.   Grignard reagents add 1,4 to the α,β-unsaturated carbonyl system, with the lactone ring remaining intact,[11] while phenyllithium leads to ring opening and the formation of 1,1-diphenyl-2-phenacylcinnamyl alcohol.[11]   The butenolide gives reduced dilactones, on treatment with lithium aluminum hydride.[12]

1. Department of Chemistry, Illinois Institute of Technology, Chicago, Illionis 60616.
2. L. F. Somerville and C. F. H. Allen, *Org. Syntheses,* Coll. Vol. 2, 81 (1943).
3. L. J. Brady, *Anal. Chem.,* 20, 1033 (1948).
4. E. Klingsberg, *Chem. Rev.,* 54, 59 (1954).
5. W. Borsche, *Ber.,* 47, 1108, 2718 (1914).
6. F. W. Schueler and C. Hanna, *J. Am. Chem. Soc.,* 73, 3528 (1951).
7. E. Baltazzi and E. A. Davies, *Chem. Ind. (London),* 1653 (1962).
8. Y. S. Rao and R. Filler, *Chem. Ind. (London),* 280 (1964).
9. R. Filler and L. M. Hebron, *J. Am. Chem. Soc.,* 81, 391 (1959).
10. R. Filler and H. A. Leipold, *J. Org. Chem.,* 27, 4440 (1962).
11. R. Filler, E. J. Piasek, and L. H. Mark, *J. Org. Chem.,* 26, 2659 (1961).
12. R. Filler and E. J. Piasek, *J. Org. Chem.,* 28, 3400 (1963).

cis-2-BENZYL-3-PHENYLAZIRIDINE

(Aziridine, 2-benzyl-3-phenyl-, *cis*-)

$$C_6H_5CH_2 \underset{\underset{NOH}{\|}}{C} CH_2C_6H_5 \xrightarrow[\text{2. } H_2O]{\text{1. LiAlH}_4, \text{THF}}$$

$$\underset{H}{\overset{C_6H_5CH_2}{\diagdown}} \underset{\underset{H}{\overset{N}{\diagup}}}{C - C} \overset{C_6H_5}{\diagup} H \quad + \quad C_6H_5CH_2 \underset{\underset{NH_2}{|}}{CH} CH_2C_6H_5$$

$$+ \quad \text{LiOH} \quad + \quad \text{Al(OH)}_3 \quad + \quad H_2$$

Submitted by Katsumi Kotera and Keizo Kitahonoki[1]
Checked by Donald R. Strobach and R. E. Benson

1. Procedure

In a 1-l., four-necked, round-bottomed flask fitted with a sealed mechanical stirrer, a thermometer, a dropping funnel, and a reflux condenser protected from atmospheric moisture with a drying tube containing calcium chloride are placed 350 ml. of dry tetrahydrofuran (Note 1) and 3.80 g. (0.100 mole) of powdered lithium aluminum hydride (Note 2). The slurry is stirred while a solution of 11.27 g. (0.0500 mole) of dibenzyl ketoxime (Note 3) in 80 ml. of dry tetrahydrofuran is added dropwise with cooling at 20° over a 10-minute period. The contents of the flask are gradually heated to reflux (Note 4) with stirring in an oil bath at 90° (external temperature) for 3 hours (Note 5); at *ca.* 62° the color of the mixture turns from the initial pale green to a permanent, light chocolate color (reaction may be exothermic at this point). The mixture is cooled with ice water and decomposed by gradual addition of 12 ml. of water at a temperature below 20°. The precipitate is collected by filtration, washed with 100 ml. of ether, and added to 200 ml. of ether. This mixture is stirred for *ca.* 10 minutes and filtered, and the residue is washed with 100 ml.

of ether. The ethereal extracts and washings are combined with the original filtrate, dried over anhydrous sodium sulfate overnight, and concentrated with a rotary evaporator at 30° (20 mm.) to give 10.60–11.0 g. of a pale yellow oil (Note 6).

The product is dissolved in 100 ml. of petroleum ether, b.p. 30–40°, with warming, and the solution is transferred to a chromatographic column consisting of 75 g. of silica gel (Note 7). The product is eluted sequentially with (A) 300 ml. of petroleum ether, (B) 300 ml. of 3:1 (v/v) petroleum ether:benzene, (C) 300 ml. of 1:1 (v/v) petroleum ether:benzene, (D) 600 ml. of 1:3 (v/v) petroleum ether:benzene, and (E) 600 ml. of benzene. Fractions A and B are discarded (Note 8). The oil (8.50–9.15 g.) obtained by distillation of the solvent from the combined fractions C, D, and E is dissolved in 65 ml. of petroleum ether. Cooling gives 5.00–6.61 g. of colorless needles, m.p. 44–45° (Note 9). Concentration of the filtrate and cooling yield successive crops of product, m.p. 41–45°. The total yield is 7.45–8.15 g. (71–78%) (Note 9).

2. Notes

1. Tetrahydrofuran of laboratory chemical grade supplied by Fisher Scientific Co. was used without further purification by the checkers. The submitters used tetrahydrofuran purified by the method of *Org. Syntheses*, Coll. Vol. **4**, 259 (1963). [*Caution! See this volume, page 976, for a warning regarding purification of tetrahydrofuran.*]

2. Obtained from Metal Hydrides, Inc.

3. The submitters used oxime prepared from Tokyo Kasei G. R. grade dibenzyl ketone in the usual manner and recrystallized from ether-petroleum ether; m.p. 123–124° (yield 93%).[2] The checkers prepared the oxime in· the following manner. A mixture of 50 g. (0.24 mole) of 1,3-diphenyl-2-propanone (Eastman Organic Chemicals, practical grade), 50 g. (0.72 mole) of hydroxylamine hydrochloride, 250 ml. of reagent grade pyridine, and 250 ml. of ethanol was heated under reflux for 2 hours. The solvent was removed by distillation at reduced pressure, and the residue was triturated with 250 ml. of cold water. The solid was collected by filtration and washed with a small volume of cold water. Crystallization of the moist product from ethanol

gave 50.5 g. (94%) of dibenzyl ketoxime, m.p. 122–124°.

4. The internal temperature is 66°. At lower temperatures the reaction takes longer, and the yield of the aziridine is lower. The submitters found that the yield is 66% after 6 hours at a reaction temperature of 50° and 55% after 30 hours at a temperature of 20° and 44 hours at −20°.

5. The consumption of the oxime can be checked by thin-layer chromatography on silica gel G with the solvent system chloroform/methanol (95/5 v/v) and a spray reagent consisting of 5% potassium dichromate in 40% sulfuric acid. The oxime appears as an immediate dark spot and the aziridine as a yellow spot. The checkers observed identical mobilities (R_f 0.8) for both compounds.

6. The submitters found that purification of the oil by direct crystallization gives only a small amount of the pure product. Attempted purification by distillation did not give satisfactory results.

7. Silica gel, particle size 0.2–0.5 mm. (Catalog No. 7733), of E. Merck A. G. (Darmstadt) was used.

8. The fractions are tested by thin-layer chromatography on silica gel G with the solvent system and spray reagent described in Note 5.

9. The product is sufficiently pure for most purposes. The pure sample after additional recrystallizations melts at 44.7–45.1°.

3. Methods of Preparation

In addition to the present method,[3–5] 2-benzyl-3-phenylaziridine has been obtained from O-substituted dibenzyl ketoximes,[3,5] chalcone oxime[4] and 3,5-diphenyl-2-isoxazoline[6] by a reduction similar to that described here.

4. Merits of the Preparation

The present preparation illustrates the general method for the synthesis of aziridines by reduction of ketoximes[3–5] and their O-acyl and -alkyl derivatives[3,5,6] having an aromatic ring attached to carbon α or β to the oximino function and of aldoximes[4] having the aromatic ring attached to the carbon atom β to the oximono group. It has also been applied to oximes of cyclic[3,4,7] and bridged[3,8,9,10] ring ketones, such as α- and β-tetralone, 1,2,3,4-dibenzo-1,3-cycloheptadien-6-one, and bicyclo[2.2.2]octanone and its benzo analogs. Examples of aziridines prepared by this method are given in Table I; derivatives of the products are listed in Table II. Because of the accessibility of oximes the present method provides a more convenient synthesis of several types of aziridines than do other methods.[11] Furthermore, the reaction proceeds stereoselectively to give the *cis*-substituted aziridine.[3] A review[10] of the present synthetic method including mechanistic aspects[3,5,6] is available. The effect of oxime configuration (*syn* or *anti*) has been investigated.[4,9,12] The addition of N-methyl-*n*-butylamine (*in situ*) has been found to increase the reaction rate and yield of aziridine.[10]

TABLE I

AZIRIDINES PREPARED BY REDUCTION OF OXIMES
WITH LITHIUM ALUMINUM HYDRIDE

Parent Ketone or Aldehyde	Aziridine	M.P., °C	Yield, %
Acetophenone	2-Phenylaziridine	(Oil)[a]	17
Phenylacetaldehyde	2-Phenylaziridine	(Oil)[a]	34
1-Acetonaphthone	2-(α-Naphthyl)-aziridine	66–67	64
3-Phenyl-2-butanone	2-(α-Methylbenzyl)-aziridine	(Oil)[b]	38
1-Tetralone		52–53.5[c]	11

[a] Cf. F. Wolfheim, *Ber.*, **47**, 1440 (1914); S. Gabriel and J. Colman, *Ber.*, **47**, 1866 (1914); S. J. Brois, *J. Org. Chem.*, **27**, 3532 (1962); A: Hassner and C. C. Heathcock, *Tetrahedron Letters*, 1125 (1964).

[b] Along with this formation of 2,3-dimethyl-2-phenylaziridine (oil, 10%) has been reported [G. Alvernhe and A. Raurent, *Bull. Soc. Chim. France*, 3003 (1970)].

[c] Cf. G. Drefahl and K. Ponsold, *Ber.*, **93**, 519 (1960); A. Hassner and C. Heathcock, *Tetrahedron*, **20**, 1037 (1964).

TABLE II
Derivatives of Aziridines Prepared by Reduction
of Oximes with Lithium Aluminum Hydride

Aziridine	1-(Phenyl-carbamoyl) Derivative, M.P., °C	1-(*p*-Nitro-benzoyl) Derivative, M.P., °C	Derived Thiazolidine-2-thione, M.P., °C
2-Phenyl-aziridine		120–122.5	170–171[a,b] 168–169[b,c]
2-(α-Naphthyl)-aziridine	133.5–135		235–237 (dec.)
2-(α-Methylbenzyl)-aziridine		65–66 and 178–179[d]	96.5–97.5 and 165.5–166[d]
NH (fused bicyclic structure)	157–158		188.5–190.5

[a] Aziridine prepared from acetophenone.
[b] Cf. C. S. Dewey and R. A. Bafford, *J. Org. Chem.*, **30**, 491 (1965).
[c] Aziridine prepared from phenylacetaldehyde.
[d] Presumably *erythro* and *threo* isomers.

1. Shionogi Research Laboratory, Shionogi and Co., Ltd., Fukushima-Ku, Osaka, Japan.
2. Cf. J. B. Senderens, *Bull. Soc. Chim. France,* [4] 7, 645 (1910); C. H. DePuy and B. W. Ponder, *J. Am. Chem. Soc.,* 81, 4629 (1959).
3. K. Kitahonoki, K. Kotera, Y. Matsukawa, S. Miyazaki, T. Okada, H. Takahashi, and Y. Takano, *Tetrahedron Lett.,* 1059 (1965); cf. M. Y. Shandala, M. D. Solomon, and E. S. Waight, *J. Chem. Soc.,* 892 (1965).
4. K. Kotera, S. Miyazaki, H. Takahashi, T. Okada, and K. Kitahonoki, *Tetrahedron,* 24, 3681 (1968).
5. K. Kotera, Y. Matsukawa, H. Takahashi, T. Okada, and K. Kitahonoki, *Tetrahedron,* 24, 6177 (1968).
6. K. Kotera, Y. Takano, A, Matsurra, and K. Kitahonoki, *Tetrahedron Lett.,* 5759 (1968); *Tetrahedron,* 26, 539 (1970).
7. K. Kotera, M. Motomura, S. Miyazaki, T. Okada, and Y. Matsukawa, *Tetrahedron,* 24, 1727 (1968).
8. K. Kitahonoki, Y. Takano, and H. Takahashi, *Tetrahedron,* 24, 4605 (1968); J. L. M. A. Schlatmann, J. G. Korsloot, and J. Schutt, *Tetrahedron,* 26, 949 (1970).
9. K. Kitahonoki, A. Matsuura, and K. Kotera, *Tetrahedron Lett.,* 1651 (1968); K. Kitahonoki, Y. Takano, A. Matsuura, and K. Kotera, *Tetrahedron,* 25, 335 (1969).
10. K. Kotera and K. Kitahonoki, *Org. Prep. Proced.,* 1, 305 (1969).

11. Cf. P. E. Fanta, in A. Weissberger, "Heterocyclic Compounds with Three and Four-Membered Rings," Part 1, Wiley-Interscience, New York, 1964, pp. 528-541; P. A. Gempitskii, N. M. Loim, and D. S. Zhuk, *Russ. Chem. Rev.*, **35**, 105 (1966); S. Hirai and W. Nagata, The Chemistry of Aziridine, in Supplementary Issue, No. 87, "Chemistry of Heterocyclic Compounds," Part 1, Kagaku no Ryoiki, Nankodo, Tokyo, 1969; O. C. Dermer and G. E. Ham, "Ethylenimine and Other Aziridines," Academic Press, New York, 1969.
12. K. Kotera, T. Okada, and S. Miyazaki, *Tetrahedron Lett.*, 841 (1967); *Tetrahedron*, **24**, 5677 (1968).

1-BENZYLPIPERAZINE*

(Piperazine, 1-benzyl-)

$$HN \overbrace{} NH \cdot HCl + C_6H_5CH_2Cl \rightarrow$$

$$C_6H_5CH_2N \overbrace{} NH \cdot HCl + ClH \cdot HN \overbrace{} NH \cdot HCl$$

$$C_6H_5CH_2N \overbrace{} NH \cdot HCl + HCl \rightarrow C_6H_5CH_2\overset{HCl}{\underset{}{N}} \overbrace{} NH \cdot HCl$$

$$C_6H_5CH_2\overset{HCl}{\underset{}{N}} \overbrace{} NH \cdot HCl \xrightarrow{OH^-} C_6H_5CH_2N \overbrace{} NH$$

Submitted by J. CYMERMAN CRAIG and R. J. YOUNG.[1]
Checked by JAMES CASON and TAYSIR JAOUNI.

1. Procedure

A solution of 24.3 g. (0.125 mole) of piperazine hexahydrate in 50 ml. of absolute ethanol, contained in a 250-ml. Erlenmeyer flask, is warmed in a bath at 65° as there is dissolved in the solution, by swirling, 22.1 g. (0.125 mole) of piperazine dihydrochloride monohydrate (Note 1). As warming in the bath at 65° is continued, there is added during 5 minutes, with vigorous swirling or stirring, 15.8 g. (14.3 ml., 0.125 mole) of recently distilled benzyl chloride. The separation of white needles commences almost immediately. After the solution has been stirred for an additional 25 minutes at 65°, it is cooled, and the unstirred solution is kept in an ice bath for about 30 minutes. The crystals of piperazine dihydrochloride monohydrate are collected by suction filtration, washed with three 10-ml. portions of ice-cold absolute ethanol, and then dried. Recovery of the dihydrochloride is 21.5–22.0 g. (97–99%) (Note 2).

The combined filtrate and washings from the piperazine dihydrochloride are cooled in an ice bath and treated with 25 ml. of absolute ethanol saturated at 0° with dry hydrogen chloride (Note 3). After the solution has been well mixed, it is cooled for 10–15 minutes in an ice bath. The precipitated white plates of 1-benzylpiperazine dihydrochloride are collected by suction filtration, washed with dry benzene, and dried. The product, which melts at about 280° with decomposition, after sintering at about 254° (Note 4), amounts to 29.0–29.5 g. (93–95%). A solution of this salt in 50 ml. of water is made alkaline (pH > 12) with about 60 ml. of 5N sodium hydroxide, then extracted twelve times with 20-ml. portions (Note 5) of chloroform. The combined extracts are dried over anhydrous sodium sulfate, and the pale-brown oil (Note 6) remaining after removal of solvent is distilled at reduced pressure in a Claisen flask. The yield of pure 1-benzylpiperazine, b.p. 122–124°/2.5 mm., n_D^{25} 1.5440–1.5450, is 14.3–16.5 g. (65–75%).

2. Notes

1. Piperazine dihydrochloride monohydrate, which is recovered almost quantitatively in this procedure, may be purchased from K and K Laboratories, Jamaica 33, New York, or from L. Light and Co., Ltd., Poyle, Colnbrook, Bucks, England. It may be readily prepared in essentially quantitative yield from the free base by the following procedure.

A brisk stream of hydrogen chloride gas is passed for 5–8 minutes into a solution of 24.3 g. (0.125 mole) of piperazine hexahydrate in 50 ml. of absolute ethanol contained in a 250-ml. Erlenmeyer flask. A wide gas-inlet tube (about 10 mm.) is used to avoid clogging, and the flask is cooled in an ice bath to keep the temperature at about 25°. After the gas stream has been discontinued, the contents of the flask are cooled to about 0°, and the crystalline product is collected by suction filtration and washed with two 25-ml. portions of ice-cold absolute ethanol. The yield is about 22 g. (0.125 mole).

2. If the filtrate from this isolation is evaporated to dryness at reduced pressure, crude 1-benzyl-4-piperazinium chloride is left as a residue. For removal of any piperazine dihydrochloride, the chloride may be crystallized after rapidly filtering a hot solution in about 50 ml. of absolute ethanol. Concentration of the filtrate, followed by cooling, gives 12.4 g. (84%) of 1-benzyl-4-piperazinium chloride as prismatic plates, m.p. 167–168°. This salt may be converted to the dihydrochloride by treatment with ethanolic hydrogen chloride.

3. When absolute ethanol is saturated with hydrogen chloride at 0°, the resultant solution is about 10.5N in hydrogen chloride.

4. The melting point has been reported as 253° by Baltzly and co-workers.[2]

5. The checkers found continuous extraction with chloroform to be convenient.

6. The free base rapidly absorbs carbon dioxide on exposure to air and should therefore be protected during both manipulation and storage. The undistilled oil may be converted in good yield to 1-benzoyl-4-benzylpiperazine hydrochloride, m.p. 245–245.5°, by treatment with benzoyl chloride in benzene solution.

3. Methods of Preparation

1-Benzylpiperazine has been prepared[2,3] by the reaction of piperazine and benzyl chloride, followed by fractionation of piperazine, and the mono- and dibenzyl derivatives. It has also been obtained[4] by alkaline hydrolysis of 1-benzyl-4-carbethoxy-piperazine. The present method, which is a modification of that first reported by Cymerman Craig, Rogers, and Tate,[5] is simple and yields an easily purified product.

4. Merits of Preparation

The benzyl group, easily removed by hydrogenolysis, is an ideal blocking group for the preparation of 1-monosubstituted, and of 1,4-unsymmetrically disubstituted, piperazines.

Published methods for preparation of 1-benzylpiperazine involve either fractionation of mixtures of piperazine and its 1-benzyl- and 1,4-dibenzyl derivatives or the use of 1-carbethoxy-piperazine as an intermediate. The procedure here described is simple; it yields, in 30 minutes, pure 1-benzylpiperazine dihydrochloride, stable to storage, from readily available starting materials, and free of any disubstituted compound.

1. Department of Chemistry, University of Sydney, Sydney, Australia.
2. R. Baltzly, J. S. Buck, E. Lorz, and W. Schon, *J. Am. Chem. Soc.*, **66**, 263 (1944).
3. R. E. Lutz and N. H. Shearer, *J. Org. Chem.*, **12**, 771 (1947).
4. B. W. Horrom, M. Freifelder, and G. R. Stone, *J. Am. Chem. Soc.*, **77**, 753 (1955).
5. J. Cymerman Craig, W. P. Rogers, and M. E. Tate, *Australian J. Chem.*, **9**, 397 (1956).

BICYCLO[2.2.1]HEPTEN-7-ONE

(2-Norbornen-7-one)

Submitted by P. G. GASSMAN and J. L. MARSHALL[1]
Checked by WILLIAM G. DAUBEN and JAMES L. CHITWOOD

1. Procedure

Into a 250-ml. Erlenmeyer flask are placed 45.9 g. (0.298 mole) of 7,7-dimethoxybicyclo[2.2.1]heptene (Note 1), 75 ml. of 5%

aqueous sulfuric acid, and a Teflon-coated magnetic stirring bar. The flask is stoppered, and the mixture is stirred vigorously with a magnetic stirrer for 20 hours. The mixture is extracted with three 40-ml. portions of pentane, and the combined extracts are dried over anhydrous magnesium sulfate. The drying agent is removed by filtration, and the solvent is distilled through a 12-in. Vigreux column. Fractional distillation of the residual oil yields 28.9 g. (90%) of colorless bicyclo[2.2.1]hepten-7-one, b.p. 96–100° (115 mm.), n^{25}D 1.4786 (Notes 2 and 3).

2. Notes

1. The preparation of 7,7-dimethoxybicyclo[2.2.1]heptene is described on p. 424.

2. The checkers, working at half-scale, obtained an 85% yield of product, b.p. 93–97° (118 mm.).

3. This material is extremely volatile and should be handled with care.

3. Methods of Preparation

Bicyclo[2.2.1]hepten-7-one has been prepared by the oxidation of anti-7-hydroxybicyclo[2.2.1]heptene with chromic acid in acetone [2] and with aluminum t-butoxide in benzene with benzoquinone as the hydrogen acceptor.[3] The procedure described here is essentially that of Gassman and Pape.[4]

4. Merits of the Preparation

Bicyclo[2.2.1]hepten-7-one is a · useful intermediate in the synthesis of a variety of norbornane derivatives. The present procedure involves a four-step synthesis from hexachlorocyclopentadiene with a 39% overall yield. The next best method [3] involves a four-step synthesis from norbornadiene with a 15% overall yield.

1. Department of Chemistry, The Ohio State University, Columbus, Ohio 43210.
2. C. J. Norton, Ph.D. Thesis, Harvard University, 1955.
3. R. K. Bly and R. S. Bly, J. Org. Chem., 28, 3165 (1963).
4. P. G. Gassman and P. G. Pape, J. Org. Chem., 29, 160 (1964).

exo-cis-BICYCLO[3.3.0]OCTANE-2-CARBOXYLIC ACID

(1-Pentalenecarboxylic acid, octahydro-)

Submitted by R. DOWBENKO [1]
Checked by E. J. COREY and B. W. ERICKSON

1. Procedure

A. *2-(Trichloromethyl)bicyclo[3.3.0]octane.*   To a 5-l. three-necked flask equipped with a mechanical stirrer, a reflux condenser, and a thermometer are added 325 g. (3.0 moles) of *cis,cis*-1,5-cyclooctadiene (Note 1), 3 l. of chloroform (Note 2), and 14.6 g. (0.06 mole) of benzoyl peroxide. The resulting solution is stirred and refluxed (63–65°) on the steam bath (Note 3) for a total of 5 days. Four 7.3 g.-(0.03 mole-)portions of benzoyl peroxide are added, one on each consecutive day of reaction (Note 4). After a total of 5 days at reflux, the reaction mixture is cooled and washed with three 250-ml. portions of aqueous sodium bicarbonate (Note 5) and with 250 ml. of water, all the washes being discarded. The chloroform solution is dried with 30 g. of magnesium sulfate and filtered. The filtrate is distilled at atmospheric pressure using a short (8-in.) Vigreux column to collect 2760–2790 ml. of chloroform, b.p. 55–64°, which is discarded. The pressure is reduced and distillation continued to obtain two fractions: (1) b.p. 31° (47 mm.) to 65° (0.2 mm.),

300 g.; (2) b.p. 65–153° (0.2 mm.), 169 g. (Note 6). Fraction 2 is refractionated with the same Vigreux column to obtain 106–117 g. (approximately 35% based on unrecovered *cis,cis*-1,5-cyclooctadiene) of 2-(trichloromethyl)bicyclo[3.3.0]octane, b.p. 116–125° (5 mm.), $n^{25}D$ 1.5110–1.5115 (Note 7). The product is pure (by gas chromatography) (Note 8) and may be used in the next step.

B. *exo-cis-Bicyclo[3.3.0]octane-2-carboxylic acid.* A mixture of 100 g. (0.440 mole) of 2-(trichloromethyl)bicyclo[3.3.0]octane and 500 ml. of 85% phosphoric acid is put into a 1-l. three-necked flask equipped with a mechanical stirrer, a reflux condenser, and a thermometer. The mixture is stirred and heated at 150° for 16 hours, during which time it evolves hydrogen chloride and darkens. The product is then allowed to cool and is poured into a separatory funnel. One liter of water is added and the resulting mixture is extracted with four 250-ml. portions of ether. The combined ether extract is then extracted with four 250-ml. portions of 2% aqueous sodium hydroxide (Note 9), and the resulting alkaline extract is washed with 100 ml. of ether to remove any neutral material (Note 10). The alkaline extract is acidified (to pH 2–3) with concentrated hydrochloric acid, and the oil which precipitates is extracted with three 250-ml. portions of ether. The resulting ether extract is dried with 15 g. of magnesium sulfate, filtered, and evaporated at 50° (30 mm.). The residue is then distilled at reduced pressure to obtain 29–32 g. (43–47%) of *exo-cis*-bicyclo[3.3.0]octane-2-carboxylic acid, b.p. 91–96° (0.15 mm.), $n^{25}D$ 1.4839–1.4847 (Note 11).

2. Notes

1. The compound was obtained from Cities Service Research and Development Co., Petrochemical Development Department, Sixty Wall Tower, New York 5, New York. Analysis by gas chromatography showed it to be pure, and it was used without further purification.

2. Either technical or pure grade chloroform may be used.

3. It may also be refluxed with boiling chips without stirring. A heating mantle may be used in place of a steam bath.

4. The portions of peroxide may be added as such or, more conveniently and safely, as solutions in 25 ml. of chloroform over a period of 10–15 minutes.

5. It is important that all benzoic acid be removed by washing at this point because otherwise it will codistil with the product and will be difficult to separate by distillation.

6. Fraction 1 is discarded. If desired, it may be redistilled at atmospheric pressure to obtain, in addition to chloroform, 182 g. (1.68 moles) of *cis,cis*-1,5-cyclooctadiene, b.p. 145–157°.

7. The higher-boiling fraction, b.p. 129° (5 mm.) to 138° (0.2 mm.), amounts to 35–50 g. and contains at least four compounds.

8. A 2-ft. column of 20% UCON Polar 50 HB 5100 on Chromosorb W, 130°, retention time $5\frac{1}{4}$ minutes.

9. Because of the high acidity of the ether extract it is more convenient to use sodium hydroxide than sodium bicarbonate.

10. This ether wash may be combined with the main neutral fraction and distilled to obtain 29–30 g. (33–34%) of 2-(dichloro-methylene)bicyclo[3.3.0]octane, b.p. 53–56° (0.1 mm.), n^{25}D 1.5179–1.5182 (pure by gas chromatography) (column as in Note 8, 125°, retention time 4 minutes).

11. Analysis by gas chromatography shows the acid to be pure (column as in Note 8), retention time $4\frac{1}{2}$ minutes at 175°.

3. Methods of Preparation

exo-cis-Bicyclo[3.3.0]octane-2-carboxylic acid has been prepared from *cis*-bicyclo[3.3.0]-2-octanone cyanohydrin,[2] by Beckmann rearrangement of tetrahydro-*exo*-dicyclopentadiene-9-one oxime,[3] and by the present method.[4]

4. Merits of the Preparation

This two-step procedure appears to be by far the most convenient one for preparing *exo-cis*-bicyclo[3.3.0]octane-2-carboxylic acid from the readily available starting materials. The first step of the procedure is also illustrative of the method of obtaining 2-substituted bicyclo[3.3.0]octanes [4, 5] from *cis,cis*-1,5-cyclooctadiene.

1. Pittsburgh Plate Glass Co., Coatings and Resins Division, Springdale, Pa.
2. A. C. Cope and M. Brown, *J. Am. Chem. Soc.*, **80**, 2859 (1958); R. Granger, P. Nau, and J. Nau, *Trav. Soc. Pharm. Montpellier*, **18**, 142 (1958) [*C.A.*, **53**, 1699 (1959)].
3. T. H. Webb, Jr., Dissertation, Duke University, 1962.
4. R. Dowbenko, *J. Am. Chem. Soc.*, **86**, 946 (1964); *Tetrahedron*, **20**, 1843 (1964).
5. L. Friedman, *J. Am. Chem. Soc.*, **86**, 1885 (1964).

BICYCLO[2.1.0]PENTANE*

Submitted by P. G. GASSMAN and K. T. MANSFIELD [1]
Checked by G. N. TAYLOR and K. B. WIBERG

1. Procedure

A. *Diethyl 2,3-diazabicyclo[2.2.1]hept-5-ene-2,3-dicarboxylate.* In a 1-l., three-necked, round-bottomed flask equipped with a constant-pressure dropping funnel, a mechanical stirrer, and a reflux condenser is placed 174 g. (1.0 mole) of ethyl azodicarboxylate [2] in 150 ml. of ether. Freshly prepared cyclopentadiene [3] (70 g., 1.06 moles) is added dropwise over a 1-hour period to the stirred ethereal solution of diethyl azodicarboxylate. During the addition a gentle reflux is maintained by external cooling with an ice-water bath as needed. When the addition is complete, the reaction mixture is allowed to stand for 4 hours, or less if the yellow color of the azodicarboxylic acid ester disappears. The dropping funnel and condenser are replaced by a glass stopper and a short distillation head, respectively. The ether and unreacted diene are distilled off on a steam bath and the residue is transferred to a 500-ml. round-bottomed boiling flask equipped with a 30-cm. Vigreux column. After a small forerun the diethyl 2,3-diazabicyclo[2.2.1]hept-5-ene-2,3-dicarboxylate distills to give 218–228 g. (91–95%) of a colorless or very pale yellow, viscous liquid, b.p. 119–120° (0.4 mm.).

B. *Diethyl 2,3-diazabicyclo[2.2.1]heptane-2,3-dicarboxylate.* A mixture of 112 g. (0.47 mole) of diethyl 2,3-diazabicyclo[2.2.1] hept-5-ene-2,3-dicarboxylate and 125 ml. of absolute ethanol is placed in a standard Paar bottle along with 0.2 g. of 5% palladium on carbon catalyst (Note 1). The bottle is attached to the Paar hydrogenation apparatus, and shaking is begun using an initial pressure of 60 p.s.i. After 2 hours, hydrogen uptake ceases. The mixture is gravity-filtered twice and the ethanol is removed using a rotary evaporator. The entire procedure is repeated on a second batch and the crude product from the combined runs is placed in a 500-ml. round-bottomed boiling flask fitted with a 15-cm. Vigreux column. Fractional distillation gives 218–223 g. (95–97%) of diethyl 2,3-diazabicyclo[2.2.1]heptane-2,3-dicarboxylate, b.p. 107–108° (0.05 mm.), $n^{22}D = 1.4730$.

C. *2,3-Diazabicyclo[2.2.1]hept-2-ene.* A slow stream of nitrogen is bubbled through 1.2 l. of ethylene glycol (Note 2) for 20 minutes in a mechanically stirred 2-l. three-necked flask with mild heating (Note 3). The gas inlet tube is replaced with a condenser and a thermometer which reaches below the level of the ethylene glycol, and 275 g. (4.2 moles) of reagent grade potassium hydroxide pellets (85% pure) is added in four portions. A constant-pressure dropping funnel containing 223 g. (0.92 moles) of diethyl 2,3-diazabicyclo[2.2.1]heptane-2,3-dicarboxylate is connected and the reaction vessel is flushed with nitrogen. The ethylene glycol solution is heated to 125° and the diethyl 2,3-diazabicyclo[2.2.1]heptane-2,3-dicarboxylate is added as rapidly as is permitted by its viscous nature. The heating source is removed whenever the reaction temperature approaches 130°. After the addition is complete, the reaction mixture is stirred at 125° for 1 hour. The reaction mixture is allowed to cool and then poured *slowly* into a 4-l. beaker which contains 1 kg. each of ice and water and 450 ml. of concentrated hydrochloric acid (*Caution. Vigorous foaming occurs*) (Note 4). When the acidification is complete, the reaction mixture is warmed to about 40° and neutralized with 5*N* ammonium hydroxide. Half of this neutral solution is transferred to a second 4-l. beaker and subsequent operations are carried out on both batches.

The solution is stirred slowly and *ca.* 25 ml. of 2*N* cupric chloride solution is added slowly. The blue-green color of the cupric chloride is rapidly discharged and a brick red coloration occurs, followed by the precipitation of voluminous bright red crystals of the cuprous chelate of 2,3-diazabicyclo[2.2.1]hept-2-ene. The *p*H is adjusted to 5–6 by the addition of 5*N* ammonium hydroxide. Addition of 25 ml. of the cupric chloride solution followed by neutralization of the generated hydrochloric acid with 5*N* ammonium hydroxide is repeated five times. The precipitate is collected by filtration and the filtrate is again treated with 25-ml. portions of cupric chloride solution and 5*N* ammonium hydroxide. The procedure is repeated until the filtrate is clear red at *p*H 3–4 and returns to a cloudy green at *p*H 6 with no further formation of precipitate (Note 5).

The combined precipitate from the two batches is carefully washed with 500 ml. of 20% ammonium chloride solution, two 400-ml. portions of 95% ethanol, and two 300-ml. portions of cold water. The product is sucked as dry as possible in the suction funnel.

The damp product is broken up and transferred to a 1-l. flask containing a magnetic stirring bar and 400 ml. of water. A cold solution of 60 g. of sodium hydroxide in 100 ml. of water is added slowly with magnetic stirring. The stirred yellow-orange suspension is then continuously extracted with 700 ml. of pentane for 48 hours.

The pentane extract is dried over 10 g. of anhydrous potassium carbonate. After removal of the drying agent by filtration, the pentane is slowly removed from the product by distillation through a 20-cm. Hempel column packed with glass helices (Note 6). When the pentane is removed, a white crystalline residue remains which weighs 78–83 g. (88–94% yield based on the hydrogenated Diels-Alder adduct). This 2,3-diazabicyclo [2.2.1]hept-2-ene melts at 98.0–99.5° (Note 7).

D. *Bicyclo[2.1.0]pentane.* Finely powdered 2,3-diazabicyclo [2.2.1]hept-2-ene (83 g.,crude product from above) is placed in a 500-ml., one-necked, round-bottomed flask. The flask is heated at 130–140° in an oil bath to completely remove any traces of pentane. A 25-cm. unpacked Hempel column is installed and connected directly to a 100-ml. receiver flask having a side arm to which is attached a drying tube packed with silica gel. The receiver is cooled in a dry ice-acetone bath. The azobicyclic is pyrolyzed by heating the oil bath to 180–195°. At the pre-ferred rate of pyrolysis, the starting material condenses about one fourth of the way up the column. Occasional flaming of the column may be necessary to prevent plugging of the column by the solidifying starting material. At the end of the pyrolysis (8 hours) only a small, black, nonvolatile residue remains: The condensed bicyclo[2.1.0]pentane is allowed to warm to room temperature, dried over anhydrous magnesium sulfate, and the drying agent removed by filtration through glass wool into a 100-ml. distillation flask. (Caution! *Bicyclo[2.1.0]pentane is a very volatile hydrocarbon and requires appropriate handling for high yields.*) Distillation leaves a residue of about 1 g. of the starting azobicyclic and affords 53.5–55.5 g. (90.0–93.5%) of bicyclo[2.1.0]pentane, b.p. 45.5°, n^{20}D 1.4220 (Note 8).

2. Notes

1. The submitters effected the hydrogenation using a medium-capacity, rocker-type, high-pressure hydrogenator with an initial hydrogen pressure of 700 p.s.i. By employing these conditions, the reaction time is reduced to 20–30 minutes. The yield is unchanged.

2. Technical grade ethylene glycol such as that sold by Union Carbide Corp. is suitable for this purpose.

3. A large oil bath supported by a laboratory jack is used for this and subsequent operations when rapid removal of the heat source might be necessary.

4. Foaming may easily be controlled even with rapid addition of the basic solution by vigorous stirring employing the mechanical stirrer used during the reaction.

5. The precipitation of copper oxides in slightly alkaline solution should not be confused with the formation of the bright red crystals of the organocuprous complex.

Recrystallization of the crude copper complex from boiling 20% ammonium chloride (pH 4) affords lustrous brick red needles. Analytically pure material is obtained on a second recrystallization from 0.001N hydrochloric acid followed by drying over phosphorus pentoxide.

6. If the supersaturated pentane solution tends to foam toward the end of the distillation, the pot should be allowed to cool. This causes the product to crystallize. Once the crystals start to form, foaming is no longer a problem.

7. This material may be further purified (m.p. 99.5–100.0°) by recrystallization from pentane or methanol, or by sublimation at 85° (60 mm.). Owing to the unusually high vapor pressure of this product, large losses may be encountered on recrystallization or sublimation unless due care is exercised.

8. If all the pentane is removed before pyrolysis, the bicyclo [2.1.0]pentane shows no impurities on vapor phase chromatography with a 20% Dow 710 on 50/60 U Anaprep column. Analysis by n.m.r. also revealed the absence of any traces of cyclopentene in the spectrum consisting of three complex multiplets at 0.3–0.8, 1.1–1.7, and 1.9–2.4 p.p.m. (downfield from internal tetramethylsilane reference).

3. Discussion

The procedure described is a modification of that developed by Diels [4] and Criegee.[5] Bicyclo[2.1.0]pentane has been prepared by the pyrolysis of 2,3-diazabicyclo[2.2.1]hept-2-ene,[5,6] the photolysis of 2,3-diazabicyclo[2.2.1]hept-2-ene,[7] the pyrolysis of N-phenyl-2-oxo-3-azabicyclo[2.2.1]heptane,[8] and the addition of methylene to cyclobutene.[9]

The procedure described is suitable for the preparation of bicyclo [2.1.0]pentane on a large scale. The product is obtained free of impurities and the general method is relatively safe. The starting materials are readily available. The hydrolysis of the diester is very reproducible, a feature that was not true of the literature procedure.[4] The pyrolysis step is much simpler and cleaner than the published description.[5] In addition, the procedure described gives a general method of hydrazo oxidation and for the pyrolysis of azo compounds. Oxidations of the type described in this procedure have been used to prepare a wide variety of cyclic azo compounds. Highly unstable azo compounds have been isolated as the stable crystalline cupric chloride complexes.[10-12] The thermolysis (and/or photolysis) of appropriately substituted cyclic azo compounds has become a highly useful method for the preparation of strained ring systems.[13-18]

1. Department of Chemistry, The Ohio State University, Columbus, Ohio 43210.
2. N. Rabjohn, *Org. Syntheses*, Coll. Vol. **3**, 375 (1955).
3. R. B. Moffett, *Org. Syntheses*, Coll. Vol. **4**, 238 (1963).
4. O. Diels, J. Blom, and W. Koll, *Ann.*, **443**, 242 (1925).
5. R. Criegee and A. Rimmelin, *Ber.*, **90**, 414 (1957); A. Ludwig, Dissertation, Karlsruhe, 1958.
6. S. Cohen, R. Zand and C. Steel, *J. Am. Chem. Soc.*, **83**, 2895 (1961).
7. C. Steel, *J. Phys. Chem.*, **67**, 1779 (1963); T. F. Thomas and C. Steel, *J. Am. Chem. Soc.*, 87, 5290 (1965).
8. G. Griffin, N. Hepfinger, and B. Shapiro, *J. Am. Chem. Soc.*, **85**, 2683 (1963).
9. G. Wittig and F. Wingler, *Ber.*, **97**, 2146 (1964).
10. E. L. Allred, J. C. Hinshaw, and A. L. Johnson, *J. Am. Chem. Soc.*, **91**, 3382 (1969); E. L. Allred and A. L. Johnson, *J. Am. Chem. Soc.*, **93**, 1300 (1971).
11. M. Martin and W. R. Roth, *Ber.*, **102**, 811 (1969).
12. R. M. Moriarty, C.-L. Yeh, and N. Ishibi, *J. Am. Chem. Soc.*, **93**, 3085 (1971).
13. E. L. Allred and R. L. Smith, *J. Am. Chem. Soc.*, **91**, 6766 (1969) and references therein.
14. W. R. Roth and M. Martin, *Ann.*, **702**, 1 (1967).
15. W. R. Roth and M. Martin, *Tetrahedron Lett.*, 4695 (1967).
16. R. M. Moriarity, *J. Org. Chem.*, **28**, 2385 (1963).
17. E. L. Allred and J. C. Hinshaw, *J. Am. Chem. Soc.*, **90**, 6885 (1968).
18. W. Luttke and V. Schabacker, *Ann.*, **698**, 86 (1967).

2,2′-BIPYRIDINE*

$$2 \quad \text{(pyridine)} \quad \xrightarrow[\text{Raney Ni}]{\text{W7-J}} \quad \text{(bipyridine)} \quad + \quad H_2$$

Submitted by W. H. F. SASSE [1]
Checked by VICTOR A. SNIECKUS and V. BOEKELHEIDE

1. Procedure

A. *Degassed Raney nickel catalyst (W7-J). Caution! Raney nickel catalysts which have been prepared by the usual methods should not be heated in vacuum, as large quantities of heat and hydrogen may be given off suddenly, and dangerous explosions may result.*[2,3]

In a hood a 2-l., wide-mouthed Erlenmeyer flask containing 600 ml. of distilled water is placed in an empty water bath and fitted with an efficient stainless steel stirrer, so that its blades are half immersed. The stirrer is started, and 160 g. of sodium hydroxide is dissolved in the water. Then 125 g. of 1:1 aluminum-nickel alloy (Note 1) is added in portions as rapidly as possible, but at such a rate that no material is lost by frothing with the stirrer running at full speed (Note 2). When all the alloy has been added the stirrer is slowed down, and the catalyst is washed down from the sides of the flask with distilled water. As soon as the reaction has subsided the water bath is filled with boiling water, and the catalyst is slowly stirred while the volume is kept up by the occasional addition of distilled water so that the catalyst is well covered at all times. After 6 hours, stirring and heating are discontinued, and the catalyst is allowed to stand at room temperature for 12–15 hours. It is then washed by decantation with ten 250-ml. portions of distilled water and transferred to a 1-l., round-bottomed, three-necked flask by means of distilled water. The total volume of catalyst and water is adjusted to 300 ml., and the flask placed in a cold water bath which is equipped with a thermometer. One side arm is fitted with a 100-ml. dropping funnel, and the other two necks are each connected to a 3-l. Büchner flask by short lengths of thick-walled, wide-bore rubber tubing (Notes 3, 4). One of the Büchner flasks is connected to a vacuum gauge and then to an efficient water pump. The other Büchner flask is connected directly to a second, equally efficient water pump. To control the pressure inside the apparatus, a screw clamp is placed between each Büchner flask and each pump (Note 5). With these clamps completely closed both pumps are turned on fully. The pressure inside the apparatus is now gradually reduced by opening the clamps at such a rate that no excessive frothing occurs. When both clamps are fully open, the water bath is heated slowly until the water in the reaction flask begins to boil. The bath is kept at this temperature until there is no more water left in the flask (Note 6). Then, with both clamps fully opened, the temperature of the water bath is raised to 100° during 15–20 minutes and kept at this temperature for 2 hours. After this time the catalyst is allowed to cool to 50–60°; it is now ready for use.

B. *2,2'-Bipyridine.* With the apparatus set as described above, 100 ml. of pure pyridine (Note 7) is poured into the dropping funnel, and the screw clamps are completely closed. Immediately afterward about 80 ml. of the pyridine is run slowly onto the catalyst from the dropping funnel. Under no circumstances is any air allowed to enter the flask. The flask is then shaken carefully in order to wet the catalyst as much as possible with pyridine. Another 80 ml. of pyridine is added in the same way, and the flask is shaken again. Finally 40 ml. of pyridine is added, and air is allowed to enter the flask. The connections to the Büchner flask are removed, and a reflux condenser is fitted. The reaction mixture is then boiled gently under reflux (Note 8). After about 48 hours the flask is allowed to cool to about 60°, and most of the liquid is decanted (*Hood!*) and filtered through a sintered-glass funnel (Note 9) into a 500-ml. round-bottomed flask. Then 50 ml. of fresh pyridine is added to the catalyst in the reaction flask, and the mixture is heated to reflux

for 10 minutes. The flask contents are allowed to cool to about
60° and the pyridine is decanted and filtered as before. This
extraction is repeated two more times. The flask containing
the filtrates is then equipped for vacuum distillation (Note 10),
and most of the pyridine is removed on a water bath under
reduced pressure (20–30 mm.) at a bath temperature not exceeding 40°. Toward the end of the distillation the bath temperature
is raised to about 75° for 10 minutes (Note 11). The residue
from this distillation is extracted with 100 ml. of boiling petroleum
ether (Note 12), and the insoluble cream-colored material is collected by filtration and washed with three 25-ml. portions of the
boiling petroleum ether (Note 13). The filtrate is chromatographed over alumina (Note 14) using petroleum ether (60–90°)
for elution. The first 2 l. of eluate is collected, concentrated to
100 ml., and allowed to stand overnight in a refrigerator. The
crude solid (23–24 g.) is collected and recrystallized from about
80 ml. of petroleum ether (60–90°) to give 21.0 g. of 2,2'-
bipyridine as white crystals, m.p. 70–71° (Notes 15, 16).

2. Notes

1. The nickel-aluminum alloy used was supplied by British
Drug Houses (through the Ealing Corporation in the U.S.A.).
Average particle size was about 4 μ as measured by a Fisher
Sub-sieve Sizer. Alloys of a finer particle size (about 2.85 μ)
gave rise to considerable loss of catalyst during the washing and
lost hydrogen during the degassing procedure in a much more
vigorous fashion.

2. With a motor running at approximately 3500–4500 r.p.m.
the addition of the alloy is completed in less than 10 minutes.
Alcohol should not be added to the catalyst to control the
frothing.

3. The Büchner flasks are included in the apparatus to accommodate relatively large quantities of hydrogen which are sometimes given off suddenly by the catalyst during the later stages
of the degassing. Instead of Büchner flasks, strong round-
bottomed flasks may be used.

4. Rubber tubing with diameters of 9 mm. I.D. and 17 mm. O.D. was used.

5. An air leak cannot be employed to regulate the vacuum inside the reaction flask since the catalyst becomes increasingly pyrophoric as the degassing progresses.

6. The average time required for the complete removal of the water from the catalyst varies from 4 to 12 hours. It is essential to evaporate the water slowly since much nickel may be lost if the water boils too vigorously. If the temperature of the bath should rise too high, the screw clamps are closed far enough to increase the pressure slightly in the apparatus. As soon as the bath has cooled to the required temperature, the clamps are reopened fully. The maximum practical temperature for the removal of the water was found to be 25–30° at pressures between 17 and 20 mm.

7. The pyridine used must be free of pyrrole since as little as 0.001% of pyrrole will markedly decrease the yield. The checkers found it most satisfactory to use spectroquality pyridine supplied by Matheson, Coleman and Bell. The submitters purified their pyridine by distillation from potassium hydroxide. To test for the presence of pyrrole, a 0.5-ml. sample of pyridine is diluted with 2.5 ml. of water, and 2 ml. of concentrated hydrochloric acid is added followed by 0.5 ml. of a 5% solution of p-dimethyl-aminobenzaldehyde in dilute hydrochloric acid (1:10 dilution of concentrated hydrochloric acid). If pyrrole is present, a red-purple color appears. Spectroquality pyridine gives a negative result in this test.

8. Excessive bumping will occur if the reaction mixture is heated too strongly.

9. A sintered-glass funnel (diameter 9 cm.) of medium porosity was used. The filtration is conveniently carried out under slightly reduced pressure, but care must be taken to keep any nickel on the funnel damp, as the catalyst is highly pyrophoric.

10. A relatively wide capillary (about 0.5 mm. I.D.) should be used to avoid blockages toward the end of the distillation.

11. The distillate contains, besides pyridine, small quantities of pyrrole, water, and 2,2'-bipyridine.

12. A petroleum fraction, b.p. 60–90° containing 5% of aromatics, was used throughout.

13. The petroleum-insoluble material is a nickel (II) complex containing 2,2'-bipyridine and 2,2'-pyrrolylpyridine.[4] About 1.5 g. of this compound is obtained.

14. The checkers used No. 2 grade neutral alumina (Woelm) in a column 3.5 x 20 cm. If a more active alumina is used, larger quantities of petroleum ether are needed for the elution.

15. Small quantities of 2,2',2''-tripyridine are removed by this recrystallization.

16. Similarly 3- and 4-alkylpyridines give the corresponding 5,5'-dialkyl-2,2'-bipyridines and 4,4'-dialkyl-2,2'-bipyridines, respectively, in good yield when treated with W7-J nickel.[5] Somewhat lower yields of 2,2'-biquinolines are obtained with quinolines.[6,7]

3. Methods of Preparation

2,2'-Bipyridine has been prepared by the action of ferric chloride,[8,9] iodine,[10] or a nickel-alumina catalyst[10] on pyridine at temperatures ranging from 300° to 400°. It has also been obtained from the reaction of 2-bromopyridine and copper.[11] The present procedure is a modification of a previously published, general method.[3] The W7-J nickel catalyst was developed from the description of the W7 Raney nickel catalyst of Billica and Adkins.[12]

4. Merits of the Preparation

This procedure serves two purposes. It provides a synthesis for the important chelating reagent 2,2'-bipyridine, a substance of interest in several fields of chemistry, and it gives a preparation of an active, degassed Raney nickel catalyst.

1. Department of Organic Chemistry, The University of Adelaide, Adelaide, South Australia. Present address: Division of Applied Chemistry, CSIRO, P.O. Box 4331, GPO, Melbourne 3001, Australia.
2. H. A. Smith, A. J. Chadwell, and S. S. Kirslis, *J. Phys. Chem.*, **59**, 820 (1955).
3. W. H. F. Sasse, *J. Chem. Soc.*, 3046 (1959).

4. A. M. Sargeson and W. H. F. Sasse, *Proc. Chem. Soc.*, 150 (1958).
5. W. H. F. Sasse and C. P. Whittle, *J. Chem. Soc.*, 1347 (1961).
6. W. H. F. Sasse, *J. Chem. Soc.*, 526 (1960).
7. G. M. Badger and W. H. F. Sasse, *J. Chem. Soc.*, 616 (1956).
8. Fr. Hein and W. Retter, *Ber.*, **61**, 1790 (1928).
9. G. T. Morgan and F. H. Burstall, *J. Chem. Soc.*, 20 (1932).
10. H. D. T. Willink, Jr., and J. P. Wibaut, *Rec. Trav. Chim.*, **54**, 275 (1935).
11. J. P. Wibaut and J. Overhoff, *Rec. Trav. Chim.*, **47**, 761 (1928).
12. H. R. Billica and H. Adkins, *Org. Syntheses*, Coll. Vol. **3**, 176 (1955).

1,2-BIS(*n*-BUTYLTHIO)BENZENE

[Benzene, *o*-bis(butylthio)-]

$$2CH_3(CH_2)_3SH + Cu_2O \rightarrow 2CH_3(CH_2)_3SCu + H_2O$$

Submitted by ROGER ADAMS,[1] WALTER REIFSCHNEIDER,[2] and ALDO FERRETTI.[3]
Checked by WILLIAM E. PARHAM, WAYLAND E. NOLAND, and JAMES R. THROCKMORTON.

1. Procedure

A. *Cuprous n-butylmercaptide.* A mixture of 42.9 g. (0.30 mole) of freshly prepared cuprous oxide (Note 1), 61.3 g. (0.68 mole) of 1-butanethiol, and 750 ml. of 95% ethanol is heated under reflux with mechanical stirring (Note 2) until the orange or red color of the cuprous oxide is completely changed to the white color of the cuprous *n*-butylmercaptide (Note 3). The product is collected by filtration, washed several times with 95%

ethanol, and dried in a vacuum. The yield is 91.6 g., essentially quantitative (Note 4).

B. *1,2-Bis(n-butylthio)benzene*. In a 1-l., round-bottomed, three-necked flask fitted with a reflux condenser, a mechanical stirrer, and a thermometer which reaches into the reaction mixture is placed a solution of 59.0 g. (0.25 mole) of *o*-dibromobenzene in a mixture of 250 ml. of quinoline and 80 ml. of pyridine. To this solution is added 84.0 g. (0.55 mole) of cuprous *n*-butylmercaptide, and the mixture is stirred and heated under reflux (Note 5) for 3.5 hours (Note 6). Heating is stopped and the reaction mixture is allowed to cool to about 100°. It is then poured into a stirred mixture of 1500 g. of ice and 400 ml. of concentrated hydrochloric acid; occasional stirring is continued for about 2 hours. The aqueous part is then decanted from the dark brown, gummy residue and is extracted twice with 400 ml. portions of ether. The ether extract is added to the residue, and the resulting mixture is stirred for about 5 minutes. The ether solution is then decanted from the residue and is filtered. The residue is extracted twice more with 400-ml. portions of ether (Note 7). The combined ether extract is washed twice with 100-ml. portions of 10% hydrochloric acid, once with water, and twice with 100-ml. portions of concentrated ammonia (Note 8). After a final wash with water, the ether solution is dried over anhydrous potassium carbonate. The potassium carbonate is collected on a filter, and the ether is removed from the filtrate by distillation. The remaining brown oil is distilled in vacuum, giving a pale orange oil, b.p. 123–124°/0.3 mm., n_D^{25} 1.5684. The yield is 46.5–56.0 g. (73–87%) (Note 9).

2. Notes

1. Cuprous oxide was prepared according to the procedure of King.[4] A good grade of commercial cuprous oxide may also be used, but the time required to complete the conversion into mercaptide may be considerably longer (see Note 3).

2. It is not necessary to carry out the reaction under nitrogen, but it is advisable to close the condenser with a cotton plug or with a capillary tube to limit the entrance of air.

3. When freshly prepared cuprous oxide is used, a period of

about 12 hours is generally sufficient. For commercial grade cuprous oxide the time required varies between 8 and 150 hours, depending on the reactivity of the cuprous oxide.

4. The checkers obtained a yield of 97%. The submitters and checkers have found that both larger and smaller runs can be carried out without difficulty or reduction in yield.

5. The pot temperature should rise during the reaction from about 150° at the beginning of reflux to about 170° at the end of the reaction time. Pot temperatures lower than 150° and higher than 180° result in lower yields.

6. Approximately 10 minutes after the mixture starts to boil a homogeneous solution is obtained.

7. If the last ether extract is not almost colorless, one more extraction of the residue with ether should be carried out.

8. If the ammonia layer is dark blue at the second extraction, extraction with ammonia should be continued until only a pale blue extract results.

9. The present procedure has also been used by the submitters to prepare the following thioethers: 1,4-bis-(*n*-butylthio)benzene, pale yellow oil, b.p. 142°/0.5 mm., n_D^{20} 1.5726, from *p*-dibromobenzene and cuprous *n*-butylmercaptide (yield 68–74%); 1,2-bis(phenylthio)benzene, white crystals, m.p. 42.5–44.5°, b.p. 190°/1 mm., from *o*-dibromobenzene and cuprous phenylmercaptide (see below) (yield 79–83%), or from *o*-dichlorobenzene (see below) and cuprous phenylmercaptide (yield 58–71%); 1,4-bis(phenylthio)benzene, white crystals, m.p. 82–83°, from *p*-dibromobenzene and cuprous phenylmercaptide (yield 80–84%), or from *p*-dichlorobenzene (see below) and cuprous phenylmercaptide (yield 59–72%).[6] The same method can be applied to the preparation of many other thioethers.

Cuprous phenylmercaptide is prepared from cuprous oxide and benzenethiol according to the procedure given for cuprous *n*-butylmercaptide. A heating period of only 2 hours (when freshly prepared cuprous oxide is used), however, is required to obtain the yellow compound. Chloro compounds can be used instead of bromo compounds for the reaction with cuprous phenylmercaptide. However, a higher reaction temperature (210–220°) and a longer reaction time (24 hours) is required.

The necessary pot temperature is obtained by using a mixture of 350 ml. of quinoline and 8 ml. of pyridine as solvents. It is also advantageous to use a larger excess of cuprous phenylmercaptide (121 g., 0.69 mole).

Aromatic chloro compounds cannot be used for reactions with aliphatic cuprous mercaptides.

3. Methods of Preparation

1,2-Bis(n-butylthio)benzene and 1,4-bis(n-butylthio)benzene have been prepared from the corresponding dibromobenzene and cuprous n-butylmercaptide, using a mixture of quinoline and pyridine as solvent.[5,6] 1,2-Bis(phenylthio)benzene and 1,4-bis(phenylthio)benzene have been prepared from the corresponding dichloro- or dibromobenzenes and cuprous phenylmercaptide, using a mixture of quinoline and pyridine as solvent.[5] 1,4-Bis(phenylthio)benzene has also been prepared from p-dibromobenzene or p-bromophenyl phenyl sulfide and lead phenylmercaptide[7] and from diazotized 4-aminophenyl phenyl sulfide and sodium phenylmercaptide.[8]

4. Merits of Preparation

As indicated (Note 9), the present procedure can be adapted for the preparation of a wide range of aryl and vinyl sulfides.[6] This, in combination with the cleavage reaction described for the preparation of 1,2-dimercaptobenzene (p. 419), provides a convenient and general method for the preparation of aryl mercaptans.

1. University of Illionis, Urbana, Illinois.
2. Agricultural Chemical Research, The Dow Chemical Co., Midland, Michigan.
3. Via Martiri Triestini, 12, Milan, Italy.
4. A. King, "Inorganic Preparations," D. Van Nostrand Co., Inc., Princeton, New Jersey, 1936, p. 39.
5. R. Adams, W. Reifschneider, and M. D. Nair, *Croat. Chem. Acta,* **29,** 277 (1957).
6. R. Adams and A. Ferretti, *J. Am. Chem. Soc.,* **81,** 4927 (1959).
7. E. Bourgeois and A. Fouassin, *Bull. Soc. Chim. France,* [4] **9,** 938 (1911); *Rec. Trav. Chim.,* **30,** 431 (1911).
8. G. Leandri and M. Pallotti, *Ann. Chim. (Rome),* **46,** 1069 (1956).

4,4'-BIS(DIMETHYLAMINO)BENZIL

[Benzil, 4,4'-bis(dimethylamino)-]

$$2C_6H_5N(CH_3)_2 + ClCOCOCl \xrightarrow{AlCl_3}$$

$$p\text{-}(CH_3)_2NC_6H_4COCOC_6H_4N(CH_3)_2\text{-}p + 2HCl$$

Submitted by CELAL TÜZÜN, MICHAEL OGLIARUSO, AND
ERNEST I. BECKER.[1]
Checked by B. C. McKUSICK and R. J. SHOZDA.

1. Procedure

In a 3-l. three-necked flask equipped with an efficient mechanical stirrer of high torque (Note 1), a reflux condenser with a calcium chloride drying tube, a thermometer, and a dropping funnel are placed 133 g. (1.00 mole) of anhydrous aluminum chloride and 200 ml. of dry carbon disulfide. The mixture is cooled in an ice bath and stirred while 182 g. (1.50 moles) of N,N-dimethylaniline (Note 2) is added through the dropping funnel during a period of 15 minutes. The dropping funnel is rinsed with 20 ml. of carbon disulfide which is then run into the flask. Any aluminum chloride sticking to the walls of the flask is now scraped into the mixture, which is an easily stirred slurry of a white solid in a light-green liquid.

The reaction mixture is cooled to 5–10° in an ice-salt bath (Note 3), and, with continued stirring, a solution of 31.7 g. (21.3 ml., 0.250 mole) of oxalyl chloride in 200 ml. of dry carbon disulfide is added through the dropping funnel in the course of 20 minutes. After the addition is complete, the thick black reaction mixture is allowed to warm to room temperature, refluxed for 1 hour, and then cooled to 0–5° in an ice bath. The mixture is stirred throughout these steps. One hundred grams of chipped ice is added with stirring, followed by 400 ml. of cold water. Steam is then passed into the flask until the carbon disulfide and unreacted dimethylaniline are removed, and the green-black aluminum complex is decomposed to a mixture of a green solid and a blue solid; this requires 1–2 hours (Note 4). The mixture is cooled to 50°, and the solid, which is principally 4,4′-bis(dimethylamino)benzil, is collected on a Büchner funnel. In order to remove the major part of the impurity, which is somewhat soluble in water, the solid is slurried in 200 ml. of water at 50°, and the slurry is filtered. This process is repeated twice, and the crude benzil, now a green solid, is washed successively on the funnel with 200 ml. of water at 50° and with 100 ml. of cold methanol. After being dried in air, it weighs 44–55 g. and melts at 191–196°.

The crude benzil is dissolved in 500 ml. of chloroform. To remove the impurity that remains (Note 5), the solution is shaken with three 400-ml. portions of 6% aqueous hydrogen peroxide solution containing 1.0 g. of sodium hydroxide in each portion, and finally with 500 ml. of water. The aqueous layers are combined, warmed to drive off dissolved chloroform, and filtered to separate about 1.5 g. of a yellow-green solid, which is dissolved in the chloroform solution.

The chloroform layer is distilled to dryness and the residue is dissolved in 1.5 l. of acetone under reflux. The hot acetone solution is filtered and then allowed to cool in a refrigerator. Yellow 4,4′-bis(dimethylamino)benzil crystallizes from the acetone solution. It is separated by filtration and washed with 100 ml. of cold methanol. After being dried in air, it weighs 28–31 g. (38–42%); m.p. 200–202°; $\lambda_{max}^{CH_3OH}$ 371 mμ(ϵ 44,700) (Note 6).

The acetone filtrate is concentrated to 700 ml. and cooled to 0–5°. An additional 4–8 g. (6–11%) of slightly less pure benzil, m.p. 198–201°, crystallizes from solution (Note 7).

2. Notes

1. A magnetic stirrer is unsatisfactory. The submitters used a glass-blade stirrer at 300 r.p.m. The stirrer shaft must be rigidly attached to the motor because the reaction mixture becomes very thick during the addition of oxalyl chloride.

2. Eastman Kodak white label compounds used without further purification are satisfactory starting materials.

3. Cooling below −10° should be avoided because the reaction stops at that temperature and large amounts of oxalyl chloride accumulate in the flask. If this mixture is then allowed to come to room temperature, a vigorous reaction that may get out of control will take place. It is probable that, at reaction temperatures about 10°, the yield of 4,4'-bis(dimethylamino)benzil is less and some Crystal Violet is formed as an impurity, for it has been reported that aluminum chloride effects the conversion of N,N-dimethylaniline and oxalyl chloride to Crystal Violet in 92–95% yield when the reaction is allowed to proceed without cooling.[2]

4. The submitters recommend that the following purification procedure be used from this point for the preparation of 4,4'-bis(diethylamino)benzil (72% yield) and 4,4'-bis(di-*n*-propylamino)benzil (58% yield). The procedure has also been used as an alternative to the one given for 4,4'-bis(dimethylamino)benzil.

One liter of water is added, making the total volume about 2 l., and, after the solution has been cooled to room temperature, it is extracted with 1 l. of chloroform and then with 150 ml. of chloroform. The combined dark-blue extracts are washed with 550 ml. of 8.5% hydrochloric acid, then with 200 ml. of water and dried over anhydrous sodium sulfate. The chloroform solution is distilled until the volume is 250 ml., and it is then passed through an 8.5 x 25-cm. column (300 g.) of Alcoa F-20 alumina. The adsorbate is eluted with 1 l. of chloroform. The eluate is evaporated to a volume of 250 ml., washed with 500 ml. of 10% sodium hydroxide solution then with 100 ml. of water, and distilled to essential dryness.

The residual crude, yellow, semi-solid product is stirred and brought to a boil with 250 ml. of ethyl acetate and then allowed to cool to room temperature while stirring. Filtration affords 39–41 g. (52–55%) 4,4'-bis(dimethylamino)benzil, m.p. 201–203°. Concentration of the mother liquor to 50 ml. gives, after cooling, an additional 3.5–4.5 g. of product, m.p. 175–180°. Recrystallization of 10 g. of the combined products from 120–150 ml. of benzene gives 9.1–9.3 g. of yellow crystals, m.p. 202–203°.

5. The checkers found that at least part of the colored impurity is Crystal Violet. Alkaline hydrogen peroxide is reported to

cleave Crystal Violet to N,N-dimethylaniline and Michler's ketone.[3]

6. The product is sometimes pale green because of traces of impurities, but it is nevertheless very pure, for repeated recrystallization does not change ϵ_{max}.

7. Addition of 1 l. of cold water to the acetone filtrate from which the second crop of benzil is separated causes about 5 g. of impure benzil to precipitate. This may be added to the crude benzil of a subsequent run prior to the treatment with hydrogen peroxide and alkali.

3. Methods and Merits of Preparation

4,4'-Bis(dimethylamino)benzil has been made previously by heating a mixture of oxalyl chloride and N,N-dimethylaniline under a pressure of 300 atmospheres of carbon monoxide in a steel pressure vessel at 100°.[4] The present method is simpler and gives better yields. As 4-dimethylaminobenzaldehyde cannot be converted to the corresponding benzoin,[5] this common route to benzils cannot be used to prepare 4,4'-bis(dimethylamino)benzil.

The present procedure is reported by the submitters to be a general way of making 4,4'-bis(dialkylamino)benzils and, with a somewhat modified purification scheme (Note 4), has been used by them to prepare 4,4'-bis(diethylamino)benzil from N,N-diethylaniline and 4,4'-bis(dipropylamino)benzil from N,N-dipropylaniline.

1. Polytechnic Institute of Brooklyn, Brooklyn 1, N. Y.
2. G. v. Georgievies, *Ber.*, 38, 884 (1905).
3. I. N. Postovskii, *J. Chem. Ind. (U.S.S.R.)*, 4, 552 (1927) [C.A., 22, 957 (1928)].
4. H. Staudinger and H. Stockmann, *Ber.*, 42, 3485 (1909).
5. S. M. McElvain, "The Acyloins," *Org. Reactions*, 4, 273 (1948).

BIS(1,3-DIPHENYLIMIDAZOLIDINYLIDENE-2)

[$\Delta^{2,2'}$-Bis(1,3-diphenylimidazolidine)]

$$2 \begin{array}{c} C_6H_5 \\ | \\ NH \\ H_2C \\ | \\ H_2C \\ NH \\ | \\ C_6H_5 \end{array} \quad + \quad 2\,CH(OC_2H_5)_3 \quad \xrightarrow{-6\,C_2H_5OH}$$

$$\begin{array}{c} C_6H_5 \quad\quad C_6H_5 \\ | \quad\quad\quad | \\ H_2C \diagdown N \quad\quad N \diagup CH_2 \\ \quad\quad C = C \\ H_2C \diagup N \quad\quad N \diagdown CH_2 \\ | \quad\quad\quad | \\ C_6H_5 \quad\quad C_6H_5 \end{array}$$

Submitted by H.-W. WANZLICK [1]
Checked by D. J. LaFOLLETTE and RONALD BRESLOW

1. Procedure

In a 250-ml. round-bottomed flask equipped with a gas-inlet tube and reflux condenser 20 g. (0.094 mole) of N,N'-diphenyl-ethylenediamine (1,2-dianilinoethane) (Note 1) and 100 ml. of purified triethyl orthoformate (Note 2) are heated by an oil bath under nitrogen (Note 3) for 5 hours. The oil bath is maintained between 190° and 200°, and water is allowed to stand in the condenser. The water in the condenser begins to boil slowly, and the alcohol which is produced is allowed to escape (Note 4). The reaction product which crystallizes during the reaction is filtered after cooling and washed with ether. There is obtained 19–20 g. (91–95%) of product, m.p. 285° (dec.) (Note 5).

2. Notes

1. 1,2-Dianilinoethane, containing water of crystallization, is best dried by melting under vacuum.

2. Commercial material, distilled.

3. The nitrogen is dried by passing it through concentrated sulfuric acid. It must be nearly oxygen-free; otherwise 1,3-diphenylimidazolidinone-2 is formed, and its removal by recrystallization results in a decreased yield.

4. An air condenser may also be employed.

5. The melting range depends on the rate of decomposition during heating. The checkers observed that in an evacuated capillary there is darkening from 270° to 290°, and fairly sharp melting at 299–300°. The product is autoxidizable and is best stored under dry nitrogen. Preparations which have oxidized on standing may be purified by digesting and washing with methylene chloride.

3. Methods of Preparation

This amino olefin was first prepared by thermal elimination of chloroform from 1,3-diphenyl-2-trichlormethylimidazolidine,[2] and later by the procedure described here.[3, 4] It can also be made by treatment of 1,3-diphenylimidazolinium salts with strong bases.[5, 6]

4. Merits of the Preparation

The procedure described is the simplest one known. All other methods also employ 1,2-dianilinoethane as starting material. This method, however, converts it directly into the amino olefin in one step.

The preparative value of this compound lies in the surprising fact that bis(1,3-diphenylimidazolidinylidene-2) behaves in many reactions (*e.g.*, with aromatic aldehydes,[2, 7] and with carbon acids [2, 7–9]) as if it dissociated to form a "nucleophilic carbene." The hydrolytic cleavage of these derived imidazolidine derivatives makes possible the preparation of formyl compounds, so that the amino olefin can be considered as a potential carbonylation reagent. In many reactions it is not necessary to isolate the reagent, as it may be produced *in situ*.[10] It should be pointed out, however, that the reaction of the amino olefin with aldehydes and carbon acids does not actually involve prior dissociation to the carbene, but it is convenient, from a preparative point of view, to describe it in these terms.[6]

1. Organisch-Chemische Institut, Technische Universität Berlin, Berlin, Germany.
2. H. W. Wanzlick and E. Schikora, *Ber.*, **94**, 2389 (1961).
3. H. W. Wanzlick and H. J. Kleiner, *Angew. Chem.*, **73**, 493 (1961).
4. H. W. Wanzlick, F. Esser, and H. J. Kleiner, *Ber.*, **96**, 1208 (1963).
5. D. M. Lemal and K. I. Kawano, *J. Am. Chem. Soc.*, **84**, 1761 (1962).
6. D. M. Lemal, R. A. Lovald, and K. I. Kawano, *J. Am. Chem. Soc.*, **86**, 2518 (1964).
7. H. W. Wanzlick, *Angew. Chem. Intern. Ed. Engl.*, **1**, 75 (1962).
8. H. W. Wanzlick, and H. J. Kleiner, *Ber.*, **96**, 3024 (1963).
9. H. W. Wanzlick and H. Ahrens, *Ber.*, **97**, 2447 (1964).
10. H. W. Wanzlick, B. Lachmann, and E. Schikora, *Ber.*, **98**, 3170 (1965).

3-BROMOACETOPHENONE*
(Acetophenone, 3-bromo-)

$$C_6H_5\overset{\overset{\displaystyle CH_3}{|}}{C}{=}O + AlCl_3 \rightarrow C_6H_5\overset{\overset{\displaystyle CH_3}{|}}{C}{=}OAlCl_3$$

$$C_6H_5\overset{\overset{\displaystyle CH_3}{|}}{C}{=}OAlCl_3 + Br_2 \xrightarrow{AlCl_3} 3\text{-}BrC_6H_4\overset{\overset{\displaystyle CH_3}{|}}{C}{=}OAlCl_3 + HBr$$

$$3\text{-}BrC_6H_4\overset{\overset{\displaystyle CH_3}{|}}{C}{=}OAlCl_3 \xrightarrow[H^+]{H_2O} 3\text{-}BrC_6H_4\overset{\overset{\displaystyle CH_3}{|}}{C}{=}O + Al^{3+} + 3Cl^-$$

Submitted by D. E. PEARSON, H. W. POPE, and W. W. HARGROVE.[1]
Checked by B. C. McKUSICK and D. W. WILEY.

1. Procedure

The apparatus consists of a 1-l. three-necked flask equipped with a condenser, a dropping funnel, and a stirrer terminating in a stiff, crescent-shaped Teflon polytetrafluoroethylene paddle. The stirrer motor must have good torque (Note 1). The assembled apparatus, which is protected from moisture by means of drying tubes in the condenser and funnel, is preferably pre-dried. About 216–224 g. (1.62–1.68 moles) of powdered anhydrous aluminum chloride is added to the apparatus with as little exposure to the moisture of the air as possible (Note 2). While the free-flowing catalyst is stirred (Note 3), 81 g. (0.67 mole) of acetophenone is added from the dropping funnel in a slow stream over a period of 20–30 minutes. Considerable heat is evolved, and, if the drops of ketone are not dispersed, darkening or charring occurs. When about one-third of the acetophenone has been added, the mixture becomes a viscous ball-like mass that is difficult to stir. Turning of the stirrer by hand or more rapid addition of ketone is necessary at this point. The addition of ketone, however, should not be so rapid as to produce a temperature above 180°. Near the end of the addition, the mass becomes molten and can be stirred easily without being either heated or cooled. The molten mass, in which the acetophenone is complexed with aluminum chloride, ranges in color from tan to brown.

Bromine (128 g., 0.80 mole) is added dropwise to the well-stirred mixture over a period of 40 minutes (Note 4). After all the bromine has been added, the molten mixture is stirred at 80–85° on a steam bath for 1 hour, or until it solidifies if that happens first (Note 5). The complex is added in portions to a well-stirred mixture of 1.3 l. of cracked ice and 100 ml. of concentrated hydrochloric acid in a 2-l. beaker (Note 6). Part of the cold aqueous layer is added to the reaction flask to decompose whatever part of the reaction mixture remains there, and the resulting mixture is added to the beaker. The dark oil that settles out is extracted from the mixture with four 150-ml. portions of ether. The extracts are combined, washed consecutively with 100 ml. of water and 100 ml. of 5% aqueous sodium bicarbonate solution, dried with anhydrous sodium sulfate, and transferred to a short-necked distillation flask. The ether is removed by distillation at atmospheric pressure, and crude 3-bromo-acetophenone is stripped from a few grams of heavy dark residue by distillation at reduced pressure. The colorless distillate is carefully fractionated in a column 20 cm. long and 1.5 cm. in diameter that is filled with Carborundum or Heli-Pak filling. The combined middle fractions of constant refractive index are taken as 3-bromoacetophenone; weight, 94–100 g. (70–75%); b.p. 75–76°/0.5 mm.; n_D^{25} 1.5738–1.5742; m.p. 7–8° (Notes 7 and 8).

2. Notes

1. Among satisfactory motors are the Sargent Cone Drive and the Waco.

2. Exposure of the aluminum chloride to air is conveniently avoided by introducing the entire contents of two 4-ounce bottles of anhydrous resublimed aluminum chloride of the Baker and Adamson Company directly into the reaction flask.

3. If the paddle width is so small as to leave isolated, un-agitated portions of aluminum chloride, it should be moved near the surface to disperse the ketone rapidly. If the ketone is not dispersed, condensation to dypnone occurs. Tars found in the stripping process are believed to originate from improper addition of the ketone to the aluminum chloride.

4. The rate of addition is regulated by the rate of evolution of hydrogen bromide. The yield of product is essentially the same whether the reaction mixture is held at 80–85° or at room temperature.

5. If the reaction mixture does not solidify during the heating, it is well to work it up at once while it can still be poured from the flask. Otherwise the work-up can be postponed to the next day. If the reaction mixture is too difficult to remove from the flask, the acid-ice slurry can be added *all at once* to the reaction flask immersed in ice. The vigorous surface decomposition is thus partly quenched. However, the cake is seldom difficult to remove unless polyhalogenation has occurred.

6. The acid prevents the formation of insoluble aluminum salts that make separation of ether-water layers difficult. It is helpful in this regard to stir the mixture of water, ketone, and acid for an hour or so before extracting the ketone with ether.

7. The present procedure has been used by the submitters to prepare the following 3-bromoacetophenones and benzaldehydes

in the indicated yields:[2] 3-bromopropiophenone, m.p. 40–41°, 60%; 3-bromo-4-methylacetophenone, m.p. 42–43°, 56%; 3,4-dibromoacetophenone, m.p. 89–90°, 55%; 3-bromo-4-*tert*-butyl-acetophenone, b.p. 92°/0.1 mm., 30%; 3,5-dibromo-4-methylace-tophenone, m.p. 102–103°, 57%; 3-bromobenzaldehyde, b.p. 105–106°/2 mm., 59%; 3-bromo-4-tolualdehyde, m.p. 48–49°, 44%.

8. The same procedure can be used to prepare 3-chloroaceto-phenones and benzaldehydes. The apparatus is modified by re-placing the dropping funnel with a gas-inlet tube that permits chlorine to be introduced under the surface of the molten com-plex of acetophenone and aluminum chloride. For a run with 81 g. (0.67 mole) of acetophenone, 31 ml. (48 g., 0.67 mole) of liquid chlorine is condensed in a trap cooled with solid carbon dioxide and acetone. The gas is passed consecutively through a safety trap, a bubble counter containing concentrated sulfuric acid, and the inlet tube into the stirred complex. The rate of addition is controlled by gradually lowering the cooling bath surrounding the liquid chlorine trap. The internal temperature of the reaction mixture rises just above room temperature and the color of the complex changes from light brown to deep red-brown. The addition of chlorine is complete in 10–14 hours; with a faster rate of addition, some chlorine escapes. Stirring is continued for another hour, and the reaction mixture is worked up. The submitters have prepared the following in this way:[2] 3-chloroacetophenone, b.p. 61–63°/0.5 mm., 54%; 3-chlorobenz-aldehyde, b.p. 93–96°/15 mm., 43%; 2,3,5,6-tetrachloro-4-methylacetophenone, m.p. 98.5–99.5°, 67%.

3. Methods of Preparation

Nuclear halogenation of acetophenone depends on formation of the aluminum chloride complex. If less than one equivalent of aluminum chloride is used, side-chain halogenation occurs.[3] 3-Bromoacetophenone has been prepared from 3-aminoaceto-phenone by the Sandmeyer reaction.[4,5] The synthesis described here has been taken from work of the submitters,[2] who have used it to prepare many 3-bromo- and 3-chloroacetophenones and benzaldehydes, as well as more highly halogenated ones (Notes 7 and 8).

1. Department of Chemistry, Vanderbilt University, Nashville, Tennessee.
2. D. E. Pearson, H. W. Pope, W. W. Hargrove, and W. E. Stamper, *J. Org. Chem.,* 23, 1412 (1958).
3. R. M. Cowper and L. H. Davidson, *Org. Syntheses,* Coll. Vol. 2, 480 (1943).
4. L. A. Elson, C. S. Gibson, and J. D. A. Johnson *J. Chem. Soc.,* 1128 (1930).
5. C. S. Marvel, R. E. Allen, and C. G. Overberger, *J. Am. Chem. Soc.,* 68, 1089 (1946).

2-BROMOALLYLAMINE

(Allylamine, 2-bromo-)

$$CH_2\!\!=\!\!\overset{\overset{\displaystyle Br}{|}}{C}\!\!-\!\!CH_2Br + C_6H_{12}N_4 \xrightarrow{\text{CHCl}_3}$$

$$CH_2\!\!=\!\!\overset{\overset{\displaystyle Br}{|}}{C}\!\!-\!\!CH_2\overset{+}{N}C_6H_{12}N_3\ \overset{-}{Br} \xrightarrow{\text{4HCl, 6H}_2\text{O}}$$

$$6CH_2O + 3NH_4Cl + CH_2\!\!=\!\!\overset{\overset{\displaystyle Br}{|}}{C}\!\!-\!\!CH_2\overset{+}{N}H_3\ \overset{-}{Cl} \xrightarrow{\text{NaOH}}$$

$$CH_2\!\!=\!\!\overset{\overset{\displaystyle Br}{|}}{C}\!\!-\!\!CH_2NH_2 + H_2O + NaCl$$

Submitted by ALBERT T. BOTTINI, VASU DEV, and JANE KLINCK.[1]
Checked by A. S. PAGANO and W. D. EMMONS.

1. Procedure

Caution!   Contact with 2-bromoallylamine can cause severe eye and skin irritation.   This preparation should be carried out in a good hood, and the operator should wear protective goggles and rubber gloves.

A. *2-Bromoallylhexaminium bromide.*   A 2-l. three-necked flask fitted with a Hershberg stirrer,[2] a dropping funnel, and a condenser is charged with a solution of 154 g. (1.10 moles) of hexamethylenetetramine (Note 1) in 1250 ml. of chloroform. The solution is stirred and heated under reflux while 200 g. (1.00 mole) of 2,3-dibromopropene (Note 2) is added dropwise over a period of 1 hour.   Precipitation of the product is noted soon after the first addition of 2,3-dibromopropene. After the addition is complete, the reaction mixture is stirred under reflux for 3 hours and allowed to stand overnight. The mixture is cooled in an ice bath, and the salt is collected by suction filtration. After air-drying, the crude yellow 2-bromoallylhexaminium bromide weighs 292–308 g. (86–91%) and melts at 183–186°.

B. *2-Bromoallylamine*. Crude 2-bromoallylhexaminium bromide (204 g., 0.60 mole) is dissolved in a warm solution prepared from 400 ml. of water, 2 l. of ethanol, and 480 ml. (5.8 moles) of 12N hydrochloric acid. A white precipitate of ammonium chloride forms within an hour. The reaction mixture is allowed to stand for 24 hours, and the precipitate is removed by suction filtration. The mother liquor is concentrated to a volume of 600 ml. (Note 3), and the precipitate (Note 4) is removed by suction filtration. The mother liquor is evaporated to dryness (Note 5), and the residue is dissolved in 300 ml. of water. The solution is cooled in an ice bath and made strongly alkaline (pH 13) with 6N sodium hydroxide solution.

The two-phase mixture is placed in a separatory funnel, and the heavy red-brown oil is separated. The aqueous phase is extracted with 100 ml. of ether. The oil and the ether extract are combined, washed with 50 ml. of saturated sodium chloride, and dried over potassium carbonate. The drying agent is removed by filtration, and the filtrate is distilled. Colorless 2-bromoallylamine is collected at 65–68°/100 mm.; weight 49–59 g. (59–72%); n_D^{25} 1.5075–1.5085 (Note 6).

2. Notes

1. The submitters used hexamethylenetetramine obtained from Matheson, Coleman and Bell.

2. The 2,3-dibromopropene was obtained from Columbia Organic Chemicals Co., Columbia, South Carolina, and was redistilled before use. The preparation of 2,3-dibromopropene is described in an earlier volume of this series.[3]

3. The submitters divided the mother liquor into 6 equal portions and concentrated each to a volume of 100 ml. at a pressure of 25 mm. in a 1-l. round-bottomed flask on a rotary film evap-

orator. The rotary film evaporator used was obtained from Cenco Scientific Co., Santa Clara, California.

4. The precipitate is ammonium chloride that contains virtually no 2-bromoallylamine hydrochloride.

5. The submitters used a rotary film evaporator to evaporate the mother liquor at a pressure of 25 mm. in a water bath heated to 90°.

6. 2-Bromoallylamine discolors slowly even when stored at 0° in a dark container. The refractometer to be used for determination of the refractive index should be placed in a good hood.

3. Methods of Preparation

2-Bromoallylamine has been prepared by heating N-(2-bromoallyl)-phthalimide with hydrazine in methanol; [4] by treatment of 2,3-dibromopropylamine hydrochloride with excess alcoholic potassium hydroxide; [5] by treatment of 1,2,3-tribromopropane with alcoholic ammonia at 100°; [6] and by the present procedure.[7]

4. Merits of the Preparation

This method gives better yields than other methods of preparation of 2-bromoallylamine, and it is the most convenient method for the preparation of large quantities of the compound. The procedure illustrates a reaction, the so-called Delépine reaction, that has been used for the preparation of many primary aliphatic amines.[8-13] It is especially useful in the preparation of derivatives of phenacylamine.[14-16] A number of primary aliphatic amines have been prepared by this method without isolation of the intermediate hexaminium salt.[11] Several preparations of aliphatic aldehydes via the hexaminium salt have been described in earlier volumes of this series.[17]

1. Department of Chemistry, University of California, Davis, California.
2. P. S. Pinkney, *Org. Syntheses,* Coll. Vol. 2, 116 (1943).
3. R. Lespieau and M. Bourguel, *Org. Syntheses,* Coll. Vol. 1, 209 (1932).
4. J. A. Lamberton, *Australian J. Chem.,* 8, 289 (1955).
5. C. Paal, *Ber.,* 21, 3190 (1888).
6. P. Galewsky, *Ber.,* 23, 1067 (1890).

7. A. T. Bottini and V. Dev, *J. Org. Chem.*, **27**, 968 (1962).
8. A. Wohl, *Ber.*, **19**, 1840 (1886).
9. M. Delepine, *Compt. Rend.*, **120**, 501 (1895); **124**, 292 (1897); *Bull. Soc. Chim. France*, [3] **17**, 290 (1897).
10. M. Delepine and P. Jaffeux, *Bull. Soc. Chim. France*, [4] **31**, 108 *(1922)*.
11. A. Galat and G. Elion, *J. Am. Chem. Soc.*, **61**, 3585 (1939).
12. K. E. Schulte and M. Goes, *Arch. Pharm.*, **290**, 118 (1957).
13. N. L. Wender, *J. Am. Chem. Soc.*, **71**, 375 (1949).
14. H. E. Baumgarten and J. M. Petersen, this volume, p. 909 and references 5–11 therein.
15. L. M. Long and H. D. Troutman, *J. Am. Chem. Soc.*, **71**, 2473 (1949).
16. M. C. Rebsock, C. D. Stratton, and L. L. Bambas, *J. Am. Chem. Soc.*, **77**, 24 (1955).
17. K. B. Wiberg, *Org. Syntheses*, Coll. Vol. 3, 811 (1955); S. J. Angyal, J. R. Tetaz, and J. G. Wilson, *Org. Syntheses*, Coll. Vol. 4, 690 (1963); E. Campaigne, R. C. Bourgeois, and W. C. McCarthy, *Org. Syntheses*, Coll. Vol. 4, 918 (1963).

N-(2-BROMOALLYL)ETHYLAMINE

(Allylamine, 2-bromo-N-ethyl-)

$$CH_2{=}CCH_2Br + 2C_2H_5NH_2 \rightarrow$$
$$\overset{|}{Br}$$

$$CH_2{=}CCH_2NHC_2H_5 + C_2H_5\overset{+}{N}H_3 \ \ Br^-$$
$$\overset{|}{Br}$$

Submitted by ALBERT T. BOTTINI and ROBERT E. OLSEN [1]
Checked by THOMAS H. LOWRY and E. J. COREY

1. Procedure

Caution! This preparation should be carried out in a hood to avoid exposure to ethylamine, 2,3-dibromopropene, and the product. 2,3-Dibromopropene is a strong lachrymator. The operator should wear rubber gloves and protective goggles because some 2-haloallyl-amines have caused severe skin and eye irritation.

A 1-l. three-necked flask is fitted with a sealed mechanical stirrer, a dropping funnel, and a dry ice condenser charged with an ice-salt mixture (Note 1). Three hundred milliliters (240 g., 3.7 moles) of aqueous 70% ethylamine solution (Note 2) is placed in the flask, the stirrer is started, and 200 g. (1.00 mole) of 2,3-dibromopropene (Note 3) is added dropwise over a period of 1 hour. After the addition is complete, the reaction mixture is stirred for 3 hours. Ether (300 ml.) is added, and the mixture is cooled in an ice bath. Sodium hydroxide (100 g.) is added with stirring and cooling. The cold mixture is transferred to a separatory funnel, and the phases are separated. The organic layer is dried in two stages over 25-g. portions of sodium hydroxide. The organic layer and the small amount of water that separates during the second stage of drying are decanted into a separatory funnel, and the phases are separated. Most of the ether and unreacted ethylamine are removed from the organic layer by distillation through a 250-mm. x 13-mm. column packed with glass helices, and the residue is distilled through the same column at reduced pressure under nitrogen to give 115–128 g. (70–78%) of N-(2-bromoallyl)ethylamine; b.p. 53–55° (27 mm.), 79–81° (75 mm.) (Note 4); n^{25}D 1.4765–1.4770.

2. Notes

1. The checkers used an inner-spiral water condenser. The cooling water was chilled to about 0° by prior passage through a short copper coil immersed in ice.

2. The aqueous 70% ethylamine solution used was the practical grade obtained from Eastman Organic Chemicals.

3. The 2,3-dibromopropene used was obtained from Columbia Organic Chemicals Co., Columbia, South Carolina, and was redistilled before use. The preparation of 2,3-dibromopropene is described in an earlier volume of this series.[2]

4. The reported boiling point of N-(2-bromoallyl)ethylamine is 148–153°.[3] It is strongly recommended that the product and other 2-haloallylamines be distilled at reduced pressure under nitrogen, for the submitters have noted two instances when a 2-haloallylamine polymerized with considerable evolution of heat during slow distillation at atmospheric pressure.

3. Methods of Preparation

This method is essentially that described by Pollard and Parcell.[3] No other procedure appears to have been used to prepare

N-(2-bromoallyl)ethylamine. A number of N-(2-haloallyl)alkyl-amines have been prepared by treatment of a 2,3-dihalopropene with a primary alkylamine in water,[3,4] ether,[3,4] or benzene.[5]

4. Merits of the Preparation

The method described here has been used for the preparation of a number of N-(2-haloallyl)alkylamines from a water-soluble amine and the corresponding 2,3-dihalopropene.[3,4]

Treatment of an N-(2-bromoallyl)alkylamine with sodium amide in liquid ammonia yields the N-alkylallenimine together with a small amount of the N-alkylpropargylamine.[3-7] Similar treatment of an N-(2-chloroallyl)alkylamine yields only the N-alkylpropargylamine.[4,6]

1. Department of Chemistry, University of California, Davis, California.
2. R. Lespieau and M. Bourguel, *Org. Syntheses*, Coll. Vol. 1, 209 (1941).
3. C. B. Pollard and R. F. Parcell, *J. Am. Chem. Soc.*, **73**, 2925 (1951).
4. A. T. Bottini, B. J. King, and R. E. Olsen, *J. Org. Chem.*, **28**, 3241 (1963).
5. J. V. Braun, M. Kuhn, and J. Weismantel, *Ann.*, **449**, 254 (1926).
6. A. T. Bottini and J. D. Roberts, *J. Am. Chem. Soc.*, **79**, 1462 (1957).
7. A. T. Bottini and R. E. Olsen, this volume, p. 541.

BROMOCYCLOPROPANE*

(Cyclopropane, bromo-)

$$2 \; \triangleright\!\!-CO_2H + HgO + 2Br_2 \longrightarrow 2 \; \triangleright\!\!-Br + 2CO_2 + HgBr_2 + H_2O$$

Submitted by JOHN S. MEEK and DAVID T. OSUGA.[1]
Checked by F. S. FAWCETT and B. C. McKUSICK.

1. Procedure

Twenty-four grams (0.11 mole) of red mercuric oxide (Note 1) and 60 ml. of freshly distilled 1,1,2,2-tetrachloroethane are placed in a 250-ml. three-necked flask equipped with a dropping funnel, a reflux condenser, and a stirrer. A solution of 32.2 g. (0.20 mole) of bromine and 17.2 g. (0.20 mole) of cyclopropanecarboxylic acid in 50 ml. of tetrachloroethane is added dropwise to the stirred suspension of mercuric oxide over a period of 45 minutes, the flask being kept in a water bath at 30–35° (Note 2). The mixture is stirred after the addition of the reactants until the evolution of carbon dioxide ceases.

The flask is then cooled in ice water, and the contents are filtered with as little suction as possible (Note 3). The filter cake is pressed dry and washed with three 15-ml. portions of tetrachloroethane first used to rinse out the flask. The combined filtrates are dried with a little calcium chloride. Sometimes the solution contains a little bromine; it is removed by adding allyl alcohol dropwise until the bromine color is discharged (usually 0.5–1.0 ml. suffices).

The solution is decanted into a 200-ml. round-bottomed flask containing a carborundum chip. The material is distilled through a 20-cm. column of glass helices or a 30-cm. spinning-band column. The fore-run boiling below 75°/760 mm. is bromocyclopropane pure enough for most purposes; weight 9.8–11.2 g. (41–46%); n_D^{25} 1.455–1.459; d_4^{26} 1.506 (Note 4). Redistillation of this product gives pure bromocyclopropane, b.p. 69°/760 mm., n_D^{25} 1.4570, with but slight loss (Note 5).

2. Notes

1. The mercuric oxide used was Mallinckrodt or Baker powdered red mercuric oxide, analytical reagent grade. Old mercuric oxide gives variable results and may lower the yield. The 1,1,2,2-tetrachloroethane used was a technical grade and was distilled to make sure no low-boiling impurities were present. Reagent-grade solvent has been used without distillation. The vapors of this chlorinated hydrocarbon are toxic, and its distillation as well as the reaction should be carried out in a hood. Suitable cyclopropanecarboxylic acid [2] is obtainable from Aldrich Chemical Company.

2. The reaction starts spontaneously and is mildly exothermic. Moderating the temperature by use of a water bath diminishes the amount of bromine and product carried off by the carbon dioxide evolved. The reaction can be followed by use of a tetrachloroethane bubbler, and at the end of the reaction the solvent

in the bubbler can be used to wash the mercuric bromide. The checkers followed the reaction with a wet test meter presaturated with carbon dioxide; 52–60% of the theoretical amount of carbon dioxide was evolved.

3. The checkers used a sintered glass pressure filter (Corning Glass Works, Cat. No. 34020) rather than a suction filter in order to minimize evaporation losses. An ordinary water aspirator can cause the mixture to boil at room temperature. The flask and filter can be cleaned readily with a little acetone, which dissolves mercuric bromide rapidly.

4. Once the boiling point starts to rise, it goes up quite rapidly. The fractions collected between 75° and 90° contain a little product and can be reworked if a second distillation is carried out.

5. After publication of this procedure, the submitters increased the yield to 53–65% by the following modification. A Barrett distilling receiver and thermometer are added to the described apparatus. A mixture of 60 ml. of undistilled technical 1,1,2,2-tetrachloroethane, 27.5 g. (0.125 mole) of mercuric oxide, and 21.5 g. (0.25 mole) of cyclopropanecarboxylic acid is heated to remove 10 ml. of solvent and water; during the distillation the mercuric oxide dissolves to form a solution of mercuric cyclopropanecarboxylate. The solution is cooled to about 70°, the distilling receiver is removed, and 40 g. (0.25 mole) of bromine is added dropwise with swirling over a period of 15–20 minutes. Mercuric bromide starts to precipitate when 50–75% of the bromine has been added and may cause frothing if the last of the bromine is added too rapidly. The evolution of carbon dioxide tapers off and ceases about 15 minutes after addition is complete; the reaction temperature is about 55° at this point. Any bromine color is discharged with allyl alcohol as described above. The condenser is replaced by one of the distillation columns specified above, and the reaction mixture is distilled without filtration. Stirring helps, but is not essential, for the finely divided mercuric bromide does not cause bumping. The yield of bromocyclopropane having the properties given above is 16–17 g. (53–56%).

3. Methods of Preparation

Bromocyclopropane has been prepared by the Hunsdiecker reaction by adding silver cyclopropanecarboxylate to bromine in dichlorodifluoromethane at −29° (53% yield) or in tetrachloroethane at −20° to −25° (15–20% yield).[3] Decomposition of the peroxide of cyclopropanecarboxylic acid in the presence of carbon tetrabromide gave bromocyclopropane in 43% yield.[4] An attempt to prepare the bromide via the von Braun reaction was unsuccessful.[3] Ten percent yields are reported for the photobromination of cyclopropane[5] and the photochemical rearrangement of allyl bromide.[6] Treatment of 1,1,3-tribromopropane with methyllithium prepared from methyl bromide furnishes a 60–65% yield of bromocyclopropane.

4. Merits of the Preparation

The present procedure is substantially simpler and quicker than the best previous procedure[3] based on using cyclopropanecarboxylic acid, which requires 4 days instead of 4 hours. It is also safer, for no explosions have been encountered; whereas care must be taken to prevent explosion of the intermediate hypobromite when the Hunsdiecker method is used,[3] and one detonation has been reported.[8]

Although the treatment of 1,1,3-tribromobutane with methyllithium is a safe procedure offering an attractive yield, it suffers from the fact that the tribromide is not commercially available while cyclopropanecarboxylic acid is.

A recent illustration of the generality of this method for preparing alkyl bromides from acids is provided by an *Organic Syntheses* procedure for 1-bromo-3-chlorocyclobutane.[10] To aid in isolating higher boiling or solid products, solvents such as carbon tetrachloride and cyclohexane can be used.[9] In preparing a solid, mercuric bromide can be removed by extraction with 5% potassium iodide. It should also be noted that mercuric bromide can be converted back to mercuric oxide easily with alkali

1. Department of Chemistry, University of Colorado, Boulder, Colorado.
2. C. M. McCloskey and G. H. Coleman, *Org. Syntheses*, Coll. Vol. 3, 221 (1955).
3. J. D. Roberts and V. C. Chambers, *J. Am. Chem. Soc.*, **73**, 3176, 5030 (1951).
4. E. Renk, P. R. Shafer, W. H. Graham, R. H. Mazur, and J. D. Roberts, *J. Am. Chem. Soc.*, **83**, 1987 (1961).
5. E. L. Dedio, P. J. Kozak, S. N. Vinogradov, and H. E. Gunning, *Can. J. Chem.*, **40**, 820 (1962).
6. S. J. Cristol and G. A. Lee, *J. Am. Chem. Soc.*, **91**, 7554 (1964).
7. W. Kirmse and B. G. v. Wedel, *Ann.*, **676**, 1 (1964).
8. J. W. Rowe, Master's Thesis, University of Colorado, 1952.
9. S. J. Cristol and W. C. Firth, Jr., *J. Org. Chem.*, **26**, 280 (1961).
10. G. M. Lampman and J. C Aumiller, *Org. Syntheses*, **51**, 106 (1971).

p-BROMODIPHENYLMETHANE

[Methane, (*p*-bromophenyl)phenyl-]

$$C_6H_5CHO + CHCl_3 \xrightarrow{\text{KOH}} C_6H_5\underset{\underset{CCl_3}{|}}{C}HOH \xrightarrow[\text{H}_2\text{SO}_4]{C_6H_5Br}$$

$$p\text{-}BrC_6H_4\underset{\underset{CCl_3}{|}}{C}HC_6H_5 \xrightarrow{\text{KOH}} p\text{-}BrC_6H_4CH_2C_6H_5$$

Submitted by A. B. Galun[1] and A. Kalir[2]
Checked by R. Breslow and H. T. Bozimo

1. Procedure

A. *1-Phenyl-2,2,2-trichloroethanol.* In a 1-l. round-bottomed-flask fitted with a mechanical stirrer, a thermometer, and a powder funnel is placed a solution of 212 g. (2.00 moles) of freshly distilled benzaldehyde in 400 g. (270 ml., 3.35 moles) of chloroform. The mixture is cooled in an ice bath, and 123 g. of commercial powdered potassium hydroxide is added with stirring at such a rate that the temperature of the solution does not exceed 45° (1–1.5 hours). The reaction mixture is stirred and kept at 40–50° for an additional hour and then poured into a solution of 60 ml. of sulfuric acid in 3 l. of water. The resulting two-phase mixture is transferred to a separatory funnel and extracted with three 250-ml. portions of chloroform (a small amount of insoluble, black resinous material is discarded). The combined organic layers are washed with three 100-ml. portions of aqueous 10% sodium carbonate, dried over anhydrous magnesium sulfate, and filtered into a 1-l. flask. The solvent is removed under reduced pressure on a hot water bath. The residue is transferred to a 250-ml. flask and distilled under reduced pressure to give 1-phenyl-2,2,2-trichloroethanol, b.p. 155–165° (26 mm.), 90–100° (0.5 mm.) (Notes 1 and 2). The yield is 170–180 g. (38–40%).

B. *1-p-Bromophenyl-1-phenyl-2,2,2-trichloroethane.* In a 500-ml. round-bottomed flask fitted with a mechanical stirrer, a dropping funnel, and a thermometer are placed 136 g. (0.60 mole) of 1-phenyl-2,2,2-trichloroethanol and 120 g. (81 ml., 0.77 mole) of bromobenzene. The flask is cooled in an ice-water bath, and a mixture of 120 ml. of concentrated sulfuric acid and 50 ml. of oleum (20% SO_3) is added with stirring at such a rate that the temperature of the reaction mixture does not exceed 10° (*ca.* 45 minutes) (Note 3). The mixture is stirred for another 30 minutes at 10° and for 4–5 hours at room temperature. It is then poured with manual stirring onto 1 kg. of cracked ice, and the mixture is allowed to stand overnight. The precipitate (Note 4) is filtered, washed with water, and recrystallized from 300 ml. of ethanol (Note 5). The yield is 129–162 g. (59–74%), m p. 95–96° (Note 6).

C. *p-Bromodiphenylmethane.* A 2-l. three-necked flask fitted with a distillation condenser, a thermometer, and an efficient mechanical stirrer is charged with 1.1 l. of diethylene glycol (Note 7) and a solution of 190 g. of potassium hydroxide in 100 ml. of water. The mixture is stirred, and water is distilled until the internal temperature reaches 180°. The resulting solution is allowed to cool to 100° or below, and 146 g. (0.40 mole) of 1-*p*-bromophenyl-1-phenyl-2,2,2-trichloroethane (Note 8) is added. The condenser is set for reflux, and the mixture is stirred and heated to boiling for 5 hours (Note 9). The hot solution is then poured onto 3 kg. of cracked ice, and the mixture is allowed to stand overnight. The oily layer is separated and dissolved in ether (any insoluble material is discarded), and the aqueous layer is extracted with 250 ml. of ether. The combined ethereal solution and extracts are dried over calcium chloride and filtered. The ether is removed under reduced pressure on a hot water bath. The product is distilled under reduced pressure; b.p. 120–130° (3 mm.), 155–163° (13 mm.) (Note 10), n^{24}D 1.6028, d_{24}^{24} 1.342. The yield is 74–79 g. (75–80%) (Note 11).

2. Notes

1. The purpose of the distillation is to separate the product from tars. Therefore no fractionation is required, and the distillation may be carried out rapidly.

2. The carbinol, which has a tendency to supercool, may crystallize overnight; m.p. 38°.

3. Solid material is sometimes deposited on the walls of the reaction flask.

4. In some cases the organic layer separates as an oil; it is then obtained in crystalline form by trituration with 200 ml. of cold methanol, which dissolves the excess of bromobenzene.

5. Wet material may require larger amounts of ethanol.

6. Trituration of the crude precipitate with methanol gives a 90% yield of material, m.p. 90–93°.

7. Eastman Organic Chemicals white label 2,2-oxydiethanol was used.

8. The material should be thoroughly freed of alcohol, preferably over phosphorus pentoxide under reduced pressure, before use. Even traces of alcohol may reduce the yield to 60%.

9. The temperature of the refluxing solution should be above 165°. Efficient stirring is essential; otherwise the precipitating potassium carbonate entrains much material, causing reduction of yield.

10. Good fractionation is not required.

11. Runs on a fourfold scale give the same yield.

3. Methods of Preparation

The procedure for the preparation of 1-phenyl-2,2,2-trichloroethanol is based on the work of Bergmann, Ginsburg, and Lavie.[3] 1-Phenyl-2,2,2-trichloroethanol has also been prepared from phenylmagnesium bromide and chloral.[4]

p-Bromodiphenylmethane has been reported as a product of the reduction of p-bromobenzophenone with hydriodic acid and red phosphorus in a sealed tube at 160°.[5] The present method is a modification of the synthesis published by Galun, Kaluszyner, and Bergmann.[6]

4. Merits of the Preparation

In this method inexpensive, commercially available chemicals are used as starting materials. The operations are simple, the yields acceptable, and the final products are free of isomers.

This procedure is especially suited for preparing variously substituted diarylmethanes.[6] The 1,1-diaryl-2,2,2-trichloro-ethanes may be converted to the corresponding benzophenones via the 1,1-diaryl-2,2-dichloroethylenes [7] and to 1,1-diarylacetic acids.[8]

1. "Zion" Chemical Products, Ltd., Yavne, Israel.
2. Israel Institute for Biological Research, Ness-Ziona, Israel.
3. E. D. Bergmann, D. Ginsburg, and D. Lavie, *J. Am. Chem. Soc.*, **72**, 5012 (1950).
4. P. Hébert, *Bull. Soc. Chim. France*, [4] **27**, 45 (1920); Z. I. Iotsich, *Zh. Russ. Fiz.-Khim. Obshch.*, **34**, 96 (1902).
5. J. H. Speer and A. J. Hill, *J. Org. Chem.*, **2**, 139 (1937).
6. A. B. Galun, A. Kaluszyner, and E. D. Bergmann, *J. Org. Chem.*, **27**, 1426 (1962).
7. O. Grummitt, A. Buck, and A. Jenkins, *J. Am. Chem. Soc.*, **67**, 155 (1945).
8. O. Grummitt, A. Buck, and R. Egan, *Org. Syntheses*, Coll. Vol. **3**, 270 (1955).

1-BROMO-2-FLUOROBENZENE*

(Benzene, 1-bromo-2-fluoro-)

Submitted by K. G. RUTHERFORD and W. REDMOND.[1]
Checked by M. PAULSHOCK and B. C. McKUSICK.

1. Procedure

A. *o-Bromobenzenediazonium hexafluorophosphate.* A solution of 95 ml. of 12N hydrochloric acid in 650 ml. of water is added with stirring to 60 g. of o-bromoaniline (0.35 mole; Note 1) in a 2-l. three-necked flask equipped with stirrer and thermometer. Solution is effected by heating the mixture on a steam bath (Note 2). A solution of 29 g. (0.42 mole) of sodium nitrite in 75 ml. of water is added with stirring while the mixture is maintained at $-5°$ to $-10°$ by means of a bath of ice and salt or of dry ice and acetone. At the end of the addition there is an excess of nitrous acid, which can be detected with starch iodide paper. Seventy-four milliliters (134 g., 0.60 mole) of 65% hexafluorophosphoric acid (Note 3) is added in one portion, with vigorous stirring, to the cold solution of the diazonium salt. Cooling and slow stirring are continued for an additional 30 minutes, and the precipitated diazonium hexafluorophosphate is then collected on a Büchner funnel. The diazonium salt is washed on the funnel with 300 ml. of cold water and with a solution of 80 ml. of methanol in 320 ml. of ether (Note 4). The salt is partly dried by drawing air through the funnel for 2 hours. It is then transferred to a pile of several filter papers, powdered with a spatula, and dried at about $25°/1$ mm. for at least 12 hours. The dried o-bromobenzenediazonium hexafluorophosphate is cream-colored; weight 108–111 g. (94–97%); m.p. 151–156° (dec.) (Note 5).

B. *1-Bromo-2-fluorobenzene. Caution! This step should be carried out in a hood because the PF₅ evolved on thermal decomposition of the diazonium salt is poisonous.* The apparatus consists of a 1-l., three-necked, round-bottomed flask equipped with a thermometer, a condenser, a magnetic stirrer (optional), and a 250-ml. Erlenmeyer flask that is attached by means of a short rubber Gooch connecting tube. The dry powdered hexafluorophosphate salt is placed in the Erlenmeyer flask, and 300 ml. of heavy mineral oil is placed in the round-bottomed flask. The mineral oil is heated to 165–170° by means of an oil bath or electric heating mantle and maintained at this temperature while the salt is added rapidly in portions over a period of 30 minutes. The flask is cooled rapidly to room temperature, the side flask is removed, and 400 ml. of 10% aqueous sodium carbonate is added slowly through the condenser. The mixture is steam-distilled until no more oil is visible in the distillate.

The oil, which is heavier than water, is separated, and the aqueous layer is extracted with three 50-ml. portions of methylene chloride. The oil and extracts are combined, dried over anhydrous sodium sulfate, and distilled from a Claisen flask with an indented neck. Colorless 1-bromo-2-fluorobenzene is collected at 58–59°/17 mm. or 156–157°/760 mm.; weight 45–47 g. (73–75% based on o-bromoaniline); n_D^{25} 1.5320–1.5325.

2. Notes

1. o-Bromoaniline obtained from Eastman Kodak and used without redistillation is satisfactory.

2. The amine is dissolved to ensure its complete conversion to the hydrochloride. The amine hydrochloride may partly crystallize as the solution is cooled, but it redissolves as diazotization proceeds.

3. The 65% hexafluorophosphoric acid (density 1.81) was obtained from the Ozark-Mahoning Company, Tulsa, Oklahoma. A graduated polypropylene (Nalgene®) cylinder was used to contain the measured quantity of the acid. Rubber gloves should be worn as a precautionary measure against burns. Working in a hood prevents any contact of exposed parts of the body with fumes. *In the event of accidental contact of the acid with the skin, the affected place should be immediately washed well with running water and then treated with a paste of magnesium oxide and glycerol* [2] *or soaked in ice water.*

4. The methanol-ether filtrate has a slight yellow color. It is not known what impurity is removed by this solvent pair. However, the submitters found that this treatment improved the yield of several aryl fluorides prepared according to the present procedure.

5. The checkers had o-bromobenzenediazonium hexafluorophosphate examined in laboratories of the Du Pont Co. Explosives Department to see if it could be detonated. It was found sensitive to neither shock nor static electricity, and to decompose but not detonate when rapidly heated to 250°. Hence it probably does not present an explosion hazard, but it should be kept away from heat, especially if in a closed container.

3. Methods of Preparation

1-Bromo-2-fluorobenzene has been prepared in 37% yield by the Schiemann reaction from *o*-bromoaniline, nitrous acid, and fluoboric acid.[3,4] The present procedure [5] is a modification of the Schiemann reaction.

4. Merits of the Preparation

This procedure is a general way of converting arylamines to aryl fluorides, for it has been used to make fifteen other aryl fluorides. It generally gives better yields than the Schiemann reaction.

1-Bromo-2-fluorobenzene is used to prepare the highly reactive intermediate, benzyne.[6]

1. Department of Chemistry, Essex College, Assumption University of Windsor, Windsor, Ontario, Canada.
2. D. T. Flood, *Org. Syntheses,* Coll. Vol. 2, 297 (1943).
3. E. Bergmann, L. Engel, and S. Sandor, *Z. Physik. Chem. (Leipzig),* **10B**, 117 (1930).
4. M. S. Kharasch, H. Pines, and J. H. Levine, *J. Org. Chem.,* 3, 347 (1938).
5. K. G. Rutherford, W. Redmond, and J. Rigamonti, *J. Org. Chem.,* 26, 5149 (1961).
6. G. Wittig and L. Pohmer, *Ber.,* 89, 1334 (1956).

1-BROMO-2-FLUOROHEPTANE

(Heptane, 1-bromo-2-fluoro-)

$$CH_3(CH_2)_4CH=CH_2 + CH_3CONHBr + HF \rightarrow$$
$$CH_3(CH_2)_4CHFCH_2Br + CH_3CONH_2$$

Submitted by F. H. Dean, J. H. Amin, and F. L. M. Pattison [1]
Checked by P. B. Sargeant and B. C. McKusick

1. Procedure

Caution! Hydrogen fluoride is very hazardous. All operations must be carried out in a hood, and the precautions outlined in Note 1 should be followed.

A 1-l. polyethylene bottle is fitted with a three-holed rubber stopper.   Three lengths of 0.25-in. (6-mm.) stiff polyethylene tubing extend through the stopper into the bottle.   One length of tubing serves as an inlet tube for dry nitrogen, which is monitored by first bubbling through mineral oil.   The second tube, the gas-outlet tube, carries a polyethylene drying tube packed with indicating Drierite®.   These tubes extend only about 2 cm. into the bottle.   The third tube serves as an inlet for anhydrous hydrogen fluoride and extends halfway into the bottle; it is connected to a hydrogen fluoride cylinder by a short length of Tygon® tubing secured to the cylinder outlet by copper wire. The bottle contains a Teflon®-covered magnetic stirring bar.

The bottle is flushed with dry nitrogen, and a slow stream of nitrogen passes through the bottle during all subsequent operations to ensure the exclusion of atmospheric moisture (Note 2). N-Bromoacetamide (80 g., 0.58 mole) is added (Note 3).   The bottle is cooled in a mixture of dry ice and acetone, and 250 ml. of anhydrous ether is added with efficient magnetic stirring. About 100 g. (100 ml., 5 moles) of anhydrous hydrogen fluoride is allowed to condense into the bottle with magnetic stirring (Note 4).   This requires about 2 hours.

1-Heptene (49 g., 0.50 mole) (Note 3) is mixed with 50 ml. of anhydrous ether.   The solution is added during 20 minutes through what was originally the nitrogen inlet; the hydrogen fluoride inlet now serves as the nitrogen inlet.   The reaction mixture is stirred in the dry ice bath for an additional 4 hours.   The dry ice bath is replaced by an ice bath, and the mixture is stirred for 40 minutes. It is then allowed to stand overnight in a mixture of dry ice and acetone in a Dewar flask.

A solution of 690 g. (5.0 moles) of potassium carbonate in 2 l. of distilled water is prepared in a 4-l. polyethylene beaker or pail, and 500 g. of crushed ice and 300 ml. of ether are added. The cold reaction mixture is cautiously added to the carbonate solution with stirring.   The pH of the aqueous layer becomes about 9.   The ether layer is separated, and the aqueous layer is extracted with three 200-ml. portions of ether.   The ether solutions are combined and washed with three 100-ml. portions of water.   The ether solution is dried over anhydrous sodium sulfate, and the ether is removed by distillation.   The oily residue is fractionated through a 15-cm. Vigreux column under reduced pressure. There is a fore-run of about 0.5 ml., and then 59–75 g. (60–77%) of 1-bromo-2-fluoroheptane is collected at 70–78° (15 mm.); n^{25}D 1.4408–1.4420. According to vapor phase chromatography, it is about 90% pure (Note 5).

2. Notes

1. Because of the hazardous nature of anhydrous hydrogen fluoride, *adequate precautions should be taken to protect the head, eyes, and skin*. Use of rubber gloves, an apron, and a plastic face mask is strongly recommended. All operations should be carried out in a hood. After completion of the reaction, all equipment should be washed with liberal quantities of water. A bottle containing magnesium oxide paste in glycerin should be available in case of emergency. *Note! Burns caused by hydrogen fluoride may not be noticed for several hours, by which time serious tissue damage may have occurred.*

The checkers recommend that, if hydrogen fluoride comes in contact with the skin, the contacted area be thoroughly washed with water and then immersed in ice water while the patient is taken to a physician.

2. Moisture or inefficient stirring reduces the yield considerably.

3. Satisfactory sources of chemicals are: N-bromoacetamide, Arapahoe Chemicals, Boulder, Colorado; 1-heptene, Aldrich Chemical Co.; hydrogen fluoride, Matheson Co.

4. The amount of hydrogen fluoride is not critical. The amount of hydrogen fluoride may be estimated by condensing in enough to increase the volume of the reaction mixture by 100 ml.

5. According to vapor phase chromatography in a 6-ft. column at 150° over silicone grease, the product contains about 8% of one impurity and 2% of another. It is sufficiently pure for conversion to 2-fluoroheptanoic acid.[2]

3. Methods of Preparation

1-Bromo-2-fluoroheptane[3] has been prepared only by the present procedure, which is similar to one described by Bowers and co-workers.[4]

4. Merits of the Preparation

The method is general for forming *vic*-bromofluorides, which in turn are useful intermediates; this is exemplified in their conversion to 2-fluoroalkanoic acids.[2] The procedure can be applied, with minor modification, to many types of alkenes.[3]

1. Department of Chemistry, The University of Western Ontario, London, Ontario, Canada.
2. F. H. Dean, J. H. Amin, and F. L. M. Pattison, this volume, p. 580.
3. F. L. M. Pattison, D. A. V. Peters, and F. H. Dean, *Can. J. Chem.*, **43**, 1689 (1965).
4. A. Bowers, Ł. C. Ibáñez, E. Denot, and R. Becerra, *J. Am. Chem. Soc.*, **82**, 4001 (1960).

2-BROMO-4-METHYLBENZALDEHYDE

(*p*-Tolualdehyde, 2-bromo-)

Submitted by S. D. JOLAD and S. RAJAGOPAL[1]
Checked by A. G. SZABO and PETER YATES

1. Procedure

A. *Formaldoxime.* A mixture of 11.5 g. (0.38 mole) of para-formaldehyde and 26.3 g. (0.38 mole) of hydroxylamine hydrochloride in 170 ml. of water is heated until a clear solution is obtained. Then there is added 51 g. (0.38 mole) of hydrated sodium acetate, and the mixture is boiled gently under reflux for

15 minutes to give a 10% solution of formaldoxime.

B. *2-Bromo-4-methylbenzenediazonium chloride.* A mixture of 46.0 g. (0.25 mole) of 2-bromo-4-methylaniline [2] and 50 ml. of water is placed in a 1-l. three-necked flask equipped with an efficient stirrer, a dropping funnel, and a thermometer. The stirrer is started, and 57 ml. of concentrated hydrochloric acid is added slowly. The mixture is cooled to room temperature, 100 g. of ice is added, and the temperature of the mixture is maintained at $-5°$ to $+5°$ by means of an ice-salt bath. To the stirred mixture there is added, dropwise, a solution of 17.5 g. (0.25 mole) of sodium nitrite in 25 ml. of water. After completion of the addition, the stirring is continued for a period of 15 minutes. The stirred solution of the diazonium salt is made neutral to Congo red by the addition of a solution of hydrated sodium acetate (22 g.) in water (35 ml.) (Note 1).

C. *2-Bromo-4-methylbenzaldehyde.* A 3-l. three-necked flask is equipped with an efficient stirrer, a dropping funnel (Note 2), and a thermometer. The aqueous 10% formaldoxime prepared in step A is placed in the flask, and to it are added 6.5 g. (0.026 mole) of hydrated cupric sulfate, 1.0 g. (0.0079 mole) of sodium sulfite, and a solution of 160 g. of hydrated sodium acetate in 180 ml. of water. The solution is maintained at 10–15° by means of a cold-water bath and stirred vigorously. The neutral diazonium salt solution prepared in step B is slowly introduced below the surface of the formaldoxime solution (Notes 3 and 4). After the addition of the diazonium salt solution is complete, the stirring is continued for an additional hour and then the mixture is treated with 230 ml. of concentrated hydrochloric acid. The stirrer and the dropping funnel are replaced by stoppers, and the mixture is gently heated under reflux for 2 hours. The flask is set up for steam distillation, and the reaction product is steam-distilled. The distillate is saturated with sodium chloride, extracted with three 150-ml. portions of ether, and the ethereal extracts are washed successively with three 20-ml. portions of a saturated sodium chloride solution, three 20-ml. portions of an aqueous 10% sodium bicarbonate solution, and again with three 20-ml. portions of a saturated sodium chloride solution.

The ether is distilled and to the residue there is added, with cooling, 90 ml. of an aqueous 40% sodium metabisulfite solution,

previously heated to 60°.   The mixture is shaken for 1 hour and
allowed to stand overnight.   The solid addition product is
filtered, washed twice with ether, and then suspended in 200 ml.
of water in a 500-ml. flask, and 40 ml. of concentrated sulfuric
acid is slowly added with cooling.   The mixture is gently boiled
under reflux for 2 hours, cooled, and extracted with three 100-ml.
portions of ether.   The ethereal extract is washed with three
15-ml. portions of a saturated sodium chloride solution and dried
over anhydrous sodium sulfate.   The ether is evaporated, and
the product is distilled under reduced pressure.   2-Bromo-
4-methylbenzaldehyde distills at 114–115° (5 mm.) as a colorless
oil, yield 17.5–22.5 g. (35–45%), which crystallizes in the receiver, m.p. 30–31°.

2. Notes

1. Exact neutralization of the diazonium salt solution is
necessary in order to minimize coupling.

2. The stem of the dropping funnel should extend a little
below the surface of the solution in the three-necked flask.

3. Addition of the diazonium salt solution sometimes results
in the formation of a pasty mass which prevents further stirring;
the mixture is then allowed to stand for a further period of 1 hour.

4. The checkers found it preferable to transfer the diazonium
salt solution by siphoning under slight nitrogen pressure.

3. Methods of Preparation

The preparation of this aldehyde is based on the reaction due
to Beech[3] for the conversion of an aromatic amine to the corresponding aldehyde and has been described earlier by Jolad
and Rajagopal.[4]

4. Merits of the Preparation

This method of preparation of a halobenzaldehyde is of wide
application and has been used for the preparation of the following
substituted benzaldehydes: 2-bromo-5-methyl-,[4] 2,3-dichloro-
and 2,4-dichloro-,[5] 2-chloro-4-methyl-,[6] 2-methyl-4-bromo- and

3-methyl-4-bromo-,[7] 2-methyl-5-chloro- and 2-methyl-5-bromo-,[8] *p*-iodo-, *p*-fluoro-, 2-iodo-4-methyl-, and 6-iodo-3-methyl-.[9]

1. Department of Chemistry, Karnatak University, Dharwar, S. India.
2. J. R. Johnson and L. T. Sandborn, *Org. Syntheses*, Coll. Vol. **1**, 111 (1951).
3. W. F. Beech, *J. Chem. Soc.*, 1297 (1954).
4. S. D. Jolad and S. Rajagopal, *J. Sci. Ind. Res. (India)*, **21B**, 359 (1961) [*C.A.*, **56**, 1381 (1962)].
5. N. Gudi, S. Hiremath, V. Badiger, and S. Rajagopal, *Arch. Pharm.*, **295**, 16 (1962).
6. S. D. Jolad and S. Rajagopal, *Naturwiss.*, **48**, 645 (1961).
7. S. S. Vernekar, S. D. Jolad, and S. Rajagopal, *Monatsh.*, **93**, 271 (1962).
8. S. D. Jolad and S. Rajagopal, *Chimia*, **16**, 196 (1962).
9. S. D. Jolad and S. Rajagopal, unpublished results.

2-BROMONAPHTHALENE *

(Naphthalene, 2-bromo-)

Submitted by J. P. Schaefer, Jerry Higgins, and P. K. Shenoy [1]
Checked by James P. Nelson, Wayland E. Noland, and William E. Parham

1. Procedure

A 500 ml., three-necked, round-bottomed flask is equipped with a Trubore stirrer, a pressure-compensating dropping funnel, and a reflux condenser with drying tube. The flask is charged with 144 g. of triphenylphosphine (0.55 mole) (Note 1) and 125 ml. of acetonitrile (Notes 2 and 3). The solution is stirred and cooled in an ice bath and 88 g. of bromine is added dropwise over a period of 20–30 minutes (Note 4). After the addition of the bromine (Note 5) is complete, the ice bath is removed and 72 g.

(0.50 mole) of β-naphthol (Note 6) in 100 ml. of acetonitrile (Note 7) is added in one portion and the reaction mixture is heated to 60–70° for at least 30 minutes (Note 8). The flask is now fitted for a simple distillation, stirring is discontinued, and the acetonitrile is distilled (Note 9) under aspirator pressure until the oil bath temperature reaches 110° (Note 10). After all the acetonitrile has been removed, the condenser is replaced by a short, large glass tube (Note 11) connected to a 500-ml. flask half-filled with water, and the oil bath is replaced by a Wood's metal bath. The bath temperature is now raised to 200–220° and kept at this temperature until all the solid has melted (Note 12). The mixture is stirred and the bath temperature is raised to 340° (Note 13) and held at this temperature until evolution of hydrogen bromide ceases (approximately 20–30 minutes). The Wood's metal bath is removed and the reaction mixture is cooled to approximately 100° and then poured into a 1-l. beaker and cooled to room temperature. Pentane (300 ml.) (Note 14) is added and the solid is broken into a fine precipitate (Note 15). The solid is filtered by suction and washed thoroughly with two 300-ml. portions of pentane. The pentane filtrates are combined, washed with 200 ml. of 20% sodium hydroxide, and dried over anhydrous magnesium sulfate. The pentane extract is then passed through a 25 mm. diameter column filled to 35 cm. in depth with alumina; distillation of the pentane at reduced pressure gives 72–81 g. (70–78%) (Note 16) of 2-bromonaphthalene, a white solid melting at 45–50° (reported:[2] 55–56.4°) (Notes 17, 18).

2. Notes

1. Triphenylphosphine was obtained from M and T Chemicals Inc. and was used without further purification.

2. The acetonitrile was *cautiously* distilled from phosphorus pentoxide. The solid phosphorus pentoxide may cause bumping during the distillation of the acetonitrile. This can be avoided if the solution is stirred during the distillation.

3. The triphenylphosphine is only partially dissolved.

4. The temperature is kept below 40° and the reaction mixture thoroughly stirred during this addition.

5. If a small amount of free bromine remains, as evidenced by color, then a sufficient amount of triphenylphosphine is added to take up the bromine.

6. Practical grade β-naphthol was obtained from Matheson Coleman and Bell and distilled at atmospheric pressure before use.

7. Warming the acetonitrile is necessary to dissolve the β-naphthol.

8. All the precipitate dissolves at this point. The checkers heated the mixture for 2 hours. The solids appeared to dissolve; then there was some reprecipitation. All solids had not dissolved after 2 hours at 70°.

9. The checkers encountered severe bumping during distillation of the acetonitrile. The acetonitrile may be removed with only moderate bumping if no external heat is applied.

10. The checkers did not remove all the acetonitrile at 110°, a factor which contributed to foaming later (Note 12).

11. If a tube smaller than $\frac{1}{2}$ inch in diameter is used, it may become plugged later in the reaction.

12. If temperature is higher than 220°, initial foaming may become troublesome. The checkers heated the solid to 240–270°, since melting did not occur at 200–220°; foaming was encountered.

13. Evolution of hydrogen bromide begins at about 280°.

14. The checkers used petroleum ether (b.p. 65–67°).

15. The checkers obtained a somewhat tarry precipitate.

16. The submitters obtained 85–87 g. (82–86% yield).

17. The checkers' product retained solvent. The product was heated for a short period at 120° (20 mm.) and a crystalline solid, free of solvent, was obtained.

18. Purity was checked by analytical vapor phase chromatography (98–99%). This product can be used for most reactions without further purification. If further purification is desired, 2-bromonaphthalene can be recrystallized from aqueous methanol (95% recovery) to give a product melting at 53–55°.

3. Discussion

2-Bromonaphthalene has been prepared from 2-aminonaphthalene by the reaction of mercuric bromide with the diazonaphthalene.[2, 3] The reaction described in this preparation appears to be fairly general and provides a useful alternative method for introducing bromine into the aromatic nucleus. Using conditions similar to those outlined, the following have been prepared from the corresponding aryl alcohols:[4, 5] α-bromonaphthalene (72%), 3-bromopyridine (76%), 2-bromopyridine (61%), 8-bromoquinoline (48)%, o-bromotoluene (72%), p-chlorobromobenzene (90%), p-nitrobromobenzene (60%), and p-methoxybromobenzene (59%). The use of the triphenylphosphine-halogen complex to convert alcohols to alkylhalides is described elsewhere in this series.[6]

1. Department of Chemistry, The University of Arizona, Tucson, Arizona.
2. M. S. Newman and P. H. Wise, *J. Am. Chem. Soc.*, **63**, 2847 (1941).
3. H. W. Schwechten, *Ber.*, **65**, 1605 (1932).
4. J. P. Schaefer and J. Higgins, *J. Org. Chem.*, **32**, 1607 (1967).
5. G. A. Wiley, R. L. Hershkowitz, B. M. Rein, and B. C. Chung, *J. Am. Chem. Soc.*, **86**, 964 (1964).
6. J. P. Schaefer, J. G. Higgins, and P. K. Shenoy, this volume, p. 249.

3-BROMOPHTHALIDE

[1(3H)-Isobenzofuranone, 3-bromo-]

Submitted by I. A. KOTEN and ROBERT J. SAUER.[1]
Checked by R. C. JUOLA, MARJORIE C. CASERIO, and JOHN D. ROBERTS.

1. Procedure

Ten grams (0.075 mole) of phthalide (Note 1), 13.3 g. (0.075 mole) of N-bromosuccinimide (Note 1), and 200 ml. of dry carbon tetrachloride (Note 1) are refluxed for 30 minutes in a 500-ml. flask carrying a reflux condenser equipped with a drying tube containing Drierite. The reaction mixture is exposed to the light of an ordinary 100-watt unfrosted light bulb placed 6–8 in. from the flask. The end of the reaction is indicated by the disappearance of N-bromosuccinimide from the bottom of the flask and accumulation of succinimide at the top of the reaction mixture. The succinimide is removed by filtration and the filtrate concentrated under atmospheric pressure to 15–20 ml. Cooling of this concentrate followed by filtration gives 12–13 g. (75–81%) of crude 3-bromophthalide, m.p. 74–80°. The crude material, when recrystallized from cyclohexane, gives colorless plates, m.p. 78–80° (Notes 2–4).

2. Notes

1. The phthalide used was obtained from Aldrich Chemical Co. It was also prepared by the method of Gardner and Naylor, *Org. Syntheses Coll. Vol.* **2**, 526 (1943). The N-bromosuccinimide was obtained from Arapahoe Chemicals, Inc.

The carbon tetrachloride used is dried over Drierite and filtered or distilled.

2. About 150 ml. of cyclohexane is necessary to recrystallize 12–13 g. of product, and the temperature of the solvent should be kept below 70° to avoid oiling of undissolved material. The recovery is 11–12 g.

3. When pure 3-bromophthalide is allowed to stand, its melting point is depressed, owing apparently to some decomposition. It may, therefore, be desirable to prepare the compound in smaller quantities than specified here. A sample of 3-bromophthalide, prepared by using 20 g. of phthalide and 26.6 g. of N-bromosuccinimide, amounted to 29.8 g. (93.4%) of crude product. Hydrolysis of the crude material[2] gave phthalaldehydic acid, m.p. 96–98°.

4. Since one of the checkers developed a serious allergy to 3-bromophthalide, suitable precautions should be taken to avoid its inhalation and contact with the skin.

3. Methods and Merits of Preparation

3-Bromophthalide has previously been prepared by direct bromination of phthalide over a period of 10–13 hours in yields of 82–83%.[2] The procedure above, a modification of the Wohl-Ziegler method, appears to be preferable since it may be completed in 3–4 hours, is applicable to the preparation of small samples, and gives comparable yields.

1. Department of Chemistry, North Central College, Naperville, Ill. This work vas supported by a grant from Research Corporation of New York City.
2. R. L. Shriner and F. J. Wolf, *Org. Syntheses*, Coll. Vol. 3, 737 (1955).

3-BROMOPYRENE

(Pyrene, 1-bromo-)

Submitted by W. H. Gumprecht [1]
Checked by Melvin S. Newman and Stephen Havlicek

1. Procedure

Caution! Many pyrenes are carcinogens. Contact of the skin with these materials should be avoided.

In a 500-ml., three-necked, round-bottomed flask fitted with a stirrer, a reflux condenser, and a dropping funnel are placed 8.08 g. (0.040 mole) of pyrene (Note 1) and 80 ml. of carbon tetrachloride (Note 2). A solution of 2.0 ml. of bromine (6.24 g., 0.039 mole) (Note 3) in 30 ml. of carbon tetrachloride is added dropwise over a period of 2–3 hours. The resulting orange solution is stirred overnight, washed with three 100-ml. portions of water, and dried over anhydrous calcium chloride. The solvent is removed under reduced pressure, the pale yellow solid residue is dissolved in 10 ml. of benzene, and the benzene solution is treated with a small amount of activated carbon. The filtrate is diluted with 120 ml. of absolute ethanol, and the solution is distilled until about 80–90 ml. of solvent remains and then cooled. The bromopyrene crystallizes as pale yellow flakes, m.p. 93–95°. Additional material of similar melting point is obtained from the mother liquor on concentration. The total yield is 8.5–9.5 g. (78–86%) (Note 4). On recrystallization from benzene-alcohol a colorless product, m.p. 94.5–95.5°, is obtained with little loss.

2. Notes

1. Pyrene, m.p. 151–153°, obtained from Chemicals Division, Union Carbide Chemicals Corp., was used by the submitter. A commercial pyrene obtained from Germany was used by the checkers.

2. A c.p. solvent was used.

3. Reagent grade material was used. Excess bromine is to be avoided, as dibromopyrene can be formed.[2]

4. The submitter has run this preparation on 80 g. of pyrene with no change in yield.

3. Methods of Preparation

This procedure is described by Lock;[2] a modification using a small amount of phenol has been published.[3] The patent literature discloses the use of a tertiary amine, such as pyridine, and its combination with other solvents for the monobromination of pyrene with elemental bromine.[4] Brominating agents, such as N-bromosuccinimide[5] and N-bromohydantoins,[6] have also been used.

4. Merits of the Preparation

3-Bromopyrene is a precursor of 3-hydroxypyrene.[7]

1. Contribution No. 334 from the Organic Chemicals Department, E. I. du Pont de Nemours and Company, Wilmington, Delaware 19899.
2. G. Lock, *Ber.*, **70**, 926 (1937).
3. H. Hock and F. Ernst, *Ber.*, **92**, 2732 (1959).
4. A. Wolfram and W. Schnurr, U.S. Patent 2,094,227 (1937) [*C.A.*, **31**, 8550 (1937)].
5. Ng. Ph. Buu-Hoi and J. Lecocq, *Compt. Rend.*, **226**, 87 (1948).
6. J. F. Salellas and O. O. Orazi, *Anales Asoc. Quim. Arg.*, **39**, 175 (1951) [*C.A.*, **47**, 2708 (1953)]; R. A. Corral, O. O. Orazi, and J. D. Bonafede, *Anales Asoc. Quim. Arg.*, **45**, 151 (1957) [*C.A.*, **53**, 342 (1959)].
7. W. H. Gumprecht, this volume, p. 632.

3-BROMOTHIOPHENE*

(Thiophene, 3-bromo-)

Submitted by S. GRONOWITZ and T. RAZNIKIEWICZ [1]
Checked by MAX TISHLER, ARTHUR J. ZAMBITO, and RONALD B. JOBSON

1. Procedure

A 5-l., three-necked, round-bottomed flask is equipped with an efficient stirrer (Note 1), a reflux condenser, and a dropping funnel. Water (1850 ml.) is added, stirring is begun and continued throughout the procedure, and 783 g. (12.0 moles) of zinc dust (Note 2) and 700 ml. of acetic acid are added. The mixture is heated to reflux, the heating mantle is removed, and 1283 g. (4.00 moles) of 2,3,5-tribromothiophene (Note 3) is added dropwise at such a rate that the mixture continues to reflux. The addition is complete in about 70 minutes. Heat is applied, and the mixture is refluxed for 3 hours. A condenser is arranged for downward distillation, and the mixture is distilled until no more organic substance distills with the water (Note 4). The heavier organic layer is separated, washed successively with 50 ml. of 10% sodium carbonate solution and 100 ml. of water, dried over calcium chloride (Note 5), and fractionated through a vacuum-mantled Dufton column (Note 6). A 19-g. fore-run, b.p. 78–159°, consists mainly of thiophene and 3-bromothiophene. 3-Bromothiophene is collected at 159–160°; n^{20}D 1.5919–1.5928; weight 580–585 g. (89–90%) (Notes 7 and 8).

2. Notes

1. The submitters used a Teflon® paddle-type stirrer sealed with rubber tubing lubricated by glycerol and driven by a powerful motor. The checkers used a Trubore® stirrer.

2. Mallinckrodt zinc powder (analytical reagent grade) is used.

3. 2,3,5-Tribromothiophene is conveniently prepared by the method of Troyanowsky.[2] Thiophene (1125 g., 13.4 moles) and 450 ml. of chloroform are charged into a 5-l. three-necked flask equipped with a stirrer, a dropping funnel, and an outlet for the hydrogen bromide evolved. The flask is in a deep pan through which cold tap water passes. Bromine (6480 g., 40.6 moles) is added dropwise to the stirred mixture over a period of 10 hours. After the mixture has stood overnight, it is heated at 50° for several hours, washed with $2N$ sodium hydroxide solution, refluxed for 7 hours with a solution of 800 g. of potassium hydroxide in 1.5 l. of 95% ethanol, and poured into water. The organic layer is separated, washed with water, dried over calcium chloride, and fractionated to give 3200–3650 g. (75–85%) of 2,3,5-tribromothiophene; b.p. 123–124° (9 mm.); m.p. 25–27°.

4. About half of the volume is distilled over. The temperature of the vapor rises during the distillation from 95° to 101°.

5. The checkers washed the drying agent with ether and combined the wash with the filtrate.

6. A Dufton column was not available to the checkers. In its place a 2.5-cm. x 38-cm. column packed with glass helices was used. This column was heated by a 4.5-cm. concentric glass jacket wrapped with Nichrome ribbon. A 6.5-cm. concentric glass jacket surrounded the whole column and served to insulate it.

7. In several experiments on one-fifth the scale, the yields were 89–92%.

8. Infrared analysis shows that the 3-bromothiophene contains about 0.5% of 2-bromothiophene, as measured by 2-bromothiophene's characteristic absorption peak at 10.26 μ. The traces of this lower-boiling isomer can easily be removed by fractionation through a more efficient column.

3. Methods of Preparation

The procedure described is a modification of that described by Gronowitz.[3]   3-Bromothiophene has been obtained more tediously and in lower yields by removal of the α-bromines of 2,3,5-tribromothiophene through the Grignard entrainment method [4] with ethyl bromide as the auxiliary halide, or by halogen-metal interconversion with *n*-butyllithium [5] followed by hydrolysis of the organometallic compounds.   It has also been obtained from 4,5-dibromo-2-thiophenecarboxylic acid through simultaneous debromination and decarboxylation.[6]

4. Merits of the Preparation

3-Bromothiophene is a key intermediate for the synthesis of 3-substituted thiophenes.[7]

1. Department of Organic Chemistry, Chemical Institute, University of Uppsala, Sweden.
2. C. Troyanowsky, *Bull. Soc. Chim. France*, 1424 (1955).
3. S. Gronowitz, *Acta Chem. Scand.*, **13**, 1045 (1959).
4. S. Gronowitz, *Arkiv Kemi*, **7**, 267 (1955).
5. S. O. Lawesson, *Arkiv Kemi*, **11**, 373 (1957).
6. R. Motoyama, S. Nishimura, E. Imoto, Y. Murakami, K. Hari, and J. Ogawa, *Nippon Kagaku Zasshi*, **78**, 950 (1957).
7. S. Gronowitz, *Arkiv Kemi*, **13**, 295 (1959).

7-*t*-BUTOXYNORBORNADIENE*

(2,5-Norbornadiene, 7-*tert*-butoxy-)

$$\text{norbornadiene} + C_6H_5CO_3C(CH_3)_3 \xrightarrow[\text{Benzene}]{\text{CuBr}} \text{7-}t\text{-butoxynorbornadiene} + C_6H_5CO_2H$$

Submitted by PAUL R. STORY and SUSAN R. FAHRENHOLTZ [1]
Checked by ERIC BLOCK and E. J. COREY

1. Procedure

A 2-l. three-necked flask, fitted with stirrer, condenser, dropping funnel, and an arrangement for maintenance of an inert atmos-

phere, is charged with a mixture of 300 g. (3.26 moles) of norbornadiene (Note 1) and 0.650 g. (4.53 mmoles) of cuprous
bromide (Note 2) in 500 ml. of benzene (Note 3). Dry nitrogen
is introduced continuously, and, after the flask contents are
brought to reflux, 245 g. (1.26 mole) of *t*-butyl perbenzoate
(Note 4), dissolved in 100 ml. of benzene, is added over approximately 1 hour to the stirred mixture. The solution immediately
becomes blue or blue-green. The modest heat of reaction requires the application of some heat throughout the reaction
period in order to maintain reflux. After the addition is completed, the solution is heated at the reflux temperature for an
additional 30 minutes (*Caution! Note 5*).

The mixture is cooled to room temperature, transferred to a
4-l. separatory funnel, and washed with three 300-ml. portions
of saturated brine, to remove copper salts, and three 300-ml.
portions of 10% aqueous sodium hydroxide, to remove benzoic
acid (Note 6). The benzene solution is then washed with 150 ml.
of brine and is dried over anhydrous sodium sulfate.

The dried solution is transferred to a 2-l. flask fitted with a
Claisen head attached to a 30-cm. Vigreux column. Benzene is
removed fairly rapidly at about 200 mm. so that the benzene
boils at 45–50° (Note 7). After removal of most of the benzene,
the pressure is slowly lowered to 10–15 mm.; a negligible
amount of fore-run is obtained before the product begins to distil
at about 65° at 15 mm. The product is collected until the temperature reaches 80–85° at the same pressure (Note 8). The
yield of 7-*t*-butoxynorbornadiene is 42–51 g. (20–25%, based on
t-butyl perbenzoate) (Note 9). Gas-phase chromatographic analysis of the product shows it to be about 95% pure. Greater
purity can be obtained, if required, by fractionation through a
spinning band column; the pure product is collected at 70–72°
(14 mm.) (Note 10).

2. Notes

1. Norbornadiene as supplied by Shell Chemical Co. is used
without further purification. Distillation of the norbornadiene
immediately before use gives no change in yield of 7-*t*-butoxy-
norbornadiene.

2. The cuprous bromide was used as obtained from E. H. Sargent Co. One instance of an ineffective batch of cuprous bromide from another source has been reported to the submitters. Cuprous bromide is only slightly soluble in the benzene solution. Greater amounts of catalyst have no effect on the yield of product.

3. Baker and Adamson or Merck reagent grade benzene was used.

4. *t*-Butyl perbenzoate was used as received from Lucidol Division of Wallace and Tiernan Co., Buffalo, New York.

5. Normally, after this time, the *t*-butyl perbenzoate is completely reacted. It is advisable, however, to check for its presence because distillation of a crude product containing some perester can lead to an explosion. *t*-Butyl perbenzoate absorbs strongly in the infrared at 5.65–5.70 μ, and examination of the infrared spectrum of the benzene solution is a sufficiently sensitive test. No difficulty has ever been encountered during reactions with norbornadiene. However, unreacted *t*-butyl perbenzoate has caused a minor explosion with another, less reactive olefin.

6. Unless the copper salts are removed first, the basic wash frequently produces a thick emulsion which requires considerable time to settle.

7. The benzene is removed at reduced pressure to minimize heating of the product which consists chiefly of high-boiling esters. The last trace of benzene and norbornadiene is removed at a lower pressure.

8. 7-*t*-Butoxynorbornadiene is collected over this wide temperature range since the last portion of the distillate is superheated because of the large quantity of high-boiling materials remaining in the pot. No other reaction products boil in the same range, and GPC analysis has shown the last fraction of distillate to be nearly as pure as the middle cut.

9. This reaction has been conducted under a variety of conditions, but the yield of 7-*t*-butoxynorbornadiene has never been less than 20% or greater than 26%.

10. Infrared (μ, CCl_4): 6.48 (w), 7.20 (m), 7.35 (s), 7.61 (m), 9.05 (s), 13.7 (s). The absorptions at 6.48 (double-bond stretch) and 13.7 (*cis* double bond C—H out-of-plane deformation) are

very characteristic of the norbornadiene nucleus. N.m.r. spectrum (CCl_4): τ = 3.56 (m), 3.70 (m), 6.77 (m), 6.38 (m), 8.94 (s).

3. Methods of Preparation

The only method reported [2] for the preparation of 7-t-butoxynorbornadiene is that described here. The general reaction of t-butyl perbenzoate with various olefins has been described by many investigators.[3] When benzoyl peroxide is used in place of t-butyl perbenzoate under similar conditions, 7-benzoxynorbornadiene is obtained in 38% yield; it is said to be more easily hydrolyzed to the alcohol than 7-t-butoxynorbornadiene.[4]

4. Merits of the Preparation

7-t-Butoxynorbornadiene provides a convenient route to 7-substituted norbornenes and norbornadienes including *anti*-7-norbornenol,[2] 7-norbornadienyl acetate,[2] 7-norbornadienol,[2] 7-chloronorbornadiene,[5] 7-methylnorbornadiene,[6] and 7-phenylnorbornadiene.[6] These compounds are useful in studies of the nature of chemical bonding.[5,7]

1. Bell Telephone Laboratories, Inc., Murray Hill, New Jersey.
2. P. R. Story, *J. Org. Chem.*, **26**, 287 (1961).
3. M. S. Kharasch, G. Sosnovsky, and N. C. Yang, *J. Am. Chem. Soc.*, **81**, 5819 (1959); D. B. Denney, D. Z. Denney, and G. Feig, *Tetrahedron Letters*, No. 15, 19 (1959); J. K. Kochi, *Tetrahedron*, **18**, 483 (1962).
4. H. Tanida and T. Tsuji, *Chem. Ind.* (*London*), 211 (1963).
5. P. R. Story and M. Saunders, *J. Am. Chem. Soc.*, **84**, 4876 (1962).
6. P. R. Story and S. R. Fahrenholtz, Unpublished results.
7. S. Winstein and C. Ordronneau, *J. Am. Chem. Soc.*, **82**, 2084 (1960).

t-BUTYL ACETOACETATE*

(Acetoacetic acid, *tert*-butyl ester)

$$CH_3\!\!-\!\!\underset{\underset{CH_3}{|}}{\overset{\overset{CH_3}{|}}{C}}\!\!-\!\!OH \; + \; \underset{CH_2\!-\!C=O}{\overset{CH_2=C\!-\!\!-O}{|\qquad\quad|}} \quad \xrightarrow{\;CH_3CO_2Na\;}$$

$$\underset{}{CH_3\overset{O}{\overset{\|}{C}}\!\!-\!\!CH_2\!\!-\!\!\overset{O}{\overset{\|}{C}}\!\!-\!\!O\!\!-\!\!\underset{\underset{CH_3}{\diagdown}}{\overset{\overset{CH_3}{\diagup}}{C}}\!\!-\!\!CH_3}$$

Submitted by Sven-Olov Lawesson, Susanne Gronwall, and Rune Sandberg.[1]
Checked by William G. Dauben and Richard Ellis.

1. Procedure

Caution! This preparation should be conducted in a hood to avoid exposure to diketene, which is toxic and which may irritate mucous tissues such as those of the eyes; the use of safety goggles is recommended.

A 500-ml. three-necked flask is equipped with a sealed mechanical stirrer, a dropping funnel, and a two-armed addition tube, one arm of which bears a reflux condenser and the other arm of which is fitted with a thermometer. *t*-Butyl alcohol (79 g., 1.07 moles) (Note 1) is added to the flask and the thermometer arranged so that its bulb is immersed in the liquid but out of the path of the stirrer. The flask is heated by means of an electric mantle until the temperature of the liquid is 80–85°, and the mantle then is removed. Anhydrous sodium acetate (0.4 g., 4.8 mmoles) is added with stirring, and then 96 g. (1.14 moles) of diketene (Note 2) is added dropwise over a period of 2.5 hours. The temperature of the solution drops to 60–70° during the first 15 minutes and then increases slowly to 110–115°. When all the diketene is added, the reaction subsides and, after the resulting brown-black solution is stirred for an additional 30 minutes, the product is distilled immediately under reduced pressure through a short column. After a small fore-run, the yield of t-butyl acetoacetate, b.p. 85°/20 mm. (Note 3), n_D^{20} 1.4200–1.4203, is 127–135 g. (75–80%) (Note 4).

2. Notes

1. Eastman Kodak white label grade is used without further purification.

2. The submitters used material directly as supplied by Dr. Theodor Schuchardt and Co., Munich, Germany. The checkers used material directly as supplied by Aldrich Chemical Co., Milwaukee, Wisconsin.

3. The still residue is dehydroacetic acid.

4. In a run five times the size described, the submitters report that the reaction goes in the same manner and in 85–92% yield.

3. Discussion

t-Butyl acetoacetate has been prepared by self-condensation of t-butylacetate.[2,3] The described procedure is based upon the method of Treibs and Hintermieier.[4]

The present preparation employs a method of considerable scope and is illustrative of a general method of preparing esters of acetoacetic acid; it gives much better yields, is considerably less laborious than other methods for the preparation of t-butyl acetoacetate, and appears to be the most convenient as starting materials are easily accessible. The title compound is of specific interest since the t-butoxy carbonyl group may be removed simply by heating the compound with catalytic amounts of p-toluenesulfonic acid. For instance, a new method has been developed for the preparation of acyloins by introducing the benzoyloxy group into t-butyl acetoacetate, followed by t-butoxy carbonyl-elimination and hydrolysis.[5] Using a similar technique and starting from t-butyl acetoacetate, levulinic esters,[6] δ-ketonitriles,[7] 1-methylcyclohexene-1-on-3,[8] α,β-unsaturated ketones,[9] and piperiton and related compounds[10] have been prepared.

1. Present address: Department of Chemistry, University of Aarhus, 8000 Aarhus C, Denmark.
2. N. Fisher and S. M. McElvain, *J. Am. Chem. Soc.*, 56, 1766 (1934).
3. W. B. Renrow and G. D. Walker, *J. Am. Chem. Soc.*, 70, 3957 (1948).
4. A. Treibs and K. Hintermeier, *Ber.*, 87, 1163 (1954)
5. S.-O. Lawesson, S. Grönwall, and M. Andersson, *Arkiv Kemi*, 17, 457 (1961).
6. S.-O. Lawesson, M. Dahlén, and C. Frisell, *Acta Chem. Scand.*, 16, 1191 (1962).
7. G. Näslund, A. Senning, and S.-O. Lawesson, *Acta Chem. Scand.*, 16, 1324 (1962).
8. G. Näslund, A. Senning, and S.-O. Lawesson, *Acta Chem. Scand.*, 16, 1329 (1962).
9. S.-O. Lawesson, E. H. Larsen, G. Sundström, and H. J. Jakobsen, *Acta Chem. Scand.*, 17, 2216 (1963).
10. S.-O. Lawesson, E. H. Larsen, H. J. Jakobsen, and C. Frisell, *Rec. Trav. Chim. Pays-Bas*, 83, 464 (1964).

t-BUTYL AZIDOFORMATE*

(Formic acid, azido-, *tert*-butyl ester)

$$NH_2NHCO_2C(CH_3)_3 \xrightarrow{\text{HONO}} N_3CO_2C(CH_3)_3$$

Submitted by LOUIS A. CARPINO, BARBARA A. CARPINO,
PAUL J. CROWLEY, CHESTER A. GIZA,
and PAUL H. TERRY [1]
Checked by VIRGIL BOEKELHEIDE and S. J. CROSS

1. Procedure

In a 1-l. round-bottomed flask fitted with a mechanical stirrer are placed 82 g. (0.62 mole) of *t*-butyl carbazate,[2] 72 g. of glacial acetic acid, and 100 ml. of water. The solution is cooled in an ice bath, and 47.0 g. (0.68 mole) of solid sodium nitrite is added over a period of 40–50 minutes, the temperature being kept at 10–15° (Note 1). The mixture is allowed to stand in the ice bath for 30 minutes, 100 ml. of water is added, and the floating oil is extracted into four 40-ml. portions of ether. The combined ether extracts are washed twice with 50-ml. portions of water and with 40-ml. portions of 1M sodium bicarbonate solution until no longer acidic (about three washings are required). The solution is dried over magnesium sulfate, and the ether is removed by distillation from a water bath maintained at 40–45°; water aspirator pressure of 140–150 mm. is used. The pressure is then lowered to 70 mm., and the water bath temperature is raised to 90–95°. The azide is distilled (*Caution! Note 2*) using a Claisen flask and is collected at 73–74° (70 mm.), n^{24}D 1.4227, after a few drops of fore-run. The yield is 57–72.8 g. (64–82%) (Notes 3 and 4).

2. Notes

1. The sodium nitrite may be added as a concentrated aqueous solution.

2. It is recommended that the distillation be carried out behind a safety shield. The submitters have distilled this compound several hundred times without incident under the conditions given on a scale up to 300–400 g. per run. On the other hand, Prof. P. G. Katsoyannis (University of Pittsburgh School of Medicine, Pittsburgh, Pennsylvania) has reported that an explosion took place in the receiving flask while the compound was being distilled under conditions previously used without incident. The reason for the explosion could not be traced. According to Prof. R. Schwyzer (Ciba, Ltd., Basel, Switzerland) tests at a Swiss Federal Institute showed that the compound could not be exploded by mere heating: it simply decomposes. For explosion, one must apply a primary explosive such as lead azide or silver azide. An attempt by the submitters to distil the azide at atmospheric pressure resulted in vigorous carbonization, but no explosion occurred. In view of the potential hazard some investigators prefer not to distil the azide; they use the crude material after removal of solvent. High yields of carbo-*t*-butoxy derivatives may be obtained in this way.

3. When freshly distilled, the azide is water-white. When the azide is allowed to stand for several weeks, it slowly develops a light yellow color; however, this does not appear to affect its reactivity as an acylating agent.[3]

4. The azide should be handled with adequate ventilation. Careless inhalation of the substance was accompanied by development of a painful throbbing headache or a sensation of giddi-

ness or both. These effects disappeared within several hours upon exposure to fresh air.

3. Discussion

t-Butyl azidoformate has been prepared by a variety of procedures,[3-12] of which the present procedure and that described elsewhere in this series[12] appear most satisfactory.

Because of the instability of *t*-butyl chloroformate a number of carbonic acid derivatives have been prepared and studied as reagents for the introduction of the carbo-*t*-butoxy group. A listing of these reagents and references to their preparation may be found in reference 13. In spite of some disadvantages the most widely used reagent is still *t*-butyl azidoformate, although *t*-butyl 2,4,5-trichlorophenyl-carbonate appears to be another potentially useful reagent. *t*-Butyl azidoformate is a convenient reagent for the acylation of amines, hydrazines, and similar compounds.[3]

The acylation product of hydroxylamine, *t*-butyl N-hydroxy-carbamate,[5] is a valuable intermediate in the synthesis of O-sub-stituted hydroxylamines such as O-acyl- and O-sulfonylhydroxyl-amines, many of which are valuable aminating agents and have not be obtained in any other way.[14,15]

1. Department of Chemistry, University of Massachusetts, Amherst, Massachusetts.
2. L. A. Carpino, D. Collins, and S. Göwecke, this volume, p. 166.
3. L. A. Carpino, *J. Am. Chem. Soc.*, 79, 4427 (1957).
4. L. A. Carpino, *J. Am. Chem. Soc.*, 79, 98 (1957).
5. L. A. Carpino, C. A. Giza, and B. A. Carpino, *J. Am. Chem. Soc.*, 81, 955 (1959).
6. K. P. Polzhofer, *Chimia (Aarau)*, 23, 298 (1969).
7. H. Yajima and H. Kawatani, *Chem. Pharm. Bull. (Tokyo)*, 16, 182 (1968).
8. M. Itoh and D. Morino, *Experientia*, 24, 101 (1968).
9. Y. A. Kiryushkin and A. I. Miroshnikov, *Experientia*, 21, 418 (1965).
10. K. Inouye, M. Kanayama, and H. Otsuka, *Nippon Kagaku Zasshi*, 85, 599 (1964).
11. D. S. Tarbell, *Accounts Chem. Res.*, 2, 296 (1969).
12. M. A. Insalaco and D. S. Tarbell, *Org. Syntheses*, 50, 9 (1970).
13. L. A. Carpino, K. N. Parameswaran, R. K. Kirkley, J. W. Spiewak, and E. Schmitz, *J. Org. Chem.*, 35, 3291 (1970).
14. L. A. Carpino, *J. Am. Chem. Soc.*, 82, 3133 (1960).
15. L. A. Carpino, *J. Am. Chem. Soc.*, 85, 2144 (1963).

t-BUTYL AZODIFORMATE

(Formic acid, azodi-, di-$tert$-butyl ester)

$$N_3CO_2C(CH_3)_3 + NH_2NHCO_2C(CH_3)_3 \xrightarrow[-HN_3]{C_5H_5N}$$

$$(CH_3)_3CO_2CNHNHCO_2C(CH_3)_3 \xrightarrow[C_5H_5N]{(CH_2CO)_2NBr}$$

$$(CH_3)_3CO_2CN=NCO_2C(CH_3)_3$$

Submitted by Louis A. Carpino and Paul J. Crowley [1]
Checked by Virgil Boekelheide and S. J. Cross

1. Procedure

A. *t-Butyl hydrazodiformate.* A solution of 28.6 g. (0.2 mole) of t-butyl azidoformate [2] and 26.4 g. (0.2 mole) of t-butyl carbazate [3] in 60 ml. of pyridine (Note 1) is allowed to stand at room temperature for 1 week and is then diluted with 500 ml. of water. The snow-white microcrystalline powder that separates is removed by filtration and is washed with two 50-ml. portions of water. The yield of air-dried hydrazide, m.p. 124–126°, is 35.5–37 g. (77–80%). The product is pure enough for most applications but may be purified by recrystallization from a 1:1 mixture of benzene and ligroin (60–90°) from which it separates as small white needles, m.p. 124–125.5°. The recovery is 92%.

B. *t-Butyl azodiformate.* In a 500-ml. Erlenmeyer flask is placed a solution of 23.2 g. (0.1 mole) of crude t-butyl hydrazodiformate (m.p. 124–126°) in 175 ml. of methylene chloride and 7.9 g. (0.1 mole) of pyridine (Note 1). The solution is cooled by a stream of running tap water while 18.2 g. (0.102 mole) (Note 2) of N-bromosuccinimide (Note 1) is added during 6–7 minutes with swirling. The resulting solution is allowed to stand at room temperature for 5 minutes and is washed with two 100-ml. portions of water followed by one 100-ml. portion of 10% sodium hydroxide. The solution is dried for 30 minutes over magnesium sulfate, filtered into a large evaporating dish, and allowed to evaporate. The yellow-orange crystalline residue, m.p. 90–91.5°, which amounts to 20.7–21.8 g. (90–94.5%), is recrystallized by covering the dry solid with 35–40 ml. of petroleum ether (b.p. 30–60°) and adding ligroin (b.p. 60–90°) to the boiling solution until the solid dissolves. *t*-Butyl azodiformate, 19.8–20.0 g., m.p. 90.7–92°, separates from the cooled solution as large lemon-yellow crystals. Evaporation of the filtrate gives an additional amount of yellow solid which is recrystallized as before, yielding 0.4–0.7 g. of pure material, m.p. 90–92°. The total yield is 20.2–20.7 g. (88–90%).

2. Notes

1. The pyridine was a pure product, b.p. 113–115°, obtained from Mallinckrodt Chemical Company and used as supplied. The methylene chloride (technical grade) and N-bromosuccinimide (practical grade) were obtained from the Matheson Company and used as received.

2. Unless a 2% excess of N-bromosuccinimide is used, the azo compound is contaminated by an impurity, possibly the original hydrazo compound, which separates along with the desired product in the form of long, easily distinguished needles. The two substances cannot be separated by crystallization from ligroin.

3. Methods of Preparation

t-Butyl hydrazodiformate has been prepared by acylation of *t*-butyl carbazate by means of *t*-butyl azidoformate [4] or *t*-butyl cyanoformate.[5] *t*-Butyl azodiformate has been prepared only by oxidation of the hydrazo compound.[6]

4. Merits of the Preparation

The potassium salt of *t*-butyl hydrazodiformate can be easily alkylated and thus used in the synthesis of acyclic and cyclic 1,2-disubstituted hydrazines.[7]

Ethyl azodiformate is a well-known and useful dienophile in the Diels-Alder reaction.[8] *t*-Butyl azodiformate behaves similarly, although it is somewhat less reactive.[6] *t*-Butyl azodiformate does, however, provide products with ester groups that are easily cleaved. Monosubstituted hydrazines may be prepared by the addition of Grignard reagents to azoformates followed by cleavage.[6]

1. Department of Chemistry, University of Massachusetts, Amherst, Massachusetts.
2. L. A. Carpino, B. A. Carpino, P. J. Crowley, C. A. Giza, and P. H. Terry, this volume, p. 157, or M. A. Insalaco and D. S. Tarbell, *Org. Syntheses*, 50, 9 (1970). This reagent is also available from Aldrich Chemical Company, Inc.
3. L. A. Carpino, D. Collins, S. Göwecke, J. Mayo, S. D. Thatte, and F. Tibbetts, this volume, p. 166.
4. L. A. Carpino, *J. Am. Chem. Soc.*, 79, 4427 (1957).
5. L. A. Carpino, *J. Am. Chem. Soc.*, 82, 2725 (1960).
6. L. A. Carpino, P. H. Terry, and P. J. Crowley, *J. Org. Chem.*, 26, 4336 (1961).
7. L. A. Carpino, *J. Am. Chem. Soc.*, 85, 2144 (1963).
8. O. Diels, J. H. Blom, and W. Koll, *Ann.*, 443, 242 (1925).

t-BUTYL CARBAMATE *

(Carbamic acid, *tert*-butyl ester)

$$(CH_3)_3COH \xrightarrow[CF_3COOH]{NaCNO} (CH_3)_3COCNH_2$$

Submitted by BERNARD LOEV, MINERVA F. KORMENDY,
and MARJORIE M. GOODMAN [1]
Checked by DAVID C. ARMBRUSTER and WILLIAM D. EMMONS

1. Procedure

Caution! *Because of the acrid nature of trifluoroacetic acid and the possibility of the evolution of toxic fumes the reaction should be carried out in a hood.*

A solution of 14.8 g. (0.20 mole) of *t*-butyl alcohol in 125 ml. of benzene (Note 1) is placed in a 500-ml. three-necked flask equipped with a stirrer, a thermometer, and an addition funnel, and 26.0 g. (0.40 mole) of sodium cyanate (Note 2) is added. The suspension is stirred as *slowly* as possible (*ca.* 120 r.p.m.; Note 3) while 48.0 g. (31.2 ml., 0.42 mole) of trifluoroacetic acid is added dropwise at a rapid rate. The temperature slowly rises to about 37° after three-quarters of the trifluoroacetic acid has been added (*ca.* 7 minutes). At this point (Note 4) the mixture is cooled to 33–35° by brief immersion in an ice-water bath, then the addition is continued. When the addition of the acid is completed (10–12 minutes total time), the temperature slowly rises to 40° and then gradually subsides. Slow stirring is continued overnight (Note 5) at room temperature.

The mixture is treated with 35 ml. of water (Note 6) and stirred vigorously for a few minutes. The benzene layer is decanted, and the aqueous slurry is rinsed with two 125-ml. portions of benzene (Note 7). The combined organic extracts are washed once with 100 ml. of aqueous 5% sodium hydroxide (Note 8) and with 100 ml. of water, dried over anhydrous magnesium sulfate, and filtered. The solvent is removed by distillation under reduced pressure, preferably on a rotary evaporator, from a water bath kept at 30° (Note 9) to give 17.7–22.0 g. (76–94%) of *t*-butyl carbamate as white needles, m.p. 104–109° (Note 10). The product may be recrystallized from hexane (Note 11); m.p. 107–109° (Note 12).

2. Notes

1. The reagents should not be dried, as traces of moisture catalyze the reaction. The choice of solvent for this type of reaction markedly affects the yield; for most alcohols the use of benzene or methylene chloride gives yields superior to those obtained in other solvents.

2. Sodium cyanate *cannot* be replaced by other cyanates (potassium, ammonium, etc.), for the yields are then drastically lowered.

3. Vigorous agitation markedly lowers the yield; stirring rates of 40–120 r.p.m. are optimum.

4. The temperature may rise to 40°; within the range 20–50° the temperature has little effect on the yield.

5. A contact time of 3–4 hours is sufficient, but it is convenient to stir the reaction mixture overnight. The yield is slightly higher after this additional time.

6. Only a limited amount of water is added at this point because *t*-butyl carbamate has some solubility in the resulting

aqueous slurry. With water-insoluble carbamates the amount of water added is immaterial.

7. The checkers found that quantitative recovery of the benzene layer by decantation was impossible, so that in the final benzene rinse the mixture was poured into a graduated cylinder, and the benzene layer was quantitatively removed by a syringe.

8. The alkaline wash serves to hydrolyze a small amount of t-butyl N-trifluoroacetylcarbamate which occasionally forms. It is not clear why this by-product forms on some occasions but not on others under apparently identical conditions. The checkers found in every case that upon standing the alkaline wash deposited 1–2 g. (after drying) of white crystals which was shown to be identical with the t-butyl carbamate obtained as the main crop. This amount is included in the yield.

9. Most carbamates, including those of high molecular weight, are volatile. They are generally thermally unstable until they are purified.

10. The melting range varies markedly with the rate of heating, the temperature at which the sample is put into the bath, the solvent used, and the crystal form of the product. The compound at this stage is analytically pure and gives a single spot on thin-layer chromatography.

11. The carbamate may also be recrystallized from water in somewhat lower recovery. With either solvent, extensive heating should be avoided since a considerable amount of product is lost by volatilization. The checkers found that a relatively large volume of hexane was required for recrystallization and therefore used a 1:1 benzene-hexane or 1:1 benzene-ligroin solvent system for the recrystallization.

12. The reported melting points range from 108° to 110°.[2–5]

3. Methods of Preparation

Although numerous methods are known for the synthesis of carbamates of primary and secondary alcohols,[6] they are not satisfactory for the preparation of carbamates of tertiary alcohols.[7, 8] t-Butyl carbamate was first obtained by reaction of sodium t-butoxide with phosgene and thionyl chloride at −60°, followed by reaction with concentrated aqueous ammonia; the overall yield was less than 20%.[2] This procedure, however, was found to be unsuitable for the preparation of carbamates of other tertiary alcohols.[8] Carbamates have been prepared by the reaction of phenyl chloroformate (prepared from phenol and phosgene at −60°) with a tertiary alcohol in pyridine, followed by treatment with liquid ammonia.[8] A variation of this procedure involves hydrazinolysis of phenyl *t*-butyl carbonate, prepared as described above, conversion to the azide, and ammonolysis.[3, 4] *t*-Butyl carbamate has also been prepared by a four-step procedure that starts with the preparation of *t*-butyl ethyl oxalate from ethoxalyl chloride. This mixed ester was converted to *t*-butyl oxamate, which was dehydrated to *t*-butyl cyanoformate, and this was treated with ammonia.[4]

The carbamates of tertiary acetylenic alcohols have also been made by reaction of these alcohols with sodium cyanate in trifluoroacetic acid.[9] The yields by this procedure are significantly lower than those obtained by the present modification, which is essentially that described by Loev and Kormendy.[5]

4. Merits of the Preparation

This one-step procedure is a convenient and general method for the preparation of carbamates. It is substantially simpler, quicker, and safer than the multistep methods hitherto used for the preparation of carbamates of tertiary alcohols. This procedure is applicable to the preparation of carbamates of primary, secondary, and tertiary alcohols and mercaptans, polyhydric alcohols, acetylenic alcohols, phenols, and oximes. It has also been extended to the preparation of carbamyl derivatives (*i.e.*, ureas) of inert (non-basic) amines.[10]

1. Research and Development Division, Smith, Kline, and French Laboratories, Philadelphia, Pennsylvania 19101.
2. A. R. Choppin and J. W. Rogers, *J. Am. Chem. Soc.*, **70**, 2967 (1948).
3. L. A. Carpino, *J. Am. Chem. Soc.*, **79**, 98 (1957); **79**, 4427 (1957).
4. L. A. Carpino, *J. Am. Chem. Soc.*, **82**, 2725 (1960).
5. B. Loev and M. F. Kormendy, *J. Org. Chem.*, **28**, 3421 (1963).
6. S. Petersen, in E. Müller, "Methoden der Organischen Chemie (Houben-Weyl)," 4th ed., Vol. 8, Georg Thieme Verlag, Stuttgart, 1952, p. 137.
7. L. Yoder, *J. Am. Chem. Soc.*, **45**, 475 (1923).
8. W. M. McLamore, S. Y. P'an, and A. Bavley, *J. Org. Chem.*, **20**, 1379 (1955).

9. P. G. Marshall, J. H. Barnes, and P. A. McCrea, U. S. Patent 2,814,637 (1957) [*C.A.*, **52**, 2901 (1958)].

10. G. J. Durant, *Chem. Ind. (London)*, 1428 (1965).

t-BUTYL CARBAZATE*

(Carbazic acid, *tert*-butyl ester)

Method I

$$CH_3SCOCl + (CH_3)_3COH \xrightarrow{C_5H_5N}$$

$$CH_3SCO_2C(CH_3)_3 \xrightarrow{NH_2NH_2} NH_2NHCO_2C(CH_3)_3$$

Submitted by Louis A. Carpino, David Collins, Siegfried Göwecke, Joe Mayo, S. D. Thatte, and Fred Tibbetts [1]
Checked by Fred G. H. Lee and Virgil Boekelheide

1. Procedure

Caution! Methyl chlorothiolformate has an obnoxious odor. All operations should be conducted in a well-ventilated hood.

A. *t-Butyl S-methylthiolcarbonate.* In a 5-l. round-bottomed flask, fitted with mechanical stirrer, reflux condenser, and dropping funnel are placed 430 ml. (422 g., 5.33 moles) of pyridine, 508 ml. (395 g., 5.33 moles) of *t*-butyl alcohol, and 1.6 l. of chloroform (Note 1). The solution is stirred while 536 g. (4.85 moles) of methyl chlorothiolformate (Note 2) is added dropwise over a period of 30–40 minutes, and the solution is then stirred and heated at the reflux temperature for 24 hours. The resulting solution is divided into three equal portions of about 1 l., and each portion is washed in a 2-l. separatory funnel with two 500-ml. portions of water, three 275-ml. portions of 5% hydrochloric acid, and finally with 350 ml. of 1*M* sodium bicarbonate solution. The combined chloroform solutions are dried over anhydrous magnesium sulfate for 5 hours, and the solvent is removed by distillation from a water bath at atmospheric pressure followed by the use of a water aspirator. Distillation (Note 3) of the residue from a 1-l. Claisen flask by means of a water or oil bath gives 419–497 g. (58–69%) of a colorless liquid, b.p. 62–65° (24 mm.), n^{25}D 1.4525. This product is sufficiently pure for use in the preparation below but may be purified by distillation through a 30-cm. helices-packed column which gives 375–447 g. (52–62%) of the ester, b.p. 60–63° (24 mm.).

B. *t-Butyl carbazate.* In a 2-l. round-bottomed flask set up in a hood and fitted with an efficient mechanical stirrer and a reflux condenser are placed 500 g. (3.47 moles) of *t*-butyl S-methylthiolcarbonate and 186 g. (3.71 moles) of 64% hydrazine solution (Note 4). The contents of the flask are heated in an oil bath at 105–110° (external temperature) with mechanical stirring for 24 hours under a reflux condenser. The resulting mixture is diluted with 650 ml. of methylene dichloride, and anhydrous magnesium sulfate is added until the aqueous layer becomes nearly solid and nonflowing. The upper layer is decanted and dried over fresh anhydrous magnesium sulfate and the solvent removed by distillation from a water bath, the last portions being removed with the aid of a water aspirator. The residual liquid solidifies on cooling and stirring to give 322–385 g. (72–86%) of a snow-white solid, m.p. 37–40°. This product is pure enough for most purposes but can be purified by distillation, a 1-l. Claisen flask with a water or oil bath at 80° being used. After 1 or 2 drops of forerun the carbazate is collected at 55–57° (0.4 mm.). The oil solidifies on cooling to give 312–358 g. (70–80%) of snow-white crystalline solid, m.p. 40–42° (Note 5).

2. Notes

1. The pyridine was a pure product, b.p. 113–115°, obtained from the Mallinckrodt Chemical Company. *t*-Butyl alcohol, m.p. 24.5–25.5°, and chloroform (U.S.P. or reagent grade) were obtained from the Matheson Company. All reagents were used as supplied.

2. The methyl chlorothiolformate, b.p. 110–111°, was used as supplied by the Stauffer Chemical Company.

3. This distillation is accompanied by foaming which is very difficult to prevent. The checkers recommend carrying out the

distillation in two separate batches to allow greater free space in the distillation flask.

4. Hydrazine hydrate (64% hydrazine) was used as supplied by the Fairmount Chemical Company.

5. Further purification can be effected by recrystallization with 90% recovery from a 50-50 mixture of low- (b.p. 30–60°) and high-boiling (b.p. 60–70°) ligroin. This procedure gives white needles, m.p. 41–42°.

Method II

$$C_6H_5OCOCl + (CH_3)_3COH \xrightarrow{C_9H_7N}$$

$$C_6H_5OCO_2C(CH_3)_3 \xrightarrow{NH_2NH_2} NH_2NHCO_2C(CH_3)_3$$

Submitted by Louis A. Carpino, Barbara A. Carpino, Chester A. Giza, Robert W. Murray, Arthur A. Santilli, and Paul H. Terry [1]
Checked by Virgil Boekelheide and S. J. Cross

1. Procedure

A. *t-Butyl phenyl carbonate.* In a 2-l. round-bottomed flask fitted with thermometer, dropping funnel, and mechanical stirrer are placed 248 g. (3.35 moles) of *t*-butyl alcohol, 430 g. (3.33 moles) of quinoline, and 500 ml. of methylene dichloride (Note 1). The solution is stirred while 520 g. (3.32 moles) of phenyl chloroformate (Note 2) is added dropwise over a period of 4 hours. During the addition the temperature is maintained at 28–31° (Note 3) by cooling the flask, as needed, by a stream of tap water. The solution is allowed to stand overnight and is then treated with 800 ml. of water to dissolve the precipitated quinoline hydrochloride (Note 4). The mixture is shaken well in a separatory funnel; the organic layer is separated and washed with two 200-ml. portions of water and three or four 200-ml. portions of 5% hydrochloric acid. After the extract has dried over anhydrous magnesium sulfate for 5 hours, the solvent is removed by distillation, a water aspirator being used to remove the last portions of the methylene dichloride. Distillation of the residue from a 1-l. Claisen flask by means of an air bath maintained at 125–135°

gives 460–495 g. (71–76%) (Note 5) of a colorless oil, b.p. 74–78° (0.5 mm.), n^{24}D 1.4832. This product is sufficiently pure for use in the preparation below.

B. *t-Butyl carbazate*. In a 1-l. Erlenmeyer flask are placed 388.4 g. (2.0 moles) of phenyl *t*-butyl carbonate and 120.2 g. (2.4 moles) of a 64% hydrazine solution (Note 6). The mixture is swirled by hand and heated on a hot plate. When the internal temperature reaches 75–80°, it then rises rapidly and spontaneously to 104–110°, the two layers forming a clear solution. The solution is allowed to cool overnight. The mixture is then diluted with 500 ml. of ether and transferred to a separatory funnel in which it is shaken vigorously for about 10 minutes with a solution prepared from 160 g. (4.0 moles) of sodium hydroxide and 1.2 l. of water. The resulting two layers are placed in a 2-l. continuous extractor and extracted for 48 hours with ether. The ether solution is dried over magnesium sulfate, and the ether is removed by distillation from a water bath. The last portions of ether are removed with the aid of a water aspirator. The residual oil is then distilled using a Claisen flask with an air bath maintained at 115–125°. After 1 or 2 drops of fore-run the carbazate is collected at 61–65° (1.2 mm.), n^{24}D 1.4518. The yield is 235–256 g. (89–97%) (Note 7).

2. Notes

1. The *t*-butyl alcohol and methylene dichloride were used directly as received from the Matheson Company. The quinoline was a practical grade material (Eastman Kodak Company) which was redistilled [b.p. 100–102° (25 mm.)].

2. Phenyl chloroformate was prepared by the method of Strain *et al.*,[2] except that methylene dichloride or chloroform was used in place of benzene as the solvent. The yield was 85–95% [b.p. 74–76° (15 mm.), n^{24}D 1.5125]. The checkers used commercial phenyl chloroformate (Eastman Kodak).

3. The reaction may be run without cooling by adding the acid chloride at a rate to maintain the temperature at 39–43°. On the scale indicated this requires about 5 hours. The yield is substantially the same, although more high-boiling material is formed.

4. Occasionally the quinoline hydrochloride does not separate;

this does not affect the yield, however. If it is desired to recover the quinoline, the salt may be filtered at this point, dissolved in water, and converted to the free base.

5. The carbonate decomposes on attempted distillation of large amounts at water aspirator pressure (20–25 mm.).

6. Hydrazine hydrate (64% hydrazine) was used as supplied by the Olin-Mathieson Company.

7. The carbazate is sufficiently pure for most applications and is conveniently handled as a liquid. When the product is cooled in an ice box, it crystallizes as a waxy mass; such samples remain tacky after several recrystallizations from petroleum ether, however. A pure sample may be obtained by extracting an ether solution of the carbazate with dilute sodium hydroxide to remove any phenol, followed by distillation and recrystallization of the product from petroleum ether (b.p. 30–60°). White needles are obtained which melt at 41–42°.

3. Methods of Preparation

t-Butyl S-methylthiolcarbonate has been prepared from sodium *t*-butoxide, carbonyl sulfide, and methyl iodide [3] and from methyl chlorothiolformate and *t*-butyl alcohol.[4] *t*-Butyl phenyl carbonate has been prepared from phenyl chloroformate and *t*-butyl alcohol.[5, 6]

t-Butyl carbazate has been prepared by reaction of hydrazine with *t*-butyl phenyl carbonate,[5, 6] *t*-butyl S-methylthiolcarbonate,[3] *t*-butyl-*p*-nitrophenyl carbonate,[7] and N-*t*-butyloxycarbonyl-imidazole.[8]

4. Merits of the Preparation

Method I is easily adapted to larger-scale operation and provides a product which crystallizes readily. Method II provides a product which is difficult to crystallize; however, the procedure obviates the use of methyl chlorothiolformate and affords higher yields of product.

t-Butyl carbazate is a useful reagent for preparing 1,1-disubstituted hydrazines.[6] In turn, the 1,1-disubstituted hydrazines can undergo an elimination of nitrogen followed by radical coupling of the two substituent groups. This reaction is promoted either by direct oxidation of the 1,1-disubstituted hydrazine [9-11] or by base-catalyzed elimination of benzenesulfinic acid from the corresponding benzenesulfonhydrazide.[6] For example, 1,1-dibenzyl-2-benzenesulfonhydrazide is converted to bibenzyl in 85% yield.

t-Butyl carbazate is also a key intermediate in the synthesis of *t*-butyl azidoformate,[5, 6, 12] *t*-butyl hydrazodiformate,[6] *t*-butyl azodiformate,[13] and *t*-butyl N-hydroxycarbamate,[12, 14] all of which are valuable synthetic intermediates.

1. Department of Chemistry, University of Massachusetts, Amherst, Massachusetts.
2. F. Strain, W. E. Bissinger, W. R. Dial, H. Rudoff, B. J. DeWitt, H. C. Stevens, and J. H. Langston, *J. Am. Chem. Soc.*, **72**, 1254 (1950).
3. L. A. Carpino, *J. Am. Chem. Soc.*, **82**, 2725 (1960).
4. L. A. Carpino, *J. Org. Chem.*, **28**, 1909 (1963).
5. L. A. Carpino, *J. Am. Chem. Soc.*, **79**, 98 (1957).
6. L. A. Carpino, *J. Am. Chem. Soc.*, **79**, 4427 (1957).
7. F. Eloy and C. Moussebois, *Bull. Soc. Chim. Belges*, **68**, 409 (1959).
8. W. Klee and M. Brenner, *Helv. Chim. Acta*, **44**, 2151 (1961).
9. C. G. Overberger and B. S. Marks, *J. Am. Chem. Soc.*, **77**, 4104 (1955).
10. M. Busch and B. Weiss, *Ber.*, **33**, 2701 (1900).
11. C. G. Overberger, J. G. Lombardino, and R. G. Hiskey, *J. Am. Chem. Soc.*, **79**, 6430 (1957).
12. L. A. Carpino, C. A. Giza, and B. A. Carpino, *J. Am. Chem. Soc.*, **81**, 955 (1959).
13. L. A. Carpino, P. H. Terry, and P. J. Crowley, *J. Org. Chem.*, **26**, 4336 (1961).
14. L. A. Carpino, *J. Am. Chem. Soc.*, **82**, 3133 (1960).

t-BUTYL CYANOACETATE*

(Cyanoacetic acid, *tert*-butyl ester)

$$NCCH_2CO_2H + PCl_5 \rightarrow NCCH_2COCl + POCl_3 + HCl$$
$$NCCH_2COCl + (CH_3)_3COH + C_6H_5N(CH_3)_2 \rightarrow$$
$$NCCH_2CO_2C(CH_3)_3 + C_6H_5N(CH_3)_2 \cdot HCl$$

Submitted by ROBERT E. IRELAND and MICHAEL CHAYKOVSKY.[1]
Checked by MAX TISHLER and ARTHUR J. ZAMBITO.

1. Procedure

In a 5-l., three-necked, round-bottomed flask equipped with a rubber- or mercury-sealed mechanical stirrer and a reflux condenser carrying a drying tube are placed 340 g. (4 moles) of cyanoacetic acid (Note 1) and 2 l. of anhydrous ether. To the stirred solution, 834 g. (4 moles) of phosphorus pentachloride is added in portions through the third neck of the flask, which is sealed between additions. The mixture is cooled occasionally with an ice bath to prevent excessive refluxing, and, after the addition is complete, stirring is continued for 0.5 hour or until the phosphorus pentachloride dissolves completely. The reflux condenser is removed and replaced with apparatus for downward distillation (Note 2), and the ether is distilled from a water bath at 50–60° (Note 3), after which most of the phosphorus oxychloride is removed at reduced pressure (20–25 mm. with a bath temperature of 55–65°) (Note 4), the receiver being cooled in an ice-salt bath. The red, oily residue is dissolved in 200 ml. of benzene and the benzene and residual phosphorus oxychloride distilled under reduced pressure. This operation is repeated with 200 ml. of fresh benzene to ensure complete removal of phosphorus oxychloride (Note 5). The residue is then cooled to room temperature (Note 6) and is transferred to a 500-ml. pressure-equalized dropping funnel for immediate use in the following step.

The same 5-l. flask used in the preceding step is used again, without washing; it is fitted with a reflux condenser carrying a drying tube, a sealed mechanical stirrer, and the dropping funnel containing the acid chloride. In the flask are placed 296 g. (4 moles) of dry t-butyl alcohol (Note 7) and 484 g. (4 moles) of dimethylaniline in 600 ml. of anhydrous ether (Note 8). The acid chloride is added dropwise to the stirred solution, the mixture being cooled occasionally with an ice bath to prevent excessive refluxing. After the addition is complete, the reaction mixture is refluxed for 2 hours and then stirred gently at room temperature for 15 hours. Two liters of water is added with stirring, and the mixture is filtered with suction through a Büchner funnel fitted with a matting of glass wool (Note 9). The matting is washed with three 250-ml. portions of ether (Note 10). After separation of the combined ethereal layers, the aqueous layer is extracted twice with 250-ml. portions of ether. The combined ether solutions are washed with successive portions of 2N sulfuric acid (a total of 1 l.) until free of dimethylaniline, then with two 200-

ml. portions of 2N sodium carbonate solution, and dried over sodium carbonate. After removal of the ether by distillation, the residue is transferred to an alkali-washed flask and distilled at reduced pressure through a 20-cm. alkali-washed Vigreux column (Note 11). The yield of colorless product is 355–378 g. (63–67%) boiling at 67–68°/1.5 mm. (90°/10 mm., 54–56°/0.3 mm), $n_D^{20} = 1.4198$.

2. Notes

1. Cyanoacetic acid of 98% purity, obtained from Kay-Fries Chemicals, Inc., was used.

2. A Claisen head with a condenser leading into a flask with a suction arm connected to a drying tube is suitable. Ground-glass joints are recommended.

3. A large bucket containing water and placed on a steam bath serves as a suitable water bath. The temperature is easily controlled between the limits mentioned.

4. The ether is removed from the receiving flask before the phosphorus oxychloride is distilled. A drying tube should be placed between the suction arm of the flask and the water pump, which serves as the source of suction. The reaction mixture may be stirred during the distillation of the oxychloride, or the stirrer may be removed and replaced with a capillary ebulliator tube to which is attached a balloon filled with dry nitrogen.

5. The checkers found that the distillation with benzene ensures a more complete removal of phosphorus oxychloride which, if still present, interferes in the subsequent step and a lower yield of product results.

6. The submitters found that on several occasions, when the residue was not cooled before transfer, it began to generate considerable heat while standing in the dropping funnel and resulted in the total carbonization of the acid chloride.

7. The submitters dried the *t*-butyl alcohol by refluxing it over calcium hydride overnight and distillation in a moisture-free apparatus. The checkers found that stirring the *t*-butyl alcohol at 60–70° over calcium hydride for several hours and then distilling the alcohol, using an air condenser, is a satisfactory procedure. When the *t*-butyl alcohol is refluxed, the alcohol

vapors condense and solidify (m.p. 24–25°) in the reflux condenser and cause clogging.

8. These reagents should be weighed out beforehand in order to prevent delay in commencing with this step.

9. The filtration removes some tarry resinous material which would otherwise interfere in the separation of the layers. The checkers found that unless the filtrate is recycled through the same matting several times, to remove practically all the tarry residue, the separation of layers and the subsequent extractions prove troublesome owing mainly to emulsion formation.

10. The checkers found that a considerable amount of product is withheld by the residue on the glass-wool matting. The product is extracted by placing the matting in a beaker, stirring with ether, and filtering. This procedure is repeated twice, and the ether extracts are combined with the original filtrate.

11. The distilling flask and Vigreux column to be used should be washed with 25% aqueous sodium hydroxide solution, rinsed three times with water, and then dried. Alternatively, about 1 g. of anhydrous potassium carbonate may be added to the residue before distillation.

3. Methods of Preparation

t-Butyl cyanoacetate has been prepared from t-butyl bromoacetate and potassium cyanide in methanol[2] and from t-butyl chloroacetate and potassium cyanide in methyl Cellosolve.[3] The present method is an adaptation of that of Beech and Pigott[4] and is similar to the *Organic Syntheses* preparation of t-butyl acetate.[5]

4. Merits of Preparation

The present preparation employs a method of considerable scope which gives much better yields and is considerably less laborious than other methods for the preparation of t-butyl cyanoacetate. The compound is of specific interest since, for example, it may be used in any reaction where ethyl cyanoacetate is used (condensation reactions, etc.), but it has the added advantage that the carbo-t-butoxy group, which may serve in conjunction with the α-cyano group to activate the α-hydrogens (for cyanoethylations, etc.), may be later removed simply by pyrolysis of the compound.[6]

1. University of Michigan, Ann Arbor, Mich.
2. B. Abramovich and C. R. Hauser, *J. Am. Chem. Soc.*, **64**, 2271 (1942).
3. Private communication, W. S. Johnson, University of Wisconsin.
4. W. F. Beech and H. A. Pigott, *J. Chem. Soc.*, 423 (1955).
5. C. R. Hauser, B. E. Hudson, B. Abramovitch, and J. C. Shivers, *Org. Syntheses*, Coll. Vol. 3, 142 (1955).
6. S.-O. Lawesson, E. H. Larsen, and H. J. Jakobsen, *Arkiv Kemi*, **23**, 453 (1965).

trans-4-*t*-BUTYLCYCLOHEXANOL*

(Cyclohexanol, 4-*t*-butyl, *trans*-)

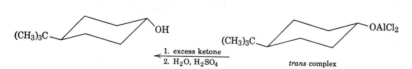

Submitted by E. L. Eliel, R. J. L. Martin, and D. Nasipuri [1]
Checked by E. J. Corey and Barbara Kaski

1. Procedure

In a 3-l. three-necked flask, equipped with a mercury-sealed Hershberg stirrer, a dropping funnel, and a reflux condenser protected with a calcium chloride tube, is placed 67 g. (0.5 mole) of powdered anhydrous aluminum chloride. The flask is cooled in an ice bath, 500 ml. of dry ether is slowly added from the dropping funnel, and the mixture is stirred for a few minutes. Meanwhile 5.5 g. of powdered lithium aluminum hydride (Note 1) is placed in a 1-l. flask fitted with a condenser, and 140 ml. of dry ether is added slowly from the top of the condenser with caution while the flask is kept cooled in ice. The mixture is gently refluxed for 30 minutes to effect, as far as possible, solution of the hydride. It is then cooled, and the resulting slurry (which consists of a suspension of lithium aluminum hydride in its solution) is transferred to the dropping funnel of the previous setup and is added to the ethereal solution of aluminum chloride with stirring within 10 minutes. After the addition is complete, the reaction mixture is stirred for an additional 30 minutes without cooling to complete the formation of the "mixed hydride."

A solution of 77.2 g. (0.5 mole) of 4-t-butylcyclohexanone (Note 2) in 500 ml. of dry ether is then placed in the dropping funnel and slowly added to the "mixed hydride" solution without much cooling so that gentle refluxing is maintained (Note 3) After addition of the ketone over a period of 45–60 minutes the reaction mixture is refluxed for 2 hours more to complete the reduction. The excess hydride is destroyed by the addition of 10 ml. of dry t-butanol, and the mixture is refluxed for an additional 30 minutes. 4-t-Butylcyclohexanone (3 g.) in 20 ml. of dry ether is then added to the reaction mixture, which is refluxed for 4 hours more and allowed to stand overnight (Notes 4, 5). The reaction mixture is cooled in an ice bath and decomposed by successive additions of 100 ml. of water and 250 ml. of 10% aqueous sulfuric acid. The ethereal layer is separated and the aqueous layer extracted once with 150 ml. of ether. The combined ether extracts are washed once with water and dried over anhydrous magnesium sulfate. The extract is filtered from magnesium sulfate and the ether removed by distillation on a steam bath. The residue, weighing 85–87 g., solidifies in the flask and on gas chromatographic analysis (see Note 5) is found to contain 96% *trans* alcohol, 0.8% *cis* alcohol, and 3.2% ketone.

The crude white product is dissolved in 150 ml. of hot petroleum ether (b.p. 60–70°). On cooling, it forms a solid cake which is transferred to a Buchner funnel and rinsed with small portions of cooled petroleum ether. The yield of product, m.p. 75–78°

(Note 6), is 57–61 g. (73–78%). It has an approximate composition of 99.3% *trans* alcohol, 0.3% *cis* alcohol, and 0.4% ketone. A further crop of 12 g. is obtained by concentration of the mother liquor. It contains less than 1% of the *cis* alcohol and the ketone and is sufficiently pure for most preparative purposes (Note 7).

2. Notes

1. The lithium aluminum hydride was obtained from Metal Hydrides Incorporated and was more than 95% pure. For calculation of the quantity of hydride required it was assumed that the purity was 95%.

2. 4-*t*-Butylcyclohexanone was supplied by the Dow Chemical Company.

3. The continuance of gentle refluxing as the last portion of the ketone is added assures that there is an excess of "mixed hydride" present.

4. It is not necessary to allow the reaction mixture to stand overnight, and it may be decomposed at this stage without any loss in purity.

5. At this stage the attainment of equilibrium can be checked by removing a 5-ml. aliquot from the reaction product and working it up in the same way as described in the preparation. The product is then analyzed by gas-liquid chromatography using a 20% Carbowax 20M on firebrick column at 150°. The features to note in the chromatogram are almost complete absence (less than 1%) of the *cis* isomer (second peak, disregarding initial solvent and *t*-butanol peaks) and the presence of some 4-*t*-butylcyclohexanone (first peak). The *trans* isomer constitutes the third peak with longest retention time.

6. The melting point of a highly purified sample[2] of *trans* alcohol is 82.5–83°.

7. Since the alcohol has a relatively high solubility in petroleum ether, the yield from the crystallization depends on the volume of solvent used. However, by concentrating the mother liquor the overall yield from the first and second crops of crystals varies from 74% to 94%.

3. Methods of Preparation

4-*t*-Butylcyclohexanol has been prepared from *p*-*t*-butylphenol by reduction under a variety of conditions.[3, 4] Winstein and Holness [5] prepared the pure *trans* alcohol from the commercial alcohol by repeated crystallization of the acid phthalate followed by saponification of the pure *trans* ester. Eliel and Ro [6] obtained 4-*t*-butylcyclohexanol containing 91% of the *trans* isomer by lithium aluminum hydride reduction of the ketone. Hückel and Kurz [7] reduced *p*-*t*-butylphenol with platinum oxide in acetic acid and then separated the isomers by column chromatography.

4. Merits of the Preparation

The procedure employs a readily available starting material and produces the pure *trans* isomer in high yield. The method described is an improvement on that used by Eliel and Rerick [2] in that it is not necessary to use a clear solution of lithium aluminum hydride in ether for the preparation of the "mixed hydride." It is not necessary to know the precise amount of lithium aluminum hydride used so long as a slight excess is present. The excess hydride is destroyed by adding *t*-butanol; the excess *t*-butanol has no effect on the subsequent equilibration and purification. The equilibration of the 4-*t*-butylcyclohexanol is effected by adding a small amount of 4-*t*-butylcyclohexanone.

The method is useful in the preparation of other equatorial alcohols.[2, 8]

1. Department of Chemistry, University of Notre Dame, Notre Dame, Indiana.
2. E. L. Eliel and M. N. Rerick, *J. Am. Chem. Soc.*, **82**, 1367 (1960).
3. G. Vavon and M. Barbier, *Bull. Soc. Chim. France*, [4] **49**, 567 (1931).
4. H. Pines and V. Ipatieff, *J. Am. Chem. Soc.*, **61**, 2728 (1939).
5. S. Winstein and N. J. Holness, *J. Am. Chem. Soc.*, **77**, 5562 (1955).
6. E. L. Eliel and R. S. Ro, *J. Am. Chem. Soc.*, **79**, 5992 (1957).
7. W. Hückel and J. Kurz, *Ann.*, **645**, 194 (1961).
8. E. L. Eliel, *Record Chem. Progr. (Kresge-Hooker Sci. Lib.)*, **22**, 129 (1961); M. N. Rerick, in R. L. Augustine, "Reduction," Marcel Dekker, Inc., New York, 1968, pp. 32–34.

t-BUTYL DIAZOACETATE

(Acetic acid, diazo-, *tert*-butyl ester)

$$p\text{-}CH_3C_6H_4SO_2Cl + NaN_3 \rightarrow p\text{-}CH_3C_6H_4SO_2N_3$$

$$CH_3COCH_2CO_2C(CH_3)_3 + p\text{-}CH_3C_6H_4SO_2N_3 \xrightarrow{(C_2H_5)_3N}$$
$$CH_3COCN_2CO_2C(CH_3)_3 + p\text{-}CH_3C_6H_4SO_2NH_2$$

$$CH_3COCN_2CO_2C(CH_3)_3 \xrightarrow[CH_3OH]{CH_3ONa}$$
$$N_2CHCO_2C(CH_3)_3 + CH_3CO_2CH_3$$

Submitted by Manfred Regitz, Jürgen Hocker,
and Annemarie Liedhegener [1]
Checked by C. John Blankley and Herbert O. House

1. Procedure

Caution! Diazoacetic esters are toxic and potentially explosive and must be handled with caution. This preparation should be carried out in a hood, and the distillation of t-butyl diazoacetate should be conducted behind a safety shield.

A. *p-Toluenesulfonyl azide.*[2,3] A solution of 71.5 g. (1.10 moles) of sodium azide (Note 1) in 200 ml. of water is placed in a 2-l. Erlenmeyer flask and diluted with 400 ml. of 90% aqueous ethanol (Note 2). To this solution is added with stirring a warm (45°) solution of 190.5 g. (1.00 mole) of *p*-toluenesulfonyl chloride (Note 3) in 1 l. of 99% ethanol (Note 2). During this addition, sodium chloride separates, and the reaction mixture takes on a light brown color. After the reaction mixture has been stirred at room temperature for 2.5 hours, most of the solvent is removed at 35° (15 mm.) with a rotary evaporator (Note 4). The residue is mixed with 1.2 l. of water in a separatory funnel, and the oily *p*-toluenesulfonyl azide is separated. This oil is washed with two 100-ml. portions of water and dried over anhydrous sodium sulfate. Filtration with suction gives 160–170 g. (81–86%, based on *p*-toluenesulfonyl chloride) of pure, colorless *p*-toluenesulfonyl azide which completely crystallizes on standing at 5°.

B. *t-Butyl α-diazoacetoacetate.* In a 2-l., wide-mouthed, Erlenmeyer flask are placed 118.5 g. (0.75 mole) of *t*-butyl acetoacetate (Note 5), 1 l. of anhydrous acetonitrile, and 75.8 g. (0.75 mole) of previously distilled triethylamine (b.p. 88.5–90.5°). The temperature of the mixture is adjusted to 20°, and 148 g. (0.75 mole) of *p*-toluenesulfonyl azide is added dropwise with vigorous stirring over 10–15 minutes. The addition causes the reaction mixture to warm to 38–40° and assume a yellow color. After the mixture has been stirred at room temperature for 2.5 hours, the solvent is evaporated at 35° (12 mm.). The partially crystalline residue is triturated with 1 l. of ether, and the mixture, including the insoluble residue, is placed in a 2-l. separatory funnel. The mixture is washed successively with a solution of 45 g. of potassium hydroxide in 500 ml. of water, a solution of 7.5 g. of potassium hydroxide in 250 ml. of water, and 250 ml. of water (Note 6). The yellow-orange ethereal phase is dried over anhydrous sodium sulfate, and the solvent is evaporated at 35° (15 mm.) until the residue has attained a constant weight. The yellow-orange diazo ester weighs 130–135 g. (94–98%) (Note 7).

C. *t-Butyl diazoacetate.* Into a 1-l. three-necked flask fitted with a stirrer, a dropping funnel, and a thermometer is placed a solution of 92.6 (0.50 mole) of *t*-butyl α-diazoacetoacetate in 150 ml. of methanol. After this solution has been cooled to 2–3° in an ice bath, a solution of sodium methoxide, prepared from 11.5 g. (0.50 g. atom) of sodium and 150 ml. of methanol, is added dropwise with stirring at such a rate that the reaction mixture remains within the temperature range 0–5° (about 30 minutes is required for the addition). After the addition is completed, the mixture is stirred in the ice bath for an additional 30 minutes. The red reaction solution is poured into 1 l. of ice water, and the resulting mixture is extracted with 500 ml. of ether. The aqueous phase is saturated with sodium chloride and extracted with two 500-ml. portions of ether (Note 8). The combined ethereal extracts are washed with 500 ml. of water and dried over anhydrous sodium sulfate. After the mixture has been filtered and the residue has been washed with ether, the bulk of the solvent is removed from the combined ethereal filtrates at 30° and water aspirator pressure with a rotary evaporator (Note 9). The remaining ether is removed by distillation under slightly reduced pressure while the stillpot is heated with a water bath at 50°. The residual red oil is distilled. (*Caution! See above.*) (Note 10). After a small forerun the diazo ester distills during which time the temperature of the water bath is raised from 60° to 75°. The yield is 48–50 g. (68–70%) of yellow-orange liquid, b.p. 51–53° (12 mm.), n^{20}D 1.4551, $R_f = 0.56$ (chloroform) (Note 11).

2. Notes

1. The submitters used sodium azide obtained from Dr. F. Raschig, GmbH, 67 Ludwigshafen, Rhein, Germany. The checkers used material from Eastman Organic Chemicals.

2. The checkers found 95% ethanol denatured with methanol to be a satisfactory substitute.

3. The submitters used *p*-toluenesulfonyl chloride obtained from Badische Anilin- und Soda-Fabrik, 67 Ludwigshafen, Rhein, Germany. Very impure *p*-toluenesulfonyl chloride can be purified by recrystallization from ether. The checkers used material from Matheson, Coleman and Bell without further purification.

4. In order to prevent foaming, the concentration is begun with the water bath at *ca.* 10°, and the bath is warmed slowly to 35°.

5. *t*-Butyl acetoacetate may be prepared from *t*-butyl alcohol and diketene.[4] The checkers obtained this material from Eastman Chemical Products, Inc.

6. Acidification of the aqueous potassium hydroxide phase with 6N hydrochloric acid gives *p*-toluenesulfonamide. After being dried at 85° (50 mm.) the sample weighs 110–120 g. (86–94%) and melts at 132–134°.

7. If desired, the α-diazo β-keto ester can be purified by a low-temperature crystallization. The diazo ester (10 g.) is cooled to −70° to −75° in a dry ice-acetone bath, and crystallization is initiated by rubbing. (*Caution! The rubbing should not be continued after crystallization has been initiated.*) This material is treated with 5 ml. of anhydrous ether which has been previously cooled, and the mixture is filtered with suction. The residue

from the filtration is placed in a flask, and the residual ether is removed by evaporation at 35° (15 mm.) to give 5–6 g. of the yellow diazo ester.[5]

8. If the ethereal phase contains a small amount of insoluble material, the mixture should be filtered to avoid difficulty in separating the phases.

9. The distillate is light yellow and contains some t-butyl diazoacetate.

10. This distillation has been conducted with the usual precautions (safety glasses, safety shield) with no explosions up to the present time.

11. The thin-layer chromatogram was obtained on "DC-Fertiplatte Merck Kieselgel F_{254}" purchased from E. Merck A. G., 61 Darmstadt, Germany. Employing an Eastman Chromatoplate K301R1 (silica without indicator) with chloroform as eluent, the checkers found an R_f value of 0.72.

3. Methods of Preparation

t-Butyl diazoacetate has been prepared by the present method, by alkaline decomposition of t-butyl N-nitroso-N-acetylglycinate,[6] and by diazotization of t-butyl glycinate.[7]

4. Merits of the Preparation

The transformation of an active CH compound into the corresponding diazo derivative with p-toluenesulfonyl azide has been designated a "diazo transfer reaction"[8] and possesses a variety of preparative uses. The method has been useful for the syntheses of diazo derivatives of cyclopentadiene,[3, 9] 1,3-dicarbonyl compounds,[5, 10, 11] 1,3-disulfonyl compounds,[12] 1,3-ketosulfonyl compounds,[13–15] 1,3-ketophosphono compounds,[16] 1,3-ketophosphinyl compounds,[17] ketones,[18–21] carboxylic acid esters[18,21] and β-keto imines.[22] Further reaction of these diazo intermediates can lead to zao compounds,[11,13] 1,2,3-triazoles,[22,23] 1,2,3-thiadiazoles,[24] and pyrazolinones.[25] This and related diazo transfer reactions have been reviewed.[26]

1. Institut für Organische Chemie der Universität des Saarlandes, 66 Saarbrücken, Germany.

2. T. Curtius and G. Kraemer, *J. Prakt. Chem.*, [2] **125**, 303 (1930).

3. W. von E. Doering and C. H. DePuy, *J. Am. Chem. Soc.*, **75**, 5955 (1953).

4. S.-O. Lawesson, S. Gronwall, and R. Sandberg, this volume, p. 155.

5. M. Regitz and A. Liedhegener, *Ber.*, **99**, 3128 (1966).

6. H. Reimlinger and L. Skatteböl, *Ber.*, **93**, 2162 (1960).

7. E. Müller and H. Huber-Emden, *Ann.*, **660**, 54 (1962).

8. M. Regitz, *Angew. Chem.*, **79**, 786 (1967); *Angew. Chem. Intern. Ed. Engl.*, **6**, 733 (1967).

9. M. Regitz and A. Liedhegener, *Tetrahedron*, **23**, 2701 (1967).

10. M. Regitz, *Ann.*, **676**, 101 (1964).

11. M. Regitz and D. Stadler, *Ann.*, **687**, 214 (1965).

12. F. Klages and K. Bott, *Ber.*, **97**, 735 (1964).

13. M. Regitz, *Ber.*, **98**, 36 (1965).

14. A. M. van Leusen, P. M. Smid, and J. Strating, *Tetrahedron Lett.*, 337 (1965).

15. M. Regitz and W. Bartz, *Ber.*, **103**, 1477 (1970).

16. M. Regitz, W. Anschütz, and A. Liedhegener, *Ber.*, **101**, 3734 (1968).

17. M. Regitz and W. Anschütz, *Ber.*, **102**, 2216 (1969).

18. M. Regitz, *Ber.*, **98**, 1210 (1964); *Tetrahedron Lett.*, 1403 (1964).

19. M. Rosenberger, P. Yates, J. B. Hendrickson, and W. Wolf, *Tetrahedron Lett.*, 2285 (1964).

20. M. Regitz and J. Rüter, *Ber.*, **101**, 1263 (1968).

21. M. Regitz and F. Menz, *Ber.*, **101**, 2622 (1968).

22. M. Regitz and H. Schwall, *Ann.*, **728**, 99 (1969).

23. M. Regitz and A. Liedhegener, *Ber.*, **99**, 2918 (1966).

24. M. Regitz, *Ann.*, **710**, 118 (1967).

25. M. Regitz and H. J. Geelhaar, *Ber.*, **101**, 1473 (1968).

26. M. Regitz, *Synthesis*, 351 (1972).

t-BUTYL HYPOCHLORITE

WARNING

It has been reported[1] that, during the preparation of *t*-butyl hypochlorite according to the directions published in this series,[2] an explosion occurred and caused moderate physical damage and minor injury to the operator. The cause of the accident has been attributed to lack of proper temperature control during addition of chlorine. It is strongly recommended that the reaction vessel be fitted with a thermometer that dips into the reaction mixture and that the rate of flow of chlorine be regulated so that the temperature of the reaction mixture never exceeds 20°.

1. C. P. C. Bradshaw and A. Nechvatal, *Proc. Chem. Soc.*, 213 (1963).
2. *Org. Syntheses*, **32**, 20 (1952); Coll. Vol. **4**, 125 (1963).

t-BUTYL HYPOCHLORITE*

$$(CH_3)_3COH + NaOCl + CH_3COOH \rightarrow$$
$$(CH_3)_3COCl + CH_3COONa + H_2O$$

Submitted by M. J. MINTZ [1] and C. WALLING [2]
Checked by LOIS A. ABLIN and HENRY E. BAUMGARTEN

1. Procedure

Caution! This preparation should be carried out in a hood to avoid exposure to the hypochlorite produced. To avoid vigorous decomposition the product should be handled only in dim light and should not be heated above its boiling point or be exposed to rubber.

In a 1-l. Erlenmeyer or round-bottomed flask equipped with a mechanical stirrer is placed 500 ml. of a commercial household bleach solution (Note 1). The flask is placed in a pail of ice and rapidly stirred until the temperature drops below 10°. At this point the lights in the vicinity of the apparatus should be turned off (Note 2). A solution of *t*-butyl alcohol (37 ml., 0.39 mole) and glacial acetic acid (24.5 ml., 0.43 mole) (Note 3) is added in a single portion to the rapidly stirred bleach solution, and stirring is continued for about 3 minutes (Note 4).

The entire reaction mixture is poured into a 1-l. separatory funnel. The lower aqueous layer (Note 5) is discarded, and the oily yellow organic layer is washed first with a 50-ml. portion of 10% aqueous sodium carbonate and then with 50 ml. of water. The product is dried over 1 g. of calcium chloride and filtered. The yield of *t*-butyl hypochlorite, 99–100% pure, is 29.6–34 g. (70–80%) (Notes 1, 6). The product can be stored conveniently in a freezer or refrigerator over calcium chloride in amber glass bottles (Note 7).

2. Notes

1. Both the submitters and checkers used the commercial household bleach solution, Clorox (Proctor and Gamble Co.). This solution is stated to be 5.25% sodium hypochlorite (NaOCl). The submitters found it to be 0.75–0.80M by iodometric titration for total oxidant (assumed to be NaOCl). Thus 500 ml. of this solution would contain 0.375–0.400 mole of NaOCl. The checkers found that as little as 440 ml. of fresh Purex (Purex Corp., Ltd., stated to be 6% sodium hypochlorite) gave the stated yield. However, samples from different bottles from one case of "Purex" gave consistently lower yields, 57–70%. Probably the lower yield was due to a lowering of the hypochlorite concentration on standing. The submitters and checkers recommend either discarding bleach solution over 6 months old or checking the titer before use. Presumably other household bleaches will give comparable results with possible small variations in yield.

2. Whereas the inorganic hypochlorite is rather stable to photodecomposition, *t*-butyl hypochlorite is much more readily decomposed. It is not necessary to work in a totally darkened room, but the incidence of strong light should be avoided—both for reasons of safety and to ensure that hypochlorite of high purity will be isolated.

3. The *t*-butyl alcohol was a commercial product obtained from Matheson Coleman and Bell, and the glacial acetic acid a commercial product obtained from Union Carbide.

4. The submitters have carried out runs using up to 4 l. of the commercial bleach solution (3 moles)—as the largest scale conveniently run in the laboratory—and found no change in the reaction behavior.

5. The checkers observed that the aqueous layer was colorless when Clorox was used and was yellow when Purex was used.

6. The purity of the hypochlorite may be determined by iodometric titration. This titration is run conveniently by weighing out a small portion of the hypochlorite (<0.5 g.) in a 4-ml. vial and then dropping the vial and its contents into an iodine flask containing 20 ml. of glacial acetic acid, 10 ml. of water, and 3 g. of potassium iodide. The titration is then conducted in the

usual fashion.

7. The product isolated by this procedure is sufficiently pure for almost any purpose. It was found that distillation did not change the product purity and often led to product of lower purity.

3. Discussion

t-Butyl hypochlorite has been prepared by treatment of an alkaline solution of *t*-butyl alcohol with chlorine,[3-7] and a recent warning [8, 9] cautions against allowing the temperature to rise above 20° during this reaction. *t*-Butyl hypochlorite has been prepared in solution by shaking a solution of the alcohol in carbon tetrachloride,[10] fluorotrichloromethane (Freon 11), and other solvents [12] with aqueous hypochlorous acid. It has also been prepared by the action of chlorine on an aqueous *t*-butyl alcohol suspension of calcium carbonate,[13] and by the action of chlorine monoxide on a carbon tetrachloride solution of the alcohol.[11]

The procedure described here has previously been reported by Mintz [14] and was adapted from work by Geneste and Kergomard,[15] Kergomard,[16] Sumner,[17] and Clark.[18] It eliminates the dangers in working with a compressed gas (chlorine) and the danger from explosion due to poor temperature control during the addition of chlorine.[3, 8, 9] The availability and low cost of the commercial bleach solution (NaOCl), the simplicity of the equipment needed, the short time involved, and the high purity of the *t*-butyl hypochlorite produced confer additional merit on this preparation. The submitters have also prepared benzyldimethylcarbinyl hypochlorite, cumyl hypochlorite, and isopropyl hypochlorite by this procedure. The checkers have used essentially the same procedure (with twice as much sodium hypochlorite solution and acetic acid) to prepare N,N-dichloro-*t*-butylamine.

1. E. C. Britton Research Laboratory, The Dow Chemical Company, Midland, Michigan 48640.
2. Department of Chemistry, Columbia University, New York 10027.
3. H. Teeter and E. Bell, *Org. Syntheses*, Coll. Vol. 4, 125 (1963).
4. F. Chattaway and O. Backeberg, *J. Chem. Soc.*, 2999 (1923).
5. R. Deanesly, U. S. Patent 1,938,175 [*Chem. Abstr.*, 28, 1053 (1934)].
6. C. Irwin and G. Hennion, *J. Am. Chem. Soc.*, 63, 858 (1941).

7. H. Teeter, R. Bachmann, E. Bell, and J. Cowan, *Ind. Eng. Chem.*, **41**, 849 (1949).
8. *Org. Syntheses*, **44** (1964); sheet to be inserted in *Org. Syntheses*, Coll. Vol. **4**, 125 (1963).
9. C. P. C. Bradshaw and A. Nechvatal. *Proc. Chem. Soc.*, 213 (1963).
10. M. Taylor, R. MacMullin, and C. Gammal, *J. Am. Chem. Soc.*, **47**, 395 (1925).
11. M. Anbar and I. Dostrovsky, *J. Chem. Soc.*, 1105 (1954).
12. R. Fort and L. Denivelle, *Bull. Soc. Chim. France*, 1109 (1954).
13. W. Hanby and H. Rydon, *J. Chem. Soc.*, 114 (1946).
14. M. J. Mintz, Ph.D. Thesis, Columbia University, 1965, p. 66.
15. J. Geneste and A. Kergomard, *Bull. Soc. Chim. France*, 470 (1963).
16. A. Kergomard, *Bull. Soc. Chim. France*, 2360 (1961).
17. G. Sumner, Ph.D. Thesis, Massachusetts Institute of Technology, 1934.
18. B. F. Clark Jr., *Chem. News*, **143**, 265 (1931).

2-*n*-BUTYL-2-METHYLCYCLOHEXANONE

(Cyclohexanone, 2-butyl-2-methyl-)

Submitted by S. Boatman, T. M. Harris, and C. R. Hauser [1]
Checked by William G. Dauben, Michael H. McGann, and Noel Vietmeyer

1. Procedure

Caution! This preparation should be carried out in a hood to avoid exposure to ammonia.

In a 3-l. three-necked flask fitted with a calcium chloride drying tube, a nitrogen-inlet tube, and a sealed mechanical stirrer are placed 54.0 g. (1.00 mole) of commercial, anhydrous sodium methoxide (Note 1) and 2 l. of anhydrous ether. The flask is purged with dry nitrogen and cooled in an ice bath. The inlet tube is replaced by an addition funnel containing a solution of 123 g. (1.10 moles) of 2-methylcyclohexanone (Note 2) and 81.4 g. (1.10 moles) of ethyl formate (Note 3). The solution is added rapidly, dropwise, and at the end of the addition the funnel is replaced by the nitrogen-inlet tube. After 15 minutes the ice bath is removed, and the mixture is stirred for 12 hours at room temperature. The thick suspension is filtered by suction, and the filter cake is washed with anhydrous ether, care being taken to protect the product from atmospheric moisture (Note 4). The solid salt is dried in a vacuum oven at *ca.* 70°, powdered (Note 5), and stored in a tightly capped bottle. Sodio-2-formyl-6-methylcyclohexanone, a cream-colored powder, is obtained in 80–85% (130–138 g.) yield (Note 6).

In a 1-l. three-necked flask equipped with a dry ice-acetone condenser and a sealed mechanical stirrer is placed 700 ml. of commercial, anhydrous, liquid ammonia. To the stirred ammonia is added a small piece of potassium metal. (*Caution! Care should be exercised in handling potassium metal, since it is extremely reactive and it ignites on contact with water, atmospheric moisture, or alcohol. It should be manipulated under toluene or xylene, and blotted with filter paper before addition.*) After the appearance of a blue color a few crystals of ferric nitrate hydrate (*ca.* 0.1 g.) are added, followed by small pieces of freshly cut potassium metal until 7.0 g. (0.18 g. atom) has been added. After all the potassium has been converted to the amide (Note 7), 24.9 g. (0.154 mole) of sodio-2-formyl-6-methylcyclohexanone is added carefully through a powder funnel (Note 8). After 1 hour a solution of 28.2 g. (0.21 mole) of *n*-butyl bromide (Note 9) in 50 ml. of anhydrous ether is added dropwise from an addition funnel. The mixture is stirred for 3 hours, and then the dry ice-acetone condenser is replaced by a water condenser. A steam bath is placed under the flask, and the ammonia is evaporated (*Caution!*) as 400 ml. of anhydrous ether is added. When the ammonia has been removed and the ether has refluxed for 5 minutes, 100 g. of ice is added, followed by 300 ml. of water. When the solid has dissolved, the layers are separated, and the ethereal layer is extracted twice with cold water. The combined aqueous extracts are placed in a 1-l. round-bottomed flask, and 6.4 g. of sodium hydroxide is added. The flask is warmed briefly to remove dissolved ether from the solution. The flask is equipped with an efficient condenser, and the mixture is refluxed until an enol test is no longer obtained (6–8 hours) (Notes 10 and 11). The mixture is cooled and extracted with three 200-ml. portions of ether. The combined ethereal extracts are washed with dilute hydrochloric acid and dried over anhydrous magnesium sulfate. The ether is evaporated, and the residue is distilled under reduced pressure to give 14–19 g. (54–74%) of 2-*n*-butyl-2-methylcyclohexanone, b.p. 116–118° (20 mm.) (Note 12).

2. Notes

1. Sodium methoxide was obtained from Matheson, Coleman and Bell. The best results were obtained with material from freshly opened bottles.

2. Eastman Organic Chemicals "Eastman grade" 2-methylcyclohexanone was distilled; b.p. 56° (20 mm.).

3. Eastman Organic Chemicals practical grade ethyl formate was shaken for 30 minutes with anhydrous sodium carbonate and for 30 minutes with anhydrous magnesium sulfate, and distilled; b.p. 54°.

4. A rubber dam was fastened tightly over the top of the Buchner funnel by means of rubber bands. It was pulled down onto the surface of the filter cake by the vacuum.

5. The solid should be powdered to allow complete formation of the dianion in the following reaction. This is most readily accomplished if the solid is ground before it is completely dry (*i.e.*, when it appears to be dry but is still cool). The fine powder is then replaced in the oven to complete the drying.

6. The checkers, working at one-quarter scale, obtained a yield of 86–88%.

7. Conversion is indicated by discharge of the deep blue color. This generally requires about 20 minutes. When conversion is completed, the stirrer should be speeded up or the contents of the flask swirled so that potassium splattered on the upper part of the flask is converted to amide; this should be done until all traces of blue color are gone.

8. The escaping ammonia will blow away some of the fine powder unless this is done carefully.

9. Eastman Organic Chemicals "Eastman grade" *n*-butyl bromide was distilled; b.p. 101–102°.

10. The enol test is performed with about 0.5 ml. of solution, which is neutralized with dilute hydrochloric acid and treated with 3–5 drops of 10% ethanolic ferric chloride. A reddish brown color denotes the presence of unhydrolyzed formyl ketone.

11. An alternative procedure is steam distillation of the basic, aqueous solution until no further organic material distills. This may be done either instead of, or after, the refluxing of the aqueous solution. The steam distillate is extracted with ether, and the ether is removed by distillation.

12. A higher-boiling fraction consisting of 2-formyl-6-*n*-butyl-6-methylcyclohexanone, b.p. 201–203° (20 mm.), is obtained if hydrolysis is not complete.

3. Methods of Preparation

This procedure is an adaptation of one described by Boatman, Harris, and Hauser.[2]

4. Merits of the Preparation

The present method affords 2-*n*-butyl-2-methylcyclohexanone uncontaminated by the isomeric 2-*n*-butyl-6-methylcyclohexanone.

2,2-Dimethylcyclohexanone and 2-benzyl-2-methylcyclohexanone have been prepared similarly in yields of 60% and 55%, respectively.[2] The procedure has been extended to the synthesis of 9-methyl-, 9-*n*-butyl-, and 9-benzyl-1-decalone from the dianion of 2-formyl-1-decalone in yields of 55%, 48%, and 58% respectively.[2]

1. Department of Chemistry, Duke University, Durham, North Carolina 27706.
2. S. Boatman, T. M. Harris, and C. R. Hauser, *J. Am. Chem. Soc.*, **87**, 82 (1965).

2-*t*-BUTYL-3-PHENYLOXAZIRANE

(Oxaziridine, 2-*t*-butyl-3-phenyl-)

A. $C_6H_5CHO + (CH_3)_3CNH_2 \rightarrow C_6H_5CH{=}NC(CH_3)_3 + H_2O$

B. $C_6H_5CH{=}NC(CH_3)_3 + C_6H_5CO_3H \rightarrow$

$$\overset{\displaystyle O}{\overset{\displaystyle \diagup \diagdown}{C_6H_5CH\text{——}NC(CH_3)_3}} + C_6H_5CO_2H$$

Submitted by W. D. Emmons and A. S. Pagano [1]
Checked by G. Ryan and Ronald Breslow

1. Procedure

Caution! The preparation and distillation of the oxazirane, like that of any active oxygen compound, should be carried out behind a safety screen.

A. *N-t-butylbenzaldimine.* A 1-l. three-necked flask equipped with stirrer, thermometer, and condenser for downward distillation is charged with 109.5 g. (1.5 moles) of *t*-butylamine (Note 1). Benzaldehyde (106 g., 1.0 mole) is then added in four increments to the stirred solution over a 20-minute period. A mild exotherm is noted which raises the temperature to 40–50°. Benzene (150 ml.) is then added and the solution is heated until distillation commences. Solvent (a mixture of amine, water, and benzene) is removed by distillation until a pot temperature of 110° is reached. The product mixture is then cooled to room temperature, dried over magnesium sulfate, and stripped free of solvent at aspirator pressure. Distillation of the yellow liquid so obtained yields 120–151 g. (78–94%) of colorless N-*t*-butylbenzaldimine, b.p. 59–63° (1 mm.), n^{26}D 1.5174, n^{20}D 1.5260 (Note 2).

B. *2-t-Butyl-3-phenyloxazirane.* A 1-l. three-necked flask equipped with an addition funnel, stirrer, and condenser is charged with 68.0 g. (0.422 mole) of N-*t*-butylbenzaldimine (Note 3) and 50 ml. of benzene. The stirred solution is cooled in an ice bath and a solution of 61 g. (0.445 mole) of perbenzoic acid in 315 ml. of benzene is added dropwise over a 40-minute period. After one additional hour the stirrer is stopped, and the reaction mixture is allowed to stand overnight with the concurrent melting of the ice bath. The light blue benzene solution is then filtered to remove the precipitated benzoic acid and is washed sequentially with three 100-ml. portions of sodium carbonate, 100 ml. of 5% hydrochloric acid, 100 ml. of saturated sodium bisulfite solution, and finally with 100 ml. of water. The solution is dried over magnesium sulfate, and the solvent is evaporated at room temperature (Note 4) at aspirator pressure. There is obtained 46–60 g. (60–78%) of crude oxazirane, n^{26}D 1.5065. This product assays 96–98% purity by iodimetric titration (Note 5) and is sufficiently pure for many purposes. Distillation of the crude product through a short Vigreux column yields, after a few drops of forerun, 42–55 g. (56–74%) of pure oxazirane, n^{26}D 1.5062, n^{20}D 1.5144, b.p. 55–58° (0.05 mm.), 74–76° (0.2 mm.). Iodimetric assay of this product indicates a purity of 99–100%.

2. Notes

1. Eastman Kodak white label reactants are satisfactory. The benzaldehyde should be freshly distilled before use.

2. The checkers handled and stored this material under nitrogen.

3. The charge of N-*t*-butylbenzaldimine is adjusted according to the amount of perbenzoic acid available. The perbenzoic acid in benzene is prepared by the procedure of Silbert, Siegel, and Swern,[2] and a 5% excess of this reagent is employed in the oxidation. In one attempt with commercial *m*-chloroperbenzoic acid instead, the checkers obtained only a 34% yield of oxazirane.

4. A rotary evaporator is very convenient for this operation.

5. A 0.200–0.300 g. sample of the oxazirane is weighed into a stoppered flask to which is added 15 ml. of glacial acetic acid and 2 ml. of saturated aqueous sodium iodide solution. After 5–10 minutes 25 ml. of deionized water is added, and the liberated iodine is titrated with 0.1N sodium thiosulfate with freshly pre-

pared starch as indicator. Each milliliter of thiosulfate solution is equivalent to 0.00885 g. of 2-*t*-butyl-3-phenyloxazirane.

3. Discussion

This procedure is an adaptation of that described by Emmons for the preparation of oxaziranes from imines using peracetic acid.[3] Other procedures which may be more useful for oxazirane preparation in specific instances are the oxidation of imines with *m*-chloroperbenzoic acid [4] and the reaction of aldehydes or ketones with hydroxylamine *O*-sulfonic acid in alkaline solution.[5] 2-*t*-Butyl-3-phenyloxazirane has also been prepared by photolysis of α-phenyl-N-*t*-butylnitrone [6] (a general reaction of considerable theoretical interest since it represents direct conversion of electromagnetic energy to chemical energy) and in low yields by ozonolysis of N-*t*-butylbenzaldimine.[7]

Oxaziranes are in a real sense active oxygen compounds and exhibit many reactions grossly analogous to those of organic peroxides. Thus they undergo one electron transfer reaction with ferrous salts and on pyrolysis they are converted to amides. Oxaziranes are also useful synthetic intermediates since in appropriate cases they may be isomerized to aromatic nitrones which are a convenient source of N-alkylhydroxylamines.[3] The reaction of oxaziranes with peracids also provides a source of nitrosoalkanes and is in many instances the method of choice for preparation of these compounds.[8]

1. Rohm and Haas Company, Spring House, Pa.
2. L. S. Silbert, E. Siegel, and D. Swern, this volume, p. 904.
3. W. D. Emmons, *J. Am. Chem. Soc.*, **79**, 5739 (1957).
4. R. G. Pews, *J. Org. Chem.*, **32**, 1628 (1967).
5. E. Schmitz, R. Ohme, and D. Murawski, *Agnew. Chem.*, **73**, 708 (1961).
6. J. S. Splitter and M. Calvin, *J. Org. Chem.*, **23**, 651 (1958).
7. A. H. Riebel, R. F. Erickson, C. J. Abshire, and P. S. Bailey, *J. Am. Chem. Soc.*, **82**, 1801 (1960).
8. W. D. Emmons, *J. Am. Chem. Soc.*, **79**, 6522 (1957).

D,L-10-CAMPHORSULFONIC ACID (REYCHLER'S ACID)*

Submitted by PAUL D. BARTLETT and L. H. KNOX [1]
Checked by JOHN D. ROBERTS and DINSHAW PATEL

1. Procedure

In a 3-l., three-necked, round-bottomed flask fitted with a powerful slow-speed stirrer having a Teflon® blade, a 500-ml. dropping funnel, and a thermometer arranged to dip into the liquid is placed 588 g. (366 ml., 6 moles) of concentrated sulfuric acid. The flask is surrounded by an ice-salt mixture, the stirrer started, and 1216 g. (1170 ml., 12 moles) of acetic anhydride (Note 1) is added at such a rate that the temperature does not rise above 20° (Note 2). The separatory funnel is removed and 912 g. (6 moles) of coarsely powdered D,L-camphor is added (Note 3). The flask is then closed with a stopper and stirring is continued until the camphor is dissolved. The stirrer is replaced by a stopper, the ice bath allowed to melt, and the mixture left to stand for 36 hours (Note 4). The camphorsulfonic acid is collected on a suction filter and washed with ether (Note 5). After being dried in a vacuum desiccator at room temperature, the nearly white crystalline product weighs 530–580 g. (38–42%). It melts at 202–203° with rapid decomposition and is relatively pure (Note 6).

2. Notes

1. If the acetic anhydride is of a good commercial grade, it need not be redistilled.

2. When the temperature is allowed to rise above 20°, the acetic-sulfuric anhydride mixture acquires a yellow to orange color from which discolored crystals are subsequently deposited. The addition, which must be slow at first, requires 1–1.5 hours

depending on the efficiency of the cooling bath.

3. The camphor employed is of the synthetic variety supplied by Howe and French, Boston.   If an optically active product is desired, active natural camphor may be used.

4. The yields vary with the length of the crystallization period. After 16 hours the yield is 470 g. (34%).   When the crystallization period is extended to 2 weeks, the yield is 615–655 g. (44–47%).

5. The checkers found the product to be very hygroscopic, in fact deliquescent, in a reasonably humid atmosphere.   In such circumstances, it was preferable to decant the mother liquor from the crystals in the flask and to wash the solid by stirring it up with four 250-ml. portions of anhydrous ether, each washing being removed by decantation.   The well-drained residual solid can then be transferred to a crystallizing dish and the ether removed by pumping under reduced pressure before the final drying in a vacuum desiccator over sulfuric acid.

6. The product can be purified with some loss by recrystallization from glacial acetic acid.   About 60 g. of crude product dissolves in 90 ml. of acetic acid at 105° and gives a recovery of about 40 g. of purified material.

3. Method of Preparation

The procedure described is that of Reychler.[2]

4. Merits of the Preparation

D,L-10-Camphorsulfonic acid is used for the preparation of the corresponding chloride (p. 196). The optically active acid has been used widely for the resolution of basic compounds into optical antipodes.

1. Converse Memorial Laboratory, Harvard University, Cambridge, Massachusetts.   Preparation was submitted November 1, 1939.
2. A. Reychler, *Bull. Soc. Chim.*, [3] **19**, 120 (1898); see also H. E. Armstrong and T. M. Lowry, *J. Chem. Soc.*, 1441 (1902); B. Rewald, *Ber.*, **42**, 3136 (1909); F. Girault, *J. Pharm. Chim.*, [8] **20**, 207 (1934) [*C.A.*, **29**, 144 (1935)]; Y. Asahina and K. Yamguti, *Proc. Imp. Acad.* (Tokyo), **13**, 38 (1937) [*C.A.*, **31**, 4305 (1937)]; R. Poggi and A. Polverini, *Ann. Chim. Appl.*, **30**, 284 (1940) [*C.A.*, **35**, 1395 (1941)]; and R. Poggi and A. Pasquarelli, *Ann. Chem. Appl.*, **37**, 321 (1947) [*C.A.*, **42**, 8788 (1948)].

D,L-10-CAMPHORSULFONYL CHLORIDE

$$CH_2SO_3H \qquad \xrightarrow{PCl_5} \qquad CH_2SO_2Cl$$

Submitted by PAUL D. BARTLETT and L. H. KNOX [1]
Checked by JOHN D. ROBERTS

1. Procedure

In a 2-l., three-necked, round-bottomed flask (Note 1) fitted with a sealed stirrer having a Teflon® blade and, on the two side necks, with gas-outlet tubes connected by rubber or plastic tubing to an efficient hydrogen chloride absorption trap (Note 2), 464 g. (2 moles) of D,L-10-camphorsulfonic acid (Note 3) is mixed with 416 g. (2 moles) of phosphorus pentachloride (Note 4). The flask is immersed in ice water and, as soon as the mixture has liquefied sufficiently, the stirrer is started but must be run slowly at first because of lumps. When the vigorous reaction has subsided, the cooling bath is removed and stirring continued until the chloride is completely dissolved (Note 5). The mixture is then allowed to stand for 3 or 4 hours. It is poured (*Hood!*) onto 500 g. of crushed ice contained in a 2-l. beaker. This mixture is immediately poured into a second beaker containing a similar quantity of crushed ice. The mixture is then poured back and forth between the two beakers until all evidence of reaction has disappeared (Note 6). The fine white product is collected on a suction filter and washed several times with cold water. The yield is essentially quantitative (500 g.) of moist sulfonyl chloride which is pure enough to be used for the preparation of D,L-ketopinic acid (p. 55). When carefully dried, the crude material has m.p. 81–83° (Notes 7 and 8) and may be preserved in a desiccator.

2. Notes

1. The checker found it expedient to carry on the reaction in a 2-l. Pyrex® reaction kettle (Corning 6947), the large closure making the initial mixing of the solid reactant and the removal of the product much simpler.

2. The type of trap described in *Org. Syntheses*, Coll. Vol. **2**, p. 4, is particularly useful.

3. The D,L-10-camphorsulfonic acid employed is the unrecrystallized product described on p. 194.

4. The initial mixing is conveniently made by turning the stirrer back and forth by hand.

5. The mixture does not usually become a clear solution because the product begins to crystallize. It is not difficult, however, to recognize yellow lumps of unreacted phosphorus pentachloride.

6. 10-Camphorsulfonyl chloride is rather rapidly hydrolyzed by warm water. The procedure here provides for complete hydrolysis of phosphorus oxychloride and excess phosphorus pentachloride without local heating and loss of product due to hydrolysis. For best results, the whole hydrolysis operation should be carried out quickly and steadily. It is well to have additional quantities of crushed ice on hand because the mixture may become quite hot if all of the ice added initially melts.

7. If the crude moist sulfonyl chloride is to be preserved, it must be thoroughly and reasonably rapidly dried. The checker found it very convenient to use a "freeze-drying" apparatus to remove the bulk of the moisture.

8. The submitters report that crystallization of the crude product from ligroin produces material of m.p. 83–84°. The sulfonyl chloride from (+)-camphorsulfonic acid has m.p. 67–68°.

3. Methods of Preparation

The procedure described here is adapted from that of Reychler.[2] The chloride may also be made from treatment of the acid with thionyl chloride.[3]

4. Merits of the Preparation

D,L-10-Camphorsulfonyl chloride may be oxidized to ketopinic acid (p. 690). The optically active forms of the sulfonyl chloride are useful for resolving alcohols and amines into optical antipodes.

1. Converse Memorial Laboratory, Harvard University, Cambridge, Massachusetts. Preparation was submitted November 1, 1939.
2. A. Reychler, *Bull. Soc. Chim.*, [3] **19**, 120 (1898).
3. S. Smiles and T. P. Hilditch, *J. Chem. Soc.*, **91**, 519 (1907); H. Sutherland and R. L. Shriner, *J. Am. Chem. Soc.*, **58**, 62 (1936); J. Read and R. A. Storey, *J. Chem. Soc.*, 2761 (1930).

2-CARBETHOXYCYCLOOCTANONE

(2-Oxocyclooctanecarboxylic acid, ethyl ester)

$$\text{(cyclooctanone)} \quad + \quad C_2H_5O\overset{O}{\overset{||}{C}}OC_2H_5 \quad \xrightarrow[\text{Benzene}]{\text{NaH}} \quad \text{(product with } CO_2C_2H_5\text{)}$$

Submitted by A. PAUL KRAPCHO, JOSEPH DIAMANTI,
CHARLES CAYEN, and RICHARD BINGHAM [1]
Checked by WILLIAM G. DAUBEN and CHARLES DALE POULTER

1. Procedure

A 2-l. two-necked, round-bottomed flask equipped with a magnetic stirrer (Note 1) is fitted with a 250-ml. pressure-equalizing constant-rate dropping funnel and a condenser, the top of which is connected to a mercury trap to prevent the entrance of air during the reaction and for the detection of gas evolution. The dropping funnel is removed, and 35 g. (0.85 mole) of sodium hydride dispersed in mineral oil is added (Note 2). The mineral oil is removed by washing the dispersion four times with 100-ml. portions of benzene (Note 3). The benzene is removed with a pipet after the sodium hydride is allowed to settle (Note 4).

After most of the mineral oil has been removed, 400 ml. of benzene is added to the sodium hydride, followed by 71 g. (0.6 mole) of diethyl carbonate (Note 5). This mixture is heated to reflux, and a solution of 38 g. (0.3 mole) of cyclooctanone (Note 6) in 100 ml. of benzene is added dropwise from the dropping funnel over a period of 3–4 hours. After the addition is complete, this mixture is allowed to reflux until the evolution of hydrogen ceases (15–20 minutes).

When the reaction mixture has cooled to room temperature, 60 ml. of glacial acetic acid is added dropwise, and a heavy, pasty solid separates. Ice-cold water (about 200 ml.) is added dropwise, and the stirring is continued until all the solid material has gone into solution (Note 7). The benzene layer is separated, and the aqueous layer is extracted three times using 100-ml. portions of benzene. The combined benzene extracts are washed three times with 100-ml. portions of cold water. The benzene is removed by distillation at atmospheric pressure, and the excess diethyl carbonate is removed under water-pump pressure with gentle heating. The residual material is transferred to a 100-ml. distillation flask, and the fraction boiling at 85–87° (0.1 mm.) is collected. The yield of 2-carbethoxycyclooctanone is 54–56 g. (91–94%), n^{25}D 1.4795–1.4800.

2. Notes

1. The checkers found that the agitation of the reaction mixture required later in this reaction is better achieved by use of a sealed mechanical stirrer.

2. The sodium hydride was obtained as a 58.6% dispersion in mineral oil from Metal Hydrides, Inc., Beverly, Massachusetts.

3. The benzene (Fisher certified reagent, thiophene free) was dried over potassium hydroxide and distilled from sodium metal.

4. By this procedure about 80–85% of the mineral oil was removed. Because some sodium hydride is lost in the pipetting procedure, an excess is initially employed.

5. The product supplied by Matheson, Coleman and Bell was used as received. Lower yields were obtained when a molar equivalent of diethyl carbonate was utilized, possibly because of self-condensation of the ketone.

6. The cyclooctanone was obtained from the Aldrich Chemical Co. and was utilized as received.

7. At this point the aqueous layer should be acidic, or more acetic acid should be added.

3. Methods of Preparation

The reaction of cyclooctanone with diethyl oxalate, followed by decarbonylation of the resulting glyoxylate, has been reported to yield 32% of 2-carbethoxycyclooctanone.[2] The reaction of cyclooctanone with sodium amide in ether, followed by the addition of diethyl carbonate, provided the product in 70% yield.[3]

The preparation of several medium- and large-sized 2-carbomethoxycycloalkanones has been accomplished by treatment of the cycloalkanone with sodium triphenylmethyl, followed by carbonation with dry ice, and esterification with diazomethane.[4] The yields are good but the procedure is laborious. The synthesis of 2-carbomethoxycyclooctanone via the Dieckmann cyclization of dimethyl azelate with sodium hydride yields 48% of this product when the procedure is carried out over a 9-day period.[5]

4. Merits of the Preparation

The reaction described is of general synthetic utility for the preparation of a variety of cyclic β-keto esters from the corresponding ketones. Using this procedure the 2-carbethoxycycloalkanones have been prepared from cyclononanone, cyclodecanone, and cyclododecanone in yields of 85%, 95%, and 90%, respectively. The procedure is simpler and gives much higher yields than other synthetic routes to these systems.

This procedure has been patterned after the method by which the carbethoxy group is introduced into a few alicyclic ketones [6] and several cyclic ketones. Cyclohexanone has been reported to yield 50% of 2-carbethoxycyclohexanone when treated with sodium hydride and diethyl carbonate using ether as the solvent.[7] The preparation of 2-carbethoxycycloheptanone using potassium t-butoxide and diethyl carbonate in benzene has been reported in 40% yield.[8] Jacob and Dev report an 80% yield of the latter compound using sodium hydride as the base.[9]

1. Department of Chemistry, University of Vermont, Burlington, Vermont. Supported by the National Science Foundation Undergraduate Research Participation Grant during the summer of 1964 and National Science Foundation Grant No. G-19490.

2. A. C. Cope, E. Ciganek, and J. Lazar, *J. Am. Chem. Soc.*, **84**, 2591 (1962).
3. P. LaFont and Y. Bonnet, Fr. Patent 1,281,926 (1962) [*C.A.*, **58**, 1373 (1963)].
4. V. Prelog, L. Ruzicka, P. Barman, and L. Frenkiel, *Helv. Chim. Acta*, **31**, 92 (1948) [*C.A.*, **42**, 4180 (1948)].
5. F. F. Blicke, J. Azuara, N. J. Doorenbos, and E. B. Hotelling, *J. Am. Chem. Soc.*, **75**, 5418 (1953).
6. S. B. Soloway and F. B. LaForge, *J. Am. Chem. Soc.*, **69**, 2677 (1947); N. Green and F. B. LaForge, *J. Am. Chem. Soc.*, **70**, 2287 (1948); Yuh-Lin Chen and W. F. Barthel, *J. Am. Chem. Soc.*, **75**, 4287 (1953).
7. M. D. Banus and A. A. Hinckley, *Advan. Chem. Ser.*, No. 19, "Handling and Uses of the Alkali Metals," American Chemical Society, 1957, p. 106.
8. H. Christol, M. Mousseron, and F. Plenat, *Bull. Soc. Chim. France*, 543 (1959).
9. T. M. Jacob and S. Dev, *Chem. Ind.* (*London*), 576 (1956).

1,1'-CARBONYLDIIMIDAZOLE *

(Imidazole, 1,1'-carbonyldi-)

Submitted by HEINZ A. STAAB and KURT WENDEL[1]
Checked by A. C. MACKAY and PETER YATES

1. Procedure

Caution! This preparation must be carried out in a hood to avoid exposure to phosgene.

Anhydrous benzene (*ca.* 200 ml.) (Note 1) is poured into a calibrated, 500-ml., standard-taper dropping funnel equipped with a gas-inlet tube containing a fritted-glass filter; the dropping funnel is stoppered and weighed accurately. The funnel is protected with a calcium chloride tube, and 15–20 g. of phosgene is introduced at room temperature over a period of *ca.* 1 hour; this quantity corresponds to an increase in volume of 12–16 ml. (Note 2). The calcium chloride tube is removed, and the funnel is immediately restoppered and reweighed (Note 3). The amount of imidazole corresponding to the increase in weight observed (*e.g.*, 16.55 g., 0.167 mole, of phosgene) is calculated on the basis of a phosgene:imidazole molar ratio of 1:4 (Note 4).   The funnel is placed on a 1-l., three-necked, round-bottomed flask, that contains a solution of the imidazole (here, 45.60 g., 0.669 mole) in 500 ml. of anhydrous tetrahydrofuran (Note 5) and is equipped with a sealed mechanical stirrer and a calcium chloride tube.   The flask is cooled with cold water, and the solution of phosgene in benzene is added with stirring from the dropping funnel over a period of 15–30 minutes.   The reaction mixture is stirred for an additional 15 minutes and then allowed to stand for 1 hour at room temperature.   The precipitate of imidazolium chloride is removed by suction filtration with exclusion of atmospheric moisture by the use of a standard-taper fritted-glass filter funnel (Note 6).   The filtrate is evaporated to dryness under reduced pressure on a water bath at 40–50° (Note 7).   The yield of color-less crystalline 1,1′-carbonyldiimidazole is 80–94% (here, 24.8 g.; 91%).   The product obtained in this way sinters at 110° and melts between 112° and 117° (here, 114–115°).   This material can be used without further purification for most reactions, *e.g.*, ester, peptide, and aldehyde syntheses [2] (Note 8).   The purity of a product of m.p. 113–117° was $98\pm2\%$.[3]   The quality and yield of 1,1′-carbonyldiimidazole are not reduced when the scale is doubled.

1,1′-Carbonyldiimidazole may be kept for a long period of time in either a desiccator over phosphorus pentoxide or in a sealed tube.   It is hydrolyzed by water to give carbon dioxide and imidazole.

2. Notes

1. The benzene is heated under reflux over sodium with benzo-phenone until a permanent blue coloration develops and then is distilled with exclusion of atmospheric moisture.

2. Use of a calibrated dropping funnel permits approximate estimation of the amount of phosgene absorbed; a volume in-crease of 1 ml. corresponds to about 1.3 g. of phosgene.   The stream of phosgene is led through a wash bottle containing con-centrated sulfuric acid and should not be too fast in order to avoid loss of solvent by evaporation.

3. The checkers used a fritted-glass inlet tube and drying tube incorporated in a standard-taper adapter that fitted the neck of

the dropping funnel. Owing to small losses of solution on with-drawal of the adapter, the weight of phosgene was slightly underestimated.

4. Technical grade imidazole (from Badische Anilin- und Soda-Fabrik, Ludwigshafen, Rhein, Germany) was recrystallized from benzene containing 1.0–1.5% ethanol; m.p. 90°. The checkers used imidazole obtained from Aldrich Chemical Co. without further purification.

5. Technical grade tetrahydrofuran was predried for a few days over sodium hydroxide. It was then heated under reflux over sodium wire with benzophenone until it developed a per-manent blue color and distilled with exclusion of atmospheric moisture. [*Caution! See p. 976 of this volume for a warning regarding purification of tetrahydrofuran.*]

6. By working quickly, the imidazolium chloride may be removed by suction filtration through a Buchner funnel. How-ever, the precipitate should not be freed of solvent completely because imidazolium chloride is extremely hygroscopic. If the moist precipitate is washed with 50–100 ml. of anhydrous tetra-hydrofuran, the yield of 1,1'-carbonyldiimidazole may be slightly increased; however, there is some danger of the introduction of too much moisture into the reaction solution.

7. The checkers used an antifoaming head for the solvent evaporation.

8. In order to obtain a purer product the crude material may be recrystallized from hot anhydrous tetrahydrofuran with careful exclusion of moisture. After this operation the yield is reduced to 65–75%; the m.p. is then between 114° and 118°; *e.g.*, recrystallization of 24.8 g. (91%) of 1,1'-carbonyldiimidazole from 60 ml. of anhydrous tetrahydrofuran yielded 19.9 g. (73%); m.p. 116–118°.

3. Methods of Preparation

1,1'-Carbonyldiimidazole has been prepared by the reaction of imidazole and phosgene in anhydrous benzene and anhydrous tetrahydrofuran.[3-5] It has also been obtained by the reaction of 1-(trimethylsilyl)imidazole and phosgene in anhydrous benzene,[6] but that method offers no advantages that justify the more extensive preparative effort required.

4. Merits of the Preparation

1,1'-Carbonyldiimidazole has been used for the preparation of such compounds as esters, anhydrides, amides, peptides, ketones, ethers, and isocyanates.[2] The present procedure provides a convenient method for its preparation in good yield.

1. Department of Chemistry, University of Heidelberg, Heidelberg, Germany.
2. H. A. Staab, *Angew. Chem.*, **74**, 407 (1962); *Angew. Chem. Intern. Ed. Engl.*, **1**, 351 (1962).
3. R. Paul and G. W. Anderson, *J. Am. Chem. Soc.*, **82**, 4596 (1960); G. W. Anderson and R. Paul, *J. Am. Chem. Soc.*, **80**, 4423 (1958).
4. H. A. Staab, *Ann.*, **609**, 75 (1957).
5. H. A. Staab and K. Wendel, *Ber.*, **93**, 2910 (1960).
6. L. Birkofer, P. Richter, and A. Ritter, *Ber.*, **93**, 2804 (1960).

α-CHLOROACETYL ISOCYANATE
(Isocyanic acid, anhydride with chloroacetic acid)

$$\text{ClCH}_2-\overset{\overset{\displaystyle O}{\|}}{\text{C}}-\text{NH}_2 + \text{ClCOCOCl} \rightarrow$$

$$\text{ClCH}_2-\overset{\overset{\displaystyle O}{\|}}{\text{C}}-\text{NCO} + \text{CO} + 2\text{HCl}$$

Submitted by A. John Speziale and Lowell R. Smith [1]
Checked by Leif A. Hoffmann and V. Boekelheide

1. Procedure

In a 250-ml. round-bottomed flask fitted with a magnetic stirrer (Note 1), a thermometer, and a condenser carrying a calcium chloride tube (Note 2) are placed 46.7 g. (0.5 mole) of α-chloroacetamide (Note 3) and 100 ml. of ethylene dichloride. The mixture is chilled in an ice bath to about 2° and stirred while 76.2 g. (0.6 mole) of oxalyl chloride (Note 4) is added all at once. The mixture is removed from the ice bath, stirred for 1 hour, and then heated to reflux at 83° with stirring for 5 hours (Note 5).

The solution is chilled in an ice bath to 0–10°, the condenser is replaced by a 120-mm. distillation column packed with glass helices, and the solvent is removed at 70 mm. pressure with stirring. The ice bath is removed after the solvent boils without foaming and is replaced by a heating mantle or oil bath. Distillation gives 39 g. (65%) of α-chloroacetyl isocyanate, b.p. 68–70° (70 mm.), as a colorless oil, n^{25}D1.4565.

2. Notes

1. For larger-scale preparations mechanical stirring is recommended.

2. Moisture must be rigorously excluded from the reaction mixture and the product.

3. The α-chloroacetamide was obtained from Eastman Kodak Co. and used without purification.

4. The oxalyl chloride was obtained from Aldrich Chemical Co. and used without purification. Oxalyl chloride vapor is irritating and toxic, and therefore manipulations must be carried out in a hood.

5. Because a large amount of hydrogen chloride is evolved, the reaction must be carried out in a hood.

3. Methods of Preparation

The only preparation reported for α-chloroacetyl isocyanate is that described by the submitters.[2]

4. Merits of the Preparation

The procedure may be adapted for the preparation of other acyl isocyanates (*i.e.*, dichloroacetyl, trichloroacetyl, phenylacetyl, diphenylacetyl, benzoyl, etc.) and is generally more convenient than the reaction of acid chlorides with silver cyanate.[3, 4] Acyl isocyanates react with amines, alcohols, and mercaptans to yield acyl ureas, carbamates, and thiocarbamates, and have been shown to undergo a variety of interesting reactions.[5]

1. Research Department, Agricultural Division, Monsanto Company, St. Louis 66, Missouri.
2. A. J. Speziale and L. R. Smith, *J. Org. Chem.*, **27**, 3742 (1962); **28**, 1805 (1963).
3. O. C. Billeter, *Ber.*, **36**, 3213 (1903).
4. A. J. Hill and W. M. Degnan, *J. Am. Chem. Soc.*, **62**, 1595 (1940).
5. L. R. Smith, A. J. Speziale, and J. E. Fedder, *J. Org. Chem.*, **34**, 633 (1969).

9-CHLOROANTHRACENE*

(Anthracene, 9-chloro-)

$$+ Cu_2Cl_2 + HCl$$

Submitted by D. C. NONHEBEL.[1]
Checked by R. B. GREENWALD and E. J. COREY.

1. Procedure

In a dry, 1-l., two-necked flask, equipped with a mechanical stirrer and a reflux condenser fitted with a drying tube, are placed 17.8 g. (0.100 mole) of anthracene (Note 1), 27.2 g. (0.202 mole) of anhydrous cupric chloride (Note 2), and 500 ml. of carbon tetrachloride (Note 3). The reaction mixture is stirred and heated under reflux for 18–24 hours. The brown cupric chloride is gradually converted to white cuprous chloride, and hydrogen chloride is gradually evolved. At the end of the reaction the cuprous chloride is removed by filtration, and the carbon tetrachloride solution is passed through a 35-mm. chromatographic column filled with 200 g. of alumina (Note 4). The column is eluted with 400 ml. of carbon tetrachloride. The combined eluates are evaporated to dryness to give 19–21 g. (89–99%) of 9-chloroanthracene as a lemon-yellow solid, m.p. 102–104° (Note 5). Crystallization of the product from petroleum ether (b.p. 60–80°) gives 16–17 g. (75–80%) of 9-chloroanthracene as yellow needles, m.p. 104–106°.

2. Notes

1. Anthracene, B. D. H. (blue fluorescence), was used. Traces of ethylene glycol, glycerol, ethanol, or water considerably retard the reaction and lead to unsatisfactory results.

2. Anhydrous cupric chloride is dried in an oven at 110–120° for several hours and stored in a desiccator or over phosphorus pentoxide before use.

3. Chlorobenzene or *sym*-tetrachlorethane may be used instead of carbon tetrachloride as solvent, in which case the reaction is complete as soon as the mixture has reached reflux. The product is liable to be contaminated by a small amount of 9,10-dichloroanthracene.

4. Merck alumina or Spence Type H alumina was used.

5. The 9-chloroanthracene at this stage usually contains a small amount of unreacted anthracene.

3. Discussion

9-Chloroanthracene has been prepared by the action of chlorine,[2] *t*-butyl hypochlorite,[3] 1,3-dichloro-5,5-dimethylhydantoin,[4] or phosphorus pentachloride[5] on anthracene. The present method is a one-step synthesis giving a high yield of 9-chloroanthracene from readily available starting materials.

The method outlined can be applied to the preparation in better than 90% yield of the 10-chloro derivatives of 9-alkyl-,[6] 9-aryl-,[6] and 9-halogenoanthracenes.[7] For the less reactive substrates chlorobenzene should be used as solvent. This is the only satisfactory procedure for the preparation of 9-bromo-10-chloroanthracene.[7] Other methods of chlorination lead to mixtures of the desired compound and 9,10-dichloroanthracene. Pyrene can likewise be converted to 1-chloropyrene (90% yield).[8] Analogous procedures with cupric bromide lead to the brominated compounds in similar high yields.

1. Chemistry Department, Royal College of Science and Technology, Glasgow, Scotland.
2. Y. Nagaki and M. Tanabe, *Kogyo Kagaku Zasshi*, **60**, 294 (1957) [C.A., **53**, 8087 (1959)].
3. J. W. Engelsma, E. Farenhorst, and E. C. Kooyman, *Rec. Trav. Chim.*, **73**, 884 (1954).
4. O. O. Orazi, J. F. Salellas, M. E. Fondovila, R. A. Corral, N. M. I. Mercere, and E. C. Rakunas de Alvarez, *Anales Asoc. Quim. Arg.*, **40**, 61 (1952) [C. A., **47**, 3244 (1953)].
5. B. M. Mikhailov and M. Sh. Promyslov, *J. Gen. Chem. (USSR)*, **20**, 338 (1950) [English Language Edition, Consultants Bureau, p. 359].
6. D. Mosnaim, D. C. Nonhebel, and J. A. Russell, *Tetrahedron*, **25**, 3485 (1969).
7. D. Mosnaim and D. C. Nonhebel, *Tetrahedron*, **25**, 1591 (1969).
8. D. C. Nonhebel, *Proc. Chem. Soc.*, 307 (1961).

N-CHLOROCYCLOHEXYLIDENEIMINE

(Cyclohexanimine, N-chloro-)

Submitted by G. H. ALT[1] and W. S. KNOWLES[2]
Checked by P. M. BURKE and PETER YATES

1. Procedure

A. *N,N-Dichlorocyclohexylamine* (Note 1).   In a 300-ml. three-necked flask fitted with stirrer, addition funnel, thermometer, and calcium chloride tube are placed 9.92 g. (0.10 mole) of cyclohexylamine (Note 2) and 50 ml. of dry benzene (Note 3), and the mixture is cooled to 0–5° by an ice bath.   A solution of 24 g. (0.22 mole) of *t*-butyl hypochlorite[3] in 50 ml. of dry benzene is added dropwise at such a rate that the temperature of the mixture does not exceed 10°.   The mixture is allowed to come to room temperature and is then stirred for 1 hour, giving a solution of N,N-dichlorocyclohexylamine suitable for use in the next step.

B. *N-Chlorocyclohexylideneimine.* In a 500-ml. three-necked flask fitted with stirrer, addition funnel, thermometer, and reflux condenser fitted with a calcium chloride tube are placed 15 g. (0.15 mole) of potassium acetate (Note 4) and 100 ml. of absolute ethanol. The mixture is heated to reflux temperature, and, when the potassium acetate has dissolved, the N,N-dichlorocyclohexylamine solution is added at such a rate as to maintain reflux (Note 5). The reaction mixture is heated under reflux for an additional 3 hours, during which time potassium chloride precipitates. The mixture is cooled to room temperature, 200 ml. of ether and 100 ml. of water are added, and the resulting mixture is transferred to a 1-l. separatory funnel. The aqueous layer is separated and discarded. The ethereal layer is washed with three 100-ml. portions of water, three 50-ml. portions of 2N hydrochloric acid, and an additional three 100-ml. portions of water, the washings being discarded. The ethereal solution is dried over anhydrous calcium sulfate, and the solvent is removed at room temperature with a rotary evaporator and water aspirator. The residue is transferred to a 25 ml. distilling flask and fractionally distilled at reduced pressure through a short, vacuum-jacketed Vigreux column equipped with a Claisen type still head and a condenser through which ice water is circulated (Note 6). N-Chlorocyclohexylideneimine, b.p. 53–54° (3 mm.) (*Caution!* *Note 7*), $n^{25}D$ 1.506, is obtained in 48–69% yield (6.3–9.1 g.) (Note 8).

2. Notes

1. This method is essentially that of Baumgarten and Petersen.[4]
2. Eastman Organic Chemicals cyclohexylamine, white label grade, was redistilled prior to use.
3. Dried by azeotropic distillation; the first 10% of distillate is discarded.
4. Baker and Adamson, reagent grade.
5. The checkers found it preferable to maintain some external heating; otherwise the rate of addition had to be very rapid to maintain reflux.

6. Ice water is essential, and cooling of the receiver is recommended; otherwise considerable losses by evaporation occur.

7. The pot temperature should not be allowed to rise above 70° (the submitters used a hot-water bath at 75°), as a fume-off which may proceed with explosive violence is likely to occur. A nitrogen bubbler may be used to eliminate bumping. The distillation should be carried out behind a safety shield.

8. The compound decomposes slowly even under refrigeration and should be used within 24 hours of preparation. Analytically pure material, b.p. 36° (1.5 mm.), may be obtained by redistillation.

3. Methods of Preparation

N-Chlorocyclohexylideneimine has been prepared by the treatment of N,N-dichlorocyclohexylamine with triethylamine,[5] potassium hydroxide,[5] or potassium acetate [6] and by reaction of chloramine with cyclohexanone [7] or N-cyclohexylideneaniline.[8]

4. Merits of the Preparation

This method, which is an adaptation of that of Alt and Knowles,[6] obviates the need to isolate the N,N-dichlorocyclohexylamine.

N-Chlorocyclohexylideneimine is of theoretical interest, being isoelectronic with the oxime tosylate. On treatment with 1 mole of base the imine undergoes a Neber-type rearrangement to the α-amino ketone [6] and has been shown to be an intermediate in the rearrangement of N,N-dichlorocyclohexylamine to 2-aminocyclohexanone.[4, 6]

1. Agricultural Research Laboratory, Monsanto Chemical Company, St. Louis, Missouri.
2. Research Department, Organic Division, Monsanto Chemical Company, St. Louis, Missouri.
3. (a) H. M. Teeter and E. W. Bell, *Org. Syntheses*, Coll. Vol. 4, 125 (1963). *Caution!* See Warning, this volume, p. 183. (b) M. J. Mintz and C. Walling, this volume, p. 184.
4. H. E. Baumgarten and J. M. Petersen, *J. Am. Chem. Soc.*, **82**, 459 (1960); see this volume, p. 909.
5. S. L. Reid and D. B. Sharp, *J. Org. Chem.*, **26**, 2567 (1961).
6. G. H. Alt and W. S. Knowles, *J. Org. Chem.*, **25**, 2047 (1960).

7. B. Rudner (to W. R. Grace and Co.), U.S. Patent 2,894,028 (1959) [*C.A.*, **53**, 19924 (1959)].
8. E. Schmitz, *Angew. Chem.*, **73**, 23 (1961).

CHLORODIISOPROPYLPHOSPHINE

(Phosphinous chloride, diisopropyl-)

$$2(CH_3)_2CHMgCl + PCl_3 \rightarrow [(CH_3)_2CH]_2PCl + 2MgCl_2$$

Submitted by W. Voskuil and J. F. Arens [1]
Checked by Hugh D. Olmstead, James E. Oliver, and Herbert O. House

1. Procedure

Caution! Because of the sensitivity of the reagents and product to moisture and oxygen, all manipulations must be performed in an anhydrous, inert atmosphere (Note 1).

A 500-ml., four-necked, round-bottomed flask is equipped with an efficient stirrer, a reflux condenser, a 250-ml. dropping funnel, and a low-temperature thermometer (Note 2). In the flask are placed 34.4 g. (21.8 ml., 0.25 mole) of phosphorus trichloride (Note 3) and 150 ml. of anhydrous ether. A solution of 0.50 mole of isopropylmagnesium chloride in about 150 ml. of ether (Notes 4 and 5) is placed in the dropping funnel.

The flask is cooled in a dry ice-acetone bath, and the Grignard reagent solution is added dropwise with rapid stirring at such a rate that the temperature of the reaction mixture remains between $-25°$ and $-30°$ with a bath temperature of $-45°$; this addition requires about 1.5 hours. After the addition has been completed, the cooling bath is removed, and the mixture is allowed to warm to room temperature. Finally, the reaction mixture is heated to reflux with continuous stirring for 30 minutes.

After the reaction mixture has cooled to room temperature, it is filtered with suction (Notes 6 and 7), and the residual salts are washed thoroughly with three 100-ml. portions of anhydrous ether. The combined ethereal filtrates are concentrated under reduced pressure at room temperature, and the residual liquid is fractionally distilled through a 15-cm. Vigreux column. After a small forerun has been collected, the product is obtained as a clear, colorless liquid, b.p. 46–47° (10 mm.), n^{20}D 1.4752 (Note 8). The yield is 21–23 g. (55–60%); practically no residue remains in the distillation pot.

2. Notes

1. The submitters used nitrogen purified by passage through B.T.S. catalyst (B.A.S.F., Ludwigshafen, Germany). The checkers used commercial prepurified nitrogen without further treatment.

2. The checkers used a three-necked flask, one neck of which was fitted with an adapter to accommodate the thermometer. They also used a pressure-equalizing dropping funnel so that a static nitrogen atmosphere several millimeters above atmospheric pressure could be maintained in the flask.

3. The submitters used Merck reagent grade phosphorus trichloride. The checkers used material from Baker and Adamson.

4. It is essential to use the Grignard reagent prepared from isopropyl chloride. From phosphorus trichloride and isopropylmagnesium bromide, bromodiisopropylphosphine is obtained because of a halogen exchange reaction between the initially formed chlorophosphine and magnesium bromide. The checkers used both the Grignard reagent prepared from isopropyl chloride and a commercial solution of isopropylmagnesium chloride available from Matheson, Coleman and Bell.

5. The concentration of the Grignard reagent should be estimated by titration. If an excess or less than the stoichiometric amount of the organometallic reagent is added, the yield is lower and the product is less pure. The checkers found the titration procedure of Watson and Eastham [2] to be most convenient. In a typical titration, performed under a nitrogen atmosphere, a 5.00-ml. aliquot of the Grignard reagent was added to a solution of about 2 mg. of o-phenanthroline in 10 ml. of anhydrous benzene. The resulting purple solution was titrated with a standard solution (0.999M) of sec-butyl alcohol in xylene until the purple color of the o-phenanthroline-Grignard reagent charge transfer

complex was just discharged. In this procedure the number of millimoles of *sec*-butyl alcohol added is equal to the number of millimoles of alkylmagnesium chloride present in the aliquot of Grignard reagent.

6. The checkers performed this filtration and subsequent washing of the precipitate by replacing the dropping funnel in the reaction flask by a sintered-glass filter stick. A slight positive nitrogen pressure was applied in the reaction flask, and the pressure was reduced in the flask that served as a receiver for the filtrate passing through the sintered-glass filter.

7. The checkers found it necessary to dislodge and break up the cake of magnesium salts that formed on the walls of the reaction flask. If this precaution was not observed, a substantial amount of product occluded in the salt cake was not recovered during the washing process.

8. The checkers verified the absence of dichloroalkylphosphine and trialkylphosphine contaminants in this product by obtaining acceptable elemental analytical results and by measuring the mass spectrum of the product, which exhibits a molecular ion peak at m/e 152 (^{35}Cl) with abundant fragment peaks at m/e 110, 43, and 41.

3. Methods of Preparation

Chlorodiisopropylphosphine has been prepared by the reduction of the diisopropyltrichlorophosphorus-aluminum chloride complex with antimony;[3, 4] this is a general method and the reduction can be performed with other reagents.[5] Other general methods for the preparation of chlorodialkylphosphines are reaction of dialkylphosphines with phosgene [6, 7] and the cleavage of N,N-dialkylaminodialkylphosphines with hydrogen chloride [8-10] or phosphorus trichloride.[11]

4. Merits of the Preparation

Chlorodialkylphosphines are important synthetic intermediates in organophosphorus chemistry. In the chemical literature there is a widespread view that the simple one-step Grignard method is not suitable for the preparation of these compounds because of dominant trisubstitution and the formation of difficultly separable mixtures.[12] Although this is true for the *n*-alkyl compounds, the present preparation demonstrates that in the case of branched primary alkyl compounds and secondary and tertiary alkyl compounds the method can be very convenient and can give pure products. The submitters have prepared [13] chloro-diisobutylphosphine (45–50%), chlorodi-*sec*-butylphosphine (75–80%), chlorodi-*t*-butylphosphine (65–70%), and chlorodicyclo-hexylphosphine (60–65%) in analogous manner.

With *t*-butylmagnesium chloride the substitution of only one chlorine atom of the phosphorus trichloride is possible, giving dichloro-*t*-butylphosphine (65–70%).

1. Laboratory of Organic Chemistry of the University of Utrecht, Netherlands.
2. S. C. Watson and J. F. Eastham, *J. Organometal. Chem.*, **9**, 165 (1967).
3. J. L. Ferron, *Nature*, **189**, 916 (1961).
4. J. L. Ferron, *Can. J. Chem.*, **39**, 842 (1961).
5. I. P. Komkov, K. V. Karavanov, and S. Z. Ivin, *Zh. Obshch. Khim.*, **28**, 2963 (1958); *J. Gen. Chem. USSR (Engl. Transl.)*, **28**, 2992 (1958).
6. R. Rabinowitz and J. Pellon, *J. Org. Chem.*, **26**, 4623 (1961).
7. W. A. Henderson, Jr., S. A. Buckler, N. E. Day, and M. Grayson, *J. Org. Chem.*, **26**, 4770 (1961).
8. A. B. Burg and P. J. Slota, Jr., *J. Am. Chem. Soc.*, **80**, 1107 (1958).
9. K. Issleib and W. Seidel, *Ber.*, **92**, 2681 (1959).
10. W. Voskuil and J. F. Arens, *Rec. Trav. Chim.*, **81**, 993 (1962).
11. Monsanto Co., Brit. Patent 1,068,364 (1967) [*C.A.*, **67**, 54260 (1967)].
12. Cf. K. Sasse, in E. Müller, "Methoden der Organischen Chemie (Houben-Weyl)," 4th ed., Vol. 12, Part I, Georg Thieme Verlag, Stuttgart, 1963, p. 203.
13. W. Voskuil and J. F. Arens, *Rec. Trav. Chim.*, **82**, 302 (1963).

2-CHLORO-1-FORMYL-1-CYCLOHEXENE

(2-Chloro-1-cyclohexenealdehyde)

$$\underset{CH_3}{\overset{CH_3\quad O}{N-C-H}} + POCl_3 \xrightarrow{ClCH=CCl_2} \underset{CH_3}{\overset{CH_3\quad H}{N=C}}\overset{+}{\quad}\underset{Cl}{\quad} OPOCl_2{}^-$$

$$\xrightarrow{\quad} \overset{CH_3}{\underset{Cl}{CH=N^+}} \cdot HPO_3Cl^- \xrightarrow[NaOCOCH_3]{H_2O,} \overset{CHO}{\underset{Cl}{}}$$

Submitted by L. A. PAQUETTE,[1] B. A. JOHNSON,[2] and F. M. HINGA[2]
Checked by WILLIAM E. PARHAM and ROBERT W. GRADY

1. Procedure

To a 12-l. three-necked flask (Note 1) fitted with a stirrer, thermometer, reflux condenser, dropping funnel, nitrogen inlet, and calcium chloride drying tube are added 310 g. (4.24 moles) of dimethylformamide and 800 ml. of trichloroethylene (Note 2). The stirred solution is cooled to 5° with an external ice bath, and the system is blanketed with nitrogen (Note 3). Phosphorus oxychloride (460 g., 3.0 moles) is added during approximately 1 hour through the dropping funnel, the temperature of the stirred reaction mixture being maintained below 10°. The mixture is then allowed to warm to room temperature.

A solution of 320 g. (3.26 moles) of cyclohexanone in 800 ml. of trichloroethylene is prepared and is added to the stirred reaction mixture at such a rate that the temperature does not rise above 60° (Note 4). When the addition is completed, the mixture is heated at 55–60° for 3 hours.

The solution is cooled to below 35° by use of an ice bath, and a solution of 1.2 kg. of anhydrous sodium acetate in 2.8 l. of water is cautiously added through the dropping funnel (Notes 5 and 6). The organic layer is separated and is washed twice with 1.5-l. portions of saturated aqueous salt solutions and once with 1.5 l. of deoxygenated water. The organic solution is dried over anhydrous sodium sulfate.

To the dried solution is added 10 g. of anhydrous sodium acetate. The solvent is evaporated under reduced pressure on a water bath heated to 50–60° (Note 7). The concentrate is distilled under nitrogen through a 14-in. vacuum-jacketed Vigreux column. The distillate is collected in receivers containing 1 g. of anhydrous sodium acetate per 100 ml. of flask capacity (Note 8). There is obtained 230–320 g. (53–74%) of colorless liquid, b.p. 86–88° (10 mm.), n^{20}D 1.5198 (Notes 9 and 10).

2. Notes

1. The checkers carried out this procedure using a 5-l. flask and employed five-twelfths of the quantities of reagents specified.

2. Du Pont extraction grade of trichloroethylene was employed throughout the course of this work.

3. A nitrogen atmosphere is maintained over the reaction mixture and the product at all times when possible.

4. Approximately 1.5 hours is required to complete the addition.

5. The temperature is maintained below 35° during this addition, which is of approximately 1-hour duration.

6. The resulting two-phase mixture appears to be stable and may be allowed to stand overnight or for several days at room temperature.

7. The concentrate may be conveniently stored at −45° or below before distillation.

8. The aldehyde is quite unstable and tends to decompose with some violence on standing at room temperature. However, when treated with 1 g. of anhydrous sodium acetate per 100 ml. of distilled product, the compound has remained stable for 2 weeks when stored in this condition at room temperature. It may be stored quite indefinitely in this condition at −45°.

9. Gas chromatography is a convenient method of monitoring the distillation. Early fractions contain trichloroethylene and an unidentified reaction by-product.

10. The product obtained by the checkers was pale yellow in color. The color was not removed by redistillation.

3. Methods of Preparation

2-Chloro-1-formyl-1-cyclohexene has been prepared only by the action of phosphorus oxychloride (or phosgene) and di-methylformamide on cyclohexanone.[3-5] 2-Bromo-1-formyl-1-cyclohexene has been synthesized by a method analogous to the above by the use of phosphorus oxybromide or phosphorus tribromide.[6]

4. Merits of the Preparation

The described procedure is useful for the conversion of ketones to chloroalkene aldehydes. Methyl ethyl ketone,[3, 4] phenyl ethyl ketone,[3, 4] cyclobutanone,[7] cyclopentanone,[3-5] cyclooctanone,[3-5] α-tetralone,[5] and benzosuberone[5] are illustrative of the wide variety of ketones which have been so treated. The yields are reported generally to be 65–80%.

The chlorovinyl aldehydes, although still a relatively new class of compounds, show great promise as useful synthetic intermediates.[5, 7, 8]

1. Department of Chemistry, The Ohio State University, Columbus, Ohio.
2. Research Laboratories of The Upjohn Co., Kalamazoo, Michigan.
3. Z. Arnold and J. Zemlicka, *Proc. Chem. Soc.*, 227 (1958).
4. Z. Arnold and J. Zemlicka, *Coll. Czech. Chem. Commun.*, **24**, 2385 (1959).
5. W. Ziegenbein and W. Lang, *Ber.*, **93**, 2743 (1960).
6. Z. Arnold and A. Holy, *Coll. Czech. Chem. Commun.*, **26**, 3059 (1961).
7. J. Zemlicka and Z. Arnold, *Coll. Czech. Chem. Commun.*, **26**, 2852 (1961).
8. W. Ziegenbein and W. Franke, *Angew. Chem.*, **71**, 628 (1959).

bis(CHLOROMETHYL) ETHER[1]

HAZARD NOTE

Very high carcinogenic activity has been reported for bis-(chloromethyl) ether when administered to rats by inhalation and by subcutaneous injection. This compound should be handled with great care.

Reported by B. L. Van Duuren, A. Sivak, B. M. Goldschmidt, C. Katz, and S. Melchionne, *J. Nat. Cancer Inst.*, **43**, 481 (1969).

1. *Org. Syntheses*, Coll. Vol. **4**, 101 (1963).

CHLOROMETHYLPHOSPHONOTHIOIC DICHLORIDE

$$10 \ ClCH_2POCl_2 + P_4S_{10} \rightarrow 10 \ ClCH_2PSCl_2 + P_4O_{10}$$

Submitted by R. Schmutzler [1]
Checked by M. D. Hurwitz and W. D. Emmons

1. Procedure

A 500-ml. three-necked flask is provided with a mechanical stirrer, thermometer, and reflux condenser equipped with a drying tube. The flask is flushed with dry nitrogen and charged under nitrogen with 502 g. (3 moles) of chloromethylphosphonic dichloride (Note 1) and 160 g. (0.36 mole) of tetraphosphorus decasulfide (Note 2). The reaction mixture is heated under reflux with stirring for 6 hours, the liquid temperature being 180–190° (Note 3). The nearly black reaction mixture is then allowed to cool to room temperature and is distilled under reduced pressure.   Material distilling between 70° (40 mm.) and 150° (20 mm.) is collected (Note 4).   The yield is 364–396 g. (66–72%). There is no impurity in the material thus obtained which is detectable by gas chromatography (Note 5).   The product may be redistilled if desired, although in most cases this is superfluous; b.p. 64–65° (10 mm.); n^{25}D 1.5730–1.5741 (Note 6).   The P[31] n.m.r. spectrum of the product shows a peak at −74.2 p.p.m. relative to external phosphoric acid.

2. Notes

1. Chloromethylphosphonic dichloride is used as obtained from Stauffer Chemical Co.   Alternatively it may be prepared from the reaction of phosphorus trichloride with paraformaldehyde.[2]

2. Technical tetraphosphorus decasulfide (Stauffer Chemical Co.) is employed.   The product is weighed under nitrogen protection.

3. In order to prevent contact of the boiling reaction mixture with air, nitrogen is passed through a T-tube on top of the drying tube on the reflux condenser.

4. Toward the end of the distillation a thick residue is formed, and this makes the distillation difficult.   After the contents of the distillation flask are cooled to room temperature, this residue may be disposed of by careful continuous rinsing with water under a well-ventilated hood.

5. An F&M 500 Program-Temperature Unit (8 ft., 20% silicon rubber on 60–80 Super Support) was used for the VPC work: program 11°/min., flow 55 ml./min.   Chloromethylphosphonothioic dichloride and chloromethylphosphonic dichloride, a potential impurity, are separated cleanly under these conditions.

6. Literature[3] values are: b.p. 89° (30 mm.), n^{25}D 1.5741, d_{25}^{25} 1.5891.

3. Methods of Preparation

Chloromethylphosphonothioic dichloride has been prepared by the reaction of chloromethylphosphonic dichloride with tetraphosphorus decasulfide [3-5] or with thiophosphoryl chloride under autogenous pressure.[3, 5]

4. Merits of the Preparation

The reaction of chloromethylphosphonic dichloride with tetraphosphorus decasulfide [3-5] or with thiophosphoryl chloride [3] are the only methods of preparation for this compound reported. The method is applicable more generally, and the syntheses of methyl-, trichloromethyl-, ethyl-, propyl-, cyclohexyl-, phenyl-, and p-chlorophenylphosphonothioic dichloride from the corresponding phosphonic dichlorides have been reported.[4, 5] Phosphinic chlorides of varying structures could also be converted to the corresponding thiono compounds by comparable procedures.[4, 5] The present method is preferable to the thiophosphoryl chloride procedure [3, 5] in that it does not require working under pressure.

Chloromethylphosphonothioic dichloride is a reactive and useful intermediate in organophosphorus chemistry.[3, 4, 6] Of special interest is its desulfurization by trivalent phosphorus compounds such as phenylphosphonous dichloride leading to the formation of chloromethylphosphonous dichloride.[3]

1. Explosives Department, E. I. duPont deNemours and Company, Wilmington, Delaware.
2. R. A. B. Bannard, J. R. Gilpin, G. R. Vavasour, and A. F. McKay, *Can. J. Chem.* **31**, 976 (1953).
3. E. Uhing, K. Rattenbury, and A. D. F. Toy, *J. Am. Chem. Soc.*, **83**, 2299 (1961).
4. M. I. Kabachnik and N. N. Godovikov, *Dokl. Akad. Nauk SSSR*, **110**, 217 (1956).
5. K. Rattenbury, U. S. Patent 2,993,929 (1961).
6. R. Schmutzler, *J. Inorg. Nucl. Chem.*, **25**, 335 (1963).

p-CHLOROPHENOXYMETHYL CHLORIDE*

(Anisole, *p*-α-dichloro-)

Submitted by H. GROSS and W. BÜRGER [1]
Checked by J. LONGANBACH and K. B. WIBERG

1. Procedure

This preparation must be carried out in an efficient hood.

A 250-ml. round-bottomed flask with a side arm is equipped with a distillation head and condenser. The receiving flask is attached to the condenser with an adapter, and the exit from the flask goes to a bubble counter containing high-boiling petroleum ether. A thermometer is inserted in the side arm of the distillation flask, and it reaches to the bottom. With exclusion of moisture, 147 g. (0.704 mole) of phosphorus pentachloride and 100 g. (0.704 mole) of *p*-chloroanisole are added to the flask. The flask is heated in an oil bath; the reaction begins when the inside temperature reaches 120° and occurs rapidly at 140°. The temperature is raised to 160° over a period of 2 hours, thereby distilling the phosphorus trichloride (Note 1). After the gas evolution subsides, the reaction mixture is heated to 175° for a short time. About 73–75 g. of phosphorus trichloride is collected.

The residue is distilled through a 30-cm. column packed with glass beads giving 10 g. of a fraction, b.p. 85–105° (10 mm.), containing mainly *p*-chloroanisole, and 85–99 g. (68–80%) of *p*-chlorophenoxymethyl chloride, b. p. 105–108° (10 mm.), $n^{23}D = 1.5496$, m.p. 28–29° (Notes 2, 3). The n.m.r. spectrum [CCl$_4$, (CH$_3$)$_4$Si reference] had bands at δ 5.6 (*s*, 2H) and 6.6–7.3 p.p.m.

2. Notes

1. Some phosphorus pentachloride may solidify in the upper part of the condenser. This may be removed by rotating the condenser.

2. The literature values are b.p. 120–124 (18 mm.), m.p. 29–30°.[2]

3. Gas chromatography analysis using a silicone gum column indicated the product to be 97% pure.

3. Discussion

Aryloxymethyl chlorides may be prepared by the reaction of sodium aryloxymethanesulfonates with phosphorus pentachloride.[2, 3] The chlorination of anisole does not, as previously reported,[4] give phenoxymethyl chloride, but rather a mixture of p- and o-chloroanisoles.[5] Similarly, anisole and other unsubstituted methyl aryl ethers undergo ring chlorination with phosphorus pentachloride and chlorine,[6] whereas ring-chlorinated anisoles, such as p-chloroanisole, undergo chlorination at the methyl group with chlorine at 190–195° in the presence of a catalytic amount of phosphorus pentachloride.[6] Ring-nitrated aryloxymethyl chlorides may be obtained by the aluminum chloride-catalyzed decarbonylation of the corresponding aryloxyacetyl chlorides.[7,8]

The present method is simple, proceeds easily and in good yield to give a single product. It is applicable to other cases, such as the preparation of 2,4-dichlorophenoxymethyl chloride (89–92%). The chlorination of p-chloroanisole with chlorine and phosphorus pentachloride gives considerable amounts of p-chlorophenoxydichloromethane which is difficult to separate from the desired compound by distillation.

The reaction of chloromethyl aryl ethers with nucleophilic reagents has been described by Barber et al.[2] Thus, by reaction with thiourea, potassium thiocyanate, or sodium cyanide, there are obtained aryloxyalkylisothiouronium salts, aryloxyalkyl thiocyanates, and aryloxyalkylacetonitriles, respectively.[2] With silver sulfonates the sulfonic acid esters of aryloxymethanols may be obtained.[8] The reaction of chloromethyl aryl ethers with butyllithium leads to an aryloxycarbene which on re-action with olefins gives aryloxycyclopropanes.[3] The ethers react with triphenylphosphine and a base to give phenoxy-methylene ylides which are useful in converting carbonyl compounds to aromatic enol ethers.[9] The reaction of the chloro ethers with trialkylphosphites gives aryloxymethane-phosphonates.[10] Most of these reactions have been studied with phenoxymethyl chloride and the *p*-methyl derivative; they also proceed well and in good yield with the readily obtainable *p*-chlorophenoxymethyl chloride.[10]

1. Institute für Organische Chemie der Deutschen Akademie der Wissenschaften zu Berlin, DDR-1199 Berlin-Adlershof.
2. H. J. Barber, R. F. Fuller, M. B. Green, and H. T. Zwartouw, *J. Appl. Chem.*, **3**, 266 (1953).
3. U. Schöllkopf, A. Lerch, and J. Paust, *Ber.*, **96**, 2266 (1963).
4. C. S. Davis and G. S. Lougheed, *Org. Syntheses,* **47**, 23 (1967).
5. M. Shamma, L. Novak, and M. G. Kelly, *J. Org. Chem.* **33**, 3335 (1968).
6. H. J. Barber, R. F. Fuller, and M. B. Green, *J. Appl. Chem.*, **3**, 409 (1953).
7. M. H. Palmer and G. J. McVie, *Tetrahedron Lett.*, 6405 (1966).
8. H. Böhme and P. H. Meyer, *Synthesis,* 150 (1971).
9. G. Wittig, W. Böll and K.-H. Krück, *Ber.*, **95**, 2514 (1962).
10. H. Gross and W. Bürger, unpublished results.

p-CHLOROPHENYL ISOTHIOCYANATE

(Isothiocyanic acid, *p*-chlorophenyl ester)

$$p\text{-ClC}_6\text{H}_4\text{NH}_2 \xrightarrow[-\text{H}_2\text{O}]{\text{CS}_2,\ \text{NH}_4\text{OH}} p\text{-ClC}_6\text{H}_4\text{NHCS}_2\text{NH}_4 \xrightarrow[-\text{NaCl}]{\text{ClCH}_2\text{CO}_2\text{Na}}$$

$$p\text{-ClC}_6\text{H}_4\text{NHCS}_2\text{CH}_2\text{CO}_2\text{NH}_4 \xrightarrow[-\text{Zn(SCH}_2\text{CO}_2\text{NH}_4)_2]{\text{ZnCl}_2,\ \text{NaOH}} p\text{-ClC}_6\text{H}_4\text{NCS}$$

Submitted by G. J. M. van der Kerk, C. W. Pluygers, and G. de Vries [1]
Checked by W. S. Wadsworth, Jr., and William D. Emmons

1. Procedure

Caution! p-Chlorophenyl isothiocyanate may cause severe derma-titis if allowed to come in contact with the skin. This preparation should be carried out in a good hood, and rubber gloves should be worn throughout.

In a 250-ml. round-bottomed flask fitted with mechanical stirrer, reflux condenser, and thermometer are placed 38.3 g. (0.30 mole) of *p*-chloroaniline (Note 1), 41 ml. (0.6 mole) of concentrated aqueous ammonia (sp. gr. 0.9), and 21 ml. (0.35 mole) of carbon disulfide. The mixture is stirred vigorously, and when it is heated to 30° the reaction starts. The temperature is maintained at 30–35° by external cooling (Note 2). The reaction mixture turns into a deep-red turbid solution within a few minutes, and then suddenly a heavy yellow precipitate of ammonium *p*-chlorophenyldithiocarbamate separates. To the mixture 15 ml. of water is added, and stirring is continued for 1 hour. The mixture is filtered with suction, and the residue is washed with two 30-ml. portions of a 3% aqueous solution of ammonium chloride and with two 15-ml. portions of 96% ethanol.

The ammonium *p*-chlorophenyldithiocarbamate obtained is transferred immediately to a 1-l. beaker fitted with an efficient mechanical stirrer. Water (250 ml.) is added, and the temperature is raised to 30°. A solution of 28.4 g. (0.30 mole) of chloroacetic acid in 30 ml. of water is neutralized with sodium carbonate [18.6 g. (0.15 mole) of $Na_2CO_3 \cdot H_2O$ in 70 ml. of water] and is added to the well-stirred dithiocarbamate suspension over a 10-minute period (Note 3). In the beginning the suspension gradually becomes less viscous, but at the end of the addition it rapidly turns into a creamy mass. Another 250 ml. of water is added to facilitate stirring, which is continued for 1 hour after the addition at about 30°.

The creamy suspension is allowed to cool to room temperature, and the electrodes of a pH meter are inserted (Note 4). A solution of 20.5 g. (0.15 mole) of zinc chloride (Note 5) in 75 ml. of water is added dropwise with vigorous stirring over a period of 45 minutes, while the pH is maintained at 7 by the simultaneous dropwise addition of a 4*N* aqueous solution of sodium hydroxide (Note 6). The mixture is stirred for 1 hour and is then filtered with suction; the solid product is dried under reduced pressure over phosphorus pentoxide. The dry material is slurried with 200 ml. of petroleum ether (b.p. 30–60°), and the solvent is decanted. This process is repeated five times, and the combined extract is evaporated at reduced pressure. The yield of almost pure *p*-chlorophenyl isothiocyanate, obtained as a readily crystallizing oil with a pleasant anise-like odor, is 33–35 g. (65–68%), m.p. 44–45°. The product can be recrystallized from the minimum amount of ethanol at 50°.

2. Notes

1. A commercial grade (Eastman Organic Chemicals, white label) of *p*-chloroaniline was used without further purification.

2. The reaction is conveniently started by dipping the flask in a hot-water bath. The reaction temperature is easily maintained by occasional dipping of the flask in a cold-water bath.

3. Any free chloroacetic acid leads to the formation of N-*p*-chlorophenylrhodanine.

4. A Beckman pH meter (Model N) was used.

5. The anhydrous zinc chloride used was obtained from Baker and Adamson, reagent grade.

6. The pH must not drop below 7, although a slightly higher pH does no harm; addition of the zinc chloride in a shorter time lowers the yield.

3. Methods of Preparation

The procedure given here is essentially that described previously by the submitters.[2] *p*-Chlorophenyl isothiocyanate has been prepared from *sym*-di-*p*-chlorophenyl thiourea with iodine in alcoholic solution,[3] from ammonium *p*-chlorophenyldithiocarbamate and lead nitrate [4] [cf. also *Org. Syntheses*, Coll. Vol. 1 447 (1932)], by the action of thiophosgene on *p*-chloroaniline [5] and from *p*-chloroaniline with thiocarbonyl tetrachloride in the presence of stannous chloride.[6]

4. Merits of the Preparation

The present method has the advantage that the whole process can be carried out in aqueous medium at low temperatures. The procedure is also attractive because of the reagents used and the relatively simple isolation procedure employed. The only restriction observed is that the formation of the aromatic dithio-carbamate must be possible.

Other isothiocyanates obtained by this method are: phenyl isothiocyanate (65%), *p*-phenylene diisothiocyanate (71%), *p*-acetylaminophenyl isothiocyanate (73%), *p*-ethoxyphenyl isothiocyanate (64%), and *p*-bromophenyl isothiocyanate (55%).

1. Organisch Chemisch Instituut T.N.O., Utrecht, Nederland.
2. G. J. M. van der Kerk, C. W. Pluygers, and G. de Vries, *Rec. Trav. Chim.*, **74,** 1262 (1955).
3. S. M. Losanitsch, *Ber.*, **5,** 156 (1872).
4. F. B. Dains, R. Q. Brewster, and C. P. Olander, *Univ. Kansas Sci. Bull.*, **13,** 1 (1922) [*C.A.*, **17,** 543 (1923)].
5. G. M. Dyson, *Org. Syntheses*, Coll. Vol. **1,** 165 (1932).
6. J. M. Connolly and G. M. Dyson, *J. Chem. Soc.*, 679 (1935).

CHLOROSULFONYL ISOCYANATE*

(Isocyanic acid, anhydride with chlorosulfonic acid)

$$SO_3 + ClCN \rightarrow ClSO_2NCO + ClSO_2OSO_2NCO + N$$

$$ClSO_2OSO_2NCO + ClCN \rightarrow 2ClSO_2NCO$$

$$\rightarrow ClSO_2NCO + ClCN$$

Submitted by RODERICH GRAF [1]
Checked by JEROME F. LEVY and WILLIAM D. EMMONS

1. Procedure

Caution! Cyanogen chloride is extremely toxic. Sulfur trioxide and chlorosulfonyl isocyanate are highly corrosive materials. This preparation should be carried out in an efficient hood, and rubber gloves should be worn throughout.

A 200-ml. four-necked flask is fitted with a mechanical stirrer (Note 1), a thermometer, a Claisen-type adapter bearing a dry ice reflux condenser and a dropping funnel, and a gas-inlet tube consisting of a length of 6-mm. glass tubing extending almost to the bottom of the flask. This gas-inlet tube is connected through a stopcock to a safety trap and then to a cylinder of cyanogen chloride (Note 2). The dropping funnel, protected with a calcium chloride drying tube, is charged with 64 g. (0.80 mole) of liquid sulfur trioxide (Note 3). The outlet from the dry ice condenser is connected to a trap cooled in dry ice.

The flask is charged with approximately 36.9 g. (0.60 mole) of cyanogen chloride while being cooled in a dry ice-methylene chloride slush bath (Note 4). The flask is allowed to warm to $-5°$ to melt the cyanogen chloride, and then the liquid sulfur trioxide is added over a period of 0.75–1.25 hours. The reaction is very exothermic. During the addition the temperature is gradually decreased from $-5°$ to $-15°$ (Note 5). After addition is completed, the reaction mixture is checked for unreacted sulfur trioxide by adding approximately 1–2 g. of cyanogen chloride and noting if a temperature rise takes place. If necessary, this test is repeated until no more unreacted sulfur trioxide is left. The reaction mixture at this point is a pulpy, stirrable mass containing some chlorosulfonyl isocyanate, some chloropyrosulfonyl isocyanate, and much precipitated 2,6-dichloro-1,4,3,5-oxathiadiazine-4, 4-dioxide (Note 6).

The adapter bearing the dry ice condenser and dropping funnel is removed and replaced by a 16 cm. x 2 cm. distillation column packed with glass helices (Note 7) and connected to an efficient air-cooled condenser. The condenser has a 100-ml. receiver and is connected to a trap cooled in dry ice and protected by a drying tube to condense unreacted cyanogen chloride. The reaction flask is heated for about 1 hour while the temperature is gradually increased to 110–115°. At this point, cyanogen chloride is bubbled into the reaction mixture at the rate of about 0.010 mole/min. (Note 8). The temperature of the flask is raised to 120–130°, whereupon chlorosulfonyl isocyanate begins to distil at a head temperature of 90–105°. When the distillation rate begins to slacken, and after most of the contents of the flask has distilled, the temperature of the flask is raised to 130–150°. When the residue in the flask is only 3–5 ml., the cyanogen chloride flow is discontinued and the distillation is stopped. This part of the reaction (from the start of the cyanogen chloride feed) requires about 0.4–0.6 mole of cyanogen chloride and takes about 0.75–1 hour.

The crude product, which may contain dissolved cyanogen chloride, is redistilled at a pressure of 100 mm. through the helices-packed column. Heating is done very slowly at first to allow the unreacted cyanogen chloride to distil and be condensed in the dry ice trap. The product is collected at 54–56° (100 mm.); d_4^{20} 1.626, weight 67.7–69.9 g. (60–62%) (Notes 9–11).

2. Notes

1. The lubricant for the ground-glass sleeve of the stirrer may be silicone oil or mineral oil; however, Teflon® oil is preferred. Glycerin should not be used.

2. The checkers used cyanogen chloride supplied in a metal cylinder by the American Cyanamid Co., Bound Brook, New Jersey. The submitter prepared cyanogen chloride beforehand [2] and either charged it as a liquid or allowed it to distil in as is required later in the reaction.[3]

3. The checkers used Sulfan®, a stabilized liquid form of sulfur trioxide which is commercially available from Baker and Adamson, General Chemical Division, Allied Chemical Corp., Morristown, New Jersey. The submitter distilled sulfur trioxide from 65% oleum directly into the reaction flask, a procedure which is described elsewhere.[3]

4. Chlorosulfonyl isocyanate reacts violently with water. For safety reasons, therefore, it is recommended that either air or dry ice mixtures be used for all cooling condensers and cooling baths. The dry ice may be mixed with methylene chloride. Acetone is not recommended, as chlorosulfonyl isocyanate may react with it.

5. Too much cooling in the early stages of the reaction may cause the cyanogen chloride to crystallize. Furthermore, if the

reaction mixture is cooled to substantially lower than $-15°$, *e.g.*, to $-30°$ or $-40°$, the rate of reaction will decrease to the extent that there is danger of an uncontrollable delayed reaction.

6. At this point the reaction mixture may be stored protected from atmospheric moisture for an unlimited length of time before converting it to chlorosulfonyl isocyanate.

7. It is highly desirable to heat the distillation column with an electrical heating tape to compensate for heat loss. A 6 ft. x ½ in., 288-watt heating tape available from Briscoe Manufacturing Co., Columbus, Ohio, was used by the checkers. This should not be necessary when the reaction is conducted on a larger scale.

8. A flowmeter calibrated for use with air was used. Although this introduces some degree of error, it is adequate for the preparation.

9. The submitter conducted the reaction on ten times the scale indicated here and obtained yields of 88–93%. The checkers, however, on a scale of 0.80 mole, reproducibly obtained the lower yields indicated.

10. For storage over a short time, glass bottles sealed with rubber stoppers that are covered with polyethylene sheet are adequate. Ground-glass stoppers, even if thoroughly coated with silicone grease, will soon become frozen. For storage over moderate periods of time (several weeks) low-pressure polyethylene may be used. If traces of sulfur trioxide are present, the walls of the polyethylene vessel will soon become black; if more than 2% of cyanogen chloride is present, the polyethylene is attacked without a change in color, and its surface is converted to a crumbly mass. For storage over a long period of time, Teflon® FEP bottles available from the Nalge Co., Inc., Rochester, New York, or sealed-glass ampoules may be used.

11. Chlorosulfonyl isocyanate is a colorless, fluid liquid which fumes slightly in moist air. The vapors have a tussive effect. The compound shows an extraordinarily violent, almost explosive-like reaction with water. The contact of a small amount of the compound with the skin has no deleterious effect if it is rapidly removed by rinsing with plenty of water. Contacts which last longer than a few seconds may result in severe burns.

Cotton fabrics will char immediately on contact with the compound and produce a dense smoke. A specific toxic effect other than the purely cauterizing effect of the compound has not been observed by the submitter during the past 10 years.

3. Methods of Preparation

The present procedure corresponds to the method described earlier by Graf.[3]

4. Merits of the Preparation

The cycloaddition of chlorosulfonyl isocyanate to olefins, followed by removal of the N-sulfonyl chloride group of the resulting β-lactam-N-sulfonyl chloride, offers a convenient synthesis of a large number of β-lactams unsubstituted on nitrogen.[4-6] Also produced in the reaction with olefins are unsaturated carboxamide-N-sulfonyl chlorides, which, like the β-lactam-N-sulfonyl chlorides, may be worked up in various ways to give a variety of products.[5,6] The reagent has also been added to 1,3-dienes,[7] cycloheptatriene,[8] and acetylenes.[9] Photochemical reactions with olefins have also been reported.[10] Chlorosulfonyl isocyanate reacts with aldehydes, e.g., benzaldehyde, to give imine-N-sulfonyl chlorides which will undergo cycloaddition reactions with ketene or dimethyl ketene to give, after removal of the sulfonyl chloride group, β-lactams also.[5]

Compounds containing active hydrogens react with chlorosulfonyl isocyanate first at the isocyanate group to give N-substituted sulfamyl chlorides which may react further with more active hydrogen compound at the sulfonyl chloride group.[3, 7-10]

The reactions of chlorosulfonyl isocyanate have been reviewed.[15,16]

1. Farbwerke Hoechst AG, Frankfurt on the Main, Germany.
2. G. Brauer, ed., "Handbook of Preparative Inorganic Chemistry," 2nd ed., Vol. 1, Academic Press, New York, 1963, pp. 662-665; see also G. H. Coleman, R. W. Leeper, and C. C. Schulze, *Inorg. Syntheses,* **2**, 90 (1946); H. Schroder, *Z. Anorg. Allgem. Chem.,* **297**, 296 (1958).
3. R. Graf, *Ber.,* **89**, 1071 (1956).

4. R. Graf, this volume, p. 673.
5. R. Graf, *Ann.*, **661**, 111 (1963).
6. H. Hoffmann and H. J. Diehr, *Tetrahedron Lett.*, 1875 (1963).
7. P. Goebel and K. Clauss, *Ann.*, **722**, 122 (1969).
8. E. J. Moriconi, C. F. Hummel, and J. F. Kelly, *Tetrahedron Lett.*, 5325 (1969).
9. E. J. Moriconi, J. G. White, R. W. Franck, J. Jansing, J. F. Kelly, R. A. Salomone, and Y. Shimakawa, *Tetrahedron Lett.*, 27 (1970).
10. D. Gunther and F. Soldan, *Ber.*, **103**, 663 (1970).
11. R. Graf, *Ber.*, **92**, 509 (1959).
12. R. Graf, *Ber.*, **96**, 56 (1963).
13. German Patent 931, 225 [C. A., 50, 7861 (1956)].
14. R. Appel and W. Senkpiel, *Ber.*, **91**, 1195 (1958).
15. R. Graf, *Angew. Chem.*, **80**, 179 (1968); *Angew. Chem. Intern. Ed. Engl.*, **7**, 172 (1968).
16. E. J. Moriconi, *Intra-science Chem. Rep.*, **3**, 131 (1968).

2-CHLOROTHIIRANE 1,1-DIOXIDE

(Thiirane, 2-chloro-, 1,1-dioxide)

Submitted by Leo A. Paquette and Lawrence S. Wittenbrook [1]
Checked by John J. Miller and William D. Emmons

1. Procedure

Caution! *Because of the toxic nature of chlorine and diazomethane and the lachrymatory properties of chloromethanesulfonyl chloride, both steps of this preparation should be carried out in a well-ventilated hood. Diazomethane is also explosive; follow the directions for its handling given in earlier volumes.*[2, 3]

A. *Chloromethanesulfonyl chloride.* A slurry of 210 g. (1.52 moles) of s-trithiane (Note 1) in a mixture of 1 l. of glacial acetic acid and 210 ml. of water is prepared in a 2-l., three-necked, round-bottomed flask equipped with an efficient mechanical stirrer, a thermometer, a coarsely fritted gas inlet tube (Note 2), and an exit tube by which excess fumes are carried to the rear of the hood. The flask is immersed in an ice bath, and the stirrer is started. A stream of chlorine is introduced at such a rate (Note 3) that the temperature of the mixture is maintained between 40° and 50° by the mildly exothermic reaction. After 1–2 hours a yellow solution results. To this solution is added 300 ml. of water, at which point the temperature rises to *ca.* 60°.

Chlorine is again introduced, and the stirred reaction mixture is cooled to maintain the temperature initially in the vicinity of 40°. During 3 hours the temperature slowly returns to that of the surroundings, and the stream of chlorine is stopped. The yellow solution is allowed to stand overnight at room temperature and is then transferred to a 4-l. Erlenmeyer flask and diluted with 1.5 l. of ice water. The flask is stoppered and placed in a refrigerator for 2–3 hours. The aqueous phase is decanted from the denser, organic layer that has separated (Note 4) and is extracted with four 300-ml. portions of methylene chloride. The methylene chloride extracts are combined with the original organic layer, dried over anhydrous magnesium sulfate, filtered, and evaporated on a rotary evaporator at 20–30°. The material that remains is distilled through a 15-cm. Vigreux column. Chloromethanesulfonyl chloride is collected as a colorless, lachrymatory liquid, b.p. 80–81° (25 mm.), n^{26}D 1.4840–1.4850; yield 135–220 g. (20–32%) (Note 5).

B. *2-Chlorothiirane 1,1-dioxide.* In a 500-ml., three-necked, round-bottomed flask fitted with an efficient mechanical stirrer, a thermometer, and two pressure-equalizing addition funnels is placed an ethereal solution of 4.6 g. (0.11 mole) of diazomethane (Note 6). The system is blanketed with nitrogen, the stirrer is started, and the solution is cooled to −10° with an ice-methanol bath. A solution of 14.9 g. (0.100 mole) of chloromethanesulfonyl chloride in 40 ml. of ether and a solution of 10.0 g. (0.099 mole) of triethylamine in 40 ml. of ether are simultaneously added dropwise from the two addition funnels. The addition requires about 45 minutes. The insoluble triethylamine hydrochloride is separated by filtration and washed with 25 ml. of cold ether. The combined filtrate and washings are evaporated at reduced pressure below 25° to give white crystalline 2-chlorothiirane 1,1-dioxide, m.p. 49–51° (Note 7); yield 10.0–10.5 g. (80–84%). The product may be further purified by recrystallization from ether-hexane at −70°; m.p. 53–54° (8.9 g. after two recrystallizations) (Notes 8 and 9).

2. Notes

1. *s*-Trithiane obtained from Eastman Organic Chemicals was used without further purification.

2. The gas inlet tube must be sufficiently long to allow the chlorine to enter near the bottom of the flask, and sufficiently coarse to prevent the pores from becoming clogged by the suspended *s*-trithiane. A 6-mm. glass tube with a slightly constricted orifice has been found to be equally satisfactory.

3. The rate of addition of chlorine appears to be important. With rates adequate to maintain the temperature between 40° and 50°, addition times of 1–2 hours are required.

4. This separation is most conveniently achieved by first decanting as much water as possible from the 4-l. Erlenmeyer flask. The remaining mixture is then placed in a 4-l. separatory funnel, and the lower layer is collected. The separatory funnel is then used in the ensuing extractions.

5. The yield is based on the assumption that 3 molecules of chloromethanesulfonyl chloride arise from each molecule of *s*-trithiane. The checkers obtained a yellow product which had to be redistilled to provide material with the reported refractive index.

6. The ethereal diazomethane is prepared by the method of Arndt.[4] The checkers employed undistilled material as described in Note 3 of the preparation cited.

7. Use of undistilled diazomethane solution gives a less pure product, m.p. 38–42°, which can be recrystallized as described.

8. The characteristic infrared maxima of 2-chlorothiirane

1,1-dioxide (Nujol) occur at 3.24, 7.53, and 8.56 μ. Its n.m.r. spectrum (CDCl$_3$) shows a doublet of doublets at δ 3.17 ($J = 9.5$ and 5.5 Hz), a triplet at δ 3.75 ($J = 9.5$ Hz), and a doublet of doublets at δ 4.85 ($J = 9.5$ and 5.5 Hz).

9. A characteristic property of thiirane 1,1-dioxides is the ease with which such molecules fragment into sulfur dioxide and the related olefin on standing for several hours at room temperature. The title compound is no exception; however, the rate of decomposition may be reduced substantially by storage under an inert atmosphere in a freezing compartment (ca. $-5°$). Under such conditions the product may be kept for many months.

3. Discussion

2-Halothiirane 1,1-dioxides are known to be intermediates in the Ramberg-Bäcklund rearrangement of α,α-dihalo sulfones.[5, 6, 9, 10] These three-membered cyclic sulfones are not isolable from such reactions, however, because they are not stable under the conditions of the rearrangement and they undergo further transformations. The present procedure represents the only means presently available for the preparation of halogen-substituted thiirane dioxides.[5-7] The addition of halosulfenes to diazoalkanes is a convenient and general synthesis which may be extended to the preparation of a variety of thiirane 1,1-dioxides with relative ease.[8]

1. Department of Chemistry, The Ohio State University, Columbus, Ohio 43210.
2. T. J. de Boer and H. J. Backer, Org. Syntheses, Coll. Vol. 4, 250 (1963).
3. J. A. Moore and D. W. Reed, this volume, p. 351.
4. F. Arndt, Org. Syntheses, Coll. Vol. 2, 165 (1943).
5. L. A. Paquette and L. S. Wittenbrook, Chem. Commun., 471 (1966).
6. L. A. Paquette, L. S. Wittenbrook, and V. V. Kane, J. Am. Chem. Soc., 89, 4487 (1967).
7. L. A. Carpino and R. H. Rynbrandt, J. Am. Chem. Soc., 88, 5682 (1966).
8. G. Opitz and K. Fischer, Z. Naturforsch., 186, 775 (1963); Angew. Chem., 77, 41 (1965); Angew. Chem. Intern. Ed. Engl., 4, 70 (1965); N. Fischer and G. Opitz, Org. Syntheses, 48, 106 (1968).
9. L. A. Paquette, J. Am. Chem. Soc., 86, 4089 (1964).
10. L. A. Carpino and L. V. McAdams, III, J. Am. Chem. Soc., 87, 5804 (1965).

1-CHLORO-1,4,4-TRIFLUOROBUTADIENE

(Butadiene, 1-chloro-1,4,4-trifluoro-)

$$CF_2{=}CFCl \;+\; CH_2{=}CHOCOCH_3 \longrightarrow$$

$$\xrightarrow[-CH_3CO_2H]{\text{Heat}}$$

$$\longrightarrow CFCl{=}CHCH{=}CF_2$$

Submitted by R. E. Putnam, B. C. Anderson, and W. H. Sharkey.[1]
Checked by R. D. Birkenmeyer, M. A. Rebenstorf, and F. Kagan.[2]

1. Procedure

A. *2-Chloro-2,3,3-trifluorocyclobutyl acetate* (Note 1). A mixture of 1.0 g. of hydroquinone, 3 drops of a terpene inhibitor (Note 2), and 140 g. (1.63 moles) of inhibited redistilled vinyl acetate (Note 3) is placed in a 400-ml. high-pressure shaker tube lined with stainless steel (Note 4). The shaker tube is closed, cooled in a mixture of solid carbon dioxide and acetone, evacuated, and charged with 47 g. (0.40 mole) of chlorotrifluoroethylene (Note 5). The shaker tube is heated with agitation to 215° in a period of about 1 hour and is then heated at 215° for 3 hours. The shaker tube is cooled to room temperature and is bled slowly to remove excess chlorotrifluoroethylene. The black, viscous reaction mixture (Note 6) is transferred to a distillation flask and heated on a steam bath. After a fore-run of dichloro-hexafluorocyclobutane and vinyl acetate is collected at atmospheric pressure, a receiver cooled in solid carbon dioxide and acetone is attached, and crude 2-chloro-2,3,3-trifluorocyclobutyl acetate is rapidly distilled by gradually reducing the pressure to about 10 mm. (Note 7). Redistillation through a 30-cm. column packed with glass helices provides 22–30 g. (27–37%) (Note 8) of the acetate, b.p. 60–65°/100 mm., n_D^{25} 1.3916–1.3921.

B. *1-Chloro-1,4,4-trifluorobutadiene*. The apparatus is similar to that described in a previous volume.[12] It consists of a "Vycor" glass reaction tube, 60 cm. long by 25 mm. outside diameter, mounted vertically in an electric furnace about 35 cm. long (Note 9). Attached to the top of the tube is a graduated dropping funnel. A thermocouple well extending to the center of the heated section is inserted through the bottom of the tube. The heated section of the tube is packed with quartz tubing (8 mm. outside diameter), cut into 0.5-cm. lengths, and held in place by indentations in the tube. Ten centimeters from the bottom of the tube is a side arm leading successively to two traps cooled with solid carbon dioxide and acetone, an inlet tube for nitrogen, a manometer, and a vacuum pump.

The system is evacuated to a pressure of 5–10 mm., and the tube is heated to 700°, measured at the center of the heated zone. 2-Chloro-2,3,3-trifluorocyclobutyl acetate is admitted at the rate of 10–20 g. per hour. From 70 g. (0.35 mole) of the cyclobutyl acetate there is obtained 62–68 g. of mixed solid and liquid condensate (Note 10). Fractionation through a 30-cm. column packed with glass helices affords 30–35 g. (60–70%) of 1-chloro-1,4,4-trifluorobutadiene (Note 11), b.p. 50–51°, n_D^{25} 1.3870; 18–22 g. of acetic acid; and 7–18 g. of recovered 2-chloro-2,3,3-trifluoro-cyclobutyl acetate (Note 12).

2. Notes

1. The exact structure of the cyclobutane is not known. Any of the possible isomers would undergo pyrolysis to give 1-chloro-1,4,4-trifluorobutadiene. 2-Chloro-2,3,3-trifluorocyclobutyl acetate is now favored as the structure of the cycloadduct rather than 3-chloro-2,2,3-trifluorocyclobutyl acetate as originally proposed.[3] The basis for this preference is mass spectral data. Ions of m/e 64 [$(CF_2{=}CH_2)^+$, relative abundance 1.2%] and 138 [$(CFCl{=}CHOCOCH_3)^+$, relative abundance 0.96%] were much more abundant than ions of m/e 80 [$(CFCl{=}CH_2)^+$, relative

abundance 0.10%] and 122 $[(CF_2=CHOCOCH_3)^+$, relative abundance 0.026%].

2. The purpose of the terpene is to inhibit polymerization of the fluoroölefin. Terpenes that are effective include dipentene and terpinolene.

3. Ordinary commercial-grade vinyl acetate is redistilled. One gram of hydroquinone per 100 g. of vinyl acetate is added to inhibit polymerization of the latter, which is then stored at 0–4° until needed.

4. A shaker tube equipped with a 1200-atm. rupture-disk assembly was used by the submitters. The checkers used a 1270-ml. stainless steel rocking autoclave fitted with a thermocouple well that extended into the reaction mixture and a stainless steel 5000-p.s.i. rupture disk. The agitation rate was 58 cycles per second. Attempts to use a magnetically stirred autoclave were unsuccessful.

5. Chlorotrifluoroethylene is available in 1-lb. and 5-lb. cylinders from the Matheson Company, East Rutherford, New Jersey.

6. The checkers found the reaction mixture to be dark but not viscous. In experiments in which a magnetically stirred autoclave was used, dark viscous reaction mixtures were obtained, but no product.

Difficulties encountered by the checkers when they used a magnetically stirred autoclave led the submitters to re-examine the reaction. It was found that certain batches of vinyl acetate gave very poor yields. In these cases, 27–37% yields were obtained by heating to 175° in 1 hour followed by heating at 175° for 16 hours.

7. The quantity of fore-run depends on the amount of polymerization of vinyl acetate. Distillation of the product through a packed column goes more smoothly, with less heat having to be applied to the distillation flask, if the product has been separated from high-boiling material by a quick preliminary distillation.

8. Up to 25% by weight of the product is ethylidene diacetate. The diacetate can be detected by gas chromatographic analysis using a column of the diglyceride of 6,6,6-trifluorohexanoic acid on firebrick at 120°.[10] The checkers obtained yields ranging from 24% in a 0.75-scale experiment to 47% on a three-fold in-

crease in scale.

9. A standard tube furnace such as the 120-volt "Multiple Unit" electric furnace manufactured by the Hevi Duty Electric Company was used.

10. Care must be taken to prevent plugging of the first cold trap by solid acetic acid because the back-pressure produced leads to greatly reduced yields and appreciable carbonization. Should plugging occur, the cooling bath is removed and the plug is melted with warm acetone. Some diene will distil into the second trap during this process.

11. 1-Chloro-1,4,4-trifluorobutadiene is a mixture of equal amounts of *cis* and *trans* isomers. This has been demonstrated by gas chromatographic analysis of the mixture on a packed column of high efficiency using Dow-Corning silicone 703 oil or 200 oil on firebrick, or on a capillary gas chromatographic column using squalane as the partitioning liquid.

12. The checkers did not isolate any recovered 2-chloro-2,3,3-trifluorocyclobutyl acetate.

3. Methods of Preparation

The procedure for chlorotrifluorocyclobutyl acetate [3] is a modification of one used by Coffman, Barrick, Cramer, and Raasch [4] for the preparation of tetrafluorocyclobutanes from tetrafluoroethylene.

The method for the pyrolysis of chlorotrifluorocyclobutyl acetate to chlorotrifluorobutadiene is that of Anderson, Putnam, and Sharkey.[3]

4. Merits of the Preparation

The synthesis of 2-chloro-2,3,3-trifluorocyclobutyl acetate illustrates a general method of preparing cyclobutanes by heating chlorotrifluoroethylene, tetrafluoroethylene, and other highly fluorinated ethylenes with alkenes. The reaction has recently been reviewed.[11] Chlorotrifluoroethylene has been shown to form cyclobutanes in this way with acrylonitrile,[5] vinylidene chloride,[6] phenylacetylene,[7] and methyl propiolate.[3] A far greater number of cyclobutanes have been prepared from tetrafluoroethylene and alkenes;[4,11] *when tetrafluoroethylene is used, care must be exercised because of the danger of explosion.* The fluorinated cyclobutanes can be converted to a variety of cyclobutanes, cyclobutenes, and butadienes.

The synthesis of chlorotrifluorobutadiene illustrates a general method that has been used to make tetrafluorobutadiene [3, 8] and substituted fluorodienes.[3, 8, 9] The same procedure can be used to transform fluorocyclobutenes and chlorofluorocyclobutenes to the isomeric dienes; 2-methyl-1,1,4,4-tetrafluorobutadiene, 2-chloro-1,1,4,4-tetrafluorobutadiene, and 1-chloro-1,4,4-trifluoro-2-phenylbutadiene have been made thus.[3]

1. Contribution No. 597 from the Central Research Department, Experimental Station, E. I. du Pont de Nemours and Company, Wilmington, Delaware.
2. Upjohn Co., Kalamazoo, Michigan.
3. J. L. Anderson, R. E. Putnam, and W. H. Sharkey, *J. Am. Chem. Soc.,* 83, 382 (1961).
4. D. D. Coffman, P. L. Barrick, R. D. Cramer, and M. S. Raasch, *J. Am. Chem. Soc.,* 71, 490 (1949).
5. A. L. Barney and T. L. Cairns, *J. Am. Chem. Soc.,* 72, 3193 (1950).
6. M. S. Raasch, R. E. Miegel, and J. E. Castle, *J. Am. Chem. Soc.,* 81, 2678 (1959).
7. E. J. Smutny and J. D. Roberts, *J. Am. Chem. Soc.,* 77, 3420 (1955).
8. J. L. Anderson and K. L. Berry, U.S. Patent pending.
9. J. L. Anderson, U.S. Patent 2,754,323 (1956) [C.A., 51, 2026 (1957)].
10. J. F. Harris, Jr., and F. W. Stacey, *J. Am. Chem. Soc.,* 83, 844 (1961).
11. J. D. Roberts and C. M. Sharts, *Org. Reactions,* 12, 1 (1962).
12. R. E. Benson and R. C. McKusick, *Org. Syntheses,* Coll. Vol. 4, 746 (1963).

3-CHLORO-2,2,3-TRIFLUOROPROPIONIC ACID

(Propionic acid, 3-chloro-2,2,3-trifluoro-)

$$CFCl{=}CF_2 + NaCN + 2H_2O \rightarrow ClCHFCF_2CO_2Na + NH_3$$

$$ClCHFCF_2CO_2Na + H_2SO_4 \rightarrow ClCHFCF_2CO_2H + NaHSO_4$$

Submitted by D. C. ENGLAND and L. R. MELBY.[1]
Checked by MAX TISHLER and W. J. JONES.

1. Procedure

Caution! This is a strongly exothermic reaction. The reaction should be carried out in a hood. A protective shield should be placed between the operator and the reaction bottle.

A modified Parr low-pressure hydrogenation apparatus is used for this preparation.[2]   The bottle is fitted with a two-holed rubber stopper through which is passed a thermocouple well made of 5-mm. glass tubing and a gooseneck made of 8-mm. heavy-walled glass tubing. The thermocouple well extends into the bottle within about 2 cm. of the bottom. The gooseneck extends 1 cm. into the bottle, and the other end is connected directly to a manifold system with heavy-walled pressure tubing using screw clamps. Also attached to the manifold system through needle valves are a vacuum line, a storage cylinder of chlorotrifluoroethylene, a pressure gauge, and a bleed line.

The bottle is charged with 52 g. (1.0 mole) of 95% sodium cyanide, 100 ml. of water, and 100 ml. of acetonitrile, giving a two-phase liquid system. After the bottle is clamped in the metal cage of the shaking apparatus, the bottle is evacuated and filled to a gauge pressure of 10 lb. (0.68 atm.) with chlorotrifluoroethylene. This procedure is repeated twice to purge the system of air. Finally, the bottle is pressured to 40 lb. (2.7 atm.), chlorotrifluoroethylene leaving the valve open to maintain this pressure (Note 1). The thermocouple is fitted into place and shaking is started. The temperature steadily increases, and in 10–15 minutes it is about 75° and rising more rapidly. It is kept at 75–80° by cooling and/or slowing the rate of chlorotrifluoroethylene addition. *Caution! Careful control of the reaction is mandatory. If the reaction rate cannot be controlled, it is imperative to shut the gas addition valve and to stop the agitation immediately.* Cooling is accomplished by packing ice inside the wire cage holding the bottle and/or pouring ice water on the bottle. When cooling is used, it is possible to complete the reaction in 1 hour. When the rate of chlorotrifluoroethylene absorption is negligible and the temperature is about 30°, pressure is released through the bleed line and the bottle is removed. The gain in weight is 130–135 g. (about 1.1 moles).

The product is a dark-colored solution which is poured slowly into sulfuric acid (100 ml. of concentrated sulfuric acid in 100 ml. of water) with cooling (Note 2). *This operation should be conducted in a hood because some hydrogen cyanide may be evolved from unreacted sodium cyanide.* The mixture is then extracted four times with 100-ml. portions of ether (Note 3). The first 100 ml. of ether yields about 260 ml. of organic material, and succeeding portions about 95 ml. each. The extract is dried over about 20 g. of magnesium sulfate and distilled (Note 4). After removal of low-boiling materials, there is obtained 124–128 g. (76–79%) of crude 3-chloro-2,2,3-trifluoropropionic acid, b.p. 70–85°/30 mm.; the bulk of this fraction distills at 83°/30 mm., n_D^{25} 1.3708–1.3717. This product is of sufficient purity for most purposes (Note 5). About 25–30 g. of high-boiling residue remains which is chiefly chlorotrifluoropropionamide (Note 6).

2. Notes

1. The chlorotrifluoroethylene cylinder is fitted with a pressure-reduction valve which is set at a maximum pressure of 40 lb. (2.7 atm.).

2. The amount of product was not changed by cooling, but the acidification reaction is very exothermic and cooling is necessary before extraction with ether.

3. The checkers encountered troublesome emulsions in the extraction step. It was found helpful to filter the acidified reaction mixture through a sintered-glass funnel before ether extraction.

4. A precision distillation column is not necessary. The submitters have used an 80-cm. spinning-band column with a 10 mm. inside diameter or a 25-cm. helix-packed column of about 10 mm. inside diameter.

5. The neutral equivalent of this material is 169–173 (theory is 162.5). It is readily purified by mixing it thoroughly with 10% of its weight of phosphorus pentoxide and redistilling. The purified acid boils at 82–83°/30 mm. (159–160°/760 mm.), n_D^{25} 1.3695, N.E. 163.5. Recovery from the crude is 92%.

6. The higher-boiling material is largely 3-chloro-2,2,3-trifluoropropionamide, and some fractions crystallize at room temperature. The amide can be purified by redistillation, b.p. 111–112°/30 mm. (198–200°/760 mm.) or recrystallization from carbon tetrachloride. The pure amide melts at 41–42°. The amount of amide formed is not changed by running the above charge at a lower temperature. However, when the amount of water is reduced to 50 ml. and the temperature kept below 50°, the yield of crude acid is 38% and of amide, 21%. When the same charge is run at 80° maximum temperature, the yield of

crude acid is 78% and of amide, 7%.

3. Methods of Preparation

3-Chloro-2,2,3-trifluoropropionic acid has been prepared by permanganate oxidation of 3-chloro-2,2,3-trifluoropropanol [3] which is one of the telomerization products of chlorotrifluoroethylene with methanol. The present procedure is a modification of one reported earlier [4] and is undoubtedly the method of choice for making propionic acids containing 2–4 fluorine atoms, i.e., 2,2,3,3-tetrafluoropropionic acid, 3,3-dichloro-2,2-difluoropropionic acid, and 3-bromo-2,2,3-trifluoropropionic acid. When preparing 2,2,3,3-tetrafluoropropionic acid from tetrafluoroethylene, it is desirable to use an additional 50 ml. of acetonitrile and externally applied heat to initiate the reaction.

1. Contribution No. 554 from the Central Research Department, Experimental Station, E. I. du Ponte de Nemours and Company.
2. R. Adams and V. Voorhees, *Org. Syntheses,* Coll. Vol. 1, 66 (Fig. 6) (1941). Later models of this apparatus use a larger-mouthed bottle which is more easily adapted to this preparation.
3. R. M. Joyce, U. S. pat. 2,559,628 (1951) [C.A., 46, 3063 (1952)].
4. D. C. England, L. R. Melby, and R. V. Lindsey, Jr., *J. Am. Chem. Soc.,* 80, 6422 (1958).

CHOLANE-24-AL

Submitted by JOHN G. MOFFATT [1]
Checked by ROBERT FAIRWEATHER and RONALD BRESLOW

1. Procedure

Cholane-24-ol (1.033 g., 3 mmoles) (Note 1) is dissolved by gentle warming in 10 ml. of anhydrous benzene (Note 2) in a

50-ml. flask, and 10 ml. of rigorously dried dimethyl sulfoxide (Note 3) is added. To the clear solution are added 0.24 ml. (3.0 mmoles) of anhydrous pyridine (Note 4), 0.12 ml. (1.5 mmoles) of distilled trifluoroacetic acid, and 1.85 g. (9 mmoles) of dicyclohexylcarbodiimide (Note 5), in that order. The flask is tightly stoppered and left at room temperature for 18 hours (Note 6). Benzene (30 ml.) is then added, and the crystalline dicyclohexylurea is removed by filtration (Note 7) and washed with benzene. The combined filtrates and washings are extracted three times with 50-ml. portions of water (Note 8) to remove the dimethyl sulfoxide. The organic layer is dried with sodium sulfate and evaporated to dryness under reduced pressure. There is obtained 2.12 g. of syrup which partially crystallizes. Thin-layer chromatography of this material (Note 9) shows a very intense spot of cholane-24-al, traces of starting material, and two compounds near the solvent front as well as excess carbodiimide (Note 9).

The crude product is dissolved in benzene-hexane (1:1) and applied to a column containing 125 g. of silicic acid (Note 10). Elution with the same solvent gives traces (less than 5 mg. each) of the two fast-moving components in fractions 2 and 4 (125-ml. fractions) and chromatographically pure cholane-24-al in fractions 5–8 (Note 11). Evaporation of the pooled fractions yields 870 mg. (84%) of the pure crystalline aldehyde, m.p. 102–104°. Recrystallization from 5 ml. of acetone raises the melting point to 103–104° (Note 12).

The compound gives a crystalline 2,4-dinitrophenylhydrazone, m.p. 163–164°, from ethanol.

2. Notes

1. Available from Aldrich Chemical Company.

2. Dried by storage over calcium hydride.

3. Dried by distillation under reduced pressure and storage for several days over Linde Molecular Sieves Type 4A.

4. Dried by distillation from, and storage over, calcium hydride.

5. The dry, crystalline material may be obtained from Aldrich Chemical Company. If the reagent is at all oily at room temper-

ature, it should be distilled under reduced pressure, b.p. 140° (5 mm.).

6. Crystalline dicyclohexylurea (m.p. 234°) starts to separate after a short time. The checkers found a decrease in yield if this is allowed to run longer; a yield of 54% was found in a 22-hour reaction time.

7. Roughly 0.6–0.8 g. of the urea is usually obtained, m.p. 232–234°. The excess dicyclohexylcarbodiimide remains in the benzene. The oxidation is generally less satisfactory if less than 2.5 molar equivalents of carbodiimide is used.

8. Some further dicyclohexylurea tends to separate at the interface during the first extraction, and a clean separation of the layers near the interface is aided by mild centrifugation.

9. On Merck Silica G using benzene as the solvent and 5% ammonium molybdate in 10% sulfuric acid followed by brief heating at 150° to develop the spots. Under these conditions cholane-24-al has an R_f of 0.76 while cholane-24-ol has R_f 0.19; dicyclohexylcarbodiimide streaks between 0.3 and 0.5.

10. Merck silica gel with 0.05–0.20 mm. particles obtained from Brinkman Instruments Inc. and packed in a 3-cm. diameter column under benzene-hexane (1:1).

11. The fractions were examined by thin-layer chromatography of 25–50 μl. aliquots as in Note 9. The checkers found that cholane-24-al is found in fractions 4–7.

12. The compound is very soluble in most organic solvents. In order to get a high recovery, it is necessary to complete the crystallization in the deep freeze. From aqueous ethanol the aldehyde crystallized in high yield as the hemihydrate, m.p. 95°.

3. Methods of Preparation

Cholane-24-al has not been previously synthesized by other methods.

4. Merits of the Preparation

The oxidation reaction described is a very general one that may be used for the preparation of both aldehydes and ketones [2] in high yield. The reaction conditions are extremely mild and only slightly acidic, thus allowing the preparation of otherwise very unstable compounds.[2] Of particular merit is the fact that the oxidation of primary alcohols stops selectively at the aldehyde and gives no traces of acidic products. Among the many different acids that have been examined as the proton source for this type of reaction,[2] pyridinium trifluoroacetate consistently gives the best results.

1. Syntex Institute for Molecular Biology, Palo Alto, California.
2. K. E. Pfitzner and J. G. Moffatt, *J. Am. Chem. Soc.*, **87**, 5661, 5670 (1965).

CHOLESTANYL METHYL ETHER

(Cholestane, 3β-methoxy-)

Submitted by M. Neeman [1,2] and William S. Johnson.[1]
Checked by F. Kaplan and John D. Roberts.

1. Procedure

To a solution of 0.200 g. (0.515 mmole) of dry dihydrocholesterol (Note 1) in 10 ml. of methylene chloride contained in a 50-ml. Erlenmeyer flask is added 0.3 ml. of a catalyst stock solution containing 0.0016 ml. (0.018 mmole) of concentrated fluoboric acid (Note 2) in 3:1 anhydrous diethyl ether-methylene chloride (Note 3). The solution is swirled, and a 0.45M solution of diazomethane (Note 4) in dry methylene chloride is added from a buret (Note 5) at a rate of about 2 ml. per minute. The yellow color of diazomethane disappears rapidly on contact with the reaction mixture and nitrogen is vigorously evolved. When about 3 ml. of diazomethane solution has been added, the reaction becomes sluggish. The yellow color persists for several minutes after the total amount of 3.9 ml. of diazomethane solution (1.76 mmoles) has been added (Note 6). After 1 hour the reaction mixture is filtered to remove a small amount of amorphous poly-methylene, which is washed with methylene chloride. The wash-ings are combined with the methylene chloride solution, washed with 5 ml. of saturated aqueous sodium bicarbonate, with three 5-ml. portions of water, and dried over anhydrous sodium sulfate. The solvent is removed on a steam bath in a stream of nitrogen and finally at reduced pressure. The crystalline residue of 0.207 g. (Note 7) is recrystallized in a 10-ml. conical flask from 1 ml. of acetone. When the flask has cooled to room temperature, 0.5 ml. of methanol is added, and the flask is chilled to $+2°$ for 2 hours. The crystals are collected on a tared Hirsch funnel of 40-mm. diameter, washed on the funnel with two 0.5-ml. portions of ice-cold methanol, and dried for 2 hours at $40°/2$ mm. The first crop of cholestanyl methyl ether thus obtained forms large color-less glistening plates, m.p. 85.5–86° (Note 8). An additional 0.002 g. of pure methyl ether adheres to the flask and spatula and is collected by washing with acetone. The total first crop material (0.197 g.) represents a 95% yield of methyl ether (Note 9).

2. Notes

1. Satisfactory material of melting point 143–143.5° is prepared as already described,[3] and dried for 2 hours at $110°/2$ mm.

2. Commercial 50% fluoboric acid is evaporated at 50–60°/5 mm. to afford a residue of about $11N$ total acidity, which is satisfactory for use as a catalyst.

3. The catalyst stock solution should be freshly prepared by placing 19 ml. of anhydrous diethyl ether in a 25-ml. volumetric flask cooled to 0° and adding 0.133 ml. of concentrated fluoboric acid (Note 2). The volume is made up to 25 ml. with methylene chloride.

4. Diazomethane solution in dry methylene chloride may be prepared from N-nitroso-N-methyl-N′-nitroguanidine by a pro-cedure based on McKay's method.[4] Methylene chloride is sub-stituted for diethyl ether used in the original procedure. A satisfactory solution of diazomethane is obtained, without distil-

ling, by separating the methylene chloride layer from the reaction mixture, drying it for 2 hours over potassium hydroxide pellets, and decanting through a funnel plugged loosely with cotton. The diazomethane solution is kept in a loosely stoppered test tube immersed in a Dewar flask containing Dry Ice during the drying period and prior to use. *All handling of the highly toxic diazomethane should be done in an efficiently exhausted hood.* Attention is called to other precautions;[5] see also pp. 16–17.

Rigorous drying and exclusion of moisture are not necessary. The concentration of diazomethane solutions is determined by analysis,[5] using about 0.12 g. of benzoic acid per milliliter of solution and assuming a concentration of about $0.8M$ as in McKay's method.[4b] Solutions approximately $0.45M$ are obtained by appropriate dilution.

5. Burets with ground-glass stopcocks should not be used, as leaking is caused by polymethylene formed preferentially on the ground surfaces. A buret such as "Ultramax F and P," having a stopcock of plastic material, is satisfactory. The buret should be filled immediately before commencement of the reaction to keep the diazomethane solution cool and thus to minimize polymerization. The technique used is very similar to that of a titration, and a number of methylations of prepared batches can be quickly performed with one filling of the buret.

6. Addition of a drop of catalyst stock solution after addition of diazomethane solution is complete causes rapid disappearance of the yellow color. The yield is not affected.

7. The crude reaction product is slightly yellow and has a very faint ammoniacal odor. It may be dissolved in acetone; on slow evaporation to dryness, the solution leaves large glistening transparent plates of good-quality cholestanyl methyl ether, m.p. 83–85°.

8. All melting points are corrected for stem exposure. Reported [6] melting point 83°.

9. The mother liquor may be evaporated to dryness and the slightly colored residue recrystallized in a 3-ml. conical flask from 6 drops of 1:1 acetone-methanol. The resulting large plates are easily transferred to a small Hirsch funnel and washed with 5 drops of methanol. This second crop of colorless methyl ether amounts to 0.010 g., m.p. 78.5–79.5°.

3. Methods of Preparation

Cholestanyl methyl ether has been prepared by catalytic hydrogenation of cholesteryl methyl ether [6, 7] and of cholest-4-en-3-one in methanolic hydrobromic acid,[7] and by methylation of cholestanol with methyl iodide in the presence of "activated" silver oxide and sodium hydroxide.[8] The reported [9] formation of cholestanyl methyl ether from *epi*cholestanol in 96% yield by refluxing with "molecular" potassium in benzene and subsequent treatment with methyl iodide stands unconfirmed.[10] Methanolysis of *epi*cholestanyl tosylate afforded a 23% yield of cholestanyl methyl ether.[11] The procedure described here,[12] with slight changes in the molar proportions of the reactants, also gave a 98% yield of *epi*cholestanyl methyl ether from *epi*cholestanol, and a 95% yield of cholesteryl methyl ether from cholesterol.

4. Merits of Preparation

The present procedure is illustrative of the utility of the general method for preparation of methyl ethers from diazomethane and alcohols with fluoboric acid as catalyst.[12]

1. Department of Chemistry, University of Wisconsin, Madison, Wis.
2. On leave from Technion Israel Institute of Technology, Haifa, Israel.
3. W. F. Bruce and J. O. Ralls, *Org. Syntheses,* Coll. Vol. 2, 191 (1943).
4. *(a)* A. F. McKay, *J. Am. Chem. Soc.,* 70, 1974 (1948); *(b)* A. F. McKay et al., *Can. J. Research,* 28B, 683 (1950).
5. F. Arndt, *Org. Syntheses,* Coll. Vol. 2, 165 (1943).
6. Th. Wagner-Jaurregg and L. Werner, *Z. Physiol. Chem.* 213, 119 (1932).
7. J. C. Babcock and L. F. Fieser, *J. Am. Chem. Soc.,* 74, 5472 (1952).
8. J. L. Dunn, I. M. Heilbron, R. F. Phipers, K. M. Samant, and F. S. Spring, *J. Chem. Soc.,* 1576 (1934).
9. J. H. Benyon, I. M. Heilbron, and F. S. Spring, *J. Chem. Soc.,* 406 (1937).
10. J. R. Lewis and C. W. Shoppee, *J. Chem. Soc.,* 1375 (1955), have obtained *epi*cholestanyl methyl ether under these conditions.
11. H. R. Nace, *J. Am. Chem. Soc.,* 74, 5937 (1952).
12. M. C. Caserio, J. D. Roberts, M. Neeman, and W. S. Johnson, *J. Am. Chem. Soc.,* 80, 2584 (1958); M. Neeman, M. C. Caserio, J. D. Roberts, and W. S. Johnson, *Tetrahedron,* 6, 36 (1959).

CINNAMYL BROMIDE*

(Benzene, 3-bromopropenyl-)

$$(C_6H_5)_3P \ + \ Br_2 \ \longrightarrow \ (C_6H_5)_3PBr_2$$

$$\xrightarrow{(C_6H_5)_3PBr_2}$$

$$+ \ (C_6H_5)_3PO \ + \ HBr$$

Submitted by JOHN P. SCHAEFER, J. G. HIGGINS,
and P. K. SHENOY [1]
Checked by R. BRESLOW and J. T. GROVES

1. Procedure

A 1-l., three-necked, round-bottomed flask equipped with a Trubore stirrer, a pressure-equalizing dropping funnel, and a reflux condenser with a drying tube is charged with 350 ml. of acetonitrile (Note 1) and 106.4 g. (0.41 mole) of triphenylphosphine (Note 2). The flask is cooled in an ice-water bath (Note 3), and 64 g. (0.40 mole) of bromine is added dropwise over a period of *ca.* 15–20 minutes (Notes 4 and 5). The ice-water bath is removed, and a solution of 54 g. (0.40 mole) of cinnamyl alcohol in 50 ml. of acetonitrile is added in portions over a period of 5–10 minutes with continued stirring (Note 6). The solvent is removed by distillation with the use of a water aspirator (30–40 mm.) and an oil bath until the bath temperature reaches 120°. The water aspirator is replaced by a vacuum pump and the water-cooled condenser with an air condenser, and the distillation is continued with rapid stirring (Notes 7, 8, and 9). Most of the product (Note 10) distills at 91–98° (2–4 mm.), and about 59 g. of product crystallizes in the receiving flask (63–75% yield) (Note 11).

The product is dissolved in 200 ml. of ether, and the solution is washed with 75 ml. of saturated aqueous sodium carbonate, dried over anhydrous magnesium sulfate, and distilled to give 47–56 g. (60–71%) of product, b.p. 66–68° (0.07 mm.), 84–86° (0.8 mm.); m.p. 29°.

2. Notes

1. The acetonitrile was distilled from phosphorus pentoxide.

2. Triphenylphosphine was obtained from M and T Chemicals, Inc., and used without further purification.

3. The triphenylphosphine is only partially dissolved.

4. If a slight excess of bromine persists after addition, a small amount of triphenylphosphine should be added until the color of bromine disappears.

5. The solid triphenylphosphine disappears, but at the same time the adduct, $(C_6H_5)_3PBr_2$, precipitates as a white solid.

6. This addition is mildly exothermic, and the temperature rises to 50–60°. All the precipitate should dissolve at this point; warming by external heat may be necessary.

7. To protect the vacuum pump from damage a dry ice-acetone trap and two liquid nitrogen traps are necessary to condense and solidify the hydrogen bromide evolved.

8. The receiving flask is placed in an ice-water bath.

9. The distillation is continued until the triphenylphosphine oxide solidifies and no more product distills. The oil bath is maintained at 130–140° during the distillation.

10. Some product is carried over by the hydrogen bromide in the initial stages of the distillation.

11. When this distillation was replaced by a procedure in which the acetonitrile was removed with a rotary evaporator and steam bath, and the product was extracted from the triphenylphosphine oxide with small portions of acetonitrile totaling *ca.* 250 ml., the checkers obtained an improved yield (79%) of cinnamyl bromide.

3. Methods of Preparation

Cinnamyl bromide has been prepared from cinnamyl alcohol by the action of hydrogen bromide in cold acetic acid [2] and of phosphorus tribromide in boiling benzene.[3] It has also been prepared by the action of N-bromosuccinimide on 3-phenyl-propene [4] and on 1-phenylpropene.[5]

4. Merits of the Preparation

The method described is general for converting alcohols to alkyl halides and is stereospecific.[6]

1. Department of Chemistry, The University of Arizona, Tucson, Arizona 85721.
2. J. von Braun and Z. Kohler, *Ber.*, **51**, 79 (1918).
3. M. Gredy, *Bull. Soc. Chim. France*, [5] **3**, 1098 (1936).
4. M. Lora-Tomayo, F. Martin-Panizo, and R. P. Ossorio, *J. Chem. Soc.*, 1418 (1950); E. A. Braude and E. S. Waight, *J. Chem. Soc.*, 1116 (1952).
5. O. A. Orio, *Rev. Fac. Cienc. Quim., Univ. Nacl. La Plata*, **30**, 21 (1957) [*C.A.*, **54**, 12062 (1960)].
6. L. Horner, H. Oediger, and H. Hoffmann, *Ann.*, **626**, 26 (1959); G. A. Wiley, R. L. Hershkowitz, B. M. Rein, and B. C. Chung, *J. Am. Chem. Soc.*, **86**, 964 (1964); J. P. Schaefer and D. S. Weinberg, *J. Org. Chem.*, **30**, 2635 (1965).

COUMARONE *

(Benzofuran)

A. (reaction of o-hydroxybenzaldehyde with $ClCH_2COOH$ / $NaOH$ to give o-formylphenoxyacetic acid)

B. (o-formylphenoxyacetic acid with Ac_2O, $HOAc$ / $NaOAc$ to give benzofuran) $+ CO_2 + H_2O$

Submitted by ALBERT W. BURGSTAHLER and LEONARD R. WORDEN [1]
Checked by WAYLAND E. NOLAND, WILLIAM E. PARHAM, and CAROL WONG

1. Procedure

A. *o-Formylphenoxyacetic acid.* A solution of 80.0 g. (2 moles) of sodium hydroxide pellets in 200 ml. of distilled water is added to a mixture of 106 ml. (122 g., 1 mole) of salicylaldehyde (Note 1), 94.5 g. (1 mole) of chloroacetic acid (Notes 1 and 2), and 800 ml. of water. The mixture is stirred slowly and heated to boiling. The resulting black solution (Note 3) is heated under reflux for 3 hours (Note 4). The solution is acidified with 190 ml. of concentrated hydrochloric acid (sp. gr. 1.19) and is steam-distilled to remove unchanged salicylaldehyde (40.0–40.5 g.) (Note 5). The residual acidic mixture is cooled to 20°, and the precipitated product is collected on a Büchner funnel and rinsed with water. The light tan solid when dry weighs 99–100 g. (82–83% based on recovered salicylaldehyde), m.p. 130.5–133.0° (Note 6).

B. *Coumarone.* A mixture of 90.0 g. (0.5 mole) of crude (Note 7), dry o-formylphenoxyacetic acid, 180 g. of anhydrous, powdered sodium acetate, 450 ml. of acetic anhydride, and 450 ml. of glacial acetic acid (Note 8) in a 2-l. flask is heated under gentle reflux with stirring for 8 hours. The hot black solution (total volume *ca.* 1.2 l.) (Note 3) is poured into 2.5 l. of ice water and extracted with one 600-ml. portion of ether (Note 9). The ether layer is washed with one 600-ml. portion of water and then with several portions of cold dilute 5% sodium hydroxide solution (Note 10) until the aqueous layer is basic. The ether layer is washed successively with water and saturated sodium chloride solution and is partially dried over anhydrous granular sodium sulfate. The ether is removed at water-bath temperature and the product is distilled, b.p. 166.5–168.0° (735 mm.) or 97.5–99.0° (80 mm.). The water-white benzofuran weighs 37.5–40.0 g. (63.5–67.8%, 52–56% overall from salicylaldehyde), n^{20}D 1.5672; λ_{max} 245 (log ϵ 4.08), 275 (3.45), and 282 mμ (3.48).

2. Notes

1. Matheson, Coleman and Bell practical grade material was used.

2. The yield is not increased by use of bromoacetic acid or 2 moles of chloroacetic acid and an additional mole of sodium hydroxide.

3. At no time did the checkers observe a black solution. The color of the solution changed from yellow to red-brown.

4. The yield is not increased by longer reflux periods.

5. Removal of unchanged salicylaldehyde by steam distillation (followed conveniently by testing the distillate with 2,4-dinitrophenylhydrazine reagent) provides a product sufficiently pure for use in the next step. Also, the recovered salicylaldehyde can be used again without further purification.

6. Three crystallizations of 36 g. of the crude o-formylphenoxyacetic acid from 360 ml. of water with 10 g. of activated carbon give 18 g. of glistening colorless plates, m.p. 133.0°–133.5°.

7. Use of purified o-formylphenoxyacetic acid increases the yield in this step by only 11%.

8. If no acetic acid is used, benzofuran is formed in only 30–31% yield, and coumarilic acid, m.p. 194–196°, is isolated in about 45% yield.

9. An additional extraction does not increase the yield appreciably.

10. About 250 ml. of this solution is required.

3. Methods of Preparation

o-Formylphenoxyacetic acid has been prepared previously in 46% yield by alkylation of salicylaldehyde with chloroacetic acid [2, 3] and in unspecified yield by alkylation with ethyl bromoacetate followed by hydrolysis.[4]

Benzofuran is found in coal tar.[5] It has been prepared in 40–46% overall yield from coumarin by bromination, conversion of the resulting 3,4-dibromocoumarin to coumarilic acid, and then decarboxylation,[6] and also by passage of coumarin vapor through an iron tube at 860°.[7] The method given here is a variation of that described by Rössing,[2] who omitted the addition of acetic acid (see Note 8 above). Benzofuran also has been prepared by the cyclization of ω-chloro-o-hydroxystyrene [8] or phenoxyacetaldehyde [9] in unspecified low yields and by the cyclization of phenoxyacetaldehyde diethyl acetal in 9% yield.[10] High-temperature catalytic dehydrocyclization of o-ethylphenol affords benzofuran in as much as 59% yield after recycling unchanged o-ethylphenol.[11]

4. Merits of the Preparation

Although the high-temperature catalytic dehydrocyclization of o-ethylphenol [11] gives benzofuran in fair yield, these conditions are not convenient in the laboratory and cannot be applied easily to functionally substituted o-formylphenoxyacetic acids. The other methods of preparation give unsatisfactory yields, are unnecessarily lengthy, or require expensive starting materials. The method of Rössing,[2] on which the present procedure is based, gives good yields in its original form only in the case of o-acylphenoxyacetic acids; o-formylphenoxyacetic acids give principally the corresponding coumarilic acids. Often these can be decarboxylated only in very poor yield.[12] In the preparation described here, benzofuran is obtained directly in fair overall yield from readily available and inexpensive starting materials without the necessity of a separate decarboxylation step.

1. Department of Chemistry, University of Kansas, Lawrence, Kansas. This investigation was supported in part by a Public Health Service Fellowship, GPM-13, 681-03, from the Division of General Medical Sciences, U.S. Public Health Service.
2. A. Rössing, Ber., 17, 2988 (1884).
3. H. Cajar, Ber., 31, 2803 (1898); A. Zubrys and C. O. Siebenmann, Can. J. Chem., 33, 11 (1955).
4. H. Dumont and St. v. Kostanecki, Ber., 42, 911 (1909).
5. G. Kraemer and A. Spilker, Ber., 23, 78 (1890).
6. R. Fittig and G. Ebert, Ann., 216, 162 (1883); A. L. Mndzhoian, A. A. Aroian, and N. Kh. Khachatrian, "Synthesis of Heterocyclic Compounds," Vols. 3 and 4, Consultants Bureau, New York, 1960, p. 83.
7. N. A. Orlow and W. W. Tistschenko, Ber., 63, 2948 (1930).
8. G. Komppa, Ber., 26, 2968 (1893).
9. R. Stoermer, Ber., 30, 1700 (1897).
10. R. Stoermer, Ann., 312, 237 (1900).
11. C. Hansch and G. Helmkamp, J. Am. Chem. Soc., 73, 3080 (1951), and earlier papers cited therein; B. B. Corson, H. E. Tiefenthal, J. E. Nickels, and W. J. Heintzelman, J. Am. Chem. Soc., 77, 5428 (1955).
12. Cf. R. Aneja, S. K. Mukerjee, and T. R. Seshadri, Tetrahedron, 2, 203 (1958).

γ-CROTONOLACTONE

(Δ$^{\alpha,\beta}$-Butenolide)

$$\underset{O}{\overset{CH_2-CH_2}{\underset{CH_2\quad CO}{\big|\qquad\big|}}} + Br_2 \xrightarrow{P} \underset{O}{\overset{CH_2-CHBr}{\underset{CH_2\quad CO}{\big|\qquad\big|}}} \xrightarrow{(C_2H_5)_3N} \underset{O}{\overset{CH=CH}{\underset{CH_2\quad C=O}{\big|\qquad\big|}}}$$

Submitted by CHARLES C. PRICE and JOSEPH M. JUDGE [1]
Checked by RICHARD F. ATKINSON and E. J. COREY

1. Procedure

Caution! Contact with α-bromo-γ-butyrolactone can cause severe eye and skin irritation. This preparation should be carried out in a good hood, and the operator should wear protective goggles and rubber gloves.

A. *α-Bromo-γ-butyrolactone.* In a 1-l., three-necked, round-bottomed flask equipped with a dropping funnel, sealed stirrer, and an efficient reflux condenser (Note 1) are placed 100 g. (1.16 moles) of redistilled γ-butyrolactone and 13.4 g. (0.43 g. atom) of red phosphorus. Over a half-hour interval, 195 g. (66.5 ml., 1.22 moles) of bromine is added, the mixture being stirred moderately and cooled by an ice bath.

This mixture is heated to 70° and an additional 195 g. (66.5 ml., 1.22 moles) of bromine added over a half-hour interval. After the bromine addition, the temperature is raised to 80° and the mixture held at that temperature for 3 hours. Air is blown into the cooled reaction until the excess bromine and hydrogen bromide are removed (Note 1). This process usually requires one hour (Note 2).

The aerated reaction mixture is heated to 80° and 25 ml. of water is added cautiously, with stirring. A vigorous reaction ensues, and upon cessation of the reaction an additional 300 ml. of water is added.

The reaction mixture of two layers and some solid residue is heated under reflux for 4 hours. Upon cooling, two layers again appear. The product is extracted with two portions of ether (200 ml. each), and the extracts are dried over magnesium sulfate (Note 3). *Care should be taken since the α-bromolactone is a vesicant.*

The dried crude material is distilled, b.p. 125–127° (13 mm.), n^{25}D 1.5030, yield 105 g. (55%).

B. *$\Delta^{\alpha,\beta}$-Butenolide.* In a 500-ml. three-necked flask fitted with a mechanical stirrer, a reflux condenser, and a 250-ml. dropping funnel containing a solution of 61 g. (84.5 ml., 0.6 mole) of triethylamine in 70 ml. of dry diethyl ether, a solution of 83 g. (0.5 mole) of α-bromo-γ-butyrolactone and 200 ml. of dry diethyl ether is heated to reflux, with stirring. The amine solution is added, slowly, during 5 hours and the stirring under reflux continued for an additional 24 hours. The brown precipitate (40 g.) is removed by filtration. Most of the solvent is removed from the filtrate by evaporation, and the additional precipitate (8 g.) is removed. This precipitate is predominantly triethylamine hydrobromide. The liquid residue is distilled under reduced pressure and the $\Delta^{\alpha,\beta}$-butenolide is collected at 107–109° (24 mm.); yield 25 g. (60%, 33% overall), m.p. 5° [2-4] (Note 4).

2. Notes

1. A trap to catch the resulting bromine-hydrogen bromide vapors is desirable.

2. Plieninger [5] reports that the product at this stage is α,γ-dibromobutyryl bromide.

3. Extraction with ether is necessary to separate the bromolactone efficiently.

4. The infrared spectrum of γ-crotonolactone shows two bands in the carbonyl region at 5.60 and 5.71 μ in carbon tetrachloride (5%) [shifted to 5.61 and 5.71 μ in chloroform (5%)] and carbon-carbon stretching absorption at 6.23 μ. The nuclear magnetic resonance spectrum shows olefinic peaks centered at 2.15τ (pair of triplets) and 3.85τ (pair of triplets), each due to one proton, and a two-proton triplet centered at 5.03τ (in CCl₄).

In the ultraviolet, γ-crotonolactone shows end absorption at 205 mμ (ε *ca.* 11,000) and no maximum at higher wavelength.

Oxidation of this product by potassium permanganate affords 2,3-dihydroxy-4-butyrolactone.[2]

3. Methods of Preparation

The original preparation of γ-crotonolactone by Lespieau involved a five-step sequence from epichlorohydrin and sodium cyanide.[2] A recent detailed study of this procedure reported an overall yield of 25% for the lactone.[3] Glattfeld[4] used a shorter route from glycerol chlorohydrin and sodium cyanide; hydrolysis and distillation of the intermediate dihydroxy acid yielded γ-crotonolactone in 23% yield and β-hydroxy-γ-butyrolactone in 28% yield.[4] The formation of γ-crotonolactone in 15% yield has also been reported from pyrolysis of 2,5-diacetoxy-2,5-dihydrofuran at 480–500°.[6] The lactone has been prepared in 37% overall yield from propynol by carboxylation and hydrogenation.[7]

The formation of α-bromo-γ-butyrolactone has been reported in 70% yield by uncatalyzed reaction of bromine at 160–170°,[8] as well as by the catalyzed procedure used here.[3]

4. Merits of the Preparation

γ-Crotonolactone is the simplest example of the butenolide ring system, which occurs in many natural products. In view of the availability of butyrolactone, the present procedure represents the most convenient method of synthesis of the unsaturated lactone.

The dehydrohalogenation by a tertiary amine illustrates the utility of such amines for dehydrohalogenations which produce a double bond normally activated for attack by many bases.[9]

1. Department of Chemistry, University of Pennsylvania, Philadelphia, Pennsylvania.
2. M. Lespieau, *Compt. Rend.*, **138**, 1050 (1904); *Bull. Soc. Chim. France*, [4] **1**, 1113 (1907).
3. R. Rambaud, S. Ducher, A. Broche, M. Brini-Fritz, and M. Vessiere, *Bull. Soc. Chim. France*, 877 (1955); see also R. Rambaud and S. Ducher, *Bull. Soc. Chim. France*, 466 (1956).
4. J. W. E. Glattfeld, G. Leavell, G. E. Spieth, and D. Hutton, *J. Am. Chem. Soc.*, **53**, 3168 (1931).
5. H. Plieninger, *Ber.*, **83**, 265 (1950).
6. N. Clauson-Kaas and N. Elming, *Acta Chem. Scand.*, **6**, 560 (1952).
7. R. J. D. Smith and R. N. Jones, *Can. J. Chem.*, **37**, 2092 (1959).

8. G. Bischoff, Swiss Patent 264,598 (1950) [C.A., 45, 1622d (1951)].
9. C. Grundmann and E. Kober, *J. Am. Chem. Soc.*, 77, 2332 (1955).

CROTYL DIAZOACETATE

(Acetic acid, diazo-, *trans*-2-butenyl ester)

$$p\text{-}CH_3C_6H_4SO_2NHNH_2 + (HO)_2CHCO_2H \rightarrow$$
$$p\text{-}CH_3C_6H_4SO_2NHN\!\!=\!\!CHCO_2H$$

$$p\text{-}CH_3C_6H_4SO_2NHN\!\!=\!\!CHCO_2H + SOCl_2 \rightarrow$$
$$p\text{-}CH_3C_6H_4SO_2NHN\!\!=\!\!CHCOCl$$

$$p\text{-}CH_3C_6H_4SO_2NHN\!\!=\!\!CHCOCl + trans\text{-}CH_3CH\!\!=\!\!CHCH_2OH \xrightarrow{(C_2H_5)_3N}$$
$$[trans\text{-}CH_3CH\!\!=\!\!CHCH_2OCOCH\!\!=\!\!NNHSO_2C_6H_4CH_3\text{-}p] \xrightarrow{(C_2H_5)_3N}$$
$$trans\text{-}CH_3CH\!\!=\!\!CHCH_2OCOCHN_2 + p\text{-}CH_3C_6H_4SO_2^- \;\; (C_2H_5)_3\overset{+}{N}H$$

Submitted by C. John Blankley, Frederick J. Sauter,
and Herbert O. House [1]
Checked by J. H. Ham and R. E. Ireland

1. Procedure

A. *Glyoxylic acid p-toluenesulfonylhydrazone.* A solution of 46.3 g. (0.50 mole) of 80% glyoxylic acid (Note 1) in 500 ml. of water is placed in a 1-l. Erlenmeyer flask and warmed on a steam bath to approximately 60°. This solution is then treated with a warm (approximately 60°) solution of 93.1 g. (0.50 mole) of *p*-toluenesulfonylhydrazide (Note 2) in 250 ml. (0.63 mole) of aqueous 2.5*M* hydrochloric acid. The resulting mixture is heated on a steam bath with continuous stirring until all the hydrazone, which initially separates as an oil, has solidified (about 5 minutes is required). After the reaction mixture has been allowed cool to room temperature and then allowed to stand in a refrigerator overnight, the crude *p*-toluenesulfonylhydrazone is collected on a filter, washed with cold water, and allowed to dry for 2 days (Note 3). The crude product (110–116 g., m.p. 145–149° dec.) is dissolved in 400 ml. of boiling ethyl acetate, filtered to remove any insoluble material, and then diluted with 800 ml. of carbon tetrachloride and allowed to cool. After the mixture has been allowed to stand overnight in a refrigerator, the *p*-toluenesulfonylhydrazone is collected and washed with cold mixture of ethyl acetate and carbon tetrachloride (1:2 by volume). The yield is 92.4–98.5 g. (76–81%) of the hydrazone as white crystals, m.p. 148–154° dec. (Note 4).

B. *The p-toluenesulfonylhydrazone of glyoxylic acid chloride. Caution! Since hydrogen chloride and sulfur dioxide are liberated during this reaction, it should be conducted in a hood.* To a suspension of 50.2 g. (0.21 mole) of glyoxylic acid *p*-toluenesulfonylhydrazone in 250 ml. of benzene is added 30 ml. (49 g. or 0.42 mole) of thionyl chloride (Note 5). The reaction mixture is heated under reflux with stirring until vigorous gas evolution has ceased and most of the suspended solid has dissolved (about 1.5–2.5 hours is required, Note 6). The reaction mixture is then cooled immediately (Note 6) and filtered through a Celite mat on a sintered-glass funnel. After the filtrate has been concentrated to dryness under reduced pressure, the residual solid is mixed with 40–50 ml. of anhydrous benzene, warmed, and the solid mass is broken up to give a fine suspension. This suspension is cooled and filtered with suction. The crystalline product is washed quickly with two portions of cold benzene to remove most of the residual colored impurities, and then the remaining crude acid chloride is transferred to a flask for recrystallization.

The combined benzene filtrates from this initial washing procedure are concentrated under reduced pressure, and the washing procedure with benzene is repeated to give a second crop of the crude acid chloride which is transferred to a flask for recrystallization.

For recrystallization each crop of the crude acid chloride is dissolved in a minimum volume of boiling anhydrous benzene (about 100 ml. is required for the first crop) and petroleum ether (b.p. 30–60°; about 50 ml. is required for the first crop) is added to the hot solution. Crystallization begins on cooling. After the mixture has cooled to room temperature, it is allowed to stand overnight at room temperature and the acid chloride is collected on a filter and washed with a small portion of cold benzene. The yield of recrystallized acid chloride from the first crop of crude acid chloride is 27.6–33.4 g. (50–61%) of pale yellow prisms, m.p. 101–112° (Note 4). The product obtained from crystallization of the second crop of acid chloride amounts to 3.3–3.6 g. (6–7%), m.p. 104–108°.

C. *Crotyl diazoacetate.* A solution of 10.0 g. (0.038 mole) of the *p*-toluenesulfonylhydrazone of glyoxylic acid chloride in 100 ml. of methylene chloride is cooled in an ice bath. Crotyl alcohol (2.80 g. or 0.038 mole) (Note 7) is added to this cold solution, and then a solution of 7.80 g. (0.077 mole) of redistilled triethylamine (b.p. 88.5–90.5°) in 25 ml. of methylene chloride is added to the cold reaction mixture dropwise and with stirring over a 20-minute period. During the addition a yellow color develops in the reaction mixture and some solid separates near the end of the addition period. The resulting mixture is stirred at 0° for 1 hour and then the solvent is removed at 25° under reduced pressure with a rotary evaporator. A solution of the residual dark orange liquid in approximately 200 ml. of benzene is thoroughly mixed with 100 g. of Florisil (Note 8) and then filtered. The residual Florisil, which has adsorbed the bulk of the dark colored by-products, is washed with two or three additional portions of benzene of such size that the total volume of the combined benzene filtrates is 400–500 ml. This yellow benzene solution of the diazoester is concentrated under reduced pressure at 25° with a rotary evaporator, and the residual yellow liquid is distilled under reduced pressure. (*Caution! This distillation should be conducted in a hood behind a safety shield*) (Note 9). The diazo ester is collected as 2.20–2.94 g. (42–55%) of yellow liquid, b.p. 30–33° (0.15 mm.), n^{24}D 1.4853 − 1.4856 (Note 10).

2. Notes

1. The submitters used a practical grade of material containing 80% glyoxylic acid purchased from Eastman Organic Chemicals.

2. The submitters used *p*-toluenesulfonylhydrazide prepared as described in *Organic Syntheses*.[2] This material may be purchased from Eastman Organic Chemicals.

3. Difficulty was encountered in the subsequently described recrystallization if the crude product was not dry.

4. The broad melting range is presumably due to the presence of a mixture of stereoisomers.

5. If the thionyl chloride is not taken from a freshly opened bottle, it should be redistilled before use.

6. The heating period is critical to the success of this reaction. After a heating period of 45–90 minutes the initially white suspension begins to turn yellow and the color gradually deepens as heating is continued. The correct heating period is normally reached about 10 minutes after vigorous gas evolution ceases; at this time the color of the reaction mixture is yellow-orange to orange. At this point the mixture is not clear, but relatively little suspended material separates when the stirrer is stopped for a short period. If heating is stopped too soon, a large amount of acid is lost during the filtration and the product is difficult to crystallize. If heating is continued too long, the product is contaminated with a brown-colored impurity and is difficult to crystallize.

7. The submitters used material purchased from the Aldrich Chemical Company.

8. This material was purchased from Fisher Scientific Company.

9. Although this distillation has been performed repeatedly without incident, the product is potentially explosive and the operator should take suitable precautions including surrounding the distillation apparatus with a hood and a safety shield and wearing an effective face shield.

10. A pot residue amounting to 2–3 g. of orange liquid remains at the end of this distillation.

3. Discussion

Crotyl diazoacetate has been prepared by the procedure described here[3] and by the reaction of diazomethane with crotyl chloroformate.[3] The lower homolog, allyl diazoacetate, has been prepared by the reaction of allyl glycinate with nitrous acid[4] and by the successive conversion of allyl chloroacetate to the corresponding azide, iminophosphorane, and, finally, the diazo ester.[5]

Other methods for the preparation of diazoacetic acid esters include the diazotization of glycine esters,[6] the thermal or base-catalyzed decomposition of N-acyl-N-nitrosoglycine esters,[7] the base-catalyzed cleavage of α-diazo-β-keto acetates,[8] the reaction of carboalkoxymethylenephosphoranes with arylsulfonyl azides,[9] the acid-catalyzed decomposition of acetic esters with α-aryltriazene substituents,[10] and the reaction of chloroformate esters with diazomethane.[3, 11] The present procedure is unique in that a diazoacetic ester may be prepared in one step by reaction of the desired alcohol with a relatively stable, solid acylating agent which may be prepared in quantity and stored (in a desiccator). Consequently, the method is of particular value for alcohols which are not readily available or for alcohols containing other functional groups which would be incompatible with the reaction conditions required in other diazoacetate syntheses.

Although the present procedure illustrates the formation of the diazoacetic ester without isolation of the intermediate ester of glyoxylic acid p-toluenesulfonylhydrazone, the two geometric isomers of this hydrazone can be isolated if only one molar equivalent of triethylamine is used in the reaction of the acid chloride with the alcohol.[3] The extremely mild conditions required for the further conversion of these hydrazones to the diazo esters should be noted. Other methods for decomposing arylsulfonylhydrazones to form diazocarbonyl compounds have included aqueous sodium hydroxide,[12] sodium hydride in dimethoxyethane at 60°,[13] and aluminum oxide in methylene chloride or ethyl acetate.[14] Although the latter method competes in mildness and convenience with the procedure described here, it was found not to be applicable to the preparation of aliphatic diazoesters such as ethyl 2-diazopropionate. Hence the conditions used in the present procedure may offer a useful complement to the last-mentioned method when the appropriate arylsulfonylhydrazone is available.

1. Department of Chemistry, Massachusetts Institute of Technology, Cambridge, Massachusetts 02139.
2. L. Friedman, R. L. Litle, and W. R. Reichle, *Org. Syntheses,* **40**, 93 (1960); cf. this volume, p. 1055.
3. H. O. House and C. J. Blankley, *J. Org. Chem.,* **33**, 53 (1968).
4. W. Kirmse and H. Dietrich, *Ber.,* **98**, 4027 (1965).
5. L. Solomon, Ph.D. Dissertation, Columbia University, 1964 [*Diss. Abstr.,* **26**, 101 (1965)].

6. E. B. Womack and A. B. Nelson, *Org. Syntheses*, Coll. Vol. **3**, 392 (1955); N. E. Searle, *Org. Syntheses*, Coll. Vol. **4**, 424 (1963).
7. E. H. White and R. J. Baumgarten, *J. Org. Chem.*, **29**, 2070 (1964); H. Reimlinger, *Angew. Chem.*, **72**, 33 (1960).
8. M. Regitz, J. Hocker, and A. Liedhegener, this volume, p. 179.
9. G. R. Harvey, *J. Org. Chem.*, **31**, 1587 (1966).
10. R. J. Baumgarten, *J. Org. Chem.*, **32**, 484 (1967).
11. J. Shafer, P. Baronowsky, R. Laursen, F. Finn, and F. H. Westheimer, *J. Biol. Chem.*, **241**, 421 (1966); H. Chaimovich, R. J. Vaughan, and F. H. Westheimer, *J. Am. Chem. Soc.*, **90**, 4088 (1968).
12. M. P. Cava, R. L. Litle, and D. R. Napier, *J. Am. Chem. Soc.*, **80**, 2257 (1958).
13. E. J. Corey and A. M. Felix, *J. Am. Chem. Soc.*, **87**, 2518 (1965).
14. J. M. Muchowski, *Tetrahedron Lett.*, 1773 (1966).

1-CYANOBENZOCYCLOBUTENE

(Bicyclo[4.2.0]octa-1,3,5-triene-7-carbonitrile)

Submitted by J. A. Skorcz and F. E. Kaminski [1]
Checked by V. Z. Williams and K. B. Wiberg

1. Procedure

A 2-l. three-necked flask is thoroughly dried and fitted with a large dry-ice condenser, a mechanical stirrer, a nitrogen inlet, and a powder funnel in an efficient hood. With nitrogen flowing through the system, 62.5 g. (1.60 moles) of commercial sodium amide (Note 1) is added rapidly. (*Caution! Sodium amide is corrosive and readily decomposes in the presence of moisture.*) The funnel is replaced by a gas-inlet tube, the condenser is filled with a mixture of dry ice and acetone, and *ca.* 400 ml. of liquid ammonia is introduced from a cylinder. The gas-inlet tube is replaced by an addition funnel, stirring is commenced, and 66.3 g. (0.400 mole) of *o*-chlorohydrocinnamonitrile (Note 2) is added over a 10-minute period. The last traces of the nitrile are washed into the flask with small amounts of anhydrous ether.

The dark green reaction mixture is stirred vigorously for 3

hours and then is treated carefully with 96 g. (1.2 moles) of solid ammonium nitrate (Note 3). All the fittings are removed from the flask, and the ammonia is allowed to evaporate (Note 4). Water (300 ml.) is added cautiously to the residue. (*Caution! Traces of undecomposed sodium amide may adhere to the upper walls of the flask.*) The organic layer is taken up in two 160-ml. portions of chloroform, and the solutions are combined and washed twice with 100 ml. of 5% hydrochloric acid and once with 100 ml. of water. (*Caution! The extraction procedure and subsequent chloroform distillation should be conducted in a hood because some hydrogen cyanide is usually evolved.*) The chloroform solution is dried over anhydrous sodium sulfate, and the chloroform is removed by distillation. The residual liquid is distilled under reduced pressure through an insulated, 5-in. Vigreux column. The forerun, b.p. 95–100° (3 mm.), weighs *ca.* 1 g.; the product boils at 100–101° (3 mm.); n^{25}D 1.5451. The yield of 1-cyanobenzocyclobutene is 33–34 g. (64–66%) (Notes 5 and 6).

2. Notes

1. The sodium amide was obtained from Farchan Research Laboratories and was approximately 90% pure.

2. The submitters prepared *o*-chlorohydrocinnamonitrile by the following procedure. Ethyl cyanoacetate (3040 g., 27 moles) was added to a solution of 140 g. (6.1 g. atoms) of sodium in 4 l. of absolute ethanol, followed by 970 g. (6 moles) of *o*,α-dichloro-toluene (Eastman Organic Chemicals), to afford 890 g. (63%) of ethyl 2-(*o*-chlorobenzyl)cyanoacetate,[2] b.p. 117–123° (0.03 mm.). Hydrolysis of this material in 2 l. of 10% aqueous sodium hydroxide at room temperature gave a quantitative yield (790 g.) of 2-(*o*-chlorobenzyl)cyanoacetic acid, m.p. 129–132° without recrystallization. Decarboxylation of 750 g. of the acid in 750 ml. of refluxing dimethylformamide gave 550 g. (93%) of *o*-chloro-hydrocinnamonitrile,[3] b.p. 82–85° (0.3 mm.), n^{25}D 1.5362. The checkers carried out this preparation starting with 8 moles of ethyl cyanoacetate and obtained comparable yields.

3. Other ammonium salts, such as ammonium chloride, are equally satisfactory.

4. Overnight evaporation at room temperature is convenient.

5. The submitters carried out the reaction on 1-molar and 3-molar scales and obtained yields of 62–64% and 67%, respectively.

6. This procedure has also been used to obtain 1-cyano-5-methoxybenzocyclobutene from 2-bromo-4-methoxyhydrocinnamonitrile.[4]

3. Methods of Preparation

1-Cyanobenzocyclobutene has been prepared from sodium cyanide and 1-bromobenzocyclobutene,[5] formed by reaction of benzocyclobutene with N-bromosuccinimide,[6] and by ring closure of o-chlorohydrocinnamonitrile with potassium amide in liquid ammonia.[7] The present procedure is a modification of the latter method and was previously described by one of the submitters.[8]

4. Merits of the Preparation

Cyclization by addition of a side-chain carbanion to an aryne bond has been proposed as the method of choice for synthesis of the versatile 1-substituted benzocyclobutene system.[7] This general procedure now has been modified to permit convenient large-scale preparations utilizing a commercially available base, a minimum amount of liquid ammonia, and distillation for isolation of the product.

1. Lakeside Laboratories, Division of Colgate-Palmolive Company, Milwaukee, Wisconsin 53201.
2. P. E. Gagnon, J. L. Boivin, and J. Giguère, *Can. J. Res.*, **28B**, 352 (1950).
3. A. D. Grebenyuk and I. P. Tsukervanik, *Zh. Obshch. Khim.*, **25**, 286 (1955); *J. Gen. Chem. USSR (Engl. Transl.)*, **25**, 269 (1955).
4. J. A. Skorcz and J. E. Robertson, *J. Med. Chem.*, **8**, 255 (1965).
5. M. P. Cava, R. L. Litle, and D. R. Napier, *J. Am. Chem. Soc.*, **80**, 2257 (1958).
6. M. P. Cava and D. R. Napier, *J. Am. Chem. Soc.*, **80**, 2255 (1958).
7. J. F. Bunnett and J. A. Skorcz, *J. Org. Chem.*, **27**, 3836 (1962).
8. J. A. Skorcz, J. T. Suh, C. I. Judd, M. Finkelstein, and A. C. Conway, *J. Med. Chem.*, **9**, 656 (1966).

7-CYANOHEPTANAL

(Heptanal, 7-cyano-)

$$\text{cyclooctene} + \text{NOCl} \xrightarrow{\text{HCl}} \text{2-chlorocyclooctanone oxime (NOH·HCl, Cl)} \xrightarrow[\text{Et}_3\text{N}]{\text{CH}_3\text{OH}}$$

$$\text{(NOH, OCH}_3) \xrightarrow[\text{2. H}_2\text{O}]{\text{1. PCl}_5} \text{NC-(CH}_2)_7\text{-CHO}$$

Submitted by MASAJI OHNO, NORIO NARUSE, and ISAO TERASAWA [1]
Checked by E. J. COREY and I. VLATTAS

1. Procedure

A. *2-Chlorocyclooctanone oxime hydrochloride.* In a 2-l. three-necked, round-bottomed flask, fitted with a mechanical stirrer, a gas inlet tube, and a tube fitted with a thermometer and a calcium chloride tube, is placed 55 g. (0.50 mole) of freshly distilled cyclooctene and 600 ml. of trichloroethylene. The solution is cooled with ice water to 5°, and 36 g. (0.55 mole) of nitrosyl chloride (Note 1) and excess of hydrogen chloride gas (about 400–600 ml. per minute) are bubbled into the solution, keeping the reaction temperature between 5–10°. The solution gradually becomes light reddish brown. The addition of nitrosyl chloride should be carried out in about 1.5 hours. After completion of the addition of nitrosyl chloride, hydrogen chloride gas is bubbled in for another 15 minutes. A light brown oily material is obtained after evaporation of the solvent under an aspirator pressure below 35° (Note 2) by using an efficient rotatory evaporator. On cooling this product in a refrigerator, 107.2 g. (*ca.* 100%) of crude 2-chlorocyclooctanone oxime hydrochloride is obtained as a solid.

B. *2-Methoxycyclooctanone oxime.* In a 500-ml., three-necked, round-bottomed flask, fitted with a mechanical stirrer, a dropping funnel, and a reflux condenser equipped with a calcium chloride tube, is placed a solution of 53.5 g. (0.252 mole) of crude 2-chlorocyclooctanone oxime hydrochloride in 250 ml. of methanol. While cooling the vessel with running water, 60.7 g. of triethylamine (0.60 mole) is added dropwise during 40 minutes. The reaction temperature is kept below 50° and the reaction is continued for 30 minutes with stirring. After removal of methanol under reduced pressure using an efficient rotatory evaporator, a light brown semisolid is obtained; it is treated with 200 ml. of ether and 200 ml. of water to effect complete solution. The ether layer is separated and the aqueous layer is further extracted twice with ether. The combined ether solution is washed with saturated sodium chloride and dried over sodium sulfate. Removal of ether affords 42.8 g. of crude 2-methoxycyclooctanone oxime (Note 3) as a brown oil.

C. *Beckmann fission of 2-methoxycyclooctanone oxime.* In a 500-ml., three-necked, round-bottomed flask equipped with a mechanical stirrer, a dropping funnel, and a calcium chloride tube is placed a suspension of 62.5 g. (0.30 mole) of phosphorus pentachloride (Note 4) in 150 ml. of absolute ether, and the reaction vessel is cooled with ice. A solution of 42.8 g. of crude 2-methoxycyclooctanone oxime (0.25 mole) in 100 ml. of absolute ether is added over 30 minutes with vigorous stirring and the reaction is continued for 50 minutes at 5°. The reaction mixture, which becomes a transparent reddish brown solution (Note 5), is poured with mechanical stirring into 500 g. of ice in a 2-l. beaker. Stirring is continued for 1.5 hours at 5° (Note 6). The ether layer is separated and the aqueous layer is extracted with methylene chloride three times. The combined organic extracts are neutralized with dilute sodium carbonate solution and dried over sodium sulfate (Note 7). Removal of the solvent below 40° affords a reddish brown oil which is distilled to give 29.6 g. (85.2%) of 7-cyanoheptanal (Note 8), b.p. 109–115° (0.3 mm.) n^{26}D = 1.4456. The 2,4-dinitrophenylhydrazone has m.p. 74–75° after recrystallization from ethanol.

2. Notes

1. Solid nitrosyl chloride stored in a dry-ice box is quickly melted by warming, and as rapidly as possible the liquid nitrosyl chloride is weighed into a flask contained in a hood. Nitrosyl chloride is simply allowed to volatilize into the reaction from this flask under ambient conditions; rapid addition of nitrosyl chloride causes a decrease of the yield of α-chlorooxime. It may sometimes be necessary to control the rate of addition by cooling the nitrosyl chloride container with ice water.

2. 2-Chlorocyclooctanone oxime hydrochloride is unstable to heat. Therefore the temperature during removal of methanol should be kept below 35°.

3. The methanol-triethylamine reagent is superior to the previously used [2] methanolic sodium methoxide, and the crude 2-methoxycyclooctanone oxime thus obtained can be used for the Beckmann fission reaction without further purification. However, it is easily purified by distillation, b.p. 101° (0.7 mm.).

4. Thionyl chloride can also be used as the reagent for the Beckmann fission.

5. A very small amount of excess of phosphorus pentachloride is sometimes observed at the bottom of the reaction vessel.

6. If necessary, the temperature is kept at 5–10° by adding ice occasionally.

7. If the solution is acidic, the yield of ω-cyanoaldehyde is diminished by the occurrence of aldo condensation.

8. Although this distilled product, a pale yellow oil, is pure enough to use for most purposes, pure 7-cyanoheptanal, a colorless oil, is obtained by redistillation, b.p. 85–87° (0.013 mm.), n^{26}D = 1.4451.

3. Discussion

The only preparation reported for 7-cyanoheptanal is that described by the submitters.[3] The present procedure starting from 2-methoxycyclooctanone oxime is superior to modifications employing 2-alkylamino- or 2-ethylthiocyclooctanone oxime in the Beckmann cleavage step.

ω-Cyanoaldehydes are not easily accessible by other routes but are interesting synthetic intermediates,[4] since the two terminal function groups are in different oxidation states which readily allow separate modification or elaboration.[5,6] The general applicability of the method described herein allows the synthesis of a wide variety of ω-cyanoaldehydes from available cycloolefins.

1. Basic Research Laboratories, Toray Industries, Inc., Kamakura, Japan.
2. M. Ohno, N. Naruse, S. Torimitsu, and M. Okamoto, *Bull. Chem. Soc. Japan,* **39**, 1119 (1966).
3. M. Ohno, N. Naruse, S. Torimitsu, and I. Terasawa, *J. Am. Chem. Soc.,* **88**, 3168 (1966).
4. M. Ohno and I. Terasawa, *J. Am. Chem. Soc.,* **88**, 5683 (1966).
5. E. J. Corey, N. H. Andersen, R. M. Carlson, J. Paust, E. Vedejs, I. Vlattas, and R. E. K. Winter, *J. Am. Chem. Soc.,* **90**, 3245 (1968).
6. E. J. Corey, I. Vlattas, N. H. Anderson, and K. Harding, *J. Am. Chem. Soc.,* **90**, 3247 (1968).

2-CYANO-6-METHYLPYRIDINE*

(6-Methylpicolinonitrile)

Submitted by WAYNE E. FEELY, GEORGE EVANEGA, and ELLINGTON M. BEAVERS.[1]
Checked by WILLIAM E. PARHAM, STUART W. FENTON, and WILLIAM W. HENDERSON.

1. Procedure

Caution! All the operations should be carried out in a well-ventilated hood because of the toxic natures of dimethyl sulfate, hydrogen cyanide, and cyanide solutions.

A. *1-Methoxy-2-methylpyridinium methyl sulfate.* In a 1-l. three-necked flask equipped with a Hirshberg stirrer, a thermometer which extends deep into the flask, and a 250-ml. pressure-equalizing, dropping funnel fitted with a calcium chloride drying tube is placed 109 g. (1.0 mole) of dry powdered 2-picoline-1-oxide (Note 1). The stirrer is started at a slow rate, and 126 g. (1.0 mole) of dimethyl sulfate (Note 2) is added dropwise at a rate such that the temperature of the reaction mixture slowly rises to between 80° and 90° and remains in this range throughout the addition (Note 3). When the addition is about two-thirds complete, gentle heating with a steam bath is necessary to maintain this temperature. After complete addition (about 1 hour), the mixture is heated for an additional 2 hours on a steam bath at 90–100°. The molten salt is then poured into a large evaporating dish and placed in a vacuum desiccator under partial vacuum to cool. The salt is obtained as a white crystalline mass in essentially quantitative yield (235 g.) (Notes 4 and 5).

B. *2-Cyano-6-methylpyridine.* In a 2-l., three-necked, round-bottomed flask equipped with a Hershberg stirrer, a 500-ml. pressure-equalizing, dropping funnel without a stopper, and a thermometer-gas inlet adapter (Note 6) fitted with a thermometer which reaches deep into the flask is placed a solution of 147 g. (3.0 mole) of sodium cyanide dissolved in 400 ml. of water. The stirrer is started and the apparatus is flushed with prepurified nitrogen for 1 hour (Note 7). The solution in the flask is then cooled to 0° with an ice bath, and a solution of 235 g. (1.0 mole) of 1-methoxy-2-methylpyridinium methyl sulfate dissolved in 300 ml. of water is added dropwise over a period of 2 hours. The dropping funnel and the thermometer-adapter are then quickly removed and replaced by stoppers, and the flask is allowed to stand in a refrigerator overnight (12–16 hours). The flask, containing needles of the crude nitrile (Note 8), is removed from the refrigerator and the contents stirred at room temperature for 6 hours. After addition of 200 ml. of chloroform, the contents of the flask are transferred to a large separatory funnel and the layers separated. Extraction of the aqueous phase is repeated twice with 100-ml. portions of chloroform, and the combined extracts are dried over anhydrous magnesium sulfate. After removal of the drying agent by filtration, the filtrate is concentrated on a steam bath to remove chloroform, and the residual crude cyanopicoline (90–110 g.) is transferred, while hot, to a distilling flask. Distillation under reduced pressure (30 mm.) (Note 9) gives three fractions: Fraction I, b.p. 99–106°, weighs 15–20 g.; Fraction II, b.p. 106–124°, weighs 5–10 g.; and Fraction III, b.p. 125–131°, weighs 60–70 g. (Note 10). Fraction III is dissolved in 1 l. of hot 10% ethyl alcohol, treated with 0.5 g. of activated carbon, filtered, and the filtrate is allowed to cool slowly to room temperature. The 2-cyano-6-methylpyridine separates as white prismatic needles, m.p. 71–73°, and weighs 48–54 g. (40–46% based on 2-picoline-1-oxide) (Notes 11 and 12).

2. Notes

1. The preparation of 2-picoline-1-oxide is described by Boekelheide and Linn.[2] The oxide is hygroscopic, and best results are obtained if it is redistilled just before use. The submitters used 2-picoline-1-oxide, obtained from the Reilly Tar and Chemical Company, Indianapolis, Indiana, which was freshly redistilled and boiled at 118–120°/10 mm.

2. Eastman Kodak Company practical grade was used. Dimethyl sulfate is toxic and must be handled with caution. Provision should be made for containing the contents should breakage occur. Ammonia is a specific antidote for dimethyl sufate and should be at hand to destroy any accidentally spilled.

3. The submitters have observed that, when 1-methoxypyridinium methyl sulfate salts are heated above about 140–150°, violent explosions usually result.

4. The salt is very hygroscopic. Aqueous solutions of the salt slowly hydrolyze upon standing to di(1-methoxy-2-methylpyridinium) sulfate but may be used in the subsequent step without adverse effects.

5. The salt may be recrystallized from anhydrous acetone, giving colorless prisms, m.p. 67–70°.[3]

6. A thermometer adapted with a gas-addition tube may be purchased from Ace Glass Inc., Vineland, New Jersey (Cat. No. 5266).

7. The presence of small amounts of air during the formation of the nitrile rapidly darkens the reaction mixture.

8. The crude 2-cyano-6-methylpyridine which has separated (40–50 g.) may be recrystallized from dilute ethyl alcohol to yield 35–45 g. of pure product.

9. The distillation is conveniently performed in a Claisen flask with a fractionating side arm. The checkers used a heat lamp to prevent solidification of product in the condenser.

10. Fraction I, b.p. 99–106°/30 mm., is mostly 4-cyano-2-methylpyridine and is best purified by redistillation.[4] Fraction II, b.p. 106–125°/30 mm., contains a mixture of the two nitriles and may be further purified by redistillation.

11. Physical constants reported for 2-cyano-6-methylpyridine are b.p. 135–136°/38 mm.,[5] m.p. 69–71°,[5] m.p. 72–74°.[6,7]

12. This general method has been used to prepare 2- and 4-cyanopyridine from pyridine-1-oxide in 32% and 49% yields, respectively; 2-cyano-4,6-dimethylpyridine (73%) from 4,6-dimethylpyridine-1-oxide; 2-cyanoquinoline (93%) from quinoline-1-oxide; and 1-cyanoisoquinoline (95%) from isoquinoline-2-oxide.[3]

3. Methods of Preparation

The present method is essentially that given by Feely and Beavers.[3] 2-Cyano-6-methylpyridine also has been prepared by the fusion of sodium 6-methylpyridine-2-sulfonate with potassium cyanide.[5] In addition, this nitrile has been prepared from 2-chloro-6-methylpyridine [6] (no yield stated) and from a catalytic reaction of 2,6-lutidine with air and ammonia in low yield.[6,7]

4. Merits of Preparation

This preparation describes a convenient and general method for preparing cyano derivatives of pyridine, quinoline, and isoquinoline from the corresponding, and readily available, amine oxides.

1. Research Laboratories, Rohm and Haas Co., Philadelphia, Pennsylvania.
2. V. Boekelheide and W. J. Linn, *J. Am. Chem. Soc.*, 76, 1286 (1954).
3. W. E. Feely and E. M. Beavers, *J. Am. Chem. Soc.*, 81, 4004 (1959).

4. E. Ochiai and I. Suzuki, *Pharm. Bull. (Tokyo)*, **2**, 247 (1954).
5. I. Suziki, *Pharm. Bull. (Tokyo)*, **5**, 13 (1957).
6. G. Mayurnik, A. F. Moschetto, H. S. Block, and J. V. Scudi, *Ind. Eng. Chem.*, **44**, 1630 (1952).
7. Pyridium Corp., Brit. pat. 671,763 [C.A. **47**, 1746 (1953)].

CYCLOBUTYLAMINE*

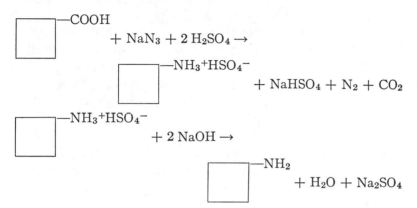

$$\text{—COOH} + NaN_3 + 2\,H_2SO_4 \rightarrow$$

$$\text{—NH}_3{}^+\text{HSO}_4{}^- + NaHSO_4 + N_2 + CO_2$$

$$\text{—NH}_3{}^+\text{HSO}_4{}^- + 2\,NaOH \rightarrow$$

$$\text{—NH}_2 + H_2O + Na_2SO_4$$

Submitted by NEWTON W. WERNER and JOSEPH CASANOVA, JR.[1]
Checked by DONALD BARTH and KENNETH B. WIBERG

1. Procedure

Caution! The reaction should be carried out in a good hood because hydrazoic acid is very toxic. Care should also be taken in handling sodium azide.

In a 1-l. three-necked, round-bottomed flask equipped with a mechanical stirrer, reflux condenser, and powder funnel are placed 180 ml. of reagent grade chloroform, 16.0 g. (0.16 mole) of cyclobutanecarboxylic acid (Note 1), and 48 ml. of concentrated sulfuric acid. The flask is heated in an oil bath to 45–50°, and 20.0 g. (0.31 mole) of sodium azide (Note 2) is added over a period of 1.5 hours (Note 3). After the addition of sodium azide is complete, the reaction mixture is heated at 50° for 1.5 hours. The flask is cooled in an ice bath, and approximately

200 g. of crushed ice is added slowly. A solution of 100 g. of sodium hydroxide in 200 ml. of water is prepared, cooled to room temperature, and then added slowly to the reaction mixture until the pH of the mixture is approximately 12–13. The mixture is poured into a 2-l. three-necked, round-bottomed flask, the flask is set up for steam distillation, and about 2 l. of distillate is collected in a cooled receiver containing 90 ml. of $3N$ hydrochloric acid (Note 4). The water and chloroform are removed by distillation under reduced pressure (Note 5), and the amine hydrochloride is transferred to a 50 ml. round-bottomed flask with a few milliliters of water. A straight condenser is connected to the flask, and the flask is cooled in an ice bath. A slush is prepared by grinding potassium hydroxide pellets in a mortar and then adding a minimum volume of water. The slush is added in portions through the top of the condenser. After the mixture has become sufficiently basic, the amine appears as a separate phase. More potassium hydroxide pellets are added to dry the amine phase. The condenser is replaced by a heated, vacuum-jacketed Vigreux column equipped with a soda-lime tube, and the fraction having a boiling point of 79–83° is collected. The distillate is dried over potassium hydroxide pellets for 2 days. The liquid is decanted into a distilling flask containing a few potassium hydroxide pellets and distilled through the apparatus described above to give 7–9 g. (60–80%) of cyclobutylamine, b.p. 80.5–81.5°, n^{25}D 1.4356 (Notes 6, 7).

2. Notes

1. Cyclobutanecarboxylic acid was purchased from the Aldrich Chemical Co., Milwaukee, Wisconsin. A synthesis of the acid is described in *Org. Syntheses*, Coll. Vol. **3**, 213 (1955).

2. Eastman practical grade was used.

3. The sodium azide is added at such a rate that a gentle reflux of vapors in the powder funnel is maintained. After somewhat more than the theoretical amount of azide has been added, the rate of addition may be much more rapid.

4. The distillation should be carried out carefully at first until all the chloroform has distilled. A distilling adapter dipping just below the surface of the acid solution should be used in order to

minimize loss of cyclobutylamine.   Care must be taken that the basic solution in the distillation flask which still contains sodium azide does not come in contact with the hydrochloric acid solution in the receiver.

5. A water aspirator is sufficient.

6. Contact of the amine with the atmosphere should be avoided since the amine reacts with carbon dioxide.

7. The purity of the product was checked by vapor phase chromatography on a polyethylene glycol on Teflon column at 72°, 15 p.s.i., and a flow rate of 102 ml. of helium per minute. The sample appeared to be homogeneous, but, since the amine tails badly on the column, it is not possible to detect the presence of a small amount of water (less than 3%).

An n.m.r. spectrum of cyclobutylamine in carbon tetrachloride showed no resonance signals at less than 1 p.p.m. from tetramethylsilane.   This suggests that no cyclopropylcarbinyl-amine was formed by rearrangement during the reaction.

3. Methods of Preparation

The preparation of cyclobutylamine from cyclobutanecarboxylic acid and hydrazoic acid has been reported previously.[2, 3] Cyclobutylamine has also been prepared by the Hofmann-type rearrangement of cyclobutanecarboxamide.[4-7]   More recently it has been prepared in 82–87% overall yield from cyclobutanecarboxamide by oxidative rearrangement with lead tetraacetate or iodosobenzene diacetate.[8]

4. Merits of the Preparation

This procedure permits the synthesis of cyclobutylamine from cyclobutanecarboxylic acid in one step and in high yield.   The procedures involving the Hofmann rearrangement[4-7] require the preparation of the amide from the acid and afford lower yields of the amine.

The interest in the synthesis of compounds containing the cyclobutyl ring system is due to the observation that reactions which are thought to proceed through cationic intermediates give rise to rearrangement products.   For example, deamination of cyclobutylamine in aqueous solution gives cyclopropylcarbinol and allylcarbinol as well as cyclobutanol.[9] Recent investigations have been concerned with the exact nature of these cationic intermediates.[10,11]

1. Department of Chemistry, California State College at Los Angeles, Los Angeles, California.
2. G. B. Heisig, *J. Phys. Chem.*, **43**, 1207 (1939).
3. D. C. Iffland, G. X. Criner, M. Koral, F. J. Lotspeich, Z. B. Papanastassiou, and S. M. White, *J. Am. Chem. Soc.*, **75**, 4044 (1953).
4. M. Freund and E. Gudeman, *Ber.*, **21**, 2692 (1888).
5. W. H. Perkin, Jr., *J. Chem. Soc.*, **65**, 950 (1894).
6. N. Zelinsky and J. Gutt, *Ber.*, **40**, 4744 (1907).
7. N. J. Demjanow and Z. I. Shuikina, *J. Gen. Chem. (USSR) (Eng. Transl.)*, **5**, 1213 (1935).
8. H. L. Smith, Ph.D. Thesis, University of Nebraska-Lincoln, June, 1970, p. 89, 118.
9. N. J. Demjanow, *Ber.*, **40**, 4393, 4961 (1907).
10. J. D. Roberts and R. H. Mazur, *J. Am. Chem. Soc.*, **73**, 2509 (1951).
11. R. H. Mazur, W. N. White, D. A. Semenow, C. C. Lee, M. S. Silver, and J. D. Roberts, *J. Am. Chem. Soc.*, **81**, 4390 (1959).

CYCLODECANONE*

Submitted by R. D. BURPITT and J. G. THWEATT [1]
Checked by WILLIAM G. DAUBEN, MICHAEL H. McGANN, and
NOEL VIETMEYER

1. Procedure

To a 500-ml. round-bottomed flask fitted with a 25- to 30-cm. column packed with glass helices to which is attached a water separator [2] filled with hexane (Note 1) are added 126 g. (1.00 mole) of cyclooctanone (Note 2), 100 g. (1.4 moles) of pyrrolidine, 100 ml. of xylene, and 0.5 g. of p-toluenesulfonic acid. The solution is heated under reflux until the separation of water ceases (Note 3). The water separator is replaced by a distillation head, and the reaction mixture is distilled through the column under reduced pressure to remove solvent and unreacted starting materials. When the head temperature reaches 50° (1 mm.), distillation is stopped, and the residue of almost pure N-(1-cyclo-octen-1-yl)pyrrolidine (152–161 g.) is used in the next step without further purification (Note 4).

The crude enamine is dissolved in 450 ml. of ether, and the solution is transferred to a 1-l. three-necked flask equipped with a sealed stirrer, a 250-ml. dropping funnel, and a two-necked adapter fitted with a calcium chloride tube and a thermometer immersed in the solution. A solution of 71–76 g. (0.85–0.90 mole) (Note 5) of methyl propiolate (*Caution! Methyl propiolate is a severe lachrymator and should be handled only in the hood.*) in 150 ml. of ether is added dropwise. During the addition the temperature of the mixture is maintained at 25–30° by periodic cooling of the reaction flask in a dry ice-acetone bath. When the addition is almost complete, a white solid begins to separate. The mixture is stirred at 25–30° for an additional hour, cooled to 0°, and filtered to remove the solid. This is dissolved in 700 ml. of 6% hydrochloric acid (Note 6), the acidic solution is warmed at 55–60° for 1 hour, and the mixture is cooled and extracted with two 100-ml. portions of ether. The ether is removed on a steam bath, and the residue of crude methyl 10-oxocyclodec-2-ene-1-carboxylate is dissolved in 300 ml. of methanol and hydrogenated over 5 g. of 5% palladium-on-alumina catalyst at 40 p.s.i. pressure and room temperature.

The catalyst is filtered, 200 g. (155 ml.) of 25% aqueous sodium hydroxide is added to the filtrate, and the mixture is heated under reflux for 1 hour. The condenser is replaced by a short Vigreux column and distillation head, and the heating is continued until most of the methanol has distilled. The two-phase residue is cooled and extracted with two 100-ml. portions of ether. The ether is removed on a steam bath, and the residue is distilled through a 20-cm. Vigreux column to yield 68–77 g. (44–50%) of cyclodecanone, b.p. 94–98° (10 mm.), m.p. 20–22° (Note 7).

2. Notes

1. If hexane is not used in the trap, an excessive amount of pyrrolidine is lost in the aqueous layer.

2. Cyclooctanone from Aldrich Chemical Co., methyl propiolate from Farchan Research Laboratories, and pyrrolidine from Eastman Organic Chemicals were used as received.

3. The reaction is usually complete after 3–6 hours at reflux. Owing to dissolved pyrrolidine, the aqueous layer amounts to 35–45 ml., and thus its volume is not a good measure of the extent of reaction.

4. Pure N-(1-cycloocten-1-yl)pyrrolidine, b.p. 76–78° (1 mm.), may be isolated by distillation through a Vigreux column.

5. The amount used should be adjusted to be equimolar with the amount of crude enamine.

6. This solid intermediate is reasonably stable to storage under nitrogen; however, the yield in the acid hydrolysis step is better when freshly prepared material is hydrolyzed immediately.

7. The same reaction sequence may be used to convert cyclododecanone to cyclotetradecanone. Preparation of the pyrrolidine enamine of cyclododecanone requires 2–3 days at reflux, and reaction of the enamine with methyl propiolate is best carried out in refluxing hexane. The enamine-propiolate reaction may also be used to convert cycloheptanone to cyclononanone. In this case the procedure must be modified to provide for partial hydrogenation of the intermediate amino ester without prior hydrolysis.[3] The reduced intermediate is saponified as described in the present procedure.

3. Methods of Preparation

Cyclodecanone has been obtained together with other products in the pyrolysis of the thorium or yttrium salts of nonanedioic acid.[4] It has also been prepared by reduction of sebacoin with zinc and hydrochloric acid,[5, 6] by dehydration of sebacoin followed by catalytic hydrogenation,[7] by ring enlargement of cyclononanone with diazomethane [8, 9] and of cyclooctanone with diazomethane in the presence of a Lewis acid catalyst,[9] by hydroboration of 1,2-cyclodecadiene followed by oxidation of the organoborane,[10] and by the present procedure.[3]

4. Merits of the Preparation

The chief merits of this preparation are its simplicity and the high purity of the product. Although the synthesis involves several steps, each step is a simple operation, and all intermediates may be used in the subsequent steps without purification. The purity of even the crude product is high, and any impurities which may be present are readily removed by a simple distillation.

The overall yield of cyclodecanone is comparable to the overall yield obtained by conversion of dimethyl sebacate to sebacoin [11] and subsequent reduction to cyclodecanone.[6] In addition, the present procedure does not require the use of a high-speed stirrer, the rigorous exclusion of air, and the high dilution that are necessary in preparing sebacoin.

1. Research Laboratories, Tennessee Eastman Company, Division of Eastman Kodak Company, Kingsport, Tennessee 37662.
2. S. Natelson and S. Gottfried, *Org. Syntheses*, Coll. Vol. **3**, 381 (1955).
3. K. C. Brannock, R. D. Burpitt, V. W. Goodlett, and J. G. Thweatt, *J. Org. Chem.*, **29**, 818 (1964).
4. L. Ruzicka, M. Stoll, and H. Schinz, *Helv. Chim. Acta*, **9**, 249 (1926); **11**, 670 (1928).
5. V. Prelog, L. Frenkiel, M. Kobelt, and P. Barman, *Helv. Chim. Acta*, **30**, 1741 (1947).
6. A. C. Cope, J. W. Barthel, and R. D. Smith, *Org. Syntheses*, Coll. Vol. **4**, 218 (1963).
7. M. Stoll, *Helv. Chim. Acta*, **30**, 1837 (1947).
8. E. P. Kohler, M. Tishler, H. Potter, and H. T. Thompson, *J. Am. Chem. Soc.*, **61**, 1057 (1939).
9. E. Muller, M. Bauer, and W. Rundel, *Tetrahedron Lett.*, No. 13, 30 (1960).
10. D. Devaprabhakara and P. D. Gardner, *J. Am. Chem. Soc.*, **85**, 1458 (1963).
11. N. L. Allinger, *Org. Syntheses*, Coll. Vol. **4**, 840 (1963).

cis-CYCLODODECENE*

Submitted by MASAJI OHNO and MASARU OKAMOTO [1]
Checked by FREDERICK J. SAUTER and HERBERT O. HOUSE

1. Procedure

In a 2-l., three-necked, round-bottomed flask equipped with a mechanical stirrer, an efficient condenser, and an air inlet tube (Note 1) are placed 60.0 g. (0.370 mole) of *cis,trans,trans*-1,5,9-cyclododecatriene (Note 2), 224.4 g. (7.00 moles) of 95% hydrazine (Note 3), 350 ml. of 95% ethanol, and 3.0 g. (0.012 mole) of copper(II) sulfate pentahydrate (Note 4). Air is bubbled through the reaction mixture (Note 5) with vigorous stirring for 8–12 hours or longer until the reaction mixture contains primarily the desired *cis*-monoolefin (Note 6). During the early stages in the reaction a considerable amount of heat is generated and the temperature of the reaction mixture rises to 50–60°.[19]

When the reaction has progressed to the desired stage (Note 6), the flow of air is stopped and the mixture is filtered. After the filtrate has been extracted with two 350-ml. portions of petroleum ether (b.p. 30–60°), the combined hydrocarbon extracts are washed successively with two 100-ml. portions of 2N hydrochloric acid and three 100-ml. portions of water. The petroleum ether is distilled from the solution, heated in a water bath, through a 60-cm. Vigreux column, and the residual liquid is distilled under reduced pressure. The fraction, b.p. 64–65° (1.0 mm.) or 132–134° (35 mm.), is collected as 39.5–52.0 g. (64–85%) of colorless liquid, n^{25}D 1.4846–1.4850. This distillation fraction contains (Note 6) 80–90% of the *cis*-cyclododecene (51–76%) accompanied by 10–20% of a mixture of cyclododecane and *cis,trans*-1,5-cyclododecadiene (Note 7). If desired, the *cis*-cyclododecene may be further purified by preparative chromatography or separation of the silver nitrate-olefin addition complex (Note 8).

2. Notes

1. An air inlet tube with a sintered-glass disk or cylinder at the end immersed in the solution is recommended.

2. The submitters used material available from Hüls Company in Germany. This material was contaminated with 1–3% of the more easily reduced *trans,trans,trans*-1,5,9-cyclododecatriene. The checkers purchased the starting triene from Aldrich Chemical Company, Inc. The gas chromatogram (see Note 6) of this material exhibited no peak corresponding to the all-*trans*-triene, an indication that less than 1% of this contaminant was present.

3. The submitters had specified the use of either hydrazine hydrate (aqueous 85% hydrazine) or aqueous 80% hydrazine. The checkers observed only partial reduction of the triene and intermediate diene under these conditions, apparently because sufficient water was present in the reaction mixture to prevent adequate partitioning of the olefins between the hydrocarbon layer and the aqueous ethanolic layer containing the diimide. The checkers avoided this difficulty by use of hydrazine containing less than 5% water (95+ % hydrazine) available from Olin Mathieson Chemical Company or from Eastman Organic Chemicals. This difficulty could probably also be avoided by use of absolute ethanol rather than 95% ethanol.

4. Copper(II) acetate can also be used.

5. The rate of air flow, measured with a precalibrated mercury flow meter in the gas inlet tube, was adjusted to 400–450 ml. per minute.

6. In order to stop the reaction when the amount of monoolefinic product in the reaction mixture is highest, aliquots of the reaction mixture are removed at intervals and analyzed by infrared spectrometry or by gas chromatography. In the infrared spectrum the relative intensities of bands at 965 cm.$^{-1}$ (*trans-*

CH=CH) and 702 cm.$^{-1}$ (*cis*-CH=CH) are observed in successive aliquots. The reaction is stopped when the band at 965 cm.$^{-1}$, attributable to the *trans* double bonds of the starting triene, has almost completely disappeared and the band at 702 cm.$^{-1}$ (*cis*-olefin) remains.

Gas chromatographic analyses are obtained at about 120° with a 2 m. x 7 mm. column packed with a suspension of 5% (by weight) of silver nitrate and 15% (by weight) of Carbowax 6000 (polyethylene glycol) on either Chromosorb P or Celite 545. With this column the relative retention times of the various possible components in the reaction mixture are: cyclododecane, 1.00; *trans,trans,trans*-1,5,9-cyclododecatriene, 1.20; *trans*-cyclododecene, 1.13; *cis*-cyclododecene, 1.33; *cis,trans*-1,5-cyclododecadiene, 1.51; *cis,trans-trans*-1,5,9-cyclododecatriene, 1.72. The reaction should be stopped when the rate of reduction of *cis,trans*-1,5-cyclododecadiene to *cis*-cyclododecene has become approximately equal to the rate of conversion of the *cis*-monoolefin to cyclododecane.

7. The submitters reported obtaining a product after a 60–72 hour reaction period which contained 91%–95% of the *cis*-monoolefin and 5–9% of cyclododecane with no *trans*-monoolefin being detected. The checkers found the maximum amount of *cis*-monoolefin was present in the reaction mixture after a reaction period of 8–12 hours. At this time the resulting distilled product had the approximate composition: cyclododecane, 8%; *trans*-cyclododecene, 3%; *cis*-cyclododecene, 80%; and *cis,trans*-1,5-cyclododecadiene, 9%. The use of longer reaction times resulted in the further reduction of the *cis*-monoolefin to cyclododecane more rapidly than it was produced from the residual *cis,trans*-diene.

8. The conversion of the *cis*-monoolefin to its silver nitrate complex [16] was accomplished by adding 1.66 g. (0.010 mole) of the distilled reaction product to a solution of 1.70 g. (0.010 mole) of silver nitrate in 50 ml. of boiling methanol. The resulting solution, when cooled, deposited the complex as white needles, m.p. 79° dec.; recrystallization from methanol separated 1.0 g. of the complex, m.p. 80° dec. After this complex had been partitioned between water and ether, the ether phase was separated,

dried over magnesium sulfate, and concentrated. Distillation of the residual liquid in a short path still separated 0.45 g. of the pure (Note 6) *cis*-cyclodecene, b.p. 70° (1.0 mm.), n^{25}D 1.4852.

3. Discussion

Cyclododecene may be prepared from 1,5,9-cyclododecatriene by the catalytic reduction with Raney nickel and hydrogen diluted with nitrogen,[2] with nickel sulfide on alumina,[3] with cobalt, iron, or nickel in the presence of thiophene,[4] with palladium on charcoal,[5] with palladium chloride in the presence of water,[6] with palladium on barium sulfate,[7] with cobalt acetate in the presence of cobalt carbonyl,[8] and with cobalt carbonyl and tri-*n*-butyl phosphine.[9] It may also be obtained from the triene by reduction with lithium and ethylamine,[10] by disproportionation,[11, 12] by epoxidation followed by isomerization to a ketone and Wolff-Kishner reduction,[13] and from cyclododecanone by the reaction of its hydrazone with sodium hydride.[14]

These methods generally afford a mixture of *cis-* and *trans*-cyclododecene. *cis*-Cyclododecene has also been prepared by the reduction of cyclododecyne with Lindlar catalyst,[15, 16] and from 1,5-cyclododecadiene [17] or from 1,2-dichlorocyclododecane.[18] The *cis*-olefin is usually obtained as a minor product from the Hofmann degradation of cyclododecyltrimethylammonium hydroxide [15] and from the pyrolysis of cyclododecyl acetate.

The procedure described is based on the selective reduction with diimide described by Ohno and Okamoto [19] and by Nozaki and Noyori.[20] It illustrates the generation of diimide from the air oxidation of hydrazine and the use of diimide for the selective reduction of the *trans* double bond in *cis,trans,trans*-1,5,9-cyclododecatriene, the product of trimerization of butadiene.[21]

The use of diimide provides a particularly convenient and general method for the selective reduction of *trans* double bonds of medium ring systems.[22] The *cis*-cyclododecene produced in this selective reduction is thermodynamically less stable than the corresponding *trans*-isomer.[23]

1. Basic Research Laboratories, Toray Industries, Inc., Kamakura, Japan.
2. Fr. Patent 1,357,114 (1964) [*Chem. Abstr.*, **61**, 5536 (1964)].

3. Belg. Patent 634,763 (1964) [*Chem. Abstr.*, **61**, 13214 (1964)].
4. Ger. Patent 1,226,568 (1966) [*Chem. Abstr.*, **66**, 10657 (1967)].
5. Neth. Appl. 6,412,540 (1964) [*Chem. Abstr.*, **63**, 11390 (1965)].
6. Neth. Appl. 6,507,159 (1965) [*Chem. Abstr.*, **64**, 19448 (1966)].
7. L. I. Zakharkin and V. V. Korneva, *Zh. Organ. Khim.*, **1**, 1608 (1965) [*Chem. Abstr.*, **64**, 611 (1966)].
8. U.S. Patent 3,308,177 [*Chem. Abstr.*, **67**, 21504 (1967)].
9. A. Misono and I. Ogata, *Bull. Chem. Soc. Japan*, **40**, 2718 (1967).
10. U.S. Patent 3,173,964 (1965) [*Chem. Abstr.*, **63**, 515 (1965)].
11. U.S. Patent 3,182,093 (1965) [*Chem. Abstr.* **63**, 515 (1965)].
12. Fr. Patent 1,389,362 (1965) [*Chem. Abstr.*, **63**, 514 (1965)].
13. W. Stumpf and K. Rombusch, *Ann.*, **687**, 136 (1965).
14. A. P. Krapcho and J. Diamanti, *Chem. Ind.* (*London*), 847 (1965).
15. V. Prelog and M. Speck, *Helv. Chim. Acta*, **38**, 1786 (1955).
16. M. Svoboda and J. Sicher, *Chem. Ind.* (*London*), 290 (1959).
17. U.S. Patent 3,294,853 (1966) [*Chem. Abstr.*, **66**, 46145 (1967)].
18. W. Ziegenbein and W. M. Schneider, *Ber.*, **98**, 824 (1965).
19. M. Ohno and M. Okamoto, *Tetrahedron Lett.*, 2423 (1964); M. Ohno, M. Okamoto, and S. Torimitsu, *Bull. Chem. Soc. Japan*, **39**, 316 (1966).
20. H. Nozaki and R. Noyori, *J. Org. Chem.*, **30**, 1652 (1965).
21. G. Wilke, *Angew. Chem.*, **75**, 10 (1963).
22. J. G. Traynham, G. R. Franzen, G. A. Knesel, and D. J. Northington, Jr., *J. Org. Chem.*, **32**, 3285 (1967).
23. A. C. Cope, P. T. Moore, and W. R. Moore, *J. Am. Chem. Soc.*, **82**, 1744 (1960).

1,3-CYCLOHEXADIENE *

Submitted by JOHN P. SCHAEFER and LELAND ENDRES [1]
Checked by R. J. CRAWFORD and PETER YATES

1. Procedure

In a 3-l. three-necked, round-bottomed flask fitted with a mechanical stirrer and set up for a simple vacuum distillation are placed 500 ml. of triethylene glycol dimethyl ether (Note 1) and 300 ml. of isopropyl alcohol. Mechanical stirring is started, and 53.5 g. (2.23 moles) of sodium hydride in a mineral oil suspension is added in small portions. After the addition is complete, the remaining neck of the flask is fitted with a Y-tube to which is connected a two-holed rubber stopper containing a thermometer which reaches into the flask below the liquid level and a piece of glass tubing which is connected to a nitrogen tank. A pressure-equalizing dropping funnel containing 242 g. (1.00 mole) of 1,2-dibromocyclohexane (Note 2) is placed in the other arm of the Y-tube.

The temperature of the reaction flask is raised to 100–110°, and the receiving flask is cooled in a dry ice-isopropyl alcohol bath as a rapid stream of nitrogen is passed through the system. After most of the isopropyl alcohol has been removed by distillation (Note 3), the receiver is changed, and the system is evacuated by a water aspirator (Note 4). Dropwise addition of 1,2-dibromocyclohexane is begun, and the rate of addition is adjusted so that the temperature of the reaction mixture is maintained at 100–110° without external heating. The addition requires about 30 minutes; the reaction is terminated when distillation becomes very slow.

The distillate is washed four times with 200-ml. portions of water, and the organic layer is dried with anhydrous magnesium sulfate. The yield of 1,3-cyclohexadiene is 56 g. (70%) (Note 5). The diene can be separated from higher-boiling contaminants by a simple distillation at atmospheric pressure under nitrogen; b.p. 78–80°, yield 28–32 g. (35–40%) (Note 6).

2. Notes

1. 1,2-Bis(methoxyethoxy)ethane (triethylene glycol dimethyl ether) was obtained from Matheson, Coleman and Bell and used without further purification.

2. The 1,2-dibromocyclohexane was prepared by the method of Snyder and Brooks.[2] If the cyclohexene is cooled to *ca.* $-30°$ with a dry ice-isopropyl alcohol bath and the bromine is not diluted, it is possible to run this preparation on a threefold scale in one-third of the recorded time. The product was always purified by the recommended procedure.

3. If the flow of nitrogen is rapid, the distillation can be completed in about 1 hour; otherwise the distillation is very slow.

4. A dry ice-isopropyl alcohol trap was inserted before the

aspirator to catch any uncondensed product. The checkers also inserted a manometer between this trap and the aspirator, and maintained the pressure during the reaction at 130–170 mm. by careful adjustment of the regulator valve of the nitrogen cylinder.

5. If the temperature rises too high or the vacuum is not sufficient to flash out the diene as it forms, the product will be contaminated with small amounts of cyclohexene, benzene, and 1,4-cyclohexadiene.

6. The checkers found that distillation without the use of a nitrogen atmosphere gave 43–44 g. (54–55%) of product, b.p. 80–83°, of excellent purity as shown by n.m.r. spectroscopy.

3. Methods of Preparation

1,3-Cyclohexadiene has been prepared by dehydration of cyclohexen-3-ol,[3] by pyrolysis at 540° of the diacetate of cyclohexane-1,2-diol,[4] by dehydrobromination with quinoline of 3-bromocyclohexene,[5] by treating the ethyl ether of cyclohexen-3-ol with potassium bisulfate,[6, 7] by heating cyclohexene oxide with phthalic anhydride,[8] by treating cyclohexane-1,2-diol with concentrated sulfuric acid,[9] by treatment of 1,2-dibromocyclohexane with tributylamine,[10] with sodium hydroxide in ethylene glycol,[10] and with quinoline,[6] and by treatment of 3,6-dibromocyclohexene with sodium.[6]

4. Merits of the Preparation

Because of its convenience and simplicity this procedure is the method of choice for laboratory preparation of 1,3-cyclohexadiene. This olefin is an intermediate of some importance because it offers a route via the Diels-Alder reaction to a variety of bicyclic compounds.[4, 7, 10]

1. Department of Chemistry, The University of Arizona, Tucson, Arizona.
2. H. R. Snyder and L. A. Brooks, *Org. Syntheses*, Coll. Vol. **2**, 171 (1943).
3. G. Clement and J. Balaceanu, U. S. Patent 3,096,376 (1963) [*C.A.*, **59**, 13842 (1963)].
4. W. J. Bailey and W. B. Lawson, *J. Am. Chem. Soc.*, **79**, 1444 (1957).
5. N. A. Domnin and M. A. Larionova, *Zh. Obshch. Khim.*, **26**, 1398 (1956).
6. N. A. Domnin and A. S. Beletskaya, *Zh. Obshch. Khim.*, **24**, 1636 (1954).
7. R. Seka and O. Tramposch, *Ber.*, **75**, 1379 (1942).

8. P. Bedos and A. Ruyer, *Compt. Rend.*, **188**, 962 (1929).
9. J. B. Senderens, *Compt. Rend.*, **177**, 1183 (1923).
10. J. Hine, J. A. Brown, L. H. Zalkow, W. E. Gardner, and M. Hine, *J. Am. Chem. Soc.*, **77**, 594 (1955).

1,4-CYCLOHEXANEDIONE *

$$2 \begin{array}{c} CH_2CO_2C_2H_5 \\ | \\ CH_2CO_2C_2H_5 \end{array} \quad \xrightarrow[\text{2. } H_2SO_4]{\text{1. } NaOC_2H_5, -2C_2H_5OH}$$

Submitted by ARNOLD T. NIELSEN and WAYNE R. CARPENTER[1]
Checked by WILLIAM G. DAUBEN and E. JOHN DEVINY

1. Procedure

A. *2,5-Dicarbethoxy-1,4-cyclohexanedione.* A solution of sodium ethoxide is prepared by adding small pieces of sodium (92 g., 4 g. atoms) as rapidly as possible to 900 ml. of commercial absolute ethanol contained in a 3-l., three-necked, round-bottomed flask equipped with two stoppers and a reflux condenser fitted with a drying tube packed with calcium chloride and soda lime. The reaction is completed by heating the mixture under reflux for 3–4 hours (Note 1). To the hot solution is added diethyl succinate (348.4 g., 2 moles) (Note 2) in one portion (*Caution! Exothermic reaction*), and the mixture is heated under reflux by maintaining the original bath temperature for 24 hours. A thick pink-colored precipitate is formed almost immediately and remains throughout the reaction.

At the end of the 24-hour period, the ethanol is removed under reduced pressure on a steam bath. A 2N sulfuric acid solution (2 l.) is added to the warm residue, and the mixture is stirred vigorously for 3–4 hours (Note 3). The solid is removed by suction filtration and washed several times with water. The air-dried product is a pale-buff powder weighing 180–190 g., m.p. 126–128°. The solid is added to 1.5 l. of ethyl acetate, the mixture is heated to boiling and is filtered rapidly while hot (Note 4). The filtrate is chilled, and it yields cream to pink-cream colored crystals of 2,5-dicarbethoxy-1,4-cyclohexanedione, 160–168 g., m.p. 126.5–128.5°. The filtrate is concentrated to one-tenth of its original volume in order to obtain a second crop of crystals, 5–7 g., m.p. 121–125°. The total yield is 165–175 g. (64–68%).

B. *1,4-Cyclohexanedione.* The purified 2,5-dicarbethoxy-1,4-cyclohexanedione (170 g., 0.66 mole) (Note 5) and 170 ml. of water are placed in a glass liner (vented) of a steel pressure vessel of 1.5-l. capacity (fitted with a pressure-release valve). The vessel is sealed, heated as rapidly as possible to 185–195°, and kept at this temperature for 10–15 minutes (Note 6). The reaction vessel is immediately removed from the heater, placed in a large tub of ice water, and cooled to room temperature. The gas pressure then is carefully released. The resulting yellow to orange liquid is transferred to a distillation flask with the aid of a minimum volume of ethanol, and most of the water and ethanol is removed under reduced pressure by means of a rotary evaporator. The flask is attached to a short heated column fitted with a short air condenser. The remainder of the water and ethanol is removed under reduced pressure, and the 1,4-cyclohexanedione is distilled, b.p. 130–133° (20 mm.). The product solidifies to a white to pale-yellow solid, m.p. 77–79°, yield 60–66 g. (81–89% yield from 2,5-dicarbethoxy-1,4-cyclohexanedione). The compound may be conveniently recrystallized from carbon tetrachloride (7 ml. per gram of dione); the purified product is obtained as white plates, m.p. 77–79° (90% recovery).

2. Notes

1. A heating bath containing a liquid heat exchanger such as hydrogenated cottonseed oil should be used. Employment of an electric heating mantle may cause extreme charring in the later stages of the reaction.

2. The diethyl succinate was obtained from Eastman Organic Chemicals and used without purification.

3. The lumps of the sodium salt of 2,5-dicarbethoxy-1,4-cyclohexanedione should be completely reacted before the filtration step. If desired, the mixture may be stirred overnight at this point. The checkers found that in some runs a rock-like precipitate persisted on the bottom of the flask, and it had to be broken up manually by using a spatula with care.

4. A large fluted filter paper and a heated funnel are recommended for the filtration. The dark insoluble material which is removed by this process quickly fills the pores of the filter paper; more than one filter paper may be required. If a large amount of material remains in the filter, the material should be treated with additional ethyl acetate, the mixture filtered, and the filtrate combined with the first filtrate.

5. Use of unpurified ester results in a much lower yield of 1,4-cyclohexanedione.

6. An electrically heated pressure bomb, 4.5 in. in diameter, of 1.5-l. capacity, was employed (American Instrument Company, Model E 1143, cold-tested to 23,000 p.s.i.). About 90 minutes was required to raise the temperature from 25° to 185°.

3. Methods of Preparation

2,5-Dicarbethoxy-1,4-cyclohexanedione has been prepared by the self-condensation of diethyl succinate by use of sodium or sodium ethoxide catalyst (with or without a solvent)[2-10] and by reaction of ethyl 4-bromo-3-ketobutanoate[11] or ethyl 4-chloro-3-ketobutanoate[12,13] with sodium ethoxide in ethanol.

1,4-Cyclohexanedione has been prepared by hydrolysis and decarboxylation of 2,5-dicarbethoxy-1,4-cyclohexanedione by using concentrated sulfuric acid,[14] aqueous alcoholic phosphoric acid,[15] or water at 195-200°,[7,8,16] and by peroxyvanadic acid oxidation of cyclohexanone.[17]

4. Merits of the Preparation

The present procedure is simpler than others previously described and gives equally good yields. It is easily adapted to the preparation of large quantities of either the diester or the diketone. It can be extended to the preparation of various alkylated 1,4-cyclohexanediones[18] and bicyclic diketodicarboxylic esters such as diethyl bicyclo[2.2.2]octane-2,5-dione-1,4-dicarboxylate.[19,20] 1,4-Cyclohexanedione is a useful intermediate for the preparation of 1,4-substituted cyclohexanes such as the dioxime,[21] diamine,[22] 1,4-dichloro-1,4-dinitrosocyclohexane,[23] and 1,4-dinitrocyclohexane.[24] It is also the precursor of 7,7,8,8-tetracyanoquinodimethan.[25]

1. Organic Branch, Chemistry Division, Naval Weapons Center, China Lake, California.
2. H. Ebert, *Ann.*, **229**, 52 (1885).
3. A. Jeanrenaud, *Ber.*, **22**, 1282 (1889).
4. A. Piutti, *Gazz. Chim. Ital.*, **20**, 167 (1890).
5. H. Liebermann, *Ann.*, **404**, 272 (1914).
6. A. E. Uspenskii and I. Turin, *J. Russ. Phys.-Chem. Soc.*, **51**, 263 (1920) [*C.A.*, **18**, 1484 (1924)].
7. J. R. Vincent, A. F. Thompson, and L. I. Smith, *J. Org. Chem.*, **3**, 603 (1938).
8. H. Musso and D. Dopp, *Ber.*, **97**, 1147 (1964).
9. D. S. Deorha and S. K. Mukerji, *J. Indian Chem. Soc.*, **41**, 604 (1964).
10. C. R. Hauser and B. E. Hudson, Jr., *Org. Reactions*, **1**, 283 (1942).
11. W. Mewes, *Ann.*, **245**, 74 (1888).
12. V. M. Rodionov and M. A. Gubareva, *Zh. Obshch. Khim.*, **23**, 1830 (1953) [*C.A.*, **49**, 926 (1955)].
13. M. Sommelet and P. Couroux, *Bull. Soc. Chim. France*, [4] **29**, 403 (1921).
14. A. Baeyer, *Ann.*, **278**, 90 (1894).
15. W. von E. Doering and A. A.-R. Sayigh, *J. Org. Chem.*, **26**, 1365 (1961).
16. H. Meerwein, *Ann.*, **398**, 248 (1913).
17. J. Vene, *Bull. Soc. Sci. Bretagne*, **23**, 123 (1948) [*C.A.*, **44**, 6395 (1950)].
18. A. Baeyer, *Ber.*, **25**, 2122 (1892); **26**, 232 (1893).
19. P. C. Guha, *Ber.*, **72**, 1359 (1939).
20. J. D. Roberts, W. T. Moreland, Jr., and W. Frazer, *J. Am. Chem. Soc.*, **75**, 637 (1953).
21. A. Baeyer and W. A. Noyes, *Ber.*, **22**, 2168 (1889).
22. K. Hosino, *J. Chem. Soc. Japan*, **62**, 190 (1941) [*C.A.*, **36**, 5140 (1942)].
23. O. Piloty and H. Steinbock, *Ber.*, **35**, 3101 (1902).
24. A. T. Nielsen, *J. Org. Chem.*, **27**, 1993 (1962).
25. D. S. Acker and W. R. Hertler, *J. Am. Chem. Soc.*, **84**, 3370 (1962).

CYCLOHEXANONE DIALLYL ACETAL

$$C_6H_{10}O + 2CH_2{=}CHCH_2OH + (CH_3)_2C(OCH_3)_2 \xrightarrow{H^+}$$
$$C_6H_{10}(OCH_2CH{=}CH_2)_2 + (CH_3)_2CO + 2CH_3OH$$

Submitted by W. L. HOWARD and N. B. LORETTE.[1]
Checked by E. J. COREY and R. A. E. WINTER.

1. Procedure

A solution of 294 g. (3 moles) of cyclohexanone, 343 g. (3.3 moles) of acetone dimethyl acetal, 418 g. (7.2 moles) of allyl alcohol, 1 l. of benzene, and 0.2 g. of *p*-toluenesulfonic acid monohydrate (Note 1) is distilled using a good fractionating column until the acetone and the benzene-methanol azeotrope are completely removed (Note 2). The solution is cooled below the boiling point, and a solution of 0.5 g. of sodium methoxide in 20 ml. of methanol is added all at once with stirring (Note 3). Distillation is resumed, and unreacted allyl alcohol and benzene are removed at atmospheric pressure and then at reduced pressure (Note 4). Distillation is continued at a pressure in the range 5–20 mm. to remove forerun (on the order of 100 ml.) (Note 5). The cyclohexanone diallyl acetal, b.p. 84°/5 mm., 98°/10 mm., 114°/20 mm., n_D^{25} 1.4600, is then collected. The yield is 382–435 g. (65–74%). A small amount of higher-boiling residue remains.

2. Notes

1. Commercial acetone dimethyl acetal and allyl alcohol from The Dow Chemical Company and cyclohexanone from Eastman Kodak Company were used without further treatment.

2. These reaction products distil within a narrow range. The head temperature was maintained in the range 56–59°. About 750 ml. of distillate is collected, depending on the efficiency of fractionation. The combined amount of methanol and acetone may be estimated by washing an aliquot of the distillate with 2

volumes of water and taking the difference between the original volume and that of the residual benzene as the volume of methanol-acetone. Usually this is about 450 ml. The distillation should be as rapid as possible to avoid the formation of by-product 2-allylcyclohexanone. A 1.9 x 120 cm. vacuum-jacketed, silvered column packed with 0.25-in. glass helices and fitted with a vapor-dividing head controlled by a timed relay was used.

The checkers used a 1.3 x 92 cm. vacuum-jacketed, silvered column packed with 0.25-in. glass ring chains. With this column it was necessary to carry out the distillation of benzene-acetone-methanol using reflux ratios varying from 2:1 initially to 11:1 at the conclusion. The use of a shorter column is not satisfactory.

3. Other soluble, non-volatile bases may be used to neutralize the acid. The reactants may be kept at room temperature safely after addition of base.

4. An azeotrope of benzene and allyl alcohol distils at about 77°, followed by benzene. When the temperature in the boiling flask reaches 120–130°, the pressure is reduced and the remaining benzene is taken to a cold trap.

5. The forerun contains some acetone diallyl acetal and about 35–40 g. of 2-allylcyclohexanone, b.p. 78°/10 mm.

3. Methods of Preparation

Cyclohexanone diallyl acetal has been prepared from cyclohexanone and allyl orthosilicate[2] and by the above procedure.[3]

4. Merits of Preparation

The preparation given here is operable for a large number of ketone acetals, including those formed from both primary and secondary alcohols and from alcohols and ketones containing other functional groups which are stable under the conditions used.[3]

1. The Dow Chemical Company, Texas Division, Freeport, Texas.
2. B. Helferich and J. Hausen, *Ber.*, 57B, 795 (1924) [*C.A.*, 18, 2869 (1924)].
3. N.B. Lorette and W. L. Howard, *J. Org. Chem.*, 25, 521 (1960).

2-CYCLOHEXENONE*

$$\text{(structure with O, OC}_2\text{H}_5\text{)} + 2[H] \rightarrow \text{(structure with OH, OC}_2\text{H}_5\text{)}$$

$$\text{(structure with OH, OC}_2\text{H}_5\text{)} \xrightarrow{\text{H}_3\text{O}^{\oplus}} \text{(structure with O)} + \text{C}_2\text{H}_5\text{OH}$$

Submitted by WALTER F. GANNON and HERBERT O. HOUSE.[1]
Checked by WILLIAM E. PARHAM, ALLAN M. HUFFMAN,
GEORGE J. MEISTERS, and WAYLAND E. NOLAND.

1. Procedure

In a dry 500-ml. three-necked flask, equipped with a reflux condenser, a mechanical stirrer, and a dropping funnel and protected from atmospheric moisture with drying tubes, are placed 6.0 g. (0.16 mole) of lithium aluminum hydride and 200 ml. of anhydrous ether. A solution of 43 g. (0.307 mole) of 3-ethoxy-2-cyclohexenone (Note 1) in 50 ml. of anhydrous ether is added, dropwise and with stirring, to the reaction flask at a rate which maintains gentle refluxing of the solvent (Note 2). After the addition is complete, the reaction solution is boiled under reflux for an additional 30 minutes and then allowed to cool. The complex is hydrolyzed and the excess lithium aluminum hydride is destroyed by the cautious addition, dropwise and with stirring, of 15 ml. of water (Note 3). The resulting reaction mixture is poured into 500 ml. of cold aqueous 10% sulfuric acid. The ether layer which forms is separated, and the residual aqueous phase is extracted with three 300-ml. portions of ether. The combined ether solutions are washed successively with one 100-ml. portion of water and one 100-ml. portion of saturated, aqueous sodium bicarbonate solution and then dried ɔver magnesium sulfate. The ether is removed by distillation through a 50-cm. Vigreux column, and the residue is distilled under reduced pressure through a 40-cm. spinning-band column (Note 4). The yield of 2-cyclo-hexenone (Note 5), b.p. 56–57.5°/10 mm. or 96–97°/72 mm., n_D^{27} 1.4858, is 18.2–22.1 g. (62–75%).

2. Notes

1. The preparation of 3-ethoxy-2-cyclohexenone is described elsewhere in this volume.[2]   (It has been reported that 3-isobutoxy-2-cyclohexenone can be prepared in somewhat higher yield than the 3-ethoxy compound by the same proce-dure and that the 3-isobutoxy compound is just as useful for the present preparation (private communication from D. A. H. Taylor).

2. This addition requires approximately 1.5 hours.

3. The addition of water is accompanied by foaming, and care must be taken to avoid excessive loss of the solvent.

4. The 2-cyclohexenone obtained by an ordinary distillation at this point is contaminated with lower-boiling impurities (see Note 5), primarily ether and ethanol.

5. The purity of the 2-cyclohexenone may be assayed by gas chromatography on an 8 mm. x 215 cm. column heated to 125° and packed with di-(2-ethylhexyl) sebacate suspended on ground firebrick. This method of analysis indicates that the 3-cyclo-hexenone in the product amounts to no more than 3%. The fore-run from this fractional distillation contains substantial amounts of 2-cyclohexenone accompanied by ether, ethanol, and minor amounts of other lower-boiling impurities. Additional quantities of pure 2-cyclohexenone can be recovered by redistilla-tion of this fore-run. The preparation of 2-cyclohexenone has been run on twice the scale described with no loss in yield. The ultraviolet spectrum of an ethanol solution of the 2-cyclohexenone obtained has a maximum at 226 mμ (ε = 10,400).

3. Methods of Preparation

2-Cyclohexenone has been prepared by dehydrohalogenation of

2-bromocyclohexanone,[3,4] by the hydrolysis and oxidation of 3-chlorocyclohexene,[5] by the dehydration of α-hydroxycyclohexanone,[6] by the oxidation of cyclohexene with chromic acid [7] or hydrogen peroxide in the presence of a vanadium catalyst,[8] by the addition of acrolein to ethyl acetoacetate followed by cyclization, hydrolysis, and decarboxylation,[9] by the reduction of N,N-dimethylaniline with sodium and ethanol in liquid ammonia followed by hydrolysis,[10] by the reduction of anisole with lithium in liquid ammonia,[11] and by the reduction of 3-alkoxy-2-cyclohexanones with lithium aluminum hydride followed by acid-catalyzed hydrolysis and rearrangement.[12]

The procedure described illustrates a general method for the preparation of α,β-unsaturated aldehydes and ketones from the enol ethers of β-dicarbonyl compounds.[12–14]

1. Department of Chemistry, Massachusetts Institute of Technology, Cambridge, Massachusetts.
2. See p. 539, this volume.
3. E. A. Braude and E. A. Evans, *J. Chem. Soc.*, 607 (1954).
4. I. N. Nazarov, L. D. Bergel'son, I. V. Torgov, and S. N. Ananchenko, *Izvest. Akad. Nauk S. S. S. R., Otdel. Khim. Nauk*, 1953, 889; C.A., 49, 1082 (1955).
5. C. Courtot and J. Pierron, *Bull. Soc. Chim. France*, 45 (4), 286 (1929).
6. P. D. Bartlett and G. F. Woods, *J. Am. Chem. Soc.*, 62, 2933 (1940).
7. F. C. Whitmore and G. W. Pedlow, Jr., *J. Am. Chem. Soc.*, 63, 758 (1941).
8. W. Treibs, G. Franke, G. Leichsenring, and H. Roder, *Ber.*, 86, 616 (1953).
9. M. Mousseron, R. Jacquier, A. Fontaine, and R. Zagdoun, *Bull. Soc. Chim. France*, 1954, 1246.
10. A. J. Birch, *J. Chem. Soc.*, 593 (1948)
11. A. L. Wilds and N. A. Nelson, *J. Am. Chem. Soc.*, 75, 5360 (1953).
12. H. Born, R. Pappo, and J. Szmuskovicz, *J. Chem. Soc.*, 1779 (1953); M. Stiles and A. L. Longroy, *J. Org. Chem.*, 32, 1095 (1967).
13. P. Seifert and H. Schinz, *Helv. Chim. Acta*, 34, 728 (1951).
14. H. Favre, B. Marinier, and J. C. Richer, *Can. J. Chem.*, 34, 1329 (1956).

CYCLOHEXYLIDENECYCLOHEXANE

(Bicyclohexylidene)

Submitted by NICHOLAS J. TURRO,[1] PETER A. LEERMAKERS,[2] and GEORGE F. VESLEY [2]
Checked by ALEX G. FALLIS and PETER YATES

1. Procedure

A. *Dispiro[5.1.5.1]tetradecane-7,14-dione.* Cyclohexanecarbonyl chloride (Note 1) (30.0 g., 0.205 mole) and 250 ml. of dry benzene are placed in a three-necked, round-bottomed flask equipped with a stirrer, condenser, and dropping funnel. A nitrogen atmosphere is maintained in the system. Dry triethylamine (35.0 g., 0.35 mole) is slowly added, and the mixture is heated under reflux overnight. The amine hydrochloride is then filtered, and the filtrate is washed with dilute hydrochloric acid and with water. Solvent is removed on a steam bath, and the residue is recrystallized from ligroin-ethanol; yield 11–13 g. (49–58%), m.p. 161–162°.

B. *Cyclohexylidenecyclohexane.* In a Hanovia 450-watt immersion photochemical reactor (Note 2), equipped with a side arm attachment to monitor gas evolution, is placed 15 g. (0.068 mole) of dispiro[5.1.5.1]tetradecane-7,14-dione dissolved in 150 ml. of methylene chloride. The sample is irradiated, and carbon monoxide starts to evolve rapidly after a few minutes. Irradiation is continued until gas evolution has ceased, usually about 8–10 hours (Note 3). After the irradiation most of the solvent is removed on a steam bath. The residual oil is transferred to a sublimator. The sublimator, with the cold finger removed, is placed in a vacuum desiccator, and the system is evacuated to remove any remaining methylene chloride. The semisolid residue is then sublimed at 45° (1 mm.) to yield 7 g. (63%) of crude cyclohexylidenecyclohexane. The product after recrystallization from methanol weighs 5.5 g. (49%), m.p. 53–54°.

2. Notes

1. Cyclohexanecarbonyl chloride was obtained from Eastman Organic Chemicals.

2. The reactor, manufactured by the Hanovia Division of Engelhard Industries, consists of a water-jacketed Vycor well through which a stream of water is continuously passed. Since wavelengths shorter than 3000 Å are not needed, the immersion well may be made of Pyrex instead. Within the well is a No. 679A-36 450-watt medium-pressure mercury lamp, also manufactured by Hanovia, and a cylindrical Pyrex filter which surrounds the lamp. The well is placed in an appropriately shaped flask containing the solution to be irradiated. The flask is essentially cylindrical and is equipped with a side arm near the top through which gas can escape and be bubbled through a container of water. The flask is so designed that the liquid level is above the top of the lamp. The reaction vessel is quite similar to that shown in Fig. 1 (p. 65).

The same synthesis could be carried out in an ordinary flask using one or two sunlamps or sunlight, but the irradiation time would necessarily be much longer.

3. The system should be relatively free of oxygen during irradiation. Oxygen apparently combines with a photochemical intermediate to form cyclohexanone.[3] Under the conditions recommended in the procedure, oxygen is prevented from entering the system by the water trap which also serves as a monitor for gas evolution.

3. Methods of Preparation

Ethyl 1-bromocyclohexanecarboxylate, when treated with magnesium in anhydrous ether-benzene with subsequent addition of cyclohexanone, yields ethyl 1-(1-hydroxycyclohexyl)cyclohexanecarboxylate. Dehydration and saponification give rise to 1-(1-cyclohexenyl)cyclohexanecarboxylic acid, which upon decarboxylation at 195° yields cyclohexylidenecyclohexane in 8% overall yield, m.p. 54°.[4] This olefin has also been prepared by the debromination of 1,1'-dibromobicyclohexyl with zinc in acetic acid.[5]

The preparation of the dispiro[5.1.5.1]tetradecane-7,14-dione intermediate is essentially that of Walborsky and Buchman.[6]

4. Merits of the Preparation

The most obvious features of this synthesis are its simplicity and overall yield, which appear to be superior to those of any other published report. An important merit lies in the generality of the reaction, and the fact that it is an example of a reasonably large-scale photochemical preparation. Tetramethylethylene is readily produced from commercially available tetramethyl-1,3-cyclobutanedione by an identical route.[7]

1. Department of Chemistry, Columbia University, New York, New York 10027.
2. Hall Laboratory of Chemistry, Wesleyan University, Middletown, Connecticut.
3. P. A. Leermakers, G. F. Vesley, N. J. Turro, and D. C. Neckers, *J. Am. Chem. Soc.*, **86**, 4213 (1964).
4. J. Jacques and C. Weidmann-Hattier, *Bull. Soc. Chim. France*, 1478 (1958).
5. R. Criegee, E. Vogel, and H. Höger, *Ber.*, **85**, 144 (1952).
6. H. M. Walborsky and E. R. Buchman, *J. Am. Chem. Soc.*, **75**, 6339 (1953).
7. N. J. Turro, G. W. Byers, and P. A. Leermakers, *J. Am. Chem. Soc.*, **86**, 955 (1964).

CYCLOHEXYL ISOCYANIDE*

$$2 \underset{\substack{CH_2CH_2 \\ CH_2 \\ CH_2CH_2}}{\diagdown} CHNHCHO + POCl_3 + 4C_5H_5N \rightarrow$$

$$2 \underset{\substack{CH_2CH_2 \\ CH_2 \\ CH_2CH_2}}{\diagdown} CHN{\equiv}C + 3C_5H_5N \cdot HCl + C_5H_5N \cdot HPO_3$$

Submitted by Ivar Ugi, Rudolf Meyr, Martin Lipinski,
Ferdinand Bodesheim, and Friedrich Rosendahl.[1]
Checked by B. C. McKusick and M. E. Hermes.

1. Procedure

*Caution! Isocyanides should be prepared in a hood since they
have pungent odors and some are known to be toxic.*

A solution consisting of 127 g. (1.00 mole) of N-cyclohexylformamide (Note 1), 500 ml. (490 g., 6.2 moles) of pyridine, and
300 ml. of petroleum ether (b.p. 40–60° or 30–60°) is charged into
a 2-l., three-necked, round-bottomed flask equipped with a Hershberg stirrer,[2] dropping funnel, reflux condenser, and thermometer.
The flask is immersed in an ice bath, and 92 g. (0.60 mole) of
phosphorus oxychloride is added from the dropping funnel to the
stirred mixture in the course of 30–40 minutes. The mixture is
stirred under reflux for 10 minutes after all the phosphorus oxychloride is added. The mixture is then cooled to 0–5°; this converts it to a heavy slurry. Ice water (800 ml.) is gradually added
with stirring, and stirring of the cold mixture is continued until
all solid material has dissolved. The organic phase is separated
in a separatory funnel. The aqueous phase is extracted with three
60-ml. portions of petroleum ether, and the extracts are combined
with the organic phase, which is then extracted with three 100-ml.
portions of water, dried over 20 g. of magnesium sulfate, and
distilled through a 40-cm. vacuum-jacketed Vigreux column

(Note 2).   The petroleum ether is rapidly removed under slightly reduced pressure from a bath at a temperature not exceeding 50–60°.   Cyclohexyl isocyanide, a colorless foul-smelling liquid (Note 3), is collected at 56–58°/11 mm.; weight 73–79 g. (67–72%); n_D^{25} 1.4488–1.4501.

2. Notes

1. The checkers prepared N-cyclohexylformamide by slowly adding 260 g. (3.52 moles) of ethyl formate with stirring to 396 g. (4.00 moles) of cyclohexylamine in a flask immersed in an ice bath.   After the exothermic reaction ceased, the solution was refluxed for 2 hours and distilled through a 25-cm. Vigreux column to give 403 g. (90%) of N-cyclohexylformamide, b.p. 137–138°/10 mm., n_D^{25} 1.4849.[3]

2. The checkers used a 50-cm. spinning-band column.[4]   In order to minimize resinification of the cyclohexyl isocyanide, distillation should be as rapid as possible and the temperature in the still pot should not exceed 90°.

3. The disagreeable odor of cyclohexyl isocyanide can be removed from the equipment used in this preparation by washing it with 5% methanolic sulfuric acid solution.

3. Discussion

A variety of methods has been employed in the synthesis of cyclohexyl isocyanide[5] but the dehydration of N-cyclohexylformamide is the most favorable method.

Of the numerous dehydrating agent/base systems which have been used in the preparation of isonitriles, the phosgene/tertiary amine system seems to afford the best yields. Examples of the phosgene procedure may be found in reference 5. The disadvantage of phosgene is its extreme toxicity and the difficulty with which it is handled by the novice. The present procedure is therefore the best combination of convenience and safety for the preparation of aliphatic isocyanides boiling above ethyl isocyanide. (Methyl and ethyl isocyanides may be prepared by using high-boiling amines like quinoline.) It has been applied to the synthesis of the following isocyanides:[5] isopropyl (38%), n-butyl (61%), t-butyl (68%), and benzyl (56%). In preparing isopropyl isocyanide or t-butyl isocyanide, the petroleum ether should be of boiling point 30–35°, as otherwise it is difficult to separate these low-boiling isocyanides in the indicated yield, and, even then, substantial amounts of isocyanide are found in the petroleum ether fraction.

Aromatic isocyanides can also be prepared conveniently by the dehydration of the corresponding formamides by phosphorous oxychloride, but much better results are obtained if the reaction is done in the presence of potassium t-butoxide rather than pyridine.[7] The preparation of methyl isocyanide by the dehydration of N-methylformamide with p-toluenesulfonyl chloride and quinoline is described elsewhere in this volume.[8]

1. Institute of Organic Chemistry, University of Munich, Munich, Germany.
2. P. S. Pinkney, *Org. Syntheses*, Coll. Vol. 2, 117 (1943).
3. R. Wietzel, German pat. 454,459 (1928) [*chem. Zentr.*, 99, I, 2540 (1928)].
4. R. G. Nester, *Anal. Chem.*, 28, 278 (1956).
5. P. Hoffman, G. Gokel, D. Marquarding, and I Ugi, in I. Ugi, "Isonitrile Chemistry," Academic Press, New York, N.Y., 1971, p. 9.
6. I. Ugi and R. Meyr, *Chem. Ber.*, 93, 239 (1960).
7. I. Ugi and R. Meyr, *this volume*, p. 1060.
8. R. E. Schuster, J. E. Scott, and J. Casanova, Jr., this volume, p. 772.

2-CYCLOHEXYLOXYETHANOL

[Ethanol, 2-(cyclohexyloxy)-]

$$\text{Cyclohexanone} =O + HOCH_2CH_2OH \xrightarrow[-H_2O]{H^+} \text{1,4-dioxaspiro[4.5]decane}$$

$$\xrightarrow[\text{2. } H_2O,\ H^+]{\text{1. LiAlH}_4,\ AlCl_3} \text{cyclohexyl} —OCH_2CH_2OH$$

Submitted by RONALD A. DAIGNAULT and E. L. ELIEL [1]
Checked by J. R. EDMAN and B. C. McKUSICK

1. Procedure

A. *1,4-Dioxaspiro[4.5]decane.* A 1-l. round-bottomed flask is charged with 118 g. (1.20 moles) of cyclohexanone, 82 g. (1.32 moles) of 1,2-ethanediol, 250 ml. of reagent grade benzene, and 0.05 g. of *p*-toluenesulfonic acid monohydrate. The flask is attached to a water separator[2] under a reflux condenser fitted with a drying tube. A heating mantle is placed under the flask, and the reaction mixture is refluxed until close to the theoretical amount of water (21.6 ml.) has collected in the trap; this requires about 6 hours. The reaction mixture is cooled to room temperature, extracted successively with 200 ml. of 10% sodium hydroxide solution and five 100-ml. portions of water, dried over anhydrous potassium carbonate, and distilled through a 20-cm. Vigreux column. 1,4-Dioxaspiro[4.5]decane is obtained as a colorless liquid, b.p. 65–67° (13 mm.), weight 128–145 g. (75–85%), $n^{25}D$ 1.4565–1.4575.

B. *2-Cyclohexyloxyethanol.* A well-dried, 3-l. three-necked, round-bottomed flask is equipped with a stirrer, a pressure-equalizing dropping funnel, and a condenser to whose top is attached a calcium chloride drying tube. The flask is charged with 242 g. (1.81 moles) of anhydrous aluminum chloride powder and is immersed in an ice-salt bath. Anhydrous ether (25–50 ml.) is added dropwise through the dropping funnel, stirring is begun as soon as possible, and an additional 450–475 ml. of ether is added rapidly (total volume of ether added: 500 ml.). The mixture is stirred for approximately 30 minutes and becomes a light gray solution. During this period a mixture of 16.7 g. (0.44 mole) of lithium aluminum hydride and 500 ml. of anhydrous ether is vigorously stirred in a 1-l. round-bottomed flask under a nitrogen atmosphere (Note 1). The resulting suspension is added to the ethereal aluminum chloride solution through the dropping funnel. The resulting mixture, a gray slurry, is stirred for at least 30 minutes.

A solution of 125 g. (0.88 mole) of 1,4-dioxaspiro[4.5]decane in 200 ml. of anhydrous ether is added at a rate to cause gentle refluxing. The ice-salt bath is replaced by a steam bath, and the reaction mixture is refluxed for 3 hours. The calcium chloride drying tube is removed, and the steam bath is replaced by an ice bath. The excess hydride is carefully destroyed by adding water dropwise until hydrogen is no longer evolved; about 12 ml. of water is needed. This is followed by the more rapid addition of 1 l. of 10% sulfuric acid and then 400 ml. of water. This combination dissolves all the inorganic salts formed and results in the formation of two clear layers. The ether layer is separated in a 3-l. separatory funnel, and the aqueous layer is extracted with three 200-ml. portions of ether. The combined ethereal extracts are washed successively with 200 ml. of saturated sodium bicarbonate solution and 200 ml. of saturated brine. The ethereal solution is dried overnight over anhydrous potassium carbonate, filtered through a fluted filter paper, and concentrated by distillation on a steam bath. The residue, a pale yellow liquid weighing about 130 g., is distilled through a 20-cm. Vigreux column under reduced pressure. 2-Cyclohexyloxyethanol is obtained as a colorless liquid, b.p. 96–98° (13 mm.), weight 105–119 g. (83–94%), n^{25}D 1.4600–1.4610.

2. Note

1. Most of the lithium aluminum hydride is in solution, but some is in suspension. When the humidity is below 35%, lithium

aluminum hydride can be weighed in air; otherwise the weighing should be done in a dry box. Although some workers pulverize lithium aluminum hydride before dissolving or suspending it in a liquid, *the checkers recommend that this not be done because it has led to several explosions in their laboratory.* The present procedure gives a fine suspension that generally passes through the stopcock of the dropping funnel without plugging it. A wooden stick or copper wire should be in readiness to clear the stopcock if it plugs up.

Twice the theoretical amount of lithium aluminum hydride is used, but this is necessary for the best yields.

3. Methods of Preparation

The method of preparing 1,4-dioxaspiro[4.5]decane is that of Salmi.[3] The methods used by Lorette and Howard [4] to prepare ketals are convenient for preparing 1,4-dioxaspiro[4.5]decane.

The present method of preparing 2-cyclohexyloxyethanol has been described before,[5] but on a smaller scale. Other β-hydroxy ethers [5] and β-hydroxy thio ethers [6] can be prepared by the same method. Hydrogenolysis of the C—O bond in acetals has also been reported [7] with diisobutylaluminum hydride; for example, 2-cyclohexyloxyethanol was obtained in 91% yield in this manner.

2-Cyclohexyloxyethanol has also been prepared by reduction of cyclohexyloxyacetic acid with lithium aluminum hydride [8] and by decomposition of cyclohexanone methanesulfonylhydrazone with sodium in ethylene glycol.[9]

4. Merits of the Preparation

The method described is more convenient than earlier methods of preparing 2-cyclohexyloxyethanol. It may be adapted to the preparation of other β-hydroxyethyl and γ-hydroxypropyl ethers[5] and the corresponding thio ethers.[6,10] Although ketals are resistant to reduction by lithium aluminum hydride alone, the presence of a Lewis acid facilitates C—O cleavage, presumably via an oxocarbonium ion,[11] as the procedure demonstrates.

1. Department of Chemistry, University of Notre Dame, Notre Dame, Indiana.
2. S. Natelson and S. Gottfried, *Org. Syntheses*, Coll. Vol. **3**, 381 (1955).
3. E. J. Salmi, *Ber.*, **71B**, 1803 (1938).
4. N. B. Lorette and W. L. Howard, *J. Org. Chem.*, **25**, 521, 525, 1814 (1960); *Org. Syntheses*, **42**, 34 (1962).
5. E. L. Eliel, V. G. Badding, and M. N. Rerick, *J. Am. Chem. Soc.*, **84**, 2371 (1962).
6. E. L. Eliel, L. A. Pilato, and V. G. Badding, *J. Am. Chem., Soc.*, **84**, 2377 (1962).
7. L. I. Zakharkin and I. M. Khorlina, *Izv. Akad. Nauk. SSSR, Otd. Khim. Nauk*, 2255 (1959) [*C.A.*, **54**, 10837 (1960)].
8. M. Mousseron, R. Jacquier, M. Mousseron-Canet, and R. Zagdoun, *Bull. Soc. Chim. France*, [5] **19**, 1042 (1952).
9. J. W. Powell and M. C. Whiting, *Tetrahedron*, **7**, 305 (1959).
10. M. N. Rerick in R. L. Augustine, "Reduction," Marcel Dekker, Inc., New York, 1968, pp. 46-52.
11. B. E. Leggetter and R. K. Brown, *Can. J. Chem.*, 41, 2671 (1963).

1,2-CYCLONONADIENE

+(CH₃)₃COH + KBr

Submitted by L. Skattebøl and S. Solomon [1]
Checked by L. S. Keller and K. B. Wiberg

1. Procedure

A. *9,9-Dibromobicyclo[6.1.0]nonane.* A dry 3-l. three-necked flask is fitted with a glass stopper, stirrer, and condenser. The flask is kept under a positive nitrogen pressure by means of a gas-trap arrangement connected to the top of the condenser (Note 1). The flask is quickly charged with 2 l. of anhydrous *t*-butyl alcohol (Note 2) and 73 g. (1.87 g. atoms) of potassium metal. (*Caution! See earlier volume* [2] *for handling of this metal.*) The flow of nitrogen is stopped and the mixture is stirred and boiled under reflux until the potassium has reacted, hydrogen being liberated through the trap. The condenser is arranged for distillation by means of an adapter. The glass stopper is replaced by a pressure-equalized dropping funnel with the nitrogen inlet connected to the top. About 1.5 l. of *t*-butyl alcohol (Note 3) is then distilled into a predried flask under an atmosphere of nitrogen. A water pump is then connected, the nitrogen inlet is closed, and the distillation is continued under reduced pressure while the three-necked flask is gradually heated to 150° in an oil bath. Finally, the water pump is replaced by an oil pump and the white remaining solid is heated at 150° under a pressure of 0.1-1 mm. for 2 hours. The connection to the vacuum system is closed, the oil bath removed, and nitrogen again introduced. The condenser with adapter is replaced by a glass stopper, and the flask is cooled in an ice-salt bath.

Freshly distilled *cis*-cyclooctene, 178 g. (214 ml., 1.62 moles) (Note 4) and 200 ml. of sodium-dried pentane (Note 5) are introduced to the flask, and the dropping funnel is charged with 420 g. (148 ml., 1.66 moles) of bromoform (Note 6). The bromoform is added dropwise to the stirred slurry over a period of 6–7 hours, the color of the reaction mixture changing gradually from light yellow to brown. When the addition is complete, the reaction mixture is allowed to warm to room temperature and left stirring overnight. Water (400 ml.) is added to the reaction mixture followed by enough 10% aqueous hydrochloric acid to neutralize the slightly basic solution. The reaction mixture is transferred to a separatory funnel and the organic layer is separated. The aqueous layer is extracted with three 50-ml. portions of pentane, and the combined pentane solutions are washed with three 50-ml. portions of water. The pentane solution is dried over anhydrous magnesium sulfate, filtered, and stripped of solvent on a rotary evaporator. Distillation of the residue yields 237–299 g. (52–65%) of 9,9-dibromobicyclo[6.1.0]nonane, b.p. 62° (0.04 mm.), n^{23}D 1.5493–1.5507 (Note 7).

B. *1,2-Cyclononadiene.* A dry 2-l. three-necked flask is equipped with mechanical stirrer, pressure-equalized dropping funnel, and a nitrogen inlet tube connected to a gas-trap arrangement (Note 1). To the flask are added 187 g. (116 ml., 0.66 mole) of 9,9-dibromobicyclo[6.1.0]nonane and 100 ml. of anhydrous ether. The dropping funnel is charged with 450 ml. of 1.9M ether solution of methyllithium (0.85 mole) (Note 8). The flask is cooled by means of an acetone-dry ice bath maintained at $-30°$ to $-40°$, and the methyllithium is added dropwise with stirring during 1 hour (Note 9). After the addition is complete, the reaction mixture is stirred for 30 minutes, and excess lithium reagent is decomposed by dropwise addition of 100 ml. of water. An additional 400 ml. of water is then added, and the ether layer is separated. The aqueous layer is extracted with three 30-ml. portions of ether. The combined ether solutions are washed with 30-ml. portions of water until neutral and dried over magnesium sulfate. The latter is filtered and the ether is distilled through a 40-cm Vigreux column. Distillation of the residue (Note 10) yields 66–73 g. (81–91%) of 1,2-cyclononadiene, b.p. 62–63° (16 mm.), n^{20}D 1.5060 (Note 11).

2. Notes

1. A suitable gas-trap has been described.[3] Mercury can conveniently be replaced by paraffin oil.

2. Reagent grade t-butyl alcohol distilled from calcium hydride was used.

3. The t-butyl alcohol thus recovered can be used for a second preparation without further purification.

4. cis-Cyclooctene was obtained from Columbia Organic Chemicals or Aldrich Chemical Co. It was distilled from sodium and a fraction, b.p. 81–82° (95 mm.), n^{25}D 1.4682, was used. Gas chromatography showed 98% purity, the impurity being mainly cyclooctane.

5. Pentane is added as a diluent in order to obtain an easily stirred slurry. Amounts varying from 100 to 250 ml. per mole of olefin have been used with no appreciable change in yield of product.

6. Reagent grade bromoform was used without further purification.

7. The submitters have also used commercially available dry potassium t-butoxide with varying success in this reaction; with a sample purchased from M.S.A. Research Corporation a 65%

yield of product was obtained. The submitters reported a 65–76% yield range for this step.

8. An ethereal solution of methyllithium was either prepared from lithium metal and methyl bromide or purchased from Alfa Inorganics, Inc. Concentrations of 0.5–2M were used with no change in result.

9. Solid methyllithium and lithium halide occasionally separate out on the tip of the dropping funnel, probably owing to the low temperature, and this may cause plugging. It can be avoided by using a faster rate of addition.

10. The submitters used a 40-cm. spinning band column. Owing to polymerization of the product, the checkers obtained consistently low yields when this column was used. Distillation through a 40-cm. Vigreux column gave the indicated yield without a significant decrease in product purity.

11. The product is more than 99% pure as shown by gas chromatography.

3. Discussion

Cyclic allenes have previously been obtained only admixed with the isomeric acetylenes.[4] The present two-step synthesis is a practical method for the preparation of cyclic allenes, and at the same time it describes a general method for the preparation of allenes.[5, 6] It is based on the original work of Doering and co-workers.[7] Examples of the reaction sequence above are known in which allenes are not produced,[8] or they represent only a part of the reaction products.[9] A one-step synthesis of 1,2-cyclononadiene has been reported.[10]

R-(+)-1,2-Cyclononadiene and S-(−)-1,2-cyclononadiene have been prepared from R-(−)- and S-(+)-*trans*-cyclooctene, respectively.[11] Optically active 1,2-cyclononadiene has also been obtained when the reaction of the dibromo bicyclo intermediate with methyllithium is carried out in the presence of an optically active amine.[12] Reduction of 1,2-cyclononadiene with sodium in liquid ammonia gives *cis*-cyclononadiene in almost quantitative yield.[13]

1. Union Carbide Research Institute, Tarrytown, New York.

2. W. S. Johnson and W. P. Schneider, *Org. Syntheses*, Coll. Vol. **4**, 132 (1963).
3. L. F. Fieser, "Experiments in Organic Chemistry," 3rd ed., D. C. Heath and Co., Boston, 1957, p. 267; A. L. Vogel, "A Textbook of Practical Organic Chemistry," 3rd ed., John Wiley & Sons, New York, 1962, p. 69.
4. A. T. Blomquist, L. H. Liu, and J. C. Bohrer, *J. Am. Chem. Soc.*, **74**, 3643 (1952).
5. L. Skattebøl, *Tetrahedron Lett.*, 167 (1961); *Acta Chem. Scand.*, 17, 1683 (1963).
6. W. R. Moore and H. R. Ward, *J. Org. Chem.*, **27**, 4179 (1962).
7. W. von E. Doering and A. K. Hoffmann, *J. Am. Chem. Soc.*, **76**, 6162 (1954); W. von E. Doering and P. M. LaFlamme, *Tetrahedron*, **2**, 75 (1958).
8. W. R. Moore and H. R. Ward, *J. Org. Chem.*, **25**, 2073 (1960); W. R. Moore, H. R. Ward, and R. F. Merritt, *J. Am. Chem. Soc.*, **83**, 2019 (1961).
9. L. Skattebøl, *J. Org. Chem.*, 31, 2789 (1966); *Tetrahedron*, 23, 1107 (1967); L. Skattebøl, *Tetrahedron Lett.*, 2361 (1970); A. C. Cope, W. R. Moore, K. G. Taylor, P. Muller, S. S. Hall, and Z. L. F. Gaibel, *Tetrahedron Lett.*, 2365 (1970); W. R. Moore and J. B. Hill, *Tetrahedron Lett.*, 4343, 4553 (1970); M. S. Baird, *Chem. Commun.*, 1145 (1971).
10. K. G. Untch, D. J. Martin, and N. T. Castellucci, *J. Org. Chem.*, 30, 3572 (1965).
11. A. C. Cope, W. R. Moore, R. D. Bach, and H. J. S. Winkler, *J. Am. Chem. Soc.*, **92**, 1243 (1970); W. R. Moore, R. D. Bach, and T. M. Ozreitch, *J. Am. Chem. Soc.*, **91**, 5918 (1969); W. R. Moore, H. W. Anderson, S. D. Clark, and T. M. Ozreitch, *J. Am. Chem. Soc.*, **93**, 4932 (1971); W. R. Moore and and R. D. Bach, *J. Am. Chem. Soc.*, **94**, 3148 (1972); R. D. Bach, U. Mazur, R. N. Brummel, and L.-H. Liu, *J. Am. Chem. Soc.*, **93**, 7120 (1971).
12. H. Nozaki, T. Aratani, T. Toraya, and R. Noyori, *Tetrahedron*, **27**, 905 (1971).
13. P. D. Gardner and M. Narayana, *J. Org. Chem.*, **26**, 3518 (1961); D. Devaprabhakara and P. D. Gardner, *J. Am. Chem. Soc.*, **85**, 648 (1963).

CYCLOÖCTANONE*

Submitted by E. J. Eisenbraun [1]
Checked by E. J. Corey and Ernest Hamanaka

1. Procedure

The chromic acid oxidizing reagent is prepared by dissolving 67 g. of chromium trioxide in 125 ml. of distilled water. To this solution is added 58 ml. of concentrated sulfuric acid (sp. gr. 1.84), and the salts which precipitate are dissolved by addition of a minimum quantity of distilled water; the total volume of the solution usually does not exceed 225 ml.

A solution of 64 g. (0.5 mole) of cycloöctanol (Note 1) in 1.25 l. of acetone (Note 1) is added to a 2-l. three-necked flask fitted with a long-stem dropping funnel, a thermometer, and a powerful mechanical stirrer (Note 2). The vigorously agitated solution is cooled in a water bath to about 20°. The chromic acid oxidizing reagent is added from the dropping funnel as a slow stream, and the rate of addition is adjusted so that the temperature of the reaction mixture does not rise above 35° (Note 3). The addition is continued until the characteristic orange color of the reagent persists for about 20 minutes (Notes 4 and 5). The volume of reagent added is about 120 ml.

The stirrer is removed, the mixture is decanted into a 2-l. round-bottomed flask, and the residual green salts are rinsed with two 70-ml. portions of acetone. The rinsings are added to the main acetone solution and additional oxidizing agent is added, if necessary, to ensure complete reaction. The stirrer is replaced and isopropyl alcohol is added dropwise until the excess chromic acid is destroyed (Note 6). In small portions and with caution there is added 63 g. of sodium bicarbonate, and the suspension is stirred vigorously until the pH of the reaction mixture tests neutral (Note 7). The suspension is filtered and the filter cake is washed with 25 ml. of acetone. The filtrate is concentrated by distillation through a 75-cm. length of Vigreux column until the pot temperature rises to 80° and a water film begins to develop in the lower portions of the distillation column (Note 8). The cooled pot residue (about 110 ml.) is transferred to a 1-l. separatory funnel, 500 ml. of saturated sodium chloride solution is added, and the mixture is extracted with two 150-ml. portions of ether. The ether extracts are combined, washed with a total of 25 ml. water in several portions, dried over anhydrous magnesium sulfate, filtered, and the ether distilled at atmospheric pressure. The pot residue is distilled under reduced pressure, b.p. 76–77° (10 mm.) (Note 9). The yield of cycloöctanone is 58–60 g. (92–96%), m.p. 40–42°.

An additional 2.2 g. (4%) of cycloöctanone may be obtained by addition of 250 ml. of water to the green salts formed during the reaction (Note 10), extraction of the mixture with ether, distillation of the ether, and addition of 12 ml. of acetone. To the acetone solution there is added sufficient chromic acid oxidizing reagent to permit the orange color of the reagent to persist (Note 11), and the mixture is processed as above.

2. Notes

1. Cycloöctanol is available from Aldrich Chemical Company, Inc. A redistilled solvent grade of acetone is satisfactory.

2. The submitter has also carried out this preparation starting from 2 moles of cycloöctanol. An 8-l. Pyrex® bottle, Corning No. 1595, is ideally suited for this scale. A round-bottomed flask is less desirable because it is necessary to see into the reaction vessel. Vigorous stirring is essential; a Lightnin Model L stirrer fitted with two 2-in., three-blade propellers is adequate for the larger-scale run. A cold-water bath for the 8-l. bottle may be conveniently constructed from an open-top 5-gallon solvent can by cutting a 1.5-cm. hole 5 cm. from the bottom and a 2.8-cm. hole 5 cm. from the top. These holes are respectively fitted with a rubber inlet tube ($\frac{11}{16}$ in. O.D. by $\frac{3}{8}$ in. I.D.) and a rubber outlet tube ($1\frac{1}{4}$ in. O.D. by 1 in. I.D.). The rubber tubing fits directly in the holes without adapter or nipples.

3. The temperature is kept below 35° to avoid the use of a condenser.

4. The characteristic end point orange color can be demonstrated by addition of a slight excess of the chromic acid oxidizing reagent to a few milliliters of acetone containing a few drops of isopropyl alcohol.

5. The course of the reaction can conveniently be followed by gas chromatography. A sample of the reaction mixture is withdrawn at intervals, neutralized with solid sodium bicarbonate, dried over magnesium sulfate, and injected directly into a gas chromatography column consisting of 15% phenyldiethanolamine succinate (PDEAS) substrate coated on 60/80 mesh, acid-washed fire brick contained in a $\frac{1}{4}$ in. by 5 ft. spiral-shaped copper

tube. A Wilkens Instrument and Research, Inc., gas chromatography apparatus, Model A-90-P, operating at column temperature of 155°, 80 ml. per min. helium flow, was used. Complete separation of peaks (5.9 minutes for cycloöctanone, 7.0 minutes for cycloöctanol) is observed, and the reaction is considered complete when a peak for cycloöctanol can no longer be observed in the gas chromatogram.

6. The reaction mixture must be slightly acidic for the oxidation to proceed. On one occasion it was necessary to add a few drops of sulfuric acid to consume the oxidizing agent completely.

7. Calcium carbonate has also been used to remove residual acid.

8. If additional runs are contemplated, the recovered acetone may be used again.

9. A heat lamp may be used to prevent solidification during distillation.

10. The chromium salts formed during the oxidation are quite sticky and tend to occlude product as well as starting material.

11. The material freed from the chromium salts should be checked for completeness of reaction by gas chromatographic analysis to ensure the absence of starting material.

3. Methods of Preparation

Cycloöctanone has been prepared by distilling the calcium and thorium salts of azelaic acid,[2] by heating azelaic acid with barium oxide in the presence of iron,[3] by the action of nitrous acid on 1-(aminomethyl)-cycloheptanol,[4] by Dieckman cyclization of azelaic acid dimethyl ester[4] and diethyl ester,[5] and by ring expansion of cycloheptanone with diazomethane.[6, 7]

4. Merits of the Preparation

This preparation illustrates a general and convenient way of oxidizing secondary alcohols to ketones in high yield. This procedure, usually called the Jones oxidation or oxidation by use of the Jones reagent,[8] offers the advantage of almost instantaneous oxidation of the alcohol under mild conditions. The reagent rarely attacks unsaturated centers; using this procedure an 81% yield of 2-cyclohexenone can be obtained from 2-cyclohexenol. The present example illustrates how this reagent can be utilized for a large-scale preparation. The major limitation of the reaction is the low solvent power of acetone. Another example of the Jones oxidation is given on p. 863 of this volume.

An attractive alternative to the Jones oxidation is oxidation with chromic acid in the two-phase system, water-ether, the details of which were reported recently.[9] By this procedure cyclooctanone was obtained in 93% yield (as determined by gas-liquid chromatography). Although the yield of isolated yields of ketones from other secondary alcohols were very good, particularly when a 100% excess of chromic acid was used at 0°.

1. Department of Chemistry, Oklahoma State University, Stillwater, Oklahoma. Work done at Aldrich Chemical Co., Milwaukee, Wisconsin.
2. L. Ruzicka and W. Brugger, *Helv. Chim. Acta*, **9**, 339 (1926).
3. A. I. Vogel, *J. Chem. Soc.*, 721 (1929).
4. F. F. Blicke, J. Azuara, N. J. Doorenbos, and E. B. Hotelling, *J. Am. Chem. Soc.*, **75**, 5418 (1953).
5. N. J. Leonard and C. W. Schimelpfenig, *J. Org. Chem.*, **23**, 1708 (1958).
6. E. P. Kohler, M. Tishler, H. Potter, and H. T. Thompson, *J. Am. Chem. Soc.*, **61**, 1057 (1939).
7. S. J. Kaarsemaker and J. Coops, *Rec. Trav. Chim.*, **70**, 1033 (1951).
8. A. Bowers, T. G. Halsall, E. R. H. Jones, and A. J. Lemin, *J. Chem. Soc.*, 2548 (1953).
9. H. C. Brown, C. P. Garg, and K.-T. Liu, *J. Org. Chem.*, **36**, 387 (1971).

trans-CYCLOOCTENE

A.

B.

C.

1. AgNO₃
2. NH₄OH

Submitted by ARTHUR C. COPE [1] and ROBERT D. BACH [2]
Checked by A. DeMEIJERE and K. B. WIBERG

1. Procedure

A. *N,N,N-Trimethylcyclooctylammonium iodide.* To a 2-l., three-necked, round-bottomed flask equipped with a stirrer, condenser, drying tube, and pressure-equalizing dropping funnel are added 155.3 g. (1 mole) of N,N-dimethylcyclooctylamine (Note 1) and 700 ml. of reagent grade methanol. To the stirred solution is added 170.3 g. (1.2 moles) of iodomethane (Note 2) dropwise over a 30-minute period. The flask is cooled intermittently with an ice bath to keep the reaction temperature at approximately 25° (Note 3). After 1 hour the bath is removed and the reaction mixture is allowed to stir at 25° for an additional 3 hours.

The light yellow solution is transferred to a 2-l. round-bottomed flask, and the solvent is removed under reduced pressure (Note 4) with slight warming. The solid product is triturated with 500 ml. of diethyl ether, filtered, and washed with three 200-ml. portions of diethyl ether. The white solid (291–296 g.) is dried under reduced pressure, m.p. 269–270° dec. (Note 5).

B. *N,N,N-Trimethylcyclooctylammonium hydroxide.* To a 1-l. round-bottomed flask equipped with a stirrer are added 100 g. (0.34 mole) of N,N,N-trimethylcyclooctylammonium iodide, 76 g. of silver oxide (Note 6), and 350 ml. of distilled water. The suspension is stirred at room temperature for 5 hours and is filtered through a Buchner funnel. The filter cake is washed with four 35-ml. portions of distilled water. The light yellow filtrate is transferred to a 1-l. round-bottomed flask and the volume is reduced to approximately 90 ml. employing a rotary evaporator and a 40° water bath. The viscous N,N,N-trimethylcyclooctylammonium hydroxide solution is transferred (Note 7) to a 200-ml. dropping funnel, with a pressure-equalizing side arm, for use in the next step in the synthesis (Note 8).

C. *trans-Cyclooctene.* A 500-ml., three-necked, round-bottomed flask is equipped with a nitrogen inlet capillary tube (Note 9), a short (10–20 cm.) unpacked column (Note 10), and a pressure-equalizing dropping funnel. The round-bottomed flask is connected by the unpacked column to a 100-ml. trap cooled in an ice bath. This trap is then connected to a 200-ml. trap cooled in dry ice-acetone (Note 11). The flask is heated in an oil bath to 110–125°, and the apparatus is evacuated to a pressure of *ca.* 10 mm. under a constant sweep of nitrogen. The hydroxide solution is added dropwise at approximately the rate of decomposition of the quaternary ammonium hydroxide (Note 12).

The combined distillates from the cold traps are allowed to come to room temperature (Note 13) and are placed in a 1-l. separatory funnel with 200 ml. of 5% hydrochloric acid solution. The mixture of *cis-* and *trans-*cyclooctenes (Note 14) is extracted with 200 ml. of *n*-pentane and subsequently with two 50-ml. portions of *n*-pentane. The *n*-pentane extracts are combined and washed with 170 ml. of 5% sodium bicarbonate solution.

To a 1-l. separatory funnel is added *ca.* 500 ml. of 20% aqueous silver nitrate solution (100 g. of Mallinckrodt C.P. crystals to *ca.* 500 ml. of water). The pentane solution is added to the separatory funnel in five approximately equal portions, with intermittent shaking until all the silver nitrate complex has gone into solution (Note 15).

The silver nitrate solution is extracted as described above with three portions of *n*-pentane to remove *cis*-cyclooctene (Note 16). The aqueous silver nitrate solution is added slowly to 300 ml. of concentrated ammonium hydroxide containing cracked ice. The hydrocarbon that separates is extracted with 300 ml. of *n*-pentane as described above and the pentane solution is dried over anhydrous magnesium sulfate and the pentane distilled through a 23 x 250 mm. column packed with glass beads.

The product is distilled under reduced pressure through a short (8 cm.) Vigreux column (Note 17) and has b.p. 75° (78 mm.), 44° (23 mm.), n^{25}D $= 1.4741$, $d_4^{25} = 0.8456$. The yield of pure *trans*-cyclooctene is 15.0 − 15.3 g. (40%) (Notes 18, 19, 20, 21).

2. Notes

1. Cyclooctylamine was purchased from Aldrich Chemical Co. It was converted to N,N-dimethylcyclooctylamine in 74% yield using a procedure analogous to that for β-phenylethyldimethylamine.[3]

2. Fischer reagent grade methanol and Eastman Organic Chemicals iodomethane were used.

3. The molar ratio of iodomethane to N,N-dimethylcyclooctylamine may be reduced if precautions are taken to prevent loss of iodomethane due to vaporization.

4. It is convenient to use a rotary evaporator for removal of the solvents.

5. After one recrystallization from an acetone-methanol mixture, the compound melts at 273–275° dec. The compound is sufficiently pure for the next step in the synthesis without recrystallization.

6. Mallinckrodt purified silver oxide powder was used. The reaction flask should be protected from direct sunlight with a suitable wrapping.

7. The flask may be rinsed with a minimum of water and transferred to the dropping funnel. The total volume of hydroxide solution at this point should not exceed 100 ml.

8. The conversion of the quaternary ammonium iodide to the hydroxide may also be carried out using a strongly basic ion exchange resin.[4]

9. The decomposition should be carried out under a constant sweep of nitrogen. The nitrogen may be introduced through the pressure-equalizing dropping funnel if that is more convenient.

10. The unpacked column should be wrapped with a heating tape, or Nichrome heating wire, and kept at *ca.* 110° throughout the decomposition.

11. The reaction is stopped and the trap, cooled in dry ice-acetone, is emptied when the reaction is *ca.* one half finished to prevent plugging by ice. Most of the olefinic products are found in the first trap. The second trap contains mostly trimethylamine and water.

12. About 3 hours is required to add the hydroxide solution. The rate of addition may be increased, but considerable foaming occurs during the decomposition, and caution should be taken that the hydroxide does not foam over into the traps.

13. This part of the experiment should be carried out in a hood because trimethylamine is evolved.

14. The decomposition of N,N,N-trimethylcyclooctylammonium hydroxide forms a mixture of *cis*- and *trans*-cyclooctenes which contains *ca.* 60% of the *trans*- and 40% of the *cis*-isomer (see Note 19). The mixture is separated by extraction of the *trans*-isomer with aqueous silver nitrate.[5]

15. If the pentane solution is added to the silver nitrate solution too rapidly, the *trans*-cyclooctene forms a dark precipitate that is difficult to get into solution. This situation can, however, be remedied by the addition of more silver nitrate solution and continued shaking.

16. The pentane is removed and *ca.* 11 g. of *cis*-cyclooctene is obtained on distillation, b.p. 65° (59 mm.); $n^{25}D = 1.4684$.

17. Considerable foaming occurs during distillation of *trans*-cyclooctene. The distillation may therefore be facilitated by use of a distilling adapter with a foam trap. The distilling adapter (5225) may be purchased from Ace Glass Incorporated, Vineland,

New Jersey. The bath temperature should be kept below 100° owing to the possibility of isomerization to *cis*-olefin and polymerization.[6] The distillation should be carried out as rapidly as possible because the condensed product evaporates under prolonged exposure to reduced pressure.

18. The submitters carried out this preparation on a 1.0-mole scale and obtained 49–51 g. (45–46%) of *trans*-cyclooctene.

19. The purity of the *trans*-cyclooctene may be determined by infrared spectroscopy[5] or by gas chromatography using an NMPN (3-nitro-3-methylpimelonitrile) column.[7] A low injection port temperature is desirable (<200°).

20. *trans*-Cyclooctene is stable for at least 1 year if kept under refrigeration and if a free radical inhibitor is used (*e.g.* di-*t*-butyl-resorcinol). The compound has a very disagreeable odor.

21. Since this procedure was submitted the submitter has found that higher overall yields of *trans*-cyclooctene may be obtained from a stirred suspension of trimethylcyclooctylammonium iodide in anhydrous dimethyl sulfoxide using commercially available (sublimed) potassium *t*-butoxide as the base at 25°. The *cis*/*trans* ratio under these conditions is 1/4.6 (private communication from R. D. Bach).

3. Discussion

trans-Cyclooctene has been prepared by the Hofmann elimination of N,N,N-trimethylcyclooctylammonium hydroxide,[5, 8] the present method; by the treatment of N,N,N-trimethylcyclooctylammonium bromide with phenyllithium, methyllithium, and potassium amide:[9] and by the treatment of *trans*-1,2-cyclooctene thiocarbonate with triisooctyl phosphite.[10]

This procedure illustrates a general method for preparing olefins by the elimination of an amine and a β-hydrogen atom.[11] The present method is more convenient for adaptation to large-scale laboratory preparation than is the Wittig modification, which utilizes liquid ammonia; both methods give essentially the same overall yield of *trans*-cyclooctene.

The preparation of olefins *via* their thiocarbonate[10] is a stereospecific elimination reaction which may be used to advantage when a mixture of *cis*- and *trans*-olefins is difficult to separate.

However, all the reagents required to prepare the thiocarbonate are not readily available.

1. Deceased, June 4, 1966; formerly of Massachusetts Institute of Technology, Cambridge, Massachusetts.
2. Massachusetts Institute of Technology, Cambridge, Massachusetts.
3. R. N. Icke, B. B. Wisegarver and G. A. Alles, *Org. Syntheses*, Coll. Vol. 3, 723 (1955); cf. R. D. Bach, *J. Org. Chem.*, 33, 1647 (1968).
4. J. L. Coke and M. C. Mourning, *J. Am. Chem. Soc.*, 90, 5561 (1968).
5. A. C. Cope, R. A. Pike, and C. F. Spencer, *J. Am. Chem. Soc.*, 75, 3212 (1953).
6. A. C. Cope and B. A. Pawson, *J. Am. Chem. Soc.*, 87, 3649 (1965).
7. A. C. Cope, C. R. Ganellin, H. W. Johnson, T. V. Van Auken, and H. J. S. Winkler, *J. Am. Chem. Soc.*, 85, 3276 (1963).
8. K. Ziegler and H. Wilms, *Ann.*, 567, 1 (1950).
9. G. Wittig and R. Polster, *Ann.*, 612, 102 (1957).
10. E. J. Corey, F. A. Carey, and R. A. E. Winter, *J. Am. Chem. Soc.*, 87, 934 (1965).
11. A. C. Cope and E. R. Trumbull, *Org. Reactions*, 11, 317 (1960).

CYCLOPENTANECARBOXALDEHYDE*

$$\text{(cyclohexene)} + 2HgSO_4 \xrightarrow{H_2SO_4} C_6H_{10}\cdot2HgSO_4 \xrightarrow[-H_2SO_4,\, -Hg_2SO_4]{H_2O,\, \Delta} \text{(cyclopentane-CHO)}$$

Submitted by OLIVER GRUMMITT, JOHN LISKA, and GERHARD GREULL[1]
Checked by WILLIAM E. PARHAM and GERALD E. STOKKER

1. Procedure

In a 5-l. three-necked flask fitted with reflux condenser, mechanical stirrer, thermometer, and nitrogen gas inlet is placed a solution of 80.0 g. (43.5 ml., 0.82 mole) of concentrated sulfuric acid in 3 l. of water. The solution is stirred under nitrogen, and 740.0 g. (2.49 moles) of reagent mercuric sulfate is added to form a suspension of deep-yellow, basic mercuric sulfate. The mixture is stirred and heated to 55° under nitrogen, and 82.0 g. (101 ml., 1.0 mole) of cyclohexene (Note 1) is added at once. A temperature of 55–65° (Note 2) is maintained for 1 hour. During this time the color of the reaction mixture changes from a deep yellow to the cream color of the cyclohexene-mercuric sulfate complex.

At the end of 1 hour the condenser is set for distillation. The temperature of the reaction mixture is raised (Note 3), the mixture is stirred while a slow current of nitrogen is continued, and 300 ml. of a mixture of crude cyclopentanecarboxaldehyde and water is distilled over a period of approximately 2 hours. The crude product is removed in a separatory funnel from the aqueous layer, which is extracted with three 50-ml. portions of ether. The extracts are combined with the product and dried over anhydrous sodium sulfate. The solution is filtered into a 250-ml. Claisen flask set for vacuum distillation, the pressure is gradually reduced to 100 mm. to distil ether, and the cyclopentanecarboxyaldehyde is distilled rapidly (Note 4) at 74–78° (100 mm.). The yield of aldehyde (n^{20}D 1.4420–1.4428) is 45–52 g. (46–53%).

Unless the aldehyde is to be used immediately, it is stored in a brown bottle at 0° after the addition of 0.1 g. of hydroquinone and a blanket of nitrogen (Note 5).

2. Notes

1. Cyclohexene from an unopened bottle or freshly distilled material (b.p. 82–84°) is used.

2. This is the optimal temperature range to form the cyclohexene-mercuric sulfate complex.[2, 3]

3. The complex undergoes oxidation-reduction at about 100° to give cyclopentanecarboxyaldehyde, mercurous sulfate, and some mercury. If desired, the mercury products can be regenerated to mercuric sulfate. However, approximately 40–140 g. of mercuric sulfate will be lost in the various filtrates in the recovery operation. Best current practice dictates that these filtrates be collected and disposed of as toxic chemicals. Concentration of the filtrates is not advisable since acetone, nitric acid, and sulfuric acid are present along with the mercury compounds. If the mercuric sulfate is not regenerated, the water and sludge of mercurous sulfate and mercury, which remain after the distillation of cyclopentanecarboxyaldehyde, should be disposed of with due regard to its toxic nature.

The regeneration procedure is as follows:

The water and sludge of mercurous sulfate and mercury, which remain after the distillation of cyclopentanecarboxyaldehyde, are filtered with suction, washed with three 100-ml. portions of boiling water, three 100-ml. portions of acetone, and finally with three 100-ml. portions of boiling water. The gray-black solid is sucked dry.

To the solid in a 3-l. Erlenmeyer flask is added 270 ml. of water, then 90 g. (65 ml., 1.0 mole) of concentrated nitric acid *slowly* (*Hood!*). The contents of the flask are swirled and allowed to stand until frothing and evolution of reddish brown oxides of nitrogen subsides. Dow-Corning Antifoam A helps to control frothing.

The mixture is heated *cautiously* on a hot plate to avoid excess frothing. When the frothing has almost subsided, additional concentrated nitric acid is added, 90 g. (65 ml., 1.0 mole) at a time, with swirling and intermittent heating to control the vigorous reaction. The solid changes from gray-black to a cream color after six 90-g. portions of concentrated nitric acid have been used.

Six hundred milliliters of concentrated nitric acid is then added with heating to form a clear, deep-orange solution of mercuric nitrate. This mixture is allowed to cool and is then filtered with suction through sintered glass to remove a small amount of solid.

Four hundred and sixty grams (250 ml., 4.7 moles) of concentrated sulfuric acid is added to the filtrate to precipitate mercuric sulfate. The mixture is boiled under the hood for 1 hour, cooled to 10–25°, and filtered with suction through sintered glass.

The solid mercuric sulfate is washed with three 100-ml. portions of approximately 40% aqueous sulfuric acid solution (110 ml. of concentrated sulfuric acid mixed with 300 ml. of water). The solid is sucked dry, transferred to an evaporating dish, broken up, and dried in the hood under a heat lamp.

The yield of recovered mercuric sulfate is 600–700 g. This material plus fresh mercuric sulfate to give 740 g. can be used in a subsequent preparation of cyclopentanecarboxaldehyde without affecting the yield.

4. Cyclopentanecarboxyaldehyde may trimerize if heating is prolonged; hence a fast, simple distillation is done. When the distillation residue is cooled, a solid may appear. This solid can be distilled above 78° (100 mm.) as a clear liquid which solidifies when allowed to stand. Recrystallization of this material from 95% ethanol gives a white solid melting at 122–124°. This

product was shown to be cyclopentanecarboxaldehyde trimer by a mixed melting-point determination with trimer prepared from cyclopentanecarboxyaldehyde and 85% phosphoric acid.[4]

5. During storage there is a slow formation of cyclopentane-carboxaldehyde trimer.

3. Methods of Preparation

Cyclopentanecarboxaldehyde has been prepared by the procedure described above;[2,3] by the reaction of aqueous nitric acid and mercuric nitrate with cyclohexene;[5] by the action of magnesium bromide etherate[6] or thoria[7] on cyclohexene oxide; by the dehydration of trans-1,2-cyclohexanediol over alumina mixed with glass helices;[8] by the dehydration of divinyl glycol over alumina followed by reduction;[9] by the reaction of cyclopentene with a solution of $[HFe(CO)_4]^-$ under a carbon monoxide atmosphere;[10] and by the reaction of cyclopentadiene with dicobalt octacarbonyl under a hydrogen and carbon monoxide atmosphere.[11]

4. Merits of the Preparation

This procedure uses readily available starting materials and in one operational step generally gives higher yields of cyclopentane-carboxaldehyde than other preparations described in the literature. Because mercuric sulfate is an expensive reactant, a method of regenerating the mercury products is given. Cyclopentane-carboxaldehyde is a useful intermediate for many cyclopentane derivatives.

1. Department of Chemistry, Western Reserve University, Cleveland, Ohio.
2. H. L. Yale and G. W. Hearne (Shell Development Co.), U.S. Patent 2,429,501 (1947) [C. A., 42, 1966 (1948)].
3. J. English, Jr., J. D. Gregory, and J. R. Trowbridge, II, J. Am. Chem. Soc., 73, 615 (1951).
4. A. G. Brook and G. F. Wright, Can. J. Chem., 29, 308 (1951).
5. D. A. Shearer and G. F. Wright, Can. J. Chem., 33, 1002 (1955).
6. S. M. Naqvi, J. P. Horwitz, and R. Filler, J. Am. Chem. Soc., 79, 6283 (1957).
7. P. Bedos and A. Ruyer, Compt. Rend., 188, 962 (1929).
8. C. C. Price and G. Berti, J. Am. Chem. Soc., 76, 1211 (1956).
9. E. Urion, Ann. Chim. (Paris), (11) 1, 5 (1934).
10. H. W. Sternberg, R. Markby, and I. Wender, J. Am. Chem. Soc., 78, 5704 (1956).
11. H. Adkins and J. L. R. Williams, J. Org. Chem., 17, 980 (1952).

2-CYCLOPENTENE-1,4-DIONE

Submitted by GARY H. RASMUSSON,[1] HERBERT O. HOUSE,[1]
EDWARD F. ZAWESKI,[2] and CHARLES H. DEPUY.[2]
Checked by WILLIAM G. DAUBEN, PHILIP E. EATON, and
RICHARD SCHNEIDER.

1. Procedure

In a 2-l. three-necked flask equipped with a thermometer, a mechanical stirrer, and a dropping funnel (Note 1) is placed a mixture of 45.1 g. (0.45 mole) of a dihydroxycyclopentene mixture (Note 2), 200 ml. of water, and 300 ml. of methylene chloride. After this mixture has been cooled to −5° to 0° by means of an external cooling bath (Note 3), the addition of a solution of 100 g. (1.0 mole) of chromium trioxide and 160 ml. of concentrated sulfuric acid in 450 ml. of water is begun. The solution of the oxidant is added dropwise, and with stirring, at such a rate that the temperature of the reaction mixture remains between −5° and 0°. After the addition is complete (Note 4), the mixture is stirred at −5° to 0° for 1 hour and then 200 ml. of chloroform is added. The resulting mixture is stirred for 10 minutes, the organic layer is separated, and then the aqueous layer is extracted with two 200-ml. portions of a mixture (1:1 by volume) of chloroform and methylene chloride. The combined organic extracts are washed with 100 ml. of water, dried over anhydrous magnesium sulfate, and concentrated under reduced pressure at room temperature. The yield of 2-cyclopentene-1,4-dione (Note 5), which crystallizes as yellow plates melting at 30.0–32°, is 17–22 g. (39–50% based on the dihydroxycyclopentene mixture) (Note 6).

2. Notes

1. All glassware must be washed with acid before use.

2. The dihydroxycyclopentene mixture was prepared from cyclopentene and peracetic acid.[3] The mixture contains approximately 70% of 2-cyclopentene-1,4-diol.

3. The submitters found a cooling bath composed of a Dry Ice and methanol-water mixture (1:3 by volume) to be convenient.

4. At this point in the preparation, an excess of the oxidant should be present. The presence of excess oxidant may be established by diluting 2 drops of the aqueous phase from the reaction mixture with 2 ml. of water and then adding 1 drop of a 0.4% solution of sodium diphenylaminesulfonate in water. A deepening in color is observed if excess oxidant is present.

5. This product is sufficiently pure for most applications. Further purification may be achieved either by sublimation of the product at 30–40°/0.1 mm. or by recrystallization of the dione from diethyl ether at Dry Ice temperatures. The dione decomposes rapidly at temperatures above 40°. An ethanol solution of pure dione, m.p. 35–36°, exhibits a maximum in the ultraviolet at 222 mμ (log ϵ 4.16).

6. Starting with pure 2-cyclopentene-1,4-diol, the submitters obtained the dione in 67–79% yield.

3. Methods of Preparation

2-Cyclopentene-1,4-dione has been prepared by oxidation of 2-cyclopentene-1,4-diol with chromium trioxide in aqueous acetic acid [4,5] or in aqueous acetone,[4] and with silver chromate.[6] The present method eliminates the tedious removal of large amounts of acetic acid and gives a higher yield.

4. Merits of Preparation

2-Cyclopentene-1,4-dione is a very reactive dienophile in the Diels-Alder reaction and thus provides access to a variety of compounds containing the reactive β-dicarbonyl grouping in a five-membered ring.[4] Also, its multiple functionality makes it a versatile starting material for other types of reactions as well.

1. Department of Chemistry, Massachusetts Institute of Technology, Cambridge, Massachusetts.
2. Department of Chemistry, Iowa State University, Ames, Iowa.
3. M. Kovach, D. R. Nielsen, and W. H. Rideout, this volume, p. 414.
4. C. H. DePuy and E. F. Zaweski, *J. Am. Chem. Soc.,* **81,** 4290 (1959).
5. V. F. Kucherov and L. I. Ivanova, *Doklady Akad. Nauk S.S.S.R.,* **131,** 1077 (1960); [*C.A.,* **54,** 21021 (1960)].
6. E. Y. Gren and G. Vanags, *Doklady Akad. Nauk S.S.S.R.,* **133,** 588 (1960); [*C.A.,* **54,** 2442 (1960)].

2-CYCLOPENTENONE*

(2-Cyclopentene-1-one)

Submitted by CHARLES H. DePuy and K. L. EILERS.[1]
Checked by WILLIAM G. DAUBEN, ROBERT A. FLATH, and GILBERT H. BEREZIN.

1. Procedure

In a 250-ml. round-bottomed flask fitted for vacuum distillation with a short path distilling head (Note 1), a condenser, and a 250-ml. receiving flask is placed 100 g. (1.0 mole) of a mixture of cyclopentenediols.[2] A few Carborundum boiling chips are added. The receiver is cooled in ice and the mixture heated to 50–55° (Note 2). At this time, the flask is opened momentarily and 1–2 g. of *p*-toluenesulfonic acid monohydrate is added. The flask is immediately closed and the pressure is reduced to 10–15 mm. Careful heating is continued, and a mixture of 2-cyclopentenone and water begins to distil with the temperature in the distilling head rising from 45° to 60° (Note 3). The temperature of the flask is gradually increased as necessary to maintain a reasonably rapid distillation rate.

The reaction is complete when approximately 10% of the original material remains in the distilling flask. The distillation normally requires 30–60 minutes.

The distillate, containing 2-cyclopentenone, water and varying amounts of cyclopentenediols, is dissolved in 150 ml. of methylene chloride and dried over anhydrous sodium sulfate. The solvent is carefully removed through a Vigreux column and the residue purified by distillation. After a forerun (b.p. 50–150°), there is collected 44–49 g. (53–60%) of pure 2-cyclopentenone, b.p. 151–154°. Cyclopentenediols may be recovered from the pot residue by distillation at 0.1 mm. The forerun contains appreciable amounts of cyclopentenone and should be added to a succeeding preparation before final distillation.

2. Notes

1. The distilling head should be short and unobstructed, for any attempt at fractionation at this stage leads to resinification of the 2-cyclopentenone by the acid. No capillary bleed is used, since the product is extremely sensitive to oxygen.

2. The submitters have found a 250-watt infrared heat lamp controlled by a Variac to be the most convenient source of heat. Occasionally, the reaction may become rapid and exothermic, and it is important to remove the heat source as quickly as possible. If an oil bath is used, the temperature is gradually increased until it approaches 150° at the end of the reaction.

3. If the temperature of the distillate rises much above 60° at this pressure, considerable amounts of diols co-distil and the yield of 2-cyclopentenone is diminished. If a reasonably rapid distillation does not occur with a head temperature below 60°, an additional gram of acid should be added after lowering the temperature of the distillation flask below 50°.

3. Methods of Preparation

Previous preparations of 2-cyclopentenone have involved the elimination of HCl from 2-chlorocyclopentanone [3] or its ketal.[4] The oxidation of 3-chloro- [5] or 3-hydroxycyclopentene [6] has been utilized as well as the direct oxidation of cyclopentene with H_2O_2.[7] Cyclopentenone has also been prepared from 1-dicyclopentadienol.[8]

4. Merits of Preparation

The α,β-unsaturated ketone system in 2-cyclopentenone makes possible a wide variety of reactions of the Michael and Diels-Alder type. Thus 2-cyclopentenone is a versatile starting material for preparing compounds containing a five-membered ring. The availability of the dihydroxycyclopentene mixture (p. 414) makes the present procedure the method of choice for its preparation.

1. Department of Chemistry, Iowa State University, Ames, Iowa.
2. M. Kovach, D. R. Nielsen, and W. H. Rideout, this volume, p. 414.
3. E. J. Corey and K. Osugi, *Pharm. Bull. (Tokyo)*, 1, 99 (1953).
4. H. Wanzlick, G. Gollmer, and M. Milz, *Ber.*, 88, 69 (1955).
5. K. Alder and F. H. Flock, *Ber.*, 89, 1732 (1956).
6. E. Dane and K. Eder, *Ann.*, 539, 207 (1939).
7. W. Treibs, G. Franke, G. Leichsenring, and H. Roder, *Ber.*, 86, 616 (1953).
8. M. Rosenblum, *J. Am. Chem. Soc.*, 79, 3179 (1957).

CYCLOPROPYLBENZENE *

(Benzene, cyclopropyl-)

$$C_6H_5CH_2CH_2CH_2Br + \begin{array}{c} CH_2-C \\ | \\ CH_2-C \end{array} \begin{array}{c} O \\ \nearrow \\ N-Br \\ \searrow \\ O \end{array} \xrightarrow{(C_6H_5CO_2)_2}$$

$$C_6H_5CHBrCH_2CH_2Br \xrightarrow{Zn-Cu} C_6H_5CH \begin{array}{c} CH_2 \\ | \\ CH_2 \end{array}$$

Submitted by THOMAS F. CORBIN, ROGER C. HAHN.[1]
and HAROLD SHECHTER [2]
Checked by WILLIAM G. DAUBEN and PAUL LAUG

1. Procedure

Caution! N-Bromosuccinimide is a skin irritant.
A. *1,3-Dibromo-1-phenylpropane.* In a 3-l. three-necked flask

fitted with a sealed stirrer and two efficient reflux condensers are placed 199 g. (1.0 mole) of 1-bromo-3-phenylpropane (Note 1), 187 g. (1.05 moles) of N-bromosuccinimide (Note 2), 3 g. of benzoyl peroxide, and 1.2 l. of carbon tetrachloride. The mixture is heated cautiously with a flame to reflux until a spontaneous reaction starts; ice-bath cooling is then applied if necessary (Note 3). When the spontaneous reaction subsides, the stirring is stopped; if more than a negligible amount of N-bromosuccinimide remains in the bottom of the flask (succinimide rises to the surface of the solvent), heating and stirring are continued until an evolution of hydrogen bromide is noted. The mixture is cooled, and the solids are removed by suction filtration and washed with carbon tetrachloride. The washings are combined with the original filtrate, and the bulk of the carbon tetrachloride is removed (Note 4) by distillation at water aspirator pressure and a bath temperature of 40–50° (Note 5). The remainder of the solvent is removed at the same bath temperature and at 0.1 mm. pressure (Note 6). The orange-yellow residue (nearly 100% of the theoretical yield of 1,3-dibromo-1-phenylpropane) is used without further purification (Note 7) in the next step.

B. *Cyclopropylbenzene.* In a 1-l. three-necked flask equipped with a stirrer and a thermometer extending into the flask but free from the stirrer are placed 500 ml. of redistilled dimethylformamide and zinc-copper couple prepared from 131 g. (2 g. atoms) of zinc (Note 8). The mixture is cooled to 7° in an ice bath, and 1,3-dibromo-1-phenylpropane is added to the stirred mixture at a rate sufficient to maintain the reaction temperature at 7–9° (Note 9). The mixture is stirred for 30 minutes after the addition is completed, poured into 1 l. of water, and then steam-distilled until the condensate is homogeneous or 1 l. of water has been collected. The organic layer is separated from the distillate, and the aqueous layer is extracted with three 100-ml. portions of ether. The combined organic portions are washed with four 50-ml. portions of water and dried over anhydrous potassium carbonate. The ether is removed by distillation at atmospheric pressure at water bath temperature. The residue is distilled to give 88–100 g. (75–85%) of cyclopropylbenzene, b.p. 170–175° (Note 10), n^{26}D 1.5306–1.5318.

2. Notes

1. The 1-bromo-3-phenylpropane was obtained from Columbia Organic Chemicals Co., Inc., Columbia, South Carolina, and from Aldrich Chemical Co., Inc., Milwaukee, Wisconsin. Redistillation of the commercial material does not noticeably affect yields.

2. N-Bromosuccinimide was obtained from Arapahoe Chemicals, Inc., Boulder, Colorado, and from Coleman and Bell, Norwood, Ohio. The material utilized by the checkers was shown to be 98.6% pure by iodometric analyses.

3. This reaction may become vigorously exothermic; two condensers and a highly mobile setup, allowing quick (5 seconds) removal of heat and application of cooling, are then necessary to contain it. *Caution* must be taken to control but not stop the reaction.

4. Any bromine present at this point is entrained by the carbon tetrachloride; the separated carbon tetrachloride may be purified by shaking with a small quantity of sodium bisulfite, drying over anhydrous potassium carbonate, and distilling.

5. Higher bath temperatures cause darkening of the residue with evolution of hydrogen bromide.

6. An efficient dry ice trap is essential to protect the vacuum pump.

7. Attempts to distil the residue usually cause evolution of large amounts of hydrogen bromide.

8. Zinc powder, obtainable from Mallinckrodt Chemical Works, St. Louis, Missouri, and Merck and Co., Rahway, New Jersey, is placed in a beaker and is washed consecutively and *rapidly* ($\sim$10 seconds) with three 100-ml. portions of 3% hydrochloric acid, two 100-ml. portions of water, two 200-ml. portions of 2% aqueous copper sulfate (until blue color disappears), two 200-ml. portions of water, two 100-ml. portions of acetone, two 100-ml. portions of dimethylformamide, and is washed into the reaction vessel with dimethylformamide. This procedure is a modification of one described by Hennion and Sheehan.[3]

9. This *highly exothermic* reaction often has an induction period the end of which is characterized by a rapid temperature rise dependent on the amount of dibromide already added. At the first sign of reaction (*watch the thermometer closely*), addition of

dibromide should be stopped and should be resumed only after the temperature has stopped rising. Careful purification of the dimethylformamide appears to minimize the induction period.

10. Analysis of this product by gas liquid chromatography (QF-1 coated column, 130°) showed it to be >98.5% pure. The boiling point of a sample collected by chromatography was 169–171°.

3. Methods of Preparation

Cyclopropylbenzene has been prepared by decomposition of 5-phenylpyrazoline,[4] addition of hydrogen bromide to cinnamyl bromide followed by cyclization with zinc,[5] decarboxylation of 1-phenylcyclopropanecarboxylic acid,[6] reaction of magnesium with 3-bromo-3-phenyl methyl ether followed by decomposition of the intermediate Grignard reagent,[7] reaction of styrene with methylene iodide and zinc-copper couple,[8] reaction of sodium amide with 3-phenylpropyltrimethylammonium iodide in liquid ammonia,[9] decarbonylation of 1-phenylcyclopropanecarboxaldehyde,[10] and reaction of sodium hydroxide with (1-phenylcyclopropyl)-diphenylphosphine oxide.[11]

4. Merits of the Preparation

Because of the unique properties of the cyclopropane ring, cyclopropylbenzene is a compound of considerable interest. Only one of the alternative methods [9] for the preparation of this compound has been reported to give more than 32% yield; the procedure described affords an olefin-free product without a relatively laborious purification process. By its utilization of readily available starting materials, and by its applicability to the preparation of large quantities of product, this method of synthesis provides easy access to many cyclopropylbenzene derivatives.[12]

1. State University of South Dakota, Vermillion, South Dakota.
2. The Ohio State University, Columbus, Ohio.
3. G. F. Hennion and J. J. Sheehan, *J. Am. Chem. Soc.*, **71**, 1964 (1949).
4. N. Kizhner, *J. Russ. Phys.-Chem. Soc.*, **45**, 949 [*C. A.*, **7**, 3964 (1913)].
5. M. Lespieau, *Compt. Rend.*, **190**, 1129 (1930).
6. F. H. Case, *J. Am. Chem. Soc.*, **56**, 715 (1934).
7. J. T. Gragson, K. W. Greenlee, J. M. Derfer, and C. E. Boord, *J. Org. Chem.*, **20**, 275 (1955).

8. H. E. Simmons and R. D. Smith, *J. Am. Chem. Soc.*, **80**, 5323 (1958).
9. C. L. Bumgardner, *J. Am. Chem. Soc.*, **83**, 4420 (1961).
10. D. I. Schuster and J. D. Roberts, *J. Org. Chem.*, **27**, 51 (1962).
11. L. Horner, H. Hoffman, and V. Toscano, *Ber.*, **95**, 536 (1962).
12. T. F. Corbin, R. C. Hahn, and H. Shechter, Unpublished work.

m-CYMENE*

(*m*-Isopropyltoluene)

Submitted by D. E. PEARSON, ROBERT D. WYSONG, and J. M. FINKEL [1]
Checked by RICHARD A. HAGGARD and W. D. EMMONS

1. Procedure

Caution! It is necessary to carry out the entire operation including the workup in a well-ventilated hood. Rubber gloves and safety glasses should be worn (Note 1).

Anhydrous hydrogen fluoride (135 ml., approx. 7 moles) is liquefied by passing the gas through an 8-ft. spiral of ¼-in. I.D. copper tubing surrounded by an isopropyl alcohol-dry ice bath (Note 2). The liquid is delivered to a 500-ml. polyethylene squeeze bottle (Note 3) containing a magnetic stirring bar via a polyethylene tube inserted through the screw cap of the bottle. The squeeze bottle is contained in a 2-l. beaker and is surrounded by powdered dry ice (Note 4). After the hydrogen fluoride is collected, the cap and delivery tube are removed, and *p*-cymene (terpene-free, 67 g., 0.5 mole), precooled to −50° to −60°, just above the slush point, is added to the hydrogen fluoride. The cap and polyethylene tube assembly is now attached to deliver boron trifluoride. The delivery tube in this case dips below the surface of the hydrogen fluoride, the bottom layer. Boron trifluoride is bubbled through while the mixture is efficiently stirred with a magnetic stirrer (Note 5). A light orange color develops immediately, and the two layers become one in about 30 minutes (Note 6). Additional powdered dry ice must be added to the beaker during the boron trifluoride addition. The volume of the complex increases about 25%. After homogeneity is effected (in 30 minutes), a somewhat slower stream of boron trifluoride is added for an additional 30 minutes (Note 7). The delivery tube is replaced by a cap on the polyethylene bottle, the drying tube is removed, and the original side-arm tube is lowered to the bottom of the container. The cold reaction mixture is squirted, by squeezing the bottle, in a continuous small stream into a 4-l. beaker half-filled with cracked ice, vigorously hand-stirred. The bottle is rinsed, and the contents of the beaker are placed in a large separatory funnel. The upper colorless layer is separated, and the aqueous phase is extracted 3 times with 50-ml. portions of hexane. The combined organic layers are washed with three 50-ml. portions of water and dried overnight with anhydrous sodium sulfate under refrigeration (Note 8). The hexane solution at this point contains *m*-cymene with about 8% disproportionation impurities including toluene. The solution is fractionated through a 1-ft. helices-packed column, with the *m*-cymene at boiling point 173–176°, 50–54 g. (75–80%) being collected (Note 9).

2. Notes

1. In the event of accidental contact of hydrogen fluoride with the skin, the affected area must be washed immediately and thoroughly with cold water. Additional treatment has been described.[2]

2. If one half the given quantity of hydrogen fluoride is used, all other factors being kept the same, the *m*-cymene formed contains as much as 5% of *p*-cymene.

3. A Nalgene "15–500" polyethylene bottle (Nalge Co., Rochester, New York) was used. The side arm coming off the shoulder of the bottle is kept well above the liquid level, the constricted tip removed, and the end of the tube connected to a drying tube containing clay plate chips impregnated with concentrated sulfuric acid.

A squeeze bottle can be simply made from a 500-ml. narrow-mouthed polyethylene bottle and polyethylene tubing. Holes in the bottle cap and shoulder are made with a sharp cork borer of the appropriate size to ensure a tight fit with the inserted tubing.

4. A polyethylene or copper foil loop 1 in. wide is placed between the squeeze bottle and the side of the beaker in such a position as to exclude the dry ice from the space and to provide a window to permit one to see that the liquid hydrogen fluoride fills the bottle to a premarked level. The frost on the beaker must be scraped off to allow inspection through the window.

5. The stirring motor is housed in a polyethylene bag to protect it from acid fumes.

6. Larger quantities of boron trifluoride are evolved from the drying tube at this point. During the first 30 minutes, boron trifluoride is added at a rate which gives a slow emanation of fuming vapor from the drying tube. The checkers found that the product was contaminated by p-cymene when inefficient stirring and slow boron trifluoride addition rates were employed (Note 7).

7. The checkers used a flowmeter to monitor boron trifluoride addition and found that an indicated addition rate of 1800 ml./min. (calibrated with air) for the first 5 minutes followed by an average rate of 600 ml./min. gave homogeneity in the pre-scribed time. A rate of 150 ml./min. was used for the second 30 minutes.

8. All glassware is rinsed with water immediately after use to prevent etching.

9. The product is analyzed by vapor phase chromatography using a 6-ft., $\frac{1}{4}$-in. O.D. copper tube, packed with 5% Bentone-34 (Wilkins Instrument Co.) and 0.5% XF-1150 (General Electric Silicone Products) on Diatoport-S (80–100 mesh) (F and M Co.); flow rate of helium 60 ml./min., oven temperature 85°. This column separates m-cymene (retention time 12 minutes) from

p-cymene (retention time 10 minutes) but does not resolve the *ortho* isomer. The purity of the distilled *m*-cymene is above 98%.

3. Methods of Preparation

m-Cymene has been prepared from the Grignard reagent of *m*-bromotoluene and acetone followed by conversion of the carbinol to the chloride and reduction with sodium in liquid ammonia.[3] It also has been prepared from *m*-toluoyl chloride and excess methylmagnesium bromide followed by catalytic reduction of the olefin formed.[4] The best set of physical properties for the isomeric cymenes appears to be that of Birch and co-workers.[4] Many examples of Friedel-Crafts alkylation of toluene with propylene are described; apparently the best of them gives a 90% yield of cymenes containing 65–70% *m*-cymene.[5]

The method of preparation in this procedure is adapted from that of McCauley and Lien by which they obtained *m*-cymene in unstated yields.[6] The procedure has been altered to operate at $-78°$ rather than $-20°$.

4. Merits of the Preparation

Aromatic hydrocarbons substituted by alkyl groups other than methyl are notorious for their tendency to disproportionate in Friedel-Crafts reactions. This tendency has previously limited the application of the isomerization of *para*-(or *ortho*-)dialkylbenzenes to the corresponding *meta* compounds. At the lower temperature of the present modification, disproportionation can be minimized.

1. Department of Chemistry, Vanderbilt University, Nashville, Tennessee 37203.
2. G. A. Olah and S. J. Kuhn, this volume, p. 66.
3. P. E. Verkade, K. S. deVries, and B. M. Wepster, *Rec. Trav. Chim.*, **82**, 637 (1963).
4. S. F. Birch, R. A. Dean, F. A. Fidler, and R. A. Lowry, *J. Am. Chem. Soc.*, **71**, 1362 (1949).
5. B. L. Kozik, I. S. Vol'fson, M. B. Vol'f, and L. I. Germash, *Khim. i Tekhnol. Topliva i Masel*, **6**, (10) 9 (1961); see G. A. Olah, "Friedel-Crafts and Related Reactions," Vol. II, Part I, Interscience Publishers, New York, 1964, p. 149 (summary of the alkylation of toluene by propylene).
6. D. A. McCauley and A. P. Lien, U.S. Patent 2,741,647 [*C.A.*, **50**, 11658 (1956)].

DEAMINATION OF AMINES. 2-PHENYLETHYL BENZOATE* VIA THE NITROSOAMIDE DECOMPOSITION

(Benzoic acid, 2-phenylethyl ester)

$$C_6H_5CH_2CH_2NH_2 + C_6H_5COCl \xrightarrow{C_5H_5N}$$

$$C_6H_5CH_2CH_2NHCOC_6H_5$$

$$C_6H_5CH_2CH_2NHCOC_6H_5 + N_2O_4 \xrightarrow{CH_3CO_2Na}$$

$$C_6H_5CH_2CH_2\underset{\underset{NO}{|}}{N}COC_6H_5$$

$$C_6H_5CH_2CH_2\underset{\underset{NO}{|}}{N}COC_6H_5 \xrightarrow{\Delta} C_6H_5CH_2CH_2OCOC_6H_5 + N_2$$

Submitted by EMIL WHITE [1]
Checked by WILLIAM G. DAUBEN and WILLIAM C. SCHWARZEL

1. Procedure

Caution! Dinitrogen tetroxide is extremely toxic and should only be handled in an excellent hood.

A. *N-(2-Phenylethyl)benzamide.* To a solution of 12.1 g. (0.10 mole) of 2-phenylethylamine and 7.9 g. (0.10 mole) of pyridine in a 250-ml. Erlenmeyer flask immersed in an ice bath is added, slowly with stirring, 15.5 g. (0.11 mole) of benzoyl chloride. The resulting crystalline mixture is extracted with chloroform and the chloroform solution washed with water, 5% hydrochloric acid, 5% sodium hydroxide, water, and dried. The solvent is removed under reduced pressure to yield 20–22 g. (89–98%) of the crude amide, m.p. 100–110°, which is of sufficient purity for use in the next step. If desired, however, the crude product may be recrystallized from 95% ethanol, m.p. 115–116°.

B. *N-Nitroso-N-(2-phenylethyl)benzamide.* A solution of 10.4 g. (0.046 mole) of the crude N-(2-phenylethyl)benzamide, 7.36 g. (0.09 mole) of anhydrous sodium acetate, and 50 ml. of glacial acetic acid is placed in a 250-ml. Erlenmeyer flask equipped with a drying tube, and the mixture is cooled to the crystallization point of the acetic acid (Note 1). A solution of dinitrogen tetroxide (Notes 2, 3) in glacial acetic acid (85 ml. of a solution approximately $1M$ in N_2O_4) is then added with stirring. The reaction mixture is allowed to warm to about 15° (15 minutes), and then it is poured into a mixture of ice and water. The yellow solid nitroso derivative is dissolved in 75 ml. of carbon tetrachloride, and this solution is washed with 5% sodium bicarbonate, water, and dried. The solution is used directly in the next step.

If the nitroso derivative is desired, the yellow solid is separated by filtration, washed with water, 5% sodium bicarbonate solution, water, and dried under reduced pressure at room temperature. The solid is recrystallized from ether to give yellow needles of pure nitrosoamide; yield 7.5 g. (64%), m.p. 57–58° (dec.).

C. *2-Phenylethyl benzoate.* The carbon tetrachloride solution of N-nitroso-N-(2-phenylethyl)benzamide (Note 4) and 0.1 g. of sodium carbonate (Note 5) are placed in a 200-ml. round-bottomed flask equipped with a condenser, and the mixture is heated under reflux for 24 hours. The evolution of nitrogen ceases, and the yellow color of the nitrosoamide weakens or disappears near the end of this period. The solution is washed with 5% sodium hydroxide solution, water, and dried. The solvent is removed under reduced pressure and the 2-phenylethyl benzoate distilled; b.p. 139-142° (1 mm.), yield 5.8-6.1g. [56-59% based on N-(2-phenylethyl)benzamide].

2. Notes

1. Carbon tetrachloride, methylene chloride, and other solvents may be used in this reaction. In these cases it is profitable to cool the reaction mixture to $-40°$ or lower and then allow the mixture to warm to 10° after the dinitrogen tetroxide has been added.

2. Dinitrogen tetroxide (nitrogen dioxide) is available from the Matheson Company, Inc., East Rutherford, New Jersey.

3. Dinitrogen tetroxide is a *poisonous gas* and should only be handled in a well-ventilated hood. The boiling point of dinitrogen tetroxide is 21°, and it is convenient to condense a given volume (or weight, density = 1.5 g./ml. at 0°) from a cylinder of dinitrogen tetroxide and to pour the liquid into the required amount of solvent, or into the reaction mixture directly at temperatures below *ca.* −20°. Impure dinitrogen tetroxide, which is green because of the presence of lower oxides of nitrogen, may also be used.

4. Any nonreactive solvent may be used, but excessive temperatures favor the concurrent elimination reaction.

5. The sodium carbonate may be omitted if it is desired to titrate the acid formed in the reaction. The carbonate prevents denitrosation (observed in a few cases).

3. Methods of Preparation

The nitrosation of amides may also be carried out with nitrosyl chloride.[2] Related methods of deamination of aliphatic amines are the triazene[3] and nitrous acid methods.[4]

4. Merits of the Preparation

Dinitrogen tetroxide is the most versatile of the nitrosating reagents and, in addition, it is readily available. The nitrosoamide method of deamination gives far superior yields and much less skeletal isomerization than the nitrous acid method (which is essentially limited to aqueous media), and it leads to a greater retention of optical activity than the triazene method.[3] In general, the nitrosoamide decomposition proceeds with retention of configuration.[5, 6]

The method outlined here works well for amides of primary carbinamines. For amides of secondary carbinamines, lower temperatures must be used for the nitrosation step (∼0°), and solvents such as methylene chloride are used in place of the acetic acid (the amide need not be completely soluble in the solvent); the procedure of White and Aufdermarsh[5] used for a trimethylacetamide is recommended in such a case. Nitrosoamides of tertiary carbinamines are very unstable, and the "salt method" of preparation is suggested for these compounds.[6]

1. Department of Chemistry, The Johns Hopkins University, Baltimore, Maryland 21218.
2. H. France, I. M. Heilbron, and D. H. Hey, *J. Chem. Soc.*, 369 (1940).
3. E. H. White and H. Scherrer, *Tetrahedron Lett.*, 758 (1961).
4. F. C. Whitmore and D. P. Langlois, *J. Am. Chem. Soc.*, **54**, 3441 (1932).
5. E. H. White and C. A. Aufdermarsh, Jr., *J. Am. Chem. Soc.*, **83**, 1174 (1961).
6. E. H. White and J. E. Stuber, *J. Am. Chem. Soc.*, **85**, 2168 (1963).

DEOXYANISOIN*

[Acetophenone, 4'-methoxy-2-(p-methoxyphenyl)-]

$$CH_3O-C_6H_4-CHOH-CO-C_6H_4-OCH_3 + Sn + 2HCl \rightarrow$$

$$CH_3O-C_6H_4-CH_2CO-C_6H_4-OCH_3 + SnCl_2 + H_2O$$

Submitted by P. H. CARTER, J. CYMERMAN CRAIG, RUTH E. LACK, and M. MOYLE.[1]
Checked by MAX TISHLER and E. M. CHAMBERLIN.

1. Procedure

A 500-ml. round-bottomed flask, equipped with a reflux condenser, is charged with 40 g. (0.33 mole) of powdered tin (Notes 1 and 2), 52 g. (0.19 mole) of anisoin (Note 3), 52 ml. of concentrated hydrochloric acid (Note 4), and 60 ml. of 95% alcohol. After the mixture is refluxed for 24 hours (Note 5), the boiling solution is decanted from undissolved tin, cooled to 0°, and the white crystals are filtered by suction. The filtrate is heated to boiling and then used to wash the tin by decantation. The combined washings are cooled to 0°, and the crystalline solid is collected by suction filtration. Recrystallization of the combined solids from 450 ml. of boiling 95% ethanol (Notes 6 and 7) gives, on cooling to 0°, colorless crystals of deoxyanisoin melting at 108–111°. The yield is 42–45 g. (86–92%).

2. Notes

1. A reduction in the amount of tin lowers the yield. Best results are obtained using powdered tin of between 100 and 200 mesh size. Use of tin coarser than 100 mesh results in the presence of unchanged anisoin, while tin finer than 200 mesh tends to conglomerate, causing lower yields. The checkers used tin obtained from E. H. Sargent Co., Chicago, labeled 200 mesh.

2. No advantage is gained by using amalgamated tin.

3. This procedure is generally applicable to the preparation of symmetrical deoxybenzoins. The submitters have prepared (a) deoxybenzoin (m.p. 56–58°) in 80–84% yield from 53 g. of benzoin (0.25 mole), 53 ml. of concentrated hydrochloric acid, 50 ml. of 95% alcohol, and 53 g. of powdered tin (0.44 mole), recrystallizing from 160 ml. of boiling 95% alcohol and cooling to 0°; (b) deoxypiperoin (m.p. and mixed m.p. 112–114°) in 89% yield from 14.3 g. of piperoin (0.048 mole), 13 ml. of concentrated hydrochloric acid, 30 ml. of 95% alcohol, and 10 g. of powdered tin (0.083 mole), recrystallizing from 140 ml. of boiling 95% alcohol and cooling to 0°.

4. A reduction in the amount of hydrochloric acid lowers the yield.

5. Lower yields are obtained by using reflux periods of 16 or 18 hours.

6. The checkers recommend washing the product with cold 95% alcohol; otherwise the product tends to discolor on standing.

7. Addition of water to the filtrate does not yield any further crystalline products.

3. Methods of Preparation

The synthesis of deoxybenzoin from phenacetyl chloride and benzene by the Friedel-Crafts reaction has been described.[2] For symmetrically substituted deoxybenzoins, direct reduction of the readily accessible benzoin is a more convenient method. Reduction of benzoin by zinc dust and acetic acid,[3] and by hydrochloric acid and granulated tin [4,5] or amalgamated powdered tin [6] has been reported. The present method is based on a publication of the authors.[7]

1. Dyson Perrins Laboratory, Oxford, England.

2. C. F. H. Allen and W. E. Barker, *Org. Syntheses,* Coll. Vol. 2, 156 (1943).
3. E. P. Kohler and E. M. Nygaard, *J. Am. Chem. Soc.,* 52, 4133 (1930).
4. J. S. Buck and S. S. Jenkins, *J. Am. Chem. Soc.,* 51, 2163 (1929).
5. I. Allen and J. S. Buck, *J. Am. Chem. Soc.,* 52, 310 (1930).
6. O. A. Ballard and W. M. Dehn, *J. Am. Chem. Soc.,* 54, 3970 (1932).
7. P. H. Carter, J. C. Craig, R. E. Lack, and M. Moyle, *J. Chem. Soc.,* 476 (1959).

4,4'-DIAMINOAZOBENZENE*

(Aniline, 4,4'-azodi-)

Submitted by Pasco Santurri, Frederick Robbins,
   and Robert Stubbings.[1]
Checked by Max Tishler, Earl M. Chamberlin,
   and William Harrison.

1. Procedure

In a 1-l. three-necked round-bottomed flask equipped with an efficient stirrer, a reflux condenser, and a thermometer (Note 1) are placed 500 ml. of glacial acetic acid (Note 2), 29.0 g. (0.19 mole) of *p*-aminoacetanilide (Note 3), 40 g. (0.26 mole) of sodium perborate tetrahydrate, and 10 g. (0.16 mole) of boric acid. The mixture is heated with stirring to 50–60° and held at this temperature for 6 hours. Initially the solids dissolve but, after heating for approximately 40 minutes, the product begins to separate. At the end of the reaction period, the mixture is cooled to room temperature and the yellow product is collected on a Büchner funnel. It is washed with water until the washings are neutral to pH paper (Note 4) and then dried in an oven at 110°. The yield of 4,4'-bis(acetamido)azobenzene, m.p. 288–293° (dec.), is 16.5 g. (57.7%). It is used as such for the hydrolysis step (Note 5).

In a 500-ml. round-bottomed flask equipped with a reflux condenser and a magnetic stirrer (Note 6) are placed 150 ml. of methanol, 150 ml. of 6N hydrochloric acid, and the total yield of 4,4'-bis(acetamido)azobenzene. The mixture is heated under reflux for 1.5 hours. The reaction mixture is cooled and the violet solid collected on a Büchner funnel (Note 7). The damp product is suspended in 500 ml. of water in a 1-l. beaker equipped with a stirrer, and the mixture is slowly neutralized by the addition of 2.5N sodium hydroxide. In the course of the neutralization, the salt dissolves and the free base separates. The 4,4'-diaminoazobenzene is collected on a Büchner funnel, washed with water, and dried under reduced pressure. The yield of yellow product, m.p. 238–241° (dec.), is 11–12 g. The over-all yield from p-aminoacetanilide is 52–56%.

2. Notes

1. The submitters carried out the reaction in a 1-l. beaker. The checkers found that considerable evaporation of acetic acid occurs when a beaker is used.

2. The use of dilute acetic acid decreases the yield.

3. Technical grade p-aminoacetanilide (m.p. 158–160°) obtained from Eastman Kodak Company (T-13) was used.

4. If the product is not washed well, the dried material will turn violet, indicating unreacted p-aminoacetanilide.

5. Witt and Kopetschni [2] report a melting point of 295–296° (dec.). This compound may be recrystallized from glacial acetic acid or ethanol.

6. The submitters did not stir the hydrolysis mixture. The checkers found that, if stirring was omitted, bumping occurred during the reflux period.

7. This product is probably the monohydrochloride salt. The dihydrochloride is reported to be red.[3]

3. Methods of Preparation

4,4'-Diaminoazobenzene was reported by Nietzki[4] to have been prepared by diazotizing *p*-nitroaniline and coupling the product with aniline. The resulting 4-nitrodiazoaminobenzene[5] is rearranged and the nitro group reduced. The submitters tried several times to carry out this procedure but were unsuccessful. 4,4'-Diaminoazobenzene has been prepared by the oxidation of *p*-nitroaniline with potassium persulfate[6] followed by the reduction of the nitro groups.[7]

The general method used by the submitters has been reported by others[8] for the preparation of other azo compounds.

4. Use of 4,4'-Diaminoazobenzene

The product is useful as a model compound for studies of the chemistry of derivatives of colored diamines. Specifically, the submitters used the compound for the preparation of colored diisocyanates.

1. Leather Research Institute, Lehigh University, Bethlehem, Pennsylvania.
2. O. N. Witt and E. Kopetschni, *Ber.*, 45, 1134 (1912).
3. F. Kehrmann and S. Hempel, *Ber.*, 50, 867 (1917).
4. R. Nietzki, *Ber.*, 17, 343 (1884).
5. E. Noelting and F. Binder, *Ber.*, 20, 3015 (1887).
6. A. H. Cook and D. G. Jones, *J. Chem. Soc.*, 1309 (1939).
7. J. N. Ashley, H. J. Barber, A. J. Ewins, G. Newbery, and A. D. H. Self, *J. Chem. Soc.*, 103 (1942).
8. S. M. Mehta and M. V. Vakilwala, *J. Am. Chem. Soc.*, 74, 563 (1952).

DIAMINOMALEONITRILE*
(HYDROGEN CYANIDE TETRAMER)

(Maleonitrile, diamino-)

$$p\text{-}CH_3C_6H_4SO_3^- \overset{+}{N}H_3CH(CN)_2 + NaCN \rightarrow$$

$$\underset{H_2N}{\overset{NC}{\diagdown}} C{=}C \underset{NH_2}{\overset{CN}{\diagup}} + p\text{-}CH_3C_6H_4SO_3Na$$

Submitted by J. P. Ferris and R. A. Sanchez[1]
Checked by O. W. Webster and R. E. Benson

1. Procedure

Caution! The preparation should be carried out in a hood because hydrogen cyanide may be evolved.

To a cooled (0°), stirred suspension of 10.0 g. (0.0395 mole) of aminomalononitrile p-toluenesulfonate[2] in 20 ml. of water is added 10.0 g. (0.204 mole) of sodium cyanide in 30 ml. of ice water. One minute (Note 1) after the addition of the sodium cyanide the precipitated product is collected by filtration and washed with 20 ml. of ice water. The solid is immediately dissolved (Note 1) in 30 ml. of boiling isobutyl alcohol, and the solution is stirred with 0.4 g. of Darco activated carbon (Note 2). The mixture is filtered rapidly through 10 g. of Celite filter aid, and the filter cake is washed with 10 ml. of hot isobutyl alcohol. The product that crystallizes on cooling is collected by filtration and washed with 10 ml. of isobutyl alcohol to give 0.95–1.1 g. (22–26%) of white needles, m.p. 181–183° (dec.).

2. Notes

1. The product darkens on long standing.
2. Hydrogen cyanide tetramer is strongly adsorbed on activated carbon; no more than the recommended amount of carbon

should be used, and it should be added carefully to avoid frothing.

3. Methods of Preparation

The present procedure is a modification of the original synthesis.[3] Hydrogen cyanide tetramer can be prepared directly from hydrogen cyanide.[4]

4. Merits of the Preparation

This is a convenient laboratory preparation of hydrogen cyanide tetramer that avoids the hazards in using hydrogen cyanide itself. Hydrogen cyanide tetramer is a useful intermediate for the synthesis of heterocycles such as imidazoles [5, 6] and thiadiazoles.[7]

1. The Salk Institute for Biological Studies, San Diego, California [Present address (J.P.F.): Department of Chemistry, Rensselaer Polytechnic Institute, Troy, New York 12181].
2. J. P. Ferris, R. A. Sanchez, and R. W. Mancuso, this volume, p. 32.
3. J. P. Ferris and L. E. Orgel, *J. Am. Chem. Soc.*, **88**, 3829 (1966); **87**, 4976 (1965).
4. H. Bredereck, G. Schmötzer, and E. Oehler, *Ann.*, **600**, 81 (1956).
5. H. Bredereck and G. Schmötzer, *Ann.*, **600**, 95 (1956).
6. J. P. Ferris and L. E. Orgel, *J. Am. Chem. Soc.*, **88**, 1074 (1966).
7. M. Carmack, L. M. Weinstock, and D. Shew, Abstracts, 136th National Meeting of the American Chemical Society, Atlantic City, N. J., Sept. 1959, p. 37P; D. Shew, *Dissertation Abstr.*, **20**, 1593 (1959).

2,3-DIAMINOPYRIDINE *

(Pyridine, 2,3-diamino-)

Submitted by B. A. Fox[1] and T. L. Threlfall[2]
Checked by Max Tishler, G. A. Stein, G. Lindberg, and M. Ryder

1. Procedure

Caution! The bromination and nitration steps should be carried out in a well-ventilated hood.

A. *2-Amino-5-bromopyridine.* In a 2-l. three-necked flask equipped with stirrer, dropping funnel, and condenser is placed a solution of 282 g. (3.0 moles) of 2-aminopyridine (Note 1) in 500 ml. of acetic acid. The solution is cooled to below 20° by immersion in an ice bath, and 480 g. (154 ml., 3.0 moles) of bromine dissolved in 300 ml. of acetic acid is added dropwise with vigorous stirring over a period of 1 hour. Initially the temperature is maintained below 20°, but after about half the bromine solution has been added it is allowed to rise to 50° to delay as long as possible the separation of the hydrobromide of 2-amino-5-bromopyridine. At 50° the hydrobromide usually begins to crystallize when about three-quarters of the bromine has been added. When addition of bromine is completed, the mixture is stirred for 1 hour and is then diluted with 750 ml. of water to dissolve the hydrobromide. The contents of the flask are transferred to a 5-l. beaker and are neutralized, with stirring and cooling, by the addition of 1.2 l. of 40% sodium hydroxide solution.

The precipitated 2-amino-5-bromopyridine, contaminated with some 2-amino-3,5-dibromopyridine, is collected by filtration and, after washing with water until the washings are free of ionic bromide, is dried at 110° (Note 2). The 2-amino-3,5-dibromo-pyridine is removed from the product by washing with three 500-ml. portions (Note 3) of hot petroleum ether (b.p. 60–80°). The yield of 2-amino-5-bromopyridine, m.p. 132–135°, sufficiently pure for use in the next stage, is 320–347 g. (62–67%) (Notes 4 and 5).

B. *2-Amino-5-bromo-3-nitropyridine.* A 1-l. three-necked flask immersed in an ice bath and equipped with stirrer, dropping funnel, condenser, and thermometer is charged with 500 ml. of sulfuric acid (sp. gr. 1.84), and 86.5 g. (0.5 mole) of 2-amino-5-bromopyridine is added at such a rate that the temperature does not exceed 5°. Twenty-six milliliters (39 g., 0.57 mole) of 95% nitric acid is added dropwise with stirring at 0°, and the mixture is stirred at 0° for 1 hour, at room temperature for 1 hour, and at 50–60° for 1 hour. The contents of the flask are cooled and poured onto 5 l. of ice and neutralized with 1350 ml. of 40% sodium hydroxide solution. The yellow precipitate of 2-amino-5-bromo-3-nitropyridine is collected by filtration and washed with water until the washings are free of sulfate, slightly acidulated water being used at the end to prevent colloidal breakthrough. The product is dried at room temperature to constant weight. The yield of 2-amino-5-bromo-3-nitropyridine, m.p. 204–208°, sufficiently pure for the next stage, is 85–93 g. (78–85%) (Notes 6 and 7).

C. *2,3-Diamino-5-bromopyridine* (Note 8). A 100-ml. flask fitted with a reflux condenser is charged with 10.9 g. (0.05 mole) of 2-amino-5-bromo-3-nitropyridine, 30 g. of reduced iron, 40 ml. of 95% ethanol, 10 ml. of water, and 0.5 ml. of concentrated hydrochloric acid (Notes 9 and 10). The mixture is heated on a steam bath (Note 11) for 1 hour, and at the end of this period the iron is removed by filtration and is washed three times with 10-ml. portions of hot 95% ethanol. The filtrate and washings are evaporated to dryness, and the dark residue is recrystallized from 50 ml. of water, 1 g. of activated carbon being used and the mixture being filtered while hot. The charcoal is washed with hot ethanol to avoid losses. 2,3-Diamino-5-bromopyridine crystallizes as colorless needles, m.p. 163°. The yield is 6.5–7.1 g. (69–76%).

D. *2,3-Diaminopyridine* (Note 12). In an apparatus for catalytic hydrogenation (Note 13) 56.4 g. (0.3 mole) of 2,3-diamino-5-bromopyridine suspended in 300 ml. of 4% sodium hydroxide solution is shaken with hydrogen in the presence of 1.0 g. of 5% palladized strontium carbonate (Note 14). When absorption of hydrogen is completed, the catalyst is removed by filtration, and, after saturation with potassium carbonate (about 330 g. is required), the resulting slushy mixture is extracted continuously with ether until all the precipitate completely disappears (usually about 18 hours, but this depends on the efficiency of the extraction apparatus). The ether is removed by distillation, and the residue of crude 2,3-diaminopyridine is recrystallized from benzene (about 600 ml. is required) using 3 g. of activated charcoal and filtering rapidly through a preheated Büchner funnel. The yield of 2,3-diaminopyridine, obtained as colorless needles, m.p. 115–116°, pK_a 6.84, is 25.5–28.0 g. (78–86%) (Note 15).

2. Notes

1. The checkers used a pure grade of 2-aminopyridine (m.p. 58–60°) obtained from Matheson, Coleman and Bell.

2. The checkers dried their product at room temperature to constant weight in order to avoid loss of product due to its high volatility at 110°. It was found that 95% of an aliquot had sublimed during drying for 24 hours at 110° and atmospheric pressure. The residue analyzed high in bromine, indicating that the monobromo derivative is more volatile than the dibromo derivative.

3. The checkers washed their crude product by first refluxing its suspension in 600 ml. of petroleum ether (b.p. 60–71°) for about 20 minutes. The product, obtained by filtration, was slurry-washed on the funnel with two 600-ml. portions of boiling petroleum ether followed by air-drying to constant weight.

4. The checkers' yield, for reasons outlined in Note 2, were appreciably higher. In two runs using one-half the quantities of reactants they obtained 211 g. and 224 g. (81–86%), of product, m.p. 132–133.5° and 133.5–135°, respectively; water content (K.F.) in both products was 0.2%.

5. If required, the 2-amino-5-bromopyridine may be recrystallized from benzene as colorless prisms, m.p. 137°.

6. The checkers' yield was 85.3 g. (78.2%), m.p. 202–204°.

7. Pure 2-amino-5-bromo-3-nitropyridine, yellow needles, m.p. 210°, may be obtained by recrystallizing the product from ethyl methyl ketone.

8. The method is essentially that of Petrow and Saper.[3]

9. Attempts to reduce larger quantities of the amino-nitro compound by this method usually give lower yields. For larger quantities several reductions may be carried out simultaneously, and the filtrates may be combined for isolation of the diamine.

10. The checkers reduced a double batch and obtained 12.8 g. (68%), m.p. 159.5–160°. In this run, heating time was doubled and charcoal was extracted repeatedly (by recycling mother liquors) to assure complete extraction.

11. The checkers employed a mechanical stirrer.

12. This is essentially the procedure of Leese and Rydon.[4]

13. The apparatus described in *Org. Syntheses*, Coll. Vol. **1**, 61 (1941), or a commercial equivalent of it, is suitable.

14. The palladized strontium carbonate is prepared as follows. Suspend 33 g. of strontium carbonate in 350 ml. of water at 70°. Add 2 g. of palladium chloride dissolved in 10 ml. of concentrated hydrochloric acid, and stir the mixture at 70° for 15 minutes. Filter the mixture, wash the product thoroughly with hot water, and dry the product at 110°.

15. The checkers' yields were 74.8%–84.7% of analytically pure material giving a negative Beilstein test.

3. Methods of Preparation

2,3-Diaminopyridine has been prepared by reduction of 2-amino-3-nitropyridine with iron and aqueous acidified ethanol,[3] tin and hydrochloric acid,[5] or stannous chloride and hydrochloric acid,[5] by catalytic reduction of 3-amino-2-nitropyridine,[6] by reduction of 3-amino-2-nitropyridine,[7] 2-amino-5-chloro-3-nitropyridine,[8] or 2-amino-5-bromo-3-nitropyridine[4] with sodium hydroxide solution and an aluminum nickel alloy, and by catalytic reduction of 2-amino-5-bromo-3-nitropyridine.[4] Amination of 3-aminopyridine with sodamide[9] and of 3-amino-2-chloropyridine with concentrated aqueous ammonia [10] have also been employed.

4. Merits of the Preparation

By this method of preparation 2,3-diaminopyridine is obtained in 26–43% yield from the readily available 2-aminopyridine. The intermediates 2-amino-5-bromopyridine and 2-amino-5-bromo-3-nitropyridine are prepared in higher yields than previously recorded.

Methods of preparation of 2,3-diaminopyridine which involve the reduction of 2-amino-3-nitropyridine are laborious. The material is obtained in yields of less than 10% by nitration of 2-aminopyridine, and its separation from 2-amino-5-nitropyridine, which is the major product of the nitration, is tedious and inconvenient. Reduction of amino-nitro or amino-halo-nitro compounds with sodium hydroxide solution and an aluminum nickel alloy gives variable yields of an inferior product, and the method can be used only for preparing comparatively small quantities of 2,3-diaminopyridine. Catalytic reduction of 3-amino-2-nitropyridine gives a good yield of 2,3-diaminopyridine, but preparation of the amino-nitro compound is a difficult and time-consuming process. The method of Schickh, Binz, and Schulz,[10] which involves chlorination of 3-aminopyridine to 3-amino-2-chloropyridine and amination of the latter by heating for 20 hours at 130° with concentrated aqueous ammonia, suffers from the disadvantage that 3-aminopyridine is less readily available than is 2-aminopyridine. Furthermore the yields obtained in the amination stage are somewhat erratic, and the yields obtained by the submitters never approached the 57% reported.

1. Science and Food Technology Department, Salford Technical College, Salford 5, England.
2. Department of Chemistry, Color Chemistry and Dyeing, Huddersfield College of Technology, Huddersfield, England.
3. V. Petrow and J. Saper, J. Chem. Soc., 1390 (1948).
4. C. L. Leese and H. N. Rydon, J. Chem. Soc., 4039 (1954).
5. T. Takahashi and S. Yajima, J. Pharm. Soc. Japan, 64, (7a) 30 (1944).
6. J. W. Clark-Lewis and M. J. Thompson, J. Chem. Soc., 442 (1957).
7. H. M. Curry and J. P. Mason, J. Am. Chem. Soc., 73, 5046 (1951).
8. J. B. Ziegler, J. Am. Chem. Soc., 71, 1891 (1949).

9. A. Konopnicki and E. Plazek, *Ber.*, **60**, 2045 (1927).
10. O. V. Schickh, A. Binz, and A. Schulz, *Ber.*, **69**, 2593 (1936).

DIAZOMETHANE

(Methane, diazo-)

$$\text{(diazomethane precursor)} + 2NaOH \rightarrow 2CH_2N_2 + \text{(terephthalate)} + 2H_2O$$

Submitted by JAMES A. MOORE and DONALD E. REED.[1]
Checked by D. J. PASTO and E. J. COREY.

1. Procedure

Caution! Diazomethane is toxic and explosive. The operation must be carried out in a good hood with an adequate shield (see Note 1).

An efficient condenser (60 cm. or longer) is fitted with an adapter to which is sealed a length of 9-mm. tubing extending nearly to the bottom of a 5-l. round-bottomed flask, which serves as the distillation receiver (Notes 2 and 3). The adapter should be connected to the receiver with a two-hole stopper carrying a drying tube if anhydrous diazomethane is desired. The receiver is placed in a well-mixed ice-salt mixture, and sufficient anhydrous ether (about 200 ml.) is added to cover the tip of the adapter.

In a 5-l. round-bottomed flask are placed 3 l. of U.S.P. solvent grade ether, 450 ml. of diethylene glycol monoethyl ether (Note

4), and 0.6 l. of 30% aqueous sodium hydroxide solution (Note 5). The mixture is chilled in an ice-salt bath to 0° (Note 6), and 180 g. (0.5 mole) of N,N'-dimethyl-N,N'-dinitrosoterephthalamide (70% in mineral oil) (Note 7) is added in one portion. The flask is immediately transferred to a heating mantle and connected by a gooseneck to the condenser. The yellow color of diazomethane appears in the receiver almost immediately. About 2 l. of ether is distilled in 2–2.5 hours (Note 8); the distilling ether is practically colorless at this point. The tip of the adapter should be kept just below the surface of the distillate during the distillation. The distillate contains 0.76–0.86 mole (76–86%) (Notes 9 and 10) of diazomethane as determined by titration.[2] When the apparatus has been protected with a drying tube, the diazomethane is suitable for reaction with an acid chloride without further drying.

2. Notes

1. Diazomethane is not only toxic but also potentially explosive. Hence one should wear heavy gloves and goggles and work behind a safety screen or a hood door with safety glass, as is recommended in the preparation of diazomethane described by De Boer and Backer.[3] As is also recommended there, ground joints and sharp surfaces should be avoided. Thus all glass tubes should be carefully fire-polished, connections should be made with rubber stoppers, and separatory funnels should be avoided, as should etched or scratched flasks. Furthermore, at least one explosion of diazomethane has been observed at the moment crystals (sharp edges!) suddenly separated from a supersaturated solution. Stirring by means of a Teflon-coated magnetic stirrer is greatly to be preferred to swirling the reaction mixture by hand, for there has been at least one case of a chemist whose hand was injured by an explosion during the preparation of diazomethane in a hand-swirled reaction vessel.

It is imperative that diazomethane solutions not be exposed to direct sunlight or placed near a strong artificial light because light is thought to have been responsible for some of the explosions that have been encountered with diazomethane. Particular caution should be exercised when an organic solvent boiling higher than ether is used. Because such a solvent has a lower vapor pressure than ether, the concentration of diazomethane in the vapor above the reaction mixture is greater and an explosion is more apt to occur.

Most diazomethane explosions occur during its distillation. Hence diazomethane should not be distilled unless the need justifies it. An ether solution of diazomethane satisfactory for many uses can be prepared as described by Arndt,[2] where nitrosomethylurea is added to a mixture of ether and 50% aqueous potassium hydroxide and the ether solution of diazomethane is subsequently decanted from the aqueous layer and dried over potassium hydroxide pellets (not sharp-edged sticks!). When distilled diazomethane is required, the alternative procedure of De Boer and Backer [3] is particularly good because at no time is much diazomethane present in the distilling flask.

Both the toxicity and explosion hazards associated with diazomethane are discussed by Gutsche.[4]

2. If it is desired to determine the yield of diazomethane by titration, the receiver should be calibrated so that the volume of the distillate can be measured without the necessity of transferring to a graduated vessel.

3. The submitters have used equipment having all connections made with ungreased 29/42 ground-glass joints. This is contrary to previously recommended practice (see Note 1). The submitters feel that ground-glass joints do not represent an added hazard, and that their use expedites the completion of consecutive runs. In the course of many preparations, however, a film of polymethylene was found to accumulate on the joints and prevent a tight fit. This film can be removed by a brief treatment with hot concentrated alkali and vigorous rubbing.

In some forty preparations made by the submitters, one explosion occurred which was attributed to the cracking of the adapter tube during the distillation. The adapter and the drying tube were disintegrated, but the receiver and the contents of the distilling flask were not affected, indicating a local detonation that was not sustained.

The checkers did not use glassware with ground-glass joints. New unmarked flasks and condenser were used which were connected together with fire-polished glass tubing and rubber stoppers.

4. Practical grade 2-(2-ethoxyethoxy)ethanol (Matheson, Coleman and Bell) can be used without further treatment. In a few preparations, the submitters encountered difficulty with the formation of a very stiff gel of disodium terephthalate in the flask during distillation. In one case, this difficulty was traced to the use of an old bottle of 2-(2-ethoxyethoxy)ethanol from another source.

This relatively large volume of cosolvent was found to give optimum yields. The submitters have found that the evolution of diazomethane from a stirred suspension of the reagent in ether and 40% aqueous sodium hydroxide is extremely slow and incomplete.

5. The use of more concentrated solutions of potassium hydroxide gave somewhat lower yields.

6. *Caution! It is extremely important that the flask contents be cooled to at least 0°. The reaction is rapid and a considerable amount of diazomethane is generated at this temperature.*

7. This material is available from Eastman Organic Chemicals and Aldrich Chemical Company. It is also available from E. I. du Pont de Nemours and Company, who use the trade name Nitrosan for it. The 30% white mineral oil acts as a stabilizer. The material may be stored indefinitely at room temperature. It sometimes turns green on long standing, but this does not affect the yield of diazomethane (private communication from B. C. McKusick).

8. The yield of diazomethane is slightly lower if the distillation is carried out more slowly.

9. The average yield in some thirty runs was over 80%; yields as high as 95% have been obtained. It is probable that a second receiver in series would permit the recovery of a small additional amount of diazomethane.

10. The checkers decomposed the small amount of diazomethane remaining in the reaction flask by careful addition of 100 ml. of acetic acid before disposal.

3. Methods of Preparation

Diazomethane has been prepared by the action of base on nitrosomethylurea,[2] nitrosomethylurethane,[5] N-nitroso-β-meth-

ylaminoisobutyl methyl ketone,[6] *p*-tolylsulfonylmethylnitros-
amide,[3] and N-nitroso-N-methyl-N'-nitroguanidine.[7]

4. Merits of Preparation

The great advantages of the present method are the availability, moderate cost, and high stability of the nitrosamide,
and the suitability for large-scale preparations. The procedure
is rapid and simple, and the yields are consistently higher than in
any other method tried by the submitters.

1. Department of Chemistry, University of Delaware, Newark, Del.
2. F. Arndt, *Org. Syntheses,* Coll. Vol. 2, 165 (1943).
3. Th. J. De Boer and H. J. Backer, *Org. Syntheses,* Coll. Vol. 4, 250 (1963).
4. C. D. Gutsche, *Org. Reactions,* 8, 391–394 (1954).
5. W. D. McPhee and E. Klingsberg, *Org. Syntheses,* Coll. Vol. 3, 119 (1955).
6. C. E. Redemann, F. O. Rice, R. Roberts, and H. P. Ward, *Org. Syntheses,* Coll. Vol. 3, 244 (1955).
7. (a) A. F. McKay, *J. Am. Chem. Soc.,* 70, 1974 (1948); (b) A. F. McKay et al., *Can. J. Research,* 28B, 683 (1950).

DI-*t*-BUTYL NITROXIDE

(Nitroxide, di-*tert*-butyl)

$$(CH_3)_3CNO_2 \xrightarrow[\text{2. H}_2\text{O}]{\text{1. Na}} [(CH_3)_3C]_2NO$$

Submitted by A. K. Hoffmann, A. M. Feldman,
E. Gelblum, and A. Henderson [1]
Checked by R. A. Haggard and William D. Emmons

1. Procedure (Note 1)

Thirteen hundred milliliters of 1,2-dimethoxyethane (glyme)
is distilled from lithium aluminum hydride (*Caution! Under no
circumstances should distillation be carried to dryness since explosive decomposition of the residual hydride may occur.*) (Note 2)
directly into a 2-l., nitrogen-flushed, three-necked, Morton flask
equipped with a nitrogen inlet, an outlet, and a high-speed
stirrer with a stainless steel propeller-type blade. The flask is
charged with 89.7 g. (0.87 mole) of *t*-nitrobutane (Note 3) and

19.9 g. (0.87 g. atom) of sodium cut into pea-sized pieces (Note 4). The stirrer is started and initially operated at a speed just adequate to draw some of the sodium through the blade. The onset of reaction is signaled when the solution is pale lavender and the sodium surface is clearly etched and colored bright gold (Note 5). The temperature of the reaction mixture is maintained at 25–30° (Note 6) by directing an air blast at the sides of the flask and by controlling the rate at which sodium is drawn through the blades of the stirrer. As the reaction progresses, colorless solid is formed, and at the end of the reaction (*ca.* 24 hours) the reaction mixture consists of solid and a colorless glyme solution. Most of the glyme is removed by evaporation under reduced pressure at room temperature with a water bath at 20–25° to leave a thick, colorless slurry. To the slurry under nitrogen (Note 7) is added 270 ml. of water, and the reddish brown organic layer is separated. The aqueous layer (Note 8) is extracted with several portions of pentane until the extract is colorless. The organic layer and pentane extracts are combined, cooled to 0°, and washed rapidly and thoroughly with two 70-ml. portions of ice-cold 0.25N hydrochloric acid to remove hydroxylamine impurities (Note 9). The pentane solution of the product is washed immediately with 70 ml. of cold water followed by 70 ml. of cold, aqueous 0.2N sodium hydroxide. The combined, cold, aqueous acidic washings are extracted with small portions of pentane until colorless. This pentane extract is used to extract the aqueous sodium hydroxide layer and is then washed with water and combined with the initial pentane extract.

The pentane solution is dried over anhydrous magnesium sulfate and fractionated with an efficient spinning-band column (Note 10). After foreruns of pentane and glyme containing *t*-nitrosobutane are removed (Note 11), 26–27 g. (42–43%) of red di-*t*-butyl nitroxide, b.p. 59–60° (11 mm.), is obtained.

2. Notes

1. The submitters obtained similar results using a preparative scale 7.5 times that described here; yield 36%.

2. The 1,2-dimethoxyethane (Ansul Chemical Co.) was pre-dried for several days over calcium hydride, filtered, and stored

over lithium aluminum hydride prior to its distillation at atmospheric pressure immediately before use. For a larger-scale preparation it is expeditious to distil simultaneously from two 5-l. flasks rather than from a single large one. Under these conditions, distillation of the glyme can be completed in 8–10 hours.

3. The *t*-nitrobutane employed was prepared by the procedure of Kornblum, Clutter, and Jones.[2] This method is essentially the same as that previously reported in *Organic Syntheses*.[3]

4. Throughout all transfers, air must be rigorously excluded from the flask by the nitrogen blanket.

5. When great care has not been taken to ensure the absence of moisture, induction times as long as several minutes are observed before the onset of reaction.

6. The temperature of the reaction mixture must never be allowed to exceed 30°; above this temperature drastic diminution of yield occurs.

7. A nitrogen blanket is used here to prevent ignition of hydrogen resulting from traces of unreacted sodium.

8. At this point during one run the checkers obtained a considerable amount of water-insoluble, colorless solid; however, the product yield was not changed.

9. It is essential to keep the reaction mixture and acid cold since otherwise substantial decomposition of product results.

10. The pot temperature should not exceed 100° until the di-*t*-butyl nitroxide fraction is collected.

11. In several runs it was noted that, despite the acid extraction, small amounts of N,N-di-*t*-butylhydroxylamine crystallized in the cooler parts of the fractionating column head. In such cases, repetition of the acid extraction procedure is required before fractionation.

3. Methods of Preparation

The procedure described is that of Hoffmann, Feldman, Gelblum, and Hodgson [4] and is the only one known at this time for the preparation of substantial amounts of di-*t*-butyl nitroxide.

4. Merits of the Preparation

The method is specific for the preparation of di-*t*-butyl nitroxide, a liquid member of a group of stable free radicals useful for the inhibition of a variety of reactions proceeding by radical chain mechanisms as well as for providing standards for e.s.r. measurements.

1. American Cyanamid Company, Stamford Research Laboratories, 1937 West Main Street, Stamford, Connecticut 06904.
2. N. Kornblum, R. J. Clutter, and W. J. Jones, *J. Am. Chem. Soc.*, **78**, 4003 (1956).
3. N. Kornblum and W. J. Jones, this volume, p. 845.
4. A. K. Hoffmann, A. M. Feldman, E. Gelblum, and W. G. Hodgson, *J. Am. Chem. Soc.*, **86**, 639 (1964).

2,2′-DICHLORO-α,α′-EPOXYBIBENZYL

(Bibenzyl, α,α′-epoxy-, 2,2′-dichloro-)

Submitted by V. Mark [1]
Checked by G. A. Frank and W. D. Emmons

1. Procedure

To a solution of *o*-chlorobenzaldehyde (56.2 g., 0.4 mole) (Note 1) in 50 ml. of benzene in a 250-ml. three-necked flask equipped with a stirrer, thermometer, dropping funnel, and reflux condenser, there is added a solution of hexamethylphosphorous triamide (37.9 g., 0.232 mole) in 20 ml. of dry ether [2] at such a rate that the temperature remains between 24° and 36°. The ensuing exothermic reaction is controlled readily by immersing the flask in a water bath (Note 2). After completion of the addition, which requires 30–50 minutes, the clear solution is maintained at 50° for 15 minutes. The solvent is removed on a rotary evaporator, and the oily residue is triturated with 100 ml. of water and then with 150 ml. of pentane. At this point only a small portion of the product is left undissolved (Note 3). The aqueous layer is extracted with 150 ml. of pentane (Note 4). The combined pentane solution is washed with two 100-ml. portions of water and concentrated to dryness to give 46–50 g. of a light yellow solid. Recrystallization from 100 ml. of methanol yields 37.5–43.1 g. of white crystals (71–81%), composed of a mixture of the *trans* epoxide (about 50–55%) and the *cis* epoxide (about 45–50%) (Notes 5, 6, 7).

2. Notes

1. The commercial product was distilled before use. The checkers found the undistilled material equally satisfactory.

2. The solvent can be omitted, but more efficient cooling is then required to control the reaction.

3. Pentane dissolves the epoxide and water dissolves the co-product, hexamethylphosphoric triamide. The insoluble, thick, yellow syrup sometimes found is the betaine 1:1 adduct of the aldehyde and amide.[3] The checkers found no insoluble portion in their preparations.

4. Filtration through a sintered-glass funnel readily breaks up the emulsion which is formed occasionally.

5. The simplest and most accurate way to determine the composition of the product is by proton n.m.r. spectroscopy. The ratio of the oxirane hydrogen atoms (*cis* 4.48 p.p.m. and *trans* 3.97 p.p.m. downfield from internal tetramethylsilane reference, determined in carbon tetrachloride or deuteriochloroform solution)[3] gives directly the ratio of the isomers. Infrared spectroscopy, although it readily distinguishes between the isomers, gives a less accurate quantitative relationship.

6. Chromatography over silica gel (60–200 mesh), using benzene as eluent, yields pure *trans*-epoxide (m.p. 76–77°; δ 4.08) in the first fractions, and the *cis*-epoxide (m.p. 94–95°, after recrystallization from hexane; δ 4.56) in the last fractions. The assignment was also confirmed by analysis of the ^{13}C-satellite bands (*cis,* J(^{13}C-H) 181, J(H-H) 4.4, and

trans, J(^{13}C-H) 182, J(H-H) 1.9 Hz.). In all of the cases studied the *trans* epoxide was eluted first and had lower δ and J(H-H) values than its *cis* isomer.

7. Hexaethylphosphorous triamide[2] may be substituted for the methyl homolog without adverse effect on the quality and yield of the product.

3. Methods of Preparation

2,2'-Dichloro-α,α'-epoxybibenzyl has been prepared only by the present procedure.[3]

4. Merits of the Preparation

The reaction of aldehydes with hexaalkylphosphorous triamides to yield the corresponding epoxides is a synthetic procedure of considerable scope (Table I) and represents a new and simple, one-step method of forming symmetrical and unsymmetrical epoxides.[3] In contrast to the most widely used epoxide synthesis, *i.e.,* from olefins with peroxides or peracids, the present pro-

TABLE I

Synthesis of Symmetrical Expoxides

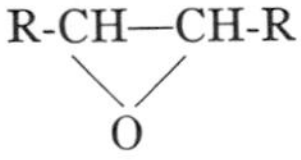

R-CH—CH-R
\ /
O

R	% Yield	Composition	
		% *trans*	% *cis*
o-Bromophenyl	90–95	59	41
m-Bromophenyl	45–50	72[a]	28
o-Fluorophenyl	90–95	60	40
3,4-Dichlorophenyl	88–95	60	40
m-Nitrophenyl	75–80	74[b]	26
p-Cyanophenyl	90–95	57	43
p-Formylphenyl	75–80	53	47
1-Naphthyl	83–87	53	47
2-Thienyl[c]	60–65	53	47
2-Pyridyl	85–90	75[d]	25

[a]M.p. 84–86°.
[b]M.p. 156–158°.
[c]Hexaethylphosphorous triamide was used.
[d]M.p. 95–97°.

cedure may be used to obtain epoxides having structural features (*e.g.*, thiophene or pyridine rings) which would not survive the more drastic peroxide route. The procedure does not, however, afford stereochemically unique products. The yields of the epoxides from the corresponding aldehydes are usually high, and new members of the underpopulated class of aromatic and heterocyclic epoxides become readily accessible. Application of this method to certain aromatic dialdehydes yielded the first examples of cyclic aromatic epoxides.[4]

1. Hooker Chemical Corporation, Niagara Falls, New York.
2. | V. Mark, this volume, p. 602.
3. V. Mark, *J. Am. Chem. Soc.*, **85**, 1884 (1963).
4. M. S. Newman and S. Blum, *J. Am. Chem. Soc.*, **86**, 5598 (1964).

DICHLOROMETHYLENETRIPHENYLPHOSPHORANE

[Phosphorane, (dichloromethylene) triphenyl-1]
AND
β,β-DICHLORO-*p*-DIMETHYLAMINOSTYRENE

$$(C_6H_5)_3P + HCCl_3 \xrightarrow[-KCl, -(CH_3)_3COH]{(CH_3)_3COK} (C_6H_5)_3P{=}CCl_2$$

$$(C_6H_5)_3P{=}CCl_2 + (CH_3)_2N{-}\bigcirc{-}CHO \longrightarrow$$

$$(CH_3)_2N{-}\bigcirc{-}CH{=}CCl_2 + (C_6H_5)_3P{=}O$$

Submitted by A. J. Speziale, K. W. Ratts, and D. E. Bissing [1]
Checked by William E. Parham and L. Dean Edwards

1. Procedure

A. *Potassium t-butoxide.* To 500 ml. of *t*-butyl alcohol (Note 1) in a 3-l. three-necked flask equipped with an efficient sealed stirrer, a nitrogen inlet (Note 2), a 500-ml. dropping funnel with a pressure-equalizing side tube (Note 3), and a reflux condenser there is added 20 g. (0.5 g. atom) of clean potassium metal. After the potassium has reacted, the condenser is replaced by a 12-in.

distillation column and the excess t-butyl alcohol is removed by distillation until crystals begin to form in the solution. There is added 2 l. of dry heptane and the distillation is continued until the head temperature reaches 98° (Notes 4 and 5). The residual mixture is adjusted to a 1.5-l. volume by addition of dry heptane and the resulting slurry of potassium t-butoxide in heptane is cooled to 0–5° in an ice bath (Note 6).

B. *Dichloromethylenetriphenylphosphorane.* In one portion 131 g. (0.5 mole) of triphenylphosphine (Note 7) is added to the cooled suspension of potassium t-butoxide in heptane, and to the well-stirred mixture a solution of 59.5 g. (0.5 mole) of chloroform in 500 ml. of dry heptane is added dropwise over a period of 1 hour, maintaining the temperature below 5° and an atmosphere of purified nitrogen. The resulting stirred suspension is concentrated to a 750-ml. volume at reduced pressure and at 15–20° (Note 8).

C. *β,β-Dichloro-p-dimethylaminostyrene.* To the heptane suspension of the phosphorane there is added over a period of 30 minutes 74.5 g. (0.5 mole) of p-dimethylaminobenzaldehyde in six equal portions; the reaction temperature is maintained below 10°. The mixture is stirred for 2 hours in an ice bath, for an additional 5 hours at room temperature, and is then allowed to stand overnight. The precipitated phosphine oxide is filtered and the solvent is removed from the filtrate at 45–50° using a rotary evaporator. The resulting brown solid is recrystallized from methanol to yield 74–85 g. (68–79%) of crude olefin, m.p. 56–60°. The major impurity is unreacted triphenylphosphine.

The crude product is dissolved in absolute ethanol (10 ml. per gram of material), and a saturated solution of mercuric chloride (1 g. per 5 g. of crude olefin) in absolute ethanol is added (Note 9). The precipitate is filtered (Note 10) and washed with absolute ethanol. The filtrate is concentrated to half of its original volume (Note 11) and cooled in an ice bath. The yield of olefin is 42–60 g. (39–56%), m.p. 71–72°.

2. Notes

1. The t-butyl alcohol should be distilled from metallic sodium before use, care being taken to exclude moisture.

2. The nitrogen was purified by passing it through two wash bottles containing Fieser's solution [2] and single wash bottles containing concentrated sulfuric acid and solid anhydrous calcium chloride, respectively.

3. If available, it is more convenient to use a flask which also accommodates a thermometer extending into the reaction mixture.

4. It may be necessary to add more heptane during the distillation, as the slurry of potassium t-butoxide in heptane becomes very difficult to stir if the total volume is less than 1 liter.

5. About 2 hours is required for removal of all the excess t-butyl alcohol.

6. The potassium t-butoxide prepared in this manner is a 1:1 complex with t-butyl alcohol; neutralization equivalent calculated for $(CH_3)_3COH \cdot (CH_3)_3COK$, 186. Found: 184, 182. The complex can be isolated by simply removing the solvent at 20–25 mm. pressure on a steam bath. It can be stored for several months under a nitrogen atmosphere.

7. Triphenylphosphine was used as supplied by Eastman Organic Chemicals.

8. It is desirable to remove the t-butyl alcohol formed during the generation of dichlorocarbene because the t-butyl alcohol reacts with the phosphorane, thus lowering the yield of olefin. The evaporation is best accomplished with a vacuum pump (e.g., a Langdon pump) since the removal of t-butyl alcohol and heptane by water aspiration is very slow at this temperature. It is imperative that this step be accomplished as rapidly as possible and that the temperature be maintained below 20°. Although the suspension of phosphorane in heptane can be stored overnight under a nitrogen atmosphere, it is better to use it immediately.

9. Mercuric chloride forms with triphenylphosphine a double salt which is insoluble in ethanol.

10. It is necessary to use a fine or ultra-fine sintered-glass funnel or Whatman No. 1 filter paper because the precipitate is finely divided.

11. The checkers obtained better results by reducing the volume to one-third the original volume.

3. Methods of Preparation

Dichloromethylenetriphenylphosphorane has been prepared by the direct reaction of triphenylphosphine with carbon tetrachloride.[3]   1,1-Dichloroethylenes have been prepared by dehydrochlorination of 1,1,1-trichloro compounds [4-6] or by specialized methods applicable only to specific compounds.[7, 8]

4. Merits of the Preparation

The procedure described illustrates a general method for the preparation of 1,1-dichloroethylenes.   Dichloromethylenetriphenylphosphorane has been treated with a variety of aldehydes and ketones including p-nitrobenzaldehyde, 2,6-dichlorobenzaldehyde, cinnamaldehyde, lauraldehyde, acetaldehyde, cyclohexanone, and benzophenone to give the corresponding 1,1-dichloroethylene in good yield.[9]

Chlorofluoromethylenetriphenylphosphorane   has   been utilized in an extension of this method to prepare chlorofluoroethylenes.[10]

1. Research Department, Agricultural Division, Monsanto Company, St. Louis 66, Missouri.
2. L. F. Fieser, "Experiments in Organic Chemistry," D. C. Heath and Co., Boston, 1955, p. 299.
3. R. Rabinowitz and R. Marcus, *J. Am. Chem. Soc.*, **84**, 1312 (1962).
4. R. Rumschneider and W. Cohnen, *Ber.*, **89**, 2702 (1956).
5. R. E. Dunbar and A. O. Geiszler, *Science*, **122**, 241 (1955).
6. H. Bader, W. A. Edmiston, and H. H. Rosen, *J. Am. Chem. Soc.*, **78**, 2590 (1956).
7. R. Kh. Freidlina, N. A. Semenov, and A. N. Nesmeyanov, *Izv. Akad. Nauk SSSR, Otd. Khim. Nauk*, 652 (1959).
8. P. Tarrant and M. R. Lilyquist, *J. Am. Chem. Soc.*, **77**, 3640 (1955).
9. A. J. Speziale and K. W. Ratts, *J. Am. Chem. Soc.*, **84**, 854 (1962).
10. H. Yamanaki, T. Ando, and W. Funasaka, *Bull. Chem. Soc. Japan*, **41**, 756 (1968).

DICHLOROMETHYL METHYL ETHER*

$$HCOOCH_3 + PCl_5 \longrightarrow Cl_2CHOCH_3 + POCl_3$$

Submitted by H. Gross, A. Rieche, E. Höft, and E. Beyer [1]
Checked by G. N. Taylor and K. B. Wiberg

1. Procedure

In a 2-l. three-necked flask equipped with a stirrer, a reflux condenser, and a dropping funnel (Note 1) 832 g. (4.0 moles) of phosphorus pentachloride is stirred with 250 ml. of phosphorus oxychloride (Note 2). To this is added with stirring 264 g. (272 ml., 4.4 moles) of methyl formate (Note 3). During the addition the reaction vessel is cooled in an ice bath to maintain a reaction temperature of 10–20°. The addition requires about 1.75 hours. When the addition is complete, the solution is stirred at a temperature under 30° until all the phosphorus pentachloride has dissolved (about 1 hour). Then the stirrer is removed, the reflux condenser is replaced by a distilling head, and the reaction mixture is distilled under a pressure of 80–120 mm. with a bath temperature of 50–65° (Note 4). During the distillation the receiver is cooled to −10° to −20° by an ice-salt bath.

The material which is collected is redistilled through a 90-cm. vacuum-jacketed column packed with glass beads (5 mm.) using a 1:10 reflux ratio. The fraction, b.p. 80–100°, is redistilled through the same column to give 353–386 g. (77–84%) of dichloromethyl methyl ether, b.p. 82–85.5°, n^{20}D 1.4303 (Note 5).

2. Notes

1. The reflux condenser and dropping funnel must be provided with calcium chloride tubes.

2. Phosphorus oxychloride serves only as a suspension medium for phosphorus pentachloride and makes possible a homogeneous reaction. The phosphorus oxychloride obtained during workup

may be recycled in this preparation.

3. Commercial methyl formate was dried over sodium sulfate and used without special purification.

4. If it is not first distilled under reduced pressure, extensive decomposition will occur during fractional distillation.

5. The product must be protected from moisture when stored.

3. Methods of Preparation

Dichloromethyl methyl ether has been prepared by the chlorination of chlorodimethyl ether in the liquid [2-4] or gas phase,[5] by the reaction of chlorodimethyl ether with sulfuryl chloride and benzoyl peroxide,[6, 7] and by the treatment of methyl formate with phosphorus pentachloride.[8-10]

4. Merits of the Preparation

Dichloromethyl methyl ether may be employed preparatively in various ways. Thus it effects the replacement of carbonyl and hydroxyl oxygens by chlorine,[11] and may be used in the preparation of α-acetochlorosugars [12] and acid chlorides, particularly those derived from acetylated monocarboxylic acid sugars [12, 13] and acetylated amino acids.[14] In addition, the *ortho* derivatives of formic acid may be prepared from dichloromethyl methyl ether.[15] With aromatic compounds, dichloromethyl methyl ether reacts under Friedel-Crafts conditions followed by hydrolysis to give the corresponding aromatic aldehydes.[10, 16, 17]

1. Institute for Organic Chemistry of the German Academy of Sciences at Berlin, Berlin-Adlershof, Germany.
2. P. L. Salzberg and J. H. Werntz, U.S. Patent 2,065,400 (1936) [*C.A.*, **31**, 1046 (1937)].
3. L. R. Evans and R. A. Gray, *J. Org. Chem.*, **23**, 745 (1958).
4. A. Rieche and H. Gross, *Chem. Tech.* (Berlin), **10**, 515 (1958).
5. H. Gross, *Chem. Tech.* (*Berlin*), **10**, 659 (1958).
6. H. Böhme and A. Dörries, *Ber.*, **89**, 723 (1956).
7. H. Laato, *Suomen Kemistilehti*, **35B**, 90 (1962).
8. H. Laato, *Suomen Kemistilehti*, **32B**, 67 (1959).
9. H. Gross, A. Rieche, and E. Höft, *Ber.*, **94**, 544 (1961).
10. A. Rieche, H. Gross, and E. Höft, *Ber.*, **93**, 88 (1960).
11. A. Rieche and H. Gross, *Ber.*, **92**, 83 (1959).
12. H. Gross and I. Farkas, *Ber.*, **93**, 95 (1960).
13. R. Bognar, I. Farkas, I. F. Szabo, and G. D. Szabo, *Ber.*, **96**, 689 (1963).

14. K. Poduska and H. Gross, *Ber.*, **94**, 527 (1961).
15. H. Gross and A. Rieche, *Ber.*, **94**, 538 (1961).
16. H. Gross, A. Rieche, and G. Matthey, *Ber.*, **96**, 308 (1963).
17. A. Rieche, H. Gross, and E. Höft, this volume, p. 49.

2,6-DICHLORONITROBENZENE

(Benzene, 1,3-dichloro-2-nitro-)

Submitted by A. S. PAGANO and W. D. EMMONS [1]
Checked by V. Z. WILLIAMS, JR. and K. B. WIBERG

1. Procedure

Caution! The preparation and handling of peroxytrifluoroacetic acid should be carried out behind a safety screen. Precautions to be observed with 90% hydrogen peroxide are described in Note 3 and should be carefully followed.

A 300-ml. three-necked flask equipped with a Trubore stirrer, dropping funnel, and reflux condenser protected with a calcium chloride drying tube is charged with 100 ml. of methylene chloride (Note 1). To this solvent is added without stirring 5.4 ml. (0.20 mole) of 90% hydrogen peroxide (Notes 2, 3, 4, 5). The hydrogen peroxide is not miscible with the solvent and separates as the lower layer at the bottom of the flask. The flask is then cooled in an ice bath, and the stirrer is started. To this cold solution over a 20-minute period is added 34.0 ml. (0.24 mole) of trifluoroacetic anhydride. After addition is complete, the ice bath is removed and the solution is stirred at room temperature for 30 minutes.

A solution is then prepared from 8.1 g. (0.05 mole) of 2,6-dichloroaniline (Note 6) and 80 ml. of methylene chloride. This solution is added dropwise over a 30-minute period to the previously prepared peroxytrifluoroacetic acid reagent (Note 7).

During this addition the exothermic reaction causes the mixture to reflux. After addition is complete, the mixture is heated under reflux for 1 hour. It is then cooled and poured into 150 ml. of cold water. The organic layer is separated, washed with 100 ml. of water, with two 100-ml. portions of 10% sodium carbonate solution (Note 8), and finally with 50 ml. of water. The organic extract is treated with activated charcoal and anhydrous magnesium sulfate. After standing overnight, the volatile solvent is removed at aspirator pressure with the aid of a warm water bath. There is obtained 8.6–8.8 g. (89–92%) of yellow 2,6-dichloronitrobenzene, m.p. 63–68°. The product is recrystallized from a minimum volume (12–15 ml.) of ethanol and washed on the filter with 10 ml. of cold ethanol to give 5.7–7.0 g. (59–73%) of a slightly off-white product, m.p. 69–70° (reported,[4] 70.5°).

2. Notes

1. Reagent grade methylene chloride was used without further purification.

2. Available from Becco Chemical Division, Food, Machinery and Chemical Co., Buffalo, New York.

3. The precautions to be observed with 90% hydrogen peroxide have been described in detail.[2] In essence, it is important to prevent contact of this reagent with any easily oxidizable substrate such as wood, alcohols, and sugars and with heavy metal salts since these substances catalyze its decomposition. Storage of hydrogen peroxide in the laboratory should be arranged in such a way that, even if the bottle containing the reagent breaks, the hydrogen peroxide will not come into contact with any material of this kind. Small samples of 90% hydrogen peroxide are regularly shipped in vented glass bottles provided with a protective outside metal container, and it is desirable to use this container while storing the reagent in the laboratory. In the event that spillage of the reagent occurs, dilution with at least several volumes of water is recommended. In weighing out 90% hydrogen peroxide it is good practice never to return excess reagent to the stock bottle; rather, it should be diluted with water and discarded to avoid any possibility that the stock bottle will be contaminated.

4. It is convenient to measure out the hydrogen peroxide by a 10-ml. graduate or by a 10-ml. pipet actuated by a glass syringe connected via a ground-glass joint.

5. The procedure described here for the preparation of peroxytrifluoroacetic acid in methylene chloride has been carried out by the submitters several hundred times without incident and is believed to be the best available. However, it has been pointed out that suspensions of 90% hydrogen peroxide in methylene chloride can be detonated by impact under certain conditions.[3] Accordingly, the use of the recommended safety screen is imperative, and the preparation should not be scaled up without special precautions. The homogeneous solution of peroxytrifluroacetic acid, once obtained, is undoubtedly much safer to handle than the suspension of hydrogen peroxide in methylene chloride. The latter suspension is not transferred, however, and exists for only a brief time period during the preparation.

6. Available from Aldrich Chemical Company, Inc.

7. Addition of the peracid solution to the aniline invariably resulted in a poor-quality product in low yield.

8. The sodium carbonate extracts are quite dark.

3. Discussion

2,6-Dichloronitrobenzene has been prepared by deamination of 3,5-dichloro-4-nitroaniline [4] and of 2,4-dichloro-3-nitroaniline.[5] This procedure is an example of the rather general oxidation of anilines to nitrobenzenes with peroxytrifluoroacetic acid.[6, 7] Use of this reagent is frequently the method of choice for carrying out this transformation, and it is particularly suitable for oxidation of negatively substituted aromatic amines. Conversely, those aromatic amines, such as *p*-anisidine and β-naphthylamine, whose aromatic nuclei are unusually sensitive to electrophilic attack give intractable mixtures with this reagent.[6] This is not a serious limitation, however, and many of the nitrobenzenes which are available from this procedure have in the past required tedious multistep syntheses.

1. Rohm and Haas Company, Spring House, Pennsylvania.
2. E. S. Shanley and F. P. Greenspan, *Ind. Eng. Chem.*, **39**, 1536 (1947).

3. J. D. McClure and P. H. Williams, *J. Org. Chem.*, **27**, 627 (1962).
4. C. B. Kremer and A. Bendich, *J. Am. Chem. Soc.*, **61**, 2658 (1939).
5. G. Körner and A. Contardi, *Atti Accad. Nazl. Lincei*, **18**, I, 100 (1909).
6. W. D. Emmons, *J. Am. Chem. Soc.*, **76**, 3470 (1954).
7. L. I. Klmel'nitskii, T. S. Novikova, and S. S. Novikov, *Izv. Akad. Nauk SSSR, Otd. Khim. Nauk*, 516 (1962) [*Chem. Abstr.* **57**, 14979 (1962)].

DI-(*p*-CHLOROPHENYL)ACETIC ACID[1]

CORRECTION

In the first line of Note 5, p. 271, 1,1-di-(*p*-chlorophenyl)-ethylene should read

1,1-di-(*p*-chlorophenyl)-2,2-dichloroethylene.

Reported by B. Stavric and G. A. Neville, *J. Ass. Offic. Anal. Chem.*, **53**, 1270 (1970):

1. *Org. Syntheses*, Coll. Vol. 3, 271 (1955).

3,4-DICHLORO-1,2,3,4-TETRAMETHYLCYCLOBUTENE

(Cyclobutene, 1,2,3,4-tetramethyl-3,4-dichloro-)

$$2CH_3—C\equiv C—CH_3 \ + \ Cl_2 \xrightarrow{-20°}$$

Submitted by R. Criegee [1]
Checked by G. Brown and V. Boekelheide

1. Procedure

A 400-ml. three-necked flask equipped with a mechanical stirrer, a gas dispersion tube, and a thermometer is charged with

216 g. (4 moles) of 2-butyne, 14 ml. of boron trifluoride etherate, and 1 ml. of water (Note 1). The mixture is stirred vigorously while the flask is partially immersed in a dry ice-acetone bath to maintain an internal temperature of -20°. Chlorine (195 g., 2.75 moles) is then added (Note 2) gradually over a period of 17–20 hours.

When all the chlorine has been added, the flask and contents (Note 3) are cooled to −78° and held at this temperature for 30 minutes. The white crystalline product which separates is collected on a sintered-glass funnel; 160–170 g. (45–48%) of crude crystals, m.p. 53–55°, is obtained. The crude product when stored at −20° takes on a reddish or blue color after a few days (Note 4).

Purification of the crude material is accomplished by dissolving it in petroleum ether (30–50°) or methylene chloride, shaking this solution three times with water, and passing the organic layer through a fluted filter paper. After the filtrate has been dried over anhydrous sodium sulfate, it is concentrated to a small volume and cooled to −78°. The perfectly white, crystalline product which separates is collected yielding 110–120 g. (30–33%) of crystals, m.p. 57–58° (Note 5).

2. Notes

1. The submitter reports that he and his colleagues have found since the procedure was submitted and checked that the yield and purity of the product is the same whether the preparation is carried out in the presence or absence of boron trifluoride etherate and water. Recommends that the 14 ml. of boron trifluoride etherate and 1 ml. of water *not* be added.

2. The chlorine is most conveniently added by liquefying the required amount in a gas trap. The gas trap is placed in an empty Dewar flask and covered with glass wool. While the liquid chlorine slowly warms up, the resulting gaseous chlorine is passed through a sulfuric acid wash bottle into the reaction mixture. The addition generally takes 17–20 hours.

3. The final reaction mixture is slightly yellow, but at times it can be reddish. During the reaction some white crystalline product collects at the walls of the reaction vessel.

4. It is important to purify the crude product as soon as possible since it decomposes readily in the impure state. Once decomposition has set in, purification is difficult.

5. A still purer product is obtained if the dried petroleum ether solution is evaporated to dryness with a water aspirator and the residual crude product distilled through a Vigreux column. After a small fore-run, pure tetramethyl-3,4-dichlorocyclobutene distils at 59–60° at 12 mm. The product melts at 58° and is more stable at room temperature than the recrystallized but undistilled material.

3. Methods of Preparation

The method used is that of Criegee and Moschel.[2] Smirnow-Samkow[3] made the same substance by the reaction of 2-butyne with sulfuryl chloride in 10–15% yield.

4. Merits of the Preparation

3,4-Dichloro-1,2,3,4-tetramethylcyclobutene is an unusually versatile intermediate.[4] The tertiary and allylic chlorine atoms undergo ready solvolysis. With lithium aluminum hydride the chlorine atoms are replaced by hydrogen. The resulting cis,trans-tetramethylcyclobutenes are starting materials for numerous transformations, e.g., thermolysis leads to stereoisomeric tetramethylbutadienes.[5] Lithium amalgam in ether results in the formation of octamethyltricyclooctadiene. With nickel carbonyl the nickel chloride complex of tetramethylcyclobutadiene is formed.[6] Ammonia transforms the dichloride into tetramethylpyrrole.[7] Other reactions have been reported.[8]

This method of preparation is simpler, more reproducible, and gives considerably better yields than the original one of Smirnow-Samkow.

1. Technische Hoschschule Karlsruhe, Institut für Organische Chemie, Karlsruhe, Germany.
2. R. Criegee and A. Moschel, *Ber.*, **92**, 2181 (1959).
3. J. W. Smirnow-Samkow, *Dokl. Akad. Nauk SSSR*, **83**, 869 (1952); J. W. Smirnow-Samkow and N. A. Kostromina, *Ukr. Khim. Zh.*, **21**, 233 (1955).
4. R. Criegee, *Angew. Chem.*, **74**, 703 (1962).
5. R. Criegee and K. Noll, *Ann.*, **627**, 1 (1959).

6. R. Criegee and G. Schröder, *Ann.*, **623**, 1 (1959).
7. R. Criegee and M. Krieger, *Ber.* **98**, 387 (1965).
8. R. Criegee, J. Dekker, W. Engel, P. Ludwig, and K. Noll, *Ber.*, **96**, 2362 (1963).

DIETHYL ACETAMIDOMALONATE*

(Malonic acid, acetamido-, diethyl ester)

$$CH_2(CO_2C_2H_5)_2 + NaNO_2 + CH_3CO_2H \rightarrow$$

$$HON{=}C(CO_2C_2H_5)_2 + CH_3CO_2Na + H_2O$$

$$HON{=}C(CO_2C_2H_5)_2 + 2Zn + 3CH_3CO_2H + (CH_3CO)_2O \rightarrow$$

$$CH_3CONHCH(CO_2C_2H_5)_2 + 2Zn(C_2H_3O_2)_2 + H_2O$$

Submitted by ARTHUR J. ZAMBITO and EUGENE E. HOWE.[1]
Checked by JOHN C. SHEEHAN and ALMA M. BOSTON.

1. Procedure

A. *Diethyl isonitrosomalonate.* In a 500-ml. three-necked, round-bottomed flask, equipped with a mechanical stirrer and thermometer, is placed 50 g. (47.4 ml., 0.312 mole) of diethyl malonate. The flask is cooled in an ice bath, and a mixture of 57 ml. of glacial acetic acid and 81 ml. of water is added with stirring. With the temperature at about 5°, a total of 65 g. of sodium nitrite (Note 1) (0.944 mole) is added in portions over a period of 1.5 hours, the temperature being maintained around 5° during the addition. After all the sodium nitrite is added, the ice bath is removed, and the stirring is continued for 4 hours (Note 2). During this time, the temperature reaches a maximum of 34–38° within 2 hours and falls to about 29° by the end of the stirring period. Gases which escape during the reaction (mostly oxides of nitrogen) are led to the hood.

The reaction mixture is transferred to a 300-ml. separatory funnel and is extracted with two 50-ml. portions of ether. The combined ethereal solution of diethyl isonitrosomalonate is used in the next step immediately or, if desired, may be used after storage in a refrigerator overnight (Note 3).

B. *Diethyl acetamidomalonate.* The solution of diethyl iso-nitrosomalonate described above, 86 g. (0.842 mole) of acetic anhydride, and 225 ml. (3.95 moles) of glacial acetic acid are placed in a 1-l., three-necked, round-bottomed flask fitted with a mechanical stirrer, a thermometer, and a dropping funnel. With vigorous stirring 78.5 g. (1.20 moles) of zinc dust is added in small portions over a period of 1.5 hours in such a manner that the temperature of the reaction is maintained at 40–50°. The reaction is markedly exothermic during most of the zinc addition, and intermittent cooling (water bath) is required. After all the metal has been added, the mixture is stirred for an additional 30 minutes.

The reaction mixture is filtered with suction and the cake is washed thoroughly with two 200-ml. portions of glacial acetic acid (Note 4). The combined filtrate and washings are evaporated under reduced pressure on the steam bath until a thick oil, which generally partially crystallizes, remains. To purify the crude product, 100 ml. of water is added, and the flask is warmed on a steam bath until the solid melts. The mixture of water and oil is stirred rapidly in an ice bath, and diethyl acetamidomalonate crystallizes as a fine white product. After cooling in an ice bath for an additional hour, the product is collected by filtration, washed once with cold water, and dried in air at 50°. A second crop is obtained by concentrating the mother liquor under reduced pressure. The yield of diethyl acetamidomalonate, m.p. 95–97° (Note 5), is 52–53 g. (77–78%) based on malonic ester.

2. Notes

1. Owing to the instability of sodium nitrite solutions, the addition of the solid salt is preferred.

2. Prolonging the stirring to 24 hours has no effect on the yield of diethyl acetamidomalonate.

3. No attempt has been made to purify diethyl isonitroso-malonate. This product has been known to explode during distillation.

4. The zinc cake is very heavy and may be washed by slurrying on a sintered-glass funnel or, if a standard Büchner funnel is

used, by removing and slurrying in a beaker.

5. The diethyl acetamidomalonate obtained is of high purity. If a product of inferior quality is obtained, it may be recrystallized from hot water, using 2.5 cc. per g. Upon cooling, the product separates first as an oil. With rapid stirring, it is converted to fine white crystals which are easily washed with cold water. Diethyl acetamidomalonate may be recrystallized in this manner with 97% recovery. The first crop amounts to 91% and the mother liquors may be concentrated to yield an additional 6%.

3. Methods of Preparation

Diethyl acetamidomalonate was first reported by Cherchez [2] in 1931, when in an attempt to carry out a carbon alkylation of diethyl aminomalonate with acetyl chloride he obtained a quantitative yield of diethyl acetamidomalonate. This method of preparation, however, is not practical since diethyl aminomalonate is unstable and is made in relatively poor yields.

Snyder and Smith [3] prepared diethyl acetamidomalonate in 40% yield by reduction of diethyl isonitrosomalonate in ethanol over palladium on charcoal followed by direct acetylation of diethyl aminomalonate in the filtrate with acetic anhydride. Ghosh and Dutta [4] used zinc dust instead of palladium. A modification using Raney nickel is described by Akabori et al. [5] Shaw and Nolan [6] reported a 98% yield by conversion of diethyl oximinomalonate-sodium acetate complex.

4. Use of Diethyl Acetamidomalonate

Diethyl acetamidomalonate is useful in the synthesis of α-amino acids by alkylation, as, for example, histidine and tryptophan.

1. Merck Sharp and Dohme Research Laboratories, Division of Merck and Co., Rahway, New Jersey.
2. V. Cherchez, Bull. Soc. Chim., 49, 45 (1931).
3. H. R. Snyder and C. W. Smith, J. Am. Chem. Soc., 66, 350 (1944).
4. T. N. Ghosh and S. Dutta, J. Indian Chem. Soc., 32, 20 (1955).
5. S. Akabori et al., Japan pat. 274 (January 20, 1954).
6. K. N. F. Shaw and C. Nolan, J. Org. Chem., 22, 1668 (1957).

DIETHYL AMINOMALONATE HYDROCHLORIDE*

(Malonic acid, amino-, diethyl ester, hydrochloride)

$$HON\!\!=\!\!CH(CO_2C_2H_5)_2 \xrightarrow[\text{50-60 lb.}]{\text{Pd, H}_2} NH_2\!\!-\!\!CH(CO_2C_2H_5)_2$$

$$NH_2\!\!-\!\!CH(CO_2C_2H_5)_2 \xrightarrow[\text{dry ether}]{\text{HCl}} (HCl)NH_2\!\!-\!\!CH(CO_2C_2H_5)_2$$

Submitted by WALTER H. HARTUNG, JAN H. R. BEAUJON,
and GEORGE COCOLAS.[1]
Checked by JOHN C. SHEEHAN and ROBERT W. PARSONS, JR.

1. Procedure

A. *Diethyl aminomalonate.* An ethereal solution (about 150 ml.) of diethyl isonitrosomalonate prepared from 50 g. of diethyl malonate (Note 1) is washed with 80-ml. portions of 1% sodium bicarbonate solution until the final washing has a distinct yellow color (Note 2). The ethereal solution is dried over 40 g. of anhydrous sodium sulfate in a refrigerator overnight and then filtered into a tared round-bottomed flask. The solvent is removed under reduced pressure at a temperature below 30° (water bath). The weight of the residue in one case was 59.6 g. Assuming complete conversion of the 0.312 mole of diethyl malonate to diethyl isonitrosomalonate, a 0.1-mole aliquot (19.1 g.) of the residue is placed in a 500-ml. reduction bottle provided for the Parr Hydrogenator (Note 3). To this is added 100 ml. of absolute alcohol and 3 g. of 10% palladium-on-charcoal catalyst (Note 4). The bottle is placed in a hydrogenator, and the system is flushed three or four times with hydrogen. With the initial reading on the pressure gauge at 50–60 lb., the bottle is shaken until no further drop in pressure is observed (about 15 minutes).

The catalyst is removed by filtration, using an absolute alcohol wash, and the clear filtrate is concentrated under reduced pressure at a temperature below 50° (water bath). As diethyl aminomalonate is not so stable as its salts, the crude product is converted directly to diethyl aminomalonate hydrochloride (Note 5).

B. *Diethyl aminomalonate hydrochloride.* The crude diethyl aminomalonate is diluted with 80 ml. of dry ether and filtered to remove a small amount of white solid. The filtrate is collected in a 250-ml. Erlenmeyer flask and cooled in an ice bath. Dry hydrogen chloride is passed just over the solution while it is being stirred mechanically (Note 6). The fine white crystals which precipitate are collected by suction filtration and washed three times with a total of 60 ml. of dry ether (Note 7). The filtrate and washings are treated again with hydrogen chloride, and a second crop of diethyl aminomalonate hydrochloride is collected and washed as before. This process is repeated until no further precipitation results from passing hydrogen chloride into the solution. A total of 16.5–17.4 g. (78–82% yield based on diethyl malonate) of diethyl aminomalonate hydrochloride, m.p. 162–163°, is obtained. Recrystallization from alcohol-ether affords a purer product, 164–165°.

2. Notes

1. It is convient to use the ether solution of diethyl isonitrosomalonate described by Zambito and Howe.[2]

2. About six washings are required. It may be necessary to add a total of 50 ml. of ether during the first three washings and 20 ml. during the final washing to facilitate breaking of the interphase emulsions. In each case, after partial separation of phases has occurred, ether is added and the separatory funnel is swirled gently until the interphase clears. The washing process requires about 1.5 hours.

3. As diethyl isonitrosomalonate may decompose with explosive violence on heating, further purification by distillation is not recommended.

According to the submitters, reductions using as much as 0.3 mole of diethyl isonitrosomalonate were carried out.

4. The checkers used 10% palladium-on-charcoal catalyst obtained from Baker and Company, Inc., 113 Astor Street, Newark, New Jersey.

5. According to the submitters, diethyl aminomalonate may

be purified by distillation, b.p. 116–118°/12 mm. or 122–123°/16 mm.; $n_D^{16} = 1.4353$; $d_{14}^{16} = 1.100$.

6. Hydrogen chloride is dried by passage through a train of two gas washing bottles containing concentrated sulfuric acid. A 10-mm. tube through which the hydrogen chloride is passed is placed just over the stirring liquid, instead of under the surface, to prevent clogging of the tube by the bulky precipitate which is formed.

The checkers found magnetic stirring satisfactory. More ether may have to be added to prevent the heavy slurry from stopping the stirrer.

7. A medium-porosity sintered-glass funnel was used by the checkers.

3. Methods of Preparation

Diethyl isonitrosomalonate has been reduced catalytically, over palladium on charcoal,[3] Raney nickel,[4] and chemically by aluminum amalgam[5] or hydrogen sulfide.[6]

4. Use of Diethyl Aminomalonate Hydrochloride

Diethyl aminomalonate is a useful intermediate, lending itself to N-acylation;[3,7] the N-acyl derivatives may be alkylated by procedures as established for syntheses via malonic ester.

1. Medical College of Virginia, Richmond, Virginia.
2. See p. 373, this volume.
3. H. R. Snyder and C. W. Smith, *J. Am. Chem. Soc.*, **66**, 350 (1944).
4. C. E. Redemann and M. S. Dunn, *J. Biol. Chem.*, **130**, 341 (1939).
5. R. Locquin and V. Cherchez, *Compt. Rend.*, **186**, 1360 (1926).
6. T. B. Johnson and B. H. Nicolet, *J. Am. Chem. Soc.*, **36**, 352 (1914).
7. J. H. R. Beaujon and W. H. Hartung, *J. Am. Pharm. Assoc., Sci. Ed.*, **41**, 578 (1952); G. H. Cocolas and W. H. Hartung, *J. Am. Chem. Soc.*, **79**, 5203 (1957).

DIETHYL [O-BENZOYL]ETHYLTARTRONATE

(Malonic acid, ethylhydroxy-, diethyl ester, benzoate)

$$C_2H_5CH(CO_2C_2H_5)_2 \xrightarrow[-H_2]{NaH, C_6H_6}$$

$$Na^{\oplus}[C_2H_5C(CO_2C_2H_5)_2]^{\ominus} \xrightarrow[-C_6H_5CO_2Na]{(C_6H_5\overset{O}{\overset{\|}{C}}-O)_2}$$

Submitted by E. H. Larsen and S.-O. Lawesson [1]
Checked by M. R. Michalewich and William D. Emmons

1. Procedure

Caution! This reaction should be carried out behind a safety screen. The solvent removal and product distillation steps should also be carried out behind a screen to minimize trouble if the product is contaminated with undetected peroxides. Benzoyl peroxide should be handled with caution because it is impact-sensitive.

To a 1-l., three-necked, round-bottomed flask is added 7.2 g. (0.15 mole) of a 50% dispersion of sodium hydride in mineral oil (Note 1). The sodium hydride is washed several times by decantation with dry ether and is then covered with 300 ml. of dry benzene (Note 2). The flask is equipped with dropping funnel, stirrer, and reflux condenser. Diethyl ethylmalonate (28.2 g., 0.15 mole) (Note 1) is added dropwise over a 5-minute period, and the reaction mixture is stirred for 2 hours until a clear solution forms. The solution is cooled in an ice bath, and 24.2 g. (0.1 mole) of benzoyl peroxide (Note 3) in 300 ml. of dry benzene is added dropwise over a 1-hour period with continuous stirring. After another 30 minutes, a peroxide test (Note 4) is made to ensure that all the peroxide has reacted.

The porridge-like mixture is then poured into 300 ml. of water and vigorously shaken in a 1-l. separatory funnel. The benzene phase is separated, and the water phase is extracted three times with 100-ml. portions of ether. The combined extracts are washed until neutral and are dried over anhydrous sodium sulfate. The volatile solvents are evaporated at aspirator pressure, and the residue (Note 5) is distilled through a short Vigreux column. After a fore-run of diethyl ethylmalonate 23.3–24.1 g. (75–78%) of diethyl [O-benzoyl]ethyltartronate is obtained, b.p. 132° (0.1 mm.); $n^{20}\text{D}$ 1.4885.

2. Notes

1. The sodium hydride is obtained from Metal Hydrides Inc., Beverly, Massachusetts; the diethyl ethylmalonate from Eastman Organic Chemicals.

2. Reagent grade benzene was dried over calcium hydride prior to use.

3. The benzoyl peroxide is recrystallized from chloroform and methanol at room temperature. The checkers used the 96% purity commercial grade available from the Lucidol Division of Wallace and Tiernan without further purification.

4. A few drops of the reaction mixture are added to a dilute solution of sodium iodide in glacial acetic acid; if a brown ring is not formed, all peroxides have reacted.

5. A peroxide test on the residue is recommended before the distillation is begun.

3. Methods of Preparation

The present procedure is essentially that described by one of the submitters.[2]

4. Merits of the Preparation

The reaction described is of considerable general utility for the preparation of benzoyloxy derivatives of β-carbonyl compounds. Thus O-benzoyl tartronates have been prepared, from which routes to diethyl tartronates and tartronic acids have been developed.[2] Ethyl benzoyloxy cyanoacetates have similarly been prepared and are of potential interest in connection with the chemistry of amino acid precursors.[3] Similarly the benzoyloxy group has been introduced into β-keto esters [4, 5] and β-diketones.[6] Also a new method for the preparation of acyloins was found.[5] An extension of the method has led to certain types of benzoyloxy γ-keto esters[7] and benzoyloxy δ-ketonitriles.[8] The method has been reviewed.[9]

1. Department of Chemistry, University of Åarhus, 8000 Åarhus C, Denmark.
2. S.-O. Lawesson, T. Busch, and C. Berglund, *Acta Chem. Scand.*, **15**, 260 (1961).
3. S.-O. Lawesson and C. Frisell, *Arkiv Kemi*, **17**, 393 (1961).
4. S.-O. Lawesson, M. Andersson, and C. Berglund, *Arkiv Kemi*, **17**, 429 (1961).
5. S.-O. Lawesson, S. Grönwall, and M. Andersson, *Arkiv Kemi*, **17**, 457 (1961).
6. S.-O. Lawesson, P.-G. Jönsson, and J. Taipale, *Arkiv Kemi*, **17**, 441 (1961).
7. S.-O. Lawesson, M. Dahlen, and C. Frisell, *Acta Chem. Scand.*, **16**, 1191 (1962).
8. G. Näslund, A. Senning, and S.-O. Lawesson, *Acta Chem. Scand.*, **16**, 1324 (1962).
9. S.-O. Lawesson, C. Frisell, D. Z. Denney, and D. B. Denney, *Tetrahedron*, **19**, 1229 (1963).

DIETHYL BIS(HYDROXYMETHYL)MALONATE*

(Malonic acid, bis(hydroxymethyl)-, diethyl ester)

$$CH_2(CO_2C_2H_5)_2 + 2HCHO \rightarrow (HOCH_2)_2 C(CO_2C_2H_5)_2$$

Submitted by PAUL BLOCK, JR.[1]
Checked by MELVIN S. NEWMAN and REINHARDT STEIN.

1. Procedure

Formaldehyde solution equivalent to 60 g. of formaldehyde (2 moles) (Note 1) and 8 g. of potassium bicarbonate (small crystals) are placed in an 800-ml. beaker standing in a water bath at 20° in a hood. Mechanical stirring is started and 160 g. (1 mole) of diethyl malonate (Note 2) is added dropwise during about 40–50 min., at such a rate that the temperature of the reaction mixture is held at 25–30°. Stirring is continued for 1 hour. The reaction mixture is transferred to a separatory funnel. A saturated solution of ammonium sulfate (320 ml.) (Note 3) is added and the mixture is extracted with 320 ml. of ether. The ethereal extract, dried for 1 hour with anhydrous sodium sulfate (20 g.), is filtered into a 1-l., three-necked flask through a fluted filter paper. The sodium sulfate and paper are washed with 50 ml. of anhydrous ether. The flask (fitted with a thermometer reaching to the bottom, a condenser set for downward distillation; third neck closed) is placed in a suitable heating bath (Note 4). Boiling chips are introduced and the ether is distilled until the temperature of the liquid has risen to 45-50°. The heating bath is then removed and the distillation assembly is replaced by a glass tube (about 4 mm. I.D.) reaching to the bottom of the flask and closed by a piece of rubber tubing and a screw clamp. An aspirator is now connected to the third neck of the flask. Vacuum is applied and volatile material is removed until the pressure falls to 20–30 mm. The temperature of the contents of the flask is brought to 40° and maintained there until crystallization begins and for an additional 30 minutes (Note 5). Isopropyl ether (500 ml.) (Note 6) is added. The mixture is warmed to 50° and swirled until the product (crystalline and oily) dissolves. The solution is transferred to an Erlenmeyer flask, and cooled in ice water with stirring until a thick suspension of crystals results. The suspension is refrigerated for 1 hour, filtered with suction (rubber dam), and the crystals are dried overnight at room temperature, and then in a vacuum desiccator over sulfuric acid. The yield of colorless crystals, m.p. 48–50°, is 158–166 g. (72–75%). This product may be recrystallized from isopropyl ether (3.5 volumes) with an 85% recovery to yield material with m.p. 50–52°. Melting points of 52–53° and 52° are reported.[2,3]

2. Notes

1. Exactly 3 ml. of formaldehyde solution is assayed by the method of U.S.P. XIII, and the result calculated to grams of formaldehyde per milliliter of solution.

2. The diethyl malonate used boiled over a 2° range.

3. Ammonium sulfate (175 g.) is added to 235 ml. of water, warmed until it is dissolved, and cooled.

4. A heating mantle or liquid bath may be used.

5. The screw clamp may be adjusted so that there will be a spattering on the upper, cool part of the flask, while the pressure is still maintained below 30 mm. Under these conditions, the initiation of crystallization is speeded.

6. A practical grade is adequate, but it should be peroxide-free.

3. Methods of Preparation

The only reported method of preparation of diethyl bis(hydroxymethyl)malonate is by the reaction described here.[2,3]

4. Use of Diethyl Bis(hydroxymethyl)malonate

Diethyl bis(hydroxymethyl)malonate is a useful intermediate for preparing substituted malonic esters, acrylic esters, and isobutyric esters.[2,4]

1. Department of Chemistry, University of Toledo, Toledo, Ohio.
2. K. N. Welch, *J. Chem. Soc.*, 257 (1930).
3. H. Gault and A. Roesch, *Bull. Soc. Chim. France,* 5 (4), 1410 (1937).
4. A. F. Ferris, *J. Org. Chem.*, 20, 780 (1955).

DIETHYL β-KETOPIMELATE

(Pimelic acid, β-oxo-, diethyl ester)

$$\begin{array}{c} CH_3 \\ | \\ CO \\ / \\ CH_2 \\ \backslash \\ CO_2C_2H_5 \end{array} + Na + ClCO(CH_2)_3CO_2C_2H_5 \rightarrow$$

$$C_2H_5O_2C(CH_2)_3COCH \begin{array}{c} CH_3 \\ | \\ CO \\ / \\ \\ \backslash \\ CO_2C_2H_5 \end{array}$$

$$C_2H_5O_2C(CH_2)_3COCH \begin{array}{c} CH_3 \\ | \\ CO \\ / \\ \\ \backslash \\ CO_2C_2H_5 \end{array} + NH_3 \rightarrow$$

$$C_2H_5O_2C(CH_2)_3COCH_2CO_2C_2H_5 + CH_3CONH_2$$

Submitted by MAYA GUHA and D. NASIPURI.[1]
Checked by WILLIAM G. DAUBEN and RICHARD ELLIS.

1. Procedure

A. *Diethyl α-acetyl-β-ketopimelate.* In a 2-l. three-necked flask equipped with a mercury-sealed Hershberg stirrer, a dropping funnel, and a reflux condenser protected with a calcium chloride tube are placed 11.5 g. (0.5 g. atom) of finely powdered sodium (Note 1) and 500 ml. of dry ether. The flask is placed in an ice bath, and 65.0 g. (63.5 ml., 0.5 mole) of freshly distilled ethyl acetoacetate in 150 ml. of dry ether is slowly added from the dropping funnel with stirring (approximate time for addition is 30–40 minutes). The mixture is stirred overnight, then it is cooled in an ice bath, and 89.0 g. (0.5 mole) of γ-carbethoxy-butyryl chloride (Note 2) in 200 ml. of dry ether is added gradually over the course of 1 hour. The reaction is first stirred overnight at room temperature, then gently refluxed by heating in a water bath for 30 minutes. The mixture is cooled in an ice bath, and a cold solution of 20 ml. of concentrated sulfuric acid in 300 ml. of water is added cautiously with vigorous stirring. The stirring is continued until two clear layers form when the stirring is stopped. The ethereal layer is separated and the aqueous layer extracted once with 100 ml. of ether. The two organic layers are combined, washed once with water, and dried over anhydrous sodium sulfate. After removal of the sodium sulfate by filtration, the solvent is removed by heating the ethereal solution on a water bath held at about 50–60°. The residual light-brown liquid is transferred to a 150 ml. Claisen flask and distilled under reduced pressure. The fraction boiling at 142–147°/0.4 mm. or 158–162°/2.5 mm. is collected (Note 3). The yield is 84–91 g. (61–66%), n_D^{28} 1.4649–1.4655.

B. *Diethyl β-ketopimelate.* In a 250-ml. distillation flask fitted with an inlet tube reaching near the bottom of the flask and a soda-lime drying tube on the side-arm is placed a solution of 50 g. (0.18 mole) of diethyl α-acetyl-β-ketopimelate in 75 ml. of dry ether. The solution is cooled by placing the flask in an ice-salt bath, and then a slow stream of ammonia gas is passed through the inlet tube. The solution becomes turbid during the first few minutes and soon becomes clear again. The gas stream is continued for 45–50 minutes, and the yellow liquid is allowed to stand at room temperature overnight with due protection from atmospheric moisture. Most of the ether is then removed by passing a stream of dry air through the solution, and the residue is transferred to a separatory funnel with the aid of 50 ml. of ether. The ethereal solution is washed with three 70-ml. portions of cold 3N hydrochloric acid, each extraction being shaken vigorously for 10 minutes. The ethereal layer is set aside, and the acid washings are extracted twice with 50-ml. portions of ether. The combined ethereal extracts are washed once with water and dried over anhydrous sodium sulfate. After removal of the sodium sulfate by filtration, the solvent is removed by heating the solution on a water bath. The residue is transferred to a Claisen flask with a short Vigreux column, and the fraction boiling at 130–132°/0.5 mm. or 120–121°/0.2 mm. is collected (Notes 4 and 5). The yield is 21–25 g. (50–59%), n_D^{28} 1.4400, $n_D^{31.5}$ 1.4376, $n_D^{36.5}$ 1.4338.

2. Notes

1. Clean pieces of sodium are melted under xylene and powdered by vigorous shaking. When cold, the xylene is decanted and the sodium powder is washed by decantation with a few milliliters of dry ether and then washed into the reaction flask with dry ether.

2. The γ-carbethoxybutyryl chloride (b.p. 100–101°/5–6 mm. or 108–110°/15 mm.) was prepared by the method of Bachmann, Kushner, and Stevenson.[2]

3. The distillate contains mostly C-acyl ester with a little of O-acyl ester. Separation of these two esters by means of a carbonate solution in which only the C-acyl ester is soluble[3,4] is possible. This separation is unnecessary in the present procedure for the O-acyl derivative gives rise to ethyl acetoacetate during decomposition with ammonia. This low-boiling ester is removed during the distillation of diethyl β-ketopimelate.

4. A small fore-run, b.p. 60–80°/5 mm., is collected which gives a positive test with ferric chloride reagent. Presumably this fraction consists mostly of ethyl acetoacetate.

5. The checkers used an 18-in. Vigreux column with a heated jacket and observed a boiling point of 126–128°/2 mm.

3. Methods of Preparation

The described method of preparing diethyl β-ketopimelate is a modification of that described by Bouveault[5] and is essentially the same as that reported by Bardhan and Nasipuri.[6] This ester has also been prepared by condensation of γ-carbethoxybutyryl chloride with ethoxymagnesiummalonic ester and cleavage of the resulting acylated malonic ester by β-naphthalenesulfonic acid[7] or by acetic or propionic acids containing a trace of concentrated sulfuric acid.[8]

4. Merits of Preparation

The present method offers a more convenient synthesis with

appreciably higher yields of diethyl β-ketopimelate. It is reported to be useful for the preparation of dimethyl β-ketoadipate[3,9] and diethyl β-ketosuberate.[4]

1. Department of Chemistry, University of Calcutta, Calcutta 9, India.
2. W. E. Bachmann, S. Kushner, and A. C. Stevenson, *J. Am. Chem. Soc.*, 64, 974 (1942).
3. J. C. Bardhan, *J. Chem. Soc.*, 1848 (1936).
4. S. Archer and G. Pratt, *J. Am. Chem. Soc.*, 66, 1656 (1944).
5. L. Bouveault, *Compt. Rend.*, 131, 45 (1900).
6. J. C. Bardhan and D. Nasipuri, *J. Chem. Soc.*, 350 (1956).
7. J. H. Hunter and J. A. Hogg, *J. Am. Chem. Soc.*, 71, 1922 (1949).
8. H. J. E. Loewenthal, *J. Chem. Soc.*, 3962 (1953); R. E. Bowman, *J. Chem. Soc.*, 322 (1950).
9. J. Korman, *J. Org. Chem.*, 22, 849 (1957).

N,N-DIETHYL-1,2,2-TRICHLOROVINYLAMINE

(Vinylamine, 1,2,2-trichloro-N,N-diethyl-)

$$Cl_3CCON(C_2H_5)_2 + (C_4H_9)_3P \rightarrow$$

$$Cl_2C{=}C\begin{smallmatrix} Cl \\ \\ N(C_2H_5)_2 \end{smallmatrix} + (C_4H_9)_3PO$$

Submitted by A. J. SPEZIALE and R. C. FREEMAN.[1]
Checked by B. C. MCKUSICK and H. D. HARTZLER.

1. Procedure

A. *N,N-Diethyl-2,2,2-trichloroacetamide.* A 1-l. three-necked flask equipped with a stirrer and dropping funnel is charged with 73 g. (1.00 mole) of diethylamine, 500 ml. of ether, and a solution of 40 g. (1.00 mole) of sodium hydroxide in 160 ml. of water. The mixture is stirred and maintained at a temperature of $-10°$ to $-15°$ by a bath of Dry Ice and acetone while 200 g. (1.10 moles) of trichloroacetyl chloride is added in the course of 1 hour. The cooling bath is removed, the temperature is allowed to rise to 10°, and the organic layer is separated. The aqueous layer is extracted with two 50-ml. portions of ether. The ether extracts are combined, washed with 50 ml. of 5% hydrochloric acid, two 50-ml. portions of 5% sodium bicarbonate solution, and 50 ml. of water, and dried over magnesium sulfate. The ether is removed by distillation at atmospheric pressure. The residue is distilled through a short indented Claisen still head at reduced pressure. N,N-Diethyl-2,2,2-trichloroacetamide is collected at 77–79°/1.5 mm.; n_D^{25} 1.4902–1.4912; weight 183–200 g. (84–92%).

B. *N,N-Diethyl-1,2,2-trichlorovinylamine.* The reaction is carried out in a 500-ml. three-necked flask equipped with an efficient mechanical stirrer, a thermometer, a reflux condenser to which a drying tube containing calcium chloride is attached, and a 250-ml. dropping funnel with a pressure-equalizing tube. The flask is charged with 219 g. (2.00 moles) of N,N-diethyl-2,2,2-trichloroacetamide. A gas-inlet tube is attached to the dropping funnel, and dry nitrogen (Note 1) is passed through the apparatus for 5 minutes with stirring. The gas-inlet tube is removed briefly, 202 g. (2.00 moles) of tri-*n*-butylphosphine (Note 2) is placed in the dropping funnel, the gas-inlet tube is replaced, and nitrogen is passed through the apparatus in a slow stream; the slow flow of nitrogen is continued all during the reaction. The phosphine is added at such a rate that a temperature of 85–90° is reached in 30 minutes (Note 3). The rate of addition is then slowed in order to maintain the temperature within this range. The total addition time is 45–55 minutes.

After all the phosphine has been added, the water bath is replaced by a heating mantle, and the reaction mixture is held at 85–95° for one additional hour and cooled to room temperature. The nitrogen-inlet tube, dropping funnel, and reflux condenser are removed, and the reaction flask is fitted with a 15 x 150-mm. Vigreux column for distillation under reduced pressure. The reaction mixture is then distilled (Note 4). The pot temperature rises from 94° to 150° during the distillation, and the crude N,N-diethyl-1,2,2-trichlorovinylamine, weight 151–164 g., is collected at 73–120°/8–11 mm. Redistillation of the crude vinylamine through a 20 x 400-mm. column packed with glass helices affords 140–150 g. (69–74%) of pure N,N-diethyl-1,2,2-trichlorovinylamine, b.p. 78–79°/18 mm., n_D^{25} 1.4857–1.4867 (Notes 5 and 6).

2. Notes

1. Tri-*n*-butylphosphine reacts exothermically with atmospheric oxygen to form tri-*n*-butylphosphine oxide.

2. Tri-*n*-butylphosphine obtainable from Westvaco Mineral Products, 161 East Forty-second St., New York City, can be used without further purification.

3. Because this reaction is very exothermic, the phosphine should be added cautiously.

4. The reaction and the initial distillation should be carried out on the same day.

5. As N,N-diethyl-1,2,2-trichlorovinylamine reacts rapidly with atmospheric moisture, it should be stored under nitrogen, preferably in a refrigerator.

6. If desired, the tri-*n*-butylphosphine oxide can be recovered by continuing the distillation. Crude tri-*n*-butylphosphine oxide distils at 115–118°/1–2 mm. (pot temperature 125–135°). The pure phosphine oxide [2] distils at 94–95°/0.03 mm.; m.p. 64.6–66.6°; yield 135–159 g. (62–73%). *Caution! The reaction mixture should not be distilled to dryness.* There should be a residue of about 40–50 ml.

3. Methods of Preparation

N,N-Diethyl-1,2,2-trichlorovinylamine has been prepared by the action of trimethyl, triethyl, or triisopropyl phosphite or triphenylphosphine on N,N-diethyl-2,2,2-trichloroacetamide.[3] These methods require a reaction temperature of 150–160° and several distillations in order to obtain a pure product. Consequently, the yields of the vinylamine are lower than by the present procedure.[3]

4. Merits of Procedure

The procedure has also been applied to the synthesis of N,N-dimethyl-1,2,2-trichlorovinylamine from trichloroacetamide (60% yield),[3] and it probably is a general means of preparing N,N-dialkyl-1,2,2-trichlorovinylamines. The reaction is an unusual one involving reduction of the amide and halogen migration and is of theoretical interest.

N,N-Diethyl-1,2,2-trichlorovinylamine undergoes certain reactions which involve the 1-chlorine atom. Acids and alcohols are converted to their respective chlorides. Aniline converts the vinylamine to N,N-diethyl-N'-phenyl-2,2-dichloroacetamidine.[3]

1. Organic Chemicals Division, Monsanto Chemical Co., St. Louis, Mo.
2. G. M. Kosolapoff, *J. Am. Chem. Soc.*, 72, 5508 (1950).
3. A. J. Speziale and R. C. Freeman, *J. Am. Chem. Soc.*, 82, 903 (1960).

β,β-DIFLUOROSTYRENE

(1,1-Difluoro-2-phenylethylene)

$$C_6H_5CHO + (C_6H_5)_3P + ClF_2CCO_2Na \longrightarrow$$

$$C_6H_5CH{=}CF_2 + CO_2 + NaCl + (C_6H_5)_3PO$$

Submitted by Samuel A. Fuqua,[1] Warren G. Duncan,[2] and Robert M. Silverstein[2]
Checked by John J. Miller, Herbert Aschkenasy, and William D. Emmons

1. Procedure

In a 250-ml. two-necked flask fitted with a reflux condenser, a drying tube, a magnetic stirrer, and a heated dropping funnel with a pressure-equalizing side arm (Note 1) are placed 23.1 g. (0.088 mole) of triphenylphosphine, 8.5 g. (0.081 mole) of benzaldehyde, and 10 ml. of anhydrous 2,2'-dimethoxydiethyl ether (diglyme) (Note 2). A solution of 18.3 g. (0.12 mole) of dry sodium chlorodifluoroacetate (Note 3) is prepared by stirring the finely divided salt in 50 ml. of anhydrous diglyme at 70° for about 5 minutes. This warm solution is placed in the dropping funnel which is heated to 60°. The system is purged with dry nitrogen. The solution in the flask is stirred and heated in an oil bath held at 160°, while the contents of the dropping funnel are added dropwise over a period of 1.5–2 hours (Note 4). The diglyme and product are flash-distilled at 1 mm. and a bath temperature of 100° into a receiver cooled with dry ice. The distillate is fractionated through a spinning-band column (18 in. × 6 mm. I.D.); the yield of product collected at a head temperature of 52–54° (40 mm.) is 7.6–8.9 g. (67–79%) (Note 5), $n^{20}D$ 1.4939 (Note 6).

2. Notes

1. All glassware is oven-dried. The dropping funnel is wrapped with heating tape, and a thermometer is inserted between the funnel and the tape.

2. Triphenylphosphine is available from M and T Chemicals, Inc. Benzaldehyde is distilled immediately before use. Diglyme is refluxed for 4 hours over calcium hydride and distilled under reduced pressure.

3. Sodium chlorodifluoroacetate is prepared from chlorodifluoroacetic acid (K & K Laboratories) as follows: To a cooled, stirred solution of 60.7 g. (1.52 moles) of sodium hydroxide in 700 ml. of methanol is slowly added a solution of 198 g. (1.52 moles) of chlorodifluoroacetic acid in 300 ml. of methanol, the temperature being kept below 40°. The methanol is removed under reduced pressure at 40°. The salt, which is pulverized and dried overnight at room temperature at 1 mm., is obtained in essentially quantitative yield. The salt is again dried in the same way immediately before use.

4. During the development of this procedure, evolution of carbon dioxide was monitored with a wet-test meter. At the bath temperature given (160°), sodium chlorodifluoroacetate eliminates carbon dioxide as rapidly as it is added over a period of 1.5–2 hours. If the bath temperature is allowed to drop, there is danger of buildup of sodium chlorodifluoroacetate followed by violent exothermic decomposition. Addition of the sodium chlorodifluoroacetate solution should not be started until the flask contents are equilibrated with the oil bath.

It is quite feasible to run the reaction at a bath temperature of 90–95° by adding all reagents to the flask initially; a quantitative evolution of carbon dioxide occurs over a period of about 18 hours. The reaction can also be carried out in refluxing 1,2-dimethoxy-ethane (Arapahoe Chemicals, Inc.) over a period of about 50 hours (yield 40–55%), or in triethylene glycol dimethyl ether (Ansul Chemical Company) at a bath temperature of 160° over a period of 2 hours (yield 64%).

5. The distilled product gave a single symmetrical peak on gas chromatography under the following conditions: 25% LAC on Chromosorb W, 6 ft. $\times$ $\frac{1}{4}$ in., 110°, helium flow 41 ml./min., elution time 14.2 minutes. The checkers used LB 5-50 on Fluoropak 80 and obtained a single peak. Gas chromatography of the flash distillate before fractionation showed an actual yield of 10.6 g. (95%). The product fumes in moist air, and some etching of glass containers was noted. This is presumably due to elimination of hydrogen fluoride. Samples in open glass containers deposit a small amount of solid on standing; the solid is probably a product of the glass-hydrogen fluoride reaction.

6. Care must be taken to clean and dry the refractometer prisms before and after use in order to prevent etching of the prisms.

3. Methods of Preparation

The literature preparation [3] of β,β-difluorostyrene consists of seven steps from sodium difluoroacetate, the last step involving pyrolysis at 600°; the overall yield was 5%.

4. Merits of the Preparation

The method described is a general synthesis for compounds containing the —CH=CF$_2$ moiety. There is no other simple general route to such compounds. Aromatic, aliphatic, and heterocyclic aldehydes to which this procedure has been applied are: p-fluorobenzaldehyde (65%), p-methoxybenzaldehyde (60%), heptanal (43–51%), and furfural (75%).[4] The method is also applicable to ketones[5] and to α-perfluoroketones.[6] Substitution of lithium chlorodifluoroacetate for the sodium salt has been advocated.[7]

1. Deceased.
2. Stanford Research Institute, Menlo Park, California.
3. M. Prober, *J. Am. Chem. Soc.*, **75**, 968 (1953).
4. S. A. Fuqua, W. G. Duncan, and R. M. Silverstein, *J. Org. Chem.*, **30**, 1027 (1965).

5. S. A. Fuqua, W. G. Duncan, and R. M. Silverstein, *J. Org. Chem.*, **30**, 2543 (1965).
6. F. E. Herkes and D. J. Burton, *J. Org. Chem.*, 32, 1311 (1967).
7. R. C. Slagel, *Chem. Ind. (London)*, 848 (1968).

2,2-DIFLUOROSUCCINIC ACID*

(Succinic acid, 2,2-difluoro-)

A. $CH_2\!=\!CCl_2 + CF_2\!=\!CClF \rightarrow$
$$\begin{array}{c} CH_2\text{—}CCl_2 \\ | \qquad | \\ CF_2\text{—}CClF \end{array}$$

B.
$$\begin{array}{c} CH_2\text{—}CCl_2 \\ | \qquad | \\ CF_2\text{—}CClF \end{array} + Zn \rightarrow \begin{array}{c} CH_2\text{—}CCl \\ | \qquad \| \\ CF_2\text{—}CF \end{array} + ZnCl_2$$

C.
$$3\begin{array}{c} CH_2\text{—}CCl \\ | \qquad \| \\ CF_2\text{—}CF \end{array} + 4KMnO_4 + 8NaOH \xrightarrow{\;H_2SO_4\;} 3\begin{array}{c} CF_2CO_2H \\ | \\ CH_2CO_2H \end{array}$$

Submitted by M. S. Raasch and J. E. Castle.[1]
Checked by R. H. Uloth and R. R. Covington.[2]

1. Procedure

Caution! In the absence of toxicity data, the fluorine compounds should all be treated as though they were toxic materials.

A. *1,1,2-Trichloro-2,3,3-trifluorocyclobutane.* In a 1-l. rocker bomb are placed 350 g. (3.6 moles) of 1,1-dichloroethylene (Note 1) and 1 g. of hydroquinone. The bomb is cooled in a mixture of Dry Ice and acetone and evacuated. The vacuum is released with nitrogen and the bomb is again evacuated. The bomb is then charged with 300 g. (2.6 moles) of chlorotrifluoroethylene (Note 1) from a cylinder. The bomb is heated at 180° for 7 hours behind a barricade (Note 2) and is then cooled, vented, and unloaded. The solid polymer (about 45 g.) is removed by filtration and is rinsed with 50 ml. of ether. The combined filtrate and rinse are concentrated and then distilled through a 30-cm. packed column to give 242–262 g. (44–48%) of 1,1,2-trichloro-2,3,3-trifluorocyclobutane, b.p. 120–121°, n_D^{25} 1.4139–1.4141.

B. *1-Chloro-2,3,3-trifluorocyclobutene.* A 1-l. three-necked flask fitted with mercury-sealed stirrer, dropping funnel, reflux condenser, and heater is charged with 150 ml. of absolute ethyl alcohol and 76 g. (1 mole) of 95% zinc dust. The alcohol is heated to boiling, and 235 g. (1.1 moles) of 1,1,2-trichloro-2,3,3-trifluorocyclobutane is added through the dropping funnel during 40 minutes. After the reaction has started, external heating is decreased. The mixture is heated under reflux for 1 hour after the end of the addition. It is then cooled below reflux temperature, 15 g. (0.2 mole) more of zinc powder is added, and the heating under reflux is continued for 30 minutes more. The mixture is again cooled below reflux temperature, and a simple still head arranged for downward distillation is attached in place of the condenser. Distillation is carried out until 165 ml. of distillate has been collected. The still head reaches a temperature of about 90°. The distillate is washed with two 250-ml. portions of water and dried by shaking gently for 5 minutes with 5 g. of 8-mesh calcium chloride. The product is decanted from the calcium chloride and distilled through a 30-cm. packed column to give 107–113 g. (68–72%) of 1-chloro-2,3,3-trifluorocyclobutene, b.p. 52–53°, n_D^{25} 1.3614–1.3619.

C. *2,2-Difluorosuccinic acid.* In a 3-l. three-necked flask fitted with stirrer, thermometer, and dropping funnel, 80 g. (2 moles) of sodium hydroxide is dissolved in 2 l. of water and 158 g. (1 mole) of potassium permanganate is then added. The mixture is cooled to 15–20° with an ice-salt bath, and 107 g. (0.75 mole) of 1-chloro-2,3,3-trifluorocyclobutene is added through the dropping funnel during 1 hour while the permanganate solution is stirred and maintained at 15–20°. After the solution has been stirred for 2 hours more at this temperature, the manganese dioxide is removed by filtration and rinsed with three 300-ml. portions of water. The combined filtrate and washings are concentrated to a volume of 500 ml. by evaporation on a steam bath (Note 3). The solution is then cooled and 85 ml. of concentrated sulfuric acid is added slowly with stirring. The cold solution is extracted with four 250-ml. portions of ether (Note 4). Drying of the ether extract is accomplished by agitating it for 5 minutes with 30 g. of anhydrous magnesium sulfate. When the drying agent has been removed by filtration and rinsed with ether, the filtrate and the ether washings are combined and concentrated to give 91–97 g. (79–84%) of 2,2-difluorosuccinic acid. The acid is recrystallized by dissolving it in hot nitromethane (1.25 ml. per g.), filtering the solution through a layer of filter aid if necessary, and cooling the solution to 3°. The crystals are collected by suction filtration and rinsed with 30 ml. of cold nitromethane. After drying, this gives 85–92 g. (74–80%) of 2,2-difluorosuccinic acid, m.p. 144–146°.

2. Notes

1. The checkers employed 1,1-dichloroethylene supplied by Dow Chemical Co., Midland, Michigan, and chlorotrifluoroethylene supplied by the Matheson Company, Joliet, Illinois.

2. This reaction has been carried out many times without incident in a 1-l. stainless-steel rocker bomb. A pressure gauge was attached during two runs and recorded a maximum of 750 p.s.i. However, an attempt in another laboratory to scale up the reaction using a 3-l. autoclave resulted in a bulged vessel. Uncontrolled polymerization may be hazardous.

3. Evaporation may be carried out in porcelain or glass dishes, but the fluoride present will cause some etching.

4. The amounts removed in the third and fourth extractions are about 5.4 g. and 1.5 g., respectively.

3. Methods of Preparation

The procedure described is the method of Raasch [3, 4] and is the only one published so far.

4. Merits of Preparation

The first step illustrates a very general reaction, the addition of fluoroalkenes to alkenes to give fluorocyclobutanes.[5] The subsequent steps illustrate the synthetic possibilities of fluorocyclobutanes as intermediates. 1,1,2-Trichloro-2,3,3-trifluorocyclobutane may also be converted into chlorotrifluorosuccinic acid, trifluorosuccinic acid, fluoromaleic acid, fluorofumaric acid, difluoromaleic acid, and difluorofumaric acid.[4, 6]

1. Central Research Department, Experimental Station, E. I. du Pont de Nemours and Company, Wilmington, Delaware.
2. Mead Johnson Research Center, Evansville, Indiana.
3. M. S. Raasch, U. S. pat. 2,824,888 [*C.A.*, 52, 12901 (1958)].
4. M. S. Raasch, R. E. Miegel, and J. E. Castle, *J. Am. Chem. Soc.*, 81, 2678 (1959).
5. D. D. Coffman, P. L. Barrick, R. D. Cramer, and M. S. Raasch, *J. Am. Chem. Soc.*, 71, 490 (1949); J. D. Roberts and C. M. Sharts, *Org. Reactions*, 12, 1 (1962).
6. M. S. Raasch, U. S. pat. 2,891,968 [*C. A.*, 54, 1323 (1960)].

α,α-DIFLUOROTOLUENE

(Toluene, α,α-difluoro-)

AND BENZENESULFINYL FLUORIDE

$$C_6H_5SF_3 + C_6H_5\overset{\overset{\displaystyle O}{\|}}{C}H \rightarrow C_6H_5\overset{\overset{\displaystyle O}{\|}}{S}F + C_6H_5CHF_2$$

Submitted by William A. Sheppard [1]
Checked by E. S. Glazer and John D. Roberts

1. Procedure

Caution! Phenylsulfur trifluoride is toxic, and this reaction should be carried out in a good hood. The reagent should not be allowed to come in contact with the skin.

Phenylsulfur trifluoride [2] (16.6 g., 0.10 mole) is placed in a two-necked 50-ml. flask equipped with a dropping funnel and connected to a dry distillation column (Note 1). The flask is heated to 50–70° in an oil bath, and 10.6 g. (0.10 mole) of benzaldehyde is added in small portions over 30 minutes. A mild exothermic reaction occurs. After the addition is completed, the reaction flask is heated to 100° with an oil bath, and the pressure on the column is reduced until α,α-difluorotoluene distills. The major portion of product distills at 68° (80 mm.), but a small final cut, b.p. 45° (15 mm.), is obtained. The yield of α,α-difluorotoluene is 9.2–10.2 g. (71–80%) (Note 2). The pressure is reduced and the distillation is continued. An intermediate cut of 1–2 g., b.p. 45° (15 mm.) to 60° (2.5 mm.), is discarded, and benzenesulfinyl fluoride, 11.7–13.2 g. (81–91%), b.p. 60° (2.5 mm.), is collected. Since the benzenesulfinyl fluoride slowly attacks glass and may be unstable to storage at room temperature, it is recommended that this product be stored at −80°.

2. Notes

1. A 45-cm. spinning band column was employed by the submitter, but any distillation column with a low holdup may be used. Since the products have widely different boiling points, careful fractionation during distillation is not needed. Because the phenylsulfur trifluoride and benzenesulfinyl fluoride slowly attack glass, all equipment should be rinsed with water and acetone *immediately* after use to minimize etching.

2. The product attacks glass slowly on standing, and a moderate increase in pressure takes place. The product can be stored for a period of several days in a polyethylene bottle, but it is best to prepare the material shortly before use. If prolonged storage is required, a stainless steel cylinder or a bottle fabricated from Teflon ® polytetrafluoroethylene resin is suggested.

3. Methods of Preparation

α,α-Difluorotoluene has been prepared by the reaction of α,α-dichlorotoluene with antimony(III) fluoride,[3] by the hydrogenation of α-chloro-α,α-difluorotoluene,[4] by the action of sulfur tetrafluoride on benzaldehyde,[5] and by the present method.[6, 7]

4. Merits of the Preparation

Sulfur tetrafluoride provides an inexpensive method for selectively converting a carbonyl to a difluoromethyl group. However, the reactions involving sulfur tetrafluoride, in general, require pressure equipment constructed of fluorine-resistant material such as "Hastelloy-C" bombs.[8] Phenylsulfur trifluoride may be used to advantage for the same reaction, where small amounts are involved, since the reaction may be run at atmospheric pressure in glass, polyethylene, or metal containers.

Although the sulfur trifluoride compounds are generally useful as selective agents for conversion of carbonyl and carboxyl groups to difluoromethylene and trifluoromethyl groups, variations in reaction conditions are often necessary.[7] Thus the reaction of aromatic ketones requires heating at 150°. Since the reaction with aliphatic aldehydes and ketones is exothermic, it is advantageous to run it in a solvent such as methylene chloride or acetonitrile containing a small amount of sodium fluoride powder (with ketones an induction period of several hours may be observed). Reactions with carboxylic acids should be carried out in a container resistant to hydrogen fluoride, and they require heating at 120° to 150°.

1. Contribution No. 669 from the Central Research Department, Experimental Station, E. I. du Pont de Nemours and Company.
2. W. A. Sheppard, this volume, p. 959.
3. T. Van Hove, *Bull. Classe Sci. Acad. Roy. Belg.*, 1074 (1913).
4. F. Swarts, *Bull. Classe Sci. Acad. Roy. Belg.*, 410 (1920).
5. W. R. Hasek, W. C. Smith, and V. A. Engelhardt, *J. Am. Chem. Soc.*, 82, 543 (1960).
6. W. A. Sheppard, *J. Am. Chem. Soc.*, 82, 4751 (1960).
7. W. A. Sheppard, *J. Am. Chem. Soc.*, 84, 3058 (1962).
8. W. R. Hasek, this volume, p. 1082.

9,10-DIHYDROANTHRACENE*

(Anthracene, 9,10-dihydro-)

Submitted by K. C. Bass.[1]
Checked by Virgil Boekelheide and S. T. Young.

1. Procedure

In a 2-l., three-necked, round-bottomed flask fitted with a rubber-tube sealed mechanical glass stirrer, a reflux condenser

(Note 1), and a thermometer reaching to the bottom of the flask are placed 50 g. (0.28 mole) of anthracene (Note 2) and 750 ml. of commercial absolute ethanol. The suspension obtained is stirred and heated (Note 3) to 50°, and 75 g. (3.25 g. atom) of freshly cut sodium is added in quantities of about 10 g. each to the stirred mixture over a period of 5 minutes. The reaction mixture boils vigorously (Note 4) and stirring is continued for 15 minutes longer. The reaction mixture is then cooled and carefully diluted with 1 l. of water. The white-yellow solid which separates is a mixture of 9,10-dihydroanthracene and anthracene, and it is collected on a Büchner funnel, washed with 400 ml. of water, and dried in air.

The dry white-yellow solid is suspended in 500 ml. of commercial absolute ethanol in a 1-l., three-necked, round-bottomed flask fitted with a rubber-tube sealed mechanical glass stirrer, a reflux condenser (Note 1), and a thermometer reaching to the bottom of the flask. The suspension is stirred and heated (Note 3) to 50°, and 50 g. (2.17 g. atom) of freshly cut sodium is added in quantities of about 10 g. each to the stirred mixture over a period of 5 minutes. The reaction mixture boils vigorously (Note 4), and stirring is continued for an additional 15 minutes. The reaction mixture is then cooled and carefully diluted with 750 ml. of water. The white solid which separates is 9,10-dihydroanthracene, and it is collected on a Büchner funnel, washed with 300 ml. of water, and dried in air. It is recrystallized from ethanol (about 250–300 ml. of solvent is required), and the crystals are collected on a Büchner funnel, washed with 20 ml. of cold ethanol, and dried in air. The yield of dry 9,10-dihydroanthracene in the form of broad, colorless needles, m.p. 108–109°, is 38–40 g. (75–79%) (Note 5).

2. Notes

1. An efficient 12-in., double-surface, all-glass condenser should be used with an outlet tube carrying the evolved hydrogen into a good hood vent.

2. A purified grade of anthracene (blue fluorescence, m.p. 216°) should be used.

3. An electric heating mantle should be used. No free flames should be present anywhere near the reaction flask.

4. The reaction may be controlled by removing the heat source or slowing down the rate of stirring or both.

5. The 9,10-dihydroanthracene may be purified further by steam distillation from an aqueous suspension followed by recrystallization of the dried product from ethanol.

3. Methods of Preparation

The procedure described is adapted from the preparation outlined by Wieland.[2]

4. Merits of Preparation

9,10-Dihydroanthracene has been used as one of the hydrogen transfer reagents in a series of homolytic hydrogen transfer reactions by Braude, Jackman, and Linstead [3] and as a hydrogen donor for the hydrogenation of thiyl radicals to form thiols.[4]

1. Department of Chemistry, The City University, London, England.
2. H. Wieland, Ber. 45, 492 (1912).
3. E. A. Braude, L. M. Jackman, and R. P. Linstead, J. Chem. Soc., 3548 (1954).
4. A. F. Bickel and E. C. Kooyman, Nature, 170, 211 (1952).

1,4-DIHYDROBENZOIC ACID

(2,5-Cyclohexadiene-1-carboxylic acid)

Submitted by M. E. KUEHNE and B. F. LAMBERT.[1]
Checked by LOUISE KUDA and V. BOEKELHEIDE.

1. Procedure

Ten grams (0.082 mole) of benzoic acid is added to 100 ml. of anhydrous ethanol in a 2-l. three-necked flask equipped with a mechanical stirrer and with loose cotton plugs in the side necks.

After the benzoic acid has dissolved, 600 ml. of liquid ammonia (Note 1) is added to the stirred solution. Then 6.2 g. (0.27 g. atom) of sodium is added in small pieces. When about one-third of the sodium has been added, the white sodium salt of the acid precipitates, and there is strong foaming of the reaction mixture. After all the sodium has been consumed, as evidenced by the disappearance of the blue color, 14.6 g. (0.27 mole) of ammonium chloride is added cautiously. The mixture is stirred for an additional hour and then allowed to stand until the ammonia has evaporated.

The residue is dissolved in 300 ml. of water. The solution is poured onto 200 g. of ice and acidified to a pH of about 4 by addition of 75 ml. of 10% hydrochloric acid. The resulting mixture is extracted with four 100-ml. portions of peroxide-free ether, and the combined extracts are washed with 50 ml. of a saturated aqueous solution of sodium chloride and dried over 2 g. of anhydrous magnesium sulfate (Note 2). The ether solution is separated from the drying agent and concentrated at room temperature under reduced pressure. The residual oil is distilled from a 25-ml. Claisen flask with an indented neck. 1,4-Dihydrobenzoic acid is obtained as a colorless oil; weight 9.0–9.7 g. (89–95%); b.p. 80–98°/0.01 mm.; n_D^{24} 1.5011. This material is sufficiently pure for most purposes. However, by a careful redistillation, a small fore-run (b.p. 80–90°/0.01 mm.; n_D^{24} 1.5000) can be separated, and the remainder of the material (b.p. 91–97°/0.01 mm.; n_D^{24} 1.5019) solidifies on cooling; m.p. 15–17° (Note 3). It is stored under nitrogen in a closed vessel (Note 4).

2. Notes

1. Arrangements for cooling or condensing the ammonia can be made, but are not necessary. Most simply, the liquid ammonia can be passed directly from a cylinder into the reaction vessel through heavy rubber tubing.

2. 1,4-Dihydrobenzoic acid has a very penetrating, repulsive odor, and care should be taken to avoid contamination of hands or clothing.

3. Samples of the 1,4-dihydrobenzoic acid, after both the first and the second distillations, are transparent in the ultraviolet

region between 220 mμ and 300 mμ, indicating the absence of benzoic acid or conjugated dihydrobenzoic acids. The refractive index cited in Reference 3 is in error.

4. In the presence of air, 1,4-dihydrobenzoic acid slowly gives benzoic acid and hydrogen peroxide.[2]

3. Method of Preparation

Apparently, 1,4-dihydrobenzoic acid has been prepared only by the Birch reduction of benzoic acid, as illustrated by the present procedure.[2,3]

4. Merits of the Preparation

This procedure is illustrative of the general method of reduction of aromatic compounds by alkali metals in liquid ammonia known as the Birch reduction. The theoretical and preparative aspects of the Birch reduction have been discussed in excellent reviews,[4-6] and there is another example of a Birch reduction in *Organic Syntheses*.[7] Of particular interest in the present procedure is the effect of having a group that forms a stable anion with the alkali metal. For both simple aromatic acids and amides, a Birch reduction gives the corresponding 1,4-dihydro derivative. The same is true when o-alkyl or o-methoxyl groups are present. However, with p-alkyl or m-methoxyl substituents, the corresponding tetrahydro derivatives are formed. p-Methoxyl or p-acetamino groups, which can form stable anionic fragments, are lost during such reductions.

The following examples may be cited to illustrate these generalizations. p-Toluic acid under conditions of the Birch reduction essentially as given in this procedure yields mainly 1,2,3,4-tetrahydro-p-toluic acid (*cis* and *trans*) plus minor amounts of 1,4-dihydro-p-toluic acid (*cis* and *trans*).[3] o-Toluic acid gives 1,4-dihydro-o-toluic acid in 73% yield;[8] m-methoxybenzoic acid gives 1,4,5,6-tetrahydro-3-methoxybenzoic acid in 32% yield;[9] o-methoxybenzoic acid gives crude 1,4-dihydro-2-methoxybenzoic acid in 80% yield;[10] 3,4,5-trimethoxybenzoic acid gives 1,4-dihydro-3,5-dimethoxybenzoic acid in 87% yield;[3] 4-acetaminobenzoic acid gives 1,4-dihydrobenzoic acid in 75% yield;[3]

benzamide gives 1,4-dihydrobenzamide in 69% yield;[3] *m*-methoxybenzamide gives 1,4-dihydro-3-methoxybenzamide in 30% yield;[3] 3,4,5-trimethoxybenzamide gives 1,4-dihydro-3,5-dimethoxybenzamide in 73% yield;[3] and 3,5-dimethoxybenzamide gives 1,4-dihydro-3,5-dimethoxybenzamide in 59% yield.[3] Thus the present example of the Birch reduction illustrates a useful and general synthetic method for preparing dihydro aromatic derivatives.

1. Ciba Pharmaceutical Products Inc., Summit, New Jersey.
2. H. Plieninger and G. Ege, *Angew. Chem.*, 70, 505 (1958).
3. M. E. Kuehne and B. F. Lambert, *J. Am. Chem. Soc.*, 81, 4278 (1959).
4. G. W. Watt, *Chem. Rev.*, 46, 317 (1950).
5. A. J. Birch, *Quart. Rev. (London)*, 4, 69 (1950).
6. A. J. Birch and H. F. Smith, *Quart. Rev. (London)*, 12, 17 (1958).
7. C. D. Gutsche and H. H. Peter, *Org. Syntheses*, Coll. Vol. 4, 887 (1963).
8. A. J. Birch, *J. Chem. Soc.*, 1551 (1950).
9. A. J. Birch, P. Hextall, and S. Sternhell, *Australian J. Chem.*, 7, 256 (1954).
10. M. E. McEntee and A. R. Pinder, *J. Chem. Soc.*, 4419 (1957).

2,5-DIHYDRO-2,5-DIMETHOXYFURAN*

(Furan, 2,5-dihydro-2,5-dimethoxy-)

$$+ \; Br_2 + 2CH_3OH \;\rightarrow$$

Submitted by D. M. BURNESS.[1]
Checked by ALAN E. BLACK and HENRY E. BAUMGARTEN

1. Procedure

In a 3-l., three-necked flask equipped with a stirrer, thermometer, drying tube filled with Drierite, and a dropping funnel are placed 500 ml. of methanol, 500 ml. of benzene, 380 g. of anhydrous sodium carbonate, and 155 ml. (2.15 moles) of distilled furan (Note 1). The slurry is cooled to < -5° with a dry ice-acetone bath (maintained at -10 to -15° during the addition), and an ice-cold solution of 106 ml. (2.0 moles) of bromine in 1.0 l. of methanol (Note 2) is added dropwise or at such a rate as to maintain a reaction temperature of 0 to -5°. This requires about one hour; the mixture is stirred for an additional 2 hours at < 0° and filtered. The solids collected are washed in the funnel with two 100-ml. portions of benzene. The filtrate is stirred for 30 minutes with 100 g. of anhydrous magnesium sulfate and refiltered. The solvent is removed at reduced pressure (Note 3) to a volume of about 300 ml. Filtration and washing with benzene is repeated (Note 4), the filtrate is stirred for 15–20 minutes with anhydrous potassium carbonate, filtered, and distilled through a short packed column to give 195–205 g. (75–79%) of a clear oil, b.p. 80–82° (50 mm.), n_{D}^{25} 1.4333 (Notes 5 and 6).

2. Notes

1. The furan was distilled from anhydrous potassium carbonate. Use of undistilled furan gives a lower yield (8–10%).

2. The methanol is cooled to about 0° before adding the bromine, and the solution is kept cold during the addition.

3. An aspirator, water bath, and short packed column are used, and distillation is continued until the temperature of the distillate begins to rise about 30°. Good stirring is needed here to prevent bumping and local superheating. The pot temperature never exceeds 25–30°.

4. If an additional 500 ml. of methanol is used in lieu of the recommended 500 ml. of benzene, a small, heavier second phase, consisting mostly of methanol, water, and salts, may separate at this point. This is removed before drying the filtrate. Later, in the early stages of distillation of the product, if drying has not been thorough enough, water may appear in the distillate. This necessitates further drying, preferably by azeotroping with benzene.

The checkers observed a separation into two layers in some runs even in the absence of added methanol. The layers were filtered together and any solids in the funnel were washed with two 100-ml. portions of benzene. The combined filtrates were dried with a second 100 g. of anhydrous magnesium carbonate.

5. The product from this procedure is yellow, due to maleic dialdehyde, and occasionally gives a positive Beilstein test for bromine, although its purity is better than 99.5% by glpc. A colorless product is reported[2] to be obtained by treatment with sodium methoxide before the final distillation. Activated alumina treatment of the distillate is also effective.

6. If the product is not used promptly a peroxidation inhibitor should be used. *t*-Butyl catechol (0.01%) is recommended.[2]

3. Discussion

Numerous 2,5-dialkoxy-2,5-dihydrofurans have been prepared by electrolytic oxidation of furan and substituted furans in alcoholic ammonium bromide[3-5] or by bromine[6-7] or chlorine[8] oxidation in the appropriate alcohol. The present method is a modification of the halogen oxidation procedures cited. It is more convenient and gives better yields (~5%) than with chlorine, and halogen-containing byproducts are eliminated in the distillation.

The 2,5-dialkoxy-2,5-dihydrofurans are cyclic acetals of unsaturated dicarbonyl compounds and as such serve as sources thereof. They are also valuable intermediates in the synthesis of numerous types of heterocyclic compounds including those of the tropinone series.[3]

Hydrogenation of the described product to tetrahydro-2,5-dimethoxyfuran is best accomplished with Raney nickel catalyst at 80–100° and 1700–2000 p.s.i. in 45–60 minutes.

1. Eastman Kodak Co., Rochester, N.Y. 14650.
2. F. A. Senour, private communication.
3. N. Elming, in R. A. Raphael, E. C. Taylor, and H. Wynberg, "Advances in Organic Chemistry," Vol. 2, Interscience Publishers, New York, 1960, pp. 67–115.
4. N. L. Weinberg and H. R. Weinberg, *Chem. Rev.*, 68, 449 (1968).
5. S. D. Ross and M. Finkelstein, *J. Org. Chem.*, 34, 1018 (1969).
6. N. Clauson-Kaas, F. Limborg, and J. Fakstorp, *Acta Chem. Scand.*, 2, 109 (1950).
7. J. Fakstorp, D. Raleigh, and L. E. Schniepp, *J. Am. Chem. Soc.*, 72, 869 (1950).

8. D. G. Jones, U.S. Pat., 2,475,097, July 5, 1949.
9. J. F. Stenberg and R. W. Ryan, private communication.

1,3-DIHYDROISOINDOLE

(Isoindoline)

Submitted by J. Bornstein, J. E. Shields,
and A. P. Boisselle [1]
Checked by William G. Dauben and Harold B. Morris

1. Procedure

In a 1-l. round-bottomed flask are placed 36.0 g. (0.132 mole) of 2-(p-tolylsulfonyl)dihydroisoindole,[2] 36.0 g. (0.38 mole) of phenol, 270 ml. of 48% hydrobromic acid (Note 1), and 45 ml. of propionic acid. A few boiling chips are added, and the flask is fitted with a reflux condenser in the top of which is placed a T-tube connected to a source of low-pressure nitrogen and to a mercury bubbler. The mixture is heated under reflux for 2 hours in an atmosphere of nitrogen. The deeply colored reaction mixture is cooled to room temperature, transferred to a 1-l. separatory funnel, and washed with two 200-ml. portions of ether (Note 2). The aqueous phase is then added dropwise over a 1-hour period to a vigorously stirred (Note 3) solution of 200 g. of sodium hydroxide in 600 ml. of water in a 2-l. Erlenmeyer flask immersed in an ice bath. The solution is transferred to a 3-l. separatory funnel and extracted with five 300-ml. portions of ether. The ethereal extracts are combined, dried over anhydrous potassium carbonate (Note 4), and filtered. The solvent is distilled, and the dark residual oil is transferred to a distillation flask and distilled through a low-holdup, semimicro column (Note 5). After removal of 1 or 2 drops of forerun, colorless 1,3-dihydroisoindole is collected at 96 97° (10 mm.) or 55–56° (2 mm.), n^{25}D 1.5686, d_4^{20} 1.081. The yield is 9.9–11.2 g. (63–71%) (Note 6).

2. Notes

1. The hydrobromic acid should be colorless. Reagent grade 47–49% hydrobromic acid, obtained from J. T. Baker Chemical Co., was used as supplied. A technical grade of the constant-boiling acid is suitable if purified by distillation from stannous chloride.

2. Detection of the water-ether interface may prove troublesome; backlighting of the separatory funnel by an intense light source is recommended. The same volume of ether must be used for each washing, even when the preparation is carried out on a smaller scale, e.g., one-half or one-third the scale described here.

3. Stirring is most conveniently accomplished with a magnetic stirrer.

4. Washing of the ethereal extract with water decreases the yield of product.

5. The submitters used a 7 × 300-mm. externally heated column packed with a helix of Chromel wire and fitted with a partial reflux head.[3]

6. Since the product slowly darkens on exposure to air, it should be stored under nitrogen in a refrigerator. The compound solidifies on cooling; m.p. 16.0–16.5°. Nuclear magnetic resonance spectrum (neat, tetramethylsilane internal standard): singlets at δ 7.00 (aromatic protons), 3.93 (CH_2), and 2.24 p.p.m. (NH).

3. Methods of Preparation

1,3-Dihydroisoindole has been prepared from phthalimide by electrolytic reduction[4] and by reduction with lithium aluminum hydride.[5] Other methods that have been used are reduction of 1-chlorophthalazine with zinc and hydrochloric acid[6] and hydrogenolysis of 2-benzyl-1,3-dihydroisoindole.[7] The present method is essentially that of Bornstein, Lashua, and Boisselle.[8]

4. Merits of the Preparation

This procedure illustrates a general method for the preparation of amines by reductive cleavage of sulfonamides by hydrobromic acid in the presence of phenol.[9] The present synthesis makes 1,3-dihydroisoindole readily accessible and is superior in certain respects to the other two practical methods of preparation. Thus the method here described is shorter and gives a higher overall yield than the three-step synthesis of Neumeyer,[7] and obviates the special apparatus and careful control required by the electrochemical process of Dunet, Rollet, and Willemart.[4]

1. Department of Chemistry, Boston College, Chestnut Hill, Massachusetts 02167.
2. J. Bornstein and J. E. Shields, this volume, p. 1064.
3. C. W. Gould, Jr., G. Holzman, and C. Niemann, *Anal. Chem.*, **20**, 361 (1948).
4. A. Dunet, J. Rollet, and A. Willemart, *Bull. Soc. Chim. France*, [5] **17**, 877 (1950).
5. A. Uffer and E. Schlittler, *Helv. Chim. Acta*, **31**, 1397 (1948).
6. S. Gabriel and A. Neumann, *Ber.*, **26**, 521 (1893).
7. J. L. Neumeyer, *J. Pharm. Sci.*, **53**, 981 (1964).
8. J. Bornstein, S. C. Lashua, and A. P. Boisselle, *J. Org. Chem.*, **22**, 1255 (1957).
9. S. Searles and S. Nukina, *Chem. Rev.*, **59**, 1077 (1959).

1,3-DIHYDRO-3,5,7-TRIMETHYL-2H-AZEPIN-2-ONE

(2H-Azepin-2-one, 1,3-dihydro-3,5,7-trimethyl-)

Submitted by Leo A. Paquette [1]
Checked by Klaus Herbig and B. C. McKusick

1. Procedure

Caution! Because obnoxious fumes are liberated during the reaction with chloramine, the apparatus should be set up in a well-ventilated hood.

Five hundred forty-five grams (4.00 moles) of 2,4,6-trimethyl-phenol (Note 1) is placed in a 1-l. three-necked flask fitted with mechanical stirrer, thermometer, 90-cm. Vigreux column, and dropping funnel (not of the pressure-equalizing variety). The Vigreux column must extend sufficiently into the top of a well-ventilated hood to entrain effectively the fumes that will be generated later in the operation. The phenol is melted with the aid of an external oil bath or Glascol® heating mantle and heated to about 100°. The heating bath is removed, and 27.6 g. (1.20 g. atoms) of sodium in cubes about 1 cm. on a side or smaller is added to the stirred mixture at such a rate that the temperature does not exceed 150–160°. The molten mass gradually becomes dark red in color as the sodium dissolves. While the addition and solution of the sodium is proceeding, a cold solution of about 0.25 mole of chloramine in 250 ml. of ether is prepared (Note 2).

When all the sodium has dissolved, the phenoxide-phenol mixture is heated to 150°. With the oil bath or heating mantle still surrounding the flask, *and with a protective shield between the reaction vessel and the operator* (Note 3), the cold ethereal chloramine solution is added with rapid stirring in a thin stream from the dropping funnel at such a rate that the temperature of the reaction mixture does not drop below 125°. Best results are obtained if the thin stream of ether solution can be added directly to the molten mass without first touching the walls of the flask.

When the addition is completed, the heat source is removed and the dark-colored contents are allowed to cool until another 0.25 mole of ethereal chloramine has been prepared and is ready for use; a wait of 1.5–2 hours between chloramine additions is convenient but not essential to the success of the experiment. The cooled reaction mixture is then reheated to 150°, and the process is repeated. This sequence is repeated until a total of four 0.25-mole portions of chloramine are added.

The reaction mixture is allowed to cool. The dropping funnel, thermometer, and Vigreux column are replaced with a stopper and short-path distillation head. The mixture is stirred while the excess phenol is removed by distillation at water-aspirator pressure; b.p. 105–110° (14 mm.). When the temperature begins to rise above 110° (14 mm.), the distillation is stopped and the residue is allowed to cool (Note 4).

Water (500 ml.) and 500 ml. of ether are added to the residue, and the mixture is well stirred. The mixture is transferred to a 2-l. separatory funnel, and the two layers are carefully separated. The aqueous layer is extracted with two additional 250-ml. portions of ether. The combined organic layers are washed twice with 5% sodium hydroxide solution and then with water, dried over anhydrous magnesium sulfate, filtered, and evaporated on a rotary evaporator. The dark residue is transferred to a distillation flask and distilled through a 30-cm. Vigreux column to yield a crystalline fraction, b.p. 130–155° (13 mm.) (Note 5). Recrystallization of this distillate from ligroin gives 68–80 g. (45–53%, based on chloramine added) of 1,3-dihydro-3,5,7-trimethyl-2H-azepin-2-one as a fluffy white solid, m.p. 131–132°.

2. Notes

1. Suitable material is obtainable from the Aldrich Chemical Co., Milwaukee, Wisconsin.

2. The ethereal chloramine solution is conveniently prepared in this quantity according to the precise directions of Coleman and Johnson.[2] It is essential to the success of this reaction that their procedure be followed exactly.

3. The protective shield is recommended despite the fact that no fire or explosion has been observed in well over fifty such experiments by the submitter.

4. The phenol recovered at this stage is reusable in subsequent preparations.

5. It is not always necessary to distil the residue. The checkers obtained a tan crystalline residue that was recrystallized from about 2 l. of heptane to give 68 g. of colorless azepinone, m.p. 130–132°. An additional 12 g. of the azepinone with the same appearance and melting point was obtained by concentrating and cooling the heptane filtrate.

3. Methods of Preparation

1,3-Dihydro-2H-azepin-2-one has been prepared in a lengthy

five-step sequence by Vogel and Erb.[3] The present method,[4] the reaction of sodium 2,6-dialkylphenoxides with chloramine, easily affords the corresponding dihydroazepinones in good yield.

4. Merits of the Preparation

This reaction can be generally applied with equal success to other 2,6-dialkylphenols,[4] many of which are commercially available. Although the procedure cannot be extended to phenol or *o*-monosubstituted phenols (aminophenols result [5]), it represents a facile synthetic method for obtaining a ring system heretofore relatively unavailable. The dihydroazepinones in turn are excellent starting materials for the preparation of other novel heterocyclic systems such as 2,3-dihydro-1H-azepines,[6] 2-substituted-3H-azepines,[7] and derivatives of 2-azabicyclo[3.2.0]hept-6-ene.[8]

1. Research Laboratories of the Upjohn Company, Kalamazoo, Michigan. Present address: Department of Chemistry, The Ohio State University, Columbus, Ohio.
2. G. H. Coleman and H. L. Johnson, *Inorg. Syntheses,* 1, 59 (1939).
3. E. Vogel and R. Erb, *Angew. Chem.,* 74, 76 (1962); E. Vogel, R. Erb, G. Lenz, and A. A. Bothner-By, *Ann.,* 682, 1 (1965).
4. L. A. Paquette, *J. Am. Chem. Soc.,* 84, 4987 (1962); 85, 3288 (1963); L. A. Paquette and W. C. Farley, *J. Am. Chem. Soc.,* 89, 3595 (1967).
5. W. Theilacker and E. Wegner, *Angew. Chem.,* 72, 127 (1960).
6. L. A. Paquette, *Tetrahedron Lett.,* 2027 (1963); *J. Am. Chem. Soc.,* 86, 4096 (1964).
7. L. A. Paquette, *J. Am. Chem. Soc.,* 85, 4053 (1963); 86, 4096 (1964).
8. L. A. Paquette, *J. Am. Chem. Soc.,* 86, 500 (1964).

3,3′-DIHYDROXYBIPHENYL

(m,m′-Biphenol)

Submitted by J. F. W. McOmie and D. E. West [1]
Checked by J. E. Hiatt and K. B. Wiberg

1. Procedure

3,3′-Dimethoxybiphenyl [2] (8 g., 0.037 mole) is dissolved in 120 ml. of methylene chloride in a 250-ml. conical flask, and the flask is placed in an acetone-dry ice bath at −80°. The flask is fitted with an air condenser. A solution of 15.9 g. (6.0 ml., 0.063 mole) of boron tribromide (Notes 1, 2) in 40 ml. of methylene chloride (Notes 3, 4) is added carefully to the stirred solution through the condenser. When the addition is complete, a calcium chloride tube is fitted to the top of the air condenser in order to protect the reaction mixture from moisture. As the solution of boron tribromide is added, a white precipitate is formed. The reaction mixture is allowed to attain room temperature overnight with stirring, when a clear, brownish yellow solution is obtained. The reaction mixture is then hydrolyzed by careful shaking with 130 ml. of water, thus precipitating a white solid which is dissolved by the addition of 500 ml. of ether. The organic layer is separated and extracted with 240 ml. of 2N sodium hydroxide; the alkaline extract is neutralized with dilute hydrochloric acid, extracted with 300 ml. of ether, and the ether extract is dried over anhydrous magnesium sulfate. On removal of the ether under reduced pressure, a brownish yellow oil remains which soon crystallizes to give an off-white solid; this is recrystallized twice from hot benzene, the first time with the addition of charcoal, and gives 3,3'-dihydroxybiphenyl as white needles with a pinkish tint, m.p. 126–127° (Note 5). The yield is 5.4–6.0 g. (77–86%).

2. Notes

1. Boron tribromide of 99.9% purity, from Koch-Light Laboratories Ltd., Colnbrook, Bucks, England, was used.

2. Boron tribromide is a heavy, colorless liquid ($d = 2.6$) when pure but begins to decompose on exposure to light, liberating free bromine. It fumes vigorously in air, being rapidly hydrolyzed to boric acid, with the evolution of considerable heat.

3. Demethylation reactions proceed equally well using dry n-pentane or dry methylene chloride as the solvent for both the ether and the boron tribromide; methylene chloride, having by far the more powerful solvent action, is to be preferred.

4. When making up the solution of boron tribromide in methylene chloride, it has been found best to stand the vessel containing the solvent in an acetone-dry ice bath at −80° and to add the required amount (it is difficult to measure accurately) to the methylene chloride as rapidly as possible.

5. In order to obtain a perfectly white product, recrystallization from water is necessary;[3, 5] prismatic needles several centimeters long are obtained. The compound is moderately soluble in boiling water and slightly soluble in cold water.

3. Discussion

The above preparation of 3,3'-dihydroxybiphenyl is a good example of the utility of boron tribromide for the cleavage of aryl methyl ethers; it is based on the method of McOmie, Watts, and West.[4] 3,3'-Dihydroxybiphenyl has been prepared previously by diazotization [3] of 3,3'-diaminobiphenyl and subsequent boiling with water, by the fusion of biphenyl-3,3'-disulfonic acid with potassium hydroxide,[6] and by heating 3,3'-dimethoxybiphenyl with hydriodic acid.[5]

Almost all the methods previously employed [7] for the demethylation of aromatic methyl ethers have involved fairly high temperatures, e.g., hydrogen halides in water or acetic acid at reflux temperature, whereas the present method is effective at, or well below, room temperature although in a few cases it has been found necessary to boil the solution (b.p. of methylene chloride, 40°). Boron tribromide does not effect cleavage of methylenedioxy groups nor of diphenyl ethers. It can be used for the demethylation of aryl methyl ethers in the presence of many functional groups without affecting these.[4] It is especially valuable for the demethylation of iodinated ethers [4] and of methoxy biphenylenes [8] where the usual reagents are either ineffective or else cause decomposition. Boron tribromide was the reagent of choice for the final (demethylation) step in the synthesis of the naturally occurring macrolide, Zearalenone.[9]

1. Department of Organic Chemistry, The University, Bristol, England.
2. N. Kornblum, *Org. Syntheses*, Coll. Vol. **3**, 295 (1955).
3. R. Adams and N. Kornblum, *J. Am. Chem. Soc.*, **63**, 188 (1941).
4. J. F. W. McOmie and M. L. Watts, *Chem. Ind.* (*London*), 1658 (1963); J. F. W. McOmie, M. L. Watts, and D. E. West, *Tetrahedron*, **24**, 2289 (1968).
5. C. Haeussermann and H. Teichmann, *Ber.*, **27**, 2107 (1894).
6. G. Schultz and W. Kohlhaus, *Ber.*, **39**, 3341 (1906).
7. J. F. W. McOmie, *Adv. Org. Chem.*, **3**, (1963).
8. J. M. Blatchly, D. V. Gardner, J. F. W. McOmie, and M. L. Watts, *J. Chem. Soc.* (C), 1545 (1968).
9. D. Taub, N. N. Girotra, R. D. Hoffsommer, C. H. Kuo, H. L. Slates, S. Weber, and N. L. Wendler, *Tetrahedron*, **24**, 2443 (1968).

DIHYDROXYCYCLOPENTENE

(Cyclopentenediol)

Submitted by M. KORACH, D. R. NIELSEN, and W. H. RIDEOUT.[1]
Checked by WILLIAM G. DAUBEN and CLIFTON ASHCRAFT.

Caution! Reactions run with peracetic acid should be conducted with due regard to the properties of this reagent (Note 1).

1. Procedure

In a 1-l. three-necked flask equipped with a dropping funnel, a thermometer, and an efficient fractionation column fitted with either a vapor- or liquid-splitting head (Note 2) is placed 400 ml. of mineral oil. The oil is heated to 240–270° and dicyclopentadiene (Note 3) is added at the rate of 5–10 ml. per minute. The reflux ratio and the rate of addition of dicyclopentadiene are adjusted to maintain the distillation head temperature at 40°. The cyclopentadiene is collected in a Dry Ice-acetone receiver (Note 4).

In a 1-l. three-necked flask fitted with a sealed stirrer, a thermometer, and a connecting tube with parallel side arm to which is attached an addition funnel and a reflux condenser are placed 56 g. (0.81 mole) of 96% cyclopentadiene, 106 g. (1.0 mole) of anhydrous sodium carbonate, and 500 ml. of methylene chloride. A solution of 2 g. of sodium acetate in 76 g. of 40% peracetic acid (0.40 mole) (Note 5) is added, with stirring, over a period of 30–45 minutes, and the temperature is maintained at 20° by intermittent external cooling. The resulting mixture is stirred for an additional hour at room temperature (Note 6). The solid in the reaction is removed by suction filtration, and the filter cake is washed three times with 75 ml. of methylene chloride. The combined filtrate and washings are added, with rapid stirring, over a period of 1 hour, to 250 ml. of cold distilled water (maintained at 5–10° by external cooling) contained in a 2-l. three-necked flask fitted with a condenser, a sealed stirrer, and an addition funnel (Note 7). The stirring is continued for 1 hour as the temperature is allowed to rise to room temperature, the layers are separated, and the lower methylene chloride layer is extracted twice with 25 ml. of distilled water. The aqueous extracts are combined with the aqueous phase obtained from the hydrolysis reaction, and the combined solution is distilled at approximately 30 mm. pressure to remove the water (Note 8). The residue is distilled at reduced pressure to give 26–28 g. of colorless oil, b.p. 82–105°/ 1 mm. (Note 9). The yield of mixed cyclopentenediols is 65–70% based on the quantity of peracetic acid used.

The mixed cyclopentenediols can be separated by distillation through a 60-cm. spinning band column. The yield of pure 3-cyclopentene-1,2-diol, b.p. 65–68°/1 mm., n_D^{26} 1.4941–1.4951, is

4.5–5.5 g., and the yield of pure 2-cyclopentene-1,4-diol, b.p. 92–95°/1 mm., n_D^{26} 1.5000–1.5010, is 17–20 g. (Note 10).

2. Notes

1. When peracetic acid is used in organic synthesis the following precautions are recommended by the FMC Corporation, Inorganic Chemicals Division (Bulletins 4 and 69). All laboratory reactions and subsequent operations should be run behind a safety shield. The peracetic acid should be added to the organic material, never the reverse, and the rate of addition should be slow enough so that the peracid reacts as rapidly as it is added and no unreacted excess is allowed to build up. The reaction mixture should be stirred efficiently while the peracid is being added, and cooling facilities should be provided since most of the reactions of peracetic acid are exothermic. New or unfamiliar reactions, particularly those run at high temperatures, should be tried first on a very small scale. Reactions products should never be recovered from the final reaction mixture by distillation until all residual active oxygen compounds have been destroyed. This includes the hydrogen peroxide present in commercial 40% peracetic acid as well as the unreacted peracid. Decomposition may be readily accomplished by catalystic quantities of activated carbon or ferric sulfate added at 25–50°. The course of the decomposition may be followed by the titration described in Note 5. (See also Note 1, p. 904).

A serious accident has occurred when the reaction described in this procedure was run on a scale ten times that described and some of the above precautions were not observed (G. Benoy, private communication).

2. The checkers used a 1 x 24 in. column packed with glass helices.

3. Commercially available dicyclopentadiene of 95% or 70% purity may be used. The higher-purity material yields a less colored product.

4. Any ice present in the product can be removed by filtration of the cyclopentadiene through glass wool. Cyclopentadiene can be stored at Dry Ice temperatures in a tightly capped bottle for

several weeks without serious loss due to dimerization. Purification of a stored sample can be effected by distillation through a short Vigreux column at 10–30 mm. pressure and collection of the product in a receiver cooled by a Dry Ice-acetone bath until the temperature in the distilling flask rises to 10°. The distillate contains approximately the same concentration of cyclopentadiene as the freshly purified material (95–96%).

5. Peracetic acid (40%) is available commercially. Since the epoxycyclopentene reacts rapidly with water, it is desirable to keep the water content of the peracetic acid solution as low as possible. This is the reason for the use of the concentrated peracetic acid solution.

The peracetic acid content of the reagent may be determined by adding an aliquot to an equal volume of ice and water and titrating first with ceric sulfate solution until the orange color of the ceric ion remains (to eliminate hydrogen peroxide) and then adding potassium iodide and titrating with standard sodium thiosulfate solution.

6. The extent of peracetic acid consumption can be determined by titration of an aliquot as described in Note 4.

7. Distilled epoxycyclopentene can be hydrolyzed under identical conditions. However, distillation of the crude epoxide has occasionally resulted in a rapid, highly exothermic reaction when the pot temperature rises above 60°. The safest method for isolating the epoxide from the crude product is to remove the methylene chloride by distillation at atmospheric pressure until the pot temperature reaches 50°, then strip off crude epoxycyclopentene at reduced pressure and at a temperature below 50°. The apparatus should be shielded and the temperature in the distillation flask should be monitored. The crude distillate of epoxide and methylene chloride can be safely redistilled at 75–100 mm. pressure.

8. A rotatory evaporator has been found to be quite suitable for the rapid removal of water.

9. In some cases the distillate is yellow. The color can be removed by redistillation in the presence of 0.1% anhydrous sodium carbonate.

10. The purity of the distillation fractions can be determined by gas-liquid chromatography, using a 5 ft. x ¼ in.

column containing 20% Carbowax at a temperature of 200°.

3. Methods of Preparation

Cyclopentenediol isomers have previously been prepared by hydrolysis of acetates produced by reaction of dibromocyclopentene with potassium acetate in acetic acid; [2] by reaction of cyclopentene with selenium dioxide in acetic anhydride; [3] or by reaction of cyclopentadiene with phenyl iodosoacetate,[4] with lead tetraacetate,[5] or with peracetic acid in the absence of base.[6] Preparation of cyclopentenediol without intermediate formation of acetates has been accomplished by reaction of cyclopentadiene with hydrogen peroxide in the presence of osmium tetroxide in tert-butanol,[7] and by reaction of cyclopentadiene with peracetic acid in a methylene chloride suspension of anhydrous sodium carbonate, followed by hydrolysis of the resulting epoxycyclopentene.[8]

4. Merits of Preparation

The present method of preparation utilizes inexpensive, readily available, non-toxic reagents, is less laborious than previous methods, produces an easily purified product, and results in improved yields (65–70% vs. 10–50% for the older methods). It is a useful starting material for a variety of compounds as illustrated by the preparations of 2-cyclopentenone (p. 326) and 2-cyclopentene-1, 4-dione (p. 324).

1. Columbia-Southern Chemical Corp., subsidiary of Pittsburgh Plate Glass Co., P. O. Box 4026, Corpus Christi, Texas.
2. A. T. Blomquist and W. G. Mayes, *J. Org. Chem.*, 10, 134 (1945).
3. E. Dane, J. Schmitt, and C. Rautenstrauch, *Ann.*, 532, 29 (1937).
4. R. Criegee and H. Beucker, *Ann.*, 541, 218 (1939).
5. W. G. Young, H. K. Hall, Jr., and S. Winstein, *J. Am. Chem. Soc.*, 78, 4338 (1956).
6. L. N. Owen and P. N. Smith, *J. Chem. Soc.*, 4035 (1952).
7. N. A. Milas and L. S. Maloney, *J. Am. Chem. Soc.*, 62, 1841 (1940).
8. M. Korach, D. R. Nielsen, and W. H. Rideout, J. Am. Chem. Soc., 82, 4328 (1960).

1,2-DIMERCAPTOBENZENE

(*o*-Benzenedithiol)

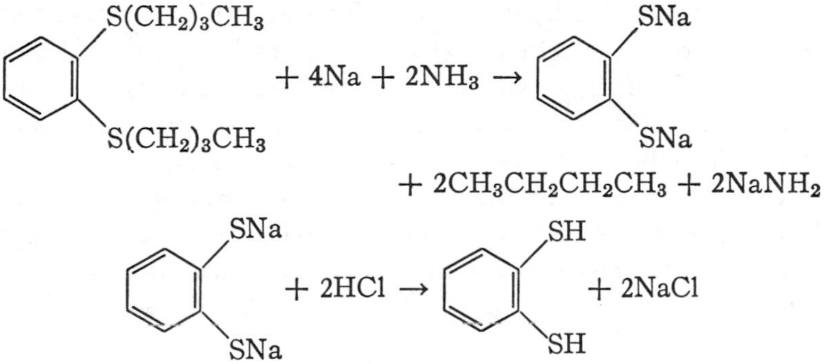

$$
\begin{array}{c}
\text{S(CH}_2\text{)}_3\text{CH}_3 \\
\text{S(CH}_2\text{)}_3\text{CH}_3
\end{array}
+ 4\text{Na} + 2\text{NH}_3 \rightarrow
\begin{array}{c}
\text{SNa} \\
\text{SNa}
\end{array}
$$

$$
+ 2\text{CH}_3\text{CH}_2\text{CH}_2\text{CH}_3 + 2\text{NaNH}_2
$$

$$
\begin{array}{c}
\text{SNa} \\
\text{SNa}
\end{array}
+ 2\text{HCl} \rightarrow
\begin{array}{c}
\text{SH} \\
\text{SH}
\end{array}
+ 2\text{NaCl}
$$

Submitted by Aldo Ferretti.[1]
Checked by William E. Parham, Wayland E. Noland, and James R. Throckmorton.

1. Procedure

A 200-ml. two-necked flask is fitted with an efficient Dry Ice-isopropyl alcohol condenser connected to a soda-lime tube, a magnetic stirrer, and a gas inlet tube. Isopropyl alcohol and Dry Ice are added to the condenser while the flask and condenser are flushed with dry nitrogen. The flask is immersed in a Dry Ice-isopropyl alcohol bath, and a vigorous stream of dry ammonia is introduced into the system. When about 80 ml. of liquid ammonia is condensed, the gas inlet tube is replaced with a ground-glass stopper. The cooling bath is removed, stirring is started, and 5.1 g. (0.020 mole) of 1,2-bis(*n*-butylthio)benzene (Note 1) is quickly introduced (Note 2).

Sodium is now added in small pieces; the solution is allowed to decolorize before each successive piece is added. A water bath is placed occasionally under the flask to ensure continuous ebullition of ammonia. The blue color will persist for at least 15 minutes after 1.6 g. (0.070 g. atom) of sodium has been added. The excess sodium is then destroyed by *cautious* addition of 6 g. (0.11 mole) of ammonium chloride, with stirring. Cooling and stirring are stopped, and a slow stream of argon is passed in for a period of about 12–15 hours. The white solid residue is transferred to a beaker, and 300 g. of ice water is added, together with sufficient pellets of sodium hydroxide to make the solution alkaline. The alkaline solution is then extracted twice with ether and the ether extracts discarded. The solution is then acidified to Congo red with cold 1:1 (by volume) hydrochloric acid and extracted three times with ether. The ether extracts are combined, washed with water, and dried over anhydrous sodium sulfate. The ether is evaporated and the 1,2-dimercaptobenzene is distilled under reduced pressure under an atmosphere of nitrogen, giving a product which boils at 95°/5 mm. and usually solidifies after distillation (Note 3). The yield is 1.6–2.4 g. (56–85%) (Note 4).

2. Notes

1. The preparation of 1,2-bis(n-butylthio)benzene is described elsewhere in this volume.[2]

2. 1,2-Bis(n-butylthio)benzene is only slightly soluble in liquid ammonia. Stirring must be very efficient during the addition and subsequently during the reaction to prevent the drops of 1,2-bis(n-butylthio)benzene from collecting as a solid mass. If this happens, the time necessary for completion of the reaction, and the quantity of sodium necessary for dealkylation, must be increased. The checkers found that the 1,2-bis(n-butylthio)benzene invariably collected as several solid masses and that it was always necessary to add additional sodium, about 0.9 g.

3. The reported melting point is 27–28°.[3]

4. The reported yield is 2.0 g. (70%) of an oil, b.p. 102°/6.5 mm.[3] A similar procedure has been used by the submitter to prepare 1,4-dimercaptobenzene, 2,5-dimercaptotoluene, 1,3,5-trimercaptobenzene, 2,4,6-trimercaptomesitylene, and 4,4'-dimercaptobiphenyl.[3]

3. Methods of Preparation

The present procedure is that of Adams and Ferretti.[3] Another method is the reduction of 1,2-benzenedisulfonyl chloride with zinc powder.[4] In a third method [5] a 2-aminomercaptobenzene is diazotized and converted to an intermediate xanthate and then to the corresponding mercaptosulfonic acid. The latter can be converted to the dimercaptan either by: (1) oxidation to a disulfonic acid, conversion to the disulfonyl chloride, and reduction to the dimercaptan, or (2) mild oxidation to the corresponding disulfide, conversion to the sulfonyl chloride disulfide, and reduction to the dimercaptan.

4. Merits of the Preparation

The present procedure, when combined with the accompanying preparation of aryl sulfides (p. 107), provides a convenient and general method for preparing aryl mercaptans from aromatic halides.

1. Via Martiri Triestini, 12, Milan, Italy.
2. R. Adams, W. Reifschneider, and A. Ferretti, this volume, p. 107.
3. R. Adams and A. Ferretti, *J. Am. Chem. Soc.*, 81, 4939 (1959).
4. W. R. H. Hurtley and S. Smies, *J. Chem. Soc.*, 1821 (1926).
5. P. C. Guha and M. N. Chakladar, *J. Indian Chem. Soc.*, 2, 318 (1925).

DIMESITYLMETHANE

(Methane, dimesityl-)

$$2 \quad \text{(mesitylene)} \quad + \frac{1}{x}\,(HCHO)_x \xrightarrow{\;HCO_2H\;}$$

$$CH_3\text{—}\text{(ring)}\text{—}CH_2\text{—}\text{(ring)}\text{—}CH_3 + H_2O$$

Submitted by JOHN H. CORNELL, JR., and MORTON H. GOLLIS.[1]
Checked by WILLIAM E. PARHAM and JAMES TOGEAS.

1. Procedure

Into a 5-l. round-bottomed flask fitted with stirrer, thermometer, and reflux condenser are introduced 165 g. (5 moles) of 91% paraformaldehyde (Note 1) and 1250 g. (24 moles) of 88% formic acid (Note 2). The mixture is heated to 80° with stirring and is stirred until the paraformaldehyde has dissolved. To the stirred mixture is rapidly added 1.8 kg. (15 moles) of mesitylene and the whole heated under reflux for 6 hours (Note 3).

On cooling to room temperature, a large mass of dirty-yellow crystals separates. The liquid layers are decanted from the yellow solid, and the aqueous (lower) layer is separated and discarded. The solid is washed in the reaction flask by stirring with 500 ml. of benzene. This slurry of solid in benzene is filtered and the solid sucked dry on a Büchner funnel. This filtrate is combined with the upper organic layer from the original reaction mixture, and the combined benzene solution is washed with 500 ml. of water, 500 ml. of 2–3% aqueous sodium carbonate (Note 4), and 200 ml. of saturated sodium chloride solution. Benzene and water are removed from this solution by distillation at atmospheric pressure. The still residue is cooled to room temperature, and precipitated solid is removed by filtration and added to the large crop of solid from the original reaction mixture. The combined solids are washed twice with 300 ml. of water, once with 400 ml. of 2–3% aqueous sodium carbonate, and once with 300–400 ml. of water and sucked dry on a Büchner funnel.

The yield is 779 g. of crude dimesitylmethane (62% of theoretical) melting at 128.5–131°, uncor.; its purity as determined by vapor-phase chromatography is 99.9 mole per cent (Note 5).

2. Notes

1. The checkers used 150 g. (5 moles) of paraformaldehyde obtained from Eastman Organic Chemicals.

2. Contact with formic acid and inhalation of its vapors should be avoided.

3. When a smaller ratio of mesitylene to formaldehyde was used, a considerable amount of polymeric residue was formed and the yield was very much reduced.

4. It is advisable to add the sodium carbonate solution cautiously and with good agitation to avoid a violent evolution of carbon dioxide.

5. The crude product is pure enough for most purposes. However, for catalytic reduction to bis(2,4,6-trimethylcyclohexyl)-methane, the residual acid must be removed by dissolving the solid in hot benzene and stirring or shaking with dilute aqueous sodium carbonate solution until the washings are basic; this is followed by a water wash and drying.

The solid can be recrystallized from boiling benzene and precipitated with about 0.15 part of boiling methanol to give white platelets (68%), m.p. 133–135°, plus a second, less pure crop (22%) melting at 128–133°. Reported for dimesitylmethane,[2] m.p. 134.4–135.4°, b.p. 212–213°/21 mm.

The reaction has been scaled up tenfold using a 50-l. flask without changes in procedure and in the same yield.

3. Methods and Merits of Preparation

Substituted diarylmethanes have been prepared from formaldehyde and a variety of its derivatives. Sulfuric acid is a common catalyst. [2,3] The procedure described is based on the general method of Gordon, May, and Lee.[4] Formic acid is preferable to sulfuric acid as a catalyst because it is capable of acting as a solvent as well, thus eliminating troublesome emulsions. Side reactions such as sulfonation are avoided.

1. Monsanto Chemical Company, Special Projects Department, Boston, Mass.
2. C. M. Welch and H. A. Smith, *J. Am. Chem. Soc.*, 73, 4391 (1951).
3. I. G. Matveev, D. A. Drapkina, and R. L. Globus, *Trudy Vsesoyuz. Nauch.-Issledovatel. Inst. Khim. Reaktivov*, No. 21, 83, (1956) [*C. A*, 52, 15474 (1958)].
4. L. B. Gordon, P. D. May, and R. J. Lee, *Ind. Eng. Chem.*, 51, 1275 (1959).

7,7-DIMETHOXYBICYCLO[2.2.1]HEPTENE

(2-Norbornen-7-one dimethyl acetal)

Submitted by P. G. GASSMAN and J. L. MARSHALL [1]
Checked by WILLIAM G. DAUBEN and JAMES L. CHITWOOD

1. Procedure

Caution! Most polychlorinated compounds show some toxicity. These compounds should be handled in a hood.

A. *5,5-Dimethoxy-1,2,3,4-tetrachlorocyclopentadiene.* In a 3-l. three-necked flask fitted with a condenser (Note 1), an addition funnel, and a mechanical stirrer (Note 2) are placed 254 g. (0.93 mole) of hexachlorocyclopentadiene (Note 3) and 800 ml. of methanol (Note 4). The stirrer is started, and a solution of 120 g. (2.14 moles) of potassium hydroxide in 600 ml. of methanol is added dropwise over a period of 2 hours (Note 5). The reaction mixture is stirred for an additional 2 hours and then poured over 3 l. of chopped ice. After the ice has melted, the mixture is extracted with three 250-ml. portions of dichloromethane. The combined extracts are dried over anhydrous magnesium sulfate and concentrated to a yellow syrup on a rotary evaporator (Note 6). The residue is distilled through a 12-in. vacuum-jacketed Vigreux column to yield 187–189 g. (76–77%) of 5,5-dimethoxy-1,2,3,4-tetrachlorocyclopentadiene as a viscous, yellow-tinted oil, b.p. 79–84° (0.6 mm.) (Note 7).

B. *7,7-Dimethoxy-1,2,3,4-tetrachlorobicyclo[2.2.1]hept-2-ene.* A large Pyrex gas washing bottle with fritted-glass inlet (Note 8) is fitted with a condenser and a drying tube. In the bottle is placed 189 g. (0.72 mole) of 5,5-dimethoxy-1,2,3,4-tetrachlorocyclopentadiene, and a slow stream of nitrogen and ethylene is passed through the fritted-glass inlet (Note 9). The bottle is heated to 180–190° by means of an oil bath. The color of the liquid changes from yellow to reddish brown as ethylene is bubbled through the reaction mixture at this temperature for 6 hours (Note 10). The reaction mixture is cooled and distilled through a 12-in. vacuum-jacketed Vigreux column to yield 155–165 g. (73–78%) of a yellow syrup, b.p. 70–75° (0.15 mm.) (Note 11).

C. *7,7-Dimethoxybicyclo[2.2.1]heptene.* A 3-l. three-necked flask is equipped with a sealed Hershberg stirrer,[2] a condenser fitted with a nitrogen inlet to maintain a slight positive pressure, and a pressure-equalizing dropping funnel. The flask is placed in a heating mantle, and into it are placed 1.5 l. of tetrahydrofuran, 130 g. (5.7 g. atoms) of sodium chopped into 5-mm. cubes, and 190 ml. (150 g., 2.0 moles) of *t*-butyl alcohol. This mixture is stirred vigorously and brought to gentle reflux (Note 12). As soon as refluxing occurs, 106 g. (0.36 mole) of 7,7-dimethoxy-1,2,3,4-tetrachlorobicyclo[2.2.1]hept-2-ene is added dropwise over a 2-hour period (Note 13). The mixture is heated under reflux for 38 hours, cooled to room temperature, and filtered through a wire screen to remove the unreacted sodium. The dark filtrate is refiltered by suction through Celite in a Buchner funnel (Note 14). The filtrate is mixed with 2 l. of chopped ice and 500 ml. of ether. The aqueous phase is separated (Note 15), and the organic phase is washed with 500-ml. portions of saturated aqueous sodium chloride until the washings are clear. The ethereal solution is dried over anhydrous magnesium sulfate and concentrated to a dark oil by removal of the ether by fractional distillation. The oil is fractionally distilled through a 6-in. Vigreux column to yield 17–24 g. (31–43%) of colorless liquid, b.p. 58–68° (17 mm.), n^{25}D 1.4598 (Notes 16 and 17).

2. Notes

1. If the directions are carefully followed, the condenser will not be utilized since it serves mainly as a safety device in case the reaction should become too exothermic.

2. A Hershberg nichrome wire stirrer [2] is well suited for this reaction.

3. The hexachlorocyclopentadiene was used as obtained from Matheson, Coleman and Bell.

4. Commercial grade methanol was used.

5. This reaction mixture should not be cooled initially because an uncontrollable exothermic reaction will occur if a large concentration of alkoxide builds up.

6. Concentration at 100° (30 mm.) is necessary for removal of most of the dichloromethane.

7. The submitters have obtained an 86% yield of product, b.p. 79–91° (0.6 mm.). They have also found that the reaction may be scaled up fivefold if 4 hours is taken for the addition of the methanolic base. No danger exists if the temperature is maintained between 50° and 60°.

8. Pyrex gas washing bottle, Corning No. 31750, was used.

9. The checkers found that best results were obtained when the slowest detectable nitrogen flow was used with a fairly rapid ethylene flow (about 1 in. of foam in the gas washing bottle at the reaction temperature).

10. The course of the reaction is readily followed by n.m.r. spectroscopy. The spectrum of the starting material has a singlet at δ 3.30 p.p.m., whereas that of the product has two singlets at δ 3.50 and 3.55 p.p.m. The time required for complete

reaction depends on the flow rate of ethylene and nitrogen. The reaction should be allowed to continue until all the starting material is consumed.

11. The distillation should be conducted carefully since the yield in the next step depends on the purity of the material used.

12. Occasionally the pieces of sodium may start to fuse together. This difficulty may be avoided by bringing the mixture to a gentle reflux and stirring vigorously. Once the addition of the chlorinated compound is started, fusing of the sodium ceases. When the reaction is finished, the sodium pieces often fuse into a single large chunk.

13. Unless the reaction mixture is heated *before* the addition process is started, there may be an initial induction period that may cause the reaction to become extremely vigorous once the mixture heats to reflux temperature.

14. The checkers found this last filtration to be a very time-consuming and cumbersome operation. They found it preferable to omit it; they cautiously added methanol [3] to decompose any traces of sodium and poured the resulting solution directly onto the chopped ice.

15. In the separation of the organic and aqueous phases it is often initially very difficult to discern the phase separation because of the dark color of the reaction mixture. The submitters found that, if phase separation cannot be detected under ordinary light, it can usually be seen by the use of an ultraviolet scanning lamp.

16. The submitters report that 65% yields of product, b.p. 61–71° (18 mm.), can be obtained by workers with experience with this reaction.

17. The product is rather volatile, and care should be taken in its handling and storing.

3. Methods of Preparation

The procedures described for the preparation of 5,5-dimethoxy-1,2,3,4-tetrachlorocyclopentadiene and 7,7-dimethoxy-1,2,3,4-tetrachlorobicyclo[2.2.1]hept-2-ene are essentially those of Newcomer and McBee [4] and of Hoch,[5] respectively. The dechlorination is a modification of an analogous dechlorination carried out by Bruck, Thompson, and Winstein.[3] The overall procedure is that of Gassman and Pape.[6]

4. Merits of the Preparation

The reactions include an unusual Diels-Alder reaction and a very useful synthetic method, the dechlorination of polychlorinated compounds. At the present time this procedure is the best one available for the removal of chlorine from an organic molecule. The end product, 7,7-dimethoxybicyclo[2.2.1]heptene, is an interesting and useful intermediate in bicyclic chemistry; it has a reactive double bond and a protected carbonyl group in the 7-position.

1. Department of Chemistry, The Ohio State University, Columbus, Ohio 43210.
2. P. S. Pinkney, *Org. Syntheses*, Coll. Vol. **2**, 116 (1943).
3. P. Bruck, D. Thompson, and S. Winstein, *Chem. Ind.* (*London*), 405 (1960).
4. J. S. Newcomer and E. T. McBee, *J. Am. Chem. Soc.*, **71**, 946 (1949).
5. P. E. Hoch, *J. Org. Chem.*, **26**, 2066 (1961).
6. P. G. Gassman and P. G. Pape, *J. Org. Chem.*, **29**, 160 (1964).

trans-4,4′-DIMETHOXYSTILBENE *

(*trans*-Stilbene, 4,4′-dimethoxy-)

Submitted by J. W. A. FINDLAY and A. B. TURNER [1]
Checked by R. E. IRELAND and G. BROWN

1. Procedure

A solution of 100 mg. (0.41 mmole) of 4,4′-dimethoxybibenzyl (Note 1) in 1.5 ml. of anhydrous dioxane (Note 2) was placed in a 10-ml. round-bottomed flask. To this was added 103 mg. (0.45 mmole) of 2,3-dichloro-5,6-dicyano-1,4-benzoquinone (DDQ; Note 3) dissolved in 1.5 ml. of anhydrous dioxane. The flask was fitted with a reflux condenser and heated in an oil bath at 105° for 18 hours. The solution, which was initially deep green, became pale yellow as the hydroquinone crystallized out. The mixture was cooled, and the solid was filtered off. It was washed with 1 ml. of warm benzene followed by 6 ml. of warm chloroform (Note 4), and dried at 100° to give 95 mg. (91%) of pure 2,3-dichloro-5,6-dicyanohydroquinone (Note 5). The filtrate and washings were combined and evaporated under reduced pressure. The semisolid residue was dissolved in 5 ml. of ethyl acetate and passed through a short column of neutral alumina (2.0 g.; Note 6). The column was eluted with 100 ml. of ethyl acetate (Note 7). Evaporation of the solvent under reduced pressure left the crude product, which was recrystallized from 35 ml. of ethanol to give 82–84 mg. (83–85%) of *trans*-4,4'-dimethoxystilbene as colorless plates, m.p. 212–213.5°.

2. Notes

1. The starting bibenzyl was prepared from *p*-methoxybenzyl chloride by a modified Wurtz reaction.[2] The checkers found the procedure described by Buu-Hoï and Lavit [2] inadequate and used the copper(I) chloride-catalyzed coupling of *p*-methoxybenzylmagnesium chloride in its place.

2. Dioxane was purified by the method of Vogel [3] and stored over molecular sieves.

3. DDQ was obtained from Koch-Light Laboratories, Ltd. It can be recrystallized from benzene if required.

4. Washing with chloroform is necessary to dissolve some of the stilbene which crystallizes out with the hydroquinone. In many reactions, washing the hydroquinone with hot benzene is sufficient, as the dehydrogenation products crystallize to a limited extent from dioxane.

5. The amount of precipitated hydroquinone is a convenient measure of the extent of hydrogen transfer. DDQ is readily regenerated in good yield from the hydroquinone by oxidation with nitric acid.[4]

6. Woelm neutral alumina, activity grade 1.

7. The volume of ethyl acetate required to elute the product can be reduced considerably for more soluble products.

3. Discussion

DDQ was first introduced for the dehydrogenation of hydro-aromatic compounds, such as tetralin and bibenzyl, which yield naphthalene and stilbene, respectively.[5] A benzene ring or an olefinic bond provides sufficient activation, although it is sometimes difficult to force the reaction to completion. This high-potential quinone has since found wide application,[6] particularly in the steroid field, and its scope has been extended by the dehydrogenation of carbonyl compounds (ketones and lactones) and alcohols. DDQ is also useful for preparing stable cations and radicals. These reactions are commonly carried out in refluxing benzene or dioxane, and the procedure described here is a general one. An alternative workup procedure involves washing with alkali.

A number of compounds react rapidly with DDQ at room temperature. They include allylic and benzylic alcohols, which can thus be selectively oxidized, and enols and phenols,[7] which undergo coupling reactions or dehydrogenation, depending on their structure. Rapid reaction with DDQ is also often observed in compounds containing activated tertiary hydrogen atoms.[8] The workup described here can be used in all these cases.

A number of side products can arise with this quinone. They include Diels-Alder adducts (DDQ is a powerful dienophile) and Michael adducts derived from the hydroquinone.

General methods for the preparation of *trans*-stilbenes have been covered previously.[9] 4,4'-Dimethoxystilbene has been prepared from deoxyanisoin and n-propylmagnesium iodide,[10] by treatment of thiophenol with 2-bromo-1,1-di-*p*-methoxy-phenylethane,[11] and by the action of nitrous acid on the corresponding amine.[12]

1. Chemistry Department, University of Aberdeen, Scotland.
2. N. P. Buu-Hoi and D. Lavit, *Bull. Soc. Chim. France*, 292 (1958).

3. A. I. Vogel, "Practical Organic Chemistry," 3rd ed., John Wiley & Sons, Inc., New York, 1956, p. 177.
4. D. Walker and T. D. Waugh, *J. Org. Chem.*, **30**, 3240 (1965).
5. E. A. Braude, A. G. Brook and R. P. Linstead, *J. Chem. Soc.*, 3569 (1954).
6. D. Walker and J. D. Hiebert, *Chem. Rev.*, **67**, 153 (1967).
7. J. W. A. Findlay and A. B. Turner, *J. Chem. Soc.* (C), 23 (1971).
8. W. Brown and A. B. Turner, *J. Chem. Soc.* (C), 2057 (1971).
9. R. L. Shriner and A. Berger, *Org. Syntheses*, Coll. Vol. 3, 786 (1955).
10. J. Cymerman-Craig, D. Martin, M. Moyle, and P. C. Wailes, *Australian J. Chem.*, **9**, 373 (1956).
11. W. Tadros and G. Aziz, *J. Chem. Soc.*, 2684 (1961).
12. R. Quelet, J. Hoch, C. Borgel, M. Mansouri, R. Pineau, E. Tchiroukine, and N. Vinot, *Bull. Soc. Chim. France*, 26 (1956).

6-(DIMETHYLAMINO)FULVENE

$$(CH_3)_2NCH\overset{O}{\|} + (CH_3O)_2SO_2 \longrightarrow \underset{H_3C}{\overset{H_3C}{>}}\overset{+}{N}=C\underset{OCH_3}{\overset{H}{<}} \quad CH_3OSO_3^-$$

$$\xrightarrow{C_5H_5Na} \quad \text{(cyclopentadienyl)}=C\underset{N(CH_3)_2}{\overset{H}{<}} \quad + \quad CH_3OSO_3Na \quad + \quad CH_3OH$$

Submitted by K. Hafner, K. H. Vöpel, G. Ploss,
and C. König [1]
Checked by S. S. Olin and Ronald Breslow

1. Procedure

A. *N,N-Dimethylformamide-dimethyl sulfate complex.* In a 500-ml. four-necked flask equipped with mechanical stirrer, reflux condenser with calcium chloride drying tube, dropping funnel, and thermometer is placed 73 g. (1.0 mole) of dimethylformamide, and 126 g. (1.0 mole) of dimethyl sulfate is added dropwise with stirring at 50–60° (Note 1). After the addition is complete, the mixture is heated for another 2 hours at 70–80°. The dimethylformamide complex forms as a viscous, colorless or pale yellow ether-insoluble oil.

B. *6-(Dimethylamino)fulvene.* A 1-l. four-necked flask is equipped with mechanical stirrer, dropping funnel with calcium chloride drying tube, thermometer, and nitrogen delivery apparatus (Note 2). The flask is flushed with dry nitrogen, and in it is placed 1.0 mole of cyclopentadienylsodium [2] in 700 ml. of tetrahydrofuran (Note 3). The dimethylformamide-dimethyl sulfate complex prepared above is transferred to the dropping funnel and added slowly with stirring under nitrogen to the cyclopentadienylsodium at $-10°$ (ice-salt bath). During the addition the temperature is kept below $-5°$. After the addition is complete, the mixture is stirred at $20°$ for 2 hours. The solution is filtered (with suction) from the precipitated sodium methyl sulfate, which is washed with another 200 ml. of tetrahydrofuran, and the combined tetrahydrofuran solutions are concentrated under reduced pressure. The residue is a dark brown oil which solidifies on cooling.

The crude product is crystallized after treatment with activated carbon from *ca.* 1.5 l. of petroleum ether (b.p. 60–80°) or 800 ml. of cyclohexane. From the orange-yellow solution 84 g. (69%) of 6-(dimethylamino)fulvene separates in yellow leaflets, m.p. 67–68° (Note 4). Concentration of the filtrate and further recrystallization of the residue from petroleum ether or cyclohexane gives an additional 8 g. of product. The combined yield is 92 g. (76%).

2. Notes

1. Dimethylformamide and dimethyl sulfate must be purified by distillation in the absence of moisture.

2. The nitrogen delivery apparatus has been completely described.[2]

3. Air and moisture must be carefully excluded from the reactants during the course of this preparation.

4. 6-(Dimethylamino)fulvene is light-sensitive and is stored in brown bottles.

3. Methods of Preparation

N,N-Dimethylaminoethoxymethylium fluoborate [3] can be used instead of N,N-dimethylaminomethoxymethylium methyl sulfate [4] to prepare 6-(dimethylamino)fulvene.[5] The same fulvene is also obtained from the condensation of cyclopentadiene with diethoxy(dimethylamino)methane.[6]

4. Merits of the Preparation

This procedure illustrates formylation by N,N-dimethylamino-methoxymethylium methyl sulfate, a compound which can be produced readily by reaction of easily available materials. 6-(Dimethylamino)fulvene is a useful intermediate for the synthesis of various fused-ring nonbenzenoid aromatic compounds.

1. Institute for Organic Chemistry, University of Munich, Munich, Germany.
2. K. Hafner and H. Kaiser, this volume, p. 1088.
3. H. Meerwein, P. Borner, O. Fuchs, H. J. Sasse, H. Schrodt, and J. Spille, *Ber.*, **89**, 2060 (1956).
4. H. Bredereck, F. Effenberger, and G. Simchen, *Ber.*, **96**, 1350 (1963).
5. K. Hafner, K. H. Vöpel, G. Ploss, and C. König, *Ann.*, **661**, 52 (1963).
6. H. Meerwein, W. Florian, N. Schön, and G. Stopp, *Ann.*, **641**, 1 (1961).

N,N-DIMETHYLAMINOMETHYLFERROCENE METHIODIDE

{Iron, cyclopentadienyl[(dimethylaminomethyl)cyclo-pentadienyl]-, methiodide}

Submitted by DANIEL LEDNICER and CHARLES R. HAUSER.[1]
Checked by B. C. McKUSICK, W. A. SHEPPARD,
R. D. VEST, and H. F. MOWER.

1. Procedure

Caution! Bis(dimethylamino)methane is a potent lachrymator, so it should be handled only in a hood.

Ferrocene (46.4 g., 0.250 mole) (Note 1) is added to a well-stirred solution of 43.2 g. (0.422 mole) of bis(dimethylamino)-methane (Note 2) and 43.2 g. of phosphoric acid in 400 ml. of acetic acid in a 2-l. three-necked round-bottomed flask equipped with a condenser, a nitrogen inlet, and a mechanical stirrer (Note 3). The resulting suspension is heated on a steam bath under a slow stream of nitrogen (Note 4) for 5 hours (Note 5). The reaction mixture, a dark-amber solution, is allowed to cool to room temperature and is diluted with 550 ml. of water. The unreacted ferrocene is removed by extracting the solution with three 325-ml. portions of ether. The aqueous solution is then cooled in ice water and made alkaline by the addition of 245 g. of sodium hydroxide pellets. The tertiary amine separates from the alkaline solution as an oil in the presence of some black tar (Note 6). The mixture is extracted with three 500-ml. portions of ether. The organic solution is washed with water and dried over sodium sulfate. Crude dimethylaminomethylferrocene is obtained as a dark-red mobile liquid when the solvent is removed at the water pump (Note 7).

To a gently swirled solution of the crude amine in 54 ml. of methanol is added 54 ml. (123 g., 0.87 mole) of methyl iodide. The solution is heated on a steam bath for 5 minutes and cooled to room temperature, and 800 ml. of ether is added. The methiodide, which separates as an oil, crystallizes on being scratched. The solid is collected on a Büchner funnel, washed with ether, and dried at 20–50 mm. for several hours at room temperature to yield 65–78 g. (68–81%; Note 8) of N,N-dimethylaminomethylferrocene methiodide as an orange powder, m.p. 200° (dec.) (Note 9).

2. Notes

1. The ferrocene may be prepared by Wilkinson's procedures [2] or it may be purchased from Matheson, Coleman and Bell, East Rutherford, New Jersey, and other companies.

2. The amine, under the name N,N,N′,N′-tetramethylmethylenediamine, may be purchased from Ames Laboratories, South Norwalk, Connecticut. The checkers prepared it by the following procedure. A solution of 60.7 g. (0.75 mole) of 37% aqueous formaldehyde solution is placed in an 800-ml. beaker equipped with a mechanical stirrer and thermometer, and cooled in an ice bath. Two hundred seventy-one grams (1.50 moles) of a 25% aqueous solution of dimethylamine is added to this solution at a rate such that the reaction temperature is kept below 15°. The solution is stirred for 30 minutes after the addition is complete, and potassium hydroxide pellets (approximately 150 g.) are added in portions until the reaction mixture separates into two layers. The upper layer is separated, dried over potassium hydroxide pellets overnight, and distilled to give 59–64 g. (77–83%) of bis(dimethylamino)methane, b.p. 83–84°.[3]

3. The mixing of the amine and acids is exothermic. It is necessary to add the amine dropwise to the solution of acids with stirring and cooling in an ice bath.

4. Since ferrocene and many of its derivatives are easily oxidized by air in the presence of acids, nitrogen is passed in at a rate sufficient to exclude air from the system.

5. It was found that the reaction is essentially complete after 5 hours. Further heating does not affect the yield.

6. The checkers found that the solution turned into a gel after the addition of sodium hydroxide. They obtained satisfactory results by adding 200 ml. of water to the gel, which made it fluid enough to be extractable with ether.

7. Distillation of the tertiary amine before methiodide formation does not significantly affect the yield or purity of the quaternary salt. The amine boils at 91–92°/0.45 mm.

8. The submitters report yields as high as 89%.

9. The melting point of this compound is ill-defined by reason of considerable darkening and shrinking that start at 175°. The product thus obtained is sufficiently pure for use in further reactions.[4]

3. Methods of Preparation

This procedure is based on the method of Lindsay and Hauser [3] as modified slightly by Osgerby and Pauson.[5] N,N-dimethyl-aminomethylferrocene methiodide has also been prepared by heating formylferrocene with dimethylamine and hydrogen in the presence of Raney nickel catalyst to give dimethylamino-methylferrocene, which was quaternized with methyl iodide.[6]

Essentially the present procedure converted 1-methylindole to 1-methyl-3-(N,N-dimethylaminomethyl)indole [7] and α-methyl-styrene to α-(N,N-dimethylaminoethyl)styrene.[8]

1. Department of Chemistry, Duke University, Durham, North Carolina. The work was supported by the Office of Ordnance Research, U. S. Army.
2. G. Wilkinson, *Org. Syntheses,* Coll. Vol. 4, 473 (1963).
3. J. K. Lindsay and C. R. Hauser, *J. Org. Chem.,* 22, 355 (1957).
4. D. Lednicer, J. K. Lindsay, and C. R. Hauser, *J. Org. Chem.,* 23, 653 (1958).
5. J. M. Osgerby and P. L. Pauson, *J. Chem. Soc.,* 656 (1958).
6. P. J. Graham, K. V. Lindsey, G. W. Parshall, M. L. Peterson, and G. M. Whitman, *J. Am. Chem. Soc.,* 79, 3416 (1957).
7. H. R. Snyder and E. L. Eliel, *J. Am. Chem. Soc.,* 70, 1703 (1948).
8. G. F. Hennion, C. C. Price, and V. C. Wolff, Jr., *J. Am. Chem. Soc.,* 77, 4633 (1955); C. J. Schmidle and R. C. Mansfield, *ibid.,* 77, 4636 (1955).

α-N,N-DIMETHYLAMINOPHENYLACETONITRILE

(Glycinonitrile, N,N-dimethyl-2-phenyl-)

$$C_6H_5CHO \xrightarrow{\text{NaHSO}_3} C_6H_5\underset{\underset{OH}{|}}{C}HSO_3Na \xrightarrow[\text{NaCN}]{\text{(CH}_3)_2\text{NH}} C_6H_5\underset{\underset{N(CH_3)_2}{|}}{C}HCN$$

Submitted by HAROLD M. TAYLOR and CHARLES R. HAUSER.[1]
Checked by W. BRUCE KOVER and JOHN D. ROBERTS.

1. Procedure

A mixture of 1.5 l. of water and 624 g. (6.00 moles) of sodium bisulfite in a 5-l. beaker equipped with a mechanical stirrer is stirred until solution is complete. Benzaldehyde (Note 1) (636 g., 6.00 moles) is added and the mixture is stirred for 20 minutes, during which time a slurry of the benzaldehyde-bisulfite addition product is formed. A 25% aqueous solution of dimethylamine (1100 g.) containing 275 g. (6.13 moles) of the amine is run in, and stirring is continued as most of the addition compound dissolves. The beaker is immersed in an ice bath, and 294 g. (6.00 moles) of sodium cyanide (*Caution! Toxic*) is added over a period of 20–25 minutes.

The ice bath is removed after addition of the sodium cyanide, and the mixture is stirred for 4 hours. The organic layer is separated, and the aqueous layer is extracted with three 500-ml. portions of ether. The combined ethereal extracts and organic layer are washed with two 100-ml. portions of cold water and dried over anhydrous magnesium sulfate. The ethereal solution is filtered, and the ether is removed at atmospheric pressure. The residue is transferred to a vacuum distillation system and distilled under reduced pressure (*Caution! See Note 2*). The yield of α-dimethylaminophenylacetonitrile boiling at 88–90°/1.9–2.1 mm. is 842–844 g. (87–88%) (Notes 3 and 4).

2. Notes

1. Eastman Kodak benzaldehyde (white label grade) was used without further purification.

2. Occasionally the odor of hydrogen cyanide can be detected during the distillation, even when a trap filled with sodium hydroxide pellets precedes the usual trap cooled in dry ice and acetone to protect the pump. For safety, the vacuum pump should be placed in a hood, or provision should be made for the pump exhaust to be vented into a hood or out-of-doors during the distillation.

3. Anhydrous dimethylamine has been used by the submitters in a slightly different procedure to give yields up to 95% of the theory.

4. The checkers carried out the preparation with one-half of the specified quantities without any decrease in the yield.

3. Methods of Preparation

The procedure described above is a modification of that of Hauser, Taylor, and Ledford [2] and of Luten [3] which avoids use of anhydrous dimethylamine. It is related to the procedure of Goodson and Christopher [4] that employs benzaldehyde, aqueous dimethylamine hydrochloride, and potassium cyanide.

The product can also be prepared from benzaldehyde, dimethylamine, and potassium cyanide in cold acetic acid and aqueous ethanol.[5]

4. Merits of the Preparation

The method can be used to prepare a number of α-aminonitriles from aliphatic or aromatic aldehydes and ketones and secondary aliphatic amines.[6]

The nitrile group of α-N,N-dimethylphenylacetonitrile can generally be replaced by an alkyl or aryl group of a Grignard reagent to form the corresponding tertiary amines.[4,7] The α-hydrogen of the aminonitrile can be alkylated,[2,7] and the resulting alkylation product can be converted to enamines [2] or to ketones.[7]

1. Department of Chemistry, Duke University, Durham, North Carolina.
2. C. R. Hauser, H. M. Taylor, and T. G. Ledford, *J. Am. Chem. Soc.*, 82, 1786 (1960).
3. D. B. Luten, Jr., *J. Org. Chem.*, 3, 588 (1939).
4. L. H. Goodson and H. Christopher, *J. Am. Chem. Soc.*, 72, 358 (1950).
5. T. S. Stevens, J. M. Cowan, and J. MacKinnon, *J. Chem. Soc.*, 2568 (1931).
6. See V. Migrdichian, "The Chemistry of Organic Cyanogen Compounds," Reinhold Publishing Company, New York, 1947, pp. 198-217.
7. H. M. Taylor and C. R. Hauser, *J. Am. Chem. Soc.*, 82, 1960 (1960).

DIMETHYL CYCLOHEXANONE-2,6-DICARBOXYLATE

(Cyclohexanone-2,6-dicarboxylic acid, dimethyl ester)

$$(CH_3O)_2Mg \ + \ CO_2 \longrightarrow CH_3OMgO_2COCH_3$$

Submitted by S. N. Balasubrahmanyam and M. Balasubramanian [1]
Checked by Frederick J. Sauter and Herbert O. House

1. Procedure

Caution! Since hydrogen is liberated in the first step of this reaction, it should be conducted in a hood. A dry 2-l. three-necked flask is fitted with a Trubore mechanical stirrer, an Allihn condenser, and a 1-l. pressure-equalizing dropping funnel, the top of which is fitted with a gas inlet tube. After 40.0 g. (1.64 g. atoms) of clean, dry magnesium ribbon (Note 1) has been placed in the flask, the system is flushed with nitrogen and 600 ml. of anhydrous methanol (Note 2) is added. As soon as the vigorous reaction begins, the nitrogen flow is stopped; if necessary, the reaction may be moderated by external cooling with wet towels. When the hydrogen evolution has ceased, a slow stream of nitrogen is passed through the reaction system and the condenser is replaced by a total condensation-partial take-off distillation head. The nitrogen flow is stopped, and the bulk of the methanol is distilled from the solution under reduced pressure (Note 3) with stirring while the reaction flask is heated to 50–55° with a water bath. This distillation is stopped when stirring of the pasty suspension of magnesium methoxide is no longer practical. Nitrogen is readmitted to the system, and the outlet from the distillation head is attached to a small trap containing mineral oil so that the volume of gas escaping from the reaction system can be estimated. Anhydrous dimethylformamide (700 ml.; Note 4) is added to the reaction flask, and the resulting suspension is stirred vigorously while a stream of anhydrous carbon dioxide (Note 5) is passed into the reaction vessel through the gas inlet tube attached to the dropping funnel. The dissolution of the carbon dioxide is accompanied by an exothermic reaction with the suspended magnesium methoxide to form a solution. When the absorption of carbon dioxide has stopped (Note 6), the colorless solution is heated with a mantle under a slow stream of anhydrous carbon dioxide gas until the temperature of the liquid distilling from the flask reaches 140°, indicating that the residual methanol has been removed from the reaction mixture. The flow of carbon dioxide is stopped, and a slow stream of nitrogen is passed through the system while the resulting solution (Note 7) is cooled below 100° with a water bath. Cyclohexanone (20.0 g., 0.204 mole) (Note 8) is added to the reaction mixture, and the solution is heated under reflux for 2 hours while a slow stream of nitrogen (2–3 bubbles per second) is passed over the reaction mixture. The resulting solution is cooled first to room temperature with a water bath and then to −5° with a dry ice-acetone bath (Note 9).

Meanwhile, 700 ml. of anhydrous methanol is placed in a 2-l. flask fitted with a gas inlet tube extending approximately 5 mm. below the surface of the methanol and a calcium chloride tube to protect the contents of the flask from atmospheric moisture. The methanol is cooled with an external cooling bath prepared from ice and calcium chloride (Note 10) and saturated with anhydrous hydrogen chloride (Note 10) (290–300 g. of hydrogen chloride is required). This solution is transferred to the dropping funnel by means of a gooseneck adapter and the methanolic hydrogen chloride solution is then added, dropwise and with stirring, to the reaction flask, the temperature of the reaction being maintained at $0° \pm 5°$ by means of a cooling bath. This addition is accompanied by the vigorous evolution of carbon dioxide and the separation of a white solid. After the addition is complete, the reaction mixture is allowed to stand overnight at room temperature and then the bulk of the methanol is removed from the solution by distillation under reduced pressure with stirring. During the distillation the temperature of the reaction mixture is kept below 55°. The remaining suspension is poured into a 4-l. beaker containing 1 kg. of crushed ice. The crude product separates as shiny white flakes which are collected on a filter and washed with water. A small second crop of the crude material may be obtained by cooling the aqueous filtrates to 0° overnight. The total crude product (25–26 g.) is dissolved in 250 ml. of boiling methanol, and this solution is concentrated to 125–150 ml. and allowed to cool. The keto diester separates as flat white needles, m.p. 128–132° (Note 11), yield 19.3–19.7 g. (44–45%). Concentration of the mother liquors affords an additional 2.2–2.5 g. of crude product, m.p. 122–128° (Note 11).

2. Notes

1. Magnesium ribbon is conveniently cleaned by immersion in aqueous 10% hydrochloric acid, rinsing the ribbon with distilled water and with acetone, and drying it in an oven at 120°.

2. To ensure a rapid reaction with the magnesium, the methanol should be heated to reflux over magnesium methoxide for 12 hours and then distilled and transferred to the reaction vessel with a siphon or a large pipet. If necessary, a crystal of iodine may be added to initiate the reaction of methanol with magnesium.

3. If a water aspirator is used, a calcium chloride tube or tower

should be included in the line connecting the distillation head and the aspirator.

4. The dimethylformamide may be purified by distillation at atmospheric pressure. The checkers distilled material purchased from the J. T. Baker Company and used the fraction collected at 153–155°.

5. Carbon dioxide obtained from a cylinder of the compressed gas was passed through a tube packed with calcium chloride and either activated silica gel or Drierite (containing an indicator) to remove water.

6. If the gas flow is turned off while carbon dioxide is still being absorbed, the pressure inside the flask falls below atmospheric pressure. This pressure change is readily observed with the mineral oil-filled trap fitted to the gas exit tube of the system.

7. The submitters found that this solution of methyl magnesium carbonate in dimethylformamide could be stored for long periods in a well-stoppered bottle without loss of potency.

8. The checkers employed cyclohexanone purchased from Eastman Organic Chemicals and distilled before use, b.p. 155–157°. The ratio of cyclohexanone to methylmagnesium carbonate is fairly critical; a proportion of ketone larger than the ratio 1:8 ketone:magnesium salt specified yields a pasty product presumably contaminated with monocarboxylated material. A smaller proportion of ketone lowers the yield.

9. The checkers measured the temperature of this solution by sliding a thermometer through the distillation head so that the thermometer bulb was immersed in the reaction mixture.

10. The checkers used a cooling bath prepared from ice and sodium chloride and dried the hydrogen chloride obtained from a compressed-gas cylinder by passing the gas through a trap filled with concentrated sulfuric acid.

11. The submitters reported that the addition of eight volumes of warm (35°) water to a warm solution of the keto diester in ten volumes of methanol followed by gradual cooling to 0° separated, on one occasion, a product, m.p. 142–143°, which was presumably one pure isomer. The checkers found that the product recrystallized readily from methanol, aqueous methanol, or benzene in good crystalline form, but invariably with a wide melting range (129–136°) which varied with the rate of heating. It would

appear that the checkers invariably obtained the product as a mixture of two or more of the three readily interconvertible forms: keto *cis*-diester, keto *trans*-diester, and enol diester. The thin-layer chromatogram of the recrystallized product, determined on a plate coated with silica gel and eluted with a mixture of carbon tetrachloride and ethyl acetate (1:1 v/v), indicated the absence of monocarbomethoxycyclohexanone in the product. Also, the elemental composition of the product and the mass spectrum of the product are consistent with the idea that the product contains only stereoisomeric and tautomeric forms. The mass spectrum exhibits a molecular ion peak at m/e 214 with abundant fragment peaks at m/e 182, 154, 126, 95, 67, and 55 but exhibits a peak of relatively low abundance at m/e 156, the mass of the molecular ion derived from 2-carbomethoxycyclohexanone. An ethanol solution of the recrystallized keto diester product initially exhibits an ultraviolet maximum at 255 mμ with a molecular extinction coefficient within the range 6000–11,000; on the addition of excess sodium hydroxide the ultraviolet maximum is shifted to 287.5 mμ (ϵ 12,600). A chloroform solution of the recrystallized product has infrared absorption at 1750 (strong), 1712 (medium), 1675 (weak), and 1610 (weak) cm.$^{-1}$; an ethanol solution of the product, when treated with ferric chloride, gives no immediate color, but a brown color develops after the solution is allowed to stand for 10 to 20 minutes. These observations suggest that the crystalline product obtained in this preparation is primarily a mixture of the *cis*- and *trans*-isomers of the keto tautomer.

3. Discussion

Dimethyl cyclohexanone-2, 6-dicarboxylate has been prepared by the alkylation of dimethyl acetonedicarboxylate with trimethylene dibromide [2] and by the carboxylation of cyclohexanone.[3] The present preparation gives a general procedure for carboxylation of active methylene compounds.[3–6] The method has been used for carboxylation of methylene groups activated by ketones,[3–5] nitro groups,[3, 4] and certain amide functions.[6] The success of the procedure is attributed to the formation of a magnesium enolate which is stabilized by chelation with an adjacent carboxylate anion.[4, 6]   In certain cases [3, 6] the magnesium enolate has been alkylated in the original reaction mixture, thereby avoiding the necessity for isolating an intermediate ester. Although the present example illustrates the fact that when two equivalent active methylene groups are present both positions may be carboxylated, the submitters were unsuccessful in obtaining a pure keto diester when the procedure was applied to cyclopentanone.

1. Department of Organic Chemistry, Indian Institute of Science, Bangalore, India.
2. P. C. Guha and N. K. Seshadriengar, *Ber.*, **69**, 1207 (1936); F. F. Blicke and F. J. McCarty, *J. Org. Chem.*, **24**, 1069 (1959).
3. M. Stiles and H. L. Finkbeiner, *J. Am. Chem. Soc.*, **81**, 505 (1959); M. Stiles, *J. Am. Chem. Soc.*, **81**, 2598 (1959).
4. M. Stiles, *Ann. N.Y. Acad. Sci.*, **88**, 332 (1960); H. L. Finkbeiner and M. Stiles, *J. Am. Chem. Soc.*, **85**, 616 (1963).
5. S. W. Pelletier, R. L. Chappell, P. C. Parthasarathy, and N. Lewin, *J. Org. Chem.*, **31**, 1747 (1966).
6. H. Finkbeiner, *J. Am. Chem. Soc.*, **86**, 961 (1964); **87**, 4588 (1965); *J. Org. Chem.*, **30**, 3414 (1965).

2,7-DIMETHYL-2,7-DINITROÖCTANE

(Octane, 2,7-dimethyl-2,7-dinitro-)

$$O_2NC(CH_3)_2CH_2CH_2CO_2CH_3 + KOH \rightarrow$$
$$O_2NC(CH_3)_2CH_2CH_2CO_2K + CH_3OH$$

Anode Reactions

$$2O_2NC(CH_3)_2CH_2CH_2COO^- \rightarrow$$
$$2O_2NC(CH_3)_2CH_2CH_2 \cdot + 2CO_2 + 2 \text{ electrons}$$
$$2O_2NC(CH_3)_2CH_2CH_2 \cdot \rightarrow$$
$$O_2NC(CH_3)_2CH_2CH_2CH_2CH_2C(CH_3)_2NO_2$$

Cathode Reactions

$$2K^+ + 2 \text{ electrons} + 2CH_3OH \rightarrow 2CH_3OK + H_2$$

Submitted by W. H. Sharkey and C. M. Langkammerer.[1]
Checked by Masaaki Takahashi, Marjorie C. Caserio, and
John D. Roberts.

1. Procedure

A. *4-Methyl-4-nitrovaleric acid.* In a 2-l. three-necked flask equipped with a stirrer and reflux condenser is placed a solution prepared from 118 g. (1.78 moles) of 85% potassium hydroxide pellets and 500 ml. of water. A thermometer may be so placed in the third neck that the bulb extends below the surface of the solution. To this solution is added 300 g. (1.71 moles) of methyl 4-methyl-4-nitrovalerate.[2] The mixture is stirred and gently heated (Notes 1 and 2). After hydrolysis has started, as indicated by changes in appearance of the hazy reaction mixture, the external source of heat is removed. The reaction is complete when the cloudy mixture has changed to a clear solution. This requires about 20–25 minutes (Note 3).

The reaction mixture is cooled to room temperature, and a saturated solution of potassium permanganate is added in an amount sufficient for a violet color to persist for about 1 minute before turning green (Note 4). About 100–110 ml. is required. Manganese dioxide is removed by filtration, and the filtrate is extracted with methylene chloride to remove non-acidic organic material. The aqueous layer is acidified with 18% hydrochloric acid, whereupon a pale yellow or green oil separates as a bottom layer. The oil is removed, and the aqueous layer is washed with methylene chloride. The washings are added to the oil, and the combined product is dried with anhydrous magnesium sulfate. Methylene chloride is removed by distillation, and the residual oil is distilled under reduced pressure through a 6-in. Vigreux column (Note 5). The fraction boiling at 125–135° (0.9–1.5 mm.) amounts to 217–225 g. (79–82%) and is a colorless oil that crystallizes on standing or when seeded to give a solid with a melting point of approximately 35–45°. This fraction is redistilled through a 24-in. spinning-band column and gives 166–180 g. (60–65%) of colorless 4-methyl-4-nitrovaleric acid, b.p. 122–124°/0.70–0.90 mm., m.p. 45–46° (Note 6).

B. *2,7-Dimethyl-2,7-dinitroöctane.* A 200-mm. desiccator is adapted for use as an electrolysis cell. A lid is prepared from a glass plate ground to fit the desiccator; holes are bored in the lid to accommodate electrodes, a thermometer, and a stirrer, and to provide an opening for making additions to the cathode compartment. The cathode (Note 7) is placed in a porous ceramic cup having outside dimensions 4 in. deep, 6 in. long, and 2 in. wide (Note 8). This cup is placed on the glazed ceramic plate of the desiccator under the holes cut in the lid for the cathode and for making additions. The glazed ceramic plate is notched so that it can be slipped down over the supports in the bottom of the desiccator. The notches should be large enough for the ceramic plate to be about halfway between its normal position and the bottom of the desiccator (Note 9). The cathode is a 26-gauge stainless-steel plate 3 in. square (Note 10), and the anode consists of two platinum plates each 0.01 in. thick and 3 in. square (Note 11) placed as close as possible, one on either side, to the ceramic cup without touching the sides of the cup.

A solution made by dissolving 161 g. (1 mole) of 4-methyl-4-nitrovaleric acid and 33.0 g. (0.5 mole) of potassium hydroxide in 2.7 l. of methanol is added to the cell. The ceramic cup is filled with 200 ml. of 2N potassium hydroxide in methanol. The cell contents are cooled to 20° (Note 12), and current is passed through the solution (Note 13). When the current is turned on, vigorous evolution of hydrogen takes place inside the porous cup. Potassium hydroxide (2N) in methanol is added as needed to maintain the original volume inside the cup (Note 14). Carbon dioxide is evolved at the anodes. After about 5 hours, white crystals of the dinitroöctane form in the anode solution. As electrolysis proceeds, the resistance of the cell increases, and after several hours the cell is operated at 3–5 amperes and 60–80 volts. If a current of 3–5 amperes cannot be maintained, it is desirable to add 2N potassium hydroxide in methanol to the anode compartment (Notes 15 and 16). After 8 hours of operation, the current is shut off, the ceramic cup removed and cleaned, and the anode solution cooled to about −30°. The anode solution is filtered to remove crystals of 2,7-dimethyl-2,7-dinitroöctane. The yield of dry product, m.p. 98–100°, amounts to 50–65.5 g. (43–56%), which is sufficiently pure for most uses (Note 17). One recrystallization from methanol (Note 18) gives pure 2,7-dimethyl-2,7-dinitroöctane, m.p. 100–101° (lit.[3] 101.5–102°).

2. Notes

1. If stirring is not used, hydrolysis does not start until the reaction mixture is heated almost to the reflux temperature. Under these conditions the reaction, which is exothermic, proceeds so rapidly that the reflux condenser is flooded.

2. Although no explosions have yet occurred, it is advisable to conduct this reaction behind a shield. The hydrolysis, once started, proceeds quite rapidly.

3. The time required varies with the initial temperature of the hydrolysis solution. With warm solutions, e.g., freshly made potassium hydroxide solution, external heating is not needed. With such a solution, the checkers found that a cloudy mixture having an initial internal temperature of 40° became clear within 2–5 minutes of mixing and during this time the temperature rose to 70°. Stirring was continued for 20 minutes to ensure complete hydrolysis.

4. Addition of potassium permanganate removes colored impurities that otherwise persist through subsequent distillations of

the acid. The maximum concentration of potassium permanganate in water at room temperature is about 5%. It may be more convenient to use a concentrated solution of sodium permanganate, which is very soluble in water. However, the end point is difficult to observe with sodium permanganate.

5. During this first distillation, nitrogen oxides are evolved in appreciable amounts. It is suggested this operation be done behind a shield in a well-ventilated area. A trap cooled with a mixture of solid carbon dioxide and acetone or with liquid air should be inserted between the column and the vacuum pump. Pumps used in this distillation did not appear to suffer corrosion damage. A dry oil was used and changed after each distillation.

6. It may be necessary in this second distillation to apply a source of heat to the fraction-collecting system to prevent crystallization of the product prior to entry of the distillate into the receivers.

7. The cathode is the electrode connected to the negative terminal of the current source.

8. A satisfactory alumina cup may be purchased from the Norton Company, Worcester, Mass., Catalog No. 44805, RA-98.

9. This is necessary because the top of the porous cup must be below the lip of the desiccator. The volume of the desiccator up to the top of the porous cup should be 4 l.

10. The considerable heat generated inside the cup leads to rapid evaporation of the methanol with which the cup is filled. The cell can be operated by constant replenishment of methanol. However, loss of methanol can be minimized by use of a water-cooled cathode. Such an electrode is conveniently prepared from two 3-in.-square stainless-steel plates and a strip of stainless steel about 0.25 in. wide, which are welded to form a stainless-steel box. Nipples serving as water inlet and outlet are welded to the top of the box.

11. The 3-in.-square platinum plates are welded to a heavy platinum wire that protrudes through one of the holes cut in the glass lid. Heavy copper wire (No. 10) is used to connect the electrodes to the source of current.

12. This temperature is easily maintained by surrounding the cell with a methanol bath cooled with solid carbon dioxide. About 100 lb. of solid carbon dioxide is required for 8 hours' operation.

Yields are severely reduced if the methanol is allowed to boil. Temperatures lower than 20° can be used but are less satisfactory.

13. The initial current is usually 6–8 amperes at 50–60 volts. The resistance of the cell constantly changes because of depletion of electrolyte and deposition of potassium methoxide in the pores of the ceramic cup. The preferred source of current is a rectifier capable of delivering 10–15 amperes at 30–90 volts. Lead storage batteries connected in series are also satisfactory but require frequent recharging.

14. In a typical run, 30 ml. of 2N potassium hydroxide in methanol was added at the end of each hour of operation. Total addition was 210 ml.

15. The amount added must be less than that required to bring the pH to 7. When the pH of the anode solution is greater than 7, undesirable reactions occur.

16. The checkers found that after 5.5 hours the current dropped below 3 amperes at 75 volts. Addition of 25 ml. of 2N potassium hydroxide in methanol to the anode solution increased the current to 6 amperes. Subsequent additions of 10–15 ml. of base were made approximately every 30 minutes. The total volume of base added was 80 ml.

17. Some unchanged 4-methyl-4-nitrovaleric acid can be recovered from the spent electrolysis solution by alkaline extraction followed by acidification and distillation of the water-insoluble oil. However, under ordinary circumstances, recovery of this material is not worth the effort expended.

18. About 500 ml. of refluxing methanol and 58 g. of crude product afford about 53 g. (95% recovery) of recrystallized material.

3. Methods of Preparation

2,7-Dimethyl-2,7-dinitroöctane has been prepared by nitration of 2,7-dimethyloctane.[3]

4. Merits of Preparation

Kolbe electrolysis is generally useful for the formation of hydrocarbons from monocarboxylic acids and for the preparation of many difunctional compounds as well. A specific illustration is the synthesis of esters of long-chain dicarboxylic acids from monoesters of appropriate dicarboxylic acids (see p. 463). A number of these syntheses are discussed by Fichter.[4] In the present preparation, a two-compartment cell is employed to avoid, or at least greatly reduce, undesired reduction of the nitro group at the cathode. It seems likely that the procedure could be adapted to the preparation of other difunctional compounds containing groups that are easily reduced.

1. Contribution No. 516 from the Central Research Department, Experimental station, E. I. du Pont de Nemours and Co., Wilmington, Del.
2. R. B. Moffett, *Org. Syntheses,* Coll. Vol. 4, 652 (1963).
3. M. Konowalow, *J. Russ. Phys-Chem. Soc.,* **38**, 124 (1906) [*Chem. Zentr.,* 1906, II, 314].
4. F. Fichter, *Organische Elektrochemie,* p. 17, Theodor Steinkopff, Dresden and Leipzig, 1942.

2,6-DIMETHYL-3,5-DIPHENYL-4H-PYRAN-4-ONE

(4H-Pyran-4-one, 2,6-dimethyl-3,5-diphenyl-)

$$C_6H_5CH_2COCH_2C_6H_5 \quad + \quad 2\,CH_3COOH$$

Polyphosphoric acid

$+ \ 3\,H_2O$

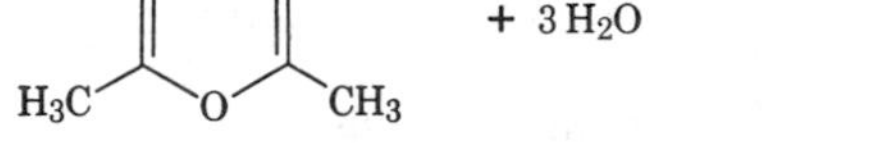

Submitted by Thomas L. Emmick and Robert L. Letsinger [1]
Checked by Donald J. MacGregor and Peter Yates

1. Procedure

A mixture of 400 g. of polyphosphoric acid and 250 ml. of glacial acetic acid is heated to reflux in a 2-l. round-bottomed flask equipped with a stirrer, reflux condenser, and thermometer. Dibenzyl ketone (42.0 g., 0.200 mole) (Note 1) is then added, and the reaction mixture is heated at reflux (130–135°) for 1.5 hours. The solution is cooled to 30° in an ice water bath, and 1 l. of water is added slowly with stirring. The brown precipitate which forms is collected by filtration, washed with 1 l. of water, and dissolved in 1 l. of hot benzene. The hot benzene solution is treated with 2 g. of activated carbon, filtered hot through a pad of diatomaceous earth, dried with 10 g. of magnesium sulfate, decanted from the magnesium sulfate, and concentrated to 500 ml. On addition of 450 ml. of hexane and cooling to 5–10°, tan crystals of the crude pyranone separate. Filtration affords 25–27 g. (45–49%) of product melting at 202–206°. For purification this material is dissolved in 500 ml. of hot benzene, treated with 1 g. of activated carbon as before, and precipitated from solution by the addition of 250 ml. of hexane and cooling of the mixture to 5–10°. On filtration 19–21 g. (34–38%) of 2,6-dimethyl-3,5-diphenyl-4*H*-pyran-4-one is obtained. This material melts sharply at 207–209° (Note 2). An additional quantity (3–4 g., 5–7%) of somewhat less pure product (m.p. 204–206°) may be recovered by evaporation of the filtrate and recrystallization of the residue from 200 ml. of benzene-hexane (50% benzene by volume).

2. Notes

1. For this preparation Matheson, Coleman and Bell practical grade dibenzyl ketone was recrystallized once from anhydrous ether at −70°. It melted at 33–34°. Practical grade dibenzyl ketone may be used directly; however, the yield of the pyranone is somewhat lower.

2. The corrected melting point is 209.5–210.0°. Melting points were obtained with a Fisher-Johns apparatus. The recrystallized sample retained a pale tan shade.

3. Methods of Preparation

This procedure is a modification of that of Letsinger and Jamison.[2] The pyranone has also been prepared by treatment of dibenzyl ketone with acetic anhydride-perchloric acid or acetyl chloride-aluminum chloride.[3]

4. Merits of the Preparation

This procedure represents a simple and unique route to certain pyran-4-ones. The reaction can be applied also to benzyl methyl ketone and diethyl ketone; the corresponding pyran-4-ones are obtained in yields of 48% and 26%, respectively.

1. Northwestern University, Evanston, Illinois.
2. R. L. Letsinger and J. D. Jamison, *J. Am. Chem. Soc.*, **83**, 193 (1961).
3. A. T. Balaban, G. D. Mateescu, and C. D. Nenitzescu, *Acad. Rep. Populare Romine, Studii Cercetari Chim.*, **9**, 211 (1961) [*C.A.*, **57**, 15065 (1962)].

4,6-DIMETHYL-1-HEPTEN-4-OL

(1-Hepten-4-ol, 4,6-dimethyl-)

$$(C_6H_5)_3SnCH_2CH=CH_2 + C_6H_5Li \rightarrow$$

$$(C_6H_5)_4Sn + CH_2=CHCH_2Li$$

$$CH_2=CHCH_2Li + CH_3\overset{\overset{\text{O}}{\|}}{C}CH_2CH(CH_3)_2 \rightarrow$$

$$CH_2=CHCH_2-\overset{\overset{\text{OLi}}{|}}{\underset{\underset{CH_3}{|}}{C}}-CH_2CH(CH_3)_2 \xrightarrow{H_2O}$$

$$CH_2=CHCH_2\overset{\overset{\text{OH}}{|}}{\underset{\underset{CH_3}{|}}{C}}CH_2CH(CH_3)_2$$

Submitted by DIETMAR SEYFERTH and MICHAEL A. WEINER.[1]
Checked by MELVIN S. NEWMAN and CLIFFORD Y. PEERY.

1. Procedure

Caution! See. p. 976.

A solution of 50 g (0.127 mole) of allyltriphenyltin (Notes 1 and 2) in 200 ml. of diethyl ether (Note 3) is prepared in a 1-l.

three-necked flask fitted with a reflux condenser, a motor-driven glass sleeve-type stirrer, nitrogen-inlet tube, and a 250-ml. dropping funnel with pressure-equalizing side arm. After the system has been flushed thoroughly with prepurified nitrogen, 113 ml. of a 1.13N solution of phenyllithium (0.127 mole) in diethyl ether (Note 4) is added rapidly to the stirred allyltriphenyltin solution. Precipitation of tetraphenyltin occurs immediately, and the reaction mixture is stirred for 30 minutes in an atmosphere of prepurified nitrogen. Through the dropping funnel is then added 12.0 g. (0.12 mole) of 4-methyl-2-pentanone (Note 5) in 25 ml. of diethyl ether at such a rate that moderate reflux is maintained. Subsequently the reaction mixture is refluxed for 1 hour, allowed to cool to room temperature, and hydrolyzed by adding 100 ml. of distilled water (Note 6). The solid tetraphenyltin is filtered (53.5 g. = 98% yield), and the filtrate is transferred to a separatory funnel. The aqueous layer is separated and extracted with three 30-ml. portions of ether. The ethereal extracts and the organic layer are combined and dried over anhydrous magnesium sulfate. After removal of the ether by distillation at atmospheric pressure, the residue is filtered through a sintered-glass funnel into a 250-ml. distilling flask and fractionally distilled at reduced pressure using a vacuum-jacketed Vigreux column equipped with a still head of the total-condensing, partial-takeoff type. 4,6-Dimethyl-1-hepten-4-ol, b.p. 70–71°/20 mm., n_D^{20} 1.4403, is obtained in 70–75% yield (12.0–12.8 g.) (Note 7).

2. Notes

1. Allyltriphenyltin is prepared as follows. To a 3-l. three-necked flask fitted with a reflux condenser, a motor-driven stirrer, nitrogen-inlet tube, and a 1-l. dropping funnel with pressure-equalizing side arm are added 50 g. (2.1 g. atom) of magnesium turnings and 800 ml. of diethyl ether (Mallinckrodt reagent grade). The dropping funnel is charged with a solution of 120 g. (1.0 mole) of allyl bromide (Eastman Kodak white label) and 250 g. (0.65 mole) of triphenyltin chloride (Metal Thermit Corporation) in 600 ml. of tetrahydrofuran (Electrochemicals Department, E. I. du Pont de Nemours and Company, Inc.)

which has been freshly distilled from lithium aluminium hydride. This solution is added to the vigorously stirred, refluxing magnesium suspension during 7 hours. After the addition is complete, 500 ml. of dry benzene is added and the reaction mixture is refluxed overnight at 60°. It is then hydrolyzed by careful addition of 150 ml. of a saturated ammonium chloride solution. The organic phase is decanted from the solids, and the latter are washed twice with ether. The combined organic layer and ethereal extracts are evaporated at reduced pressure with the aid of a rotary evaporator. The solid residue is recrystallized from 350 ml. of ligroin. The yield of product, m.p. 73–74°, is 190–205 g. (75–80%).

Allyltriphenyltin can also be prepared by using the reaction of preformed allylmagnesium bromide with triphenyltin chloride.[2] However, the submitters prefer the simpler procedure described above for large-scale preparations of allyltin compounds.

2. Tetraallyltin, triallylphenyltin, and diallyldiphenyltin may be used in place of allyltriphenyltin.

3. The diethyl ether used is Mallinckrodt reagent grade and is distilled from lithium aluminum hydride before use.

4. Ethereal phenyllithium, prepared from lithium and bromobenzene,[3] may be standardized by adding an aliquot to water and titrating with standard sulfuric acid.

5. The 4-methyl-2-pentanone used is Eastman Kodak white label grade.

6. The first few milliliters of water should be added dropwise.

7. The reported [4] physical constants for 4,6-dimethyl-1-hepten-4-ol are b.p. 68–69°/20 mm., n_D^{20} 1.4402.

3. Methods of Preparation

This procedure is essentially that reported previously.[5] 4,6-Dimethyl-1-hepten-4-ol has also been prepared by the reaction between allylmagnesium bromide and 4-methyl-2-pentanone,[4] by the treatment of magnesium with a mixture of allyl bromide and 4-methyl-2-pentanone,[6,7] and by the treatment of zinc with a mixture of allyl iodide and 4-methyl-2-pentanone.[8]

4. Merits of Preparation

The present synthesis of 4,6-dimethyl-1-hepten-4-ol is an example of the preparation and use of allyllithium. The same general procedure may be used to prepare vinyllithium from ether solutions of any of the compounds $(CH_2=CH)_n$-$Sn(C_6H_5)_{4-n}$ $(n = 1-4)$,[9] and benzyllithium from any of the $(C_6H_5CH_2)_n Sn(C_6H_5)_{4-n}$ $(n=1-4)$ compounds.[10]

Allyllithium is of particular value in the preparation of allylmetal derivatives.[5] Allyllithium may also be prepared by transmetalation between phenyllithium and tetraallyltin[11] or by cleavage of allyl phenyl ether with lithium in tetrahydrofuran.[12]

1. Department of Chemistry, Massachusetts Institute of Technology, Cambridge, Mass. Supported by the National Science Foundation under Grant G7325.
2. H. Gilman and J. Eisch, *J. Org. Chem.*, **20**, 763 (1955).
3. K. Ziegler and H. Colonius, *Ann.*, **479**, 135 (1930). See also H. Gilman, E. A. Zoellner, and W. M. Selby, *J. Am. Chem. Soc.*, **54**, 1957 (1932).
4. H. R. Henze, B. B. Allen, and W. B. Leslie, *J. Org. Chem.*, **7**, 326 (1942).
5. D. Seyferth and M. A. Weiner, *J. Org. Chem.* **24**, 1395 (1959).
6. A. Knorr, Ger. pat. 544,388 (1930) [*C.A.*, **26**, 2466 (1932)].
7. F. Bodroux and F. Taboury, *Bull. Soc. Chim. France*, [4] **5**, 812 (1909).
8. D. Marko, *J. Prakt. Chem.*, [2] **71**, 258 (1905).
9. D. Seyferth and M. A. Weiner, *Chem. & Ind. (London)*, 402 (1959).
10. H. Gilman and S. D. Rosenberg, *J. Org. Chem.* **24**, 2069 (1959); D. Seyferth and C. R. Sabet, unpublished.
11. D. Seyferth and M. A. Weiner, *J. Org. Chem.*, **26**, 4797 (1961).
12. J. J. Eisch and A. M. Jacobs, *J. Org. Chem.*, **28**, 2145 (1963).

DIMETHYLKETENE β-LACTONE DIMER

(3-Pentenoic acid, 3-hydroxy-2,2,4-trimethyl-, β-lactone)

$$(CH_3)_2C—C=O \quad \xrightarrow[C_6H_5Cl]{AlCl_3} \quad (CH_3)_2C—C=O$$

$$O=C—C(CH_3)_2 \qquad\qquad (CH_3)_2C=C—O$$

Submitted by Robert H. Hasek,[1] R. Donald Clark,[1]
and Gerald L. Mayberry [2]
Checked by V. Boekelheide, J. Witte, and G. Singer

1. Procedure

Caution! Dimethylketene β-lactone dimer is a mild but deceptively persistent lachrymator.

In a 500-ml. three-necked flask equipped with a thermometer, a mechanical stirrer, and a reflux condenser are placed 200 g. (1.43 moles) of tetramethyl-1,3-cyclobutanedione (Note 1) and 50 g. of chlorobenzene (Note 2). The mixture is heated to 135° with stirring while a total of 1.8 g. of reagent grade anhydrous aluminum chloride is added in 0.3-g. portions over a 3-hour period (Note 3). After the addition is complete, heating is continued for an additional 5 hours (Note 4).

The reaction mixture is then cooled to 35–40° and poured into a stirred solution of 230 g. of sodium chloride and 6.0 g. of sodium acetate in 600 ml. of water at 40°. The mixture is stirred for 15 minutes and then is transferred to a separatory funnel; the layers are separated, and the lower, aqueous layer is discarded. The crude product is distilled at reduced pressure through a stainless steel spinning-band column (Note 5). The yield of the β-lactone dimer of dimethylketene is 122–132 g. (61–67%); b.p. 69–71.5° (14 mm.) (Note 6). The product may be redistilled and the fraction boiling at 119.5–120° (150 mm.), n^{20}D 1.4380, collected.

2. Notes

1. Tetramethyl-1,3-cyclobutanedione is available from East-man Organic Chemicals. It melts at 115 116° and is typically 99% pure by vapor-phase chromatography.

2. More highly chlorinated aromatic solvents such as 1,2,4-trichlorobenzene can be used with similar results.

3. No special drying precautions are required; however, traces of water can be conveniently removed before the catalyst is added by distilling a small portion of the solvent until cloudiness disappears. The checkers found that the distillation of part of the solvent as indicated was necessary in order to obtain satisfactory yields. When the catalyst is added incrementally, no appreciable exotherm is observed.

4. The isomerization is usually complete in 5 hours and can easily be followed by vapor-phase chromatography. Heating periods up to 20 hours are not detrimental. The only failure among numerous preparations occurred when tetramethyl-1,3-cyclobutanedione contaminated with 4% of isobutyric acid was used. In case of partial conversion after 5 hours, additional increments (0.5 g.) of aluminum chloride should be added to complete the reaction.

5. The product may be distilled directly from the crude reaction mixture after addition of sodium acetate. The results are similar.

6. The distilled product is 99% pure by vapor-phase chromatography.

3. Methods of Preparation

The β-lactone dimer of dimethylketene can be prepared by pyrolysis of its polyester, which is formed by the base-catalyzed polymerization of dimethylketene.[3-5] In addition to the rearrangement of the normal dimer described above,[6] the direct dimerization of dimethylketene in the presence of aluminum chloride [3] or trialkyl phosphites [7] leads to the β-lactone dimer.

4. Merits of the Preparation

The β-lactone dimer of dimethylketene reacts with alcohols, phenols, mercaptans, and amines to form derivatives of 2,2,4-trimethyl-3-oxovaleric acid.[3] In this respect it is a more powerful acylating reagent than the normal dimer, tetramethyl-1,3-cyclobutanedione. The preparation of 2,2,4-trimethyl-3-oxovaleranilide, for example, is accomplished easily with the lactone dimer, but is extremely difficult with the normal dimer.[8]

In the presence of catalytic amounts of sodium methoxide, dimethylketene β-lactone dimer is polymerized at moderate temperature to a polyester.[3] At higher temperatures (above 100°), disproportionation to the cyclic trimer, hexamethyl-1.3,5-cyclohexanetrione, takes place.[9] Addition of a stoichiometric amount of sodium methoxide to the lactone dimer generates the sodium enolate of methyl 2,2,4-trimethyl-3-oxovalerate. This reaction provides a convenient entry into certain ester anion chemistry that formerly required the use of a strong base like tritylsodium.[10]

Although these reactions can be duplicated in most cases with the normal dimer of dimethylketene,[11] the more reactive lactone dimer is the preferred reagent. The liquid form of this dimer is convenient to handle. A distinct difference in behavior of the dimethylketene dimers is noted when they are pyrolyzed. The normal dimer is dissociated at 600° to dimethylketene,[12] but the lactone dimer is decarboxylated almost quantitatively at 450° to tetramethylallene.[13]

1. Research Laboratories, Tennessee Eastman Co., Kingsport, Tennessee 37662.
2. Organic Chemicals Division, Tennessee Eastman Co., Kingsport, Tennessee 37662.
3. R. H. Hasek, R. D. Clark, E. U. Elam, and J. C. Martin, *J. Org. Chem.*, **27**, 60 (1962).
4. G. Natta, G. Mazzanti, G. Pregaglia, M. Binaghi, and M. Peraldo, *J. Am. Chem. Soc.*, **82**, 4742 (1960).
5. G. Natta, G. Mazzanti, G. Pregaglia, and M. Binaghi, *Makromol. Chem.*, **44–46**, 537 (1961).
6. R. D. Clark, U. S. Patent 3,062,837 (1962) [*C.A.*, **58**, 8911 (1963)].
7. E. U. Elam, *J. Org. Chem.*, **32**, 215 (1967).
8. H. Staudinger and J. Mayer, *Ber.*, **44**, 528 (1911).
9. R. H. Hasek, R. D. Clark, E. U. Elam, and R. G. Nations, *J. Org. Chem.*, **27**, 3106 (1962).
10. Cf. C. R. Hauser and B. E. Hudson, Jr., *Org. Reactions*, **1**, 266 (1942).
11. R. H. Hasek, E. U. Elam, J. C. Martin, and R. G. Nations, *J. Org. Chem.*, **26**, 700 (1961).

12. W. E. Hanford and J. C. Sauer, *Org. Reactions*, **3**, 136 (1946).
13. J. C. Martin, U. S. Patent 3,131,234 (1964) [*C. A.*, **61**, 2969 (1964)].

DIMETHYL
3-METHYLENECYCLOBUTANE-1,2-DICARBOXYLATE

(3-Methylenecyclobutane-1,2-dicarboxylic acid, dimethyl ester)

Submitted by H. B. STEVENSON, H. N. CRIPPS, and J. K. WILLIAMS.[1]
Checked by R. D. BIRKENMEYER, W. E. RUSSEY, and F. KAGAN.[2]

1. Procedure

A. *3-Methylenecyclobutane-1,2-dicarboxylic anhydride.* A 2-l. stainless steel autoclave equipped with stirrer, pressure gauge, and thermocouple is charged with 500 g. (5.1 moles) of maleic anhydride, 645 ml. of benzene, and 0.25 g. of hydroquinone. The autoclave is closed, cooled to $-70°$ with stirring, and evacuated to a pressure of about 20 mm. Allene[3] (100 g., 2.5 moles) (Note 1) is sucked into the autoclave, and the mixture is heated with stirring for 8–10 hours at 200–210°. During this time, the pressure drops from 23 atm. to 15 atm. The vessel is cooled to 25°, and unreacted allene (6–13 g.) is vented into a cold trap (Note 2).

The benzene solution is decanted, and about 500 ml. of acetone is added to the autoclave and stirred until the dark viscous residue goes into solution. The benzene and acetone solutions are combined, filtered, and distilled through a 19-mm. x 1.8-m. Nester spinning-band still.[4] When the pot temperature reaches 170°, the pressure is reduced to 40 mm., and up to 250 g. of maleic anhydride, b.p. 110–115°/40 mm., is recovered. Finally 119 g. of crude anhydride mixture, b.p. 70–125°/3 mm. (Note 3), is collected.

The crude anhydride is carefully fractionated through a 13-mm. x 1.2-m. Nester still at a pressure of 25 mm. (Note 4) and a reflux ratio of at least 10:1. After a fore-run of maleic anhydride, b.p. 50–100°/25 mm., and a small intermediate fraction, there is obtained 75–90 g. (22–26%) of 3-methylenecyclobutane-1,2-dicarboxylic anhydride; b.p. 155–159°/25 mm.; n_D^{25} 1.4935–1.4952 (Note 5). This material is of sufficient purity for most uses, but it contains approximately 2–5% of propargylsuccinic anhydride. Redistillation through the Nester still gives 65–80 g. (19–23%) of 3-methylenecyclobutane-1,2-dicarboxylic anhydride; b.p. 155°/25 mm.; n_D^{25} 1.4946–1.4955.

By continuing the distillation after removal of the cyclobutane anhydride, there is obtained 25–30 g. (7–9%) of propargylsuccinic anhydride; b.p. 162–168°/25 mm.; m.p. 63–68°. The melting point is raised to 69–70° by one recrystallization from 100 ml. of benzene (80% recovery of purified product).

B. *Dimethyl 3-methylenecyclobutane-1,2-dicarboxylate.* One liter of methanol is added cautiously with occasional shaking to 276 g. (2.00 moles) of 3-methylenecyclobutane-1,2-dicarboxylic anhydride (n_D^{25} 1.4946–1.4955; Note 6) and 5 g. of *p*-toluenesulfonic acid in a 2-l. three-necked flask fitted with a thermometer, a condenser, and a dropping funnel. Refluxing starts after about two-thirds of the methanol has been added. The remainder is added at a rate that maintains vigorous boiling. The solution is refluxed for 30–40 hours with the pot temperature increasing from 67° to 68° (Note 7). The mixture is cooled to 15°, and methanol and water are removed by distillation under reduced pressure at temperatures below 15°, using a large receiver cooled with a mixture of solid carbon dioxide and acetone. When the pressure goes below 1 mm., the temperature is increased to 50°

until the distillation is completed. One liter of methanol (Note 8) is added to the residue, and the solution is heated under reflux for an additional 30–40 hours, during which time the pot temperature increases from 67° to 67.5°. The solution is cooled to 15°, 1.7 g. of finely powdered anhydrous sodium carbonate is added to neutralize the p-toluenesulfonic acid, and the methanol and water are removed as before. Crude dimethyl 3-methylene-cyclobutane-1,2-dicarboxylate is distilled rapidly at 65–85°/1 mm. through a 30-cm. Vigreux column (Note 9). The ester can be purified by redistillation through a 13-mm. x 1.2-m. Nester still, with the main fraction boiling at 134–137°/25 mm.; weight 297–338 g. (81–92%); n_D^{25} 1.4624–1.4630.

2. Notes

1. Freshly distilled allene should be used. It should be free of 2-chloropropene, usually present in allene prepared by zinc dehalogenation of 2,3-dichloropropene,[3] to avoid formation of chlorine-containing products that liberate hydrogen chloride on distillation.

2. The impurities present in the original allene are concentrated in the recovered material. If recovered allene is to be reused, it should be fractionated first.

3. The checkers isolated 167 g. of crude anhydride mixture boiling at 70–125°/3 mm. The large tarry residue contains allene polymers and small amounts of 1,2,3,4,5,6,7,8-octahydronaphthalene-2,3,6,7-tetracarboxylic dianhydride, which can be recovered by diluting the residue with benzene and filtering.

4. Pot temperatures above 175°, which result from use of pressures above 25 mm., cause formation of high-boiling by-products.

5. Collection of the product fraction should begin after a few milliliters of an intermediate fraction has been collected at 155°/25 mm. This material has a low index of refraction.

6. The checkers found that the use of anhydride with n_D^{25} 1.4937–1.4945 led to a product with a low index of refraction (n_D^{25} 1.4616).

7. The temperature rises because of disappearance of methanol by conversion to the methyl ester. Attainment of equilibrium is signified by the pot temperature reaching a constant temperature.

8. The second treatment with methanol increases the yield from 60% to 90%.

9. Rapid distillation from the neutralized catalyst results in much smaller loss of ester than is encountered in the more usual procedure that includes washing with water and drying.

3. Methods of Preparation

The procedure used is essentially that described by Cripps, Williams, and Sharkey.[5] The anhydride has been prepared in a similar manner by Alder and Ackermann.[6] No other methods have been described for the preparation of these materials.

4. Merits of the Preparation

The first step of this procedure illustrates a general reaction, the addition of allenes to alkenes to form methylenecyclobutanes. The reaction has been reviewed recently.[7]

Since 3-methylenecyclobutane-1,2-dicarboxylic anhydride is easily converted to 3-methyl-2-cyclobutene-1,2-dicarboxylic acid,[8] it is an intermediate to a variety of cyclobutenes. The dimethyl ester of 3-methylenecyclobutane-1,2-dicarboxylic acid is also a versatile compound; on pyrolysis it gives the substituted allene, methyl butadienoate,[9] and on treatment with amines it gives a cyclobutene, dimethyl 3-methyl-2-cyclobutene-1,2-dicarboxylate.[8]

1. Contribution No. 567 from the Central Research Department, Experimental Station, E. I. du Pont de Nemours and Co., Wilmington, Delaware.
2. The Upjohn Company, Kalamazoo, Michigan.
3. H. N. Cripps and E. F. Kiefer, this volume, p. 22.
4. R. G. Nester, *Anal. Chem.,* **28**, 278 (1956).
5. H. N. Cripps, J. K. Williams, and W. H. Sharkey, *J. Am. Chem. Soc.,* **81**, 2723 (1959).
6. K. Alder and O. Ackermann, *Ber.,* **90**, 1697 (1957).
7. J. D. Roberts and C. M. Sharts, *Org. Reactions,* **12**, 1 (1962).
8. H. N. Cripps, J. K. Williams, V. Tullio, and W. H. Sharkey, *J. Am. Chem. Soc.,* **81**, 4904 (1959).
9. J. J. Drysdale, H. B. Stevenson, and W. H. Sharkey, *J. Am. Chem. Soc.,* **81**, 4908 (1959); H. B. Stevenson and W. H. Sharkey, this volume, p. 734.

DIMETHYL OCTADECANEDIOATE

(Octadecanedioic acid, dimethyl ester)

$$2CH_3O_2C(CH_2)_8CO_2^- \xrightarrow{\text{Electrolysis}}$$

$$CH_3O_2C(CH_2)_{16}CO_2CH_3 + 2CO_2$$

Submitted by SHERLOCK SWANN, JR., and W. E. GARRISON, JR.[1]
Checked by JAMES CASON, JOHN A. CARLSON, and STANLEY WOOD.

1. Procedure

To 500 ml. of absolute methanol (Note 1) in a 1-l. electrolytic (tall form) beaker is added 1.1 g. (0.05 g. atom) of clean sodium metal. After solution of the sodium, 216 g. (1.0 mole) of methyl hydrogen sebacate (Note 2) is dissolved in the sodium methoxide solution. A magnetic stirring bar is placed in the beaker which is then fitted with a large neoprene stopper (Note 3) holding a platinum sheet anode, 12 cm.^{2} in area; and two platinum sheet cathodes, approximately 5.3 cm.^{2} in area, spaced equidistantly on either side of the anode at a distance of approximately 1.5 cm. (Note 4). The stopper is also provided with a stoppered entry tube and an efficient reflux condenser (Note 5).

The electrodes are connected to a suitable variable source of direct current (Note 6), the magnetic stirrer is started, and a potential of about 50 volts is applied. This results in a current flow of 1–2 amperes. The solution soon comes to boiling; the voltage is then regulated so that a rapid reflux is maintained (Note 7).

Completion of the run, after 30–40 hours, is indicated when a few drops of the solution show an alkaline reaction to phenolphthalein. No harm is done if the electrolysis is carried a few hours beyond this point; however, after excessively long periods, formation of polymeric material lowers the yield and renders purification of the product rather troublesome.

Upon completion of the reaction (Note 8) the solution is acidified with acetic acid, and the solvent removed under reduced pressure. The residue is dissolved in about 1.4 l. of ether and filtered into a 2-l. separatory funnel through fluted paper. After the ether solution has been washed with two 300-ml. portions of 5% aqueous sodium bicarbonate solution (Note 9), the ether is removed on a steam bath. The residue is dissolved in about 1.5 l. of warm methanol, and the solution is allowed to cool to room temperature. The crystallized product is collected by suction filtration on a Büchner funnel and pressed well. The product is rather waxy and is best washed by transferring it to a beaker and stirring thoroughly with about 150 ml. of cold methanol. The white crystals are then collected and pressed well again. If the filtrate is colored, the crystals are washed again with a smaller quantity of methanol. The combined filtrate and washings are concentrated to one-half the original volume and chilled in ice to yield a few grams of additional product of the same melting point. The combined lots are dried in a vacuum desiccator. The yield amounts to 116–126 g. (68–74%) of white, microcrystalline dimethyl octadecanedioate, m.p. 57–58°.

2. Notes

1. Analytical reagent absolute methanol is sufficiently pure for this purpose.

2. The methyl hydrogen sebacate used by the submitters was prepared by the method described by Pattison and co-workers,[2] which is a modification of the method described in *Organic Syntheses*, Coll. Vol. **2**, 276 (1943), for ethyl hydrogen sebacate. The methyl rather than the ethyl ester is preferred because of the greater ease of purification of methyl hydrogen sebacate and dimethyl octadecanedioate. The checkers prepared methyl hydrogen sebacate by the convenient procedure which has been described for methyl hydrogen hendecanedioate.[3] Absence of diacid in the half ester is imperative, for each molecule of diacid (in low concentration) will couple with two molecules of half ester.

3. A rubber stopper will suffice, but neoprene is more durable.

4. Convenient dimensions are 3 x 4 cm. for the anode, and 2.3 x 2.3 cm. for the cathode. The bottom of the anode should

extend to about 3 cm. from the bottom of the beaker, and the centers of the cathodes should be lined up with the center of the anode. The submitters and checkers used 0.002-in. platinum sheet. Thinner material is easily distorted by the action of the stirrer and may thus develop a short circuit. Platinum wire may be attached to the sheet by heating both parts to redness and hammering them together. The wire may then be sealed in a piece of Pyrex tubing, which is passed through a properly located hole in the stopper, and mercury is used to make contact with the lead-in wire.

In a run one-half the size here described, the checkers obtained similar results by using the same geometry of electrodes with the same concentrations of reactants in a 700-ml. tall-form beaker, with one-half the electrolysis time.

5. It is most convenient to insert a 24/40 outer joint in the stopper to accommodate the condenser. A 14/20 outer joint makes a convenient entry tube for withdrawing samples in order to determine when alkalinity has been reached. This joint must be stoppered during the electrolysis, which is under reflux.

6. A variable field d-c motor generator was used by the submitters. In lieu of such equipment, the checkers used a 120-volt d-c source with a heavy-duty resistor in the circuit. Occasionally, a film of polymeric material may form on the anode and reduce the current flow. This condition may be corrected by reversing the current flow for a period of 5–10 seconds. A suitable control circuit is diagrammed in Fig. 1.

7. It is usually necessary to increase the applied voltage toward the end of the electrolysis in order to keep the current flow at a high rate. The submitters usually finished the electrolysis at 140 volts; however, the checkers found that a limit of 120 volts resulted in no significant delay in completion of the electrolysis.

8. The electrodes should be removed promptly from the warm reaction mixture, for the solution sets to a solid mass on cooling. It is most convenient to proceed with work-up of the warm solution.

9. An insignificant amount of half ester is recovered by acidification of the bicarbonate washings, followed by extraction with ether.

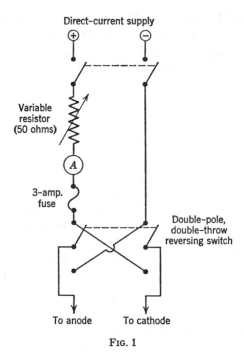

Direct-current supply

Variable resistor (50 ohms)

A

3-amp. fuse

Double-pole, double-throw reversing switch

To anode   To cathode

FIG. 1

3. Methods of Preparation

Dimethyl and diethyl octadecanedioate have been prepared by the electrolysis of the sodium or potassium salts of methyl or ethyl hydrogen sebacate in water [5] or in methanol.[6]

4. Merits of Preparation

The advantages of this preparation of dimethyl octadecane-dioate over that described earlier [5] are purity of product, no foaming of the electrolyte, higher yields, and elimination of the use of large quantities of the salt of the acid ester as supporting electrolyte. The procedure is regarded as near the optimum for a Kolbe electrolysis.

1. Department of Chemistry, University of Illinois, Urbana, Ill. This work was supported by a grant from the National Science Foundation for polymer research.
2. F. L. M. Pattison, W. C. Howell, A. J. McNamara, J. C. Schneider, and J. F. Walker, *J. Org. Chem.*, **21**, 739 (1956).
3. L. J. Durham, D. J. McLeod, and J. Cason, *Org. Syntheses*, Coll. Vol. 4, 635 (1963).

4. S. Swann, Jr., in A. Weissberger, "Technique of Organic Chemistry," Vol. 2, 2nd ed., Interscience Publishers, New York, 1950, p. 400.
5. S. Swann, Jr., R. Oehler, and P. S. Pinkney, *Org. Syntheses*, Coll. Vol. 3, 401 (1955); D. A. Fairweather, *Proc. Roy. Soc. Edinburgh*, **45**, 283 (1925); S. Shiina, *J. Soc. Chem. Ind. Japan*, **40**, Suppl. binding 324 (1937) [*C.A.*, **32**, 499 (1938)]; N. L. Drake, H. W. Carhart, and R. Mozingo, *J. Am. Chem. Soc.*, **63**, 617 (1941).
6. W. S. Greaves, R. P. Linstead, B. R. Shephard, S. L. S. Thomas, and B. C. L. Weedon, *J. Chem. Soc.*, 3326 (1950); W. Fuchs and E. Dickersbach-Boronetsky, *Fette Seifen Anstrichmittel*, **57**, 675 (1955) [*C.A.*, **51**, 1843 (1957)].

2,7-DIMETHYLOXEPIN

(Oxepin, 2,7-dimethyl-)

Submitted by L. A. Paquette and J. H. Barrett [1]
Checked by Jon Malmin and Ronald Breslow

1. Procedure

A. *1,2-Dimethyl-1,4-cyclohexadiene. Caution! This step should be conducted in a hood to avoid exposure to ammonia fumes.* A 5-l.

three-necked flask, cooled in a dry ice-isopropyl alcohol bath, is fitted with an efficient stirrer and a dry ice condenser. The flask is charged with approximately 2.5 l. of liquid ammonia, the stirrer is started, and 450 g. of anhydrous diethyl ether, 460 g. (10 moles) of absolute ethanol, and 318.5 g. (3.0 moles) of o-xylene (Note 1) are added slowly in that order (Note 2). Then 207 g. (9.0 g. atoms) of sodium is added in small pieces over a 5-hour period (Note 3). The ammonia is allowed to evaporate overnight. The flask is now equipped with a reflux condenser, and approximately 800 ml. of ice water is slowly added with stirring to dissolve the salts (Note 4). The two layers which form are separated and the upper organic layer is washed three times with 800-ml. portions of water and dried over anhydrous magnesium sulfate. The liquid is separated from the drying agent and is distilled through a 20-cm. Vigreux column. The fraction boiling at 70–72° (48 mm.) is collected and weighs 250–300 g. (77–92%). The 1,2-dimethyl-1,4-cyclohexadiene is sufficiently pure for the epoxidation reaction (Note 5).

B. *1,2-Dimethyl-1,2-epoxycyclohex-4-ene.* A 2-l. three-necked flask equipped with an efficient stirrer, a reflux condenser, and a dropping funnel is charged with 41 g. (0.38 mole) of 1,2-dimethyl-1,4-cyclohexadiene. Over a period of 2 hours 80 g. (0.46 mole of 85% assay) of m-chloroperbenzoic acid (Note 6) dissolved in 1 l. of chloroform is added with vigorous stirring. The mixture is heated to reflux on a steam bath for 3 hours and kept overnight. The contents of the flask are cooled in an ice bath and the precipitated m-chlorobenzoic acid is removed by filtration. The organic layer is washed with 25 ml. of 20% sodium bisulfite solution, three 100-ml. portions of 10% sodium bicarbonate solution (Note 7), and 100 ml. of saturated sodium chloride solution, in that order. The organic layer is dried over anhydrous magnesium sulfate, filtered, concentrated under reduced pressure, and distilled through a 30-cm. glass bead-packed column (Note 8) to afford 32.3–36.8 g. (68–78%) of the epoxide, b.p. 55–57° (15 mm.); n^{20}D 1.4642–1.4650 (Note 9).

C. *4,5-Dibromo-1,2-dimethyl-1,2-epoxycyclohexane.* Into a 1-l. three-necked flask fitted with an efficient stirrer, an alcohol thermometer, a dropping funnel, and a drying tube are placed 32 g. (0.26 mole) of the epoxide and 500 ml. of an anhydrous chloroform-methylene chloride mixture (1:1). The solution is cooled to $-65°$ and 34 g. (0.21 mole) of bromine in 50 ml. of the same solvent is added dropwise while maintaining the temperature below $-60°$ (Note 10). When the addition is complete, the reaction mixture is stirred for 30 minutes and the solvent is removed at room temperature under reduced pressure. The resulting oil (or solid) is recrystallized from a minimum amount of *n*-hexane to give 47–51 g. (80–86%) of lustrous white needles, m.p. 82–83°.

D. *2,7-Dimethyloxepin.* In a 1-l. three-necked flask fitted as above, except that the dropping funnel is replaced by a 125-ml. Erlenmeyer flask connected to the reaction flask by means of Gooch tubing, is placed a solution of 42.1 g. (0.15 mole) of the purified dibromoepoxide in 500 ml. of anhydrous ether. The solution is cooled to 0° and 33.2 g. (0.30 mole) of potassium *t*-butoxide (Note 11) is added portionwise through the Gooch tubing over a period of 1 hour while maintaining the temperature below 5°. The resulting mixture is stirred for 30 minutes and filtered. The ether is removed under reduced pressure, and the residual liquid is distilled to give 9.7–12.2 g. (52–67%) of 2,7-dimethyloxepin as an orange oil, b.p. 49–50° (15 mm.), n^{27}D 1.5010 (Note 12).

2. Notes

1. Eastman Organic Chemicals, white label grade, was used without further purification.

2. It is advisable to precool these reagents before their addition to minimize excessive boiling of the liquid ammonia.

3. Only five or six pieces of sodium should be added at one time in order to avoid an almost uncontrollable exothermic reaction. The solution turns blue and then white as the sodium is consumed. When the solution turns white, another portion of sodium may be added. The last 50 g. of sodium may be added without waiting between portions because the reaction is much slower at this point.

4. Because dissolution of the salts is a highly exothermic process, the water should be added slowly. A stream of nitrogen may be passed through the reaction during the addition of the

water to ensure that no fire is started by bits of sodium that may be adhering to the upper walls of the flask.

5. The product is readily analyzed by vapor phase chromatography. Since the only impurity is o-xylene (conversions range from 80% to 100%), the percentage of reduction product was calculated from the gas chromatogram and this value was used to determine the amount of m-chloroperbenzoic acid to be used in the epoxidation.

6. m-Chloroperbenzoic acid was obtained from Aldrich Chemical Company, Milwaukee, Wisconsin.

7. The separatory funnel must be vented frequently because of the large volume of carbon dioxide liberated at this point.

8. It appears necessary to effectively remove the residual o-xylene during this distillation in order that it not interfere (by liberation of hydrogen bromide) with the subsequent bromination of the epoxide. The checkers used a spinning-band column for this distillation.

9. The submitters worked at four times this scale with similar yields and purity.

10. Best yields are obtained if the bromination mixture is never allowed to become orange in color. If a calculated amount of bromine is added to the epoxide, the yields of dibromide are greatly diminished.

11. Potassium t-butoxide may be obtained from MSA Research Corporation, Callery, Pennsylvania.

12. 2,7-Dimethyloxepin is stable for long periods when stored under nitrogen at 0–5°.

3. Discussion

The procedure described is patterned after the method suggested by Vogel, Schubart, and Böll,[2] and it illustrates a general method of preparing oxepins. Furthermore, the first step represents an example of the Birch reduction of an aromatic hydrocarbon.[3] The second step is illustrative of the selective epoxidation of a diene system.[3]

Oxepins themselves are interesting examples of cyclic conjugated molecules with $4n$ π-electrons.

1. Department of Chemistry, The Ohio State University, Columbus, Ohio 43210.
2. E. Vogel, R. Schubart and W. A. Böll, *Angew. Chem. Intern. Ed. Engl.*, **3**, 510 (1964).
3. W. Hückel and U. Wörffel, *Ber.*, 88, 338 (1955).

α,α-DIMETHYL-β-PHENETHYLAMINE *

(Phenethylamine, α,α-dimethyl)

$$C_6H_5CH_2C(CH_3)_2OH + HCN \xrightarrow{H_2SO_4}$$

$$C_6H_5CH_2C(CH_3)_2NHCHO \xrightarrow[-HCOONa]{NaOH} C_6H_5CH_2C(CH_3)_2NH_2$$

Submitted by JOHN J. RITTER [1] and JOSEPH KALISH [2]
Checked by WILLIAM G. DAUBEN and ALAN KRUBINER

1. Procedure

Caution! This preparation should be conducted in a hood because poisonous hydrogen cyanide may be evolved.

A. *N-Formyl-α,α-dimethyl-β-phenethylamine.* To a 2-l., three-necked, round-bottomed flask fitted with a sealed stirrer carrying a crescent-shaped blade, a thermometer, an addition funnel, and a reflux condenser connected to a trap containing 20% sodium hydroxide solution is added 500 ml. of glacial acetic acid (Note 1). The contents of the flask are cooled to 20° by means of an ice bath, the addition funnel is temporarily replaced by a stopper, and to the stirred acetic acid is added 110 g. (2 moles) of 95% sodium cyanide in small portions. The temperature of the mixture is maintained around 20°, and the addition requires about 20 minutes (Note 2). The addition funnel is replaced, and a previously prepared and cooled solution of 500 g. (272 ml., 4.9 moles) of concentrated sulfuric acid in 250 ml. of glacial acetic acid is added slowly, with stirring, the temperature of the mixture being kept at about 20° by means of an ice bath (Note 3). The ice bath is removed, and 300 g. (2 moles) of α,α-dimethyl-β-phenethyl alcohol (Note 4) is added over a 20-minute period during which the temperature of the mixture rises slowly to 35–

45°. The reaction mixture is stirred for an additional 90 minutes (Note 5) and allowed to stand overnight. The amber-colored mixture containing some solid sodium sulfate is aerated with nitrogen for 2 hours (*Caution! In a hood*), poured into 3 l. of ice water, and the supernatant oil separated. The aqueous phase is neutralized with sodium carbonate and extracted with 600 ml. of ether. The ethereal extract is combined with the original oily supernatant, neutralized with sodium carbonate, and dried over anhydrous sodium sulfate. The solvent is removed under reduced pressure, and the residue is distilled to yield 230–248 g. (65–70%) of product, b.p. 137–141° (2 mm.). Redistillation of the ether-containing fore-run yields up to an additional 14 g. of material.

B. *α,α-Dimethyl-β-phenethylamine*. In a 3-l., three-necked, round-bottomed flask equipped with a reflux condenser and a sealed stirrer are placed 246 g. (1.39 moles) of N-formyl-α,α-dimethyl-β-phenethylamine and 2.1 l. of 20% sodium hydroxide solution. The mixture is heated under reflux with vigorous stirring for 2.5 hours or until a test portion of the oily layer dissolves completely in cold 5% hydrochloric acid. The reaction mixture is cooled, 750 ml. of benzene is added, the mixture is stirred, and the benzene layer is separated. The benzene solution is shaken with a saturated sodium chloride solution, the benzene removed by distillation at atmospheric pressure, and the product distilled at reduced pressure to yield 155–165 g. (75–80%) of α,α-dimethyl-β-phenethylamine, b.p. 80–82° (10 mm.).

2. Notes

1. The reaction may be conducted in other solvents (e.g., dibutyl ether) or in the absence of solvent with some alteration in the order of mixing the reagents. The submitters find for this and a large number of similar preparations that acetic acid generally is most convenient.

2. Most of the sodium cyanide does not dissolve.

3. Care must be taken during the first part of the addition because the reaction is very exothermic.

4. Methallylbenzene or isobutenylbenzene may be used in place of the carbinol with practically identical results.

5. The temperature may continue to rise during the initial

portion of this period, but it is controlled by means of an ice bath to limit the temperature of the mixture to 45–50°.

3. Methods of Preparation

α,α-Dimethyl-β-phenethylamine has been prepared from benzaldehyde and 2-nitropropane [3] and from isobutyrophenone by a series of steps involving alkylation with benzyl bromide, alkali cleavage, and Hofmann bromamide degradation.[4]

4. Merits of the Preparation

The present method is shorter and less laborious than previously described methods, and it gives better yields of material. The method, now known as the Ritter reaction, is one of considerable scope,[5] having been used with fair to excellent success with many tertiary alcohols or the corresponding alkenes, with benzyl alcohol, and with some secondary alcohols. It has also been used with alkanes, alkadienes, alicyclic and spiro alcohols, alkyl chlorides, glycols, aldehydes, chlorohydrins, N-methylolamides, ethers, carboxylic acids, esters, ketones, and ketoximes. The Ritter reaction has been reviewed.[5] Another example of this reaction is given elsewhere in this volume.[6]

1. Evans Research and Development Corporation, 250 East 43 Street, New York, New York.
2. Drug and Cosmetic Industry, 101 West 31 Street, New York, New York.
3. R. S. Shelton and M. C. Van Campen, Jr., U.S. Patent 2,408,345 (Sept. 1946) [*C.A.*, 41, 567 (1947)]; B. L. Zenitz, E. B. Macks, and M. L. Moore, *J. Am. Chem. Soc.*, 70, 955 (1948).
4. L. L. Abell, W. F. Bruce, and J. Seifter, U.S. Patent 2,590,079 (March 1952) [*C.A.*, 46, 10200 (1952)].
5. L. I. Krinsen and D. J. Cota, *Org. Reactions,* 17, 213 (1969).
6. C. L. Parris, this volume, p. 73.

2,4-DINITROBENZENESULFENYL CHLORIDE *

(Benzenesulfenyl chloride, 2,4-dinitro-)

$$O_2N-\langle\rangle-Cl + HSCH_2C_6H_5 \xrightarrow{C_5H_5N}$$

$$O_2N-\langle\rangle-SCH_2C_6H_5 + C_5H_5N \cdot HCl$$

$$O_2N-\langle\rangle-SCH_2C_6H_5 \xrightarrow{SO_2Cl_2}$$

$$O_2N-\langle\rangle-SCl + ClCH_2C_6H_5 + SO_2$$

Submitted by Norman Kharasch and Robert B. Langford [1]
Checked by D. C. Dittmer and B. C. McKusick

1. Procedure

Caution! Both steps should be carried out in a good hood.

A. *2,4-Dinitrophenyl benzyl sulfide.* The apparatus consists of a 1-l., three-necked, round-bottomed flask equipped with a sealed mechanical stirrer and a reflux condenser. In it are placed 202 g. (1.00 mole) of 2,4-dinitrochlorobenzene (m.p. 50–52°) (*Caution! A skin irritant*), 400 ml. of methanol, 124 g. (1.00 mole) of benzyl mercaptan, and 87 g. (85 ml., 1.10 moles) of pyridine. The mixture is heated at the reflux temperature with stirring for 16 hours or more (Note 1) and cooled to 0°. The 2,4-dinitrophenyl benzyl sulfide that precipitates is separated by filtration, washed with two 250-ml. portions of ice-cold methanol, and dried at 60–80°.

The sulfide, a yellow crystalline solid that melts at 128–129°, weighs 235–250 g. (81–86%) (Note 2). It may be used in the next step without further purification.

B. *2,4-Dinitrobenzenesulfenyl chloride.* Dry 2,4-dinitrophenyl benzyl sulfide (232 g., 0.80 mole) and 400 ml. of dry ethylene chloride are placed in a 2-l., one-necked, round-bottomed flask equipped with a stirrer (Note 3). Sulfuryl chloride (119 g., 0.88 mole) (Note 4) is added to the resulting suspension at room temperature. A mildly exothermic reaction causes the solid to dissolve quickly, usually within 1 to 2 minutes, with a temperature rise of 10–15° (Note 5). The resulting clear yellow solution is concentrated to an oil by heating under aspirator vacuum on a steam bath (Note 6). (*Caution! Do not heat with gas or electricity because the product, like many nitro compounds, can explode if overheated.*) The residual oil is cooled to 50–60°, and 3–4 volumes of dry petroleum ether (b.p. 30–60°) are added with vigorous hand-swirling. The oil quickly crystallizes. The mixture is cooled to room temperature and filtered to separate 2,4-dinitrobenzenesulfenyl chloride as a yellow crystalline solid. The sulfenyl chloride is washed well with dry petroleum ether and dried at 60–80° (Note 7); weight 150–170 g. (80–90%); m.p. 95–96° (Notes 8, 9).

2. Notes

1. After 2 or 3 hours, solid product usually appears in the reaction mixture.

2. When practical grade 2,4-dinitrochlorobenzene (m.p. 46–47°) is substituted, a product of equally good quality (m.p. 128–129°) is obtained, but the yield is only 70–75%.

3. All materials and equipment used in Step B of this procedure must be completely dry to avoid loss of product by hydrolysis. The checkers found, however, that the reaction may be carried out open to the air without loss of yield.

4. Practical grade sulfuryl chloride, obtained from Matheson, Coleman and Bell, gives satisfactory results.

5. The 2,4-dinitrophenyl benzyl sulfide normally undergoes cleavage at room temperature without the addition of a catalyst. If the reaction does not occur spontaneously, the mixture may be warmed gently and/or one drop of dry pyridine may be added to

initiate the reaction.

6. Rotary or other distillation equipment with metal parts should not be used in concentrating the reaction mixture because not only will the corrosive vapors damage the equipment, but also the resulting metal salts will discolor and partially decompose the product. The solution should not be heated any longer than is necessary to concentrate it; excessive heating gives a dark-colored product.

7. The product should not be dried longer than is necessary for it to reach constant weight, or there may be partial decomposition.

8. The product obtained by this procedure is pure enough for most purposes. Its melt, however, is faintly cloudy. A product of high purity, giving a clear melt, can be obtained by recrystallization from about 15 ml. of dry carbon tetrachloride per gram of sulfenyl chloride. When stored in a sealed brown bottle with a plastic cap (no metal!), the sulfenyl chloride is stable for years.

9. 2,4-Dinitrobenzenesulfenyl bromide may be similarly prepared by refluxing 2,4-dinitrophenyl benzyl sulfide with the equivalent amount of bromine in 5 parts of dry carbon tetrachloride. As it is less stable than the chloride, losing bromine if overheated, it should be concentrated on a 40° water bath under vacuum. When worked up in the same manner as the chloride, the product usually contains some *bis*-(2,4-dinitrophenyl) disulfide. Because the disulfide is insoluble in carbon tetrachloride, the sulfenyl bromide may be readily purified by recrystallization; yield 75–80%, m.p. 102–104°.

3. Methods of Preparation

2,4-Dinitrophenyl benzyl sulfide has been prepared by the reaction of benzyl chloride with 2,4-dinitrothiophenol [2] or *bis*-(2,4-dinitrophenyl) disulfide [3] and by the condensation of 2,4-dinitrochlorobenzene with benzyl mercaptan.[4]

2,4-Dinitrobenzenesulfenyl chloride has been obtained by the chlorinolysis of 2,4-dinitrophenyl thiolbenzoate,[5] 2,4-dinitrothiophenol,[6] or *bis*-(2,4-dinitrophenyl) disulfide, [7,11] and by the present procedure.[8]

4. Merits of the Preparation

2,4-Dinitrobenzenesulfenyl chloride is a versatile analytical reagent for the characterization of a wide variety of organic compounds, including alcohols, mercaptans, ketones, olefins, amines, aromatic compounds, olefin oxides, and hydroxysteroids. Review articles summarize these applications.[9, 10]

The chlorinolysis of 2,4-dinitrophenyl benzyl sulfide gives a good yield of product which is satisfactory for most purposes without recrystallization. Only simple equipment and inexpensive materials are needed, only 2 or 3 hours of the operator's time are required, and the entire procedure can be completed within 24 hours.

The best previous method of preparation, the chlorinolysis of *bis*-(2,4-dinitrophenyl) disulfide by sulfuryl chloride in the presence of pyridine,[11] requires much more time and effort with results that are uncertain, even for experienced operators.

1. Department of Chemistry, University of Southern California, Los Angeles, California.
2. C. Willgerodt, *Ber.*, **18**, 331 (1885).
3. E. Fromm, H. Benzinger, and F. Schäfer, *Ann.*, **394**, 335 (1912).
4. R. W. Bost, J. O. Turner, and R. D. Norton, *J. Am. Chem. Soc.*, **54**, 1985 (1932).
5. K. Fries and W. Buchler, *Ann.*, **454**, 258 (1927).
6. G. W. Perold and H. L. F. Snyman, *J. Am. Chem. Soc.*, **73**, 2379 (1951).
7. M. H. Hubacher, *Org. Syntheses*, Coll. Vol. **2**, 456 (1943); N. Kharasch, G. I. Gleason, and C. M. Buess, *J. Am. Chem. Soc.*, **72**, 1796 (1950).
8. N. Kharasch and R. B. Langford, *J. Org. Chem.*, **28**, 1903 (1963).
9. N. Kharasch, *J. Chem. Educ.*, **33**, 585 (1956).
10. R. B. Langford and D. D. Lawson, *J. Chem. Educ.*, **34**, 510 (1957).
11. D. D. Lawson and N. Kharasch, *J. Org. Chem.*, **24**, 858 (1959).

2,4-DINITROIODOBENZENE *

(Benzene, 1-iodo-2,4-dinitro-)

$$\text{(structure with Cl, NO}_2\text{, NO}_2\text{)} + \text{NaI} \rightarrow \text{(structure with I, NO}_2\text{, NO}_2\text{)} + \text{NaCl}$$

Submitted by J. F. BUNNETT and R. M. CONNER.[1]
Checked by VIRGIL BOEKELHEIDE and JAMES S. TODD.

1. Procedure

Caution! 2,4-Dinitrochlorobenzene causes severe skin irritation to some individuals. Sensitive persons are advised to wear rubber gloves.

To 200 ml. of redistilled technical grade dimethylformamide (Note 1) in a 1-l. round-bottomed flask are added 150 g. (1 mole) of sodium iodide (Note 2) and 40.5 g. (0.2 mole) of 2,4-dinitrochlorobenzene (Note 3). The flask is fitted with a reflux condenser, and the red-brown mixture is refluxed for 15 minutes over a free flame arranged to impinge somewhat off-center (Note 4).

The hot reaction mixture is poured with stirring over about 0.75 l. of crushed ice in a 2-l. beaker. The beaker is filled with water, and the mixture is stirred to dissolve inorganic salts (Note 5). The insoluble red-brown solid is collected on a suction filter. This crude product, even while damp, is transferred to a 2-l. round-bottomed flask, and 500 ml. of a mixture of 75% (375 ml.) of petroleum ether (b.p. 90–100°) and 25% (125 ml.) of benzene is added. The flask is provided with a reflux condenser, and the mixture is heated at reflux for 15 minutes by means of an electric mantle (Note 6). The resulting solution is decanted into a second 2-l. flask, leaving in the first flask some water and a red-brown solid residue. To the slightly cooled liquor in the second flask is added cautiously 7 g. of powdered activated carbon. The carbon is dispersed by swirling, and the mixture is heated for an additional 5 minutes. The mixture is then filtered through a fluted filter into a 1-l. Erlenmeyer flask. This flask is stoppered and chilled to cause crystallization of the product (Note 7).

The product is collected on a suction filter. The yield of air-dried product, as yellow-orange crystals, m.p. 87–89°, is 38–42 g. (65–71%) (Note 8). A purer, lemon-yellow product, m.p. 88.5–90°, is obtained by an additional recrystallization from 1 l. of petroleum ether (b.p. 90–100°) with use of carbon; the yield after this second crystallization is 28–34 g. (48–58%).

2. Notes

1. Dimethylformamide is available from E. I. du Pont de Nemours and Co., Inc., Wilmington, Delaware.

2. Reagent grade sodium iodide was used.

3. Eastman Kodak white label 2,4-dinitrochlorobenzene, which had a light tan color, was used without further purification.

4. The burner should be so arranged that the flame impinges upon the side of the flask in such a way as to dissolve the sodium iodide from the top downward. The flask should be shaken or swirled frequently during the onset of boiling. A small amount of white solid remains undissolved even at reflux.

5. At this point there may be a lumpy, dark-brown material mixed with the orange-brown crude precipitated product. There appears to be no need to break up the lumps.

6. Alternatively, the flask can be heated on an efficient steam bath with swirling until the crude product melts. After an additional 15 minutes of heating, the solution can be decanted as described.

7. The submitters placed the flasks of warm filtrate in the refrigerator for storage overnight or in a freezing cabinet at −20° for at least 1 hour.

8. From a run of five times the scale described, the submitters obtained 219 g. (71%) of an orange-yellow crystalline product, m.p. 88–90°.

3. Methods of Preparation

2,4-Dinitroiodobenzene has been prepared by the nitration of
o- or *p*-nitroiodobenzene,[2] by treatment of 2,4-dinitrobenzenedi-
azonium sulfate with potassium iodide,[3] and by the reaction of
sodium iodide with 2,4-dinitrochlorobenzene in refluxing ethylene
glycol.[4] The present procedure is a modification [5] of the last-
mentioned one.

1. Department of Chemistry, University of North Carolina, Chapel Hill, North
   Carolina.
2. W. Körner, *Gazz. Chim. Ital.*, 4, 323 (1874).
3. A. L. Beckwith, J. Miller, and G. D. Leahy, *J. Chem. Soc.*, 3556 (1952).
4. G. M. Bennett and I. H. Vernon, *J. Chem. Soc.*, 1783 (1938).
5. J. F. Bunnett and R. M. Conner, *J. Org. Chem.*, 23, 305 (1958).

3,5-DINITRO-*o*-TOLUNITRILE

(Benzonitrile, 2-methyl-3,5-dinitro-)

$$HNO_3 \;+\; HF \;+\; 2\,BF_3 \;\longrightarrow\; O_2N^+BF_4^- \;+\; H_2O \cdot BF_3$$

$$+ \; 2\,O_2N^+BF_4^- \;\longrightarrow\; + \; 2\,HF \;+\; 2\,BF_3$$

Submitted by George A. Olah [1] and Stephen J. Kuhn [2]
Checked by J. Lazar and B. C. McKusick

1. Procedure

A. *Nitronium tetrafluoroborate (Note 1).* *Caution!* *Hydrogen
fluoride is very hazardous.* *Caution is also called for in the use of
boron trifluoride.* *All operations must be carried out in a hood, and
the precautions outlined in Note 2 should be followed.* A 1-l. three-
necked polyolefin flask (Note 3) is provided with a short inlet
tube for nitrogen, a long inlet tube for gaseous boron trifluoride,
a drying tube, and a magnetic stirring bar (Note 4). The flask
is immersed in an ice-salt bath and flushed with dry nitrogen.

Under a gentle stream of nitrogen and with stirring, the flask is charged with 400 ml. of methylene chloride, 41 ml. (65.5 g., 1.00 mole) of red fuming nitric acid (95%), and 22 ml. (22 g., 1.10 moles) of cold, liquid, anhydrous hydrogen fluoride (Note 5).

Gaseous boron trifluoride (136 g., 2.00 moles) from a cylinder mounted on a scale is bubbled into the stirred, cooled reaction mixture (Note 6). The first mole is passed in rather quickly (in about 10 minutes). When approximately 1 mole has been absorbed, copious white fumes begin to appear at the exit, and the rate of flow is diminished so that it takes about 1 hour to pass in the second mole; even at this slow rate, there is considerable fuming at the exit. After all the boron trifluoride has been introduced, the mixture is allowed to stand in the cooling bath under a slow stream of nitrogen for 1.5 hours. The mixture is swirled, and the suspended product is separated from the supernatant liquid by means of a medium-porosity, sintered-glass Buchner funnel (Note 7). The gooey solid remaining in the flask is transferred to the funnel with the aid of two 50-ml. portions of nitromethane. The solid on the funnel, nitronium tetrafluoroborate, is washed successively with two 100-ml. portions of nitromethane and two 100-ml. portions of methylene chloride. In order to protect the salt from atmospheric moisture during the washing procedure, suction is always stopped while the salt is still moist. The moist salt is transferred to a round-bottomed flask and dried by evaporating the solvent (Note 8). At the end of the procedure the flask can be gently heated to 40–50° (Note 9). The yield of colorless nitronium tetrafluoroborate is 85–106 g. (64–80%) (Notes 10, 11, 12). It is stored in a wide-mouthed polyolefin bottle with a screw cap. The edge of the cap is sealed with paraffin wax after it is screwed on.

B. *3,5-Dinitro-o-tolunitrile.* A 500-ml. four-necked flask is equipped with a mechanical stirrer, a dropping funnel, a thermometer, and an inlet for dry nitrogen (Note 13). It is baked thoroughly by means of a Bunsen flame and allowed to cool to room temperature with a slow stream of dry nitrogen passing through it. The flask is charged, preferably in a dry box, with 335 g. of tetramethylene sulfone (Note 14) and 73.1 g. (0.55 mole) of nitronium tetrafluoroborate. The thermometer is adjusted so that the bulb is immersed in the liquid. The reaction mixture is stirred well and kept at 10–20° by means of an ice bath while 58.5 g. (0.50 mole) of freshly distilled *o*-tolunitrile [3] is added dropwise. The nitronium tetrafluoroborate only partially dissolves in the tetramethylene sulfone (Note 15), but through good stirring a homogeneous suspension can be obtained. As the dissolved nitronium salt reacts with the nitrile, more and more salt dissolves until all of it is in solution. The addition of *o*-tolunitrile requires 25–35 minutes.

After the addition is complete, the cooling bath is removed and stirring is continued for 15 minutes at 35°. The dropping funnel is removed, 74.5 g. (0.56 mole) of nitronium tetrafluoroborate is added, and the opening of the flask is closed with a glass stopper. The well-stirred reaction mixture is heated by an electric heating mantle to 100° in 15 minutes and kept at 100–115° for 1 hour. The reaction mixture is allowed to cool to room temperature with continued stirring and is then poured into 800 g. of ice water. Crude 3,5-dinitro-*o*-tolunitrile separates on top of the aqueous mixture as a dark oil that solidifies after standing a few minutes. The solid is collected on a Buchner funnel. It is triturated on the funnel with five 50-ml. portions of cold water and with 40 ml. of ice-cold ethanol. After being dried in a vacuum desiccator, the crude nitrile weighs 75–84 g., m.p. 60–65°. Recrystallization from about 110 ml. of hot methanol gives 50–55 g. (48–53%) of 3,5-dinitro-*o*-tolunitrile, m.p. 82–84° (Note 16).

2. Notes

1. Nitronium tetrafluoroborate is commercially available from the Ozark-Mahoning Co., Tulsa, Oklahoma.

2. *Because of the hazardous nature of anhydrous hydrogen fluoride, adequate precautions should be taken to protect the head, eyes, and skin.* Rubber gloves, an apron, and a plastic face mask are strongly recommended. All operations should be carried out in a hood. If hydrogen fluoride comes in contact with the skin, the contacted area should be thoroughly washed with water and then immersed in ice water while the patient is taken to a physician. After completion of the reaction, all equipment should be washed with liberal quantities of water. *Note! Burns caused by*

hydrogen fluoride may not be noticed for several hours, by which time
serious tissue damage may have occurred.

3. All operations involving liquid hydrogen fluoride must be carried out with equipment resisting hydrogen fluoride (fused silica, polyolefin, etc.).

4. An egg-shaped stirrer seems to work best. As the reaction proceeds, the precipitating nitronium tetrafluoroborate prevents the stirring bar from operating. This is not serious if the reaction mixture is shaken occasionally.

5. It is convenient to condense anhydrous hydrogen fluoride, b.p. 19.5°, from a cylinder into a small calibrated polyolefin flask immersed in a mixture of dry ice and acetone. As hydrogen fluoride is very hygroscopic, it should be carefully protected from atmospheric moisture, preferably by maintaining an atmosphere of dry nitrogen over it, otherwise by means of a drying tube. The hydrogen fluoride is then simply poured into the reaction flask.

6. The temperature of the reaction is not critical, but the reaction is slower at higher temperatures because of the lower solubility of boron trifluoride in the solvent.

7. Since free hydrogen fluoride is no longer present, filtration can be carried out with glass or porcelain equipment. However, commercially available polyolefin Buchner funnels and filter flasks are preferred.

8. Kel-F grease is recommended for ground-glass joints. Nitronium tetrafluoroborate slowly attacks silicone stopcock grease, causing air to enter the flask.

9. Nitronium tetrafluoroborate is thermally stable up to 170°. Above this temperature it starts to dissociate into nitryl fluoride and boron trifluoride.

10. Nitronium tetrafluoroborate is very hygroscopic. It is stable as long as it is anhydrous, but it is decomposed by moisture, and all transfers should be in a dry box. Its purity can be checked by conventional elemental analysis. However, because of the hygroscopic nature of the salt, the submitters have found it convenient to use neutron activation analysis (B, F, N, O) of samples sealed into polyolefin sample holders. Lange's method [4] for the determination of BF_4^- as the nitron salt gives good results but requires considerable care to achieve reproducibility.

11. The last part of the procedure can be used to purify nitronium tetrafluoroborate that has picked up water on standing. The impure salt is washed twice with nitromethane, twice with methylene chloride, and is dried under reduced pressure.

12. Nitronium tetrafluoroborate slowly attacks polyethylene and polypropylene, but apparatus made of these materials will last for several preparations of the salt.

13. The entire operation should be carried out in an atmosphere of dry nitrogen. If dry nitrogen is not available, rigorously anhydrous conditions should be maintained with the help of a drying tube.

14. Tetramethylene sulfone is commercially available from the Shell Chemical Company and the Phillips Petroleum Company.

15. A saturated solution at 25° contains 7 g. of nitronium tetrafluoroborate per 100 g. of tetramethylene sulfone.

16. This is pure enough for most purposes. An analytical sample melted at 86.5–88.4°.

3. Methods of Preparation

Nitronium tetrafluoroborate has been prepared by interaction of nitric acid, hydrogen fluoride, and boron fluoride in nitromethane.[5] However, mixtures of nitric acid and nitromethane are extremely explosive.[6, 7] The present modification of the procedure, in which the medium is methylene chloride instead of nitromethane, was developed to avoid this hazard. It has not been published before.

The preparation of 3,5-dinitro-o-tolunitrile is based on previously published work.[8] The nitration of o-tolunitrile using fuming nitric acid has been reported by Candea and Macovski.[9]

4. Merits of the Preparation

Nitration of aromatic rings by nitronium tetrafluoroborate is a general method. Fifty-seven arenes, haloarenes, nitroarenes, arenecarboxylic esters, arenecarbonyl halides, and arenecarbonitriles have been nitrated in high yield by this reagent.[8] The method is particularly convenient for nitrating aromatic compounds susceptible to acid-catalyzed hydrolysis. For example, although mononitration of arenecarbonitriles is easily accomplished by conventional nitrating agents, dinitration is not. The reason is that the forcing conditions required for dinitration (strongly acid media and higher temperatures) bring about hydrolysis (and oxidation) of the nitrile group. In contrast, nitrations with nitronium tetrafluoroborate can be carried out in nonaqueous acid-free systems, where the only acid originates from proton elimination during nitration. In the basic solvent used, this acid concentration generally is not sufficient to cause any detectable hydrolysis (or oxidation).

1. Western Reserve University, Cleveland, Ohio.
2. Dow Chemical of Canada, Limited, Sarnia, Ontario.
3. H. T. Clarke and R. R. Read, *Org. Syntheses*, Coll. Vol. 1, 514 (1941).
4. W. Lange, *Ber.*, **59**, 2107, 2432 (1926).
5. S. J. Kuhn, *Can. J. Chem.*, **40**, 1660 (1962).
6. D. W. Coillet and S. D. Hamann, *Trans. Faraday Soc.*, **57**, 2232 (1961).
7. J. A. Herickes, J. Ribovich, G. H. Damon, and R. W. Van Dolah, "Shock-Sensitivity Studies of Liquid Systems," Preprint, Second Conference on Explosives Sensitivity, Washington, D. C., September 16–17, 1957.
8. S. J. Kuhn and G. A. Olah, *J. Am. Chem. Soc.*, **83**, 4564 (1961); G. A. Olah, S. Kuhn, and A. Mlinko, *J. Chem. Soc.*, 4257 (1956).
9. C. Candea and E. Macovski, *Bull. Soc. Chim. France*, [5] **5**, 1350 (1938).

1,6-DIOXO-8a-METHYL-1,2,3,4,6,7,8,8a-OCTAHYDRONAPHTHALENE

(1,6-Naphthalenedione, 1,2,3,4,6,7,8,8a-octahydro-8a-methyl-)

Submitted by S. Ramachandran and Melvin S. Newman.[1]
Checked by Max Tishler, G. A. Stein, and G. Lindberg.

1. Procedure

A mixture of 63.1 g. (0.5 mole) of 2-methyl-1,3-cyclohexane-dione (Note 1), 52.6 g. (0.75 mole) of methyl vinyl ketone (Note 2), about 0.25 g. (3 pellets) of potassium hydroxide, and 250 ml. of absolute methanol is placed in a 500-ml. round-bottomed flask fitted with a reflux condenser and a drying tube (Note 3). The mixture is heated under reflux for 3 hours, and the dione gradually goes into solution. At the end of this period, methanol and the excess methyl vinyl ketone are removed by distillation under reduced pressure (Notes 4 and 5). The residual liquid is dissolved in 250 ml. of benzene, a Dean-Stark phase-separating head is attached, and 20 ml. of solvent is removed by distillation at atmospheric pressure to remove traces of water and methanol. The solution is cooled well below the boiling point, 3 ml. of pyrrolidine is added (Note 6) and the mixture held at reflux for about 30 minutes, during which time about 9 ml. of water collects in the trap. Refluxing is continued for an additional 15 minutes after the separation of water ceases. The water collected is removed, and then 50 ml. of solvent is distilled. The reddish reaction mixture is cooled to room temperature and diluted with 150 ml. of ether. This solution is washed with 100 ml. of distilled water containing 15 ml. of 10% hydrochloric acid and 100 ml. of water. The aqueous extracts are extracted with 50 ml. of ether (Note 7), and the combined organic layers are washed with three 100-ml. portions of water, then with saturated salt solution and dried over magnesium sulfate. The solvents are then removed, and on distillation of the residue (82–85 g.) (Note 8) at 0.5–1.0 mm. (Note 9) the material, b.p. 117–145°, is collected and diluted with 5 ml. of ether. The distillate is placed overnight in a refrigerator, the resulting crystals are then collected by rapid filtration and washed with about 25 ml. of cold ether (Notes 10 and 11). The first crop of diketone weighs 50–53 g. and is colorless. The combined mother liquors are redistilled to obtain a further 4–6 g. of crystalline product. A yield of 56–58 g. (63–65% based on dione) of 1,6-dioxo-8a-methyl-1,2,3,4,6,7,8,8a-octahydronaphthalene, m.p. 47–50°, suitable for most other purposes, is obtained (Note 12).

2. Notes

1. The 2-methyl-1,3-cyclohexanedione was prepared by the method described by Mekler et al., this volume, p. 743.

2. Technical grade methyl vinyl ketone supplied by Matheson, Coleman and Bell Co., Cincinnati, Ohio, was used without further purification.

3. The checkers found that stirring during reflux and concentration and in the cyclization step was advantageous.

4. The submitters report that the intermediate 2-methyl-2-(3'-oxobutyl)-1,3-cyclohexanedione can be isolated in 85% yield [2] at this point if desired.

5. The checkers removed methanol and methyl vinyl ketone at 650 mm. pressure with a water bath at 70°.

6. On adding pyrrolidine an exothermic reaction occurs rapidly. Cooling is needed to prevent too rapid a reaction. The submitters report that piperidine may be used in place of pyrrolidine.

7. The checkers extracted the aqueous extracts twice with 75-ml. portions of ether.

8. The checkers found the residual weights to amount to 87–100 g.

9. The checkers used a saddle-packed 5-in. column and found b.p. 137–150°/0.6–0.7 mm.; 123–150°/0.2–0.5 mm.; and 132–141°/0.5–0.8 mm.

10. A successful crystallization yields relatively large crystals which may be rapidly filtered and washed with ether with little loss. If fine crystals are obtained, it is preferable to redissolve and allow the material to crystallize again.

11. Because of the high solubility in ether (1 g. per 2.5 ml. at room temperature), the checkers washed the product with hexane (b.p. 60–71°).

12. Purer product, m.p. 48.6–50.0°, may be obtained by crystallization from ether.

3. Methods of Preparation

1,6-Dioxo-8a-methyl-1,2,3,4,6,7,8,8a-octahydronaphthalene has been obtained through the reaction of 2-methyl-1,3-cyclohexanedione with acetonedicarboxylic acid and formaldehyde,[2] 4-diethylamino-2-butanone methiodide,[3] pyridine and 4-diethylamino-2-butanone,[4] triethylamine and 4-diethylamino-2-butanone,[5] and by cyclization of 2-methyl-2-(3-oxobutyl)-1,3-cyclohexanedione using either aluminum *tert*-butoxide or piperidine phosphate as catalyst.[6, 7]

4. Merits of Preparation

1,6-Dioxo-8a-methyl-1,2,3,4,6,7,8,8a-octahydronaphthalene has been employed as an intermediate in the synthesis of terpenes [8, 9] and in the projected synthesis of steroids.[4, 10]

1. Department of Chemistry, Ohio State University, Columbus, Ohio.
2. C. B. C. Boyce and J. S. Whitehurst, *J. Chem. Soc.*, 2022 (1959).
3. P. Wieland and K. Miescher, *Helv. Chim. Acta,* 33, 2215 (1950).
4. S. Swaminathan and M. S. Newman, *Tetrahedron,* 2, 88 (1958).
5. M. S. Newman and A. B. Mekler, *J. Am. Chem. Soc.*, 82, 4039 (1960).
6. N. L. Wendler, H. L. Slates, and M. Tishler, *J. Am. Chem. Soc.*, 73, 3816 (1951).
7. I. N. Nazarov, S. I. Zav'yalov, M. S. Burmistrova, I. A. Gurvich, and L. I.

Shmonina, *Zhur. Obshchei. Khim.*, **26**, 441 1956; *J. Gen. Chem. U.S.S.R.*, **26**, 465 (1956) [*C.A.*, **50**, 13847c (1956)].

8. F. Sondheimer and D. Elad, *J. Am. Chem. Soc.*, **79**, 5542 (1957).
9. J. D. Cocker and T. G. Halsall, *J. Chem. Soc.*, 3441 (1957).
10. C. A. Friedman and R. Robinson, *Chem. & Ind. (London)*, 777 (1951).

DIPHENALDEHYDE

(Biphenyl, 2,2′-diformyl-)

Submitted by PHILIP S. BAILEY and RONALD E. ERICKSON.[1]
Checked by MARJORIE C. CASERIO and JOHN D. ROBERTS.

1. Procedure

A finely divided suspension (Notes 1 and 2) of 10.0 g. (0.056 mole) of phenanthrene (Note 3) in 200 ml. of dry methanol (Note 4) is placed in a standard ozonolysis vessel (Note 5). The reaction mixture is cooled in a Dewar flask by an acetone-Dry Ice mixture to about −30° (Notes 6 and 7), and ozone (Note 8) is passed through at a rate of about 20 l. per hour (Note 9) until all the phenanthrene has reacted (Note 10).

To the cooled reaction mixture are added 25–30 g. (roughly 1.5 times the theoretical 0.112 mole) of sodium or potassium iodide and 30 ml. of glacial acetic acid (Note 11). After the addition, the reaction mixture is allowed to stand at room temperature for 30 minutes to 1 hour. The released iodine is reduced with 10% sodium thiosulfate solution, after which the reaction mixture is placed immediately under an air blast (Note 12). As the methanol evaporates, the product begins to crystallize (Note 13). The crystallization should be well advanced by the time most of the methanol has evaporated. Water is then added, and the solid is removed by filtration and dried. The yield of crude product, softening at about 54° and melting at 59–62°, is 9.2–11.4 g. (78–96%). The crude product may be recrystallized by dissolving it in the minimal amount (40–50 ml.) of dry ether and slowly adding about 150 ml. of ligroin (Note 14). Small crystals separate halfway through the addition, and crystallization is completed by cooling the mixture in an acetone-Dry Ice bath. An 80–90% recovery of pale yellow crystals melting at 62–63° is obtained. A second recrystallization from 70% aqueous ethanol gives nearly colorless crystals melting at 62.5–63.5° (Note 15).

2. Notes

1. This is produced by dissolving the phenanthrene in the refluxing solvent and cooling rapidly.

2. The finely divided suspension is necessary in order for the phenanthrene to go into solution and react readily during the ozonolysis.

3. Eastman white label 599, m.p. 99–100°, was used.

4. Commercial methanol reagent containing 0.1% or less of water is satisfactory.

5. The usual long, cylindrical, gas-absorption-type vessel with an inlet tube extending to near the bottom is satisfactory.[2] The total volume of the vessel should be at least twice that of the reaction solution. More elaborate reaction vessels equipped with a stirrer [3] are very useful in reactions such as this in which the reactant is suspended in the solvent. However, the commercially available vessels of this type are not large enough for the reaction mixture described here.

6. The temperature of the reaction mixture should not be allowed to rise above −20°, because at higher temperatures ozone tends to react with the solvent and the reactions shown below also occur. Compound III is not readily reduced to the dialdehyde.

7. Compound II may precipitate during ozonolysis at −30° or below. This is in no way detrimental.

8. A Welsbach T23 ozonator was used by the submitters.

Oxygen dried by a Pittsburg Laboratory-Lectrodryer to a dew point of $-60°$ was passed through the ozonator, which was set to produce a 5–6% by weight concentration of ozone.[4] Following the ozonation flask were a potassium iodide trap and a wet-test meter.[4] The checkers used a simple ozonator capable of producing 3.8% by weight of ozone at a flow rate of 20 l. per hour from oxygen dried by passage through a 30-cm. column of silica gel.

9. The rate should be sufficiently great to cause considerable agitation of the suspended phenanthrene. As the reaction proceeds, the reaction vessel should be shaken frequently in order to maintain good contact between the phenanthrene and ozone. For smaller runs a reaction vessel that includes a stirrer is advantageous (Note 5).

The checkers found it convenient to use leads of Tygon tubing of sufficient length to allow the reaction flask to be withdrawn at intervals from the Dewar flask and shaken manually.

10. Unreacted ozone starts passing through to the potassium iodide trap toward the end of the reaction. However, it is best to continue the reaction until all the suspended phenanthrene has disappeared. This usually requires a total of 1.1–1.3 mole-equivalents of ozone. Unless all the phenanthrene has reacted, difficulty is encountered in the crystallization and/or recrystallization of the dialdehyde.

11. The reduction may be carried out in the ozonolysis flask, or the reaction mixture may be transferred first to an Erlenmeyer flask or beaker. The iodide and acetic acid should be added simultaneously. The reaction of peroxides with iodide ion is exothermic. The temperature of the reaction mixture should be kept below $-20°$ while the sodium iodide and acetic acid are added, after which it may be allowed to rise slowly to room temperature.

12. It seems to be detrimental to the crystallization and re-crystallization of the product to postpone the evaporation of the reaction mixture, probably because the product becomes contaminated with sulfur if the reduced reaction mixture is allowed to stand.

13. Sometimes difficulty is encountered in starting the crystallization, since the product may separate as a yellow oil. It is helpful to induce crystallization by rubbing the sides of the vessel with a stirring rod and seeding the solution with any crystals that form on the sides of the vessel during the evaporation.

14. The ligroin used was Skellysolve B.

15. About 30 ml. of warm absolute ethanol readily dissolves 8–9 g. of product. Addition of 15 ml. of water and cooling effect crystallization with about 90% recovery of product.

3. Methods and Merits of Preparation

Previously, diphenaldehyde has been made by the Ullman coupling of o-iodobenzaldehyde,[5,6] by bromination of o,o'-bitolyl and hydrolysis of the resulting tetrabromo compound,[7] and by lithium aluminum hydride reduction of the N-methylanilide of diphenic acid.[8] These methods involve more steps and give poorer yields than ozonolysis of phenanthrene.

The present method is based on the earlier described ozonolysis of phenanthrene in methanol.[9] The reduction of the peroxidic reaction mixture with trimethyl phosphite to give diphenalde-hyde, isolated as the di-p-nitrophenylhydrazone, in quantitative yield has been described recently.[10] The disadvantage of this method is that the dialdehyde cannot be isolated in the free state in high yield. Diphenaldehyde has also been obtained by sodium iodide reduction of peroxidic products from ozonolysis of phenanthrene in solvents that do not react[2] with the zwitter-ion intermediate.[11,12] The yields are inferior to those obtained by the present method. The aldehyde has been obtained in 91% yield using dimethyl sulfide as the reducing agent.[13] An 81% yield of diphenaldehyde has been obtained from the ozonolsis in acqueous t-butyl alcohol followed by distillation of the solvent at pH 7.5.[14] Hydrogen peroxide is a by-product.

1. Department of Chemistry, University of Texas, Austin, Tex.
2. P. S. Bailey, *Chem. Revs.,* 58, 985 (1958).
3. A. Maggiolo, *Organic Ozone Reactions and Techniques,* The Welsbach Corp., Ozone Processes Division, Philadelphia, Pa., Third Revision, 1956, pp. 21–22.
4. *Basic Manual of Applications and Laboratory Ozonization Techniques,* The Welsbach Corp., Ozone Processes Division, Philadelphia, Pa., First Revision, pp. 18–25.
5. F. Mayer, *Ber.,* 44, 2304 (1911); 45, 1107 (1912).
6. W. S. Rapson and R. G Shuttleworth, *J. Chem. Soc.,* 489 (1941).
7. J. Kenner and E. G. Turner, *J. Chem. Soc.,* 99, 2112 (1911).
8. F. Weygand, G. Eberhardt, H. Linden, F. Schafer, and I. Eigen, *Angew. Chem.,* 65, 525 (1953).
9. P. S. Bailey, *J. Am. Chem. Soc.,* 78, 3811 (1956).
10. W. S. Knowles and Q. E. Thompson, *J. Org. Chem.,* 25, 1031 (1960).
11. P. S. Bailey and S. B. Mainthia, *J. Org. Chem.,* 23, 1089 (1958).
12. W. J. Schmitt, E. J. Moriconi, and W. F. O'Connor, *J. Am. Chem. Soc.,* 77, 5640 (1955).
13. J. J. Pappas, W. P. Keaveney, E. Gancher, and M. Berger, *Tetrahedron Lett.,* 4273 (1966).
14. M. B. Sturrock, E. L. Cline, and K. R. Robinson, *J. Org. Chem.,* 28, 2340 (1963).

DIPHENALDEHYDIC ACID

(2-Biphenylcarboxylic acid, 2′-formyl-)

Submitted by PHILIP S. BAILEY and RONALD E. ERICKSON.[1]
Checked by CAROLE L. OLSON and JOHN D. ROBERTS.

1. Procedure

A. *3,8-Dimethoxy-4,5,6,7-dibenzo-1,2-dioxacyclooctane.* The ozonolysis of 10 g. (0.0562 mole) of phenanthrene in dry methanol is carried out as described in the diphenaldehyde preparation (p. 489). The reaction mixture is not reduced, however, but is acidified with 1–3 drops of concentrated hydrochloric acid (Note

1) and allowed to stand at room temperature for an hour and then in the refrigerator for several hours or overnight. Suction filtration yields 11.5–12.5 g. (75–82%) of crystals melting at 178–181°. Trituration with methyl ethyl ketone gives a 90–95% recovery of colorless crystals melting at 180–181° (Note 2).

B. *Diphenaldehydic acid.* A mixture of 10 g. (0.0368 mole) of 3,8-dimethoxy-4,5,6,7-dibenzo-1,2-dioxacyclooctane, 50 ml. of 10% sodium hydroxide solution, and 200 ml. of 95% ethyl alcohol is heated under reflux for 15 minutes, during which time the solid dissolves (Note 3). The solution is cooled, acidified with concentrated hydrochloric acid, and diluted to the cloud point with water. Crystallization is induced by rubbing the side of the vessel with a stirring rod (Note 4). More water is then added slowly until crystallization is complete. Filtration yields 6.7–7.3 g. (81–88%) of colorless to yellowish crystals melting at 130–132°. Recrystallization from 100 ml. of 1:1 methanol-water gives an 80–95% recovery of diphenaldehydic acid melting at 134–135° (Notes 5, 6, and 7).

2. Notes

1. The acid catalyzes formation of the dimethoxy compound from the initial ozonolysis products (see Note 6 of the diphenaldehyde preparation, p. 490). Compound I forms only slowly in the absence of the hydrochloric acid.

2. The trituration is carried out at room temperature, but the mixture is cooled before filtering. The product can be recrystallized from methyl ethyl ketone, but this requires large volumes of the solvent and is unnecessary.

3. After solution has occurred, 1 ml. of the solution is acidified and tested with sodium or potassium iodide. If no iodine is released, the reaction is complete.

4. If the solution resists crystallization, it can be evaporated one-half or two-thirds of its volume and cooled further. The checkers found that, if product was allowed to oil out and solidify, the subsequent purification was rendered more difficult.

5. Often recrystallization is unnecessary since the first crystalline product melts at 134–135°. The yields then are 81–84%.

6. If sodium hydroxide is omitted in this preparation and the reaction mixture is refluxed until it no longer gives a positive

peroxide test with iodide ion (Note 3) (about 2 hours), the product is the methyl ester of diphenaldehydic acid in 91–98% yield (m.p. 50–51°).[2]

7. If, in the reaction mixture described, twice the volume of 10% sodium hydroxide solution and 25 ml. of 30% hydrogen peroxide are employed and the reaction mixture is refluxed until it no longer gives a positive peroxide test with iodide ion (about 30 minutes to 1 hour), the product is diphenic acid (m.p. 220–223°) in 73–85% yield.

3. Methods and Merits of Preparation

The method here described is based on the reported ozonolysis of phenanthrene in methanol, followed by conversion of the initial ozonolysis product to diphenaldehyde (p. 489), diphenaldehydic acid, methyl diphenaldehydate (Note 6), and diphenic acid (Note 7).[2] Diphenaldehydic acid has previously been made in low yields by oxidative decomposition of the monohydrazide of diphenic acid.[3,4] The presently described method is far superior, not only in yield, but also in simplicity.

Diphenic acid has been prepared by the reduction of diazotized anthranilic acid with cuprous ion,[5] Ullman coupling of potassium o-bromobenzoate,[6] and oxidation of phenanthrene or phenanthrenequinone with various oxidizing agents.[7] The latter methods have been reviewed recently.[7] The ozonolysis method has also been carried out in solvents [8] that do not react with the zwitterion intermediate.[9]

Of the various routes to diphenic acid, the present method and the peracetic acid oxidation of phenanthrene [7] seem to be the simplest. The yields are equally good.

1. Department of Chemistry, University of Texas, Austin, Tex.
2. P. S. Bailey, *J. Am. Chem. Soc.*, 78, 3811 (1956).
3. J. W. Cook, G. T. Dickson, J. Jack, J. D. Loudon, J. McKeown, J. MacMillan, and W. F. Williamson, *J. Chem. Soc.*, 139 (1950).
4. E. F. M. Stephenson, *J. Chem. Soc.*, 2354 (1954).
5. *Org. Syntheses*, Coll. Vol. 1, 222 (1941).
6. W. R. H. Hurtley, *J. Chem. Soc.*, 1870 (1929).
7. W. F. O'Connor and E. J. Moriconi, *Ind. Eng. Chem.*, 45, 277 (1953).
8. P. S. Bailey, *Chem. Rev.*, 58, 986 (1958).
9. W. F. O'Connor, W. J. Schmitt, and E. J. Moriconi, *Ind. Eng. Chem.*, 49, 1701 (1957).

DIPHENYL-*p*-BROMOPHENYLPHOSPHINE

[Phosphine, (*p*-bromophenyl)diphenyl]

$$Br\!-\!\!\langle\bigcirc\rangle\!-\!Br + Mg \longrightarrow Br\!-\!\!\langle\bigcirc\rangle\!-\!MgBr$$

$$Br\!-\!\!\langle\bigcirc\rangle\!-\!MgBr + ClP(C_6H_5)_2 \longrightarrow$$

$$Br\!-\!\!\langle\bigcirc\rangle\!-\!P(C_6H_5)_2 + ClMgBr$$

Submitted by G. P. Schiemenz [1]
Checked by V. Z. Williams, Jr., and K. B. Wiberg

1. Procedure

A dry 1-l. round-bottomed flask with five outlets is equipped with a sealed stirrer, a 500-ml. dropping funnel, a reflux condenser attached to a calcium chloride tube,. an inlet for dry nitrogen (a weak stream of which is maintained through all the reaction until the hydrolysis step), and a thermometer reaching close to the bottom. In the flask are placed 9.0 g. (0.38 g. atom) of magnesium turnings, a crystal of iodine, and about 25 ml. of dry ether. With stirring, about 15 ml. of a solution of 88.5 g. (0.38 mole) of *p*-dibromobenzene (Note 1) in 500 ml. of dry ether (Note 2) is added at once. When the reaction has started, the remaining ether solution is added at a rate which maintains rapid refluxing. After the *p*-dibromobenzene has been added, the mixture is stirred at room temperature for 1.5 hours.

The mixture is then cooled by means of an ice-sodium chloride bath. When the internal temperatures has reached $-7°$, a solution of 71.8 g. (0.33 mole) of chlorodiphenylphosphine (Note 3) in 100 ml. of dry ether is added at such a rate that the internal temperature does not exceed $+10°$. The addition requires about 1.25 hours. The cooling bath is then removed and stirring continued for 1.5 hours.   The flask is then again immersed in an ice-sodium chloride bath, and 150 ml. of a cold saturated aqueous ammonium chloride solution is added slowly.   The ether is decanted and the remainder acidified with hydrochloric acid and extracted three times with 125 ml. of benzene each (Note 4). From the combined ether and benzene solutions, the solvents are evaporated and the residue is distilled at reduced pressure. After the low-boiling material, some *p*-dibromobenzene distills and crystallizes in the distillation bridge.   At 2×10^{-2} mm., heating is continued until the phosphine reaches the stillhead.   At this stage the distillation is interrupted, the stillhead and condenser containing *p*-dibromobenzene replaced by a clean, short distillation bridge without condenser, and the phosphine distilled at 2×10^{-2} to 10^{-3} mm., no forerun being taken (Note 5).   The main bulk distills at 180–185° (2×10^{-2} mm.).   The colorless, oily distillate begins to crystallize in the receiving flask during or shortly after the distillation (Note 6) and weighs 81 83 g. (73–77% yield) (Note 7), m.p. 64–71°.   This material is sufficiently pure for further reactions, *e.g.*, Grignard reaction.   A sample may be recrystallized from methanol to give colorless needles, m.p. 79–80°.

2. Notes

1. A commercial product, m.p. 88–89°, was used without purification.

2. The *p*-dibromobenzene may be dissolved by heating the ether to reflux.   If substantially less ether is used, part of the compound will crystallize out at room temperature.

3. A commercial product from Aldrich Chemical Company was used without purification.

4. When a larger excess of Grignard reagent was used, a polymer insoluble in either phase was observed.

5. Dividing the distillate into a forerun and a constant-boiling main fraction did not improve the melting point of the latter, 66 g. (62%) of phosphine being obtained.   The forerun likewise consisted mainly of the phosphine.

6. The distillation apparatus should be taken apart and

to a hard glass which blocks the joints and can hardly be removed from the flask.

7. No improvement of the yield was obtained when a 50% excess of Grignard reagent was used.

3. Discussion

This preparation [2] is an example of the general and versatile synthesis of t-phosphines of Michaelis [3] which, however, is usually not applicable for aromatic phosphines substituted with $-M$ substituents. The synthesis is an interesting case of the Grignard reaction in that it includes the addition of a Grignard reagent to an "inorganic" single bond and makes use of the mono-Grignard reagent of a dihalogen compound with two equivalent halogen atoms. Similarly, from the mono-Grignard reagents of m-dibromobenzene in ether [4] and of p-dichlorobenzene in tetrahydrofuran,[5] diphenyl-m-bromophenyl-[4] and diphenyl-p-chlorophenylphosphine [2] were prepared in yields of 58 and 84%, respectively.

A slightly higher yield of diphenyl-p-bromophenylphosphine has been reported using more expensive reagents (tetrahydrofuran and butyllithium rather than ether and magnesium turnings).[6] An alternative route consists of a Friedel-Crafts type of reaction of bromobenzene with phosphorus trichloride and reaction of the resulting dichloro-p-bromophenylphosphine with phenylmagnesium bromide. The submitter found this sequence less convenient, and the overall yield is given as only 21%.[7-9] In addition, this path fails for the *meta* isomer, and with other substituents the first step yields a mixture of isomers.[10] On the other hand, some phosphines containing $-M$ substituents were prepared by making use of the second step.[11, 12] A more facile synthesis of such phosphines starts from the title compound [2] or its *meta* isomer,[4] the key step being a second Grignard reaction with subsequent carbonation to give the diphenylphosphinobenzoic acids [2, 4] which are also accessible by several other, apparently less convenient and more expensive, routes.[8, 11-14] p-Diphenylphosphinobenzoic acid has been used in place of triphenylphosphine in a modification of the Wittig olefination, giving rise to a phosphine oxide which is scarcely soluble in organic solvents and easily soluble in aqueous carbonate solution, and therefore facilitates separation of the olefin from the phosphine oxide.[15]

1. Institut für Organische Chemie der Universität Kiel, Germany.
2. G. P. Schiemenz, *Ber.*, **99**, 504 (1966).
3. G. M. Kosolapoff, "Organophosphorus Compounds," John Wiley & Sons, New York, 1950.
4. G. P. Schiemenz and H.-U. Siebeneick, *Ber.*, **102**, 1883 (1969).
5. H. E. Ramsden, A. E. Balint, W. R. Whitford, J. J. Walburn and R. Cserr, *J. Org. Chem.*, **22**, 1202 (1957).
6. R. A. Baldwin and M. T. Cheng, *J. Org. Chem.*, **32**, 1572 (1967).
7. A. Michaelis, *Ann.*, **293**, 193 (1896).
8. H. Gilman and G. E. Brown, *J. Am. Chem. Soc.*, **67**, 824 (1945).
9. H. Goetz, F. Nerdel, and K. -H. Wiechel, *Ann.*, **665**, 1 (1963). Compare, however, the comment of Professor Goetz quoted in ref. 15.
10. H. Schindlbauer, *Monatsh.*, **96**, 1936 (1965).
11. E. N. Tsvetkov, D. I. Lobanov, and M. I. Kabachnik, *Teor. i Eksperim. Khim.*, **1**, 729 (1965).
12. E. N. Tsvetkov, D. I. Lobanov, and M. I. Kabachnik, *Teor. i. Eksperim. Khim.*, **2**, 458 (1966).
13. H. Schindlbauer, *Monatsh.*, **96**, 1021 (1965).
14. R. A. Baldwin, M. T. Cheng, and G. D. Homer, *J. Org. Chem.*, **32**, 2176 (1967).
15. G. P. Schiemenz and J. Thobe, *Ber.*, **99**, 2663 (1966).

1,4-DIPHENYL-1,3-BUTADIENE *

(1,3-Butadiene,1,4-diphenyl-)

$$C_6H_5CH{=}CHCH_2Cl + (C_6H_5)_3P \rightarrow$$

$$C_6H_5CH{=}CHCH_2P^+(C_6H_5)_3; \; Cl^-$$

$$C_6H_5CH{=}CHCH_2P^+(C_6H_5)_3; \; Cl^- + C_6H_5CHO + LiOC_2H_5 \rightarrow$$

$$C_6H_5CH{=}CHCH{=}CHC_6H_5 + (C_6H_5)_3PO + C_2H_5OH + LiCl$$

Submitted by RICHARD N. McDONALD and TOD W. CAMPBELL.[1]
Checked by M. S. NEWMAN, R. MARSHALL, and W. N. WHITE.

1. Procedure

A. *Triphenylcinnamylphosphonium chloride.* A mixture of 40 g. (0.26 mole) of (3-chloropropenyl)benzene (Note 1) and 92 g. (0.35 mole) of triphenylphosphine (Note 2) in 500 ml. of xylene is heated at reflux for 12 hours with stirring. The mixture is allowed to cool to about 60°, and the colorless crystalline product is filtered, washed with 100 ml. of xylene, and dried in a vacuum oven at about 20 mm. pressure and 60° to constant weight. The yield is 99–101 g. (91–93%), m.p. 224–226° (Note 3).

B. *1,4-Diphenylbutadiene*. To a solution of 60.0 g. (0.145 mole) of triphenylcinnamylphosphonium chloride and 16.4 g. (0.155 mole) of benzaldehyde in 200 ml. of ethanol (Note 4) is added 760 ml. of 0.2M lithium ethoxide in ethanol (Notes 5 and 6). After allowing this mixture to stand 30 minutes, 700 ml. of water is added (Note 7) and the colorless crystals are filtered, washed with 150 ml. of 60% ethanol, and dried in the vacuum oven at 65°. The yield of crystalline product, m.p. 153–156°, is 17.9–19.9 g. (60–67%) (Note 8). The product is the *trans-trans* isomer and is pure enough for most purposes (Note 9). Recrystallization from cyclohexane gives a product with m.p. 154–156°.

2. Notes

1. Eastman Organic Chemicals, white label grade, used without purification.

2. Commercial triphenylphosphine was used without further purification. Metal and Thermit Corp., Rahway, New Jersey, now offers this reagent for sale at a modest price.

3. The phosphonium salt can be recrystallized to analytical purity by dissolving in a small amount of boiling ethanol, adding ether at the boil until cloudy, and allowing the salt to crystallize in a refrigerator.

4. Commercial anhydrous ethanol was used throughout without further purification.

5. The lithium ethoxide solution is prepared by dissolving 1.40 g. of lithium wire in 1 l. of anhydrous ethanol.

6. A transient orange color is immediately formed, and it is replaced by crystallization of the product in about 1 minute.

7. Triphenylphosphine oxide is soluble in 60% aqueous ethanol; therefore it remains in the filtrate and affords no difficulty.

8. The yield can probably be increased by carrying out the reaction in an ether solvent with an alkyllithium as base, but the simplicity and relative ease of the conditions described appear to make the possible yield advantage secondary.

9. This procedure has been applied successfully to the synthesis of substituted bistyryls, i.e., 1-(p-tolyl)-4-phenylbutadiene (76%), 1-(4-methoxyphenyl)-4-phenylbutadiene (63%), and 1-(4-acetamidophenyl)-4-phenylbutadiene (61%), by using the corresponding substituted benzaldehydes.

3. Methods of Preparation

1,4-Diphenylbutadiene has been obtained from phenylacetic acid and cinnamaldehyde with lead oxide,[2] by the dehydrogenation of 1,4-diphenyl-2-butene with butyllithium,[3] and by the coupling reaction of benzenediazonium chloride and cinnamylideneacetic acid.[4] The present method [5] gives better yields than those previously reported, is adaptable to the preparation of a variety of substituted bistyryls, and is relatively easy to carry out.

1. Pioneering Research Division, Textile Fibers Department, E. I. du Pont de Nemours and Co., Wilmington, Delaware.
2. B. B. Corson, *Org. Syn.*, Coll. Vol. 2, 229 (1943). This reference includes previous preparative methods.
3. H. Gilman and C. W. Bradley, *J. Am. Chem. Soc.*, **60**, 2333 (1938).
4. C. F. Koelsch and V. Boekelheide, *J. Am. Chem. Soc.*, **66**, 412 (1944).
5. T. W. Campbell and R. N. McDonald, *J. Org. Chem.*, **24**, 1241. (1959).

DIPHENYLCARBODIIMIDE

(Carbodiimide, diphenyl-)

Method I

$$2C_6H_5NCO \longrightarrow C_6H_5N{=}C{=}NC_6H_5 \ + \ CO_2$$

Submitted by T. W. CAMPBELL [1] and J. J. MONAGLE.
Checked by W. S. WADSWORTH and W. D. EMMONS.

1. Procedure

A 250-ml. four-necked flask is fitted with a sealed mechanical stirrer, a condenser protected by a drying tube, a thermometer, and a gas inlet. The flask is swept with a slow stream of nitrogen

(Note 1) and dried by flaming. One hundred milliliters (108 g., 0.91 mole) of phenyl isocyanate (Note 2) is pipetted into the flask. One gram (0.052 mole) of 3-methyl-1-phenylphospholene 1-oxide [2] is added (Note 3), and the reaction mixture is heated at 50° under nitrogen for 2.5 hours (Note 4); at this point only a faint test for carbon dioxide is obtained when the off-gas is passed through saturated calcium hydroxide solution. The reaction mixture is cooled and rapidly transferred to a Claisen flask. Distillation yields 72–82 g. (82–93%) of diphenylcarbodiimide, obtained as a clear water-white oil, b.p. 110–112°/0.2 mm., n_D^{25} 1.6360–1.6362 (Note 5).

2. Notes

1. Commercial nitrogen is dried by passage through concentrated sulfuric acid.

2. Best results were obtained with material obtained from Eastman Kodak Company. Either freshly distilled material or material from a freshly opened bottle may be used. Material obtained from several other sources gave variable results even after redistillation.

3. Since the phosphine oxides are very hygroscopic and the reaction rate is sensitive to traces of moisture, the catalyst can be conveniently stored and added to the reaction mixture in long-necked, thin-walled glass ampoules. The catalyst may be dried by distillation (b.p. 168–170°/1.4 mm.) into a receiver containing the inverted ampoules. When sufficient catalyst has distilled to fill the ampoules, nitrogen is bled into the receiver, forcing the catalyst into the ampoules. An ampoule about 15 mm. in diameter will hold about 1 g. of catalyst. A small air space should be left to facilitate crushing the ampoule.

4. Use of more catalyst or higher temperature leads to an increasingly vigorous evolution of carbon dioxide.

5. Diphenylcarbodiimide can be stored for several weeks at 0°. At room temperature it gradually solidifies to a mixture of trimer and polymer. The monomer can be separated from the solid by vacuum distillation.

3. Methods of Preparation

Carbodiimides have been prepared by desulfurization of thioureas by metal oxides,[3] by sodium hypochlorite,[4] or by ethyl chloroformate in the presence of a tertiary amine; [5] by halogenation of ureas or thioureas followed by dehydrohalogenation of the N,N'-disubstituted carbamic chloride; [6] and by dehydration of disubstituted ureas using p-toluenesulfonyl chloride and pyridine.[7] The method described above is a modification of that of Campbell and Verbanc.[8]

4. Merits of the Preparation

This method may be applied to the synthesis of a variety of aryl and alkyl carbodiimides.[9] Other catalysts may also be used,[10] but the especially active one described here is the one most easily obtained. The method is superior to other methods reported in that it provides pure products under very simple and mild conditions, allows the use of readily available isocyanates with or without the use of solvent, and offers extremely easy work-up.

1. Textile Fibers Department and Organic Chemicals Department, E. I. du Pont de Nemours and Co., Wilmington, Delaware.
2. W. B. McCormack, this volume, p. 786.
3. S. Hünig, H. Lehmann, and G. Grimmer, *Ann.*, 579, 77 (1953).
4. E. Schmidt and M. Seefelder, *Ann.*, 571, 83 (1951).
5. R. F. Coles and H. A. Levine, U.S. Patent 2,942,025 (1960) [*C. A.,* 54, 24464a (1960)].
6. H. Eilingsfeld, M. Seefelder, and H. Weidinger, *Angew. Chem.*, 72, 836 (1960).
7. G. Amiard and R. Heymès, *Bull. Soc. Chim. France*, 1360 (1956).
8. T. W. Campbell and J. Verbanc, U.S. Patent 2,853,473 (1958) [*C. A.,* 53, 10126e (1959)].
9. T. W. Campbell, J. J. Monagle, and V. S. Foldi, *J. Am. Chem. Soc.*, 84, 3673 (1962).
10. J. J. Monagle, *J. Org. Chem.*, 27, 3851 (1962).

DIPHENYLCARBODIIMIDE

(Carbodiimide, diphenyl)

Method II

A. $C_6H_5NH_2 \xrightarrow{SOCl_2} C_6H_5N{=}S{=}O$

B. $C_6H_5CH{=}NOH \xrightarrow{Cl_2} C_6H_5C(Cl){=}NOH$

$C_6H_5C(Cl){=}NOH \xrightarrow{(C_2H_5)_3N} C_6H_5C{\equiv}N{\longrightarrow}O$

C. $\overset{\cdot}{C_6H_5}C{\equiv}N{\longrightarrow}O$ + $C_6H_5N{=}S{=}O \longrightarrow$

$$\left[\begin{array}{c} \text{structure} \end{array} \right]$$

$\xrightarrow[SO_2]{\Delta}$ $C_6H_5N{=}C{=}NC_6H_5$

Submitted by P. Rajagopalan, B. G. Advani, and C. N. Talaty [1]
Checked by A. Eschenmoser, R. Scheffold, and P. Mayer

1. Procedure

Caution! All the following operations should be carried out in a well-ventilated hood.

A. *N-Sulfinylaniline* (Note 1). A solution of 82.5 g. (0.69 mole) of pure thionyl chloride (Note 2) in 100 ml. of anhydrous benzene is added slowly to a solution of 46.5 g. (0.5 mole) of freshly distilled aniline in 250 ml. of anhydrous benzene contained in a 1-l. round-bottomed flask, with swirling and occasional cooling in an ice bath (if necessary). An immediate precipitation of aniline hydrochloride occurs. After the addition of the thionyl chloride solution is complete, the mixture is heated to reflux, protected from moisture, on a heating mantle until a clear solution is obtained (2–5 hours). The solvent and excess thionyl chloride are evaporated under reduced pressure (Note 3) at 50° and the residual brownish yellow liquid is distilled under vacuum to yield 63–65 g. (91–94%) of yellow N-sulfinylaniline, b.p. 88–95° (17–20 mm.), n^{21}D 1.6253.

B. *Benzohydroxamoyl chloride* (Note 4). A four-necked flask equipped with a rubber-sealed stirrer, a thermometer, an inlet tube, and an outlet tube attached to a calcium chloride tube (Note 5) and containing a solution of 50 g. (0.41 mole) of benzal-doxime (Note 6) in 450 ml. of pure chloroform (Note 7) is cooled in a dry ice-acetone bath (Note 8). When the temperature of the solution reaches −2°, stirring is started and a stream of chlorine gas (Note 9) is passed through at such a rate as to maintain the temperature below 2°. After 1 hour the passage of chlorine is stopped and the greenish yellow solution is transferred to a 1-l. round-bottomed flask which is then connected to an aspirator to remove most of the dissolved chlorine. The light yellow solution thus obtained is stripped of the solvent at 40° under reduced pressure (Note 3). The almost colorless residual liquid is dissolved in 150 ml. of petroleum ether (b.p. 40–60°) and cooled, with scratching, in a dry ice-acetone bath, whereupon a colorless crystalline solid starts separating. The cooling is continued for 30 minutes and the solid is then filtered, washed with 50 ml. of cold petroleum ether (b.p. 40–60°), pressed to remove most of the adhering mother liquor, and dried over a filter paper. The yield of benzohydroxamoyl chloride, m.p. 48–52°, which is pure enough (Note 10) for the next step, is 33–38 g. (51–59%).

C. *Diphenylcarbodiimide* (Note 11). A solution of 15.6 g. (0.1 mole) of benzohydroxamoyl chloride in 300 ml. of anhydrous benzene contained in a 500-ml. wide-mouthed Erlenmeyer flask is cooled to 5°, agitated vigorously, and treated with 10.1 g. (0.1 mole) of freshly distilled triethylamine added in one portion. The mixture is shaken continuously for 3 minutes while being cooled in an ice bath and then filtered rapidly through a Buchner funnel into a filter flask cooled in an ice bath. The residue is washed with 50 ml. of anhydrous benzene, pressed to remove as much of the adhering solution as possible, dried in an oven at 60°, and weighed (Note 12). The yield of the triethylamine hydrochloride, m.p. 254–256°, is 13.4–13.6 g. (97–99%).

The combined filtrates containing benzonitrile oxide are transferred to a 1-l. round-bottomed flask, treated immediately with

13.9 g. (0.1 mole) of N-sulfinylaniline added in one portion, with swirling, and set aside protected from moisture, while the temperature reaches a maximum of 33–34° (usually 15 minutes). The mixture is then heated to reflux, protected from moisture, in a temperature-controlled oil bath for 3–5 hours. Continuous evolution of sulfur dioxide takes place during this period at the end of which the mixture is cooled and evaporated under reduced pressure (Note 3) at 70–80° to remove the solvent. The residual dark brown liquid is transferred to a 50-ml., pear-shaped distilling flask (Note 13) and heated, protected from moisture, at 110° for 30 minutes to complete the decomposition. It is then cooled and distilled under high vacuum (Note 14). Unchanged N-sulfinylaniline (2.0–2.5 g.) distills over at 45–50° (0.1–0.2 mm.). A second fraction (1.2–1.5 g.) is collected until the temperature reaches 112° (Note 15); then diphenyl carbodiimide is collected at 114–117° (0.1–0.2 mm.) as a clear yellow liquid; yield 10.5–10.8 g. (54–56%) (Note 16); n^{23}D 1.6355; $\nu_{max.}^{CHCl_3}$ 2140 cm.$^{-1}$ (very strong), 2110 cm.$^{-1}$ (medium), and 1480 cm.$^{-1}$ (medium) (Note 17).

2. Notes

1. This method is essentially that described by Kresze and co-workers [2] which is a modification of the original procedure of Michaelis.[3]

2. The yield of the product depends on the purity of the thionyl chloride. Thionyl chloride obtained from Riedel-Haen (Hannover, West Germany) was used as such.

3. A rotary evaporator equipped with a constant-temperature water bath is ideal for this purpose.

4. This method is essentially that of Werner and Buss.[4]

5. The calcium chloride tube is, in turn, attached to a rubber tube which is either led out of a ventilator or connected to a water pump through which water is adjusted to flow gently.

6. Purum grade α-benzaldoxime, m.p. 32°, obtained from Fluka AG (Buchs, Switzerland) was used most of the time. When out of stock, it was prepared in the usual manner and distilled before use.

7. Chloroform distilled over phosphorus pentoxide was used.

8. An ice-salt mixture is not adequate to regulate the temperature, as it rises steeply when chlorine is let in.

9. It is better to lead the gas through a drying tower containing small lumps of calcium chloride before passing it through the reaction mixture.

10. Although it is not necessary, the product can be recrystallized from petroleum ether (b.p. 40–60°) without much loss. The melting point of the recrystallized product is 51–52°.

11. This is a modification of the new general method for the preparation of carbodiimides by the thermolysis of 5-substituted 4-aryl-1,2,3,5-thiaoxadiazole-1-oxides described recently by Rajagopalan and Advani,[5] whereby the 4,5-diphenyl-1,2,3,5-thiaoxadiazole-1-oxide, which is formed in this reaction, is not isolated but decomposed *in situ*.

12. The drying and weighing of triethylamine hydrochloride should be carried out only after N-sulfinylaniline has been added to the solution of benzonitrile oxide, as otherwise the latter, not being very stable in the free state, would dimerize resulting in the reduction in yield of the carbodiimide.

13. At this point it is best to use the flask that is going to be employed subsequently for the distillation of the carbodiimide to avoid unnecessary loss in transferring from one flask to the other.

14. The temperature of the bath used in this distillation should not exceed 160°. A short-path distillation apparatus should be used.

15. Most of this fraction is comprised of diphenylcarbodiimide.

16. The yield, on the basis of average recovered N-sulfinylaniline, is 64–66%.

17. The infrared spectrum was determined in a Perkin-Elmer Infracord 337 spectrophotometer.

3. Discussion

Diphenylcarbodiimide can be prepared by the removal of the elements of hydrogen sulfide from N,N'-diphenylthiourea by mercuric oxide,[6] lead oxide,[7] sodium hypochlorite,[8] or phosgene;[9] by heating phenylisocyanate in a sealed tube [10] or in the presence of catalysts such as phospholenes [11] or phosphonates;[12] by the pyrolysis of N,N',N''-triphenylguanidine,[13] 3-phenyl-4-phenyl-imino-1,3-thiazetidin-2-one (carbonythiocarbanilide),[14] and 1,5-diphenyl tetrazole;[15] and by heating phenylisocyanide dichloride with aniline hydrochloride in an inert solvent.[16]

Although this procedure offers no advantage over that of Hünig, Lehmann, and Grimmer,[7] it effectively illustrates a new method for the synthesis of symmetrical and unsymmetrical carbodiimides.[5] The generality of this procedure is limited only by the number of substituted benzohydroxamoyl chlorides that can be made without difficulty, as a variety of N-sulfinylamines is easily accessible.[2]

N-Sulfinylaniline, the procedure for the preparation of which is described in Part A, is a versatile intermediate in the synthesis of heterocyclic compounds.[2, 5, 17] Benzohydroxamoyl chloride, the method for the preparation of which is given in Part B, is the precursor of the highly reactive benzonitrile oxide, the diverse dipolar addition reactions of which have been thoroughly investigated.[18] A wide array of heterocyclic compounds can be prepared starting with benzonitrile oxide.[18]

1. Ciba Research Centre, Goregaon, Bombay 63, India.
2. G. Kresze, A. Maschke, R. Albrecht, K. Bederke, H. P. Patzschke, H. Smalla, and A. Trede, *Angew. Chem.*, **74**, 135 (1962).
3. A. Michaelis, *Ber.*, **24**, 745 (1891).
4. A. Werner and H. Buss, *Ber.*, **27**, 2197 (1894).
5. P. Rajagopalan and B. G. Advani, *J. Org. Chem.*, **30**, 3369 (1965).
6. W. Weith, *Ber.*, **7**, 10 (1874); R. Rotter, *Monatsh.*, **47**, 353 (1926).
7. S. Hünig, H. Lehmann, and G. Grimmer, *Ann.*, **579**, 77 (1953).
8. S. Senoh, N. Yano, and T. Yamada, Japan patent 11,865 [*Chem. Abstr.*, **65**, 16910 (1966)].
9. A. A. R. Sayigh and H. Ulrich, U.S. Patent 3,301,895 [*Chem. Abstr.*, **66**, 85611 (1967)].
10. R. Stolle, *Ber.*, **41**, 1125 (1908).
11. W. J. Balon, U.S. Patent 2,853,518 [*Chem. Abstr.*, **53**, 5202 (1959)]. T. W. Campbell and J. J. Monagle, this volume, p. 501.
12. J. J. Monagle and H. R. Nace, U.S. Patent 3,056,835 [*Chem. Abstr.*, **58**, 3362 (1963)].
13. W. Weith, *Ber.*, **7**, 1303 (1874).
14. W. Will, *Ber.*, **14**, 1485 (1881).
15. T. Bacchetti and A. Alemagna, *Rend. Ist. Lombardo Sci. Letters*, **94A**, 351 (1960) [*C.A.*, **55**, 16527 (1961)].
16. E. Kuehle, Ger. Patent, 1,149,712 [*Chem. Abstr.*, **59**, 12704 (1963)].
17. R. Huisgen, R. Grashey, M. Seidel, H. Knupfer, and R. Schmidt, *Ann.* **658**, 169 (1962).
18. For a review of the chemistry of nitrile oxides, see C. Grundmann, *Fortschr. Chem. Forsch.*, **7**, 62 (1966).

1,1-DIPHENYLCYCLOPROPANE

(Cyclopropane, 1,1-diphenyl-)

$$(C_6H_5)_2CO + (C_2H_5O)_2\overset{O}{\overset{\|}{P}}CH_2CO_2C_2H_5 \xrightarrow{\text{NaH}}$$

$$(C_6H_5)_2C{=}CHCO_2C_2H_5 \xrightarrow[\text{2. } H_2SO_4]{\text{1. KOH, } H_2O}$$

$$(C_6H_5)_2C{=}CHCO_2H \xrightarrow{\text{LiAlH}_4} (C_6H_5)_2C\overset{CH_2}{\overset{\triangle}{\underset{}{{-}{-}}}}CH_2$$

Submitted by M. J. Jorgenson and A. F. Thacher [1]
Checked by John J. Miller and William D. Emmons

1. Procedure

A. *β-Phenylcinnamic acid.* A suspension of 8.8 g. (0.20 mole) of a 55.1% sodium hydride dispersion in mineral oil (Note 1) and 300 ml. of 1,2-dimethoxyethane (Note 2) are added to a dry, 1-l., three-necked flask equipped with a stirrer, a condenser with drying tube, and a pressure-equalizing dropping funnel. The flask is immersed in an ice bath, and 44.8 g. (0.20 mole) of triethyl phosphonoacetate (Note 3) is added through the dropping funnel over a 20-minute period. After the addition the solution is stirred at room temperature for 30 minutes, and then 36.4 g. (0.20 mole) of benzophenone is added in one portion. The solution is heated at reflux for 4 days. The two-phase reaction mixture is cooled, and the flask is filled with water. The ester is extracted from the solution with three 100-ml. portions of ether. The ethereal extracts are combined, dried over anhydrous sodium sulfate, and evaporated to as small a volume as possible on the steam bath. The resulting mixture of ester and unreacted benzophenone is added to 140 ml. of water, 45 ml. of dioxane, and

26.4 g. (0.40 mole) of 85% potassium hydroxide in a 500-ml. one-necked flask, and the mixture is refluxed overnight (17 hours). The solution is cooled and extracted with 50 ml. of ether to remove benzophenone and any unhydrolyzed ester. The aqueous fraction is acidified with 10N sulfuric acid, and the resulting solid product is isolated by filtration, washed with two 50-ml. portions of water, and dried. Recrystallization from 200 ml. of benzene gives 28.3–33.9 g. (63–75%) of β-phenylcinnamic acid, m.p. 161–163° (Notes 4 and 5).

B. *1,1-Diphenylcyclopropane.* A 500-ml. three-necked flask is equipped with a nitrogen-inlet tube, a condenser, and a dropping funnel. The flask is charged with 100 ml. of dry tetrahydrofuran (Note 6) and 5.13 g. (0.135 mole) of lithium aluminum hydride (Note 1), and the system is purged with dry nitrogen. To the stirred solution is added dropwise over a period of 15–20 minutes a solution of 20 g. (0.089 mole) of β-phenylcinnamic acid (Note 7) in 100 ml. of tetrahydrofuran. The addition results in an exothermic reaction which causes the solution to reflux. The refluxing is allowed to subside after addition is complete, and then the mixture is carefully reheated to reflux. A mild exotherm occurs a few minutes after reflux is reached. This is easily controlled by removing the heat source or by cooling with ice. After the occurrence of this exothermic reaction, which is accompanied by a color change from rust to deep red, the mixture is heated at the reflux temperature for an additional 2 hours. The mixture is then cooled in an ice bath, and sufficient 10% sulfuric acid (*ca.* 5 ml.) is carefully (*Caution! Vigorous evolution of hydrogen.*) added dropwise to destroy excess lithium aluminum hydride. Another 100 ml. of 10% sulfuric acid is added after decomposition of the hydride is complete to dissolve the aluminum salts (Note 8). The acidified solution is worked up immediately (Note 9) by dilution with 200 ml. of water and extraction with one 100-ml. portion and three 50-ml. portions of ether. The combined ethereal extracts are washed with three 50-ml. portions of saturated aqueous sodium bicarbonate. The bicarbonate extracts are combined and washed with 50 ml. of ether, and all the ethereal extracts are combined. The ethereal solution is dried over anhydrous sodium sulfate, and the ether is removed at atmospheric pressure. The residue is distilled under reduced pressure to give

10.0–10.7 g. (57–62%) of diphenylcyclopropane (Note 10), b.p. 132–134° (10 mm.), n^{20}D 1.590.

2. Notes

1. Sodium hydride dispersions in mineral oil and lithium aluminum hydride are available from Metal Hydrides, Inc.

2. Before use, 1,2-dimethoxyethane (ethylene glycol dimethyl ether) was partially dried over anhydrous calcium chloride and then distilled from lithium aluminum hydride. It was stored over sodium ribbon.

3. Triethyl phosphonoacetate is available from Aldrich Chemical Co.

4. About half of the solid is insoluble in the hot benzene and is removed by filtration.

5. The submitters obtained 28.2 g. (63%) of β-phenylcinnamic acid, m.p. 162–163°.

6. Tetrahydrofuran is best dried according to the procedure described on page 976 of this volume. (*Caution! Note warning of danger in drying tetrahydrofuran containing peroxides.*) The checkers employed a peroxide-free grade of anhydrous tetrahydrofuran which is available from Fisher Scientific Co.

7. In the preparation of other phenylcyclopropanes by this method, esters rather than acids were employed. Use of the ester in the present preparation gave a poor yield of product. Since the acid can easily be purified, its use here is also more expedient.

8. The checkers found this amount of acid insufficient to dissolve all of the salts, and an additional 5 ml. of concentrated sulfuric acid was added.

9. In general it has been found that yields of cyclopropanes are lowered if the acidic solutions are permitted to stand before workup.

10. Fractionation is not essential because the main contaminant is high-boiling polymer. The checkers obtained two fractions: b.p. 138–142° (11 mm.), 10.0–10.2 g., 99% pure by vapor-phase chromatography; and b.p. 142–148° (11 mm.), 0.6–0.9 g., 96% pure.

3. Methods of Preparation

1,1-Diphenylcyclopropane has been prepared in 24% yield by the Simmons-Smith reaction,[2] in 78% yield by treatment of 3,3-diphenylpropyltrimethylammonium iodide with sodium or potassium amide,[3] in 61% yield by reaction of 1,1-diphenylethylene with dimethylsulfonium methylide,[4] and in unspecified yields from 1,1-diphenylethylene by reaction with diazomethane followed by pyrolysis of the resulting pyrazoline or by reaction with ethyl diazoacetate followed by distillation of the corresponding acid over calcium oxide.[5]

β-Phenylcinnamic acid has been prepared previously by a variety of methods, the best of which appear to be the dehydration of ethyl β-hydroxy-β,β-diphenylpropionate by treatment with sodium acetate in acetic acid [6] and the reaction of 1,1-diphenylethylene with oxalyl chloride.[7]

4. Merits of the Preparation

This procedure illustrates a general method for the preparation of phenylcyclopropanes from cinnamic acids, esters, aldehydes, or alcohols.[8] It complements the Simmons-Smith reaction as a general method for the preparation of such cyclopropanes. It offers advantages over the Simmons-Smith method in cases in which electron-withdrawing substituents in the benzene ring or steric crowding around the double bond lead to low yields in the Simmons-Smith reaction. Also, in the case of possible stereoisomerism in the starting material or product, the present method leads stereospecifically to a single cyclopropane. It has the disadvantage that reducible substituents on the benzene ring do not survive the reductive treatment. The present method is an exceptionally simple, one-step preparative process employing starting materials that are commercially available, or readily accessible from aldehydes and ketones via the phosphono ester addition procedure of Wadsworth and Emmons.[9] Cyclopropane formation is particularly facile when electron-withdrawing substituents are present on the aromatic ring or when the β-position is substituted by another aryl group. Alkyl substitution on the double bond also facilitates cyclopropane formation. The examples recorded in Table I illustrate these effects.   An excess of hydride is necessary for producing good yields of cyclopropanes.

TABLE I

PREPARATION OF PHENYLCYCLOPROPANES FROM CINNAMIC ESTERS

Substituents	Solvent	Reflux Time, Hours	Yield of Cyclopropane, %
R = C_6H_5, R' = CH_3 Ar = C_6H_5	Tetrahydrofuran	10	78
R = R' = CH_3 Ar = m-$CF_3C_6H_4$	Tetrahydrofuran	10	54
R = R' = CH_3 Ar = C_6H_5	Dimethoxyethane	48	52
R = R' = H Ar = C_6H_5[a]	Dimethoxyethane	320	66[b]
R = R' = CH_3 Ar = 3,4-$Cl_2C_6H_3$	Tetrahydrofuran	1	45[b]

[a] Aldehyde rather than ester was employed; the yield was 28% after 3 days.
[b] From Reference 8.

1. Department of Chemistry, University of California, Berkeley, California 94720.
2. H. E. Simmons and R. D. Smith, *J. Am. Chem. Soc.*, **81**, 4256 (1959).
3. C. L. Bumgardner, *J. Am. Chem. Soc.*, **83**, 4420 (1961).
4. E. J. Corey and M. Chaykovsky, *J. Am. Chem. Soc.*, **87**, 1353 (1965).
5. H. Wieland and O. Probst, *Ann.*, **530**, 274 (1937).
6. W. R. Dunnavant and C. R. Hauser, this volume, p. 564.
7. F. Bergmann, M. Weizmann, E. Dimant, J. Patai, and J. Szmuskowicz, *J. Am. Chem. Soc.*, **70**, 1612 (1948).
8. M. J. Jorgenson and A. W. Friend, *J. Am. Chem. Soc.*, **87**, 1815 (1965).
9. W. S. Wadsworth, Jr., and W. D. Emmons, this volume, p. 547.

DIPHENYLCYCLOPROPENONE

$$C_6H_5CH_2COCH_2C_6H_5 + 2Br_2 \longrightarrow C_6H_5CHBrCOCHBrC_6H_5 + 2HBr$$

$$C_6H_5CHBrCOCHBrC_6H_5 \xrightarrow{(C_2H_5)_3N}$$

Submitted by R. Breslow and J. Posner [1]
Checked by E. J. Corey and M. F. Semmelhack

1. Procedure

Caution! See Note 1.

A. *α,α'-Dibromodibenzyl ketone.* To a solution of 70 g. (0.33 mole) of commercial dibenzyl ketone in 250 ml. of glacial acetic acid in a 2-l. one-necked flask fitted with a magnetic stirrer a solution of 110 g. (0.67 mole) of bromine in 500 ml. of acetic acid is added through a dropping funnel over a 15-minute period. After addition is complete, the mixture is stirred for an additional 5 minutes and is then poured into 1 l. of water. Solid sodium sulfite is added in small portions until the initial yellow color of the solution is discharged, and the mixture is allowed to stand for 1 hour. The slightly yellow dibromoketone is then collected by filtration and air-dried. Recrystallization from 1 l. of ligroin yields 97 g. of white needles, m.p. 79–87°; an additional 11 g., m.p. 79–83°, is obtained by concentrating the mother liquors, and the two crops are combined (Note 1).

B. *Diphenylcyclopropenone.* A solution of 100 ml. of triethylamine (Note 2) in 250 ml. of methylene chloride is magnetically stirred in a 2-l. one-necked flask while 108 g. (0.29 mole) of the above dibromoketone in 500 ml. of methylene chloride is added dropwise over 1 hour. The mixture is stirred for an additional 30 minutes and then extracted with two 150-ml. portions of 3N hydrochloric acid; the aqueous extracts are dis-

carded. The red organic solution is transferred to a 2-l. Erlenmeyer flask and cooled in an ice bath. While this solution is
swirled, a cold solution of 50 ml. of concentrated sulfuric acid in
25 ml. of water is slowly added. A slightly pink precipitate of
diphenylcyclopropenone bisulfate gradually separates (Note 3).
This is collected on a sintered-glass funnel and washed with two
100-ml. portions of methylene chloride. The solid is then returned to the flask (Note 4) along with 250 ml. of methylene
chloride and 500 ml. of water, and 5 g. of solid sodium carbonate
is added in small portions. The organic layer is collected and the
aqueous solution extracted with two 150-ml. portions of methylene chloride. The combined organic layers are dried over
magnesium sulfate and evaporated to dryness. The impure
diphenylcyclopropenone is recrystallized by repeated extraction
with boiling cyclohexane (total 1.5 l.), the solution being decanted
in each case from a reddish oily impurity. On cooling, the solution deposits 29 g. of white crystals, and an additional 1 g. can
be obtained by concentrating the mother liquors to 150 ml.
The combined 30 g., m.p. 119–120°, represents an overall yield
of 44% based on dibenzylketone.

2. Notes

1. Care should be taken to prevent either the dibromoketone
or the cyclopropenone from coming into contact with the skin,
as allergic reactions have been observed in several cases. The
use of gloves is recommended especially for the bromoketone.
The latter product has a wide melting range because it is a mixture of the *meso-* and *d,l*-compounds.

2. For best results the commercial triethylamine (Matheson,
b.p. 89–90°) should be purified to remove primary and secondary
amines and water, either by distillation from acetic anhydride
and then from barium oxide, or by reaction with phenylisocyanate.[2, 3]

3. If the white solid fails to separate after 15–30 minutes,
concentrated sulfuric acid is added in 4-ml. portions to the cooled
solution with swirling until the white solid appears.

4. Since some of the white solid adheres to the walls of the
flask, it is convenient to use the same flask for the neutralization

after rinsing it with methylene chloride.

3. Methods of Preparation

Diphenylcyclopropenone has also been prepared by the action of phenylchlorocarbene on phenylketene acetal[4] and by the reaction of dihalocarbene with diphenylacetylene.[5] The present procedure[6] is the most convenient on a preparative scale.

4. Merits of the Preparation

Diphenylcyclopropenone is the first stable molecule prepared which has a carbonyl group in a three-membered ring. In a very real sense the compound has aromatic character and is fairly stable.[4] An interesting cycloaddition reaction of enamines with diphenylcyclopropenone has been reported.[7]

1. Department of Chemistry, Columbia University, New York, New York 10027.
2. A. Weissberger, "Technique of Organic Chemistry," Vol. VII, 2nd ed., Interscience Publishers Inc., New York, 1955, p. 445.
3. J. C. Sauer, *Org. Syntheses*, Coll. Vol. 4, 561 (1963).
4. R. Breslow, R. Haynie, and J. Mirra, *J. Am. Chem. Soc.*, **81**, 247 (1959).
5. M. E. Volpin, Yu. D. Koreshokov, and D. N. Kursanov, *Izv. Akad. Nauk SSSR*, 560 (1959).
6. R. Breslow, J. Posner, and A. Krebs, *J. Am. Chem. Soc.*, **85**, 234 (1963).
7. J. Ciabattoni and G. A. Berchtold, *J. Org. Chem.*, **31**, 1336 (1966).

DIPHENYLDIACETYLENE*

(Butadiyne, diphenyl-)

$$\text{C}_6\text{H}_5\text{—C}\equiv\text{CH} + 2\text{Cu}(\text{OCOCH}_3)_2 \xrightarrow[\text{CH}_3\text{OH}]{\text{Pyridine}}$$

$$\text{C}_6\text{H}_5\text{—C}\equiv\text{C—C}\equiv\text{C—C}_6\text{H}_5 + 2\text{CuOCOCH}_3 + 2\text{CH}_3\text{CO}_2\text{H}$$

Submitted by I. D. CAMPBELL and G. EGLINTON [1]
Checked by JOANNE GROVES and VIRGIL BOEKELHEIDE

1. Procedure

To a saturated solution of 5.5 g. (0.028 mole) of finely powdered cupric acetate monohydrate (Note 1) in 20 ml. of a 1:1 by volume pyridine-methanol mixture (Notes 2, 3, 4, and 5) contained in a 50-ml. round-bottomed flask fitted with a reflux condenser is added 2.0 g. (0.0196 mole) of phenylacetylene (Note 6). The deep-blue suspension becomes green when heated under reflux. After 1 hour of heating, the solution is cooled (Note 7) and added dropwise to 60 ml. of 18N sulfuric acid, with stirring and external cooling in an ice-salt freezing mixture (Note 8). The resulting white suspension is extracted with three 25-ml. portions of ether, and the combined ethereal extracts are washed with 15 ml. of aqueous ethanolic silver nitrate solution (Note 9) to remove any unchanged phenylacetylene. The ether solution is then washed twice with water and dried over anhydrous magnesium sulfate. When the dried ether solution is concentrated under reduced pressure, a brown oil (1.81 g.) remains which solidifies on cooling.

The crude solid is purified by dissolving it in 50 ml. of petroleum ether (b.p. 40–60°) and introducing it on a short alumina column (15 g., Brockmann Activity 1 or an equivalent chromatographic alumina). The column is then eluted with 300 ml. of a 1:9 mixture of ether-petroleum ether (b.p. 40–60°). Concentration of the eluate leaves a solid which is recrystallized from aqueous ethanol to give 1.4–1.6 g. (70–80%) of diphenyldiacetylene as large colorless needles, m.p. 87–88° (Note 10).

2. Notes

1. Commercially available crystalline cupric acetate monohydrate was used. A large excess of cupric acetate does not improve the yield. Small catalytic amounts can be used if the cupric salt is continually regenerated by passage of oxygen through the reaction mixture, but the procedure is much slower.

2. A good grade of commercial pyridine was used. The reaction can also be carried out under anhydrous conditions (anhydrous cupric acetate, anhydrous methanol); then the pyridine is distilled from potassium hydroxide pellets. The yields are similar, and, in fact, water may be added as co-solvent if desired.

The solubility of anhydrous cupric acetate is ca. 2.3 g. per 100 ml. of pyridine, and that of the hydrate is ca. 1.6 g. per 100 ml. of pyridine. The solubility is much improved by the addition of methanol (solubility ca. 8.6 g. per 100 ml. of a 1:1 mixture of pyridine-methanol).

For high-dilution experiments, for example, the cyclization of α,ω-diynes, about 4 volumes of ether per volume of reagent solution can be added as entraining solvent without precipitation of the copper salt. A lower reaction temperature results.

3. Commercial grade methanol was used. Methanol is best avoided in experiments involving esters, as methanolysis has been encountered.[2]

4. It is apparently not essential that all the cupric acetate be in solution. Large volumes of solvent ensure complete solution but are inconvenient during isolation of the product.

5. Other solvent systems have been investigated. A base appears to be essential to remove the acetic acid formed; otherwise insoluble yellow precipitates of the cuprous derivative are obtained, which are only slowly oxidized to the required product.

6. Redistilled commercial phenylacetylene, titrating as 98% with silver nitrate-sodium hydroxide,[3] was used.

7. The reaction can be followed by adding an aliquot to ethanolic silver nitrate solution (Note 9). The reaction is complete when no precipitate of the silver derivative is obtained. Also the

disappearance of the infrared absorption band at 3300 cm.$^{-1}$ (3.03 μ) (ethynyl vCH) can be followed with carbon tetrachloride extracts of aliquots.

An 89% yield of diphenyldiacetylene was obtained when the reaction was allowed to proceed for 24 hours at room temperature (20°).

8. This is more convenient than removing the pyridine and the methanol by distillation.

9. The reagent is made by dissolving 3.5 g. of silver nitrate in 5 ml. of water and adding 10 ml. of ethanol.

10. The product may be contaminated by traces of the corresponding eneyne, *trans*-1,4-diphenyl-but-1-en-3-yne, formed by Straus coupling.[4] This compound, however, has an ultraviolet absorption spectrum which differs markedly from that of diphenylbutadiyne.[5]

3. Methods of Preparation

This compound has been prepared by air oxidation of the preformed cuprous salt[6] and through use of aqueous cuprous chloride-ammonium chloride and an oxidant (e.g., oxygen).[3] More recently the compound has been prepared in 95% yield by oxidation of phenylacetylene in pyridine solution containing cuprous chloride[7] and in 97% yield by oxidation in tetramethylenediamine solution containing 5 mole percent cuprous chloride.[8] These and other coupling procedures have been reviewed.[9-11]

4. Merits of the Preparation

The reaction, "Glaser oxidative coupling," is a general one,[9] but this particular technique is recommended for the more water-insoluble ethynyl compounds, and also for the cyclization of α,ω-diynes,[2,12] where controlled dilution is required.

The cupric acetate-pyridine reagent provides a homogenous and basic reaction medium. The yields are high, and there is seldom precipitation of the cuprous derivative which may slow down the cuprous chloride-oxygen procedure.[3]

The rate of oxidative coupling is said to decrease with decreasing acidity of the ethynyl hydrogen.[13] Thus oct-1-yne underwent only limited reaction after being heated with the reagent under reflux for 24 hours.

It is to be noted that cupric acetate has been used to oxidize other systems, for example, α-ketols, phenols, thiols, and nitro-alkanes.

1. The Chemistry Department, The University of Glasgow, Glasgow, Scotland.
2. G. Eglinton and A. R. Galbraith, *J. Chem. Soc.*, 889 (1959).
3. R. A. Raphael, "Acetylenic Compounds in Organic Syntheses," Butterworths, London, 1955, pp. 205–207.
4. F. Straus, *Ann.*, 342, 225 (1905).
5. H. K. Black, D. H. S. Horn, and B. C. L. Weedon, *J. Chem. Soc.*, 1704 (1954).
6. C. Glaser, *Ann.*, 154, 137 (1870); *Ber.*, 2, 422 (1869).
7. L. Brandsma, "Preparative Acetylene Chemistry," Elsevier Publishing Co., Amersterdam, 1971, p. 156.
8. A. S. Hay, *J. Org. Chem.*, 27, 3320 (1962).
9. G. Eglinton and W. McCrae, *Advan. Org. Chem.*, 4, 225 (1963).
10. T. F. Rutledge, "Acetylenic Compounds," Reinhold Book Co., New York, 1968.
11. P. Cadiot and W. Chodkiewicz in H. G. Viehe, "Chemistry of Acetylenes," Marcel Dekker, New York, 1969, p. 597.
12. J. Carnduff, G. Eglinton, W. McCrae, and R. A. Raphael, *Chem. Ind. (London),* 559 (1960); O. M. Behr, G. Eglinton, A. R. Galbraith, and R. A. Raphael, *J. Chem. Soc.*, 3614 (1960).
13. A. L. Klebanskii, I. V. Grachev, and O. M. Kuznetsova, *Zh. Obshch. Khim.*, 27, 2977 (1957).

DIPHENYL KETIMINE

(Diphenylmethylenimine)

$$C_6H_5Br \xrightarrow{Mg} C_6H_5MgBr \xrightarrow{C_6H_5CN}$$

$$(C_6H_5)_2C{=}NMgBr \xrightarrow{CH_3OH} (C_6H_5)_2C{=}NH + CH_3OMgBr$$

Submitted by P. L. PICKARD [1] and T. L. TOLBERT [2]
Checked by C. L. DICKINSON, H. D. HARTZLER, and B. C. McKUSICK

1. Procedure

The apparatus consists of a 1-l. three-necked flask equipped with a mechanical stirrer, a 250-ml. dropping funnel, and a Fried-richs reflux condenser fitted with a calcium chloride drying tube.

Magnesium turnings (13.4 g., 0.55 g. atom) and 200 ml. of anhydrous diethyl ether are put in the flask (Note 1). Slow stirring
is started, and 4 ml. of bromobenzene (Note 2) is added from the
funnel. After reaction has started (Note 3), the stirring rate is
increased, and moderate reflux is maintained by addition of
80.5 g. of bromobenzene (making a total of 86 g. or 0.55 mole)
in 100 ml. of ether. The solution is refluxed for 30–45 minutes
after the addition and is cooled to room temperature. Stirring
is continued while a solution of 51.5 g. (0.50 mole) of benzonitrile
(Note 2) in 100 ml. of ether is added slowly enough (Note 4)
to maintain only a gentle reflux. On completion of the addition,
the reaction mixture of pale-yellow liquid and white solid is refluxed with stirring for 4–6 hours. The stirred mixture is cooled
to room temperature, and the Grignard-nitrile complex is decomposed by cautious addition of 120 ml. (3 moles) of anhydrous
methanol (Note 5).

On completion of the methanol addition, the mixture is stirred
for 30 minutes and filtered. Low-boiling material is stripped
from the filtrate, and the residue is distilled through a 45-cm.
Vigreux column at reduced pressure. There is a fore-run, b.p.
120–127° (3.5 mm.), that weighs about 5 g. Then 55–73 g. (61–
81%) of diphenyl ketimine is collected at 127–128° (3.5 mm.) or
151–152° (8 mm.); $n^{20}D$ 1.6180–1.6191 (Note 6). The product
should be stored under nitrogen to prevent yellowing.

2. Notes

1. Freshly opened commercial (Baker and Adamson) anhydrous ether is suitable. The checkers found it more convenient
to use commercial phenylmagnesium bromide than to prepare it.
They obtained 80 g. (88%) of the ketimine by charging the flask
with 175 ml. (0.525 mole) of $3N$ phenylmagnesium bromide
(Arapahoe Chemicals, Boulder, Colorado), then adding 51.5 g.
of benzonitrile as described.

2. Both bromobenzene and benzonitrile (white label grade,
Eastman Kodak Company) were distilled before use.

3. If the reaction does not start spontaneously, a crystal of
iodine may be added and the mixture may be warmed.

4. Care should be taken to prevent a buildup of unreacted

nitrile that could result in uncontrolled reaction.

5. The methanol should be added as fast as possible. A quantity of gummy material will form as the decomposition progresses, but with continued addition of methanol it will be rapidly converted to crystalline methoxymagnesium bromide.

6. Gas chromatographic analysis of the product from three consecutive preparations showed less than 0.1% impurity. Similar results were obtained on 0.005-ml. samples in an F. and M. 202 Temperature Programed Gas Chromatograph using two columns: a 12-foot column of 10% HiVac grease and 5% Marlex-50 on 100–140 mesh Gas Chrom A, at a constant temperature of 275°, with a helium flow rate of 120 ml. per minute; and a 20-foot column of 20% GE-SE 30 on 100–140 mesh Gas Chrom A, programed at 3.3° per minute from 250° to 300°, with a helium flow rate of 120 ml. per minute.

3. Methods of Preparation

This procedure is a modification of the method employed by Moureu and Mignonac,[3] who first reported the preparation of ketimines via Grignard-nitrile complexes. The use of methanol in the decomposition step results in higher yields and extends the method to the less stable ketimines.[4] The preparation of diphenyl ketimine by the thermal decomposition of benzophenone oxime has been described in *Organic Syntheses*.[5]

4. Merits of the Preparation

The procedure is general and is often the best way to make ketimines.

1. Celanese Chemical Company, Clarkwood, Texas.
2. Chemstrand Research Center, Inc., Durham, North Carolina.
3. C. Moureu and G. Mignonac, *Compt. Rend.*, **156**, 1801 (1913).
4. P. L. Pickard and T. L. Tolbert, *J. Org. Chem.*, **26**, 4886 (1961).
5. A. Lachman, *Org. Syntheses*, Coll. Vol. **2**, 234 (1943).

1,1-DIPHENYLPENTANE

(Pentane, 1,1-diphenyl-)

$$(C_6H_5)_2CH_2 \xrightarrow[\text{NH}_3]{\text{NaNH}_2} (C_6H_5)_2CHNa \xrightarrow[(C_2H_5)_2O]{n\text{-}C_4H_9Br} (C_6H_5)_2CHC_4H_9$$

Submitted by William S. Murphy, Phillip J. Hamrick, and Charles R. Hauser[1]

Checked by Prithipal Singh and Peter Yates

1. Procedure

Caution! This preparation should be carried out in a hood to avoid exposure to ammonia.

A suspension of sodium amide (0.275 mole) (Note 1) in liquid ammonia is prepared in the following manner in a 1-l. three-necked flask equipped with an air condenser (Note 2), a sealed mechanical stirrer, and a dropping funnel. Commercial anhydrous liquid ammonia (600 ml.) is introduced by pouring from an Erlenmeyer flask (Note 3). To the stirred liquid ammonia is added a small piece of sodium. After the appearance of a permanent blue color (Note 4) a few crystals of ferric nitrate hydrate (*ca.* 0.1 g.) are added, followed by small pieces of freshly cut sodium (Note 5) until 6.32 g. (0.275 g. atom) has been added. After all the sodium is converted to the amide (Note 6), 42.0 g. (0.250 mole) of diphenylmethane (Note 7) in 20 ml. of anhydrous ether is added (Note 8). The deep red suspension is stirred for 15 minutes. *n*-Butyl bromide (37.6 g., 0.274 mole) (Note 7) in 20 ml. of anhydrous ether is then added dropwise with stirring. The ammonia is allowed to evaporate (Note 9) from the resulting gray suspension. Water (100 ml.) is added carefully (Note 10), then 100 ml. of ether. The ethereal layer is separated, and the aqueous layer is extracted with two further 100-ml. portions of ether. The combined ethereal extracts are dried over Drierite and filtered, and the solvent is removed. The resulting liquid

(54.5 g., 97%) is essentially pure 1,1-diphenylpentane (Notes 11 and 12). The liquid is distilled with the use of a Claisen distillation head without a fractionating column. The fraction, b.p. 138–139° (1.5 mm.), n^{26}D 1.5501, weighs 51.6 g. (92%) (Note 13).

2. Notes

1. A 10% excess of sodium amide and n-butyl bromide with respect to diphenylmethane was adopted.

2. The checkers used a dry-ice condenser in place of the air condenser.

3. Dry commercial liquid ammonia is conveniently transferred from the cylinder via an Erlenmeyer flask without cooling and without the use of a condenser.

4. A permanent blue color may not remain after the addition of one pellet of sodium because of the presence of traces of moisture. Another pellet is added if necessary.

5. Sodium is cut into small pellets in the atmosphere but weighed under dry benzene or toluene.

6. Conversion is indicated by the discharge of the blue color (*ca*. 30 minutes). The addition of another portion of ferric nitrate hydrate will catalyze the conversion.

7. Freshly distilled diphenylmethane and n-butyl bromide were used.

8. The checkers found it important to add the diphenylmethane slowly (*ca*. 20 minutes); fast addition caused the reaction to get out of control.

9. The ammonia is allowed to evaporate overnight. A steam bath may be employed with care to facilitate the evaporation.

10. In the event of the presence of traces of unreacted sodium on the flask, water is added initially with special care.

11. The purity of the 1,1-diphenylpentane is attested by vapor-phase chromatography on a 5-ft. column of 10% Apiezon L on Celite at 200°.

12. Although 1,1-diphenylpentane undergoes air oxidation,[2] it appears to be stable in a stoppered flask under an inert atmosphere.

13. The checkers observed b.p. 127–129° (1.5 mm.).

3. Methods of Preparation

This procedure is an adaptation of one described by Hauser and Hamrick.[3] 1,1-Diphenylpentane has been prepared by the catalytic hydrogenation of 1,1-diphenyl-l-pentene [4-6] and in low yield from the reaction of diphenylmethyl bromide with di-*n*-butylmercury.[7] More recently 1,1-diphenylpentane was prepared by allowing lithium diphenylmethide to react with tri-*n*-butyl orthophosphate.[8]

4. Merits of the Preparation

This procedure illustrates a process which is general for 1,1-diphenyl substituted hydrocarbons. Diphenylmethane has been alkylated [3] with benzyl chloride, benzhydryl chloride, α-phenyl-ethyl chloride, β-phenylethyl chloride, isopropyl chloride, 2-ethylbutyl bromide, and *n*-octyl bromide in yields of 99, 96, 97, 88, 86, 96, and 99%, respectively.

The present method is superior to earlier ones in that it is shorter, the chemicals are readily available, and high yields are obtained. The Gilman method [8] affords a 74% yield but a longer reaction time (1–2 days) and less readily available starting materials make it less convenient.

1. Department of Chemistry, Duke University, Durham, North Carolina 27706. This work was supported by the National Science Foundation.
2. M. S. Eventova, P. P. Borisov, and M. V. Chistyakova, *Vestn. Mosk. Univ. Ser. Mat., Mekhan., Astron., Fiz. i Khim.*, **12**, No. 3, 185 (1957) [*C.A.*, **52**, 6275 (1958)].
3. C. R. Hauser and P. J. Hamrick, Jr., *J. Am. Chem. Soc.*, **79**, 3142 (1957).
4. K. T. Serijan and P. H. Wise, *J. Am. Chem. Soc.*, **73**, 4766 (1951).
5. G. Benoit and F. Eliopoulo, *Bull. Soc. Chim. France*, [5] **18**, 892 (1951).
6. G. Wittig and E. Stahnecker, *Ann.*, **605**, 69 (1957).
7. F. C. Whitmore and E. N. Thurman, *J. Am. Chem. Soc.*, **51**, 1500 (1929).
8. H. Gilman and B. J. Gaj, *J. Am. Chem. Soc.*, **82**, 6326 (1960).

α,β-DIPHENYLPROPIONIC ACID

(Propionic acid, 2,3-diphenyl-)

$$C_6H_5\text{—}CH_2\text{—}CO_2H \xrightarrow[\text{Liq. NH}_3]{2NaNH_2} C_6H_5\text{—}\underset{\underset{Na}{|}}{C}HCO_2Na \xrightarrow[\text{Et}_2O]{C_6H_5CH_2Cl}$$

$$C_6H_5\text{—}\underset{\underset{C_6H_5\text{—}CH_2}{|}}{C}H\text{—}CO_2Na \xrightarrow{H^+} C_6H_5\text{—}\underset{\underset{C_6H_5\text{—}CH_2}{|}}{C}H\text{—}COOH$$

Submitted by CHARLES R. HAUSER and W. R. DUNNAVANT.[1]
Checked by VIRGIL BOEKELHEIDE and P. WARRICK.

1. Procedure

Caution! This preparation should be carried out in a hood to avoid exposure to ammonia.

A solution of sodium amide (0.226 mole) in liquid ammonia is prepared in a 1-l. three-necked flask equipped with a condenser, a ball-sealed mechanical stirrer, and a dropping funnel. Commercial anhydrous liquid ammonia (500 ml.) is introduced from a cylinder through an inlet tube. To the stirred ammonia is added a small piece of sodium. After the appearance of a blue color, a few crystals of ferric nitrate hydrate (about 0.25 g.) are added, followed by small pieces of freshly cut sodium until 5.2 g. has been added. After all the sodium has been converted to the amide (Note 1), 14.2 g. (0.104 mole) of phenylacetic acid (Note 2) is added and the dark-green suspension is stirred for 15 minutes. To the green suspension is added rapidly 13.2 g. (0.104 mole) of benzyl chloride (Note 3) in 25 ml. of anhydrous ether, and the mixture is then stirred for 1 hour. The mixture is then evaporated to near dryness on a steam bath, 200 ml. of ether added, and evaporation to dryness effected. Another 200 ml. of ether is added, followed by evaporation to dryness. The resulting solid is then dissolved in 300 ml. of water and washed with three 200-ml. portions of ether. The aqueous solution is filtered through a layer of Celite to remove the slight brown coloration, and the filtrate is acidified with hydrochloric acid. A colorless oil forms which, when cooled for a few minutes in an ice bath, becomes a white crystalline solid. This is collected by filtration and washed with three 100-ml. portions of hot water (Note 4) and dried. The yield of crude α,β-diphenyl-propionic acid, m.p. 92.0–93.5°, is 19.85 g. (84.5–88%). The crude solid is recrystallized from 60 ml. of petroleum ether (60–90°) to yield 18.80 g. (80–84%) (Note 5) of α,β-diphenyl-propionic acid, m.p. 95.5–96.5° (Notes 6 and 7).

2. Notes

1. Conversion is indicated by the disappearance of any blue color. This generally requires about 20 minutes.

2. Phenylacetic acid as supplied by the Eastman Kodak Company was used without purification.

3. Eastman Kodak Company "Practical Grade" benzyl chloride was distilled before use, the fraction with b.p. 63–64°/12 mm. being used.

4. The hot water effectively removes unreacted phenylacetic acid, m.p. 76.7°, the presence of which hinders the purification of the product. The water should be hot but not boiling, since α,β-diphenylpropionic acid has some solubility in boiling water and may be recrystallized from water.

5. An identical preparation using potassium amide instead of sodium amide gave α,β-diphenylpropionic acid in 57% yield.

6. Miller and Rohde [7] report that α,β-diphenylpropionic acid exists in three crystalline modifications melting at 82°, 88–89°, and 95–96°. Although the product obtained by the submitters melted at 95.5–96.5°, corresponding to the high-melting form, the sample obtained by the checkers melted sharply at 88–89° corresponding to the crystalline form of intermediate melting point.

7. Under comparable conditions the corresponding alkylations of phenylacetic acid with α-phenylethyl chloride and benzhydryl chloride have been effected to form α,β-diphenylbutyric acid and

α,β,β-triphenylpropionic acid in yields of 74% and 51%, respectively.[2]

3. Methods of Preparation

This procedure is an adaption of one described by Hauser and Chambers.[2] Previous preparations include the benzylation of diethyl phenylmalonate followed by hydrolysis,[3] the benzylation of phenylacetonitrile followed by hydrolysis,[4,5] the benzylation of phenylacetic acid through the Ivanov reagent,[6] and the reduction of α-phenylcinnamic acid using sodium amalgam.[7]

The present procedure illustrates the use of dianions for alkylation and gives α,β-diphenylpropionic acid more conveniently and in better yield than previous preparations.

The alkylation of ethyl phenylacetate is also described in this volume (see p. 559). The submitters of this alternative procedure reported that it gave an 85% yield of ethyl a,β-diphenylpropionate.

1. Department of Chemistry, Duke University, Durham, North Carolina. Work supported by the Office of Ordnance Research.
2. C. R. Hauser and W. J. Chambers, *J. Am. Chem. Soc.*, **78**, 4942 (1956).
3. W. Wislicenus and K. Goldstein, *Ber.*, **28**, 818 (1895).
4. A. Meyer, *Ber.*, **21**, 1311 (1888).
5. C. R. Hauser and W. R. Brasen, *J. Am. Chem. Soc.*, **78**, 494 (1956).
6. T. Cohen and G. F. Wright, *J. Org. Chem.*, **18**, 432 (1953).
7. W. v. Miller and G. Rohde, *Ber.*, **25**, 2017 (1892).

cis- AND trans-1,2-DIVINYLCYCLOBUTANE

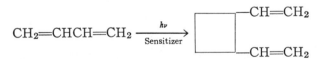

Submitted by CHARLES D. DeBOER,[1] NICHOLAS J. TURRO,[2] and GEORGE S. HAMMOND[1]
Checked by WILLIAM G. DAUBEN and JAMES H. SMITH

1. Procedure

Caution! This reaction should be carried out in an explosion-proof room behind a safety shield because it involves a glass vessel under pressure.

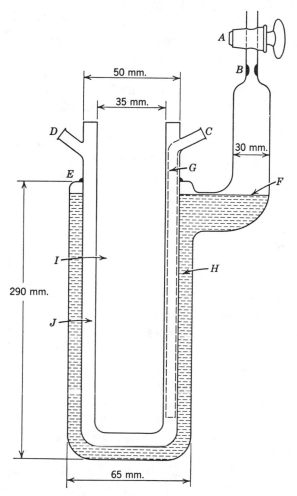

FIG. 1. *A*, 4-mm. stopcock; *B*, thickened for easy sealing; *C*, water inlet; *D*, water outlet; *E*, this seal can be replaced by a ground glass joint for higher-boiling materials than butadiene; *F*, filling level; *G*, the cooling water tube indicated by the dotted lines will permit a higher flow rate if shaped as an oval or rectangle; *H*, reaction well; *I*, lamp well; *J*, cooling water jacket.

A reaction vessel as shown in Fig. 1 is made from Pyrex tubing (Note 1). The vessel is evacuated and the stopcock closed. A 500-ml. round-bottomed, two-necked flask equipped with a gas inlet and a cold finger condenser containing dry ice is charged with 1 g. of *finely powdered* 4,4′-bis(dimethylamino)benzophenone (Michler's ketone) (Notes 2, 3). A butadiene tank is connected to the gas inlet, and about 250 ml. (160 g., 3.0 moles) of butadiene is collected in the flask (Note 4). The butadiene is cooled to dry ice temperature. The reaction vessel is immersed to the filling level shown in Fig. 1 in a slurry of dry ice and acetone. A 4-in. length of Tygon tubing is attached between the inlet of the reaction vessel and one neck of the flask containing the butadiene. The other neck of the flask is stoppered, and the flask is tipped to fill the connecting tube and the neck of the stopcock with butadiene. At the same time the stopcock is opened. The flask is shaken to ensure that the suspended Michler's ketone will be swept into the reaction vessel by the butadiene. After the vessel is filled with butadiene the stopcock is closed, the connecting tube removed, and the vessel transferred to a Dewar flask filled with liquid nitrogen. When the butadiene is frozen, the vessel is evacuated with a high-vacuum pump and sealed off below the stopcock with a torch (Note 5). The reaction vessel is fitted with cooling water hoses and a 450-watt Hanovia medium-pressure mercury arc lamp, and then the butadiene is allowed to thaw and come to room temperature (Note 6). The mixture is irradiated for 72 hours, the water jacket dried, and the vessel weighed (*Caution!* Note 6). The reaction vessel is then frozen in a Dewar flask containing a dry ice-acetone mixture, and the seal is cautiously broken. The cooling bath is removed, and the reactor is allowed to come to room temperature. The reaction mixture is removed, the vessel cleaned (Note 7), and weighed again to determine the amount of starting material. The reaction mixture is distilled and the fraction boiling between 109° and 111° (uncor.) is collected (Note 8). The yield is 96–104 g. (60–65%) of 99% pure (by gas chromatography) *trans*-1,2-divinylcyclobutane, n^{25}D 1.4429–1.4431, the impurities being butadiene and 1,5-cyclooctadiene (Note 9).

2. Notes

1. A reaction vessel, as shown in Fig. 1, is useful for many photochemical reactions because virtually all the light produced can be captured by the reagents. It can be constructed from

either Pyrex or quartz tubing, depending on the absorption spectra of the reagents.

2. The choice of a sensitizer for butadiene dimerization depends on three things: the energy of the triplet-singlet transition, the intersystem crossing efficiency, and the absorption spectrum of the sensitizer.[3]   Michler's ketone has a sufficiently high triplet energy to transfer energy at a diffusion-controlled rate to both *cis* and *trans* forms of butadiene, thus reducing the yield of 4-vinylcyclohexene produced.[4]   Furthermore it has a high intersystem crossing efficiency and a high extinction coefficient at 3660 Å, which is the most intense line from a medium-pressure mercury arc lamp.   For photochemical reactions where energy transfer is not diffusion-controlled, the lifetime of the triplet may be an important factor in the choice.

3. The amount of sensitizer is not critical since only enough is needed to absorb all the light.   One gram of Michler's ketone will not be totally dissolved in 250 ml. of butadiene, but will be totally dissolved after the reaction is finished.   Michler's ketone as obtained from Eastman Organic Chemicals was used without purification.

4. 1,3-Butadiene obtained from Matheson, Coleman and Bell was used without purification.

5. The butadiene should not be degassed by freeze-thaw cycles, because the presence of a small amount of oxygen reduces the amount of polymer formed on the walls of the vessel.

6. All experimental manipulations with the reaction vessel while it is sealed and under pressure should be carried out behind a safety shield.

7. The polymer can be conveniently removed by filling the flask two-thirds full with concentrated nitric acid and warming gently on a steam bath in a hood behind a safety shield.

8. The *cis*-1,2-divinylcyclobutane in the reaction mixture rearranges rapidly under reflux to the higher-boiling 1,5-cyclooctadiene.

9. If *cis*-1,2-divinylcyclobutane is desired, it can be isolated in 7–8% yield from the reaction mixture by preparative gas chromatography with the Beckman Megachrom instrument, using columns packed with Apiezon J.

3. Methods of Preparation

The *trans* isomer of 1,2-divinylcyclobutane may be isolated in low yield from the mixture formed by thermal dimerization of butadiene.[5] The *cis* isomer has been prepared by a sequence of reactions.[6]

4. Merits of the Preparation

Essentially the same procedure may be used to produce mixtures of cyclodimers from isoprene,[4] 1,3-cyclopentadiene,[4] and 1,3-cyclohexadiene.[7] Separation of all products is somewhat difficult in most cases but has always been possible by preparative vapor phase chromatography. Despite the problems that may be involved in separation of desired products in some instances, photocyclization frequently is the method of choice for preparation of 1,2-dialkenylcyclobutanes if they can be made major products of photoreactions. Starting materials are readily available, and the preparations are easily carried out on the scale described. There is little doubt that the method is the best for preparation of *trans*-1,2-divinylcyclobutane.

1. Contribution No. 3131 from the Gates and Crellin Laboratories of Chemistry, California Institute of Technology, Pasadena, California.
2. Department of Chemistry, Columbia University, New York, New York.
3. W. G. Herkstroeter, A. A. Lamola, and G. S. Hammond, *J. Am. Chem. Soc.*, **86**, 4537 (1964).
4. G. S. Hammond, N. J. Turro, and R. S. H. Liu, *J. Org. Chem.*, **28**, 3297 (1963).
5. E. Vogel, *Ann.*, **615**, 1 (1958).
6. H. W. B. Reed, *J. Chem. Soc.*, 685 (1951).
7. D. Valentine, N. J. Turro, and G. S. Hammond, *J. Am. Chem. Soc.*, **86**, 5202 (1964).

DOCOSANEDIOIC ACID *

$$2 \quad + \quad ClCO(CH_2)_8COCl \quad \xrightarrow[-2HCl]{2(C_2H_5)_3N} \quad$$

$$\xrightarrow{HCl/H_2O} \quad 2 \quad O\diagdown NH \cdot HCl \quad + \quad$$

$$\xrightarrow{2NaOH} \quad NaOCO(CH_2)_5CO(CH_2)_8CO(CH_2)_5CO_2Na \quad \xrightarrow{H_2NNH_2} \quad HOCO(CH_2)_{20}CO_2H$$

Submitted by S. Hünig, F. Lücke, and W. Brenninger.[1]
Checked by F. E. Mumford, E. A. LaLancette,
W. J. Middleton, and B. C. McKusick.

1. Procedure

A. *2,2'-Sebacoyldicyclohexanone.* A solution of 167 g. (1.00
mole) of 1-morpholino-1-cyclohexene [2] and 101 g. (139 ml., 1.00
mole) of anhydrous triethylamine in 500 ml. of dry chloroform
(Note 1) is put in a 5-l., three-necked, round-bottomed flask
equipped with a mechanical stirrer, a dropping funnel, and a reflux condenser. Tubes of calcium chloride are inserted in the
open ends of the dropping funnel and reflux condenser. The
reaction flask is immersed in a water bath at 35°, and a solution of
120 g. (0.50 mole) of sebacoyl chloride (Note 2) in 200 ml. of dry
chloroform is added to the well-stirred reaction mixture over a
period of about 1.5 hours. The reaction mixture gradually as-

sumes an orange to red color, and a solid precipitates. The reaction mixture is stirred for an additional 3 hours at 35°, 500 ml. of 20% hydrochloric acid is added, and the mixture is boiled under reflux for 5 hours with vigorous stirring. The reaction mixture is cooled to room temperature, and the chloroform layer is separated and extracted with six 150-ml. portions of water. The washings and the aqueous phase are combined, adjusted to a pH of 5–6 with 25% sodium hydroxide solution, and extracted with five 100-ml. portions of chloroform. The chloroform extracts are combined with the chloroform layer, and the chloroform is removed by distillation on a steam bath. The residue gradually congeals to an oily solid on standing at room temperature under a pressure of 10–50 mm. The yield of crude 2,2′-sebacoyldicyclohexanone is 181–192 g. (100–106%) (Note 3).

B. *Disodium 7,16-diketodocosanedioate.* A mixture of 120 g. (3.00 moles) of sodium hydroxide and 1.4 l. of commercial absolute ethanol is refluxed with mechanical stirring in a 5-l. round-bottomed flask until all the sodium hydroxide is dissolved (about 2 hours). The solution is cooled to room temperature, and a warm solution of the crude 2,2′-sebacoyldicyclohexanone from Step A in 300 ml. of absolute ethanol is added. The mixture is brought to a boil on a water bath or steam bath in the course of about 15 minutes and is then refluxed for 1 hour. Colorless disodium 7,16-diketodocosanedioate separates during the heating. The reaction mixture, now a thick mush, is cooled to room temperature, and the salt is collected on a 25-cm. Büchner funnel and pressed as dry as possible, preferably with the aid of a rubber dam. The moist salt is suspended in 1 l. of absolute ethanol with mechanical stirring and is then collected on the Büchner funnel as before. After being dried in air, the crude colorless disodium 7,16-diketodocosanedioate weighs 248–255 g. (112–115%, based on 1-morpholino-1-cyclohexene). It is pure enough for the following reduction to docosanedioic acid (Note 4).

C. *Docosanedioic acid.* All the crude disodium 7,16-diketodocosanedioate of Step B is added to 1 l. of triethanolamine in a 5-l. round-bottomed flask equipped with a reflux condenser, a thermometer, a mechanical stirrer, and a deep oil bath. The mixture is heated under reflux until all the salt dissolves (about 15 minutes). The solution is cooled to 130°, 610 ml. (10 moles) of 82% hydrazine hydrate is added through the reflux condenser, and the mixture is refluxed for 4 hours (Note 5).

Potassium hydroxide (168 g., 3.0 moles) is dissolved in 400 ml. of triethanolamine by heating the mixture to boiling in a 1-l. Erlenmeyer flask (about 15 minutes is required). At the end of the reflux period, the hot reaction mixture is transferred to a good hood if it is not already in one, the condenser is removed, and the hot potassium hydroxide solution is added cautiously but rapidly to the stirred reaction mixture (Note 6). The open reaction mixture is at least two-thirds immersed in the oil bath to help prevent foaming over and is heated strongly and rapidly in order to drive off water and excess hydrazine hydrate. After 2–3 hours the temperature inside the flask reaches about 140°, and decomposition of the bis-hydrazone begins, with evolution of nitrogen and considerable foaming. Foaming over is prevented by judicious regulation of the heating, good stirring, and occasional addition of a little silicone oil, which is a good antifoaming agent (Note 7). The temperature is raised as rapidly as possible to 195° (about 2 hours is needed) and held at this temperature for 6 hours. The final oil bath temperature is 200–220°.

The reaction mixture is cooled to about 100° (Note 5), washed out of the flask with 5 l. of hot water (Note 8), and acidified to a pH between 2 and 3 with 1.4 l. of 12N hydrochloric acid. The mixture is cooled to room temperature, and the docosanedioic acid that has precipitated is collected on a 25-cm. Büchner funnel and pressed as dry as possible (Note 9). The filter cake is suspended in 5 l. of water with mechanical stirring and collected on the Büchner funnel as before. The moist filter cake is dissolved in 700 ml. of hot 2-methoxyethanol, the hot solution is filtered through a fluted paper in a heated funnel, and the filtrate is gradually cooled to 0–5°. The docosanedioic acid that crystallizes out is separated on a Büchner funnel, pressed as dry as possible, and suspended in 500 ml. of 95% ethanol with mechanical stirring. The acid is collected on a Büchner funnel, washed with a little 95% ethanol, dried in air, and pulverized. The colorless docosanedioic acid thus obtained weighs 127–133 g. (69–72%, based on 1-morpholino-1-cyclohexene) and is nearly pure; m.p. 124–126°; neutralization equivalent 181–184 (calculated, 185) (Notes 10 and 11).

2. Notes

1. Satisfactory chloroform is obtained by washing 2 l. of commercial chloroform with two 100-ml. portions of $2N$ sodium carbonate solution and two 200-ml. portions of water and distilling it until no more water codistils and the boiling point is 61°. The material remaining in the distillation pot is used without distillation.

2. Satisfactory sebacoyl chloride can be purchased from the Eastman Kodak Co., Rochester, New York. The submitters prepared it as follows. A mixture of 150 g. (0.74 mole) of sebacic acid and 150 ml. of thionyl chloride is heated in a water bath at 60°. The acid gradually goes into solution with evolution of sulfur dioxide and hydrogen chloride. When gas evolution ceases, the mixture is distilled as rapidly as possible under reduced pressure. The yield of sebacoyl chloride, b.p. 171–175°/15 mm., is about 140 g. (79%). *Caution! Toward the end of the distillation, spontaneous decomposition of the residue with formation of a voluminous black foam frequently occurs.*

3. The tetraketone can be obtained in a pure form by recrystallizing it first from ether with the addition of decolorizing carbon and then from *n*-butanol; yield 50–58%; m.p. 68–72°.

4. The submitters obtained pure 7,16-diketodocosanedioic acid by the following procedure. A solution of 300 ml. of $12N$ hydrochloric acid in 3 l. of water is stirred into a warm solution of 250 g. of the crude disodium 7,16-diketodocosanedioate in 3 l. of water. The resultant suspension of the diketo acid is boiled for a few minutes to make the acid easier to filter, then cooled to room temperature and collected on a Büchner funnel. The filter cake is suspended in 3 l. of water with mechanical stirring and collected on a Büchner funnel, and this procedure is repeated. The moist well-pressed filter cake is recrystallized from 600 ml. of 2-methoxyethanol. The recrystallized acid is suspended in 500 ml. of 95% ethanol, separated on a Büchner funnel, and dried in air. About 120 g. (61%) of pure 7,16-diketodocosanedioic acid is ob-

tained; m.p. 127–129°; equivalent weight 196 (—CO$_2$H), 200 ($>$C$=$O) (calculated, 199).

5. One may interrupt the procedure at this point and allow the mixture to stand overnight at room temperature.

6. It is essential for good results that the procedure not be interrupted from the time that the potassium hydroxide solution is added until the time that the 6-hour heating at 195° is completed.

7. Additional security against foaming over is provided by a glass tube that projects into the neck of the flask and is attached to a water pump. The checkers found it helpful to use a Hershberg stirrer with two wire blades; the upper blade was adjusted so that its ends extended above the surface of the reaction mixture and into the foam.

8. The aqueous mixture is not clear because the sodium salt is sparingly soluble in water.

9. The precipitate can be more rapidly separated by means of a centrifuge.

10. Very pure docosanedioic acid can be obtained by another recrystallization from about 450 ml. of 2-methoxyethanol. The recrystallized acid is collected on a Büchner funnel, and the well-pressed filter cake is suspended in 200 ml. of 95% ethanol, refiltered, and dried in air; weight 112 g. (61%); m p. 126–127°; neutralization equivalent, 185–187.

11. The checkers found it slightly more convenient to recrystallize the moist crude docosanedioic acid from 1 l. of methyl ethyl ketone. The hot solution is filtered and cooled, and the acid is collected on a Büchner funnel, washed with methyl ethyl ketone, and dried in air.

3. Methods of Preparation

Docosanedioic acid has been prepared by Wolff-Kishner reduction of 6,17-diketodocosanedioic acid, formed by reaction of the half-ester acid chloride of adipic acid with the α,ω-cadmium derivative of decane (26% overall yield).[3] Reduction of ω-[5-(ω-carboxy-n-octyl)-2-thenoyl]caprylic acid by the Wolff-Kishner method, followed by simultaneous reduction and desulfurization with Raney nickel of the 2,5-bis(ω-carboxyoctyl)thiophene produced, is reported to yield docosanedioic acid in 68% overall yield.[4] Other routes to docosanedioic acid include electrolysis of the monomethyl ester of dodecanedioic acid (43% yield);[5] oxidative coupling of 10-undecynoic acid to docosa-10,12-diyne-dioic acid (90% yield) and reduction of this intermediate with a palladium catalyst;[6] and reaction of α,ω-diiodoeicosane with potassium cyanide followed by hydrolysis of the dinitrile produced.[7]

The present method of making docosanedioic acid has been described by Hünig and Lücke.[8] The Wolff-Kishner reduction of the diketonic intermediate is an application of the modification of Gardner, Rand, and Haynes.[9]

4. Merits of the Preparation

The present procedure has been used to convert suberoyl chloride to eicosanedioic acid (44%), and it is probably a general method for increasing by twelve carbon atoms the chain length of dicarboxylic acids whose chain length is eight or more carbon atoms.[8] A variant of the method, in the first step of which the ester chloride of a dicarboxylic acid is condensed with 1-morpholino-1-cyclohexene, has also been used to prepare dicarboxylic acids. Thus the mono-ester acid chloride of succinic acid has been converted to sebacic acid (48%), that of suberic acid to tetradecanedioic acid (34%), and that of sebacic acid to hexadecanedioic acid (32%).[8] A general method of increasing the chain length of a carboxylic acid by six carbon atoms is to employ a monoacyl chloride in the present procedure; overall yields of acids from nonanoic to tetracosanoic are 42–48%.[10] 1-Morpholino-1-cyclopentene [11] can be used in the same sort of syntheses as 1-morpholino-1-cyclohexene; thus, by starting with it and lauroyl chloride, heptadecanoic acid can be obtained in 60% yield.[12] Similarly, an enamine of cyclododecanone has been used to lengthen the chain of monocarboxylic acids by twelve carbon atoms, and of dicarboxylic acids by twenty-four; for instance, stearic acid has been converted to triacontanoic acid (70%), and suberic acid to pentatricontanedioic acid (40%).[13]

How to decide whether the enamine method or some other method is better for preparing a given mono- or dicarboxylic acid is discussed in two papers.[8,10]

1. University of Marburg, Marburg, Germany.
2. S. Hünig, E. Lücke, and W. Brenninger, this volume, p. 808.
3. A. Kreuchunas, *J. Am. Chem. Soc.*, 75, 3339 (1953).
4. N. P. Buu-Hoï, M. Sy, and N. D. Xuong, *Bull. Soc. Chim. France*, 1955, 1583.
5. R. Signer and P. Sprecher, *Helv. Chim. Acta*, 30, 1001 (1947).
6. A. Seher, *Ann.*, 589, 222 (1954).
7. S. Shiina, *J. Soc. Chem. Ind. Japan*, Suppl., 42, 147B (1939).
8. H. Hünig and E. Lücke, *Ber.*, 92, 652 (1959).
9. P. D. Gardner, L. Rand, and G. R. Haynes, *J. Am. Chem. Soc.*, 78, 3425 (1956).
10. S. Hünig, E. Lücke, and E. Benzing, *Ber.*, 91, 129 (1958).
11. E. D. Bergmann and R. Ikan, *J. Am. Chem. Soc.*, 78, 1485 (1956).
12. S. Hünig and W. Lendle, *Ber.*, 93, 909, 915 (1960).
13. S. Hünig and S. J. Buysch, unpublished results.

3-ETHOXY-2-CYCLOHEXENONE*

(Dihydroresorcinol monoethyl ether)

$$+ C_2H_5OH \xrightarrow{p\text{-}CH_3C_6H_4SO_3H} \qquad + H_2O$$

Submitted by WALTER F. GANNON and HERBERT O. HOUSE.[1]
Checked by WILLIAM E. PARHAM, WAYLAND E. NOLAND,
GEORGE MEISTERS, and ALLAN M. HUFFMAN.

1. Procedure

In a 2-l. flask fitted with a total-reflux, variable-take-off distillation head is placed a solution of 53 g. (0.472 mole) of dihydroresorcinol (Note 1), 2.3 g. of *p*-toluenesulfonic acid monohydrate and 250 ml. of absolute ethanol in 900 ml. of benzene. The mixture is heated to boiling and the azeotrope composed of benzene, alcohol, and water is removed at the rate of 100 ml. per hour. When the temperature of the distilling vapor reaches 78° (Note 2), the distillation is stopped and the residual solution is washed with four 100-ml. portions of 10% aqueous sodium hydroxide which have been saturated with sodium chloride. The resulting organic solution is washed with successive 50-ml. portions of water until the aqueous washings are neutral and then concentrated under reduced pressure. The residual liquid is distilled under reduced pressure. The yield of 3-ethoxy-2-cyclohexenone (Note 3), b.p. 66–68.5°/0.4 mm. or 115–121°/11 mm., n_D^{29} 1.5015, is 46.6–49.9 g. (70–75%).

2. Notes

1. The preparation of dihydroresorcinol was described in an earlier volume of this series.[2]

2. This distillation requires 6–8 hours.

3. The product may be analyzed by gas chromatography on an 8 mm. x 215 cm. column heated to 220–240° and packed with Dow-Corning Silicone Fluid No. 550 suspended on 50–80 mesh ground firebrick. The chromatogram obtained with this column exhibits a single major peak. The ultraviolet spectrum of an ethanol solution of the product has a maxium at 250 mμ (ε = 17,200).

3. Methods of Preparation

3-Ethoxy-2-cyclohexenone has been prepared by reaction of the silver salt of dihydroresorcinol with ethyl iodide [3] and by the reaction of dihydroresorcinol with ethyl orthoformate, ethanol and sulfuric acid.[4] The acid-catalyzed reaction of dihydroresorcinol with ethanol in benzene solution utilized in this preparation is patterned after the procedure of Frank and Hall.[4,5]

This procedure can also be used to prepare other 2-alkoxy-2-cyclohexenones. For example, it has been reported that 2-isobutoxy-2-cyclohexenone can be prepared easily in high yield (91%) from dihydroresorcinol and isobutyl alcohol.[6] The use of isobutyl alcohol[7] often gives better yields of enol ethers.[8,9]

4. Use of 3-Ethoxy-2-cyclohexenone

The 3-alkoxy-2-cyclohexenones are useful intermediates in

the synthesis of certain cyclohexenones. The reduction of 3-ethoxy-2-cyclohexenone with lithium aluminum hydride followed by hydrolysis and dehydration of the reduction product yields 2-cyclohexenone.[10] Similarly, the reaction of 3-alkoxy-2-cyclohexenones with organometallic reagents followed by hydrolysis and dehydration of the addition product affords a variety of 3-substituted 2-cyclohexenones.[3,5,6,8]

1. Department of Chemistry, Massachusetts Institute of Technology, Cambridge, Massachusetts.
2. R. B. Thompson, *Org. Syntheses,* Coll. Vol. 3, 278 (1955).
3. G. F. Woods and I. W. Tucker, *J. Am. Chem. Soc.,* 70, 2174 (9148).
4. E. G. Meek, J. H. Turnbull, and W. Wilson, *J. Chem. Soc.,* 811 (1953).
5. R. L. Frank and H. K. Hall, Jr., *J. Am. Chem. Soc.,* 72, 1645 (1950).
6. J. J. Panouse and C. Sannie, *Bull. Soc. Chim. France,* 1272 (1956).
7. B. H. Chase and J. Walker, *J. Chem. Soc.,* 3518 (1953).
8. A. Eschenmoser, J. Schreiber, and S. A. Julia, *Helv. Chim. Acta,* 36, 482 (1953).
9. D. A. H. Taylor, private communication.
10. W. F. Gannon and H. O. House, this volume, p. 294.

N-ETHYLALLENIMINE

(Aziridine, 1-ethyl-2-methylene-)

$$\text{H}_2\text{C}=\overset{|}{\underset{\text{Br}}{\text{C}}}\text{CH}_2\text{NHC}_2\text{H}_5 \xrightarrow[\text{Liq. NH}_3]{\text{NaNH}_2} \text{H}_2\text{C}=\overset{\text{CH}_2}{\underset{\underset{\text{C}_2\text{H}_5}{|}}{\text{C}}}$$

Submitted by ALBERT T. BOTTINI and ROBERT E. OLSEN [1]
Checked by THOMAS H. LOWRY and E. J. COREY

1. Procedure

Caution! This preparation should be carried out in a good hood to avoid exposure to ammonia. The operator should wear rubber gloves and protective goggles because 2-haloallylamines and ethylenimines can cause severe skin and eye irritation.

A 2-l. three-necked flask is fitted with a sealed mechanical stirrer, a gas-inlet tube, and a dry ice condenser protected from the air by a soda-lime drying tube (Note 1). The system is flushed thoroughly with dry ammonia, and 32.8 g. (0.84 mole) of sodium amide (Note 2) is added to the flask. The system is again flushed with ammonia, the condenser is provided with dry ice covered by acetone, and 1.2 l. of liquid ammonia is condensed in the flask. The gas-inlet tube is replaced with a dropping funnel, the stirrer is started, and 118 g. (0.72 mole) of N-(2-bromoallyl)ethylamine [2] is added dropwise in 20–30 minutes; during the addition, the ammonia boils vigorously, and the color of the slurry changes from gray to black. Stirring is continued for 3 hours, and the dry ice is then allowed to evaporate. The condenser is provided with an ice-salt mixture, and the ammonia is allowed to evaporate until the volume is reduced to about 800 ml. (Note 3). Ethanol-free ether (200 ml.) is added rapidly through the dropping funnel, and the reaction is stopped by the slow, dropwise addition (*Caution!*) of 5 ml. of water. The ammonia is allowed to evaporate overnight. Water (150 ml.) and 100 ml. of ether are added to the residue, and the mixture is stirred for 2 minutes in order to dissolve the precipitated salts. The resulting mixture, which consists of aqueous and ethereal solutions, is separated, and the aqueous phase is extracted with 75 ml. of ether. The ether solutions are combined, dried over sodium hydroxide (Note 4), and distilled through an *efficient* low-holdup column (Note 5). The fraction with b.p. 77–80°, n^{25}D 1.4260–1.4268, which is 96–97% N-ethylallenimine (Note 6), weighs 30–34 g. (48–55%). Pure (>99.5%) N-ethylallenimine has b.p. 77–79°, n^{25}D 1.4281–1.4284 (Notes 7 and 8).

2. Notes

1. The glassware should be dried in an oven before use, and water must be rigorously excluded from the reaction mixture.

2. The sodium amide was obtained from Roberts Chemical Co., Nitro, West Virginia.

3. About 1.5–2.5 hours is required; stirring is continued and ice is prevented from forming on the outside of the flask. The checkers used an inner-spiral condenser cooled by ice water.

4. The ether solution and fractions taken during the subse-

quent distillation may be assayed by gas-liquid partition chromatography on a 0.8-cm. x 200-cm. column heated at 120° and
packed with nonyl phthalate supported on ground firebrick.

5. The submitters concentrated the dry ether solution to a
volume of 80–100 ml. by distillation through a 1.0-cm. x 40-cm.
column packed with glass helices and equipped with a total-
reflux head. p-Xylene (10 ml.) was added to the residue, and
this solution was fractionated through a 0.8-cm. x 30-cm. Pod-
bielniak-type column fitted with a total-reflux head. The submitters recommend that, during distillation of the concentrated
solution, a slow stream of nitrogen be passed through the boiling
liquid to minimize the formation of dark, tarry products. The
checkers used a 1-cm. x 100-cm. spinning-band column (Nester
and Faust Co.) for the distillation of N-ethylallenimine and were
able to obtain material of 99% purity (Note 7) directly.

6. This fraction is 2–3% ether, 96–97% N-ethylallenimine,
and 1–2% N-ethylpropargylamine. The product from the reaction consists of 80–90% N-ethylallenimine and 10–20% N-ethylpropargylamine. N-Ethylpropargylamine has b.p. 100–102° (760
mm.), n^{25}D 1.4314–1.4316.

7. The submitters obtained essentially pure (>99.5%)
N-ethylallenimine by redistilling 30 g. or more of the 96–97%
pure product through the Podbielniak column and rejecting the
first 10–20% and the last 20% of the distillate. The yield of pure
N-ethylallenimine is 18–21 g. (30–35%). Pure N-ethylallenimine
has also been obtained in comparable yields by (a) distilling the
combined concentrated solution from the equivalent of three
runs through a 1.3-cm. x 100-cm. column packed with glass
helices and equipped with a total-reflux head, and (b) treating
the crude distillate with lithium aluminum hydride as described
for the purification of N-propylallenimine.[8]

8. Samples of pure N-ethylallenimine and other allenimines
have been stored at 0° for well over a year with no significant
deterioration. *Caution! N-Alkylallenimines, even as dilute solutions in aqueous ethanol, are rapidly destroyed by acid.*[4,5] Therefore
concentrated solutions of N-alkylallenimines should not be allowed to come in contact with acid because of the possibilities of
violent decomposition.

3. Methods of Preparation

The method described is essentially that of Pollard and Parcell.[4] N-Ethylallenimine has been prepared by treating N-(2-bromoallyl)ethylamine in liquid ammonia with sodium amide,[4,6] lithium amide,[6] or potassium amide.[6]

4. Merits of the Preparation

This is a general method for making N-alkylallenimines, and the following ones have been made in this way: N-methyl-,[6] N-propyl-,[6] N-isopropyl-,[4] N-butyl-,[4] N-hexyl-,[6] and N-(3,5,5-trimethylhexyl)-.[4] N-t-Butylallenimine[6] and 1-(1-allenimino)-2-hydroxy-3-butene[7] have also been prepared by this method, but with sodium amide/2-bromoallylamine mole ratios of 1.75 and 2.1, respectively. This method has been used for the preparation of pure N-alkylpropargylamines from 2-chloroallylamines.[6,7] The optimum sodium amide/2-chloroallylamine ratio for the preparation of N-alkylpropargylamines is 2.1.

1. Chemistry Department, University of California, Davis, California.
2. A. T. Bottini and R. E. Olsen, this volume, p. 124.
3. A. T. Bottini and R. E. Olsen, J. Am. Chem. Soc., 84, 196 (1962).
4. C. B. Pollard and R. F. Parcell, J. Am. Chem. Soc., 73, 2925 (1951).
5. A. T. Bottini and J. D. Roberts, J. Am. Chem. Soc., 79, 1462 (1957).
6. A. T. Bottini, B. J. King, and R. E. Olsen, J. Org. Chem., 28, 3241 (1963).
7. A. T. Bottini and V. Dev, J. Org. Chem., 27, 968 (1962).

ETHYL AZODICARBOXYLATE

WARNING

A report has been received that a sample of ethyl azodicarboxylate [Org. Syntheses, 28, 59 (1948); Coll. Vol. 3, 375 (1955)] decomposed upon attempted distillation with sufficient violence to shatter the distillation apparatus.

It is possible that the explosion may have been due to over-chlorination or to insufficient washing of the product with sodium bicarbonate solution.

It is recommended that ethyl azodicarboxylate be distilled only behind a safety shield, and protected from direct sources of light.

Additional precautions on the handling of ethyl azodicarboxylate may be found in the most recent procedure published for its preparation in this series.[1]

1. *Org. Syntheses,* Coll. Vol. 4, 421 (1963).

ETHYL γ-BROMOBUTYRATE *

(Butyric acid, γ-bromo-, ethyl ester)

$$C_2H_5OH \ + \ \text{[lactone]} \ + \ HBr \ \xrightarrow{0°} \ Br(CH_2)_3COOC_2H_5 \ + \ H_2O$$

Submitted by J. LAVETY and G. R. PROCTOR [1]
Checked by IAN MORRISON and VIRGIL BOEKELHEIDE

1. Procedure

A solution of 200 g. (2.33 moles) of γ-butyrolactone (Note 1) in 375 ml. of absolute ethanol is cooled to 0° in an ice bath while a stream of dry hydrogen bromide (Note 2) is introduced. The passage of gas is discontinued 1 hour after hydrogen bromide is seen to pass through the reaction mixture unchanged (Note 3).

The alcoholic solution of the product is kept for 24 hours at 0° and then poured into 1 l. of ice-cold water. The oily layer is separated, and the aqueous layer is extracted with two 100-ml. portions of ethyl bromide (Note 4). The combined oil and extracts is washed with ice-cold 2% potassium hydroxide solution, then with very dilute hydrochloric acid, and finally with water.

The organic layer is dried over sodium sulfate, the solvent is removed under reduced pressure, and the residual crude ester is purified by distillation. The yield of ethyl γ-bromobutyrate, obtained as a colorless oil, b.p. 97–99° (25 mm.), n^{25}D 1.4543, is 350–380 g. (77–84%) (Notes 5 and 6).

2. Notes

1. Technical grade γ-butyrolactone was employed.

2. The hydrogen bromide is made by burning together hydrogen and bromine. The apparatus is essentially that described previously.[2, 3] The submitters found that standard ground-glass fittings can be used throughout, connected where necessary by Neoprene® tubing. The combustion tube is made by "butt-joining" two ground-glass sockets. The checkers used commercial hydrogen bromide from a cylinder which was connected to the reaction flask through a safety trap.

3. The time varies from 6 to 8 hours, and the increase in weight from 450 g. to 480 g. Using the commercial hydrogen bromide, the checkers found the time for saturation to be 3.5–4 hours.

4. The aqueous layer may be kept and used again in repeating the reaction.

5. In smaller runs, yields as high as 95% have been obtained.[4]

6. The product is best stored in dark bottles at 0°.

3. Methods of Preparation

The ester has been made from the corresponding acid which was obtained from the nitrile,[5] but the present method is the more practicable.[4, 6] This procedure is an adaptation of the method used for the preparation of ethyl 2-bromocyclopentane-acetate.[7]

4. Merits of the Preparation

This is a simple procedure for preparing ethyl γ-bromobutyrate in good yield. Ethyl γ-bromobutyrate is used for adding chains of four carbons and has been particularly useful in syntheses of benzazepines.[3, 6, 8, 9]

1. Royal College of Science and Technology, Glasgow, C. 1., Scotland.
2. A. I. Vogel, "Textbook of Practical Organic Chemistry," Longmans, Green and Co., London-New York, 1951, p. 177.
3. J. R. Ruhoff, R. E. Burnett, and E. E. Reid, *Org. Syntheses*, Coll. Vol. **2**, 338 (1943).
4. G. R. Proctor and R. H. Thomson, *J. Chem. Soc.*, 2302 (1957).
5. E. A. Prill and S. M. McElvain, *J. Am. Chem. Soc.*, **55**, 1233 (1933).
6. J. T. Braunholtz and F. G. Mann, *J. Chem. Soc.*, 3383 (1958).
7. R. P. Linstead and E. M. Meade, *J. Chem. Soc.*, 943 (1934).
8. G. R. Proctor, *J. Chem. Soc.*, 3989 (1961).
9. B. D. Astill and V. Boekelheide, *J. Am. Chem. Soc.*, **77**, 4080 (1955).

ETHYL CYCLOHEXYLIDENEACETATE

(Δ^1-α-Cyclohexaneacetic acid, ethyl ester)

$$(C_2H_5O)_2P(O)CH_2COOC_2H_5 + \text{⬡}=O + NaH \rightarrow$$

$$\text{⬡}=CHCOOC_2H_5 + (C_2H_5O)_2P(O)ONa + H_2$$

Submitted by W. S. WADSWORTH, JR., and WILLIAM D. EMMONS [1]
Checked by WILLIAM E. PARHAM, R. M. DODSON, W. L. SALO, and J. N. WEMPLE.

1. Procedure

A dry, 500-ml., three-necked flask equipped with stirrer, thermometer, condenser, and dropping funnel is purged with dry nitrogen and charged with 16 g. (0.33 mole) of a 50% dispersion of sodium hydride in mineral oil (Note 1) and 100 ml. of dry benzene (Note 2). To this stirred mixture is added dropwise over a 45–50 minute period 74.7 g. (0.33 mole) of triethyl phosphonoacetate (Note 3). During the addition period the temperature is maintained at 30–35°, and cooling is employed if necessary (Note 4). Vigorous evolution of hydrogen is noted during this portion of the reaction. After addition of triethyl phosphonoacetate is completed, the mixture is stirred for 1 hour at room temperature to ensure complete reaction (Note 5). To this nearly clear solution is added dropwise over a 30–40 minute period 32.7 g.

(0.33 mole) of cyclohexanone (Note 6). During the addition the temperature is maintained at 20–30° by appropriate cooling with an ice bath. After approximately one-half of the ketone is added, a gummy precipitate of sodium diethyl phosphate forms, which in some instances makes agitation difficult. The mixture is then heated at 60–65° for 15 minutes, during which time it is stirred without difficulty. The resulting product is cooled to 15–20°, and the mother liquor is decanted from the precipitate. This gummy precipitate is washed well by mixing it at 60° with several 25-ml. portions of benzene and decanting at 20° (Note 7). Benzene is distilled from the combined mother and wash liquors at atmospheric pressure. The product is distilled through a 20-cm. Vigreux column, and the mineral oil remains as pot residue after distillation is completed. Ethyl cyclohexylideneacetate (37–43 g., 67–77% yield) is collected at 48–49° (0.02 mm.), n^{25}D 1.4755 (Note 8).

2. Notes

1. Sodium hydride, 50–51% in mineral oil, was supplied by Metal Hydrides Inc., Beverly, Massachusetts.

2. Reagent grade benzene is filtered from sodium hydride just before use.

3. Triethylphosphonoacetate may be obtained from the Aldrich Chemical Company and is distilled before use; b.p. 140° (10 mm.).

4. Temperatures above 40–50° are detrimental to the anion and must be avoided.

5. Approximately the stoichiometric quantity of hydrogen (7.4 l.) is evolved during preparation of the anion.

6. The cyclohexanone was obtained from Eastman Organic Chemicals and redistilled before use.

7. The submitters obtained the yields indicated by washing the gummy precipitate with two 25-ml. portions of benzene; the checkers observed that additional extraction at this point is required, and they recommend extraction with a total of four 25-ml. portions of warm benzene.

8. If an excess of sodium hydride has been used, the product contains varying amounts of the β,γ-isomer, ethyl cyclohexenyl-acetate. To ensure against the occurrence of this side reaction, a 5–10% excess of the phosphonate ester can be used.

3. Methods of Preparation

Esters of cyclohexylideneacetic acid have been prepared by the Reformatsky reaction followed by acylation and pyrolysis,[2] a laborious procedure giving low yields. The phosphonate carbanion procedure would appear to be the method of choice for preparation of these esters.

4. Merits of the Preparation

The phosphonate carbanion method is generally applicable to the synthesis of a wide variety of olefins.[3] The synthesis complements the Wittig reaction in that the latter procedure is often unsatisfactory for preparation of olefins having an electron-withdrawing group adjacent to the double bond.[4] Generally, any ketone or aldehyde can be used in the phosphonate carbanion synthesis, and yields of olefins comparable to that obtained with cyclohexanone are obtained. Although a variety of alkyl phosphonates can be employed, the present procedure is specific for phosphonates containing an electron-withdrawing group. The synthesis can be performed at room temperature or below, and product isolation is facilitated by simplified removal of the by-products, virtues which make this procedure of practical value.

1. Rohm and Haas Company, Philadelphia, Pennsylvania.
2. H. Schmid and P. Karrer, *Helv. Chim. Acta*, **31**, 1067 (1948).
3. W. S. Wadsworth and W. D. Emmons, *J. Am. Chem. Soc.*, **83**, 1733 (1961). Other workers have investigated modifications of this reaction: L. Horner, H. Hoffman, H. G. Wippel, and G. Klahre, *Ber.*, **92**, 2499 (1959); H. Pommer, *Angew. Chem.*, **72**, 911 (1960); E. J. Seus and C. V. Wilson, *J. Org. Chem.*, **26**, 5243 (1961); J. Wolinsky and K. L. Erickson, *J. Org. Chem.*, 30, 2208 (1965); H. Takahashi, K. Fujiwara, and M. Ohta, *Bull. Chem. Soc. Japan*, 35, 1498 (1962).
4. G. Wittig and W. Haag, *Ber.*, 88, 1654 (1955).

ETHYL 6,7-DIMETHOXY-3-METHYLINDENE-2-CARBOXYLATE

(Indene, 6,7-dimethoxy-3-methyl-2-carbethoxy-)

Submitted by JOHN KOO.[1]
Checked by JOHN C. SHEEHAN and J. IANNICELLI.

1. Procedure

To 300 g. of polyphosphoric acid (Note 1) precooled to 5° is added 28 g. (0.1 mole) of ethyl α-acetyl-β-(2,3-dimethoxyphenyl)-propionate[2] contained in a 400-ml. beaker. The mixture is stirred thoroughly (Note 2) with a strong spatula for 15 minutes. The temperature, which is around 10° at the beginning, increases rapidly and is kept between 20° and 25° by occasionally cooling the beaker in an ice-water bath (Note 3). The deep-yellow reaction paste is then poured immediately into 600 ml. of ice water with thorough stirring and trituration (Note 4), and the beaker is rinsed with small portions of cold water. The product, which separates from the cold water as a colorless precipitate, is extracted with several portions of chloroform (first with 400 ml., then twice with 200 ml., and finally with 100 ml. of this solvent). The combined chloroform extracts are washed successively with 100 ml. of cold water, then once with 100 ml., twice with 50 ml. of 10% sodium bicarbonate solution (Note 5), and finally with 50 ml. of cold water. The chloroform solution is dried over magnesium sulfate and filtered. The chloroform is removed by distillation on a steam bath first at atmospheric pressure and finally under reduced pressure. The residual pale-yellow oil soon solidifies and is practically pure. The yield of ethyl 6,7-dimethoxy-3-methylindene-2-carboxylate is 21.5–22.5 g. (82–86%), m.p. 81–83° (Note 6).

2. Notes

1. Obtained from Victor Chemical Works, Chicago, Illinois.

2. Since the mixture is so viscous, continuous and vigorous hand stirring is necessary.

3. After the reaction starts, the temperature should be checked every few minutes and kept between 20° and 25°, which is the most favorable temperature.

4. Careful stirring and trituration are necessary to ensure the complete decomposition of every drop of the yellow paste.

5. According to the submitter, acidification of the combined sodium bicarbonate washings with 10% hydrochloric acid yields a colorless precipitate of 6,7-dimethoxy-3-methylindene-2-carboxylic acid, which is collected by filtration, washed, and dried; yield, 3.4–3.9 g. (14–16%); m.p. 216–218°.

6. Recrystallization from 75 ml. of 60% ethanol gives a 90% recovery of colorless long needles, m.p. 83–85°.

3. Methods of Preparation

The method described here is based on the analogous preparation of some other indenes.[3]

1. Present address: National Drug Company, Research Laboratories, Philadelphia, Pennsylvania. Work done in the Laboratory of Chemical Pharmacology, National Cancer Institute, National Institutes of Health, Bethesda, Maryland.
2. E. C. Horning, J. Koo, M. S. Fish, and G. N. Walker, *Org. Syntheses,* Coll. Vol. 4, 408 (1963).
3. J. Koo, *J. Am. Chem. Soc.,* 75, 1891 (1953).

ETHYL p-DIMETHYLAMINOPHENYLACETATE

(Acetic acid, p-dimethylaminophenyl-, ethyl ester)

$$O_2N-\langle\ \rangle-CH_2CO_2C_2H_5 \xrightarrow[\text{Pd/C}]{\text{CH}_2\text{O, H}_2}$$

$$(CH_3)_2N-\langle\ \rangle-CH_2CO_2C_2H_5$$

Submitted by MICHAEL G. ROMANELLI and ERNEST I. BECKER [1]
Checked by ROY A. SIKSTROM, DOUGLAS R. JOHNSON,
WILLIAM E. PARHAM, and WAYLAND E. NOLAND

1. Procedure

In a 400-ml. Parr bottle are placed 41.8 g. (0.20 mole) of ethyl p-nitrophenylacetate, 40 ml. of 40% aqueous formaldehyde solution, 200 ml. of 95% ethanol, and 2.0 g. of 10% palladium on charcoal (Note 1). The bottle is then placed on a Parr hydrogenation apparatus. The sample is evacuated and filled with hydrogen, this process being repeated three times. The tank and bottle are then filled with hydrogen to 55 p.s.i. The shaker is started, and the hydrogenation is allowed to proceed until the pressure drop corresponds to 1.0 mole of hydrogen (Notes 2, 3). The time required for hydrogenation is approximately 2.5 hours (Note 4). After venting the hydrogen from the bottle safely (Note 1), the ethanol solution is filtered and the catalyst washed carefully (Note 5) with 20 ml. of ethanol.

The filtrate is transferred to a flask which is placed on a rotary evaporator. The ethanol is then removed under reduced pressure on a steam bath. Using ether as a washing solvent, the residue is transferred to a small distilling flask and the ether distilled. The ethyl p-dimethylaminophenylacetate is then distilled (Notes 6, 7) at reduced pressure, affording 27.7–31.8 g. (67–77%) of colorless product, b.p. 122–124° (0.4 mm.), n^{23}D 1.5358.

2. Notes

1. Care must be taken in weighing out and transferring the catalyst as it can ignite mixtures of air and flammable vapors. The operation of the Parr apparatus and appropriate safety precautions in its use have been described in detail.[2]

2. The hydrogenation is exothermic, and care must be taken in order to prevent the reaction from getting out of control. The submitters have not experienced this difficulty, but exothermic hydrogenations require supervision.

3. With the apparatus used, a pressure drop of 85 lb. corresponds to 1.0 mole of hydrogen. Either the particular apparatus used can be calibrated or the hydrogenation allowed to proceed until the pressure ceases to drop.[2]

4. The time required for the hydrogenation will depend on several factors, such as the speed of shaking, activity and particle size of the catalyst. In the experiments run on ethyl *p*-nitrophenylacetate the submitters have found that the time required varied from about 2 to 4 hours.

5. The filtration was by suction. The catalyst must not be allowed to dry out with a stream of air passing through it, as it can then readily ignite.

6. Before the actual distillation could be carried out, the flask was heated to approximately 95° and the residual formaldehyde removed at the aspirator. Only after the formaldehyde was removed could the pressure be reduced to that required for the distillation.

7. The checkers observed that rather rapid decomposition of the product occurs unless precautions were taken. A short path distillation using a 50-ml. distilling flask equipped with a capillary nitrogen bubbler was employed.

3. Methods of Preparation

Ethyl *p*-dimethylaminophenylacetate has been previously prepared in this laboratory by Fischer esterification of *p*-dimethylaminophenylacetic acid, the acid in turn being prepared by the reductive hydrolysis of *p*-dimethylaminomandelonitrile.[3]

4. Merits of the Preparation

Besides being a convenient preparation for ethyl p-dimethyl-aminophenylacetate, the procedure for reductive alkylation can be generalized.[4, 5]   Table I lists the results obtained by the submitters.

TABLE I

REDUCTIVE ALKYLATION OF NITRO COMPOUNDS

Nitro Compound	% Yield
p-Nitrophenylacetic acid	84–91
p-Nitrobenzoic acid	87
m-Nitrobenzoic acid	95
p-Nitrotoluene	90
o-Nitrotoluene	76

1. Department of Chemistry, University of Massachusetts, Boston, Massachusetts. This work was carried out at the Polytechnic Institute of Brooklyn, Brooklyn, New York.
2. R. Adams and V. Voorhees, *Org. Syntheses*, Coll. Vol. **1**, 61 (1941).
3. Celal Tüzün and E. I. Becker, unpublished results.
4. R. E. Bowman and H. H. Stroud, *J. Chem. Soc.*, 1342 (1950).
5. W. S. Emerson, *Org. Reactions*, **4**, 174 (1948).

1-ETHYL-3-(3-DIMETHYLAMINO)PROPYLCARBODIIMIDE* HYDROCHLORIDE AND METHIODIDE

(Carbodiimide, [3-(dimethylamino)propyl]ethyl-, hydrochloride and Ammonium iodide, [3-[[(ethylimino)methylene]amino]propyl]-trimethyl-)

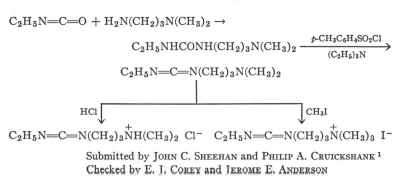

$$C_2H_5N{=}C{=}O + H_2N(CH_2)_3N(CH_3)_2 \rightarrow$$

$$C_2H_5NHCONH(CH_2)_3N(CH_3)_2 \xrightarrow[\text{(C}_2\text{H}_5)_3\text{N}]{p\text{-CH}_3\text{C}_6\text{H}_4\text{SO}_2\text{Cl}}$$

$$C_2H_5N{=}C{=}N(CH_2)_3N(CH_3)_2$$

HCl $\downarrow$ CH$_3$I $\downarrow$

$$C_2H_5N{=}C{=}N(CH_2)_3\overset{+}{N}H(CH_3)_2 \ Cl^- \quad C_2H_5N{=}C{=}N(CH_2)_3\overset{+}{N}(CH_3)_3 \ I^-$$

Submitted by John C. Sheehan and Philip A. Cruickshank [1]
Checked by E. J. Corey and Jerome E. Anderson

1. Procedure

A. *1-Ethyl-3-(3-dimethylamino)propylcarbodiimide.* A solution of 100 g. (1.41 moles) of ethyl isocyanate (Note 1) in 750 ml. of methylene chloride is prepared in a 5-l. three-necked flask equipped with a mechanical stirrer, an immersion thermometer, and a 500-ml., pressure-equalizing, addition funnel (Note 2). The flask and its contents are cooled to 5° in an ice bath, and a solution of 144 g. (1.41 moles) of N,N-dimethyl-1,3-propanedi-amine in 250 ml. of methylene chloride is added through the addition funnel at a rate such that the reaction temperature does not exceed 10° (Note 3). On completion of this addition 500 ml. of triethylamine is added to the flask, and a solution of 300 g. (1.6 moles) of p-toluenesulfonyl chloride in 300 ml. of methylene chloride is placed in the addition funnel and added to the reaction mixture, again at a rate such that the temperature does not exceed 10° (Note 4). After completion of the second addition, the ice bath is replaced with a heating mantle and the addition funnel with a reflux condenser; the reaction mixture is then heated under gentle reflux for 3 hours. Anhydrous sodium carbonate (400 g.) is added to the cooled reaction mixture, followed by 3.5 l. of ice water. The mixture is stirred vigorously for 30 minutes, after which the phases are allowed to separate, and the lower, organic phase is drawn off through a tube into a 2-l. suction flask. The aqueous phase is then extracted with three 500-ml. portions of methylene chloride (Note 5). The original methylene chloride phase and the extracts are combined and dried over anhydrous magnesium sulfate. The solvents are removed under reduced pressure, and the dark brown residue is distilled under reduced pressure through a 15-cm. Vigreux column to give 1-ethyl-3-(3-dimethylamino)propylcarbodiimide, b.p. 52–55° (0.3–0.4 mm.), n^{25}D 1.4591. The yield is 110–118 g. (50–54%) (Note 6).

B. *1-Ethyl-3-(3-dimethylamino)propylcarbodiimide hydrochloride.* A suspension of 34.6 g. (0.300 mole) of pyridine hydrochloride (Note 7) in 280 ml. of methylene chloride is prepared in a 1-l. Erlenmeyer flask. To this is slowly added 46.5 g. (0.300 mole) of 1-ethyl-3-(3-dimethylamino)propylcarbodiimide. The resulting solution is diluted with anhydrous ether (Note 8) and stored at 0–5° for 16–20 hours. The crystalline product is collected by filtration in a *dry* atmosphere (Note 9), washed with a little anhydrous ether, and dried under reduced pressure over phosphorus pentoxide. The yield is 50.5–55.5 g. (88–96.5%), m.p. 104–109° (Notes 10 and 11). This material is sufficiently pure for most purposes.

C. *1-Ethyl-3-(3-dimethylamino)propylcarbodiimide methiodide.* A solution of 30.0 g. (0.193 mole) of 1-ethyl-3-(3-dimethylamino)-propylcarbodiimide in 750 ml. of anhydrous ether is prepared in a 2-l. Erlenmeyer flask. To this is slowly added from an addition funnel a solution of 30.0 g. (0.21 mole) of methyl iodide in 100 ml. of ether. The mixture is stored in the dark for 48 hours, after which the crystalline product is collected by filtration, washed with ether, and dried. The product, m.p. 94–95°, weighs 50.5–52.5 g. (88–91.5%) (Note 12).

2. Notes

1. Available from Eastman Organic Chemicals.

2. The assembled apparatus was dried at 120° for 18 hours.

3. The addition time was *ca.* 2 hours.

4. The addition time was *ca.* 3 hours.

5. Extractions were carried out in the reaction flask; after separation, the lower, methylene chloride phase was drawn off by suction.

6. The submitters, working on a twofold scale, obtained a yield of 60–65%.

7. Pyridine hydrochloride is extremely hygroscopic; the material used must be the anhydrous crystalline form.

8. Vigorous boiling of solvent occurs during addition of the carbodiimide and during spontaneous crystallization of the product. Approximately 250 ml. of ether was used.

9. The product is hygroscopic and care must be taken to protect it from atmospheric moisture at all times.

10. A sample of analytical purity had m.p. 113.5–114.5°.

11. The submitters, working on a 4.5-fold scale, obtained 92% of product, m.p. 108–112°.

12. The submitters, working on a twofold scale, obtained 95.5% of product, m.p. 93–95°, after one recrystallization from chloroform-ether.

3. Methods of Preparation

This procedure for the preparation of 1-ethyl-3-(3-dimethyl-amino)propylcarbodiimide and its salts is a modification of one that has been published.[2] Unsymmetrical carbodiimides have also been prepared by desulfurization of the corresponding thio-ureas with mercuric oxide[3] or by dehydration of the correspond-ing ureas with *p*-toluenesulfonyl chloride in pyridine.[4] Unsym-metrical 1,3-disubstituted ureas are best prepared by the reaction of isocyanates with primary or secondary amines[5] or by the action of carbamoyl chlorides on primary or secondary amines.[6]

4. Merits of the Preparation

Carbodiimides are, in general, useful compounds for effecting certain dehydrative condensations, *e.g.*, in the formation of amides, esters, and anhydrides. These two crystalline water-soluble carbodiimides are especially useful in the synthesis of peptides and in the modification of proteins. The excess of reagent and the co-product (the corresponding urea) are easily separated from products with limited solubility in water. The hydrochloride is best employed in nonaqueous solvents (methylene chloride, acetonitrile, dimethylformamide). The methiodide is relatively stable in neutral aqueous systems, and thus is recommended for those media.

Preparation of carbodiimides by dehydration of the corresponding ureas is of general applicability and is well adapted to the laboratory preparation of substantial quantities. The intermediates for this particular preparation are commercially available at moderate cost.

1. Research Institute for Medicine and Chemistry, Cambridge, Massachusetts 02142.
2. J. C. Sheehan, P. A. Cruickshank, and G. L. Boshart, *J. Org. Chem.*, **26**, 2525 (1961).
3. H. G. Khorana, *Chem. Rev.*, **53**, 145 (1953).
4. G. Amiard and R. Heymès, *Bull. Soc. Chim. France*, 1360 (1956).
5. V. Papesch and E. F. Schroeder, *J. Org. Chem.*, **16**, 1879 (1951).
6. E. N. Abrahart, *J. Chem. Soc.*, 1273 (1936).

ETHYL 2,4-DIPHENYLBUTANOATE

(Butyric acid, 2,4-diphenyl-, ethyl ester)

$$C_6H_5CH_2CO_2C_2H_5 \xrightarrow[NH_3]{NaNH_2}$$

$$C_6H_5\overset{-}{C}HCO_2C_2H_5 \ Na^+ \xrightarrow{C_6H_5CH_2CH_2Br}$$

$$C_6H_5CH_2CH_2\overset{\overset{\displaystyle H}{|}}{\underset{\underset{\displaystyle C_6H_5}{|}}{C}}CO_2C_2H_5$$

Submitted by Edwin M. Kaiser, William G. Kenyon,
and Charles R. Hauser [1]
Checked by Joseph G. Pfeiffer and Kenneth B. Wiberg

1. Procedure

Caution! This preparation should be carried out in a hood to avoid exposure to ammonia.

A suspension of sodium amide [2] (0.1 mole) in liquid ammonia is prepared in a 500-ml. three-necked, round-bottomed flask fitted with a West condenser, a ball and socket glass mechanical stirrer (Note 1), and a dropping funnel. In the preparation of this reagent a small piece of clean sodium metal is added to 350 ml. of commercial anhydrous liquid ammonia. After the appearance of a blue color, a few crystals of hydrated ferric nitrate are added, whereupon the blue color is discharged. The remainder of the 2.3 g. (0.1 mole) of sodium (Note 2) is then rapidly added as small pieces. After all the sodium has been converted to sodium amide (Note 3), a solution of 16.4 g. (0.1 mole) of ethyl phenylacetate (Note 4) in 35 ml. of anhydrous ethyl ether is added dropwise over a 2-minute period, and the mixture is stirred for 20 minutes. To the dark green suspension is added over an 8-minute period a solution of 18.5 g. (0.1 mole) of (2-bromoethyl)benzene (Note 4) dissolved in 35 ml. of anhydrous ethyl ether. The mixture is stirred for 3 hours and is then neutralized by the addition of 5.35 g. (0.1 mole) of ammonium chloride. After addition of 150 ml. of dry ethyl ether, the ammonia is evaporated with stirring by use of a warm water bath (Note 5). The mixture is then cooled to 0° by an ice bath and hydrolyzed by the dropwise addition of 100 ml. of $3N$ hydrochloric acid. After stirring for 15 minutes, the mixture is allowed to warm to room temperature, and the layers are separated. The aqueous layer is extracted with two 50-ml. portions of ethyl ether. The combined ether extracts are then washed with two 50-ml. portions of saturated aqueous sodium bicarbonate followed by two 50-ml. portions of saturated sodium chloride. Drying is accomplished over magnesium sulfate. After filtration and solvent removal, the crude product is purified by vacuum distillation to give 20.6–21.8 g. (77–81%) of ethyl 2,4-diphenylbutanoate, b.p. 168–169° (3.5 mm.). Vapor phase chromatography shows the presence of one peak (Note 6).

2. Notes

1. Stirrers with Teflon paddles should not be used.
2. The sodium was weighed in toluene and then rinsed in anhydrous ethyl ether.
3. Conversion is indicated by the discharge of the blue color.
4. Ethyl phenylacetate and (2-bromoethyl)benzene as supplied by Eastman Organic Chemicals were used without further purification.
5. Alternatively, the ammonia may be evaporated after the ammonium chloride and ethyl ether have been added by allowing the flask to stand overnight with stirring.
6. A 5-ft. Apiezon L column at 225° was used.

3. Methods of Preparation

The procedure described is a modification of that given by Kenyon, Meyer, and Hauser.[3] No other methods appear to have been used to prepare ethyl 2,4-diphenylbutanoate. A number of alkylations of ethyl phenylacetate have previously been effected with alkyl halides by means of other basic reagents, but the yields generally have not been very satisfactory.[4, 5]

4. Merits of the Preparation

The method described is successfully used for the alkylation and aralkylation of ethyl and *t*-butyl phenylacetate.[3] The alkylation of ethyl phenylacetate with methyl iodide, *n*-butyl bromide, benzyl chloride, and α-phenylethyl chloride affords the corresponding pure monoalkylation products in 69%, 91%, 85%, and 70% (*erythro* isomer) yields, respectively. The alkylation of *t*-butyl phenylacetate with methyl iodide, *n*-butyl bromide, α-phenylethyl chloride, and β-phenylethyl bromide gives the corresponding pure monoalkylated products in 83%, 86%, 72–73%, and 76% yields, respectively.

Certain of the monoalkylated ethyl phenylacetates have been further alkylated with alkyl and aralkyl halides to produce the corresponding disubstituted phenylacetic esters.[6] Ethyl 2-phenylpropanoate has been alkylated by methyl iodide to give pure ethyl 2-methyl-2-phenylpropanoate in 81% yield. Similarly, the alkylations of ethyl 2-phenylhexanoate with methyl iodide, *n*-butyl bromide, and benzyl chloride gave the corresponding pure dialkylated products in 73%, 92%, and 72% yields, respectively.

Butylation of ethyl phenylacetate, *t*-butyl phenylacetate, and ethyl 2-phenylhexanoate has also been accomplished with *n*-butyl bromide and sodium hydride in refluxing monoglyme in 64%, 66%, and 56% yields, respectively.[6] In contrast to the sodium amide reactions above, however, careful fractionation of the crude products was required to obtain pure products.

1. Department of Chemistry, Duke University, Durham, North Carolina. This work was supported by the U. S. Army Research Office — Durham.
2. C. R. Hauser, F. W. Swamer, and J. T. Adams, *Org. Reactions*, 8, 122 (1954).
3. W. G. Kenyon, R. B. Meyer, and C. R. Hauser, *J. Org. Chem.*, 28, 3108 (1963).
4. A. C. Cope, H. L. Holmes, and H. O. House, *Org. Reactions*, 9, 284 (1957).
5. A. L. Mndzhoyan, O. L. Mndzhoyan, E. R. Bagdasaryan, and V. A. Mnatsakanyan, *Dokl. Akad. Nauk Arm. SSR*, 30, 97 (1960) [*C.A.* 55, 3508 (1961)].
6. W. G. Kenyon, E. M. Kaiser, and C. R. Hauser, *J. Org. Chem.*, 30, 2937 (1965).

ETHYLENE SULFIDE*

$$
\begin{array}{c}
\text{CH}_2\text{O} \\
| \quad\quad\ \ \text{C}{=}\text{O} + \text{KCNS} \rightarrow \\
\text{CH}_2\text{O}
\end{array}
\quad
\begin{array}{c}
\text{CH}_2 \\
| \quad\quad\ \ \text{S} + \text{KCNO} + \text{CO}_2 \\
\text{CH}_2
\end{array}
$$

Submitted by SCOTT SEARLES, EUGENE F. LUTZ, HIGH R. HAYS, and
HARLEY E. MORTENSEN.[1]
Checked by MELVIN S. NEWMAN and BERNARD C. REAM.

1. Procedure

Into a 500-ml., two-necked, round-bottomed flask equipped with a thermometer (Note 1) and leading to a condenser equipped with a distillation head and a receiver is placed 145 g. (1.5 mole) of potassium thiocyanate (Note 2). The system is evacuated to about 1 mm., and the flask is heated with a free flame until the temperature of the molten salt is in the 165–175° range (Note 3). After the flask has been heated for 15 minutes, it is cooled to room temperature and 88 g. (1.0 mole) of ethylene carbonate (Note 4) is added. The apparatus is reconnected and the receiver protected with a calcium chloride tube.

The reaction flask is slowly heated by means of a sand bath, and the receiver is cooled in a Dry Ice-acetone bath. When the temperature in the fused potassium thiocyanate layer reaches 95°, reaction occurs. Ethylene sulfide distils and is collected in the receiver. Heating is continued for about 3 hours at 95–99° (Note 6). The distillate amounts to 41–45 g. (68–75%) (Note 7) and is sufficiently pure to be used directly (Note 8).

2. Notes

1. The thermometer may be inserted through a neoprene stopper in one neck or into a suitably designed thermometer well.
2. J. T. Baker's analytical grade was used.

3. The purpose of this operation is to ensure dryness, as potassium thiocyanate is hygroscopic. The presence of even small amounts of water is detrimental to the yield. The initially colorless salt melts to a blue liquid. On cooling, this solidifies to a colorless solid.

It has been reported that this procedure is more easily reproduced when the potassium thiocyanate is heated to *ca.* 100–110° at 0.1 mm. so that the salt does not melt, and a larger surface is available for reaction (Nelson N. Schwartz, private communication).

4. Ethylene carbonate obtained from the Eastman Kodak Company or the Jefferson Chemical Company was vacuum distilled and the fraction, b.p. 125°/10 mm., was used.

5. It has been reported that an oil bath may be used to provide greater uniformity of heating (Nelson N. Schwartz, private communication).

6. Differences in the rate of heating and time of heating cause small changes in the yield of ethylene sulfide obtained. If the time of heating is reduced and the rate of heating increased, the yield drops somewhat. For many purposes the saving in time offsets the higher yield obtained under optimum conditions. The submitters have obtained yields in the 81–87% range.

7. This product can be stored at room temperature for several weeks without polymerization. On distillation, pure ethylene sulfide, b.p. 54.0–54.5°, n_D^{20} 1.4960, is obtained.

8. For example, this product is suitable for reaction with amines.[2]

3. Methods of Preparation

Ethylene sulfide has been prepared by the reaction of ethylene oxide with aqueous potassium thiocyanate;[2,3] by the reaction of 2-chloroethyl mercaptan with aqueous sodium bicarbonate;[4] by the reactions of 2-chloroethylthiocyanate[5] and 1,2-dithiocyanoethane[6] with alcoholic sodium sulfide; and by the thermal decomposition of monothiolethylene carbonate.[7]

4. Merits of Preparation

The advantage of the present procedure is the easy availability at low expense of the starting material, ethylene carbonate.[8] Its advantages over the other methods are high yields and degree of purity of the product, combined with greater simplicity of procedure.

1. Department of Chemistry, Kansas State University, Manhattan, Kansas.
2. H. R. Snyder, J. M. Steward, and J. B. Ziegler, *J. Am. Chem. Soc.*, **69**, 2672 (1947).
3. G. I. Braz, *Shur. Obshchei Khim.*, **21**, 688 (1951) [*C. A.*, **45**, 9453 (1951)].
4. W. Coltof, U. S. pat. 2,183,860 (1940) [*C. A.*, **34**, 2395 (1940)]. Also N. V. de Bataafsche Pet. Maatschappij, Brit. pat. 508,932 (1939); Dutch pat. 47,835 (1940).
5. M. M. Delepine, *Bull. Soc. Chim. France*, **27**, 740 (1920).
6. M. M. Delepine and S. Eshenbrenner, *Bull. Soc. Chim. France*, **33**, 703 (1923).
7. D. D. Reynolds, *J. Am. Chem. Soc.*, **79**, 4951 (1957).
8. S. Searles and E. F. Lutz, *J. Am. Chem. Soc.*, **80**, 3168 (1958).

ETHYL β-HYDROXY-β,β-DIPHENYLPROPIONATE

(Hydracrylic acid, 3,3-diphenyl, ethyl ester)

$$CH_3CO_2C_2H_5 \xrightarrow[\text{C}_6\text{H}_5\text{COC}_6\text{H}_5]{\text{LiNH}_2,\ \text{NH}_3}$$

$$\underset{\overset{|}{C_6H_5}}{\overset{\overset{C_6H_5}{|}}{LiO-C}}-CH_2CO_2C_2H_5 \xrightarrow{\text{NH}_4\text{Cl}} \underset{\overset{|}{C_6H_5}}{\overset{\overset{C_6H_5}{|}}{HO-C}}-CH_2CO_2C_2H_5$$

Submitted by W. R. DUNNAVANT and CHARLES R. HAUSER [1]
Checked by WILLIAM E. PARHAM and J. E. BURCSU

1. Procedure

Caution! This preparation should be carried out in a hood to avoid exposure to ammonia.

A suspension of lithium amide (0.25 mole) (Note 1) in liquid ammonia is prepared in a 1-l. three-necked flask equipped with a condenser, a ball-sealed mechanical stirrer, and a dropping funnel. In the preparation of this reagent commercial anhydrous liquid ammonia (500 ml.) is introduced from a cylinder through an inlet tube. To the stirred ammonia is added a small piece of

clean lithium metal. After the appearance of a blue color a few
crystals of ferric nitrate hydrate (about 0.25 g.) are added, followed by small pieces of freshly cut lithium metal (Note 2) until
1.73 g. has been added. After all the lithium has been converted
to the amide (Note 3), 17.6 g. (0.2 mole) of ethyl acetate (Note 4)
is added, and the gray suspension is stirred for about 30 seconds.
To the gray suspension is added 36.4 g. (0.2 mole) of benzophenone (Note 4) dissolved in 100 ml. of anhydrous ether. The
mixture is stirred for 30 minutes and is then neutralized by the
addition of 13.4 g. (0.25 mole) of ammonium chloride. The liquid
ammonia is then removed by use of a steam bath while 200–300
ml. of ether is being added (Note 5). When the ammonia has
been removed, 200 ml. of cold water is added. The ether layer
is separated, and the aqueous layer is further extracted with two
100-ml. portions of ether. The combined ether extract is dried
over magnesium sulfate and filtered, and the solvent is evaporated. The residue is dissolved in 50 ml. of hot 95% ethanol,
treated with Norit®, filtered, and allowed to cool. The yield of
ethyl β-hydroxy-β,β-diphenylpropionate, obtained as colorless,
needle-like crystals melting at 85–86°, is 40.5–45.5 g. (75–84%).
The filtrate, on reduction in volume and cooling, yields small
amounts of benzophenone, m.p. 46–47°.

2. Notes

1. This preparation may be accomplished by using one molecular equivalent of lithium amide; special reaction procedures must
be employed, however, and the yields are not reproducible.[2] The
preparation may also be accomplished (with reduced yield) by
using sodium amide, but only under special reaction conditions.[3]

2. The lithium wire or ribbon is cut in about 0.25-g. pieces,
stored under kerosene, and blotted with filter paper before
addition.

3. Conversion is indicated by the discharge of the blue color.

4. Ethyl acetate and benzophenone as supplied by the Eastman Kodak Company were used without further purification.

5. The checkers added ether and permitted the ammonia to
evaporate overnight. If a steam bath is employed, care must be
exercised to prevent charring of the product.

3. Methods of Preparation

This procedure is an adaptation of ones described by Dunnavant and Hauser.[2,4] Ethyl β-hydroxy-β,β-diphenylpropionate has been prepared previously using the Reformatsky reaction by condensing ethyl α-bromoacetate with benzophenone by means of zinc metal.[5]

4. Merits of the Preparation

This procedure illustrates the use of lithio esters for the preparation of β-hydroxy esters. Isopropyl and t-butyl β-hydroxy-β,β-diphenylpropionate may be prepared in approximately 80% yields by using isopropyl or t-butyl acetates in place of ethyl acetate.[2] This procedure is generally more convenient than the Reformatsky reaction for the preparation of such esters. Under similar conditions ethyl acetate may conveniently be condensed with various aldehydes or ketones to give the corresponding β-hydroxy esters.[4]

1. Department of Chemistry, Duke University, Durham, North Carolina. This work was supported by the Army Research Office (Durham).
2. W. R. Dunnavant and C. R. Hauser, *J. Org. Chem.*, **25**, 1693 (1960).
3. C. R. Hauser and W. R. Dunnavant, *J. Org. Chem.*, **25**, 1296 (1960).
4. W. R. Dunnavant and C. R. Hauser, *J. Org. Chem.*, **25**, 503 (1960).
5. H. Rupe and E. Busolt, *Ber.*, **40**, 4537 (1907).

ETHYL INDOLE-2-CARBOXYLATE*

(Indole-2-carboxylic acid, ethyl ester)

Submitted by WAYLAND E. NOLAND and FREDERIC J. BAUDE.[1]
Checked by E. J. COREY and RONALD J. MCCAULLY.

1. Procedure

A. *Potassium salt of ethyl o-nitrophenylpyruvate.* Anhydrous ether (300 ml.) is placed in a 5-l., three-necked, round-bottomed flask fitted with a 500-ml. dropping funnel, a motor-driven stirrer (with seal), and a reflux condenser protected with a calcium chloride tube. Freshly cut potassium (39.1 g., 1.00 g. atom) is added. *Caution! Follow the precautions for handling potassium described in an earlier volume.*[2]

A slow stream of dry nitrogen is passed through the flask above the surface of the stirred liquid, and a mixture of 250 ml. of commercial absolute ethanol and 200 ml. of anhydrous ether is added from the dropping funnel just fast enough to maintain mild boiling. When all the potassium has dissolved (Note 1), the nitrogen is shut off. The solution is allowed to cool to room temperature, and 2.5 l. of anhydrous ether is added. Diethyl oxalate (146 g., 1.00 mole) is added with stirring, followed after 10 minutes by 137 g. (1.00 mole) of *o*-nitrotoluene. Stirring is discontinued after an additional 10 minutes, and the mixture is poured, with the aid of a connecting tube, into a 5-l. Erlenmeyer flask. The flask is stoppered and set aside for at least 24 hours. The lumpy deep-purple potassium salt of ethyl o-nitrophenylpyruvate is separated by filtration (Note 2) and washed with anhydrous ether until the filtrate remains colorless. The yield of the air-dried salt is 204–215 g. (74–78%).

B. *Ethyl indole-2-carboxylate.* Thirty grams (0.109 mole) of the potassium salt is placed in a 400-ml. hydrogenation bottle and dissolved by addition of 200 ml. of glacial acetic acid, producing a yellow, opaque solution (Note 3). Platinum catalyst [3] (0.20 g.) is added, the bottle is placed in a Parr low-pressure hydrogenation apparatus, and the system is flushed several times with hydrogen. With the initial reading on the pressure gauge about 30 p.s.i., the bottle is shaken until hydrogen uptake ceases and then for an additional 1–2 hours (Note 4). The catalyst is removed by filtration and washed with glacial acetic acid. The filtrate is placed in a 4-l. beaker, and 3 l. of water is added slowly with stirring. Ethyl indole-2-carboxylate precipitates as a yellow solid. It is separated by filtration, washed with five 100-ml. portions of water, and dried over calcium chloride in a desiccator. It weighs 13.2–13.6 g. (64–66%; 47–51% based on o-nitrotoluene); m.p. 118–124°.

The dried ester can be further purified by treatment with charcoal and recrystallization from a mixture of methylene chloride and light petroleum ether (b.p. 60–68°). This gives 11.3–11.7 g. (41–44% based on o-nitrotoluene) of ethyl indole-2-carboxylate in the form of white needles, m.p. 122.5–124° (Note 5).

2. Notes

1. Complete solution of the potassium takes 1.5–2 hours with stirring and 2.5–3 hours without stirring.

2. Salt that sticks to the sides of the Erlenmeyer flask may be loosened with a piece of 10-mm. glass tubing that has not been fire-polished.

3. On addition of the acetic acid, a small amount of black solid settles out, but this dissolves when the solution is swirled for several minutes. The potassium salt of ethyl o-nitrophenyl-

pyruvate, although it undergoes no apparent change in color, does not keep indefinitely in the dry state. After 3 weeks of storage at room temperature, the salt still produced a yellow solution when dissolved in acetic acid, but, after 3 months of storage, the dry salt produced a deep-red solution from which an oil, rather than crystalline ester, was obtained after catalytic hydrogenation.

4. The hydrogen pressure-drop corresponds to 0.325–0.335 mole (99–102%). When the hydrogen pressure drops below about 15 p.s.i., the hydrogen should be replenished in the reservoir tank to bring the pressure back up to about 30 p.s.i. The checkers found a reduction period of 4–6 hours sufficient; the submitters routinely used a 24-hour reduction period.

5. The reported melting points [4, 5–13] range from 119° [6] to 125–126°. [7, 8]

3. Methods of Preparation

The potassium salt of ethyl o-nitrophenylpyruvate is prepared essentially according to the method of Wislicenus and Thoma. [14] However, the isolation of ethyl o-nitrophenylpyruvate has been eliminated by liberating the ester from its potassium salt in the acetic acid used as solvent for the hydrogenation. Catalytic hydrogenation of the ester is carried out essentially by the procedure of Brehm. [5]

Ethyl o-nitrophenylpyruvate [4, 14] and o-nitrophenylpyruvic acid [14–21] have been prepared by condensation of o-nitrotoluene with diethyl oxalate in the presence of potassium ethoxide, [4, 14] sodium ethoxide, [15–20] or sodium methoxide. [21] Sodium ethoxide is less reactive, however, and cannot be substituted successfully for potassium ethoxide in the present procedure, as it gives a very poor yield and poor quality of precipitated sodium salt. With sodium ethoxide the reaction does not appear to go to completion even under the conditions of refluxing ethanol usually employed, [15–21] which are considerably more severe than the room temperature conditions employed with potassium ethoxide in the present procedure. o-Nitrophenylpyruvic acid has also been prepared by hydrochloric acid hydrolysis of o-nitro-α-acetamino-cinnamic azlactone. [4]

Ethyl indole-2-carboxylate [5, 13] and the corresponding carboxylic acid [4,17,19,22,23] have been prepared by reductive cyclization of ethyl o-nitrophenylpyruvate and o-nitrophenylpyruvic acid, both in the presence of reducing agents such as zinc and acetic acid,[4,13] ferrous sulfate and ammonium hydroxide,[17,19,23] and sodium hydrosulfite,[17,22] and by platinum-catalyzed hydrogenation.[5] The ethyl ester has also been prepared by esterification [9,19] of the acid in the presence of sulfuric [6] and hydrochloric [12] acid catalysts, by the Fischer indole synthesis from ethyl pyruvate phenylhydrazone catalyzed by polyphosphoric acid,[11] sulfuric acid and acetic acid,[11,17] or zinc chloride,[24-26] and by stannous chloride reduction of ethyl 1-hydroxyindole-2-carboxylate.[7] Indole-2-carboxylic acid has also been prepared by the Fischer indole synthesis from pyruvic acid phenylhydrazone catalyzed by zinc chloride,[24] by the Madelung synthesis from potassium oxalyl-o-toluidine,[27] by zinc and acetic acid reduction of 1-hydroxy- and 1-methoxyindole-2-carboxylic acids,[28] by cyclizative demethanolation of o-amino-α-methoxycinnamic acid,[29] by reductive cyclization and hydrolysis of o-nitrobenzalrhodanine,[12] by alkaline hydrolysis and decarboxylation of dimethyl indole-2,3-dicarboxylate,[30] and by fusion of 2-methylindole with potassium hydroxide in the presence of air.[31]

4. Merits of the Preparation

The procedure employs the least expensive commercially available starting materials and requires the minimum number of reaction steps.

Alkaline hydrolysis of ethyl indole-2-carboxylate yields indole-2-carboxylic acid,[4,5,7,11,24,25] which can be decarboxylated to indole by heating at 230°.[24,25] The acid or its ester serves as a readily accessible indole capable of electrophilic substitution at the 3-position,[6,22] and as a precursor for the synthesis of indole-2-acylamino derivatives of interest as model compounds in the study of alkaloid synthesis [5,23,32] and as a degradation product of the mold metabolite, gliotoxin.[4,33-35] Reduction of the ester with lithium aluminum hydride yields indole-2-methanol,[5] which can be oxidized to indole-2-carboxaldehyde by potassium permanganate in acetone.[10]   Reduction of the acid chloride with lithium aluminum tri-*tert*-butoxy hydride [36] is a convenient synthesis of indole-2-carboxaldehyde.[37]

1. School of Chemistry, University of Minnesota, Minneapolis, Minnesota.
2. W. S. Johnson and W. P. Schneider, *Org. Syntheses,* Coll. Vol. 4, 132 (1963).
3. R. Adams, V. Voorhees, and R. L. Shriner, *Org. Syntheses,* Coll. Vol. 1, 463 (1941).
4. J. R. Johnson, R. B. Hasbrouck, J. D. Dutcher, and W. F. Bruce, *J. Am. Chem. Soc.,* 67, 423 (1945).
5. W. J. Brehm, *J. Am. Chem. Soc.,* 71, 3541 (1949).
6. H. Fischer and K. Pistor, *Ber.,* 56B, 2213 (1923).
7. S. Gabriel, W. Gerhard, and R. Wolter, *Ber.,* 56B, 1024 (1923).
8. F. Millich and E. I. Becker, *J. Org. Chem.,* 23, 1096 (1958).
9. W. J. Brehm and H. G. Lindwall, *J. Org. Chem.,* 15, 685 (1950).
10. W. I. Taylor, *Helv. Chim. Acta,* 33, 164, 781 (1950); 34, 787 (1951).
11. R. Andrisano and T. Vitali, *Gazz. Chim. Ital.,* 87, 949 (1957).
12. C. Gränacher, A. Mahal, and M. Gerö, *Helv. Chim. Acta,* 7, 579 (1924).
13. H. Maurer and E. Moser, *Z. Physiol. Chem.,* 161, 131 (1926).
14. W. Wislicenus and E. Thoma, *Ann. Chem.,* 436, 42 (1924).
15. A. Reissert, *Ber.,* 30, 1030 (1897).
16. F. Mayer and G. Balle, *Ann.* 403, 188 (1914).
17. J. Elks, D. F. Elliot, and B. A. Hems, *J. Chem. Soc.,* 629 (1944).
18. F. J. DiCarlo, *J. Am. Chem. Soc.,* 66, 1420 (1944).
19. E. T. Stiller, U.S. Patent 2,380,479 (1945) [*C. A.,* 40, 367 (1946)].
20. V. Rousseau and H. G. Lindwall, *J. Am. Chem. Soc.,* 72, 3047 (1950).
21. E. L. May and E. Mosettig, *J. Org. Chem.,* 11, 437 (1946).
22. R. H. Cornforth and R. Robinson, *J. Chem. Soc.,* 680 (1942).
23. W. O. Kermack, W. H. Perkin, Jr., and R. Robinson, *J. Chem. Soc.,* 119, 1625 (1921).
24. E. Fischer, *Ann.,* 236, 141 (1886); *Ber.,* 19, 1563 (1886).
25. Farbwerke Höchst, German Patent 38,784 (1886) [*Fortschritte der Teerfarbenfabrickation,* 1, 154 (1888)].
26. Gesellschaft für Teerverwertung, German Patent 238,138 (1911) [*Fortschritte der Teerfarbenfabrickation,* 10, 333 (1913)].
27. W. Madelung, *Ber.,* 45, 3521 (1912).
28. A. Reissert, *Ber.,* 29, 655 (1896).
29. K. G. Blaikie and W. H. Perkin, Jr., *J. Chem. Soc.,* 125, 334 (1924).
30. O. Diels and J. Reese, *Ann.,* 511, 179 (1934).
31. G. Ciamician and C. Zatti, *Ber.,* 21, 1929 (1888); *Gazz. Chim. Ital.,* 18, 386 (1888).
32. T. Nográdi, *Monatsh. Chem.,* 88, 1087 (1958).
33. J. R. Johnson, A. A. Larsen, A. D. Holley, and K. Gerzon, *J. Am. Chem. Soc.,* 69, 2364 (1947).
34. J. A. Elvidge and F. S. Spring, *J. Chem. Soc.,* S135 (1949).
35. J. D. Dutcher and A. Kjaer, *J. Am. Chem. Soc.,* 73, 4139 (1951).
36. H. C. Brown and B. C. Subba Rao, *J. Am. Chem. Soc.,* 80, 5377 (1958).
37. Y. Satō and Y. Matsumoto, *Ann. Rep. Takamine Lab.,* 11, 33 (1959).

α-ETHYL-α-METHYLSUCCINIC ACID *

(Succinic acid, α-ethyl-α-methyl-)

$$CH_3CCH_2CH_3 + NCCH_2CO_2C_2H_5 \xrightarrow[-H_2O,\ -CH_3COOK]{KCN,\ CH_3CO_2H}$$

with $CH_3CCH_2CH_3$ bearing $\|$ and O below.

$$\left[NC-\underset{\underset{CH_3}{|}}{\overset{\overset{C_2H_5}{|}}{C}}----\underset{\underset{H}{|}}{\overset{\overset{CO_2C_2H_5}{|}}{C}}-CN \right] \xrightarrow[-2NH_4Cl,\ -C_2H_5OH,\ -CO_2]{HCl,\ H_2O,\ \Delta}$$

$$HO_2C-\underset{\underset{CH_3}{|}}{\overset{\overset{C_2H_5}{|}}{C}}-CH_2CO_2H$$

Submitted by F. S. PROUT, V. N. AGUILAR, F. H. GIRARD, D. D. LEE, and J. P. SHOFFNER [1]
Checked by WILLIAM G. DAUBEN and DALE L. WHALEN

1. Procedure

Potassium cyanide (71.6 g., 1.1 moles, U.S.P.) and 100 ml. of 95% ethanol are placed in a 2-l. round-bottomed flask having a ground joint and arranged with a Hershberg stirrer [2] (Note 1). A solution of 113 g. (106 ml., 1 mole) of ethyl cyanoacetate, 79 g. (98 ml., 1.1 moles) of 2-butanone, and 66 ml. of glacial acetic acid is added to the stirred solution over a period of 1 hour. The mixture is stirred for an additional hour, the stirrer is removed, and the mixture is allowed to stand at room temperature for 7 days (Note 2).

Concentrated hydrochloric acid (500 ml.) is added to the semi-solid reaction mixture, a reflux condenser is placed on the flask, and the mixture is heated under vigorous reflux for a period of 4 hours (Note 3). An additional 500 ml. of hydrochloric acid is added, and the boiling under reflux is continued for an additional 4 hours.

The cooled reaction mixture is extracted (Note 4) with four

portions of ether (400 ml., 250 ml., 200 ml., 200 ml.) (Note 5). The ether extracts are filtered and combined, and about two-thirds of the ether is distilled. The ethereal solution is transferred to a 500-ml. Erlenmeyer flask, and the remaining ether is removed. The residue (about 160 g.) is dissolved in 200 ml. of 24% hydrochloric acid (1 part water, 2 parts concentrated hydrochloric acid) and the solution distilled until the boiling point reaches 108° (Note 6). The solution is cooled and allowed to stand at 5° for about 20 hours. The product is collected by vacuum filtration and dried in a vacuum desiccator containing both concentrated sulfuric acid and potassium hydroxide pellets. The yield of α-ethyl-α-methylsuccinic acid is 65–75 g. (41–47%), m.p. 91–97°. Concentration of the mother liquor to 125 ml. gives an additional 8–9 g. of acid, m.p. 85–91° (Note 7).

2. Notes

1. The reaction can be run in an open flask because only a small amount of gas escapes. See Note 3. Sodium cyanide can be substituted for potassium cyanide if 2 g. of β-alanine is also employed as a catalyst.

2. Heating the reaction for shorter periods gave erratic results. At this point the semisolid mixture can be diluted with 200 ml. of water, extracted with benzene, and the benzene extract fractionally distilled to give ethyl 2,3-dicyano-3-methylpentanoate, b.p. 146.0–147.5° (2.5 mm.), n^{27}D 1.4429 (highly purified ester has b.p. 138.5–141.5° (2 mm.), n^{25}D 1.4432). The overall yield of α-ethyl-α-methylsuccinic acid is decreased by about 5% when the dicyano intermediate is isolated.

3. During the reflux period, gases are continuously evolved; these apparently are hydrogen chloride, carbon dioxide, ethyl acetate, and possibly ethyl chloride. The reaction should be run in a hood, or the gases should be trapped.[3]

4. If no layer separates on addition of the ether, add 200 ml. of water.

5. This extraction, designed to remove organic acids from inorganic salts, may also be effected with a lighter-than-water Kutscher-Steudel extractor.[4]

6. The distillate consists of low-boiling solvents.

7. The acid can be purified further by dissolving 50 g. of it in 100 ml. of benzene. The solution is filtered, diluted with 100 ml. of hexane, and cooled to 5°. The yield of acid is 45.0 g., m.p. 97–102° (lit.[5] m.p. 101–102°).

3. Methods of Preparation

The one-step condensation to convert 2-butanone, ethyl cyanoacetate, and hydrocyanic acid to ethyl 2,3-dicyano-3-methylpentanoate is a modification of the procedure described by Smith and Horowitz [5] in which pyridine acetate was employed as the catalyst. Higson and Thorp [6] employed a two-step procedure in which butanone was converted to its cyanohydrin, which in turn was condensed with ethyl cyanoacetate.

α-Ethyl-α-methylsuccinic acid also has been prepared by the sulfuric acid hydrolysis of ethyl α-ethyl-α-methyl-β-carbethoxysuccinate,[7] the action of 80% sulfuric acid on 1-ethoxy-3-ethyl-3-methyl-1,2-cyclopropanedioic acid,[8] and the dichromate oxidation of β-ethyl-β-methylbutyrolactone.[9]

4. Merits of the Preparation

This procedure illustrates a process which should be general for many α,α-disubstituted succinic acids. It is more convenient than those previously employed because the reaction sequence is carried out in one step.

1. Department of Chemistry, DePaul University, Chicago, Illinois.
2. See *Org. Syntheses,* Coll. Vol. 2, 116 (1943).
3. See *Org. Syntheses,* Coll. Vol. 2, 4 (1943); also J. Cason and H. Rapoport, "Laboratory Text in Organic Chemistry," 3rd. Ed., Prentice-Hall, Inc., Englewood Cliffs, N.J. (1970), p. 95.
4. L. C. Craig and D. Craig, in A. Weissberger, "Technique of Organic Chemistry," Vol. III, Part I, 2nd Ed., Interscience Publishers, New York (1956), p. 230. See also *Org. Syntheses,* Coll. Vol. 3, 539 (1955).
5. P. A. S. Smith and J. P. Horowitz, *J. Am. Chem. Soc.,* **71,** 3418 (1949).
6. A. Higson and J. F. Thorpe, *J. Chem. Soc.,* **89,** 1467 (1906).
7. K. Auwers and R. Fritzweiler, *Ann.,* **298,** 166 (1897).
8. B. Singh and J. F. Thorpe, *J. Chem. Soc.,* **123,** 113 (1923).
9. S. S. G. Guha-Sircar, *J. Chem. Soc.,* 898 (1928).

N-ETHYLPIPERIDINE*

(Piperidine, 1-ethyl-)

$$\text{(piperidine)} \quad + \quad CH_2{=}CH_2 \quad \longrightarrow \quad \text{(N-ethylpiperidine)}$$

Submitted by J. WOLLENSAK and R. D. CLOSSON.[1]
Checked by C. D. VER NOOY and B. C. McKUSICK.

1. Procedure

A 1-l. three-necked flask equipped with a reflux condenser, an inlet for dry nitrogen, and a mechanical stirrer is flushed with dry nitrogen. It is then charged with 340 g. (4.00 moles) of piperidine (Note 1), 4.4 g. (0.19 g. atom) of sodium, and 5.0 g. (5.1 ml., 0.063 mole) of pyridine (Note 2). While a slow stream of nitrogen continues to pass through the flask, the solution is heated under reflux with high-speed stirring for approximately 10 minutes. During this time most of the sodium reacts without evolution of hydrogen, and the dispersion darkens. The dispersion, which contains some finely divided solids, is cooled and transferred to a 2-l. stirred autoclave (Note 3) under nitrogen. An additional 85 g. (1.00 mole) of piperidine is used to wash the last portions of the dispersion into the autoclave.

The autoclave is pressured with ethylene (Note 2) to 400 lb./in.^{2} with stirring (Note 4). It is then heated to 100° with stirring, which causes the pressure to rise to about 555 lb./in.^{2}. It is maintained at 100° with stirring until a gradual drop in pressure ceases; this usually takes about 2.5 hours, but it may take as long as 10 hours (Note 5). The autoclave is cooled to room temperature, and the excess ethylene is vented. The reaction mixture is transferred to a 1-l. round-bottomed flask, the autoclave is rinsed with 100 ml. of methanol that is added to the flask, and the mixture is fractionated through a 90-cm. column packed with glass helices. After a fore-cut of 50–100 g. distilling at 55–129°, 434–468 g. (77–83%) of N-ethylpiperidine is collected; b.p. 129–130.5°; n_D^{20} 1.443–1.444 (Note 6).

2. Notes

1. Piperidine obtained from Eastman Kodak was fractionated through a 90-cm. column packed with glass helices, and the fraction distilling at 105° was used for this work. This material contained approximately 0.36 wt.% of pyridine as indicated by vapor phase chromatography and ultraviolet analysis.

2. Sodium from Ethyl Corporation, pyridine from Eastman Kodak, and c.p. ethylene from Matheson are suitable.

3. The kind of stirrer is not important. The submitters obtained similar results with a three-blade propeller turning at 600 r.p.m. and a paddle stirrer turning at 78 r.p.m. They believe that a rocking autoclave could be substituted for a stirred one.

4. Over half the ethylene pressured into the autoclave dissolves in the piperidine. It is essential to agitate the piperidine during the pressuring operation so that the piperidine will become saturated with ethylene, for otherwise there will not be enough ethylene for the reaction.

5. The checkers found that it shortened the reaction time appreciably to repressure the autoclave to 400–500 lb./in.² whenever the pressure dropped below 350 lb./in.²

6. The checkers got the same results using half the quantities of reactants in a 1-l. stirred autoclave.

3. Methods of Preparation

The described procedure is essentially the method of Closson, Kolka, and Ligett.[2] Since N-ethylpiperidine was first prepared by Cahours by reaction of piperidine with ethyl iodide,[3] a large number of synthetic methods have been used for its preparation. Reductive alkylation of pyridine with ethanol over Langenback or Raney nickel catalyst gives N-ethylpiperidine in high yield.[4] The compound may similarly be prepared by catalytic hydrogenation of N-ethylpyridinium chloride with platinum oxide as catalyst [5] and by the alkylation of piperidine using ethanol and

Raney nickel catalyst under hydrogenating conditions.[6] Other methods that have been used are electrolytic reduction of N-ethylglutarimide,[7] interaction of pentamethylene oxide and ethylamine at high temperature over aluminum oxide,[8] interaction of ethylamine and 1,5-dibromopentane,[9] and reduction of 1-acetylpiperidine with lithium aluminum hydride.[10]

4. Merits of the Preparation

The procedure is illustrative of a general method of ethylating amines, wherein one reacts the amine with ethylene using an alkali-metal salt of the amine as catalyst.[2] Di-*n*-butylamine and *n*-hexylamine have been thus ethylated at 130–160°, aniline, *o*-toluidine, and N-methylaniline at 240–275°. In general, higher olefins add to amines only sluggishly.[2]

1. Ethyl Corporation Research Laboratories, Detroit, Michigan.
2. R. D. Closson, J. P. Napolitano, G. G. Ecke, and A. J. Kolka, *J. Org. Chem.*, 22, 646 (1957); R. D. Closson, A. J. Kolka, and W. B. Ligett, U.S. Patent 2,750,417 (1956).
3. A. Cahours, *Ann. Chim. Phys.*, (3) 38, 96 (1853).
4. A. Majrich, Z. Nerad, and A. Klouda, *Chem. Listy*, 50, 2038 (1956) [*C. A.*, 51, 5070 (1957)].
5. T. S. Hamilton and R. Adams, *J. Am. Chem. Soc.*, 50, 2260 (1928).
6. C. F. Winans and H. Adkins, *J. Am. Chem. Soc.*, 54, 306 (1932).
7. B. Sakurai, *Bull. Chem. Soc. (Japan)*, 13, 482 (1938).
8. Yu. K. Yurgev, E. Ya. Pervova, and V. A. Sazonova, *J. Gen. Chem. (USSR)*, 9, 590 (1939) [*C. A.*, 33, 7779 (1939)].
9. J. von Braun, *Ber.*, 42, 2052 (1909).
10. V. M. Mićović and M. L. Mihailović, *J. Org. Chem.*, 18, 1190 (1953).

FERROCENYLACETONITRILE *

{Iron, [(cyanomethyl)cyclopentadienyl]cyclopentadienyl-}

Submitted by Daniel Lednicer and Charles R. Hauser.[1]
Checked by B. C. McKusick and R. D. Vest.

1. Procedure

Caution! This preparation should be carried out in a hood since trimethylamine is evolved.

A solution of 57 g. (0.88 mole) of potassium cyanide in 570 ml. of water is placed in a 1-l. three-necked flask equipped with a stirrer and a reflux condenser. Fifty-eight grams (0.15 mole) of N,N-dimethylaminomethylferrocene methiodide[2] is added and the mixture is heated to boiling with good stirring. As the mixture is brought to boiling, the solid goes into solution. Within a few minutes of the onset of boiling, evolution of trimethylamine begins and a steam-volatile oil starts to separate from the solution. The reaction mixture is boiled vigorously with stirring for 2 hours and then is allowed to cool to room temperature. During the cooling the oil that has separated solidifies.

The solid is separated by filtration and the filtrate is extracted with three 150-ml. portions of ether. (*Caution! Gloves should be worn when handling this solution because of the large amount of cyanide it contains.*) The solid is dissolved in ether and this solution is combined with the extracts. The combined ethereal solutions are washed with water and dried over 5 g. of sodium sulfate.

Removal of the solvent by distillation leaves crude ferrocenyl-acetonitrile as a solid or as an oil that crystallizes on being scratched. The nitrile is dissolved in about 200 ml. of boiling technical grade hexane. The hot solution is decanted from a small amount of insoluble black tar and is cooled to room temperature. Ferrocenylacetonitrile is deposited as bright yellow crystals, m.p. 79–82° (Note 1). The yield of the nitrile is 24-26 g. (71-77%) (Note 2).

2. Notes

1. The pure nitrile melts at 81–83° after further recrystallization from hexane.[3]

2. The yield is directly dependent on the quality of the methiodide employed. Yields as high as 95% have been obtained.[3]

3. Methods of Preparation

This method is that described by Lednicer, Lindsay, and Hauser.[3] No other procedure appears to have been employed to prepare this compound.

Essentially the present procedure converted 1-methylgramine to 1-methyl-3-indoleacetonitrile,[4] but it failed to convert benzyl-dimethylphenylammonium chloride to phenylacetonitrile.[5]

1. Department of Chemistry, Duke University, Durham, North Carolina. The work was supported by the Office of Ordnance Research, U.S. Army.
2. See p. 434, this volume.
3. D. Lednicer, J. K. Lindsay, and C. R. Hauser, *J. Org. Chem.*, 23, 653 (1958).
4. H. R. Snyder and E. L. Eliel, *J. Am. Chem. Soc.*, 70, 1857 (1948).
5. H. R. Snyder and J. C. Speck, *J. Am. Chem. Soc.*, 61, 668 (1939).

2-FLUOROHEPTANOIC ACID

(Heptanoic acid, 2-fluoro-)

$$CH_3(CH_2)_4CHFCH_2Br + NaOCOCH_3 \xrightarrow[\text{HCON(CH}_3)_2]{\text{NaI}}$$
$$CH_3(CH_2)_4CHFCH_2OCOCH_3$$

$$\xrightarrow[\text{CH}_3\text{COOH}]{\text{HNO}_3} CH_3(CH_2)_4CHFCOOH$$

Submitted by F. H. Dean, J. H. Amin, and F. L. M. Pattison [1]
Checked by Michelle Moran and B. C. McKusick

1. Procedure

A. *2-Fluoroheptyl acetate.* A 1-l. two-necked flask is fitted with a thermometer reaching close to the bottom and a reflux condenser that has a calcium chloride tube in its end. It is charged with 29.6 g. (0.150 mole) of 1-bromo-2-fluoroheptane,[2] 22.5 g. (0.150 mole) of dry sodium iodide, 24.6 g. (0.30 mole) of anhydrous sodium acetate, and 600 ml. of dimethylformamide previously dried over anhydrous calcium sulfate. The mixture is stirred by a magnetic bar for 40 hours while being maintained at 120–130° by means of an electric heating mantle. The mixture is cooled to room temperature, diluted with 750 ml. of water, and extracted with 900 ml. of ether. The aqueous layer is extracted with two 150-ml. portions of ether, and the combined ether extracts are washed with five 150-ml. portions of water (Note 1). The ethereal solution is dried over anhydrous sodium sulfate. The ether is removed by distillation, and the residue is transferred to a Claisen flask with a 15-cm. indented neck. Fractionation under reduced pressure gives, after a small fore-run, 16.6–20.6 g. (63–78%) of 2-fluoroheptyl acetate, b.p. 83–87° (10 mm.), n^{25}D 1.4101.

B. *2-Fluoroheptanoic acid* (*Note 2*). A 250-ml. two-necked flask is fitted with a thermometer and a condenser that has an outlet tube to carry oxides of nitrogen to a gas absorption trap [3] or the back of a hood. It is charged with 17.6 g. (0.100 mole) of 2-fluoroheptyl acetate, 50 ml. of glacial acetic acid, and 60 ml. of 16 N nitric acid. The mixture is heated at 48–50° for 25 hours by means of an electric heating mantle. It is diluted with 500 ml. of water and crushed ice and extracted with one 300-ml. portion and five 100-ml. portions of ether (Note 3). The combined ether extracts are added carefully to a slurry of 125 g. of sodium bicarbonate in 500 ml. of water in a 2-l. beaker. The aqueous alkaline solution is extracted with two 100-ml. portions of ether, which are discarded. The aqueous solution is neutralized to a pH of approximately 4 with about 300 ml. of 10% hydrochloric acid and is extracted with one 300-ml. portion and five 100-ml. portions of ether. The combined extracts are washed with four 100-ml. portions of water to remove traces of acetic acid and dried over anhydrous sodium sulfate. The ether is removed by distillation, leaving 12–13.5 g. of crude 2-fluoroheptanoic acid that soon solidifies. It is purified by distillation under reduced pressure through a short Claisen still-head with a short condenser that can be heated by steam, a burner, or a heat lamp when the acid solidifies in it. 2-Fluoroheptanoic acid, b.p. 62–64° (0.15 mm.), 78–80° (0.7 mm.), is obtained as a moist, waxy, pale yellow solid that smells faintly of acetic acid. After being dried on a porous plate, it is odorless and nearly colorless; weight 11.5–12.5 g. (78–84%), m.p. 38–39°.

2. Notes

1. Thorough washing with water is necessary to remove residual dimethylformamide.

2. 2-Fluoroheptanoic acid required no more than the usual precautions accorded organic compounds, for it and its precursors have $LD_{50} > 100$ mg./kg. in mice. The relatively low toxicity of this and other 2-fluoroalkanoic acids is in contrast to the high toxicity of the ω-fluoro acids $F(CH_2)_nCOOH$ with n an odd number (if $n = 5$, $LD_{50} = 1.3$ mg./kg. in mice).[4]

3. The solubility of the fluoro acid in water is sufficient to require thorough ether extraction.

3. Methods of Preparation

2-Fluoroheptanoic acid[5] has been prepared only by the present procedure.

4. Merits of the Preparation

Bromofluorination[2] followed by the present procedure is a general way to convert 1-alkenes to 2-fluoroalkanoic acids; similar results have been obtained with ethylene, propylene, 1-butene, 1-hexene, 1-octene, 1-decene, and methyl 10-undecenoate.[5] It is an easy and convenient way to make 2-fluoroalkanoic acids, for it requires only conventional apparatus and readily available intermediates.

1. Department of Chemistry, University of Western Ontario, London, Ontario, Canada.
2. **F. H. Dean, J. H. Amin, and F. L. M. Pattison, this volume, p. 136.**
3. C. F. H. Allen, *Org. Syntheses*, Coll. Vol. **2**, 4 (1943).
4. F. L. M. Pattison, S. B. D. Hunt, and J. B. Stothers, *J. Org. Chem.*, **21**, 883 (1956).
5. F. L. M. Pattison, R. L. Buchanan, and F. H. Dean, *Can. J. Chem.*, **43**, 1700 (1965).

FORMAMIDINE ACETATE*

$$HC(OC_2H_5)_3 + CH_3COOH + 2NH_3 \rightarrow$$
$$[HC(=NH_2)NH_2]^+ \quad CH_3COO^- + 3C_2H_5OH$$

Submitted by EDWARD C. TAYLOR, WENDELL A. EHRHART, and M. KAWANISI[1]
Checked by JOHN J. MILLER and WILLIAM D. EMMONS

1. Procedure

In a 500-ml. three-necked flask equipped with a reflux condenser, a gas-inlet tube (Note 1) reaching to the bottom of the flask, a thermometer, and a magnetic stirrer is placed a mixture of 90.0 g. of triethyl orthoformate (Note 2) and 49.2 g. of glacial acetic acid. The flask is immersed in an oil bath maintained at 125–130° (Note 3). When the internal temperature of the mixture reaches 115°, a moderate stream of ammonia is introduced.

As the temperature decreases gradually, vigorous refluxing is observed (Note 4). Formamidine acetate starts to crystallize from the boiling mixture after 20–30 minutes. The ammonia flow is continued until no further decrease in temperature is observed (Note 5). The mixture is cooled to room temperature, the precipitate collected by filtration and washed thoroughly with 50 ml. of absolute ethanol. The yield of colorless formamidine acetate is 53.0–55.8 g. (83.8–88.2%), m.p. 162–164° (Note 6). Evaporation of the mother liquor under reduced pressure followed by chilling gives a small additional amount of product (1.0–2.2 g.) (Note 7).

2. Notes

1. An open-end gas-inlet tube should be used rather than a fritted glass inlet because the latter becomes clogged.

2. Commercial triethyl orthoformate, b.p. 50–52° (20 mm.) (Matheson, Coleman and Bell) is used without further purification. It has been reported that it is essential to this procedure that the triethyl orthoformate be slightly wet. Commercial triethyl orthoformate as available in the USA appears to fulfill this requirement, but the anhydrous reagent fails to react. If anhydrous triethyl orthoformate is used, 3 drops of water should be added to ensure a slightly wet reagent (private communication from P. R. H. Speakman).

3. If the temperature is higher than 140°, the product is colored and the yield is lower.

4. This temperature decrease serves as a useful indication of the progress of the reaction.

5. The final temperature of the reaction mixture is usually 72–73°. Total working time is 60–70 minutes.

6. Recrystallization from ethanol does not change the melting point.

7. This material is usually slightly colored and not so pure as the first crop.

3. Methods of Preparation

This method is a modification of the procedure described by

Taylor and Ehrhart.[2] Formamidine has previously been prepared (as its hydrochloride) from hydrogen cyanide via the formimino ether, which is then treated with ammonia,[3] or by desulfurization of thiourea in the presence of ammonium chloride.[4] The methosulfate salt of formamidine has been reported to be formed by reaction of formamide with dimethyl sulfate.[5]

4. Merits of the Procedure

Because formamidine hydrochloride is extremely deliquescent, considerable care must be exercised in its preparation if satisfactory results are to be achieved. Furthermore, formamidine hydrochloride cannot be used directly in most condensation reactions; it must be treated first with a mole of base to liberate free formamidine. The same restriction applies to the methosulfate salt of formamidine; in addition, complications in synthesis may be anticipated in this latter case because methyl hydrogen sulfate itself is an effective methylating agent.[6]

By contrast, formamidine acetate is not hygroscopic and no particular care need be taken to protect it from atmospheric moisture. Furthermore, formamidine acetate can be used directly without prior treatment with base in syntheses requiring free formamidine.[2, 7-10] Finally, this preparation of formamidine is by far the simplest and most convenient yet reported; it obviates the necessity of using either toxic (hydrogen cyanide) or cumbersome (Raney nickel) reagents, and the method can be adapted to the preparation of N,N'-disubstituted formamidines by substitution of primary amines for ammonia.[11]

1. Department of Chemistry, Princeton University, Princeton, New Jersey.
2. E. C. Taylor and W. A. Ehrhart, *J. Am. Chem. Soc.*, **82**, 3138 (1960).
3. A. Pinner, *Ber.*, **16**, 352, 1643 (1883).
4. D. J. Brown, *J. Appl. Chem.*, **2**, 202 (1952).
5. H. Bredereck, R. Gompper, H. Rempfer, K. Klemm, and H. Keck, *Ber.*, **92**, 329 (1959).
6. **E. H. Rodd, ed., "Chemistry of Carbon Compounds," 1st ed., Vol. 1A, Elsevier Publishing Co., New York, 1951, p. 339.**
7. H. Bredereck, F. Effenberger, G. Rainer, and H. P. Schosser, *Ann.*, **659**, 133 (1962).
8. H. Bredereck, F. Effenberger, and G. Rainer, *Ann.*, **673**, 82, 88 (1964).
9. E. C. Taylor and R. W. Hendess, *J. Am. Chem. Soc.*, **87**, 1995 (1965).
10. E. C. Taylor and E. E. Garcia, *J. Org. Chem.*, **29**, 2121 (1964).
11. E. C. Taylor and W. A. Ehrhart, *J. Org. Chem.*, **28**, 1108 (1963).

3-(2-FURYL)ACRYLONITRILE*

(2-Furanacrylonitrile)

$$\text{[furfural]} + CH_2(CN)CO_2H \xrightarrow[C_5H_5N]{CH_3CO_2NH_4}$$

$$\text{[furyl]} CH{=}CH{-}CN + H_2O + CO_2$$

Submitted by JOHN M. PATTERSON.[1]
Checked by MELVIN S. NEWMAN and HERBERT BODEN.

1. Procedure

A mixture of 105.6 g. (1.1 moles) of freshly distilled furfural, 87.0 g. (1.0 mole) of 98% cyanoacetic acid (Note 1), 3.0 g. of ammonium acetate, 200 ml. of toluene, and 110 ml. of pyridine is placed in a 1-l. round-bottomed flask equipped with a Stark and Dean water trap and reflux condenser. The mixture is boiled under reflux for 2 days. The theoretical quantity of water is collected in the trap within 1 hour. Upon completion of the reflux period, the solvent is removed under reduced pressure by heating on a water bath. The residue, distilled through a 15-cm. Vigreux column at 11 mm. pressure, yields 88.6–93.3 g. (74.5–78%) of colorless liquid boiling at 95–97°, n_D^{25} 1.5823–1.5825.

2. Note

1. Cyanoacetic acid was obtained from Distillation Products Industries, Rochester, New York, and used without further purification.

3. Methods of Preparation

The method described is a modification of the procedure used by Ghosez[2] to synthesize cinnamonitrile. 3-(2-Furyl)acrylonitrile has been prepared by catalytic condensation of furfural with acetonitrile in the vapor phase at 320°,[3] by dehydration of the corresponding amide over phosphorus pentachloride,[4] and by decarboxylation of 3-(2-furyl)-2-cyanoacrylic acid.[5]

1. University of Kentucky, Lexington, Kentucky.
2. J. Ghosez, *Bull. Soc. Chim. Belges.*, **41**, 477 (1932).
3. M. M. Brubaker, U. S. pat. 2,341,016 (1944) [*C. A.*, **38**, 4272 (1944)].
4. H. Gilman and A P. Hewlett, *Iowa State Coll. J. Sci.*, **4** (1), 27 (1929).
5. R. Heuck, *Ber.*, **27**, 2624 (1894).

GLYCINE *t*-BUTYL ESTER

$$ClCH_2COOC(CH_3)_3 \xrightarrow{\text{NaN}_3}$$

$$N_3CH_2COOC(CH_3)_3 \xrightarrow{\text{H}_2/\text{Pd-C}} H_2NCH_2COOC(CH_3)_3$$

Submitted by A. T. Moore and H. N. Rydon [1]
Checked by William G. Dauben and John A. Hennings

1. Procedure

A. *t-Butyl azidoacetate.* In a 300-ml. round-bottomed flask fitted with a reflux condenser are placed 30 g. (0.2 mole) of *t*-butyl chloroacetate (Note 1), 24 g. (0.37 mole) of sodium azide, and 90 ml. of 60% (v./v.) acetone-water. The heterogeneous mixture (two liquid phases and a solid phase) is heated under reflux on a steam bath for 18 hours, the acetone distilled, and 15 ml. of water added (Note 2). The mixture is transferred to a separatory funnel, the layers separated, and the lower aqueous layer extracted twice with 25-ml. portions of ether. The ethereal extracts are added to the original upper layer, and the solution is dried over anhydrous sodium sulfate. The ether is distilled, and the residual oil is fractionated under reduced pressure (Note 3), the fraction boiling from 33–41° (1 mm.) being collected; yield 29 g. (92%), n^{20}D 1.4356 (Note 4).

B. *Glycine t-butyl ester.* In the center neck of a 500-ml. suction filtration flask is placed a gas-inlet tube which is connected to a nitrogen cylinder, and on the side arm of the flask there is attached an exit tube leading to a suitable ventilation duct. The flask is placed on a magnetic stirrer, and a solution of 28.9 g. (0.18 mole) of *t*-butyl azidoacetate in 150 ml. of methanol and 0.7 g. of 5% palladium-on-charcoal catalyst is added to the flask. A stream of nitrogen is swept over the surface of the stirred suspension for 5 minutes, the nitrogen cylinder is replaced by a hydrogen cylinder, and hydrogen is passed over the magnetically stirred mixture for 10 hours. The hydrogen is displaced from the flask by a sweeping with nitrogen, the catalyst is removed by filtration and is washed with 5 ml. of methanol. The filtrate is transferred to a 500-ml. Erlenmeyer flask, 15 g. (0.18 mole) of phosphorous acid is added, and the mixture is warmed gently to dissolve the phosphorous acid. The solution is cooled to room temperature (Note 5), 150 ml. of ether is added slowly, and the solution is cooled at 0° for 12 hours. The precipitated glycine *t*-butyl ester phosphite is filtered, washed with ether, and dried in a vacuum oven at 70°, yield 29–32 g. (75–82%), m.p. 144–147° (dec.) (Notes 6 and 7).

To 50 ml. of a well-cooled 6*N* sodium hydroxide solution is added, with stirring, 32 g. (0.15 mole) of the phosphite salt. The stirring is continued until all the solid has dissolved. The solution is transferred to a 125 ml. separatory funnel, extracted with three 20-ml. portions of ether, and the combined extracts dried over anhydrous sodium sulfate. The drying agent is removed by filtration, the solvent removed under reduced pressure, and the glycine *t*-butyl ester distilled, b.p. 65–67° (20 mm.), n^{20}D 1.4237, yield 14 g. (72%, based on phosphite salt). The overall yield from *t*-butyl chloroacetate is 50–55%.

2. Notes

1. The *t*-butyl chloroacetate was prepared from chloroacetyl chloride and *t*-butanol following the procedure of Baker.[2]

2. The water is added to dissolve any inorganic salts which are still not in solution.

3. Owing to the possibly explosive nature of the ester, the distillation was conducted behind a safety screen, using a water

bath for the heat source and keeping the pressure as low as convenient.

4. The submitters reported a boiling point of 63–64° (5–6 mm.), n^{20}D 1.4348. The literature values are b.p. 72–73° (13 mm.) and n^{25}D 1.4332.[3] The submitters also report that the reaction has been run safely on a 200-g. scale.

5. If the mixture sets solid upon cooling, the lumps of phosphite salt should be broken up during the addition of the ether.

6. The crystallization of the crude product from methanol-isopropyl ether gave pure phosphite salt, m.p. 154–157° (dec.).

7. Some t-butyl azidoacetate can be recovered by evaporation of the mother liquor. After removal of the methanol from the filtrate, the residual oil is dissolved in ether, washed with distilled water, the ether removed, and the residue fractionally distilled under reduced pressure (using proper precautions).

3. Methods of Preparation

This method is a modification of that developed by Vollmar and Dunn.[3] Glycine t-butyl ester also has been prepared by the acid-catalyzed addition of N-benzyloxycarbonylglycine to isobutene, followed by catalytic hydrogenolysis of the resulting N-benzyloxycarbonylglycine t-butyl ester.[4] The esters of other amino acids have been prepared directly by the isobutene method.[5]

4. Merits of the Preparation

Glycine t-butyl ester is a valuable intermediate for the preparation of peptides of glycine, since the labile t-butyl group can readily be removed by acid under conditions which do not affect the blocked amino grouping. The present method using t-butyl chloroacetate is superior to that using the bromo derivative,[3] since chloride is cheaper to prepare, less lachrymatory and more easily separated, by fractional distillation, from the t-butyl azidoacetate. The method is also less cumbersome than the procedure using isobutene.

1. Department of Chemistry, University of Exeter, Exeter, England.
2. R. H. Baker, *Org. Syntheses*, **24**, 21 (1944); Coll. Vol. **3**, 144 (1955).

3. A. Vollmar and M. S. Dunn, *J. Org. Chem.*, **25**, 387 (1960).
4. G. W. Anderson and F. M. Callahan, *J. Am. Chem. Soc.*, **82**, 3359 (1960).
5. R. W. Roeske, *Chem. Ind.* (London), 1121 (1959).

GUANIDINE NITRATE

WARNING

It is strongly recommended that our procedure [1] not be used to prepare guanidine nitrate. Mixtures of ammonium nitrate and organic materials not much different from the mixture in the procedure are now used extensively as commercial explosives. The aqueous mixture of Note 10 [1] is similar to some aqueous mixtures used in sizable quantities for rock blasting; a confined mixture of this sort is especially hazardous. Only a few laboratories devoted to explosives research have the barricadcs and remote control devices needed to run this preparation of guanidine nitrate without risk.

Guanidine nitrate can be bought from Eastman Organic Chemicals and other suppliers.

[1] *Org. Syntheses*, Coll. Vol. **1**, 302 (1941).

4-HEPTANONE *

$$2\,CH_3CH_2CH_2COOH + Fe \longrightarrow (CH_3CH_2CH_2COO)_2Fe + H_2$$

$$(CH_3CH_2CH_2COO)_2Fe \xrightarrow{\Delta} (CH_3CH_2CH_2)_2CO + FeO + CO_2$$

Submitted by ROBERT DAVIS, CHARLES GRANITO,
and HARRY P. SCHULTZ [1]
Checked by WILLIAM G. DAUBEN and RICHARD J. SHAVITZ

1. Procedure

A mixture of 370 ml. (4 moles) of *n*-butyric acid (Note 1) and 123 g. (2.2 moles) of hydrogen-reduced iron powder (Note 2) is refluxed for 5 hours in a 1-l. flask equipped with a condenser

(Note 3). The apparatus is converted for downward distillation while an atmosphere of nitrogen is maintained. The nitrogen sweep is then stopped, the flask is strongly heated, and the entire distillate collected.

The crude product is washed with two 20-ml. portions of 10% sodium hydroxide solution and with one 20-ml. portion of water. The 4-heptanone is dried over 5 g. of anhydrous sodium sulfate, filtered, and distilled. The yield of 4-heptanone, b.p. 142–144°, n^{25}D 1.4031–1.4036, is 157–171 g. (69–75%).

2. Notes

1. n-Butyric acid, b.p. 162–164°, from Eastman Organic Chemicals was redistilled before use.

2. Hydrogen-reduced iron powder from Fisher Scientific Company was used.

3. Severe foaming may force brief cessations of heating during the first hour. Boric acid (0.1 g.) somewhat diminishes the extent of foaming.

3. Methods of Preparation

The present procedure is that of Davis and Schultz.[2] 4-Heptanone has also been synthesized by virtually every general method known and listed for ketones in "Chemistry of Carbon Compounds," including liquid or vapor phase decarboxylation of n-butyric acid or its salts, oxidation of 4-heptanol, and hydration of 3-heptyne.[3]

4. Merits of the Preparation

This method is illustrative of a general method of preparing simple ketones from normal aliphatic carboxylic acids. It is especially useful because the starting materials are easily accessible, the yields good, and the procedure very simple.

1. Chemistry Department, University of Miami, Coral Gables, Florida 33124.
2. R. Davis and H. P. Schultz, *J. Org. Chem.*, **27**, 854 (1962).
3. J. G. Buchanan, N. A. Hughes, F. J. McQuillin, and G. A. Swan in S. Coffey, "Rodd's Chemistry of Carbon Compounds," Elsevier Publishing Company, New York, 1965, p. 53.

HEXAHYDROGALLIC ACID AND HEXAHYDROGALLIC ACID TRIACETATE

(Cyclohexanecarboxylic acid, 3,4,5-triol and triacetate)

Submitted by ALBERT W. BURGSTAHLER and ZOE J. BITHOS.[1]
Checked by R. P. LUTZ and JOHN D. ROBERTS.

1. Procedure

A. *Hexahydrogallic acid.* A solution of 50 g. (0.266 mole) of recrystallized gallic acid monohydrate (Note 1) in 225 ml. of 95% ethanol (Note 2) is placed in a 1-l. high-pressure hydrogenation bomb (Note 3) with 8 g. of 5% rhodium-alumina catalyst (Note 4). The bomb is then closed, hydrogen admitted at full tank pressure (2200 lb., Note 5), and the temperature raised to 90–100° (Note 6) while agitation is commenced. When the hydrogen uptake is complete (8–12 hours), heating is discontinued and the bomb is allowed to cool. The residual hydrogen is bled off, and the contents of the bomb are rinsed out with two 40-ml. portions of warm distilled water and then heated to boiling on the steam bath for 5 minutes to dissolve any product which has crystallized on the catalyst. After removal of the catalyst by suction filtration (Note 7), the colorless filtrate (Note 8) is concentrated on the steam bath under reduced pressure (preferably using a rotary evaporator). The viscous residue which may have begun to deposit crystals is diluted with 75–100 ml. of ethyl acetate and the product allowed to crystallize at 0° for several hours or overnight. The product is collected on a 9-cm. Büchner funnel and washed with 75 ml. of cold 3:1 ethyl acetate-absolute ethanol and finally with 100 ml. of 30–40° petroleum ether. When dry, it weighs 21–24 g. (45–51%); an additional 2–4 g. can usually be obtained by concentration of the mother liquors and crystallization from ethyl acetate. Recrystallization is achieved by dissolution of the combined products in the minimum amount of boiling water (Note 9), suction filtration if necessary to remove suspended matter (Note 10), addition of hot ethanol to bring the volume of the solution to about 110 ml., and finally addition of about 35 ml. of acetone, sufficient to produce a faint cloudiness. The solution is allowed to cool slowly to room temperature and is then stored at 0° overnight. The fine, colorless crystals are collected on a 7-cm. Büchner funnel and washed with 80 ml. of cold 5:3 absolute ethanol-acetone, then with 100 ml. of 30–40° petroleum ether. The product when dry weighs 18–20 g. (38–43%). The yield may be increased somewhat by concentration of the combined mother liquor and washings, and treatment as before with ethanol and acetone. The melting point is not a useful criterion of purity, since the hexahydrogallic acid decomposes on heating (Note 11). The product is apparently substantially the all-*cis* isomer.[2]

B. *Hexahydrogallic acid triacetate.* A suspension of 10 g. (0.057 mole) of the dry, recrystallized hexahydrogallic acid in 40 ml. of acetic anhydride is treated with 1 drop of concentrated sulfuric acid, which initiates the reaction (Note 12). Most of the solid then goes into solution with some evolution of heat. The reaction is completed on a steam bath for 30 minutes. The acetic acid and most of the excess anhydride are then removed on the steam bath under reduced pressure with the aid of an oil pump. Twenty-five milliliters of water is added, and the mixture is shaken and heated on the steam bath for 10 minutes in order to hydrolyze residual acetic anhydride and the mixed anhydride of the product and acetic acid. Most of the solvent is then removed under reduced pressure on the steam bath; the product usually crystallizes during this process. About 15 ml. of water is added, and the mixture is heated on the steam bath until the solids dissolve. The solution is first allowed to cool slowly to room temperature and then stored at 0° for several hours to complete crystallization. The colorless crystals are collected, washed rapidly with 10–15 ml. of cold water, and dried at 60° or in a vacuum desiccator at room temperature. The yield is 15–17 g., m.p. 152–154°. The product may be recrystallized by dissolution in 25–30 ml. of hot acetone and addition of 50 ml. of 30–40° petroleum ether. The colorless crystalline granules are collected by suction filtration and washed with a small amount of 2:1 petroleum ether(30–40°)-acetone. The recrystallized product when dried amounts to 13.5–14.5 g. (78–84%), m.p. 155–156° (Note 13).

2. Notes

1. Gallic acid is conveniently recrystallized from water (heated to boiling, then cooled to 0°) with treatment with decolorizing carbon if necessary.

2. Absolute alcohol leads to partial esterification of the product; a higher percentage of water deactivates the catalyst.

3. A glass liner may be helpful in preventing poisoning of the catalyst. Stainless-steel vessels usually require one run to "condition" the surfaces before the reported yields can be obtained.

4. The catalyst is available from Englehardt Industries, Inc., Chemical Division, Newark, New Jersey. The activity appears to vary slightly with different lots. Other catalysts, such as palladium, platinum, and ruthenium on various supports, or Raney nickel, were found to be much less satisfactory or completely ineffective.

5. Pressures lower than 1800 lb. usually lead to incomplete reduction.

6. Reduction is inconveniently slow at lower temperatures; temperatures higher than 125° tend to favor esterification and other by-product formation.

7. The recovered catalyst (along with fresh catalyst) can be reused several times for further reductions.

8. The ferric chloride test is negative, and the filtrate remains colorless when hydrogenation is complete. A deep blue color (due to the presence of gallic acid or dihydro products) appears when it is not, but the hexahydro acid can usually be isolated in good yield in spite of this.

9. A hot plate equipped with a magnetic stirrer is especially convenient for this operation.

10. Any excess water used to transfer the solution or to wash the filter must be evaporated; otherwise the recovery is smaller.

11. Decomposition usually begins at about 190°, with melting at 198–200°. Melting points as high as 203–204° have been observed by the submitters and checkers. The purity of successively recrystallized products may be compared by immersing the samples enclosed in capillary tubes of uniform dimensions in a melting-point bath maintained at a constant temperature of 200° and noting the times required for complete melting.

12. Acetylation in pyridine is comparatively less satisfactory and considerably more inconvenient.

13. Using a Kofler melting-point block fitted with a microscope, the checkers observed a crystal transition at about 140° with final, moderately sharp, melting at 157–158°.

3. Methods of Preparation

The all-*cis* diasteroisomer of hexahydrogallic acid has been prepared from gallic acid in 13–19% over-all yield by a two-stage reduction, first with Raney nickel in basic solution to form the somewhat difficultly isolated dihydro intermediate, and then with a platinum catalyst to complete the reduction.[3] The present procedure is based on a published preparation.[2]

The acetylation procedure described here is based on that which has already been published.[3]

4. Merits of Preparation

The direct reduction of gallic acid described here illustrates the virtue of the rhodium-on-alumina catalyst to achieve the perhydrogenation of polyhydroxylated aromatic compounds with minimal attendant hydrogenolysis. A closely related hydrogenation, that of pyrogallol, to yield a dihydro intermediate,[3] and also the direct reduction of pyrogallol with palladium-on-strontium carbonate to afford the all cis-pyrogallitol (1,2,3-cyclohexanetriol) have been reported.[4]

1. Department of Chemistry, University of Kansas, Lawrence, Kansas.
2. A. W. Burgstahler and Z. J. Bithos, *J. Am. Chem. Soc.*, 82, 5466 (1960).
3. W. Mayer, R. Bachmann, and F. Kraus, *Ber.*, 88, 316 (1955).
4. W. R. Christian, C. J. Gogek, and C. B. Purves, *Can. J. Chem.*, 29, 911 (1951); cf. S. J. Angyal and D. J. McHugh, *J. Chem. Soc.*, 3682 (1957).

HEXAHYDROXYBENZENE

(Benzenehexol)

Submitted by A. J. Fatiadi and W. F. Sager.[1]
Checked by B. C. McKusick and J. K. Williams.

1. Procedure

One hundred grams (0.44 mole) of stannous chloride dihydrate is added to a boiling solution of 10 g. (0.058 mole) of tetrahydroxyquinone[2] in 200 ml. of 2.4N hydrochloric acid contained in a 1.5-l. beaker. The initial deep-red color disappears, and grayish crystals of hexahydroxybenzene precipitate. Two hundred fifty milliliters of 12N hydrochloric acid is added, and the mixture is heated to boiling with constant stirring. The beaker is removed from the hot plate, an additional 600 ml. of 12N hydrochloric acid is added, and the solution is cooled in a refrigerator. The hexahydroxybenzene is collected on a Büchner funnel fitted with a sintered-glass disk (Note 1) and sucked dry.

The crude hexahydroxybenzene is dissolved in 450 ml. of hot 2.4N hydrochloric acid containing 3 g. of hydrated stannous chloride and 1 g. of decolorizing carbon. The solution is filtered while hot, and the carbon is rinsed with 75 ml. of boiling water that is combined with the filtrate. One liter of 12N hydrochloric acid is added, and the mixture is cooled in a refrigerator. The snow-white crystals of hexahydroxybenzene that separate are collected under carbon dioxide or nitrogen (Note 2) on a Büchner funnel fitted with a sintered-glass disk. The hexahydroxybenzene is washed with 100 ml. of a cold 1:1 mixture of ethanol and 12N hydrochloric acid and dried in a vacuum desiccator over sodium hydroxide pellets; yield 7.1–7.8 g. (70–77%). It fails to melt on a hot plate at 310° (Note 3).

2. Notes

1. Filter paper cannot be used because it is attacked by strong hydrochloric acid.

2. By rapid manipulation it is possible to obtain a product of fair quality. The moist product is susceptible to air oxidation, as is shown by a development of pink coloration on the crystals. The filtration is best carried out under a blanket of carbon dioxide or nitrogen obtained by inverting a funnel attached to a source of carbon dioxide or nitrogen over the Büchner funnel.

3. The decomposition point of hexahydroxybenzene is not a good criterion of purity. If the product is light in color, there can be no significant amount of oxidized material in it, for even traces of tetrahydroxyquinone cause intense coloration. Decomposition of a sample with nitric acid followed by evaporation and ignition of the residue should give a negligible amount of tin oxide. The product can be characterized as the hexaacetate, m.p. 202–203°, by treating it with acetic anhydride and sodium acetate.[3]

3. Methods of Preparation

The present procedure is a modification of the procedure of Anderson and Wallis.[4] Hexahydroxybenzene can also be prepared by acidic hydrolysis of potassium carbonyl [3] or by nitration and oxidation of diacetyl hydroquinone.[5]

4. Merits of the Procedure

This is the most convenient synthesis of hexahydroxybenzene, and the present procedure gives better yields than reported by Anderson and Wallis.[4] Hexahydroxybenzene is of interest as the most highly hydroxylated member of the polyhydroxybenzene family.

It has been used as a source of the biologically important *myo*-inositol [6,7] (1235/46 isomer) by hydrogenation over palladium and of *cis*-inositol (123456 isomer) by hydrogenation over palladium-on-carbon.

1. Department of Chemistry, The George Washington University, Washington, D. C.
2. A. J. Fatiadi and W. F. Sager, this volume, p. 1011.
3. B. Nietzki and T. Benckiser, *Ber.*, 18, 1834 (1885).
4. R. C. Anderson and E. S. Wallis, *J. Am. Chem. Soc.*, 70, 2931 (1948).
5. B. Nietzki and T. Benckiser, *Ber.*, 18, 500, 1842 (1885).
6. H. Wieland and R. S. Wishart, *Ber.*, 47, 2082 (1914).
7. S. J. Angyal and D. J. McHugh, *Chem. & Ind. (London)*, 947 (1955).

2,3,4,5,6,6-HEXAMETHYL-2,4-CYCLOHEXADIEN-1-ONE

(2,4-Cyclohexadien-1-one, 2,3,4,5,6,6-hexamethyl-)

Submitted by HAROLD HART, RICHARD M. LANGE,
and PETER M. COLLINS [1]
Checked by A. S. PAGANO and WILLIAM D. EMMONS

1. Procedure

Caution! The preparation and handling of peroxytrifluoroacetic acid should be carried out behind a safety screen. Precautions to be observed with 90% hydrogen peroxide are described in Note 3 and should be followed carefully.

To a 300-ml. three-necked flask equipped with a glass Trubore stirrer and two loose-fitting ground-glass stoppers are added 46.2 g. (0.22 mole) of trifluoroacetic anhydride and 50 ml. of methylene chloride (Note 1). The stirred solution is cooled in an ice bath, and 5.40 ml. (0.20 mole) of 90% hydrogen peroxide is added from a 10-ml. graduated cylinder in *ca.* 1-ml. portions over a period of 10 minutes (Notes 2, 3, and 4). When the mixture has become homogeneous, it is allowed to warm to room temperature for a few minutes and is then cooled once more in an ice bath to 0°.

A solution of 24.2 g. (0.15 mole) of hexamethylbenzene in 300 ml. of distilled methylene chloride is prepared in a 1-l. three-necked flask equipped with two ice-jacketed addition funnels (Note 5) and a thermometer. The solution is cooled to 5° in an ice-ethanol bath and is agitated by a magnetic stirrer. The cold peroxytrifluoroacetic acid solution is added at a constant rate to the hexamethylbenzene solution from one of the ice-jacketed addition funnels at the same time that 63.3 ml. of technical (48%) boron trifluoride etherate is added from the second addition funnel. The additions require *ca.* 45 minutes, and as far as possible they should be completed at the same time. During this period the temperature of the reaction mixture is maintained between 0° and 5° (Notes 6 and 7).

The mixture is stirred at 0–5° for 1 hour after addition is complete and then is hydrolyzed with 100 ml. of water, which is added quickly. The reaction mixture is extracted consecutively with two 100-ml. portions of water, three 100-ml. portions of saturated aqueous sodium bicarbonate, one 75-ml. portion of aqueous 5% sodium hydroxide, and two 75-ml. portions of water. The organic phase is dried over anhydrous magnesium sulfate, and the solvent is removed on a rotary evaporator. The residue, a mobile yellow oil, is distilled through a 6-in. Vigreux column under reduced pressure to give pure (Note 8) 2,3,4,5,6,6-hexamethyl-2,4-cyclohexadien-1-one, b.p. 85–87° (1.0 mm.). The yield is 22–24 g. (82–90%) (Notes 9 and 10).

2. Notes

1. Excess anhydride is used to remove water introduced during the addition of 90% hydrogen peroxide.

2. Available from FMC Corp., Inorganic Chemicals Division, 808 Gwynne Building, Cincinnati 2, Ohio.

3. The precautions to be observed with 90% hydrogen peroxide have been described in detail.[2] In essence it is important to prevent contact of this reagent with any easily oxidizable substrates, such as wood, alcohols, and sugars and with heavy-metal salts, since the latter catalyze its decomposition. Storage of hydrogen peroxide in the laboratory should be arranged in such a way that, even if the bottle containing the reagent breaks, the hydrogen peroxide does not come into contact with any materials of this kind. Small samples of 90% hydrogen peroxide are regularly shipped in vented glass bottles provided with a protective outside metal container, and it is desirable to use this container while storing the reagent in the laboratory. If spillage of the reagent occurs, dilution with at least several volumes of water is recommended. In weighing out 90% hydrogen peroxide it is good

practice never to return excess reagent to the stock bottle; rather it should be diluted with water and discarded to avoid any possibility that the stock bottle will be contaminated.

4. The hydrogen peroxide may be added in one portion, but then an appreciable exotherm is noted.

5. A simple procedure for the construction of a jacketed addition funnel has been described.[3]

6. Boron trifluoride gas may be used in place of the etherate. In this case a fritted-glass gas-dispersion tube that extends below the liquid surface replaces the second addition funnel. Boron trifluoride gas (0.20 mole, 4.48 l.) is passed through the solution as the peroxytrifluoroacetic acid is added. The boron trifluoride may be metered into the mixture through a calibrated flowmeter containing carbon tetrachloride as the indicator liquid. Alternatively, a premeasured quantity of boron trifluoride may be displaced by carbon tetrachloride from a gas bulb. The yield is approximately the same regardless of the source of boron trifluoride.

7. If boron trifluoride is omitted as a reactant, the yield falls to about 67%.

8. This material should be at least 98% pure by vapor-phase chromatography (SE-30 column at 180–200°). It usually crystallizes if stored in a refrigerator. Unreacted hexamethylbenzene, present if insufficient oxidant is used, can best be removed by column chromatography on alumina with pentane as eluant.

9. The reaction can be used to prepare hexaethyl-2,4-cyclohexadienone, m.p. 44–45°, in 82% yield from hexaethylbenzene and 3,4,6,6-tetramethyl-2,4-cyclohexadienone from durene in over 80% yield.

10. One can use a solution of 90% hydrogen peroxide (*Caution!* see Note 3) in acetic-sulfuric acid as the oxidant in place of peroxytrifluoroacetic acid-boron trifluoride. The reaction is less exothermic, therefore easier to control. The generality of the procedure for other polyalkylbenzenes has not been tested. A typical procedure is as follows: To a well-stirred slurry of 100 g. (0.618 mole) of hexamethylbenzene in 200 ml. of methylene chloride, 400 ml. of glacial acetic acid, and 300 ml. of concentrated sulfuric acid maintained at 0–10° there is added, during 30 minutes, a solution of 28 ml.

(0.75 mole) of 90% hydrogen peroxide in 45 ml. of glacial acetic acid and 30 ml. of concentrated sulfuric aicd. After addition, the mixture is stirred at 0° for 4 hours before work-up, which is essentially as described in the original procedure. The yield of hexamethyl-2,4-cyclohexadienone is 95 g. (88%). The procedure is derived from that of Hart, Collins, and Warning[9] (private communication from H. Hart).

3. Methods of Preparation

The method described is that of Waring and Hart.[4] Dienones of this type have not been available by any previously described synthetic route.

4. Merits of the Preparation

Dienones of this class are useful starting materials for the preparation of bicyclic compounds via Diels-Alder reactions[4] and for the synthesis of small ring compounds.[5] The 2,4-dienone can be converted quantitatively to the 2,5-isomer by treatment with fuming sulfuric acid and subsequent hydrolysis.[6] The oxidation procedure is also applicable to the conversion of mesitylene to mesitol or of isodurene to isodurenol,[7] and can be used to convert tetramethylethylene quantitatively and directly to pinacolone.[8] A review which includes references to many applications of the procedure described here is available.[10]

1. Michigan State University, East Lansing, Michigan 48823.
2. E. S. Shanley and F. P. Greenspan, *Ind. Eng. Chem.*, **39**, 1536 (1947).
3. R. Graf, this volume, p. 673.
4. A. J. Waring and H. Hart, *J. Am. Chem. Soc.*, **86**, 1454 (1964).
5. H. Hart and A. J. Waring, *Tetrahedron Letters*, 325 (1965).
6. H. Hart and D. W. Swatton, *J. Am. Chem. Soc.*, **89**, 1874 (1967).
7. H. Hart and C. A. Buehler, *J. Org. Chem.*, **29**, 2397 (1964).
8. H. Hart and L. R. Lerner, *J. Org. Chem.*, **32**, 2669 (1967).
9. H. Hart, P. M. Collins, and A. J. Waring, *J. Am. Chem. Soc.*, **88**, 1005 (1966).
10. H. Hart, *Accounts Chem. Res.*, **4**, 337 (1971).

HEXAMETHYLPHOSPHOROUS TRIAMIDE*

(Phosphorous triamide, hexamethyl-)

$$PCl_3 + 6(CH_3)_2NH \rightarrow [(CH_3)_2N]_3P + 3(CH_3)_2NH_2{}^+Cl^-$$

Submitted by V. Mark [1]
Checked by G. A. Frank and W. D. Emmons

1. Procedure

A solution of phosphorus trichloride (137.3 g., 1.0 mole) in 1.5 l. of dry ether (Note 1) is added to a 3-l., three-necked, round-bottomed flask equipped with an efficient stirrer, thermometer, a gas-inlet tube (Note 2), and a reflux condenser vented through a nitrogen reservoir (a T-tube under slight positive nitrogen pressure) into a well-functioning hood (Note 3). The flask is cooled in an ice bath to 0–5°, and an excess of anhydrous dimethylamine (Note 4) is introduced at such a rate that the temperature does not exceed 15° The addition requires about 3–4 hours. At the end of this period the flask contains the white stirrable slurry of the amine hydrochloride and the ethereal solution of the phosphorous triamide (Note 5). The reaction mixture is allowed to warm to room temperature overnight while still being protected by nitrogen. Filtration of the slurry and thorough washing of the filter cake with three 100-ml. portions of dry ether afford dimethylamine hydrochloride, quantitatively (Note 6). The clear filtrate is concentrated on a rotary evaporator connected to a water aspirator in a bath not exceeding 40° to give 152–154 g. (94–95%) of hexamethylphosphorous triamide as a light yellow oil. The product can be purified by distillation at atmospheric pressure, b.p. 162–4°, or under reduced pressure, b.p. 49–51° (12 mm.), n^{25}D 1.4636 (Note 7). Hexamethylphosphorous triamide is best stored in a nitrogen atmosphere (Note 8).

2. Notes

1. An equal volume of a hydrocarbon solvent (pentane, hexane, benzene) can be substituted for ether without affecting the yield of the triamide.

2. The lower part of the gas-inlet tube, which reaches below the surface of the liquid, should be wide enough that it will not be clogged by the amine hydrochloride. A 12-mm. I.D. glass tube was found satisfactory.

3. Carbon dioxide is not satisfactory because it reacts with hexamethylphosphorous triamide.

4. Available from the Matheson Company. The checkers used Rohm & Haas anhydrous dimethylamine.

5. The water extract of the clear solution should give, after acidification with dilute nitric acid, no white precipitate with silver nitrate. When free of chloride ion, the water extract gives only a dark coloration or precipitate.

6. The use of a large (9.5 cm. in diameter, 8 cm. high or larger), coarse grade, sintered-glass funnel, which permits the slurrying and thorough rinsing of the filter cake, is recommended. Since the conversion of the phosphorus trichloride to the triamide is quantitative, the major cause of lower yields is the retainment of the liquid product by the salt cake.

7. As a safety precaution the exposure of the hot material in the flask to air should be avoided. The checkers recovered 134 g. (82%) of distillate from 153 g. of crude product.

8. Essentially the same procedure can be used to obtain the higher alkyl homologs of hexamethylphosphorous triamide. Since the higher dialkylamines are liquid at room temperature, the gas-inlet tube is replaced by an addition funnel. Alternatively, the mode of addition may be reversed (*i.e.*, phosphorus trichloride may be added to the amine) without affecting the subsequent workup or yield. The higher homologs of hexamethylphosphorous triamide such as the ethyl, *n*-propyl, and *n*-butyl can also be prepared in 95–100% conversion when a slight excess (5–10%) of the amine is employed.

3. Methods of Preparation

The described procedure is a modification of the method of Carmody and Zletz [2] and of Burg and Slota.[3] The higher homologs were reported by Stuebe and Lankelma.[4]

4. Merits of the Preparation

This is a general method of preparing hexaalkylphosphorous triamides from the corresponding dialkylamines. The procedure is simple, and the yields are high. Hexaalkylphosphorous triamides are powerful nucleophiles.[5] This feature can be used in a rather unique way to synthesize epoxides directly from aldehydes.[5,6] More recently the methyl and ethyl derivatives have been used in desulfurization reactions to prepare thiolactones[7] and thietanes.[8]

1. Hooker Chemical Corporation, Niagara Falls, New York.
2. D. R. Carmody and A. Zletz, U.S. Patent 2,898,732 (1959).
3. A. B. Burg and P. J. Slota, Jr., *J. Am. Chem. Soc.*, **80**, 1107 (1958).
4. C. Stuebe and H. P. Lankelma, *J. Am. Chem. Soc.*, **78**, 976 (1956).
5. V. Mark, *J. Am. Chem. Soc.*, **85**, 1884 (1963).
6. V. Mark, this volume, p. 358.
7. J. H. Markgraf, C. l. Heller, and N. L. Avery, *J. Org. Chem.*, **35**, 1588 (1970).
8. D. N. Harpp and J. G. Gleason, *J. Org. Chem.*, **35**, 3359 (1970).

HEXAPHENYLBENZENE *

(Benzene, hexaphenyl-)

Submitted by Louis F. Fieser [1]
Checked by Chester E. Ramey and V. Boekelheide

1. Procedure

A 100-ml., round-bottomed, ground-glass flask containing 40 g. of benzophenone is heated over a free flame to melt the bulk of the solid, and then 8.0 g. of tetraphenylcyclopentadienone (0.021 mole) (Note 1) and 8.0 g. of diphenylacetylene (0.043 mole) (Note 2) are introduced through a paper cone so that no material lodges on the neck or walls. An air condenser is attached, and the mixture is heated over a microburner so that it refluxes briskly but without flooding the condenser (the temperature of the liquid phase is 301–303°). Carbon monoxide is evolved, the purple color begins to fade in 15–20 minutes, and the color changes to a reddish brown in 25–30 minutes. When no further lightening in color is observed (after about 45 minutes), the burner is removed and 8 ml. of diphenyl ether is added to prevent subsequent solidification of the benzophenone. The crystals that separate are brought into solution by reheating, and the solution is let stand for crystallization at room temperature. The product is collected and washed free of brown solvent with benzene to give 9.4 g. (84%) of colorless plates, m.p. 454–456° (sealed capillary) (Note 3). A satisfactory solvent for recrystallization is diphenyl ether, using 7 ml. per gram of product.

2. Notes

1 A synthesis of tetraphenylcyclopentadienone is described in *Org. Syntheses*, Coll. Vol. **3**, 806 (1955). However, the following procedure is more convenient. A 250-ml. Erlenmeyer flask is charged with 21 g. of benzil, 21 g. of dibenzyl ketone (Eastman Organic Chemicals, practical grade), and 100 ml. of triethylene glycol, a thermometer is introduced, and the mixture is heated over a free flame until the solid has dissolved. A 10-ml. portion of a 40% solution of benzyltrimethylammonium hydroxide in methanol is made ready, the temperature of the reaction mixture is adjusted to 100°, the basic catalyst is added, and the mixture is swirled once for mixing and let stand. Within 15–20 seconds the liquid sets to a stiff paste of purple crystals and the temperature rises to 115°. After the temperature has dropped to 80°, the mixture is cooled, thinned by stirring in 50 ml. of methanol, and the product is collected and washed with methanol until the filtrate is purple, not brown. The yield of product, m.p. 219–220°, is 35.5 g. (93%). Recrystallization can be accomplished

with 92% recovery by dissolving the ketone in triethylene glycol (10 ml./gram) at 220°.

2. Improvements in the preparation and dehydrohalogenation of *meso*-stilbene dibromide [*Org. Syntheses*, Coll. Vol. **3**, 350 (1955)] are as follows. *trans*-Stilbene (20 g.) is heated with 400 ml. of acetic acid on the steam bath until dissolved, 40 g. of pyridinium bromide perbromide is added, and the mixture is heated on the steam bath and swirled for 5 minutes. *meso*-Stilbene dibromide separates at once in pearly white plates. The mixture is cooled to room temperature, and the product is collected and washed with methanol. The yield of dibromide, m.p. 236–237°, is 32.4 g. (86%). A 250-ml., round-bottomed, ground-glass flask is charged with 32.4 g. of the dibromide, 65 g. of potassium hydroxide pellets, and 130 ml. of triethylene glycol. A 15 x 125 mm. test tube containing enough of the same solvent to cover the bulb of a thermometer is inserted in the flask. The flask is supported in a clamp, which is used as a handle for swirling the flask over a free flame to mix the contents and bring the temperature to 160°, when potassium bromide begins to separate. By intermittent heating and swirling the mixture is kept at 160–170° for 5 minutes more to complete the reaction. The test tube is then removed, dipped into 500 ml. of water in a beaker, and the adhering organic material is rinsed into the beaker with 95% ethanol. The hot reaction mixture is poured into the beaker, and the flask is rinsed alternately with water and with ethanol. After cooling, the crude product is collected, washed with water, and air-dried (16.5 g.). The brown solution of this material in 50 ml. of 95% ethanol is filtered from a little dark residue, reheated, and let stand for crystallization. A first crop of diphenylacetylene (11.8 g.) separates in large colorless spars, m.p. 61.5–62.5°. Concentration of the mother liquor yields an additional 2.4 g. of crystals, m.p. 58–59°.

The checkers used diphenylacetylene provided by Aldrich Chemicals.

3. The melting point is determined conveniently with a Mel-Temp apparatus and a 90–510° thermometer designed for use with it (Laboratory Devices, Post Office Box 68, Cambridge 39, Massachusetts). An evacuated capillary containing a sample is

sealed close to the sample to prevent sublimation, and repeated determinations are made with the same sample. The figure 456° is the average of two determinations of the temperature of melting; 454° is the average of two observations of the point of solidification. When the amount of diphenylacetylene was reduced to 1.2 times the theory, the yield was the same but the melting point was 450–452°.

3. Methods of Preparation

Hexaphenylbenzene has been prepared by heating tetraphenylcyclopentadienone and diphenylacetylene without solvent [2] and by trimerization of diphenylacetylene with bis-(benzonitrile)-palladium chloride and other catalysts.[3]

4. Merits of the Preparation

Hexaphenylbenzene can be prepared satisfactorily by strong heating of a mixture of 0.5 g. each of tetraphenylcyclopentadienone and diphenylacetylene in a test tube, but the method is unsatisfactory on a larger scale because of the high melting point of the product and the poor heat transfer in a flask. The present procedure demonstrates use of benzophenone as solvent for a Diels-Alder reaction requiring a temperature of about 300°. When the reaction is completed, addition of a small amount of diphenyl ether lowers the melting point of benzophenone sufficiently to prevent this solvent from solidifying.

Other solvents tried and the liquid temperatures of the refluxing mixtures are: stearic acid (340–365°), di-n-butyl phthalate (320–325°), phenyl salicylate (290°). The first two solvents are unsatisfactory because of side reactions consuming some of the tetraphenylcyclopentadienone, the third because the addition reaction is too slow.

Note 1 describes an improvement in the preparation of the starting dienone involving use of a medium of higher solvent power and higher boiling point than ethanol and of a basic catalyst more convenient than potassium hydroxide because it is miscible with the solvent employed. Note 2 reports two improvements in the preparation of diphenylacetylene. The yield in the conversion of *trans*-stilbene to the *meso* dibromide is increased by use of the highly stereoselective reagent pyridinium bromide perbromide. In the dehydrohalogenation step the reaction time is reduced substantially and the yield increased by use of a high-boiling alcohol in place of ethanol.

1. Department of Chemistry, Harvard University, Cambridge, Massachusetts, 02138.
2. W. Dilthey and G. Hurtig, *Ber.*, **67**, 2007 (1934).
3. A. T. Blomquist and P. M. Maitlis, *J. Am. Chem. Soc.*, **84**, 2329 (1962).

1,3,5-HEXATRIENE*

A. $CH_2{=}CHCH_2Br + Mg \xrightarrow{\text{Ether}} CH_2{=}CHCH_2MgBr$

$CH_2{=}CHCH_2MgBr + CH_2{=}CHCHO \xrightarrow[\text{H}_2\text{SO}_4]{\text{H}_2\text{O}}$

$CH_2{=}CHCH_2CHOHCH{=}CH_2$

B. $CH_2{=}CHCH_2CHOHCH{=}CH_2 + PBr_3 \rightarrow C_6H_9Br$

$C_6H_9Br + (CH_3)_2NCH_2C_6H_5 \rightarrow C_6H_9\overset{\oplus}{N}(CH_3)_2CH_2C_6H_5\ Br^{\ominus}$

$C_6H_9\overset{\oplus}{N}(CH_3)_2CH_2C_6H_5\ Br^{\ominus} \xrightarrow[\text{Heat}]{\substack{\text{Aqueous}\\ \text{NaOH}}}$

$CH_2{=}CHCH{=}CHCH{=}CH_2 + C_6H_5CH_2N(CH_3)_2$

Submitted by JESSE C. H. HWA and HOMER SIMS.[1]
Checked by VIRGIL BOEKELHEIDE and E. A. CARESS.

1. Procedure

A. *1,5-Hexadien-3-ol* (Note 1). In a 5-l. three-necked flask fitted with a stirrer, a dropping funnel and an ice-water condenser are placed 153.0 g. (6.28 g. atoms) of magnesium turnings, 360 ml. of anhydrous ether (Note 2), and a few crystals of iodine. A solution of 351.0 g. (2.90 moles) of allyl bromide (Note 3) in 2.6 l. of ether is added in small portions until the reaction begins, and then at such a rate as to maintain gentle refluxing of the ether. The addition requires about 3 hours, after which the reaction mixture is refluxed on a steam bath for an additional hour. Acrolein (Note 4) (104.0 g., 1.86 moles) is added during 2 hours, and this causes gentle refluxing. After an additional hour at room temperature the reaction mixture is poured slowly into 2 l. of ice water. The precipitate is dissolved by adding slowly a solution of 120 ml. of concentrated sulfuric acid (sp. gr. 1.84) in 400 ml. of water (Note 5). The organic layer is separated and the water layer extracted with three 200-ml. portions of ether. The combined oil and ether extracts are dried over 8–10 g. of anhydrous magnesium sulfate. After removal of the ether, the residue is distilled through a 6-in. column packed with glass helices to yield 104–108 g. (57–59%, based on acrolein) of 1,5-hexadien-3-ol; b.p. 62–65/50 mm., n_D^{25} 1.4440.

B. *1,3,5-Hexatriene.* In a 500-ml., three-necked, round-bottomed flask fitted with a mechanical stirrer, a thermometer, and a graduated dropping funnel are placed 114 g. (0.42 mole) of phosphorus tribromide (Note 6) and 2 drops of 48% hydrobromic acid. As the contents of the flask are stirred and maintained at 10–15° by means of an ice-water bath, 98 g. (1.00 mole) of 1,5-hexadien-3-ol is added in the course of 1.5 to 1.75 hours. The mixture is allowed to stir at 10–15° for 40 minutes and then to stand at room temperature overnight. The flask is cooled in an ice-salt bath for 20 minutes, and the upper organic layer is decanted from the residue while still cold. The organic layer is successively washed with three 40-ml. portions each of ice water, 5% sodium bicarbonate, and water. The crude bromohexadiene weighs 147–153 g. (91–95%) (Note 7).

In an assembly similar to that used for the previous reaction, 90 g. (0.67 mole) of dimethylbenzylamine (Note 8), 0.13 g. of hydroquinone, and 500 ml. of water are stirred and heated at 50°. The crude bromohexadiene (107 g., 0.67 mole) is added in the course of 20–40 minutes, and stirring and heating are maintained at 50° for 2–2.5 more hours. The flask is then fitted for downward distillation, and the mixture is distilled at about 40–50° and 30 mm. until no more oil distils with the water. A total of 133–200 ml. of distillate is collected. This is discarded.

A solution of sodium hydroxide (106 g., 2.7 moles) in 535 ml. of water is placed in a 2-l. flask equipped with a sealed mechanical stirrer and an outlet arranged for downward distillation into an ice-cooled receiver. The aqueous solution of the quaternary bromide is added dropwise to the boiling solution of sodium hydroxide during a period of 2.5–4 hours (Note 9). The hexatriene and dimethylbenzylamine which form are distilled with the water. Distillation is continued for 10–15 minutes after the final addition of quaternary bromide solution. The clear upper layer of the distillate is separated, cooled to 5–10°, washed with three 170-ml. portions each of cold $2N$ hydrochloric acid and water, and dried over anhydrous sodium sulfate. The oil is then distilled, a spinning-band column being used to give 32.0–34.0 g. (54–60%) of 1,3,5-hexatriene; b.p. 80–80.5°, n_D^{20} 1.5103–1.5119 (Note 10).

2. Notes

1. This method is essentially that of Butz, Butz, and Gaddis except for modified charge ratios to increase the output per batch at some sacrifice in per cent yield.

2. Baker and Adamson, reagent grade.

3. Matheson, Coleman and Bell, b.p. 70–71°.

4. Shell Chemical Co., commercial grade, inhibited.

5. When the magnesium complex is dissolved, the solution may be decanted from the excess magnesium metal.

6. Dow Chemical Co., practical grade.

7. The moist crude bromohexadiene is quaternized in water without further purification. The submitters report that, if desired, the crude mixture may be dried over anhydrous calcium chloride and fractionally distilled through a 10-in. stainless-steel-packed column at reduced pressure. Crude bromohexadiene (220–230 g.) from 147 g. (1.50 moles) of 1,5-hexadiene-3-ol was found to give the fractions listed in Table I. The yield of the

TABLE I

Fraction	Wt., g.	B.P./mm.	n_D^{20}	Product
1	43.2	55–56°/34	1.4829	3-Bromo-1,5-hexadiene
2	31.7	57–72°/34	1.4923	Mixture of the 3- and 1-bromo isomers
3	70.7	72–73°/36	1.4981	1-Bromo-2,5-hexadiene
4	10.1	56–59°/18	1.4996	Mostly 1-bromo-2,5-hexadiene
5	19.2	64–103°/14	1.5196	—

total distillate is 174.9 g. of which fractions 1–4 amount to 155.7 g. (64.5%). The 1-bromo isomer has been reported in the literature,[3,4] and the infrared spectrum suggests that fraction 1 is 3-bromo-1,5-hexadiene. Both isomers when treated separately in this procedure yield hexatriene.

8. The amine should be freshly distilled.

9. This time, although not critical, was chosen to prevent accumulation of unreacted quaternary base or of hexatriene in the reaction vessel.

10. The faintly yellow product obtained before distillation is quite pure. The infrared absorption spectra of the liquid before and after distillation are identical and have, in addition to all the bands shown in the published spectrum of Woods and Schwartzman,[5] weak absorption bands, notably at 820, 989, 1187, and 1452 cm^{-1}. The differences between the two spectra are due to various ratios of *cis* and *trans* isomers in these hexatriene samples. The ratio of *cis*- to *trans*-1,3,5-hexatriene in this preparation is estimated at 3:7. This is based on studies using vapor-phase chromatography, ultraviolet absorption spectra, and refractive indices.[6] The hexatriene is stored under nitrogen at $-5°$. Although no visible change is observed after 1 week of storage, the liquid is partially polymerized to the consistency of glycol after 3 weeks. Hexatriene can be removed from the thin syrup by distillation at 40 mm., and when redistilled at atmospheric pressure under nitrogen has b.p. 80.5°, n_D^{20} 1.5101. Its infrared absorption spectrum is then identical with that of freshly prepared hexatriene.

3. Methods and Merits of Preparation

1,3,5-Hexatriene has been prepared by many workers. The more successful methods are the catalytic pyrolyses (alumina, 260–325°) of 1,3-hexadien-5-ol [5,7,8] and 2,4-hexadien-1-ol.[9] Other methods which give hexatriene of questionable purity or involve less convenient laboratory methods are dehydration of 1,5-hexadien-3-ol by sodium bisulfate at 170°,[10] or by phthalic anhydride at 160–200°,[2] and by catalytic hydrogenation of divinylacetylene.[11] Additional methods are listed in footnote 5. The present procedure is a practical laboratory method of preparing pure hexatriene in satisfactory yields.

1. Rohm and Haas Company, Philadelphia, Pa.
2. L. W. Butz, E. W. J. Butz, and A. M. Gaddis, *J. Org. Chem.*, 5, 171 (1940).
3. M. Lora-Tamayo, F. Martin-Panizo, and R. Ossorio, *J. Chem. Soc.*, 1418 (1950).
4. P. Karrer and S. Perl, *Helv. Chim. Acta*, 33, 36 (1950).
5. G. F. Woods and L. Schwartzman, *J. Am. Chem. Soc.*, 70, 3394 (1948).
6. J. C. H. Hwa, P. DeBenneville, and H. Sims, *J. Am. Chem. Soc.*, 82, 2537 (1960).
7. K. Alder and H. Von Brachel, *Ann.*, 608, 208 (1957).
8. E. Lippincott, C. White, and J. Sibilia, *J. Am. Chem. Soc.*, 80, 2926 (1958).
9. G. F. Woods, N. Bolgiano, and D. Duggan, *J. Am. Chem. Soc.*, 77, 1800 (1955).
10. O. Kiun-Houo, *Ann. Chim.*, 13, 175 (1940) [*C.A.*, 34, 4377 (1940)].
11. A. Klebanskii, L. Popov, and N. Tsukarman, *J. Gen. Chem. U.S.S.R.*, 16, 2083 (1946) [*C.A.*, 42, 857 (1948)].

HOMOPHTHALIC ACID*

Submitted by P. A. S. SMITH and R. O. KAN [1]
Checked by MELVIN S. NEWMAN and BERNARD DARRE

1. Procedure

Ten grams (0.056 mole) of 2*a*-thiohomophthalimide [2] and a solution of 30 g. of potassium hydroxide in 125 ml. of water are placed in a 300-ml., one-necked, round-bottomed flask (Note 1). The mixture is refluxed for 48 hours, filtered, and acidified with 12*N* hydrochloric acid. The solid that forms on cooling is collected by filtration and recrystallized from a mixture of 25 ml. of water and as much acetic acid (about 7 ml.) as necessary to dissolve the solid in the boiling solution, with addition of a little

activated carbon. The yield of homophthalic acid, m.p. 181°
(Note 2), is 6.1–7.5 g. (60–73%) (Note 3).

2. Notes

1. Because base can attack glass vessels, possibly introducing
difficultly removable silicates into the reaction mixture, a copper
flask is recommended for routine operations.

2. The melting point depends on the rate of heating. When
the solid is heated slowly, the melting range can be as low as
172–174°.

3. An alternative procedure involves 3 days of refluxing in a
mixture of 75 ml. of glacial acetic acid, 50 ml. of 12N hydrochloric
acid, and 30 ml. of water. The product separates on cooling in a
slightly lower yield (48%).

3. Methods of Preparation

Homophthalic acid may be obtained by the oxidation of indene,[3,4] the reduction of phthalonic acid,[5,6] and the hydrolysis
of o-carboxyphenylacetonitrile.[7] Other methods are listed in an
earlier volume.[8]

4. Merits of the Preparation

This is a general method for converting 2a-thiohomophthalim-
ides to homophthalic acids. Since 2a-thiohomophthalimides
are readily obtained from phenylacetyl chlorides,[2] this is a convenient method for preparing homophthalic acids.

1. Department of Chemistry, University of Michigan, Ann Arbor, Michigan.
2. P. A. S. Smith and R. O. Kan, this volume, p. 1051.
3. O. Grummitt, R. Egan, and A. Buck, *Org. Syntheses*, Coll. Vol. **3**, 449 (1955).
4. W. F. Whitmore and R. C. Cooney, *J. Am. Chem. Soc.*, **66**, 1239 (1944).
5. K. Miescher and J. R. Billeter, *Helv. Chim. Acta*, **22**, 608 (1939).
6. W. Davies and H. G. Poole, *J. Chem. Soc.*, 1617 (1928).
7. C. C. Price, *Org. Syntheses*, **22**, 61 (1942).

3-HYDROXYGLUTARONITRILE

(Glutaronitrile, 3-hydroxy-)

$$\text{CH}_2\text{—CH—CH}_2\text{—Cl} \quad + \quad 2\text{KCN} \quad + \quad \text{H}_2\text{O} \longrightarrow$$

$$\text{NCCH}_2\text{CHOHCH}_2\text{CN} \quad + \quad \text{KCl} \quad + \quad \text{KOH}$$

Submitted by F. JOHNSON and J. P. PANELLA [1]
Checked by A. G. ANASTASSIOU and B. C. McKUSICK

1. Procedure

A mixture of 493 g. (2.00 moles) of magnesium sulfate hepta-hydrate and 700 ml. of tap water is stirred for 5 minutes and filtered into a 2-l. three-necked flask equipped with a mechanical stirrer and an alcohol thermometer that dips into the solution. The flask is immersed in a cooling bath (Note 1), the stirrer is started, and the solution is cooled to 10°. To the solution is added, in one portion, 143 g. (2.20 moles) of potassium cyanide (*Caution! Toxic*), and the stirring is continued for 45 minutes at 8–12° (Note 2). The solution is maintained at this temperature while 102 g. (1.10 moles) of epichlorohydrin (Note 3) is added dropwise with stirring over a period of 1 hour (Note 4). The mixture is allowed to come to room temperature and is stirred for an additional 24 hours at this temperature.

The dark red-brown reaction mixture is stirred and extracted continuously with 1 l. of ethyl acetate for 48 hours (Note 5). The extract is dried over anhydrous magnesium sulfate for 18 hours (Note 6), filtered, and the filtrate is concentrated under reduced pressure on a steam bath. The residual dark-brown oil (about 90 g.) is distilled from a Claisen flask; the distillation must be rapid to minimize decomposition. About 20 g. of fore-run consisting of 4-chloro-3-hydroxybutyronitrile and 4-hydroxycro-tononitrile is collected at 90–115° (0.4 mm.). 3-Hydroxyglu-taronitrile is collected at 155–160° (0.4 mm.), yield 65–75 g.

(54–62%), n^{23}D 1.4634. This pale yellow distillate is sufficiently pure for most purposes. Further purification can be effected with only 3–5% loss by distillation of the material through a 15-cm. Vigreux column and collection of the portion boiling at 154–156° (0.2 mm.), n^{23}D 1.4632 (Note 7).

2. Notes

1. The bath contained a mixture of trichloroethylene and solid carbon dioxide kept at −20°.

2. At this point the mixture has an opaque milky-white appearance caused by precipitation of a little magnesium hydroxide.

3. Epichlorohydrin (white label brand) supplied by Eastman Organic Chemicals was used without further purification.

4. The reaction is exothermic. If the temperature of the reaction mixture is allowed to rise above 30°, the reaction is likely to get out of control.

5. The checkers found that a stirred extractor [2] was much more efficient than an unstirred one for this operation.

6. This extensive drying period is necessary to allow precipitation of traces of basic salts that have been carried over during the extraction procedure. Failure to remove these salts results in extensive decomposition of the product during the distillation step.

7. The submitters have obtained the same yield working on 10 times this scale.

3. Methods of Preparation

3-Hydroxyglutaronitrile has been prepared by the action of potassium cyanide on 1,3-dichloro-2-propanol [3–5] or on 4-chloro-3-hydroxybutyronitrile.[6, 7] More recently it has been prepared from epichlorohydrin using essentially the present method.[8]

4. Discussion

This is a much more convenient and satisfactory synthesis of 3-hydroxyglutaronitrile than earlier ones.[3–7] The method can be applied to other epichlorohydrins; 2-methylepichlorohydrin and 2-ethylepichlorohydrin have been converted to the corresponding hydroxydinitriles in 71% and 77% yields, respectively.[8] The hydroxydinitriles undergo cyclizations to heterocyclic compounds not easily prepared in other ways. Thus hydrogen bromide at 0° converts 3-hydroxyglutaronitrile to 2-amino-6-bromopyridine in 70% yield.[9]

The reaction probably proceeds as follows:[8]

$$CH_2\text{---}CHCH_2Cl \xrightarrow{CN^-} \left[NCCH_2\underset{O^-}{CH}CH_2Cl \right] \longrightarrow NCCH_2CH\text{---}CH_2$$

$$\xrightarrow[H_2O]{CN^-} NCCH_2CHOHCH_2CN$$

1. The Dow Chemical Company, Eastern Research Laboratory, Wayland, Massachusetts.
2. G. Billek, this volume p. 627.
3. M. Simpson, *Ann.*, **133**, 74 (1865).
4. O. Morgenstern and E. Zerner, *Monatsh.*, **31**, 777 (1910).
5. G. Braun, *J. Am. Chem. Soc.*, **52**, 3167 (1930).
6. R. Lespieau, *Bull. Soc. Chim. France*, [4] **33**, 725 (1923).
7. R. Legrand, *Bull. Soc. Chim. Belges*, **53**, 166 (1944).
8. F. Johnson, J. P. Panella, and A. A. Carlson, *J. Org. Chem.*, **27**, 2241 (1962).
9. F. Johnson, J. P. Panella, A. A. Carlson, and D. H. Hunneman, *J. Org. Chem.*, **27**, 2473 (1962).

2-HYDROXYISOPHTHALIC ACID

(Isophthalic acid, 2-hydroxy-)

$$\text{(2-hydroxy-3-methylbenzoic acid)} + 2KOH \rightarrow \text{(potassium salt)} + 2H_2O$$

$$\text{(potassium salt)} + KOH + 3PbO_2 \rightarrow$$

$$\text{(tripotassium salt)} + 3PbO + 2H_2O$$

$$\text{(tripotassium salt)} + 3HCl \rightarrow \text{(2-hydroxyisophthalic acid)} + 3KCl$$

Submitted by David Todd [1] and A. E. Martell.[2]
Checked by B. C. McKusick and S. Andreades.

1. Procedure

In an 800-ml. stainless-steel flanged beaker are placed 240 g. of potassium hydroxide pellets and 50 ml. of water. When the mush has cooled, 40.0 g. (0.263 mole) of 2-hydroxy-3-methyl-benzoic acid (Note 1) is added, and the slurry is stirred with a long glass rod. The beaker is placed firmly in a clamp (Note 2) and set in a cold oil bath (Note 3) in a hood, and 240 g. (1.00 mole) of lead dioxide is stirred in all at once (Note 4). The oil bath is now heated by a flame, and when the bath temperature reaches 200° steady manual stirring by means of the glass rod is begun. (*Caution! Goggles and a rubber glove should be worn to protect eyes and the stirring hand against spattering.*) The temperature is raised, with steady stirring, until the bath reaches 238–240°, when the flame is moderated to maintain this bath temperature. Several minutes after 240° is reached, the mixture boils steadily and the lumpy brown mass turns quickly to a bright-orange melt containing heavy crystals of lead monoxide. The bath is held at 240° for another 15 minutes. It is then brought briefly to 250°, the flame is removed, and the beaker is lifted out of the bath.

About 5 minutes later, the liquid contents of the beaker are poured cautiously into a 2-l. glass beaker, and this is tipped and rotated slowly so as to spread the congealing mass in a thin film on the beaker walls. When the material in the beakers has cooled, a total of 1 l. of water is poured into the beakers and the water is stirred well for at least 1 hour (Note 5). The cold suspension is filtered with suction to separate 200–210 g. of an insoluble mixture of lead monoxide and red lead, which is washed on the filter with six 50-ml. portions of water.

The alkaline filtrate and washings are combined and partially neutralized by the addition of 150–175 ml. of concentrated hydrochloric acid. Sufficient sodium sulfide solution is added to precipitate all the lead ion present (Note 6). The suspension is brought to a gentle boil to coagulate the lead sulfide, allowed to cool somewhat, and filtered with suction. The filtrate is placed in a 2-l. beaker set in an ice bath and acidified (*Caution! in the hood*) with about 150 ml. of concentrated hydrochloric acid to precipitate crude 2-hydroxyisophthalic acid monohydrate (Note 7). The suspension is cooled to 0–5° and filtered to separate the crude acid, which weighs 35–49 g. after being dried in a vacuum oven at 110°/50–150 mm. for 5 hours (Note 8).

In order to remove 1–3 g. of contaminating 2-hydroxy-3-methylbenzoic acid, the crude acid is ground in a mortar and refluxed briefly with 100 ml. of chloroform, and the suspension is filtered hot (Note 9). The separated solid is dried in air and added to 1 l. of boiling water. The mixture is boiled gently for 15 minutes and filtered by gravity to remove a small amount of gray sludge. The clear light-orange filtrate rapidly deposits needles of 2-hydroxyisophthalic acid monohydrate. After cooling to 0–5°, the acid is collected and dried at 110°/50–150 mm. for 20 hours. The anhydrous acid, which ranges in color from pale pink to tan, weighs 22–29 g. (46–61%) (Note 10) and melts at 243–255°, depending on the rate of heating and the apparatus used.

2. Notes

1. Eastman Kodak technical grade material is satisfactory.

2. A convenient clamp can be fashioned by bending a loop at the end of a 40-cm. length of iron rod 6 mm. (1/4 in.) in diameter so that the beaker can just slip up to its flange through the loop. A pinch clamp is used to hold the flange firmly to the loop.

3. The checkers used a bath wax (flash point 325°) supplied by the Fisher Scientific Co. The operation should be carried out in a hood because fumes from oil baths at high temperatures are injurious to health. A fire extinguisher should be close at hand in case the oil bath catches fire. A high-boiling silicone heat exchange oil, although more expensive, would be less of a fire hazard. Alternatively, a sodium nitrite-sodium nitrate-potassium nitrate mixture such as "Hitec" heat transfer salt sold by the du Pont Company (useful range 150–450°) can be used. Such baths will not burn, but they have the disadvantage of being oxidizing agents, so precautions should be taken not to let any organic material get into them lest a flash oxidation accompanied by spattering of hot salt take place. Hot salt baths should be well shielded to guard against spattering.

4. If the calculated amount (190 g.) of lead dioxide is used, the yield is lowered by about 20%.

5. At least an hour's contact with water is necessary in order to dislodge all the solid from the walls.

6. The amount of sodium sulfide nonahydrate needed varies from 12 to 25 g. from one run to another.

7. The acid is precipitated from aqueous solution as the monohydrate, which is soluble in cold dilute hydrochloric acid to the extent of about 6 g. per l.

8. An additional 3–5 g. (6–10%) of crude product can be obtained by concentrating the filtrate to about 800 ml., cooling the concentrate to 0°, and filtering the solid that separates. This solid must be washed well with cold water to remove coprecipitated potassium chloride.

9. The starting material is moderately soluble in hot chloroform, while 2-hydroxyisophthalic acid is quite insoluble. Fractional crystallization from water, an alternative method suggested for the separation of starting material,[3] has been found by the submitters to be unsuccessful.

10. An additional 2–5 g. (4–10%) of product can be obtained by concentrating the filtrate to one-third its volume, adding 25 ml. of concentrated hydrochloric acid, cooling the mixture, separating crude acid by filtration, and recrystallizing the acid from water.

3. Methods of Preparation

2-Hydroxyisophthalic acid has been prepared by oxidizing 2-hydroxy-3-methylbenzoic acid with lead dioxide,[3–5] by cleaving the ether group of 2-methoxyisophthalic acid with hydriodic acid,[6] and by hydrolyzing 2-iodoisophthalic acid with alcoholic sodium hydroxide.[7]

The lead dioxide-alkali method has also been applied successfully by Graebe and Kraft [4] to the three cresols, the three toluic acids, and 2,4-dimethylphenol. For the preparation of 2-hydroxyisophthalic acid, it is the only one-step method that starts from readily obtainable materials.

In general, this method is a one-step procedure for the oxidation of a cresol type of molecule to the corresponding phenolic acid. The vigorous reaction conditions clearly limit the type of functional groups that may be present in the molecule. There is no evidence that the reaction has been applied to polynuclear or heterocyclic alkylphenols.

1. Worcester Polytechnic Institute, Worcester, Massachusetts.
2. Clark University, Worcester, Massachusetts.
3. W. S. Benica and O. Gisvold, *J. Am. Pharm. Assoc., Sci. Ed.*, 34, 42 (1945).
4. C. Graebe and H. Kraft, *Ber.*, 39, 799 (1906).
5. A. Moshfegh, S. Fallab, and H. Erlenmeyer, *Helv. Chim. Acta*, 40, 1157 (1957).
6. G. R. Sprengling and J. H. Freeman, *J. Am. Chem. Soc.*, 72, 1984 (1950).
7. C. W. James, J. Kenner, and W. V. Stubbings, *J. Chem. Soc.*, 117, 775 (1920).

HYDROXYMETHYLFERROCENE *

{Iron, cyclopentadienyl[(hydroxymethyl)cyclopentadienyl]-}

$$
\text{Fe}\overset{+}{\big(}CH_2\overset{+}{N}(CH_3)_3\ I^- \quad + \ NaOH \longrightarrow \quad \text{Fe}\big(CH_2OH \quad + \ NaI + (CH_3)_3N
$$

Submitted by DANIEL LEDNICER, T. ARTHUR MASHBURN, JR., and CHARLES R. HAUSER.[1]
Checked by B. C. McKUSICK, H. F. MOWER, and G. N. SAUSEN.

1. Procedure

Caution! This preparation should be conducted in a hood because trimethylamine is evolved.

A solution of 10.0 g. (0.25 mole) of sodium hydroxide in 250 ml. of water is prepared in a 1-l. round-bottomed flask equipped with a reflux condenser and a mechanical stirrer. Twenty-five grams (0.065 mole) of N,N-dimethylaminomethylferrocene methiodide [2] is added to the solution. The resulting suspension is heated to reflux temperature with stirring. At this point the solid is in solution. Within 5 minutes oil starts to separate from the solution and trimethylamine starts to come off. At the end of 3.5 hours, at which time the evolution of the amine has virtually ceased, the reaction mixture is allowed to cool to room temperature. The oil generally crystallizes during the cooling. The mixture is stirred with 150 ml. of ether until the oil or solid is all dissolved in the ether. The ether layer is separated in a separatory funnel and the aqueous layer is extracted with two additional 150-ml. portions of ether. The combined ether extracts are washed once with water and dried over sodium sulfate.

The oil that remains when the solvent is removed from the extract crystallizes when cooled to room temperature. This orange solid is recrystallized from 150 ml. of hexane (Note 1) to yield 9.5–12.5 g. (68–89%) of hydroxymethylferrocene, m.p. 74–76°. One more recrystallization from the same solvent affords 8.2–11.0 g. (59–79%) (Note 2) of good-quality alcohol as golden needles, m.p. 76–78° (Note 3).

2. Notes

1. Eastman Kodak Company practical grade hexane is suitable.
2. The yield of this reaction is directly dependent on the purity of the quaternary salt employed. If the salt is prepared from redistilled N,N-dimethylaminomethylferrocene, the yield of alcohol may be as high as 90%.[3]
3. The pure alcohol melts at 81–82°.

3. Methods of Preparation

Hydroxymethylferrocene has been made by condensing ferrocene with N-methylformanilide to give ferrocenecarboxaldehyde, and reducing the latter with lithium aluminum hydride,[4] sodium borohydride,[5] or formaldehyde and alkali.[5] The present procedure is based on the method of Lindsay and Hauser.[3] A similar procedure has been used to convert gramine methiodide to 3-hydroxymethylindole,[6] and the method could probably be used to prepare other hydroxymethyl aromatic compounds.

1. Department of Chemistry, Duke University, Durham, North Carolina. The work was supported by the Office of Ordnance Research, U. S. Army.
2. See p. 434, this volume.
3. J. K. Lindsay and C. R. Hauser, *J. Org. Chem.*, 22, 355 (1957).
4. P. J. Graham, R. V. Lindsey, G. W. Parshall, M. L. Peterson, and G. M. Whitman, *J. Am. Chem. Soc.*, 79, 3416 (1957).
5. G. D. Broadhead, J. M. Osgerby, and P. L. Pauson, *J. Chem. Soc.*, 650 (1958).
6. E. Leete and L. Marion, *Can. J. Chem.*, 31, 775 (1953).

2-HYDROXY-3-METHYLISOCARBOSTYRIL

(Isocarbostyril, 2-hydroxy-3-methyl-)

Submitted by EMIL J. MORICONI and FRANCIS J. CREEGAN [1]
Checked by BARBARA A. ALEXANDER, HERMANN ERTL,
T. HOEKEMEIJER, and PETER YATES

1. Procedure

A solution of 8.0 g. (7.5 ml., 0.055 mole) of 2-methyl-1-indanone (Note 1) in 100 ml. of toluene is prepared in a 500-ml., three-necked, round-bottomed flask equipped with a thermometer, a dropping funnel, and a magnetic stirrer. The flask is immersed in a freezing mixture of sodium chloride and ice. When the solution temperature reaches 0°, 70 ml. (0.21 mole) of $3N$ hydrochloric acid in ethyl acetate (Note 2) is added slowly (Note 3). To this mixture 8.0 ml. (0.068 mole) of freshly prepared n-butyl nitrite (Note 4) in 25 ml. of toluene is added with stirring over a 10-minute period (Note 5). The mixture is stirred for 1 hour at 0° and for an additional hour at room temperature. The two layers are separated, and the upper layer (toluene) is concentrated to one-half volume. Both solutions are refrigerated at −20° for 4 days. The precipitated orange product is collected by filtration from each layer. Further concentration of each filtrate under reduced pressure to one-half volume gives additional crude product (Note 6).

The various fractions are combined, washed with 20 ml. of cold ether, and dried. Recrystallization from methylene chloride-ether (Note 7) gives 6.0–6.6 g. (62–69%) of 2-hydroxy-3-methylisocarbostyril as light orange plates, m.p. 175–180°. Sublimation of this material at 100–110° (0.5 mm.) gives a white product, m.p. 182–184°, with softening at 174° (Note 7).

2. Notes

1. The 2-methyl-1-indanone, b.p. 65–66° (0.6 mm.) [lit.[2] b.p. 120° (15 mm.)], was prepared by the following method, described by Colonge and Weinstein.[2]

To 15.0 g. (0.50 mole) of paraformaldehyde (Eastman Organic Chemicals) and 100 g. (0.75 mole) of propiophenone (Eastman Organic Chemicals) in a 250-ml. Erlenmeyer flask, 10 ml. of $1N$ alcoholic potassium hydroxide solution was added with stirring. After a few minutes a clear solution formed, and the temperature rose to 35° and then fell slowly. The yellow solution was stirred for 5.5 hours at room temperature, during which time the solution became turbid. The turbid solution was poured into 150 ml. of water, and the mixture was acidified with concentrated hydrochloric acid (Congo red indicator).

The mixture was extracted with two 150-ml. portions of benzene, and the combined organic extracts were washed with two 150-ml. portions of water, two 150-ml. portions of 10% aqueous sodium carbonate, and two 150-ml. portions of water. The benzene extracts were dried over anhydrous sodium sulfate, and the solvent was removed. The yellow residue was distilled under reduced pressure to give a forerun consisting of 45 g. (0.34 mole) of unconsumed propiophenone followed by 32–36 g. (39–44%) of 3-hydroxy-2-methylpropiophenone, b.p. 108–110° (0.55 mm.) [lit.[2] b.p. 158–162° (17 mm.)]; infrared band (neat) at 5.94 μ (C=O).

3-Hydroxy-2-methylpropiophenone (30 g., 0.183 mole) was added slowly to 150 ml. of concentrated sulfuric acid with stirring. The temperature rose and the solution turned dark brown. The temperature remained at 80° for 10 minutes and then slowly fell. After 1 hour the dark solution was poured onto 200 g. of cracked ice. The mixture was extracted with two 100-ml. portions of ether. The ethereal solution was washed with two 100-ml. portions of water, two 100-ml. portions of saturated aqueous sodium bicarbonate, and again with two 100-ml. portions of water. It was dried over anhydrous potassium carbonate, and the solvent was removed. The residue was distilled to give 18–19 g. (67–71%) of a pale yellow liquid, b.p. 65–66° (0.6 mm.) [lit[2]. b.p. 120° (15 mm.)], n^{20}D 1.5510 (lit.[2] n^{23}D 1.5511); infrared band (neat) at 5.80 μ (C=O).

2. Prepared by dissolving 17.5 ml. of 12N hydrochloric acid in 52.5 ml. of ethyl acetate.

3. Two phases are obtained; this heterogenous mixture is vigorously stirred during the addition of n-butyl nitrite.

4. The n-butyl nitrite must be refrigerated after preparation[3] and used as soon as possible thereafter. The use of commercially available n-butyl nitrite invariably led to lower yields of the isocarbostyril.

5. With lower hydrochloric acid concentration and reversal of the mode of addition, $i.e.$, acid to indanone-nitrite mixture, the intermediate 2-methyl-2-nitroso-1-indanone may also be isolated as its dimer. This can be isomerized to the isocarbostyril rapidly in refluxing methanolic sodium methoxide and more slowly in concentrated hydrochloric acid.[4]

6. To determine whether all the isocarbostyril has been isolated from the filtrates, a small aliquot of the filtrate is treated with excess aqueous ferric chloride. The appearance of a deep purple color indicates the necessity for further concentration under reduced pressure and precipitation of product.

7. The checkers used methylene chloride alone as the solvent for recrystallization; sublimation gave a product, m.p. 178–180°.

3. Methods of Preparation

2-Hydroxy-3-methylisocarbostyril has been prepared by the present method,[4] and in 12–15% yield by the ozonization of 3-methylisoquinoline-2-oxide.[5]

4. Merits of the Preparation

This simple, one-step ring expansion is the only available method for the preparation of 2-hydroxy-3-alkylisocarbostyrils in good yield from the corresponding 2-alkyl-1-indanones. Table I lists five new hydroxyisocarbostyrils prepared in this manner.

TABLE I

SYNTHESES OF 2-HYDROXY-3-ALKYLISOCARBOSTYRILS

2-Alkyl Substituent	Yield, %	M.P., °C
Ethyl	65	154–155
Propyl	64	139–141
Isopropyl	49	107–108
Butyl	45	108–109
t-Butyl	20	104–106

Direct reduction of the 2-hydroxy-3-alkylisocarbostyrils gives 3-alkylisocarbostyrils and provides a useful synthesis of these compounds.

1. Contribution No. 842 from the Department of Chemistry, Fordham University, New York, N. Y. 10458. This work was supported by the Directorate of Chemical Sciences, Air Force Office of Scientific Research, under Grant AF-AFOSR-62-18 and 488-64.
2. J. Colonge and G. Weinstein, *Bull. Soc. Chim. France*, [5] **19**, 462 (1952).
3. W. A. Noyes, *Org. Syntheses*, Coll. Vol. **2**, 108 (1943).
4. E. J. Moriconi, F. J. Creegan, C. K. Donovan, and F. A. Spano, *J. Org. Chem.*, **28**, 2215 (1963); E. J. Moriconi and F. J. Creegan, *J. Org. Chem.*, **31**, 2090 (1966).
5. E. J. Moriconi and F. A. Spano, *J. Am. Chem. Soc.*, **86**, 38 (1964).

p-HYDROXYPHENYLPYRUVIC ACID *

[Pyruvic acid, *p*-hydroxyphenyl-]

Submitted by GERHARD BILLEK.[1]
Checked by MAX TISHLER and ARTHUR J. ZAMBITO.

1. Procedure

A. *5-(p-Hydroxybenzal)hydantoin.*   An intimate mixture of
6.11 g. (0.050 mole) of *p*-hydroxybenzaldehyde (Note 1) and 5.5 g.
(0.055 mole) of hydantoin (Note 2) is placed in a 250-ml. round-
bottomed flask.   Dry piperidine (10 ml.) is added, a reflux condenser protected by a calcium chloride tube is fitted to the flask,
and the flask is immersed in an oil bath so that the level of the
reaction mixture is the same as the oil level of the bath.   The oil
bath is heated slowly to 130° and is held at this temperature for
30 minutes; foaming and gentle boiling occur.   The reaction mixture is cooled, and 200 ml. of water at about 60° is added.   The
contents of the flask are stirred by means of a glass rod until a
clear red solution is obtained (Note 3).   Any traces of tarry material are removed by filtration.   The solution is cooled to room
temperature, transferred to an Erlenmeyer flask, and acidified by
dropwise addition of 20 ml. of 12N hydrochloric acid.   The mixture stands at room temperature a few hours, and then the yellow

precipitate of 5-(*p*-hydroxybenzal)hydantoin is collected on a Büchner funnel and washed thoroughly with cold water (Note 4). After being dried in a vacuum desiccator over potassium hydroxide, the crude hydantoin weighs 8.5–8.8 g. (83–86%). It melts at 310–315° (dec.) and is sufficiently pure for the next step and other preparative purposes (Note 5).

B. *p-Hydroxyphenylpyruvic acid.* Crude 5-(*p*-hydroxybenzal)hydantoin (8.5 g., 0.042 mole) and a few chips of porous plate are placed in a 500-ml., three-necked, round-bottomed flask fitted with a reflux condenser, a dropping funnel, and a gas-inlet tube (Note 6). A slow stream of oxygen-free nitrogen (Note 7) is introduced. As soon as the air has been swept out of the apparatus, 240 ml. of 20% aqueous sodium hydroxide solution (w/v) is added through the dropping funnel. The mixture is boiled for 3 hours in an oil bath at 170–180° (Note 8). The 5-(*p*-hydroxybenzal)hydantoin dissolves rapidly, and a clear orange solution is obtained that becomes less deeply colored during the reaction. The reaction mixture is cooled by replacing the oil bath by a bath of cold water. Without interrupting the stream of nitrogen, 100 ml. of 12N hydrochloric acid is added through the dropping funnel at such a rate that foaming and heating of the mixture are not excessive. The flask is disconnected, and 5 g. of sodium bicarbonate is dissolved in the mixture (Note 9).

The liquid is transferred to a continuous extractor (Note 10) and extracted with ether until the supernatant layer of ether remains colorless (about 2 hours). The ethereal extract is discarded (Note 11). The aqueous solution is transferred to a 1-l. beaker and acidified by the cautious addition of 60 ml. of 12N hydrochloric acid (Note 12). The solution is returned to the extractor, which is attached to a tared round-bottomed flask. The solution is extracted with ether until no more *p*-hydroxyphenylpyruvic acid is obtained (Note 13). The undried ether solution is evaporated to dryness on a boiling water bath to give crude *p*-hydroxyphenylpyruvic acid as a pale-yellow crystalline mass. The mass is broken up with a spatula, and the flask is kept over potassium hydroxide in a vacuum desiccator until its weight is constant. The yield of crude acid is 6.9–7.2 g. (92–96%). It melts at 210–215° (dec.) (Note 14).

Twelve milliliters of water for each gram of the crude acid

(83–86 ml.) is added to the flask, which is then attached to a reflux condenser and immersed in an oil bath at 150°. After 10–20 minutes of boiling, a clear pale-yellow solution is obtained. This is filtered through a fluted filter into an Erlenmeyer flask. After crystallization has started, 8.3–8.6 ml. of 12*N* hydrochloric acid (1.2 ml. of acid for each gram of crude acid) is added, and the mixture is allowed to cool slowly to room temperature, during which period it is occasionally agitated. Crystallization is completed by keeping the flask in a refrigerator for at least 10 hours. The product is separated by suction filtration and washed with a small amount of ice water. The purified *p*-hydroxyphenylpyruvic acid weighs 4.4–4.7 g. (59–63%) and melts at 216–218° (dec.) (Notes 14 and 15).

2. Notes

1. The *p*-hydroxybenzaldehyde used was a commercial product (practical grade) melting at 114–117°.

2. Hydantoin can be prepared in a variety of ways, notably from glycine [2] or ethyl aminoacetate [3] and potassium cyanate. The checkers used Eastman Kodak "white label" hydantoin.

3. Because of the viscous nature of the reaction mixture, which sometimes shows a tendency to crystallize, this is a slow process, but a mechanical stirrer is not required.

4. The checkers found that three or four cold-water washes are sufficient to wash the precipitate to neutral pH.

5. Further purification may be achieved by crystallization from acetic acid (50 ml. per g.). A product melting at 315° (dec.) is obtained.

6. The inlet tube, preferably in the center neck, is placed in such a way that it nearly touches the bottom of the flask. Thereby nitrogen bubbles effect some agitation of the reaction mixture and prevent bumping of the boiling solution.

7. *p*-Hydroxyphenylpyruvic acid is rapidly oxidized in alkaline solution. Commercially available compressed nitrogen may be used if the gas is further purified by passage through an alkaline solution of pyrogallol (45 g. of pyrogallol dissolved in 300 ml. of 50% sodium hydroxide solution).

8. This should be done in a hood because ammonia is evolved.

9. The purpose of the first extraction is to remove phenolic impurities. Care should be taken to adjust to the proper pH range (6–7). At this pH the solution changes in color from orange to yellow. A small amount of a flocculent precipitate is formed, but, to avoid longer contact of the solution with air, it is not filtered off.

10. A convenient type of extractor is described in *Organic Syntheses*.[4] The submitter improved the efficiency of the extractor by stirring the aqueous phase with a magnetic stirrer. The inner tube had no filter on the lower end of it and was suspended so that this end was about 1 cm. above the magnetic stirring bar. Tests show that magnetic stirring of the aqueous phase increases the speed of extraction by a factor of 2 to 3. It is convenient to use an extractor of such size that the same apparatus can be used for both extractions. A greater volume of solution must be handled in the second extraction than in the first because of the addition of hydrochloric acid.

11. Evaporating the ethereal solution yields not more than 0.3 g. of brown tarry material consisting mostly of impure *p*-hydroxybenzaldehyde.

12. Carbon dioxide and fumes of ether are evolved during the addition of the acid. The solution is stirred by means of a glass rod until the foaming ceases.

13. This is a slow process, and the extraction time depends on the type of extractor used. With stirring as described in Note 10, practically quantitative extraction of *p*-hydroxyphenylpyruvic acid can be achieved within 6 hours. Extremely long extraction times may cause decomposition of the product.

14. The checkers observed a decomposition point of 198–202° for the crude acid, 211–214° for the purified acid.

15. A second crystallization from 10 parts of water raises the melting point of the acid to 220°. Any prolonged contact of the hot solution with air will cause some decomposition, notably the formation of traces of *p*-hydroxybenzaldehyde. The checkers preferred crystallization from 10 parts of glacial acetic acid and 10 parts of 12*N* hydrochloric acid,[5] from which solvent a white product was recovered in 75% yield; m.p. 220° (dec.).

The purity of the *p*-hydroxyphenylpyruvic acid may be checked by paper chromatography. By the ascending method on

Schleicher and Schüll paper No. 2043b and *n*-butanol-acetic acid-water (4:1:1) as solvent, the following R_f values are obtained: *p*-hydroxyphenylpyruvic acid, 0.71; *p*-hydroxybenzaldehyde, 0.85. Sprays: 2,4-dinitrophenylhydrazine (0.2% in 2N hydrochloric acid) and Folin-Denis reagent.[6]

3. Methods of Preparation

p-Hydroxyphenylpyruvic acid has been prepared by alkaline hydrolysis of the azlactone of α-benzoylamino-*p*-acetoxycinnamic acid [7] and by a two-step hydrolysis of the azlactone of α-acetamino-*p*-acetoxycinnamic acid.[8] *p*-Hydroxyphenylpyruvic acid has also been prepared by alkaline hydrolysis of 5-(*p*-hydroxybenzal)-3-phenylhydantoin.[9] The procedure described here is adapted from published directions for the preparation of *p*-hydroxyphenylpyruvic-3-C^{14} acid.[5] 5-(*p*-Hydroxybenzal)hydantoin is prepared according to the method of Boyd and Robson.[10]

4. Merits of the Preparation

p-Hydroxyphenylpyruvic acid plays an important role in the biogenesis of compounds with a phenylpropane skeleton, and it has been used as substrate in several enzyme studies. Published procedures for its preparation are unsatisfactory in many ways. The alkaline hydrolysis of the azlactone of α-benzoylamino-*p*-acetoxycinnamic acid [7] makes necessary a tedious separation of the resulting benzoic acid, and the yield is only 34% based on *p*-hydroxybenzaldehyde. The hydrolysis of 5-(*p*-hydroxybenzal)-3-phenylhydantoin [9] requires a separation of phenylurea. Finally, the two-step cleavage of the azlactone of α-acetamino-*p*-acetoxycinnamic acid [8] does not proceed easily, and impure products are obtained. In applying this procedure to the synthesis of a carboxyl-labeled *p*-hydroxyphenylpyruvic acid, the overall yield was only 9%.[11] It must be kept in mind that any prolonged isolation procedure will cause some decomposition of this sensitive compound.

The method of preparing *p*-hydroxyphenylpyruvic acid presented here has the advantage that only volatile by-products, ammonia and carbon dioxide, are formed. Therefore the compound can be obtained in high purity and good yield. There are no difficulties in decreasing the amounts of starting materials to the millimole scale, as shown by the application of this procedure to the preparation of labeled p-hydroxyphenylpyruvic acid.[5]

Finally, this method is of general utility, for alkaline cleavage of analogously substituted hydantoins has given a series of substituted phenylpyruvic acids.[12]

1. Organisch-Chemisches Institut der Universität Wien, Austria.
2. E. C. Wagner and J. K. Simons, *J. Chem. Educ.*, 13, 265 (1936).
3. C. Harries and M. Weiss, *Ber.*, 33, 3418 (1900).
4. N. Weiner, *Org. Syntheses,* Coll. Vol. 2, 376 (1943).
5. G. Billek and E. F. Herrmann, *Monatsh. Chem.*, 90, 89 (1959).
6. O. Folin and W. Denis, *J. Biol. Chem.*, 12, 239 (1912).
7. O. Neubauer and K. Fromherz, *Z. Physiol. Chem.*, 70, 339 (1911).
8. J. A. Saul and V. M. Trikojus, *Biochem. J.*, 42, 80 (1948).
9. M. Bergmann and D. Delis, *Ann.*, 458, 76 (1927).
10. W. J. Boyd and W. Robson, *Biochem. J.*, 29, 542 (1935).
11. S. N. Acerbo, W. J. Schubert, and F. F. Nord, *J. Am. Chem. Soc.*, 80, 1990 (1958).
12. G. Billek, *Monatsh. Chem.*, 92, 343, 352 (1961).

3-HYDROXYPYRENE

(1-Pyrenol)

$$+ \text{ NaOH} \xrightarrow[\text{Cu, Cu}_2\text{O}]{\text{H}_2\text{O, }\Delta} + \text{ NaBr}$$

Submitted by W. H. Gumprecht[1]
Checked by William G. Dauben, John R. Wiseman, and Michael H. McGann

1. Procedure

Caution! Many pyrene derivatives are carcinogens. Contact of the skin with these materials should be avoided.

In a 100-ml. pressure vessel (Note 1) are placed 14.0 g. (0.050 mole) of 3-bromopyrene (Note 2), 0.5 g. of copper bronze powder (Note 3), 1.5 g. of cuprous oxide (Note 4), and 60 ml. of 10% aqueous sodium hydroxide (Note 5). The vessel is sealed, heated rapidly with shaking to 275–280°, and maintained at this temperature for 3 hours. The vessel is allowed to cool and is opened, and the contents are poured into a 500-ml. beaker containing 200 ml. of water (Note 6). The mixture is filtered, and the filter cake is washed with water until the washings become neutral to pH paper. The combined filtrate and washings, which show a blue fluorescence, are made acid to Congo red paper with 20% aqueous sulfuric acid. The precipitate is collected by filtration, washed free of acid with water, and dried in an oven at 100° to give *ca.* 9 g. of a gray solid.

The solid is boiled under reflux with *ca.* 175 ml. of benzene for 1 hour, and the benzene-insoluble material is removed from the hot mixture by filtration through a medium-grade sintered-glass funnel (Note 7). Small portions of activated alumina are added to the hot filtrate until a strong green fluorescence develops; 20–25 g. is needed (Note 8). The mixture is boiled (Note 9) under reflux for 15 minutes and filtered hot. On concentration and cooling, the fluorescent, lemon-yellow filtrate yields 3-hydroxy-pyrene as light yellow needles, m.p 179–181°. An additional quantity of equally pure product is obtained by further concentration of the mother liquor. The total yield is 5.5–6.0 g. (50–55%) (Note 10).

2. Notes

1. Stainless steel and Hastelloy-C vessels were used with equivalent results.

2. This material, m.p. 93–95°, is readily prepared in high yields (78–86%) from commercial pyrene by the method of the submitter,[2] who used pyrene, m.p. 151–153°, obtained from Chemicals Division, Union Carbide Corp.

3. Copper bronze, type 3310, obtained from U. S. Bronze Powder Works, Inc., Flemington, New Jersey, was used. The use of some grades of copper powder leads to a considerably lower yield.

4. Technical grade cuprous oxide obtained from Baker and Adamson Products, General Chemical Division, Allied Chemical Corp., was used.

5. The amount of sodium hydroxide used does not affect the yield, provided that it is present in a quantity well in excess of that required by the stoichiometry.

6. The sodium salt of 3-hydroxypyrene is somewhat insoluble in the reaction mixture. Dilution before filtration ensures its removal from the copper residues.

7. The funnel should be preheated to prevent crystallization of the product in its pores. The filtration process can be accelerated by scraping the muddy cake from the funnel surface.

8. The use of a larger quantity reduces the yield with no significant improvement in quality.

9. The presence of the alumina causes the mixture to bump violently. Agitation of the boiling mixture with a magnetically driven stirring bar helps to alleviate this problem. The vessel should be securely clamped.

10. The submitter obtained 25–29 g. (57–67%) of product when the reaction was run with 56.2 g. (0.20 mole) of the bromide in a 400-ml. pressure vessel.

3. Methods of Preparation

This procedure is based on the method of Smith, Opie, Wawzonek, and Prichard [3] for the preparation of 2,3,6-trimethylphenol. 3-Hydroxypyrene has been prepared by fusion of pyrene-3-sulfonic acid with sodium hydroxide [4] and by desulfonation of 3-hydroxypyrene-5,8,10-trisulfonic acid with hot, dilute sulfuric acid.[5]

4. Merits of the Preparation

It has been shown [6] that two mechanisms, elimination-addition (benzyne) and S_N2 displacement, are operative in the liquid-phase hydrolysis of halogenated aromatic compounds. The formation of isomeric phenols as a result of the availability of the benzyne route makes the reaction of limited synthetic value. The incorporation of the copper-cuprous oxide system suppresses reaction via the benzyne route, so that the present method has general utility for the preparation of isomer-free phenols. For example, p-cresol is the only cresol formed from p-bromotoluene under the conditions of this preparation.

The methods previously reported for the preparation of 3-hydroxypyrene have been found to be unsatisfactory, because of both very poor yields and difficulties in operation.

1. Contribution No. 300 from the Organic Chemicals Department, E. I. du Pont de Nemours and Company, Wilmington, Delaware 19899.
2. W. H. Gumprecht, this volume. p. 147.
3. L. I. Smith, J. W. Opie, S. Wawzonek, and W. W. Prichard, *J. Org. Chem.*, **4**, 318 (1939).
4. H. Vollmann, H. Becker, M. Corell, and H. Streeck, *Ann.*, **531**, 1 (1937); W. Kern, U. S. Patent 2,018,792 (1935) [*C.A.*, **30**, 112 (1936)].
5. E. Tietze and O. Bayer, *Ann.*, **540**, 189 (1939).
6. A. T. Bottini and J. D. Roberts, *J. Am. Chem. Soc.*, **79**, 1458 (1957).

3-HYDROXYQUINOLINE

(3-Quinolinol)

A. $CH_3COCO_2H + SO_2Cl_2 \rightarrow ClCH_2COCO_2H + SO_2 + HCl$

B.

C.

Submitted by EDWARD J. CRAGOE, JR., and CHARLES M. ROBB.[1]
Checked by JAMES CASON and JAMES D. WILLETT.

1. Procedure

Caution! The preparation of chloropyruvic acid should be carried out in a fume hood, as should the purification of 3-hydroxycinchoninic acid and its decarboxylation.

A. *Chloropyruvic acid*. In a 1-l. four-necked flask (Note 1) fitted with a sealed mechanical stirrer, dropping funnel, thermometer, and reflux condenser protected with a calcium chloride tube is placed 249 g. (2.83 moles) of pyruvic acid (Note 2). The stirrer is started and 394 g. (2.92 moles) of sulfuryl chloride (Note 3) is added dropwise over a period of 2 hours. During the addition the temperature is maintained at 25–30° by cooling with a water bath.

The mixture is stirred at room temperature for an additional 60 hours (Note 4), during which time the calcium chloride tube may become spent and need replacement. The viscous, light-yellow liquid product is transferred to a large crystallizing dish and dried in a vacuum desiccator over soda-lime for about 24 hours (Note 5).

The yield of light-yellow chloropyruvic acid is 333–340 g. (96–98%) (Note 6).

B. *3-Hydroxycinchoninic acid*. A 3-l., four-necked flask (Note 1) is equipped with a sealed mechanical stirrer, gas inlet tube, gas outlet consisting of a 1-mm. capillary (Note 7), and thermometer. The flask is charged with a freshly prepared solution containing 448 g. (8 moles) of reagent grade (85% minimum assay) potassium hydroxide and 900 ml. of water. The solution (hot from dissolution of potassium hydroxide) is stirred and 147 g. (1 mole) of isatin (Note 8) is introduced. The solid quickly dissolves to give an orange-yellow solution.

After replacement of the gas outlet the flask is flushed with nitrogen, and a nitrogen atmosphere is maintained during the remaining operations (Note 9). The temperature is maintained at 20–25° by cooling when necessary. The solution is stirred vigorously as 168.5 g. (1.375 moles) of chloropyruvic acid (part A) is added gradually over a period of 2 hours (Note 10). After stirring has been continued for an additional hour, the introduction of nitrogen and the stirring are terminated, the flask is stoppered, and the mixture is allowed to stand at room temperature for 6 days.

At the end of the standing period the reaction mixture is cooled with stirring and maintained at 15–18°. A solution containing 34 g. of sodium bisulfite in 60 ml. of water (Note 11) is added, and the mixture is made acid to Congo red paper by the dropwise addition of reagent grade concentrated hydrochloric acid (Note 12) (approximately 480 ml.). The yellow product that precipitates is separated by suction filtration on a large (at least 15-cm.) Büchner (or sintered-glass) funnel and washed with water saturated with sulfur dioxide. Drainage of the filter cake, whose consistency is that of putty, is slow. Pressing with a large cork or use of a rubber dam is helpful. The pressed solid is suspended in 1.3 l. of water previously saturated with sulfur dioxide, and the mixture is mechanically stirred for 30 minutes.

After the product has been collected as before, the filter cake is pressed well, suspended in 800 ml. of water, and dissolved by stirring and adding the minimum quantity of reagent grade concentrated aqueous ammonia (approximately 60 ml.). A small amount of insoluble material is removed by filtration. A saturated solution of 8 g. of sodium bisulfite is added to the filtrate. The orange-yellow solution is stirred mechanically and made acid to Congo red paper by the dropwise addition of reagent grade concentrated hydrochloric acid (approximately 80 ml.).

The product is again collected by filtration, washed with water, resuspended in 225 ml. of water, collected, and pressed as dry as possible. The filter cake is thoroughly dispersed in 160 ml. of absolute alcohol, then filtered, air-dried, and finally dried in a vacuum desiccator over concentrated sulfuric acid. The bright-yellow solid is pulverized and redried. The yield is 115–135 g. (60–71%). When this product is inserted in a bath preheated to 210° and the temperature is increased at a rate of 1° per 10 seconds, decomposition with evolution of gas occurs at 219–220° (cor.) (Note 13).

C. *3-Hydroxyquinoline*. A 1-l. beaker is fitted with a thermometer and mechanical stirrer and clamped firmly on an efficient electric heater (Note 14). Diethyl succinate (400 ml.) (Note 15) is placed in the beaker and heated to boiling (215–220°) with stirring. 3-Hydroxycinchoninic acid (part B) (94.6 g., 0.5 mole) is added in portions to the boiling solution by means of a metal spoon or Scoopula. Care is taken to prevent too vigorous evolution of carbon dioxide. The addition requires 2–3 minutes, during which time a temperature drop is noted unless good heating is maintained.

Stirring and boiling are continued until complete solution is effected. This requires about 6 minutes (Note 16). The stirrer is withdrawn and the beaker is removed from the hot plate for a few minutes. Finally, the solution is stirred and cooled first in a warm water bath and then in an ice bath. After 30 minutes the gray-brown solid is collected by suction filtration and washed with hexane. The product is suspended in 250 ml. of hexane, filtered, and washed with hexane. After drying, the crude gray-colored 3-hydroxyquinoline weighs 57–63 g. (79–87%), m.p. 175–191° (cor.).

The crude product is suspended in 190 ml. of water and dissolved by the addition of the minimum quantity (31–35 ml.) of concentrated hydrochloric acid. The solution is filtered in order to remove a small amount of insoluble material. The filtrate is treated with decolorizing carbon (about 3.5 g.), allowed to stand for 30 minutes, and filtered. The filtrate from the charcoal is stirred and treated dropwise with concentrated aqueous ammonia (25–29 ml.) until precipitation is complete. The precipitate is removed by filtration, washed with water (two 30-ml. portions), and dried. The yield at this stage is 48–55 g. of material melting at 185–195° (cor.).

The reprecipitated product is pulverized, dissolved in a boiling mixture of methanol (about 420 ml.) and water (about 360 ml.), and treated with decolorizing charcoal for about 10 minutes. The boiling mixture is filtered through a fluted filter paper placed on a large Pyrex funnel resting on a wide-mouthed Erlenmeyer flask containing a little boiling solvent of the same composition. The filtrate is concentrated to incipient precipitation (about 600 ml. volume) and allowed to cool. After drying, the yield of tan-colored crystalline 3-hydroxyquinoline is 44–47 g. (61–65%), m.p. 199–200° (cor.) (Note 17).

2. Notes

1. If suitable dual outlets are used, a three-necked flask is satisfactory.

2. Pyruvic acid from Matheson, Coleman and Bell was distilled just before use. Material boiling at 46–47° at 4 mm. was employed.

3. Technical grade sulfuryl chloride from Matheson, Coleman

and Bell was found satisfactory.

4. The stirring prevents foaming and promotes the evolution of the gases.

5. It is advisable to change the desiccant at least once during the drying period.

6. The chloropyruvic acid prepared in this manner is satisfactory for use in the next reaction without purification. It often crystallizes to form a waxy solid or semisolid which is quite hygroscopic. The pure anhydrous material is reported [2] to melt at 45°, while the monohydrate obtained by other methods [3-5] melts at 57–58°. The chloropyruvic acid is normally used immediately, but it has been stored successfully in a desiccator at room temperature for a few days or for longer periods in an airtight container in the refrigerator. Material which has been stored for long periods gives poorer yields in the Pfitzinger reaction (part B) than that which has been freshly prepared.

7. When solids are added to the flask, the gas outlet is replaced by a powder funnel and nitrogen flow is increased slightly. If a separatory funnel is used (cf. Note 10), a dual outlet is needed.

8. Commercial isatin from Eastman Kodak or Matheson, Coleman and Bell has been used, but poorer yields are obtained (about 10% less) than when purified material is employed. Purification by reprecipitation [6] or by recrystallization from glacial acetic acid [6] is equally satisfactory.

9. Maintaining an atmosphere of nitrogen minimizes the darkening of the reaction mixture due to air oxidation.

10. If the chloropyruvic acid remains essentially as a viscous liquid, it may be introduced via a dropping funnel containing a large-bore stopcock. If the material has set up to a waxy solid, it must be introduced in portions through a powder funnel.

11. If no precautions are observed, the reaction mixture rapidly darkens after acidification when exposed to air. The sulfur dioxide generated upon acidification of the sodium bisulfite largely prevents this discoloration; however, the precipitated product should be collected *without delay* of more than a few hours. The sulfur dioxide used in the wash water also protects the product.

12. At about the midpoint in the addition of acid, frothing tends to raise the precipitate out of the flask. Addition of a 1-ml. portion of ether controls the frothing. A second portion of ether

may be required later, but the frothing subsides as the addition proceeds.

13. 3-Hydroxycinchoninic acid of this purity is adequate for decarboxylation. A sample recrystallized from dimethylformamide or 5N hydrochloric acid decomposes at 224° when observed as described before.

14. It is advantageous to use a mechanical stirrer, but successful reactions have been carried out using manual stirring. A run in which no stirring was employed gave acceptable results.

15. Commercial diethyl succinate from Carbide and Carbon Chemicals Co. or from Eastman Organic Chemicals was found satisfactory. Nitrobenzene has been used successfully a number of times; however, it is considered a less desirable solvent to handle.

16. The heating time is kept to a minimum in order to reduce the darkening of the solution, which increases as the heating time is extended.

17. Once recrystallized, 3-hydroxyquinoline is pure enough for most purposes. One or two more recrystallizations are required to produce white crystals, m.p. 201–202° (cor.).

3. Methods of Preparation

The method of synthesis described for chloropyruvic acid is essentially that reported.[2] This procedure affords the product in excellent yields from readily available materials by a short, convenient route. Other less acceptable methods involve chlorination of pyruvic acid with sulfur dichloride [7] or hypochlorous acid [8] and the treatment of ethyl chloro(1-hydroxyheptyl)- or (α-hydroxybenzyl)oxalacetate γ-lactone with 50% hydrochloric acid.[3-5]

The procedure described for the preparation of 3-hydroxycinchoninic acid is adapted from that reported.[9] This synthesis is successful when bromopyruvic acid or its ethyl ester is substituted for chloropyruvic acid.[9] The reaction of isatin with chloropyruvic acid to produce 3-hydroxycinchoninic acid has been reported; [10] however, no details or physical properties were given. This method offers a decided advantage over the method involving diazotization of the difficultly accessible 3-aminocinchoninic acid.[11]

Until recent years the only syntheses of 3-hydroxyquinoline involved multistep processes, the last step of which consisted of the conversion of 3-aminoquinoline to 3-hydroxyquinoline via the diazonium salt.[12-14] Small quantities of quinoline have been oxidized to 3-hydroxyquinoline in low yields by using oxygen in the presence of ascorbic acid, ethylenediaminetetraacetic acid, ferrous sulfate, and phosphate buffer.[15] The decarboxylation of 3-hydroxycinchoninic acid in boiling nitrobenzene has been reported.[9, 11] The procedure described involves a simplified modification of this method.

1. Merck Sharp and Dohme Research Laboratories, West Point, Pennsylvania.
2. M. Garino and I. Muzio, *Gazz. Chim. Ital.*, 52 II, 226 (1922).
3. H. Gault, J. Suprin, and R. Ritter, *Compt. Rend.*, 226, 2079 (1948).
4. H. Gault, J. Suprin, and R. Ritter, *Compt. Rend.*, 230, 1408 (1950).
5. R. Ritter, *Ann. Chim.*, 6 (12), 247 (1951).
6. C. S. Marvel and G. S. Hiers, *Org. Syntheses*, Coll. Vol. 1, 327 (1951).
7. J. Parrod and M. Rahier, *Bull. Soc. Chim. France*, 1947, 109.
8. E. A. Shilov and A. A. Yasnikov, *Ukrain. Khim. Zhur.*, 18, 611 (1952).
9. E. J. Cragoe, Jr., C. M. Robb, and M. D. Bealor, *J. Org. Chem.*, 18, 552 (1953).
10. Kracker, Luce, and Fitzky, Office of the Publication Board, Department of Commerce, P. B. Report 58,847, frames 782-786 (July 25, 1947).
11. K. C. Blanchard, E. A. Dearborn, and E. K. Marshall, Jr., *Bull. Johns Hopkins Hosp.*, 88, 181 (1951).
12. W. H. Mills and W. H. Watson, *J. Chem. Soc.*, 97, 753 (1910).
13. G. Bargellini and M. Settimj, *Gazz. Chim. Ital.*, 53, 601 (1923).
14. R. Kuhn and O. Westphal, *Ber.*, 73B, 1105 (1940).
15. B. B. Brodie, J. Axelrod, P. A. Shore, and S. Udenfriend, *J. Biol. Chem.*, 208, 741 (1954).

2-HYDROXYTHIOPHENE

[Thiophene-2-ol and 2(5H)-thiophenone]

Submitted by C. Frisell and S.-O. Lawesson.[1]
Checked by R. M. Scribner, C. G. McKay and B. C. McKusick.

1. Procedure

A. *2-t-Butoxythiophene.* A dry 1-l. three-necked flask is fitted with a mechanical stirrer (Note 1), a reflux condenser having a take-off attachment, and a 250-ml. dropping funnel with a pressure-equalizing side tube.[2] A nitrogen-inlet tube is connected to the top of the condenser, and a T-tube branch of this is led to a mercury valve. The latter consists of a U-tube the bend of which is just filled with mercury.

Ten grams (0.41 g. atom) of magnesium turnings (Note 2) is placed in the flask and covered with 200 ml. of dry ether. Ten milliliters of a solution of 65.2 g. (0.40 mole) of 2-bromothiophene in 60 ml. of dry ether is added, the Grignard reaction is started by gently warming the reaction flask, and the remainder of the solution is added dropwise during 45 minutes. The mixture is stirred for 3.5 hours, the last 15 minutes under reflux. The Grignard reagent is cooled to 0–5° by immersing the flask in ice water. t-Butyl perbenzoate (62 g., 56 ml., 0.32 mole) (Note 3) in 100 ml. of dry ether is added dropwise during 45 minutes to the stirred ice-cooled mixture. The reaction mixture is stirred overnight, poured into ice water, and acidified with concentrated hydrochloric acid. The two phases are separated, and the water phase is twice extracted with ether. The combined ether solutions are extracted with three 60-ml. portions of $2N$ sodium hydroxide solution (Note 4), washed until neutral with water, dried over anhydrous sodium sulfate, and transferred to a distillation flask equipped with a Vigreux column. *Caution! The ether extract should not be distilled unless a test shows that peroxides are absent (Note 5).* The ether is distilled off at atmospheric pressure, and the residual oil is distilled under reduced pressure to give 35–38 g. (70–76%) of 2-t-butoxythiophene, b.p. 64–66°/13 mm., n_D^{20} 1.4991.

B. *2-Hydroxythiophene.* The 2-t-butoxythiophene obtained in Step A is placed in a distillation flask equipped with a short Vigreux column and a capillary inlet for nitrogen, and 0.1 g. of p-toluenesulfonic acid is added. The apparatus is placed in an oil bath at 155°. Decomposition begins immediately. After 5–10 minutes the oil bath is removed, and the distillation assembly is connected to a water pump (Note 6). 2-Hydroxythiophene is distilled under reduced pressure, nitrogen gas being drawn through the capillary during the whole procedure (Note 7). 2-Hydroxythiophene is collected at 91–93°/13 mm.; yield 20–23 g. (89–94%); n_D^{20} 1.5613 (Note 8).

2. Notes

1. Although a mercury seal is preferable, a rubber tube lubricated with glycerol is an adequate seal.

2. Common laboratory magnesium is as satisfactory as extremely pure sublimed magnesium.

3. t-Butyl perbenzoate is supplied by Lucidol Division, Wallace and Tiernan, Inc., Buffalo, New York, and Light and Co., Colnbrook, Bucks, England.

4. Acidification of the basic solution gives 29–32 g. (80–88%) of benzoic acid if the reaction has proceeded properly.

5. To make a peroxide test, place a few milligrams of sodium iodide, a trace of ferric chloride, and 2–3 ml. of glacial acetic acid in a test tube and carefully add 1–2 ml. of the ether solution. When unconsumed perbenzoate is present, a yellow ring is immediately formed between the two phases. *If this test indicates the presence of peroxide, the extract should not be concentrated and distilled until it has been extracted first with a solution of potassium iodide in acetic acid to remove peroxide and then with aqueous sodium thiosulfate to remove iodine.*

6. The decomposition is considered to be complete when the pressure is constant.

7. 2-Hydroxythiophene resinifies on prolonged exposure to air.

8. It has recently been shown that 2-hydroxythiophene exists mainly as 2(5H)-thiophenone at room temperature.[3]

3. Methods of Preparation

The procedure described is essentially that of Lawesson and Frisell.[4] 2-Hydroxythiophene has been prepared in low yields by Hurd and Kreuz [5] from 2-thienylmagnesium bromide and oxygen in the presence of excess isopropylmagnesium bromide.

4. Merits of the Preparation

The first step of the procedure illustrates a general way of preparing aryl *t*-butyl ethers.[4,6] The second step is the best way to prepare 2-hydroxythiophene, inasmuch as the yield is good and *t*-butyl perbenzoate is a readily available perester that is relatively stable. The same procedure has been used to convert several other haloaromatic compounds to hydroxy-aromatic compounds in good yield[4] and is probably quite general.

1. Department of Chemistry, University of Aarhus, 8000 Aarhus C, Denmark.
2. K. B. Wiberg, "Laboratory Technique in Organic Chemistry," McGraw-Hill Book Company, Inc., New York, 1960, p. 207.
3. S. Gronowitz and R. A. Hoffman, *Arkiv Kemi,* 15, 499 (1960).
4. S.-O. Lawesson and C. Frisell, *Arkiv Kemi,* 17, 393 (1961).
5. C. D. Hurd and K. L. Kreuz, *J. Am. Chem. Soc.,* 72, 5543 (1950).
6. C. Frisell and S.-O. Lawesson, this volume, p. 549.

HYDROXYUREA*

(Urea, hydroxy-)

$$NH_2CO_2C_2H_5 + NH_2OH \cdot HCl \xrightarrow{NaOH}$$
$$NH_2CONHOH + NaCl + C_2H_5OH$$

Submitted by R. DEGHENGHI.[1]
Checked by MELVIN S. NEWMAN and JOHN EBERWEIN.

1. Procedure

To a solution of 20.8 g. (0.3 mole) of hydroxylamine hydrochloride and 20.6 g. (0.5 mole) of sodium hydroxide (98%) in 100 ml. of water is added 22.26 g. (0.25 mole) of ethyl carbamate. After 3 days at room temperature the solution is cooled in an ice bath and carefully neutralized with concentrated hydrochloric acid (Note 1). If necessary (Note 2), the solution is filtered and then extracted with ether; the aqueous phase is evaporated on a water bath under reduced pressure as rapidly as possible at a temperature not above 50–60°.

The dry residue is extracted by boiling with 100 ml. of absolute ethanol, and the solution is filtered through a heated funnel. On cooling, a first crop (6–8 g.) of hydroxyurea crystallizes.

The saline residue on the filter is extracted once again with 50 ml. of boiling absolute ethanol. On concentrating the filtrate from the second extraction and the mother liquor from the first crystallizate to a small volume, a second crop (4–6 g.) of product is obtained. The yield of the hydroxyurea is 10–14 g. (53–73%) of white crystals, m.p. 137–141° (dec.).

The product may be purified by recrystallization of 10 g. from 150 ml. of absolute ethanol. The rate of solution is slow (15–30 minutes is required), and the yield of hydroxyurea, m.p. 139–141° (dec.), is about 8 g. (Note 3).

2. Notes

1. Neutralization to phenolphthalein is satisfactory, but a glass electrode might give better results. Hydroxyurea is decomposed very rapidly in aqueous acidic medium, whereas its metallic salts (sodium or the copper complex salts) are stable.

2. Insoluble matter is sometimes present if a commercial grade of reactants is employed.

3. It is preferable to store the crystals in a cool, dry place. Some decomposition may occur after a few weeks.

3. Methods of Preparation

The improved method herein described is adapted from the procedure of Runti and Deghenghi.[2] Hydroxyurea has been prepared from potassium cyanate and hydroxylamine hydrochloride.[3-5] A lower melting isomeric substance, m.p. 71°, has been described.[4-6] The structure $NH_2CO_2NH_2$ has been proposed [5,6] for this low-melting substance.

1. Department of Medical Research, University of Western Ontario, Ontario, Canada.
2. C. Runti and R. Deghenghi, *Ann. Triest. Cura Univ. Trieste, sez.* 2, 22, 185 (1953); *C.A.,* 49, 1568 (1955).
3. W. F. C. Fresler and R. Stein, *Ann.,* 150, 242 (1869).
4. L. Francesconi and A. Parrozzani, *Gazz. Chim. Ital,* 31 II, 334 (1901).
5. H. Kofod, *Acta Chem. Scand.,* 7, 274, 938 (1953).
6. O. Exner, *Chem. Lisly,* 50, 2025 (1956).

2-INDANONE*

Submitted by J. E. Horan and R. W. Schiessler.[1]
Checked by William E. Parham, Wayland E. Noland,
and Abdel-Moneim M. Makky.

1. Procedure

In a 2-l. three-necked flask fitted with stirrer, dropping funnel, and thermometer are placed 700 ml. of formic acid (88%) and 140 ml. of hydrogen peroxide (30%, 1.37 moles). While the temperature is kept at 35–40° (Note 1), 116.2 g. (117.3 ml., 1.00 mole) of indene (98%) (Note 2) is added dropwise, with stirring, over a period of 2 hours. An additional 100 ml. of formic acid is used to rinse the last of the indene from the dropping funnel into the reaction flask. The reaction solution is stirred at room temperature for 7 hours to ensure complete reaction (Note 3). The solution is transferred to a 2- or 3-l. Claisen flask, and the formic acid is removed under aspirator pressure (b.p. 35–40°/20–30 mm.), care being taken to maintain the boiler temperature below 60° (Note 4). The residue, after being cooled to room temperature, is a yellowish brown crystalline solid (Note 5), the color being due to contamination by a small amount of brownish oil.

In a 5-l. flask fitted with a long condenser (about 40 cm.) connected to an ice-cooled receiver is placed 2 l. of 7% (by volume) sulfuric acid. The solution is heated to boiling, and the crude monoformate of 1,2-indanediol is added. Steam is introduced and the mixture is steam distilled, while external heat is applied with a flame in order to maintain the boiler contents at a constant volume of 2 l. The steam distillation is carried out at the rate of about 1 l. per hour until 5–6 l. of distillate have been collected and the 2-indanone has stopped distilling (Note 6). The dark-brown oily residue becomes semisolid at room temperature.

The cold distillate is filtered with suction, and the white crystalline solid is sucked thoroughly dry on the filter (Note 7). The crystals are dried further in a vacuum desiccator (at about 1 mm.) at room temperature or below for about 12 hours. The melting point of the 2-indanone is 57–58° (Note 8). The yield is 90–107 g. (69–81%).

2. Notes

1. This is the best temperature at which to control the reaction. At higher temperatures, the reaction becomes too vigorous. The stirrer must be sufficiently powerful to thoroughly mix the phases or a very low yield will result. A Hershberg stirrer has been found to be very effective in this preparation (private communication from H. E. Baumgarten).

2. The indene ($n_D^{25.5}$ 1.5698, b.p. 74–76°/24 mm.) was obtained from Rütgerswerke, A. G., West Germany, through Terra Chemicals, Inc., 500 Fifth Avenue, New York 36, N. Y. It was faintly yellow, but was used without distillation, since distillation, although it removed the yellow color, did not change the refractive index. When Matheson, Coleman and Bell technical grade indene was used (redistilled, $n_D^{25.5}$ 1.5606, b.p. 177–179°), a 45% yield of 2-indanone was obtained.

3. The reaction mixture can be left overnight at this point with no adverse effect upon yield.

4. A higher temperature should be avoided at the start to prevent boilover, and later to reduce side reactions and eliminate danger from possible residual peroxides.

5. The reaction can be interrupted at this point and the formate ester stored for several weeks, if desired.

6. The rate of flow of cooling water through the condenser must be regulated so that the condenser does not become clogged

with 2-indanone. The end point of the distillation can be recognized by lack of turbidity in the condensate or of solidification in the condenser when cold water is passed rapidly through the condenser.

7. The steam distillate, with the 2-indanone under water, can be kept for as long as a week in the refrigerator. In the dry state, 2-indanone is unstable to air at room temperature but can be kept in a closed vessel for several days at room temperature and for longer periods (several weeks or more) in a refrigerator.

8. The ultraviolet spectrum in the 300–350 mμ region showed that less than 1% of 1-indanone was present. If the 2-indanone darkens on standing, it can be repurified by steam distillation or by crystallization from ethanol.

3. Methods of Preparation

2-Indanone was first prepared by distillation of the calcium salt of o-phenylenediacetic acid [2,3] and, more recently, by the action of acetic anhydride on its potassium salt.[4] It has been obtained by the dilute sulfuric acid-catalyzed hydrolysis and decarboxylation of 2-iminoindan-1-carboxylate [5] and ethyl 2-indanone-1-carboxylate [6] 2-Indanone is commonly obtained by acid-catalyzed dehydration of an indene glycol, [7,8] as illustrated in this preparation. Indene glycol has been obtained from indene via the bromohydrin.[9-12] The most recent preparation of 2-indanone is by Curtius degradation of 2-indenecarboxylic acid.[13]

1. Pennsylvania State University, University Park, Pa.
2. P. Schad, *Ber.*, **26**, 222 (1893).
3. J. Wislicenus and H. Benedikt, *Ann.*, **275**, 351 (1893).
4. H. Waldmann and P. Pitschak, *Ann.*, **527**, 183 (1937).
5. C. W. Moore and J. F. Thorpe, *J. Chem. Soc.*, 186 (1908).
6. W. H. Perkin, Jr., and A. F. Titley, *J. Chem. Soc.*, 1562 (1922).
7. F. Heusler and H. Schieffer, *Ber.*, **32**, 28 (1899).
8. M. Mousseron, R. Jacquier, and H. Christol, *Compt. Rend.*, **236**, 927 (1953).
9. J. Read and E. Hirst, *J. Chem. Soc.*, 2550 (1922).
10. L. S. Walters, *J. Soc. Chem. Ind. (London)*, **46**, 150 (1927).
11. H. D. Porter and C. M. Suter, *J. Am. Chem. Soc.*, **57**, 2022 (1935).
12. P. Pfeiffer and T. Hesse, *J. Prakt. Chem.*, **158**, 315 (1941).
13. B. P. Sen, A. Chatterjee, S. K. Gupta, and B. K. Bhattacharyya, *J. Indian Chem. Soc.*, **35**, 751 (1958).

INDAZOLE *

(Benzopyrazole)

$o\text{-}CH_3C_6H_4NH_2 + (CH_3CO)_2O \rightarrow$

$$o\text{-}CH_3C_6H_4NHCOCH_3 + CH_3CO_2H$$

$2\ o\text{-}CH_3C_6H_4NHCOCH_3 + N_2O_3 \rightarrow$

$$2\ o\text{-}CH_3C_6H_4N\underset{\diagdown COCH_3}{\overset{\diagup NO}{}} + H_2O$$

Submitted by ROLF HUISGEN and KLAUS BAST.[1]
Checked by W. E. PARHAM, WAYLAND E. NOLAND, and
JOHN W. DRENCKPOHL.

1. Procedure

Ninety grams (90.2 ml., 0.839 mole) of o-toluidine is slowly added to a mixture of 90 ml. of glacial acetic acid and 180 ml. (1.90 mole) of acetic anhydride contained in a 750-ml. two-necked flask equipped with a thermometer and a two-hole cork stopper for a gas inlet tube (Note 1). Acetylation occurs with evolution of heat. The mixture is cooled in an ice bath (Note 2) and nitrosated by rapid admission of a stream of nitrous gases

(Note 3). The nitrous gases are obtained by the action of nitric acid (density 1.47) on sodium nitrite (Note 4). Technical grade, large-grain sodium nitrite (180 g.) is placed in a 1-l. suction flask to which a dropping funnel is fitted by means of a rubber stopper (Note 5). The acid (a total of about 250 ml. is used) is added dropwise from the dropping funnel. The rate of the gas evolution should be such that the temperature of the reaction mixture is kept between $+1°$ and $+4°$ (Note 6). The gas is passed through a wash bottle (with inlet and outlet positions reversed) containing some glass wool and positioned between the generating flask and the reaction flask. After about 6 hours the nitrosation is complete, and the solution exhibits a permanent black-green color due to excess N_2O_3 (Note 7).

The solution of N-nitroso-o-acetotoluidide is poured onto a mixture of 400 g. of ice and 200 ml. of ice water in a beaker, covered loosely with a watch glass, and allowed to stand in an ice bath for 2 hours. The oil which separates is transferred to a separatory funnel and extracted by shaking with several portions of benzene (total volume 500 ml.). The combined extract is washed with three 100-ml. portions of ice water and, after shaking with 30 ml. of methanol to remove remaining acetic anhydride, is allowed to stand, lightly covered, in an ice bath for 1 hour. Next, the mixture is washed with three 100-ml. portions of ice water, and the cold benzene solution is allowed to stand, loosely covered, over calcium chloride in the refrigerator overnight (Note 8). The brown solution is decanted from the drying agent into a 3-l. Erlenmeyer flask, and the calcium chloride is washed (by decantation) with 800 ml. of benzene. The combined benzene layer and washings are warmed to 35° in a large water bath and maintained at this temperature for 1 hour (internal temperature, Notes 9 and 10), and then at 40–45° for 7 hours. These temperatures must be strictly adhered to; otherwise overheating can occur (Note 11).

After the completion of the decomposition, the solution is boiled for a short time by heating on a steam bath. The cooled solution is transferred to a separatory funnel and extracted with 200 ml. of 2N hydrochloric acid and then with three 50-ml. portions of 5N hydrochloric acid. The combined acid extracts are treated with excess ammonia, at which point the indazole precipitates. The mixture is kept in the refrigerator for 2 hours, and the solid is then collected on a Büchner funnel, washed with water, placed in a beaker covered with a piece of paper, and dried overnight at 100–105°. The yield of crude, light brown indazole, m.p. 144–147°, is 36–46 g. (36–47%) (Note 12). For purification, vacuum distillation in a Claisen flask, modified for distillation of solids, is suitable. This gives 33–43 g. of colorless indazole (b.p. 167–176°/40–50 mm.) with a melting point of 148° (Note 13).

2. Notes

1. The inlet tube should not be too narrow and should dip far enough into the reaction mixture to permit agitation of the reaction mixture by the gas stream.

2. Crystallization of o-acetotoluidide sometimes occurs at this stage and is allowed to go to completion. This is indicated by a decrease in the evolution of heat of crystallization. Addition of nitrous acid is not begun until the reaction mixture has reached 3°.

3. No harm is done by the separation of o-acetotoluidide, which sometimes occurs at this point.

4. The acid (density 1.47) is obtained by diluting 200 ml. of fuming nitric acid with 70 ml. of concentrated nitric acid.

The amount of N_2O_3 in the nitrous gases depends on the density of the nitric acid used.

Density 1.40 corresponds to 13 vol. % N_2O_3 or its equivalent.
Density 1.43 corresponds to 23 vol. % N_2O_3 or its equivalent.
Density 1.45 corresponds to 41 vol. % N_2O_3 or its equivalent.
Density 1.47 corresponds to 78 vol. % N_2O_3 or its equivalent.

5. The stopper is covered with a thin layer of paraffin. At the rubber tubing connections, the ends of the glass tubes should be in contact.

6. This is accomplished with an ice bath (without added salt).

7. Should the gas evolution become sluggish after 3–4 hours, the generator may be replaced by a new one containing one-half as much sodium nitrite.

8. It is necessary to begin the preparation early in the morning in order to bring it to this stage in one day.

9. If the water bath is heated by a hot plate controlled by a thermoregulator, it is only necessary to set the temperature at the thermoregulator.

10. At the beginning, the internal temperature is about 5–10° higher than the bath temperature because of the exothermic character of the indazole formation.

11. The use of a large bath simplifies dissipation of the heat of reaction.

12. The submitters regularly obtained yields as high as 55–61% of crude indazole and 52–58% of pure indazole.

13. This pressure range is used in order to give a boiling point sufficiently above the melting point of the indazole.

3. Methods and Merits of Preparation

The preceding method is that of Huisgen and Nakaten [2] and is based on the preparation of indazole (ca. 40% yield) from N-nitroso-o-benzotoluidide discovered by Jacobson and Huber.[3] Mechanistic studies [2] showed this reaction to be an intramolecular azo coupling with an initial acyl shift as the determining step. The yield of indazole from N-nitroso-o-benzotoluidide can be made almost quantitative.[2] Since the low solubility of o-benzotoluidide makes large quantities of acetic acid and acetic anhydride necessary, the method using the N-acetyl compound described here is more convenient.

With respect to time and cost, this method is superior to the five-step synthesis [4] from anthranilic acid. Older literature sources have been cited in an earlier volume of *Organic Syntheses*.[4] Indazole has also been prepared recently by the reaction of 2-hydroxymethylenecyclohexanone with hydrazine and dehydrogenation of the 5,6,7,8-tetrahydro derivative.[5]

1. Institut für Organische Chemie, Universität München, München 2, Germany.
2. R. Huisgen and H. Nakaten, *Ann.*, 586, 84 (1954).
3. P. Jacobson and L. Huber, *Ber.*, 41, 660 (1908).
4. E. F. M. Stephenson, *Org. Syntheses*, Coll. Vol. 3, 475 (1955).
5. C. Ainsworth, *Org. Syntheses*, Coll. Vol. 4, 536 (1963).

INDOLE-3-ACETIC ACID*

Submitted by HERBERT E. JOHNSON and DONALD G. CROSBY [1]
Checked by W. W. PRICHARD and B. C. McKUSICK

1. Procedure

A 3-l. stainless steel, rocking autoclave (Note 1) is charged with 270 g. (4.1 moles) of 85% potassium hydroxide and 351 g. (3.00 moles) of indole (Note 2), and then 360 g. (3.3 moles) of 70% aqueous glycolic acid is added gradually (Note 3). The autoclave is closed and rocked at 250° for about 18 hours (Note 4). The reaction mixture is cooled to below 50°, 500 ml. of water is added, and the autoclave is rocked at 100° for 30 minutes to dissolve the potassium indole-3-acetate. The aqueous solution is cooled to 25° and removed from the autoclave, the autoclave is rinsed out well with water, and water is added until the total volume of solution is 3 l. The solution is extracted with 500 ml. of ether (Note 5). The aqueous phase is acidified at 20–30° with 12N hydrochloric acid and then is cooled to 10° (Note 6). The indole-3-acetic acid that precipitates is collected on a Büchner funnel, washed with copious amounts of cold water, and dried in air or a vacuum desiccator out of direct light (Note 7); weight 455–490 g. (87–93%); m.p. 163–165° (dec.).

The indole-3-acetic acid, which is cream-colored, is of high purity. If further purification is desired, it may be done conveniently by recrystallization from water. One liter of water is used for 30 g. of acid, with 10 g. of decolorizing carbon added. Recovery is about 22 g. of a nearly colorless product, m.p. 164–166° (dec.).

2. Notes

1. A stirred autoclave is just as satisfactory. The scale is not critical, for the checkers got equally good results on one-third the scale; they used a 1-l. rocking autoclave.

2. Indole from the Union Carbide Olefins Company, Institute, West Virginia, is satisfactory.

3. If the reactants are added in this order, with the glycolic acid being introduced over a 5–10 minute period, there is no violent heating because the heat of neutralization is used to melt the indole. An equivalent amount of anhydrous glycolic acid may be used, but this offers no special advantage.

4. These limits are not critical, but they are probably optimum. Reaction times of 24–30 hours are not particularly detrimental, and high yields of product can be obtained within 12 hours. The temperature can range from 230° to 270° with but slight effect on the yield of product.

5. This extraction may be omitted. It does, however, remove traces of neutral material and consequently gives a product with greater color stability.

6. This operation is most conveniently conducted in a flask equipped with a stirrer.

7. The product dries slowly, and several days in air or 24 hours in a vacuum desiccator is usually required. Considerable coloration will result if this is done in direct light. Drying in a heated oven or removing the water as a benzene azeotrope is not satisfactory because of some decarboxylation to skatole. The product should be stored in a dark bottle away from direct sunlight.

3. Methods of Preparation

Indole-3-acetic acid has been prepared by the Fischer indole synthesis,[2] by hydrolysis of indoleacetonitrile,[3] from the reaction of gramine-type compounds with cyanide ion under conditions which hydrolyze the nitrile,[4] by the reaction of indole with ethyl diazoacetate followed by hydrolysis,[5] through oxidation of indole-pyruvic acid,[6] and by ultraviolet irradiation of tryptophan.[7]

4. Merits of the Preparation

This is the most convenient method of preparing indole-3-acetic acid if an agitated autoclave is available. The method can be used to prepare other indole-3-acetic acids from α-hydroxy acids. For example, α-methylindole-3-acetic acid has been prepared by condensing indole with lactic acid.

Indole-3-acetic acid is a natural plant auxin and is used as a control in research on plant growth.

1. Union Carbide Chemicals Company, South Charleston, West Virginia.
2. A. Ellinger, *Ber.*, **37**, 1801 (1904); Z. Tanaka, *J. Pharm. Soc. Japan*, **60**, 74, 219 (1940) [*C. A.*, **34**, 3735, 5446 (1940)]; S. W. Fox and M. W. Bullock, *J. Am. Chem. Soc.*, **73**, 2756, 5155 (1951); S. W. Fox and M. W. Bullock, U.S. Patents 2,701,250 and 2,701,251 (1955) [*C. A.*, **50**, 1922 (1956)]; V. V. Feofilaktov and N. K. Semenova, *Akad. Nauk S. S. S. R., Inst. Org. Khim., Sintezy Org. Soedin., Sb.*, **2**, 63 (1952) [*C. A.*, **48**, 666 (1954)].
3. R. Majima and T. Hoshino, *Ber.*, **58**, 2042 (1925); J. Thesing and F. Schülde, *Ber.*, **85**, 324 (1952); E. L. Eliel and N. J. Murphy, *J. Am. Chem. Soc.*, **75**, 3589 (1953).
4. H. R. Snyder and F. J. Pilgrim, *J. Am. Chem. Soc.*, **70**, 3770 (1948); C. Runti and G. Orlando, *Ann. Chim. (Rome)*, **43**, 308 (1953) [*C. A.*, **49**, 3940 (1955)].
5. S. S. Nametkin, N. N. Mel'nikov, and K. S. Bokarev, *Zh. Prikl. Khim.*, **29**, 459 (1956) [*C. A.*, **50**, 13867 (1956)].
6. J. A. Bently, K. R. Farrar, S. Housley, G. F. Smith, and W. C. Taylor, *Biochem. J.*, **64**, 44 (1956).
7. A. Berthelot and G. Amoureux, *Compt. Rend.*, **206**, 699 (1938); A. Berthelot, G. Amoureux, and S. Deberque, *Compt. Rend. Soc. Biol.*, **131**, 1234 (1939).

INDOLE-3-CARBONITRILE

Submitted by H. M. BLATTER, H. LUKASZEWSKI, and G. DE STEVENS.[1]
Checked by WAYLAND E. NOLAND and KENT R. RUSH.

1. Procedure

A mixture of 1.44 g. (0.0099 mole) of indole-3-carboxaldehyde,[2] 7.0 g. (0.053 mole) of diammonium hydrogen phosphate, 30 g.

(30 ml., 0.34 mole) of 1-nitropropane, and 10 ml. of glacial acetic acid is refluxed for 12.5 hours. During the reflux period the pale-yellow mixture becomes dark red. The volatile reactants and solvent are removed under reduced pressure, and an excess of water is then added to the dark residue. After a short time, crude indole-3-carbonitrile precipitates rapidly. It is separated by filtration and dried under reduced pressure; weight 1.20–1.34 g. (85–95%). Crystallization from acetone-hexane, with decolorization by activated carbon, yields 0.68–0.89 g. (48–63%) of fairly pure indole-3-carbonitrile, m.p. 179.5–182.5° (Note 1).

2. Note

1. The checkers obtained pure indole-3-carbonitrile, m.p. 182–184°, by subliming the product at a pressure of 1.5 mm. (bath temperature 165–170°) and recrystallizing the sublimate from a mixture of acetone and light petroleum ether. The recovery was 84%.

3. Methods of Preparation

Indole-3-carbonitrile has been prepared by the dehydration of indole-3-carboxaldehyde oxime,[3-5] indole-3-glyoxalic acid oxime,[4,6] or indole-3-carboxamide;[3] by the action of cyanogen chloride on indolylmagnesium iodide;[6] by the reaction of isoamyl formate with o-aminobenzyl cyanide in the presence of metallic sodium;[7,8] by mild basic hydrolysis of 1-acetylindole-3-carbonitrile;[7] and by the present method.[9]

4. Merits of the Preparation

This synthetic process is applicable to the preparation of other aromatic nitriles from aldehydes. The submitters have used it to prepare 5-bromoindole-3-carbonitrile, 7-azaindole-3-carbonitrile, p-chlorobenzonitrile, 3,4,5-trimethoxybenzonitrile, and p-N,N-dimethylaminobenzonitrile.[9] There are several advantages to its use. They include (a) readily available and inexpensive reagents, (b) a simple, time-saving procedure, and (c) fair to good yields of nitrile obtained by a *one-step* method.

1. Research Department, Division of Chemistry, CIBA Pharmaceutical Company, Division of CIBA Corporation, Summit, New Jersey.

2. P. N. James and H. R. Snyder, *Org. Syntheses,* Coll. Vol. 4, 539 (1963).
3. F. P. Doyle, W. Ferrier, D. O. Holland, M. D. Mehta, and J. H. C. Nayler, *J. Chem. So.,* 2853 (1956).
4. K. N. F. Shaw, A. McMillan, A. G. Gudmundson, and M. D. Armstrong, *J. Org. Chem.,* 23, 1171 (1958).
5. S. Swaminathan, S. Ranganathan, and S. Sulochana, *J. Org. Chem.,* 23, 707 (1958).
6. R. Majima, T. Shigematsu, and T. Rokkaku, *Ber.,* 57, 1453 (1924).
7. R. Pschorr and G. Hoppe, *Ber.,* 43, 2549 (1910).
8. N. I. Gavrilov, *Izvest. Petrov. Selskochos, Akad. (Russia),* 1-4, 14 (1919) [*C. A.,* 19, 505 (1925)].
9. H. M. Blatter, H. Lukaszewski, and G. de Stevens, *J. Am. Chem. Soc.,* 83, 2203 (1961).

IODOSOBENZENE

(Benzene, iodoso-)

$$C_6H_5I(OCOCH_3)_2 + 2NaOH \rightarrow$$
$$C_6H_5IO + 2NaOCOCH_3 + H_2O$$

Submitted by H. Saltzman and J. G. Sharefkin.[1]
Checked by Melvin S. Newman and Narinder Gill.

1. Procedure

Caution! This compound explodes if heated to 210°.

Finely ground iodosobenzene diacetate [2] (32.2 g., 0.10 mole) is placed in a 250-ml. beaker, and 150 ml. of 3N sodium hydroxide is added over a 5-minute period with vigorous stirring. The lumps of solid that form are triturated with a stirring rod or spatula for 15 minutes, and the reaction mixture stands for an additional 45 minutes to complete the reaction. One hundred milliliters of water is added, the mixture is stirred vigorously, and the crude, solid iodosobenzene is collected on a Büchner funnel. The wet solid is returned to the beaker and triturated in 200 ml. of water. The solid is again collected on the Büchner funnel, washed there with 200 ml. of water, and dried by maintaining suction. Final purification is effected by triturating the dried solid in 75 ml. of chloroform in a beaker. The iodosobenzene is separated by filtration (Note 1) and air-dried; weight

18.7–20.5 g. (85–93%); m.p. 210° (*Caution! Explodes!*). Iodometric titration[3] shows the product to be more than 99% pure (Note 2).

2. Notes

1. The filtrate yields unreacted diacetate on evaporation.
2. The purity of the iodosobenzene depends on the purity of the diacetate used.

3. Methods of Preparation

Iodosobenzene has been prepared by the action of sodium or potassium hydroxide solution on iodobenzene dichloride[3,4] and by addition of water to the dichloride.[5]

4. Merits of the Preparation

This method of preparing iodosobenzene is preferable to older ones based on iodosobenzene dichloride because iodosobenzene diacetate[2] is more stable and more conveniently prepared than the dichloride[3] and the overall yield is greater (75% versus 54%).

The procedure seems to be a general way of preparing iodosoarenes with electron-donating substituents, for the submitters have used it to obtain good yields of *o*-, *m*- and *p*-iodosotoluene, 2- and 4-iodoso-*m*-xylene, 2-iodoso-*p*-xylene, *o*-iodosophenetole, and 4-iodosobiphenyl.

Iodosoarenes are useful in the preparation of iodonium salts, $Ar_2I^+X^-$.[6]

1. Department of Chemistry, Brooklyn College of the City University of New York, Brooklyn, New York.
2. J. G. Sharefkin and H. Slatzman, this volume, p. 660.
3. H. J. Lucas, E. R. Kennedy, and M. W. Formo, *Org. Syntheses,* Coll. Vol. 3, 483 (1955).
4. C. Willgerodt, *Ber.,* **25**, 3494 (1892); **26**, 357, 1802 (1893); P. Askenasy and V. Meyer, *Ber.,* **26**, 1354 (1893); C. Hartmann and V. Meyer, *Ber.,* **27**, 502 (1894).
5. C. Willgerodt, *Ber.,* **26**, 357 (1893); G. Ortoleva, *Chem. Zentr.,* 1900, 722.
6. F. M. Beringer, R. A. Falk, M. Karniol, I. Lillien, G. Masullo, M. Mausner, and E. Sommer, *J. Am. Chem. Soc.,* **81**, 343 (1959) C. Hartmann and V. Meyer, *Ber.,* **27**, 426, 504 (1894).

IODOSOBENZENE DIACETATE *

(Benzene, iodoso-, diacetate)

$$C_6H_5I + CH_3CO_3H + CH_3CO_2H \rightarrow$$
$$(C_6H_5IOCOCH_3)^+(CH_3COO)^- + H_2O$$

Submitted by J. G. SHAREFKIN and H. SALTZMAN.[1]
Checked by J. DIEKMANN and B. C. McKUSICK.

1. Procedure

Caution! Avoid inhaling the vapor of peracetic acid or allowing the liquid to come into contact with the skin. The reaction is best carried out in a hood (Note 1).

The apparatus consists of a 200-ml. beaker equipped with a magnetic stirrer or any other type suitable for stirring a small volume of liquid. The flask is charged with 20.4 g. (0.10 mole) of iodobenzene [2] and is immersed in a water bath maintained at 30° (Note 2). Thirty-six grams (31 ml., 0.24 mole) of commercial 40% peracetic acid (Note 3) is added dropwise to the well-stirred iodobenzene over a period of 30–40 minutes. Stirring is continued for another 20 minutes at a bath temperature of 30°, during which time a homogeneous yellow solution is formed. Crystallization of iodosobenzene diacetate may begin during this period.

The beaker is chilled in an ice bath for 1 hour. The crystalline diacetate that separates is collected on a Büchner funnel and washed with three 20-ml. portions of cold water. After drying for 30 minutes on the funnel with suction, the diacetate is dried overnight in a vacuum desiccator containing calcium chloride (Note 4). The dried diacetate weighs 26.7–29.3 g. (83–91%) and melts at 158–159° with decomposition. The purity of the diacetate, determined by the titration method of Lucas, Kennedy, and Formo,[3] is 97–98%, which is good enough for most purposes. The purity can be increased to 99–100% by a recrystallization from $5M$ acetic acid.

2. Notes

1. Rubber gloves should be worn when handling vessels containing peracetic acid, for traces of the liquid can cause severe irritation. Skin that has come into contact with peracetic acid should be washed immediately and treated with sodium bicarbonate. Details for the safe handling of peracetic acid are found in Bulletin 4 supplied by Buffalo Electrochemical Corp.

2. Appreciable amounts of iodoxybenzene are formed if the temperature of the bath is allowed to go above 30° or if the addition of peracetic acid is faster than indicated.

3. Satisfactory 40% peracetic acid is obtainable from Buffalo Electrochemical Corp., Food Machinery and Chemical Corp., Buffalo, New York. The specifications given by the manufacturer for its composition are: peracetic acid, 40%; hydrogen peroxide, 5%; acetic acid, 39%; sulfuric acid, 1%; water, 15%. Its density is 1.15 g. per ml.

A fresh sample of this 40% peracetic acid contains about 1.54 equivalents, or 0.77 mole, of peroxide per 100 ml. of solution, corresponding to 1.34 equivalents per 100 g. The concentration can be determined by treating the peroxide solution with potassium iodide and titrating the liberated iodine with standard sodium thiosulfate. The concentration of peroxide in peracetic acid decreases somewhat on long standing and should be checked before the peracetic acid is used. The yield of diacetate is lowered if the concentration of the peroxide is less than 1.0 equivalent per 100 g. of peracetic acid. The total amount of peroxide used should be 2.4 moles, or 4.8 equivalents, for each mole of iodobenzene.

4. The surface of the diacetate may become yellow during the drying, but this does not affect its usefulness for most purposes.

3. Methods of Preparation

Willgerodt[4] prepared iodosobenzene diacetate by adding chlorine to iodobenzene and hydrolyzing the dichloride to iodosobenzene, which was then reacted with acetic acid. Pausacker[5] used this method to synthesize a number of analogs but found it inferior to his modification of the method of Böeseken and Schneider [6] in which iodobenzene is treated with 30% hydrogen peroxide and acetic anhydride. Arbusov [7] obtained the diacetate in 79% yield by the action of a mixture of peracetic and acetic acids on iodobenzene. Quantitative yields of the diacetate have been claimed for the reaction of iodobenzene dichloride with lead tetraacetate in glacial acetic acid containing 10% acetic anhydride, followed by precipitation of the lead as the chloride.[8]

4. Merits of the Preparation

Iodosobenzene diacetate is best prepared by the action of peracetic acid and acetic acid on iodobenzene. The present procedure is superior to earlier ones [5–8] because it uses inexpensive, commercially available peracetic acid, is faster, and gives higher yields. The procedure seems general for aryl iodides with electron-releasing substituents, for the submitters have obtained good yields of diacetates from o-, m- and p-iodotoluene, 2- and 4-iodo-m-xylene, 2-iodo-p-xylene, o-iodophenetole, and 4-iodobiphenyl.

Iodosobenzene diacetate is used as a reagent for the preparation of glycol diacetates from olefins,[9] for the oxidation of aromatic amines to corresponding azo compounds,[10] for the ring acetylation of N-arylacetamides,[11] for oxidation of some phenols to phenyl ethers,[12] and as a coupling agent in the preparation of iodonium salts.[13] Its hydrolysis to iodosobenzene constitutes the best synthesis of that compound.[14]

1. Department of Chemistry, Brooklyn College of the City University of New York, Brooklyn, New York.
2. H. J. Lucas and E. R. Kennedy, *Org. Syntheses,* Coll. Vol. 2, 351 (1943).
3. H. J. Lucas, E. R. Kennedy, and M. W. Formo, *Org. Syntheses,* Coll. Vol. 3, 483 (1955).
4. C. Willgerodt, *Ber.,* 25, 3495 (1892).
5. K. H. Pausacker, *J. Chem. Soc.,* 107 (1953).
6. J. Böeseken and G. C. C. C. Schneider, *J. Prakt. Chem.,* 131, 285 (1931).
7. B. A. Arbusov, *J. Prakt. Chem.,* 131, 351 (1931).
8. R. Neu, *Ber.,* 72B, 1505 (1939).
9. R. Criegee and H. Beucker, *Ann.* 541, 218 (1939).
10. K. H. Pausacker, *J. Chem. Soc.,* 1989 (1953).
11. G. B. Barlin and N. V. Riggs, *J. Chem. Soc.,* 3125 (1954).
12. K. H. Pausacker and A. R. Fox, *J. Chem. Soc.,* 295 (1957).

13. F. M. Beringer, R. A. Falk, M. Karniol, I. Lillien, G. Masullo, M. Mausner, and E. Sommer, *J. Am. Chem. Soc.*, 81, 342 (1959).
14. H. Saltzman and J. G. Sharefkin, this volume, p. 658.

N-IODOSUCCINIMIDE*

(Succinimide, N-iodo-)

Submitted by W. R. BENSON,[1] E. T. McBEE,[2] and L. RAND.[3]
Checked by B. C. McKUSICK and T. J. KEALY.

1. Procedure

Twenty grams (0.079 mole) of iodine and 90 ml. of dried dioxane (Note 1) are placed in a wide-mouthed, screw-cap, brown bottle of 150–200 ml. capacity. Most of the iodine dissolves. Eighteen grams (0.087 mole) of thoroughly dried N-silver succinimide (Note 2) is added, and the bottle is shaken vigorously for several minutes. The mixture is occasionally shaken in the course of an hour and then is warmed in a water bath at 50° for 5 minutes. It is now filtered hot through a Büchner funnel into a 500-ml. filter flask well wrapped with black paper or aluminum foil. The silver iodide that is collected is washed with a 10-ml. portion of warm dioxane. Carbon tetrachloride (200 ml.) is added to the combined filtrates in the filter flask, and the solution is chilled overnight at −8° to −20°. N-Iodosuccinimide separates as colorless crystals. It is collected on a Büchner funnel with as little exposure to light as possible, washed with 25 ml. of carbon tetrachloride, and dried with suction. After being dried overnight in the dark at 25° (1 mm.) the N-iodosuccinimide weighs 14.3–15.1 g. (81–85% yield) and melts with decomposition at 193–199° (Note 3).

2. Notes

1. The dioxane is purified only by the use of sodium strips and distillation.[3] The checkers used a newly opened bottle of "Spectroquality Reagent" dioxane (Matheson, Coleman and Bell) without further treatment.

2. N-Silver succinimide was prepared by the method of Djerassi and Lenk.[4] The checkers rapidly added a solution of 64 g. (1.6 moles) of sodium hydroxide in 300 ml. of water dropwise to a stirred solution of 249 g. (1.47 moles) of silver nitrate in 700 ml. of water at room temperature. The silver oxide that formed was separated on a Büchner funnel and washed with water. The moist oxide was added in one portion to a boiling solution of 133 g. (1.34 moles) of succinimide in 4 l. of water. The reaction vessel was wrapped with aluminum foil in order to exclude as much light as possible. After 45 minutes, the suspension was filtered through a heated Büchner funnel into a filter flask also wrapped with aluminum foil. The filtrate was allowed to stand at room temperature overnight, during which time N-silver succinimide crystallized. The N-silver succinimide was separated on a Büchner funnel, dried in air under suction, and ground to a powder. After being dried in a vacuum oven for 1 hour at 110°, it weighed 128 g. (47%). N-Silver succinimide should be stored in a brown bottle.

3. This material is pure enough to use in the preparation of α-iodoketones. The checkers found that, after one recrystallization from a mixture of dioxane and carbon tetrachloride, the N-iodosuccinimide melted with decomposition at 195–200°. Pure N-iodosuccinimide is reported to melt at 200–201°.[4]

3. Methods of Preparation

N-Iodosuccinimide has been prepared by the action of iodine on N-silver succinimide,[4,5] and by the action of iodine monochloride on the sodium salt of succinimide.[6] The present procedure is essentially that of Djerassi and Lenk,[4] with the modification that dioxane is the reaction medium instead of acetone; dioxane gives a better yield without formation of a lachrymatory by-product.

4. Merits of Preparation

N-Iodosuccinimide reacts with enol acetates derived from ketones to give α-iodoketones, and the reaction has found application in the steroid field.[4,6] The iodination of the enol acetates seems to proceed by an ionic mechanism, and preliminary work indicates that N-iodosuccinimide is not capable of at least some of the radical-chain iodinations analogous to radical-chain brominations brought about by N-bromosuccinimide.[4] N-Iodosuccinimide has also been used for the iodination of purine nucleosides[7] and of fluorenone.[8] Iodination of 2-hydroxy-3-(β-alkylvinyl)-1,4-naphthoquinones results in an unusual replacement of a vinyl hydrogen.[9]

1. Chemistry Department, Colorado State University, Fort Collins, Colo.
2. Department of Chemistry, Purdue University, Lafayette, Ind.
3. L. Fieser, *Experiments in Organic Chemistry,* 3rd ed., D. C. Heath and Co., Boston, 1955, p. 284.
4. C. Djerassi and C. T. Lenk, *J. Am. Chem. Soc.,* 75, 3494 (1953).
5. N. Bunge, *Ann.,* 7 (suppl.), 117 (1870).
6. C. Djerassi, J. Grossman, and G. H. Thomas, *J. Am. Chem. Soc.,* 77, 3826 (1955).
7. D. Nowotny and D. Lipken, *Monatsh. Chem.,* 96, 125 (1965).
8. F. Dewhurst and P. Shah, *J. Chem. Soc. C,* 1503 (1969).
9. K. H. Dudley and H. W. Miller, *Tetrahedron Lett.,* 571 (1968).

IODOXYBENZENE

(Benzene, iodoxy-)

$$C_6H_5I + 2CH_3CO_3H \rightarrow C_6H_5IO_2 + 2CH_3CO_2H$$

Submitted by J. G. Sharefkin and H. Saltzman.[1]
Checked by E. J. Corey and C. P. Lillya.

1. Procedure

Caution! Avoid inhaling the vapor of peracetic acid or allowing the liquid to come into contact with the skin. The reaction is best carried out in a hood (Note 1). Iodoxybenzene explodes if heated to 230°.

A 500-ml. three-necked flask fitted with reflux condenser, stirrer, and dropping funnel and containing 20.4 g. (0.10 mole) of iodobenzene [2] is immersed in an oil bath maintained at 35°. Seventy-five grams (65 ml., 0.50 mole) of commercial 40% peracetic acid (Note 1) is added with vigorous stirring over a 30-minute period. Solid may begin to form before all the peracetic acid has been added, but, although this may slow down the stirring, it does not decrease the yield or cause a rise in temperature.

After all the peracetic acid has been added, the reaction mixture is diluted with 80 ml. of water and heated from 35° to 100° over a 20-minute period (Note 2). It is then kept at 100° for 45 minutes. The flask is cooled to 0–5° in an ice bath, and the solid iodoxybenzene is collected on a Büchner funnel and air-dried with suction for 1 hour. Additional material is obtained by concentrating the filtrate to one-fourth of its volume (*Caution! Note 3*). The two crops of crude iodoxybenzene are combined and dried overnight in a desiccator; weight 19.6–20.5 g.; m.p. 230° (*Caution! Explodes!*). Iodometric titration [3] shows the purity to be about 94% (Note 4).

Purification of the crude iodoxybenzene is effected by grinding it to a powder in a mortar, macerating it with 70 ml. of chloroform, and separating the solid by filtration. The chloroform extraction is repeated and the solid is dried; weight 17–19 g. (72–80%); purity 99.0–99.9% by iodometric titration.[3]

2. Notes

1. For a source and the specifications of 40% peracetic acid and precautions in handling it, see Notes 3 and 1 under the preparation of iodosobenzene diacetate, p. 661.

2. If the temperature of the bath is not raised slowly, foaming is difficult to control. Although the gradual rise in temperature causes considerable foaming, the reaction mixture remains within the flask.

3. The filtrate must not be evaporated to dryness because iodoxybenzene explodes when heated.

4. The major by-products in this reaction are iodobenzene and iodosobenzene diacetate. An excess of 20 ml. of peracetic acid over the 65 ml. recommended results in an increase in the amount

of iodobenzene.   Both impurities are removed from the product by washing with chloroform.

3. Methods of Preparation

Iodoxybenzene has been prepared by the disproportionation of iodosobenzene,[4-6] by oxidation of iodosobenzene with hypochlorous acid or bleaching powder,[7] and by oxidation of iodobenzene with hypochlorous acid or with sodium hydroxide and bromine.[8]   Other oxidizing agents used with iodobenzene include air,[3] chlorine in pyridine,[9] Caro's acid,[10,11] concentrated chloric acid,[12] and peracetic acid solution.[13]   Hypochlorite oxidation of iodobenzene dichloride has also been employed.[14]

4. Merits of the Preparation

This one-step method of preparing iodoxybenzene is preferable to earlier methods because it is simpler and the yield is substantially higher.   The procedure seems general for iodoxyarenes, at least those with electron-releasing substituents, for the submitters have used it to obtain good yields of *o*-, *m*- and *p*-iodoxytoluene, 2- and 4-iodoxy-*m*-xylene, 2-iodoxy-*p*-xylene, *o*-iodoxyphenetole, 4-iodoxybiphenyl, and *o*-iodoxybenzoic acid.

Iodoxyarenes are useful in the preparation of iodonium salts, $Ar_2I^+X^-$.[15]

1. Department of Chemistry, Brooklyn College of the City University of New York, Brooklyn, New York.
2. H. J. Lucas and E. R. Kennedy, *Org. Syntheses,* Coll. Vol. 2, 351 (1943).
3. C. Willgerodt, *Ber.,* 25, 3500 (1892); 26, 358 (1893).
4. C. Willgerodt, *Ber.,* 26, 1307, 1806 (1893).
5. P. Askenasy and V. Meyer, *Ber.,* 26, 1356 (1893).
6. H. J. Lucas and E. R. Kennedy, *Org. Syntheses,* Coll. Vol. 3, 485 (1955).
7. C. Willgerodt, *Ber.,* 29, 1568 (1896).
8. C. Willgerodt, *Ber.,* 29, 1571 (1896).
9. G. Ortoleva, *Chem. Zentr.,* 1900, 723.
10. E. Bamberger and A. Hill, *Ber.,* 33, 534 (1900).
11. I. Masson, E. Race, and F. E. Pounder, *J. Chem. Soc.,* 1678 (1935).
12. R. L. Datta and J. K. Choudhury, *J. Am. Chem. Soc.,* 38, 1085 (1916).
13. B. A. Arbusov, *J. Prakt. Chem.,* 131, 357 (1931).
14. M. W. Formo and J. R. Johnson, *Org. Syntheses,* Coll. Vol. 3, 486 (1955).
15. C. Hartman and V. Meyer, *Ber.,* 27, 504 (1894).

ISOPHTHALALDEHYDE*

Submitted by J. H. Ackerman and A. R. Surrey [1]
Checked by Kenneth H. Brown, Wayland E. Noland,
and William E. Parham

1. Procedure

A solution of 272 g. (261 ml., 2.00 moles) of α,α'-diamino-m-xylene (Note 1), 1.00 kg. (7.1 moles) of hexamethylenetetramine, 480 ml. of concentrated hydrochloric acid, and 3.2 l. of 50% aqueous acetic acid in a 12-l. flask is stirred and heated at the reflux temperature for 2.5 hours. The hot amber reaction mixture is then poured into a large battery jar in a well-ventilated hood, and a solution prepared from 298 g. of sodium hydroxide and 3.85 l. of water is added slowly with stirring (Note 2). The mixture is covered and allowed to stand overnight at about 5°. The product, which separates as long needles, is collected, washed on a Buchner funnel with 100 ml. of cold water, and then dried to constant weight under vacuum (Note 3) over calcium chloride. There is obtained 158–166 g. (59–62%) of almost colorless needles of isophthalaldehyde, m.p. 88–90° (Note 4).

2. Notes

1. α,α'-Diamino-m-xylene was obtained from California Chemical Company and Aldrich Chemical Company.

2. The sodium hydroxide solution is added to neutralize most of the acetic acid present. Better yields are obtained using this neutralization procedure than by merely cooling the reaction mixture.

3. The checkers observed that house vacuum removed only 50% of the water after 48 hours.

4. One lot of α,α'-diamino-m-xylene from Aldrich Chemical Company gave the isophthalaldehyde as pale pink, long needles, m.p. 88–90°. When 12.0 g. of this material was recrystallized from 500 ml. of water, there was obtained 10.9 g. (91%) of pale cream, long needles, m.p. 89–91°.

3. Methods of Preparation

The procedure described is a modification of the general procedure of Angyal[2] for the preparation of aldehydes from benzylamines by the Sommelet reaction. Isophthalaldehyde has been prepared from m-xylene by preparation of the tetrachloro derivative and hydrolysis,[3] from isophthaloyl chloride by the Rosenmund reaction,[4] from α,α'-dibromo-m-xylene by the Sommelet reaction,[5] and from isophthaloyl chloride by reduction with lithium tri-t-butoxyaluminumhydride.[6]

4. Merits of the Preparation

Isophthalaldehyde is a valuable intermediate. Although the yields obtained by some of the other reported methods of preparation are better than the yield obtained here, the availability of starting material and the simplicity of reaction make this method attractive.

This appears to be the first reported case of the Sommelet reaction starting with a diamine.

1. Sterling-Winthrop Research Institute, Rensselaer, New York.
2. S. J. Angyal, *Org. Reactions*, 8, 197 (1954).
3. A. Colson and H. Gautier, *Bull. Soc. Chim. France*, 45, 509 (1886).
4. K. W. Rosenmund, F. Zetzsche, and C. Flütsch, *Ber.*, 54, 2888 (1921).
5. K. F. Jennings, *J. Chem. Soc.*, 1172 (1957).
6. H. C. Brown and B. C. Subba Rao, *J. Am. Chem. Soc.*, 80, 5377 (1958).

3-ISOQUINUCLIDONE

(2-Azabicyclo[2.2.2]octan-3-one)

Submitted by W. M. Pearlman [1]
Checked by Peter Campbell and Ronald Breslow

1. Procedure

A. *cis- and trans-4-Aminocyclohexanecarboxylic acid.* A mixture of 27.4 g. (0.20 mole) of *p*-aminobenzoic acid (Note 1), 200 ml. of water, and 2 g. of 10% rhodium-0.1% palladium on carbon catalyst (Note 2) is placed in a pressure bottle and hydrogenated at 50 p.s.i. When 0.6 mole of hydrogen has been absorbed (Note 3), the mixture is filtered and concentrated under reduced pressure until crystals start to form (Note 4). The mixture is diluted with 200 ml. of dimethylformamide and cooled to 5°, filtered, washed with dimethylformamide, then methanol, and dried, giving 19.4–20.3 g. (68–71%) of *cis-* and *trans*-4-amino-cyclohexanecarboxylic acid, m.p. 292–296° (Note 5).

B. *3-Isoquinuclidone.* A mixture of 4.53 g. of *cis-* and *trans*-4-aminocyclohexanecarboxylic acid and 30 ml. of Dowtherm A (Note 6) is heated as rapidly as possible to reflux temperature. Heating is continued for 20 minutes during which time the water formed is allowed to distill away; at the end of this time, solution has taken place. The solution is allowed to cool to room temperature and is diluted with 100 ml. of isooctane. The solution is extracted three times with 50-ml. portions of water. The combined water extracts are treated with charcoal, filtered, and concentrated to dryness under reduced pressure. The residue is crystallized from cyclohexane giving 3.20–3.33 g. (81–84%) of 3-isoquinuclidone, m.p. 197–198° (Note 5).

2. Notes

1. Purchased from B. L. Lemke and Co., Inc., 199 Main Street, Lodi, New Jersey. The checkers used material from Eastman Organic Chemicals Department.

2. A mixture of 5.26 g. of rhodium chloride trihydrate, 0.34 g. of palladium chloride, 18 g. of carbon (Darco G-60), and 200 ml. of water is rapidly stirred and heated to 80°. Lithium hydroxide hydrate (2.7 g.) dissolved in 10 ml. of water is added all at once and the heating stopped. The mixture is stirred overnight, filtered, and washed with 100 ml. of 0.5 v/v% aqueous acetic acid. The product is dried under reduced pressure at 65°, giving 20.6–21 g. of the catalyst. One gram of this catalyst consumes 0.0022–0.0028 mole of hydrogen in aqueous suspension.[2]

3. The checkers found that the reduction requires 4–5 days, whereas the submitter reported the reaction requires 24 hours. Fresh catalyst is added whenever the rate of hydrogen uptake significantly decreases. When fresh catalyst is added to the reaction vessel, it is important that it first be wet with solvent and that the hydrogen be well evacuated. Opening the mixture to the atmosphere without careful evacuation will produce a hydrogen-oxygen mixture which may explode on contact with fresh catalyst.

4. It is necessary to concentrate the solution to one-fifth volume before crystals form.

5. The submitters report the preparation scaled up by fiftyfold with similar yields and purities.

6. Purchased from Dow Chemical Co., Midland, Michigan.

3. Discussion

Ferber and Brückner[3] reduced p-aminobenzoic acid using Adams catalyst (PtO₂) at atmospheric pressure, and Schneider and Dillman[4] reduced p-aminobenzoic acid using 10% ruthenium on carbon at 140 atm. and 70°. 3-Isoquinuclidone has been prepared, by the previously mentioned investigators,[3,4] by heating the dry 4-aminocyclohexane carboxylic acid at elevated temperatures.

The described method of preparation of 3-isoquinuclidone has the following advantages: The isolation of the *cis* form of 4-aminocyclohexane carboxylic acid is not required in order to obtain a good yield; the amount of 3-isoquinuclidone that can be prepared at a time is limited only by the size of available equipment; the yield is excellent and the workup is easy and straightforward.

3-Isoquinuclidone has been found to be an excellent substitute for camphor for molecular weight determinations.[2]

1. Research Laboratories, Parke, Davis & Company, Ann Arbor, Michigan.
2. W. M. Pearlman, *Tetrahedron Lett.,* 1663 (1967).
3. E. Ferber and H. Brückner, *Ber.,* **76**, 1019 (1943).
4. W. Schneider and R. Dillmann, *Ber.,* **96**, 2377 (1963).

β-ISOVALEROLACTAM-N-SULFONYL CHLORIDE AND β-ISOVALEROLACTAM

(2-Azetidinone-4,4-dimethyl-1-sulfonyl chloride and 2-azetidinone-4,4-dimethyl)

$(CH_3)_2C{=}CH_2 + ClSO_2N{=}C{=}O \rightarrow$

[structure: 2-azetidinone-4,4-dimethyl ring with CH₃, CH₃ substituents, N—SO₂Cl, and =O]

$+ CH_2{=}\overset{\underset{\displaystyle CH_3}{|}}{C}{-}CH_2\overset{\underset{\displaystyle H}{\displaystyle N}}{\underset{}{\overset{\displaystyle O}{\overset{\|}{C}}}}{-}SO_2Cl$

[structure: ring with CH₃, CH₃, N—SO₂Cl, O] $\xrightarrow[\substack{-Na_2SO_4 \\ -NaCl \\ -H_2O}]{3NaOH}$ [structure: ring II with CH₃, CH₃, N—H, O]

Submitted by RODERICH GRAF [1]
Checked by JEROME F. LEVY and WILLIAM D. EMMONS

1. Procedure

Caution! Chlorosulfonyl isocyanate is highly corrosive and may be contaminated with cyanogen chloride. This preparation should be carried out in a good hood, and rubber gloves should be worn.

A. *β-Isovalerolactam-N-sulfonyl chloride.* A 200-ml. four-necked flask fitted with a mechanical stirrer, a dry ice-jacketed dropping funnel (Note 1), and a thermometer is cooled with a dry ice-methylene chloride slush bath while 67 ml. of sulfur dioxide (Note 2) is condensed into the flask. Both the dry ice condenser and the dropping funnel are protected with drying tubes containing anhydrous calcium sulfate. With the liquid sulfur dioxide at $-20°$, the flask is charged with 0.3 g. of finely powdered potassium chloride (Note 3) and 47.1 g. (0.33 mole) of chlorosulfonyl isocyanate.[2] Then 19.5 g. (0.35 mole) of isobutylene, previously condensed in a cold trap, is added to the dropping funnel. The temperature of the flask is lowered to $-40°$ to $-50°$, and isobutylene is added dropwise over a 20-minute period (Notes 4, 5). After the isobutylene addition is completed, the cooling bath under the flask is removed, and the reaction mixture is allowed to warm up until the solvent begins to reflux (approximately $-6°$). The colorless contents of the flask are then poured into 125 ml. of water contained in a 400-ml. beaker over a period of 1 minute with vigorous agitation provided by a mechanically driven paddle stirrer. Sulfur dioxide is evolved while β-isovalerolactam-N-sulfonyl chloride precipitates as a gritty, crystalline, white solid (Note 6). The major portion of dissolved sulfur dioxide is removed by impinging a vigorous stream of air on the surface of the liquid in the beaker until the temperature of the mixture rises again after falling to 0–4°. The precipitate is removed by suction filtration and is washed three times with 33-ml. portions of ice water. The yield of moist product containing 10–20% water is 52–56 g. (Note 7). The product in this form is more suitable for subsequent conversion to the free β-lactam than if it is anhydrous or in a more coarsely crystalline form.

To prepare the anhydrous compound, the solid is dissolved in methylene chloride, whereupon the water separates as the upper phase. The organic layer is dried over anhydrous sodium sulfate, and the solvent is removed under reduced pressure at room temperature to give a colorless, crystalline mass, m.p. 75–77°. Small amounts of the pure compound may be obtained in the form of long needles by recrystallization from ether, m.p. 77–78°. The yield is 43–46 g. (65–70%) in anhydrous form (Note 8).

B. *β-Isovalerolactam.* A 200-ml. beaker is provided with a combination pH electrode, a mechanically driven paddle stirrer, a thermometer, and a syringe or dropping funnel (Note 9). An amount of water just sufficient for immersion of the pH electrode (20 ml.) is introduced, and with vigorous stirring the first portion (about one quarter) of the 52–56 g. (about 0.20–0.22 mole) of moist β-isovalerolactam-N-sulfonyl chloride is added. The liberated acid is neutralized by dropwise addition of approximately 10N sodium hydroxide solution (Note 10) to maintain the pH of the mixture between 2 and 8, and preferably in the range 5–7 (Note 11). The temperature is kept at 20–25° with cooling supplied by an ice bath as necessary. More lactam-N-sulfonyl chloride is added as the hydrolysis proceeds.

Hydrolysis is very sluggish at first, especially if anhydrous or coarsely crystalline sulfonyl chloride is used. Hydrolysis of the first one quarter to one half of the sulfonyl chloride requires 1–3 hours, and by that time the increasing salt concentration of the solution will cause the separation of the β-lactam as a second liquid phase. The sulfonyl chloride is significantly more soluble in this water-containing lactam phase, and as a result the rate of hydrolysis increases markedly (Note 12). Hydrolysis of the remainder of the sulfonyl chloride can be completed in about 30 minutes to 1 hour. The pH is adjusted finally to 7, and the mixture is cooled to 10° while being repeatedly seeded with sodium sulfate decahydrate. This converts all the precipitated sodium sulfate to the decahydrate (Note 13), and the β-lactam which originally separated in oily form is dissolved. The sodium sulfate decahydrate is removed by suction filtration and is washed with 90 ml. of chloroform. Small portions of the chloroform washings are then used for repeated extraction of the lactam from the salt solution. The extracts are combined, dried over anhydrous potassium carbonate, and distilled through an 8-in. Vigreux column. Most of the chloroform is removed by distillation at atmospheric pressure, and the product is distilled under reduced pressure, b.p. 70° (1.0 mm.), n^{25}D 1.4475, freezing point 14.7° (Note 14), weight 16.7–17.3 g. (51–53% overall from chlorosulfonyl isocyanate). The product is 99.8% pure by vapor phase chromatographic analysis.

2. Notes

1. A jacket which is suitable for holding dry ice may be made easily for use with a cylindrical dropping funnel. The neck and

bottom of a narrow-mouthed polyethylene bottle are cut off, and two or three vertical slits are made at the narrow end to allow it to slip over the body of the dropping funnel and rest on the stopcock barrel. The size of the dropping funnel will determine the size of polyethylene container to be used.

2. The checkers used anhydrous sulfur dioxide supplied in cylinders by the Matheson Co., Inc., East Rutherford, New Jersey, without further purification. Traces of moisture will not interfere with the reaction, and it is sufficient if the liquid sulfur dioxide is clear and colorless. If necessary, however, the gaseous sulfur dioxide may be dried with anhydrous calcium chloride before condensing it.

3. Addition of potassium chloride may be omitted if the chlorosulfonyl isocyanate is free of sulfur trioxide. Otherwise, traces of sulfur trioxide will give rise to a yellow or brown coloration of the reaction mixture and to formation of small amounts of byproducts which, because of their emulsifying activity, may interfere with further processing.

4. The reaction may also be carried out at higher temperatures, e.g., at $-10°$, with simultaneous addition of gaseous isobutylene over a prolonged period of time. This, however, results in a reaction product of lower purity than is obtained by the present procedure.

5. The reaction is exothermic, and the rate of addition should be controlled to keep the temperature within the limits indicated.

6. The cycloaddition reaction is accompanied by another reaction giving about 30% of 3-methyl-3-butenamide-N-sulfonyl chloride which is readily hydrolyzed during the aqueous work-up. The β-lactam-N-sulfonyl chloride which is the major product of the reaction is relatively stable to hydrolysis under the conditions of its isolation.

7. The product may be stored for several days before its conversion to the free β-lactam in a polyethylene bag placed in a refrigerator or preferably under dry ice in a Dewar flask.

8. The cycloaddition reaction can also be carried out with ether as solvent, especially in small batches.[3]

9. For hydrolysis of a much larger quantity of material, a four-necked flask may be employed instead of a beaker.

10. Three moles of sodium hydroxide solution is used per mole of sulfonyl chloride. The use of base of known concentration provides a means of following the hydrolysis as well as determining the true amount of product present for calculating the yield in Part A.

11. At high pH (>10–11) saponification of the sulfonyl chloride to the sodium salt of 3-amino-3-methylbutyric acid-N-sulfonic acid will predominate, and at too low a pH (*e.g.*, pH of 0) hydrolysis to 3-hydroxy-3-methylbutyramide will prevail.

12. Care should be taken that the sulfonyl chloride is not added too rapidly, as the increased hydrolysis rate at this point will not permit adequate control of temperature and pH if a large amount of sulfonyl chloride is present. For repeat preparations a portion of the reaction mixture from a preceding batch may be introduced to achieve a more rapid hydrolysis rate sooner in the reaction. For the first preparation there are ways of increasing the initial rate of hydrolysis, or shortening the time interval before the transition from low to higher hydrolysis rate occurs. These are use of sodium sulfate solution instead of pure water, addition of a few tenths of a gram of potassium iodide, or addition of a small amount (1 ml.) of methylene chloride. However, these steps are not necessary if a reasonable amount of patience is exercised.

13. If seeding with sodium sulfate decahydrate is omitted, the unstable heptahydrate may crystallize.

14. The freezing point given was determined by the checkers as the temperature of a solid-melt equilibrium for a sample of 99.8% purity. The submitter reports the melting point at 15.3° after recrystallization from isopropyl ether and redistillation.

3. Methods of Preparation

The only methods reported[3] for the preparation of 4,4-dimethyl-2-azetidinone-1-sulfonyl chloride and 4,4-dimethyl-2-azetidinone are those described here.

Conversion of β-lactam-N-sulfonyl chlorides to the free lactams may also be accomplished by means of reducing agents,[3,4] and for β-lactam-N-sulfonyl chlorides which are hydrolytically more stable than the one in the present example this represents a method to be preferred over pH-controlled hydrolysis.[3] The N-sulfonyl chlorides are reduced to N-sulfinic acids which spontaneously decompose to β-lactams and sulfur dioxide. Among the reducing agents which may be used are thiophenol, hydrogen sulfide, zinc dust, iron power, iodide ion, and sodium sulfite.[4] Iodide ion need be used only in catalytic amounts, as the liberated iodine is reduced again to iodide ion by the sulfur dioxide split off.

4. Merits of the Preparation

This procedure illustrates a general method for preparing aliphatic and, in certain cases, aromatic β-lactams containing a free NH group and substituted in either the 4 position or in both the 3 and 4 positions of the 2-azetidinone ring. The major by-product of the cycloaddition step is a β,γ-unsaturated carboxamide-N-sulfonyl chloride which, in the case of certain aromatic olefins, may predominate. Reactions of both β-lactam-N-sulfonyl chlorides and the β,γ-unsaturated carboxamide-N-sulfonyl chlorides have been tabulated.[3]

Recently additional examples of the synthesis of 3,4-disubstituted 2-azetidinones have been reported[5,6] as well as a kinetic study of the reaction of chlorosulfonyl isocyanate with olefins.[7]

β-Lactams substituted only in the 3 position cannot be prepared by the present procedure, since the lactam formed has the nitrogen of the chlorosulfonyl isocyanate attached to the more highly substituted carbon atom of the olefinic double bond in Markownikoff fashion. 2-Azetidinones substituted in the 3 position only have been prepared by Grignard reagent-catalyzed cyclizations of esters of appropriately substituted β-amino acids.[8,9]

1. Farbwerke Hoechst AG, Frankfurt on the Main, Germany.
2. R. Graf, this volume, p. 226.
3. R. Graf, *Ann.*, **661**, 111 (1963).
4. T. Durst and M. J. O'Sullivan, *J. Org. Chem.*, **35**, 2043 (1970).
5. H. Bestian, H. Biener, K. Clauss, and H. Heyn, *Ann.*, **718**, 94 (1968).
6. H. J. Friedrich, *Tetrahedron Lett.*, 2981 (1971).
7. K. Clauss, *Ann.*, **722**, 110 (1969).

8. R. W. Holley and A. D. Holley, *J. Am. Chem. Soc.*, 71, 2129 (1949).
9. G. Cignarella, G. F. Cristiani, and E. Testa, *Ann.*, 661, 181 (1963); and previous papers in the series.

KETENE

$$CH_2 = C\text{---}O \xrightarrow{550°} 2CH_2 = C = O$$

Submitted by S. ANDREADES and H. D. CARLSON [1]
Checked by R. DAVID CLARK, JAMES J. FUERHOLZER,
and HENRY E. BAUMGARTEN

1. Procedure (Note 13)

Caution! Ketene (b.p. −41°) is a poisonous gas of the same order of toxicity as phosgene. All operations with ketene should be carried out in an efficient hood.

The pyrolysis apparatus consists of a vertical, electrically-heated Vycor® tube (25 mm. I.D.) packed with 6-mm. lengths of Pyrex® tubing (10 mm. O.D.) and mounted in an electric furnace about 45 cm. long (Notes 1 and 2). Attached to the top is a 100-ml. dropping funnel with a pressure-equalizing side arm [2] that has an inlet for nitrogen (Note 3). A thermocouple well inside the tube holds a movable thermocouple and extends to the bottom of the heated section (Note 4). The bottom of the reactor is fitted to a 500-ml. side-arm flask packed in ice. The side arm leads to two traps in series cooled in ice and to a final trap cooled in a bath of dry ice and acetone (Note 5).

The hottest part of the tube, which is near the middle of the heated section, is maintained at 550° ± 10° while dry oxygen-free nitrogen is passed successively through a flowmeter and the tube at about 150 ml. per hr. for at least 30 minutes (Note 6). The dropping funnel is charged with 56 g. (0.67 mole) of diketene (Notes 7 and 8), which is then introduced into the hot tube at a rate of about 0.5 ml. per min. while the nitrogen flow continues.

Essentially pure ketene (Note 9), yield 26–31 g. (46–55%) (Note 10), collects in the dry-ice trap as a colorless or nearly colorless liquid. The ketene is distilled directly from this trap for use in reactions.

If the ketene is not to be used at once, drying tubes should be attached to the trap, which should then be stored at $-80°$. Ketene can be kept for as long as 2 weeks in this way, although some transformation to high-boiling material occurs (Note 11). However, pure ketene can be readily obtained from a partially decomposed mixture by simple distillation from the trap (Notes 11 and 12). *Caution! Do not store ketene under pressure, as an explosion may result.*

2. Notes

1. The furnace used by the submitters and checkers was an 1870-watt hinged type manufactured by the Hevi-Duty Electric Company, Milwaukee, Wisconsin; length 18 in.; inside diameter $2\frac{3}{8}$ in.; catalog No. M-2018. With the packing described, the total surface area in the packed tube is about 2000 cm.^{2} In addition to this setup the checkers used a similar tube (20 mm. inside diameter) packed with $\frac{1}{8}$-in. I.D. single-turn Pyrex® glass helices and inserted in a 550-watt furnace manufactured by the Hoskins Mfg. Co., Detroit, Michigan; length 12 in.; inside diameter $1\frac{1}{2}$ in.; catalog No. FD303A.

2. To prevent heat loss from the ends of the furnace opening, the ends are packed with Pyrex® glass wool or Fiberglas® insulation PF-105. In addition, the checkers capped each end of the furnace with a plate constructed from $\frac{1}{4}$-in. Transite® sheet in which a hole just large enough for a loose sliding fit on the tube had been bored. The plate at the bottom of the furnace was held in place with a rubber stopper also bored to fit the tube, and the plate at the top of the furnace was held in place with an iron ring attached to the ring stand supporting the funnel.

3. The lower part of the barrel of the dropping funnel should be bent in such a way as to offset it from the path of the thermocouple to permit adjustment of the latter. The checkers used a 8-mm. glass tube bent for offset and fitted at the upper end with

a nitrogen inlet and a standard-taper ground joint to permit attachment of funnels of various sizes.

4. Attachment of dropping funnel and thermocouple well to the pyrolysis tube may be made with a rubber stopper suitably bored.

5. The first two traps may be packed loosely with glass wool to prevent mechanically entrained impurities (or aerosol) from passing through into the final trap. Omission of the glass wool may allow as much as 0.5–1.0 g. of colored material to be collected in the product. However, this colored impurity is easily removed by simple distillation. The checkers cooled the two traps in ice-ethanol.

6. The checkers passed the nitrogen through a gas absorption bottle filled to a depth of 10 cm. with concentrated sulfuric acid which had been calibrated roughly for flow rate. The rate of flow used was one bubble per 7 seconds (*ca.* 145 ml. per hr.) for the larger furnace, and one bubble per 10 seconds (*ca.* 100 ml. per hr.) for the smaller furnace.

7. Suitable diketene can be obtained from the Aldrich Chemical Company, Milwaukee, Wisconsin. If at all colored, this material should be distilled or sublimed before pyrolysis, for use of colored material may lead to a colored product. Distillation is easily carried out through use of the apparatus illustrated in Fig. 1. The impure diketene is placed in flask A and is cooled in ice and stirred with a magnetic stirrer. The Dewar trap is filled with dry ice and ethanol. Evacuation of the system with a vacuum pump is begun very carefully with appropriate upward adjustment of the pressure if necessary to prevent bumping and splashing. After the initial degassing and removal of low-boiling materials, the diketene distils smoothly (at 0.1–1.0 mm. pressure), collecting as a white solid on the cold surface of the Dewar trap. After the distillation is completed, the apparatus is partially disassembled and the dry ice and ethanol are poured out. As the diketene melts, it flows into flask B. The diketene should be stored under nitrogen in tightly stoppered brown bottles in a refrigerator or, better, below its freezing point of $-6.5°$, as otherwise it may slowly decompose.

8. Equally good results have been obtained with charges

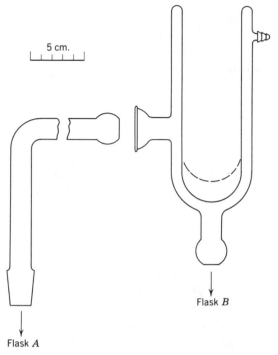

5 cm.

Flask *B*

Flask *A*

Fɪɢ. 1. Low-temperature distillation or evaporation apparatus. When solids are being collected, the Dewar flask bottom should be as indicated by the broken line.

one-fourth to twice that of the present procedure.

9. The ketene obtained by simple distillation from the final trap is essentially pure. The purity can be determined by passing a weighed sample into $1N$ sodium hydroxide solution and titrating for excess base. The checkers found it very difficult to avoid entirely the collection of minute traces of high-boiling colored impurities with the ketene. However, these impurities were easily removed by simple distillation of the ketene, just before use, from the original trap into a clean trap by warming the former in ice water and cooling the latter in dry ice and ethanol.

10. The checkers' yields were 54–66% for the larger furnace and 58–78% for the smaller furnace. The yield was found to be somewhat dependent on the rate of addition, with the rate specified in the procedure giving a good yield in a reasonable length of time. The checkers used an addition rate of 0.25–0.30

ml. per min. for the smaller furnace to obtain the yields cited.

11. Major impurities that collect on standing are all high-boiling, *e.g.*, diketene, b.p. 127°, and dehydroacetic acid, b.p. 270°. If the ketene has been stored long enough to allow a considerable portion of higher-boiling materials to accumulate, it is desirable to insert a trap cooled in an ice bath between the ketene-containing trap and the reaction vessel in order to minimize mechanical entrainment of the impurities. Repeated warming of the container to remove portions of the ketene naturally hastens transformation of the ketene to high-boiling materials.

12. The carbonaceous material that is deposited inside the pyrolysis tube is easily removed by passing a stream of oxygen through at about 550° after a thorough flushing with nitrogen.

13. The submitters no longer consider this method to be the most convenient one for preparing laboratory quantities of ketene. It has been supplanted at the Du Pont Experimental Station by pyrolysis of acetic anhydride according to the procedure of Fisher, MacLean, and Schnizer.[8] It is recommended that the wires of the preheater be glass-covered, since polymerization of the ketene at −78° seemed to occur frequently when bare wires were used, possibly because of catalysis by trace amounts of metal. Using this procedure 55 g. (1.31 moles) of ketene per hour could be produced (private communication, J. B. Sieja and H. D. Carson).

3. Methods of Preparation

Ketene can be generated conveniently by pyrolysis of acetone in a hot tube[3] or over a hot wire in a "ketene lamp,"[4] or by pyrolysis of diketene in a hot tube,[5,6] or by pyrolysis of acetic anhydride.[8] Other methods of preparation have been summarized.[3] It has been shown that diketene cracks quite cleanly to ketene,[6,7] although some allene and carbon dioxide are formed at the same time.[7]

4. Merits of the Preparation

The most convenient procedures for the preparation of ketene are the present one and the pyrolyses of acetone[4] or acetic ahnydride.[8]   The acetone procedure gives ketene at a relatively fast rate (0.45 mole per hour), but it takes considerable adjustment to get optimum conditions, and trouble is sometimes caused by the wire getting coated with carbon. Furthermore, because the efficiency of a given wire coil varies with time, passing through a maximum, frequent calibration of the apparatus is necessary. The acetic anhydride method is even faster (1.31 moles per hour) and uses a readily available chemical. It appears to be the method of choice at this time. The diketene procedure described here is relatively simple and reliable; however, it is relatively slow (0.2 mole per hour) and requires a somewhat less readily available starting material.

1. Contribution No. 884 from the Central Research Department, Experimental Station, E. I. du Pont de Nemours and Company, Wilmington 98, Delaware.
2. R. E. Benson and B. C. McKusick, *Org. Syntheses*, Coll. Vol. 4, 746 (1958).
3. C. D. Hurd, *Org. Syntheses*, Coll. Vol. 1, 330 (1941).
4. J. W. Williams and C. D. Hurd, *J. Org. Chem.*, 5, 122 (1940).
5. A. B. Boese, *Ind. Eng. Chem.*, 32, 16 (1940).
6. F. O. Rice and R. Roberts, *J. Am. Chem. Soc.*, 65, 1677 (1943).
7. J. T. Fitzpatrick, *J. Am. Chem. Soc.*, 69, 2236 (1947).
8. G. J. Fisher, A. F. MacLean, and A. W. Schnizer, *J. Org. Chem.*, 18, 1055 (1953).

KETENE DI(2-METHOXYETHYL) ACETAL

[Ketene bis(2-methoxyethyl) acetal]

$$2\,CH_3OCH_2CH_2ONa + H_2C{=}CCl_2 \longrightarrow$$
$$H_2C{=}C(OCH_2CH_2OCH_3)_2 + 2\,NaCl$$

Submitted by WILLIAM C. KURYLA and JOHN E. HYRE [1]
Checked by EARL M. LEVI and PETER YATES

1. Procedure

In a dry 1-l. five-necked flask equipped with a mechanical stirrer, a reflux condenser (Note 1), a thermometer, a nitrogen inlet, and a stoppered port (Note 2) are placed 152 g. (2.00 moles) of 2-methoxyethanol (Note 3) and 50 g. of xylene (Note 4). A constant dry nitrogen purge is maintained on the apparatus throughout the following operations (Note 5).   Metallic sodium (46.0 g., 2.00 moles) is added in small chunks through the stoppered port to the stirred reaction mixture over a 2-hour period at a temperature of 130–150°.   After all the sodium has reacted (Note 6), heating is discontinued, and 120 g. (1.24 moles) of vinylidene chloride (Note 7) is added dropwise to the stirred reaction mixture over a 20-minute period.   During the addition of vinylidene chloride the reaction mixture becomes dark, and its temperature increases rapidly from an initial 140° to a maximum of 170–175°.   It then decreases as addition continues, and concomitant precipitation of sodium chloride is noted.   Stirring is continued for an additional 10 minutes after completion of the addition of vinylidene chloride.   Anhydrous diethyl ether (100 ml.) is slowly added, serving both to reduce the viscosity and to cool the reaction mixture to about 60°.   This mixture is then filtered through a medium-grade fritted-glass funnel, and the sodium chloride cake is washed with several 20-ml. portions of fresh ether (Note 8).

The ethereal filtrate and washings are distilled under reduced pressure (Note 9) with the use of a 6 in. Vigreux column, and pure ketene di(2-methoxyethyl) acetal (Note 10) is obtained; b.p. 81–84° (2.0 mm.), $n^{25}D$ 1.4411, yield 98–132 g. (56–75%). The infrared spectrum of the product shows a very strong $C=C$ absorption band at 1640 cm.$^{-1}$.

2. Notes

1. A dry ice type of condenser has been found to be the most satisfactory because of the low boiling point (32°) of the vinylidene chloride.   An efficient water-cooled condenser is satisfactory, however.

2. An addition funnel is fitted in this port after the sodium addition is complete.

3. 2-Methoxyethanol (methyl Cellosolve) from Union Carbide Corporation, Chemicals Division, was used.

4. Xylene (analytical reagent grade) from Mallinckrodt Chemical Works was distilled from sodium before use.

5. This is essential to avoid both the excessive oxidation of the reactants and the danger of a sodium-sparked fire.

6. Small amounts of metallic sodium, such as a few *very* small spheres floating in the reaction mass, are tolerable as long as dry nitrogen is being continuously passed through the reaction flask. A slow nitrogen purge is also maintained on the apparatus during the addition of the vinylidene chloride. The checkers found that appreciable amounts of sodium remained unconsumed after 3.5 hours; they added more 2-methoxyethanol (10–12 g.) to complete the reaction (significant loss of this reagent appeared to occur from the port during addition of the sodium).

7. Vinylidene chloride (inhibited grade) from Dow Chemical Company was used.

8. The sodium chloride cake is washed with as many 20-ml. portions of ether as are required to make the filtrate essentially colorless (usually four or five).

9. The diethyl ether is collected directly into traps cooled in dry ice-acetone.

10. The directions of McElvain and Kundiger [2] regarding the storage of ketene acetals should be followed. The submitters have found that storage at 0°, in a bottle which was previously washed with a hot concentrated caustic solution, is satisfactory.

3. Methods of Preparation

Ketene di(2-methoxyethyl) acetal has been obtained by the present method with the use of diethylene glycol dimethyl ether as solvent.[3] Other methods for the preparation of ketene acetals include the dehydrohalogenation of a halo acetal with potassium *t*-butoxide [2, 4] and the reaction of an α-bromo orthoester with metallic sodium.[5]

4. Merits of the Preparation

This synthetic process is applicable to the preparation of other ketene acetal derivatives of β-alkoxy alcohols. Examples include the ketene acetal derivatives of tetrahydrofurfuryl alcohol and 1-methoxy-2-propanol.[3] There are a number of advantages in its use, including a simple, time-saving procedure, readily available and inexpensive reagents, and good yields of ketene acetal obtained by a *one-step* method.

1. Union Carbide Corporation, Technical Center, Research and Development Department, South Charleston, West Virginia.
2. S. M. McElvain and D. Kundiger, *Org. Syntheses*, Coll. Vol. **3**, 506 (1955).
3. W. C. Kuryla and D. G. Leis, *J. Org. Chem.*, **29**, 2773 (1964).
4. F. Beyerstedt and S. M. McElvain, *J. Am. Chem. Soc.*, **58**, 529 (1936).
5. P. M. Walters and S. M. McElvain, *J. Am. Chem. Soc.*, **62**, 1482 (1940).

α-KETOGLUTARIC ACID*

(Glutaric acid, 2-oxo-)

$$\begin{array}{c} CH_2CO_2C_2H_5 \\ | \\ CH_2CO_2C_2H_5 \end{array} + \begin{array}{c} CO_2C_2H_5 \\ | \\ CO_2C_2H_5 \end{array} \xrightarrow{C_2H_5ONa}$$

$$\begin{array}{c} COCO_2C_2H_5 \\ | \\ CHCO_2C_2H_5 \\ | \\ CH_2CO_2C_2H_5 \end{array} \xrightarrow{3H_2O\ (H^+)} \begin{array}{c} COCO_2H \\ | \\ CH_2 \\ | \\ CH_2CO_2H \end{array} + 3C_2H_5OH + CO_2$$

Submitted by E. M. BOTTORFF and L. L. MOORE [1]
Checked by WILLIAM G. DAUBEN and ROBERT M. COATES

1. Procedure

A. *Triethyl oxalylsuccinate.* In a 2-l. three-necked flask equipped with a sealed stirrer and a reflux condenser bearing a calcium chloride drying tube is placed 356 ml. (276 g., 6.00 moles) of anhydrous ethanol (Note 1). Sodium (23 g., 1.0 g. atom) is added in small portions at a rate sufficient to keep the ethanol boiling. External heating is required to dissolve the last portions of the metal. After all the sodium has dissolved, the excess ethanol is removed by distillation at atmospheric pressure; as the mixture becomes pasty, dry toluene is added in sufficient amounts to permit stirring and to prevent splattering of the salt. Distillation and addition of toluene is continued until all the ethanol is removed and the contents of the flask reach a temperature of 105° (Note 2). The sodium ethoxide slurry is cooled to room temperature and 650 ml. of anhydrous ether is added, followed by 146 g. (1.00 mole) of diethyl oxalate. To the yellow solution there is added 174 g. (1.00 mole) of diethyl succinate, and the mixture is allowed to stand at room temperature for at least 12 hours.

The mixture is hydrolyzed by the addition of 500 ml. of water with stirring. The layers are separated, the ether layer is washed with 150 ml. of water, and the ether layer is discarded. The combined aqueous layers are acidified with 100 ml. of 12N hydrochloric acid, and the layers are separated. The aqueous layer is extracted with three 150-ml. portions of ether, which are added to the oily layer. The ethereal solution is dried over anhydrous magnesium sulfate, and the ether is removed by evaporation under water-pump pressure at a bath temperature of 35–45°. Triethyl oxalylsuccinate, a yellow oil weighing 235–250 g. (86–91%), remains in the flask (Note 3).

B. α-*Ketoglutaric acid.* A mixture of 225 g. (0.82 mole) of triethyl oxalylsuccinate, 330 ml. of 12N hydrochloric acid, and 660 ml. of water is heated under reflux for 4 hours, and the mixture is distilled to dryness under reduced pressure at a bath temperature of 60–70° (Note 4). The liquid residue, which solidifies readily on standing, is warmed with 200 ml. of nitroethane on a steam bath until it is in solution. The warm solution is filtered, the funnel is washed with 40 ml. of nitroethane, and the filtrate is stirred at 0–10° for 5 hours. α-Ketoglutaric acid is separated by filtration and dried at 90° under reduced pressure for 4 hours. It is obtained as a tan solid; weight 88–99 g. (73–83%); m.p. 103–110° (Note 5).

2. Notes

1. Commercial absolute ethanol is dried by heating with sodium and diethyl succinate and is then distilled directly into the reaction flask.

2. If the toluene method to remove all the ethanol is not used, the yield is lower by 5–10%.

3. Triethyl oxalylsuccinate begins to decompose at 84° at 760 mm. It cannot be distilled without decomposition even at a pressure of 1 mm.

4. The color of the α-ketoglutaric acid is darker if the pot tem-

perature goes much above 90° during the evaporation and re-crystallization.

5. The product is pure enough for most purposes. Further recrystallization from nitroethane does not improve the melting point.

3. Methods of Preparation

The present procedure is a modification of one reported in an earlier volume of *Organic Syntheses*.[2] The methods used to pre-pare triethyl oxalylsuccinate and α-ketoglutaric acid are sum-marized in that volume.

4. Merits of the Preparation

The advantages of this procedure over the earlier version are the use of sodium ethoxide instead of potassium ethoxide and better reproducibility.

1. Organic Chemical Development, Eli Lilly and Company, Indianapolis, Indiana.
2. L. Friedman and E. Kosower, *Org. Syntheses*, Coll. Vol. **3**, 510 (1955).

D,L-KETOPINIC ACID

(1-Apocamphanecarboxylic acid, 2-oxo-)

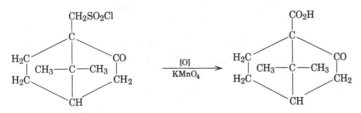

Submitted by PAUL D. BARTLETT and L. H. KNOX [1]
Checked by JOHN D. ROBERTS

1. Procedure

A 4-l. beaker containing a solution of 100 g. (0.95 mole) of anhydrous sodium carbonate in 900 ml. of water is placed on a steam bath, provision being made for efficient mechanical stirring. The stirrer is started and, when the solution is hot, one-third of a solution of 100 g. (0.63 mole) of potassium permanganate in 600 ml. of hot water is added all at once, followed by a 34-g. portion of D,L-10-camphorsulfonyl chloride (Note 1). After an interval of 5–10 minutes, half the remaining permanganate is poured in, followed by 33 g. of the chloride. After a similar interval, the remaining permanganate solution and a final 33-g. portion of the chloride are added and heating is continued for an hour.

The excess permanganate is destroyed by adding a few milliliters of an acidified solution of sodium sulfite. The reaction mixture is cooled and made strongly acidic by cautious addition (foaming may occur) of 20% sulfuric acid. The mixture is heated, and the precipitated manganese dioxide is dissolved by stirring in powdered sodium sulfite (usually 70–80 g. is required). The resulting solution is cooled and extracted with one 200-ml., two 150-ml., and one 100-ml. portions of ether. The combined ether extracts are dried over anhydrous sodium sulfate and the bulk of the ether removed by distillation from a steam bath. The residue is evaporated in a crystallizing dish (Note 2). The crude acid (38–45 g.) is recrystallized from hot water. Considerable oiling may occur and 250–400 ml. of water is usually required to give complete solution. The yield of recrystallized acid is 28–32 g. (38–43%), m.p. 233–234° (Note 3).

2. Notes

1. The camphorsulfonyl chloride is the crude product obtained as described on p. 196. If it is not carefully dried, it should be oxidized reasonably promptly after its preparation. The oxidation is conveniently carried out in 100-g. portions. Several reactions can easily be carried out in parallel.

2. The checker found it convenient to use a rotary evaporator at this point.

3. An additional small crop of crystals may be obtained by concentration of the mother liquor. The checker observed m.p. 240–242°.

3. Methods of Preparation

D,L-Ketopinic acid has been prepared by oxidation of bornyl

chloride with nitric acid at 20° [2] or with perbenzoic acid in acetic acid;[3] from 10,10-dinitrocamphan-2-ol [4] or apocamphan-2-ol-1-carboxylic acid [5] with alkaline permanganate; and from the oxidation of 10-camphorchlorosulfoxide, obtained from 10-camphorsulfonyl chloride by the action of pyridine, with potassium permanganate.[6] The present procedure represents a simplification of the latter and gives as high an overall yield.[7]

4. Merits of the Preparation

Ketopinic acid is of interest as a β-keto acid which fails to decarboxylate readily.[8] It may be converted to apocamphane-1-carboxylic acid.[7]

1. Converse Memorial Laboratory, Harvard University, Cambridge, Massachusetts. Preparation was submitted November 1, 1939.
2. H. E. Armstrong, *J. Chem. Soc.*, **69**, 1397 (1896).
3. G. Gallas and J. M. Montañés, *Anales Soc. Espan. Fis. Quim.*, **28**, 1163 (1930) [*C.A.*, **25**, 506 (1931)].
4. P. Lipp, *Ann.*, **399**, 241 (1913).
5. J. Bredt and R. May, *Chem. Ztg.*, **34**, 65 (1910) [*C.A.*, **4**, 1476 (1910)].
6. E. Wedekind, *Ber.*, **57**, 664 (1924).
7. P. D. Bartlett and L. H. Knox, *J. Am. Chem. Soc.*, **61**, 3184 (1939).
8. For references and discussion, see F. S. Fawcett, *Chem. Rev.*, **47**, 219 (1950).

18,20-LACTONE OF 3β-ACETOXY-20β-HYDROXY-5-PREGNENE-18-OIC ACID

(Pregn-5-en-18-oic acid, 3β,20β-dihydroxy, 18,20-lactone, 3-acetate)

Submitted by K. HEUSLER, P. WIELAND, and CH. MEYSTRE [1]
Checked by E. J. COREY and WILLIAM E. RUSSEY

1. Procedure

A. *3β-Acetoxy-20β-hydroxy-5-pregnene.* In a 2-l. five-necked flask fitted with a mechanical stirrer, 250-ml. dropping funnel, thermometer, nitrogen-inlet tube, and reflux condenser with calcium chloride tube is placed 750 ml. of anhydrous tetrahydrofuran (Note 1). The vessel is flushed with nitrogen, and 101.6 g. (0.4 mole) of lithium aluminum tri-*t*-butoxyhydride[2] (Note 2) is added. The suspension is cooled to about 2°, and 71.7 g. (0.2 mole) of pregnenolone acetate (Note 3) is added in one portion while stirring, the particles adhering to the wall of the flask being rinsed into the solution with an additional 50 ml. of tetrahydrofuran. The reaction mixture is stirred at 0–5° for 6 hours. A solution of 100 g. of ammonium sulfate in 150 ml. of water is added, with stirring, over a 15–20 minute period through the dropping funnel, the temperature of the reaction mixture being kept below 10° by efficient cooling with ice. A considerable quantity of hydrogen is evolved. There is added 20 g. of filter aid (Celite® or Hyflo Supercel®), the mixture is stirred for another 30 minutes, and it is finally filtered with suction through a layer of filter aid. The reaction vessel is rinsed and the filter residue thoroughly washed with 1.5 l. of tetrahydrofuran (Note 4). The filtrate is evaporated to dryness under reduced pressure. The crystalline residue is dissolved in 750 ml. of hot acetone, filtered (if necessary), and the solution is concentrated to a volume of about 200 ml. (crystallization may begin during this evaporation). The flask is kept overnight at 0° to −10° and the product isolated by suction filtration. The crystals are washed with 75 ml. of ice-cold acetone and dried at 60°. The yield of the product is 54–57 g. (75–79%), m.p. 161–164°,[3] $[\alpha]^{25}_D$ −74° (*c* 1.0, CHCl$_3$) (Note 5).

B. *3β-Acetoxy-18-iodo- and 18-hydroxy-18,20β-oxido-5-pregnene.* In a 5-l. three-necked flask fitted with a mechanical stirrer, a thermometer, and a reflux condenser are placed 3 l. of cyclohexane (Note 6), 180 g. (*ca.* 0.37 mole) of commercial lead tetraacetate containing approximately 10% acetic acid (Note 7), 24 g. (0.095 mole) of iodine and 30 g. (0.083 mole) of 3β-acetoxy-20β-hydroxy-5-pregnene. The reaction mixture is stirred and heated to the boiling point by irradiation with a 1000-watt lamp (Note 8) from underneath. When the iodine color has disappeared (usually after about 60–90 minutes) (Note 9), the reaction mixture is cooled to room temperature, filtered with suction, and the filter residue is rinsed with 600 ml. of cyclohexane. The filtrate is washed with two 500-ml. portions of 5% sodium thiosulfate solution and then with water. The combined aqueous solutions are extracted once with 500 ml. of ether. To the combined organic layers is added 6 ml. of pyridine, the solution is dried over sodium sulfate, filtered (Note 10), and the solvent evaporated under reduced pressure at a bath temperature of 35–40° (preferably by using a rotary evaporator). About 60 g. of an oily residue (Note 11), which is immediately oxidized (Note 12), is obtained.

C. *Oxidation to the 18,20-lactone of 3β-acetoxy-20β-hydroxy-5-pregnene-18-oic acid.* The above residue is dissolved in 600 ml. of acetone (Note 13), the solution is transferred to a 3-l. three-necked flask with a rigid mechanical stirrer, a dropping funnel, and a thermometer; the evaporation flask is rinsed with an additional 120 ml. of acetone. The solution is cooled to 0° to +5°, and 38.4 ml. of a chromic acid solution [4] (prepared by mixing 13.3 g. of chromium trioxide and 11.5 ml. of concentrated sulfuric acid and carefully diluting the mixture to 50.0 ml. with water while cooling) is slowly added within 10 minutes from the dropping funnel. The mixture is stirred (Note 14) for another 30 minutes at 0° to +5°, and a solution of 270 g. of crystalline sodium acetate in 780 ml. of water is added. The dark green solution is transferred to a separatory funnel, and it is extracted once with 2.4 l. and once with 600 ml. of benzene. Each extract is washed twice with 600 ml. of half-saturated sodium chloride solution, dried over sodium sulfate, and the solvent is evaporated under reduced pressure. The combined semisolid residue (48–50 g.) is triturated with 50 ml. of ether and kept overnight at 0° to −10°. The crude product is filtered, and the filter residue is washed with pentane. The yield of crystalline lactone is 14–16 g. (45–52% based on pure 3β-acetoxy-20β-hydroxy-5-pregnene). For further purification the product is dissolved in 200 ml. of hot acetone, if necessary 250 mg. of charcoal is added, the mixture is brought to the boiling point on a steam bath, filtered through a layer of filter aid, and the filter residue is washed with warm acetone. The solution is concentrated on the steam bath to a volume of about 90 ml. During this operation, crystallization begins. Then 150 ml. of hexane is added, the mixture is again concentrated to a volume of about 50–80 ml.

with swirling, and finally kept at 0° overnight. A first crop of 12.0–13.5 g. (39–43%) of pure lactone, m.p. 201–206°, is obtained, $[\alpha]^{25}$D −44° to −45° (c 1.0, CHCl₃). Concentration of the mother liquor yields a second crop of 0.3–2.7 g. of less pure lactone.

2. Notes

1. Tetrahydrofuran freshly distilled from lithium aluminum hydride should be used. A commercial product with a peroxide content giving a positive iodine test must be treated with about 0.3% of cuprous chloride (boiling for 30 minutes and distillation) before the addition of the hydride. [*Caution! See p. 976.*]

2. Obtained from Metal Hydrides Inc., Beverly, Massachusetts.

3. A commercial product, m.p. 142.5–148.5°; $[\alpha]^{20}$D +141.5° (c 1.003, CHCl₃) was used.

4. Tetrahydrofuran free of peroxide, distilled from cuprous chloride, may be used. [*Caution! See p. 976.*]

5. The first crop of the product contains only a trace of the 20α-epimer.[3] The main portion of this compound is found in the mother liquor. Use of the material of a second crop for the subsequent steps is not recommended. The residue of the mother liquor can, however, be reoxidized to pregnenolone acetate by the method described in step C. By crystallization of the oxidation product of the mother liquor residue from methanol, 13–14 g. of pure pregnenolone acetate can be recovered.

6. Commercial product, redistilled. Small amounts of cyclohexene do not interfere.

7. Obtained from Fluka A. G., Buchs, S. G., Switzerland, and Arapahoe Chemical Co., Boulder, Colorado. Dry lead tetraacetate may also be used.

8. An ordinary 1000-watt lamp was used for heating as well as irradiation. The light reduces the induction period and accelerates the reaction. It is important that the solution be at reflux during the irradiation; an aluminum foil tent may be used to prevent excessive loss of energy from the light/heat source. In smaller runs the intensity of the light should be reduced accordingly (see Note 9).

9. The disappearance of the iodine color is not indicative

of the end of the reaction, since lead tetraacetate itself reacts with iodine under the reaction conditions giving lead diacetate, carbon dioxide, and methyl iodide. Under intense irradiation the decolorization can take place very quickly. In this case the rate of the decomposition of lead tetraacetate becomes comparable to the rate of the desired hypoiodite reaction, and the intermediate 18-iodo-20-alcohol will accumulate. The latter compound is oxidized by chromic acid to the 18-iodo-20-ketone. If an accumulation of the iodo alcohol is observed, a weaker light source or a larger excess of lead tetraacetate and iodine should be used to bring the reaction to completion.

10. The filtrate contains labile iodine derivatives which in light and at room temperature give off iodine. If the solution is not concentrated immediately, it should be kept at $0°$ in the dark; but it should be processed within less than 4 hours because of the instability of the hemiacetal-type intermediate.

11. The oil contains considerable amounts of derivatives formed by reaction with the solvent, e.g., cyclohexanol acetate, bicyclohexyl, and a number of high-boiling, iodine-containing substances. These by-products are removed only after oxidation.

12. It is important to oxidize the product as soon as possible because the crude 18,20-hemiacetal is unstable. In solution, in the presence of traces of acid, bimolecular anhydro products are formed which are stable to chromic acid oxidation and greatly diminish the yield of lactone.

13. Commercial acetone, boiled with 0.05% potassium permanganate for about 2 hours and distilled from potassium carbonate, was used.

14. The chromium sulfate tends to aggregate in large lumps. The stirrer should therefore be rigid and at least an inch away from the bottom of the flask.

3. Methods of Preparation

The preparation of the title lactone has been described by a ·multistep synthesis from holarrhimine.[5] The method described in detail above is essentially an application of the "hypoiodite reaction" published by Ch. Meystre and co-workers.[6] These authors also describe the isolation of the intermediate hemiacetal in pure form. Saturated lactones epimeric at C-20 have also been obtained by chromic acid oxidation of 18,20-dihydroxy compounds [7] which were in turn prepared by treatment of 20-hydroxypregnanes with lead tetraacetate, acetolysis of the resulting 18,20β-oxides, and hydrolysis. Saturated lactones of the 20α- and 20β-series were also obtained by photolysis of the corresponding 20-nitrites, hydrolysis, and oxidation.[8]

4. Merits of the Preparation

For the substitution of the angular methyl groups in steroids five methods are known: (a) homolysis of N-chloramines [Löffler-Freytag reaction [9] (only C-18)]; (b) oxidation of alcohols with lead tetraacetate; [10] (c) photolysis of nitrite esters; [11] (d) homolysis of hypochlorites; [12] (e) the "hypoiodite reaction." [13]

Of these methods the hypoiodite cleavage appears to be the simplest and most efficient one. It leads directly to compounds which are oxidized at the angular C-18 substituent to the aldehyde stage. In common with method (b), it has the advantage that an alcohol can be used directly as starting material, which under the reaction conditions is transformed into a derivative which is then homolytically cleaved; but, in contrast to method (b), the hypoiodite method is much less susceptible to steric effects, 20β-alcohols being oxidized almost as efficiently as 20α-alcohols. Methods (a), (c), and (d) require the formation of reactive derivatives in a separate step before homolysis. No special apparatus or special light source is needed for the hypoiodite reaction. Further applications and the scope of the reaction are discussed elsewhere.[14]

The lactone described can be used as starting material for the preparation of a number of 18-oxygenated steroids. Hydrolysis and Oppenauer oxidation [5b, 6] leads to the 18,20-lactone of 3-oxo-20β-hydroxy-4-pregnene-18-oic acid. This lactone is a suitable starting material for the preparation of 18-hydroxy- and 18-oxoprogesterone.[15] On the other hand, microbiological oxidation leads to the corresponding 11α-hydroxylactone [16] which is a suitable starting material for the preparation of aldosterone.[17]

1. Research Laboratories, Pharmaceutical Division, CIBA Limited, Basel, Switzerland.
2. H. C. Brown and R. F. McFarlin, *J. Am. Chem. Soc.*, **78**, 252 (1956); **80**, 5372 (1958); for reduction of 20-oxo steroids with this reagent, cf. K. Heusler, J. Kalvoda, P. Wieland, and A. Wettstein, *Helv. Chim. Acta*, **44**, 179 (1961).
3. P. Wieland and K. Miescher, *Helv. Chim. Acta*, **32**, 1922 (1949).
4. Cf. K. Bowden, I. M. Heilbron, E. R. H. Jones, and B. C. L. Weedon, *J. Chem. Soc.*, 39 (1946); C. Djerassi, R. R. Engle, and A. Bowers, *J. Org. Chem.*, **21**, 1547 (1956).
5. L. Lábler and F. Šorm, (a) *Chem. Ind.* (*London*), 598 (1959); *Coll. Czech. Chem. Commun.*, **25**, 265 (1960); (b) *Chem. Ind.* (*London*), 935 (1960); *Coll. Czech. Chem. Commun.*, **25**, 2855 (1960).
6. Ch. Meystre, K. Heusler, J. Kalvoda, P. Wieland, G. Anner, and A. Wettstein, *Helv. Chim. Acta*, **45**, 1317 (1962).
7. B. Kamber, G. Cainelli, D. Arigoni, and O. Jeger, *Helv. Chim. Acta*, **43**, 347 (1960).
8. D. H. R. Barton, J. M. Beaton, L. E. Geller, and M. M. Pechet, *J. Am. Chem. Soc.*, **82**, 2640 (1960); **83**, 4076 (1961).
9. P. Buchschacher, J. Kalvoda, D. Arigoni, and O. Jeger, *J. Am. Chem. Soc.*, **80**, 2905 (1958); E. J. Corey and W. R. Hertler, *J. Am. Chem. Soc.*, **80**, 2903 (1958); **81**, 5209 (1959).
10. G. Cainelli, M. Lj. Mihailović, D. Arigoni, and O. Jeger, *Helv. Chim. Acta*, **42**, 1124 (1959).
11. Cf. A. L. Nussbaum and C. H. Robinson, *Tetrahedron*, **17**, 35 (1962).
12. M. Akhtar and D. H. R. Barton, *J. Am. Chem. Soc.*, **83**, 2213 (1961); J. S. Mills and V. Petrow, *Chem. Ind.* (*London*), 946 (1961).
13. Ch. Meystre, K. Heusler, J. Kalvoda, P. Wieland, G. Anner, and A. Wettstein, *Experientia*, **17**, 475 (1961).
14. K. Heusler and J. Kalvoda, *Angew. Chem.*, **76**, 518 (1964) [*Angew. Chem. Intern. Ed.*, **3**, 525 (1964)].
15. J. Kalvoda, J. Schmidlin, G. Anner, and A. Wettstein, *Experientia*, **18**, 398 (1962).
16. L. Lábler and F. Šorm, *Chem. Ind.* (*London*), 1114 (1961); *Coll. Czech. Chem. Commun.*, **27**, 276 (1962).
17. K. Heusler, P. Wieland, and A. Wettstein, *Helv. Chim. Acta*, **44**, 1374 (1961); K. Heusler, J. Kalvoda, Ch. Meystre, P. Wieland, G. Anner, A. Wettstein, G. Cainelli, D. Arigoni, and O. Jeger, *Helv. Chim. Acta*, **44**, 502 (1961).

LEVOPIMARIC ACID

Pine oleoresin + $(CH_3)_2$—C—NH_2 (with CH_2OH below) $\longrightarrow$

(structure with $CH(CH_3)_2$, $COO^{\ominus}$, $H_3\overset{\oplus}{N}$—C—$(CH_3)_2$ with CH_2OH below)

Amine salt + H_3PO_4 $\longrightarrow$

(structure with $CH(CH_3)_2$ and CO_2H)

Submitted by Winston D. Lloyd and Glen W. Hedrick [1]
Checked by William G. Dauben and Robert M. Coates

1. Procedure

Pine oleoresin [1 kg. containing 260 g. (0.86 mole) of levo-pimaric acid] (Notes 1 and 2) is dissolved in 2 l. of acetone in a 4-l. beaker. A solution of 200 g. (2.2 moles) of 2-amino-2-methyl-1-propanol (Note 3) in 200 ml. of acetone is added as rapidly as possible with stirring. The pasty precipitate which forms almost immediately is collected by suction filtration and is pressed as dry as possible using a rubber dam (Note 4). The crude moist precipitate is returned to a 2-l. beaker and is dissolved in the minimum volume (~1 l.) of boiling methanol. The methanolic solution is cooled to 5° in a refrigerator and stirred occasionally to expedite crystallization. When the crystallization is completed, the solid is collected by suction filtration. The precipitate is redissolved in a minimum volume of boiling methanol (~1 l.) (Note 5), the solution concentrated to two-thirds its original volume (Note 6), cooled to 5°, and the amine salt allowed to crystallize. The solid is filtered by suction, and the filter cake is air-dried to yield 68–78 g. (20–23% of the available levopimaric acid) of the 2-amino-2-methyl-1-propanol salt of levopimaric acid, $[\alpha]^{25}$D $-202°$ (Notes 7 and 8). The recrystallization is repeated; approximately 0.8 l. of boiling methanol is used and then concentrated, the yield of amine salt is 41–46 g. (12–14% of the available levopimaric acid), $[\alpha]^{25}$D $-210°$ (Note 9).

In a 1-l. separatory funnel there are first placed 400 ml. of ether and 75 ml. of 10% phosphoric acid (Note 10), and then the above amine salt is added (Note 11). The mixture is shaken vigorously for a few minutes, an additional 50 ml. of 10% phosphoric acid added, and the vigorous shaking continued until all the solid has disappeared. The ether layer is separated, washed twice with 100-ml. portions of water, and dried over anhydrous sodium sulfate. The drying agent is separated by filtration, the ether is removed at room temperature under reduced pressure using a rotary evaporator, and the residue dissolved in 40–60 ml. of boiling ethanol. The levopimaric acid is collected by suction filtration, yield 26–31 g. (10–12%), m.p. 147–150°, $[\alpha]^{25}$D $-265°$ (Notes 12, 13, and 14).

2. Notes

1. The longleaf pine (*Pinus palustris*) oleoresin used was analyzed by the method of Lloyd and Hedrick [2] and was found to contain a total resin acid content of 660 g. The oleoresin used by the checkers was obtained from Shelton Naval Stores Co., Valdosta, Georgia. The oleoresin can also be obtained from the following sources: K. S. Varn and Co., Hoboken, Georgia; The Langdale Co., Valdosta, Georgia; Vidalia Gum Turpentine Co., Vidalia, Georgia; Stallworth Pine Products Co., Mobile, Alabama; Filtered Rosin Products Company, Baxley, Georgia; Taylor-Lowenstein and Co., Mobile, Alabama; and Nelio Chemicals, Inc., Jacksonville, Florida.

2. If any woody material remains undissolved, it should be removed by filtration of the acetone solution.

3. The 2-amino-2-methyl-1-propanol, m.p. 25–29°, N.E. 88.5–99.0, was obtained from the Commercial Solvents Corporation

and was used without further purification. The checkers obtained their material from Matheson Coleman and Bell Co.

4. The use of a rubber dam is essential in this step to effect the separation of the residual acetone. It is also beneficial to use a rubber dam in the other suction filtrations in this process.

5. If a clear solution is not obtained, the undissolved material should be removed by filtration.

6. Concentration of the solution at this point gives a major improvement in yield. Crystallization does not occur during this concentration step unless the solution is seeded.

7. All rotations were taken with a 2% methanolic solution.

8. If the rotation is −210° or more negative, the next recrystallization may be omitted and the levopimaric acid generated directly.

9. The maximum observed rotation for the 2-amino-2-methyl-1-propanol salt of levopimaric acid is $[\alpha]^{24}D$ −218°.[3] Methanol and ethanol solutions give the same specific rotations, but methanol is the preferred solvent because the time required to effect solution in ethanol is longer. If pure levopimaric acid, m.p. 151–153°, $[\alpha]^{24}D$ −276° is desired, the salt with −210° rotation should be dissolved in 8 parts of boiling methanol, the solution concentrated to the point of incipient crystallization, cooled, and filtered. The yield in this recrystallization is about 70%.

10. The submitters find phosphoric acid more convenient than boric acid[3] or acetic acid.[4] Acid isomerization to abietic acid[3, 5] did not occur under the conditions used here.

11. After an induction period of approximately 1 week, the amine salt begins to be oxidized by air and the salt should be converted to levopimaric acid as soon as possible after it has been isolated.

12. The checkers found that their material with $[\alpha]^{25}D$ −260° had a melting point at 125–150°. The melting point is very sensitive to impurities, and a few percent of impurities can lower it drastically.

13. The yields obtained using this procedure can vary with the source of the oleoresin; the submitters report a yield of 29–34% of levopimaric acid.

14. Using slash pine (*Pinus elliotti*) oleoresin containing ap-

proximately 16% of levopimaric acid,[6, 7] the submitters found yields consistently less than their reported 29–34%. Pine "scrape," a material which crystallizes on the surface of the pine tree, has been used in a similar process [4] but gives highly variable yields and, owing to interference by oxidation products, may fail to give the desired material. To avoid the deleterious effects of oxidized materials, the use of fresh oleoresin is recommended.

3. Methods of Preparation

The described process of isolating levopimaric acid is based on the method of Summers, Lloyd, and Hedrick.[8] The procedure, a modification of the process devised by Harris and Sanderson [3] and Loeblich, Baldwin, O'Connor, and Lawrence,[4] is more convenient and gives improved yields.

4. Merits of the Preparation

In pine oleoresin, many resin acids occur. This procedure illustrates how, by the use of a specific amine, it is possible to get a specific precipitation of one resin acid from a mixture of acids.

1. Contribution of the Naval Stores Laboratory, Olustee, Florida, one of the laboratories of the Southern Utilization Research and Development Division, Agricultural Research Service, U.S. Department of Agriculture.
2. W. D. Lloyd and G. W. Hedrick, *J. Org. Chem.*, **26**, 2029 (1961).
3. G. C. Harris and T. F. Sanderson, *J. Am. Chem. Soc.*, **70**, 334 (1948).
4. V. M. Loeblich, D. E. Baldwin, R. T. O'Connor, and R. V. Lawrence, *J. Am. Chem. Soc.*, **77**, 6311 (1955).
5. D. E. Baldwin, V. M. Loeblich, and R. V. Lawrence, *J. Am. Chem. Soc.*, **78**, 2015 (1956).
6. B. L. Davis and E. E. Fleck, *Ind. Eng. Chem.*, **35**, 171 (1943).
7. D. E. Baldwin, V. M. Loeblich, and R. V. Lawrence, *J. Chem. Eng. Data*, **3**, 342 (1958).
8. H. B. Summers, Jr., W. D. Lloyd, and G. W. Hedrick, *Ind. Eng. Chem., Prod. Res. Develop.*, **2**, 143 (1963).

2-MERCAPTOPYRIMIDINE *

(2-Pyrimidinethiol)

$$
\begin{array}{c}
CH(OC_2H_5)_2 \\
| \\
CH_2 \\
| \\
CH(OC_2H_5)_2
\end{array}
\quad + \quad
\begin{array}{c}
NH_2 \\
| \\
C{=}S \\
| \\
NH_2
\end{array}
\quad
\xrightarrow[-4C_2H_5OH]{HCl}
\quad
\underset{}{\text{pyrimidine-N·HCl-SH}}
\quad
\xrightarrow{NaOH}
$$

$$
\underset{}{\text{pyrimidine-SH}}
\quad + \quad NaCl \quad + \quad H_2O
$$

Submitted by Donald G. Crosby, Robert V. Berthold, and Herbert E. Johnson.[1]
Checked by B. Bellin, J. L. Gibbs, and V. Boekelheide.

1. Procedure

A. *2-Mercaptopyrimidine hydrochloride.* Thiourea (61 g., 0.80 mole) and 600 ml. of ethyl alcohol (Note 1) are placed in a 2-l. three-necked flask equipped with a sealed mechanical stirrer, a reflux condenser, and a stopper. The stirrer is started, and 200 ml. of concentrated hydrochloric acid is added in one portion through the open neck. After several minutes, when the warm mixture has become homogeneous, 176 g. (0.80 mole) of commercial-grade 1,1,3,3-tetraethyoxypropane (Note 2) is added rapidly, the open neck is stoppered, and the yellow solution is boiled for about 1 hour with continuous stirring. During this period the reaction mixture darkens in color and the product separates (Note 3).

The reaction mixture is chilled to about 10° by immersing it in an ice bath for about 30 minutes, and the yellow crystalline precipitate is collected on a Büchner funnel. It is then washed with 100 ml. of cold alcohol and air-dried at room temperature. The yield of 2-mercaptopyrimidine hydrochloride is 71–76 g.

(60–64%). The product is pure enough for most purposes (Note 4), but it may be recrystallized by dissolving it in 12N hydrochloric acid (10 ml. per gram of solid) at about 75°, filtering the hot solution through glass wool or a sintered glass filter, chilling the filtrate in ice, and collecting the golden-yellow crystals on a sintered glass filter. Recovery is 60–65% (Note 5).

B. *2-Mercaptopyrimidine.* Crude 2-mercaptopyrimidine hydrochloride (25 g., 0.17 mole) is suspended in 50 ml. of water in a beaker and stirred rapidly while a 20% aqueous solution of sodium hydroxide (about 27 ml.) is added until the pH of the mixture is 7–8 (Note 6). The precipitated solid is collected on a Büchner funnel and washed on the funnel with 50 ml. of cold water. The damp product is dissolved by heating it in a mixture of 300 ml. of water and 300 ml. of alcohol on the steam bath, and the hot solution is filtered through a fluted paper and allowed to cool slowly to room temperature. The crystals of 2-mercaptopyrimidine are collected, washed with about 50 ml. of the aqueous alcohol, and dried either at room temperature overnight or for several hours in an oven at 110°. The yield is 15–16 g. (80–85%) of yellow needles, m.p. 218–219° (sealed tube).

2. Notes

1. Either commercial absolute alcohol or the 95% grade may be used.

2. 1,1,3,3-Tetraethoxypropane is available from Kay-Fries Chemicals, Inc., New York 16, New York, or from Distillation Products Industries, Rochester 3, New York, and may be used without further purification.

3. Longer boiling does not affect the yield but causes the product to be somewhat dark-colored. A shorter heating period or lack of mechanical stirring decreases the yield.

4. Electrometric titration shows the purity to be at least 95%. The product does not melt below 300°.

5. If concentrated sulfuric acid is substituted for the hydrochloric acid in the procedure, 2-mercaptopyrimidine bisulfate is obtained in about 50% yield. Recrystallization from aqueous acetic acid provides the bisulfate as yellow needles, m.p. 186–

186.5°. (*Anal.* Calcd. for $C_4H_6N_2O_4S_2$: C, 22.9; H, 2.9. Found:
C, 23.1; H, 2.9.)

6. pH indicator paper may be used, or the solution may be
made weakly basic to litmus. Excess base dissolves the product
and should be avoided.

3. Methods of Preparation

The synthesis of 2-substituted pyrimidines from 1,3-dicarbonyl
compounds and urea derivatives was first described by Evans [2]
and was later improved by Hunt, McOmie, and Sayer [3] for the
preparation of 2-mercapto-4,6-dimethylpyrimidine. Burness [4]
employed 3-ketobutyraldehyde acetal in this procedure to give
2-mercapto-4-methylpyrimidine.   2-Mercaptopyrimidine has
been prepared from 1,1,3,3-tetraethoxypropane and thiourea by
variations of this basic method [3,5,6] as well as by the reaction of
2-chloropyrimidine with thiourea [7] or sodium hydrosulfide. [8]

4. Merits of the Preparation

This preparation describes a convenient and general method of
synthesis of substituted pyrimidines from compounds containing
a β-dicarbonyl group, either intact or as the corresponding ketal.
The usefulness of the 2-mercaptopyrimidines is enhanced by the
ease of removal of the mercapto group by desulfurization [9]
or oxidation [10] and its replacement by other functional groups. [10]

1. Research Department, Union Carbide Chemicals Company, South Charleston,
   West Virginia.
2. P. N. Evans, *J. Prakt. Chem.*, [2] 48, 489 (1893).
3. R. R. Hunt, J. F. W. McOmie, and E. R. Sayer, *J. Chem. Soc.*, 525 (1959).
4. D. M. Burness, *J. Org. Chem.*, 21, 97 (1956).
5. T. V. Protopopova and A. P. Skoldinov, *J. Gen. Chem. (USSR)*, 27, 1276
   (1957).
6. J. W. Copenhaver and R. F. Kleinschmidt, Canadian Patent 534,307 (1956).
7. M. P. V. Boarland and J. F. W. McOmie, *J. Chem. Soc.*, 1218 (1951).
8. R. O. Roblin, Jr., and J. W. Clapp, *J. Am. Chem. Soc.*, 72, 4890 (1950).
9. G. R. Pettit and E. E. van Tamelen, *Org. Reactions*, 12, 364 (1962).
10. G. W. Kenner and A. Todd, in R. C. Elderfield, "Heterocyclic Compounds,"
    Vol. 6, John Wiley & Sons, New York, 1956, pp. 283-287.

MESITOIC ACID*

(Benzoic acid, 2,4,6-trimethyl-)

$$+ \text{ClCOCOCl} \xrightarrow[\text{(2) } H_2O^+]{\text{(1) } AlCl_3}$$

$+ CO + 2HCl$

Submitted by PHILLIP E. SOKOL [1]
Checked by MELVIN S. NEWMAN and VERN G. DEVRIES

1. Procedure

Caution! The reaction should be carried out in a hood because carbon monoxide is evolved.

The apparatus consists of a 2-l. three-necked flask fitted with a sealed stirrer, a 500-ml. addition funnel, and a condenser protected by a drying tube connected to an alkaline trap. In it are placed 146 g. (1.10 moles) of anhydrous aluminum chloride (Note 1) and 700 ml. of dry carbon disulfide. The suspension is cooled to 10–15° in an ice bath, and 139 g. (1.10 moles) of oxalyl chloride (Note 2) is added dropwise with stirring over a 30-minute period. After this addition the reaction mixture is stirred for 15 minutes. A solution of 120 g. (1.00 mole) of mesitylene (Note 3) in 200 ml. of dry carbon disulfide is added dropwise with stirring over a 1-hour period to the mixture, the temperature being maintained at 10–15°. Hydrogen chloride evolution is observed after about 5 minutes, and a red complex soon forms.

After the addition is completed, the reaction mixture is refluxed for 1 hour and is then poured very cautiously with manual stirring onto a mixture of 2 kg. of crushed ice and 300 ml. of 12N hydrochloric acid in a 4-l. beaker. The mixture thus formed is extracted with three 250-ml. portions of carbon tetrachloride. The combined organic extracts are washed with two 500-ml. portions of water, and the acid is then extracted with 500 ml. of ice-cold 10% sodium hydroxide solution. The aqueous extract is then slowly added to 250 ml. of 6N hydrochloric acid. The suspension is cooled, and the mesitoic acid is separated by filtration, washed thoroughly with water, and dried. The colorless crude acid (m.p. 149–150°) weighs 106–124 g. (65–76%) (Note 4) and is sufficiently pure for most purposes (Note 5).

2. Notes

1. Good results have been obtained with several different varieties of anhydrous aluminum chloride.

2. High-purity commercial oxalyl chloride was used without further purification.

3. Commercial mesitylene of high purity (99+%) was used.

4. Similar yields were obtained when experiments were run on a 0.10-mole scale.

5. For recrystallization, 10 g. of crude acid is dissolved in 20 ml. of 45% methanol at reflux. About 9.5 g. of mesitoic acid, m.p. 153–154° (uncor.),[2] is obtained.

3. Methods of Preparation

Mesitoic acid has been prepared by carbonation of mesityl-magnesium bromide;[2–4] by hydrolysis of its amide prepared by condensation of mesitylene with carbamyl chloride under the influence of aluminum chloride;[5] by oxidation of isodurene with dilute nitric acid;[6,7] by distillation of 2,4,6-trimethylmandelic acid (low yield);[8] by dry distillation of 2,4,6-trimethylphenyl-glyoxylic acid;[9] by oxidation of the latter with potassium permanganate;[10] and by treating 2,4,6-trimethylphenylglyoxylic acid with concentrated sulfuric acid in the cold[11] or with heating.[12]

4. Merits of the Preparation

The method described in this preparation of mesitoic acid avoids the preparation of bromomesitylene,[13] and the yield of acid is essentially the same as that from the two-step synthesis.[2,13] This procedure appears to be general and can be used to prepare such acids as α- and β-naphthoic acids,[14] cumenecarboxylic acid, 2,5-dimethylbenzoic acid, and durenecarboxylic acid. Carboxylic acids could not be obtained from benzothiophene, veratrole, p-dimethoxybenzene, and ferrocene under the conditions of this reaction. Although there has been no exhaustive study, this procedure is probably applicable to a variety of aromatic compounds, especially alkylated aromatics. Aromatic compounds which readily undergo oxidation, e.g., ferrocene, catechol, and hydroquinone, do not lend themselves to this method.

1. Northwestern University, Evanston, Illinois; present address, De Soto Chemical Coatings, Chicago, Illinois.
2. D. M. Bowen, *Org. Syntheses*, Coll. Vol. **3**, 553 (1955).
3. R. P. Barnes, *Org. Syntheses*, Coll. Vol. **3**, 555 (1955).
4. E. P. Kohler and R. Baltzly, *J. Am. Chem. Soc.*, **54**, 4015 (1932).
5. A. Michael and K. J. Oechslin, *Ber.*, **42**, 329 (1909).
6. P. Jannasch and M. Weiler, *Ber.*, **27**, 3444 (1894).
7. O. Jacobsen, *Ber.*, **15**, 1855 (1882).
8. V. Meyer and W. Molz, *Ber.*, **30**, 1273 (1897).
9. E. Feith, *Ber.*, **24**, 3544 (1891).
10. A. Claus, *J. Prakt. Chem.*, (2) **41**, 506 (1890).
11. S. Hoogewerff and W. A. Van Dorp, *Rec. Trav. Chim.*, **21**, 358 (1902).
12. M. L. van Scherpenzeel, *Rec. Trav. Chim.*, **19**, 377 (1900).
13. L. I. Smith, *Org. Syntheses*, Coll. Vol. **2**, 95 (1943).
14. L. Gattermann, *Ann.*, **244**, 56 (1888).

METHANESULFINYL CHLORIDE

Method I

$$CH_3SSCH_3 + Cl_2 \rightarrow 2CH_3SCl$$

$$CH_3SCl + Cl_2 \rightarrow CH_3SCl_3$$

$$CH_3SCl_3 + (CH_3CO)_2O \rightarrow CH_3SOCl + 2CH_3COCl$$

Submitted by IRWIN B. DOUGLASS and RICHARD V. NORTON[1]
Checked by R. D. THOMPSON and HENRY E. BAUMGARTEN

1a. Procedure

In a 500-ml., three-necked flask, fitted with an efficient sealed stirrer, a gas inlet tube extending within 1 in. of the reaction mixture but not below its surface, a gas outlet tube connected to a calcium chloride tube, and a low-temperature thermometer to register the temperature of the reaction mixture, are placed 23.55 g. (0.25 mole) of freshly distilled methyl disulfide (Note 1) and 51.05 g. (0.5 mole) of acetic anhydride (Note 2). The reaction flask is then partially immersed in a bath of acetone cooled by dry ice until the internal temperature has reached 0° to −10°. The temperature is maintained within these limits throughout the reaction (Note 3). The reaction is best carried out in a well-ventilated hood.

Chlorine is passed into the well-stirred mixture at such a rate that the temperature is controlled between 0° and −10°. The progress of the reaction can be followed by color changes. At first the mixture turns yellow, then reddish as methanesulfenyl chloride (CH$_3$SCl) accumulates but gradually the color fades, and the mixture may become colorless. The addition of chlorine is terminated when a faint greenish-yellow color indicates an excess of chlorine (Note 4).

The reaction mixture is then transferred (Note 5) with a few fresh chips of porous plate to a distilling flask having a thermometer well, and the flask is fitted to a 12- to 18-in. Vigreux column and vacuum distillation apparatus. The fraction collector of the latter is connected through a trap cooled in an acetone-dry ice bath and a safety trap to a water pump. The cold trap must have sufficient capacity to contain, without backing up, all the acetyl chloride formed in the reaction.

The pressure is gradually decreased without heating the flask. The excess chlorine rapidly escapes and at 15 mm. the acetyl chloride boils smoothly below 0°. No heat is applied until most of the acetyl chloride has been removed (Note 6). Heat is then gradually increased until at 47-48° (15 mm.) the methanesulfinyl chloride distills as a nearly colorless liquid with n_D^{25} 1.500 (lit.[2] n_D^{25} 1.5038). Its nmr spectrum shows a single peak at δ 3.33 ppm with no sign of a singlet at δ 3.64 ppm, the frequency characteristic of methanesulfonyl chloride. The yield is 41–42 g. (83–86%) (Note 7).

Method II

$$CH_3SSCH_3 + 3Cl_2 \rightarrow 2CH_3SCl_3$$

$$CH_3SCl_3 + CH_3CO_2H \rightarrow CH_3SOCl + CH_3COCl + HCl$$

Submitted by IRWIN B. DOUGLASS and BASIL SAID FARAH.[1]
Checked by MELVIN S. NEWMAN, J. T. GOLDEN, and W. N. WHITE.

1b. Procedure

In a 1-l. three-necked flask, fitted with an efficient sealed stirrer, a gas inlet tube extending within 1 in. of the reaction mixture but not below its surface, and a gas outlet tube leading through a trap cooled in a dry ice-acetone bath to a hydrogen chloride absorption system (Note 8), are placed 94.2 g. (1 mole) of methyl disulfide (Note 9) and 120 g. (2 moles) of glacial acetic acid. The reaction flask is then surrounded by a bath of acetone cooled by dry ice until the internal temperature has reached 0° to −10°. The temperature is maintained within these limits throughout the reaction (Note 10) except that, toward the end, the temperature should be kept at $-10°$ to $-15°$ to minimize the escape of hydrogen chloride which might carry with it unreacted chlorine. The entire reaction is best carried out in a well-ventilated hood.

Three moles of chlorine (212.7 g.) is condensed in a tared flask cooled by a dry ice-acetone bath (Note 11). This flask is then connected to the gas inlet tube of the chlorination apparatus, and the cooling bath surrounding the chlorine container is removed. After about 2 hours the chlorine has all been absorbed by the reaction mixture (Note 12).

When the last of the chlorine has been added, the cooling bath is removed and, while vigorous stirring is continued (Note 13), the reaction mixture is allowed slowly to warm to room temperature. During this period there is a vigorous evolution of hydrogen chloride. The flask is finally warmed to 35° to facilitate the escape of more hydrogen chloride.

The reaction mixture and contents of the cold trap are then transferred (Note 14) to a 500-ml. distilling flask attached through a short fractionating column to a water-cooled condenser which is connected in series to a receiver, a trap cooled in a dry ice-acetone bath, and a hydrogen chloride absorption trap which may later be attached to a water pump. The mixture is then distilled until the pot temperature reaches 100° and practically all of the acetyl chloride has been driven over.

The residue, consisting chiefly of methanesulfinyl chloride, is then cooled immediately to 20° or lower and transferred to a 250-ml. flask and distilled through the same equipment under reduced pressure to remove the remaining acetyl chloride and other lower-boiling impurities. As the pot temperature begins to rise, a yellow intermediate fraction is collected very slowly until the distillate and pot temperatures are within 3° of each other (Note 15). At this point the distillation is temporarily discontinued while the accumulated acetyl chloride (Note 16) is removed from the cold trap and the receiver for the main product is attached. On resuming distillation the main product should come over within a 5° boiling range, and the distillation and pot temperature should remain within 2° of each other until the major part of the product has distilled. The yield is 161–177 g. (82–92%) of straw-colored or yellow product boiling at 55–59° (40 mm.) and having n_D^{25} 1.500–1.501 (Note 7).

2. Notes

1. The submitters used methyl disulfide obtained from the Crown Zellerback Corporation, Chemical Products Division, Camas, Washington, which was dried and redistilled. The fraction boiling at 108–109° was used. The checkers used. methyl disulfide obtained from the Aldrich Chemical Co., Inc.

2. The success of the reaction is closely related to having stoichiometric quantities of reagents. An excess of disulfide will produce a colored product due to the presence of methanesulfenyl chloride or products from the decomposition of methylsulfur trichloride. An excess of acetic anhydride will give a problem of purification during distillation, or if chlorine is also in excess, will lead to the formation of methanesulfonyl chloride.[3]

3. The reaction of acetic anhydride with methylsulfur trichloride is temperature dependent. If the chlorination is carried out at too low a temperature (>-20°), solid methylsulfur trichloride separates as white crystals which may cause the reaction mixture to become semi-solid. At 0° to -10°, however, no solid trichloride is observed.

4. At the colorless stage, the weight of the reaction mixture indicates that a stoichiometric quantity of chlorine has been added. In some preparations, a colorless condition is not achieved and one depends on the color of excess chlorine to indicate the end of the reaction. Excess chlorine does not seem to affect the quality of the product. An alternative to depending on the color change to determine the end of the reaction is to condense the calculated weight of chlorine in a tube and allow it to evaporate quantitatively into the reaction mixture.

5. Because of the unpleasant nature of methanesulfinyl chloride, acetyl chloride, and chlorine, the transfer should be made in the hood.

6. It is not necessary to empty the acetyl chloride from the cold trap until after the distillation is complete. Great care must be taken, however, to provide an intervening safety trap so that no water from the pump can back up into the trap holding the acetyl chloride.

7. Pure methanesulfinyl chloride boils at 48° (22 mm.) and 59° (42 mm.) and has n_D^{25} 1.5038, d_4^0 1.4044 and d_4^{25} 1.3706. On standing at room temperature, it slowly decomposes with the liberation of hydrogen chloride. It should not be stored for a long period in a tightly sealed container.

8. The hydrogen chloride absorption system must be of such design that there is no possibility for water to suck back into the reaction flask after a sudden surge of escaping hydrogen chloride.

9. The methyl disulfide was obtained from the Crown Zeller-bach Corporation, Chemical Products Division, Camas, Washington, and redistilled. The 108–109° boiling fraction was used.

10. Crystalline acetic acid separates at first but redissolves as the reaction progresses.

11. The flask should be full of chlorine gas when the tare weight is taken. The success of this preparation depends in large measure on the use of stoichiometric quantities of all re-agents. An excess or deficiency of any one will lead to an impure product and will greatly complicate the problem of purification.[2]

12. The white solid which collects inside the upper part of the flask is methylsulfur trichloride. This must be washed down with the cold reaction mixture before the flask warms to room temperature. The progress of the chlorination is accompanied by definite color changes. When one-third of the chlorine has been added, the reaction mixture is a deep reddish orange color which gradually fades as more chlorine is added until at the end the color should be a pale golden yellow or light straw color.

13. Vigorous stirring is necessary to prevent loss of material through too rapid escape of hydrogen chloride.

14. Since both methanesulfinyl and acetyl chlorides are unpleasant materials and the reaction mixture still contains much hydrogen chloride, all transfers should be made in the hood.

15. The difference between the distillation and pot temperatures is closely related to the success of this preparation. If the specified weights of reactants and temperatures are employed a 3° temperature difference should be reached before the yellow intermediate fraction, probably containing methanesulfenyl chloride, chloromethanesulfenyl chloride, and acetic acid, has attained a volume of 10 ml. Toward the end of the distillation the pot temperature may begin to rise owing to the presence in the residue of methanesulfonyl chloride (b.p. 63°/20 mm., 72°/31 mm., 82°/48 mm.), methyl methanethiolsulfonate (b.p. 115°/15 mm.) or both. The residue at the end of the distillation should amount to less than 10 ml.

16. The acetyl chloride obtained is yellow in color, probably because of the presence of the sulfenyl chlorides mentioned above. The addition of cyclohexene will discharge the color (although a darker color develops later) and redistillation then yields a stable water-clear product. The yield of acetyl chloride varies from 60% to 85%, depending on the care with which liquids are transferred and the vapors are trapped. The amount of sulfinyl chloride which may be recovered by redistilling the acetyl chloride fraction does not justify the time required.

3. Methods of Preparation

Alkanesulfinyl chlorides have been prepared by the action of thionyl chloride on alkanesulfinic acids[4,5] and by solvolysis of alkylsulfur trichlorides with water, alcohols, and organic acids.[2] Method II is an earlier procedure of the submitters,[6] which appears to be general for the preparation of sulfinyl chlorides in either the aliphatic or the aromatic series and is based on an improvement in the solvolysis method whereby the use of inert solvent is eliminated and the reaction is carried out in a one-phase system. In the submitters' laboratory this procedure has been supplanted by Method I which is superior to any of those previously employed in that the reaction is carried out in a homogeneous medium without the formation of hydrogen chloride gas. An adaptation of the present method is to chlorinate a mixture of a thiolester and acetic anhydride.[7] The latter modification is especially effective in the preparation of α-toluenesulfinyl chloride.

The present method seems to be general for the preparation of most sulfinyl chlorides in either the aliphatic or aromatic series. Benzyl and t-butyl disulfides, however, do not yield sulfinyl chlorides by this method because chlorine appears to cleave the carbon-sulfur rather than the sulfur-sulfur bonds.

Sulfinyl chlorides are highly reactive and unstable compounds. Methanesulfinyl chloride, at room temperature, disproportionates into methanesulfonyl and methanesulfenyl chlorides. It should *not* be stored in a sealed container at room temperature for a long period.[8] *Aromatic sulfinyl chlorides should not be distilled* because they have been known to explode on heating. They can, however, be prepared by the present method.

4. Uses of Methanesulfinyl Chloride

The utility of methanesulfinyl chloride lies in its great activity as a chemical intermediate. Through its ready hydrolysis, it serves as a convenient source of methanesulfinic acid. It reacts at low temperatures with aromatic amines to form sulfinamides, and with alcohols to form sulfinate esters. When it is hydrolyzed in the presence of an equimolar quantity of a sulfenyl chloride, a triolsulfonate ester is produced.

1. University of Maine, Orono, Maine.
2. I. B. Douglass and D. R. Poole, *J. Org. Chem.*, **22**, 536 (1957).
3. I. B. Douglass and R. V. Norton, *J. Org. Chem.*, **33**, 2104 (1968).
4. J. von Braun and K. Weissbach, *Ber.*, **63**, 2836 (1930).
5. A. Meuwsen and H. Gebhardt, *Ber.*, **69**, 937 (1936).
6. I. B. Douglass and B. S. Farah, *J. Org. Chem.*, **23**, 330 (1958); *Org. Syntheses*, **40**, 62 (1960).
7. M. L. Kee and I. B. Douglass, *Org. Prep. Proced.*, **2**, 235 (1970).
8. I. B. Douglass and D. A. Koop, *J. Org. Chem.*, **29**, 951 (1964).

1-(2-METHOXYCARBONYLPHENYL)PYRROLE

[Pyrrole, 1-(2-methoxycarbonylphenyl)-]

Submitted by A. D. Josey [1]
Checked by William G. Dauben and Juraj Hostynek

1. Procedure

A solution of 90 g. (0.59 mole) of methyl anthranilate (Note 1) in 265 ml. of glacial acetic acid is placed in a 1-l. round-bottomed flask equipped with a reflux condenser and a magnetic stirrer. The stirrer is started, and 78 g. (0.59 mole) of 2,5-dimethoxy-tetrahydrofuran (Note 2) is added during 10–15 minutes (Note 3). The solution is heated under reflux for 1 hour, during which time the solution turns deep red to black in color. The heating is discontinued, the condenser is replaced with a Vigreux column, and the acetic acid is removed by distillation at aspirator pressure. The dark residue is distilled under reduced pressure through a 25-cm. column packed with glass helices, and 84–96 g. (70–80%) of slightly yellow 1-(2-methoxycarbonylphenyl)pyrrole is collected, b.p. 90–95° (2 mm.), n^{25}D 1.5729.

2. Notes

1. Methyl anthranilate from Eastman Kodak Company was used without further purification.
2. 2,5-Dimethoxytetrahydrofuran from Eastman Kodak Company was used without further purification. This material also can be prepared by catalytic hydrogenation[2] of 2,5-dihydro-2,5-dimethoxyfuran.[3]

3. The submitter reports that much heat is liberated during the addition; the checkers did not find the reaction to be markedly exothermic.

3. Methods of Preparation

1-(2-Methoxycarbonylphenyl)pyrrole has not been prepared previously. An attempt to prepare the material via the mucic acid pyrrole synthesis using methyl anthranilate was unsuccessful.[4]

4. Merits of the Preparation

The condensation of primary amines with 2,5-dialkoxytetrahydrofurans to give in one step N-substituted pyrroles is applicable to a variety of substituted aliphatic and aromatic amines.[5] The method, largely developed by Clauson-Kaas and associates, has the advantages of simplicity, mild conditions, and generally excellent yields from readily available starting materials.

The submitter has used the method to prepare the corresponding 1-pyrrolyl derivatives [6] from the following amines in the indicated yields· ethyl β-aminobutyrate 88%, methyl β-aminoglutarate 87%, β-aminopropionitrile 58%, and 2,5-diamino-3,4-dicyanothiophene 22%.

On saponification 1-(2-methoxycarbonylphenyl)pyrrole yields 1-(2-carboxyphenyl)pyrrole, m.p. 106–107°, which on reaction with polyphosphoric acid at 70° is cyclized to 9-keto-9H-pyrrolo-(1,2-*a*)indole in 28–32% yield. Through the choice of the appropriate amine and acetal components, the substituted 1-(2-methoxycarbonylphenyl)pyrroles become readily available intermediates in the preparation of a variety of derivatives of the pyrrolo(1,2-*a*)indole ring system.

1. Contribution No. 977 from the Central Research Department, Experimental Station, E.I. du Pont de Nemours and Co., Inc.
2. J. Fakstorp, D. Raleigh, and L. E. Schniepp, *J. Am. Chem. Soc.*, **72**, 869 (1950).
3. D. M. Burness, this volume, p. 403.
4. D. A. Shirley, B. H. Gross, and P. A. Roussel, *J. Org. Chem.*, **20**, 225 (1955).
5. N. Clauson-Kaas and Z. Tyle, *Acta Chem. Scand.*, **6**, 667 (1952); **6**, 867 (1952).
6. A. D. Josey and E. L. Jenner, *J. Org. Chem.*, **27**, 2466 (1962).

1-(*p*-METHOXYPHENYL)-5-PHENYL-1,3,5-PENTANETRIONE

[1,3,5-Pentanetrione, 1-(*p*-methoxyphenyl)-5-phenyl-]

Submitted by MARION L. MILES, THOMAS M. HARRIS, and CHARLES R. HAUSER [1]

Checked by VICTOR NELSON, WAYLAND E. NOLAND, and WILLIAM E. PARHAM

1. Procedure

Caution! Sodium hydride causes severe burns if brought into contact with the skin and, in the dry state, is pyrophoric. Since hydrogen is evolved during the course of the reaction, the necessary precautions against fire and explosion should be taken.

A 1-l. three-necked flask is fitted with a sealed mechanical stirrer, an addition funnel with a pressure-equalizing side arm, and a reflux condenser with a gas take-off at the upper end. The gas take-off is connected by means of rubber tubing to one arm of a glass Y-tube. The other arm of the Y-tube is connected to a source of dry nitrogen gas. The bottom of the Y-tube is immersed just beneath the surface of a little 1,2-dimethoxyethane (monoglyme) contained in a small beaker.

The flask is swept with a stream of dry nitrogen. Monoglyme (100 ml.) (Note 1) and sodium hydride (6 g., 0.25 mole) (Note 2) are placed in the flask. A solution of 8.1 g. (0.050 mole) of benzoylacetone (Note 3) and 12.5 g. (0.075 mole) of methyl anisate (Note 4) in 100 ml. of monoglyme is placed in the addition funnel. The funnel is stoppered, and the nitrogen flow rate is adjusted so that approximately 10 bubbles per minute are emitted from the bottom of the Y-tube. The sodium hydride slurry is stirred and heated on the steam bath. When reflux is obtained, the solution of benzoylacetone and methyl anisate is added slowly so that hydrogen evolution is maintained at a controllable rate. The reaction mixture is kept at the reflux temperature for 6 hours.

The reaction mixture is then cooled to room temperature, the reflux condenser is replaced with a distillation condenser equipped with a vacuum take-off, and most of the solvent is removed under reduced pressure (*ca.* 100 mm.) until a thick paste is obtained. The mixture is then cooled in an ice water bath, and 150 ml. of ether is added. Cold water (200 ml.) is placed in the addition funnel, and initially the water is added dropwise (*Caution! Vigorous evolution of hydrogen*) until the excess sodium hydride is destroyed; then the remainder of the water is added more rapidly (Note 5).

The reaction mixture is poured into a 1-l. separatory funnel, and the aqueous layer is removed. The ether layer is extracted with two 100-ml. portions of water and then with 100 ml. of aqueous 1% sodium hydroxide. The extracts are combined with the original aqueous layer, and the resulting solution is washed once with 100 ml. of fresh ether. Crushed ice (100 g.) is added to the solution, followed by 30 ml. of 12N hydrochloric acid. The product, which precipitates at this point, is removed by filtration, washed with water, and recrystallized from 450 ml. (Note 6) of 95% ethanol. The yield of 1-(*p*-methoxyphenyl)-5-phenyl-1,3,5-pentanetrione is 11.4–12.8 g. (77–86%) (Note 7), m.p. 120–121.5°.

2. Notes

1. Eastman Kodak (Eastman grade) 1,2-dimethoxyethane was dried for 24 hours over calcium hydride, distilled from sodium metal, and then stored over calcium hydride.

2. The submitters used sodium hydride obtained as a 50% dispersion in mineral oil from Ventron Corp. This material was used as received.

3. The benzoylacetone was obtained from Eastman Kodak (Eastman grade), m.p. 57–58°.

4. Eastman Kodak (Eastman grade) methyl anisate was used without further purification.

5. The. checkers observed that the thick brown oil, which separates as the water is added, may partially solidify when stored overnight. The solid was melted on a steam bath, placed in the separatory funnel, and processed with the rest of the reaction mixture as described.

6. The checkers used 480 ml. of 95% ethanol.

7. The checkers found that the product retained as much as 2 g. of ethanol after 20 minutes of air drying. The product was dried to constant weight.

3. Methods of Preparation

The method described is that of Miles, Harris, and Hauser [2] and is an improvement over the earlier procedure of Hauser and co-workers.[3, 4] In the earlier method the dianion of benzoylacetone, formed by the action of alkali amide in liquid ammonia, was treated with methyl anisate to yield 1-(p-methoxyphenyl)-5-phenyl-1,3,5-pentanetrione (61% based on the ester). This compound has also been prepared by the base-catalyzed ring opening of 2-(p-methoxyphenyl)-6-phenyl-4-pyrone; however, no yield is reported.[5]

4. Merits of the Preparation

This procedure appears to be fairly general for the aroylation of β-diketones to give 1,3,5-triketones. Using this method, the submitters [2] have aroylated benzoylacetone with methyl benzoate (87%), methyl p-chlorobenzoate (78%), and ethyl nicotinate (69%). Also, acetylacetone has been monobenzoylated with methyl benzoate to form 1-phenyl-1,3,5-hexanetrione in 75% yield or dibenzoylated with the same ester to form 1,7-diphenyl-

1,3,5,7-heptanetetraone in 56% yield.[6] Symmetrical 1,5-diaryl-1,3,5-pentanetriones can be conveniently prepared by a similar procedure[2] from acetone and two equivalents of the appropriate aromatic ester; for example, 1,5-diphenyl-1,3,5-pentanetrione and 1,5-di(p-methoxyphenyl)-1,3,5-pentanetrione are formed in yields of 82% and 77%, respectively.

1,3,5-Triketones are useful intermediates in the preparation of 4-pyrones, 4-pyridones,[3, 4] and other cyclic products.

1. Chemistry Department, Duke University, Durham, North Carolina. This research was supported by the National Science Foundation.
2. M. L. Miles, T. M. Harris, and C. R. Hauser, *J. Org. Chem.*, **30**, 1007 (1965).
3. C. R. Hauser and T. M. Harris, *J. Am. Chem. Soc.*, **80**, 6360 (1958).
4. R. J. Light and C. R. Hauser, *J. Org. Chem.*, **25**, 538 (1960).
5. G. Soliman and I. E. El-Kholy, *J. Chem. Soc.*, 1755 (1954).
6. M. L. Miles, T. M. Harris, and C. R. Hauser, *J. Am. Chem. Soc.*, **85**, 3884 (1963).

2-(p-METHOXYPHENYL)-6-PHENYL-4-PYRONE

[4H-Pyran-4-one, 2-(p-methoxyphenyl)-6-phenyl-]

Submitted by MARION L. MILES and CHARLES R. HAUSER[1]
Checked by VICTOR NELSON, WAYLAND E. NOLAND, and WILLIAM E. PARHAM

1. Procedure

In a 50-ml. Erlenmeyer flask is placed 10 ml. of concentrated (36N) sulfuric acid (Note 1), and the flask is then immersed in an ice water bath. When the temperature of the acid reaches 0°, 2.96 g. (0.010 mole) of 1-(*p*-methoxyphenyl)-5-phenyl-1,3,5-pentanetrione (Note 2) is added in small portions. As each portion is added, the flask is swirled until the triketone dissolves. After the addition is completed, the solution is kept at 0° for 1 hour and then poured into 500 ml. of cold water. To the resulting slurry is added solid sodium bicarbonate until a pH of 7–8 (Note 3) is obtained. The mixture is filtered, and the filter cake is washed with cold water and then recrystallized from 15 ml. of 95% ethanol to give 2.46–2.72 g. (88–98%) of 2-(*p*-methoxyphenyl)-6-phenyl-4-pyrone, m.p. 161–163°.

2. Notes

1. Regular commercial grade of concentrated sulfuric acid (sp. gr. 1.84) obtained from the General Chemical Division of Allied Chemical Corporation was used.

2. For the preparation of this compound see this volume, p. 718.

3. This pyrone has a tendency to form a salt in aqueous sulfuric acid. The submitters used "Hydrion" paper to check the pH.

3. Methods of Preparation

The method is an adaptation of the procedure of Light and Hauser.[2] 2-(*p*-Methoxyphenyl)-6-phenyl-4-pyrone has been prepared in 50% yield by a Claisen-type acylation of *p*-methoxyacetophenone with ethyl phenylpropiolate accompanied by cyclization.[3]

4. Merits of the Preparation

This procedure offers an extremely simple and fairly general method for the preparation of 2,6-disubstituted 4-pyrones. Pyrones which have been prepared[2] by this procedure are: 2-methyl-6-phenyl-4-pyrone (60%), 2-(*p*-chlorophenyl)-6-methyl-4-pyrone (90%), 2,6-diphenyl-4-pyrone (91%), 2-(*p*-chlorophenyl)-6-phenyl-4-pyrone (90%), 2-phenyl-6-(3-pyridyl)-4-pyrone (91%), 5,6,7,8-tetrahydroflavone (76%), 4'-methoxy-5,6,7,8-tetrahydroflavone (70%), cyclopenteno[b]-6-(p-methoxyphenyl)-4-pyrone (59%), and flavone (63%).

1. Chemistry Department, Duke University, Durham, North Carolina. This research was supported by the National Institutes of Health.
2. R. J. Light and C. R. Hauser, *J. Org. Chem.*, **25**, 538 (1960).
3. G. Soliman and I. E. El-Kholy, *J. Chem. Soc.*, 1755 (1954).

METHYL BENZENESULFINATE

(Benzenesulfinic acid, methyl ester)

$$(C_6H_5S)_2 + 3Pb(OAc)_4 + 4CH_3OH \rightarrow$$
$$2C_6H_5S(O)OCH_3 + 3Pb(OAc)_2 + 4AcOH + 2AcOCH_3$$

Submitted by LAMAR FIELD [1] and J. MICHAEL LOCKE [2]
Checked by JOHN J. MILLER and WILLIAM D. EMMONS

1. Procedure

Caution! Care should be taken to keep methyl benzenesulfinate off the skin (Note 1).

In a 5-l., three-necked, round-bottomed flask equipped with a sealed mechanical stirrer and a reflux condenser carrying a drying tube are placed 54.6 g. (0.25 mole) of diphenyl disulfide (Note 2), 450 ml. of chloroform (Note 3), and 450 ml. of methanol. To the stirred solution at the reflux temperature is added 443.4 g. (1.00 mole) of lead tetraacetate (Note 4) in 2 l. of chloroform during 8 hours. Owing to formation of lead dioxide, the initially yellow solution becomes dark brown during the addition. The mixture is kept at the reflux temperature overnight (about 12 hours), after which 2 l. of chloroform is removed by distillation at atmospheric pressure (Note 5). The mixture then is cooled to room temperature, and 330 ml. of distilled water is added with stirring to decompose any excess lead tetraacetate. Lead dioxide is removed by filtration of the entire mixture using a Celite®-coated filter paper. The chloroform layer is washed with distilled water until the washings are free of lead ions (Note 6.) The chloroform solution is dried over anhydrous magnesium sulfate and, after separation of the drying agent, is concentrated by means of a rotating-flask evaporator. The oily yellow residue is left overnight under vacuum (about 0.1 mm.) to remove any traces of hexachloroethane (Note 7). Distillation is effected through a 15-cm. Vigreux column under reduced pressure (Note 8). The yield of methyl benzenesulfinate is 48.6–53 g. (62–68%), b.p. 59–60° (0.04 mm.), 76–78° (0.45 mm.); n^{25}D 1.5410–1.5428, reported n^{20}D 1.5400,[3] 1.5440.[4]

2. Notes

1. The checkers experienced an extreme and prolonged burning sensation on contact.

2. Diphenyl disulfide, supplied by Distillation Products Industries, Rochester 3, New York, was used as received.

3. Reagent grade chloroform is satisfactory.

4. Used as received from Arapahoe Chemicals, Inc., Boulder, Colorado. This product, usually about 85–96% lead tetraacetate moist with acetic acid, is stored at about 5°. The molar amount specified is based on occasional iodometric titration (Arapahoe brochure) as follows:[5] An accurately weighed sample of about 0.5 g. is dissolved in 5 ml. of glacial acetic acid with gentle warming, and 100 ml. of an aqueous solution of 12 g. of anhydrous sodium acetate and 1 g. of potassium iodide is added. After several minutes, with occasional swirling, the flask wall is rinsed with water. Liberated iodine is titrated with 0.1N sodium thiosulfate to a starch end point. The percent of lead tetraacetate is calculated from the formula 22.17 (milliliters of thiosulfate) (normality of thiosulfate)/(weight of sample).

The submitters recommend that the lead tetraacetate be added in eight separate portions of 0.125 mole of lead tetraacetate, each in 250 ml. of chloroform, because the solution of lead tetraacetate decomposes on standing.

5. This can be done conveniently by removing the reflux condenser and replacing it with apparatus for downward distillation.

6. A solution of sodium sulfide can be used to test for the presence of lead ions in the wash liquors. The checkers found

that the yield can be improved somewhat by extraction of the initial water layer with chloroform.

7. The small amount of hexachloroethane produced during the reaction presumably is formed from chloroform by a free radical process.

8. The residue after distillation is diphenyl disulfide. It may be recovered by recrystallization from ethanol. The methyl benzenesulfinate may be pale yellow when first distilled, but if so it becomes colorless on standing. If possible, a spinning-band column should be used for distillation, and distillation should be as rapid as possible; use of a 47-cm. spinning-band column gave analytically pure ester, n^{25}D 1.5436 (cf. Field and co-workers).[6]

3. Methods of Preparation

Methyl benzenesulfinate has been prepared by the three-stage process of reduction of benzenesulfonyl chloride to benzenesulfinic acid, conversion of the acid to benzenesulfinyl chloride, and esterification of the chloride with methanol.[3, 7] It has been prepared also by ozonolysis of methyl benzenesulfenate.[4] Alkane- and arenesulfinate esters have been prepared from thiols or disulfides by the following sequence: conversion to a sulfinyl chloride by treatment with chlorine, reaction with the appropriate alcohol, treatment with an amine to remove any sulfonyl chloride, and distillation of the sulfinate.[8] The present procedure is based on one reported by Field, Hoelzel, and Locke.[6]

4. Merits of the Preparation

This procedure affords a one-step synthesis of aromatic sulfinic esters from readily available starting materials. It is successful with a variety of types of aromatic sulfinic esters.[6] The method is rather unattractive for aliphatic disulfides, however, because the nature of by-products formed makes rigorous purification of the sulfinic esters impracticable.[9]

1. Department of Chemistry, Vanderbilt University, Nashville, Tennessee. This work was partly supported by the U.S. Army Research Office, Durham, North Carolina.
2. Department of Chemistry, University of Southampton, Southampton, England.
3. H. F. Herbrandson, R. T. Dickerson, Jr., and J. Weinstein, *J. Am. Chem. Soc.*, **78**, 2576 (1956).
4. D. Barnard, *J. Chem. Soc.*, 4547 (1957).
5. O. Dimroth and R. Schweizer, *Ber.*, **56B**, 1375 (1923).
6. L. Field, C. B. Hoelzel, and J. M. Locke, *J. Am. Chem. Soc.*, **84**, 847 (1962).
7. S. Detoni and D. Hadzi, *J. Chem. Soc.*, 3163 (1955).
8. I. B. Douglass, *J. Org. Chem.*, **30**, 633 (1965); see also I. B. Douglass and R. V. Norton, *J. Org. Chem.*, **33**, 2104 (1968).
9. L. Field, J. M. Locke, C. B. Hoelzel, and J. E. Lawson, *J. Org. Chem.*, **27**, 3313 (1962).

10-METHYL-10,9-BORAZAROPHENANTHRENE

(Dibenz[c,e][1,2]azaborine, 5,6-dihydro-6-methyl-)

Submitted by M. J. S. DEWAR, R. B. K. DEWAR, and Z. L. F. GAIBEL [1]
Checked by JACK A. SNYDER and B. C. McKUSICK

1. Procedure

A. *Bis(10,9-borazarophenanthryl) ether.   Caution!   All   the
operations involving boron trichloride should be carried out in a
good hood.*

A solution of 250 g. (1.48 moles) of 2-aminobiphenyl (Note 1) in 2.0 l. of dry xylene (Note 2) is placed in a 5-l. four-necked flask equipped with a 500-ml. pressure-equalized dropping funnel, a reflux condenser fitted with a drying tube loosely packed with calcium chloride, a thermometer, and a mechanical stirrer. A solution of 250 g. (174 ml., 2.14 moles) of boron trichloride in 250 ml. of very cold xylene is placed in the dropping funnel (Note 3).

The boron trichloride solution is added dropwise to the stirred amine solution over a period of 30 minutes; a thick precipitate forms as the first third of the solution is added, but it gradually dissolves, and the reaction mixture is a clear, dark amber solution at the end of the addition. The mixture is heated under reflux for 1 hour; the temperature of the mixture gradually rises from 110° to 140° as hydrogen chloride and boron chloride are evolved (Note 4). The mixture is cooled to about 100°, the dropping funnel is replaced by a powder funnel, and 15 g. of anhydrous aluminum chloride is cautiously added. The funnel is replaced by a well-greased stopper (Note 5), and the mixture is heated under reflux for 2 hours. The mixture is cooled slightly, and an additional 5 g. of aluminum chloride is added. The reaction mixture is heated under reflux at least 16 hours, and then the reflux condenser is replaced by a Claisen head leading to a water condenser and receiver. About 80% of the solvent (1.6–1.7 l.) is distilled with vigorous stirring.

A dropping funnel with a pressure-equalizing arm is inserted in place of the stopper, and 2.5 l. of distilled water is added from it with vigorous stirring. The reaction with water is extremely exothermic at first, and the first 30 ml. of water is added not faster than 1 drop per second. The mixture is steam-distilled with heating and vigorous stirring until the head temperature reaches 100°. A fresh receiver is attached, and the distillation is continued until the distillate no longer smells of xylene; this is usually after about 300 ml. of distillate has been collected in the receiving vessel (Note 6). If necessary, water is added from time to time to maintain the liquid level in the flask. The flask is cooled to room temperature with vigorous stirring (Note 7), and crystalline 10-hydroxy-10,9-borazarophenanthrene separates.

The solid is collected by suction filtration, washed with about 750 ml. of water, and dried overnight at 75–80° in an oven under reduced pressure (200 mm. or below). Dehydration to bis(10,9-borazarophenanthryl) ether usually takes place during the drying. The ether, a tan solid, weighs 221–250 g. (80–91%) (Note 8).

B. *10-Methyl-10,9-borazarophenanthrene.*   The tan bis(10,9-borazarophenanthryl) ether from the previous step is transferred to a dry, 5-l., four-necked flask equipped with a mechanical stirrer, a nitrogen inlet, a 1-l. pressure-equalized dropping funnel, a thermometer, and a very efficient reflux condenser with a drying tube packed with silica gel at the top (Note 9). A 3.2-l. portion of anhydrous ether is placed in the flask, and the mechanical stirrer is started. The flask is cooled in an ice bath, and 700 ml. (2.1 moles) of $3M$ methylmagnesium bromide solution in ether (Note 10) is added dropwise through the funnel during 1 hour. The reaction mixture is heated under reflux overnight. The mixture is cautiously and slowly poured into 1 l. of ice water and then cautiously acidified with 385 ml. (10% excess) of $6N$ hydrochloric acid. The ether layer is separated, and the water layer is extracted with five 400-ml. portions of ether. The combined ether fractions are washed with 200 ml. of saturated sodium bicarbonate solution; the sodium bicarbonate solution is washed with 300 ml. of ether, which is added to the other ether fractions.

The combined ether fractions are dried over 50 g. of magnesium sulfate, the ether is removed at reduced pressure, and the residue is dried overnight at 60° (10 mm.). The resulting slightly oily, crystalline, brown solid is a mixture of 10-methyl-10,9-borazarophenanthrene, bis(10,9-borazarophenanthryl) ether, and tars; weight about 220 g.; m.p. 92–98°. The methyl derivative is isolated by continuous chromatography over about 500 g. of Merck basic alumina (chromatography grade) in the apparatus shown in Fig. 1 (Note 11). The crude product is placed on top of the alumina, 4 l. of petroleum ether (b.p. 30–60°) is poured into the flask, and the extractor is operated for 30–35 hours. 10-Methyl-10,9-borazarophenanthrene gradually crystallizes as fine needles on the walls of the flask.

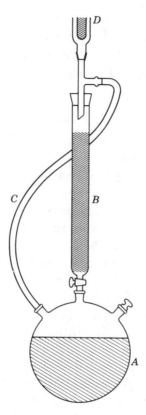

FIG. 1. *A*, 5-l. three-necked flask with electric heating mantle; *B*, alumina column 45 cm. × 3.8 cm. (Merck basic alumina); *C*, polyethylene tubing; *D*, condenser (see Note 11).

The flask is cooled to room temperature, all solvents are allowed to drain from the column, and 120–150 g. of colorless product, m.p. 99–101°, is separated by filtration and washed with petroleum ether. A second crop, weight 25–40 g., m.p. 98.5–101.0°, is obtained by concentrating the combined filtrates to about 250 ml. and cooling the concentrate to room temperature. The crops of 10-methyl-10,9-borazarophenanthrene are combined; weight 155–182 g. (54–65% based on 2-aminobiphenyl). The product is sufficiently pure for many purposes. A purer product, m.p. 103–104°,[2] may be obtained by recrystallization from petroleum ether (b.p. 35°) (Note 12).

2. Notes

1. A technical grade of 2-aminobiphenyl obtainable from Columbia Organic Chemicals Co. is satisfactory. This black material contains the carcinogenic 4-isomer; it should therefore be handled carefully to avoid contact with the skin. Purification of the amine by vacuum distillation removes the black impurities but does not improve either the yield or the quality of the final product.

2. The xylene is dried over sodium wire or sodium-lead alloy before use.

3. Boron trichloride boils at 13°. To prepare the solution, the submitters poured liquid boron trichloride into a tared beaker containing dry xylene until the weight increased by 250 g. The beaker was in a methanol-ice bath.

The checkers placed 250 ml. of dry xylene in a 500-ml. round-bottomed flask marked with a line corresponding to a volume of 424 ml. and with a gas inlet above the line. A cold finger containing a mixture of dry ice and methanol was attached to the flask, which was immersed in the same mixture. Gaseous boron trichloride (Matheson Co.) was passed in until the flask was filled to the 424-ml. line, and the cold solution was transferred to the dropping funnel.

4. During the most vigorous evolution of gas it is advisable to replace the drying tube by a tube leading to the back of the hood or to a gas absorption trap.

5. Unless the stopper is well greased, the aluminum chloride will cement it to the flask.

6. The purpose of the steam distillation is to remove all the xylene. If the crystals are collected prematurely, the last traces of xylene are hard to remove.

7. If the mixture is not vigorously stirred while cooling, an intractable cake forms at the bottom of the flask.

8. The product may be bis(10,9-borazarophenanthryl) ether, 10-hydroxy-10,9-borazarophenanthrene, or a mixture. The melting point of either product varies widely with the method of determination. The checkers placed analytical samples in a bath at 125° with the temperature rising 9–12° per minute and

observed m.p. 154–159° for the ether, m.p. 133° (with frothing) for the hydroxy compound. The infrared spectrum of the hydroxy compound has a band at 2.83 μ that is lacking in the spectrum of the ether. According to the submitters, either the ether or the hydroxy compound can be used in step B. The checkers used only the ether.

The tan product is almost pure. Extraction with petroleum ether (b.p. 60–70°) in a Soxhlet gives white crystals.

9. A West condenser and a Friedrich condenser in tandem are recommended.

10. The methylmagnesium bromide solution is obtainable from Arapahoe Chemicals, Inc., Boulder, Colorado.

11. This procedure may prove useful in other cases where impurities are strongly adsorbed on alumina; it avoids the use of enormous volumes of solvent. The heating mantle is controlled by a Variac, set so that the liquid level remains constant. The plastic tubing should not be of Tygon®, since this is attacked by the solvent; polyethylene tubing is suitable. An efficient condenser is essential; the use of a West condenser and a Friedrich condenser in tandem is recommended. A similar but more complicated apparatus is described by Meier and Fletschinger.[3]

12. Some unreacted bis(10,9-borazarophenanthryl) ether can be recovered from the column by extraction with methanol.

3. Methods of Preparation

10-Chloro-10,9-borazarophenanthrene has been obtained only by Friedel-Crafts cyclization of the adduct from 2-aminobiphenyl and boron trichloride;[2] the procedure described here is an improvement on the original process,[2] in which no solvent was used.

10-Methyl-10,9-borazarophenanthrene has been obtained by the action of methylmagnesium halides on 10-chloro-10,9-borazarophenanthrene [2] or bis(10,9-borazarophenanthryl) ether.[4] The former preparation was satisfactory when pure 2-aminobiphenyl was available, but existing grades lead to products containing intractable impurities, presumably derived from the 4-isomer. These impurities are eliminated during the hydrolysis to 10-hydroxy-10,9-borazarophenanthrene.

4. Merits of the Preparation

10,9-Borazarophenanthrene was the first representative of a new class of heteroaromatic compounds containing boron atoms in six-membered aromatic rings.[4] These compounds are of a different order of stability from previously known types of organoboron compounds, being chemically similar to "normal" aromatics, and their discovery has opened up a new field of aromatic chemistry. The procedure indicated here has been used to prepare a large number of related aromatic systems.

The cyclization can be carried out with halogenated amines, and substitution products of the new ring systems can also be obtained in the conventional way, by nitration, etc. Similar compounds can be prepared directly by using arylboron dichlorides in place of boron trichloride in the procedure indicated above. The parent borazarene derivatives, with hydrogen attached to boron, can be made from the B-hydroxy compounds with lithium aluminum hydride in the presence of aluminum chloride.[4, 5] N-Alkyl derivatives can be made either by using N-alkyl derivatives of the aminobiphenyls as starting materials, or by N-alkylation of the unsubstituted compounds via their N-lithio derivatives.[5] Apart from their inherent interest, compounds of this type can serve as intermediates in various syntheses. Thus benzocinnolines and 2,2'-dihydroxybiphenyls can be obtained from derivatives of 2-aminobiphenyl.[5-7]

1. Department of Chemistry, The University of Texas, Austin, Texas.
2. M. J. S. Dewar, V. P. Kubba, and R. Pettit, *J. Chem. Soc.*, 3073 (1958).
3. R. Meier and J. Fletschinger, *Angew. Chem.*, **68**, 373 (1956).
4. M. J. S. Dewar, R. Dietz, V. P. Kubba, and A. R. Lepley, *J. Am. Chem. Soc.*, **83**, 1754 (1961).
5. For reviews see P. M. Maitlis, *Chem. Rev.*, **62**, 223 (1962).
6. M. J. S. Dewar and P. M. Maitlis, *J. Am. Chem. Soc.*, **83**, 187 (1961).
7. M. J. S. Dewar and W. E. Poesche, *J. Chem. Soc.*, 2201 (1963).

METHYL BUTADIENOATE

(Butadienoic acid, methyl ester)

$$\text{(cyclobutane structure)} \xrightarrow{\text{Heat}} CH_2\!=\!C\!=\!CH-CO_2CH_3 \;+\; CH_2\!=\!CHCO_2CH_3$$

where the cyclobutane bears $H_2C\!=$ and CO_2CH_3 substituents, with CO_2CH_3 below.

Submitted by H. B. STEVENSON and W. H. SHARKEY.[1a]
Checked by R. D. BIRKENMEYER, W. E. RUSSEY, and F. KAGAN.[1b]

1. Procedure

A "Vycor" tube 550 mm. long and 25 mm. in outside diameter is packed for 500 mm. of its length with 6-mm. x 6-mm. quartz rings and mounted vertically so that the upper section is encased in a tube furnace 150 mm. long and the lower section is encased in a tube furnace 300 mm. long. The upper furnace, which serves as a preheater, is monitored by a thermocouple placed between the tube and the furnace heating elements. The temperature of the lower furnace is monitored by a thermocouple located in the center of the lower packed section. The upper end of the pyrolysis tube is fitted with a Y-tube carrying the thermocouple well and a graduated addition funnel with a pressure-equalizing arm. The lower end is attached through one 500-ml. trap and two 200-ml. traps, each immersed in a mixture of solid carbon dioxide and acetone, to a regulated vacuum source. One gram of hydroquinone is placed in the first trap.

The pyrolysis tube is flushed with nitrogen, the lower section is heated to 600° and the upper section to 300° (Note 1), and the pressure is regulated at 25 mm. Then 184 g. (1.00 mole) (Note 2) of dimethyl 3-methylenecyclobutane-1,2-dicarboxylate[2] is admitted over a period of 3 hours (Note 3). The product, which amounts to 172–177 g., collects in the traps. It is distilled through a 13-mm. x 1.2-m. Nester spinning-band still.[3] First there is 41–47 g. (48–55%) of methyl acrylate, n_D^{25} 1.4010, b.p. 34–36°/150 mm., then 39–43 g. (40–44%) of methyl butadienoate, n_D^{25} 1.4635 (Note 4), b.p. 59–60°/52 mm. or 48–49°/26 mm. By continuing the distillation, 20–30 g. of starting material, b.p. 125–150°/25 mm., can be recovered.

2. Notes

1. The temperature of the lower section is quite critical and should be maintained within the range 590–610°. However, the preheater section is needed only to volatilize the ester, so any temperature between 200° and 400° is satisfactory.

2. Since an estimated 3–5 g. of carbon is deposited in the tube during the pyrolysis, it is advisable to pyrolyze only 1 mole of ester at a time and to burn out the carbon with a slow stream of air at 600° between pyrolyses.

3. The space velocity is approximately 125 l. of standard gas per l. of free space per hour, and the contact time is approximately 0.3 second.

4. The checkers used a 10-mm. x 0.76-m. Nester spinning band still and obtained material having n_D^{25} 1.4620 that could not be purified by redistillation. Analysis by vapor-phase chromatography (silicone gum rubber, 20% w/w on firebrick, 120-cm. x 6-mm. outside diameter column at 125°) showed this material to be 95% methyl butadienoate contaminated by small amounts of two other materials.

3. Methods of Preparation

The method used is described by Drysdale, Stevenson, and Sharkey.[4] The methyl ester of butadienoic acid has not been described previously, but the free acid contaminated by 2-butynoic acid has been prepared by Wotiz, Matthews, and Lieb [5] by carbonation of propargylmagnesium bromide. Ethyl butadienoate has been prepared by Eglinton, Jones, Mansfield, and Whiting [6] by alkali-catalyzed isomerization of ethyl 3-butynoate prepared from 3-butynol by chromic acid oxidation and esterification.

4. Merits of the Preparation

This procedure gives a product free of acetylenic groups. It illustrates the synthesis of an olefinic compound by cracking a cyclobutane into two fragments.

1a. Contribution No. 568 from the Central Research Department, Experimental Station, E. I. du Pont de Nemours and Co., Wilmington, Delaware.
1b. Upjohn Co., Kalamazoo, Michigan.
2. H. B. Stevenson, H. N. Cripps and J. K. Williams, this volume, p. 459.
3. R. G. Nester, *Anal. Chem.*, 28, 278 (1956).
4. J. J. Drysdale, H. B. Stevenson, and W. H. Sharkey, *J. Am. Chem. Soc.*, 81, 4908 (1959).
5. J. H. Wotiz, J. S. Matthews, and J. A. Lieb, *J. Am. Chem. Soc.*, 73, 5503 (1951).
6. G. Eglinton, E. R. H. Jones, G. H. Mansfield, and M. C. Whiting, *J. Chem. Soc.*, 3197 (1954).

N-METHYLBUTYLAMINE *

(Butylamine, N-methyl-)

$$C_6H_5CHO + C_4H_9NH_2 \xrightarrow[-H_2O]{} C_6H_5CH=NC_4H_9 \xrightarrow{(CH_3O)_2SO_2}$$

$$C_6H_5CH=\overset{+}{N}(CH_3)C_4H_9 \quad CH_3OSO_3^- \xrightarrow{2H_2O}$$

$$C_6H_5CHO + CH_3OH + C_4H_9\overset{+}{N}H_2CH_3 \quad SO_3H^- \xrightarrow{2NaOH}$$

$$C_4H_9NHCH_3 + Na_2SO_4 + 2H_2O$$

Submitted by JOHN J. LUCIER, ARLO D. HARRIS, and PHILIP S. KOROSEC [1]
Checked by MAX TISHLER and M. BENNETT.

1. Procedure

A 1-l. round-bottomed flask fitted with a reflux condenser bearing a soda-lime drying tube is successively charged with 100 ml. of anhydrous benzene, 36.6 g. (0.50 mole) of *n*-butylamine, and 63.7 g. (61 ml., 0.60 mole) of benzaldehyde (Note 1). The mixture is heated under reflux for 30 minutes (Note 2). The condenser is replaced by a Claisen distillation head, and the mixture is distilled until the temperature reaches 100° (Note 3). The residue, which is mostly N-benzylidenebutylamine, is cooled, and the distillation head is replaced by the reflux condenser bearing a soda-lime drying tube.

A solution of 75.6 g. (57 ml., 0.60 mole) of dimethyl sulfate (*Toxic! Note 4*) in 200 ml. of anhydrous benzene is added through the condenser with intermittent swirling (Note 2). The mixture is then heated *gently*. After a short period (about 10 minutes) the reaction becomes mildly vigorous, and the heating is stopped (Note 5). The ebullition subsides after about 10 minutes. The mixture is heated under reflux for 30 minutes. It is then steam-distilled until the distillate becomes clear; about 500 ml. of distillate is collected (Note 6). The residue is cooled in an ice bath, and 60 g. (1.5 moles) of sodium hydroxide is added with continuous swirling. The layers are separated, and the amine layer is dried for several hours over 5 g. of sodium hydroxide. The amine layer is separated, dried over a second 5-g. portion of sodium hydroxide (Note 7), and distilled from a 50-ml. Claisen flask containing 2 g. of sodium hydroxide. N-Methylbutylamine is collected at 86–90°(745 mm.); weight 19.6–23.0 g. (45–53%); n^{20}D 1.4010–1.4020. The product contains 3–5% of impurity according to vapor-phase chromatographic analysis.

2. Notes

1. The benzaldehyde and *n*-butylamine were obtained from the Eastman Kodak Company (white label grade). The benz-aldehyde was used without further purification. The *n*-butyl-amine was redistilled before use.

The checkers dried benzene over sodium-lead alloy (dri-Na, Baker). Its water content was less than 0.1 mg. per ml. by Karl Fischer titration.

2. The checkers stirred the mixture with a magnetic stirrer.

3. This distillation is carried out to remove the water formed by the first reaction. No more water comes over after the temperature reaches 100°. Toward the end of distillation, slight bumping may occur.

4. The dimethyl sulfate was purchased from Matheson, Coleman and Bell and was used without further purification. Both the liquid and the vapors of dimethyl sulfate are *toxic*, and the compound must be handled with care.

5. An ice bath should be kept ready to restrain the reaction if necessary.

6. In a simpler alternative to steam distillation, 200 ml. of water is added to the benzene solution, and the mixture is vigorously stirred under gentle reflux for 20 minutes. The mixture is cooled to room temperature. The aqueous layer is separated, extracted with 100 ml. of ether to remove traces of benzaldehyde, and then treated with 60 g. of sodium hydroxide as in the present procedure.[2]

7. If a larger portion of sodium hydroxide is used, a semisolid mass is formed from which the product can be separated only with difficulty.

3. Methods of Preparation

Unsymmetrical secondary aliphatic amines have been prepared by reaction of alkyl halides with benzylidene amines and subsequent hydrolysis; [3,4] by reaction of alkyl halides with alkyl amines; [5] by reduction of amine-aldehyde adducts; [6-8] and by dealkylation of tertiary amines with dibenzoyl peroxide.[9]

In the present procedure, the method of Decker and Becker [3] has been modified by substitution of a dialkyl sulfate for the corresponding alkyl halide.

4. Merits of the Preparation

The procedure is a general one for the preparation of unsymmetrical aliphatic amines, for the submitters have used it to obtain good yields of N-methylpentylamine, N-methylhexylamine, N-methylheptylamine, N-ethylbutylamine, N-ethylpentylamine, and N-ethylheptylamine.

Compared to the procedure of Decker and Becker [3] and that of Wawzonek, McKillip, and Peterson in this volume,[4] the present procedure has the advantages of being simpler and using cheaper alkylating agents. It tends to give lower yields and less pure products than the procedure of Wawzonek, McKillip, and Peterson.

1. Department of Chemistry, University of Dayton, Dayton, Ohio. This work was supported by Wright Air Development Division of the United States Air Force, Air Research and Development Command, under Contract No. AF 35(616)-6607.

2. C. L. Dickinson (E. I. du Pont de Nemours and Co.), private communication.
3. H. Decker and P. Becker, *Ann.*, **395**, 362 (1913).
4. S. Wawzonek, W. McKillip, and C. J. Peterson, this volume, p. 758.
5. O. Westphal and D. Jerchel, *Ber.*, **73B**, 1002 (1940).
6. H. Henze and D. Humphreys, *J. Am. Chem. Soc.*, **64**, 2878 (1942).
7. K. Campbell, A. Sommers, and B. Campbell, *J. Am. Chem. Soc.*, **66**, 82 (1944).
8. W. S. Emerson, *Org. Reactions*, **4**, 174 (1948).
9. L. Horner and W. Kirmse, *Ann.*, **597**, 48 (1955).

1-METHYLCYCLOHEXANECARBOXYLIC ACID*

(Cyclohexanecarboxylic acid, 1-methyl-)

Submitted by W. HAAF [1]
Checked by D. M. GALE and B. C. McKUSICK

1. Procedure

Caution! Because carbon monoxide is evolved, the reaction should be carried out in a good hood.

Two hundred seventy milliliters (497 g., 4.86 moles) of 96% sulfuric acid (Note 1) is poured into a 1-l. three-necked flask equipped with a paddle stirrer driven by a powerful motor, a dropping funnel with a gas by-pass, and a thermometer that dips into the acid. The reaction mixture is stirred *vigorously* (Note 2) and maintained at 15–20° by means of a cooling bath as 3 ml. of 98–100% formic acid (Note 3) is added dropwise. Under the same conditions, a solution of 28.5 g. (0.25 mole) of 2-methyl-cyclohexanol (Note 4) in 46 g. (1.00 mole) of 98–100% formic acid is added in the course of 1 hour. The reaction mixture foams during the additions. The mixture, which is a very light cream color, is stirred for 1 hour at 15–20° and then is poured with stirring onto 1 kg. of crushed ice in a 4-l. beaker. The carboxylic acid separates as a white solid.

The acid is taken up in 200 ml. of hexane (Note 5), the hexane layer is separated, and the aqueous layer is extracted with two 150-ml. portions of hexane. The combined hexane solutions are extracted twice with a mixture of 175 ml. of 1.4N potassium hydroxide solution and 50 g. of crushed ice. The two alkaline solutions are combined and extracted with 100 ml. of hexane to remove traces of neutral oil, and then acidified to pH 2 with 12N hydrochloric acid (about 35 ml.). The liberated carboxylic acid is taken up in 150 ml. of hexane. The aqueous layer is extracted with 100 ml. of hexane, and the combined hexane layers are washed with 75 ml. of water and dried over 3 g. of anhydrous magnesium sulfate. The hexane is evaporated by warming the solution at 30–60° (15–30 mm.) overnight. The residue is 33–36 g. (93–101%) of colorless 1-methylcyclohexanecarboxylic acid, m.p. 34–36°, that is pure enough for most purposes. Distillation from a 100-ml. Claisen flask (Note 6) gives 31.5–33.5 g. (89–94%) of the acid, b.p. 132–140° (19 mm.), 79–81° (0.5 mm.); m.p. 38–39°.

2. Notes

1. Three moles of 99–100% sulfuric acid may be used in place of the 96% sulfuric acid.

2. With slow stirring there is a higher concentration of carbon monoxide and hence less rearrangement. For example, cyclohexanol in a slowly stirred reaction mixture gave 75% cyclohexanecarboxylic acid and 14% 1-methylcyclopentanecarboxylic acid; with rapid stirring the corresponding yields were 8% and 61%.[2]

3. Technical grade 85% formic acid can be substituted for 98–100% formic acid if the decrease in sulfuric acid concentration

that would result is compensated for by a suitable increase in the amount of sulfuric acid charged.

4. The checkers used "*o* Methylcyclohexanol," available from K & K Laboratories, Jamaica, New York. They redistilled it before use; b.p. 70–80° (20 mm.), n^{25}D 1.4617. 3- or 4-Methylcyclohexanol can be used in place of 2-methylcyclohexanol, or mixtures of the three can be used.

5. Normal hexane, commercial grade, from the Phillips Petroleum Co., Bartlesville, Oklahoma, was used. Other organic solvents, such as benzene, are satisfactory.

6. The checkers used a 30-cm. spinning-band column for the distillation.

3. Methods of Preparation

1-Methylcyclohexanecarboxylic acid can be prepared by carbonation of the Grignard reagent from 1-chloro-1-methylcyclohexane[3] or by Friedel-Crafts condensation of 1-chloro-1-methylcyclohexane with methyl 2-furancarboxylate followed by saponification and oxidation.[4] It can also be prepared by successive hydrogenation and saponification of the Diels-Alder adduct from butadiene and methyl methacrylate,[5] by oxidation of 1-methyl-1-acetylcyclohexane with nitric acid[6] or sodium hypobromite,[7] and by the present method of synthesis.[8]

4. Merits of the Preparation

Carboxylation by formic acid is a rapid and simple method of preparing many tertiary carboxylic acids.[8] It can be applied to primary, secondary, and tertiary alcohols as well as to olefins and other compounds equivalent to the alcohols under the reaction conditions. The reaction often proceeds with rearrangement of the carbon skeleton. The scope of the reaction is indicated by Table I, which lists 13 alcohols to which the reaction has been applied.

In a related reaction, saturated hydrocarbons with a tertiary hydrogen are carboxylated by a mixture of formic acid, *t*-butyl alcohol, and sulfuric acid.[9, 10]

TABLE I
Carbonylation of Alcohols to Acids

Alcohol		Acids Formed[a]	Yield of RCO₂H, %[b]
1- or 2-Butanol	100	2-Methylbutyric acid	36 or 43
t-Butyl alcohol	95	Trimethylacetic acid	75
1- or 2-Pentanol	80	2,2-Dimethylbutyric acid	76 or 81
	20	C_{11} acids	
2-Methyl-2-butanol	10	Trimethylacetic acid	73
	42	2,2-Dimethylbutyric acid	
	12	C_7 acids	
	36	C_{11} acids	
2,2,3-Trimethyl-2-butanol	100	2,2,3,3-Tetramethylbutyric acid	88
2,2-Dimethyl-1-propanol	100	2,2-Dimethylbutyric acid	83
Cyclopentanol	63	Cyclopentanecarboxylic acid	26
	37	*cis*-9-Decalincarboxylic acid	
Cyclohexanol	80	1-Methylcyclopentanecarboxylic acid	78
	9	Cyclohexanecarboxylic acid	
	11	C_{13} acids	
Cycloheptanol	100	1-Methylcyclohexanecarboxylic acid	91
2-Decalol	80	*cis*-9-Decalincarboxylic acid	95
	20	*trans*-9-Decalincarboxylic acid	
1-Hydroxyadamantane	100	1-Adamantanecarboxylic acid[10,11]	95

[a] The number before each acid is its volume percent in the mixture of carboxylic acids formed.

[b] Total yield of all carboxylic acids formed.

1. Max-Planck Institut für Kohlenforschung, Mülheim a. d. Ruhr, Germany.
2. W. Haaf, *Ber.*, **99**, 1149 (1966).
3. J. Gutt, *Ber.*, **40**, 2069 (1907).
4. T. Reichstein, H. R. Rosenberg, and R. Eberhardt, *Helv. Chim. Acta.*, **18**, 721 (1935).
5. V. N. Ipatieff, J. E. Germain, and H. Pines, *Bull. Soc. Chim. France*, 259 (1951).
6. H. Meerwein, *Ann.*, **396**, 235 (1913).
7. H. Meerwein and J. Schäfer, *J. Prakt. Chem.*, [2] **104**, 299, 306 (1922).
8. H. Koch and W. Haaf, *Ann.*, **618**, 251 (1958).
9. W. Haaf and H. Koch, *Ann.*, **638**, 122 (1960).
10. H. Koch and W. Haaf, this volume, p. 20.
11. H. Stetter, M. Schwarz, and A. Hirschhorn, *Ber.*, **92**, 1629 (1959).

2-METHYL-1,3-CYCLOHEXANEDIONE*

(1,3-Cyclohexanedione, 2-methyl-)

Submitted by A. B. MEKLER, S. RAMACHANDRAN,
S. SWAMINATHAN, and MELVIN S. NEWMAN.[1]
Checked by MAX TISHLER, A. J. ZAMBITO, and
W. B. WRIGHT.

1. Procedure

A freshly prepared solution of 96.0 g. (2.4 moles) of sodium hydroxide, 335 ml. of water, and 220.2 g. (2.0 moles) (Note 1) of resorcinol (Note 2) is placed in a 1.3-l. hydrogenation bomb together with 40.0 g. of finely powdered nickel catalyst (Note 3). The hydrogenation is carried out under an initial pressure of about 1900 lb. of hydrogen with agitation. The reaction is slightly exothermic, and gentle heating is applied to maintain a temperature of 45–50° (Note 4). The hydrogenation is continued until about a 10% excess of the theoretical amount of hydrogen (2.0 moles) has been absorbed (Note 5). At this point, the agitation is stopped, and the bomb is cooled to room temperature. The reaction mixture is poured into a 1-l. beaker, and the catalyst is removed by filtration with the aid of three 50-ml. portions of water for the combined operations. The filtrate and washings are transferred to a 2-l. round-bottomed flask and treated with

33.5 ml. of concentrated hydrochloric acid (for partial neutralization), 145 ml. of dioxane, and 335 g. (2.4 moles) of methyl iodide (Note 6). The reaction mixture is refluxed for a total of 12–14 hours. After 7 or 8 hours, an additional 33.5 g. (0.24 mole) of methyl iodide is added. The system is cooled several hours in an ice bath, and the 2-methyl-1,3-cyclohexanedione which crystallizes is collected by filtration (Note 7), washed with four 200-ml. portions of cold water (Note 8), and then dried in an oven at 110°. The initial crop of dione, m.p. 206–208° dec., weighs 138–142 g. (54–56%). The mother liquors are concentrated under reduced pressure to one-half of their original volume and are then cooled in an ice-salt bath to yield an additional 7–11 g. (3–5%) of slightly yellow dione which melts at 200–204° dec.

The 2-methyl-1,3-cyclohexanedione thus obtained (Note 9) can be purified by recrystallization from 95% ethanol, using about 20 ml. of ethanol for each 5 g. of product to give colorless crystals, m.p. 208–210° dec., with only minor loss of material.

2. Notes

1. The submitters have carried out runs with as much as 990 g. (9.0 moles) of resorcinol.

2. Both Merck resorcinol u.s.p. powder and resorcinol of practical grade give satisfactory results.

3. Many different nickel catalysts have been used: e.g., Grade 0140T1/8 Harshaw Chemical Co., Cleveland 6, Ohio; reduced Universal Oil Products hydrogenation catalyst pellets which are pulverized just before use; and Raney nickel catalyst W-2. Before using the Raney nickel, care must be taken to free it of aluminum by careful washing with 5% sodium hydroxide solution, followed by thorough rinsing with distilled water (private communication from Dr. Richard Weiss, van Ameringen-Haebler, Inc., New York 19, N. Y.). The checkers prepared Raney nickel catalyst W-2 according to the procedure in *Org. Syntheses*, Coll. Vol. **3**, 181 (1955).

4. The temperature should not exceed 50°; at higher temperatures complex condensation products result.

5. The length of time required for the hydrogenation varies with the size of the run and the activity of the catalyst. For a 9-mole run the hydrogenation usually proceeds for 10–12 hours before external heating is needed to bring the temperature into the 45–50° range. The system is periodically recharged with hydrogen to maintain a pressure of about 1800 lb. Reduction is continued until a total of 9.0 moles of hydrogen has been absorbed. Absorption of the calculated amount of hydrogen serves as the criterion for stopping the hydrogenation. For the 2.0-mole run described, 4–5 hours are needed.

6. The conversion of resorcinol to 2-methyl-1,3-cyclohexanedione can be effected by first isolating the dihydroresorcinol [2] and subjecting it to the methylation reaction. However, this procedure is more laborious and the yield is no better.

7. The checkers found that filtration through a coarse sintered-glass funnel is rapid and better-quality product is obtained. When a medium-porosity sintered-glass funnel or Büchner funnel is used, the filtration is exceedingly slow and may even stop owing to a small amount of gelatinous impurity which is present.

8. If all the sodium iodide is not removed by the washings with water, the product tends to become yellow.

9. This product is sufficiently pure for use in conversion to 1,6-dioxo-8a-methyl-1,2,3,4,6,7,8,8a-octahydronaphthalene.[3,4]

3. Methods of Preparation

2-Methyl-1,3-cyclohexanedione has been prepared by cyclization of ethyl 5-oxoheptanoate [5] and methyl 5-oxoheptanoate [3] with sodium ethoxide and sodium methoxide, respectively, and by methylation of dihydroresorcinol [2] employing sodium methoxide in methanol,[3,6] potassium hydroxide in aqueous methanol,[7] potassium methoxide in methanol,[8] potassium hydroxide in aqueous acetone,[9] potassium carbonate in aqueous acetone,[4] or sodium ethoxide in ethanol.[10] The present method is essentially that of Stetter,[7] except that the unnecessary isolation of the intermediary dihydroresorcinol is omitted, and this greatly enhances the ease of preparation.

4. Merits of Preparation

2-Methyl-1,3-cyclohexanedione has been used as starting material for the syntheses of several polycyclic compounds for projected syntheses of steroids and terpenoids.[11-13] It has also been used to prepare 1,6-diketo-8a-methyl-1,2,3,4,6,7,8,8a-octahydronaphthalene.[3, 4, 14, 15]

1. Department of Chemistry, Ohio State University, Columbus, Ohio.
2. R. B. Thompson, *Org. Syntheses*, Coll. Vol. 3, 278 (1955).
3. M. S. Newman and S. Swaminathan, *Tetrahedron*, 2, 88 (1958).
4. I. N. Nazarov, S. I. Zav'yalov, M. S. Burmistrova, I. A. Gurvich, and L. I. Shmonina, *J. Gen. Chem. U.S.S.R.*, 26, 465 (1956) [*C.A.*, 50, 13847 (1956)].
5. E. E. Blaise and M. Maire, *Bull. Soc. Chim., France* [4] 3, 421 (1908).
6. E. G. Meek, J. H. Turnbull, and W. Wilson, *J. Chem. Soc.*, 811 (1953).
7. H. Stetter, *Angew. Chem.*, 67, 783 (1955).
8. H. Stetter and W. Dierichs, *Ber.*, 85, 61 (1952).
9. H. Smith, *J. Chem. Soc.*, 803 (1953).
10. H. Born, R. Pappo, and J. Szmuszkovicz, *J. Chem. Soc.*, 1779 (1953).
11. F. Sondheimer and D. Elad, *J. Am. Chem. Soc.*, 79, 5542 (1957); J. D. Cocker and T. G. Halsall, *J. Chem. Soc.*, 3441 (1957).
12. A. Eschenmoser, J. Schreiber, and S. S. Julia, *Helv. Chim. Acta*, 36, 482 (1953).
13. J. J. Panouse and C. Sannie, *Bull. Soc. Chim. France*, 5(e), 1435 (1956).
14. P. Wieland and K. Miescher, *Helv. Chim. Acta*, 33, 2215 (1950).
15. N. L. Wendler, H. L. Slates, and M. Tishler, *J. Am. Chem. Soc.*, 73, 3816 (1951).

2-METHYLCYCLOPENTANE-1,3-DIONE*

(1,3-Cyclopentanedione, 2-methyl-)

Submitted by JOSEPH P. JOHN, S. SWAMINATHAN,
and P. S. VENKATARAMANI [1]
Checked by J. A. BEREZOWSKY and PETER YATES

1. Procedure

A. *2-Methyl-4-ethoxalylcyclopentane-1,3,5-trione.*   A   solution
of sodium ethoxide is prepared in a 2-l. three-necked, round-
bottomed flask fitted with a mercury-sealed stirrer, a reflux condenser carrying a drying tube, and a stopper by the addition of
69.0 g. (3 moles) of sodium to 950 ml. of absolute ethanol. The
solution is cooled to 0–5° in an ice bath and stirred.   The stopper

is replaced by a dropping funnel, and a cold mixture (5–15°) of 108 g. (1.50 moles) of freshly distilled 2-butanone and 482 g. (3.30 moles) of diethyl oxalate (Note 1) is added gradually over a period of 30 minutes. After the addition is complete, the thick, orange-red mixture is allowed to warm with continued stirring to room temperature, heated under reflux for 30 minutes, and cooled again to 0° in an ice bath. The mixture is decomposed by stirring with 165 ml. of sulfuric acid (1:1 by volume) added in portions. The sodium sulfate formed is filtered by suction and washed with ethanol (150–200 ml.) (Note 2). The washings and filtrate are combined and concentrated by evaporation at room temperature for 3–4 days in two wide-mouthed (6-in.) 1-l. crystallizing basins (Note 3). The yellowish brown product which accumulates by slow crystallization is collected by filtration, washed with small quantities of ice-cold water, and dried in air. The crude product weighs 140–150 g. Further evaporative concentration of the mother liquor followed by cooling furnishes an additional 40–50 g. of the keto ester, bringing the total yield to 180–200 g. (53–59%) (Note 2). This crude material (m.p. 120–130°) is used in the next step. A pure sample can be obtained by crystallization from ethyl acetate after treatment with Norit activated carbon, m.p. 160–162°.

B. *2-Methylcyclopentane-1,3,5-trione hydrate.* A mixture of 200 g. (0.89 mole) of the keto ester prepared above, 910 ml. of water, and 100 ml. of 85% phosphoric acid is heated under reflux for 4 hours and then cooled in an ice-salt bath to −5°. The trione mixed with oxalic acid separates and is collected by filtration and dried under reduced pressure. The dried material is extracted with boiling ether (250–300 ml.) under reflux, and the ethereal extract is separated from the undissolved oxalic acid. The original aqueous filtrate is also extracted with ether in a continuous extractor. The two extracts are combined, and ether is removed by distillation. The crude trione separates as a dark brown solid and is crystallized from *ca.* 250 ml. of hot water. The once-crystallized, faintly yellow product weighs 95–105 g. (74–82%), m.p. 70–74°. This product is used in the next step without further purification. A better specimen, m.p. 77–78°, which is almost colorless, can be obtained by recrystallization from hot water after treatment with Norit activated carbon.

C. *2-Methylcyclopentane-1,3,5-trione   5-semicarbazone.*   The above trione hydrate (144 g., 1.00 mole) is dissolved in a mixture of 500 ml. of water and 1 l. of ethanol.   A solution of 150 g. of sodium acetate in 200 ml. of water is added with stirring to raise the pH to 5–5.5, and the precipitate formed (Note 4) is filtered and washed with a little water (*ca.* 25 ml.).   The filtrate is transferred to a 4-l. beaker and warmed to 45° on a water bath. Heating is then stopped, and a solution of 112 g. (1.00 mole) of semicarbazide hydrochloride and 150 g. of sodium acetate in 250 ml. of water is added dropwise from a dropping funnel with vigorous stirring during the course of 1.5 hours (Note 5).   The stirring is continued for an additional hour, and the cream-colored monosemicarbazone is collected by filtration, washed with a little aqueous ethanol, and dried at 100°.   The dried material weighs 110–120 g. (60–66%) and does not melt below 300°.

D. *2-Methylcyclopentane-1,3-dione.*   In a 2-l. three-necked flask equipped with a reflux condenser and a stirrer are placed 115 g. (2.00 moles) of potassium hydroxide pellets and 1150 ml. of ethylene glycol.   The flask is immersed in an oil bath which is heated to 130°.   To the stirred mixture is added 12 ml. of water followed by 115 g. (0.628 mole) of the semicarbazone prepared above, added in portions over 30–40 minutes through the third neck of the flask, which is kept stoppered between additions. After the addition is complete, the bath temperature is raised to 150° and kept at this temperature for 30 minutes and then raised again to 180–185°.   After 2 hours at 180–185° the reaction mixture is cooled, and the ethylene glycol is removed under reduced pressure (preferably below 4 mm.) (Note 6).   The dry residue remaining is dissolved in 200–225 ml. of water, and the solution is cooled and carefully (Note 7) made acidic to Congo red with concentrated hydrochloric acid.   The crude dione, which separates as a brown solid, is collected by filtration and crystallized from a mixture of 250 ml. of water and 200 ml. of ethanol after treatment with Norit activated carbon.   The almost colorless crystalline product weighs 40–44 g., m.p. 206–207°.   The mother liquor is concentrated to furnish an additional 10–12 g. of product, m.p. 204–205°.   The crops are combined and recrystallized as before to give 42–47 g. (60–67%) of colorless dione, m.p. 211–212°.

2. Notes

1. Eastman Organic Chemicals or B.D.H. Laboratory reagent grade diethyl oxalate was used.

2. The checkers washed the sodium sulfate with 500 ml. of ethanol. They obtained 202–213 g. (60–63%) of product, m.p. 140–155°.

3. The combined filtrate and washings may be concentrated to about 350 ml. under reduced pressure with a bath temperature not exceeding 40° and then worked up as described. However, the final yield of the keto ester is decreased to 120–140 g.

4. This pale yellow precipitate weighs 20–30 g. and is rejected.

5. The conditions described for the preparation of the semicarbazone are critical and should be strictly observed. Otherwise, the yield of the product in the subsequent Wolff-Kishner reduction is decreased.

6. The checkers found that it is important to remove the ethylene glycol immediately upon completion of the reaction; in a run in which the reaction mixture was allowed to stand overnight, a drastic reduction in the yield of product was observed.

7. There is considerable evolution of carbon dioxide and consequent frothing during acidification. Care must be exercised so that the contents of the flask do not spill over.

3. Methods of Preparation

2-Methylcyclopentane-1,3-dione has been prepared in 15% yield by the catalytic reduction of 2-methylcyclopentane-1,3,5-trione over platinum.[2] The present method is based on the original procedure [3] of Panouse and Sannie with improvements as effected by Boyce and Whitehurst [4] and the submitters.[5]

4. Merits of the Preparation

2-Methylcyclopentane-1,3-dione has found increasing use as an intermediate in the synthesis of steroids.[6–12] The method described is the only practicable method available for the preparation of 2-methylcyclopentane-1,3-dione in large amounts.

1. Department of Organic Chemistry, University of Madras, Madras-25, India.
2. M. Orchin and L. W. Butz, *J. Am. Chem. Soc.*, **65**, 2296 (1943).
3. J. J. Panouse and C. Sannic, *Bull. Soc. Chim. France*, 1036 (1955).
4. C. B. C. Boyce and J. S. Whitehurst, *J. Chem. Soc.*, 2022 (1959).
5. S. Swaminathan, J. P. John, P. S. Venkataramani, and K. Viswanathan, *Proc. Indian Acad. Sci., Sect. A.*, **57**, 44 (1963) [*C.A.*, **59**, 1503 (1963)].
6. G. A. Hughes and H. Smith, *Chem. Ind. (London)*, 1022 (1960).
7. D. J. Crispin and J. S. Whitehurst, *Proc. Chem. Soc.*, 356 (1962).
8. S. N. Ananchenko, V. Ye. Limanov, V. N. Leonov, V. N. Rzheznikov, and I. V. Torgov, *Tetrahedron*, **18**, 1355 (1962).
9. D. J. Crispin and J. S. Whitehurst, *Proc. Chem. Soc.*, 22 (1963).
10. T. Miki, K. Hiraga, and T. Asako, *Proc. Chem. Soc.*, 139 (1963).
11. S. N. Ananchenko and I. V. Torgov, *Tetrahedron Lett.*, 1553 (1963).
12. T. B. Windholz, J. H. Fried, and A. A. Patchett, *J. Org. Chem.*, **28**, 1092 (1963).

METHYLENECYCLOHEXANE*

(Cyclohexane, methylene-)

$$(C_6H_5)_3P + CH_3Br \rightarrow [(C_6H_5)_3P-CH_3]Br \xrightarrow{C_4H_9Li}$$

$$(C_6H_5)_3P=CH_2$$

$$(C_6H_5)_3P=CH_2 + \text{[cyclohexanone]}=O \rightarrow \text{[cyclohexane]}=CH_2 + (C_6H_5)_3PO$$

Submitted by GEORGE WITTIG and U. SCHOELLKOPF.[1]
Checked by JOHN D. ROBERTS and MARTIN VOGEL.

1. Procedure

A. *Triphenylmethylphosphonium bromide.* A solution of 55 g. (0.21 mole) of triphenylphosphine dissolved in 45 ml. of dry benzene is placed in a pressure bottle, the bottle is cooled in an ice-salt mixture, and 28 g. (0.29 mole) of previously condensed methyl bromide is added (Note 1). The bottle is sealed, allowed to stand at room temperature for 2 days, and is reopened. The white solid is collected by means of suction filtration with the aid of about 500 ml. of hot benzene and is dried in a vacuum oven at 100° over phosphorus pentoxide. The yield is 74 g. (99%), m.p. 232–233°.

B. *Methylenecyclohexane* (Note 2). A 500-ml. three-necked round-bottomed flask is fitted with a reflux condenser, an addition funnel, a mechanical stirrer, and a gas inlet tube. A gentle flow of nitrogen through the apparatus is maintained throughout the reaction. An ethereal solution of *n*-butyllithium[2] (0.10 mole, about 100 ml., depending on the concentration of the solution) and 200 ml. of anhydrous ether is added to the flask. The solution is stirred and 35.7 g. (0.10 mole) of triphenylmethylphosphonium bromide is added cautiously over a 5-minute period (Note 3). The solution is stirred for 4 hours at room temperature (Note 4).

Freshly distilled cyclohexanone (10.8 g., 0.11 mole) is now added dropwise. The solution becomes colorless and a white precipitate separates. The mixture is heated under reflux overnight, allowed to cool to room temperature, and the precipitate is removed by suction filtration. The precipitate is washed with 100 ml. of ether, and the combined ethereal filtrates are extracted with 100-ml. portions of water until neutral and then dried over calcium chloride. The ether is carefully distilled through an 80-cm. column packed with glass helices. Fractionation of the residue remaining after removal of the ether through an efficient, low-holdup column (Note 5) gives 3.4–3.8 g. (35–40%) of pure methylenecyclohexane, b.p. 99–101°/740 mm., n_D^{25} 1.4470 (Note 6).

2. Notes

1. Eastman Kodak Co. white label triphenylphosphine and Matheson Co. methyl bromide were used. Triphenylphosphine is available from the Metal and Thermit Corp., Rahway, New Jersey.

2. Since this procedure was first published in this series the use of methylsulfinyl carbanion (as the base) in dimethyl sulfoxide (as solvent) has been found to give better results in many examples than the base-solvent pair described here.[3] The Wittig reaction appears to proceed more rapidly in dimethyl sulfoxide and the yields of olefin are frequently superior. For purposes of comparison with the present procedure the

procedure of Greenwald, Chaykovsky, and Corey[3] was repeated several times with the following deviations from the published procedure. Triphenylmethylphosphonium bromide from two different batches was purchased from the Aldrich Chemical Co., Inc. This material contained a small amount (*ca.* 1–2 percent) of benzene (as indicated by nmr analysis). Drying as described in part A reduced the benzene content to less than 1 percent. The apparatus was set up and the purification of cyclohexanone and dimethylsulfoxide was carried out as described in the preparation of methylenecyclohexane oxide (this volume, p. 755). With these modifications the procedure resembled that described by Monson.[4] In several experiments the *apparent* yields (75–94%) of methylenecyclohexane clustered about the 8.10 g. (84.2%) reported by Greenwald, Chaykovsky, and Corey;[3] however, analysis by nmr and glpc showed that the product was a mixture of methylenecyclohexane and benzene in the ratio of 4 to 1. Methylenecyclohexane and benzene may be separated quite easily by distillation (using a spinning-band column) or by preparative gas chromatography on any of several columns (SE-30, di-2-ethylhexyl sebacate, silver nitrate-ethylene glycol) Thus, the *true* yield of methylenecyclohexane was 60–78%. This is a substantial improvement over the yield obtained by the present procedure. The source of the benzene was not determined. Using different batches of the various reagents did not alter the results. Omitting the washing of the sodium hydride with *n*-pentane resulted in a 50–60% reduction in apparent yield (private communication from D. G. McMahan and H. E. Baumgarten).

3. If the triphenylmethylphosphonium bromide is added too rapidly, the evolution of butane causes excessive frothing of the solution.

4. The small amount of precipitate in the orange solution is not unreacted starting material but triphenylphosphinemethylene.

5. The checkers used a 6 x 350 mm. spinning-band column with a total reflux variable take-off head. The band was spun at 1100 r.p.m.

6. The methylenecyclohexane was analyzed by vapor-phase chromatography and found to be better than 99% pure.

3. Methods of Preparation

The procedure is a modification of that published.[5] Methyl-enecyclohexane has been prepared by the pyrolysis of N,N-dimethyl-1-methylcyclohexylamine oxide, N,N,N-trimethyl-1-methylcyclohexylammonium hydroxide, N,N-dimethylcyclo-hexylmethylamine oxide, and N,N,N-trimethylcyclohexylmethyl-ammonium hydroxide.[6] It has also been obtained from the pyroly-sis of cyclohexylmethyl acetate[7] and of cyclohexylideneacetic acid[8] and from the dehydrohalogenation of cyclohexylmethyl iodide.[9]

1. University of Heidelberg, Heidelberg, Germany.
2. R. G. Jones and H. Gilman, *Org. Reactions*, 6, 352 (1951).
3. R. Greenwald, M. Chaykovsky, and E. J. Corey, *J. Org. Chem.*, 28, 1128 (1963).
4. R. S. Monson, "Advanced Organic Synthesis," Academic Press, New York, 1971, p. 105.
5. G. Wittig and U. Schoellkopf, *Ber.*, 87, 1318 (1954).
6. A. C. Cope, C. L. Bumgardner, and E. E. Schweizer, *J. Am. Chem. Soc.*, 79, 4729 (1957).
7. D. H. Froemsdorf, C. H. Collins, G. S. Hammond, and C. H. DePuy, *J. Am. Chem. Soc.*, 81, 643 (1959).
8. O. Wallach, *Ann.*, 359, 287 (1908).
9. A. Faworsky and I. Borgmann, *Ber.*, 40, 4863 (1907).

METHYLENECYCLOHEXANE OXIDE

(Octane, 1-oxaspiro[2.5]-)

Submitted by E. J. Corey [1] and Michael Chaykovsky [2]
Checked by William Washburn and Ronald Breslow

1. Procedure

A. *Dimethyloxosulfonium methylide* (Note 1). In a 500-ml., three-necked, round-bottomed flask with a magnetic stirrer (Note 2) are placed 8.8 g. (0.22 mole) of sodium hydride (60% oil dispersion) (Note 3) and 150 ml. of petroleum ether (30–60°). The suspension is stirred, the hydride allowed to settle, the petroleum ether decanted (Note 4), and 250 ml. of dry dimethyl sulfoxide (Note 5) is added. The flask is immediately fitted with an inlet and outlet for nitrogen and a piece of Gooch tubing connected to a 125-ml. Erlenmeyer flask containing 50.6 g. (0.23 mole) of trimethyloxosulfonium iodide (Note 6). A gentle stream of dry nitrogen is then continuously passed through the system. With stirring, the oxosulfonium iodide is added, in portions, over a period of 15 minutes (Note 7) and stirring is then continued for an additional 30 minutes (Note 8).

B. *Methylenecyclohexane oxide.* The Gooch tubing is removed from the reaction flask and immediately replaced with a sealed, pressure-compensated dropping funnel containing 19.6 g. (0.2 mole) of cyclohexanone (Note 9), which is then added to the reaction mixture over a 5-minute period. After stirring for 15 minutes, the reaction mixture is heated to 55–60° for 30 minutes with an oil bath and then poured into 500 ml. of cold water and extracted with three 100-ml. portions of ether. The combined ether extracts are washed with 100 ml. of water, then with 50 ml. of saturated aqueous salt solution, dried over anhydrous sodium sulfate, and the ether is distilled at atmospheric pressure through a 20-cm. Vigreux column. The almost colorless residue is transferred to a 50-ml. round-bottomed flask and distilled under reduced pressure through a 5-cm. Vigreux column to yield 15–17 g. (67–76%) of the oxide as a colorless liquid, b.p. 61–62° (39 mm.); $n^{23}D$ 1.4485 (Note 10). The n.m.r. spectrum (CDCl$_3$; (CH$_3$)$_4$Si internal standard) showed a band at δ 1.58 (10H) and a sharp singlet at δ 2.53 (2H).

2. Notes

1. The reaction should be carried out in a well-ventilated hood because hydrogen is evolved.

2. The submitters used a mechanical stirrer, but the checkers found that the more convenient magnetic stirrer works as well.

3. The submitters used Alfa Inorganics Inc. sodium hydride dispersion.

4. The petroleum ether removes most of the oil from the hydride dispersion.

5. Matheson, Coleman and Bell anhydrous dimethyl sulfoxide was stirred over powdered calcium hydride overnight and then distilled under reduced pressure, b.p. 64–65° (4 mm.). Dimethyl sulfoxide should not be distilled at temperatures above 90° since at these higher temperatures appreciable disproportionation occurs producing dimethyl sulfone and dimethyl sulfide, the latter of which contaminates the distilled solvent.

6. Trimethyloxosulfonium iodide was purchased from Aldrich Chemical Co. and was recrystallized from water, crushed, and dried in a desiccator over phosphorus pentoxide before use. The

salt may be prepared by reaction of dimethyl sulfoxide with excess methyl iodide.[3]

7. The reaction is only mildly exothermic. No cooling is necessary.

8. After this time the evolution of hydrogen is essentially complete.

9. Eastman Kodak white label cyclohexanone was used without further purification.

10. Reported physical constants are b.p. 62–63° (37 mm.), n^{20}D 1.4470;[4] b.p. 66–68° (50 mm.) n^{20}D 1.4506.[5]

3. Discussion

Methylenecyclohexane oxide has been prepared by the oxidation of methylenecyclohexane with benzonitrile-hydrogen peroxide or with peracetic acid;[5] by treatment of 1-chlorocyclohexylmethanol with aqueous potassium hydroxide;[6] and by the reaction of dimethylsulfonium methylide with cyclohexanone.[7]

This reaction illustrates a general method for the conversion of ketones and aldehydes[8] into oxiranes using the methylene-transfer reagent dimethyloxosulfonium methylide. The yields of oxiranes are usually high, and the crude products, in most cases, are of sufficient purity to be used in subsequent reactions (*e.g.*, rearrangement to aldehydes) without further purification.

1. Department of Chemistry, Harvard University, Cambridge, Massachusetts.
2. Chemical Research Laboratories, Hoffman-La Roche Inc., Nutley, New Jersey.
3. R. Kuhn and H. Trischmann, *Ann.*, **611**, 117 (1958).
4. J. S. Traynham and O. S. Pascual, *Tetrahedron*, **7**, 165 (1959).
5. G. B. Payne, *Tetrahedron*, **18**, 763 (1962).
6. M. Tiffeneau, P. Weill, and B. Tchoubar, *Compt. Rend.*, **205**, 144 (1937).
7. V. Franzen and H. E. Driesen, *Ber.*, **96**, 1881 (1963).
8. E. J. Corey and M. Chaykovsky, *J. Am. Chem. Soc.*, **87**, 1353 (1965).

N-METHYLETHYLAMINE*

(Ethylamine, N-methyl-)

$$C_6H_5CHO + C_2H_5NH_2 \rightarrow C_6H_5CH=NC_2H_5 \xrightarrow{CH_3I}$$

$$C_6H_5CH=\overset{+}{N}(CH_3)C_2H_5 \; I^- \xrightarrow{H_2O}$$

$$C_6H_5CHO + C_2H_5\overset{+}{N}H_2CH_3 \; I^- \xrightarrow{NaOH}$$

$$C_2H_5NHCH_3 + NaI + H_2O$$

Submitted by S. Wawzonek, W. McKillip, and C. J. Peterson [1]
Checked by J. K. Williams, H. E. Winberg, C. L. Dickinson, and B. C. McKusick

1. Procedure

A. *N-Benzylideneethylamine.* Benzaldehyde (466 g., 4.40 moles) is placed in a 2-l. three-necked flask equipped with a mechanical stirrer and a thermometer. The flask is cooled to 5° in an ice bath, and 200 g. (4.44 moles) of anhydrous ethylamine (Note 1) is added to the stirred benzaldehyde at such a rate that the temperature remains below 15°; about 50 minutes is required for the addition. The mixture is stirred for an additional 30 minutes at room temperature and allowed to stand for 1 hour.

The condenser is arranged for downward distillation, and the water is removed from the product by codistillation with 200 ml. of benzene. The residue, N-benzylideneethylamine, is purified by distillation through a 25-cm. Fenske column; b.p. 52–53° (4.5 mm.); n^{23}D 1.5400; weight 470–523 g. (80–89%) (Note 2).

B. *N-Methylethylamine.* N-Benzylideneethylamine (133 g., 1.00 mole) is heated with 156 g. (1.10 moles) of methyl iodide (Note 3) in a 300-ml. pressure bomb at 100° for 24 hours (Note 4). The bomb is cooled to 50° (Note 5), and the dark, viscous oil is poured into a 1-l. beaker containing 200 ml. of water. The bomb is rinsed with three 50-ml. portions of water, and the washings are combined with the main solution. The resulting mixture is heated with manual stirring on a steam bath for 20 minutes and then cooled in an ice bath to room temperature. The resulting mixture is extracted with two 75-ml. portions of ether (Note 6). The ether layer is washed with two 50-ml. portions of water, and the washings are combined with the main aqueous layer, which is then heated at 100° on a steam bath for 20 minutes to remove traces of ether.

For the liberation of N-methylethylamine, a 1-l. Claisen flask is equipped with a 250-ml. separatory funnel and an efficient condenser for distillation. The receiver is cooled with a mixture of acetone and dry ice (Note 7). A solution of 100 g. (2.5 moles) of sodium hydroxide in 100 ml. of water is added to the flask and kept at about 100° by heating on a steam bath. The aqueous solution of N-methylethylamine hydriodide is added to this solution through the separatory funnel in the course of 1.5 hours. After the addition is complete, the final solution is heated for an additional 30 minutes. Crude N-methylethylamine, b.p. 30–70°, collects in the cooled receiver. It is purified by distillation from 25 g. of solid potassium hydroxide in a 250-ml. modified Claisen flask fitted with a 25-cm. Fenske column and a receiver cooled by dry ice and acetone. N-Methylethylamine is collected at 34–35°; weight 49–55 g. (83–93%); n^5D 1.3830.

2. Notes

1. The ethylamine is cooled to 5° to prevent loss by evaporation. Addition is made directly from the bottle with intermittent cooling in an ice bath.

2. The aldimine need not be distilled but can be used directly in the next step.

3. Dimethyl sulfate, when substituted for the methyl iodide, reacts vigorously with the aldimine at ice-bath temperatures and gives a 49% yield of N-methylethylamine together with considerable tar.

4. Four times these amounts have been used for N-methylbutylamine with equal success.

5. The pressure bomb is opened while still warm (50°). If the bomb is allowed to cool below this temperature, the product solidifies and removal becomes a problem.

6. The benzaldehyde may be recovered after removal of the ether.

7. Because of the low boiling point of N-methylethylamine, there must be efficient cooling or a portion of the product will be lost.

3. Methods of Preparation

This procedure is a modification of the method used for N-methylallylamine.[2]

N-Methylethylamine has been prepared by heating ethylamine with methyl iodide in alcohol at 100°;[3] by the hydrolysis of N-methyl-N-ethylarenesulfonamides,[4,5] p-nitroso-N-methyl-N-ethylaniline,[6] or methylethylbenzhydrylidene ammonium iodide;[7] by catalytic hydrogenation of ethyl isocyanate or ethyl isocyanide;[8] and by the reduction of ethyl isocyanate by lithium aluminum hydride,[9] of N-methylacetisoaldoxime by sodium amalgam and acetic acid,[10] or of a nitromethane/ethylmagnesium bromide adduct by zinc and hydrochloric acid.[11]

4. Merits of the Preparation

This preparation illustrates a general method for the synthesis of N-methylalkylamines. The submitters have used it to prepare N-methylbutylamine (Note 4) and N-methylallylamine, and the checkers have used it to prepare N-methylisopropylamine (80%), N-methylisobutylamine (67%), N-methyl-tert-butylamine (52%), and N-methyl-2-methoxyethylamine (55%). Secondary amines are useful as starting materials for the synthesis of 1,1-disubstituted hydrazines and asymmetric amine imides.

The method gives better yields, utilizes more readily available starting materials, and is much less laborious than the hydrolysis of N-methyl-N-alkylarenesulfonamides and p-nitroso-N, N-dialkylanilines, or the lithium aluminum hydride reduction of alkyl isocyanates. Compared to the closely related procedure of Lucier, Harris, and Korosec,[12] in which the N-benzylidenealkylamine is treated with dialkyl sulfate at atmospheric pressure, the present procedure tends to give higher yields and purer products, but it is less convenient because of the need for a pressure vessel.

1. Department of Chemistry, State University of Iowa, Iowa City, Iowa.
2. A. L. Morrison and H. Rinderknecht, *J. Chem. Soc.*, 1478 (1950).
3. Z. H. Skraup and D. Wiegmann, *Monatsh.*, **10**, 101 (1889).
4. O. Hinsberg, *Ann.*, **265**, 178 (1891).
5. H. Lecher and F. Graf, *Ann.*, **445**, 68 (1925).
6. J. Meisenheimer, *Ann.*, **428**, 252 (1922).
7. M. Sommelet, *Compt. Rend.*, **184**, 1338 (1927).
8. P. Sabatier and A. Mailhe, *Compt. Rend.*, **144**, 824, 955 (1907); *Bull. Soc. Chim. France*, (4) **1**, 615 (1907); *Ann. Chim. (Paris)*, (8) **16**, 70 (1909).
9. A. E. Finholt, C. D. Anderson, and C. L. Agre, *J. Org. Chem.*, **18**, 1338 (1953).
10. W. R. Dunstan and E. Goulding, *J. Chem. Soc.*, **79**, 628 (1901).
11. G. D. Buckley, *J. Chem. Soc.*, 1492 (1917).
12. J. J. Lucier, A. D. Harris, and P. S. Korosec, this volume, p. 736.

3-METHYLHEPTANOIC ACID

(Heptanoic acid, 3-methyl-)

$$CH_3CH=CHCO_2H + C_2H_5CHOHCH_3 \xrightarrow{H_2SO_4}$$

$$CH_3CH=CHCO_2\overset{\displaystyle CH_3}{\underset{\displaystyle C_2H_5}{CH}} + H_2O$$

$$CH_3CH=CHCO_2\overset{\displaystyle CH_3}{\underset{\displaystyle C_2H_5}{CH}} \xrightarrow[\text{(2) H}_2\text{O, H}^+]{\text{(1) } n\text{-C}_4\text{H}_9\text{MgBr}}$$

$$C_4H_9\underset{\displaystyle CH_3}{\overset{}{CH}}CH_2COO\overset{\displaystyle CH_3}{\underset{\displaystyle C_2H_5}{CH}}$$

$$C_4H_9\underset{\displaystyle CH_3}{\overset{}{CH}}CH_2CO_2\overset{\displaystyle CH_3}{\underset{\displaystyle C_2H_5}{CH}} \xrightarrow[\text{(2) HCl}]{\text{(1) KOH}} C_4H_9\underset{\displaystyle CH_3}{\overset{}{CH}}CH_2CO_2H$$

Submitted by JON MUNCH-PETERSEN.[1]
Checked by MELVIN S. NEWMAN and DONALD E. HARSH.

1. Procedure

A. *sec-Butyl crotonate.* In a 2-l. round-bottomed flask are placed 258 g. (3 moles) of crotonic acid (Note 1), 370 g. (5 moles) of *sec*-butyl alcohol in which has been dissolved 6–7 ml. of concentrated sulfuric acid, and 300 ml. of benzene. A few boiling chips are added, and the flask is fitted with a suitable water separator (Note 2) in the top of which is placed a reflux condenser. The mixture is heated under reflux for about 12 hours or until no further separation of aqueous phase occurs. About 65 ml. of water is collected. The cooled reaction mixture is diluted with 200 ml. of ether, washed with 10% sodium carbonate solution until neutral to litmus, washed with saturated sodium chloride solution, and finally dried over magnesium sulfate. The solvent is distilled, and the ester is fractionated under reduced pressure through a small column. The yield of *sec*-butyl crotonate, b.p. 74–75°/30 mm. or 83–84°/45 mm., n_D^{25} 1.4261, is 360–390 g. (85–90%) (Note 3).

B. *3-Methylheptanoic acid.* In a 2-l. three-necked flask fitted with a mercury-sealed stirrer, a reflux condenser carrying a calcium chloride tube, and a dropping funnel are placed 25.0 g. (1.04 g. atoms) of magnesium turnings. The flask is heated to about 100° for a few minutes and then cooled to room temperature. A solution of 178 g. (1.30 moles) of *n*-butyl bromide in 300 ml. of dry ether is prepared; and of this solution about 10 ml., together with 30 ml. of dry ether, is run into the flask. The reaction is started by heating to reflux for a few seconds, the stirrer is started, and the remainder of the bromide solution is added at such a rate as to maintain constant reflux (about 1 hour).

After the addition has been completed, the solution is heated under reflux for 10–15 minutes. The flask is now surrounded by an ice and water bath, and stirring is continued for 15 minutes (Note 4). From a graduated dropping funnel, a solution of 56.8 g. (0.4 mole) of *sec*-butyl crotonate (Note 5) in 400 ml. of dry ether is then added dropwise during a period of about 3 hours (Note 6) while the reaction mixture is effectively stirred and cooled in the ice bath. After the addition of the ester solution is complete, the reaction mixture is stirred in the ice bath for an additional 15 minutes. The ice bath is then removed, and stirring of the grayish brown solution is continued at room temperature for 1 hour.

In a 3-l. Erlenmeyer flask are placed about 500 g. of crushed ice, 110 ml. (1.3 equivalents) of concentrated hydrochloric acid, and 100 ml. of ether. This mixture is vigorously swirled and shaken while the Grignard reaction mixture is cautiously added in small portions. More ice is added to the Erlenmeyer flask as required to keep the temperature near 0°. The resulting mixture is poured into a separatory funnel and shaken thoroughly. The water layer is separated and extracted three times with 100 ml. of ether. The combined ether solutions are washed with 100 ml. of saturated sodium bicarbonate solution, and then with 100 ml. of water. The solution is dried over anhydrous magnesium sulfate, and the ether distilled on a water bath. The residue is fractionated at reduced pressure through a modified Claisen flask to yield 54–62 g. (68–78%) (Note 4) of sec-butyl 3-methylheptanoate, b.p. 92–93°/9 mm., n_D^{25} 1.4190.

A solution of 40 g. (0.2 mole) of sec-butyl 3-methylheptanoate in 100 ml. of ethanol containing 18.5 g. (0.3 mole) of potassium hydroxide and 20 ml. of water is heated under reflux for 30 minutes (Notes 7 and 8). The cooled solution is diluted with 200 ml. of water and acidified by the addition of 60 ml. of concentrated hydrochloric acid. The organic acid is extracted with three 100-ml. portions of 1:1 benzene-ether, and the combined benzene-ether extracts are washed with 50 ml. of saturated sodium chloride solution. The resulting solution is filtered by gravity through a bed of anhydrous magnesium sulfate. After removal of solvents by distillation, 26–27 g. (90–94%) of 3-methylheptanoic acid, b.p. 116–117°/10 mm., n_D^{25} 1.4242, is obtained by distillation in a modified Claisen flask (Note 9).

2. Notes

1. Eastman Organic Chemicals practical grade of crotonic acid (containing 10% water) was used by the checkers without further purification.

2. The water separator preferred by the submitter is that described by Wideqvist,[2] but any continuous water separator which will return the benzene to the reaction mixture may be used, e.g., the modified Dean-Stark water separator.[3]

3. By essentially the same procedure the submitter has prepared the following sec-butyl esters: sec-butyl acrylate, b.p. 127–129°, n_D^{20} 1.4158; sec-butyl methacrylate, b.p. 59–62°/34 mm., n_D^{25} 1.4161; sec-butyl tiglate, b.p. 84.5°/27 mm., n_D^{25} 1.4332; sec-butyl β,β-dimethylacrylate, b.p. 68–70°/13 mm., n_D^{25} 1.4379; sec-butyl cinnamate, b.p. 122°/2 mm., n_D^{25} 1.5382. With sec-butyl acrylate and methacrylate, 2–3% of hydroquinone should be added to the reaction mixtures and 0.1% of hydroquinone to the esters if stored at room temperature.

4. Recent investigations[4] by the submitter have shown that the yield of sec-butyl 3-methylheptanoate is improved to 80–85% if 1.4 g. (1.4 mole% with respect to the Grignard reagent) of cuprous chloride (commercial grade, analytically pure) is added in seven portions during the course of the addition of the ester.

5. Ethyl crotonate may be used with the same yield (70%) of addition product if cuprous chloride is present during the addition[5] (cf. Note 4). Methyl crotonate under these conditions yields methyl α,γ-di-(2-hexyl)-acetoacetate [methyl 2-(2'-hexyl)-3-keto-5-methylnonanoate], b.p. 135°/2.5 mm., n_D^{25} 1.4419, in 67% yield.[6]

6. The large excess of Grignard reagent, the dilution of the ester, and the slow addition are essential features of the procedure. If these conditions are not fulfilled the yields drop considerably, and a greater amount of high-boiling residue, di-sec-butyl α-(2-hexyl)-β-methylglutarate, b.p. 145°/1.5 mm., n_D^{25} 1.4400, is formed.[6] When the reaction is run on a 0.2-mole scale the addition time of the ester may be reduced to 1.5 hours.

7. In the case of the analogous products obtained by the addition reactions with sec-butyl tiglate (cf. Note 9) considerably more drastic conditions are necessary in order to secure complete saponification. The submitter generally employs reflux for 6–8 hours with 35 g. (0.6 mole) of potassium hydroxide dissolved in 250 ml. of 95% ethanol.

8. An alternative procedure is used by the submitter from this point to the final distillation of solvent and ester: The condenser is then set for downward distillation, and about 40 ml. of alcohol is distilled. Then 100 ml. of water is added, and an additional 100 ml. of alcohol and water is distilled. The cooled residue is diluted with 200 ml. of water and the solution freed of insoluble organic material by washing three times with 50 ml. of ether. After acidification with 40 ml. of concentrated hydrochloric acid, the organic layer is extracted with three 50-ml. portions of ether. The combined ether extracts are washed with water and dried over anhydrous magnesium sulfate.

9. The submitter has, by either cuprous chloride-catalyzed or uncatalyzed reactions, prepared a variety of 3-methyl-substituted fatty acids from the adducts of sec-butyl crotonate and other

Grignard reagents.[4-9] The uncatalyzed reaction has also been used with *sec*-butyl tiglate to obtain 2,3-dimethyl-substituted fatty acids.[7]

Methods of Preparation

3-Methylheptanoic acid has been prepared by mixed electrolysis of β-methylglutaric acid monomethyl ester and butyric acid, followed by saponification of the methyl ester,[10] and by the malonic ester synthesis from 2-bromohexane.[11] The present method [7] has the advantage of avoiding the use of secondary bromides, which are often difficult to secure entirely pure.[12]

4. Merits of Procedure

The reactions here described are of considerable general utility for the preparation of a variety of fatty acids from the addition products of Grignard reagents and α,β-unsaturated esters.[4-9, 13]

1. Department of Organic Chemistry, Polyteknisk Laereanstalt, Copenhagen, Denmark.
2. S. Wideqvist, *Acta Chem. Scand.*, 3, 303 (1949).
3. S. Natelson and S. Gottfried, *Org. Syntheses*, 23, 38 (1943); Coll. Vol. 3, 382 (1955).
4. J. Munch-Petersen and V. K. Andersen, *Acta Chem. Scand.*, 15, 271 (1961).
5. J. Munch-Petersen, *Acta Chem. Scand.*, 12, 2046 (1958).
6. J. Munch-Petersen, *J. Org. Chem.*, 22, 170 (1957).
7. J. Munch-Petersen, *Acta Chem. Scand.*, 12, 967 (1958).
8. J. Munch-Petersen, *Acta Chem. Scand.*, 23, 2007 (1958).
9. J. Munch-Petersen and V. K. Andersen, *Acta Chem. Scand.*, 15, 293 (1961).
10. S. Stallberg-Stenhagen, *Arkiv Kemi.*, 2, 95 (1950) [*C.A.*, 44, 7761 (1950)].
11. R. P. Linstead, B. R. Shephard, B. C. L. Weedon, and J. C. Lunt, *J. Chem. Soc.*, 1538 (1953).
12. J. Cason and R. H. Mills, *J. Am. Chem. Soc.*, 73, 1354 (1951).·
13. Cf. E. L. Eliel, R. O. Hutchins, and Sr. M. Knoeber, *Org. Syntheses*, 50, 38 (1970).

5-METHYL-5-HEXEN-2-ONE

(5-Methylene-2-hexanone)

$$\text{CH}_3\text{COCH}_2\text{COCH}_3 \xrightarrow[\text{K}_2\text{CO}_3,\ \text{C}_2\text{H}_5\text{OH}]{\overset{\text{CH}_3}{\underset{}{\text{CH}_2=\text{CCH}_2\text{Cl}}}} \left[\begin{array}{c} \overset{\text{CH}_3}{\underset{\mid}{}} \\ \text{CH}_2=\text{CCH}_2 \\ \mid \\ \text{CH}_3\text{COCHCOCH}_3 \end{array} \right]$$

$$\xrightarrow{\text{Cleavage}} \text{CH}_3\text{CO}_2\text{C}_2\text{H}_5 + \text{CH}_2=\overset{\text{CH}_3}{\underset{\mid}{\text{C}}}\text{CH}_2\text{CH}_2\text{COCH}_3$$

Submitted by SANDRA BOATMAN and CHARLES R. HAUSER [1]
Checked by E. J. COREY and WILLIAM E. RUSSEY

1. Procedure

In a 1-l. round-bottomed flask equipped with a condenser are placed 78.0 g. (0.56 mole) of commercial anhydrous potassium carbonate, 45.0 g. (0.50 mole) of methallyl chloride (Note 1), 55.0 g. (0.55 mole) of 2,4-pentanedione (Note 1), and 300 ml. of anhydrous ethanol (Note 2). The mixture is refluxed on a steam bath for 16 hours. The condenser is replaced by a distilling head and condenser, and about 200 ml. of ethanol is distilled from the mixture (Note 3). Ice water (600 ml.) is added to dissolve the salts, and the mixture is extracted three times with ether. The combined ether extracts are washed twice with 100 ml. of saturated sodium chloride solution, dried for 30 minutes over anhydrous magnesium sulfate, and filtered; the solvent is evaporated. The residue is distilled through a 6-in. Vigreux column using an oil bath maintained at 190° to give 26–29 g. (47–52%) of the product, b.p. 145–155° (Notes 4, 5).

2. Notes

1. Eastman Organic Chemicals practical grade methallyl chloride was distilled (b.p. 70–71°); Union Carbide Chemicals Co. 2,4-pentanedione was distilled (b.p. 134.5–135.5°).

2. Commercial grade absolute ethanol was dried over Linde 3A molecular sieves.

3. At this point most of the ethyl acetate, which is formed as a by-product of the reaction, also is removed.

4. The checkers used a 2-ft. spinning-band column at 200 mm. and observed b.p. 110–111.5°.

5. In the distillation residue (5.7–6.3 g.) remain other by-products, presumably 1,1-dimethallyl-2-propanone, 3-methallyl-2,4-pentanedione, and 3,3-dimethallyl-2,4-pentanedione (indicated by vapor phase chromatography). The checkers carried out v.p.c. analyses using an 8-ft. column of 5% silicone oil XE-60 on Diatoport S at 100° for analysis of the distillate and 175° for analysis of the residue.

3. Methods of Preparation

5-Methyl-5-hexen-2-one has been prepared by alkylation of acetoacetic ester with methallyl chloride, followed by cleavage; the overall yield in the two steps was 51%.[2]

4. Merits of the Preparation

The present procedure, which is characterized by its extreme simplicity, has been employed to prepare various ketones of type RCH_2COCH_3 from 2,4-pentanedione,[3] as indicated in Table I.

TABLE I

ALKYLATION AND CLEAVAGE OF 2,4-PENTANEDIONE

Alkyl Halide	Ketone	% Yield
Benzyl chloride	4-Phenyl-2-butanone	73
o-Bromobenzyl bromide	4-(o-Bromophenyl)-2-butanone	75
m-Bromobenzyl bromide	4-(m-Bromophenyl)-2-butanone	78
o-Chlorobenzyl chloride	4-(o-Chlorophenyl)-2-butanone	78
m-Chlorobenzyl bromide	4-(m-Chlorophenyl)-2-butanone	65
p-Chlorobenzyl bromide	4-(p-Chlorophenyl)-2-butanone	62
m-Fluorobenzyl chloride	4-(m-Fluorophenyl)-2-butanone	60
m-Nitrobenzyl chloride	4-(m-Nitrophenyl)-2-butanone	65
α-Chloromethylnaphthalene	4-(α-Naphthyl)-2-butanone	61
Phenacyl chloride	1-Phenyl-1,4-pentanedione	55
n-Butyl iodide	2-Heptanone	60

1. Department of Chemistry, Duke University, Durham, North Carolina. This work was supported by the National Science Foundation.
2. W. Kimel and A. C. Cope, *J. Am. Chem. Soc.*, **65**, 1992 (1943).
3. S. Boatman, T. M. Harris, and C. R. Hauser, *J. Org. Chem.*, **30**, 3321 (1965).

1-METHYLINDOLE*

(Indole, 1-methyl-)

Submitted by K. T. POTTS and J. E. SAXTON.[1]
Checked by W. E. PARHAM, WAYLAND E. NOLAND, and BRYCE A. CUNNINGHAM.

1. Procedure

Caution! Ammonia gas is an extreme irritant and can cause serious burns to the eyes, etc. It is necessary to carry out the entire reaction under a well-ventilated hood.

In a 1-l. three-necked flask fitted with a motor stirrer (Note. 1), gas inlet tube, dropping funnel, and a wide-bore soda-lime tube are placed 400–500 ml. of liquid ammonia and 0.1 g. of ferric nitrate nonahydrate (Note 2).

In small portions, just sufficient to maintain the blue color, 5.0 g. (0.22 gram atom) of clean, metallic sodium is added with vigorous stirring. After dissolution is complete (Note 3), a solution of 23.4 g. (0.20 mole) of indole (Note 4) in 50 ml. of anhydrous ether is added slowly and then, after an additional 10 minutes, a solution of 31.2 g. (0.22 mole) of methyl iodide in an equal volume of anhydrous ether is added dropwise. Stirring is continued for a further 15 minutes. The ammonia is allowed to evaporate (Note 5), 100 ml. of water is added, followed by 100 ml. of ether. The ether layer is separated, the aqueous phase extracted with an additional 20 ml. of ether, and the combined ether extracts washed with three 15 ml.-portions of water (Note 6) and dried over anhydrous sodium sulfate. The solvent is removed at atmospheric pressure, and the crude oil ($n_D^{18.5°}$ 1.6078) is purified by distillation under reduced pressure. 1-Methylindole is obtained as a colorless oil, b.p. 133°/26 mm., $n_D^{18.5°}$ 1.6082. In several runs the yield is 22.3–24.9 g. (85–95%).

2. Notes

1. Any sealed mechanical stirrer may be used. Those of the Hershberg [2] type were found particularly efficient during the formation of the sodium amide.

2. It was found most advantageous to run in the liquid ammonia from the commercial cylinder, laid on the floor with the foot raised slightly, and connected to the gas inlet tube with rubber tubing approximately 1 cm. in diameter. The large excess of liquid ammonia used obviates the need of a dry ice-acetone cooling bath and permits a reasonably rapid formation of sodium amide.

3. This occurs when the blue color has disappeared. The formation of the light gray sodium amide is usually complete within 20 minutes and may be observed by washing a portion of the outside of the flask with a little alcohol.

4. A commercial grade of indole is satisfactory.

5. The checkers removed the ammonia by distillation (water aspirator).

6. The checkers extracted with two 50-ml. portions of ether and three 50-ml. portions of water.

3. Methods of Preparation

1-Methylindole has been prepared from the *as*-methylphenyl-hydrazone of pyruvic acid,[3] by the action of sodium amide or sodium hydride on indole followed by methyl iodide at elevated temperatures,[4,5] by treatment of indole with methyl *p*-toluene-sulfonate and anhydrous sodium carbonate in boiling xylene,[6] and by the action of methyl sulfate on indole previously treated with sodium amide in liquid ammonia.[7] The present method is essentially that of Potts and Saxton.[8]

1-Methylindole has also been prepared by lithium aluminum hydride reduction of 1-methylindoxyl.[9] Compounds giving rise to NH absorption in the infrared (indole, skatole) can be completely removed [10] by refluxing the crude 1-methylindole over sodium for 2 days and then distilling the unreacted 1-methylindole from the sodio derivatives and tarry decomposition products.

1. Department of Organic Chemistry, University of Adelaide, Adelaide, South Australia.
2. E. B. Hershberg, *Ind. Eng. Chem., Anal. Ed.*, 8, 313 (1936).
3. E. Fischer and O. Hess, *Ber.*, 17, 559a (1884).
4. R. Weissgerber, *Ber.*, 43, 3522 (1910).
5. A. P. Gray, *J. Am. Chem. Soc.*, 75, 1253 (1953).
6. D. A. Shirley and P. A. Roussel, *J. Am. Chem. Soc.*, 75, 375 (1953).
7. H. Plieninger, *Ber.*, 87, 127 (1954).
8. K. T. Potts and J. E. Saxton, *J. Chem. Soc.*, 2641 (1954).
9. P. L. Julian and H. C. Printy, *J. Am. Chem. Soc.*, 71, 3206 (1949).
10. W. E. Noland, W. C. Kuryla, and R. F. Lange, *J. Am. Chem. Soc.*, 81, 6015 (1959).

METHYL ISOCYANIDE

$$CH_3NHCHO \ + \ CH_3-\text{⬡}-SO_2Cl \ + \ 2\,\text{⬡⬡N} \longrightarrow$$

$$CH_3N\overset{\equiv}{}C \ + \ Cl^- \ + \ CH_3-\text{⬡}-SO_2O^- \ + \ 2\,\text{⬡⬡}\underset{H^+}{N}$$

Submitted by R. E. Schuster, James E. Scott
and Joseph Casanova, Jr. [1]
Checked by John A. DuPont and William D. Emmons

1. Procedure

Caution! Methyl isocyanide should be prepared in a good hood since it is toxic and has a very unpleasant odor. The reaction and subsequent distillation of the product should be conducted behind safety shields (Note 1).

In a 2-l. four-necked flask (Note 2) equipped with a 250-ml. pressure-equalizing dropping funnel, a sealed mechanical stirrer, a thermometer, and a receiver trap (Note 3) are placed 1034 g. (8.0 moles) of quinoline and 572 g. (3.0 moles) of *p*-toluenesulfonyl chloride (Note 4). The solution is heated to 75° by an oil bath and the system evacuated to a pressure of 15 mm. The receiver is cooled in a bath of liquid nitrogen (Note 5). While the solution is vigorously stirred and maintained at this temperature, 118 g. (2.0 moles) of N-methylformamide (Note 4) is added dropwise to maintain a smooth distillation rate. The addition is complete in 45–60 minutes.

The material which collects in the receiver is distilled through a 15-cm. Vigreux column at atmospheric pressure. Methyl isocyanide, a colorless, vile-smelling liquid, is collected at 59–60°; weight 57–61 g. (69–74%) (Note 6). Analysis by gas liquid chromatography indicates that the purity exceeds 99% (Note 7).

2. Notes

1. An explosion involving methyl isocyanide has been reported.[2] For this reason, prudence dictates the use of adequate shielding in all heating operations.

2. A three-necked flask may be used by employing a suitable adapter on one of the necks. The checkers used a standard wide-bore, 75-degree side-arm adapter fitted with a long-stemmed thermometer extending into the reaction solution. A Trubore® stirrer equipped with a semicircular Teflon® paddle was also used.

3. A vapor trap having a wide-bore inlet tube and the appropriate condensate capacity is used. The checkers used a 4.8 cm. x 30.0 cm. trap having a 2.0 cm. x 18.0 cm. inlet tube. Best results are obtained when the trap is connected directly to the flask or the adapter (see Note 2) by a wide-bore tube. Ground-glass joints should be used throughout the apparatus.

4. All materials were obtained from Eastman Kodak Company. Quinoline, practical grade, b.p. 72–74° (0.2 mm.), was freshly distilled from zinc dust. If undistilled quinoline is employed, a major contaminant, which appears to be methyl isocyanate, will be formed. "White label" p-toluenesulfonyl chloride and N-methylformamide were used without further purification.

5. The checkers found that a bath of dry ice and acetone worked equally well.

6. A single transfer under high vacuum afforded a product of identical purity. The major contaminants appear to be small amounts of high-boiling starting materials.

7. When a 2-m. polypropylene glycol on firebrick column at 75° is used, the retention volume of methyl isocyanide is 55 cc. of helium. Because of an unknown factor in conditioning the column, it is advisable to perform at least two consecutive analyses.

The checkers employed a 5-ft. 20% Carbowax 20 M (terminated with terephthalic acid) on Chromosorb W (acid washed) column at 60°. Methyl isocyanide showed a retention volume of 300 cc. of helium. Only traces of lower-boiling impurities were observed.

3. Methods of Preparation

Methyl isocyanide has been prepared chiefly by minor modifications of the original method of Gautier,[3] which is the alkylation of silver cyanide by an alkyl halide.

4. Merits of the Preparation

The excellent procedures for dehydration of N-alkyl- and N-arylformamides developed by Hertler and Corey [4] and by Ugi and co-workers [5] are unsuccessful with low-molecular-weight isocyanides. This common failure is probably due to poor efficiency in extraction of these very polar substances from water. The present method also has been successfully employed for the preparation of smaller quantities of methyl (50%), ethyl (45%), s-butyl (35%) and cyclobutyl (24%) isocyanides.[6] The procedure is less laborious than that reported earlier for ethyl isocyanide.[7] For comments on procedures for the preparation of the higher alkyl isocyanides see p. 300 of this volume and reference 8.

1. Department of Chemistry, California State College at Los Angeles, Los Angeles, California.
2. A. R. Stein, *Chem. Eng. News,* **46,** Oct. 21, 1968, p. 8; cf. also M. P. Lemoult, *Compt. Rend.,* **143,** 902 (1906).
3. A. Gautier, *Ann.,* **146,** 119 (1868); *Ann. Chim. et Phys.,* [4] **17,** 103, 203 (1869).
4. W. Hertler and E. J. Corey, *J. Org. Chem.,* **23,** 1221 (1958).
5. I. Ugi and R. Meyr, *Ber.,* **93,** 239 (1960); I. Ugi, R. Meyr, and M. Lipinski, this volume, p. 300; I. Ugi and R. Meyr, this volume, p. 1060; I. Ugi, U. Fetzer, U. Eholzer, H. Knupfer, and K. Offermann, *Angew. Chem. Intern. Ed.,* **4,** 472 (1965).
6. J. Casanova, Jr., R. E. Schuster, and N. D. Werner, *J. Chem. Soc.,* 4280 (1963).
7. H. L. Jackson and B. C. McKusick, *Org. Syntheses,* Coll. Vol. **4,** 438 (1963).
8. P. Hoffman, G. Gokel, D. Marquarding, and I. Ugi, in I. Ugi, "Isonitrile Chemistry," Academic Press, New York, 1971, p. 9.

METHYL KETONES FROM CARBOXYLIC ACIDS:
CYCLOHEXYL METHYL KETONE*

(Ketone, cyclohexyl methyl)

Submitted by Thomas M. Bare and Herbert O. House [1]
Checked by A. De Meijere and K. B. Wiberg

1. Procedure

Caution! Since hydrogen is liberated in the first step of this reaction, it should be conducted in a hood. A dry, 500-ml., three-necked flask is fitted with a reflux condenser, a pressure-equalizing dropping funnel, a mechanical stirrer, and an inlet tube to maintain a static nitrogen atmosphere in the reaction vessel throughout the reaction. In the flask are placed 1.39 g. (0.174 mole) of powdered lithium hydride (Note 1) and 100 ml. of anhydrous 1,2-dimethoxyethane (Note 2). While this suspension is stirred vigorously, a solution of 19.25 g. (0.150 mole) of cyclohexane-carboxylic acid (Note 3) in 100 ml. of anhydrous 1,2-dimethoxy-ethane (Note 2) is added dropwise over a 10-minute period. The resulting mixture is heated to reflux with stirring for 2.5 hours, at which time hydrogen evolution and the formation of lithium cyclohexanecarboxylate are complete. After the resulting suspension has been cooled to approximately 10° with an ice bath, it is stirred vigorously while 123 ml. of an ethereal solution containing 0.170 mole of methyllithium (Note 4) is added dropwise over a 30-minute period. After the addition is complete, the ice bath is removed and the resulting suspension is stirred at room temperature for 2 hours. The dropping funnel is removed from the reaction flask and replaced by a rubber septum fitted with a 4-mm. O.D. glass tube of suitable dimensions to permit the reaction mixture to be siphoned from the reaction flask when a slight positive nitrogen pressure is present in the flask.

The fine suspension in the reaction flask is agitated and siphoned into a vigorously stirred mixture of 27 ml. (0.32 mole) (Note 5) of concentrated hydrochloric acid and 400 ml. of water. The reaction flask is rinsed with an additional 100 ml. of ether which is also added to the aqueous solution. After the resulting mixture has been saturated with sodium chloride, the organic phase is separated and the alkaline (Note 5) aqueous phase is extracted with three 150-ml. portions of ether. When the combined organic solutions have been dried over magnesium sulfate, the bulk of the ether is distilled from the mixture through a 40-cm. Vigreux column (Note 6) and then the residual ether and the 1,2-dimethoxyethane are distilled from the mixture through a 10-cm. Vigreux column. Distillation of the residual pale yellow liquid separates 17.1–17.7 g. (91–94%) of the methyl ketone as a colorless liquid, b.p. 57–60° (8 mm.), $n^{26}D$ 1.4488–1.4489 (Note 7).

2. Notes

1. Lithium hydride of suitable quality may be purchased from Alfa Inorganics, Inc., 8 Congress Street, Beverly, Massachusetts 01915.

2. Commercial 1,2-dimethoxyethane, b.p. 85–86°, purchased from Eastman Organic Chemicals, was distilled from lithium aluminum hydride before use.

3. The cyclohexanecarboxylic acid, m.p. 31–32°, purchased from Aldrich Chemical Company was used without further purification.

4. An ethereal solution which was 1.38M in methyllithium was purchased from Foote Mineral Company. The concentration of methyllithium in ethereal solutions may be conveniently determined by a procedure described elsewhere [2, 3] in which the lithium reagent is titrated with sec-butyl alcohol, utilizing the charge transfer complex formed from bipyridyl or o-phenanthroline and the lithium reagent as an indicator.

5. The quantity of hydrochloric acid used is normally insufficient to neutralize all the lithium hydroxide produced when the reaction mixture is quenched in the aqueous solution. As a result, any unchanged cyclohexanecarboxylic acid will be present as its lithium salt and will remain in the aqueous phase.

6. The product is sufficiently volatile that use of a rotary evaporator or an open flask to distill the bulk of the ether and 1,2-dimethoxyethane from this solution may result in loss of a significant fraction of the product.

7. The product may be analyzed by use of a gas chromatography column packed with either LAC-728 (diethylene glycol succinate) or Carbowax 20M suspended on Chromosorb P. Using a 2.5-m. LAC-728 column heated to 100°, the submitters found retention times of 9.4 and 13.0 minutes for cyclohexyl methyl ketone and cyclohexyldimethylcarbinol. Less than 1% of the carbinol by-product was present.

3. Discussion

Apart from the reaction of cyclohexanecarboxylic acid with methyllithium,[4] cyclohexyl methyl ketone has been prepared by the reaction of cyclohexylmagnesium halides with acetyl chloride or acetic anhydride [5-7] and by the reaction of methylmagnesium iodide with cyclohexanecarboxylic acid chloride.[8] Other preparative methods include the aluminum chloride-catalyzed acetylation of cyclohexene in the presence of cyclohexane,[9] the oxidation of cyclohexylmethylcarbinol,[10, 11] the decarboxylation and rearrangement of the glycidic ester derived from cyclohexanone and t-butyl α-chloropropionate,[12] and the catalytic hydrogenation of 1-acetylcyclohexene.[13, 14]

This preparation illustrates the procedure for reaction of

organolithium reagents with the lithium salts of carboxylic acids to form ketones.[15] The reaction is generally applicable to alkyl, vinyl, and aryl organolithium reagents and carboxylic acid salts which do not contain other interfering functional groups. However, the reaction has been most often used with methyllithium for the preparation of methyl ketones. The reaction is known to effect the *stereospecific* conversion of a carboxylic acid to methyl ketone and, consequently, is a useful part of the sequence illustrated in the accompanying equations for interrelating the stereochemistry of alcohols and carboxylic acids.[16, 17]

$$R-CO_2H \longrightarrow R-CO_2Li \xrightarrow{CH_3Li} \xrightarrow{H_2O} R-CO-CH_3 \xrightarrow{R'CO_3H}$$

$$R-O-COCH_3 \xrightarrow[H_2O]{OH^-} R-OH$$

The reaction is believed to proceed by the indicated conversion of a lithium carboxylate to the dilithium salt [18] which is stable

$$R-CO_2Li + CH_3Li \longrightarrow R-\overset{\overset{\displaystyle OLi}{|}}{\underset{\underset{\displaystyle OLi}{|}}{C}}-CH_3 \xrightarrow{H^+} R-CO-CH_3$$

at room temperature in the absence of proton-donating solvents or reactants. However, the rapidity with which this dilithium salt is decomposed to a ketone in the presence of proton donors, accompanied by the rapidity of the subsequent reaction of the ketone with more methyllithium, can lead to a common side reaction in which an alkyldimethyl carbinol is formed. Both of the aforementioned reactions appear to be fast compared with 0.01 second usually required to disperse components with the conventional mixing techniques. This side reaction can be almost entirely avoided by taking precautions to minimize the

$$R-\overset{\overset{\displaystyle OLi}{|}}{\underset{\underset{\displaystyle OLi}{|}}{C}}-CH_3 \xrightarrow{H^+} R-CO-CH_3 \xrightarrow{CH_3Li} \xrightarrow{H_2O} R-\overset{\overset{\displaystyle OH}{|}}{C}(CH_3)_2$$

possibility that high *local concentrations* of the geminal dialkoxide and the lithium reagent are present when a proton donor is added during either the reaction or the *subsequent quenching*. It is frequently possible to obtain only relatively small amounts of the alcohol by-product by the dropwise addition *with vigorous mixing* of 2 equivalents of the organolithium reagent *to* a solution of the carboxylic acid.   During this procedure it is not uncommon for the lithium carboxylate to separate during the addition of the first mole of organolithium reagent and then to react and redissolve as the second equivalent of organolithium reagent is added. The reverse procedure, adding the acid to a solution of the organolithium reagent, appears always to produce substantial amounts of the alcohol by-product.   The procedure used in this preparation illustrates how this mixing problem may be avoided by the initial conversion of the carboxylic acid to its lithium salt with lithium hydride.   In all cases it is important not to add a large excess of organolithium reagent and to add the final reaction mixture to the aqueous quenching bath slowly and with vigorous stirring if the formation of substantial amounts of alcohol by-product is to be avoided.

1. Department of Chemistry, Massachusetts Institute of Technology, Cambridge, Massachusetts 02139.
2. W. Voskuil and J. F. Arens, this volume, p. 211.
3. S. C. Watson and J. F. Eastham, *J. Organometallic Chem.*, **9**, 165 (1967).
4. V. Theus and H. Schinz, *Helv. Chim. Acta.*, **39**, 1290 (1956).
5. J. Rouzaud, G. Cauquil, and L. Giral, *Bull. Soc. Chim. France*, 2908 (1964).
6. B. D. Tiffany, J. B. Wright, R. B. Moffett, R. V. Heinzelman, R. E. Strube, B. D. Aspergren, E. H. Lincoln, and J. L. White, *J. Am. Chem. Soc.*, **79**, 1682 (1957).
7. C. G. Overberger and A. Lebovits, *J. Am. Chem. Soc.*, **76**, 2722 (1954).
8. G. Darzens and H. Rost, *Compt. Rend.*, **153**, 772 (1911).
9. C. D. Nenitzescu and E. Cioranescu, *Ber.*, **69B**, 1820 (1936).
10. J. Lecomte and H. Gault, *Compt. Rend.*, **238**, 2538 (1954).
11. R. B. Wagner and J. A. Moore, *J. Am. Chem. Soc.*, **72**, 974 (1950).
12. E. P. Blanchard, Jr., and G. Buchi, *J. Am. Chem. Soc.*, **85**, 955 (1963).
13. E. M. Cherkasova and G. S. Erkomaishvili, *Zh. Obshch. Khim.*, **31**, 1832 (1961) [*Chem. Abstr.*, **55**, 24593 (1961)].
14. T. Takeshima, K. Wakamatsu, and A. Furuhashi, *Bull. Chem. Soc. Japan*, **31**, 640 (1958).
15. M. J. Jorgenson, *Org. Reactions*, 18, 1 (1970).
16. W. G. Dauben and E. Hoerger, *J. Am. Chem. Soc.*, **73**, 1504 (1951); W. G. Dauben, V. M. Alhadeff, and M. Tanabe, *J. Am. Chem. Soc.*, **75**, 4580 (1953); W. G. Dauben, R. C. Tweit, and C. Mannerskantz, *J. Am. Chem. Soc.*, **76**, 4420 (1954).
17. H. O. House and T. M. Bare, *J. Org. Chem.*, **33**, 943 (1968).
18. H. F. Bluhm, H. V. Donn, and H. D. Zook, *J. Am. Chem. Soc.*, **77**, 4406 (1955).

2-METHYLMERCAPTO-N-METHYL-Δ²-PYRROLINE

(2-Pyrroline, 1-methyl-2-methylthio-)

Submitted by R. GOMPPER and W. ELSER [1]
Checked by HERMANN ERTL, IAN D. RAE, and PETER YATES

1. Procedure

Caution! Hydrogen sulfide is very poisonous. Procedure A should be conducted in a hood.

A. *N-Methyl-2-pyrrolidinethione.* Phosphorus pentasulfide (667 g., 3.00 moles) is suspended in 600 ml. of carbon disulfide in a 2-l. three-necked flask equipped with a mechanical stirrer with a segment-shaped paddle, an efficient reflux condenser, and a dropping funnel. N-Methyl-2-pyrrolidinone (300 g., 3.03 moles) is added in large portions from the dropping funnel with vigorous stirring. The reaction mixture warms up considerably and, after addition of about two-thirds of the amide, becomes so viscous that further stirring is impossible. The addition of the remainder is carried out without stirring at a rate such that the solution boils gently. The semisolid, yellow-brown mixture is boiled under reflux on a water bath for a further 10 hours.

The carbon disulfide is decanted, and 200–300 ml. of water is added to the contents of the flask. Initially the mixture reacts slowly, but after some time the reaction becomes so vigorous that it is necessary to pour off the water (Notes 1 and 2). The addition of water, followed by its removal when the reaction becomes very vigorous, is repeated until decomposition is complete. The combined aqueous solutions are extracted several times with chloroform, and the combined extracts are dried over anhydrous sodium sulfate. Distillation of the solvent gives a dark brown liquid residue which is distilled under reduced pressure to give 310.3 g. (89%) of the thioamide as a yellow liquid, b.p. 133–135° (12 mm.) (Note 3).

B. *2-Methylmercapto-N-methyl-Δ ^{1}-pyrrolinium iodide.* N-Methyl-2-pyrrolidinethione (310 g., 2.69 moles) is dissolved with stirring in 1.1 l. of anhydrous ether in a 2-l. three-necked flask equipped with a mechanical stirrer, a reflux condenser fitted with a drying tube, and a dropping funnel. To this solution is added *ca.* 5 g. of the product as seed crystals (Note 4) to prevent initial deposition of the iodide as an oil that suddenly crystallizes with considerable evolution of heat. Methyl iodide (520 g., 3.66 moles) is then added rather rapidly. The solution becomes turbid after a short time, and separation of the salt begins with heat evolution. After 12 hours the hygroscopic, crystalline paste is filtered and dried in a desiccator; yield 663 g. (96%).

This product can be used without further purification. Crystallization from mixtures of acetonitrile and ether gives colorless, felted needles, m.p. 118–120° (dec.) (Note 5).

C. *2-Methylmercapto-N-methyl-Δ ^{2}-pyrroline.* 2-Methylmercapto-N-methyl-Δ ^{1}-pyrrolinium iodide (662 g., 2.57 moles) is suspended in 1.25 l. of anhydrous ether in a 4-l. three-necked flask equipped with a mechanical stirrer and a reflux condenser with a segment-shaped paddle. Potassium *t*-butoxide (448 g., 4.0 moles) is added in one batch to this suspension with vigorous stirring. The mixture warms up a little, and later the solid becomes fine-grained and more mobile as a result of separation of potassium iodide. After being stirred for 1.5 hours at room temperature, the mixture is treated with 1.8 l. of anhydrous ether and boiled under reflux on a water bath for 5 hours.

The flask is cooled in an ice bath, and the precipitate is filtered onto a large Buchner funnel and washed well with ether (Note 6). Ether is distilled from the yellow-brown filtrate at atmospheric pressure, and t-butyl alcohol is distilled under reduced pressure at 40°. Distillation of the residue through a 10-cm. Vigreux column at the water aspirator gives, after a small forerun ($ca.$ 20 ml.), the product as a colorless liquid, b.p. 70–73° (10–12 mm.); yield 268 g. (81%) (Notes 7, 8, and 9). This becomes brown on standing.

Repeated fractional distillation gives a product of analytical purity (Note 10), b.p. 65–68° (8–10 mm.), n^{25}D 1.5222 (Note 11); yield 62%. This is an extremely disagreeable, musty liquid which on standing in the atmosphere warms up slightly and immediately turns red.

2. Notes

1. The large amount of hydrogen sulfide produced is destroyed by passage into two wash bottles containing concentrated aqueous potassium hydroxide.

2. It is also possible to moderate the reaction by the addition of chloroform.

3. The checkers, working at quarter scale, obtained 84–87% of product, b.p. 131–133° (11 mm.), 139–142° (14–15 mm.).

4. This can be prepared readily in a test tube by addition of methyl iodide to a solution of the thioamide in ether.

5. $Anal.$ Calcd. for $C_6H_{12}INS$: I, 49.36. Found: 49.34.

6. The precipitate may be so fine-grained that the filtrate is initially turbid; the turbid filtrate is then filtered again through the Buchner funnel containing the precipitate.

7. This fraction is followed by tailings (12 g.), b.p. 74–75° (10–12 mm.).

8. The checkers, working at quarter scale, obtained 75% of product, b.p. 68–71° (13 mm.), followed by tailings, b.p. 71–74° (13 mm.).

9. This product is contaminated with N-methyl-2-pyrrolidin-one, which does not impair its usefulness for most further reactions.

10. *Anal.* Calcd. for $C_6H_{11}NS$: C, 55.76; H, 8.58; N, 10.84; S, 24.82. Found: C, 55.60; H, 9.08; N, 10.84, S, 24.66.

11. The checkers found that the refractive index of the pyrroline increases rapidly on exposure to air, impairing its usefulness as a criterion of purity.

3. Methods of Preparation

The method described for the preparation of N-methyl-2-pyrrolidinethione is very similar to that of Peak and Stansfield [2] for the preparation of 4-thioacetylmorpholine. N-Methyl-2-pyrrolidinethione has also been prepared by the reaction of 2-chloro-N-methyl-Δ^1-pyrrolinium chloride with hydrogen sulfide [yield 83%; b.p. 144–145° (15 mm.)] [3] and by heating N-methyl-2-pyrrolidinone with 2 equivalents of phosphorus pentasulfide in xylene [yield not reported; b.p. 125–132° (10 mm.)].[4] General procedures for the preparation of N,N-disubstituted thioamides have been reviewed.[5, 6]

The method described for the preparation of 2-methylmercapto-N-methyl-Δ^1-pyrrolinium iodide is based on the general procedure of several authors.[2, 7–9] Its preparation has not been described previously; the corresponding methyl sulfate has been obtained as a noncrystalline, viscous mass.[4, 10]

2-Methylmercapto-N-methyl-Δ^2-pyrroline has been prepared by the present method only.[11, 12]

4. Merits of the Preparation

The procedure illustrates a general method for the preparation of ketene S,N-acetals via thioamides and their crystalline quaternary iodides; some other examples are shown in Table I. 2,2-Dialkyl-substituted ketene S,N-acetals cannot be prepared by this method because the nature of the products makes rigorous purification impractical. Ketene S,N-acetals are useful starting materials for many syntheses.[11, 13-15]

TABLE I

PREPARATION OF KETENE S,N-ACETALS

$$RCH=C\begin{array}{c} SCH_3 \\ \diagup \\ \diagdown \\ NR'R'' \end{array}$$

R	R'	R''	B.P., °C.	Yield of Crude Product, % (Purity, %)[a]
H	CH$_3$	CH$_3$	31–32 (11 mm.)	77 (92–96)[b]
H	(CH$_2$)$_4$		79–81 (10 mm.)	60 (72–88)
CH$_3$	CH$_3$	CH$_3$	38–40 (10 mm.)	68 (88)
C$_2$H$_5$	CH$_3$	CH$_3$	56 (11 mm.)	74 (92)
C$_6$H$_5$	(CH$_2$)$_2$O(CH$_2$)$_2$[c]		114–120 (10^{-3} mm.)[d]	81
(CH$_2$)$_3$		CH$_3$	75–76 (12 mm.)	69–73 (94)
(CH$_2$)$_4$		CH$_3$	86–87.5 (12 mm.)	82 (94)

[a] Determined by gas chromatography; products distilled through a spinning-band column were 94–99% pure.
[b] Yield of pure product, 52%.
[c] M. A. T. Rogers, *J. Chem. Soc.*, 3350 (1950).
[d] M.p. 44–45° (ex ethanol).

1. Institut für Organische Chemie der Universität München, München, Germany.
2. D. A. Peak and F. Stansfield, *J. Chem. Soc.*, 4067 (1952).
3. H. Eilingsfeld, M. Seefelder, and H. Weidinger, *Ber.*, **96**, 2671 (1963).
4. O.Riester and F. Bauer, German Patent 927,043 (1955) [*C.A.*, **52**, 938 (1958)].
5. A. Schöberl and A. Wagner, in E. Müller, "Methoden der Organischen Chemie (Houben-Weyl)," 4th ed., Vol. 9, Georg Thieme Verlag, Stuttgart, 1955, p. 762.
6. R. N. Hurd and G. DeLamater, *Chem. Rev.*, **61**, 45 (1961).
7. J. Renault, *Compt. Rend.*, **233**, 182 (1951).
8. P. Chabrier and S. H. Renard, *Compt. Rend.*, **228**, 850 (1949).
9. R. I. Meltzer, A. D. Lewis, and J. A. King, *J. Am. Chem. Soc.*, **77**, 4062 (1955).
10. S. Hünig and F. Müller, *Ann.*, **651**, 89 (1962).
11. W. Elser, Ph.D., Dissertation, Technische Hochschule, Stuttgart, 1965.
12. R. Gompper and W. Elser, *Tetrahedron Lett.*, 1971 (1964).
13. R. Buijle, A. Halleux, and H. G. Viehe, *Angew. Chem.*, **78**, 593 (1966); *Angew. Chem. Intern. Ed. Engl.*, **5**, 584 (1966).
14. R. Gompper and W. Elser, *Angew. Chem.*, **79**, 382 (1967); *Angew. Chem. Intern. Ed. Engl.*, **6**, 366 (1967).
15. R. Gompper, W. Elser, and H.-J. Müller, *Angew. Chem.*, **79**, 473 (1967); *Angew. Chem. Intern. Ed. Engl.*, **6**, 453 (1967).

3-METHYLPENTANE-2,4-DIONE

(2,4-Pentanedione, 3-methyl-)

$$CH_3COCH_2COCH_3 + CH_3I \xrightarrow{K_2CO_3} CH_3COCH(CH_3)COCH_3$$

Submitted by A. W. Johnson, E. Markham, and R. Price.[1]
Checked by Virgil Boekelheide and M. Kunstmann.

1. Procedure

A mixture of 65.2 g. (0.65 mole) of pentane-2,4-dione, 113 g. (0.80 mole) of methyl iodide, 84 g. of anhydrous potassium carbonate (Note 1), and 125 ml. of acetone is placed in a 500-ml. round-bottomed flask fitted with a reflux condenser and a calcium chloride guard tube. The mixture is heated under reflux for 20 hours and is then allowed to cool. The insoluble material is removed by filtration and washed with acetone (Note 2). The combined filtrate and acetone washings are concentrated on the steam bath (Note 3), and the residual oil is distilled. There is collected 56–57 g. (75–77%) of a colorless oil, b.p. 170–172°/760 mm., n_D^{24} 1.4378 (Note 4).

2. Notes

1. The potassium carbonate is dried at 100° for 2 hours before use.

2. Thorough washing of the inorganic residues is essential and requires about 200 ml. of acetone.

3. During removal of the acetone, potassium iodide is deposited and it is advisable to decant the crude 3-methylpentane-2,4-dione from this material before distillation.

4. It has been reported by A. M. Roe and J. B. Harbride (private communication) that this procedure yields a product containing 20–25% of 3,3-dimethylpentane-2,4-dione as shown by gas chromatography. The amount of the dialkylation product is said to be reduced to 5–10% when the reflux period is shortened from 20 to 4.5 hours. The impurity is not

readily removed, but it does not interfere with the preparation of 2,3,4,5-tetramethylpyrrole [this volume, p. 1022].

The same authors report that the work-up of the reaction may be improved by adding 250 ml. of petroleum ether (b.p. 40–60°) to the cold reaction mixture before filtering, and washing the solids with a 1:1 mixture of acetone and petroleum ether. With this change it is not necessary to decant the product from the precipitated potassium iodide as recommended in Note 3.

3. Methods of Preparation

3-Methylpentane-2,4-dione has been prepared by the reaction of the sodium derivative of pentane-2,4-dione with methyl iodide in a sealed tube at 140°,[2] and from the sodium [3] and potassium [4] derivatives of pentane-2,4-dione and methyl iodide in alcoholic solution. It has also been prepared by the reaction of methyl iodide and pentane-2,4-dione in the presence of potassium carbonate in alcoholic or ethereal solution [5] and in acetone solution,[6,7] and by heating 2-aminopenten-4-one with methyl iodide at 100°.[8] The present modification affords improved yields.

4. Merits of Preparation

The method presented here has also been used for the preparation of 3-ethyl- and 3-isopropylpentane-2,4-diones and is probably of general applicability in the preparation of 3-alkylpentane-2,4-diones.

1. Department of Chemistry, University of Nottingham, Nottingham, England.
2. W. R. Dunstan and T. S. Dymond, *J. Chem. Soc.*, 59, 428 (1891).
3. J. Salkind, *Chem. Zentr.*, 1905, II, 753.
4. W. H. Perkin, *J. Chem. Soc.*, 61, 848 (1892).
5. L. Claisen, *Ber.*, 27, 3184 (1894).
6. K. von Auwers and H. Jacobsen, *Ann.*, 426, 229 (1922).
7. A. M. Roe and J. B. Harbridge, *Chem. & Ind. (London)*, 182 (1965).
8. A. Combes and C. Combes, *Bull. Soc. Chim. France*, [3] 7, 785 (1892).

3-METHYL-1-PHENYLPHOSPHOLENE OXIDE

(Phospholene, 3-methyl-1-phenyl-, 1-oxide)

Submitted by W. B. McCormack.[1]
Checked by S. N. Lewis and W. D. Emmons.

1. Procedure

A. *3-Methyl-1-phenylphospholene 1,1-dichloride.* A dry 1-l. suction flask (Note 1) is charged with 179 g. (1.00 mole) of dichlorophenylphosphine (n_D^{28} 1.592; Note 2), 300 ml. (about 204 g., 3.0 moles) of commercial isoprene (Note 3), and 2.0 g. of Ionol® (Note 4). The flask is then stoppered, the side arm is sealed with tubing and a clamp, and the homogeneous solution is allowed to sit at room temperature in the back of a hood for 5–7 days. White solid is usually apparent within 2–4 hours, and after the reaction period the liquid phase is full of a white crystalline adduct, 1,1-dichloro-1-phenyl-3-methyl-1-phospha-3-cyclopentene. The granular adduct is crushed, slurried with petroleum ether, collected on a sintered glass Büchner funnel, and washed with petroleum ether; exposure to moisture of the air is minimized by covering the funnel with a clock glass (Note 5).

B. *3-Methyl-1-phenylphosphacyclopentene 1-oxide.* The adduct is hydrolyzed by stirring it into 700 ml. of ice water, and stirring is continued until essentially all of it is in solution (Note

6). The total amount of acid in the solution is determined by titrating an aliquot, and the solution is nearly neutralized by slow addition of about 93% of the theoretical amount of 30% sodium hydroxide with good stirring and sufficient ice to keep the temperature below 25°. The solution generally contains about 1.62 equivalents of acid, which calls for 150 ml. (1.50 equivalents) of 30% sodium hydroxide. The pH is then adjusted to 6.5 with sodium bicarbonate solution (Note 7). After saturation of the aqueous solution with sodium chloride, the product is extracted with three 250-ml. portions of chloroform. The chloroform extracts are combined, dried briefly over calcium sulfate, filtered, and concentrated at atmospheric pressure until the temperature of the liquid in the distillation flask is 130°. The residual liquid is fractionated through a 30-cm. packed column. A water aspirator is used initially to strip small amounts of low boilers at 100° (Note 8), and then a mechanical pump is used. There is a fore-shot of a partially solidifying oil, weight 2–3 g., b.p. 163–168°/0.6–0.7 mm. Then 110–120 g. (57–63%) of 3-methyl-1-phenyl-1-phospha-3-cyclopentene 1-oxide, b.p. 173–174°/0.7 mm., is collected. It is a viscous liquid of a very pale yellow color that solidifies to a white solid, m.p. 60–65° (Note 9). A small amount of residue remains (Note 10).

2. Notes

1. A suction flask provides a strong-walled reactor that is conveniently stored for the reaction period.

2. Commercial material (currently from Stauffer Chemical Company, Special Chemicals Division) is suitable after distillation.

3. Shell Chemical material (92% minimum purity) was used. It was noted by the checkers that redistilled isoprene gave slightly better yields. The excess isoprene serves both as a solvent and for mass action. Moreover, dichlorophenylphosphine tends to dissolve in the product, and excess isoprene, by extracting the solid, promotes completion of reaction.

4. Ionol® is a commercial antioxidant, 2,6-di-*tert*-butyl-*p*-cresol, manufactured by Shell Chemical Corp. Inhibitors appear

to minimize formation of polymeric side products, although with isoprene the effect is often small.

5. Sometimes, as with less pure reagents or on heating, a viscous oil, red to dark in color, will form instead of a white solid. At other times the solid is rather gummy. In these cases, mixing with petroleum ether as much as possible and decanting must suffice.

6. On occasion, gelatinous material is apparent; in time it usually dissolves or swells greatly.

7. The product is relatively sensitive to basic conditions, showing both polymerization and addition of water. Therefore alkaline conditions must be avoided. Neutralization serves to convert monophenylphosphinic acid (formed by hydrolysis of unreacted, unextracted dichlorophenylphosphine) to the monosodium salt, thereby preventing its subsequent extraction from water along with the phosphine oxide.

8. The crude mixtures all have strong odors of aromatic phosphines. Some of this odor presumably arises from disproportionation of monophenylphosphinic acid to phenylphosphine. It is recommended that manipulations be carried out with rubber gloves to prevent transfer of these rather durable odors to the skin, and that all equipment be washed with a bleach such as Clorox® before it is taken from the hood.

9. The distilled product can be used as a catalyst, although it usually has a relatively strong phenylphosphine odor. It is quite deliquescent, and it has not been satisfactorily recrystallized. If rigorous purification and deodorization are desired, the product is dissolved in water, a small amount of hydrogen peroxide is added to oxidize the phosphines, the solution is reneutralized, saturated with salt, and extracted with chloroform, and the product is refractionated. One cycle is normally enough. Pure product is essentially odorless, very hygroscopic, and soluble in polar solvents.

10. This residue is mostly the linear copolymer

$$\left[\begin{array}{c} O \\ \uparrow \\ P-CH_2-CH=\overset{\displaystyle CH_3}{\underset{\displaystyle }{C}}-CH_2 \\ | \\ C_6H_5 \end{array} \right]_n$$

of nearly 1:1 composition. On occasion it can be present in substantial amounts, especially if higher temperatures are used to increase the reaction rate.

3. Methods of Preparation

3-Methyl-1-phenylphosphacyclopentene 1-oxide has been prepared only as described here.[2]

4. Merits of the Preparation

The reaction given here has been described before as a general reaction,[2] and there can be a wide variety of alkyl, aryl, and halo substituents on the diene and phosphorus. Dibromophosphines are appreciably more reactive than dichlorophosphines. If a free-radical catalyst is used instead of an inhibitor, the copolymers can be made in good yield.[3] The 1,4-addition of dichlorophosphines to 1,3-dienes is of theoretical interest because of its analogy to the well-known 1,4-addition of sulfur dioxide to 1,3-dienes.

The unsaturated cyclic phosphine oxides are active catalysts for the conversion of aryl isocyanates to carbodiimides.[4] The polymeric material[3] and the saturated cyclic phosphine oxides[5] are also catalysts but are less active. The unsaturated cyclic phosphine oxides show properties analogous to those of the unsaturated cyclic sulfones from dienes and sulfur dioxide in that the double bond is quite reactive to basic reagents and relatively resistant to acidic reagents. Physically these cyclic phosphine oxides are stable to over 300°, are very powerful hydrogen bond acceptors, and are excellent solvents for polar materials.

The most recent review[6] summarizes the chemistry of these materials.

1. Contribution No. 317, Organic Chemicals Department, E. I. du Pont de Nemours and Co., Wilmington, Delaware.
2. W. B. McCormack, U.S. Patent 2,663,737 (1953) [C. A., 49, 7601a (1955)].
3. W. B. McCormack, U.S. Patent 2,671,079 (1954) [C. A., 48, 6738c (1954)].
4. T. W. Campbell and J. J. Verbanc, U.S. Patent 2,853,473 (1958) [C. A., 53, 10126e (1959)]; T. W. Campbell and J. J. Monagle, this volume, p. 504.
5. W. B. McCormack, U.S. Patent 2,663,739 (1953) [C. A., 49, 7602f (1955)].
6. F. G. Mann, "The Heterocyclic Derivatives of Phosphorus, Arsenic, Antimony, and Bismuth," 2nd Edition, Wiley-Interscience, New York, 1970, pp. 31–61.

METHYL PHENYL SULFOXIDE

$$C_6H_5SCH_3 + NaIO_4 \rightarrow C_6H_5\overset{\overset{O}{\|}}{S}CH_3 + NaIO_3$$

Submitted by Carl R. Johnson and Jeffrey E. Keiser [1]
Checked by Wayland E. Noland, Leonard J. Czuba,
and William E. Parham

1. Procedure

In a 500-ml. round-bottomed flask equipped with a magnetic stirrer are placed 22.5 g. (0.105 mole) of powdered sodium metaperiodate and 210 ml. of water. The mixture is stirred and cooled in an ice bath (Note 1), and 12.4 g. (0.100 mole) of thioanisole (Note 2) is added. The reaction mixture is stirred for 15 hours at ice-bath temperature and is then filtered through a Büchner funnel. The filter cake of sodium iodate is washed with three 30-ml. portions of methylene chloride. The water-methylene chloride filtrate is transferred to a separatory funnel, the lower methylene chloride layer is removed, and the water layer is extracted with three 100-ml. portions of methylene chloride. The combined methylene chloride extracts are treated with activated carbon (Note 3) and dried over anhydrous sodium sulfate (Note 4). The solvent is removed at reduced pressure to yield 13.6–13.9 g. of a slightly yellow oil (Note 5) which crystallizes on cooling. The crude sulfoxide is transferred to a 25-ml. distillation flask with the aid of a small amount of methylene chloride. After removal of the solvent, a pinch of activated carbon is added to the distillation flask (Note 6). Simple vacuum distillation (Note 7) of the crude product through a short path still affords 12.7–12.8 g. (91%) of pure methyl phenyl sulfoxide, b.p. 78–79° (0.1 mm.), m.p. 33–34° (Notes 8 and 9).

2. Notes

1. The periodate oxidation of aryl sulfides to sulfoxides may be carried out at room temperature; however, an ice bath is necessary to prevent overoxidation of dialkyl sulfides.

2. Thioanisole (methyl phenyl sulfide) supplied by Aldrich Chemical Company, Milwaukee, Wisconsin, was used without further purification.

3. The checkers used 1 g. of activated carbon.

4. The checkers used 3–5 g. of sodium sulfate.

5. Gas-phase chromatography shows this crude material to be sulfide-free sulfoxide containing a small amount of methylene chloride.

6. The simple technique of adding a pinch of activated carbon to the distillation pot affords a more nearly colorless distillate.

7. No fore-run is observed other than a small amount of methylene chloride which does not condense. The pot is taken nearly to dryness.

8. Methyl phenyl sulfoxide is extremely hygroscopic. The melting point is best taken by rapid transfer of the easily super-cooled oil to a melting-point capillary by means of a finely drawn pipet. The sealed capillary is then cooled to effect crystallization of the sulfoxide.

9. This procedure, with slight modifications depending on the physical properties of the sulfide and sulfoxide in question, has been used to prepare a variety of sulfoxides as illustrated by examples provided in Table I. In the case of very insoluble sulfides, co-solvents such as methanol or dioxane may be employed. Very soluble sulfoxides are best isolated by continuous extraction with chloroform or methylene chloride.

TABLE I
PREPARATION OF SULFOXIDES

Products	Yields, %	M.P. [B.P.], °C.
Methyl 4-ketopentyl sulfoxide	98	22.5–23.5 [99–101 (0.12 mm.)]
Thian 1-oxide	99	67–68
1,4-Oxathian 1-oxide	83	46–47
Bis(2-diethylaminoethyl)sulfoxide	85	Dipicrate 146–148
Acetoxymethyl methyl sulfoxide	72	[85–90 (0.1 mm.)]
1-Benzylsulfinyl-2-propanone	89	126.0–126.5
Phenylsulfinylacetic acid	99	118.0–119.5
Benzyl sulfoxide	96	135–136
Ethyl sulfoxide	65	[45–47 (0.15 mm.)]
Ethyl phenyl sulfoxide	93	[101–102 (1.5 mm.)]

3. Methods of Preparation

Methyl phenyl sulfoxide has also been prepared from thio-anisole by the action of hydrogen peroxide,[2, 3] lead tetraacetate,[4] and dinitrogen tetroxide,[5, 6] and from methanesulfinyl chloride and benzene with anhydrous aluminum chloride.[7]

4. Merits of the Preparation

The present procedure is a specific example of the method generalized by Leonard and Johnson.[8] The method employs extremely mild reaction conditions and affords high yields of sulfoxides (Note 9) free of contamination by sulfides or sulfones. Sodium periodate is easily and safely handled; however, the higher cost of this reagent in comparison to certain other oxidants, *e.g.*, hydrogen peroxide, may prohibit its use in large-scale reactions.

1. Department of Chemistry, Wayne State University, Detroit, Michigan. This work was supported in part by grant GP-1159 of the National Science Foundation.
2. C. C. Price and J. J. Hydock, *J. Am. Chem. Soc.*, **74**, 1943 (1952).
3. D. Barnard, J. M. Fabien, and H. P. Koch, *J. Chem. Soc.*, 2442 (1949)
4. H. Böhme, H. Fischer, and R. Frank, *Ann.*, **563**, 54 (1949).
5. L. Horner and F. Hübenett, *Ann.*, **579**, 193 (1953).
6. D. W. Goheen, W. H. Hearon, and J. Kamlet, U.S. Patent 2,925,442 (1960) [*C. A.*, **54**, 11994a (1960)].
7. I. B. Douglass and B. S. Farah, *J. Org. Chem.*, **23**, 805 (1958).
8. N. J. Leonard and C. R. Johnson, *J. Org. Chem.*, **27**, 282 (1962).

4-METHYLPYRIMIDINE*

(Pyrimidine, 4-methyl-)

$$CH_3COCH_2CH(OCH_3)_2 \quad + \quad 2HCONH_2 \quad \xrightarrow{NH_4Cl}$$

$$+ \quad 2CH_3OH \quad + \quad CO \quad + \quad 2H_2O$$

Submitted by H. Bredereck.[1]
Checked by Max Tishler, G. A. Stein, W. F. Jankowski, and J. ten Broeke.

1. Procedure

A 2-l. three-necked flask is equipped with a stirrer, a thermometer, an addition funnel, and a wide-bore reflux condenser (Note 1). A second condenser set downward for distillation is connected to the top of the reflux condenser by means of a head provided with a thermometer well. The thermometer well should be positioned in the connecting head in such a manner that the thermometer gives the temperature at the head of the reflux condenser. *Caution! The flask should be in a hood so that the carbon monoxide evolved cannot be a hazard.*

The three-necked flask is charged with 750 ml. of formamide, 25 ml. of water, and 50 g. of ammonium chloride (Note 2). The mixture is heated to 180–190° in an oil bath, and 400 g. (3.02 moles) of 4,4-dimethoxy-2-butanone (Note 3) is added dropwise with stirring over the course of 6 hours (Note 4). The flow of cooling water in the reflux condenser should be adjusted to a rate such that the methanol and methyl formate formed during the reaction distil out (Note 5). After all the acetal has been added, heating is continued for 1 hour (Note 6). The mixture is allowed to cool and is poured into 1 l. of $1N$ sodium hydroxide. The resultant solution is extracted with chloroform in a liquid-liquid extractor for 24 hours. The chloroform is separated, dried over sodium sulfate, and removed by distillation through a short column on a steam bath.

The residue is distilled under reduced pressure, and all the distillate boiling at 60–80°/15 mm. is collected (Notes 7 and 8). Pure 4-methylpyrimidine is obtained by redistillation through a short column at atmospheric pressure; b.p. 140–142°; n_D^{25} 1.4936; weight 153–180 g. (54–63%).

2. Notes

1. A wide-bore condenser is employed to prevent clogging of the tube by ammonium salts that may sublime from the reaction mixture (Note 5). A Liebig condenser is most suitable because it can be cleaned from the top with a rod or wire during the reaction.

2. An acidic salt must be present so that the acetal bonds will be hydrolyzed. Other salts, such as ammonium formate, may be substituted for ammonium chloride.

3. 4,4-Dimethoxy-2-butanone from Chemische Werke Huls, Huls, Germany, was used by the submitter. The checkers used 4,4-dimethoxy-2-butanone (practical grade) from Eastman Organic Chemicals Co., Rochester, New York; it was 95% pure by vapor-phase chromatography.

4. The checkers found that, during the course of the addition, the internal temperature fell from 190° to 140°.

5. The checkers found that, if the temperature at the head of the reflux condenser is kept at 50–55°, no ammonium salts collect in the condenser and, therefore, there is no problem of clogging.

6. The checkers found that the temperature of the reaction mixture remains at 140° during the additional hour of heating.

7. It is advisable to carry out a vacuum distillation prior to the final distillation because the tarry residues obtained by distillation at atmospheric pressure retain a considerable amount of product.

8. The checkers collected their product at 50–70°/30 mm.

3. Methods of Preparation

4-Methylpyrimidine has been obtained by the present method [2] and by a three-step method that begins with the condensation of acetoacetic ester with urea to give 2,6-dihydroxy-4-methyl-pyrimidine; the latter is treated with phosphorus oxychloride to give 2,6-dichloro-4-methylpyrimidine, which is reduced by zinc dust and water [3] or by catalytic hydrogenolysis.[4]

4. Merits of the Preparation

The present one-step procedure for making 4-methylpyrimidine is simpler and easier than the three-step method used in the past. The present procedure and modifications of it have been used to make a variety of 4- and 4,6-substituted pyrimidines.[2, 5]

1. Institue für Organische Chemie, Technische Hochschule, Stuttgart, Germany.
2. H. Bredereck, R. Gompper, and G. Morlock, *Ber.,* 90, 942 (1957).
3. S. Gabriel and J. Colman, *Ber.,* 32, 1534 (1899).
4. W. Pfleiderer and H. Mosthaf, *Ber.,* 90, 733 (1957).
5. H. Bredereck, R. Gompper, and G. Morlock, *Ber.,* 91, 2830 (1958); H. Bredereck, R. Gompper, and H. Herlinger, *Ber.,* 91, 2832 (1958).

1-METHYL-3-*p*-TOLYLTRIAZENE*

(Triazene, 1-methyl-3-*p*-tolyl-)

AND ITS USE IN THE ESTERIFICATION OF ACIDS

$$p\text{-CH}_3\text{C}_6\text{H}_4\text{NH}_2 \quad + \quad \text{HONO} \quad \xrightarrow[\text{HCl}]{\text{H}_2\text{O}} \quad p\text{-CH}_3\text{C}_6\text{H}_4\text{N}_2{}^+ \text{ Cl}^-$$

$$p\text{-CH}_3\text{C}_6\text{H}_4\text{N}_2{}^+ \text{ Cl}^- \quad + \quad \text{CH}_3\text{NH}_2 \quad \xrightarrow{\text{H}_2\text{O}} \quad p\text{-CH}_3\text{C}_6\text{H}_4\text{N}{=}\text{NNHCH}_3$$

$$p\text{-CH}_3\text{C}_6\text{H}_4\text{N}{=}\text{NNHCH}_3 \quad +$$

Submitted by E. H. WHITE, A. A. BAUM,
and D. E. EITEL [1]
Checked by I. KATZ and R. BRESLOW

1. Procedure

A. *1-Methyl-3-p-tolyltriazene.* *p*-Toluidine (50.2 g., 0.47 mole) is added to a 2-l. flask equipped with a 200-ml. dropping funnel and an efficient stirrer, and the flask is immersed in an ice-salt bath at *ca.* −10°. A solution of 46.8 g. (0.55 mole) of potassium nitrite in 150 ml. of water is placed in the dropping funnel, and a mixture of 250 g. of crushed ice and 140 ml. of concentrated hydrochloric acid is added to the *p*-toluidine with stirring. The potassium nitrite solution is slowly added with continued stirring during 1–2 hours until a positive starch-potassium iodide test is obtained (Note 1), and the mixture is stirred for an additional hour to ensure the reaction of all the toluidine.

The solution of *p*-toluenediazonium chloride is then brought to pH 6.8–7.2 at 0° with cold, concentrated, aqueous sodium carbonate, whereupon the solution becomes red to orange in color and a small amount of red material settles out. The cold, neutral solution is transferred to a dropping funnel and added slowly to a vigorously stirred mixture of 150 g. of sodium carbonate, 300 ml. of 30–35% aqueous methylamine (Note 2), and 100 g. of crushed ice in a 3-l. flask. The reaction mixture is kept at *ca.* −10° during the addition, which requires about 45 minutes (Note 3). The solution is extracted with three 1-l. portions of ether. The ethereal extracts are dried with anhydrous sodium sulfate and evaporated on a rotary evaporator at room temperature to give 65 g. of crude 1-methyl-3-*p*-tolyltriazene (Note 4). This is placed in a water-cooled sublimer, and the triazene is sublimed at 50° (1 mm.) (*Caution! See Note 5*); 43.3 g. (0.29 mole, 62%) of a yellow, crystalline sublimate, m.p. 77–80°, is obtained (Note 6). The sublimate can be recrystallized from hexane to give the triazene as white needles, m.p. 80.5–81.5°. More conveniently, it is dissolved in the minimum amount of ether, and the solution is diluted with 2 volumes of hexane and cooled to 0° to give flat plates with a slightly yellow cast; m.p. 79–81°. The yield of pure triazene is 33–37 g. (47–53%) (Note 7).

B. *Esterification of 3,5-dinitrobenzoic acid with 1-methyl-3-p-tolyltriazene* (Note 8). A solution of 1-methyl-3-*p*-tolyltriazene (1.05 g., 7.0 mmoles) in 10 ml. of ether is placed in a 100-ml. flask equipped with a 100-ml. dropping funnel (Note 9). A solution of 1.50 g. (7.1 mmoles) of 3,5-dinitrobenzoic acid in 25 ml. of ether is placed in the dropping funnel and is slowly added to the triazene solution; the contents of the flask are gently swirled from time to time. During the addition, nitrogen is evolved, and the solution becomes red in color. After the nitrogen evolution has ceased (*ca.* 1 hour), the ethereal solution is transferred to a separatory funnel and washed with 5*N* hydrochloric acid to remove toluidine (Note 10). It is then washed with 5% aqueous sodium carbonate and dried over anhydrous sodium sulfate. Evaporation of the ether yields methyl 3,5-dinitrobenzoate (1.11–

1.42 g., 70–90%) as light tan crystals, m.p. 106–107.5° (Note 11). Recrystallization from ether yields small, flat plates, m.p. 107–107.5°.

2. Notes

1. The individual tests with starch-potassium iodide paper should be made 1–2 minutes after the addition of potassium nitrite has been stopped.

2. The checkers used 40% aqueous methylamine supplied by Matheson, Coleman and Bell.

3. The reaction is over when a drop of solution no longer gives a red color with a solution of β-naphthol in aqueous sodium carbonate.

4. The chief impurity is 1,5-di-*p*-tolyl-3-methyl-1,4-pent-azadiene (m.p. 148°). This can be removed by fractional crystallization, but it is easier to sublime the triazene from the reaction mixture.

5. Care should be exercised during the sublimation, which should be conducted behind an appropriate shield. During the sublimation of an analog of 1-methyl-3-*p*-tolyltriazene, 1-benzyl-3-*p*-tolyltriazene, a violent explosion has occurred. In order to achieve effective sublimation, a temperature of 90–100° was required, and this elevated temperature may have been a factor in causing the explosion (private communication from D. W. Hutchinson).

6. The sublimate contains a trace of 1,3-di-*p*-tolyltriazene, as shown by thin-layer chromatography. Recrystallization yields the pure 1-methyl-3-*p*-tolyltriazene.

7. This procedure works well only with water-soluble amines. A procedure has been given elsewhere for the preparation of triazenes of water-insoluble amines.[2]

8. The ethyl, propyl, butyl, and other esters may be prepared similarly from the corresponding triazenes.

9. Many solvents may be used for this reaction; the reaction rate, however, is greater in nonpolar solvents. Less color develops in the reaction mixture if the system is flushed with nitrogen at this point.

10. The triazenes of secondary carbinamines also yield some N-alkyltoluidine.[3] Colored impurities normally enter the aqueous phase at this point in the work-up.

11. The infrared spectrum of this material is essentially identical with that of the pure, recrystallized ester.

3. Methods of Preparation

1-Methyl-3-*p*-tolyltriazene has been prepared by the reaction of methylmagnesium bromide with *p*-tolyl azide,[4] and, in unspecified yield, by the addition of methylamine to *p*-toluenediazonium chloride.[5]

4. Merits of the Preparation

This procedure represents the most convenient synthesis of 1-methyl-3-*p*-tolyltriazene. Triazenes with more complex alkyl groups may be prepared from the corresponding amine[2] or Grignard reagent.[4]

The alkylation of acids with triazenes is superior to alkylation with diazomethane and other diazoalkanes in that the triazenes are crystalline, stable materials which are easy to prepare and store. Alkylations with triazenes are unlikely to be accompanied by side reactions, such as addition to strained or conjugated double bonds, which are frequently observed in alkylations with diazoalkanes.

1. Department of Chemistry, The Johns Hopkins University, Baltimore, Maryland 21218.
2. E. H. White and H. Scherrer, *Tetrahedron Lett.*, 758 (1961).
3. E. H. White, H. Maskill, D. J. Woodcock, and M. A. Schroeder, *Tetrahedron Lett.*, 1713 (1969).
4. O. Dimroth, M. Eble, and W. Gruhl, *Ber.*, 40, 2390 (1907).
5. C. S. Rondestvedt, Jr., and S. J. Davis, *J. Org. Chem.*, 22, 200 (1957).

N-MONO- AND N,N-DISUBSTITUTED
UREAS AND THIOUREAS

Submitted by Roy G. Neville [1] and John J. McGee [2]
Checked by William E. Parham and J. Kent Rinehart

METHOD I

CYCLOHEXYLUREA

(Urea, cyclohexyl-)

$$C_6H_{11}NH_2 + Si(NCO)_4 \rightarrow$$
$$Si(NHCONHC_6H_{11})_4 \xrightarrow[-SiO_2]{H_2O} C_6H_{11}NHCONH_2$$

A solution of cyclohexylamine (39.7 g., 0.4 mole) (Note 1) in 100 ml. of anhydrous benzene (Note 2) is added slowly to a stirred solution (Note 3) of silicon tetraisocyanate (19.6 g., 0.1 mole) (Note 4) in 150 ml. of anhydrous benzene contained in a 1-l. round-bottomed flask. After the exothermic reaction has subsided, the mixture is heated at the reflux temperature for 30 minutes; the benzene is then removed using a rotary evaporator. Dilute isopropyl alcohol (200 ml.) (Note 5) is added to the residue, and the resulting mixture is heated at the reflux temperature for 30 minutes. The hot mixture is filtered through a 0.5-in. layer of Celite® contained in a coarse-grade sintered-glass funnel (Note 6). The gelatinous silica is washed with two 75-ml. portions of acetone and is then pressed and drained. The combined filtrates are evaporated to dryness on a steam bath (Note 7). The crude cyclohexylurea (m.p. 185–191°, 55.0 g., 97% yield) is recrystallized from 220 ml. of isopropyl alcohol (Note 8) to give 37 g. (65%) of product, m.p. 192–193°. Concentration of the mother liquor affords about 9 g. (16%) of additional product which is less pure (m.p. 189–192°) (Note 9).

METHOD II

2,6-DIMETHYLPHENYLTHIOUREA

[Urea, 1-(2,6-dimethylphenyl)-2-thio-]

Silicon tetraisothiocyanate (26.0 g., 0.10 mole) (Note 10) is finely ground under 100 ml. of anhydrous benzene, and the mixture is quickly transferred to a 1-l. round-bottomed flask. The mortar and pestle are washed with two 25-ml. portions of anhydrous benzene, and the washings are added to the flask. A solution of 2,6-dimethylaniline (48.5 g., 0.4 mole) (Note 1) in 100 ml. of anhydrous benzene is added to the well-stirred mixture. The reaction is mildly exothermic. The mixture is heated at the reflux temperature for 30 minutes, and the benzene is then removed using a rotary evaporator. Dilute isopropyl alcohol (200 ml.) (Note 5) is added to the residue, and the resulting mixture is heated at the reflux temperature for 30 minutes. The mixture is then processed in exactly the same manner as described above for the preparation of cyclohexylurea. The crude 2,6-dimethylphenylthiourea (m.p. 193–197°, 71.3 g., 99% yield) is recrystallized from 280 ml. of isopropyl alcohol (Note 8) to give 50 g. (72%) of product, m.p. 201–202°. Concentration of the mother liquor affords 11 g. (15%) of less pure product, m.p. 197–199° (Note 11).

2. Notes

1. Cyclohexylamine and 2,6-dimethylaniline were obtained from Eastman Organic Chemicals and were redistilled prior to use.

2. Benzene was dried over sodium wire.

3. The mixture becomes viscous; however, a good magnetic stirrer is adequate. The checkers found it convenient to decrease the viscosity of the mixture by increasing the volume of benzene from 100 ml. to 150–300 ml.

4. Silicon tetraisocyanate is prepared from silicon tetrachloride and silver cyanate or lead cyanate.[3, 4]

5. Dilute isopropyl alcohol is prepared by mixing the alcohol (180 ml.) with water (20 ml.). The use of more than about 10% water in the alcohol results in an intractable mass of gelatinous silica from which it is very difficult to separate a good yield of the urea.

6. As gelatinous silica clogs the filter when too strong a suction is applied, it is best to carry out the filtration using very gentle suction. Only when almost all the liquid has passed through the filter is strong suction applied. The checkers used Hyflo Supercel® as the filter aid, and a 600-ml. coarse-grade sintered-glass funnel.

7. An open dish or a rotary evaporator is satisfactory.

8. Isopropyl alcohol is a good solvent to employ for recrystallizing most ureas; however, occasionally a mixture of alcohol and benzene or pure benzene is superior.

9. This material is of sufficient purity for most purposes. If a purer product is required, the first crop of the cyclohexylurea (37 g.) is recrystallized from 135 ml. of isopropyl alcohol to yield 23 g. of product, m.p. 195.5–196.0°

10. Silicon tetraisothiocyanate is prepared from silicon tetrachloride and silver thiocyanate [5, 6] or, preferably, ammonium thiocyanate.[6, 7] The silicon tetraisothiocyanate used by the checkers was slightly yellow; however, this did not affect the yield of product.

11. This material is of sufficient purity for most purposes. If purer material is required, the first crop of 2,6-dimethylphenylthiourea (50 g.) is recrystallized from 250 ml. of isopropyl alcohol to yield 41 g. of product, m.p. 203.5–204.0°.

3. Methods of Preparation

Cyclohexylurea has been prepared by the reaction of cyclohexyl isocyanate with gaseous ammonia [8] or ammonium hydroxide,[9] by thermal decomposition of cyclohexyl allophanamide,[10] by treating cyclohexylamine hydrochloride with an aqueous solution of potassium cyanate,[11] by heating nitrosomethylurea with cyclohexylamine,[12] and by heating an ethanolic solution of cyclohexylamine and 3,5-dimethyl-1-carbamylpyrazole.[13]

2,6-Dimethylphenylthiourea has been synthesized by allowing 2,6-dimethylaniline hydrochloride to react with ammonium thiocyanate.[14]

4. Merits of the Preparation

These procedures are generally applicable to aliphatic, alicyclic, aralkyl, aromatic, and heterocyclic primary or secondary amines. The reactions fail or give poor yields with sterically hindered amines such as 2-trifluoromethylaniline, 2,6-dibromoaniline, and diphenylamine. In general, however, excellent (95–100%) yields of N-mono- or N,N-disubstituted ureas or thioureas can be obtained by employing these versatile reactions which are, in most cases, superior to and supplement the methods conventionally employed for the synthesis of ureas and thioureas.[15] Because of the rapidity, ease, and excellent yields of these reactions, silicon tetraisocyanate and tetraisothiocyanate (both of which are readily prepared [4, 6]) are likely to become standard reagents for the preparation of N-mono- and N,N-disubstituted ureas and thioureas. The submitters have employed these reagents to prepare large-scale (0.4M) amounts of the following compounds (yields in parentheses): benzylurea, m.p. 148° (96%); [16a] phenylurea, m.p. 147° (95%); [16b] t-butylurea, m.p. 182° (95%); [17] N-(2-benzothiazolyl)urea, m.p. >350° (95%); [18, 19] dibenzylthiourea, m.p. 140° (95%); [16c] t-butylthiourea, m.p. 168° (98%). [16d] In addition, the scope and limitation of the reactions of silicon tetraisocyanate and tetraisothiocyanate have been investigated with more than fifty alkyl, aralkyl, aromatic, and heterocyclic primary and secondary amines.[20, 21]

1. North American Aviation, Inc., Los Angeles, California.
2. Beckman Instruments, Fullerton, California.
3. G. S. Forbes and H. H. Anderson, *J. Am. Chem. Soc.*, **62**, 761 (1940).
4. R. G. Neville and J. J. McGee, *Inorg. Syntheses*, **8**, 24 (1966).
5. G. S. Forbes and H. H. Anderson, *J. Am. Chem. Soc.*, **67**, 1911 (1945).
6. R. G. Neville and J. J. McGee, *Inorg. Syntheses*, **8**, 27 (1966).
7. M. G. Voronkov and B. N. Dolgov, "Soviet Research in Organo-Silicon Chemistry, 1949–56," English Translation, Consultants Bureau, Inc., New York, Part II, pp. 344–347.
8. A. Skita and H. Rolfes, *Ber.*, **53B**, 1242 (1920).
9. S. O. Olsen and E. Enkemeyer, *Ber.*, **81**, 359 (1948).
10. J. Bougault and J. Leboucq, *Bull. Soc. Chim. France*, [4] **47**, 594 (1930).
11. O. Wallach, *Ann.*, **343**, 46 (1905).
12. J. L. Boivin and P. A. Boivin, *Can. J. Chem.*, **29**, 478 (1951).
13. F. L. Scott, D. G. O'Donovan, M. R. Kennedy, and J. Reilly, *J. Org. Chem.*, **22**, 820 (1957).
14. F. Kurzer and P. M. Sanderson, *J. Chem. Soc.*, 4461 (1957).
15. R. B. Wagner and H. D. Zook, "Synthetic Organic Chemistry," John Wiley and Sons, Inc., New York, 1953, pp. 645–652.
16. (a) I. Heilbron and H. M. Bunbury, "Dictionary of Organic Compounds," Oxford University Press, New York, 1953, Vol. 1, p. 284; (b) Vol. 4, p. 184; (c) Vol. 2, p. 65; (d) Vol. 1, p. 400.
17. L. I. Smith and O. H. Emerson, *Org. Syntheses*, Coll. Vol. **3**, 151 (1960).
18. H. P. Kaufman and P. Schulz, *Arch. Pharm.*, **273**, 31 (1935).
19. R. H. Wiley and A. J. Hart, *J. Org. Chem.*, **18**, 1368 (1953).
20. R. G. Neville, *J. Org. Chem.*, **23**, 937 (1958).
21. R. G. Neville and J. J. McGee, unpublished results.

MONOPERPHTHALIC ACID

(Phthalic monoperoxy acid)

$$C_6H_4 \underset{CO}{\overset{CO}{\diagdown\!\!\!\diagup}} O + H_2O_2 \rightarrow C_6H_4 \underset{CO_2H}{\overset{CO_2OH}{\diagdown\!\!\!\diagup}}$$

Submitted by GEORGE B. PAYNE.[1]
Checked by K. NAGARAJAN and JOHN D. ROBERTS.

1. Procedure

In a 1-l. round-bottomed flask equipped with a mechanical stirrer and cooled in an ice-salt bath is placed a solution of 62 g.

(0.5 mole) of sodium carbonate monohydrate in 250 ml. of water. This is cooled to 0°, and 69 g. (63 ml., 0.6 mole) of 30% hydrogen peroxide (Note 1) is added in one portion. With the temperature at −5 to 0°, 74 g. (0.5 mole) of phthalic anhydride (Note 1) which has been pulverized to pass a 14-mesh sieve is added (Note 2).

The reaction mixture is stirred vigorously at −5 to 0° for 30 minutes, then the resulting solution or suspension (Note 3) is poured into a 2-l. separatory funnel, shaken with 350 ml. of ether (Note 4), and carefully acidified with an ice-cold solution of 30 ml. of concentrated sulfuric acid in 150 ml. of water. The liberated monoperphthalic acid is extracted into the ether and removed completely from the water by extraction with two more 150-ml. portions of ether. The combined ether extracts are washed with two 200-ml. portions of 40% ammonium sulfate solution and dried overnight in a refrigerator over 50 g. of anhydrous magnesium sulfate.

The peracid content is determined by adding 30 ml. of 20% potassium iodide solution to 2 ml. of the peracid solution and, after 10 minutes, titrating the liberated iodine with 0.1 N thiosulfate. The yield is 71–78 g. (78–86% based on phthalic anhydride).

If crystalline monoperphthalic acid is desired, it may be prepared as described earlier.[2]

2. Notes

1. Commercial phthalic anhydride and hydrogen peroxide, both of reagent grade, are used.

2. Fieser and Fieser[3] recommend that the phthalic acid be recrystallized from benzene rather than pulverized.

3. The sodium salt of monoperphthalic acid may precipitate during the reaction.

4. The ether may be used, along with a *small amount* of cold water, to effect a quantitative transfer of the suspension from reaction vessel to separatory funnel.

3. Methods of Preparation

Monoperphthalic acid has been prepared by hydrolysis of phthalyl peroxide with sodium hydroxide,[4] by reaction of phthalic anhydride with excess alkaline peroxide solution,[5] by reaction of phthalic anhydride with hydrogen peroxide,[6] and by stirring phthalic anhydride with mildly alkaline peroxide.[7] The method described here is a slight modification of the last procedure.

The methods of preparation, properties, analysis, and safe handling of monoperphthalic acid have been reviewed.[8]

4. Merits of Preparation

The Böhme procedure [2,5] for preparing perphthalic acid from phthalic anhydride and hydrogen peroxide is sensitive to slight variations in the experimental conditions.[6,7] The present method gives reproducible yields with quite short reaction times.

1. Shell Development Company, Emeryville, Calif.
2. H. Böhme, *Org. Syntheses,* Coll. Vol. 3, 619 (1955).
3. L. F. Fieser and M. Fieser, *Reagents for Org. Synthesis,* 1, 820 (1967).
4. A. Baeyer and V. Villiger, *Ber.,* 34, 764 (1901).
5. H. Böhme, *Ber,* 70, 379 (1937); G. B. Bachman and D. E. Cooper, *J. Org. Chem.,* 9, 307 (1944).
6. E. E. Royals and L. L. Harrell, Jr., *J. Am. Chem. Soc.,* 77, 3405 (1955).
7. G. B. Payne, *J. Org. Chem.,* 24, 1354 (1959).
8. D. Swern, "Organic Peroxides," Vol. I, Wiley-Interscience, New York, 1970, p. 424, 489.

1-MORPHOLINO-1-CYCLOHEXENE*

[Morpholine, 4-(1-cyclohexenyl)-]

Submitted by S. Hünig, E. Lücke, and W. Brenninger.[1]
Checked by B. C. McKusick and F. E. Mumford.

1. Procedure

A solution of 147 g. (1.50 moles) of cyclohexanone, 157 g. (1.80 moles) of morpholine (Note 1), and 1.5 g. of p-toluenesulfonic acid in 300 ml. of toluene is heated to boiling in a 1-l. round-bottomed flask to which is attached a water separator[2] under a reflux condenser. The separation of water begins at once and ceases after 4 or 5 hours. An indented Claisen stillhead is attached to the flask, and the reaction mixture is distilled. Most of the toluene is removed at atmospheric pressure. 1-Morpholino-1-cyclohexene is collected as a colorless liquid at 118–120°/10 mm.; n_D^{25} 1.5122–1.5129 (Note 2). It weighs 180–200 g. (72–80%).

2. Notes

1. An excess of morpholine is required because the water that separates during the reaction always contains a considerable amount of it in solution.

2. 1-Morpholino-1-cyclohexene is very easily hydrolyzed. Accordingly one must be careful to keep moisture out. On long standing in a refrigerator, the compound generally becomes somewhat yellowish, but this does not affect its usefulness in subsequent reactions.

3. Methods of Preparation

The procedure is that of Hünig, Benzing and Lücke.[3] It is based on earlier work on the preparation of enamines.[4,5]

4. Merits of Preparation

This is a general method of preparing enamines from a secondary aliphatic amine and cyclohexanone or cyclopentanone. Acylation of such enamines is the first step in a general procedure for increasing the chain length of a carboxylic acid by 5 or 6 carbon atoms and of a dicarboxylic acid by 10 or 12 carbon atoms.[6] Alkylation of enamines of cyclohexanones by alkyl halides [5,7] or electrophilic olefins,[8] followed by hydrolysis, is a good route to α-monoalkylcyclohexanones. The chemistry of enamines has been reviewed.[9]

1. University of Marburg, Marburg, Germany.
2. S. Natelson and S. Gottfried, *Org. Syntheses,* Coll. Vol. 3, 381 (1955).
3. S. Hünig, E. Benzing, and E. Lücke, *Ber.,* 90, 2833 (1957).
4. M. E. Herr and F. W. Heyl, *J. Am. Chem. Soc.,* 74, 3627 (1952); 75, 1918 (1953).
5. G. Stork, R. Terrell, and J. Szmuszkovicz, *J. Am. Chem. Soc.,* 76, 2029 (1954).
6. S. Hünig, E. Lücke, and E. Benzing, this volume, p. 533; *Ber,* 91, 129 (1958); S. Hünig and E. Lücke, *Ber.,* 92, 652 (1959); S. Hünig and W. Lendle, *Ber.,* 93, 909, 913 (1960).
7. D. M. Locke and S. W. Pelletier, *J. Am. Chem. Soc.,* 80, 2588 (1958).
8. G. Stork and H. K. Landesman, *J. Am. Chem. Soc.,* 78, 5128 (1956).
9. J. Szmuszkovicz., *Advan. Org. Chem.,* 4, 1 (1963).

2,3-NAPHTHALENEDICARBOXYLIC ACID*

$$\xrightarrow{2Na_2Cr_2O_7}$$

$$+ 2Cr_2O_3 + 2NaOH + H_2O$$

$$\xrightarrow{2HCl}$$

$$+ 2NaCl$$

Submitted by LESTER FRIEDMAN.[1]
Checked by G. A. BOSWELL and B. C. McKUSICK.

1. Procedure

An autoclave (Note 1) is charged with 200 g. (1.28 moles) of 2,3-dimethylnaphthalene (Note 2), 940 g. (3.14 moles, 23% excess) of sodium dichromate dihydrate, and 1.8 l. of water. The autoclave is closed, heated to 250°, and shaken continuously at this temperature for 18 hours. The autoclave is cooled with continued agitation (Note 3), the pressure is released, and the autoclave is opened. The contents are transferred to a large vessel (Note 4). To effect complete transfer, the autoclave is rinsed with several 500-ml. portions of hot water. Green hydrated chromium oxide in the reaction mixture is separated on a large Büchner funnel and washed with warm water until the filtrate is colorless. The combined filtrates (7–8 l.) are acidified with 1.3 l. of 6N hydrochloric acid. The acidified mixture is allowed to cool to room temperature overnight. The 2,3-naphthalenedicarboxylic acid that has precipitated is collected on a large Büchner funnel, washed with water until the filtrate is colorless, and dried to constant weight in a vacuum oven at 50°/20 mm. or by long standing in air. The 2,3-naphthalenedicarboxylic acid is a white powder; m.p. 239–241°; weight 240–256 g. (87–93%).

2. Notes

1. An autoclave fitted for stirring or shaking is essential for good yields. The submitter used a hydrogenation autoclave of the type supplied by the American Instrument Company, catalog No. 406–21, having a capacity of 3.2 l. The autoclave is shaken by means of a "Bomb Shaker" in which it is placed. If a stirred autoclave or "Magne-Dash" is used, the reaction time can be shortened to 3–5 hours. At 250° the gauge pressure is about 600 lb./in.² The checkers' yield was 92% in a shaker tube, but only 75% in a rocker tube; the latter yield was raised to 82% by extending the reaction time to 40 hours.

This oxidation does not poison the autoclave for subsequent hydrogenations.

2. Material produced by Ruetgerswerke A. G. is satisfactory. This can be obtained in the United States from Terra Chemicals Inc., New York, New York; Aldrich Chemical Co., Milwaukee 10, Wisconsin; and K and K Laboratories Inc., Jamaica 33, New York.

3. It is convenient to empty the autoclave while the contents are still warm.

4. Commercially available 10-quart polyethylene pails are very satisfactory.

3. Methods of Preparation

2,3-Naphthalenedicarboxylic acid has been prepared by the present method [2] and by hydrolysis of 3-cyano-2-naphthoic acid, which is obtainable from 3-amino-2-naphthoic acid by the Sandmeyer reaction.[3]

4. Merits of the Preparation

This procedure illustrates a general method for the preparation of aromatic carboxylic acids by oxidation of the corresponding alkylarenes.[2] For example, 2-naphthoic acid (360 g., 93% yield; m.p. 184–185°) was obtained from 2-methylnaphthalene (320 g., 2.25 moles), sodium dichromate (975 g., 3.26 moles, 45% excess), and water (1.8 l.).

2,3-Naphthalenedicarboxylic acid is useful in the synthesis of linear polyacenes,[3] 3-halo-2-naphthoic acids,[4] and 3-amino-2-naphthoic acid.[4]

1. Chemistry Department, Case Institute of Technology, Cleveland, Ohio.
2. L. Friedman, D. L. Fishel, and H. Shechter, *J. Org. Chem.*, **30**, 1453 (1965).
3. H. Waldmann and H. Mathiowetz, *Ber.*, **64**, 1713 (1931).
4. L. Friedman, unpublished data.

2,6-NAPHTHALENEDICARBOXYLIC ACID*

Submitted by Bernhard Raecke and Hubert Schirp.[1]
Checked by B. C. McKusick and P. E. Aldrich.

1. Procedure

A solution of 66.5 g. (1.01 moles) of 85% potassium hydroxide in 300 ml. of water in an 800-ml. beaker is heated to 60–70°, and 100 g. (0.505 mole) of commercial 1,8-naphthalic anhydride (Note 1) is stirred in. The pH of the resultant deep-brown solution is adjusted to a value of 7 (Note 2) with 6N hydrochloric acid and 3N potassium hydroxide. It is treated with 10 g. of decolorizing carbon and filtered. This operation is repeated. The filtrate is concentrated in a 1.5-l. beaker on a steam bath to about 180 ml. The concentrate is cooled to room temperature, 800 ml. of methanol is added with vigorous stirring by hand, and the mixture is cooled to 0–5°. The precipitated dipotassium naphthalate is separated by filtration, washed with 150 ml. of methanol, and dried in a vacuum oven at 150°/150 mm. The dried cream-colored salt weighs 130–135 g. (88–92%).

A mixture of 100 g. of dipotassium naphthalate and 4 g. of

anhydrous cadmium chloride [2] is ground in a ball mill for 4 hours. The mixture is placed in a 0.5-l. autoclave (Notes 3 and 4) that can be rocked, rolled, or shaken. The autoclave is evacuated, for oxygen lowers the yield of the product. The autoclave is then filled with carbon dioxide to a pressure of about 30 atm. The agitated autoclave is heated to an internal temperature of 400–430° in the course of 1–2 hours and is maintained at this temperature for 1.5 hours. The pressure rises to about 90 atm. in the course of the heating (Note 4).

The autoclave is cooled to room temperature, and the carbon dioxide is bled off. The solid reaction product is taken from the autoclave, pulverized, and dissolved in 1 l. of water at 50–60°. Ten grams of decolorizing carbon is added, and the mixture is stirred well and filtered to remove cadmium salts and carbon. The filtrate is heated to 80–90° and acidified with concentrated hydrochloric acid to pH 1 (Note 5). 2,6-Naphthalenedicarboxylic acid precipitates. It is separated from the hot mixture by filtration. It is then suspended in 500 ml. of water at 90–95° (Note 5), separated by filtration, and washed successively with 300 ml. of water, 300 ml. of 50% ethanol, and 300 ml. of 90% ethanol. After being dried at 100–150°/150 mm. in a vacuum oven, the 2,6-naphthalenedicarboxylic acid weighs 42–45 g. (57–61%). It decomposes on a heated block at 310–313°.

2. Notes

1. Suitable 1,8-naphthalic anhydride, m.p. 274–275°, is obtainable from Coaltar Chemicals Corp., 420 Lexington Ave., New York, N. Y.

2. Bromothymol blue or commercial universal indicator pH paper (graduated in 0.2-pH units) may be used as external indicators.

3. A 150-ml. shaking Hastelloy-C autoclave, manufactured by the Haynes Stellite Division of Union Carbide Co., Kokomo, Indiana, was used by the checkers. The approximate composition of the alloy is: Cr, 15.5–17.5%; Mo, 16–18%; Fe, 4.5–7%; W, 3.7–4.75%; and the remainder, Ni. Because their autoclave had only three-tenths the capacity of that used by the sub-

mitters, the checkers used three-tenths the quantities of materials given here.

4. The line to the pressure gauge tends to become clogged during the reaction.

5. Heat conduction in the heavy slurry that is formed is poor, and bumping may occur if the mixture is overheated. Efficient mechanical stirring aids this operation.

3. Methods of Preparation

2,6-Naphthalenedicarboxylic acid has been prepared by fusing dipotassium 2,6-naphthalenedisulfonate with potassium cyanide to give the corresponding dinitrile, which is hydrolyzed;[3] by oxidation of 2-methyl-6-acetylnaphthalene with dilute nitric acid at 200°;[4] by the thermal disproportionation of potassium α- or β-naphthoate to dipotassium 2,6-naphthalenedicarboxylate and naphthalene;[5] and by the present method.[6] The present method is much more convenient than earlier methods, if a suitable autoclave is available.

The present method for preparing aromatic dicarboxylic acids has been used to convert phthalic or isophthalic acid to terephthalic acid (90–95%); 2,2'-biphenyldicarboxylic acid to 4,4'-biphenyldicarboxylic acid; 3,4-pyrroledicarboxylic acid to 2,5-pyrroledicarboxylic acid; and 2,3-pyridinedicarboxylic acid to 2,5-pyridinedicarboxylic acid.[7] A closely related method for preparing aromatic dicarboxylic acids is the thermal disproportionation of the potassium salt of an aromatic monocarboxylic acid to an equimolar mixture of the corresponding aromatic hydrocarbon and the dipotassium salt of an aromatic dicarboxylic acid. The disproportionation method has been used to convert benzoic acid to terephthalic acid (90–95%); pyridinecarboxylic acids to 2,5-pyridinedicarboxylic acid (30–50%); 2-furoic acid to 2,5-furandicarboxylic acid; 2-thiophenecarboxylic acid to 2,5-thiophenedicarboxylic acid; and 2-quinolinecarboxylic acid to 2,4-quinolinedicarboxylic acid.[7] One or the other of these two methods is often the best way to make otherwise inaccessible aromatic dicarboxylic acids. The two methods were recently reviewed.[7]

1. Henkel & Cie. GmbH, Post Box 1100, Düsseldorf, Germany.
2. J. Cason and F. S. Prout, *Org. Syntheses,* Coll. Vol. 3, 603 (1955).
3. R. Ebert and V. Merz, *Ber.,* 9, 604 (1876).
4. G. E. Tabet, U. S. pat. 2,644,841 (1953) [*C.A.,* 48, 6469 (1954)].
5. B. Raecke and H. Schirp, Ger. pat. 953,072 (1955) [*C.A.,* 53, 1289 (1959)].
6. B. Raecke, W. Stein, and H. Schirp, Ger. pat. 949,652 (1953) [*C.A.,* 52, 20106 (1958)].
7. B. Raecke, *Angew. Chem.,* 70, 1 (1958).

N-β-NAPHTHYLPIPERIDINE

(Piperidine, 1-(2-naphthyl)-)

Submitted by J. F. Bunnett, T. K. Brotherton, and S. M. Williamson.[1]
Checked by Virgil Boekelheide and F. Lind.

1. Procedure

A dry 1-l. three-necked round-bottomed flask is fitted in the center neck with a sweep-blade stirrer whose shaft passes through an airtight bearing (Note 1). One side neck is fitted with a condenser topped by a soda-lime drying tube, and the other is fitted with a solid stopper. In the flask are placed 75 ml. of piperidine (Note 2) and 15.6 g. (0.4 mole) of sodium amide (Note 3), and the mixture is heated at reflux (Note 4) for 15 minutes with good stirring. The mixture is cooled just below reflux temperature, and 46 g. (0.2 mole) of sodium β-naphthalenesulfonate (Note 5) is added, followed by an additional 75 ml. of piperidine. The mixture is then heated at reflux for 12 hours with stirring.

To the cooled reaction mixture, 200 ml. of water is added carefully with stirring. Potassium carbonate is added with continued stirring until the water layer is saturated; the mixture is now transferred to a separatory funnel and extracted three times with 60-ml. portions of ether. The combined ether extracts are dried over solid sodium hydroxide and are then transferred to a simple distillation apparatus. Distillation is commenced with a steam bath as source of heat; when most of the ether has been removed, the steam bath is replaced by a flame. Distillation is continued until most of the piperidine (b.p. 106°) has been removed. The cooled residue in the distillation flask is recrystallized from petroleum ether (boiling range 30–60°) with the use of charcoal. There is obtained 30.0 g. (71%) of N-β-naphthylpiperidine as tan crystals, m.p. 52–56°. An additional recrystallization from the same solvent gives crystals, m.p. 56–58°, with about 10% loss in weight (Note 6).

2. Notes

1. The submitters used a ball-joint bearing. A mercury seal or a Trubore bearing should also suffice.

2. Commercial piperidine was purified by 6 hours' refluxing with sodium metal followed by distillation from sodium.

3. Sodium amide from Farchan Research Laboratories, Cleveland, Ohio, was used.

4. The submitters used an electric heating mantle as a source of heat.

5. Sodium β-naphthalenesulfonate, technical grade, from Matheson, Coleman and Bell was dried in an oven and then used directly.

6. The melting point of pure β-naphthylpiperidine is 58–58.5°.[2,3] By the same procedure the submitters have obtained N-phenylpiperidine (94%) from sodium benzenesulfonate and N-α-naphthylpiperidine (23%) from sodium α-naphthalenesulfonate.

3. Methods of Preparation

N-β-Naphthylpiperidine has been prepared by the condensation of β-bromonaphthalene[2,3] or of β-naphthol[4] with piperidine at elevated temperatures; from the action of 1,5-dibromopentane on β-naphthylamine;[5] and from the action of the sodium amide-piperidine reagent on each of the eight monohalonaphthalenes [3, 6] or on methyl β-naphthyl sulfone.[6] The present procedure is adapted from that of Brotherton and Bunnett.[7]

1. Department of Chemistry, University of North Carolina, Chapel Hill, North Carolina.
2. E. Lellman and M. Büttner, *Ber.*, 23, 1383 (1890).
3. J. F. Bunnett and T. K. Brotherton, *J. Am. Chem. Soc.*, 78, 155 (1956).
4. W. Roth, *Ber.*, 29, 1175 (1896).
5. M. Scholtz and E. Wassermann, *Ber.*, 40, 856 (1907).
6. J. F. Bunnett and T. K. Brotherton, *J. Am. Chem. Soc.*, 78, 6265 (1956).
7. T. K. Brotherton and J. F. Bunnett, *Chem. & Ind. (London)*, 60, (1957).

NEOPENTYL ALCOHOL *

(2,2-Dimethyl-1-propanol)

$$(CH_3)_3CCH_2C(CH_3)=CH_2 + H_2O_2 \xrightarrow{H^+}$$

$$(CH_3)_3CCH_2C(CH_3)_2O_2H$$

$$(CH_3)_3CCH_2C(CH_3)_2O_2H \xrightarrow{H^+} (CH_3)_3CCH_2OH + CH_3COCH_3$$

Submitted by JOSEPH HOFFMAN.[1]
Checked by JOHN D. ROBERTS and J. ERIC NORDLANDER.

1. Procedure

A. *Preparation of hydroperoxide.* In a 2-l. three-necked round-bottomed flask, equipped with a mechanical stirrer (Note 1), a dropping funnel, and a thermometer, is placed 800 g. of 30% hydrogen peroxide (Note 2). The flask is surrounded by an ice bath and rapid stirring is started. In the meantime, 800 g. of 95–96% sulfuric acid is added to 310 g. of cracked ice and the solution is cooled to 10°. When the temperature of the hydrogen peroxide reaches 5–10°, the cold sulfuric acid is added slowly from the dropping funnel during a period of about 20 minutes (Note 3). The temperature of the solution should not exceed 20° during the addition. Commercial diisobutylene (224.4 g., 2 moles) is now added over a period of 5–10 minutes. The ice bath is removed and replaced by a water bath maintained at approximately 25° (Note 4). Vigorous agitation is maintained for 24 hours (Note 5). At the end of this time, mixing is discontinued, the mixture is transferred to a 2-l. separatory funnel and the two layers are allowed to separate (Note 6).

B. *Decomposition of hydroperoxide.* The upper organic layer (240–250 g.) is removed (Note 7) and added with vigorous stirring to 500 g. of 70% sulfuric acid in a 1-l. three-necked round-bottomed flask fitted with thermometer, stirrer, and dropping funnel and surrounded by an ice bath. The reaction temperature is maintained at 15–25° during addition, which requires 65–75 minutes (Note 8). Stirring is continued for 30 minutes at 5–10°, and then the reaction mixture is allowed to stand (0.5–3 hours) until the two layers are completely separated. The mixture is now transferred to a 1-l. separatory funnel and allowed to stand for about 15 minutes, after which time the lower layer is drawn off into 1 l. of water. The resulting mixture is distilled from a 3-l. flask without fractionation. The distillation is complete when 50-100 ml. of water has been collected (Note 9). The upper organic layer from the distillate (180–190 g.) is removed and dried over anhydrous magnesium sulfate (Note 10). The dried organic layer is filtered with the aid of a small amount of ether and distilled through an efficient fractionating column. The fraction which boils at 111–113° is collected (Notes 11 and 12). The yield is 60–70 g. (34–40% of theory, based upon the diisobutylene).

2. Notes

1. A heavy nichrome wire twisted into 4 loops (similar to an egg beater) was found to be very satisfactory. The two ends of the wire, extending several inches beyond the loops, were pushed into a piece of glass tubing for the stirrer shaft.

2. If the hydrogen peroxide is slightly below 30%, enough should be used to give the amount called for. The acid concentration should be maintained by increasing the sulfuric acid proportionately.

3. The hydrogen peroxide-sulfuric acid solution consists of approximately 12.5% hydrogen peroxide and 40% sulfuric acid.

4. Good results have been obtained in the temperature range 23–27°. Although the heat given off during the reaction is not great and is spread over a long period of time, the reaction vessel must nevertheless be surrounded by a water bath. If the room temperature is in the range indicated, no further regulation of the water bath temperature is required.

5. It is necessary to provide rapid and vigorous stirring in order to obtain good results.

6. The aqueous layer now contains approximately 8% hydrogen peroxide. This layer may be reused by adjusting the hydrogen peroxide percentage to 12.5 by use of either 30% or 50% hydrogen peroxide. The sulfuric acid must be readjusted to 40%. Approximately 1.9 kg. of aqueous layer is required for 2 moles of diisobutylene.

7. The submitter states that the procedure may be interrupted at this point by washing the organic layer free of acid with a saturated solution of sodium bicarbonate and that the hydroperoxide concentrate will keep for a long time, especially if refrigerated.

8. Care should be taken in decomposing the hydroperoxide. If the temperature is kept too low, decomposition takes place too slowly and hydroperoxide may accumulate. Heat is liberated during the decomposition, and after each small addition the temperature will rise. At the start, small amounts of hydroperoxide are added until the temperature rises above 15°; the rate is then adjusted to keep the temperature in the range 15–25°. The rise in temperature after each small addition is the best evidence that decomposition is actually taking place.

9. The submitter states that, when the distillation is essentially complete, there is a tendency for the residue to foam; this should be watched for and the heat should be turned back to avoid carryover into the distillate. The checkers did not experience any difficulties of this sort.

10. The checkers found that the commercial grade of "dried" magnesium sulfate was not completely effective and, even after three treatments with fresh drying agent, the subsequent distilla-

tion afforded an azeotrope of neopentyl alcohol and water of b.p. 80–85°. The organic layer of the azeotrope had to be separated from the water, dried, and redistilled to give the stated total yields.

11. The fraction between 95° and 110° should be taken off at a high reflux ratio; toward the end of the distillation the reflux ratio should again be increased. Most of the product comes over at 113°.

12. Pure neopentyl alcohol melts at about 55°. From time to time it will be necessary to circulate hot water through the take-off condenser in order to facilitate removal of the alcohol.

3. Methods of Preparation

Neopentyl alcohol has been made by lithium aluminum hydride reduction of trimethylacetic acid [2] and by treating *tert*-butyl-magnesium chloride with methyl formate.[3]

The preparation of neopentyl alcohol from diisobutylene herein described represents an example of acid-catalyzed addition of hydrogen peroxide to a branched olefin, followed by an acid-catalyzed rearrangement of the tertiary hydroperoxide formed. In addition to neopentyl alcohol, there are formed acetone and also small amounts of methanol and methyl neopentyl ketone by an alternative rearrangement of the hydroperoxide.

1. Air Reduction Co., Murray Hill, New Jersey.
2. R. F. Nystrom and W. G. Brown, *J. Am. Chem. Soc.*, 69, 2548 (1947).
3. L. H. Sommer, H. D. Blankman, and P. C. Miller, *J. Am. Chem. Soc.*, 76, 803 (1954).

NICOTINIC ANHYDRIDE

$$+ \ 2(C_2H_5)_3\overset{+}{N}H \ \ Cl^- \ + \ \ \ CO_2$$

Submitted by HEINRICH RINDERKNECHT and MORRIS GUTENSTEIN [1]
Checked by WALTER K. SOSEY, WAYLAND E. NOLAND,
and WILLIAM E. PARHAM

1. Procedure

Nicotinic acid (10 g., 0.081 mole) (Note 1) and anhydrous benzene (275 ml.) (Note 2) are placed in a 500-ml. three-necked, round-bottomed flask (Note 3) fitted with a sealed Hershberg stirrer, a dropping funnel with a pressure-equalizing tube, and a stillhead connected to a condenser. In order to remove traces of moisture introduced with the nicotinic acid the mixture is heated until about 75 ml. of benzene has distilled. The stillhead is replaced by a Claisen head fitted with a thermometer and a calcium chloride tube, and the mixture is cooled to 5° by stirring in an ice bath. To the cold suspension of nicotinic acid is added all at once 8.65 g. (0.086 mole, 5% excess) of triethylamine (Note 4). The resulting clear solution is stirred with continued cooling while 34 g. of a 12.5% solution of phosgene (0.043 mole, 5% excess) in benzene (Note 5) is added through the dropping funnel. The rate of addition is regulated so that the temperature of the reaction mixture does not exceed 7°. Triethylamine hydrochloride precipitates immediately. After the addition of phosgene the mixture is stirred at room temperature for 45 minutes, heated to the boiling point, and filtered under slightly reduced pressure (Note 6) while hot. The triethylamine hydrochloride cake (Note 7) is washed on the filter with three 25-ml. portions of warm benzene (60°). The combined filtrate and washes are transferred to a 500-ml. round-bottomed flask and evaporated to dryness on a rotary evaporator at low temperature and pressure. The dry residue is simmered with 75 ml. of anhydrous benzene (Note 2), and the mixture is filtered while hot. The triethylamine hydrochloride cake (Note 7) is washed with two 5-ml. portions of cold benzene, and the filtrate and washes are allowed to stand at 20° for 2–3 hours. The crystalline product is collected on a filter, washed with two 4-ml. portions of cold anhydrous benzene, and dried in a vacuum. The yield of nicotinic anhydride, m.p. 122–125° (Note 8), is 6.25 g. (68%). The combined filtrate and washes are evaporated to dryness on a rotary evaporator. The residue is simmered with 175 ml. of a mixture of benzene and cyclohexane (2:3) (Note 2), and a small amount of insoluble material is removed by filtration of the hot mixture (Note 6). The filtrate is stored at 5° for 18 hours (Note 9); the crystalline deposit is collected, washed with 3 ml. of cold benzene-cyclohexane mixture, and dried in a vacuum. An additional 2.4 g. (25%) of colorless product, m.p. 122–123°, is thus obtained. The total yield of nicotinic anhydride is 8.05–8.65 g. (87–93%).

2. Notes

1. Nicotinic acid supplied by Matheson, Coleman and Bell yielded a colorless anhydride; u.s.p. grade material gave a slightly buff-colored product.

2. Benzene and cyclohexane are freshly distilled and stored over sodium wire or calcium hydride.

3. Nicotinic anhydride is extremely sensitive to moisture; all glassware is therefore dried overnight in an oven at 200° before use.

4. Triethylamine is freshly distilled and stored over potassium hydroxide pellets.

5. A 12.5% solution of phosgene in benzene is available from Matheson, Coleman and Bell.

6. It is essential to carry out filtration under only slightly reduced pressure in order to minimize evaporation, cooling, and crystallization in the filter plate and funnel.

7. The yield of triethylamine hydrochloride obtained in this and subsequent extractions amounts to over 96% of that expected.

8. Melting points were determined in capillary tubes and are corrected. Reported:[2] 123–126°.

9. Rigorously anhydrous conditions are essential throughout this procedure and the flask must be air-tight.

3. Methods of Preparation

The present method is that described by Rinderknecht and Ma.[3] It is equally applicable to a variety of other heterocyclic, aromatic, and aliphatic anhydrides.[4, 5] Nicotinic anhydride was first prepared by reaction of nicotinoyl chloride with sodium nicotinate,[6, 7] and more recently by reaction of potassium nicotinate with oxalyl chloride in anhydrous benzene.[2]

4. Merits of the Preparation

The present method of preparing anhydrides is distinguished from other procedures by its simplicity and high yield. It avoids the two-phase reaction systems of older methods and the need for often inaccessible and highly sensitive acid chlorides. The only nongaseous by-product, triethylamine hydrochloride, is readily removed from the reaction mixture and leaves, in nearly quantitative yield, a solution of product suitable for further reaction or isolation.

1. Contribution No. 3176 from the Gates and Crellin Laboratories of Chemistry, California Institute of Technology, Pasadena, California. This work was supported by grant No. HD-00347 from the National Institute of Child Health and Human Development.
2. A. W. Schrecker and P. B. Maury, *J. Am. Chem. Soc.*, **76**, 5803 (1954).
3. H. Rinderknecht and V. Ma, *Helv. Chim. Acta*, **47**, 162 (1964).
4. T. Wieland and H. Bernhard, *Ann.*, **572**, 190 (1951).
5. T. K. Brotherton, J. Smith, Jr., and J. W. Lynn, *J. Org. Chem.*, **26**, 1283 (1961).
6. R. Graf, *Biochem. Z.*, **229**, 164 (1930).
7. C. O. Badgett, *J. Am. Chem. Soc.*, **69**, 2231 (1947).

o-NITROBENZALDEHYDE *

(Benzaldehyde, *o*-nitro-)

Submitted by A. Kalir [1]
Checked by John H. Sellstedt, Wayland E. Noland,
and William E. Parham

1. Procedure

A. *o-Nitrobenzylpyridinium bromide.* A 1-l. flask fitted with a reflux condenser is charged with 102 g. (0.744 mole) of *o*-nitrotoluene, 120 g. (0.675 mole) of N-bromosuccinimide, 1.0 g. of benzoyl peroxide, and 450 ml. of dry carbon tetrachloride. The mixture is heated under reflux until, after the refluxing is temporarily interrupted, all the solid is seen to float on the surface (usually 6–8 hours suffices).

The hot mixture is filtered with suction into a 1-l. round-bottomed flask through a Büchner-type sintered glass funnel provided with a ground joint (Note 1). The solid on the funnel is washed successively with two 50-ml. portions of hot carbon tetrachloride. The solvent is removed from the filtrate under reduced pressure on a water bath (Note 2). The flask is then fitted with a reflux condenser, and 400 ml. of commercial absolute ethanol and 65 ml. (0.81 mole) of good grade pyridine (Note 3) are added to the residue.

The solution is heated at the reflux temperature for 45 minutes and immediately transferred to a wide-mouthed Erlenmeyer flask. Crystallization begins at once, and, after the mixture is cooled, the crystals of nearly pure *o*-nitrobenzylpyridinium bromide are collected, washed with cold ethanol, and used in the next step (Note 4).

B. *N-(p-Dimethylaminophenyl)-α-(o-nitrophenyl)nitrone.* The wet *o*-nitrobenzylpyridinium bromide, together with 100 g. (0.536 mole) of *p*-nitrosodimethylaniline hydrochloride (Note 5) and 800 ml. of ethanol are introduced into a 2-l. three-necked flask equipped with an efficient stirrer, thermometer, and a dropping funnel, and immersed in an ice-salt bath. The stirrer is started, and a solution of 54 g. (1.35 mole) of sodium hydroxide in 500 ml. of water is added at 0–5° (Note 6). The color changes gradually from yellow to green, brown, and orange. The stirring is continued at 5–10° over a period of 1 hour. At the end of this time 500 ml. of ice-cold water is added to the flask, and the orange N-(*p*-dimethylaminophenyl)-α-(*o*-nitrophenyl)nitrone is collected on a large Büchner funnel, pressed well, and washed with cold water. The nitrone is used in the next step without further purification (Note 7).

C. *o-Nitrobenzaldehyde.* The wet crude nitrone is placed in a 3-l. beaker. A solution of approximately 6N sulfuric acid (Notes 8, 9) is then added, and the mixture is hand-stirred with a spatula or a glass rod. Crushed ice is added after 10 minutes, and the crude solid *o*-nitrobenzaldehyde is filtered, washed successively with dilute sodium bicarbonate solution and water, and dried over calcium chloride in a desiccator.

The light brown material is best purified by distillation under reduced pressure. The yellow aldehyde is collected at 120–140° (3 mm.) (Note 10) and melts at 41–44°. This material weighs 48–54 g. (47–53% overall yield based on N-bromosuccinimide) and is sufficiently pure for most uses (Notes 11, 12).

2. Notes

1. Suction filtration is necessary. The filtration is conveniently carried out through a regular Büchner funnel connected through a rubber stopper to a 1-l. suction flask. *Since o-nitrobenzyl bromide is a powerful lachrymator, the filtration should be carried out in a fume hood.*

2. The checkers used a rotary evaporator.

3. The checkers used Merck Reagent A.C.S. grade pyridine.

4. The yield of air-dried *o*-nitrobenzylpyridinium bromide is 125–135 g. (63–68%). The product melts at 206–208° (cor.).

5. Freshly prepared *p*-nitrosodimethylaniline hydrochloride [2] was used without further purification.

6. Sometimes a difficulty in stirring is encountered, and 100–200 ml. of ethanol should be added to the reaction mixture. The checkers found that the reaction mixture became a very thick paste which was quite difficult to stir. Use of a sturdy Hershberg stirrer is recommended.

7. The wet material contains about 55–65% of water. When the product is dried in a vacuum desiccator and recrystallized from ethyl acetate or acetone, it melts at 130–134°.

8. The solution is prepared by careful addition of 170 ml. of concentrated sulfuric acid to 850 ml. of water.

9. Hydrochloric acid (15%) can be substituted for sulfuric acid with equal results.

10. The checkers collected the product at 97–99° (1 mm.).

11. Very pure material can be obtained by dissolving *o*-nitrobenzaldehyde in toluene and precipitating with petroleum ether, according to earlier instructions.[3]

12. The same yields are obtained when the scale of this preparation is doubled.

3. Methods of Preparation

o-Nitrobenzaldehyde has been prepared by numerous methods.[3] The best-known and most widely used route involves the oxidation of *o*-nitrotoluene by chromium trioxide in acetic anhydride-acetic acid solution.[3] The present preparation is an example of the Kröhnke reaction.[4] It is adapted from the published directions for the synthesis of a series of halo- and nitrobenzaldehydes.[5]

4. Merits of the Preparation

The present procedure is a general method for preparing aromatic and heterocyclic aldehydes. It is of particular value in the synthesis of *o*-nitrobenzaldehydes in 100–200 g. lots. The benzaldehydes are useful starting materials for cinnamic acids, β-nitrostyrenes, etc. The manipulations are simple, the yields are reproducible, and the intermediates can be easily isolated and purified. The intermediates themselves have many synthetic uses.

The submitter has prepared the fluoro-*o*-nitrobenzaldehydes shown in Table I by application of this method.[6]

TABLE I

FLUORO-*o*-NITROBENZALDEHYDES

Position of Fluorine	M. P. of Fluoro-2-nitrobenzylpyridinium Bromide, °C	M. P. of N-(*p*-dimethylaminophenyl)-α-(fluoro-*o*-nitrophenyl)nitrone, °C	M. P. of Aldehyde, °C	Overall Yield, %
4	200–201	164–165	32–33	55–62
5	189–190	155–156	93–95	45–55
6	202–204	151–152	62–63	59–66

1. Israel Institute for Biological Research, Ness-Ziona, Israel.
2. G. M. Bennett and E. V. Bell, *Org. Syntheses*, Coll. Vol. **2**, 223 (1943).
3. S. M. Tsang, E. H. Wood, and J. R. Johnson, *Org. Syntheses*, Coll. Vol. **3**, 641 (1955).
4. F. Kröhnke, *Angew. Chem. Intern. Ed.*, **2**, 380 (1963).
5. K. Clarke, *J. Chem. Soc.*, 3807 (1957).
6. A. Kalir and D. Balderman, *Israel J. Chem.*, **6**, 927 (1968).

2-NITROCARBAZOLE

(Carbazole, 2-nitro-)

Submitted by G. David Mendenhall and Peter A. S. Smith [1]
Checked by Howard A. Harris and Kenneth B. Wiberg

1. Procedure

A. *o-Aminobiphenyl.* A Parr bottle is charged with 60 g. (0.30 mole) of *o*-nitrobiphenyl (Note 1), 3 g. of 5% palladium-on-carbon catalyst (Note 2), and 200 ml. of 95% ethanol. The mixture is shaken with hydrogen under 25–50 p.s.i. until the gas is no longer absorbed (about 70 minutes), the catalyst is filtered from the hot solution and washed with 20 ml. of ethanol, and the filtrates are poured in a thin stream into 1 l. of ice water contained in a 2-l. Erlenmeyer flask (Note 3). After standing for 20 minutes the white solid is filtered with suction, pressed to remove excess water, and allowed to dry in air. The yield of essentially pure *o*-aminobiphenyl is 48–51 g. (94–100%), m.p. 43–45.5°.

B. *o-Amino-p′-nitrobiphenyl.* Concentrated sulfuric acid (400 ml.) is placed in a 1-l. round-bottomed flask fitted with a mechanical stirrer and a thermometer. Stirring is begun, and 45.0 g. (0.27 mole) of powdered *o*-aminobiphenyl is added all at once through a powder funnel. When the amine has dissolved, the flask is placed in an ice-salt bath and its contents cooled to a temperature between 0° and −5°. A mixture of 30 ml. of concentrated sulfuric acid and 11.0 ml. of fuming nitric acid (density 1.5) is then added dropwise from a separatory funnel while the temperature is kept below 0°. The addition requires about an hour, and stirring is continued 45 minutes longer. The liquid is poured onto 1.5 kg. of ice in a 4-l. beaker and treated carefully until neutral with a solution of 580 g. (14.5 moles) of sodium hydroxide in 1.5 l. of water cooled to room temperature. The resultant hot suspension of product is allowed to cool nearly to room temperature, filtered with suction, and the orange solid is washed with 500 ml. of water. The crude material is pressed free of excess water and recrystallized from 850–1000 ml. of 95% ethanol (Note 4), giving 32–42 g. (56–74%) of orange needles, m.p. 156–158.5°.

C. *o-Azido-p′-nitrobiphenyl.* Water (100 ml.) is placed in a 1-l. round-bottomed flask equipped with a thermometer and an efficient mechanical stirrer. With stirring, 30 ml. of concentrated sulfuric acid is added, followed by 32.1 g. (0.15 mole) of recrystallized *o*-amino-*p′*-nitrobiphenyl. When all the amine has been converted to the white sulfate, 50 ml. more of water is added and the suspension is cooled to 0–5° in an ice-salt bath. A solution of 11 g. (0.16 mole) of sodium nitrite in 30 ml. of water is added dropwise over a period of 15 minutes (Note 5), and the mixture is stirred for 45 minutes longer. A thick precipitate of the sparingly soluble diazonium salt may have separated from the initially clear solution by this time. With strong stirring, a solution of 12 g. (0.17 mole) of sodium azide in 40 ml. of water is run in (Note 6), and stirring is continued for 40 minutes longer. The thick white solid is filtered with suction and washed with 200 ml. of water. After pressing free of excess water, the material is allowed to dry in air in a dark place. The yield of gray-white azide is 35.5–36 g. (99–100%), m.p. 91.5–92.5° (Note 7).

D. *2-Nitrocarbazole.* In a 2-l. round-bottomed flask fitted with a mechanical stirrer, a thermometer, and a short air condenser are placed 35.5 g. (0.15 mole) of powdered *o*-azido-*p*'-nitro-biphenyl and 1 l. of *o*-dichlorobenzene (Note 8). The stirred mixture is heated above 170° for 1 hour by means of a heating mantle, allowed to cool to room temperature, and chilled in a refrigerator (5°) for several hours. The crude product is filtered with suction, washed with 40 ml. of light petroleum, and sucked dry on the filter. There results 26–28 g. of yellow-brown crystals, m.p. 171.5–174°. The filtrate is distilled under aspirator pressure to a volume of 150–200 ml. and chilled as before, to yield an additional 2–3 g., m.p. 171–174°. The total yield is 28–30 g. (89–96%). The combined crops are dissolved in 400–450 ml. of boiling 95% ethanol with 3–4 g. of Norit® to remove impurities and filtered through a preheated Büchner funnel. The filtrate on cooling deposits bright yellow needles of product, which are filtered after standing at 5° for several hours. This crop weighs 23–25 g., m.p. 174–175.5°. Concentration of the mother liquor to a small volume (50–70 ml.) and chilling gives a second crop of lesser purity, 1–2 g., m.p. 172–175°. The total yield of recrystallized material is 24–26.5 g. (77–85%), and the overall yield from *o*-nitrobiphenyl is 40–63%.

2. Notes

1. An Eastman Kodak technical grade of *o*-nitrobiphenyl was used by the submitters. This is no longer available, and the checkers used the material supplied by K and K Laboratories. Both *o*-amino- and *o*-nitrobiphenyl are available from the Aldrich Chemical Company.

2. The Baker Co. catalyst was used.

3. This carcinogen is more easily handled in a flask than in a beaker. Contact with the skin obviously should be avoided.

4. Recrystallization is best accomplished by adding the compound to boiling ethanol and filtering. Prolonged heating should be avoided, as the substance gradually decomposes in hot solvent.

5. The sodium nitrite solution must be added carefully in order to avoid loss of material due to vigorous foaming.

6. This operation should be carried out in a hood to avoid the unpleasant effects of exposure to hydrogen azide vapors.

7. The compound may be recrystallized from a large volume of ethanol, but no increase in yield was noted using recrystallized material in the next step.

8. Eastman Kodak o-dichlorobenzene of 95% purity was used. Olefin-free kerosene or decalin may be substituted for the solvent, keeping the reaction temperature between 170° and 190°.

3. Methods of Preparation

o-Aminobiphenyl has been prepared by the reduction of the corresponding nitro compound with zinc and acetic acid,[2] zinc and hydrochloric acid,[3] iron and hydrochloric acid,[4] sodium bisulfite under pressure,[5] or hydrazine and palladium;[6] by the Hofmann reaction on o-phenylbenzamide;[7] and by pyrolysis of diazoaminobenzene.[8, 9]

o-Amino-p'-nitrobiphenyl has been made by the nitration of o-aminobiphenyl with ethyl nitrate;[10] by hydrolysis of the corresponding acetamide derivative;[11, 12] and by partial reduction of o, p'-dinitrobiphenyl with sodium bisulfite under pressure.[5]

2-Nitrocarbazole has been prepared by the dehydrogenation of 2-nitro-1,2,3,4-tetrahydrocarbazole with chloranil,[13] by the deamination of 2-nitro-3-aminocarbazole,[14] and by the thermal decomposition of o-azido-p'-nitrobiphenyl.[15] The procedure given here is a slight modification of the last-mentioned method.

4. Merits of the Preparation

The decomposition of o-azidobiphenyls is a convenient and general synthesis for a variety of carbazoles in good yield,[15] especially those not available through direct substitution of carbazole itself. Many of the required intermediates can be prepared from o-aminobiphenyl by substitution reactions. The method is also applicable to the preparation of analogs of the carbazole system in which a heterocyclic ring replaces a benzene ring, to the preparation of indoles, and to certain analogous aliphatic systems.

1. Department of Chemistry, University of Michigan, Ann Arbor, Michigan.
2. H. Hübner, *Ann.* **209**, 339 (1881).
3. F. Fichter and A. Sulzberger, *Ber.* **37**, 878 (1904).
4. T. Maki and K. Obayashi, *J. Chem. Soc. Japan, Ind. Chem. Sect.*, **55**, 108 (1952).
5. C. Finzi and G. Leandri, *Ann. Chim.* (*Rome*), **40**, 334 (1950).
6. P. M. G. Bavin, *Can. J. Chem.*, **36**, 238 (1958).
7. M. Chaix and F. de Rochebouët, *Bull. Soc. Chim. France*, [5] **2**, 273 (1935).
8. F. Heusler, *Ann.*, **260**, 227 (1890).
9. J. A. Aeschlimann, N. D. Lees, N. P. McCleland, and G. N. Nicklin, *J. Chem. Soc.*, **127**, 66 (1925).
10. C. Finzi and V. Bellavito, *Gazz. Chim. Ital.*, **64**, 335 (1934).
11. H. A. Scarborough and W. A. Waters, *J. Chem. Soc.*, 89 (1927).
12. S. Sako, *Bull. Chem. Soc. Japan*, **10**, 585 (1935).
13. B. M. Barclay and N. Campbell, *J. Chem. Soc.*, 530 (1945).
14. G. Anderson and N. Campbell, *J. Chem. Soc.*, 2904 (1950).
15. P. A. S. Smith and B. B. Brown, *J. Am. Chem. Soc.*, **73**, 2435 (1951); see also P. A. S. Smith and J. H. Hall, *J. Am. Chem. Soc.*, **84**, 480 (1962) and P. A. S. Smith, in W. Lvowski, "Nitrenes," Wiley-Interscience, New York, 1970, Chapter 4.

2-NITROETHANOL*

(Ethanol, 2-nitro-)

$$O_2NCH_3 + CH_2O \xrightarrow[\text{2. } H_2SO_4]{\text{1. KOH}} O_2NCH_2CH_2OH$$

Submitted by WAYLAND E. NOLAND.[1]
Checked by MELVIN S. NEWMAN and SURJAN S. RAWALAY.

1. Procedure

In a 5-l., three-necked, round-bottomed flask fitted with a 30-ml. dropping funnel, mechanical stirrer, and thermometer extending down into the liquid is placed a suspension of paraformaldehyde (trioxymethylene, 125 g., 4.16 moles) in freshly distilled (Note 1) nitromethane (2.5 l., 46.6 moles). The suspension is stirred vigorously, and 3N methanolic potassium hydroxide solution is added dropwise from the dropping funnel until, at an apparent pH of 6–8, but closer to pH 8 (pH paper), the paraformaldehyde begins to dissolve and the suspension assumes a clearer appearance. About 10 ml. of the alkaline solution is required, and the addition takes about 10 minutes. About 15–20 minutes after addition of the alkaline solution is complete, the paraformaldehyde dissolves completely. Shortly thereafter, the solution temperature reaches a maximum of 13–14 degrees above room temperature and then slowly drops. Stirring is continued 1 hour after addition of the alkaline solution is complete.

Stirring is continued while the added alkali is *completely* neutralized by adding concentrated (36N) sulfuric acid (1 ml.) dropwise from a medicine dropper over a period of about 3 minutes until an apparent pH of about 4 is reached (Note 2). The solution is then stirred for an hour, during which time the pH should not change (Note 3).

The precipitated potassium sulfate is filtered by passing the solution through a 12-cm. Büchner funnel. The light-yellow or yellowish green filtrate is transferred to a 5-l., one-necked, round-bottomed flask fitted with a Claisen head containing a capillary ebulliator tube and a thermometer, and connected to a water-cooled condenser. The condenser is connected through a vacuum adapter to a 3-l., one-necked, round-bottomed flask, cooled in ice, to act as a receiver. About 2.3 l. of pure, unchanged nitromethane is removed by distillation at aspirator pressure and a water-bath temperature of 40–50°. The distillation takes about 6–7 hours.

The golden-yellow residue (315–365 g.) is transferred to a 1-l., one-necked, round-bottomed flask containing an equal weight of diphenyl ether (Note 4). The flask is fitted with a Claisen head containing a capillary ebulliator tube and a thermometer, and connected to a water-cooled condenser. The condenser is connected to a 3- or 4-port fraction cutter fitted with 100–500 ml., one-necked, round-bottomed flasks, at least one of which is 500 ml. or larger to accommodate the main fraction (Note 5). The mixture is distilled under the vacuum of a good pump. The fore-run, b.p. 29–33° at about 0.10 mm., consisting of nitromethane (about 56 ml.), can be distilled at a water-bath temperature of 32–79° and usually passes directly into the Dry Ice trap protecting the vacuum pump. The temperature then rises as 2-nitroethanol and diphenyl ether codistil. The main fraction, a two-phase distillate initially richer in 2-nitroethanol than diphenyl ether, gradually changes in composition until the proportion of 2-nitroethanol becomes negligible. The main fraction of 410–425 g., b.p. 54–57° at about 0.10 mm. (or 64–66° at about 0.4 mm.), collects at a water-bath temperature of 79–88°. Care should be taken to prevent clogging of the condenser or fraction cutter with solid diphenyl ether (m.p. 27°). The distillation is continued until the proportion of 2-nitroethanol (lower layer) observed in the distillate becomes negligible, and the temperature suddenly starts to rise. At this point heating is stopped, *but the residue is cooled to room temperature or below before the vacuum is broken* (Note 6).

The two-phase main fraction of the distillate is placed in a 500-ml. separatory funnel and the lower layer of crude 2-nitroethanol (185–200 g., 146–158 ml., n_D^{25} 1.4493–1.4513, containing about 92–94 mole % 2-nitroethanol) is drawn off. The 2-nitroethanol is then extracted in a 500-ml. separatory funnel with an equal volume of light petroleum ether (b.p. 60–68°, such as Skellysolve B) or hexane, and the colorless lower layer of 2-nitroethanol (174–188 g., 46–49%, n_D^{25} 1.4432–1.4433, containing about 98 mole % 2-nitroethanol) is drawn off (Notes 7 and 8). The product turns light yellow after standing for a day or more.

2. Notes

1. Commercial nitromethane is sometimes quite acidic, and much more methanolic potassium hydroxide is required to initiate the reaction when such material is used. For safety, the nitromethane should be distilled at aspirator pressure instead of atmospheric pressure.

2. Sulfuric acid *must* be used in an amount slightly *more* than enough exactly to neutralize the alkali, and not just sufficient to make the reaction acidic. Otherwise, the metal salts of nitromethane can form explosive fulminates upon heating.

3. This is a suitable point at which to interrupt the experiment overnight.

4. 2-Nitroethanol prepared by the formaldehyde-nitromethane method should not be distilled without use of diphenyl ether as a

heat-dispersing agent. The residue, consisting of di- and tri-condensation products of formaldehyde with nitromethane, when hot and concentrated, and particularly when the vacuum is broken and air is let in on the hot distillation residue, is very likely to undergo a flash detonation, or at least a fume-off which may proceed with explosive violence. Use of diphenyl ether is a wise safety precaution in the distillation of 2-nitroethanol made by other methods as well.

5. If a fraction cutter is not used, the residue should be cooled to room temperature each time before the vacuum is broken.

6. The large amount of diphenyl ether (80–125 g.) left as the upper layer in the distilling flask has served the useful purpose, by its mass and volatility, of preventing superheating of the residue and subsequent violent decomposition, as described in Note 4.

7. This procedure has been carried out 30 times by students in the advanced organic laboratory course at the University of Minnesota. The extreme ranges of yields obtained were 32–52%, and the median yield was 46%.

8. The 2-nitroethanol obtained by this procedure is quite satisfactory for synthetic purposes, such as the preparation of nitro-ethylene. The small amount of light petroleum ether dissolved in the 2-nitroethanol can easily be removed under reduced pressure. Most of the remaining diphenyl ether can be removed by one redistillation under vacuum, since the fore-run is relatively rich in diphenyl ether. The main fraction has n_D^{25} 1.4425–1.4431. Although vacuum redistillation of 2-nitroethanol which has been freed by the present procedure from higher condensation products of formaldehyde with nitromethane is relatively safe, it is recommended that the procedure be carried out behind a safety shield or a barricade.

3. Methods of Preparation

The present procedure is that of Controulis, Rebstock, and Crooks,[2] modified to include the diphenyl ether purification method of Roy.[3] 2-Nitroethanol has been prepared by condensation of formaldehyde (usually employed in the solid form as para-formaldehyde) with a large excess of nitromethane in the presence of an alkali catalyst,[2, 4-6] as illustrated by the present procedure, or in the presence of a strongly basic ion-exchange resin.[7] Dimethoxymethane has also served as a source of formaldehyde in a reaction catalyzed by a mixture of acidic and basic ion-exchange resins.[8] 2-Nitroethanol has also been prepared by the action of silver nitrite on 2-iodoethanol (ethylene iodohydrin); [9-12] by selective catalytic hydrogenation over 5% palladium on barium sulfate in pyridine solution of halogenated derivatives, including 2-chloro-2-nitroethanol, 2,2-dichloro-2-nitroethanol, and 2-bromo-2-nitroethanol; [13] by the action of fuming nitric acid on ethylene; [14] and by the action of dinitrogen tetroxide on ethylene in the presence of oxygen [15-19] or nitric oxide,[20] or in carbon tetrachloride solution.[21]   The preparation of 2-nitroethanol from ethylene oxide by the action of aqueous solutions of barium,[22] calcium,[22] magnesium, [23] or zinc [22] nitrite, or by the action of sodium nitrite and carbon dioxide,[24] has also been reported.   The submitter has been unable to prepare 2-nitroethanol from ethylene oxide using the procedures described for barium [6] or sodium nitrite; his observation with respect to barium nitrite has been confirmed in another laboratory.[25]   More recently, the preparation of 2-nitroethanol in 50% yield has been reported by adding ethylene oxide to aqueous sodium nitrite at 20° under nitrogen in the presence of a nitrite scavenger, such as sodium hydrosulfite or phloroglucinol, at a pH of 7.1–7.3 controlled by the addition of phosphoric acid.[26]   The action of dinitrogen tetroxide on ethylene oxide in chloroform solution has been reported to yield 2-nitroethyl nitrate, from which 2-nitroethanol could be obtained by alkaline saponification.[27] This report has since been refuted with the finding that the initial product is the mononitrite mononitrate ester of ethylene glycol, which saponifies to ethylene glycol mononitrate and diethylene glycol mononitrate.[28]

4. Merits of Preparation

The present procedure has the advantage of using inexpensive, commercially available starting materials, combined with an apparently safe method of isolating the product.   2-Nitroethanol is particularly valuable as a synthetic intermediate for the preparation of nitroethylene. Nitroethylene is conveniently prepared by heating 2-nitroethanol with phthalic anhydride and allowing the nitroethylene to distil under reduced pressure.[29,30]

1. School of Chemistry, University of Minnesota, Minneapolis, Minn.
2. J. Controulis, M. C. Rebstock, and H. M. Crooks, Jr., *J. Am. Chem. Soc.*, 71, 2465 (1949); Harry M. Crooks, Jr., Parke, Davis and Co., Detroit, Mich., private communication to W. E. Noland, Jan. 8, 1954.
3. H. T. Roy, Jr. (to General Tire and Rubber Co.), U.S. pat. 2,710,830 (June 14, 1955).
4. I. M. Gorsky and S. P. Makarov, *Ber.*, 67, 996 (1934).
5. J. T. Hays, G. F. Hager, M. H. Engelmann, and H. M. Spurlin, *J. Am. Chem. Soc.*, 73, 5369 (1951).
6. W. E. Noland, H. I. Freeman, and M. S. Baker, *J. Am. Chem. Soc.*, 78, 188 (1956).
7. M. J. Astle and F. P. Abbott, *J. Org. Chem.*, 21, 1228 (1956).
8. C. J. Schmidle (to Rohm and Haas Co.), U.S. pat. 2,736,741 (Feb. 28, 1956) [*C.A.*, 50, 10761 (1956)].
9. R. Demuth and V. Meyer, *Ann.*, 256, 28 (1890).
10. L. Henry, *Rec. Trav. Chim.*, 16, 252 (1897); *Bull. Classe Sci. Acad. Roy. Belg.*, [3] 34, 547 (1897).
11. H. Wieland and E. Sakellarios, *Ber.*, 53, 201 (1920).
12. W. E. Noland and P. J. Hartman, *J. Am. Chem. Soc.*, 76, 3227 (1954).
13. R. Wilkendorf and M. Trénel, *Ber.*, 56, 611 (1923).
14. P. V. McKie, *J. Chem. Soc.*, 962 (1927).
15. A. E. Wilder Smith and C. W. Scaife (to Imperial Chemical Industries, Ltd.), U.S. pat. 2,384,048 (Sept. 4, 1945) [*C.A.*, 40, 347 (1946)].
16. A. E. Wilder Smith, C. W. Scaife, and Imperial Chemical Industries, Ltd., Brit. pat. 575,604 (Feb. 26, 1946) [*C.A.*, 41, 6893 (1947)].
17. A. E. Wilder Smith, R. H. Stanley, C. W. Scaife, and Imperial Chemical Industries, Ltd., British pat. 575,618 (Feb. 26, 1946) [*C.A.*, 41, 6893 (1947)].
18. A. E. Wilder Smith, R. H. Stanely, and C. W. Scaife (to Imperial Chemical Industries, Ltd.), U.S. pat. 2,424,510 (July 22, 1947) [*C.A.*, 41, 6893 (1947)].
19. N. Levy, C. W. Scaife, and A. E. Wilder Smith, *J. Chem. Soc.*, 1096 (1946).
20. V. L. Volkov, Russ. pat. 66,229 (April 30, 1946) [*C.A.*, 41, 2074 (1947)].
21. E. I. du Pont de Nemours and Co., Brit. pat. 603,344 (June 14, 1948) [*C.A.*, 43, 665 (1949)].
22. S. Miura (to Tanabe Chemical Industries Co.), Japan. pat. 156,256 (April 28, 1943).
23. G. V. Chelintsev and V. K. Kuskov, *Zhur. Obshchei Khim.*, 16, 1482 (1946).
24. S. Miura (to Yamanouchi Pharmaceutical Co.), Japan. pat. 6910 (Nov. 6, 1951) [*C.A.*, 48, 1412 (1954)].
25. T. E. Stevens and W. D. Emmons, *J. Am. Chem. Soc.*, 79, 6008 (1957).
26. H. N. Lee and R. W. Van House (to Parke, Davis and Co.), U.S. pat. 3,426,084 (Feb. 4, 1969) [*C.A.*, 70, 67594k (1969)].
27. G. Darzens, *Compt. Rend.*, 229, 1148 (1949).
28. G. Rossmy, *Ber.*, 88, 1969 (1955).
29. G. D. Buckley and C. W. Scaife, *J. Chem. Soc.*, 1471 (1947).
30. G. D. Buckley, C. W. Scaife, and Imperial Chemical Industries, Ltd., Brit. pat. 595,282 (Dec. 31, 1947) [*C.A.*, 42, 3773 (1948)].

N-NITROMORPHOLINE

(Morpholine, 4-nitro-)

$$\underset{(CH_3)_2\overset{\overset{\displaystyle OH}{|}}{C}-CN}{} \;+\; (CH_3CO)_2O \;+\; HNO_3 \;\longrightarrow\; \underset{(CH_3)_2\overset{\overset{\displaystyle ONO_2}{|}}{C}-CN}{} \;+\; 2CH_3CO_2H$$

$$\underset{(CH_3)_2\overset{\overset{\displaystyle ONO_2}{|}}{C}-CN}{} \;+\; 2\,O\!\!\diagup\!\!\diagdown\!NH \;\longrightarrow\; O\!\!\diagup\!\!\diagdown\!N-NO_2 \;+$$

$$O\!\!\diagup\!\!\diagdown\!N-\underset{\overset{\displaystyle |}{CN}}{C(CH_3)_2} \;+\; H_2O$$

Submitted by JEREMIAH P. FREEMAN and INELLA G. SHEPARD.[1]
Checked by C. G. BOTTOMLEY and B. C. McKUSICK.

1. Procedure

Caution! The nitrating mixture consisting of fuming nitric acid and acetic anhydride is an extremely active one, and combinations of it and organic materials are potentially explosive. The nitration should be carried out behind adequate safety shields. Acetone cyano-hydrin nitrate is moderately explosive (Note 6) and all operations with it, but particularly its distillation, should be carried out behind safety shields.

A. *Acetone cyanohydrin nitrate.* White fuming nitric acid (106 ml., 158 g., 2.3 moles) (Note 1) is added dropwise to 380 ml. (408 g., 4.00 moles) of acetic anhydride at 3–5° contained in a 2-l. three-necked flask fitted with a stirrer, a thermometer, and a dropping funnel and immersed in an ice bath. After the addition, which requires 45 minutes, the mixture is stirred at 5° for 15 minutes (Note 2). Acetone cyanohydrin (92 ml., 85 g., 1.00 mole) (Note 3) is added dropwise to the mixture at 5–10° over a 45-minute period. After the addition, the ice bath is removed and the mixture is allowed to warm to room temperature and is stirred there for 30 minutes. It is then poured into 600 g. of ice and water, and the resulting mixture is stirred for 90 minutes to dissolve the acetic anhydride.

The mixture is extracted with four 100-ml. portions of methylene chloride. The extracts are combined, washed successively with 100 ml. of water and four 100-ml. portions of 5% sodium carbonate solution (Note 4), and dried over anhydrous magnesium sulfate. The methylene chloride is removed by evaporation at 30–40° under the pressure of a water aspirator, and the residue is distilled through a 30-cm. Vigreux column to yield 85–90 g. (65–69%) (Note 5) of acetone cyanohydrin nitrate (Note 6); b.p. 62–65°/10 mm.; n_D^{20} 1.4170–1.4175 (Note 7).

B. *N-Nitromorpholine.* Morpholine (34.8 g., 0.40 mole) and 26 g. (0.20 mole) of acetone cyanohydrin nitrate are mixed in a 50-ml. round-bottomed flask equipped with a thermometer well. A condenser is attached, and the mixture is heated slowly. At about 60° an exotherm ensues that raises the temperature of the mixture to 110°. The mixture is allowed to cool to 80° and maintained there for 1 hour. It is poured into 200 ml. of 10% hydrochloric acid (*Caution! Do in a hood! Note 8*) and extracted with three 100-ml. portions of methylene chloride (Note 9). The extracts are combined, washed successively with two 100-ml. portions of water, 100 ml. of 10% hydrochloric acid, and 100 ml. of water, and dried over anhydrous magnesium sulfate. The solvent is removed by evaporation on a water aspirator at room temperature to yield a pale-yellow oil (Note 10).

The oil is dissolved in 80 ml. of absolute ethanol. The solution is cooled to 0–5°, causing white crystals of N-nitromorpholine to precipitate; weight 15–17 g. (57–64%); m.p. 52–54° (Note 11).

2. Notes

1. This is 90% nitric acid, d. 1.48–1.50. In order to remove dissolved nitrogen oxides from it, 0.5 g. of urea is added and the mixture is air-sparged for 20 minutes. The acid should be colorless before it is added to the acetic anhydride.

2. The nitrating mixture should be colorless at this point. If it is not, 0.5 g. of urea should be added and the mixture air-sparged until colorless.

3. Suitable acetone cyanohydrin can be purchased from the Rohm and Haas Co. and other commercial sources, or it can be prepared as described in *Organic Syntheses*.[2]

4. Washing with the carbonate solution should be continued until the organic layer is free of acid. Traces of acid may cause extensive decomposition during the distillation.

5. Similar yields were observed in preparations on three times this scale.

6. Acetone cyanohydrin nitrate should be regarded as a moderately explosive material and should be handled carefully and distilled behind a safety shield. For purposes of comparison, the drop-weight sensitivities on the Olin-Mathieson drop-weight tester of three materials are: propyl nitrate, 10 kg.-cm.; acetone cyanohydrin nitrate, 40 kg.-cm.; nitromethane, 60 kg.-cm.

7. The product obtained from this distillation usually contains small amounts of acetone cyanohydrin acetate, as evidenced by an ester carbonyl band at 1740 cm.$^{-1}$ in its infrared spectrum. This material does not interfere with the nitration reactions of the reagent. It may be removed by fractionation through a more efficient column.

8. This operation should be carried out in a good hood because hydrogen cyanide is evolved at this point.

9. The aqueous solution contains α-morpholinoisobutyronitrile in the form of its hydrochloride. It is formed by condensation of morpholine with the acetone and hydrogen cyanide formed in the nitration reaction. It is because of this side reaction that the excess amine is employed.

10. Occasionally this oil solidifies after removal of the last traces of solvent; in these instances it is necessary to warm the ethanol slightly to effect solution.

11. In nitrating amines other than morpholine, particularly on a larger scale, it may be desirable to carry out the reaction in acetonitrile to control the temperature better.[3]

3. Methods of Preparation

N-Nitromorpholine has been prepared by the oxidation of N-nitrosomorpholine with peroxytrifluoroacetic acid,[4] by the chloride ion-catalyzed reaction of nitric acid with morpholine,[5] by the action of nitric acid and acetic anhydride on N-formyl-morpholine,[6] by the reaction of dinitrogen pentoxide with morpholine,[7] and by alkaline nitration of morpholine with acetone cyanohydrin nitrate.[3]

4. Merits of the Preparation

This synthesis of N-nitromorpholine is representative of a rather general reaction for the preparation of both primary and secondary nitramines.[3] It represents the simplest process for obtaining both types of compounds. The reaction is unique in that a nitration is carried out under neutral or alkaline conditions. Acetone cyanohydrin nitrate may also be used for the nitration of many active methylene compounds.[8]

1. Rohm and Haas Company, Redstone Arsenal Research Division, Huntsville, Alabama. This research was carried out under Ordnance Contract W-01-021-ORD-334.
2. R. F. B. Cox and R. T. Stormont, *Org. Syntheses,* Coll. Vol. 2, 7 (1943).
3. W. D. Emmons and J. P. Freeman, *J. Am. Chem. Soc.,* 77, 4387 (1955).
4. W. D. Emmons, *J. Am. Chem. Soc.,* 76, 3468 (1954).
5. W. J. Chute, G. E. Dunn, J. C. MacKenzie, G. S. Myers, G. N. R. Smart, J. W. Suggitt, and G. F. Wright, *Can. J. Res.,* 26B, 114 (1948).
6. J. H. Robson, *J. Am. Chem. Soc.,* 77, 107 (1955).
7. W. D. Emmons, A. S. Pagano, and T. E. Stevens, *J. Org. Chem.,* 23, 311 (1958).
8. W. D. Emmons and J. P. Freeman, *J. Am. Chem. Soc.,* 77, 4391 (1955).

NITROSOMETHYLURETHANE

WARNING

Nitrosomethylurethane [1] has been reported to be a potent carcinogen by Druckrey and Preussmann.[2] These investigators suggest that nitrosomethylurethane be handled with greatest care or, preferably, be replaced whenever possible with *p*-tolyl-sulfonylmethylnitrosamide,[3] which was shown to be practically non-toxic and non-carcinogenic under conditions for which the urethane was toxic and/or carcinogenic.

1. W. W. Hartman and R. Phillips, *Org. Syntheses,* Coll. Vol. 2, 464 (1943).
2. H. Druckrey and R. Preussmann, *Nature,* 195, 1111 (1962).
3. Th. J. DeBoer and H. J. Backer, *Org. Syntheses,* Coll. Vol. 4, 943, 250 (1963).

m-NITROPHENYL DISULFIDE*

[Disulfide, bis-(*m*-nitrophenyl)]

$$2 \, m\text{-}NO_2C_6H_4SO_2Cl + 10HI \rightarrow$$
$$(m\text{-}NO_2C_6H_4)_2S_2 + 5I_2 + 2HCl + 4H_2O$$

Submitted by W. A. Sheppard.[1]
Checked by John D. Roberts and W. H. Graham.

1. Procedure

A 5-l. three-necked round-bottomed flask equipped with a re-flux condenser, a sealed mechanical stirrer, and a dropping funnel is set up in a hood and charged with 333 g. (1.50 moles) of *m*-nitrobenzenesulfonyl chloride (Note 1). The stirrer is started and 1033 ml. (7.5 moles) of 55–58% hydriodic acid (Note 2) is rapidly added dropwise over a period of 30–45 minutes (Note 3). After the addition is complete, the reaction mixture is stirred and re-fluxed on a steam bath for 3 hours. It is then cooled to room temperature, and the dropping funnel is replaced with an open powder funnel. Solid sodium bisulfite powder (Note 4) is added in portions until all the iodine has been reduced, and the reaction mixture is a suspension of pale yellow *m*-nitrophenyl disulfide in an almost colorless solution. The reaction mixture is filtered through a coarse-grade sintered-glass funnel to separate the disulfide, which is washed thoroughly with warm water to remove all inorganic salts. There is obtained 210–221 g. (91–96%) of crude *m*-nitrophenyl disulfide, m.p. 81–83°. This material is purified by dissolving it in approximately 800 ml. of boiling acetone, which is filtered hot and cooled to give 170–183 g. (74–79%) of the disulfide in the form of pale yellow prisms, m.p. 82–83°. By concentration of the mother liquor an additional 30–40 g. (13–17%) of disulfide, m.p. 82–83°, is obtained, so that the total yield of satisfactory product is 200–210 g. (86–91%).

2. Notes

1. Eastman Kodak white label *m*-nitrobenzenesulfonyl chloride was used.

2. Reagent grade hydriodic acid was generally employed, but material of lower purity may be used without decreasing the yield. The calculated amount of 45–47% hydriodic acid may also be employed [2] with only a slight diminution in yield.

3. The reaction of the hydriodic acid with *m*-nitrobenzenesulfonyl chloride is mildly exothermic, and iodine crystals precipitate as the reaction proceeds.

4. Approximately 3 lb. of sodium bisulfite is required to reduce the iodine. Technical grade bisulfite may be used satisfactorily. Caution should be observed in adding the bisulfite, since evolution of sulfur dioxide can cause excessive foaming. This foaming occurs a short time after each addition and is most noticeable when the iodine is almost neutralized. Iodine and product clinging to the upper walls of the flask and in the condenser may be conveniently rinsed into the reaction mixture with a stream of water from a wash bottle.

3. Methods of Preparation

The described method of preparation of *m*-nitrophenyl disulfide is essentially that of Foss and co-workers [2] and is a modification of that reported by Ekbom.[3] The disulfide has been prepared by reaction of potassium ethyl xanthate with *m*-nitrobenzenediazonium chloride solution, followed by hydrolysis to yield the mercaptan, which is subsequently oxidized with potassium ferrocyanide or dilute nitric acid to the disulfide.[4]

The usual method of preparing aromatic disulfides is to treat an aryl halide with Na_2S_2.[5] However, this method is limited to compounds where the halogen is strongly activated by electronegative groups (for example, *o*- or *p*-nitrochlorobenzene). The reaction of diazonium salts with xanthate is unsatisfactory for large-scale preparations because dilute solutions must be employed to reduce the hazard of explosion. Aromatic sulfonyl chlorides (not containing nitro groups) are also reduced with zinc and mineral acid to mercaptans,[6] which must be subsequently oxidized to the disulfide. The present method has been used to prepare nitronaphthalene disulfides,[7, 8] naphthalene disulfide, and phenyl disulfide [8] and should be applicable to the preparation of any symmetrical aromatic disulfides containing substituents stable to hydriodic acid.

4. Use of *m*-Nitrophenyl Disulfide

The disulfides are useful intermediates in the preparation of sulfenyl chlorides.[2, 9]

1. Contribution No. 515 from the Central Research Department, Experimental Station, E. I. du Pont de Nemours and Company, Inc., Wilmington, Delaware.
2. N. E. Foss, J. J. Stehle, H. M. Shusett, and D. Hadburg, *J. Am. Chem. Soc.*, 60, 2729 (1938).
3. A. Ekbom, *Ber.*, 24, 335 (1891).
4. R. Leuckart and W. Holtzapfel, *J. Prakt. Chim.*, 41 (2), 197 (1890).
5. M. T. Bogert and A. Stull, *Org. Syntheses*, Coll. Vol. 1, 220 (1941).
6. Th. Zinke and O. Krüger, *Ber.*, 45, 3468 (1912).
7. P. T. Cleve, *Ber.*, 20, 1534 (1887).
8. P. T. Cleve, *Ber.*, 21, 1099 (1888).
9. M. H. Hubacher, *Org. Syntheses*, Coll. Vol. 2, 455 (1943).

4-NITRO-2,2,4-TRIMETHYLPENTANE

(Pentane, 2,2,4-trimethyl-4-nitro-)

$$(CH_3)_3CCH_2\underset{\underset{NH_2}{|}}{C}(CH_3)_2 + 2KMnO_4 \rightarrow$$

$$(CH_3)_3CCH_2\underset{\underset{NO_2}{|}}{C}(CH_3)_2 + 2MnO_2 + 2KOH$$

Submitted by NATHAN KORNBLUM and WILLARD J. JONES.[1]
Checked by WILLIAM G. DAUBEN and PAUL R. RESNICK.

1. Procedure

A solution of 25.8 g. (0.20 mole) of 4-amino-2,2,4-trimethyl-pentane (*tert*-octylamine) (Note 1) in 500 ml. of C.P. acetone is placed in a 1-l. three-necked flask equipped with a "Tru-Bore" stirrer and a thermometer and is diluted with a solution of 30 g.

of magnesium sulfate (Note 2) in 125 ml. of water. Potassium permanganate (190 g., 1.20 moles) is added to the well-stirred reaction mixture in small portions over a period of about 30 minutes (Note 3). During the addition the temperature of the mixture is maintained at 25–30° (Note 4), and the mixture is stirred for an additional 48 hours at this same temperature (Note 5). The reaction mixture is stirred under water-aspirator vacuum at an internal temperature of about 30° until most of the acetone is removed (Note 6). The resulting viscous mixture is steam-distilled; approximately 500 ml. of water and a pale-blue organic layer are collected. The distillate is extracted with pentane, the extract is dried over anhydrous sodium sulfate, and the pentane is removed by distillation at atmospheric pressure. The residue is distilled through a column (Note 7) at reduced pressure to give 22–26 g. (69–82%) of colorless 4-nitro-2,2,4-trimethylpentane, b.p. 53–54°/3 mm., n_D^{28} 1.4314, m.p. 23.5–23.7°.

2. Notes

1. The *tert*-octylamine employed was redistilled commercial-grade material, b.p. 140°/760 mm., n_D^{20} 1.4240.

2. The magnesium sulfate was purified dried powder of J. T. Baker Chemical Co. This is approximately 70% magnesium sulfate and 30% water.

3. Good agitation prevents the permanganate from caking on the bottom of the flask. The formation of a cake results in local overheating and consumption of the permanganate as mentioned in Note 4.

4. If a constant-temperature bath is not available, a bucket of water, initially at 25°, serves to dissipate the heat of reaction. At higher temperatures the potassium permanganate is rapidly consumed, presumably by reaction with the acetone.

5. At the end of the reaction time there was no unreacted amine as shown by the following test: A 10-ml. aliquot was filtered through "Supercel" to remove the manganese dioxide, and the filtrate was added to a mixture of 25 ml. of benzene and 25 ml. of water. Extraction of the benzene layer with 10% hydrochloric acid, followed by the addition of sodium hydroxide, gave no oil layer or characteristic odor of the free amine.

6. If agitation becomes difficult during the concentration, 100 ml. of water can be added to give a more fluid mixture.

7. A 60-cm. x 1-cm. externally heated column packed with 4-mm. glass helices and equipped with a total-reflux variable take-off head was used.

3. Methods of Preparation

The procedure described is that of Kornblum, Clutter, and Jones.[2] 4-Nitro-2,2,4-trimethylpentane has been prepared previously, in low yield, by allowing isoöctane to react with concentrated nitric acid in a sealed tube at elevated temperature.[3]

4. Merits of the Preparation

This is a general method of preparing trialkylnitromethanes from the corresponding (trialkylmethyl)amines.[2,4] Table I lists

TABLE I

Synthesis of Trialkylnitromethanes, R_3CNO_2

Nitro Compound	Yield, %
2-Nitro-2-methylpropane	83
2-Nitro-2,3-dimethylbutane	71
2-Nitro-2,4-dimethylpentane	82
1-Nitro-1-methylcyclopentane	72
1-Nitro-1-methylcyclohexane	73
1-Nitro-1,4-dimethylcyclohexane	70
1,8-Dinitro-p-menthane	61

[a] This oxidation was carried out in water.

seven prepared in this way. The procedure is simple and reliable, and the yields of product are high. Other methods give mixtures of products and low yields of nitro compounds and are inconvenient to perform.

1. Department of Chemistry, Purdue University, West Lafayette, Indiana. This research was supported by the United States Air Force under Contract No. AF 18(600)-310 monitored by the Office of Scientific Research, Air Research and Development Command.

2. N. Kornblum, R. J. Clutter, and W. J. Jones, *J. Am. Chem. Soc.*, 78, 4003 (1956).

3. S. S. Nametkin and K. S. Zabrodina, *Doklady Akad. Nauk SSSR,* **75,** 395 (1950) [*C.A.,* **45,** 6998 (1951)].

4. N. Kornblum, *Org. Reactions,* **12,** 115 (1962).

2,4-NONANEDIONE

$$CH_3COCH_2COCH_3 \xrightarrow[\text{Liq. NH}_3]{2 \text{ NaNH}_2} \overset{Na}{NaCH_2COCHCOCH_3}$$

$$\downarrow n\text{-}C_4H_9Br$$

$$CH_3(CH_2)_4COCH_2COCH_3 \xleftarrow{H^+} \overset{Na}{CH_3(CH_2)_4COCHCOCH_3}$$

Submitted by K. Gerald Hampton, Thomas M. Harris, and Charles R. Hauser [1]

Checked by Eugene Gosselink and Peter Yates

1. Procedure

Caution! This preparation should be carried out in a hood to avoid exposure to ammonia.

A suspension of sodium amide (1.10 moles) in liquid ammonia is prepared in a 1-l. three-necked flask equipped with an air condenser (Note 1), a ball-sealed mechanical stirrer, and a glass stopper. In the preparation of this reagent commercial anhydrous liquid ammonia (800 ml.) is introduced from a cylinder through an inlet tube. To the stirred ammonia is added a small piece of sodium. After the appearance of a blue color a few crystals of ferric nitrate hydrate (about 0.25 g.) are added, followed by small pieces of freshly cut sodium until 25.3 g. (1.10 moles) has been added. After the sodium amide formation is complete (Note 2), the glass stopper is replaced by a pressure-equalizing dropping funnel containing 60.0 g. (0.600 mole) of 2,4-pentanedione (Note 3) in 40 ml. of anhydrous ether. The top of the addition funnel is fitted with a nitrogen inlet tube. The reaction flask is immersed at least 3 in. into a dry ice-acetone bath (Note 4), and simultaneously the slow introduction of dry nitrogen through the inlet tube is begun. After the reaction mixture is thoroughly cooled (about 20 minutes), 2,4-pentane-dione is added intermittently in small portions (Note 4) over 10 minutes. The cooling bath is removed. After 20 minutes the nitrogen purge is stopped, and 68.5 g. (0.500 mole) of 1-bromo-butane (Note 5) in 40 ml. of anhydrous ether is introduced drop-wise during 10–20 minutes. The addition funnel is rinsed with a small volume of anhydrous ether, which is added to the reaction mixture. After 30 minutes 400 ml. of anhydrous ether is added, and the ammonia is removed by cautious heating on the steam bath. Crushed ice (200 g.) is added causing a thick slurry to form. Next a mixture of 60 ml. of concentrated hydrochloric acid and 10 g. of crushed ice is added. The reaction mixture is stirred until all solids are dissolved and then is transferred to a separa-tory funnel, the flask being washed with a little ether and dilute hydrochloric acid. The ethereal layer is separated, and the aqueous layer (Note 6) is further extracted three times with 100-ml. portions of ether. The combined ethereal extracts are dried over anhydrous magnesium sulfate. After filtration and removal of the solvent the residue is distilled through a 12-in. Vigreux column to give 63.0–63.6 g. (81–82%) of 2,4-nonane-dione, b.p. 100–103° (19 mm.), as a colorless liquid (Note 7).

2. Notes

1. The checkers used a dry ice condenser during the introduc-tion of ammonia to the reaction flask and replaced it with an air condenser before the addition of sodium.

2. Conversion to sodium amide is indicated by the disappear-ance of the blue color. This generally requires about 20 minutes.

3. Eastman Organic Chemicals 2,4-pentanedione was dried over potassium carbonate and distilled before use, the fraction boiling at 133–135° at atmospheric pressure being used.

4. The addition of 2,4-pentanedione to liquid ammonia is a highly exothermic process. Also, ammonia vapor reacts with the β-diketone to produce an insoluble ammonium salt, which tends to clog the tip of the addition funnel. Cooling the reac-tion mixture in dry ice-acetone reduces the vigor of the reaction and minimizes the clogging of the addition funnel. The 2,4-

pentanedione should be added in spurts which fall on the surface of the reaction mixture rather than on the wall of the flask.

5. Eastman Organic Chemicals 1-bromobutane was used without purification.

6. The aqueous layer should be acidic to litmus paper. If it is basic, indicating that the ammonia was not completely removed from the reaction mixture, more hydrochloric acid should be added until an acidic test is obtained.

7. A small forerun of 2,4-pentanedione, b.p. 32–100° (19 mm.), is obtained. The purity of the product may be demonstrated by gas chromatography on a 2-ft. column packed with silicone gum rubber (F and M Scientific Co., Avondale, Pennsylvania) programmed linearly from 100° to 300°. The chromatogram obtained is a single sharp peak. The three conceivable impurities, 2,4–pentanedione, 3-butyl-2,4-pentanedione, and 6,8-tridecanedione, would have been observed under these conditions if they had been present.

3. Methods of Preparation

The method described is that of Hampton, Harris, and Hauser[2] and is an improvement over the earlier procedures[3, 4] of Hauser and co-workers, which employed potassium amide. 2,4-Nonanedione has been prepared by the condensation of ethyl caproate with acetone in the presence of sodium hydride (54–80%),[5, 6] and by the acylation of ethyl acetoacetate followed by cleavage and decarboxylation (51%).[7] Other preparations include the acetylation of 2-heptanone with ethyl acetate and sodium amide (61%),[8] the acetylation of 1-heptyne with acetic anhydride and boron fluoride (42%),[9] and the sulfuric acid-catalyzed hydration of 3-nonyn-2-one.[10] Low yields (6–7%) of 2,4-nonanedione were obtained by the acetylation of 2-heptanone with acetic anhydride and boron fluoride,[11] and by the pyrolysis of the enol acetates of 2-heptanone.[12] The last two methods afforded substantial amounts of the isomeric 3-butyl-2,4-pentanedione.

4. Merits of the Preparation

This procedure represents a novel, convenient, and fairly general method for preparing higher β-diketones. By this method the submitters have alkylated 2,4-pentanedione at the 1-position with methyl iodide to give 2,4-hexanedione (59–65%) and with n-octyl bromide to give 2,4-tridecanedione (66–79%).[2] Alkylation at the 3-position is not observed, and little or no 1,5-dialkylation occurs. By similar procedures employing potassium amide, 2,4-pentanedione has also been alkylated with benzyl chloride,[3, 4] allyl bromide,[4] n-heptyl bromide,[4] and isopropyl bromide.[4] Numerous other β-diketones have been alkylated similarly. They include benzoylacetone,[3] 6-phenyl-2,4-hexanedione,[4, 13] 2-acetylcyclopentanone,[13] 2-acetylcyclohexanone,[13] and 2,4-tridecanedione.[4] In contrast to 2,4-pentanedione, these β-diketones may be added readily to the liquid ammonia solution without employing a dry ice cooling bath.

1. Chemistry Department, Duke University, Durham, North Carolina. This research was supported in part by the National Science Foundation.
2. K. G. Hampton, T. M. Harris, and C. R. Hauser, *J. Org. Chem.*, **28**, 1946 (1963); **30**, 61 (1965).
3. C. R. Hauser and T. M. Harris, *J. Am. Chem. Soc.*, **80**, 6360 (1958).
4. R. B. Meyer and C. R. Hauser, *J. Org. Chem.*, **25**, 158 (1960).
5. F. W. Swamer and C. R. Hauser, *J. Am. Chem. Soc.*, **72**, 1352 (1950).
6. N. Green and F. B. LaForge, *J. Am. Chem. Soc.*, **70**, 2287 (1948).
7. A. Bongert, *Compt. Rend.*, **133**, 820 (1901); L. Bouveault and A. Bongert, *Bull. Soc. Chim. France*, [3] **27**, 1083 (1902); Y. K. Yur'ev and Z. V. Belyakova, *J. Gen. Chem. USSR*, **29**, 1432 (1959).
8. R. Levine, J. A. Conroy, J. T. Adams, and C. R. Hauser, *J. Am. Chem. Soc.*, **67**, 1510 (1945).
9. V. E. Sibirtseva and V. N. Belov, *Tr., Vses. Nauch.-Issled. Inst. Sintetich. i Natural'n. Dushistykh Veshchestv*, No. **5**, 42 (1961) [*C.A.*, **57**, 16379 (1962)].
10. C. Moureu and R. Delange, *Bull. Soc. Chim. France*, [3] **25**, 302 (1901).
11. J. T. Adams and C. R. Hauser, *J. Am. Chem. Soc.*, **66**, 345 (1944); **67**, 284 (1945).
12. Z. Budesinsky and V. Musil, *Collection Czech. Chem. Commun.*, **24**, 4022 (1959).
13. T. M. Harris and C. R. Hauser, *J. Am. Chem. Soc.*, **81**, 1160 (1959).

2-NORBORNANONE*

(Norcamphor)

$$3 \quad \underset{\text{O-C-H}}{\overset{\text{O}}{\underset{\|}{}}} + 4CrO_3 + 6H_2SO_4 \xrightarrow{\text{Acetone}} 3$$

$$+ 3CO_2 + 2Cr_2(SO_4)_3 + 9H_2O$$

Submitted by DONALD C. KLEINFELTER and PAUL VON R. SCHLEYER.[1]
Checked by WILLIAM E. PARHAM, WAYLAND E. NOLAND, and
LYNETTE E. CHRISTENSEN.

1. Procedure

A. *2-exo-Norbornyl formate*. Approximately 800 g. (17.4 moles) of 98–100% formic acid (Note 1) is added to 400 g. (4.25 moles) of norbornene (Note 2) in a 2-l. round-bottomed flask equipped with a condenser, and the mixture is boiled under reflux for 4 hours (Note 3). The dark solution is cooled and the flask arranged for distillation using a 30-cm. Vigreux column. The excess formic acid is removed under reduced pressure (b.p. 26–30°/21–30 mm.). Distillation of the residue then gives a forerun of about 100 ml. of a mixture of formic acid and ester followed by about 485 g. of 2-*exo*-norbornyl formate, a colorless oil, b.p. 65–67°/14–16 mm., n_D^{25} 1.4594–1.4597. Another 55–65 g. of ester is obtained by adding water to the forerun, extracting with 30–60° petroleum ether, washing the extracts with dilute sodium carbonate solution, drying over sodium sulfate, and distilling. The total yield is 540–550 g. (90.5–92.5%) (Note 4).

B. *2-Norbornanone.* A solution of 510 g. (3.64 moles) of 2-*exo*-norbornyl formate in 1.5 l. of reagent grade acetone is contained in a 5-l. three-necked flask equipped with a thermometer, stirrer, and dropping funnel containing 8*N* chromic acid solution (Note 5). The flask is cooled with an ice bath and the oxidant is added at a rate such that the reaction temperature is maintained at 20–30°. Approximately 1870 ml. of oxidant solution is required, completion of the reaction being shown by the persistence of the brownish orange color. A slight excess of oxidant is added, and the solution is stirred overnight at room temperature. Solid sodium bisulfite is added in portions to reduce the excess oxidant.

The reaction mixture is poured into a large separatory funnel. The dark green chromic sulfate sludge, which has formed during the course of the reaction, is separated either by decantation and washing or by drawing it off from the bottom of the funnel. The acetone solution is washed three times with 200–250 ml. portions of an aqueous saturated potassium carbonate solution and finally is dried over anhydrous potassium carbonate. The acetone is removed by distillation through a 30-cm. Vigreux column at atmospheric pressure; benzene may be added near the end to assist in the removal of water by azeotropic distillation. When it is observed that the distillation of solvent is complete and the considerably hotter vapors of product begin to ascend the column, the condenser is removed from the top and replaced by an adapter and collection flask immersed in ice water. The adapter is heated and maintained above 100° by a free flame until the product begins to distil (Note 6). 2-Norbornanone, 335–350 g. (83–87%), distils at 170–173° and crystallizes immediately in the collection flask. The crystals melt at 90–91° (Note 7) and are sufficiently pure for most preparative purposes (Note 8).

2. Notes

1. Baker and Adamson 98–100% formic acid was used.

2. Prepared as described in *Org. Syntheses,* Coll. Vol. 4, 738 (1963).

3. Norbornene is not soluble in cold formic acid; initially there are two layers. As heat is applied to the flask, solution occurs and the reaction becomes quite exothermic. It is recommended

that a splash trap be mounted at the top of the condenser and that an ice bath be nearby in case the refluxing becomes too rapid. *Caution! Formic acid causes severe burns!*

4. Quite pure 2-*exo*-norborneol, m.p. 127–128°, can be prepared by saponification of 2-*exo*-norbornyl formate in an aqueous ethanolic solution of potassium hydroxide. The product can be isolated in about 85% yield by distillation and boils at 178–179°.

5. The 8*N* chromic acid solution [2] is prepared by dissolving 534 g. of chromium trioxide in ice water, adding 444 ml. of conc. sulfuric acid carefully, and diluting to 2 l. with water.

6. The product solidifies readily. Care should be taken to prevent clogging of the adapter. Once begun, there is no difficulty if the distillation proceeds smoothly to completion.

7. The melting points given in the literature vary from 90–91° (Ref. 3) to 95.5° (Ref. 4). Pure 2-norbornanone, m.p. 97.2–98.0°, may be made by regeneration from its semicarbazone derivative, m.p. 196.5–197.6°.

8. Gas chromatographic analysis shows this material to have a purity of about 96%. Besides a small amount of water (up to 0.5%) there are two minor impurities. Neither 2-*exo*-norbornyl formate nor 2-*exo*-norbornanol is present, however. Oxidation of 2-*exo*-norbornanol with chromic acid, under a variety of conditions, gives 2-norbornanone contaminated with some starting material.

3. Methods of Preparation

2-Norbornanone is generally prepared from the Diels-Alder adduct of cyclopentadiene and vinyl acetate by hydrogenation, saponification, and oxidation with chromic acid in acetic acid solution.[5] The present procedure, which gives higher over-all yields in fewer steps, makes use of the superior solvent, acetone, for mild chromic acid oxidations [2] and of the observation that formate esters of secondary alcohols can be oxidized directly to ketones.[6]

4. Merits of Preparation

2-Norbornanone is a useful starting material for various bicyclic derivatives of theoretical interest. The present procedure provides a convenient method for its preparation and illustrates a general method for the oxidation of formate esters to ketones.

1. Department of Chemistry, Princeton University, Princeton, New Jersey.
2. K. Bowden, I. M. Heibron, E. R. H. Jones, and B. C. L. Weedon, *J. Chem. Soc.,* 39 (1946); P. Bladon, J. M. Fabian, H. B. Henbest, H. P. Koch, and G. W. Wood, *J. Chem. Soc.,* 2402 (1951).
3. G. Komppa and S. Beckmann, *Ann.,* 512, 172 (1934).
4. G. Becker and W. A. Roth, *Ber.,* 67, 627 (1934).
5. K. Alder and H. F. Rickert, *Ann.,* 543, 1 (1940).
6. E. J. Corey, M. Ohno, S. W. Chow, and R. A. Scherrer, *J. Am. Chem. Soc.,* 81, 6305 (1959).

NORCARANE*

(Bicyclo[4.1.0]heptane)

Submitted by R. D. SMITH and H. E. SIMMONS.[1]
Checked by WILLIAM E. PARHAM and M. D. DHAVSAR.

1. Procedure

A. *Zinc-copper couple.* In a 500-ml. Erlenmeyer flask fitted with a magnetic stirrer are placed 49.2 g. (0.75 g. atom) of zinc powder (Note 1) and 40 ml. of 3% hydrochloric acid. The mixture is stirred rapidly for 1 minute, then the supernatant liquid is decanted. In a similar manner, the zinc powder is washed successively with three additional 40-ml. portions of 3% hydrochloric acid, five 100-ml. portions of distilled water, two 75-ml. portions of 2% aqueous copper sulfate solution, five 100-ml. portions of distilled water, four 100-ml. portions of absolute ethanol, and five 100-ml. portions of absolute ether (Note 2). The couple is finally transferred to a Büchner funnel, washed with additional anhydrous ether, covered tightly with a rubber dam, and suction-dried until it reaches room temperature. The zinc-copper couple is stored overnight in a vacuum desiccator over phosphorus pentoxide and is then ready for use in the preparation of norcarane (Note 3).

B. *Norcarane.* In a 500-ml. round-bottomed flask fitted with a magnetic stirrer and a reflux condenser protected by a drying tube filled with Drierite are placed 46.8 g. (0.72 g. atom) of zinc-copper couple and 250 ml. of anhydrous ether. A crystal of iodine is added, and the mixture is stirred until the brown color has disappeared (Note 4). A mixture of 53.3 g. (0.65 mole) of cyclohexene and 190 g. (0.71 mole) of methylene iodide is added in one portion (Note 5). The reaction mixture is then heated under gentle reflux with stirring. After 30–45 minutes, a mildly exothermic reaction occurs which may require cessation of external heating. After the exothermic reaction has subsided (approximately 30 minutes), the mixture is stirred under reflux for 15 hours. At the end of this time, most of the gray couple has been converted to finely divided copper. The ether solution is decanted (Note 6) from the copper and unreacted couple, which are then washed with two 30-ml. portions of ether. The washes are combined with the bulk of the solution and shaken with two 100-ml. portions of saturated ammonium chloride solution (Note 7), 100 ml. of saturated sodium bicarbonate solution, and 100 ml. of water. The ether solution is dried over anhydrous magnesium sulfate and filtered. The ether is distilled through a 20 x 2-cm. column packed with glass helices. The residue is distilled through a 45-cm. spinning-band column[2] to give 35–36 g. (56–58%) of norcarane, b.p. 116–117°, n_D^{25} 1.4546 (Note 8).

2. Notes

1. Mallinckrodt A. R. zinc dust was found satisfactory for this preparation. The checkers used Merck zinc dust but found it necessary to start with 51 g. of zinc in order to obtain sufficient couple for the next step.

2. The washings with hydrochloric acid should be done rapidly to avoid adsorption of bubbles of hydrogen on the zinc which

make subsequent washings more difficult. The use of a magnetic stirrer greatly facilitates the washings. The absolute ethanol and absolute ether washings are decanted directly on a Büchner funnel to prevent loss of the couple.

3. This method of preparing zinc-copper couple is essentially that of Shank and Shechter.[3] An equally active couple can be prepared by reduction of cupric oxide in the presence of zinc powder.[4] Mallinckrodt A. R. wire-form cupric oxide (30 g.) is ground to a powder in a mortar and mixed with 240 g. of Mallinckrodt A. R. zinc dust. The mixture is placed in a Vycor combustion boat lined with copper foil, and a thermocouple is imbedded in the powder. The boat is placed in a Vycor tube heated by a muffle furnace. A mixed gas (hydrogen, 65 l. per hour; nitrogen, 25 l. per hour) is passed through the tube while the temperature is raised to 500° during 4 hours. The mixture is kept at 500° for 30 minutes, and the tube is then allowed to cool to room temperature in a hydrogen atmosphere. The zinc-copper couple is obtained as dark gray lumps, which are ground to a fine powder in a mortar before use. In some instances, there is also found in the mixture a small amount of material which has apparently melted and agglomerated during heating. This shiny, metallic material is easily separated from the powdered couple and is not used in the preparation of norcarane.

4. The addition of iodine appears to promote the subsequent reaction of the zinc-copper couple with methylene iodide.

5. Commercial cyclohexene (Eastman Kodak) was distilled and passed over a column of activated alumina just before use. Methylene iodide (Matheson, Coleman and Bell) was distilled under reduced pressure, b.p. 50–51°/7 mm., and was stored in a brown bottle over iron wire.

6. The checkers filtered the solution because the finely divided copper and unreacted couple did not settle completely.

7. Care must be taken when adding the ammonium chloride solution to the ether solution since considerable heat is generated.

8. About 10–12 g. of cyclohexene, b.p. 82–84°, is recovered. The intermediate fraction, b.p. 84–116°, amounts to 1.5–2.5 g., and 10–12 g. of a dark residue remains in the still pot.

3. Discussion

This method is generally applicable to the stereospecific synthesis of cyclopropane derivatives from a large variety of substituted olefins.[4]

Three methods have been employed to generate iodomethylzinc iodide, the intermediate active in cyclopropanation: (1) reaction of methylene iodide with a zinc-copper couple in an ether solvent;[4] (2) reaction of diazomethane with zinc iodide in an ether solvent;[5] (3) reaction of methylene iodide with diethylzinc in ether or hydrocarbon solvents.[6] Method (1) has been used more extensively to prepare cyclopropanes from olefins because it is generally the simplest, most convenient, and most economical variation when applicable.[7] An active zinc-copper couple has been reported that is easily prepared and is recommended for the cyclopropanation reaction.[8]

Norcarane has been prepared by the reduction of 7,7-dichloronorcarane with sodium and alcohol,[9] and by the light-catalyzed reaction of diazomethane with cyclohexene.[9] The reaction of cyclohexene with methylene iodide and zinc-copper couple represents the most convenient preparation of norcarane which is of high purity.

1. Contribution No. 622 from the Central Research Department, Experimental Station, E. I. du Pont de Nemours and Co., Wilmington, Del.
2. R. G. Nester, *Anal. Chem.*, 28, 278 (1956).
3. R. S. Shank and H. Shechter, *J. Org. Chem.*, 24, 1825 (1959).
4. H. E. Simmons and R. D. Smith, *J. Am. Chem. Soc.*, 81, 4256 (1959).
5. G. Wittig and F. Wingler, *Ann.*, 656, 18 (1962); *Ber.*, 97, 2146 (1964).
6. J. Furukawa, N. Kawakata, and J. Mishimura, *Tetrahedron*, 24, 53 (1968).
7. H. E. Simmons, T. L. Cairns, and C. M. Hoiness, *Org. Reactions*, in press.
8. E. LeGoff, *J. Org. Chem.*, 29, 2048 (1964).
9. W. von E. Doering and A. K. Hoffman, *J. Am. Chem. Soc.*, 76, 6162 (1954).

exo/endo-7-NORCARANOL

Submitted by U. Schöllkopf, J. Paust, and M. R. Patsch [1]
Checked by William G. Dauben, Michael H. McGann, and
Noel Vietmeyer

1. Procedure

A. *exo/endo*-7-*(2-Chloroethoxy)bicyclo[4.1.0]heptane.* A 2-l.,
three-necked, round-bottomed flask is equipped with a sealed
stirrer, a pressure-equalizing dropping funnel, and a condenser
fitted with a nitrogen-inlet tube (Note 1). The flask is flushed
with dry nitrogen, and to it are added 500 ml. of cyclohexene
(Note 2) and 49.0 g. (0.300 mole) of dichloromethyl 2-chloroethyl
ether (Note 3). To the stirred solution at room temperature is
added dropwise 430 ml. (0.47 mole) of a 1.1N ethereal solution of
methyllithium (Note 4) at a rate adequate to maintain gentle
reflux of the ether; the addition requires *ca.* 4 hours (Note 5).
The reaction mixture is poured into 1.5 l. of ice water, the aqueous
layer is separated, and the organic layer is extracted with four
300-ml. portions of water and dried over anhydrous sodium
sulfate. The solvents are removed by distillation through a
10-cm. Vigreux column (Note 6), and the residue is distilled
under reduced pressure to yield 21–29 g. (40–56%) of *exo/endo*-
7-(2-chloroethoxy)bicyclo[4.1.0]heptane, b.p. 98–101° (10 mm.).
This material is sufficiently pure for the next step (Note 7).

B. *exo/endo*-7-*Norcaranol.* A 500-ml. three-necked flask
equipped with a magnetic stirrer, a pressure-equalizing dropping

funnel, and a condenser fitted with a nitrogen-inlet tube (Note 1) is flushed with nitrogen, and a solution of 20.0 g. (0.115 mole) of *exo/endo*-7-(2-chloroethoxy)bicyclo[4.1.0]heptane in 150 ml. of dry ether is added. To this solution is added dropwise at room temperature 280 ml. (0.45 mole) of a 1.6N solution of *n*-butyl-lithium in hexane over a 30- to 45-minute period (Note 5). The mixture is poured into 800 ml. of ice-cold, saturated, aqueous sodium bicarbonate, and the aqueous phase is separated and extracted with four 150-ml. portions of ether. The organic solutions are combined and dried over anhydrous sodium sulfate, and the solvents are removed by distillation through a 10-cm. Vigreux column at a maximum bath temperature of 65°. The residue is distilled under reduced pressure to yield 11.6–12.3 g. (90–95%) of *exo/endo*-7-norcaranol, b.p. 80–85° (10 mm.) (Notes 8 and 9).

2. Notes

1. The nitrogen-inlet system described by Johnson and Schneider [2] is satisfactory.

2. The cyclohexene was dried over potassium hydroxide pellets and distilled from sodium before use.

3. The checkers prepared this ether in the following manner. 2-Ethoxy-1,3-dioxolane was prepared in 82% yield from ethylene glycol and ethyl orthoformate and treated with acetyl chloride to give 2-chloroethyl formate by the procedures of Baganz and Domaschke;[3] the overall yield was 56–60%. The formate was converted to dichloromethyl 2-chloroethyl ether with phosphorus pentachloride by the procedure of Gross, Rieche, and Höft,[4] and the product was distilled through a 10-cm. column containing glass helices; b.p. 107–111° (110 mm.); yield 85%.

4. The methyllithium must be prepared from methyl iodide because the presence of the iodide anion is essential. The submitters prepared methyllithium in the following manner. Methyl iodide (425.7 g., 3.00 moles) was added with stirring to 48 g. (7.0 g. atoms) of lithium in 2.5 l. of ether under nitrogen at a rate adequate to maintain gentle reflux of the ether. After 24 hours the solution of methyllithium was decanted into a storage vessel filled with nitrogen. The concentration was estimated in the

usual way by hydrolysis of an aliquot and titration with 0.1*N* hydrochloric acid.

5. The addition of the organolithium solution is continued until a positive Gilman test[5] is obtained.

6. Isopropyl 2-chloroethyl ether, b.p. 118–121°, is formed in variable amounts as a by-product.

7. The *exo/endo* ratio is ∼6:1; the *exo* and *endo* isomers show characteristic triplets in their n.m.r. spectra at δ 2.9 and 3.1 p.p.m., respectively.

8. The *exo/endo* ratio is ∼8:1; the *exo* and *endo* isomers show characteristic triplets in their n.m.r. spectra at δ 3.0 and 3.25 p.p.m., respectively.

9. In some runs, *exo*-7-norcaranol, m.p. 57–58°, crystallized in the condenser or in the receiver.

3. Discussion

This method for the preparation of *exo/endo*-7-norcaranol is an adaptation of that described by Schöllkopf, Paust, Al-Azrak, and Schumacher.[6] The method has been used by the submitters for the preparation of the following cyclopropanols: *exo/endo*-6-hydroxybicyclo[3.1.0]hexane, *exo/endo*-8-hydroxybicyclo[5.1.0] octane, *exo/endo*-9-hydroxybicyclo[6.1.0]nonane, 2,2-dimethylcyclopropanol, 2,2,3,3-tetramethylcyclopropanol, *trans*-2,3-dimethyl cyclopropanol, *cis*-2,3-dimethyl-*cis/trans*-cyclopropanol, *cis/trans*-2,2,3-trimethylcyclopropanol, and *cis/trans*-2-phenylcyclopropanol.

The principal disadvantage of this procedure is that the olefin must be used in at least three- to fourfold excess in order to obtain reasonable yields. In case of rare olefins, or of olefins containing groups such as the carbonyl group which add organolithium compounds, other methods[7, 8] might be more advantageous. The method is also limited to the preparation of secondary cyclopropanols.

The most satisfactory procedure for obtaining cyclopropanol itself is that of Cottle[7, 9] which is also recommended for the synthesis of 1-arylcyclopropanols.[7] 1-Alkylcyclopropanols are best prepared via the corresponding acetates which are obtained by the method of Freeman [10] that involves thermolysis of a 3-acetoxy-1-pyrazolin. According to DePuy,[7] cyclopropyl acetates are best cleaved to cyclopropanols by methyllithium. However, the preparation of cyclopropyl acetates is somewhat laborious. It usually involves reactions of an olefin with ethyl diazotate—in this step the olefin must be used in excess, too—followed by a Baeyer-Villiger rearrangement of the corresponding methyl cyclopropyl ketone.[7]

The cyclopropanols, the study of whose chemistry is still in its early stages,[7, 8] show promise as useful synthetic intermediates. The chemistry of their derivatives should aid in the understanding of the nature of nucleophilic substitution on three-membered rings.[7, 8, 11]

1. Organisch-Chemisches Institut der Universität, Göttingen, Germany.
2. W. S. Johnson and W. P. Schneider, *Org. Syntheses*, Coll. Vol. 4, 132 (1963).
3. H. Baganz and L. Domaschke, *Ber.*, **91**, 650, 653 (1958).
4. H. Gross, A. Rieche, and E. Höft, *Ber.*, **94**, 544 (1961).
5. H. Gilman and L. L. Heck, *J. Am. Chem. Soc.*, **52**, 4949 (1930).
6. U. Schöllkopf, J. Paust, A. Al-Azrak, and H. Schumacher, *Ber.*, **99**, 3391 (1966).
7. C. H. DePuy, *Accounts Chem. Res.*, **1**, 33 (1968); C. H. DePuy, G. M. Dappen, K. L. Eilers, and R. A. Klein, *J. Org. Chem.*, **29**, 2813 (1964).
8. U. Schöllkopf, *Angew. Chem. Intern. Ed.*, **7**, 588 (1968).
9. J. L. Magrane, Jr. and D. L. Cottle, *J. Am. Chem. Soc.*, **64**, 484 (1942); G. W. Stahl and D. L. Cottle, *J. Am. Chem. Soc.*, **65**, 1782 (1943).
10. J. P. Freeman, *J. Org. Chem.*, **29**, 1379 (1964); for a typical example see T. S. Cantrell and H. Schechter, *J. Am. Chem. Soc.*, **89**, 5868 (1967).
11. C. H. DePuy, L. G. Schnack, and J. W. Hausser, *J. Am. Chem. Soc.*, **88**, 3343 (1966); P. von R. Schleyer, G. W. Van Dine, U. Schöllkopf, and J. Paust, *J. Am. Chem. Soc.*, **88**, 2868 (1966); U. Schöllkopf, K. Fellenberger, M. Patsch, P. von R. Schleyer, G. W. Van Dine, and T. Su, *Tetrahedron Lett.*, **3639** (1967).

NORTRICYCLANOL

(Tricyclo[2.2.1.0^{2,6}]heptan-3-ol)

Submitted by J. Meinwald, J. Crandall, and W. E. Hymans [1]
Checked by J. R. Roland and B. C. McKusick

1. Procedure

A. *Nortricyclyl acetate.* A mixture of 156 g. (1.70 moles) of bicyclo[2.2.1]hepta-2,5-diene (Note 1), 105 g. (100 ml., 1.75 moles) of glacial acetic acid, and 3 ml. of boron trifluoride etherate (Note 2) is placed in a 500-ml. flask attached to a condenser equipped with a drying tube. The mixture is heated on a steam bath for 6 hours, cooled to room temperature, and diluted with 250 ml. of ether. The ethereal solution is washed successively with two 50-ml. portions of 3N ammonia and 50 ml. of water and dried over magnesium sulfate. The ether is removed by distillation through a short column of glass helices, and the dark residue is distilled under reduced pressure to give about 200 g. of a mixture of nortricyclyl acetate and bicyclo[2.2.1]hepta-5-en-2-yl acetate as a colorless liquid, b.p. 85–95° (15 mm.) (Note 3).

The acetate mixture is dissolved in 500 ml. of chloroform (analytical reagent grade) in a 2-l. Erlenmeyer flask equipped with a thermometer and a gas-inlet tube and located in a good hood. *Caution! All operations using nitrosyl chloride should be performed in a good hood.* The solution is cooled to −10° in an ice-salt bath, and nitrosyl chloride (Note 4) is bubbled into the solution with swirling at −10° ± 3° until the color of the solution changes through bright green to a brownish green that indicates excess nitrosyl chloride. A white precipitate begins to form at this point. There is added 500 ml. of 30–60° petroleum ether, and the mixture is cooled at −10° ± 3° for an additional 15 minutes. The precipitated nitrosyl chloride adduct (Note 5) is collected by suction filtration (*Caution! Hood*). The filtrate is washed successively with two 200-ml. portions of saturated sodium carbonate solution and 500 ml. of saturated sodium chloride and dried over magnesium sulfate. The solvent is removed through a short column of glass helices, and the dark residue is distilled under reduced pressure (*Caution! Note 6*) to give 132–167 g. (52–66%) of nortricyclyl acetate as a faintly green liquid, b.p. 83–85° (13 mm.), $n^{25}D$ 1.4673–1.4681 (Note 7).

B. *Nortricyclanol.* The nortricyclyl acetate obtained above is added to a solution of 0.5 g. of sodium in 500 ml. of anhydrous methanol (analytical grade reagent). The solution is heated on a steam bath, and the methanol is slowly distilled through a short column packed with glass helices (Note 8). The residue is cooled, diluted with 250 ml. of 30–60° petroleum ether, and the solution is washed with two 50-ml. portions of water and dried over magnesium sulfate. The solvent is removed by distillation through a short column packed with glass helices, finally at 25° and water-pump pressure. The crude product, which solidifies on cooling, is sublimed at 80° (2 mm.) to yield 84–107 g. (45–57% based on bicycloheptadiene; Note 9) of nortricyclanol, m.p. 108–110°. It is pure enough for most purposes. A slightly purer product is obtained by resublimation (Note 10).

2. Notes

1. Bicycloheptadiene supplied by Shell Chemical Corporation can usually be used without purification. If the material is cloudy or contains a precipitate, it should be distilled before being used.

2. The purified grade of Eastman Organic Chemicals is satisfactory. After the procedure had been checked, the submitters found that the use of only 1 ml. of this reagent gave the specified yield more consistently.

3. The exact proportion of unsaturated acetate varies slightly but is typically 10–15% as determined by vapor-phase chromatography.

4. Nitrosyl chloride from the Matheson Company is satisfactory.

5. The adduct may be recrystallized from chloroform to give a white crystalline product, m.p. 152–153°.

6. The hot residue decomposes vigorously with the evolution of irritating gases when opened to the atmosphere. Consequently, the distillation flask should be cooled to room temperature before breaking the vacuum.

7. Vapor-phase chromatographic analysis of this product showed less than 1% of isomeric material.

8. The checkers used a 1.3-cm. x 25-cm. column of helices and removed the solvent over a period of about 5 hours.

9. The yield of nortricyclanol from nortricyclyl acetate is 82–95%.

10. Vapor-phase chromatography shows no detectable impurities under conditions where <1% of isomeric material would be easily visible. A melting point of 108–109° has been reported.[2]

3. Methods of Preparation

The addition of carboxylic acids to bicyclo[2.2.1]hepta-2,5-diene has been described by several authors;[3] the method described here is a modification of these procedures. Nortricyclanol has been prepared by the hydration of bicyclo[2.2.1]hepta-2,5-diene [4] and the solvolysis of nortricyclyl [2] and bicyclo[2.2.1]hept-2-en-5-yl [5] halides, as well as by the saponification [3a, 6] and transesterification [3c] of the corresponding esters.   The described conversion of the acetate to the alcohol is patterned after a similar procedure of Hall.[3c]

4. Merits of the Preparation

The present preparation affords high-purity nortricyclanol in good yield without the necessity of tedious purification.   It illustrates a convenient way to convert olefins to alcohols and to remove olefinic impurities from alcohols.   Nortricyclanol is of current interest in studies of highly strained ring systems.   It is readily oxidized to nortricyclanone.[7]

1. Department of Chemistry, Cornell University, Ithaca, New York.
2. J. D. Roberts, E. R. Trumbull, Jr., W. Bennett, and R. Armstrong, *J. Am. Chem. Soc.*, **72**, 3116 (1950).
3. (*a*) L. Schmerling, J. P. Luvisi, and R. W. Welch, *J. Am. Chem. Soc.*, **78**, 281 (1956); (*b*) S. Winstein and M. Shatavsky, *Chem. Ind. (London)*, 56 (1956); (*c*) H. K. Hall, Jr., *J. Am. Chem. Soc.*, **82**, 1209 (1960); (*d*) H. Krieger, *Suomen Kemistilehti*, **33B**, 183 (1960).
4. G. T. Youngblood, C. D. Trivette, Jr., and P. Wilder, Jr., *J. Org. Chem.*, **23**, 684 (1958).
5. J. D. Roberts, W. Bennett, and R. Armstrong, *J. Am. Chem. Soc.*, **72**, 3329 (1950).
6. K. Alder, F. H. Flock, and H. Wirtz, *Ber.*, **91**, 609 (1958).
7. J. Meinwald, J. Crandall, and W. E. Hymans, *Org. Syntheses*, this volume, p. 866.

NORTRICYCLANONE

(Tricyclo[2.2.1.0 ^{2,6}]heptan-3-one)

Submitted by J. Meinwald, J. Crandall, and W. E. Hymans [1]
Checked by J. R. Roland and B. C. McKusick

1. Procedure

The oxidation reagent is prepared by dissolving 70 g. (0.70 mole) of chromium trioxide in 100 ml. of water in a 500-ml.

beaker. The beaker is immersed in an ice bath, and 112 g. (61 ml., 1.10 moles) of concentrated (18M) sulfuric acid followed by 200 ml. of water is added cautiously with manual stirring. The solution is cooled to 0–5°.

A solution of 110 g. (1.00 mole) of nortricyclanol (Note 1) in 600 ml. of acetone (analytical reagent grade) is cooled to 0–5° in a 2-l. three-necked flask immersed in an ice bath and equipped with an efficient mechanical stirrer, a thermometer, and a dropping funnel with a pressure-equalizing arm. The cooled oxidation reagent prepared above is poured into the dropping funnel, and the reagent is added with vigorous stirring, at a rate to maintain the temperature of the reaction mixture at about 20°. The stirring is continued for 3 hours after the addition is completed.

Sodium bisulfite is added in small portions until the brown color of chromic acid is gone from the upper layer of the two-phase mixture. The top layer is decanted, and the dense, green, lower layer is extracted with 200 ml. of 30–60° petroleum ether. Combination of this extract with the original upper layer causes a separation into two phases. The lower phase is drawn off and added to the original lower phase, which is then extracted with three 200-ml. portions of 30–60° petroleum ether (Note 2). The extracts are combined, washed successively with two 50-ml. portions of saturated sodium chloride, two 50-ml. portions of saturated sodium bicarbonate solution, and 50 ml. of saturated sodium chloride solution, and dried over magnesium sulfate. The solvent is removed by distillation through a short column containing glass helices, and the residue is distilled under reduced pressure to give 85–95 g. (79–88%) of nortricyclanone, b.p. 103–105° (77 mm.), n^{26}D 1.4873 (Notes 3 and 4).

2. Notes

1. The preparation of nortricyclanol is described in this volume, p. 863. The crude, unsublimed nortricyclanol is a satisfactory starting material.

2. Sufficient time between extraction and separation of layers must be allowed, for the organic layer separates slowly from the thick aqueous phase.

3. The product was shown to contain $<1\%$ of the starting material and no other detectable impurity by vapor-phase chromatographic analysis.

4. The reported [2] properties of this material are b.p. 78–79° (24 mm.), n^{25}D 1.4878.

3. Methods of Preparation

Nortricyclanone has been prepared by oxidation of nortricyclanol using chromic acid in acetic acid [2, 3] and using a modified Oppenauer reaction.[4] The present procedure is based on the Jones modification [5] of chromic acid oxidations.

4. Merits of the Preparation

This preparation illustrates a general and convenient way of oxidizing secondary alcohols to ketones. The novel feature of the reaction is represented by acetone solvent which affects markedly the properties of the oxidizing agent. The reaction is very rapid (if not instantaneous), and the yields are high, the reagent rarely attacking unsaturated centers. The procedure is applicable to acetylenic carbinols, allyl and other unsaturated alcohols, and saturated carbinols. The main limitation is the low solvent power of acetone. Another example of the Jones oxidation is given on p. 310 of this volume.

An attractive alternative to the Jones oxidation is oxidation with chromic acid in the two-phase system, water-ether, the details of which were reported recently;[6] however, this method has not been applied to the preparation of nortricyclanone.

The preparation illustrates the utility of the reaction in preparing the highly strained and reactive nortricyclanone, a compound of current interest in studies of strained ring systems.

1. Department of Chemistry, Cornell University, Ithaca, New York.
2. H. K. Hall, Jr., *J. Am. Chem. Soc.*, **82**, 1209 (1960).
3. J. D. Roberts, E. R. Trumbull, Jr., W. Bennett, and R. Armstrong, *J. Am. Chem. Soc.*, **72**, 3116 (1950); H. Krieger, *Suomen Kemistilehti*, **32B**, 109 (1959).
4. K. Alder, F. H. Flock, and H. Wirtz, *Ber.*, **91**, 609 (1958).
5. R. G. Curtis, I. Heilbron, E. R. H. Jones, and G. F. Woods, *J. Chem. Soc.*, 457 (1953).
6. H. C. Brown, C. P. Garg, and K.-T. Liu, *J. Org. Chem.*, **36**, 387 (1971).

$\Delta^{1(9)}$-OCTALONE-2

[2(3H)-Naphthalenone, 4,4a,5,6,7,8-hexahydro-]

Submitted by ROBERT L. AUGUSTINE and JOSEPH A. CAPUTO [1]
Checked by WILLIAM G. DAUBEN and JEFFREY N. LABOVITZ

1. Procedure

A. $\Delta^{1(9)}$-*Octalone-2* and $\Delta^{9(10)}$-*octalone-2*. In a 2-l., three-necked, round-bottomed flask equipped with a sealed stirrer, a condenser, and a dropping funnel is placed a solution of 102 g. (0.61 mole) of 1-morpholino-1-cyclohexene [2] (Note 1) in 600 ml. of purified dioxane (Notes 2 and 3). To this stirred solution is added 45 g. (0.64 mole) of freshly distilled methyl vinyl ketone at such a rate that the addition requires approximately 1 hour. The resulting solution is heated under reflux for 4 hours, after which time 750 ml. of water is added, and the heating under reflux is continued for an additional 10–12 hours. The solution is cooled to room temperature and poured into 1 l. of water. The resulting mixture is extracted four times with 500-ml. portions of ether. The combined ether extracts are washed three times with 250-ml. portions of $3N$ hydrochloric acid, twice with 100-ml. portions of a saturated aqueous sodium bicarbonate solution, once with a 250-ml. portion of water, once with a 200-ml. portion of a saturated aqueous sodium chloride solution, and dried over anhydrous magnesium sulfate. The mixture is filtered, the ethereal filtrate evaporated, and the residual octalones distilled through a short column. The yield is 54–59 g. (59–65%) of an octalone mixture (Note 4), b.p. 75–78° (0.2 mm.), 101–103° (2 mm.).

B. $\Delta^{1(9)}$-*Octalone-2*. A solution of 35 g. (0.23 mole) of the above octalone mixture in 200 ml. of 60–110° petroleum ether is cooled to $-80°$ in an acetone-dry ice bath and kept at this temperature for 1 hour. The crystalline $\Delta^{1(9)}$-octalone-2 is filtered by suction through a jacketed sintered-glass funnel kept at $-80°$. The residue is washed with 100 ml. of cold petroleum ether, removed from the funnel, and recrystallized a second time in the same way. After the second recrystallization the white crystals are removed from the funnel, allowed to melt by warming to room temperature, and distilled. The yield of purified $\Delta^{1(9)}$-octalone-2 (Note 5), b.p. 143–145° (15 mm.), is 20–25 g. (34–46% based on starting enamine). The petroleum ether mother liquors can be distilled to yield a fraction boiling at 143–145° (15 mm.) which is enriched in $\Delta^{9(10)}$-octalone-2.

2. Notes

1. The enamine should be utilized as soon as possible after distillation.

2. A suitably purified dioxane can be obtained by distillation of reagent grade dioxane from lithium aluminum hydride.

3. Absolute ethanol and dry benzene are also useful solvents for enamine reactions. In this instance, however, the use of these solvents results in a lower yield of octalones.

4. The octalone mixture contains 10–20% of the $\Delta^{9(10)}$-isomer. This mixture may be used as such for many purposes.

5. The purified octalone still contains 1–3% of the $\Delta^{9(10)}$-isomer which cannot be removed even on further crystallization.

3. Methods of Preparation

The method of preparation used here is styled after the general procedure described for the reaction of enamines with electrophilic olefins.[3] $\Delta^{1(9)}$-Octalone-2 also has been prepared by condensation of 4-diethylamino-2-butanone with cyclohexanone; by condensation of 2-diethylaminomethylcyclohexanone with ethyl acetoacetate;[4] by condensation of methyl vinyl ketone with cyclohexanone;[5] by condensation of 4-oxo-1,1-dimethylpiperidinium salts with 2-carbethoxycyclohexanone;[6] by the oxidation of α-decalones;[7] and by the reduction of 6-methoxytetralin.[8]

4. Merits of the Preparation

The present procedure is a general method for the preparation of monoalkylated ketones from enamines of aldehydes and ketones with electrophilic olefins.[3]   There are many advantages in this method of alkylation. Generally only monoalkylation occurs, even when such reactive species as acrylonitrile are used; and, when a cyclic ketone like 2-methylcyclohexanone is used, reaction occurs only at the lesser substituted center. In a general base-catalyzed reaction, substitution occurs on the more substituted center.

Another advantage of this method is that no catalyst is needed for the addition reaction; this means that the base-catalyzed polymerization of the electrophilic olefin (*i.e.*, α,β-unsaturated ketones, esters, etc.) is not normally a factor to contend with, as it is in the usual base-catalyzed reactions of the Michael type. It also means that the carbonyl compound is not subject to aldol condensation which often is the predominant reaction in base-catalyzed reactions.   An unsaturated aldehyde can be used only in a Michael addition reaction when the enamine method is employed.

1. Department of Chemistry, Seton Hall University, South Orange, New Jersey.
2. S. Hünig, E. Löcke, and W. Brenninger, this volume, p. 808.
3. G. Stork, A. Brizzolara, H. Landesman, J. Szmuszkovicz, and R. Terrell, *J. Am. Chem. Soc.*, 85, 207 (1963); M. E. Kuehne, "Enamines in Organic Synthesis," in A. G. Cook (editor), "Enamines," Marcel Dekker, Inc., New York, 1969, Chapter 8.
4. E. C. DuFeu, F. J. McQuillin, and R. Robinson, *J. Chem. Soc.*, 53 (1937).
5. E. Bergmann, R. Ikan, and H. Weiler-Feilchenfeld, *Bull. Soc. Chim. France*, 290 (1957).
6. H. M. E. Cardwell and F. J. McQuillin, *J. Chem. Soc.*, 708 (1949).
7. H. H. Zeiss and W. B. Martin, Jr., *J. Am. Chem. Soc.*, 75, 5935 (1953).
8. A. J. Birch, *J. Chem. Soc.*, 593 (1946).

OCTANAL*

$$n\text{-}C_8H_{17}I + (CH_3)_3N{\to}O \xrightarrow[\text{CHCl}_3]{} n\text{-}C_7H_{15}CHO + (CH_3)_3\overset{+}{N}H \quad I^-$$

Submitted by Volker Franzen [1]
Checked by Claibourne Smith and V. Boekelheide

1. Procedure

To a solution of 30 g. (0.4 mole) of anhydrous trimethylamine oxide (Note 1) in 100 ml. of dry chloroform placed in a 250-ml. three-necked, round-bottomed flask fitted with a reflux condenser, a stirrer, and a dropping funnel with protection against moisture is added 48.0 g. (0.2 mole) of n-octyl iodide (Note 2) dropwise with stirring. At the beginning of the addition the flask is warmed to 40–50° on a steam bath to initiate the reaction. The start of the reaction can be recognized by the evolution of heat, and the rate of addition can then be adjusted to maintain the temperature around 50°. Overall the addition requires about 20–30 minutes. When addition is complete, the solution is boiled under reflux for another 20 minutes. Then the solution is cooled, and 110 ml. of 2N aqueous sulfuric acid solution is added with stirring. The chloroform layer is separated and washed successively with water, 2N aqueous sodium carbonate solution, and again with water. After the chloroform solution has been dried over sodium sulfate, it is concentrated under reduced pressure, and then the residue is distilled. The first distillation (Note 3) is carried out at atmospheric pressure and gives 12.0–12.5 g. of a crude oil, b.p. 155–165°. Redistillation using a simple Vigreux column gives 10.6–11.0 g. (41.5–43%) of a colorless oil, b.p. 69–71° (19 mm.), n^{26}D 1.4167.

2. Notes

1. Trimethylamine oxide is normally available as a hydrate, and for the present preparation it is necessary to convert it to its anhydrous form. A convenient way of doing this is as follows.

A solution of 45.0 g. of trimethylamine oxide dihydrate (supplied by Beacon Chemicals) is dissolved in 300 ml. of warm dimethylformamide and placed in a three-necked flask set up for distillation. At atmospheric pressure the flask is heated and solvent distilled off until the boiling point reaches 152–153°. Then the pressure is reduced using a water aspirator, and the remainder of the solvent is distilled. At the end of the distillation the temperature of the bath is slowly raised to 120°. The residual anhydrous trimethylamine oxide (30 g.) can be dissolved in 100 ml. of chloroform and may remain in the same flask for use in the present preparation.

2. In place of n-octyl iodide other derivatives such as n-octyl bromide, n-octyl p-toluenesulfonate, and n-octyl chlorosulfonate can be substituted; the submitter reports that the yields of octanal in these cases are comparable.

3. At this stage the product is a waxy semisolid, presumably a trimer or polymer of octanal, and the higher temperature of an atmospheric distillation is needed to generate the monomeric octanal.

3. Methods of Preparation

Octanal has been prepared by the reduction of caprylonitrile with hydrogen chloride and stannous chloride,[2] by the passage of a mixture of caprylic acid and formic acid over titanium dioxide[3] or manganous oxide,[4] by dehydrogenation of 1-octanol over copper,[5] and by oxidation of 1-octanol.[6]

4. Merits of the Preparation

The conversion of alkyl halides to aldehydes is a synthetic step of broad utility. Earlier procedures for such conversions involving the use of dimethyl sulfoxide[7] or pyridine N-oxide[8] have worked best with activated alkyl halides, although Kornblum has described a modification of the dimethyl sulfoxide procedure for use with ordinary aliphatic halides.[9] The present procedure using trimethylamine oxide avoids some of the complicating side reactions of pyridine N-oxide and is useful with ordinary aliphatic halides.[10,11]

1. Max-Planck Institute for Medical Research, Heidelberg, Germany.
2. H. Stephen, *J. Chem. Soc.*, **127**, 1875 (1925).
3. P. Sabatier and A. Maihle, *Compt. Rend.*, **154**, 561 (1912).
4. P. Sabatier and A. Maihle, *Compt. Rend.*, **158**, 985 (1914).
5. R. H. Pickard and J. Kenyon, *J. Chem. Soc.*, **99**, 56 (1911).
6. R. P. A. Sneeden and R. B. Turner, *J. Am. Chem. Soc.*, **77**, 190 (1955).
7. N. Kornblum, J. W. Powers, G. J. Anderson, W. J. Jones, H. O. Larson, O. Levand, and W. M. Weaver, *J. Am. Chem. Soc.*, **79**, 6562 (1957).
8. W. Feely, W. L. Lehn, and V. Boekelheide, *J. Org. Chem.*, **22**, 1135 (1957).
9. N. Kornblum, W. J. Jones, and G. J. Anderson, *J. Am. Chem. Soc.*, **81**, 4113 (1959).
10. V. Franzen and S. Otto, *Ber.*, 94, 1360 (1961).
11. Cf. L. D. Bergelson and M. M. Shemyakin, *Angew. Chem. Intern. Ed. Engl.*, 3, 250 (1964).

2-OXA-7,7-DICHLORONORCARANE

(7,7-Dichloro-2-oxabicyclo[4.1.0]heptane)

$$+ Cl_3CCO_2C_2H_5 + NaOCH_3 \rightarrow$$

$$+ NaCl + \text{mixed carbonic acid esters}$$

Submitted by WILLIAM E. PARHAM, EDWARD E. SCHWEIZER, and SIGMUND A. MIERZWA, JR.[1]
Checked by WILLIAM G. DAUBEN and RICHARD ELLIS.

1. Procedure

In a 1-l. three-necked flask (Note 1) is placed 50 g. (0.92 mole) of sodium methoxide (Notes 2 and 3). The flask is temporarily stoppered and then fitted with a nitrogen inlet tube, a sealed stirrer, and a 250-ml. pressure-equalized dropping funnel carrying a calcium chloride tube.

The dropping funnel is removed, and 67.4 g. (0.8 mole) of dihydropyran (Note 4) and 600 ml. of dry, olefin-free pentane (Note 5) are added successively. The light-yellow solution is stirred for 15 minutes in an ice-water bath, and then 164.8 g. (0.86 mole) of ethyl trichloracetate (Note 6) is added from the dropping funnel over a period of 3–4 minutes. The dropping funnel is removed and replaced by a calcium chloride tube.

The reaction mixture is stirred for 6 hours (Note 7) at the ice-bath temperature and then is allowed to warm to room temperature overnight while the stirring is continued. During this period the color of the mixture changes from yellow-orange to brown.

Water (200 ml.) is added, the mixture is transferred to a 2-l. separatory funnel and shaken. The layers are separated and the aqueous layer is extracted twice with 100-ml. portions of petroleum ether (b.p. 60–68°). The organic layers are combined and dried over anhydrous magnesium sulfate.

The solvent is removed at a maximum water-bath temperature of 60° and a minimum pressure of 30 mm. (Note 8). The residual liquid is distilled through a 25-cm. Vigreux column, and the fraction boiling at 74–76/8 mm. is collected (Note 9). The yield of 2-oxa-7,7-dichloronorcarane is 91–100 g. (68–75%), n_D^{25} 1.4974–1.4983.

2. Notes

1. All the glassware used is dried in an oven at 120°. The reaction vessel is arranged so that all the steps prior to hydrolysis are carried out under a constant positive pressure of dry nitrogen.

2. The sodium methoxide was obtained from Matheson, Coleman and Bell Co. The submitters carried out all operations with this reagent in a dry-box under a stream of dry nitrogen. Sodium ethoxide and potassium *tert*-butoxide have been successfully substituted for sodium methoxide;[2] the choice of sodium methoxide is here principally one of convenience. With other olefins, the choice of alkoxide depends upon the boiling points of the dichlorocarbene adduct and the corresponding dialkyl carbonates.

3. The checkers did not use a dry-box but simply rapidly weighed the sodium methoxide on a balance which was constantly swept with a stream of dry nitrogen from a large inverted funnel and then transferred the solid directly to the reaction flask.

4. The dihydropyran (Matheson) is dried over sodium carbonate and distilled once prior to use.

5. Technical grade pentane (Eastman Kodak) is freed of olefins by five successive washes each with 100 ml. of concentrated

sulfuric acid per liter of pentane. The olefin-free pentane is then washed with an equal amount of water, dried over magnesium sulfate, distilled, and stored over sodium wire.

6. Ethyl trichloracetate (Eastman Kodak) is distilled prior to use.

7. The nitrogen flow must be slow enough to prevent significant loss of the pentane by evaporation.

8. A rotary evaporator is a convenient apparatus for this operation.

9. Distillation at significantly higher pressures results in increased decomposition.

3. Methods of Preparation

The present procedure is that described by the submitters.[3]

4. Merits of the Preparation

The generation of dichlorocarbene for addition to olefins has been realized by the use of chloroform and alkali metal alkoxides[4,5] (preferably potassium *tert*-butoxide), sodium trichloroacetate,[6] butyllithium and bromotrichloromethane,[7] phenyl(trichloromethyl)mercury,[8] and the reaction of an ester of trichloracetic acid with an alkali metal alkoxide.[2,3] The latter method, which is here illustrated by the preparation of 2-oxa-7,7-dichloronorcarane, and procedures using phenyl-(trichloromethyl)mercury generally give higher yields of adducts.

1. Department of Chemistry, University of Minnesota, Minneapolis, Minn. 55455.
2. W. E. Parham and E. E. Schweizer, *J. Org. Chem.*, 24, 1733 (1959).
3. E. E. Schweizer and W. E. Parham, *J. Am. Chem. Soc.*, 82, 4085 (1960).
4. W. E. Doering and A. K. Hoffmann, *J. Am. Chem. Soc.*, 76, 6162 (1954).
5. H. E. Winberg, *J. Org. Chem.*, 24, 264 (1959).
6. W. M. Wagner, *Proc. Chem. Soc.*, 229 (1959).
7. W. T. Miller and C. S. Y. Kim, *J. Am. Chem. Soc.*, 81, 5008 (1959).
8. D. Seyferth, J. M. Burlitch and J. K. Heeren, *J. Org. Chem.*, 27, 1491 (1962); cf. T. J. Logan, this volume, p. 969.

D-2-OXO-7,7-DIMETHYL-1-VINYLBICYCLO[2.2.1]HEPTANE

(2-Norbornanone, 7,7-dimethyl-1-vinyl-, D-)

Submitted by Nikolaus Fischer[1] and G. Opitz[2]
Checked by Hermann Ertl, Ian D. Rae, and Peter Yates

1. Procedure

Caution! Diazomethane is both explosive and poisonous, and all operations involving its preparation and use must be carried out in a hood. Follow the directions for its handling given in earlier volumes.[3,4]

In a 500-ml. three-necked flask equipped with a mechanical stirrer, a dropping funnel, and a reflux condenser fitted with a potassium hydroxide drying tube are placed 7.0 g. (0.069 mole) of triethylamine (Note 1) and a solution of 3.15 g. (0.075 mole) of diazomethane in 200 ml. of ether (Note 2). The flask is cooled in an ice bath, and a solution of 13.0 g. (0.052 mole) of D-camphor-10-sulfonyl chloride (Note 3) in 75 ml. of anhydrous ether is added dropwise over a period of 1 hour. Triethylamine hydrochloride slowly precipitates. The reaction mixture is stirred for an additional 30 minutes and then concentrated to *ca.* 150 ml. under reduced pressure (water aspirator) with continued stirring to remove the excess of diazomethane. The mixture is filtered under reduced pressure; the precipitate is washed with 50 ml. of

anhydrous ether, giving 6.7 g. (94%) of triethylamine hydrochloride. The combined filtrate and washings are freed of solvent
on a rotary evaporator at room temperature to give 10.7 g.
(90%) of crude episulfone, m.p. 76–85° (dec.) (Note 4). This is
used without purification in the next step; it can be purified by
crystallization from a little methanol at −20°. This gives colorless episulfone, m.p. 83–85° (dec.), $[\alpha]^{24}_D$ −6.72° (methanol,
$c = 3.20$); infrared bands at 3070, 1300, and 1170 cm.$^{-1}$ (Note 5).

The crude episulfone (3.0 g.) is placed in a 10-ml. round-
bottomed flask fitted with a reflux condenser and is heated at
95° for 30 minutes, when it decomposes with loss of sulfur dioxide.
The reflux condenser is replaced with a distillation head (Note 6),
and the yellow residue is distilled under reduced pressure (water
aspirator). D-2-Oxo-7,7-dimethyl-1-vinylbicyclo[2.2.1]heptane
(1.7 g., 71% based on sulfonyl chloride), b.p. 95–96° (10 mm.),
distills at a bath temperature of 110–120°. Sublimation at 60°
(0.01 mm.) gives the olefin as colorless, waxy crystals, m.p.
64–65°, $[\alpha]^{25}_D$ +16.35° (methanol, $c = 2.16$); infrared band at
1650 cm.$^{-1}$.

2. Notes

1. The triethylamine was purified by treatment with naphthyl
isocyanate and distilled; the distillate was stored over sodium
wire.

2. The ethereal diazomethane was prepared from N-nitro-
somethylurea and aqueous potassium hydroxide and dried over
potassium hydroxide pellets for 2–3 hours. The solid potassium
hydroxide was replaced once or twice to ensure complete dryness.
The checkers used the procedure of Arndt [5] for this preparation
and for the estimation of the diazomethane.

3. D-Camphor-10-sulfonyl chloride can be prepared from
commercially available D-camphor-10-sulfonic acid and phosphorus pentachloride [6] or thionyl chloride.[7] The checkers used
the following procedure. D-Camphor-10-sulfonic acid (50.0 g.)
was added slowly to 50 g. of thionyl chloride. The mixture was
boiled under reflux until homogeneous and then for a further
2 hours. The solution was cooled and poured onto ca. 500 g. of

crushed ice. The crude product (51.1 g., 95%) was filtered and crystallized twice from hexane to give the sulfonyl chloride, m.p. 65–67.5°; yield, 35.5 g. (66%).

4. The checkers obtained higher yields (94–97%) of less pure material [m.p. 50–78° (dec.)].

5. The checkers observed that the episulfone decomposed slowly at room temperature and, on one occasion, during evaporation at 15°.

6. The checkers found it advisable to use an apparatus with a wide-bore side arm (18 mm.) without a condenser.

3. Methods of Preparation

The only method reported for the preparation of D-2-oxo-7,7-dimethyl-1-vinylbicyclo[2.2.1]heptane is that of the present procedure.[8,9]

4. Merits of the Preparation

The method is of general applicability[8,9] for the synthesis of olefins. Other sulfonyl chlorides, RCH_2SO_2Cl, have been used where R = H, C_2H_5, C_6H_5, and $C_6H_5CH_2$; other diazoalkanes that have been used are diazoethane and 1-diazo-2-methylpropane. In all cases the olefins form without double-bond migration. A review[9] of the method is available.

1. Department of Chemistry, Louisiana State University, Baton Rouge, La. 70803.
2. Chemisches Institut der Universität, Tübingen, Germany.
3. T. J. de Boer and H. J. Backer, *Org. Syntheses,* Coll. Vol. 4, 250 (1963).
4. J. A. Moore and D. E. Reed, this volume, p. 351.
5. F. Arndt, *Org. Syntheses,* Coll. Vol. 2, 165 (1943).
6. P. D. Bartlett and L. H. Knox, this volume, p. 196.
7. S. Smiles and T. P. Hilditch, *J. Chem. Soc.,* 91, 519 (1907).
8. G. Opitz and K. Fischer, *Angew. Chem.,* 77, 41 (1965); *Angew. Chem. Intern. Ed. Engl.,* 4, 70 (1965) [Correction: Compounds (1f) and (2i) were obtained from 1-diazo-2-methylpropane (R^2=H, R^3=CH(CH$_3$)$_2$) and not from 2-diazobutane (R^2=CH$_3$, R^3=C$_2$H$_5$)].
9. N. H. Fisher, *Synthesis,* 393 (1970).

PALLADIUM CATALYST FOR PARTIAL REDUCTION OF ACETYLENES[1]

$$H_2PdCl_4 + H_2O + CaCO_3 \rightarrow PdO/CaCO_3 + 4HCl$$

$$PdO/CaCO_3 + HCO_2H \rightarrow Pd/CaCO_3 + H_2O + CO_2$$

$$Pd/CaCO_3 \xrightarrow{Pb(OCOCH_3)_2} \text{conditioned Pd/CaCO}_3$$

Test for selectivity:

$$C_6H_5C{\equiv}CH + H_2 \xrightarrow[\text{Quinoline}]{\text{Conditioned Pd/CaCO}_3} C_6H_5CH{=}CH_2$$

Submitted by H. Lindlar and R. Dubuis [2]
Checked by F. N. Jones and B. C. McKusick

1. Procedure

Palladous chloride (1.48 g., 0.0083 mole) (Note 1) is placed in a 10-ml. Erlenmeyer flask, and 3.6 ml. (0.043 mole) of 37% hydrochloric acid is added. The flask is shaken at about 30° until the palladous chloride is dissolved. The chloropalladous acid solution is transferred to a 150-ml. beaker with 45 ml. of distilled water (Note 2). The beaker is equipped with a pH meter and a magnetic or mechanical stirrer. The pH of the stirred solution is brought to 4.0–4.5 by slow addition of aqueous $3N$ sodium hydroxide from a buret. A precipitate may form at high local concentrations of sodium hydroxide, but it dissolves on further stirring. The solution is diluted to approximately 100 ml. in a graduated cylinder and placed in a 200-ml. or 250-ml., three-necked, round-bottomed flask equipped with a mechanical stirrer and a thermometer and partly immersed in a bath of oil or water. Precipitated calcium carbonate (18 g.) (Note 3) is added.   The well-stirred suspension is heated to 75–85° and held at this temperature until all the palladium has precipitated, as indicated by loss of color from the solution; this takes about 15 minutes.   With the mixture still at 75–85°, 6.0 ml. of sodium formate solution (about 0.7N) (Note 4) is added.   During the addition CO_2 escapes and the catalyst turns from brown to gray; rapid stirring is essential to keep the mixture from foaming over. An additional 4.5 ml. of the sodium formate solution is added, and the reduction is completed by stirring the mixture at 75–85° for 40 minutes.   The catalyst, which is now black, is separated on a 10-cm. Büchner funnel (Note 5) and washed with eight 65-ml. portions of water.

The moist catalyst is placed in a 200-ml. or 250-ml. round-bottomed flask equipped as described above.   Water (60 ml.) and 18 ml. of a 7.7% solution of lead acetate (Note 6) are added. The slurry is stirred and heated at 75–85° for 45 minutes.   The catalyst is separated on a 10-cm. Büchner funnel, washed with four 50-ml. portions of water, sucked as dry as possible, and dried in an oven at 60–70° (Note 7).   The dried catalyst, a dark gray powder, weighs 19–19.5 g. (Note 8).

To establish that the catalyst is active and selective, it is convenient to test it by quantitative hydrogenation of phenylacetylene to styrene.   The reaction flask of a low-pressure hydrogenation apparatus (Note 9) is charged with 2.04 g. (0.0200 mole) of phenylacetylene, 0.10 g. of the palladium catalyst, 1.0 ml. of quinoline (Note 10), and 15 ml. of olefin-free petroleum ether (b.p. 80–105°) or hexane (Note 11).   The apparatus is evacuated, and hydrogen is admitted to a pressure slightly above 1 atm. Stirring or shaking is started, causing rapid absorption of hydrogen.   The hydrogen pressure is kept close to 1 atm.   Absorption of the first 0.0200 mole of hydrogen requires 10–90 minutes, depending on the activity of the catalyst.   Hydrogen absorption then abruptly slows but does not stop.   In synthetic work it is desirable to stop the reaction soon after the required amount of hydrogen has been absorbed.

2. Notes

1. Palladium chloride, Engelhard, 60% Pd, was obtained from Engelhard Industries, 113 Astor Street, Newark 14, New Jersey.

2. All water used in the procedure should be distilled or deionized water or chlorine-free tap water.

3. A commercial grade of "precipitated" (not "powdered" or "prepared") calcium carbonate is satisfactory, if it can be filtered easily. The checkers used Fisher Catalog No. C-62 Calcium Carbonate, u.s.p. (Precipitated Chalk).

4. To prepare the sodium formate solution, a filtered solution of 15 g. (0.14 mole) of anhydrous sodium carbonate in 80 ml. of water is diluted to 120 ml. Approximately 4 ml. (4.9 g., 0.10 mole) of 99% formic acid (a commercial grade) is then added dropwise until the solution is weakly alkaline to phenolphthalein.

5. The checkers found a funnel with a fritted disk convenient. Thorough washing is essential.

6. The solution is prepared by dissolving 9.0 g. (0.024 mole) of a commercial grade of lead acetate, $Pb(OCOCH_3)_2 \cdot 3H_2O$, in 100 ml. of water.

7. The checkers dried the catalyst for 2 hours in a vacuum oven at 60° (1 mm.).

8. The catalyst has been stored for more than 3 years with no loss in activity.

9. The hydrogen pressure and the design of the apparatus are not critical; any apparatus in common use is satisfactory. The checkers used an apparatus having a magnetically stirred reaction flask as described by Wiberg.[3]

10. Less quinoline (as little as 5% of the weight of the catalyst) often suffices for hydrogenations of this type. The submitters report that 2,2'-(ethylenedithio) diethanol (available from Fluka AG., Chemische Fabrik, CH-9470 Buchs, Switzerland) has proved to be more effective than quinoline in most selective hydrogenations. Between 1 and 10 parts per thousand calculated on the weight of the catalyst is usually sufficient (private communication from H. Lindlar and P. P. Gutmann).

11. Benzene, toluene, or acetone may be used as solvent for substances insoluble in paraffins. Alcohols are usually unsatisfactory media.

3. Methods of Preparation

The preparation of the catalyst is a slight modification of the original procedure.[4]

4. Merits of the Preparation

This form of palladium can be used for the hydrogenation of almost any triple bond to the double bond. Reduction of doubly substituted acetylenes gives *cis* olefins.

1. Checkers' note: Among organic chemists this catalyst is commonly called "Lindlar catalyst."
2. Chemical Research Department, F. Hoffmann-LaRoche & Co. Ltd., Basel, Switzerland.
3. K. B. Wiberg, "Laboratory Technique in Organic Chemistry," McGraw-Hill Book Company, New York, 1960, pp. 228–230.
4. H. Lindlar, *Helv. Chim. Acta*, **35**, 446 (1952).

[2.2]PARACYCLOPHANE*

(Tricyclo[8.2.2.2 ^{4,7}]hexadeca-4,6,10,12,13,15-hexaene)

CH_3—⟨⟩—CH_2 Br + $(CH_3)_3$N →

CH_3—⟨⟩—CH_2—$\overset{+}{N}(CH_3)_3$ · Br⁻

2CH_3—⟨⟩—$CH_2\overset{+}{N}(CH_3)_3$ Br⁻ + Ag_2O →

2CH_3—⟨⟩—$CH_2\overset{+}{N}(CH_3)_3$ OH⁻ + 2AgBr

2CH_3—⟨⟩—$CH_2\overset{+}{N}(CH_3)_3$ OH⁻ →

CH_2—⟨⟩—CH_2 + 2$(CH_3)_3$N

CH_2—⟨⟩—CH_2 + 2H_2O

Submitted by H. E. WINBERG and F. S. FAWCETT.[1]
Checked by WILLIAM E. PARHAM, WAYLAND E. NOLAND, and THOMAS A. CHAMBERLIN.

1. Procedure

Caution! This preparation should be conducted in a hood to avoid exposure to trimethylamine and to α-bromo-p-xylene.

A. *p-Methylbenzyltrimethylammonium bromide*. In a 1-l. three-necked flask equipped with a stirrer, a reflux condenser provided with a Drierite drying tube, and a gas inlet tube about 1 cm. above the surface of the liquid are placed 600 ml. of dry ether and 100 g. (0.54 mole) of α-bromo-*p*-xylene (Note 1). The flask is cooled in an ice-water bath with stirring. A dry, weighed trap is cooled in a mush of Dry Ice-acetone, and 50 g. (0.85 mole) of liquid trimethylamine is condensed into the trap. A boiling chip is added to the cold trimethylamine, the trap is connected to the gas inlet tube and is then removed from the cooling bath. The trimethylamine is allowed to distil into the flask during a period of 2 hours, during which time *p*-methylbenzyltrimethylammonium bromide separates as a white solid (Note 2). The resulting pasty mixture is allowed to stand overnight at room temperature. The bromide is collected on a Büchner funnel, the transfer being aided by re-use of the filtrate. The bromide is washed on the filter with 200 ml. of dry ether and dried in air to give 125–130 g. (95–99%) of product (Note 3).

B. *[2.2]Paracyclophane*. *p*-Methylbenzyltrimethylammonium bromide (24.4 g., 0.10 mole) is dissolved in 75 ml. of water. Silver oxide (23 g.) (Note 4) is added, and the mixture is stirred at room temperature for 1.5 hours. The mixture is filtered, the solid is rinsed with 40 ml. of water, and the combined liquids are collected and then dried azeotropically as follows. A 500-ml. three-necked flask is equipped with a Tru-bore stirrer with a paddle of "Teflon" tetrafluoroethylene resin, a Dean-Stark water separator[2] attached to a reflux condenser, and a heating mantle. In the flask are placed 300 ml. of toluene, 0.5 g. of phenothiazine (Note 5), and the above aqueous solution containing the quaternary ammonium hydroxide. The mixture is stirred and heated under reflux during about 3 hours, the water being separated as it collects in the separator. When the water has been removed, decomposition occurs, as indicated by trimethylamine evolution and the separation of solid polymer. Heating and stirring are continued for 1.25 hours, after which time the evolution of trimethylamine has virtually ceased. The mixture is cooled, the solid is separated by filtration, and the somewhat gelatinous solid is extracted overnight in a Soxhlet apparatus (Note 6) employing the toluene used in the azeotropic drying step. After extraction there remains 5.7–6.7 g. of air-dried insoluble poly-*p*-xylylene. The toluene extract is concentrated to dryness under reduced pressure, and the solid residue is washed with three 10-ml. portions of acetone. Sublimation of the remaining solid at 0.5–1.0 mm. (temperature of oil bath 150–160°) gives a sublimate of 1.0–1.1 g. (10–11%; Note 7) of white, crystalline [2.2]paracyclophane, m.p. 284–287° (sealed capillary tube).

2. Notes

1. The α-bromo-*p*-xylene was obtained from Eastman Organic Chemicals; it melted at 35–36.5°.

2. A convenient, alternative procedure consists in slowly passing trimethylamine directly from a cylinder through a trap into the reaction mixture until an excess is present.

3. *p*-Methylbenzyltrimethylammonium bromide is not noticeably hygroscopic. The crude product, after it has been dried at 80° under reduced pressure over phosphorus pentoxide, melts at 197–199°. Recrystallization from absolute ethanol followed by similar drying gives crystals melting at 199–200°. Less thoroughly dried samples show lower and erratic melting points.

4. The silver oxide used may be the commercially available material or that freshly prepared by adding, with stirring, 8.8 g. of sodium hydroxide in 80 ml. of water to a solution of 34 g. of silver nitrate in 200 ml. of water. The precipitate is collected by filtration and washed with water to remove the bulk of the alkali. The wet cake is used directly for preparation of the quaternary hydroxide. The strongly basic quaternary hydroxide solution should be protected from excessive exposure to air because of carbon dioxide absorption.

5. The addition of a polymerization inhibitor appears to increase the amount of paracyclophane formed in the reaction.

6. The polymer is bulky when swollen by the toluene; a Soxhlet thimble 12 cm. long and 145 mm. in diameter is used. Additional toluene may be used in the extraction.

7. The checkers' yields are reported; the submitters report yields of 1.75–2.05 g. (17–19%).

3. Methods of Preparation

The above procedures are essentially those described by Winberg, Fawcett, Mochel, and Theobald.[3]   Equally good results have been obtained starting with α-chloro-p-xylene, but the hygroscopic nature of p-methylbenzyltrimethylammonium chloride makes this intermediate less convenient to use than the bromide. [2.2]Paracyclophane has been isolated from the pyrolysis of p-xylene[4] and by dimerization of p-xylylene.[5]   Paracyclophanes have been synthesized by intramolecular Wurtz reactions at high dilution.[6]

4. Merits of Preparation

This reaction, a 1,6-elimination of the Hofmann type, gives [2.2]paracyclophane from readily available starting materials without requiring complex equipment or manipulations, and, accordingly, it is probably the most convenient method of preparing [2.2]paracyclophane.   This substance is of interest because of its unusual geometrical features.[6]   The method is fairly general.   Thus in addition to the hydrocarbon [2.2]paracyclophane, heterocyclophanes have been prepared by similar procedures.[3] The thiophene derivative, 5,5'-ethylene-1,2-bis-(2-thienyl)-ethane, has been made in 19% yield and the furan derivative, 5,5'-ethylene-1,2-bis-(2-furyl)-ethane, has been made in 72% yield.   A monomeric intermediate, 2,5-dimethylene-2,5-dihydrofuran, was isolated in the latter case.

1. Contribution No. 672 from the Central Research Department, Experimental Station, E. I. du Pont de Nemours and Company, Wilmington, Delaware.
2. S. Natelson and S. Gottfried, *Organic Syntheses,* Coll. Vol. 3, 382, 1955.
3. H. E. Winberg, F. S. Fawcett, W. E. Mochel, and C. W. Theobald, *J. Am. Chem. Soc.,* 82, 1428 (1960); J. L. Anderson and H. E. Winberg, U.S. pat. 2,876,216 (March 3, 1959); H. E. Winberg, U.S. pat. 2,904,540 (Sept. 15, 1959); F. S. Fawcett, U.S. pat. 2,757,146 (July 31, 1956); T. E. Young, U.S. pat. 2,999,820 (1961).
4. C. J. Brown and A. C. Farthing, *Nature,* 164, 915 (1949); C. J. Brown, *J. Chem. Soc.,* 3265 (1953).
5. L. A. Errede, R. S. Gregorian, and J. M. Hoyt, *J. Am. Chem. Soc.,* 82, 5218 (1960).
6. D. J. Cram and H. Steinberg, *J. Am. Chem. Soc.,* 73, 5691 (1951).

2,3,4,5,6-PENTA-O-ACETYL-D-GLUCONIC ACID AND 2,3,4,5,6-PENTA-O-ACETYL-D-GLUCONYL CHLORIDE

(Gluconic acid, pentaacetyl-,D-, and Gluconyl chloride, pentaacetyl-, D-)

$$CH_2(OH)-\overset{\overset{\displaystyle H}{|}}{\underset{\underline{\qquad O \qquad}}{C}}-(CHOH)_3-C{=}O + 4Ac_2O \rightarrow$$

$$CH_2(OAc)-\overset{\overset{\displaystyle H}{|}}{\underset{\displaystyle OH}{C}}-(CHOAc)_3-\overset{\overset{\displaystyle O}{\|}}{C}\diagdown_{OH} + 4HOAc$$

$$CH_2(OAc)-\overset{\overset{\displaystyle H}{|}}{\underset{\displaystyle OH}{C}}-(CHOAc)_3-\overset{\overset{\displaystyle O}{\|}}{C}\diagdown_{OH} + Ac_2O \rightarrow$$

$$CH_2(OAc)(CHOAc)_4-\overset{\overset{\displaystyle O}{\|}}{C}\diagdown_{OH} + HOAc$$

$$CH_2(OAc)(CHOAc)_4-\overset{\overset{\displaystyle O}{\|}}{C}\diagdown_{OH} + PCl_5 \rightarrow$$

$$CH_2(OAc)(CHOAc)_4-\overset{\overset{\displaystyle O}{\|}}{C}\diagdown_{Cl} + POCl_3 + HCl$$

Submitted by CHARLES E. BRAUN and CLINTON D. COOK.[1]
Checked by W. G. WOODS, E. F. SILVERSMITH, and JOHN D. ROBERTS.

1. Procedure

A. *2,3,4,6-Tetra-O-acetyl-D-gluconic acid monohydrate.* Crushed, fused zinc chloride (20 g.) is shaken with 250 ml. of acetic anhydride in a 1-l. three-necked flask until most of the solid dissolves. The flask is then equipped with mechanical stirrer, thermometer reaching into the liquid, and a dropping funnel. As the flask is cooled in an ice bath, 50 g. (0.28 mole) of D-glucono-δ-lactone (Note 1) is added slowly with vigorous stirring. During the addition, the temperature should be kept below 65°. After an hour in the ice bath, the solution is kept at room temperature for 24 hours and is then poured into 1 l. of water and stirred until the hydrolysis of the acetic anhydride is complete (about an hour). The mixture is placed in a refrigerator until the product crystallizes completely (Note 2). The crude material is removed by filtration and washed with a small amount of ice water. The 2,3,4,6-tetra-O-acetyl-D-gluconic acid monohydrate thus obtained melts at 113–117°. The yield is 79–84 g. (74–79%).

B. *2,3,4,5,6-Penta-O-acetyl-D-gluconic acid.* Tetra-O-acetyl-D-gluconic acid monohydrate (50 g., 0.13 mole) is slowly added to a chilled (0–5°) solution of 18 g. of fused zinc chloride in 190 ml. of acetic anhydride contained in a 1-l. Erlenmeyer flask. The solution is kept in an ice bath for an hour and then allowed to stand at room temperature for 24 hours. After dilution with 1 l. of water, the solution is extracted with four 100-ml. portions of chloroform. In order to remove the chloroform, 200 ml. is distilled, 250 ml. of toluene is added, and 250 ml. of this solution is distilled. Another 250 ml. of toluene is then added and the volume is reduced to 300 ml. The product crystallizes on standing at 0° (Note 2). The solid is removed by filtration, washed with toluene and then with petroleum ether (b.p. 35–55°). A yield of 44–45 g. (83–84%) of anhydrous 2,3,4,5,6-penta-O-acetyl-D-gluconic acid, melting at 110–111°, is obtained; $[\alpha]_D^{23}$ + 11.5° (c = 4.0 in ethanol-free chloroform).

C. *2,3,4,5,6-Penta-O-acetyl-D-gluconyl chloride.* Anhydrous 2,3,4,5,6-penta-O-acetyl-D-gluconic acid (25 g., 0.062 mole) is shaken with 185 ml. of anhydrous ethyl ether in a 1-l. round-bottomed flask until most of the solid dissolves. Then 15 g. (0.072 mole) of phosphorus pentachloride is added with shaking. The flask is fitted with a calcium chloride drying tube and stored overnight at room temperature. Any solid material is removed by filtration through a fritted-glass funnel into a 1-l. round-bottomed flask, and the ethereal solution is concentrated to about one-half volume under reduced pressure at room temperature.

The concentrated solution is allowed to stand overnight at 0° or below. The mother liquor is decanted from the crystals, which are then broken up, transferred to a fritted-glass funnel, washed quickly with petroleum ether (b.p. 35–55°), and dried in a vacuum desiccator. The mother liquor is again concentrated to one-half its volume under reduced pressure and a second crop of crystals collected. The total yield of 2,3,4,5,6-penta-O-acetyl-D-gluconyl chloride, melting at 68–71° (Note 3), is 21–24 g. (80–92%), depending on the temperature of crystallization (Notes 4 and 5).

2. Notes

1. The D-glucono-δ-lactone (m.p. 153–155°; assay > 99%, Pfizer specification; water content < 0.2%, Karl Fischer titration) is obtained from Charles Pfizer & Co., Inc., 630 Flushing Avenue, Brooklyn 6, New York. Material having a water content greatly in excess of 0.2% may be dried for 48 hours at 100°. Drying at higher temperatures has in some samples produced decomposition. Independent experience with many preparations in the laboratories of one of the editors (Max Tishler, Merck Sharp & Dohme Research Laboratories) has indicated that drying is generally not necessary.

2. As long as 48 hours may be required.

3. The checkers found m.p. 69.5–74°.

4. Since 2,3,4,5,6-penta-O-acetyl-D-gluconyl chloride is appreciably soluble in ethyl ether at 0°, the yield can be improved by carrying out the crystallization at lower temperatures.

5. The submitters report $[\alpha]_D^{23°} = +2.2$ ($c = 4.0$ in ethanol-free chloroform). The checkers found $[\alpha]_D^{25°} = +2.5°$.

3. Methods of Preparation

The method followed for the preparation of the two acids is a modification of that of Major and Cook.[2] The preparation for penta-O-acetyl-D-gluconyl chloride is that of Braun and co-workers.[3] A slightly different technique has also been described.[4]

1. University of Vermont, Burlington, Vt.
2. R. T. Major and E. W. Cook, *J. Am. Chem. Soc.*, **58**, 2475, 2477 (1936).
3. C. E. Braun, S. H. Nichols, Jr., J. L. Cohen, and T. E. Aitken, *J. Am. Chem. Soc.*, **62**, 1619 (1940).
4. R. T. Major, and E. W. Cook, U.S. pat. 2,368,557 [*C.A.*, **40**, 3549 (1946)].

PENTACHLOROBENZOIC ACID

(Benzoic acid, pentachloro-)

$$\text{(hexachlorobenzene)} \xrightarrow[-2MgBr_2, \ -2CH_2=CH_2]{3Mg + 2BrCH_2CH_2Br} \text{(pentachlorophenylmagnesium chloride)}$$

$$\text{(pentachlorophenylmagnesium chloride)} \xrightarrow[\text{2. HCl, H}_2\text{O}]{\text{1. CO}_2} \text{(pentachlorobenzoic acid)}$$

Submitted by D. E. Pearson and Dorotha Cowan [1]
Checked by Virgil Boekelheide and Fred G. H. Lee

1. Procedure

Magnesium turnings (39 g., 1.6 g. atoms) and hexachlorobenzene (142.4 g., 0.5 mole, m.p. 228–229°) in 1 l. of dry ether are brought to gentle reflux in a 3-l. three-necked flask heated by a Glascol® mantle at 20 volts (Note 1). Ethylene bromide (188 g., 1.0 mole) in 200 ml. of dry benzene is added through a Hershberg funnel [2] over a period of 48 hours (about 1 drop/25 seconds) (Note 2). Efficient stirring is maintained throughout the period of addition, during which the reaction mixture turns dark brown and forms a precipitate. The mixture is cooled to room temperature, and carbon dioxide, generated from dry ice and dried by passage through anhydrous calcium chloride, is added under the surface of the stirred mixture for at least 3 hours and at such a rate as to minimize clogging of the entrance tube (Note 3). After this addition 10% aqueous hydrochloric acid is added slowly until the mixture is strongly acid. The ether and benzene are removed by distillation, and the crude pentachlorobenzoic acid left in the water is removed by filtration and is washed free of salts with water. The dark-brown damp acid is converted to the ammonium salt by repeated extraction with hot dilute ammonium hydroxide (1 part by volume of concentrated ammonium hydroxide and 2 parts of water) followed by decantation. The combined decanted solutions are treated with Norit® while still hot, filtered, and then strongly acidified while still hot with concentrated hydrochloric acid. The precipitated acid is digested for at least several hours (Note 4). After the suspension has been cooled, the crude brown-colored acid is removed by filtration, washed with cold water, and air-dried to give 113 g. (77%) of product. The crude acid is recrystallized from 900 ml. of 50% aqueous methanol to yield 95 g. (65%) of tan-colored needles, m.p. 202–206° (Note 5).

2. Notes

1. The atomic proportions of magnesium are not related to the mole quantity of hexachlorobenzene in this or any other entrainment reaction. The excess magnesium (1.1 g. atoms in this case) is used to react with ethylene bromide and leave 0.5 g. atom of clean-surfaced magnesium. Ordinarily 1 mole of entrainment reagent is used per mole of "inert" halide, but for this preparation 2 moles of entrainment reagent per mole of halide gives a better yield.

2. Little attention is needed provided that the capillary tube is fitted properly. The capillary tube of the Hershberg dropping funnel should be about 4.5 in. long, and a Band S 24 platinum wire should be inserted to fit very snugly.

Rather than a Hershberg funnel, a commercial constant addition funnel (Kontes Glass Co., Vineland, N.J.) can be used.

3. A T-tube in the carbon dioxide stream serves to bypass the gas if its rate of addition is too rapid. Also, the T-tube is large enough to permit the insertion of a plunger to dislodge particles within the mouth of the tube.

4. Without digestion the acid will contain appreciable amounts of the ammonium salt. In an alternative method of purification the crude acid is converted to the insoluble sodium salt. The sodium salt can be recrystallized from 95% ethanol to give flaky white crystals, m.p. 339–340°. Digestion of the sodium salt with

1 part of concentrated hydrochloric acid and 1 part of water yields the free acid. From 10 g. of crude acid, 7.3 g. of purified acid can be obtained from the sodium salt. The free acid is reported to crystallize well from toluene and light petroleum ether.[3]

5. The melting point is reported variously in the range from 199° to 208°.[4] The acid is colorless if purified by conversions through the sodium salt (Note 4), but the yield is lower.

3. Methods of Preparation

Pentachlorobenzoic acid has been prepared by oxidation of pentachlorotoluene with nitric acid and mercury,[3] by oxidation of pentachlorobenzaldehyde by potassium permanganate,[5] and by chlorination of tetrachlorophthalyl chloride [6] and of dichlorobenzoic acids.[7] Pentachlorobenzoic acid recently has been prepared by the exhaustive chlorination of benzoic acid in sulfuric acid containing iodine.[8] The present procedure has been adapted from that of Pearson, Cowan, and Beckler.[9]

4. Merits of the Preparation

Ethylene bromide has been demonstrated to be as efficient as ethyl bromide as an entrainment agent.[9] Its use is advantageous because a second Grignard reagent is not introduced in the reaction mixture—only magnesium bromide. An additional feature of this preparation and of most preparations involving entrainment agents is the slow rate of addition of the entrainer, which permits adequate time for the "inert" halide (in this preparation, hexachlorobenzene) to react on the bright, clean surfaces of the magnesium turnings.

Although pentachlorophenylmagnesium chloride can be made in tetrahydrofuran without the use of the entrainment method, the Grignard reagent in this solvent does not react with carbon dioxide to give pentachlorobenzoic acid in good yield.[10]

1. Department of Chemistry, Vanderbilt University, Nashville, Tennessee.
2. L. F. Fieser, "Experiments in Organic Chemistry," 3rd ed., D. C. Heath, Boston, 1955, p. 265.
3. O. Silberrad, *J. Chem. Soc.*, **127**, 2684 (1925).

4. E. H. Huntress, "Organic Chlorine Compounds," John Wiley and Sons, New York, 1948, p. 464.
5. G. Lock, *Ber.*, **72**, 303 (1939).
6. A. Kirpal and H. Kunze, *Ber.*, **62**, 2105 (1929).
7. A. Claus and A. W. Bucher, *Ber.*, **20**, 1627 (1887).
8. L. G. Zagorskaya, S. I. Burmistrov, and S. A. Yashkova, *J. Gen. Chem. USSR (Engl. Transl.)*, **32**, 2612 (1962).
9. D. E. Pearson, Dorotha Cowan, and J. D. Beckler, *J. Org. Chem.*, **24**, 504 (1959).
10. H. E. Ramsden, A. E. Balint, W. R. Whitford, J. J. Walburn, and R. Cserr, *J. Org. Chem.*, **22**, 1202 (1957).

1,2,3,4,5-PENTACHLORO-5-ETHYLCYCLOPENTADIENE

(Cyclopentadiene, 1,2,3,4,5-pentachloro-5-ethyl-)

Submitted by V. Mark, R. E. Wann, and H. C. Godt, Jr.[1]
Checked by William E. Parham, Wayland E. Noland, and G. Paul Richter.

1. Procedure

A solution of 183 g. (1.10 moles) of triethyl phosphite (Note 1) in 200 ml. of petroleum ether (b.p. 30–60°) is added to a 3-l., three-necked, round-bottomed flask equipped with a mechanical stirrer, a thermometer, a dropping funnel, and an air condenser; the open end of the condenser is connected to a drying tube filled with calcium sulfate or calcium chloride. The flask is immersed in a freezing mixture of sodium chloride and ice, and the stirrer is started. When the temperature of the phosphite solution reaches 0°, a solution of 273 g. (1.00 mole) of hexachlorocyclopentadiene (Note 2) in 100 ml. of petroleum ether (b.p. 30–60°) is added through the dropping funnel at such a rate that the temperature remains between 0° and 10°. The addition requires about 4–6 hours. After the addition is complete, the freezing mixture is removed, and the brown, clear solution is allowed to warm up to room temperature.

The air condenser is replaced by an efficient water condenser set downward for steam distillation. One liter of water is added in one portion to the stirred reaction mixture, and stirring is continued for 30 minutes (Note 3). The dropping funnel is replaced by a steam-inlet tube reaching into the liquid, and steam is passed through the mixture until first the petroleum ether, then, separately, the ethylpentachlorocyclopentadiene is completely removed. The diene is separated from the steam distillate as a pale-yellow heavy oil. The aqueous phase of the steam distillate is extracted with petroleum ether, and the extract is combined with the diene and dried over calcium sulfate (Note 4). The petroleum ether is removed by evaporation on a steam bath or through a water aspirator at room temperature, leaving 245–257 g. (92–96%) of 1,2,3,4,5-pentachloro-5-ethylcyclopentadiene as a pale-yellow oil, n_D^{25} 1.5387–1.5400. The diene can be distilled without appreciably lowering the yield; b.p. 51–53°/0.2 mm.; n_D^{25} 1.5398.

2. Notes

1. Triethyl phosphite can be obtained from Virginia Carolina Chemical Corp., Eastman Kodak Co., Aldrich Chemical Co., K and K Laboratories, and Matheson, Coleman and Bell. The presence of dialkyl hydrogen phosphite or trialkyl phosphate is not deleterious, but a correction for assay is required. Fractionation readily separates triethyl phosphite (b.p. 48–49°/11 mm.) from diethyl hydrogen phosphite (b.p. 72°/11 mm.) and triethyl phosphate (b.p. 90°/10 mm.). The presence of amines and amine

hydrochlorides may seriously interfere with the alkylation, especially in the case of trimethyl phosphite (see Table I).   The checkers redistilled triethyl phosphite obtained from Matheson, Coleman and Bell.

2. A commercial product obtained from Matheson, Coleman and Bell was used.

3. Water hydrolyzes diethyl phosphorochloridate [chlorodiethoxyphosphorus(V) oxide] readily but does not affect the diene.   Alternatively, the reaction mixture can be processed by fractionation.   Evaporation of the petroleum ether and fractionation of the residue through a 25-cm. x 2.2-cm. column of glass helices yields 170 g. (98.5%) of diethyl phosphorochloridate, b.p. 34–36°/0.2 mm., n_D^{25} 1.4210–1.4250 (the refractive index indicates that it contains 5–10% of the title compound), and 240–255 g. (90–96%) of 1,2,3,4,5-pentachloro-5-ethylcyclopentadiene, b.p. 51–53°/0.2 mm., n_D^{25} 1.5398.

The reaction mixture can also be processed by chromatography. The crude reaction mixture is poured on a 90-cm. x 4.5-cm. column of alumina (e.g., Fisher "adsorption grade") and eluted with about 2 l. of technical-grade pentane.   This yields a pale-yellow solution that is free of diethyl phosphorochloridate. Evaporation of the pentane gives 240–255 g. (90–96%) of 1,2,3,4,5-pentachloro-5-ethylcyclopentadiene.

4. Calcium chloride or sodium sulfate can also be used.

3. Methods of Preparation

1,2,3,4,5-Pentachloro-5-ethylcyclopentadiene has been prepared only by the present procedure.[2]

4. Merits of the Preparation

5-Alkyl-1,2,3,4,5-pentachlorocyclopentadienes are a novel class of compounds.[2]   The alkylation of hexachlorocyclopentadiene by trialkyl phosphites is a synthetic procedure of considerable scope (Table I) and represents a new method of forming carbon-to-carbon bonds.   The products, 5-alkylpentachlorocyclopentadienes, show the manifold reactions of the parent chlorocarbon and undergo a variety of substitution and addition reactions, including Diels-Alder reactions.

TABLE I

SYNTHESIS OF 5-ALKYLPENTACHLOROCYCLOPENTADIENES, RC_5Cl_5

R [a]	Tempera- ture, °C. [b]	B.P., °C./mm.	n_D^{25}	% Yield
Methyl	20–22	45–47/0.3	1.5465	89
Isopropyl	2–5	67–68/0.4	1.5397	95
Butyl	5–10	72–74/0.3	1.5270	84
Isobutyl	5–10	73–75/0.4	1.5254	72
sec-Butyl	20–25	90–92/0.7	1.5370	89
2-Ethylhexyl	5–10	105–107/0.3	1.5172	90
Dodecyl	5–10	130–135/0.2	1.5043	78
Allyl	20–22	63–65/0.5	1.5450	75
Methallyl	25–40	80–83/0.5	1.5385	80

[a] All the trialkyl phosphites required for the preparations listed are available from the suppliers mentioned in Note 1.
[b] Temperature range during the addition period.

1. Monsanto Chemical Co., Agricultural Chemicals Division, St. Louis, Missouri.
2. V. Mark, *Tetrahedron Lett.*, 296 (1961).

3,3-PENTAMETHYLENEDIAZIRINE

Submitted by ERNST SCHMITZ and ROLAND OHME [1]
Checked by E. J. COREY and RICHARD GLASS

1. Procedure

A. *3,3-Pentamethylenediaziridine.* A solution of 147 g. (1.5 moles) of cyclohexanone in 400 ml. of 15N aqueous ammonia (6.0 moles) in a 1-l. beaker is stirred mechanically and cooled to 0° with an ice-salt mixture. Maintaining the temperature of the solution between 0° and +10°, 124 g. (1.0 mole) of 90% hydroxylamine-O-sulfonic acid (Note 1) is added in portions of about 1 g. The addition requires about 1 hour, and the mixture is stirred for another hour at 0° and allowed to stand overnight at −15° in a refrigerator. The precipitated crystalline cake is filtered and pressed tight with a glass stopper. The solid is washed with 50-ml. portions of ice-cold ether, toluene, and finally ether. There is obtained 110–115 g. of product which is 70–90% pure (Notes 2 and 3). The product is divided into two portions, each of which is boiled briefly with a 50-ml. portion of toluene; the solutions are decanted from small salt residues and cooled to 0° for 2 hours. The precipitates are filtered with suction and washed with 50 ml. of ice-cold petroleum ether. The combined yield of 3,3-pentamethylenediaziridine is 68–78 g. (61–70%), m.p. 104–107°. The purity is 96–100% (Note 4).

B. *3,3-Pentamethylenediazirine.* *Caution! See Note 5.* A solution of 34.0 g. (0.2 mole) of silver nitrate in 100 ml. of water is treated dropwise, with shaking, with 100 ml. of 2N sodium hydroxide. The precipitate is filtered with suction and washed thoroughly with water, methanol, and lastly ether. A mixture of 10.0 g. (0.089 mole) of 3,3-pentamethylenediaziridine (Note 5) and 220 ml. of ether is warmed, the resulting solution cooled to room temperature, and within a 5-minute period the silver oxide prepared above is added in small portions, with shaking, to the cooled solution. During the addition the reaction mixture is cooled with tap water and then is shaken without cooling until an aliquot does not liberate iodine from an acidified iodide solution; the reaction is normally complete in 30–60 minutes. The mixture is filtered through a fluted filter, the solid residue washed with a small volume of ether, and the filtrate dried over potassium carbonate. The ether is distilled at a bath temperature of 45° through a 30-cm. Vigreux column. The last 20 ml. of the solvent is removed at a pressure of 30 mm. and a bath temperature of 10°. Using a protective shield, the residue is distilled at 33° (30 mm.) to yield 6.4–7.4 g. (65–75%) of 3,3-pentamethylenediazirine. In order to prevent decomposition of the product on storage, it is diluted with ether and kept in a refrigerator.

2. Notes

1. Hydroxylamine-O-sulfonic acid is prepared according to the method of Gösl and Meuwsen,[2] or of Matsuguma and Audrieth.[3] The material is available from Eastman Organic Chemicals.

Analysis of this substance, just prior to use, is carried out in the following manner. A sample is dissolved in water and treated with a solution of potassium iodide in 2N sulfuric acid. After 5 minutes the solution is titrated with thiosulfate solution; near the end of the titration the solution is boiled to ensure completeness of iodine liberation. Instead of the 90% product, a correspondingly greater amount of the 80% product can be employed.

2. For the analysis an ethanolic solution of the 3,3-pentamethylenediaziridine is treated with a solution of potassium iodide in 2N sulfuric acid. It liberates two equivalents of iodine instantaneously.

3. The crude product is recrystallized without additional drying; it undergoes partial decomposition on standing.

4. If subsequent treatment is not to be performed within a few days, it is recommended that the preparation be recrystal-

lized a second time to obtain a stable product. Smaller amounts can be advantageously purified by vacuum sublimation.

5. It is recommended that the dehydrogenation be done with small amounts if the dehydrogenation product is to be isolated. Although decomposition was never observed with the preceding procedure, *diazirines should be handled with caution. Explosions were reported when working with 3-methyldiazirine*[4] *and 3-n-pro-pyldiazirine*[4] *as well as when overheating pentamethylenediazirine.*[5] Most of the reactions of diazirines, especially the reaction with Grignard compounds,[5] can be done without purification of the diazirine.

3. Methods of Preparation

Diazirines have been prepared by dehydrogenation of diaziridines with mercuric oxide,[6] silver oxide,[5] or dichromate-sulfuric acid.[4] The present procedure corresponds to that of Schmitz and Ohme.[5] The procedure for the preparation of the 3,3-penta-methylenediaziridine has been reported by H. J. Abendroth.[8]

4. Merits of the Preparation

Diazirines are the cyclic isomers of the alphatic diazo compounds. Both the diaziridines and the diazirines are starting materials for the synthesis of alkyl hydrazines. 3,3-Pentamethyl-enediaziridine can be hydrolyzed quantitatively to hydrazine. Methylamine[9] may be substituted for ammonia in the procedure resulting in 1-methyl-3,3-pentamethylenediaziridine (m.p. 35–36°, yield 62% of theoretical) and then methyl hydrazine. Use of ethylenediamine leads to ethylene bis-hydrazine[7] via a bifunctional diaziridine (m.p. 143–144°, yield 48% of theoretical). Ammonia can also be replaced by n-propylamine[8] or cyclohexylamine[7]; cyclohexanone by acetone.

3,3-Pentamethylenediazirine and other diazirines easily add Grignard reagents to the N—N double bond. The reaction leads to N-alkyl diaziridines which can be hydrolyzed to alkyl hydrazines. Cyclohexylhydrazine (85% yield), n-propylhydrazine (88%), isopropylhydrazine (95%), and benzylhydrazine were prepared from 3,3-pentamethylenediazirine and the corresponding Grignard reagent.[10]

1. Institute for Organic Chemistry of the German Academy of Sciences, Berlin-Adlershof, East Germany.
2. R. Gösl and A. Meuwsen, *Ber.*, **92**, 2521 (1959).
3. H. J. Matsuguma and L. Audrieth, *Inorg. Syn.*, **5**, 122 (1957).
4. E. Schmitz and R. Ohme, *Ber.*, **95**, 795 (1962).
5. E. Schmitz and R. Ohme, *Ber.*, **94**, 2166 (1961).
6. S. R. Paulsen, *Angew. Chem.*, **72**, 781 (1960).
7. R. Ohme, doctoral dissertation, Humboldt University, Berlin, 1962.
8. H. J. Abendroth, *Angew. Chem.*, **73**, 67 (1961).
9. E. Schmitz, R. Ohme, and R. D. Schmidt, *Ber.*, **95**, 2714 (1962).
10. E. Schmitz and R. Ohme, *Angew. Chem.*, **73**, 220 (1961).

PERBENZOIC ACID

WARNING

It has been reported that the evaporation of a chloroform solution of peroxybenzoic acid (perbenzoic acid) according to the directions published in this series[1] has resulted in a heavy explosion. Hence suitable precautions should be observed in carrying out solvent evaporations from solutions of peroxybenzoic acid. Such precautions and other useful information are given on p. 904 of this volume.

1. *Org. Syntheses,* Coll. Vol. 1, 431 (1941).

PERCHLOROFULVALENE

(Bicyclopentadienylidene, octachloro-)

$+$ $[(CH_3)_2CHO]_3P$ $\longrightarrow$

$+$ $(CH_3)_2CHCl + [(CH_3)_2CHO]_2POCl$

Submitted by V. Mark [1]
Checked by M. Rosenberger and Peter Yates

1. Procedure

A slurry of 47.5 g. (0.100 mole) of decachlorobi-2,4-cyclo-pentadienyl (Note 1) in 200 ml. of petroleum ether (b.p. 30–60°) is prepared in a 1-l., three-necked, round-bottomed flask equipped with a Hershberg stirrer, thermometer, dropping funnel, and air condenser fitted with a drying tube. The reaction flask is immersed in a water bath at 20–24° and the stirrer is started. A solution of 25.2 g. (0.121 mole) of triisopropyl phosphite (Note 2) in 25 ml. of petroleum ether (b.p. 30–60°) is added from the dropping funnel at such a rate that the temperature of the mildly exothermic reaction remains between 20° and 25°. During the addition, which requires 50–80 minutes, the light yellow slurry of the starting material is converted to a dark blue slurry of perchlorofulvalene.

After the addition is completed, the reaction mixture is stirred for an additional period of 20 minutes. The crystalline product is filtered rapidly by suction through a sintered-glass funnel (Note 3) and rapidly washed with three 20-ml. portions of petroleum ether (b.p. 30–60°). After the last washing, the perchlorofulvalene is allowed to dry at room temperature. The yield of the dark bluish-violet, uniform crystals is 26–29 g. (65–72%). The chlorocarbon is of a high purity, as indicated by infrared and ultraviolet spectroscopy (Note 4); it can be recrystallized from benzene, cyclohexane, hexane, carbon tetrachloride, or methylene chloride without appreciable lowering of the yield.

2. Notes

1. Decachlorobi-2,4-cyclopentadienyl [bis-(pentachlorocyclopentadienyl)], m.p. 123–124°, can be obtained from Columbia Organic Chemicals Co., Inc., and Aldrich Chemical Co., Inc. The checkers used purified material, m.p. 124–126°, obtained by recrystallization of commercial material from hexane. They found that without prior purification of the commercial starting materials or correction for assay (cf. Note 2) the yield was reduced to 50%. The checked runs were carried out under nitrogen, but it was not determined whether this influenced the yield.

2. Triisopropyl phosphite can be obtained from Virginia Carolina Chemical Corporation, Eastman Organic Chemicals, Aldrich Chemical Company, Matheson Coleman and Bell, and K and K Laboratories. The presence of diisopropyl hydrogen phosphite or triisopropyl phosphate is not deleterious, but a correction for the assay is required. Fractionation readily separates triisopropyl phosphite, b.p. 60–61° (10 mm.), from diisopropyl hydrogen phosphite, b.p. 70–71° (10 mm.), and triisopropyl phosphate, b.p. 95–96° (10 mm.). The checkers used a fraction, b.p. 85–88° (33 mm.) (cf. Note 1).

Alternatively, triethyl phosphite (available from the suppliers given above) can be substituted, in equivalent amount, for triisopropyl phosphite; the yield of perchlorofulvalene is 55–60%. The use of trimethyl phosphite gives lower yields (45–48%) and a less pure product.

3. The use of a large-size (500-ml.), coarse-grade, sintered-glass funnel permits rapid filtration and the washing of the filter cake

directly in the funnel. Rapid removal of the co-product, diiso-propyl phosphorochloridate, $(C_3H_7O)_2P(O)Cl$, from perchloroful-valene is necessary in order to prevent its hydrolysis to petroleum ether-insoluble products.

4. Since perchlorofulvalene does not melt below its decomposi-tion point around 200°, infrared and ultraviolet spectroscopic analyses provide the most satisfactory method of checking the purity of the product. The characteristic infrared maxima of perchlorofulvalene occur at 6.55, 7.56, 7.97, 8.10, 8.63, 10.37, 13.03, 14.24, and 14.53 μ, and the ultraviolet and visible maxima at 386 mμ (ϵ 35,800) and 590 mμ (ϵ 505). The absence of deca-chlorobi-2,4-cyclopentadienyl in the product is indicated by the absence of its characteristic bands at 6.27, 12.35, and 14.83 μ.

3. Methods of Preparation

The present procedure is that described by the submitter.[2] Perchlorofulvalene has also been obtained by the dechlori-nation of decachlorobi-2,4-cyclopentadienyl catalytically at 500° and 0.1 mm. in 57% yield,[3] or by a mixture of iron and ferric chloride in warm tetrahydrofuran in 14% yield[4] or by stannous chloride in refluxing acetone in 36% yield.[5]

4. Merits of the Preparation

Perchlorofulvalene and the more recently prepared perbro-mofulvalene[6] are the only stable compounds known at present in which the fulvalene system alone represents all the unsaturation. The current listing of a compound as "perchlo-rofulvalene" in various chemical catalogs is based on earlier work shown to be in error;[2] this compound has been shown to be a $C_{15}Cl_{12}$ chlorocarbon with a trindane skeleton.[2]

Perchlorofulvalene is a highly reactive chlorocarbon which undergoes a variety of reactions, including dimerization and oligomerization[7] and other addition reactions, including chlorination,[5b] (2+4)-cycloadditions,[7] and reaction with nucleophiles.[7] It has a unique structure[8] which reflects a compromise between steric hindrance and conjugation. More recently its dipole moment[9] and nqr spectrum[10] have been determined.

1. General Electric, Plastics Department, Mt. Vernon, Indiana 47620.
2. V. Mark, *Tetrahedron Lett.*, 333 (1961); V. Mark, U.S. Patent 3,328,472 (1967).
3. A. E. Ginsberg, P. Raatz, and F. Korte, *Tetrahedron Lett.*, 779 (1962).
4. R. R. Hindersinn, U.S. Patent 3,475,502 (1969).
5. (a) D.C.F. Law, Ph.D. Thesis, The University of Wisconsin, 1966; (b) R.M. Smith and R. West, *J. Org. Chem.*, 35, 2681 (1970).
6. P. T. Kwitowski and R. West, *J. Am. Chem. Soc.*, 88, 4541 (1966); R. West and P. T. Kwitowski, *J. Am. Chem. Soc.*, 90, 4697 (1968).
7. V. Mark, unpublished results.
8. P. J. Wheatley, *J. Chem. Soc.*, 4936 (1961).
9. I. Agranat, H. W. Feilchenfeld, and R. J. Loewenstein, *Chem. Comm.*, 1154 (1970).
10. I. Agranat, D. Gill, M. Hayek, and R. J. Loewenstein, *J. Chem. Phys.*, 51, 2756 (1969).

PEROXYBENZOIC ACID

$$C_6H_5CO_2H + H_2O_2 \ (70\%) \xrightarrow{CH_3SO_3H} C_6H_5CO_3H + H_2O$$

Submitted by LEONARD S. SILBERT, ELAINE SIEGEL, and DANIEL SWERN.[1]
Checked by A. S. PAGANO and W. D. EMMONS.

1. Procedure

Caution! All reactions in which 50% or more concentrated hydrogen peroxide is employed must be conducted behind a safety shield. Beakers are recommended as reaction vessels to permit rapid escape of gas and avoidance of pressure build-up in the event of a rapid decomposition. (See Note 1.)

Twenty-two grams (0.45 mole) of 70% hydrogen peroxide (Note 2) is added dropwise with efficient agitation to a slurry or partial solution of 36.6 g. (0.30 mole) of benzoic acid (Note 3) in 86.5 g. (0.90 mole) of methanesulfonic acid (Note 4) in a 500-ml. tall-form beaker. The reaction temperature is maintained at 25–30° by means of an ice-water bath. The reaction is exothermic during the hydrogen peroxide addition, which requires approximately 30 minutes. During this period the benzoic acid completely dissolves.

The solution is stirred for an additional 2 hours and is then cooled to 15°. Fifty grams of chopped ice and 75 ml. of ice-cold saturated ammonium sulfate solution are cautiously added in sequence while the temperature is maintained below 25° during the dilution (Note 5). The contents of the beaker are transferred to a separatory funnel, and the peroxybenzoic acid solution is extracted with three 50-ml. portions of benzene at room temperature (Note 6). The aqueous layer is discarded, and the combined benzene extracts are washed twice with 15 ml. of cold saturated ammonium sulfate solution to ensure complete removal of methanesulfonic acid and hydrogen peroxide, dried over anhydrous sodium sulfate, and filtered. Iodometric titration of an aliquot of the benzene solution (Note 7) indicates that the conversion of benzoic to peroxybenzoic acid is 85–90%. This solution can be used directly for epoxidation or other oxidation reactions without further treatment (Note 8).

2. Notes

1. For a summary of procedures to be followed in the safe handling of hydrogen peroxide and peroxy acids see reference 2, p. 90 and 478.

2. Hydrogen peroxide of this concentration can be obtained from various commercial sources. The submitters have also used 50% and 95% hydrogen peroxide instead of the 70% concentration. With 50% hydrogen peroxide, conversion to peroxybenzoic acid is only about 75%. With 95% peroxide, the reaction proceeds more rapidly and is slightly more exothermic, and conversions of benzoic acid to peroxybenzoic acid are about 90–95% instead of 85–90%. Little advantage is seen in using the more concentrated hydrogen peroxide in the preparation of peroxybenzoic acid except when a high yield of pure crystalline material is needed (Note 7).

3. Benzoic acid of analytical reagent grade is used.

4. Methanesulfonic acid, Eastman Chemicals, practical grade, is satisfactory.

5 Dilution of the methanesulfonic acid is exothermic. Since peroxybenzoic acid has an appreciable solubility in aqueous methanesulfonic acid, dilution and washing are conducted with minimal quantities of saturated ammonium sulfate solution.

6 In the first extraction, 90% of the available peroxybenzoic acid is extracted. The second extraction removes 7%, and the third 2%. The first benzene extract is an approximately 40% solution of peroxybenzoic acid ($2.8M$).

7. A 1-ml. or 2-ml. aliquot of the benzene solution of peroxybenzoic acid is pipetted into an iodine flask, the walls of the flask are rinsed with a small quantity of chloroform, and 15 ml. of acetic acid is added. Two milliliters of a saturated aqueous solution of analytical reagent grade sodium iodide is added. After a reaction period of about 5 minutes, 50–75 ml. of water is added, and the liberated iodine is titrated with $0.1N$ sodium thiosulfate solution (starch indicator). One milliliter of $0.1N$ sodium thiosulfate is equivalent to 0.00691 g. of peroxybenzoic acid.

The analytical method described is also used in following the consumption of peroxybenzoic acid or other peroxy acids during an oxidation reaction; it has also been used in determining the conversion of other carboxylic acids to peroxy acids when solvent extraction has been used in the isolation.

8. If a solvent other than benzene *must* be used in an oxidation reaction, peroxybenzoic acid can be isolated by evaporation of the benzene in an evaporating dish in the hood under a stream of nitrogen gas, or preferably in a rotary evaporator. Evaporation of the solvent from the peroxybenzoic acid solution is preferably conducted as rapidly as possible from a water bath at a temperature below 30°. *Caution! This operation must be carried out behind a good shield. A heavy explosion once occurred during such evaporation of a chloroform solution of perbenzoic acid.*[8] Owing to the volatility of peroxybenzoic acid, some is lost during solvent evaporation; overall recovery of peroxy acid is 70–90%. The crude peroxybenzoic acid obtained as a residue is a pale-yellow mushy solid or liquid if it still contains traces of benzene. The peroxy acid should be stored in a refrigerator if it is not used immediately.

Analytically pure solid peroxybenzoic acid decomposes at the rate of about 2–3% per day at room temperature, but it can be stored for long periods in a refrigerator without significant loss of active oxygen. Crude preparations lose active oxygen more rapidly. Pure peroxybenzoic acid can be obtained readily from peroxybenzoic acid of 90–95% purity by crystallization at −20° from *olefin-free* 3:1 petroleum ether/diethyl ether cosolvent. About 4.5 ml. per gram of crude peroxy acid is needed, and the solution should be seeded at about 5°. From 15 g. of crude peroxy acid, 9–10 g. of the pure acid, m.p. 41–42°, is obtained as long white needles. To obtain reaction products containing 90–95% peroxybenzoic acid, 95% hydrogen peroxide must be used in the preparation.

3. Methods of Preparation

The methods of preparation, properties, analysis, and safe handling of peroxybenzoic acid have been reviewed.[2] Numerous methods of preparing peroxy acids are described in the literature,[2,3] and many of them have been applied to the synthesis of peroxybenzoic acid. A common way of preparing it has been by the action of sodium methoxide on benzoyl peroxide followed by acidification.[3] The present method is adapted from one in a publication of the submitters.[4]

The preparation of *m*-chloroperoxybenzoic acid is described elsewhere in this series.[5]

4. Merits of the Preparation

The present procedure for peroxybenzoic acid is easier and more reliable than earlier ones. Thus that in an earlier volume of *Organic Syntheses*[3] has been found by the submitters to be difficult to reproduce, and yields are frequently low. The modified procedure of Kolthoff, Lee, and Mairs[6] is an improvement, but it is tedious and indirect.

There are other methods for converting aliphatic acids directly to peroxy acids, but this is the first that converts aromatic acids directly to peroxy acids. With suitable modifications it is applicable to a wide variety of aliphatic and aromatic peroxy acids.[4] The methyl ester may be used in place of highly insoluble acids. Water-insoluble peroxy acids such as *p*-nitroperoxybenzoic acid (an outstanding epoxidizing agent [7]), *p-tert*-butylperoxybenzoic acid, and peroxystearic acid require 90–95% hydrogen peroxide for best results; the procedure is essentially the same *except that greater precautions are necessary with hydrogen peroxide of such high strength.*[4]

1. Eastern Regional Research Laboratory, Philadelphia, Pennsylvania.
2. D. Swern, "Organic Peroxides," Vol. I, Wiley-Interscience, New York, 1970, p. 424, 489.
3. G. Braun, *Org. Syntheses*, Coll. Vol. 1, 431 (1941).
4. L. S. Silbert, E. Siegel, and D. Swern, *J. Org. Chem.*, 27, 1336 (1962).
5. R. N. McDonald, R. N. Steppel, and J. E. Dorsey, *Org. Syntheses*, 50, 15 (1970).
6. I. M. Koltoff, T. S. Lee, and M. A. Mairs, *J. Polymer Sci.*, 2, 199 (1947).
7. B. M. Lynch and K. H. Pausacker, *J. Chem. Soc.*, 1525 (1955); M. Vilkas, *Bull. Soc. Chim. France*, 1401 (1959).
8. P. Westerhof, private communication.

PHENACYLAMINE HYDROCHLORIDE*

(Acetophenone, 2-amino-, hydrochloride)

Submitted by Henry E. Baumgarten and
James M. Petersen.[1]
Checked by William E. Parham, Norman Newman and
R. M. Dodson.

1. Procedure

In a thoroughly dry 500-ml., three-necked, round-bottomed flask fitted with a mechanical stirrer, dropping funnel, and a Y-tube containing a calcium chloride drying tube and a thermometer (Note 1) are placed 24.2 g. (26 ml., 0.20 mole) of α-phenylethylamine (Note 2) and 50 ml. of dry benzene (Note 3). The solution is cooled in an ice-salt bath to 5°, and a solution of 44.5 g. (50 ml., 0.41 mole) of tert-butyl hypochlorite [2] (Note 4) in 50 ml. of dry benzene (Note 3) is added at such a rate as to maintain the temperature below 10° (Note 5). After the addition of the tert-butyl hypochlorite solution is complete, the reaction mixture is stirred at room temperature 1–4 hours (Note 6).

The Y-tube is replaced by a reflux condenser fitted with a calcium chloride drying tube, and a freshly prepared solution of 13.8 g. (0.60 g. atom) of sodium in 140 ml. of anhydrous methanol (Note 7) is added to the benzene solution of N,N-dichloro-α-phenylethylamine at such a rate as to maintain gentle reflux (Note 8). After addition of the sodium methoxide is complete, the reaction mixture is heated under reflux until a test with acidified starch-iodide paper is negative (about 45–70 minutes) (Note 9). The reaction mixture is cooled in an ice-water bath, and the precipitated sodium chloride is removed by filtration through a Büchner funnel. The filter cake is washed with three 25-ml. portions of dry benzene. The combined filtrates are added very slowly with shaking or stirring *to* 150 ml. of 2*N* hydrochloric acid contained in a 1-l. beaker (Note 10). The layers are separated, and the benzene layer is extracted with three 50-ml. portions of 2*N* hydrochloric acid. The combined acid extracts are washed twice with 50-ml. portions of ether (Note 11). The ether extracts are discarded. The pale amber to yellow aqueous solution is evaporated to dryness at a temperature not greater than 40° (Note 12). The residue is transferred to a 1-l. round-bottomed flask fitted with a reflux condenser to which is added 400 ml. of isopropyl alcohol-hydrochloric acid solution (Note 13). The mixture is heated under reflux for at least 30 minutes and is filtered hot through a Büchner funnel. The residual solid is returned to the flask and extracted in the same manner with a 150-ml. portion of the isopropyl alcohol-hydrochloric acid solution. The solid residue (sodium chloride) is discarded (Note 14). The two extracts are cooled separately in the refrigerator overnight and then filtered on a Büchner funnel (Note 15). The nearly colorless crystals are washed on the filter with two 50-ml. portions of dry ether. Each of the filtrates is diluted with an equal volume of dry ether (400 ml. and 150 ml., respectively) and is allowed to stand in the refrigerator overnight. From these diluted filtrates additional crops of crystals are collected (Note 16). The combined yield of the three to four crops is 18.9–24.8 g. (55–72%), m.p. 185–186° dec. (Notes 17 and 18). Normally the product is sufficiently pure for use without further purification; however, the product may be recrystallized from isopropyl alcohol-hydrochloric acid solution (Note 12), using 100 ml. of the solution for each 6 g. of compound.   The recovery is about 5.5 g. per 6.0 g. of crude product.

2. Notes

1. The submitters used apparatus with ground-glass joints and dried the various pieces in the oven at 120–140° overnight before use.   The Y-tube was constructed from a 24/40 male joint by joining a short length of 8-mm. i.d. glass tubing to the unground end of the joint in such a fashion as to permit insertion of a thermometer through the joint into the flask and then joining a second short piece of 8-mm. i.d. glass tubing in such a fashion as to permit attachment of a calcium chloride tube without interfering with the thermometer opening.   If desired, a four-necked flask may be substituted for the Y-tube and three-necked flask.

2. The submitters used dl-α-phenylethylamine obtained either from the Eastman Kodak Company or Matheson, Coleman and Bell without further purification.   The preparation of dl-α-phenylethylamine has been described previously in *Organic Syntheses*.[3, 4]

3. Reagent grade dry benzene is dried by simple distillation, the first 10% of the distillate being discarded.

4. The submitters did not redistil the *tert*-butyl hypochlorite. If it is desired to avoid the use of *tert*-butyl hypochlorite, an equivalent quantity of dichloramine B (N,N-dichlorobenzene-sulfonamide, Arapahoe Chemical Co., Boulder, Colorado) may be substituted.   This material is soluble in benzene but the benzene-sulfonamide is not; therefore the reaction mixture must be filtered just before the addition of the sodium methoxide solution.   Using this technique, the submitters obtained 44–52% of phenacylamine hydrochloride.

5. The rate of addition is not critical, for the reaction is not especially exothermic.   However, at even slightly elevated temperatures the N,N-dichloroamines may begin to decompose with the formation of undesired products; therefore the addition can be carried out as rapidly as desired within the specified temperature range.   With a reasonable cooling efficiency this will be

well below 30 minutes, but no harm will be done if a longer period is required.

6. The halogenation reaction appears to be quite rapid; therefore the time of stirring is not critical but probably should not be prolonged beyond 4 hours. The submitters used this time interval to prepare the sodium methoxide solution, and the actual time lapse depended upon the time required to prepare this solution. The solution of N,N-dichloro-α-phenylethylamine should be clear yellow after the stirring period. A turbid solution or one containing a precipitate usually indicates a poor sample of *tert*-butyl hypochlorite.

7. Commercial absolute methanol is dried by heating the material under reflux over magnesium turnings for 4 hours, followed by distillation into a dried receiver. Normally 1 g. of magnesium turnings per 100 ml. of absolute methanol will be sufficient. To allow for losses during the drying and distillation, the charge of methanol should be at least twice the amount required for the preparation.

It is advantageous to dry the methanol the day before the preparation is to be carried out and to store the dried methanol in a carefully sealed, *dry* flask or to allow the methanol-magnesium mixture to reflux overnight followed by distillation just prior to use.

The submitters used the inverse addition procedure for preparing the methanolic sodium methoxide, as follows. In a thoroughly dry 500-ml., three-necked, round-bottomed flask fitted with a mechanical stirrer, dropping funnel, and a reflux condenser carrying a calcium chloride drying tube is placed 13.8 g. (0.60 g. atom) of sodium freshly cut into small pieces. To this is added slowly 140 ml. of anhydrous methanol at such a rate as to maintain vigorous reflux. If all the sodium does not dissolve during the addition of the methanol, the mixture may be heated on the steam bath until solution is effected or additional methanol (up to 25 ml.) may be added. The preparation of the solution of sodium methoxide requires about 30 minutes.

It may be advantageous to allow a slow stream of dry nitrogen to pass through the apparatus during the addition of the methanol. The submitters routinely omit this precaution and, as yet, have experienced no accidents or fires. For other precautions see Note 1, *Org. Syntheses*, Coll. Vol. **3**, 215 (1955).

8. The rate of addition probably is not critical but should not be allowed to proceed uncontrolled. The submitters added the sodium methoxide solution at such a rate as to cause vapors of the refluxing methanol to condense in the first 2–3 in. of the reflux condenser without application of external heating.

9. A positive test is the immediate formation of a dark violet or brown spot on starch-iodide paper moistened with 2*N* hydrochloric acid. A negative test may consist of a very faint beige color or complete absence of color.

10. The reverse mode of addition may lower the yield and introduce unwanted condensation products of the amino ketone, which is not stable in neutral or alkaline solution.

11. The procedure should be continued up to at least this point without stopping. After this operation the sequence may be interrupted at any time.

If the *tert*-butyl hypochlorite has been prepared in advance and if the methanol to be used has been allowed to reflux over magnesium overnight, the solvents can be distilled and the reaction carried to this point in an 8-hour day. However, it may be preferable to prepare the dry solvents the day before the reaction is to be run. The latter procedure appears to have little effect on the final yield provided that the solvents are stored in tightly sealed containers and transferred with due care.

12. The submitters strongly recommend the use of a rotating evaporator (such as the Flash-Evaporator, Laboratory Glass Supply Co., New York 31, N. Y.) with which the solution can be reduced to a syrup in about 4 hours. The further evaporation is facilitated by adding 100 ml. of commercial absolute ethanol at this point and continuing the evaporation. Total time for evaporation will be about 6 hours, and the product will be a crystalline mass. The extraction step may be carried out in the 2-l. flask normally used with the evaporator.

If a rotating evaporator is not available, the solution is poured into a large porcelain evaporating dish and is allowed to stand protected in the hood for several days. Toward the end of this time, the evaporation may be accelerated by the addition of 100 ml. of ethanol as described above. The checkers removed water by blowing air over the solution.

13. The isopropyl alcohol-hydrochloric acid solution contains 1 ml. of concentrated hydrochloric acid per 100 ml. of isopropyl alcohol. If the hydrochloric acid is omitted, the product will be impure and the yield greatly reduced. Sodium chloride is not appreciably soluble in this solution.

14. The yield of sodium chloride is usually 33–35 g. (94–100%). It is often helpful to recover and weigh the sodium chloride before discarding it. An excess over the theoretical amount indicates incomplete extraction.

15. If the reflux period has been sufficiently long, little or no precipitate will be formed at this stage in the second extracting solution, which is used to ensure efficient extraction.

16. At this stage, crops of crystals may be formed in each solution.

17. Further treatment of the filtrate normally will yield little crystalline material.

18. The submitters report that, on the basis of experience in student preparations courses, the usual percentage yields for fairly capable technicians on their first trial are in the low fifties, and on subsequent trials in the sixties. Persons with exceptionally good laboratory technique may get even greater yields than those specified (up to 78%).

This procedure may be used for the preparation of a variety of α-amino ketones as is indicated in Table I, which summarizes most of the submitters' experience with this reaction. Principal deviations from the procedure will be in the time required for a negative starch-iodide test and the nature and amount of extraction and recrystallization solvent. *It is strongly recommended that any one using the reaction for the first time carry out the preparation on α-phenylethylamine before attempting to use it on other more valuable amines.*

3. Methods of Preparation

Phenacylamine hydrochloride has been prepared by (1) the hydrolysis of the quaternary salt obtained from phenacyl bromide and hexamethylenetetramine (the Delepine reaction),[5-11] (2) the hydrolysis of N-phenacylphthalimide (the Gabriel reaction),[12-14]

(3) the reduction or catalytic hydrogenation of α-oximinoaceto-phenone,[10, 15-19] (4) the reduction of α-nitroacetophenone,[20, 21] (5) the catalytic hydrogenation of α-azidoacetophenone,[22] (6) the catalytic hydrogenation of α-benzylaminoacetophenone,[23] (7) the base-catalyzed rearrangement of the tosylate of acetophenone oxime (the Neber rearrangement),[24, 25] (8) the base-catalyzed rearrangement of acetophenone dimethylhydrazone methiodide,[26] (9) the Friedel-Crafts acylation of benzene with glycyl chloride hydrochloride,[27] as well as by other procedures of uncertain preparative value. The present procedure is adapted from those of Baumgarten and Bower [19] and Baumgarten and Petersen.[28]

TABLE I

PREPARATION OF α-AMINO KETONES

Product	Approx. Reaction Time,[a] min.	Yields, %	Recryst. Solvent	M.P., °C.[b]
Hydrochloride of				
2-Aminocyclopentanone	180	34–36	i-PrOH	146–147
2-Aminocyclohexanone	25–45	49–73	i-PrOH	156
3-Amino-2-heptanone	210	50–75	i-PrOH	134–135
p-Bromophenacylamine	70	58–73	2N HCl	275
p-Chlorophenacylamine	80–90	49–60	EtOH	270–271
p-Methoxyphenacylamine	270	62–74	EtOH	200
p-Nitrophenacylamine	60	50–56	MeOH	243
p-Methylphenacylamine	80–90	70–72	i-PrOH	206–207
p-Phenylphenacylamine	80	54–71	2N HCl	185–186
α-Aminovalerophenone	30	65–66	i-PrOH-Et₂O	156.5–158
2-Amino-1-tetralone	100	63–70	i-PrOH	201–202
2-Amino-4,4-dimethyl-1-tetralone	300	61–65	i-PrOH	212–213
Desylamine (2-amino-2-phenylacetophenone)	30	45–46	i-PrOH	233–234
Phenacylamine	45–70	55–78	i-PrOH	185–186

[a] Time for negative starch-iodide test.
[b] Usually with decomposition.

4. Merits of the Preparation

The present procedure is a specific example of a synthetic method of some generality. The procedure describes an example which is of considerable interest *per se* but, perhaps more importantly, which also serves as a model for the use of this procedure for the preparation of other α-amino ketones. In the submitters' laboratory, this specific procedure is used routinely for the training of persons who will be using this general technique or related techniques.[29]

Of the procedures cited in Section 3, procedures (1), (3), and (4) have been examined by the submitters for comparison with the present procedure. Of these, the present procedure and that based on the Delepine reaction (1) appeared to be the most satisfactory for preparative purposes. Yields by the two procedures were comparable; however, the Delepine reaction could be run somewhat more conveniently on a larger scale (provided that one was willing to accept a tedious extraction of the product from the copious quantity of ammonium salts with which it is mixed). The Delepine reaction also makes a lesser demand on the skill and technique of the operator. On the other hand, attempts in the submitters' laboratory to extend the Delepine reaction to *sec*-bromides have been unsuccessful; therefore the Delepine reaction appears to lack the generality of the present procedure, which shares such generality, apparently, with procedures (2), (3), (7), and (8). Furthermore, the Delepine reaction gives a mixture of phenacylamine hydrochloride and hydrobromide [5,10] (although the submitters have found that by careful fractional crystallization from isopropyl alcohol-hydrochloric acid solution about 50% of the pure hydrochloride can be obtained).

Under appropriate conditions each of the intermediates shown in the equation may be prepared. Suitable conditions for the preparation of many N-chloroimines are described in this volume.[30] In the event that this procedure is found unsuitable, the use of slightly more than one equivalent of sodium methoxide *at room temperature* may be an acceptable alternative.[31] Solutions of the methoxy aziridine may be obtained by carrying out the present procedure up to the point where the starch-iodide test is negative. In favorable examples the methoxy aziridines may be isolated.[31]

1. Avery Laboratory, University of Nebraska, Lincoln, Neb. This work was supported in part by grants G-1090 and G-11339 of the National Science Foundation.
2. II. M. Teeter and E. W. Bell, *Org. Syntheses,* Coll. Vol. 4, 125 (1963); M. J. Mintz and C. Walling, this volume, p. 184.
3. A. W. Ingersoll, *Org. Syntheses,* Coll. Vol. 2. 503 (1943).
4. J. C. Robinson, Jr., and H. R. Snyder, *Org. Syntheses,* Coll. Vol. 3, 717 (1955).
5. C. Mannich and F. L. Hahn, *Ber.,* 44, 1542 (1911).
6. K. H. Slotta and H. Heller, *Ber.,* 63, 1024 (1930).
7. B. Reichert and H. Baege, *Pharmazie,* 2, 451 (1947).
8. M. A. Moscosco C., *Anales Fac. Farm. y Bioquim., Univ. Nacl. Mayor San Marcos (Lima),* 5, 573 (1954).
9. N. A. Adrova, M. M. Koton, and F. S. Florinskiĭ, *Izvest. Akad. Nauk S.S.S.R., Otdel. Khim. Nauk,* 1957, 385.
10. H. O. House and E. J. Grubbs, *J. Am. Chem. Soc.,* 81, 4733 (1959).
11. M. Nagawa, R. Myokei, and Y. Murase, *Takamine Kenkyujo Nempo,* 81 (1956).
12. C. Goedeckemeyer, *Ber.,* 21, 2684 (1888).
13. S. Gabriel, *Ber.,* 41, 1132 (1908).
14. H. V. Euler, H. Hasselquist, and O. Cedar, *Ann.,* 581, 198 (1953).
15. E. Braun and V. Meyer, *Ber.,* 21, 1271 (1888).
16. H. Rupe, *Ber.,* 28, 251 (1895).
17. A. Angeli, *Gazz. Chim. Ital.,* 23, II, 349 (1893).
18. A. K. Mills and J. Grigor, *J. Chem. Soc.,* 1568 (1934).
19. H. E. Baumgarten and F. A. Bower, *J. Am. Chem. Soc.,* 76, 4561 (1954).
20. J. Thiele and S. Haeckel, *Ann.,* 325, 13 (1886).
21. L. M. Long and H. D. Troutman, *J. Am. Chem. Soc.,* 71, 2469 (1949).
22. H. Bretschneider and H. Hörmann, *Monatsh.,* 84, 1021 (1953).
23. R. Simonoff and W. H. Hartung, *J. Am. Pharm. Assoc.,* 35, 306 (1946).
24. P. W. Neber and G. Huh, *Ann.,* 515, 283 (1935).
25. S. Tatsuoka, K. Osugi, A. Minato, M. Honjo, and Y. Tokuda, *J. Pharm. Soc., Japan,* 71, 774 (1951).
26. P. A. S. Smith and E. F. Most, Jr., *J. Org. Chem.,* 22, 358 (1957).
27. H. Zinner and G. Brossman, *J. Prakt. Chem.,* [4] 5, 91 (1957).
28. H. E. Baumgarten and J. M. Petersen, *J. Am. Chem. Soc.,* 82, 459 (1960).
29. H. E. Baumgarten, J. E. Dirks, J. M. Petersen, and D. C. Wolf, *J. Am. Chem. Soc.,* 82, 4422 (1960).
30. G. H. Alt and W. S. Knowles, this volume, p. 208.
31. S. D. Carlson, Ph.D. Thesis, University of Nebraska, 1966, p. 59.

PHENOLS: 6-METHOXY-2-NAPHTHOL

(2-Naphthol, 6-methoxy-)

Submitted by R. L. KIDWELL, M. MURPHY, and S. D. DARLING [1]
Checked by R. E. IRELAND, J. W. TILLEY, and C. KOWALSKI

1. Procedure

A 2-l. three-necked flask equipped with a condenser and containing 27 g. (1.1 mole) of magnesium is flame-dried and the atmosphere replaced with nitrogen. A 200-ml. portion of tetrahydrofuran (Note 1) is added along with several lumps, totaling about 95 g., of 6-bromo-2-methoxynaphthalene (Note 2) and a small crystal of iodine. The mixture is heated to reflux until the boiling becomes spontaneous. An additional 600 ml. of tetrahydrofuran is added with more of the bromide to maintain a vigorous reflux, until 237.4 g. (1 mole) of 6-bromo-2-methoxynaphthalene has been added. After the spontaneous reflux has subsided, the dark solution is heated to reflux for 20 minutes.

A 5-l. three-necked flask fitted with a paddle stirrer, a Claisen adapter containing a thermometer well and nitrogen inlet, and a dropping funnel is flame-dried and placed under nitrogen. Into the flask are introduced 125 ml. (1.1 mole) of trimethyl borate (Note 3) and 600 ml. of tetrahydrofuran. This solution is cooled to −10° with an all-encompassing ice-salt bath or a dry ice-carbon tetrachloride bath. The dropping funnel is charged with the Grignard solution which is added over 30 minutes to the borate solution while stirring rapidly and maintaining the temperature between −10° and −5°. A white sludge separates from the solution during the addition. After stirring for an additional 15 minutes, 86 ml. (1.5 mole) of chilled acetic acid (Note 4) is added all at once. This is followed by the addition of a cold solution of 112 ml. (1.1 mole) of 30% hydrogen peroxide in 100 ml. of water, dropwise over 15 minutes, while maintaining the temperature below 0° (Note 5) and stirring vigorously.

The mixture is allowed to warm up over 20 minutes and is poured into a 2-l. separatory funnel. The purplish solution is washed with a saturated ammonium sulfate solution (about 1.5 l.) containing ferrous ammonium sulfate until the rust-brown ferric color is no longer produced. The organic layer is dried over magnesium sulfate and concentrated, leaving a dark solid. Purification of the solid by high-vacuum short path distillation gives 127–142 g. (73–81%) of a pinkish or tan-colored product, b.p. 148–150° (0.15 mm.), m.p. 145–147°. It may be further purified by sublimation, or recrystallization from benzene-hexane, m.p. 148–149°.

2. Notes

1. Reagent grade tetrahydrofuran (Mallinckrodt) has been used directly. The formation of the Grignard reagent starts readily and no precipitates are formed. Tetrahydrofuran obtained from the Quaker Oats Company in 1-gal. cans has also been used; the reaction, however, is slower to start, a cloudy precipitate is formed, and the yield is slightly lower. [*Caution!* *See p. 976.*]

2. This starting material is obtained conveniently from the bromination [2] and methylation of 2-naphthol. The procedure is modified by not removing the tin salts.

After bromination of 144 g. (1 mole) of 2-naphthol, the hot solution is poured into water and filtered. The dry precipitate is mixed with a solution of 200 ml. of concentrated sulfuric acid in 500 ml. of technical methanol and heated to vigorous reflux for 4 hours. An oily layer separates during the heating period. The hot mixture is poured into 3 l. of ice and water, and the solids are removed by filtration. The moist solid is triturated with 1 l. of hot 5% sodium hydroxide. After chilling the mixture to solidify the oil, it is filtered and the product is washed and dried. The 6-methoxy-2-bromonaphthalene is purified by distillation, b.p. 114–118° (0.2 mm.). After distillation the product is most conveniently handled by remelting and pouring it into a mold to solidify. The overall yield is 173–208 g. (73–88%), m.p. 101.5–103°.

3. Commercial trimethyl borate contains an appreciable amount of methanol. It is removed by adding anhydrous lithium chloride [3] to the bottle and allowing the mixture to stand with occasional shaking. The upper layer is decanted off and fractionated, b.p. 68–69°. The product must be protected from moisture.

4. Water added at this point hydrolyzes the arylboronic ester extremely rapidly to 2-methoxynaphthalene.

5. The reaction is exothermic. Except for a darkening of the product, no apparent harm results if occasionally the temperature rises to 10–15°.

3. Discussion

The classic caustic fusion of sulfonic acid salts has been used for preparing 2,6-dinaphthol [4] and its derivatives. Other more recent procedures have employed the direct hydrolysis of aryl bromides [5] and the oxidation of aryl Grignard reagents.[6]

The indirect oxidation of an aryl Grignard reagent through a boronic ester nearly doubles the yield of phenol obtained by direct oxidation and decreases the reaction time. Tetrahydrofuran is the preferred solvent. It facilitates the dissolution of the bromide, which is relatively insoluble in diethyl ether, solvates the Grignard reagent, and renders the oxidation reaction homogeneous.

The preparation of 6-methoxy-2-naphthol is of particular interest as the starting point in many synthetic sequences. It is readily converted to 6-methoxy-2-tetralone through a Birch reduction.[7]

1. Department of Chemistry, Southern Illinois University, Carbondale, Illinois.
2. C. F. Koelch, *Org. Syntheses*, Coll. Vol. **3**, 132 (1955); H. E. French and K. Sears, *J. Am. Chem. Soc.*, **70**, 1279 (1948).
3. H. I. Schlesinger, H. C. Brown, D. L. Mayfield, and J. R. Gilbreath, *J. Am. Chem. Soc.*, **75**, 213 (1953).
4. R. Willstatter and J. Parnas, *Ber.*, **40**, 1406 (1907).
5. M. Gates and W. G. Webb, *J. Am. Chem. Soc.*, **80**, 1186 (1958).
6. H. E. French and K. Sears, *J. Am. Chem. Soc.*, **70**, 1279 (1948).
7. N. A. Nelson, R. S. P. Hsi, J. M. Schuck, and L. D. Kahn, *J. Am. Chem., Soc.*, **82**, 2573 (1960).

PHENYLBROMOETHYNE

(Benzene, bromoethynyl)

$$C_6H_5C{\equiv}CH \xrightarrow[-NaBr, -H_2O]{NaOH + Br_2} C_6H_5C{\equiv}CBr$$

Submitted by SIDNEY I. MILLER, GENE R. ZIEGLER, and R. WIELESECK [1]
Checked by WILLIAM E. PARHAM and JAMES N. WEMPLE

1. Procedure

To a 2-l. bottle equipped with a rubber stopper and immersed in a mixture of ice and water (slush) there is added a cold (about 0°) solution containing 300 g. (7.5 moles) of sodium hydroxide (Note 1) and 800 ml. of water. The mixture is swirled or stirred while 160 g. (2 moles) of bromine is added. Phenylacetylene (84 g., 0.82 mole) (Note 2) is then added to the yellow solution, and the resulting mixture is stoppered and shaken. The rubber stopper is wired down, the bottle is covered with opaque cloth or paper, and the bottle is then placed in a mechanical shaker for 60 hours at room temperature (Note 3).

The crude oil is then separated from the aqueous phase, dried with calcium chloride (Note 4), and fractionated (Note 5) at reduced pressure under nitrogen (*Caution! Note 6*). The distillation receiver should be cooled in an ice-salt or dry ice-acetone mixture. After a small fore-run of phenylacetylene, there is obtained 109–124 g. (73–83% yield) of water-white phenylbromoethyne, b.p. 40–41° (0.1 mm.), n^{25}D 1.6075 (Note 7).

2. Notes

1. Practical grade sodium hydroxide and bromine were used.

2. Commercially available phenylacetylene can be used. The checkers used material as obtained from Columbia Organic Chemicals Co., Inc.

3. Vigorous shaking is essential. For this reaction rate = $k[C_6H_5C{\equiv}CH]$ $[OBr^-]$ $[OH^-]$ with $k = 7$ M^{-2} sec.$^{-1}$ at 25°.[12] Since the solubility of phenylacetylene at 25° is 5.1×10^{-3} M in water (2.0×10^{-3} M in $2M$ sodium chloride), efficient mixing of the reagents is of paramount importance.[12] An ordinary motor-drive stirrer proved to be inadequate. Phenylbromoethyne gradually darkens when exposed to light or air. The product is best stored under nitrogen in a refrigerator and should be distilled within a few days of its preparation.

4. The checkers observed that the calcium chloride absorbs appreciable quantities of product. The crude oil was dissolved in peroxide-free ether (about 300 ml.) prior to drying with calcium chloride, or the calcium chloride was extracted with several 50-ml. portions of dry ether after use. The ethereal extracts were concentrated under nitrogen and added to the product before distillation.

5. The checkers distilled the product from a flask equipped with a Claisen head but no column.

6. No air should be allowed to come in contact with the hot pot liquid during the distillation, for an exothermic reaction may occur; at best this may fill the apparatus with tarry material and the room with noxious fumes; at worst, pressure built up may destroy all or part of the apparatus. As a precaution, this distillation should be carried out behind a safety shield.

7. The checkers observed that the refractive index of a sample stored for 5 days in the refrigerator in a stoppered tube wrapped in aluminum foil and cloth changed from n^{25}D 1.6074 to n^{25}D 1.6082.

3. Methods of Preparation

Phenylbromoethyne has been prepared by base-catalyzed dehydrobromination of 1,1- or 1,2-dibromostyrene;[2] by the thermal decomposition of silver 1,2-dibromocinnamate;[2] from phenylethynylmagnesium Grignard reagent and bromine;[3, 4] cyanogen bromide,[5] or benzenesulfonic anhydride;[6] from phenylethynylsodium and cyanogen bromide[4] or p-toluenesulfonylbromide[7]; from phenylethynylsilver and bromine in pyridine;[8] and from phenylethynyllithium and N-bromoimides.[9] The present method is a modification of one in which the hypobromite-phenylacetylene mixture is warmed for 1.5 hours in the presence of an emulsifying agent, 1% potassium stearate[10] or soap,[11] to give 88% yield of product.

4. Merits of the Preparation

The hypohalite route to 1-chloro-, 1-bromo-, or 1-iodoalkynes is both general and convenient. The purity of the reagents does not appear to be critical.

1. Department of Chemistry, Illinois Institute of Technology, Chicago, Illinois.
2. J. V. Nef, *Ann.*, **308**, 264 (1899).
3. J. I. Iotsitch, *J. Phys. Chem. Soc. (Russia)*, **35**, 1269 (1903) [*Bull. Soc. Chim. France*, [3] **34**, 181 (1905)].
4. V. Grignard and C. Courtot, *Bull. Soc. Chim. France*, [4] **17**, 228 (1915); V. Grignard, E. Bellet, and C. Courtot, *Ann. Chim. (Paris)*, [9] **4**, 39 (1915).
5. C. Moureu and R. Delange, *Bull. Soc. Chim. France*, [3] **25**, 99 (1901).
6. L. Field, *J. Am. Chem. Soc.*, **74**, 394 (1952).
7. R. Truchet, *Ann. Chim. (Paris)*, [10] **16**, 309 (1931).
8. T. Agawa and S. I. Miller, Unpublished result.
9. V. Wolf and F. Kowitz, *Ann.*, **638**, 33 (1960).
10. F. Straus, L. Kollek, and W. Heyn, *Ber.*, **63**, 1868 (1930).
11. M. Murray and F. F. Cleveland, *J. Chem. Phys.*, **12**, 156 (1944).
12. R. R. Lii and S. I. Miller, *J. Am. Chem. Soc.*, **91**, 7524 (1969); R. R. Lii, Ph.D. Thesis, Illinois Institute of Technology, 1971.

PHENYL *t*-BUTYL ETHER*

(Ether, *tert*-butyl phenyl)

Method I

$$C_6H_5Br + Mg \xrightarrow{\text{Ether}} C_6H_5MgBr$$

$$C_6H_5MgBr + C_6H_5\overset{\displaystyle O}{\overset{\|}{C}}O_2C(CH_3)_3 \xrightarrow[\text{HCl}]{\text{H}_2\text{O}}$$

$$C_6H_5OC(CH_3)_3 + C_6H_5CO_2H$$

Submitted by CHRISTER FRISELL and SVEN-OLOV LAWESSON.[1]
Checked by WILLIAM G. DAUBEN and GILBERT H. BEREZIN.

1. Procedure

A 1-l., three-necked, round-bottomed flask equipped with a sealed mechanical stirrer, a reflux condenser, and a 500-ml. pressure-equalized dropping funnel is arranged for conducting a reaction in an atmosphere of nitrogen by fitting into the top of the condenser a T-tube attached to a low-pressure supply of nitrogen and to a mercury bubbler. The flask is dried by warming with a soft flame as a slow stream of nitrogen is passed through the system. In the cooled flask a solution of phenylmagnesium bromide is prepared from 13 g. (0.53 g. atom) of magnesium turnings, 79 g. (0.5 mole, 53.6 ml.) of bromobenzene, and 200 ml. of anhydrous ether.

After the preparation of phenylmagnesium bromide is complete, the ethereal solution is cooled in an ice bath and 200 ml. of anhydrous ether is added. A solution of 58.3 g. (0.3 mole, 56 ml.) of *t*-butyl perbenzoate (Note 1) in 120 ml. of anhydrous ether is added, dropwise, with stirring over a 30-minute period, and the stirring is continued for an additional 5 minutes.

The reaction mixture is poured carefully into a cold solution of 40 ml. of concentrated hydrochloric acid in 1 l. of water. The ethereal layer is separated, and the aqueous layer is extracted twice with 150-ml. portions of ether. The combined organic layers are extracted with three 25-ml. portions of 2*M* sodium hydroxide solution, washed with water until the washings are neutral, and then dried over anhydrous magnesium sulfate (Notes 2 and 3). The dried solution is concentrated and the product distilled under reduced pressure, b.p. 57–59°/7 mm. The yield of phenyl *t*-butyl ether is 35–38 g. (78–84%), n_D^{25} 1.4870–1.4880. This synthetic process is applicable to the preparation of other *t*-butyl ethers (Note 4).

2. Notes

1. *t*-Butyl perbenzoate is supplied by Lucidol Division, Wallace and Tiernan Inc., Buffalo 5, New York, or L. Light & Co., Ltd., Colnbrook, Bucks, England. The Lucidol product contains 98% *t*-butyl perbenzoate.

2. The ethereal solution should be tested for peroxides as follows: A few milligrams of sodium iodide, a trace of ferric chloride, and 3 ml. of glacial acetic acid are placed in a test tube and 2 ml. of the ether solution added carefully. When unconsumed perbenzoate is present, a yellow ring is formed immediately between the two phases. If a positive test is obtained, the acid and base treatments should be repeated.

3. By acidification of the sodium hydroxide solution, 29–32 g. of benzoic acid (80–90%) is obtained.

4. The same general method[2] has been used by the submitters to prepare o-tolyl *t*-butyl ether,[3] *m*-tolyl *t*-butyl ether,[3] benzyl *t*-butyl ether,[3] and *p*-anisyl *t*-butyl ether.

3. Methods of Preparation

The method presented is essentially that described by Lawesson and Yang.[4] Phenyl *t*-butyl ether has been prepared by acid-catalyzed condensation of isobutylene and phenol[5] by reaction of diphenyliodonium chloride with potassium *t*-butoxide,[6] and by the next procedure in this volume.[7]

4. Merits of Preparation

The synthesis of *t*-butyl ethers by the reaction of Grignard reagents with *t*-butyl perbenzoate appears to have considerable generality (Note 4), and the perester is a stable, readily available material.

1. Department of Chemistry, University of Uppsala, Uppsala, Sweden.
2. S.-O. Lawesson and G. Schroll, in S. Patai, "The Chemistry of Carboxylic Acids and Esters," Wiley-Interscience, London, 1969, p. 669.
3. S.-O. Lawesson and C. Frisell, *Arkiv Kemi,* 17, 287 (1961).
4. S.-O. Lawesson and N. C. Yang, *J. Am. Chem. Soc.,* 81, 4230 (1959).
5. D. R. Steven, *J. Org. Chem.,* 20, 1232 (1955).
6. F. M. Beringer, P. S. Forgione, and M. D. Yudis, *Tetrahedron,* 8, 49 (1960).
7. M. R. V. Sahyun and D. J. Cram, this volume, p. 926.

PHENYL *t*-BUTYL ETHER*

(Ether, *t*-butyl phenyl)

Method II

$$C_6H_5Br + KOC(CH_3)_3 \rightarrow C_6H_5OC(CH_3)_3 + KBr$$

Submitted by MELVILLE R. V. SAHYUN and DONALD J. CRAM [1]
Checked by WILLIAM G. DAUBEN and DAVID J. ELLIS

1. Procedure

In a loosely stoppered 1-l. round-bottomed flask are placed 37.5 g. (48 ml.) of *t*-butyl alcohol, 150 ml. of dimethyl sulfoxide (Note 1), and a Teflon®-coated magnetic stirring bar. The solution is heated in an oil bath which is placed on a combination magnetic stirrer-hotplate. When the temperature of the mixture reaches 125–130°, 75 g. (0.67 mole) of alcohol-free potassium *t*-butoxide (Notes 2 and 3) is added, the stopper is replaced loosely, and the mixture is stirred. When all the potassium *t*-butoxide is in solution, the stopper is removed, 25 g. (0.159 mole, 17 ml.) of bromobenzene is added in one portion to the hot solution, and an air condenser fitted with a drying tube is *rapidly* placed on the flask. The solution immediately turns dark brown, and an extremely vigorous, exothermic reaction occurs. After 1 minute the reaction mixture is poured into 500 ml. of water. The aqueous solution is saturated with sodium chloride and extracted with four 200-ml. portions of ether (Note 4). The ether extract is washed with three 100-ml. portions of water and dried over anhydrous potassium carbonate. The ether is distilled at atmospheric pressure on a steam bath to leave 17–18 g. of crude phenyl *t*-butyl ether (Note 5). The brown oil is distilled to yield 10–11 g. (42–46%) of pure phenyl *t*-butyl ether, b.p. 45–46° (2 mm.), m.p. −17 to −16°, n^{25}D 1.4860–1.4890 (Note 6). The ether may be hydrolyzed readily to phenol (Note 7).

2. Notes

1. "Baker Analyzed" dimethylsulfoxide, which is freshly opened and dry to Karl Fischer reagent, is used without further purification.

2. Commercial potassium *t*-butoxide is used directly as obtained from the Mine Safety Appliance Research Corp., Callery, Pennsylvania.

3. This amount of potassium *t*-butoxide is not soluble in the dimethylsulfoxide at a lower temperature. An excess of base over *t*-butyl alcohol is necessary to the reaction, and a high concentration of *t*-butyl alcohol (3.3*M*) considerably improves the yield of product desired.

4. The aqueous residue from the extraction can be acidified and extracted to yield phenol which is purified by chromatography on a silica gel column, with an eluant solution composed of 95% pentane and 5% ether by volume. The purified phenol weighs 4.3 g. (29% of the theoretical amount) and is obtained as long needles.

5. The crude material also contains some polymeric material and traces of solvent. Gas chromatography indicates that the phenyl *t*-butyl ether is 60–70% pure at this point.

6. The checkers found their product to contain 0.5–1.0% bromobenzene. Careful redistillation is required to free the product of this impurity.

7. Phenyl *t*-butyl ether is swirled with 6*N* hydrochloric acid until solution is completed. The solution is then saturated with sodium chloride and extracted with ether to yield phenol, identifiable as the tribromide.

3. Methods of Preparation

The procedure described here is based on a method outlined by Cram, Rickborn, and Knox.[2] The method is not a general one for the preparation of a substituted phenyl *t*-butyl ether because an aryne intermediate is involved. It appears that aryl fluorides undergo direct substitution to yield unrearranged aryl *t*-butyl ethers. Alternative methods for preparation of these ethers are listed on page 924.

4. Merits of the Preparation

This reaction sequence illustrates how the rates of many base-catalyzed reactions can be enhanced greatly by substitution of dimethylsulfoxide for the usual hydroxylic solvents.[3] Other examples of the enhanced reactivity of anions in dimethylsulfoxide are found in Wolff-Kishner reductions and Cope elimination reactions.[5] The present reaction illustrates the generation of an aryne intermediate from bromobenzene.[4]

1. Department of Chemistry, University of California, Los Angeles 24, California.
2. D. J. Cram, B. Rickborn, and G. R. Knox, *J. Am. Chem. Soc.*, **82**, 6412 (1960).
3. D. J. Cram, B. Rickborn, C. A. Kingsbury, and P. Haberfield, *J. Am. Chem. Soc.*, **83**, 3678 (1961).
4. D. J. Cram, M. R. V. Sahyun, and G. R. Knox, *J. Am. Chem. Soc.*, **84**, 1734 (1962).

PHENYLCYCLOPROPANE*

(Benzene, cyclopropyl-)

Submitted by R. J. PETERSEN and P. S. SKELL [1]
Checked by THOMAS R. LYNCH and PETER YATES

1. Procedure

Caution! *This reaction should be carried out behind a safety screen.*

A 1-l. three-necked flask is fitted with a reflux condenser, an addition funnel, and a thermometer. It is charged with 450 ml. of 95% ethanol, 230 ml. (236 g.) of 85% hydrazine hydrate (Note 1), and several porcelain boiling chips. The solution is brought to reflux with a heating mantle. Cinnamaldehyde (200 g., 1.51 moles) (Note 2) is added dropwise over a period of 45 minutes to the refluxing solution, while the mixture turns orange because of the formation of cinnamalazine in a side reaction (Note 3). After an additional 30 minutes at reflux, the flask is fitted with a simple takeoff head, and ethanol, water, and hydrazine hydrate are slowly removed by distillation. After approximately 3 hours the pot temperature rises to 200°, and phenylcyclopropane begins to codistil with the last of the hydrazine hydrate (Note 4). The distillate from this point is collected in a 250-ml. receiver, the main fraction coming over at 170–80°. When the pot temperature exceeds 250°, the decomposition is essentially complete (Notes 5, 6).

The crude, cloudy distillate (110–130 g.) is washed twice with 100-ml. portions of water and dried over anhydrous potassium carbonate. Distillation at reduced pressure, b.p. 60° (13 mm.), 79–80° (37 mm.), through a 12-in. Vigreux column gives phenyl-cyclopropane pure enough for most purposes; yield 80–100 g. (45–56%), n^{25}D 1.5309.

2. Notes

1. Matheson, Coleman and Bell technical grade 85% hydrazine hydrate was used.

2. Eastman Organic Chemicals cinnamaldehyde gave satisfactory results. If colorless crystals are present in the neck of the bottle or on the walls above the liquid, the cinnamaldehyde is seriously contaminated with cinnamic acid and should be distilled before use. A small amount of cinnamic acid apparently does not affect the yield of phenylcyclopropane.

3. Reversal of the addition procedure results in formation of cinnamalazine as a major product.

4. Earlier investigators employed strong bases (sodium hydroxide, potassium hydroxide) or platinum on asbestos to catalyze the decomposition of 5-phenylpyrazoline. These catalysts are not necessary and should be avoided because they also cause the reduction of cinnamalhydrazone to propenyl-benzene. Phenylcyclopropane can be freed from propenyl-benzene only with great difficulty.

5. The checkers found in a full-scale run that the pot temperature had to be raised to close to 250° before the onset of reaction, which was then very vigorous.

6. Pyrazoline vapors are known to be rather flammable. It is advisable, therefore, to cool the pyrolysis flask somewhat before dismantling the apparatus. The syrupy residue in the flask sets to a hard mass on cooling; it can be removed by heating under dimethylformamide on a steam bath overnight.

3. Methods of Preparation

Phenylcyclopropane has been prepared by the base-catalyzed decomposition of 5-phenylpyrazoline (33%),[2] by the reaction of

1,3-dibromo-1-phenylpropane with magnesium (68%),[3] and by the reaction of 3-phenylpropyltrimethylammonium iodide with sodium amide in liquid ammonia (80%).[4] However, the method frequently used at present is the reaction of styrene with the methylene iodide-zinc reagent (32%).[5]

4. Merits of the Preparation

The procedure outlined is much quicker and simpler than previous methods. Starting materials are readily available, and the preparation can be run on any scale in the length of a day. Because exclusion of a basic catalyst eliminates the Wolff-Kishner reduction of the cinnamalhydrazone, separation of the 5-phenylpyrazoline from cinnamalhydrazone, or of phenylcyclopropane from propenylbenzene, does not have to be effected. The present procedure can also be used to convert other ring-substituted cinnamaldehydes to the corresponding arylcyclopropanes.

1. Department of Chemistry, The Pennsylvania State University, University Park, Pennsylvania.
2. S. G. Beech, J. H. Turnbull, and W. Wilson, *J. Chem. Soc.*, 4686 (1952).
3. P. J. C. Fierens and J. Nasielski, *Bull. Soc. Chim. Belges*, **71**, 187 (1962).
4. C. L. Bumgardner, *J. Am. Chem. Soc.*, **83**, 4420 (1961).
5. H. E. Simmons and R. D. Smith, *J. Am. Chem. Soc.*, **81**, 4256 (1959).

R(+)- AND S(–)-α-PHENYLETHYLAMINE*

(Benzylamine, α-methyl)

$$\underset{CH_3}{\overset{\displaystyle |}{C_6H_5CH}}\!\!-\!NH_2 \quad \xrightarrow{\text{Resolution}} \quad \text{R(+)-isomer} \qquad \text{S(–)-isomer}$$

Submitted by Addison Ault [1]
Checked by Martin Gall, Elia J. Racah, and
Herbert O. House

1. Procedure

A. *S(–)-α-Phenylethylamine.* A mixture of 31.25 g. (0.208 mole) of (+)-tartaric acid and 450 ml. of methanol is placed in a 1-l. Erlenmeyer flask and heated to boiling. To the hot solution is added, cautiously to avoid foaming, 25.0 g. (26.2 ml., 0.206 mole) of racemic α-phenylethylamine (Note 1) and the resulting solution is allowed to cool. Since crystallization occurs slowly, the solution should be allowed to stand at room temperature for approximately 24 hours. The (–)-amine (+)-hydrogen tartrate salt separates as white prismatic crystals (Note 2). The product (18.1–19.3 g.) should be collected on a filter and washed with a small volume of methanol. The combined mother liquor and methanol washings should be concentrated to a volume of 175 ml. with a rotary evaporator. The resulting mixture is then heated to boiling, and the solution is allowed to cool and stand at room temperature for approximately 24 hours. In this way an additional crop (2.0–3.8 g.) of the (–)-amine (+)-hydrogen tartrate salt may be separated as white prisms (Note 2). The combined methanolic mother liquors and washings from these crystallizations are concentrated to dryness on a rotary evaporator. The crude residual salt is used for the preparation of the (+)-amine.

The combined crops of crude (−)-amine (+)-hydrogen tartrate are pulverized in a mortar and redissolved in 450–500 ml. of boiling methanol. The resulting hot solution is concentrated to 350 ml. (Note 3) and then allowed to cool and stand for 24 hours. After the initial crop (14.3–16.2 g.) of pure (−)-amine (+)-hydrogen tartrate has been collected as white prisms (Note 2) (m.p. 179–182° dec.), the mother liquors and washings are concentrated to 75 ml. and again allowed to stand for 24 hours. In this way a second crop (2.9–3.6 g.) of the pure (−) -amine salt is obtained. The total yield of the pure (−)-amine salt is 17.9–19.1 g. (64–68%).

A mixture of the pure (−)-amine salt (17.9–19.1 g.) and 90 ml. of water is treated with 15 ml. of aqueous 50% sodium hydroxide and the resulting mixture is shaken with four 75-ml. portions of ether. After the combined ether extracts have been washed with 50 ml. of saturated aqueous sodium chloride and dried over magnesium sulfate, the bulk of the ether is distilled from the mixture through a 30-cm. Vigreux column and the residual liquid is distilled under reduced pressure. The (−)-amine is collected as 6.9–7.2 g. (55–58%) of colorless liquid, b.p. 94–95° (28 mm.), n^{25}D 1.5241–1.5244, $[\alpha]^{29}$D −39.4° (neat) (Notes 4, 5).

B. *R(+)-α-Phenylethylamine.* The residual salts (approximately 35 g.) obtained by concentration of the methanolic mother liquors from the initial crystallization of the (−)-amine (+)-hydrogen tartarate are treated successively with 160 ml. of water and 25 ml. of aqueous 50% sodium hydroxide. After the resulting mixture has been extracted with ether, the extract is dried, concentrated, and distilled as previously described. The recovered amine amounts to 12.5–14.1 g. of colorless liquid, b.p. 79–80° (18 mm.), $[\alpha]^{28}$D +23.8 to +24.7° (neat). From the weight and specific rotation data for this amine sample and the reported [2] specific rotation, $[\alpha]^{25}$D +40.6° (neat), for the pure (+)-amine, the amount of excess (+)-amine present in the recovered amine sample is calculated. Typical values range from 0.06 to 0.07 mole of excess (+)-amine. A solution of this partially resolved amine in 90 ml. of 95% ethanol is heated to boiling and then treated with 180 ml. of an ethanolic solution containing a sufficient amount (0.03–0.035 mole) of concentrated sulfuric acid to convert the excess (+)-amine to its neutral sulfate salt

(Note 6). The hot solution is allowed to cool to room temperature, and the crude (+)-amine sulfate which separates as white needles (7.8–9.3 g.) is collected on a filter and washed with 95% ethanol. The combined ethanolic mother liquors and washings are concentrated and allowed to cool to separate a second crop (1.2–1.4 g.) of the crude (+)-amine sulfate. The combined crops of (+)-amine sulfate are dissolved in a minimum volume (about 45 ml. of hot water), and the resulting hot solution is diluted with acetone until it is just saturated at the boiling point. After the solution has been allowed to cool to room temperature, the pure (+)-amine sulfate which separates as white needles, m.p. 240–265° dec. (5.0–6.1 g.) is collected on a filter and washed with cold 95% ethanol. The combined mother liquors and washings are concentrated to dryness, and the residual solid is recrystallized from aqueous acetone to separate additional crops (2.6–2.8 g.) of the pure (+)-amine sulfate. The total yield of the pure amine sulfate is 7.8–8.9 g. (45–51% on the basis of the starting α-phenylethylamine).

A mixture of the pure (+)-amine sulfate (7.8–8.9 g.) and 40 ml. of water is treated with 6.0 ml. of aqueous 50% sodium hydroxide and the resulting mixture is shaken with four 75-ml. portions of ether. The combined ether extracts are washed with 50 ml. of saturated aqueous sodium chloride, dried over magnesium sulfate, and concentrated by distillation of the ether through a 30-cm. Vigreux column. The residual liquid is distilled under reduced pressure to separate 5.1–5.5 g. (41–44%) of the (+)-amine as a colorless liquid, b.p. 85–86° (21 mm.), $n^{25}D$ 1.5243–1.5248, $[\alpha]^{29}D +39.7°$ (neat) (Note 7).

2. Notes

1. A practical grade of racemic α-phenylethylamine supplied by Eastman Organic Chemicals is satisfactory. However, if the racemic amine is highly discolored, distillation before use is recommended.

2. Sometimes a salt separates in the form of white needles. The (−)-amine recovered from these needlelike crystals is not optically pure; $[\alpha]^{25}D -19°$ to $-21°$ (neat). If the product separates either partially or completely as needlelike crystals

during the crystallization, the mixture should be warmed until all the needlelike crystals have dissolved, and then the solution should be allowed to cool slowly. If possible, the solution should be seeded with the prismatic crystals. Separation of the prismatic and needlelike crystals can also be effected by taking advantage of the fact that the needles dissolve more rapidly than the prisms in warm methanol.

3. Because of the low rate of solution of the amine salt, the desired solution is obtained most rapidly by dissolving the salt in excess solvent and then concentrating the solution.

4. The literature value (d_4^{25} 0.9528)[2] for the density of α-phenylethylamine was used to calculate the specific rotation.

5. From the reported specific rotation value, $[\alpha]^{25}$D $-40.14°$ (neat),[3] $[\alpha]^{22}$D $-40.3°$ (neat),[4] the optical purity of this preparation is estimated to be 98%. The boiling point of this amine at atmospheric pressure is 186–187°.

6. For example, a 14.1-g. (0.116 mole) sample of amine, $[\alpha]^{28}$D $+23.8°$ (neat), was estimated to contain 0.0676 mole of excess (+)-amine. Therefore 3.52 g. (0.0345 mole) of concentrated sulfuric acid was added.

7. From the reported specific rotation value, $[\alpha]^{25}$D $+40.6°$ (neat),[2] the optical purity of this preparation is estimated to be 98%. The boiling point of this amine at atmospheric pressure is 186–187°.

3. Discussion

The method presented is based on the procedure of Theilaker and Winkler.[4] It makes use of (+)-tartaric acid, an inexpensive and readily available material, as the resolving agent and provides optically pure samples of both enantiomers of α-phenylethylamine.

Some other methods of resolution include the use of l-malic acid [(+)-form],[5] l- and dl-malic acids [(+)- and (−)-forms],[6] l-malic acid and d-tartaric acids [(+)- and (−)-forms],[7] d-α-bromocamphor-π-sulfonic acid [(−)-form],[8] l-quinic and d-tartaric acids [(+)- and (−)-forms],[9] 2,3,4,6-tetraacetyl-D-glucose [(+)-form],[10] and barium (−)-bornyl sulfate [(+)- and (−)-forms].[11]

The enantiomers of this amine are useful resolving agents. Some of the compounds which have been resolved with one of the optically active forms of α-phenylethylamine are: mandelic acid,[12] α-methylmandelic acid,[13] α-ethylmandelic acid,[14] 2-phenyl-propionic acid,[15] 2-(p-nitrophenyl)propionic acid,[16] 2,3-dichloro-2-methylpropionic acid,[17] 2-phenylbutyric acid,[15] 2-phenylvaleric acid,[18] 2-phenylcaproic acid,[18] α-methylhydrocinnamic acid,[19] β-methylhydrocinnamic acid,[20] benzylsuccinic acid,[21] N-formyl-phenylalanine,[22] N-acetyl-3,5-dibromotyrosine,[23] N-acetyltrypto-phan,[24] 6,6'-dinitrodiphenic acid,[25] and 3-methylcyclohexanone and β-methylcinnamaldehyde, via the amine bisulfite complexes.[26]

1. Department of Chemistry, Cornell College, Mount Vernon, Iowa.
2. A. C. Cope, C. R. Ganellin, H. W. Johnson, Jr., T. V. Van Auken, and H. J. S. Winkler, *J. Am. Chem. Soc.*, **85**, 3276 (1963).
3. R. D. Bach, Ph.D. Dissertation, Massachusetts Institute of Technology, 1967, pp. 35–38.
4. W. Theilacker and H. G. Winkler, *Ber.*, **87**, 690 (1954).
5. J. M. Loven, *J. Prakt. Chem.*, **72**, 307 (1905).
6. A. W. Ingersoll, *J. Am. Chem. Soc.*, **47**, 1168 (1925).
7. A. W. Ingersoll, *Org. Syntheses*, Coll. Vol. **2**, 506 (1943).
8. A. E. Hunter and F. S. Kipping, *J. Chem. Soc.*, **83**, 1147 (1903). C. K. Ingold and C. L. Wilson, *J. Chem. Soc.*, 1493 (1933).
9. E. Andre and C. Vernier, *Compt. Rend.*, **19?**, 1192 (1931).
10. B. Helferich and W. Portz, *Ber.*, **86**, 1034 (1953).
11. A. P. Terent'ev and U. M. Potapov, *Zh. Obshch. Khim.*, **26**, 1225 (1956) [*Chem. Abstr.*, **50**, 16709 (1956)].
12. L. Smith, *J. Prakt. Chem.*, **84**, 743 (1911); A. W. Ingersoll, S. H. Babcock, and F. B. Burns, *J. Am. Chem. Soc.*, **55**, 411 (1933).
13. L. Smith, *J. Prakt. Chem.*, **84**, 731 (1911).
14. L. Smith, *J. Prakt. Chem.*, **84**, 744 (1911).
15. K. Pettersson, *Arkiv Kemi*, **10**, 283 (1956–1957).
16. A. Fredga, *Arkiv Kemi*, **7**, 241 (1954–1955).
17. C. E. Glassick and W. E. Adcock, *Ind. Eng. Chem. Prod. Res. Develop.*, **3**, 14 (1964).
18. K. Pettersson and G. Willdeck, *Arkiv Kemi*, **9**, 333 (1956).
19. A. W. Schrecker, *J. Org. Chem.*, **22**, 33 (1957).
20. E. L. Eliel, P. H. Wilken, and F. T. Fang, *J. Org. Chem.*, **22**, 231 (1957).
21. A. Fredga, *Arkiv Kemi, Mineral. Geol.*, **26B**, No. 11 (1948) [*Chem. Abstr.*, **43**, 1747 (1949)].
22. L. R. Overby and A. W. Ingersoll, *J. Am. Chem. Soc.*, **73**, 3363 (1951).
23. H. D. DeWitt and A. W. Ingersoll, *J. Am. Chem. Soc.*, **73**, 5782 (1951).
24. L. R. Overby, *J. Org. Chem.*, **23**, 1393 (1958).
25. A. W. Ingersoll and J. R. Little, *J. Am. Chem. Soc.*, **56**, 2123 (1934).
26. R. Adams and J. D. Garber, *J. Am. Chem. Soc.*, **71**, 522 (1949).

PHENYLGLYOXAL *

(Glyoxal, phenyl-)

$$CH_3SOCH_3 \xrightarrow{(CH_3)_3COK} CH_3SOCH_2^- K^+$$

$$C_6H_5CO_2C_2H_5 + CH_3SOCH_2^- K^+ \longrightarrow [C_6H_5COCHSOCH_3]^- K^+$$

$$\xrightarrow{H_3O^+} C_6H_5CO\overset{\overset{\displaystyle OH}{\displaystyle |}}{C}HSCH_3 \xrightarrow{Cu(OAc)_2} C_6H_5COCHO$$

Submitted by GERARD J. MIKOL and GLEN A. RUSSELL [1]
Checked by WILLIAM G. DAUBEN, MICHAEL H. McGANN, and
NOEL VIETMEYER

1. Procedure

A. *Phenylglyoxal hemimercaptal.* In a 1-l. three-necked flask equipped with an all-glass mechanical stirrer, a 125-ml. dropping funnel, and a condenser fitted with a nitrogen-inlet tube are placed 90 ml. (99 g., 1.27 moles) of dry dimethyl sulfoxide (Note 1), 120 ml. of dry *t*-butyl alcohol (Note 1), and 57.4 g. (0.51 mole) of potassium *t*-butoxide (Notes 2 and 3). The mixture is warmed to 80°; when all the solid has dissolved, the heating is discontinued, and 72 ml. (75 g., 0.50 mole) of dry ethyl benzoate (Note 1) is added slowly from the dropping funnel. The reaction mixture is stirred at room temperature for 4 hours, and the solvent is removed at 80–90° under reduced pressure until the volume of the reaction mixture has been reduced to 150 ml. (Note 4). The residue is poured into 500 ml. of an ice-water slurry. The resulting aqueous solution is extracted with three 100-ml. portions of ether, and the ethereal extracts are discarded (Note 5). The aqueous solution is acidified with a solution of 190 ml. of concentrated hydrochloric acid in 675 ml. of water, and the mixture is allowed to stand at room temperature for 30 hours. The pale yellow precipitate is removed by suction filtration, washed with 500 ml. of cold water, and air-dried to yield 69–74 g. (76–81%) of phenylglyoxal hemimercaptal, m.p. 103–105°.

B. *Phenylglyoxal.* The phenylglyoxal hemimercaptal prepared as described in procedure A (69–74 g.) is dissolved in 400 ml. of warm chloroform, and 60 g. (0.30 mole) of powdered cupric acetate monohydrate is added in one portion to the well-stirred solution. The mixture is stirred at room temperature for 1 hour; the solids are removed by suction filtration and washed with two 75-ml. portions of chloroform. The combined chloroform filtrate and washings are shaken in a separatory funnel with 75 ml. of water; 20 g. of powdered sodium carbonate is added in small portions to the funnel, and the chloroform solution is shaken with the neutralized aqueous solution. (*Caution! Carbon dioxide is evolved.*) The aqueous layer is separated and extracted with four 30-ml. portions of chloroform. The chloroform solutions are combined and dried with anhydrous magnesium sulfate, and the chloroform is removed under reduced pressure. The residue is fractionally distilled under reduced pressure to yield 43–49 g. (64–73%, based on ethyl benzoate) of anhydrous phenylglyoxal as a yellow liquid, b.p. 63–65° (0.5 mm.).

2. Notes

1. The presence of water results in very rapid saponification of ethyl benzoate. Dimethyl sulfoxide (Crown Zellerbach Corp.) may be dried by stirring with calcium hydride for 4–8 hours, followed by distillation under reduced pressure at 80–90° without filtration. Commercial *t*-butyl alcohol and ethyl benzoate are conveniently dried by stirring for 2–4 hours with calcium hydride followed by filtration.

2. Potassium *t*-butoxide was obtained from Mine Safety Appliances Corp.

3. The potassium salt of dimethyl sulfoxide can also be prepared in the following manner. In a 1-l. three-necked flask equipped with an all-glass mechanical stirrer, a 125-ml. dropping funnel containing 90 ml. of dry dimethyl sulfoxide (Note 1), and a Claisen distillation head and condenser is placed 425 ml. of dry *t*-butyl alcohol (Note 1). The system is flushed with dry nitrogen, and 20 g. (0.51 g. atom) of potassium is added (Note 6). The system is closed to the atmosphere by a mineral oil bubbler through which the evolved hydrogen escapes. The mixture is stirred at 80° until the potassium has dissolved. After cooling, the unreacted alcohol is removed by distillation under reduced pressure until a thick slurry of potassium *t*-butoxide remains

(Note 7). The dimethyl sulfoxide is added from the dropping funnel, and the mixture is heated to 80–90° to dissolve all the solid. The solution is maintained at this temperature, and additional t-butyl alcohol is removed under reduced pressure until the volume of the solution is reduced to 300 ml.

4. Since the volume of the solution at this point is critical, the reaction flask should be calibrated.

5. The aqueous solution can be used to prepare 2-(methylsulfinyl)acetophenone by the following procedure. The solution is acidified to pH 1–2 (Hydrion paper) by the slow addition of concentrated hydrochloric acid with vigorous stirring and is extracted immediately with two 100-ml. portions of chloroform. The chloroform extracts are combined, washed with 75 ml. of saturated aqueous sodium carbonate and two 75-ml. portions of water, and dried over anhydrous magnesium sulfate. The chloroform is removed under reduced pressure, and the resulting solid is pulverized, slurried with 100 ml. of ether, collected by filtration, and air-dried. The 2-(methylsulfinyl)acetophenone weighs 75–77 g. (82–85%); m.p. 85–86°. It can be converted to phenylglyoxal hemimercaptal by treatment with dilute hydrochloric acid in dimethyl sulfoxide solution at room temperature (2 ml. of dimethyl sulfoxide, 2 ml. of concentrated hydrochloric acid, and 15 ml. of water per gram of the keto sulfoxide). The solution is allowed to stand at room temperature for 30 hours, after which the phenylglyoxal hemimercaptal can be isolated as described in procedure A.

6. The potassium should be free of oxide and/or hydroxide to avoid subsequent saponification of ethyl benzoate.

7. A heating mantle may be used, but care must be taken to avoid decomposition on the walls of the flask due to overheating during the later stages of the distillation.

3. Methods of Preparation

Phenylglyoxal has been prepared from isonitrosoacetophenone via the bisulfite compound [2] and by treatment with nitrosyl sulfuric acid [3] or nitrous acid [4]. It has also been prepared by oxidation of benzoylcarbinol with cupric acetate,[5] by heating or aqueous hydrolysis of 2-acetoxy-2-bromoacetophenone,[6] by selenium dioxide oxidation of acetophenone,[7] by oxidation of phenacyl bromide with dimethyl sulfoxide,[8] by oxidative bromination of phenylglyoxal diethyl mercaptal,[9] and by treatment of 2,2-dibromoacetophenone with morpholine followed by acidic hydrolysis.[10] Excellent yields of phenylglyoxal hemihydrate can be obtained on a small scale by the hydrolysis of phenylglyoxal hemimercaptal with boiling dilute hydrochloric acid [11] or in one step from 2-(methylsulfinyl)acetophenone by hydrolysis with boiling 8% phosphoric acid.

4. Merits of the Preparation

This procedure provides a convenient synthesis of phenylglyoxal from readily available starting materials. In addition, the method described appears to have general utility for the synthesis of glyoxals. It has been used for the synthesis of p-tolylglyoxal, p-methoxyphenylglyoxal, p-bromophenylglyoxal, and cyclohexylglyoxal. Since β-keto sulfoxides are readily alkylated in basic solution to yield α-alkyl β-keto sulfoxides,[12] it would appear possible to extend the scope of the reaction to yield a variety of α-diketones.

1. Department of Chemistry, Iowa State University, Ames, Iowa 50010.
2. H. von Pechmann, *Ber.*, **20**, 2904 (1887); H. Müller and H. von Pechmann, *Ber.*, **22**, 2556 (1889); A. Pinner, *Ber.*, **35**, 4131 (1902); **38**, 1531 (1905); I. Smedley, *J. Chem. Soc.*, **95**, 218 (1909).
3. C. Neuberg and E. Hofmann, *Biochem. Z.*, **229**, 443 (1930).
4. C. Neuberg and E. Hofmann, *Biochem. Z.*, **239**, 495 (1931); S. Cusmano, *Gazz. Chim. Ital.*, **68**, 129 (1938).
5. J. U. Nef, *Ann.*, **335**, 269 (1904); M. Henze, *Z. Physiol. Chem.*, **198**, 82 (1931); **200**, 232 (1931).
6. W. Madelung and M. E. Oberwegner, *Ber.*, **65**, 931 (1932).
7. H. L. Riley, J. F. Morley, and N. A. C. Friend, *J. Chem. Soc.*, 1875 (1932); H. L. Riley, Brit. Patent 354,798 (1930) [*C.A.*, **26**, 3804 (1932)]; U.S. Patent 1,955,890 (1934) [*C.A.*, **28**, 4067 (1934)]; H. A. Riley and A. R. Gray, *Org. Syntheses*, Coll. Vol., **2**, 509 (1943); R. Bousset, *Bull. Soc. Chim. France*, [5] **6**, 986 (1939).
8. N. Kornblum, J. W. Powers, G. J. Anderson, W. J. Jones, H. O. Larson, O. Levand, and W. M. Weaver, *J. Am. Chem. Soc.*, **79**, 6562 (1957).
9. F. Weygand and H. J. Bestmann, *Ber.*, **90**, 1230 (1957).
10. M. Kerfanto, *Compt. Rend.*, **254**, 493 (1962).
11. H.-D. Becker, G. J. Mikol, and G. A. Russell, *J. Am. Chem. Soc.*, **85**, 3410 (1963).
12. E. J. Corey and M. Chaykovsky, *J. Am. Chem. Soc.*, **86**, 1639 (1964); **87**, 1345 (1965).

2-PHENYLINDAZOLE

(2*H*-Indazole, 2-phenyl-)

o-NO$_2$C$_6$H$_4$CHO + C$_6$H$_5$NH$_2$ $\longrightarrow$ o-NO$_2$C$_6$H$_4$CH$=$NC$_6$H$_5$

$$\xrightarrow{\text{(C}_2\text{H}_5\text{O)}_3\text{P}}$$

Submitted by J. I. G. Cadogan and R. K. Mackie [1]
Checked by William G. Dauben, Harold B. Morris,
and Kent E. Opheim

1. Procedure

A. *o-Nitrobenzalaniline.* A mixture of 14 g. (0.15 mole) of aniline (Note 1) and 22.7 g. (0.15 mole) of *o*-nitrobenzaldehyde (Note 2) is heated in a 100-ml. round-bottomed flask on a water bath for 1 hour, allowed to cool, and dissolved in 100 ml. of ether. The ethereal solution is dried, and the ether is removed by distillation. The residue solidifies on standing (Note 3) and is recrystallized from 55 ml. of water-ethanol (1:8) to yield 29.4–31.8 g. (87–94%) of yellow *o*-nitrobenzalaniline, m.p. 64–66° (Note 4).

B. *2-Phenylindazole.* In a 200-ml. round-bottomed flask fitted with a condenser are mixed 50 g. (0.30 mole) of triethyl phosphite (Note 5) and 22.6 g. (0.10 mole) of *o*-nitrobenzalaniline. The apparatus is sealed from the atmosphere by means of a liquid paraffin bubbler that consists of a U-tube the bend of which is just filled with mineral oil. The apparatus is flushed with nitrogen, and the contents are kept under nitrogen during the reaction. The mixture is heated at 150° in an oil bath for 8 hours and cooled, and the condenser is replaced by a Claisen distillation head. Triethyl phosphite, b.p. 46–48° (10 mm.), and triethyl phosphate, b.p. 90–92° (10 mm.), are removed by distillation under reduced pressure; the volume of distillate is 48–50 ml. On cooling, the black residue solidifies. The flask is filled with glass wool, and the remaining phosphite and phosphate (1–3 g.) are removed by distillation at 30–50° (1 mm.). The residue of crude 2-phenylindazole is distilled at 10^{-4} mm.; b.p. 108–112°. The yield is 13–15 g. (67–78%) (Notes 6 and 7).

This product is crystallized from 75–100 ml. of ethanol-water (7:3) to yield pale yellow crystals, m.p. 81–82°. Additional material is obtained by dilution of the mother liquor with *ca.* 200 ml. of water and two crystallizations as before. The total yield is 10–12 g. (52–62%).

2. Notes

1. Aniline is purified by distillation from zinc dust.

2. The reagent as supplied by British Drug Houses or Eastman Organic Chemicals was used directly.

3. If the product does not solidify at room temperature, it should be cooled with dry ice.

4. *o*-Nitrobenzalaniline is very photosensitive and should be kept away from light as much as possible.

5. The reagent as supplied by Albright and Wilson, Ltd., or Matheson, Coleman and Bell was fractionally distilled from sodium and used within a few days of distillation.

6. A slightly purer sample may be obtained by chromatography on alumina. Elution with chloroform-benzene (1:4) gives a pale yellow solid which is purified further by crystallization from 70% ethanol.

7. The checkers found it more convenient to transfer the crude, black 2-phenylindazole to an apparatus for simple bulb-to-bulb distillation and not to retain the distillation head.

3. Methods of Preparation

The procedure given here is essentially that described previously by the submitters [2] and is based on the early work of Knoevenagel.[3] 2-Phenylindazole has been prepared by reduction of N-(*o*-nitrobenzyl)aniline with tin and hydrochloric acid,[4] by reduction of N-(*o*-nitrobenzyl)-N-nitrosoaniline with tin and hydrochloric acid,[5] by dehydration of 2-(phenylazo)benzyl alcohol,[6] by elimination of acetic acid from 2-(phenylazo)benzyl acetate,[7] by dehydrogenation of 3,3a,4,5,6,7-hexahydro-2-phenylindazole with sulfur,[8] and by thermal decomposition of *o*-azidobenzalaniline.[9]

4. Merits of the Preparation

Reductive cyclization of nitro compounds by triethyl phosphite is a general method for the preparation of a variety of nitrogen-containing heterocyclic systems. The submitters have synthesized the following ring systems by this method from the starting materials given in parentheses: 2-arylindoles (*o*-nitrostilbenes),[2] 2-arylindazoles (*o*-nitrobenzalanilines),[2] 2-arylbenzotriazoles (*o*-nitroazobenzenes),[2] carbazoles (*o*-nitrobiphenyls),[2] phenothiazines (*o*-nitrodiphenyl sulfides),[10, 11] and anthranils (*o*-nitrophenyl ketones).[10]

The products are isolated in good yield in a one-stage synthesis from starting materials that are readily available in the main. An alternative method involves the decomposition of the corresponding azides.[9, 12] These compounds are less readily available and are more hazardous to use than are the nitro compounds used in the present synthesis. This synthesis also gives better yields than the cyclization using ferrous oxalate,[12, 13] which is performed under much harsher conditions. The present method of synthesis has been reviewed.[14]

1. Department of Chemistry, St. Salvator's College, University of St. Andrews, St. Andrews, Fife, Scotland.
2. J. I. G. Cadogan, M. Cameron-Wood, R. K. Mackie, and R. J. G. Searle, *J. Chem. Soc.*, 4831 (1965); J. I. G. Cadogan and R. J. G. Searle, *Chem. Ind. (London)*, 1282 (1963).
3. E. Knoevenagel, *Ber.*, **31**, 2609 (1898).
4. C. Paal and F. Krecke, *Ber.*, **23**, 2634 (1890); C. Paal, *Ber.*, **24**, 959 (1891).
5. M. Busch, *Ber.*, **27**, 2897 (1894).
6. P. Freundler, *Compt. Rend.*, **136**, 1136 (1903); *Bull. Soc. Chim. France*, [3] **29**, 742 (1903).
7. P. Freundler, *Compt. Rend.*, **138**, 1425 (1904); *Bull. Soc. Chim. France*, [3] **31**, 868 (1904).
8. I. I. Grandberg, A. N. Kost, and L. S. Yaguzhinskii, *Zh. Obshch. Khim.*, **29**, 2537 (1959); *J. Gen. Chem. USSR (Engl. Transl.)*, **29**, 2499 (1959).
9. L. Krbechek and H. Takimoto, *J. Org. Chem.*, **29**, 1150 (1964).
10. J. I. G. Cadogan, R. Marshall, D. M. Smith, and M. J. Todd, *J. Chem. Soc.* (C), 2441 (1970).
11. J. I. G. Cadogan, S. Kulik, C. Thomson, and M. J. Todd, *J. Chem. Soc.* (C), 2437 (1970); J. I. G. Cadogan and S. Kulik, *J. Chem. Soc.* (C), 2621 (1971).

12. R. A. Abramovitch and B. A. Davis, *Chem. Rev.*, **64**, 149 (1964).
13. R. A. Abramovitch, Y. Ahmad, and D. Newman, *Tetrahedron Lett.*, 752 (1961).
14. J. I. G. Cadogan, *Quart. Rev.*, **22**, 222 (9168); *Synthesis*, **1**, 111 (1969).

N-PHENYLMALEIMIDE*

(Maleimide, N-phenyl-)

Submitted by M. P. Cava, A. A. Deana, K. Muth, and
M. J. Mitchell.[1]
Checked by Carole L. Olson, Marjorie C. Caserio,
and John D. Roberts.

1. Procedure

A. *Maleanilic acid.* In a 5-l. three-necked flask provided with a paddle-type stirrer, a reflux condenser, and a dropping funnel are placed 196 g. (2 moles) of maleic anhydride (Note 1) and 2.5 l. of ethyl ether (Note 2). The stirrer is started and, when all the maleic anhydride has dissolved, a solution of 182 ml. (186 g., 2 moles) of aniline (Note 3) in 200 ml. of ether (Note 2) is run in through the dropping funnel (Note 4). The resulting thick suspension is stirred at room temperature for 1 hour and is then cooled to 15–20° in an ice bath. The product is obtained by suction filtration. It is a fine, cream-colored powder, m.p. 201–202°, suitable for use in the next step without purification. The yield is 371–374 g. (97–98%).

B. *N-Phenylmaleimide.* In a 2-l. Erlenmeyer flask are placed 670 ml. of acetic anhydride (Note 5) and 65 g. of anhydrous sodium acetate. The maleanilic acid (316 g.), obtained as described above, is added, and the resulting suspension is dissolved by swirling and heating on a steam bath for 30 minutes (Note 6). The reaction mixture is cooled almost to room temperature in a cold water bath and is then poured into 1.3 l. of ice water. The precipitated product is removed by suction filtration, washed three times with 500-ml. portions of ice-cold water and once with 500 ml. of petroleum ether (b.p. 30–60°), and dried. The yield of crude N-phenylmaleimide is 214–238 g. (75–80%), m.p. 88–89°. Recrystallization from cyclohexane gives canary-yellow needles, m.p. 89–89.8° (Note 7).

2. Notes

1. Reagent grade maleic anhydride is used without purification.
2. Reagent grade anhydrous ether is used.
3. Reagent grade aniline is used without further purification.
4. The aniline solution may be run in as fast as is possible without flooding the condenser.
5. Carbide and Carbon or Baker's Analyzed technical grade acetic anhydride is used.
6. The sodium acetate fails to dissolve completely.
7. About 500 ml. of the refluxing solvent will dissolve some 58 g. of N-phenylmaleimide. The recovery of recrystallized material is approximately 93%.

3. Methods of Preparation

The procedure described here is based on a method outlined in U. S. patent 2,444,536.[2] N-Phenylmaleimide has also been prepared by the dry distillation of the aniline salt of malic acid,[3,4] by treating the aniline salt of malic acid with phosphorus pentoxide,[5] and by treating maleanilic acid with phosphorus trichloride or with phosphorus pentoxide.[6] Ring-substituted N-phenylmaleimides, viz., N-(*p*-methoxyphenyl)-, N-(*p*-ethoxy-phenyl)-, and N-(*p*-nitrophenyl)maleimide, have been prepared by treatment of the appropriate maleanilic acids with acetic anhydride and fused potassium acetate.[7]

4. Merits of Preparation

N-Phenylmaleimide is an active dienophile in the Diels-Alder reaction and usually gives crystalline adducts.

1. Chemistry Department, The Ohio State University, Columbus, Ohio.
2. N. E. Searle (to E. I. du Pont de Nemours and Co., Inc.) U.S. pat. 2,444,536 (1948) [*C.A.*, 42, 7340c (1948)].
3. A. Michael and J. F. Wing, *Am. Chem. J.*, 7, 278 (1885).
4. R. Anschutz and Q. Wirtz, *Ann.*, 239, 140, 142 (1887).
5. K. Auwers, *Ann.*, 309, 346 (1899).
6. A. E. Kretov and N. E. Kul'chitskaya, *Zhur. Obshchei Khim.*, 26, 208 (1956) [*C.A.*, 50, 13771g (1956)].
7. W. R. Roderick, *J. Am. Chem. Soc.*, 79, 1710 (1957).

2-PHENYL-5-OXAZOLONE*

(2-Phenyl-2-oxazolin-5-one)

$$\underset{\substack{|\\ \text{C}=\text{O}\\ |\\ \text{C}_6\text{H}_5}}{\text{NHCH}_2\text{COOH}} \quad \xrightarrow{\text{(CH}_3\text{CO)}_2\text{O}} \quad \underset{\substack{|\\ \text{C}\\ |\\ \text{C}_6\text{H}_5}}{\underset{\text{N}\diagdown\quad\diagup\text{O}}{\overset{\text{CH}_2-\text{C}=\text{O}}{|\qquad\quad|}}} \quad + \quad \text{CH}_3\text{COOH}$$

Submitted by G. E. VANDENBERG,[1] J. B. HARRISON,[2] H. E. CARTER,[2] and B. J. MAGERLEIN[1]
Checked by WILLIAM G. DAUBEN, NOEL VIETMEYER, and STEVEN A. SCHMIDT

1. Procedure

A mixture of 537 g. (3 moles) of hippuric acid (Note 1) and 1.6 l. (17 moles) of acetic anhydride is prepared in a 3-l. three-necked, round-bottomed flask fitted with a sealed stirrer, a reflux condenser with a drying tube, a thermometer, and a nitrogen inlet tube (Note 2).

In a nitrogen atmosphere the reaction mixture is heated to 80° *on a water bath* over a period of 40 minutes with stirring. The solids slowly dissolve, and a yellow-orange solution results (Note 3). The reaction mixture is cooled to 5°, and the reflux condenser is turned downward for distillation. The condenser is cooled with a circulating fluid maintained at −20° to −40° (Note 4). The receiver is immersed in dry ice and acetone. At 1–3 mm. pressure, 1.5 l. of distillate is collected while the distillation flask is kept in a water bath maintained at 50° (Note 5). A capillary tube is used to bubble nitrogen through the reaction mixture to ensure good agitation. The distillation residue is dissolved in 1 l. of *t*-butanol, the solution scratched to encourage crystallization, and it is refrigerated overnight. The crystals are collected by filtration (Note 2). The filter cake is washed with a minimum volume of *t*-butanol and then with Skellysolve B (Note 6) or hexane. The moist cake is slurried with Skellysolve B or hexane under a nitrogen atmosphere (Note 7), filtered, and dried to constant weight under vacuum at ambient temperature. The yield of light buff to yellow crystals, m.p. 89–92°, is 320–328 g. (66–68%).

2. Notes

1. The checkers ran the reaction on one-fifth scale.

2. Exposure to air produces a pink to red product of lower melting point and purity.

3. Temperature control is important in order to obtain a high yield of light-colored product. Prolonged heating at 80° is to be avoided.

4. Acetone, cooled by passing through a copper coil placed in a dry ice-acetone bath, may be circulated through the condenser. If a temperature lower than −40° is used, the distillate will freeze in the condenser.

5. This distillate should be collected within about 1 hour, as longer distillation time diminishes the yield.

6. A saturated hydrocarbon fraction, b.p. 60–71°, available from the Skelly Oil Company, Kansas City, Missouri.

7. Alternatively, the moist cake may be recrystallized by dissolving in hot *t*-butanol (about 800 ml.) and diluting with Skellysolve B. The yield is then about 40–50% and the melting point about the same as that of the reslurried product.

3. Methods of Preparation

This procedure is a modification of the original method of preparation of 2-phenyl-5-oxazolone [3] which has since appeared in the literature in various forms.[4-6] In addition to the use of acetic anhydride, the cyclization of hippuric acid to 2-phenyl-5-oxazolone has been described using phosphorus tribromide [7] and N,N'-dicyclohexylcarbodiimide.[8]

4. Merits of the Preparation

This procedure offers a reproducible method for the preparation of 2-phenyl-5-oxazolone, which is not commercially available. It illustrates that strict attention to detail often smooths out an erratic procedure. 2-Phenyl-5-oxazolone is, of course, an important intermediate in the synthesis of α-amino acids and related materials.[6]

1. Research Laboratories, The Upjohn Company, Kalamazoo, Michigan.
2. University of Illinois, Urbana, Illinois.
3. J. W. Cornforth, in H. T. Clarke, J. R. Johnson, and R. Robinson, "Chemistry of Penicillin," Princeton University Press, Princeton, New Jersey, 1949, p. 778.
4. M. M. Shemyakin, S. I. Lur'e, and E. I. Rodionovskaya, *Zh. Obshch. Khim*, **19**, 769 (1949) [*C.A.*, **44**, 1096 (1950)].
5. H. E. Carter, J. B. Harrison, and D. Shapiro, *J. Am. Chem. Soc.*, **75**, 4705 (1953).
6. J. M. Stewart and D. W. Woolley, *J. Am. Chem. Soc.*, **78**, 5336 (1956).
7. J. H. Hunter, J. W. Hinman, and H. E. Carter, in H. T. Clarke, J. R. Johnson, and R. Robinson, "Chemistry of Penicillin," Princeton University Press, Princeton, 1949, p. 915.
8. I. T. Strukov, *Zh. Obshch. Khim*, **29**, 2359 (1959) [*C.A.*, **54**, 9889 (1960)].

2-PHENYLPERFLUOROPROPENE

[Styrene, β,β-difluoro-α-(trifluoromethyl)-]

$$C_6H_5COCF_3 + CClF_2CO_2Na + (C_6H_5)_3P \rightarrow$$

$$C_6H_5C{=}CF_2 + (C_6H_5)_3PO + NaCl + CO_2$$
$$\underset{CF_3}{|}$$

Submitted by Frank E. Herkes and Donald J. Burton [1]
Checked by W. C. Ripka and R. E. Benson

1. Procedure

Caution! This compound is an analog of the toxic olefin, perfluoroisobutylene. Since its toxicity is unknown, proper care should be exercised in its handling.

A 1-l., three-necked, round-bottomed flask is equipped with a mechanical stirrer, a nitrogen-inlet tube, and a reflux condenser connected to a dry ice-trichloroethylene trap (Note 1) that is followed by a water trap to measure carbon dioxide evolution (Note 2). The flask is charged with 65.6 g. (0.250 mole) of triphenylphosphine (Note 3) and 43.5 g. (0.250 mole) of α,α,α-trifluoroacetophenone (Note 4) in 200 ml. of dry diglyme (Note 5). The solution is heated to a bath temperature of 140° under a nitrogen atmosphere, and the nitrogen-inlet tube is replaced by a pressure-equalizing dropping funnel (Note 6) containing a solution of 76.2 g. (0.500 mole) of sodium chlorodifluoroacetate (Note 7) in 150 ml. of dry diglyme. The solution of sodium chlorodifluoroacetate is added dropwise over a period of 1 hour (Note 8), and the reaction mixture is heated for an additional hour at 130–140° to ensure complete decarboxylation of the salt (Note 9). The colors of the reaction mixture are characteristic of ylid reactions, changing from a creamy white to a creamy orange and finally to a deep brown.

The reaction mixture is then cooled to room temperature under nitrogen. The cool mixture is steam distilled until 2 l. of distillate has been collected. The lower, organic layer is separated from the distillate, washed with cold water (Note 10) to remove diglyme, and dried over anhydrous calcium sulfate. Fractional distillation gives 26–31 g. (50–60%) of 2-phenylperfluoropropene, b.p. 58–59° (54 mm.), n^{20}D 1.4225 (Notes 11 and 12).

2. Notes

1. In addition to carbon dioxide, small amounts of fluoroformyl fluoride and chloride are formed and swept out with the carbon dioxide.

2. The rate of decarboxylation can be followed qualitatively by collecting the liberated carbon dioxide over water, *e.g.*, by the use of a wet test meter.

3. Eastman Organic Chemicals white label triphenylphosphine was used directly.

4. Available from Pierce Chemical Co., Rockford, Illinois. The ketone can also be prepared conveniently from phenyl-magnesium bromide and trifluoroacetic acid by the method of Dishart and Levine.[2]

5. Diglyme (Ansul Ether 141) was predried over calcium hydride and distilled under reduced pressure from lithium alumi-num hydride; b.p. 62–63° (15 mm.).[3] The yield of olefin is dependent on the dryness of the solvent. The formation of 2-phenyl-2H-perfluoropropane is favored by the presence of water in the solvent.

6. A dropping funnel of the type described by Benson and McKusick [4] is satisfactory.

7. Sodium chlorodifluoroacetate is prepared in quantitative yield by careful neutralization of 130.5 g. (1.00 mole) of chlorodi-fluoroacetic acid (available from Allied Chemical Corp.) in 300 ml. of ether with 53.0 g. (0.500 mole) of anhydrous sodium carbonate, removal of the solvent and water under reduced pressure, and drying over phosphorus pentoxide in a vacuum desiccator. Studies have shown that the best yield of olefin is obtained when a 100% excess of salt is used.

8. The salt is added at a rate sufficient to cause a constant evolution of carbon dioxide. The reaction is slightly exothermic (*ca.* 10° temperature rise).

9. A total of 9460 ml. (72% STP) of carbon dioxide was collected.

10. Eight washings with 50-ml. portions of water were found to remove all the diglyme.

11. The submitters used an 18-in. spinning-band column. The product was shown to be >99.9% pure by gas-liquid chromatography on Carbowax 20M.

12. The checkers used a 40-cm. spinning-band column. The product, n^{25}D 1.4237, was shown to be 98.9% pure by gas-liquid chromatography on a 6-ft. 20% fluorosilicone column. The retention time was 3.75 minutes with a flow rate of helium of 100 ml. per minute, and a column temperature of 125° with the injection port at 170°. The ^{19}F n.m.r. spectrum (56.4 MHz) consists of four lines of equal intensity centered at +3396 Hz from trichlorofluoromethane (internal) and two sets of two overlapping quartets centered at +4369 Hz and +4461 Hz, respectively. The integrated intensities of the three sets of fluorine resonances are 3:1:1.

3. Methods of Preparation

This procedure is a modification of the method previously reported by the submitters.[5] 2-Phenylperfluoropropene has been reported as a by-product of the thermal decomposition of 7,7-bis-(trifluoromethyl)-1,3,5-cycloheptatriene; however, no experimental procedure was given.[6]

4. Merits of the Preparation

The procedure illustrates a fairly general method for the preparation of β-substituted perfluoroolefins. The method has been applied to the synthesis of 2-cyclohexyl- (70%), 2-benzyl- (61%), and 2-(p-fluorophenyl)perfluoropropenes (67%), and it is probably applicable to any α-trifluoromethyl ketone. Olefins containing a perfluoroalkyl group other than trifluoromethyl can be prepared by the same procedure by the substitution of lithium chlorodifluoroacetate for sodium chlorodifluoroacetate.[7] Other routes to β-substituted perfluoroolefins are not general or convenient. Routes to perfluoroolefins generally yield the α-substituted olefin rather than the β-substituted olefin.

This method can also be utilized as a general method for the preparation of olefins with terminal difluoromethylene groups from aldehydes.[8] Also, by the substitution of tributylphosphine for triphenylphosphine in this procedure, ketones other than those containing an α-perfluoroalkyl group can be converted to terminal difluoromethylene compounds.[9]

1. Department of Chemistry, University of Iowa, Iowa City, Iowa 52240.
2. K. T. Dishart and R. Levine, *J. Am. Chem. Soc.*, **78**, 2268 (1956).
3. G. Zweifel and H. C. Brown, *Org. Reactions*, **13**, 28 (1963).
4. R. E. Benson and B. C. McKusick, *Org. Syntheses*, Coll. Vol. **4**, 747 (1963).
5. D. J. Burton and F. E. Herkes, *Tetrahedron Lett.*, 1883 (1965).
6. D. M. Gale, W. J. Middleton, and C. G. Krespan, *J. Am. Chem. Soc.*, **87**, 657 (1965).
7. D. J. Burton and F. E. Herkes, *Tetrahedron Lett.*, 4509 (1965).
8. S. A. Fuqua, W. G. Duncan, and R. M. Silverstein, *J. Org. Chem.*, **30**, 1027 (1965); this volume, p. 390.
9. S. A. Fuqua, W. G. Duncan, and R. M. Silverstein, *J. Org. Chem.*, **30**, 2543 (1965).

9-PHENYLPHENANTHRENE

(Phenanthrene, 9-phenyl-)

Submitted by FRANK B. MALLORY and CLELIA S. WOOD[1]
Checked by WILLIAM G. DAUBEN and DONALD N. BRATTESANI

1. Procedure

A solution of 2.56 g. (0.01 mole) of triphenylethylene[2] (Note 1) and 0.127 g. (0.5 mmole) of iodine in 1 l. of cyclohexane (Note 2) is placed in a 1.5-l. beaker and stirred magnetically (Note 3). A Hanovia water-cooled 19433 Vycor immersion well fitted with a

200-watt 654A-36 mercury lamp (Notes 4 and 5) is inserted into the beaker, and the lamp is started (Note 6). The irradiation is continued for about 3 hours (Note 7).

The reaction mixture is transferred to a 2-l. round-bottomed flask, and the solvent is evaporated under reduced pressure (Note 8). The residue is dissolved in 50 ml. of warm cyclohexane (Note 2), and the solution (Note 9) is poured onto a column of alumina (Note 10) 1.8 cm. in diameter and 6–7 cm. in length. The round-bottomed flask is rinsed with three 10-ml. portions of cyclohexane, and the rinsings are poured onto the column. The column is eluted with additional cyclohexane (about 100 ml.) until no appreciable amount of 9-phenylphenanthrene is obtained in the eluate. The elution of any yellow material from the column should be avoided. The total eluate is evaporated to dryness under reduced pressure (Note 8), and the residue is re-crystallized from 40–45 ml. of 95% ethanol to give 1.65–1.90 g. (65–75%) of 9-phenylphenanthrene, m.p. 103.5–104.5° (Notes 11, 12, and 13).

2. Notes

1. The triphenylethylene used by the submitters had been recrystallized from absolute ethanol, and the material melted at 68.0–68.6°.

2. Eastman Organic Chemicals practical grade cyclohexane is distilled before use.

3. A 4-cm. Teflon®-coated stirring bar gives sufficiently effective stirring.

4. This unit is sold as a Hanovia Laboratory Photochemical Reactor by the Hanovia Lamp Division, Engelhardt Industries, Inc., 100 Chestnut Street, Newark 5, New Jersey.

5. A relatively inexpensive light source and probe can be made as described below and used in place of the Hanovia unit. A 100-watt General Electric H100A4/T or H100A38-4 mercury lamp, available from the Lamp Department, General Electric Co., Nela Park, Cleveland, Ohio, is modified by cutting away the outer glass envelope and by detaching the inner quartz bulb from the screw base on which it is mounted. The two electrical

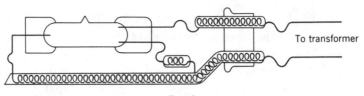

F<small>IG.</small> 2

leads from the lamp are then connected, s shown in Fig. 2, by means of insulated wire to a suitable ower supply such as a General Electric 9T64Y-3518 or 9T64 -1019 transformer. The modified mercury lamp is then inserted in a 17-mm. I.D. quartz tube which is about 30 cm. long and is sealed on one end. This tube is made from Clear Fused Quartz tubing available from the Lamp Glass Department, General Electric Co., Nela Park, Cleveland 12, Ohio.

Irradiations with this type of light source are carried out using a 1-l. Erlenmeyer flask as the reaction vessel instead of a 1.5-l. beaker. The flask is placed in a cold-water bath that is supported on a magnetic stirrer. The bath can be made from a 10-qt. polyethylene bucket with a $\frac{3}{4}$-in. hole bored about 1 in. from the top of the bucket and a piece of rubber tubing with a $\frac{5}{8}$-in. bore and $\frac{1}{8}$-in. wall inserted as a drain. A stream of 24° tap water run into the bucket at a flow rate of 5 l. per min. maintains the temperature of the reaction mixture below 33°.

6. Unfiltered light from mercury lamps is damaging to the eyes; suitable precautions, such as wearing appropriate glasses and surrounding the reaction vessel with aluminum foil, should be taken.

7. The irradiation time required depends on the type of light source used and can be determined by following the progress of the reaction by infrared spectroscopy. A 10-ml. aliquot is withdrawn from the reaction mixture and evaporated to dryness under reduced pressure; the residue is dissolved in 0.5 ml. of carbon tetrachloride, and the spectrum is obtained using 0.1-mm. sodium chloride cells. A new peak appears at 899 cm.$^{-1}$, and the ratio of the absorbance of the peak at 703 cm.$^{-1}$ to that of the peak at 727 cm.$^{-1}$ continuously decreases during the course of the reaction. Using these spectral criteria, the submitters judged the

reaction to be complete after 4 hours of irradiation with the lamp described in Note 5; however, the recrystallized products from 4-hour reactions melted about 0.4° lower than those from 5-hour reactions. The submitters found that varying the irradiation time from 4 hours to 8 hours had no significant effect on the yield of 9-phenylphenanthrene. The extent of the reaction can also be monitored by gas-liquid chromatography.

8. It is convenient to use a rotary evaporator and a water aspirator for this operation.

9. This solution may be purple in color owing to incomplete removal of iodine during the reduced-pressure evaporation.

10. The submitters used Merck 71707 aluminum oxide. The checkers used Woelm neutral alumina, Activity I.

11. A sample of 9-phenylphenanthrene that had been exhaustively purified by zone refining and by recrystallization melted at 104.1–104.7°. The melting point has been reported as 104–105°[3-5] and as 105–106°.[6, 7]

12. Reactions carried out at higher concentrations or on larger scales give slightly lower yields of less pure material. To obtain larger amounts of the product, the submitters recommend irradiating in batches on the scale specified in the procedure and combining the reaction mixtures prior to the chromatographic purification.

13. Using the apparatus described in Note 5, the submitters obtained 2.08–2.17 g. (82–85%) of 9-phenylphenanthrene.

3. Methods of Preparation

This preparation is based on a procedure published by the submitters.[8, 9] 9-Phenylphenanthrene has been prepared previously by the reaction of phenyllithium with 9-chlorophenanthrene,[10] by the high-temperature dehydrogenation with palladium on charcoal of the Diels-Alder dimer of 1-phenyl-1,3-butadiene,[11] and by the acid-catalyzed cyclization of the alcohol formed from the reaction of 2-biphenylylmagnesium iodide and 2-phenoxyacetophenone.[3]

4. Merits of the Preparation

This preparation illustrates a reasonably general method for obtaining 1-, 3-, or 9-substituted phenanthrenes in good yields from the photocyclization of the corresponding *o*-, *p*-, or *α*-substituted stilbenes.[8, 9] The submitters have obtained satisfactory results with bromo, chloro, fluoro, methoxy, methyl, phenyl, trifluoromethyl, and carboxyl substituents. *α*-Styrylnaphthalene gives chrysene, *β*-styrylnaphthalene gives benzo[*c*]phenanthrene, and 1,2-di-*α*-naphthylethylene gives picene.

The photocyclization has been found not to occur with stilbenes substituted with acetyl, dimethylamino, or nitro groups. Iodo substituents are replaced by hydrogen by photolysis in cyclohexane solution.[9, 12] *m*-Substituted stilbenes give mixtures of 2- and 4-substituted phenanthrenes which generally are difficult to separate.

1. Department of Chemistry, Bryn Mawr College, Bryn Mawr, Pennsylvania.
2. H. Adkins and W. Zartman, *Org. Syntheses*, Coll. Vol. **2**, 606 (1943).
3. C. K. Bradsher and A. K. Schneider, *J. Am. Chem. Soc.*, **60**, 2960 (1938).
4. S. Wawzonek, E. Dufek, and N. M. Sial, *J. Org. Chem.*, **21**, 276 (1956).
5. R. C. Fuson and S. J. Strycker, *J. Am. Chem. Soc.*, **79**, 2633 (1957).
6. C. F. Koelsch, *J. Am. Chem. Soc.*, **56**, 480 (1934).
7. H. E. Eschinazi and F. Bergmann, *J. Am. Chem. Soc.*, **65**, 1411 (1943).
8. F. B. Mallory, C. S. Wood, and J. T. Gordon, *J. Am. Chem. Soc.*, **86**, 3094 (1964); C. S. Wood and F. B. Mallory, *J. Org. Chem.*, **29**, 3373 (1964).
9. C. S. Wood, doctoral dissertation, Bryn Mawr College, 1963.
10. R. Huisgen, J. Sauer, and A. Hauser, *Ber.*, **91**, 2366 (1958).
11. K. Alder, J. Haydn, and W. Vogt, *Ber.*, **86**, 1302 (1953).
12. N. Kharasch, W. Wolf, T. J. Erpelding, P. G. Naylor, and L. Tokes, *Chem. Ind.* (*London*), 1720 (1962).

2-PHENYL-3-*n*-PROPYLISOXAZOLIDINE-4,5-*cis*-
DICARBOXYLIC ACID N-PHENYLIMIDE

$$n\text{-}C_3H_7\text{—CHO} \quad + \quad C_6H_5NHOH \longrightarrow$$

Submitted by INGRID BRÜNING, RUDOLF GRASHEY, HANS HAUCK,
ROLF HUISGEN, and HELMUT SEIDL [1]
Checked by ROBERT ELIASON, WAYLAND E. NOLAND,
and WILLIAM E. PARHAM

1. Procedure

N-Phenylhydroxylamine (11 g., 0.10 mole)[2] (Note 1) and N-phenylmaleimide (17.4 g., 0.10 mole)[3] are suspended in 40 ml. of ethanol contained in a 200-ml. Erlenmeyer flask. To the mixture is added immediately (Note 2) 8.98 g. (11.2 ml., 0.124 mole) of freshly distilled *n*-butyraldehyde. An exothermic reaction ensues, and the mixture spontaneously heats to the boiling point. A clear slightly yellow solution results which, upon cooling, deposits an almost colorless crystalline cake. The mixture is allowed to stand in the ice box for 1 day; it is then filtered through a Büchner funnel, and the crystals are washed twice with 25-ml. portions of ice-cold ethanol. The yield of air-dried product, m.p. 99–101°, is 31–32 g. (92–95%). For further purification the crude material is dissolved in 60 ml. of boiling ethanol on the steam bath, and the resulting solution is allowed to cool slowly to room temperature. If crystallization does not spontaneously begin in 5–10 minutes, it can then be induced by seeding. After being kept for 5 hours in the refrigerator, the solution is filtered and the colorless crystals are washed twice with 20-ml. portions of cold ethanol. The dried product weighs 29–30 g. An additional recrystallization of the air-dried product from 60 ml. of ethanol gives 26–27 g. (77–80%) of the pure isoxazolidine, m.p. 106.5–107.5°.

2. Notes

1. The phenylhydroxylamine should be free of sodium chloride. This can be easily removed by dissolution of the substance in benzene followed by filtration, and then addition of petroleum ether to precipitate the pure compound.

2. The checkers observed, in two runs, that when n-butyraldehyde is added after 10–15 minutes, the reaction is only mildly exothermic, and the white precipitate that forms does not dissolve. The infrared spectrum of the white crystalline product (19–20 g., m.p. 181–184° dec.) suggests that it may be the adduct of phenylhydroxylamine and N-phenylmaleimide formed by addition of the N—H bond of the amine to the olefinic bond of the imide; however, the structure of the product was not further examined.

3. Methods of Preparation

The preparation of 2-phenyl-3-n-propylisoxazolidine-4,5-cis-dicarboxylic acid N-phenylimide from n-butyraldehyde, N-phenylhydroxylamine, and N-phenylmaleimide is new and is described by Hauck.[4] The intermediate, C-(n-propyl)-N-phenyl-nitrone, is an unstable compound and is difficult to purify. The procedure described avoids the isolation of the nitrone by adding it *in situ* to a suitable dipolarophile.

4. Merits of the Preparation

The present procedure serves as a model for the generation and use *in situ* of unstable nitrones in 1,3-dipolar cycloaddition reactions.

1. Institut für Organische Chemie der Universität München, München, Germany.
2. O. Kamm, *Org. Syntheses*, Coll. Vol. 1, 445 (1941).
3. M. P. Cava, A. A. Deana, K. Muth, and M. J. Mitchell, this volume, p. 944.
4. H. Hauck, Dissertation, Universität München, 1963.

PHENYLSULFUR TRIFLUORIDE

(Benzenesulfenyl trifluoride)

$$(C_6H_5S)_2 + 6AgF_2 \rightarrow 2C_6H_5SF_3 + 6AgF$$

Submitted by William A. Sheppard [1]
Checked by E. S. Glazer and John D. Roberts

1. Procedure

Caution! Phenylsulfur trifluoride and by-products (e.g., hydrogen fluoride from hydrolysis) are toxic, and all manipulations should be carried out in a good hood. Silver difluoride is a powerful oxidative fluorinating agent and reacts vigorously with many organic materials. These reagents should not be allowed to come in contact with the skin.

A 1-l., four-necked, round-bottomed flask equipped with reflux condenser, sealed stirrer, thermometer, and solid addition funnel [2] and protected from atmospheric moisture with a Drierite® guard tube is carefully dried and flushed with a dry inert gas (Note 1). The flask is charged with 453 g. (3.1 moles) of silver difluoride (Note 2) and 500 ml. of 1,1,2-trichloro-1,2,2-trifluoroethane (Note 3), and phenyl disulfide (100 g., 0.458 mole) (Note 4) is weighed into the solid addition funnel. The stirrer is started, and phenyl disulfide is added to the slurry in small portions. An exothermic reaction occurs, and after the addition of several portions the reaction mixture reaches a temperature of 40° (Note 5). By intermittent use of a cooling bath and by adjusting the rate of addition of the disulfide, the reaction temperature may be maintained between 35° and 40°. The addition of the phenyl disulfide requires 45–60 minutes. On completion of the addition the suspension of black silver difluoride has been converted to yellow silver monofluoride, and the exothermic reaction gradually subsides. The reaction mixture is stirred for an additional 15–30 minutes without external cooling and then quickly heated to reflux.

The reaction mixture is filtered hot through a fluted filter paper under a blanket of dry nitrogen into a dry, 1-l., round-bottomed flask. The residue of solid silver fluoride is washed with a total of 500 ml. of boiling 1,1,2-trichloro-1,2,2-trifluoroethane in portions (Note 6). The filtrates are combined and distilled through a short Vigreux column, an oil bath not heated over 70° being used (Note 7). The residue of phenylsulfur trifluoride is transferred to a 200-ml. round-bottomed flask and distilled, b.p. 47–48° (2.6 mm.), through a Claisen-type distillation column, discarding a small fore-run. The product is obtained in a yield of 84–92 g. (55–60%) as a colorless liquid, m.p. −10° (Note 8). Since phenylsulfur trifluoride slowly attacks Pyrex® glass, it should be used immediately. It can be stored for several days in glass at −80° or in polyethylene, however, and may be stored indefinitely at room temperature in bottles of Teflon® polytetrafluoroethylene resin or aluminum (Note 9).

2. Notes

1. The equipment should be dried carefully by the techniques normally employed when preparing for a Grignard reaction. Dry nitrogen gas was normally employed to flush the apparatus, but any dry inert atmosphere, or dry air, could be employed.

2. A technical grade of silver difluoride (approximately 85%) is available from Harshaw Chemical Company. Better grades of silver difluoride are available and may be employed. It is important that the silver difluoride be a black powder. If the material is light brown and lumpy, a lower yield of product may be obtained. Normally, the contents of a 1-lb. can (approximately 435–470 g.) are employed.

3. 1,1,2-Trichloro-1,2,2-trifluoroethane (trademark "Freon-113"), b.p. 47°, is available from the Organic Chemicals Department, E. I. du Pont de Nemours and Company, Wilmington, Delaware.

4. Eastman's white label grade phenyl disulfide is suitable.

5. Caution must be exercised in the addition of the phenyl disulfide. There is a short induction period between the addition of disulfide and the exothermic reaction. If the disulfide is added too rapidly, a vigorous exothermic reaction, which is difficult to control, will result. The extensive use of a cooling bath should be avoided because the reaction rate is sufficiently slow at lower temperatures to allow buildup of reactants and the development of a vigorous, uncontrollable reaction.

6. Etching of the glass equipment is reduced to a minimum if all equipment used in the preparation and subsequent manipulation is rinsed with water and acetone *immediately* after use.

7. The Freon® solvent may be removed under reduced pressure in order to shorten the distillation time. Since phenylsulfur trifluoride attacks glass, the total time involved in the preparation and distillation in the glass equipment should be kept to a maximum of a few hours. It is recommended that the column be changed after distillation of the Freon®. If the preparation cannot be completed within a day, the Freon® solution of crude phenylsulfur trifluoride may be stored in polyethylene bottles overnight.

8. In contact with moisture of glass, phenylsulfur trifluoride develops pink, green, or bluish colors. A small amount of discoloration does not appear to affect the quality. Phenylsulfur trifluoride prepared in glass equipment always contains a few percent of phenylsulfinyl fluoride. The amount of this impurity depends on the care taken to exclude moisture during preparation and manipulation.

9. Phenylsulfur trifluoride slowly oozes through polyethylene bottles after storage for several days. However, a sample of phenylsulfur trifluoride has been stored in a bottle of Teflon® for several years without decomposition. Storage in a dry atmosphere in a well-ventilated area is recommended.

3. Methods of Preparation

Phenylsulfur trifluoride has been prepared only by the present method.[3,4]

4. Merits of the Preparation

This procedure illustrates a fairly general method for the preparation of alkyl- and arylsulfur trifluorides. The method has also been applied to the synthesis of nitrophenyl-, tolyl-, and fluorobutylsulfur trifluorides,[3,4] and it is probably applicable to any disulfide that does not contain groups reactive with silver difluoride. 2,4-Dinitrophenyl- and perfluoroalkylsulfur trifluorides have been prepared by reaction of disulfides with fluorine or by electrolytic fluorination.[5,6] These other routes to sulfur trifluoride compounds are not general or convenient, and they often give low yields.

The sulfur trifluoride compounds are useful as selective agents for conversion of carbonyl and carboxyl groups to difluoromethylene [7] and trifluoromethyl groups,[3, 4] respectively, and as intermediates for synthesis of arylsulfur pentafluorides.[3, 8]

1. Contribution No. 670 from the Central Research Department, Experimental Station, E. I. du Pont de Nemours and Company.
2. R. C. Fuson, E. C. Horning, S. P. Rowland, and M. L. Ward, *Org. Syntheses*, Coll. Vol. **3**, 549 (1955); see Note 7.
3. W. A. Sheppard, *J. Am. Chem. Soc.*, **82**, 4751 (1960).
4. W. A. Sheppard, *J. Am. Chem. Soc.*, **84**, 3058 (1962).
5. D. L. Chamberlain and N. Kharasch, *J. Am. Chem. Soc.*, **77**, 1041 (1955).
6. J. Burdon and J. C. Tatlow in M. Stacey, J. C. Tatlow, and A. G. Sharpe, "Advances in Fluorine Chemistry," Academic Press, Inc., New York, 1960, p. 151.
7. W. A. Sheppard, this volume, p. 396.
8. W. A. Sheppard, *J. Am. Chem. Soc.*, **84**, 3064 (1962).

3-PHENYLSYDNONE

(N-Phenylsydnone)

$$C_6H_5NHCH_2CO_2H \xrightarrow[\text{HCl}]{\text{NaNO}_2} C_6H_5N(NO)CH_2CO_2H$$

$$C_6H_5N(NO)CH_2CO_2H \xrightarrow{(CH_3CO)_2O} \begin{array}{c} C_6H_5N\text{------}CH \\ | \quad (\pm) \quad | \\ N_{\diagdown O \diagup}CO \end{array}$$

Submitted by CHARLES J. THOMAN and DENYS J. VOADEN [1]
Checked by WILLIAM E. PARHAM and EDWARD A. WALTERS

1. Procedure

A. *N-Nitroso-N-phenylglycine.* One hundred grams (0.66 mole) of N-phenylglycine (Note 1) is suspended in 1.2 l. of water contained in a 3-l. beaker placed in an ice-salt bath and stirred until the temperature has dropped below 0°. A solution of 50 g. (0.72 mole) of sodium nitrite in 300 ml. of water is added dropwise over a period of 40 minutes at such a rate that the temperature never exceeds 0°. The red, almost clear solution (Note 2) is filtered as quickly as possible with suction, after which 3 g. of Norit® is added and allowed to stir with the cold solution for several minutes (Note 3). The mixture is again filtered with suction. Addition of 100 ml. of concentrated hydrochloric acid to the well-stirred solution produces, after about 30 seconds, a profusion of light, fluffy crystals. The suspension is stirred for 10 minutes and is then filtered with suction and washed twice with ice-cold water. The precipitate is best dried by leaving it on the suction funnel overnight. The resulting product melts at 103–104°, weighs 96–99 g. (80–83%) (Note 4), and is off-white in color. It can be used without recrystallization.

B. *3-Phenylsydnone*. The 99 g. (0.55 mole) of N-nitroso-N-phenylglycine is dissolved in 500 ml. of acetic anhydride in a 1-l. Erlenmeyer flask fitted with a reflux condenser topped by a drying tube. The deep-red solution is heated in a boiling water bath for 1.5 hours with magnetic stirring (Note 5) and is then allowed to cool to room temperature. The cool solution is poured slowly into 3 l. of cold water which is very well stirred (Note 6); white crystals separate almost immediately. After 5 minutes of stirring, the solid is filtered with suction, washed twice with ice-cold water, and dried on the funnel with suction overnight. The dried product is cream-colored, weighs 74–75 g. (83–84%), and melts at 136–137° (Note 7). The overall yield for the two steps is 67–70%.

2. Notes

1. Eastman Organic Chemicals practical grade material, mud-brown in color, was used without purification.

2. Often a small amount of insoluble, dark-brown material remains in suspension. Filtration of this product is most difficult, since it tends to clog the filter paper. It seems advisable to change filter papers two or three times during the filtration, if necessary. This step usually requires from 30 minutes to 1.5 hours; however, the time can be shortened appreciably by the use of Hyflo Supercel®.

3. This Norit® treatment, when combined with the preceding filtration, does much to improve the purity of the N-nitroso-N-phenylglycine; though the yield of nitroso compound thereby is

lowered, the yield of the sydnone is increased correspondingly and the sydnone is much purer.

4. Earl and Mackney[2] report a tan product (96.8% yield) melting at 102–103°. They did not use the preliminary filtration or Norit® treatment described above.

5. The usual method (Earl and Mackney[2]) has been to let the solution stand at room temperature for 24 hours. Control experiments proved, however, that the procedure described above gives comparable results.[3]

6. On rare occasions a small amount of insoluble material may be present in the cool solution; the solution can be poured into the water through a funnel fitted with a plug of glass wool.

7. Earl and Mackney[2] report a very light tan product, melting at 134–135°, in 73% yield. The product can be recrystallized from boiling water to give cream-colored needles, but this does not improve the purity of the product.

3. Methods of Preparation

This procedure is a modification of preparations of 3-phenyl-sydnone described earlier.[2, 3] The dehydration of N-nitroso-N-phenylglycine has also been effected by the use of thionyl chloride and pyridine in dioxane,[4] thionyl chloride in ether,[4] trifluoroacetic anhydride in ether,[4] and diisopropylcarbodiimide in water;[5] or by reaction of the alkali metal salts of N-nitroso-N-phenylglycine with phosgene or benzenesulfonyl chloride in water[5] or with acetyl chloride in benzene.[4]

4. Merits of the Preparation

The present procedure makes possible the preparation of large quantities of very pure 3-phenylsydnone without recrystallization. The earlier procedure[2] produced a tan or brown product which lost its color only after several recrystallizations. Slight variations in this procedure can be used to prepare a variety of 3-substituted and 3,4-disubstituted sydnones.

3-Phenylsydnone is the prototype of that class of mesoionic compounds called sydnones. On acidic hydrolysis it produces phenylhydrazine, whereas basic hydrolysis regenerates N-nitroso-

N-phenylglycine. This sydnone undergoes a variety of electrophilic substitutions,[3, 4, 6-19] including mercuration [11, 13, 16] and formylation,[19] with an ease comparable to thiophene, and a number of "1,3-dipolar cycloadditions" with numerous alkenes,[15, 18] alkynes,[17] and quinones [8] to form, with loss of carbon dioxide, a variety of pyrazole derivatives.

1. Department of Chemistry, University of Massachusetts, Amherst, Massachusetts. This work was supported in part by a research grant from the National Cancer Institute, U.S. Public Health Service.

2. J. C. Earl and A. W. Mackney, *J. Chem. Soc.*, 899 (1935).

3. W. Baker, W. D. Ollis, and V. D. Poole, *J. Chem. Soc.*, 307 (1949).

4. W. Baker, W. D. Ollis, and V. D. Poole, *J. Chem. Soc.*, 1542 (1950).

5. G. Wolfrum, G. Unterstenhofer, and R. Pütter, Brit. Patent 832,001 and Ger. Patent 1,069,633 [*C.A.*, **54**, 8854b (1960)].

6. J. Kenner and K. Mackay, *Nature*, **158**, 909 (1946).

7. R. A. Eade and J. C. Earl, *J. Chem. Soc.*, 2307 (1948).

8. D. Ll. Hammick and D. J. Voaden, *Chem. Ind.* (*London*), 739 (1956).

9. J. C. Earl, *Rec. Trav. Chim.*, **75**, 1080 (1956).

10. H. Kato and M. Ohta, *Nippon Kagaku Zasshi*, **77**, 1304 (1956) [*C.A.*, **53**, 5250 (1959)].

11. K. Nakahara and M. Ohta, *Nippon Kagaku Zasshi*, **77**, 1306 (1956) [*C.A.*, **53**, 5251 (1959)].

12. M. Hashimoto and M. Ohta, *Nippon Kagaku Zasshi*, **78**, 181 (1957) [*C.A.*, **54**, 511 (1960)].

13. V. G. Yashunskiĭ, V. F. Vasil'eva, and Yu. N. Sheĭnker, *Zh. Obshch. Khim.*, **29**, 2712 (1959) [*C.A.*, **54**, 10999 (1960)].

14. V. G. Yashunskiĭ and V. F. Vasil'eva, *Dokl. Akad. Nauk SSSR*, **130**, 350 (1960) [*C.A.*, **54**, 10999 (1960)].

15. V. F. Vasil'eva, V. G. Yashunskiĭ, and M. N. Shchukina, *Zh. Obshch. Khim.*, **30**, 698 (1960) [*C.A.*, **54**, 24674 (1960)]; **31**, 1501 (1961) [*C.A.*, **55**, 22291 (1961)].

16. J. M. Tien and I. M. Hunsberger, *J. Am. Chem. Soc.*, **83**, 178 (1961).

17. R. Huisgen, R. Grashey, H. Gotthardt, and R. Schmidt, *Angew. Chem.*, **74**, 29 (1962).

18. R. Huisgen, H. Gotthardt, and R. Grashey, *Angew. Chem.*, **74**, 30 (1962).

19. Rev. C. J. Thoman, S.J., D. J. Voaden, and I. M. Hunsberger, unpublished results.

1-PHENYL-2-THIOBIURET

(Biuret, 1-phenyl-2-thio-)

$$C_6H_5NCS + H_2NC{=}NH \cdot HCl + KOH \rightarrow$$
$$\underset{\displaystyle OCH_3}{|}$$

$$C_6H_5NHCNHC{=}NH + KCl + H_2O$$
$$\underset{\displaystyle S}{\|} \quad \underset{\displaystyle OCH_3}{|}$$

$$C_6H_5NHCNHC{=}NH + HCl \rightarrow C_6H_5NHCNHCNH_2 + CH_3Cl$$
$$\underset{\displaystyle S}{\|} \; \underset{\displaystyle OCH_3}{|} \qquad\qquad \underset{\displaystyle S}{\|} \; \underset{\displaystyle O}{\|}$$

Submitted by Frederick Kurzer and W. Tertiuk.[1]
Checked by James Cason and Francis J. Schmitz.

1. Procedure

A. *1-Phenyl-2-thio-4-methylisobiuret.* Into a 500-ml. three-necked flask fitted with a Hershberg stirrer [2] and reflux condenser are introduced a solution of 23.0 g. (0.35 mole) of 85% potassium hydroxide in 75 ml. of water, followed by 38.7 g. (0.35 mole) of methylisourea hydrochloride (Note 1). The clear liquid is diluted with 150 ml. of acetone, and the resulting suspension, containing finely divided crystalline solid, is treated with 27.0 g. (24 ml., 0.2 mole) of recently distilled phenyl isothiocyanate.

After the additions have been completed, the third neck of the flask is closed and the temperature of the stirred reaction mixture is raised to its boiling point during 15–20 minutes, then heating under reflux is continued for 10–15 minutes. The contents of the flask, which first change to a greenish yellow clear solution, then later separate into two phases, are kept well mixed by rapid stirring (Note 2). The flask is next disconnected, fitted with a stillhead, and the acetone is distilled rapidly at 25–35° under reduced pressure. The residual semicrystalline suspension, or two-phase mixture containing the crude product in the upper viscous layer, is carefully stirred onto 300–400 g. of crushed ice. This yields the crude isobiuret as a very pale-yellow granular solid, which is collected by suction filtration, washed with successive small portions of water, drained well, and allowed to dry at room temperature. The dry product is dissolved in 120–150 ml.

of boiling benzene. Small quantities of suspended yellow powdery material (and possibly droplets of water) are removed by gravity filtration through a heated funnel or by suction filtration through a preheated Büchner funnel. The clear yellow filtrate deposits large prismatic crystals of 1-phenyl-2-thio-4-methylisobiuret, which are collected by suction filtration at room temperature, washed with a little benzene, and air-dried. The yield of material having a melting point in the range of 122–128° is 27–31.5 g. (65–75%) (Note 3). Further small quantities (2–4 g.) of less pure material may be obtained by partial vacuum evaporation of the combined mother liquors and washings.

B. *1-Phenyl-2-thiobiuret.* A solution of 20.9 g. (0.1 mole) of 1-phenyl-2-thio-4-methylisobiuret in 200 ml. of hot absolute ethanol is treated with 40 ml. of concentrated hydrochloric acid, and the clear liquid is heated under reflux until no more methyl chloride is evolved (6–12 minutes, Note 4). The resulting solution is stirred into 2 l. of water, and the separated crystalline precipitate is collected after storage at 0° for at least 24 hours. The dried product is dissolved in boiling absolute ethanol (5–6 ml. per g.), then the hot solution is quickly filtered by light suction and diluted with half its volume of petroleum ether (b.p. 60–80°). The separated 1-phenyl-2-thiobiuret is collected by suction filtration after storage for 12 hours at room temperature, and rinsed with small portions of a mixture of equal volumes of ethanol and petroleum ether. The yield of product, m.p. 159–161° (Note 5), is 8.8–10.5 g. (45–54%) (Note 6).

2. Notes

1. Methylisourea hydrochloride is accessible from commercially available calcium cyanamide by the method described in *Organic Syntheses.*[3]

2. The reaction is complete when a withdrawn sample of the liquid, stirred on a watch-glass in an air current, solidifies rapidly and smells only very faintly of phenyl isothiocyanate.

3. This product, though still pale yellow, is suitable for most synthetic purposes. Colorless glass-like prisms of m.p. 128–130° (cor.) are obtainable on further crystallization from benzene.

4. The top of the condenser is fitted with a short vertical piece of hard-glass tubing at the mouth of which the escaping

methyl chloride may be burned off. The completeness of the reaction is indicated when insufficient gas is evolved to support a *steady* flame. Methyl chloride will continue to diffuse out and produce a flickering flame when a match is held to the outlet. Prolonging the reaction time excessively reduces the yield.

5. Rather variable melting points have been reported for this compound, probably because the melting is accompanied by decomposition. In a bath heated at about 2° per minute, the checkers obtained capillary tube melting points for all samples in the range 149.5–152° (cor.).

6. The submitters report that partial evaporation of the mother liquors gives additional small quantities of low-melting fractions from which additional pure material may be obtained by further crystallizations. In contrast, the checkers obtained nearly one-half the yield in a second crop which had essentially the same melting point as the first crop. This somewhat different behavior may result from a difference in solvent characteristics of different samples of petroleum ether.

3. Methods of Preparation

1-Phenyl-2-thiobiuret has been prepared by the pyrolysis, at 75–90°, of 1-phenyl-2-thio-4-methylisobiuret hydrochloride,[4] and by the condensation of carbamyl isothiocyanate with aniline.[5] The method here described, which is based on the former method, is regarded as most convenient. A comprehensive review of syntheses of biurets, thiobiurets, and dithiobiurets is available.[6]

4. Merits of Preparation

This synthesis is generally applicable. For example, condensation of phenyl isothiocyanate and ethylisourea by the procedure above gives 70–80% yield of 1-phenyl-2-thio-4-ethylisobiuret,[4] which forms lustrous massive prisms, m.p. 98–99° (from benzene).

1. Royal Free Hospital School of Medicine, University of London, England.
2. P. S. Pinkney, *Org. Syntheses,* Coll. Vol. 2, 116 (1943).
3. F. Kurzer and A. Lawson, *Org. Syntheses,* Coll. Vol. 4, 645 (1963).
4. W. M. Bruce, *J. Am. Chem. Soc.,* 26, 449 (1904); F. Kurzer and S. A. Taylor, *J. Chem. Soc.,* 379 (1958).
5. L. Birckenbach and K. Kraus, *Ber.,* 71, 1492 (1938).
6. F. Kurzer, *Chem. Rev.,* 56, 95 (1956).

PHENYL(TRICHLOROMETHYL)MERCURY*

$$C_6H_5HgCl + Cl_3CCO_2Na \xrightarrow[\Delta]{(CH_3OCH_2)_2} C_6H_5HgCCl_3 + NaCl + CO_2$$

Submitted by TED J. LOGAN [1]
Checked by WILLIAM E. PARHAM and JOHN R. POTOSKI

1. Procedure

Into a 250-ml. round-bottomed flask equipped with a magnetic stirrer and reflux condenser fitted with a drying tube containing Drierite® are placed 150 ml. of dimethoxyethane (Note 1), 27.8 g. (0.15 mole) of sodium trichloroacetate (Note 2), and 31.3 g. (0.1 mole) of phenylmercuric chloride. The stirred mixture (Note 3) is heated to reflux (~85°) by use of a heating mantle. Carbon dioxide evolution, which begins shortly after heating is begun, is accompanied by the appearance of a precipitate of sodium chloride. The reactants are heated at the reflux temperature until no more carbon dioxide evolution is obvious (~1 hour), then cooled to room temperature and poured into 500 ml. of water. The resulting mixture, consisting of a dense oil layer, a solid, and an aqueous layer, is extracted with four 50-ml. portions of diethyl ether. The combined ether layers are then washed with two 50-ml. portions of water, dried over anhydrous magnesium sulfate, filtered, and the solvent removed using a rotary evaporator. The resulting white solid, which weighs 44.8 g., is dissolved in 130 ml. of hot chloroform and fractionally crystallized. The first three fractions weigh 2.3 g. and are recovered phenylmercuric chloride. Successive reduction of solvent volume and further fractional crystallization provides 25.6 g. of product (65% yield), m.p. 110° (Notes 4 and 5).

2. Notes

1. The 1,2-dimethoxyethane (monoglyme) was purchased from Matheson Coleman and Bell and purified by distillation from lithium aluminum hydride. The use of unpurified solvent had little effect on the yield of product.

2. Sodium trichloroacetate may be purchased from the Dow Chemical Company (96.4% pure by Cl analysis) or prepared by neutralizing trichloroacetic acid (Matheson Coleman and Bell) with aqueous sodium hydroxide to the phenolphthalein end point. The product is dried under vacuum for 12 hours, sieved, then dried an additional 12 hours under vacuum, all at room temperature. The salt prepared by this method and used in this preparation was 98.5% pure, based on chlorine analysis, and can be stored indefinitely without decomposition. The submitter has obtained nearly identical yields of phenyltrichloromethyl-mercury from the commercial and from the prepared salts.

3. If all the reactants are stirred for several minutes at room temperature, they dissolve to give a turbid solution. Stirring while heating then becomes unnecessary, except to promote more even heating, since the refluxing solvent and carbon dioxide evolution keep the precipitated sodium chloride in suspension.

4. Purity of the product was ascertained by quantitative X-ray fluorescence analysis for chlorine and mercury, which showed satisfactory agreement with calculated values. Compounds containing both mercury and chlorine are difficult to analyze by classical "wet" analytical procedures.

5. Yields as high as 77% have been obtained by this procedure. It is difficult to recover all the product from the mother liquor. The use of a 1:1 ratio of sodium trichloroacetate and phenyl-mercuric chloride gave yields of 39–45%, while a 1.25:1 ratio gave a 61% yield of product.

3. Methods of Preparation

This procedure is essentially identical with that previously published by the submitter.[2]

The pyrolysis of sodium trichloroacetate in 1,2-dimethoxy-ethane was originally described by Wagner.[3] Razuvaev later adapted this procedure to the synthesis of organomercurials, including the title compound.[4]

Phenyl(trichloromethyl)mercury has also been prepared by the reaction of phenylmercuric bromide with sodium methoxide and ethyl trichloroacetate[5] (62–71% yield); of phenylmercuric chloride with potassium t-butoxide and chloroform[6] (75% yield); of phenylmagnesium bromide with trichloromethylmercuric bromide[7] (24% yield); of trichloromethylmercuric bromide with diphenyldichlorotin (49%); and of trichloromethylmercuric bromide with phenylmagnesium bromide[8] (no yield given).

4. Merits of the Preparation

The main advantages of this procedure are simplicity of apparatus and technique, availability of reactants, ease of product isolation in good yield, and purity of product. The submitter has also used this method successfully for the preparation of trichloromethylmercuric chloride (from mercuric chloride), bis-(trichloromethyl)mercury (from a 2:1 ratio of sodium trichloroacetate to mercuric chloride or mercuric acetate), and trichloromethylmercuric bromide (from mercuric bromide).

Phenyl(trihalomethyl)mercurials, including the title compound, can be thermally decomposed in the presence of olefins to yield the corresponding dichlorocyclopropane derivatives.[2,9–11] Olefins such as tetrachloroethylene and ethylene, which give exceptionally low yields of dichlorocyclopropanes when treated with other reagents for generating dichlorocarbene ($:CCl_2$), give reasonable yields of dichlorocyclopropanes when heated with phenyltrichloromethylmercurials.[12]

These mercurials have also been employed in the preparation of dihalomethyl derivatives of carbon, silicon, and germanium,[13] in the conversion of carboxylic acids to dichloromethyl esters,[14] in the deoxygenation of pyridine N-oxide,[5] in the synthesis of diarylcyclopropenones from diaryl acetylenes,[17] and in numerous other applications. Leading references to these applications may be found in a recent review on the use of phenyl (trihalomethyl)mercury compounds as divalent carbon transfer reagents.[18]

1. The Procter and Gamble Company, Miami Valley Laboratories, Cincinnati, Ohio.
2. T. J. Logan, *J. Org. Chem.*, 28, 1129 (1963).
3. W. M. Wagner, *Proc. Chem. Soc.*, 229 (1959).
4. G. A. Razuvaev, N. S. Vasileiskaya, and L. A. Nikitina, *Tr. po Khim. i Khim. Tekhnol.*, 3, 638 (1960) [*C. A*, 56, 15116d (1962)].
5. E. E. Schweizer and C. J. O'Neill, *J. Org. Chem.*, 28, 851 (1963).
6. O. A. Reutov and A N. Lovtsova, *Dokl. Akad. Nauk SSSR*, 139, 622 (1961) [*C.A.*, 56, 1469b (1962)]; *Izv. Akad. Nauk SSSR, Otd. Khim. Nauk*, 1716 (1960) [*C.A.*, 55, 9319h (1961)]; cf. D. Seyferth and R. L. Lambert, Jr., *J. Organometal. Chem.*, 16, 21 (1969).
7. R. Kh. Freidlina and F. K. Velichko, *Izv. Akad. Nauk SSSR, Otd. Khim. Nauk*, 1225 (1959) [*C.A.*, 54, 1379g (1960)].
8. A. N. Nesmeyanov, R. Kh. Freidlina, and F. K. Velichko, *Dokl. Akad. Nauk SSSR*, 114, 557 (1957) [*C.A.*, 52, 296d (1958)].
9. D. Seyferth, J. M. Burlitch, and J. K. Heeren, *J. Org. Chem.*, 27, 1491 (1962).
10. D. Seyferth and J. M. Burlitch, *J. Am. Chem. Soc.*, 84, 1757 (1962); 86, 2730 (1964).
11. D. Seyferth, J. Y.-P. Mui, M. E. Gordon, and J. M. Burlitch, *J. Am. Chem. Soc.*, 87, 681 (1965).
12. D. Seyferth, R. J. Minasz, A. J. H. Treiber, J. M. Burlitch, and S. R. Dowd, *J. Org. Chem.*, 28, 1163 (1963).
13. D. Seyferth and J. M. Burlitch, *J. Am. Chem. Soc.*, 85, 2667 (1963).
14. D. Seyferth, J. Y.-P. Mui, and L. J. Todd, *J. Am. Chem. Soc.*, 86, 2961 (1964).
15. W. E. Parham and J. F. Dooley, *J. Am. Chem. Soc.*, 89, 985 (1967).
16. W. E. Parham and R. J. Sperley, *J. Org. Chem.*, 32, 926 (1967).
17. D. Seyferth and R. Damrauer, *J. Org. Chem.*, 31, 1660 (1966).
18. D. Seyferth, *Accounts Chem. Res.*, 5, 65 (1972).

N-PHTHALYL-ʟ-β-PHENYLALANINE*

(2-Isoindolineacetic acid, α-benzyl-1,3-dioxo-, ʟ-)

Submitted by AJAY K. BOSE.[1]
Checked by MAX TISHLER and GEORGE A. DOLDOURAS.

1. Procedure

In a 300-ml. flask fitted with a water separator and a reflux condenser are placed 16.5 g. (0.1 mole) of ʟ-phenylalanine (Note 1), 14.8 g. (0.1 mole) of finely ground phthalic anhydride, 150 ml. of toluene, and 1.3 ml. of triethylamine. The flask is heated on an oil bath or with an electric mantle so as to maintain a vigorous reflux (Note 2). Separation of water is rapid at the beginning but becomes slower with time and is virtually over in 1.5 hours.

After 2 hours the water separator and the reflux condenser are disconnected, and volatile material is removed from the mixture under reduced pressure and on a steam bath (Note 3). The solid residue is stirred with 200 ml. of cold water and 2 ml. of hydrochloric acid until all the lumps are broken (Note 4).

The mixture is filtered under suction, and the product is washed with three 50-ml. portions of cold water. After drying in an air oven the product weighs 27–28 g. (91.5–95%) and consists of a white crystalline powder, m.p. 179–183°, $[\alpha]_D^{25}$ −198 to −200° (in alc.) (Note 5).

This product can be recrystallized by dissolving 10 g. in 20 ml. of hot ethyl alcohol (95%), adding 14 ml. of water, and allowing the solution to cool slowly so that no oiling out takes place.

Colorless needles, m.p. 183–185°, $[\alpha]_D^{25}$ −211 to −217° (in alc.) (Note 6), are obtained (first crop, 7.5–8.5 g., 83–90% recovery; further quantities can be recovered from the mother liquor).

2. Notes

1. Material (supplied by Nutritional Biochemicals Corp., Cleveland, Ohio) of $[\alpha]_D^{25}$ −32° (c = 1.98 in water) was used.

2. An applicator stick (available from drug stores) can be used very conveniently to ensure smooth boiling. If a boiling chip is used, it should be colored (Carborundum, for example) to make its separation from the product convenient.

3. This evaporation during which a solid separates is very conveniently carried out in a rotary vacuum evaporator (manufactured by Rinco Instrument Co., Greenville, Illinois). An equally convenient alternative arrangement for solvent stripping that is in use in some laboratories is shown in Fig. 1. The splashhead A permits rapid removal of solvent under reduced pressure. Any solid carried beyond the flask B by spattering is arrested in A and can be washed down into B by introducing a low-boiling

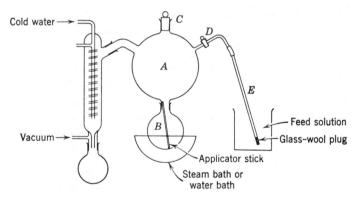

Fig. 1. Solvent stripper.

liquid like acetone through the port C. The side arm D permits continuous feeding into the "stripper." If the feed tube E is fitted with a plug of glass wool, the feed solution can be automatically filtered from suspended solids such as drying agents.

Another advantage of this stripper is that a conveniently small collecting flask B can be used for a feed solution that is very dilute and large in volume.

4. The purity of the product depends largely on the efficient breaking up of all lumps and the subsequent washing.

5. This material gives satisfactory elemental analysis and can be used without purification for further reactions.

6. The checkers observed a rotation of -211 to $-217°$, as against -207 to $-212°$ reported by the submitter; Sheehan, Chapman, and Roth [2] report $-212°$.

3. Methods of Preparation

N-Phthalyl-L-phenylalanine has been prepared by the fusion of L-alanine with phthalic anhydride.[2]

The present method, based on a recent publication,[3] ensures a low temperature of reaction which precludes racemization and is more convenient than the fusion method for large-scale operation.

The method described here can be applied to other amino acids and on a larger scale. Thus β-alanine on a 1.5-mole scale gave N-phthalyl-β-alanine in 96% yield and L-alanine gave N-phthalyl-L-alanine in 91% yield, and glycine ethyl ester hydrochloride (using more than one molar equivalent of triethylamine) gave the ethyl ester of N-phthalylglycine in 96% yield.

1. Department of Chemistry, Stevens Institute, Hoboken, New Jersey.
2. J. C. Sheehan, D. W. Chapman, and R. W. Roth, *J. Am. Chem. Soc.*, 74, 3822 (1952).
3. A. K. Bose, F. Greer, and C. C. Price, *J. Org. Chem.*, 23, 1335 (1958).

PURIFICATION OF TETRAHYDROFURAN

WARNING

It has been reported that serious explosions may occur when impure tetrahydrofuran is treated with solid potassium hydroxide or with concentrated aqueous potassium hydroxide, as has been recommended widely for the purification of tetrahydrofuran; see *Org. Syntheses*, Coll. Vol. **4**, 474, 792 (1963); Vol. **40**, 94 (1960). There is evidence that the presence of peroxides in the tetrahydrofuran being purified was causal. It is strongly recommended, therefore, that this method not be used to dry tetrahydrofuran, if the presence of peroxides is indicatéd by a qualitative or quantitative test with acidic aqueous iodide solution. Traces of peroxide can be removed by treatment with cuprous chloride; see p. 692. The safety of this operation should be checked first on a small scale (1–5 ml.). It is recommended that tetrahydrofuran containing larger than trace amounts of peroxides be discarded by flushing down a drain with tap water. It must be kept in mind that mixtures of tetrahydrofuran vapor and air are easily ignitable and explosive; purification is best carried out in a hood which is well exhausted and which does not contain an ignition source.

The best procedure for drying tetrahydrofuran appears to be distillation (under nitrogen) from lithium aluminum hydride. This operation should not be attempted until it is ascertained that the tetrahydrofuran is peroxide-free and also not grossly wet. A small-scale test can be carried out in which a small amount of lithium aluminum hydride is added to *ca.* 1 ml. of the tetrahydrofuran to determine whether a larger-scale drying operation with lithium aluminum hydride would be too vigorous for safe operation. Tetrahydrofuran so purified rapidly absorbs both oxygen and moisture from air. If not used immediately, the purified solvent should be kept under nitrogen in a bottle labeled with the date of purification. Storage for more than a few days is not advised unless 0.025% of 2,6-di-*t*-butyl-4-methyl phenol is added as an antioxidant.

There are no indications that *peroxide-free*, but moisture-containing, tetrahydrofuran cannot safely be predried over potassium hydroxide. However, even this operation should be attempted only after a test-tube scale experiment to make sure that a vigorous reaction does not occur.

A peroxide-free grade of anhydrous tetrahydrofuran (stabilized by 0.025% of 2,6-di-*t*-butyl-4-methyl phenol) is available currently (1966) from Fisher Scientific Co. in 1-lb. bottles. This product as obtained from freshly opened bottles has been found to be suitable for reactions such as the formation of Grignard reagents in which purity of solvent is critical (du Pont Co., unpublished observations). It is standard practice in at least one laboratory to use only tetrahydrofuran (Fisher) from freshly opened bottles and to discard whatever material is not used within 2–3 days.

4-PYRIDINESULFONIC ACID

Submitted by Russell F. Evans,[1,2] Herbert C. Brown,[1] and H. C. van der Plas.[3]
Checked by James Cason and Taysir Jaouni.

1. Procedure

A. *N-(4-Pyridyl)pyridinium chloride hydrochloride.* In a 2-l. round-bottomed flask equipped with a ground joint (Note 2) is placed 395 g. (5.00 moles) of dry pyridine (Note 3). As this flask is cooled by swirling in a bath of cold water (Note 4), there is added during a few minutes 1190 g. (10.0 moles) of a good commercial grade of thionyl chloride (Note 1). After completion of the addition, the flask is protected by a drying tube, and the reaction mixture is allowed to stand at room temperature under a hood for 3 days. During this period, the color of the mixture changes from deep yellow through brown to black.

The flask is fitted with a Claisen head, and excess thionyl chloride is distilled at reduced pressure (water pump) and collected in a receiver cooled in a mixture of dry ice and acetone (Note 5). The flask is heated with a water bath that is slowly raised from room temperature to about 90°, then held at that temperature until no more distillation occurs and a black residue remains.

The black residue is cooled to 0°, and 100 ml. of ice-cold ethanol is added very cautiously to react with residual thionyl chloride. An additional 400 ml. of ice-cold ethanol is added, and the solid mass left at the bottom of the flask is broken up with the aid of a rod (Note 2). The resultant light-brown powder is collected by suction filtration, preferably on a sintered glass funnel, and washed with five 150-ml. portions of ethanol. The yield of crude N-(4-pyridyl)pyridinium chloride hydrochloride is 230–257 g. (40–45%). This product is very deliquescent and should be used immediately or stored over phosphorus pentoxide.

B. *4-Pyridinesulfonic acid.* A 115-g. (0.50 mole) quantity of N-(4-pyridyl)pyridinium chloride hydrochloride is dissolved in 750 ml. of water in a 2-l. round-bottomed flask, and 378 g. (1.50 moles) of solid sodium sulfite heptahydrate is added cautiously. After the evolution of sulfur dioxide has ceased, the solution is gently heated under reflux in a nitrogen atmosphere for 24 hours. After slight cooling, 20 g. of charcoal is added to the mixture, and it is heated under reflux for an additional hour. The resultant mixture is filtered through a fluted paper, the filtrate is evaporated to dryness on a steam bath under reduced pressure, and the residue is air-dried at 100–110° (Note 6). This solid is now continuously extracted with absolute ethanol for 24 hours in a Soxhlet apparatus. The alcohol is distilled from the extract on a steam bath, and the crude sodium 4-pyridinesulfonate is dissolved in about 160 ml. of hot water. After 320 ml. of $12N$ hydrochloric acid has been added with mixing, the solution is cooled to room temperature. The precipitate of sodium chloride is filtered, and the filtrate is evaporated to dryness under reduced pressure on a steam bath. Crystallization of the residue from 600 ml. of 70% aqueous ethanol yields 27–30 g. of colorless crystals of 4-pyridinesulfonic acid, m.p. 313–315° (dec.). Concentration of the mother liquor affords about 10 g. of additional product which is less pure. The total yield is 36–40 g. (45–50%) (Note 7). Recrystallization from 70% aqueous ethanol affords a purer specimen, m.p. 317–318° (dec.). (Note 8).

2. Notes

1. Although N-(4-pyridyl)pyridinium chloride hydrochloride is formed by reaction of pyridine with thionyl chloride, followed by treatment with ethanol, the intermediates involved in the reaction have not been well established. It has been suggested [4,5] that 1 mole of thionyl chloride converts 2 moles of pyridine to the compound

This intermediate would be further oxidized by thionyl chloride and solvolyzed by ethanol to the pyridinium chloride hydrochloride. According to this reaction route, the stoichiometric ratio of pyridine to thionyl chloride for the overall process would be about 1:1. Varying ratios of thionyl chloride have been used [6-8] and varying yields of the product have been reported, ranging from 60% of crude product [7] to 48% of recrystallized product.[8] In one run in which the checkers used one-half the specified amount of thionyl chloride, the yield was unaffected. The submitters report yields in the range 58–62% by the procedure described here.

2. Thionyl chloride attacks rubber so rapidly that all-glass apparatus is highly desirable for this procedure. Since breaking up the residual product in a flask results in a high mortality of

flasks, the checkers preferred a distilling vessel with a removable top of the type used with vacuum desiccators (e.g., Corning Glass Works, No. 3480).

3. Since moisture reacts with thionyl chloride to give hydrogen chloride, which forms the salt of pyridine and thus inactivates it, the pyridine should be dried over barium oxide for 24 hours, then distilled under anhydrous conditions shortly before use.

4. Provided that this addition is carried out rapidly, ingress of moisture is not significant, and more complicated apparatus is not recommended.

5. Since thionyl chloride ruins all rubber tubing with which it comes in contact, efficient cooling of the receiver is recommended.

6. Alternatively, to decrease the time required to complete drying at 100–110°, the moist solid residue may be triturated with chloroform and the chloroform distilled from the steam bath. The checkers used a vacuum oven for drying.

7. The submitters report yields in the range 63–70%.

8. Because the sulfonic acid melts with decomposition, the value observed for the melting point is highly dependent on the rate of heating of the sample.[9–12]

3. Methods of Preparation

The preparation of N-(4-pyridyl)pyridinium chloride hydrochloride follows the procedure of Koenigs and Greiner,[6] while the preparation of the sulfonic acid is a modification of a patent procedure.[13]

4-Pyridinesulfonic acid has been prepared by oxidation of 4-pyridinethiol with hydrogen peroxide in barium hydroxide solution,[9] with hydrogen peroxide in glacial acetic acid,[10] with nitric acid-chlorine or nitric acid-chlorine-hydrochloric acid mixtures,[11] and with nitric acid alone.[10, 12, 14] The latter reaction gives a mixture of 4-pyridinesulfonic acid and other products, e.g., di-4-pyridyl disulfide dinitrate, and this has led to some confusion in the literature.[10–12,14] 4-Pyridinesulfonic acid has also been obtained from its N-oxide derivative by reduction of the N-oxide group with iron and acetic acid[15] or catalytically.[16]

Sodium 4-pyridinesulfonate has been formed by the oxidation of 4-pyridinethiol with hydrogen peroxide in sodium hydroxide solution,[17,18] and from the reaction of 4-chloropyridine with aqueous sodium sulfite.[19] The salt has been converted to the free acid by treatment with a cation-exchange resin[10,11] or with sulfuric acid.[11]

4. Merits of the Preparation

This is the most convenient preparation of 4-pyridinesulfonic acid, a useful intermediate for the synthesis of various pyridine derivatives.

1. Department of Chemistry, Purdue University, Lafayette, Indiana.
2. National Chemical Laboratory, D.S.I.R., Teddington, Middlesex, England.
3. Landbouwhogeschool, Laboratorium voor Organische Chemie, Wageningen, Holland.
4. F. Kröhnke, Angew. Chem., 65, 623 (1953).
5. D. Jerchel, H. Fischer, and K. Thomas, Chem. Ber., 89, 2921 (1956).
6. E. Koenigs and H. Greiner, Ber., 64, 1049 (1931).
7. B. Bak and D. Christensen, Acta Chem. Scand., 8, 390 (1954).
8. K. Bowden and P. N. Green, J. Chem. Soc., 1795 (1954).
9. R. F. Evans and H. C. Brown, Chem. Ind. (London), 1559 (1958).
10. A. M. Comrie and J. B. Stenlake, J. Chem. Soc., 1853 (1958).
11. J. Angulo and A. M. Municio, Chem. Ind. (London), 1175 (1958).
12. H. J. den Hertog, H. C. van der Plas, and D. J. Buurman, Rec. Trav. Chim., 77, 963 (1958).
13. A. E. Tiesler, U.S. Patent 2,330,641 (1943) [C.A., 38, 1249 (1944)].
14. E. Koenigs and G. Kinne, Ber., 54, 1357 (1921).
15. L. Thunus and J. Delarge, J. Pharm. (Belg.), 21, 485 (1966) [C.A., 66, 3736h (1967)].
16. M. Van Ammers and H. J. den Hertog, Rec. Trav. Chim., 78, 586 (1959).
17. H. King and L. L. Ware, J. Chem. Soc., 873 (1939).
18. W. Walter, J. Voss, and J. Curts, Ann., 695, 77 (1966).
19. E. Ochiai and I. Suzuki, Pharm. Bull. (Tokyo), 2, 247 (1954) [C.A., 50, 1015 (1956)].

α-PYRONE

(2-Oxo-1, 2H-pyrane-)

Submitted by HOWARD E. ZIMMERMAN, GARY L. GRUNEWALD, and ROBERT M. PAUFLER [1]

Checked by E. J. COREY, W. H. PIRKLE, and M. J. HADDADIN

1. Procedure (Note 1)

A 37.5-g. (0.266-mole) sample of coumalic acid (Note 2) is placed in a 30 x 10 cm. cylindrical flask attached horizontally to a 55 x 3 cm. oven-heated Vycor tube (Note 3) *loosely* packed with 20 g. of fine copper turnings (Note 4). Following the Vycor tube successively are two ice-cooled 50-ml. receivers and a dry ice trap. The latter is connected to an efficient vacuum pump (Note 5). The system is evacuated, and the Vycor tube is heated to 650–670°. Then the flask containing the coumalic acid is heated with a nichrome wound heating jacket to 180°, and the temperature is allowed to rise slowly to 215°. During this time coumalic acid sublimes into the Vycor tube and α-pyrone distills into the ice-cooled receivers. The pressure is held below 5 mm. (Note 6). The yield of pale yellow crude material is 18–19.3 g. (70–75%). Distillation affords 16.9–18 g. (66–70%) of colorless oily α-pyrone, b.p. 110° (26 mm.), n^{25}D 1.5270.

2. Notes

1. Since this procedure was checked, a method[1] has been developed which uses a vertical arrangement of the oven together with a powder sifter placed above the oven. This allows the use of less pure starting material but with reduced yields. Also, it allows more rapid throughput. It is required that the top portion of the tube be filled with porcelain Berl saddles onto which the powder is sifted. The saddles then pick up any tarry residue which might clog the pyrolysis tube.

1. Coumalic acid, m.p. 206–209°, was prepared by the method of Wiley and Smith[2] and recrystallized from methanol as described. Starting material prepared in this way still contains impurities but is satisfactory for the preparation of α-pyrone. A purer grade of colorless coumalic acid, m.p. 206–208.5°, may be obtained by further recrystallization and a subsequent sublimation at 180–190° (0.5 mm.) (Precision Glass Macro Sublimator No. JM7410), and its use leads to higher yields of α-pyrone (80–85%) by the procedure given here. However, the losses of coumalic acid incurred in the purification (about 50%) and the time involved render this modification unprofitable. On the other hand, when unrecrystallized coumalic acid is used, the yields are generally somewhat lower (60–65%) and the results are slightly more variable.

3. The following apparatus was used by the submitters. The furnace was a Lindberg Model CF-1R High Temperature Combustion Furnace (Fisher Scientific Co. Catalog No. 10-467-1; E. H. Sargent Co. Catalog No. S-35955). This furnace has a hot zone of $8\frac{3}{4}$ in. and a maximum temperature of 1450°C. Because of the short heating length, the Vycor tube was packed with copper turnings over its entire length to prevent condensation of coumalic acid in the cooler parts of the tube. Glass wool insulation was used at both ends of the furnace to prevent heat loss.

The heater used for the sublimation vessel was made from a length of 15-cm. Pyrex tubing. The tube was covered partially with moistened asbestos fiber strips (*ca.* 1 mm. thick) which remain in place when dry. The tube was only partially covered with asbestos to allow visual inspection of the sublimation vessel. A sufficient length of nichrome wire (depending on the resistance of the wire) was wound over the asbestos base, and more asbestos was added over that already in place to hold the wire loops apart. During use, the open end was well stuffed with glass wool.

The sublimation vessel was made from 10-cm. Pyrex tubing sealed at one end and fitted with a standard taper 34/45 female joint at the other end.

The checkers used a Hoskins tube furnace, type FD303A (Central Scientific Co.), 17 in. long. The heater for the sublimation vessel was wound in two sections with heating wire in such a way that a decreasing temperature gradient in the direction of the pyrolysis oven was maintained. The open end of the heater was closed by an asbestos end plate which could be heated independently by a small nichrome coil.

4. Copper appears to function only as a surface heat transfer agent. Broken pieces of porous plate, for example, may also be used.

5. Better yields are obtained at low pressures (preferably below 5 mm.) because of more efficient sublimation of coumalic acid. The submitters report that a water aspirator could be used with crude (unrecrystallized) coumalic acid to avoid damage to the vacuum pump by untrapped corrosive vapors, and that yields of α-pyrone averaged 45% in this modification. The checkers used a mechanical pump with an efficient sodium hydroxide trap in all runs.

6. It is important that the melting of the coumalic acid be prevented during the sublimation process; hence maintenance of the lowest possible pressure is recommended. If the material in the sublimation vessel begins to melt, resinification occurs with no further sublimation and a correspondingly lower yield of product. The temperature of the sublimator should be maintained high enough to allow a maximum rate of sublimation of coumalic acid, but not so high as to cause melting. Increasing the scale of the preparation increases the possibility of resinification of coumalic acid in the sublimator before complete reaction. However, the submitters have successfully carried out the preparation of α-pyrone on twice the scale described here using the same procedure and apparatus (except that the collection flasks were 100-ml. size). On the larger scale it is advisable to use pure sublimed coumalic acid. In the experience of the checkers, larger-scale runs are much less readily reproducible.

3. Methods of Preparation

α-Pyrone has previously been prepared in low yield by the pyrolysis of heavy metal salts of coumalic acid [3] and by the small-scale pyrolysis of α-pyrone-6-carboxylic acid over copper.[4]

4. Merits of the Preparation

This method affords α-pyrone in quantity and in good yield not achieved previously.[3] The compound has considerable possibilities in Diels-Alder reactions, such as a decarboxylative double diene synthesis.[5]

1. Department of Chemistry, University of Wisconsin, Madison, Wisconsin 53706.
2. H. E. Zimmerman, G. L. Grunewald, R. M. Paufler, and M. A. Sherwin, *J. Am. Soc.*, **91**, 2330 (1969).
3. R. Wiley and N. Smith, *Org. Syntheses*, Coll. Vol. 4, 201 (1963).
4. H. von Pechmann, *Ann.*, **264**, 272 (1891).
5. J. Fried and R. C. Elderfield, *J. Org. Chem.*, **6**, 566 (1941).
6. H. E. Zimmerman and R. M. Paufler, *J. Am. Chem. Soc.*, **82**, 1514 (1960).

p-QUINQUEPHENYL

$p\text{-}ClCH_2C_6H_4CH_2Cl + 2(C_6H_5)_3P \rightarrow$

$$p\text{-}(C_6H_5)_3P^+CH_2C_6H_4CH_2P^+(C_6H_5)_3; 2Cl^-$$

$p\text{-}(C_6H_5)_3P^+CH_2C_6H_4CH_2P^+(C_6H_5)_3; 2Cl^-$

$+ 2C_6H_5CH{=}CHCHO + 2LiOC_2H_5 \rightarrow$

$$C_6H_5CH\overset{CH-CH}{\diagup\diagdown}CHC_6H_4CH\overset{CH-CH}{\diagup\diagdown}CHC_6H_5$$

$$+ 2(C_6H_5)_3PO + 2C_2H_5OH + 2LiCl$$

$$C_6H_5CH\overset{CH-CH}{\diagup\diagdown}CHC_6H_4CH\overset{CH-CH}{\diagup\diagdown}CHC_6H_5$$

$+ 2C_2H_5O_2CC{\equiv}CCO_2C_2H_5 \xrightarrow{\quad} \xrightarrow{OH^- \ Fe(CN)_6^{3-}}$

$$p\text{-}C_6H_5{-}C_6H_4{-}C_6H_4{-}C_6H_4{-}C_6H_5$$

Submitted by Tod W. Campbell and Richard N. McDonald.[1]
Checked by Virgil Boekelheide and Richard E. Partch.

1. Procedure

A. *p-Xylylene-bis(triphenylphosphonium chloride).* A mixture of 262 g. (1.0 mole) of triphenylphosphine (Note 1) and 84 g.

(0.48 mole) of *p*-xylylene dichloride (Note 2) in 1 l. of dimethylformamide is heated at reflux with stirring for 3 hours (Note 3). The mixture is then allowed to cool to room temperature with stirring, and the white crystalline solid is collected, washed with 100 ml. of dimethylformamide followed by 300 ml. of ether, and dried in a vacuum oven at 20 mm. pressure and 80°. The dry weight is 313–329 g. (93–98%).

B. *1,4-Bis-(4-phenylbutadienyl)benzene.* To a solution of 70 g. (0.10 mole) of *p*-xylylene-bis(triphenylphosphonium chloride) and 35 g. (0.26 mole) of cinnamaldehyde in 250 ml. of ethanol (Note 4) is added a solution of 0.25*M* lithium ethoxide in ethanol (Note 5). After being allowed to stand overnight at room temperature the yellow solid is collected by filtration, washed with 300 ml. of 60% ethanol, and dried in a vacuum oven at 20 mm. and 70°. The dry weight is 29–32 g. (87–95%). The solid is then dissolved in the minimum amount (about 2 l.) of boiling xylene, treated with decolorizing charcoal, and filtered. The filtrate is reduced in volume to about 1.2 l. and digested at the boiling point with a trace of iodine for 3 hours (Note 6). After the solution has stood overnight at room temperature, the yellow plates are collected by filtration, washed with benzene, and dried in a vacuum oven at 20 mm. pressure at 70°. The weight of crystals, m.p. 285–287° (Note 7), is 23–25 g. (69–75%).

C. *p-Quinquephenyl.* A mixture of 3.40 g. (0.020 mole) of diethyl acetylenedicarboxylate [2] and 3.34 g. (0.010 mole) of 1,4-bis-(4-phenylbutadienyl)benzene is refluxed with 20 ml. of *o*-dichlorobenzene for 3 hours. It is allowed to cool to about 80°, then 100 ml. of ethanol and 5 g. of potassium hydroxide are added, and the mixture is refluxed for about 2 hours. The solvent is evaporated on a steam bath under a nitrogen atmosphere (Note 8), and the damp solid is extracted with 200 ml. of water. The intense yellow or yellow-orange aqueous layer is filtered, then extracted twice with 75-ml. portions of ether, charcoal is added and then filtered to separate water-insoluble matter. The filtrate is just neutralized with dilute hydrochloric acid (Note 9) and then made basic with 5 g. of sodium carbonate. To this is added a solution of 30 g. of potassium ferricyanide in 200 ml. of water. The mixture rapidly becomes milky and is allowed to stand overnight (Note 10). The suspended solid is centrifuged and washed with water twice by centrifugation. It is dried in a vacuum oven to give 3.1–4.2 g. of a green-tinged solid (Note 11). The combined material from five runs (18 g.) is sublimed to give 10 g. (52%) of pure quinquephenyl, m.p. 385–390°. This can be recrystallized from dimethylsulfoxide to give well-defined leaflets.

2. Notes

1. Commercial triphenylphosphine was used without further purification.

2. A sample of this compound was obtained from Hooker Electrochemical Co. and used without further purification.

3. The salt begins to precipitate after about 30 minutes.

4. Commercial anhydrous ethanol was used throughout.

5. Prepared by dissolving 1.74 g. of lithium wire in 1 l. of ethanol.

6. After the volume is reduced, a small crystal of iodine is added whereupon large yellow leaflets of product begin to separate.

7. The product can be recrystallized readily from dimethyl-formamide to give yellow leaflets, m.p. 290–293°. However, it is pure enough to be used in the next step. This synthesis has also been applied to the preparation of 1,4-bis-[4-(*p*-tolyl)buta-dienyl]benzene (100%), 1,4-bis-[4-(3-nitrophenyl)butadienyl]-benzene (56%), and 1,4-bis-(3-methyl-4-phenylbutadienyl)ben-zene (87%).

8. Nitrogen is used both for rapid removal of the solvent and to maintain an inert atmosphere.

9. The acid is added slowly with stirring until a trace of permanent precipitate is formed.

10. All the steps to this point can be completed in 1 day.

11. This procedure has been applied to the synthesis of 4,4′′′′ dimethylquinquephenyl and 2′,3′′′-dimethylquinquephenyl from 1,4-bis-[4-(*p*-tolyl)butadienyl]benzene and 1,4-bis-(3-methyl-4-phenylbutadienyl)benzene, respectively.

3. Methods of Preparation

1,4-Bis-(4-phenylbutadienyl)benzene has been obtained by condensation of cinnamaldehyde and *p*-phenylenediacetic acid with lead oxide.[3] *p*-Quinquephenyl has been prepared by the reaction of biphenyllithium with 1,4-cyclohexanedione, followed by dehydration and air oxidation of the dihydroquinquephenyl;[4] by the Gatterman coupling reaction of benzenediazonium formate with copper;[5] by the Ullmann coupling of 4-iodoterphenyl and 4-iodobiphenyl with silver;[5] by the catalytic reduction of *p*-dibromobenzene;[6] and the Friedel-Crafts reaction of cyclohexene with terphenyl followed by dehydrogenation.[7] The procedure described represents the best route to both the 1,4-bis-(4-aryl-butadienyl)benzenes and quinquephenyls that has been reported.[8] The details of the oxidative decarboxylation step for reactions of the type described here have been established by Fieser and Haddadin.[9]

1. Pioneering Research Division, Textile Fibers Department, E. I. du Pont de Nemours and Co., Wilmington, Delaware.
2. E. H. Huntress, T. E. Leslie, and J. Bornstein, *Org. Syntheses,* Coll. Vol. 4, 330, (1963).
3. G. Drefahl and G. Plotner, *Ber.,* 91, 1274 (1958).
4. E. Muller and T. Topel, *Ber.,* 72, 273 (1939).
5. O. Gerngross and M. Dunkel, *Ber.,* 57, 739 (1924).
6. M. Busch and W. Waber, *J. Prakt. Chem.,* 146, 1 (1936).
7. N. P. Buu-Hoi and P. Cagniant, *Compt. Rend.,* 216, 381 (1943).
8. R. N. McDonald and T. W. Campbell, *J. Org. Chem.,* 24, 1969 (1959).
9. L. F. Fieser and M. J. Haddadin, *J. Am. Chem. Soc.,* 86, 2392 (1964).

3-QUINUCLIDONE HYDROCHLORIDE *

(3-Quinuclidinone, hydrochloride)

Submitted by H. U. DAENIKER and C. A. GROB [1]
Checked by E. CIGANEK, W. R. HERTLER, A. D. JOSEY,
and B. C. McKUSICK

1. Procedure

A. *1-Carbethoxymethyl-4-carbethoxypyridinium bromide.* A solution of 151 g. (1.00 mole) of ethyl isonicotinate (Note 1) and 167 g. (1.00 mole) of ethyl bromoacetate in 500 ml. of ethanol is allowed to stand overnight at room temperature in a 1-l. round-bottomed flask equipped with a reflux condenser (Note 2). The mixture is then heated at the reflux temperature for 4 hours. The resulting solution of 1-carbethoxymethyl-4-carbethoxypyridinium bromide is used directly for the next step (Note 3).

B. *1-Carbethoxymethyl-4-carbethoxypiperidine.* Fifteen grams of 10% palladium on charcoal [2] is added to the solution of the pyridinium bromide. The mixture is placed in an agitated 2-l. hydrogenation autoclave and hydrogenated at 90° under an initial pressure of 100 atm. (Note 4). Slightly more than the calculated amount of hydrogen (3 moles) is absorbed within 30–60 minutes. The mixture is cooled to 25°, and the catalyst is separated by filtration and washed with 100 ml. of ethanol. The filtrate is evaporated to dryness under water-aspirator vacuum at a bath temperature of 50–60°. The residue, semicrystalline 1-carbethoxymethyl-4-carbethoxypiperidine hydrobromide, is taken up in 500 ml. of ice-cold water. The solution is added to 500 ml. of chloroform in a 5-l. beaker immersed in an ice bath, and an ice-cold solution of 150 g. of potassium carbonate in 250 ml. of water is added gradually with stirring (Note 5). After the carbon dioxide evolution has subsided, the mixture is placed in a 2-l. separatory funnel and thoroughly shaken for some time. The lower, organic layer is drawn off and washed once with 200 ml. of water. The aqueous layers are combined and washed once with 500 ml. of chloroform. The two chloroform extracts are combined and dried over anhydrous sodium sulfate. After 1 hour the sodium sulfate is separated on a Büchner funnel and washed with two 200-ml. portions of chloroform. The chloroform is removed on a steam bath, and the resulting oily residue is distilled under high vacuum through a 20-cm. Vigreux column. A fore-run, weight 4–8 g., is collected below 110° (0.20 mm.). Then 156–190 g. (64–78%) of 1-carbethoxymethyl-4-carbethoxypiperidine is collected as a colorless oil, b.p. 111–113° (0.2 mm.), d_{15}^{15} 1.057, n^{20}D 1.4585.

C. *3-Quinuclidone hydrochloride.* A 2-l. three-necked flask is fitted with a Hershberg stirrer, a pressure-equalizing addition funnel, and a condenser connected to a source of dry nitrogen. Absolute toluene (330 ml.) and 80 g. (2.05 g. atom) of potassium free of oxide crust are added. (*Caution! Directions* [3] *for the safe handling of potassium should be consulted.*) The air in the flask is replaced by an atmosphere of dry nitrogen that is maintained until the reaction mixture is decomposed. The flask is heated in an oil bath until the toluene begins to reflux gently. As soon as the potassium is molten, it is pulverized by vigorous stirring. One hundred twenty-five milliliters (98.6 g., 2.14 moles) of absolute ethanol (Note 6) is added through the addition funnel within 30 minutes while heating and stirring are continued. After disappearance of the potassium the temperature is raised to 130°, and a solution of 200 g. (0.822 mole) of 1-carbethoxymethyl-4-carbethoxypiperidine in 500 ml. of absolute toluene is added within 2 hours. The mixture is stirred and heated for an additional 3 hours.

The resulting solution is cooled to 0° and decomposed by careful addition of 500 ml. of 10N hydrochloric acid. The mixture is transferred to a separatory funnel, the aqueous phase is separated, and the toluene layer is extracted with two 250-ml. portions of 10N hydrochloric acid. The aqueous extracts are combined and heated under reflux for 15 hours to effect decarboxylation. The hot, dark-colored solution is treated with 10 g. of activated charcoal, filtered, and evaporated to dryness under reduced pressure. The residue is washed into a separatory funnel with 300 ml. of water. The solution is treated with saturated aqueous potassium carbonate solution until it is alkaline to litmus; the carbonate solution must be added very carefully to prevent excessive foaming. Solid potassium carbonate is added until a thin slurry is obtained, and the slurry is extracted with four 400-ml. portions of ether. The combined ether extracts are dried for at least 60 minutes over calcined potassium carbonate and then filtered.

The ether is removed by distillation on a steam bath through a column filled with Raschig rings. The yellowish crystalline residue is treated with 150 g. of ice and 150 g. (130 ml.) of 10N hydrochloric acid, and the solution is evaporated to dryness under reduced pressure (Note 7). The crystalline residue is dissolved in the minimum amount of hot water (about 70 ml.), and boiling isopropyl alcohol (about 1.5 l.) is added until crystalline 3-quinuclidone hydrochloride begins to separate. The mixture is cooled to 0–5°, and the solid is separated by filtration, washed with acetone, and dried. The yield of 3-quinuclidone hydrochloride, m.p. 294–296° (sealed capillary) (Note 8), is 102–109 g. (77–82%).

2. Notes

1. The checkers used ethyl isonicotinate purchased from K and K Laboratories, Inc., Jamaica, New York, or prepared by esterification of isonicotinic acid as described by La Forge [4] for nicotinic acid.

2. The quaternization is slightly exothermic.

3. The quaternary salt may be isolated by evaporation of the solution and subsequent recrystallization of the residue from isopropyl alcohol; m.p. 159° (dec.). Calcd. for $C_{12}H_{16}BrNO_4$: C, 45.30; H, 5.07; Br, 25.12. Found: C, 45.41; H, 5.14; Br, 25.28.

4. The checkers found that hydrogenation proceeded rapidly and quite exothermically at a pressure of only 7 atm. at 90°. They used 10% palladium-on-carbon powder purchased from Engelhard Industries Inc., Newark, New Jersey.

5. The evolution of carbon dioxide causes considerable foaming. Losses are easily avoided if a 5-l. beaker is used.

6. Commercial absolute alcohol was further dried by treatment with magnesium and a little iodine with subsequent redistillation, as described by Lund and Bjerrum.[5]

7. In an alternative method of isolating crude quinuclidone hydrochloride, found by the checkers to give equally good results, the dried ether solution of quinuclidone is transferred to a 2-l. round-bottomed flask equipped with a stirrer, a gas-inlet tube, and a gas-exit tube. The flask is immersed in an ice bath, and gaseous hydrogen chloride is passed into the stirred solution until it begins to bubble out, indicating that the solution is saturated. The quinuclidone hydrochloride that precipitates is collected on a Büchner funnel, washed with acetone, and dried in a vacuum desiccator. The product is then dissolved in hot water and precipitated with isopropyl alcohol as described in the procedure.

8. The melting point depends on the rate of heating and the apparatus used. The checkers observed m.p. 297–305°, 298–303°, and 301° under various conditions.

3. Methods of Preparation

Quinuclidone hydrochloride has been prepared by intramolecular condensation of 1-carbethoxymethyl-4-carbethoxypiperidine with potassium [6-8] or, as in the present procedure, with potassium ethoxide.[9] 1-Carbethoxymethyl-4-carbethoxypiperidine has been prepared by alkylating ethyl hexahydroisonicotinate with ethyl chloroacetate [6, 8] or by the present method.[7]

4. Merits of the Preparation

This is the most convenient way to prepare quinuclidone hydrochloride. The second step illustrates the conversion of an N-alkylpyridinium salt to an N-alkylpiperidine. The third step illustrates the formation of a bicyclic system by the Dieckmann condensation.

Quinuclidone can be reduced to quinuclidine.[6] Depending on the availability of starting materials, either this reduction or the dehydrative cyclization of 4-(2-hydroxyethyl)piperidine [10] is the most convenient synthesis of quinuclidine.

1. University of Basel, Basel, Switzerland.
2. R. Mozingo, *Org. Syntheses*, Coll. Vol. **3**, 687 (1955).
3. W. S. Johnson and W. P. Schneider, *Org. Syntheses*, Coll. Vol. **4**, 132 (1963).
4. F. B. La Forge, *J. Am. Chem. Soc.*, **50**, 2477 (1928).
5. H. Lund and J. Bjerrum, *Ber.*, **64**, 210 (1931).
6. G. R. Clemo and T. P. Metcalfe, *J. Chem. Soc.*, 1989 (1937).
7. L. H. Sternbach and S. Kaiser, *J. Am. Chem. Soc.*, **74**, 2215 (1952).
8. C. A. Grob and E. Renk, *Helv. Chim. Acta*, **37**, 1689 (1954).
9. E. E. Mikhlina and M. V. Rubtsov, *Zh. Obshch. Khim.*, **29**, 118 (1959) [*J. Gen. Chem. USSR (Engl. Transl.)*, **29**, 123 (1959)]; C. A. Grob and J. Zergenyi, *Helv. Chim. Acta*, **46**, 2658 (1963).
10. S. Leonard and S. Elkin, *J. Org. Chem.*, **27**, 4635 (1962).

REDUCTION OF CONJUGATED ALKENES WITH CHROMIUM(II) SULFATE: DIETHYL SUCCINATE*

(Succinic acid, diethyl ester)

$$Cr_2(SO_4)_3 + Zn(Hg) \rightarrow 2CrSO_4 + ZnSO_4$$

$$+ 2Cr(II)^{2+} + 2H^+ \longrightarrow$$

$$H_5C_2O_2CCH_2CH_2CO_2C_2H_5 + 2Cr(III)^{3+}$$

Submitted by A. ZURQIYAH and C. E. CASTRO [1]
Checked by FREDERICK J. SAUTER and HERBERT O. HOUSE

1. Procedure

A. *Chromium(II) sulfate solution.* A 3-l., three-necked flask fitted with a gastight mechanical stirrer and nitrogen inlet and outlet stopcocks is charged with 300 g. (*ca.* 0.55 mole) of hydrated chromium(III) sulfate (Note 1), 2 l. of distilled water, 75 g. (1.15 g. atoms) of mossy zinc (Note 2), and 4.0 ml. (54 g., 0.27 g. atom) of mercury (Note 3). After the flask has been flushed with nitrogen for 30 minutes, the mixture is warmed in a water bath to about 80° (Note 4) with stirring for 30 minutes under a nitrogen atmosphere to initiate reaction. Then the mixture is stirred at room temperature under a nitrogen atmosphere for an additional 30 hours, at which time the originally green reaction mixture has been converted to a clear, deep blue solution. While a nitrogen atmosphere is maintained over the reaction solution, the mechanical stirrer is removed and replaced with a nitrogen outlet (Note 5). The third neck of the reaction flask is fitted with a short adapter closed with a rubber septum (Note 6).

The solution is standardized by withdrawing 5.0-ml. aliquots into a hypodermic syringe (Note 6) fitted with a relatively wide-bore needle and flushed with nitrogen before use. The aliquots are quenched by injecting them into 10 ml. of aqueous $1M$ ferric chloride solution in an Erlenmeyer flask under a nitrogen atmosphere. After 2 minutes the flow of nitrogen is stopped, and the resulting solution is diluted with 50 ml. of water and titrated with $0.1N$ ceric sulfate to the ferrous ion-phenanthroline end point (Note 7). Solutions prepared in this fashion are usually $0.55M$ in chromium(II) species (Note 8) and are stable for years, if they are protected from reaction with oxygen.

B. *Reduction of diethyl fumarate.* A 1-l. three-necked flask is equipped with a magnetic stirring bar, an addition funnel with a pressure-equalizing tube, a stopcock connected to a mercury trap, and a rubber septum (Note 9). The addition funnel is charged with a solution of 13.87 g. (0.080 mole) of diethyl fumarate (Note 10) in 137 ml. of dimethylformamide (Note 11). A nitrogen line is connected to the top of the addition funnel and the system is thoroughly flushed with nitrogen (Note 12). With a hypodermic syringe, 318 ml. (0.175 mole) of the previously described $0.55M$ chromium(II) sulfate solution (0.175 mole) is added to the reaction flask through the rubber septum. After the stirrer has been started, the diethyl fumarate solution is added rapidly. The solution immediately turns green and the reduction is complete in 10 minutes (Note 13). The resulting solution is diluted with 100 ml. of water and 30 g. of ammonium sulfate is added. The mixture is shaken with four 150-ml. portions of ether, and the combined ether extracts are washed with three 50-ml. portions of water and then dried over magnesium sulfate. After the ether has been removed by distillation through a 60-cm. Vigreux column, the residual liquid is distilled through a short Vigreux column to separate 12.4–13.2 g. (88–94%) of diethyl succinate (Note 14); b.p. 129° (44 mm), n^{23}D 1.4194.

2. Notes

1. Mallinckrodt analytical reagent, chromium(III) sulfate crystals, $Cr_2(SO_4)_3(H_2O)_x$, were employed. Repeated preparations with this substance have indicated its average formula weight to be 542.

2. Either Baker and Adamson or Mallinckrodt reagent grades of mossy zinc have been used interchangeably.

3. Distilled mercury was employed.

4. Warming is not always essential, but a more rapid reduction occurs routinely if the reaction is initiated by warming. In some cases a longer heating period may be required.

5. In order to maintain an oxygen-free atmosphere over the solution, it is essential that all standard taper joints be adequately lubricated and that the various joints be held together with rubber bands, wire, or springs.

6. Transfers are conveniently made by maintaining a slightly positive nitrogen pressure in the reaction vessel before the aliquots are removed and using an adapter consisting of a standard taper joint sealed to a wide-bore stopcock. The short length of glass tubing above the stopcock is fitted with a securely fastened rubber septum. The rubber septum above the stopcock is pierced with the hypodermic syringe, and then the stopcock is opened to place a slightly positive nitrogen pressure in the small septum-capped chamber. This procedure forces the plunger of the syringe out and sweeps any remaining oxygen from the syringe. The syringe plunger is replaced and the syringe needle is pushed below the surface of the solution. The internal nitrogen pressure forces

liquid into the syringe until slightly more than the desired amount is obtained. The syringe is then withdrawn and inverted and the excess solution is expelled into an absorbent paper. Finally, the syringe containing the desired volume of solution is emptied into a reaction vessel under a nitrogen atmosphere.

7. The preparation of the indicator solution is described by Kolthoff and Sandell.[2] The red-brown to green end point is easily observed.

8. Solutions of higher or lower concentrations can be prepared by adjusting the amounts of reagents.

9. Though unnecessary for this reduction, it is more usually convenient for chromium(II) sulfate reductions to fit the rubber septum to a stopcock adapter of the type described in Note 6.

10. Diethyl fumarate was purchased from either Eastman Organic Chemicals, Inc., or Aldrich Chemical Company and used without purification.

11. Baker reagent grade dimethylformamide was used without further purification.

12. The submitters recommend that the system be flushed with a slow stream of nitrogen for 30 minutes.

13. The kinetics of this reduction have been reported.[3] The reaction is easily followed by withdrawing aliquots and analyzing them for chromium(II) content.

14. Diethyl succinate is the sole product of the reduction. The yield reflects the efficiency of the workup. The distilled product gives a single sharp peak on gas chromatography employing a column packed with Carbowax 20M suspended on Chromosorb P. On this column the checkers found the retention times of diethyl fumarate and diethyl succinate to be 38.8 minutes and 43.6 minutes, respectively.

3. Discussion

Aqueous solutions of chromium(II) sulfate have been prepared from chromium(III) sulfate by reduction with zinc powder [3] and from potassium dichromate by reduction with amalgamated zinc and sulfuric acid.[4] Solid chromium(II) sulfate pentahydrate can be obtained from the reaction of highly purified chromium metal with concentrated sulfuric acid.[5] The present procedure is especially simple since it avoids filtration of zinc powder and avoids the acid present in the dichromate reduction.

Chromium(II) sulfate is a versatile reagent for the mild reduction of a variety of bonds. Thus aqueous dimethylformamide solutions of this reagent at room temperature couple benzylic halides,[3, 6] reduce aliphatic monohalides to alkanes,[6] convert vicinal dihalides to olefins,[7] convert geminal halides to carbenoids,[8] reduce acetylenes to *trans*-olefins,[9] and reduce α,β-unsaturated esters, acids, and nitriles to the corresponding saturated derivatives.[10] These conditions also reduce aldehydes to alcohols.[7]

The reduction of diethyl fumarate described in this preparation illustrates the mildness of the reaction conditions for the reduction of acetylenes and α,β-unsaturated esters, acids, and nitriles.

The reduction of diethyl fumarate to diethyl succinate has also been effected with diethyl 1,4-dihydro-2,6-dimethylpyridine-3,5-dicarboxylate [11] and by catalytic hydrogenation.

1. Department of Nematology, University of California, Riverside, California, 92502.
2. I. M. Kolthoff and E. B. Sandell, "Textbook of Quantitative Inorganic Analysis," The Macmillan Co., New York, 1952, p. 475.
3. C. E. Castro, *J. Am. Chem. Soc.*, **83**, 3262 (1961).
4. J. J. Lingane and R. L. Pecsok, *Anal. Chem.*, **20**, 425 (1948).
5. H. Lux and G. Illman, *Ber.*, **91**, 2143 (1958).
6. C. E. Castro and W. C. Kray, Jr., *J. Am. Chem. Soc.*, **85**, 2768 (1963).
7. W. C. Kray, Jr., and C. E. Castro, *J. Am. Chem. Soc.*, **86**, 4603 (1964).
8. C. E. Castro and W. C. Kray, Jr., *J. Am. Chem. Soc.*, **88**, 4447 (1966).
9. C. E. Castro and R. D. Stephens, *J. Am. Chem. Soc.*, **86**, 4358 (1964).
10. C. E. Castro, R. D. Stephens, and S. Moje, *J. Am. Chem. Soc.*, **88**, 4964 (1966).
11. E. A. Braude, J. Hannah, and R. Linstead, *J. Chem. Soc.*, 3257 (1960).

REDUCTION OF ORGANIC HALIDES.
CHLOROBENZENE TO BENZENE

$$C_6H_5Cl + (CH_3)_2CHOH + Mg \longrightarrow C_6H_6 + (CH_3)_2CHOMgCl$$

Submitted by D. Bryce-Smith and B. J. Wakefield [1]
Checked by William G. Dauben and Louis E. Friedrich

1. Procedure

To a 250-ml. round-bottomed flask fitted with a glass-blade stirrer, a pressure-equalizing dropping funnel, a thermometer, and a reflux condenser equipped with a nitrogen bubbler (Note 1) are added 6.0 g. (0.25 mole) of magnesium powder (Note 2), 50 ml. of decahydronaphthalene (Note 3), and a crystal of iodine. The flask is swept with nitrogen, and a nitrogen atmosphere is maintained throughout the reaction. The mixture is heated to reflux without stirring, and from the dropping funnel there is added slowly one-fifth of a solution of 11.3 g. (0.1 mole) of chlorobenzene (Note 4) and 9.0 g. (0.15 mole) of dry 2-propanol. Reaction is almost immediately apparent in the region of the iodine crystal, and as the reaction becomes progressively more vigorous (*ca.* 15 minutes) the stirrer is started and the external heating is reduced (Note 5). The remainder of the chlorobenzene solution is added over a 30-minute period; this rate of addition causes the mixture to reflux gently without external heating. An additional 25 ml. of decahydronaphthalene is added to facilitate the stirring, and the mixture is heated under reflux for one additional hour.

To the cooled mixture $6N$ hydrochloric acid is added dropwise with stirring, until no solid remains. The organic layer is separated, washed four times with 30-ml. portions of water (Note 6), dried over powdered calcium chloride (Note 7), and distilled through a 1×15 cm. column packed with Fenske helices (Note 8). The yield of benzene is 5.5–6.5 g. (70–83%), b.p. 80–82°, n^{20}D 1.5007. The fraction boiling at 82–180° contains no unreacted chlorobenzene (Notes 9, 10, 11).

2. Notes

1. To minimize loss of volatile products such as benzene, it is advisable to employ a dry ice condenser on top of the conventional condenser.

2. Magnesium powder (Grade 4) from Magnesium Elektron, Inc., 610 Fifth Avenue, New York 20, New York, or from Magnesium Elektron Ltd., Manchester, England, was employed within six months of the date of its grinding by the manufacturer. The use of older or coarser material may lead to lengthened induction periods, particularly when chlorides are used.

3. Freshly distilled decahydronaphthalene was used. With the more easily reduced halides, and where the boiling point of the neutral reduction product was close to that of decahydronapthalene, an excess of 2-propanol was used as the reaction medium. Other hydrocarbons and secondary or tertiary alcohols may be employed for convenience in particular reductions. Diethyl ether and tetrahydrofuran were not found to be generally suitable media.

4. The checkers found it necessary to distil the chlorobenzene just before use.

5. When there is no sustained reaction after 10 minutes, initiation can often be accomplished by the addition of another crystal of iodine (no stirring) and/or a small amount of an easily reduced halide such as 1-bromobutane.

6. These washings remove the bulk of the 2-propanol.

7. This drying also removes the last traces of 2-propanol.

8. The checkers used a Nester-Faust 44-cm. spinning-band column.

9. This procedure has been used to effect the following reductions at *ca.* 150°: bromobenzene to benzene (89%), iodobenzene to benzene (95%), 1-chlorobutane to *n*-butane (95%), 2-chloro-2-methylbutane to 2-methylbutane (32%), and isopropyl chloroacetate to isopropyl acetate (63%).

10. The following reductions have been carried out at 80° with the use of an excess of 2-propanol as the reaction medium (see Note 3): carbon tetrachloride to methane (47%), 1-bromonaphthalene to naphthalene (90%), β-bromostyrene to styrene (72%), *p*-bromoaniline to aniline (61%), *p*-bromophenol to phenol (66%), and monochloroacetone to acetone (30%).

11. Certain halides, notably fluorides, are comparatively inert under these reaction conditions. In such cases the *entrainment method* can be used, and reduction can be accomplished in the presence of a reactive halide such as 1-bromonaphthalene or 1-bromobutane. Also with certain halides, such as chlorocyclohexane, the tendency for dehydrohalogenation is diminished by the use of such entraining agents.

A typical example is the following reduction of chlorocyclohexane to cyclohexane. The general procedure is employed using 8.0 g. (0.33 mole) of magnesium powder in decahydronaphthalene (50 ml. + 20 ml.) and a solution of 6.0 g. (0.05 mole) of chlorocyclohexane, 10.4 g. (0.05 mole) of 1-bromonaphthalene, and 18 g. (0.3 mole) of 2-propanol. The product fraction, b.p. 78–80°, is a mixture of 3.5 g. (83%) of cyclohexane and 0.4 g. (10%) of cyclohexene. The olefin is removed by treatment with concentrated sulfuric acid in the usual manner.

Under the foregoing conditions, fluorocyclohexane gives cyclohexane (33%), and benzotrifluoride gives toluene (10%); fluorobenzene is inert.

3. Methods of Preparation

The present procedures are based on those briefly described by the submitters in conjunction with E. T. Blues,[2] and are based on the observation that magnesium does not, under normal conditions, readily react with secondary and tertiary alcohols in the absence of a halogen or an organic halide; little or no hydrogen is evolved during the reduction. Magnesium reacts readily with primary alcohols, evolving hydrogen, and the system is much less active in the reduction of organic halides. 2-Propanol is recommended as a general-purpose alcoholic component, but other secondary and tertiary alcohols can also be employed.

4. Merits of the Preparation

Reduction with magnesium and 2-propanol provides a simple and effective procedure for the reduction of alkyl and aryl chlorides, bromides, and iodides; with an entraining agent some alkyl fluorides are attacked. Groups such as amino, phenolic hydroxyl, ester carbonyl, and ethylenic linkages have not interfered. Nitro compounds must be absent as they inhibit the reaction with magnesium. Many carbonyl compounds, for example, p-bromobenzophenone, undergo much simultaneous reduction of the carbonyl groups, but acetone was obtained in fair yield from chloroacetone.

1. Department of Chemistry, The University, Reading, England.
2. D. Bryce-Smith, B. J. Wakefield, and E. T. Blues, *Proc. Chem. Soc.*, 219 (1963).

RUTHENOCENE*

(Ruthenium, dicyclopentadienyl-)

$$2Ru + 3Cl_2 \longrightarrow 2RuCl_3$$

$$2\,C_5H_5\ominus Na\oplus + H_2$$

$$6C_5H_5\ominus Na\oplus + 2RuCl_3 + Ru \longrightarrow 3 \;\; Ru \;\; + 6NaCl$$

Submitted by D. E. Bublitz, William E. McEwen, and Jacob Kleinberg.[1]
Checked by Hans G. Essler and John H. Richards.

1. Procedure

A 500-ml. three-necked flask is equipped with a Trubore stirrer, reflux condenser, and a pressure-equalizing dropping funnel that carries an inlet for admission of nitrogen. The system is purged with nitrogen (Note 1), and 300 ml. of 1,2-dimethoxyethane (Note 2) is added, followed by 7.2 g. (0.312 g. atom) of sodium either as wire or freshly cut small pieces. The solution is stirred, and 31.0 ml. (0.376 mole) of cyclopentadiene (Note 3) is added dropwise. When the evolution of hydrogen has almost ceased, the mixture is maintained at slightly below the reflux temperature for 1–2 hours. In the event that all the sodium does not dissolve, the solution is cooled to room temperature, a few milliliters more of cyclopentadiene added, and the mixture heated again until dissolution of the sodium is complete.

A mixture of 14.6 g. (0.07 mole) of ruthenium trichloride and 2.4 g. (0.024 g. atom) of ruthenium metal (Note 4) is added, and the reaction mixture is heated and stirred under nitrogen for 80 hours (Note 5) at slightly below the reflux temperature. With the use of stirring, the solvent is removed at aspirator pressure, and the flask then refilled with nitrogen. The solid is transferred to a sublimator in a dry-box containing a nitrogen atmosphere (Note 6) and sublimed at 0.1 mm. pressure with a heating bath at 130° (Note 7). The sublimate is dissolved in benzene and passed through a 1 x 12-in. column of activated alumina. Evaporation of the benzene gives 12.2–15.1 g. (56–69%) of ruthenocene, m.p. 199–200° (Note 8).

2. Notes

1. The submitters used prepurified nitrogen, obtained from Matheson Company, Inc., East Rutherford, New Jersey, without further purification. The checkers passed Linde (H. P. Dry) nitrogen successively through chromous chloride solution, solid potassium hydroxide, Ascarite, and solid phosphorus pentoxide.

2. 1,2-Dimethoxyethane is dried over sodium wire and then distilled under nitrogen from lithium aluminum hydride just before use.

3. For preparation of cyclopentadiene from the dimer, see G. Wilkinson, *Org. Syntheses,* Coll. Vol. 4, 475 (1963). The dicyclopentadiene used as starting material was dried by passage through a 1 x 12-in. column of activated alumina prior to cracking.

4. Ruthenium trichloride was prepared by chlorination of powdered ruthenium at 650–700° [2] with the use of metal obtained from Goldsmith Bros. Smelting and Refining Co., 111. N. Wabash Ave., Chicago 2, Illinois. Complete chlorination could not be effected under these conditions, and on the average about 85%

of the metal was converted to trichloride. Consequently, in all the preparations of ruthenocene, mixtures of trichloride and metal, as obtained from the chlorination reaction, were employed. The equations given for the preparation are idealized; the submitters believe that during the course of reaction the trichloride is gradually reduced to dichloride by ruthenium metal, and that it is the dichloride which reacts with sodium cyclopentadienide.

5. Somewhat lower yields than those reported are obtained when the reaction is carried out for a shorter period of time.

6. From this point on, the solid materials are pyrophoric, especially the residual solids from the sublimation process. However, the ruthenocene obtained by sublimation is not pyrophoric. The checkers found that careful addition of the sublimation residues to water under nitrogen destroys their pyrophoric character.

7. The checkers found the use of a Dry Ice-cooled sublimation finger advantageous.

8. The yield reported here is based on the total amount of ruthenium (both Ru^{III} and Ru^0) available for formation of ruthenocene. An additional quantity of ruthenocene may be obtained by extraction of the pyrophoric residue from the sublimation step with benzene in a Soxhlet extractor under a nitrogen atmosphere. The benzene solution is filtered through activated alumina, the solvent evaporated, and the residue sublimed.

3. Methods of Preparation

Ruthenocene has been prepared in 20% yields by reaction of cyclopentadienylmagnesium bromide with ruthenium(III) acetylacetonate.[3] More recently,[4] the compound has been made in 43–52% yield by treatment of sodium cyclopentadienide with ruthenium trichloride in tetrahydrofuran or 1,2-dimethoxyethane.

4. Merits of the Preparation

Ruthenocene is an example of a stable π-bonded organometallic compound which undergoes substitution reactions similar to those displayed by ferrocene. Because ruthenocene has heretofore been relatively unavailable, its chemistry has not been extensively studied.

1. Department of Chemistry, University of Kansas, Lawrence, Kans.
2. G. Brauer, "Handbuch der präparativen anorganischen Chemie," Ferdinand Enke Verlag, Stuttgart, Germany, 1952, p. 1194.
3. G. Wilkinson, *J. Am. Chem. Soc.*, **74**, 6146 (1952).
4. E. O. Fischer and H. Grubert, *Ber.*, **92**, 2302 (1959).

SODIUM NITROMALONALDEHYDE MONOHYDRATE

WARNING

It has been reported (Bruno Camerino, private communication) that, during the preparation of sodium nitromalonaldehyde monohydrate[1] on a pilot-plant scale, two operators were so affected by the fumes evolved during the preparation that their immediate hospitalization was necessary. It was determined subsequently that hydrogen cyanide, up to approximately 1 g./kg. of mucobromic acid utilized, was formed in the reaction mixture. It is essential that precautions specified in Note 1 of the procedure be followed carefully.

1. *Org. Syntheses*, Coll. Vol. **4**, 844 (1963).

STYRYLPHOSPHONIC DICHLORIDE

(Phosphonic dichloride, styryl-)

$$C_6H_5CH\!\!=\!\!CH_2 \xrightarrow[-HCl]{2PCl_5} C_6H_5CH\!\!=\!\!CH\overset{\oplus}{P}Cl_3 \ \overset{\ominus}{P}Cl_6$$

$$\xrightarrow[-POCl_3, -2SOCl_2]{2SO_2} C_6H_5CH\!\!=\!\!CHPOCl_2$$

Submitted by R. Schmutzler [1]
Checked by William G. Dauben and David A. Cox

1. Procedure

The reaction is conducted in a 500-ml. three-necked flask equipped with a sealed mechanical stirrer, a dropping funnel, and a reflux condenser carrying a drying tube. The flask is flushed with dry nitrogen, and 104 g. (0.50 mole) of phosphorus pentachloride in 150 ml. of dry benzene is added. The mixture is cooled in an ice bath (Note 1) and stirred while a solution of 26 g. (0.25 mole) of styrene in 50 ml. of dry benzene is added through the dropping funnel during a period of 30 minutes. A dense crystalline solid begins to form immediately, and after the addition is completed the mixture is stirred for 30 minutes at room temperature. The dropping funnel is replaced by a gas-inlet tube which is connected to a cylinder of sulfur dioxide through a wash bottle containing concentrated sulfuric acid. Sulfur dioxide is bubbled through the stirred mixture until all the precipitate is dissolved. The mildly exothermic reaction is controlled by occasionally cooling the reactants with an ice bath. The benzene solvent is removed from the clear solution under reduced pressure, and the residue is distilled at reduced pressure from a Claisen flask with Vigreux indentations. The yield of styrylphosphonic dichloride is 49–52 g. (89–94%), b.p. 107–110° (0.2 mm.). The distillate solidifies during or after the distillation, m.p. 71–72°.

2. Note

1. Care must be taken not to freeze the benzene before the styrene is added.

3. Methods of Preparation

Styrylphosphonic dichloride has been prepared by the addition of phosphorus pentachloride to styrene with subsequent reaction of the adduct with phosphorus pentoxide [2] or sulfur dioxide.[3, 4]

4. Merits of the Preparation

The addition reaction of phosphorus pentachloride to styrene and its derivatives provides a convenient route to styrylphosphonic acids and their derivatives.[2-7] The styrene phosphorus pentachloride adduct also can be reduced with phosphorus to give the corresponding dichlorophosphine.[4, 8]

The behavior of phosphorus pentachloride toward carbon-carbon multiple bonds has received considerable attention, and the procedure described represents but one example of a wide variety of derivatives of unsaturated phosphonic acids which are accessible. Indene was the first olefinic compound to be reacted with phosphorus pentachloride,[9] and the reaction of phosphorus pentachloride with other unsaturated compounds has been described.[2-6, 10-13] More recent examples include the reaction of phosphorus pentachloride with vinyl ethers [14-16] and vinyl thioethers,[17] providing access to β-alkoxy- and β-alkylmercaptovinylphosphonic and phosphonothioic acid derivatives.

1. Explosives Department, E. I. du Pont de Nemours and Company, Wilmington, Delaware.
2. W. H. Woodstock, U.S. Patent 2,471,472 [*C.A.*, **43**, 7499 (1949)].
3. K. N. Anisimov, *Bull. Acad. Sci. USSR, Div. Chem. Sci. (Engl. Transl.)*, 693 (1954).
4. G. K. Fedorova and A. V. Kirsanov, *J. Gen. Chem. USSR (Engl. Transl.)*, 4006 (1960).
5. E. Bergmann and A. Bondi, *Ber.*, **63**, 1158 (1930).
6. G. M. Kosolapoff and W. F. Huber, *J. Am. Chem. Soc.*, **68**, 2540 (1946).
7. A. D. F. Toy, U.S. Patent 2,425,766 [*C.A.*, **42**, 596 (1948)].
8. E. N. Walsh, T. M. Beck, and W. H. Woodstock, *J. Am. Chem. Soc.*, **77**, 929 (1955).
9. J. Thiele, *Chemiker-Ztg.*, **36**, 657 (1912); cf. C. Harnist, *Ber.*, **63**, 2307 (1930).
10. E. Bergmann and A. Bondi, *Ber.*, **64**, 1455 (1931); **66**, 278, 286 (1933).
11. L. Anschütz, F. König, F. Otto, and H. Walbrecht, *Ann.*, **525**, 297 (1936).
12. G. B. Bachman and R. E. Hatton, *J. Am. Chem. Soc.*, **66**, 1513 (1944).
13. K. N. Anisimov and N. E. Kolobova, *Bull. Acad. Sci. USSR (Engl. Transl.)*, 943, 947 (1956).

14. K. N. Anisimov and A. N. Nesmeyanov, *Bull. Acad. Sci. USSR (Engl. Transl.)*, 521 (1954).
15. K. N. Anisimov, *Bull. Acad. Sci. USSR (Engl. Transl.)*, 693 (1954).
16. K. N. Anisimov, N. E. Kolobova, and A. N. Nesmeyanov, *Bull. Acad. Sci. USSR (Engl. Transl.)*, 685, 689 (1954).
17. K. N. Anisimov, N. E. Kolobova, and A. N. Nesmeyanov, *Bull. Acad. Sci. USSR (Engl. Transl.)*, 21 (1956).

TETRACYANOETHYLENE OXIDE

(Ethanetetracarbonitrile, 1,2-epoxy-)

$$(NC)_2C{=}C(CN)_2 + H_2O_2 \longrightarrow (NC)_2C\overset{O}{\overbrace{\qquad}}C(CN)_2 + H_2O$$

Submitted by W. J. Linn [1]
Checked by A. Eschenmoser, W. Lusuardi, and R. Scheffold

1. Procedure

Caution! Both tetracyanoethylene and tetracyanoethylene oxide slowly evolve hydrogen cyanide when exposed to water. Therefore all operations should be conducted in an efficient hood and contact with the skin should be avoided.

In a 500-ml. Erlenmeyer flask fitted with an efficient ·stirrer and thermometer are placed 25.6 g. (0.2 mole) of tetracyanoethylene (Note 1) and 150 ml. of acetonitrile (Note 2). The flask is surrounded by an ice-salt bath and the stirrer is started. When the temperature is about −4°, 21 ml. of 30% hydrogen peroxide is added from a buret at the rate of 3–5 ml. per minute. The rate is adjusted to keep the temperature at 10–12° (Note 3). Near the end of the addition, the color of the reaction mixture changes from dark amber to pale yellow. When all the peroxide has been added, the reaction mixture is stirred with efficient cooling for 3–4 minutes. Without delay the solution is then poured slowly, with very rapid stirring, into a mixture of 500 ml. of water and approximately 250 g. of crushed ice contained in 2-l. beaker (Note 4). The solid is filtered rapidly by suction through a coarse, sintered-glass funnel and washed with 200 ml. of ice water. For best results the product is dried on the funnel with continuous suction for 3–4 hours and recrystallized from 1,2-dichloroethane (10 ml. per g.) (Note 5). The yield of nearly colorless needles melting at 177–178° (sealed capillary) is 17.1–19.6 g. (59–68%) (Note 6). The infrared spectrum of the oxide (Nujol mull) is simple and useful in product identification. In addition to the strong —C≡N absorption at 4.38μ, there are bands at 7.68, 8.47, 8.66, 10.54, and 11.23μ.

2. Notes

1. Tetracyanoethylene may be purchased from the Columbia Organic Chemicals Co., the Eastman Kodak Co., or prepared by the method of Carboni.[2] This procedure has been simplified in this laboratory as follows.[3]

To 450 ml. of cold water in the apparatus of Part A[2] there is added 99 g. (1.5 moles) of molten malononitrile followed by 250 g. of ice and 158 ml. (3.05 moles at 25°) of bromine. The bromine is added during 5–10 minutes, and during the addition enough ice (about 200 g.) is added to maintain the temperature at 10–15°. The mixture is stirred at 20° for 1 hour. A heavy layer of dibromomalononitrile is separated, and the aqueous layer is extracted with two 50-ml. portions of 1,2-dichloroethane. The dibromomalononitrile and the extracts are combined, dried over magnesium sulfate, and added to 750 ml. of dry 1,2-dichloroethane in the flask of Part B. Twenty grams of copper powder is added, and the mixture is heated to gentle reflux with stirring. An exothermic reaction generally occurs; when it subsides, or after about 10 minutes, a second 20-g. portion is added, and this process is continued until 120 g. has been added. The mixture is allowed to reflux a total of 4–6 hours. The solids are separated from the hot mixture using a fluted filter paper, which is washed with a little hot 1,2-dichloroethane. The filtrate is stored overnight at 0–5°. Nearly colorless tetracyanoethylene crystallizes out. It is separated on a Buchner funnel, washed with a little 1,2-dichloroethane, and dried in a vacuum desiccator; weight 29–38 g. (30–40%). The purity of the product is over 98% as judged by the ϵ_{max} [pure tetracyanoethylene has $\lambda_{max.}^{CH_2Cl_2}$ 277 mμ

(ϵ 12,050) 267 mμ (ϵ 13,600)]. It is pure enough for most purposes including synthesis of the epoxide. If very pure material is needed, tetracyanoethylene can be recrystallized from 1,2-dichloroethane (15 ml. per g.) or sublimed at 130–140° at 1 mm.

2. Eastman Kodak Co. practical grade is sufficiently pure for the reaction.

3. The rate of addition of hydrogen peroxide is fairly fast initially but is slowed to maintain the indicated temperature. It is important to get the reaction over in a short time (5–7 minutes) for the best yield.

4. It is wise to use a mechanical or magnetic stirrer in order to induce rapid crystallization of the product. Prolonged contact of the product with water at this stage diminishes the yield markedly. The presence of anions, *e.g.*, chloride, can lead to more rapid decomposition of the product, and it is best to use distilled water and ice prepared from distilled water at this point. If the oil cannot be induced to crystallize rapidly, more ice water should be added.

5. If it is necessary to interrupt the preparation before recrystallization, the product should be stored in a desiccator with continuous evacuation until it is absolutely dry.

6. This preparation has been carried out on a 4.2-mole scale using essentially the same procedure with only a slight diminution in yield. In larger runs the crude product may be more efficiently washed by rapidly resuspending the filter cake in fresh ice water, filtering, and drying. The only problem is that of drying the product rapidly. The drying can be hastened on a large scale by heating the mass on the funnel slightly with an infrared lamp.

3. Discussion

The usual method for epoxidation of an olefin with a peracid fails when the double bond is substituted with an electron-withdrawing group.[4] This difficulty has been circumvented in certain cases by the use of a very strong peracid; *i.e.*, peroxytrifluoroacetic acid, in the presence of a buffer [5] or by the use of alkaline hydrogen peroxide.[6] In the latter case, the attack is by the hydroperoxide anion.[7] This method is normally not applicable to the synthesis of epoxynitriles because of the simultaneous conversion of the nitrile to an amide group.[8] However, the four nitrile groups of tetracyanoethylene so diminish the electron density at the double bond that it is attacked by hydrogen peroxide in the absence of any added base. There is no significant attack on the nitrile groups when the reaction is carried out rapidly in a mutual solvent for the olefin and peroxide.[9] Olefins that are somewhat less electrophilic, e.g., phenyltricyanoethylene and diethyl 1,2-dicyanoethylene-1,2-dicarboxylate, can be epoxidized by essentially the same procedure using a catalytic amount of a mild base such as pyridine.[10]

Slight variations in the procedure described above have been used to prepare tetracyanoethylene oxide. Hydrogen peroxide in ether or t-butyl hydroperoxide in benzene [11] gives the epoxide in higher yield than the present method but requires large amounts of organic solvents and is not readily adaptable to large-scale preparations. An apparent contradiction of the opening statement above is the observation that tetracyanoethylene *can* be epoxidized with a peracid.[12] This is undoubtedly due to *nucleophilic* attack by the peracid or its anion on the electron-deficient double bond, a mechanism which cannot operate with olefins containing only one or two electronegative substituents. An example of this type of epoxidation is the preparation of 1,1-dicyano-2,2-bis(trifluoromethyl)ethylene oxide.[13]

Tetracyanoethylene oxide does not undergo reactions typical of epoxides of simple hydrocarbon olefins. Entirely new types of reactions are observed; e.g., cleavage by nucleophilic reagents into the elements of dicyanomethylene and carbonyl cyanide [10] and cleavage of the carbon-carbon bond, followed by addition to a wide variety of olefins, acetylenes, and aromatic compounds.[14] For example, tetracyanoethylene oxide adds thermally to adjacent positions on the benzene ring to give 1,1,3,3-tetracyano-1,3,3a,7a-tetrahydroisobenzofuran.

1. Contribution No. 1277 from the Central Research Department, Experimental Station, E. I. du Pont de Nemours and Company, Wilmington, Delaware 19898.
2. R. A. Carboni, *Org. Syntheses* Coll. Vol. **4**, 877 (1963).
3. Based on the work of Dr. E. L. Martin and Mr. H. D. Carlson.
4. D. Swern, *Org. Reactions*, **7**, 378 (1953).
5. W. D. Emmons and A. S. Pagano, *J. Am. Chem. Soc.*, **77**, 89 (1955).

6. E. Weitz and A. Scheffer, *Ber.*, **54**, 2327 (1921).
7. C. A. Bunton and G. J. Minkoff, *J. Chem. Soc.*, 665 (1949).
8. J. V. Murray and J. B. Cloke, *J. Am. Chem. Soc.*, **56**, 2749 (1934).
9. W. J. Linn (to E. I. du Pont de Nemours and Co.), U.S. Patent 3,238,228 (1966).
10. W. J. Linn, O. W. Webster, and R. E. Benson, *J. Am. Chem. Soc.*, **87**, 3651 (1965).
11. A. Rieche and P. Dietrich, *Ber.*, **96**, 3044 (1963).
12. Unpublished observations from this laboratory.
13. W. J. Middleton, *J. Org. Chem.*, **31**, 3731 (1966).
14. W. J. Linn and R. E. Benson, *J. Am. Chem. Soc.*, **87**, 3657 (1965).

TETRAHYDROXYQUINONE *

(Quinone, tetrahydroxy-)

Submitted by A. J. FATIADI and W. F. SAGER.[1]
Checked by B. C. McKUSICK and J. K. WILLIAMS.

1. Procedure

A 5-l., three-necked round-bottomed flask is fitted with a thermometer, an air-inlet tube of 10-mm. diameter extending to within approximately 1 cm. of the bottom, and an outlet tube connected to an aspirator (Note 1). A solution of 400 g. (3.17 moles) of anhydrous sodium sulfite and 150 g. (1.79 moles) of anhydrous sodium bicarbonate (Note 2) in 3 l. of water is heated to 40–45° in the flask. Six hundred grams (480 ml., 3.11 moles) of 30% glyoxal solution (Note 3) is added, and a brisk stream of air is drawn through the solution for 1 hour without application of heat. Within a few minutes, greenish black crystals of the sodium salt of tetrahydroxyquinone begin to separate. The flask is warmed to between 80° and 90° over a period of an hour. The air current is then stopped, and the mixture is heated to incipient boiling and set aside for 30 minutes. It is then cooled to 50°

(Note 4), and the sodium salt of tetrahydroxyquinone is separated by filtration and washed successively with 50 ml. of cold
15% sodium chloride solution, 50 ml. of cold 1:1 methanol-water,
and 50 ml. of methanol. The air-dried salt weighs 20–21 g.

The salt is added to 250 ml. of 2N hydrochloric acid, and the
mixture is heated to incipient boiling. The resultant solution is
cooled in an ice bath, and the glistening black crystals of tetrahydroxyquinone that precipitate are collected on a Büchner funnel and washed with ice water to give 11–15 g. (6.2–8.4%) of
product. The quinone fails to melt on a hot plate at 320° (Note 5).

2. Notes

1. A tube of 10-mm. diameter is necessary to prevent the
clogging of the outlet that occurs if tubing of smaller diameter
is used.

2. Equivalent amounts of hydrated salt may be used.

3. Dow commercial grade 30% glyoxal solution is satisfactory.

4. It is not necessary to cool below this temperature since
crystallization is essentially complete at 50°.

5. This material is pure enough for reduction to hexahydroxybenzene [2] and most other purposes. A purer product can be
obtained by dissolving the crude tetrahydroxyquinone in acetone
and adding petroleum ether of b.p. 60–80° to precipitate it.

3. Methods of Preparation

The procedure employed for tetrahydroxyquinone is based on
an observation by Homolka.[3] Tetrahydroxyquinone may also be
prepared by treatment of the glyoxal-bisulfite addition compound
with sodium carbonate [3] or magnesium hydroxide and potassium
cyanide [4] or by treatment of 50% glyoxal with sodium hydrosulfite.[5]

4. Merits of the Preparation

Tetrahydroxyquinone is of interest because of its application in
analytical chemistry to the determination of barium and as a com-
plexing agent for many ions.[6] Moreover, it serves as a convenient

source not only of the reduction product hexahydroxybenzene [2] but also of the oxidation products rhodizonic acid and triquinoyl and of the product of catalytic reduction, *meso*-inositol.

This procedure serves as a particularly simple method for preparing tetrahydroxyquinone. The low yield is more than off-set by the simplicity of the set-up, the ease of manipulation, and the low cost and ready availability of the starting materials.

1. Department of Chemistry, The George Washington University, Washington, D. C.
2. A. J. Fatiadi and W. F. Sager, this volume, p. 595.
3. B. Homolka, *Ber.*, 54, 1393 (1921).
4. R. Kuhn, G. Quadbeck, and E. Röhm, *Ann.*, 565, 1 (1949).
5. B. Eistert and G. Bock, *Angew. Chem.*, 70, 595 (1958).
6. J. H. Yoe and L. A. Sarver, "Organic Analytical Reagents," John Wiley and Sons, Inc., New York, 1941 (out of print); S. J. Kocher, *Ind. Eng. Chem., Anal. Ed.*, 9, 331 (1937).

TETRAMETHYLAMMONIUM
1,1,2,3,3-PENTACYANOPROPENIDE

(1-Propene-1,1,2,3,3-pentacarbonitrile, tetramethylammonium salt)

$$(NC)_2C{=}C(CN)_2 + CH_2(CN)_2 + \text{(pyridine)}N \rightarrow$$

$$[C(CN)_2{=}C(CN)C(CN)_2]^- \ \text{(pyridinium)}NH + HCN$$

$$[C(CN)_2{=}C(CN)C(CN)_2]^- \ \text{(pyridinium)}NH + N(CH_3)_4Cl \rightarrow$$

$$[C(CN)_2{=}C(CN)C(CN)_2]^- \ [N(CH_3)_4]^+ + \ \text{(pyridinium)}NHCl$$

Submitted by W. J. MIDDLETON and D. W. WILEY.[1]
Checked by JAMES CASON and WILLIAM T. MILLER.

1. Procedure

Caution! This reaction must be carried out in an efficient hood because large amounts of hydrogen cyanide are evolved. It is also recommended that tetracyanoethylene not be allowed to come into contact with the skin.

A solution of 6.6 g. (0.10 mole) of malononitrile (Note 1) in 8.7 g. (0.11 mole) of pyridine and 25 ml. of water is prepared in a 125-ml. Erlenmeyer flask and stirred mechanically (no stirrer seal required) as there is added rapidly in small portions a total of 12.8 g. (0.10 mole) of powdered recrystallized tetracyanoethylene (Note 2). The resulting mixture is warmed on a hot plate as stirring is continued until complete solution occurs (5–10 minutes, Note 3). The hot dark solution is poured into a swirled solution of 12.1 g. (0.11 mole) of tetramethylammonium chloride (Note 4) in 500 ml. of water. The resultant mixture is heated almost to boiling to give a dark-red solution, which is then allowed to cool spontaneously to room temperature. After final cooling in an ice bath, the orange needles of tetramethylammonium 1,1,2,3,3-pentacyanopropenide are collected by suction filtration and washed with two 100-ml. portions of cold water. This product is dissolved in 500 ml. of hot water, decolorized with about 5 g. of activated carbon, and allowed to crystallize as described above. The yield of bright yellow-orange needles, m.p. 314–315^c (Note 5), is 19.5–20.5 g. (81–85%).

2. Notes

1. Malononitrile, m.p. 30–31°, obtained from Winthrop-Stearns Corp., New York, N. Y., is satisfactory.

2. Tetracyanoethylene is available from the Aldrich Chemical Company, Inc., or may be prepared by a simplified version of the procedure of Carboni.[2] In the simplified version, 99 g. of molten malononitrile is added to 450 ml. of cold water in the apparatus of Part A. This is followed by about 250 g. of ice and 158 ml. of bromine. The bromine is added during 5–10 minutes, and during the addition enough ice (about 200 g.) is added to maintain the temperature at 10–15°. The mixture is stirred at 20° for one hour. A heavy layer of dibromomalononitrile is separated, and the aqueous layer is extracted with two 50-ml. portions of 1,2-dichloroethane. The dibromomalononitrile and the extracts

are combined, dried over magnesium sulfate, and added to 750 ml. of dry 1,2-dichloroethane in the flask of Part B. Twenty grams of copper power is added, and the mixture is heated to gentle reflux with stirring. An exothermic reaction generally occurs; when it subsides, or after about 10 minutes, a second 20-g. portion is added, and this process is continued until 120 g. has been added. The mixture is refluxed a total of 4–6 hours. The solids are separated from the hot mixture on a fluted filter, which is washed with a little hot 1,2-dichloroethane. The filtrate is stored overnight at 0–5°. Nearly colorless tetracyanoethylene crystallizes out. It is separated on a Büchner funnel, washed with a little 1,2-dichloroethane, and dried in a vacuum desiccator; weight 29–38 g. (30–40%). It is over 98% pure as judged by its ϵ_{max} (pure tetracyanoethylene has $\lambda_{max}^{CH_2 Cl_2}$ 277 mμ (ϵ 12,050), 267 mμ (ϵ 13,600)). It is pure enough for most purposes, but can be sublimed at 1 mm. (bath 130–140°) if very pure material is needed (private communication from E. L. Martin and H. D. Carlson via B. C. McKusick).

3. The rate of solution depends upon the fineness of the tetracyanoethylene powder; however, solution should occur soon after the temperature reaches 60–70°.

4. A technical grade of tetramethylammonium chloride is satisfactory, provided old samples that have absorbed considerable water are not used.

5. In a heated block, the checkers observed melting points in the range 318–321°.

3. Methods of Preparation

Tetramethylammonium 1,1,2,3,3-pentacyanopropenide has been prepared by the base-catalyzed hydrolysis of tetracyanoethylene,[3] and by the present method, which is more economical of tetracyanoethylene.

4. Merits of Preparation

Tetramethylammonium 1,1,2,3,3-pentacyanopropenide is useful for preparation of pentacyanopropenide salts of other metal and quaternary ammonium cations by metathesis.[3] The free acid, which may be obtained by use of an ion-exchange resin,[3] has an ionization constant comparable to that of a strong mineral acid ($pK_a < -8.5$; the anion is not detectably protonated in $12M$ sulfuric acid).[4]

1. Contribution No. 482 from Central Research Department, Experimental Station, E. I. du Pont de Nemours and Co., Wilmington, Delaware.
2. R. A. Carboni, *Org. Syntheses*, Coll. Vol. 4, 877 (1963).
3. W. J. Middleton, E. L. Little, D. D. Coffman, and V. A. Engelhardt, *J. Am. Chem. Soc.*, 80, 2795 (1958).
4. R. H. Boyd, unpublished experiments.

TETRAMETHYLBIPHOSPHINE DISULFIDE *

(Diphosphine, tetramethyl-, disulfide)

$$6 \, CH_3MgBr + 2 \, PSCl_3 \rightarrow (CH_3)_2P\!\!-\!\!P(CH_3)_2 + 6 \, MgBrCl$$
$$\underset{S}{\|} \quad \underset{S}{\|}$$

Submitted by G. W. PARSHALL [1]
Checked by W. S. WADSWORTH and WILLIAM D. EMMONS

1. Procedure

A 3-l. round-bottomed flask equipped with mechanical stirrer, condenser (surmounted by a drying tube), thermometer, and addition funnel is charged with 800 ml. of $3M$ methylmagnesium bromide solution (2.4 moles) (Note 1) and 600 ml. of anhydrous ether. The solution is stirred and cooled to 0–5° while a solution of 135 g. (83 ml., 0.80 mole) of thiophosphoryl chloride (Note 2) in 85 ml. of ether is added over a period of 3 hours. A thick white precipitate forms during the course of the addition. After completion of the addition, the reaction mixture is poured onto 500 g. of ice in a 4-l. beaker. Sulfuric acid (900 ml. of 10% solution) is added over a period of 20 minutes with gentle stirring. The mixture is filtered, and the white solid product is washed with 4 l. of water and recrystallized from 2 l. of ethanol. The product is dried over phosphorus pentoxide in a vacuum desiccator to give 50–55 g. (67–74%) of white crystalline tetramethylbiphosphine disulfide, m.p. 223–227° (Note 3). Evaporation of the mother liquor to a volume of 900 ml. gives an additional 3 g. of tetramethylbiphosphine disulfide, m.p. 222–225°.

2. Notes

1. A suitable $3M$ solution of methylmagnesium bromide in diethyl ether can be purchased from Arapahoe Chemical Co., Boulder, Colorado.

2. Although practical grade thiophosphoryl chloride obtained from Eastman Organic Chemicals will serve in this reaction, a much cleaner product is obtained if the thiophosphoryl chloride is redistilled (b.p. 122–123°).

3. Tetramethylbiphosphine disulfide melts sharply at 227° when pure, but the material obtained as described above is satisfactory for most reactions.

3. Methods of Preparation

Tetramethylbiphosphine disulfide has been prepared by reaction of methylmagnesium halides with thiophosphoryl chloride.[2–4]

4. Merits of the Preparation

Tetramethylbiphosphine disulfide is an extremely versatile intermediate for the preparation of compounds containing two methyl groups on phosphorus, for example, dimethylphosphine.[5] Most other methods for the preparation of such compounds give large amounts of mono- and trimethylated by-products. Tetramethylbiphosphine disulfide has been converted in high yields to dimethylphosphinic acid,[3,4] dimethylphosphinyl chloride,[4,6] dimethylchlorophosphine,[7] and dimethylthiophosphinic bromide.[8] Other tetraalkylbiphosphine disulfides have been converted to tetraalkylbiphosphines, dialkylthiophosphoryl bromides, and dialkylphosphinic anhydrides.[9] Addition of tetramethylbiphosphine disulfide to ethylene followed by desulfurization gives tetramethylethylenediphosphine, a powerful chelating agent.[10] Other alkyl Grignard reagents also react with thiophosphoryl chloride under the conditions of the present procedure to give the corresponding tetraalkylbiphosphine disulfides in high yield.[5,11]

1. Contribution No. 582 from the Central Research Department, Experimental Station, E. I. du Pont de Nemours and Company, Wilmington, Delaware.

2. M. I. Kabachnik and E. S. Shepeleva, *Izv. Akad. Nauk SSSR, Otd. Khim. Nauk*, 56 (1949).
3. G. M. Kosolapoff and R. M. Watson, *J. Am. Chem. Soc.*, **73**, 5466 (1951).
4. H. Reinhardt, D. Bianchi, and D. Mölle, *Ber.*, **90**, 1656 (1957).
5. H. Niebergall and B. Langenfeld, *Ber.*, **95**, 64 (1962); G. W. Parshall, *Inorg. Syntheses*, **11**, 157 (1968).
6. K. A. Pollart and H. J. Harwood, *J. Org. Chem.*, **27**, 4444 (1962).
7. G. W. Parshall, *J. Inorg. Nucl. Chem.*, **12**, 372 (1960).
8. R. Schmutzler, *Inorg. Syntheses*, **12**, 287 (1970).
9. W. Kuchen and H. Buchwald, *Angew. Chem.*, **71**, 162 (1959).
10. G. W. Parshall, *J. Inorg. Nucl. Chem.*, **14**, 291 (1960).
11. K. Issleib and A. Tzschach, *Ber.*, **92**, 704 (1959).

TETRAMETHYL-*p*-PHENYLENEDIAMINE*

(*p*-Phenylenediamine, N,N,N'N'-tetramethyl)

$$H_2N-\!\!\left\langle\bigcirc\right\rangle\!\!-NH_2 \xrightarrow[\text{NaHCO}_3]{\text{(CH}_3)_2\text{SO}_4} (CH_3)_3\overset{+}{N}-\!\!\left\langle\bigcirc\right\rangle\!\!-\overset{+}{N}(CH_3)_3$$

$$2CH_3SO_4^-$$

$$(CH_3)_3\overset{+}{N}-\!\!\left\langle\bigcirc\right\rangle\!\!-\overset{+}{N}(CH_3)_3 + 2H_2N\!-\!CH_2\!-\!CH_2\!-\!OH \longrightarrow$$

$$2CH_3SO_4^-$$

$$(CH_3)_2N-\!\!\left\langle\bigcirc\right\rangle\!\!-N(CH_3)_2 + 2CH_3\overset{+}{N}H_2\!-\!CH_2\!-\!CH_2\!-\!OH$$

$$2CH_3SO_4^-$$

Submitted by S. Hünig, H. Quast, W. Brenninger, and E. Frankenfeld [1]
Checked by R. A. Schwartz and K. B. Wiberg

Caution! Tetramethyl-p-phenylenediamine may induce a painful dermatitis when brought into contact with the skin.[2] Suitable precautions should be taken to avoid such contact.

1. Procedure

In a 2-l. three-necked flask fitted with a stirrer, thermometer, and pressure-compensated dropping funnel are placed 54 g. (0.5 mole) of powdered *p*-phenylenediamine (Note 1), 310 g. (3.7 mole) of sodium bicarbonate, and 250 ml. of water. The temperature of the solution is maintained at 18–22° using an ice bath while 320 ml. (3.4 mole) of dimethyl sulfate (Note 2) is added with stirring over a 30- to 50-minute period. Carbon dioxide is evolved vigorously and a transient purple color is developed; it changes to a brown tinge later on.

When the addition of dimethyl sulfate is complete, stirring is continued for 1 hour at 20–25°. Then the temperature is raised to 60–65° during 10 minutes (Note 3) and is kept at this value until the evolution of carbon dioxide ceases. After the addition of 250 ml. of cold water, the reaction flask is cooled rapidly in an ice bath and 100 ml. of ethanolamine (Note 4) is added. The resultant crystalline slurry is removed from the flask, and the apparatus is rearranged as indicated in Note 5, using an upright condenser between the dropping funnel (Note 6) and the receiving flask.

To the reaction flask is added 200 ml. of ethanolamine, and it is heated to 140° with stirring. The slurry above is added in moderate portions over a 40- to 50-minute period (Note 7). When the heating bath is maintained at 230–240°, the addition of the slurry should provide an inner temperature at 120–140° as the water and oily product distill. After the addition is complete, the dropping funnel is rinsed with 100–150 ml. of water. As soon as the inner temperature has reached 160°, 50 ml. of ethanolamine is added and the temperature is maintained at 160–170° for 20 minutes. Water (50 ml.) is added through the dropping funnel to initiate a rapid steam distillation. Steam distillation is continued by the addition of 50-ml. portions of water at an inner temperature of 120–140° and a bath temperature of 230–240° until no more oil appears in the distillate (Note 8).

The oily product solidifies on cooling to about 20°, forming white lumps. After filtration by suction, the lumps are crushed, filtered, and washed four times with 50-ml. portions of ice water. Drying over silica gel in a vacuum gives 62–72 g. (82–88%) of white glistening scales, m.p. 51°.

2. Notes

1. A technical grade of *p*-phenylenediamine was used.
2. Dimethyl sulfate was distilled, b.p. 73–75° (13 mm.).

3. At this temperature the excess of dimethyl sulfate is destroyed.

4. Ethanolamine was distilled, b.p. 74–75° (13 mm.).

5. The apparatus shown in Fig. 1 is suitable for this step. It is essential that the condenser be very effective since the steam distillation is very rapid. If vapor is lost from the top of an internal coil condenser, cold towels placed on the outside of the condenser will provide additional cooling.

6. The dropping funnel should have a stopcock bore as large as possible.

7. Stirring and addition of only 40–50 ml. of the slurry into the dropping funnel should avoid obstruction of the stopcock. The use of a thin metal wire is sometimes helpful.

8. Transient blue colors in the distillate result from autoxidation. They do not, however, affect the purity of the final product.

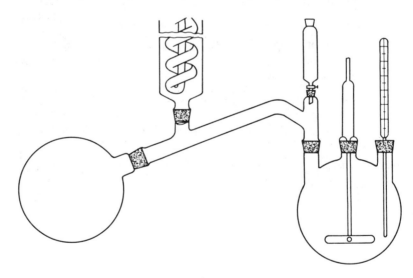

Fig. 1

3. Discussion

Tetramethyl-*p*-phenylenediamine has been obtained in low yield by the reaction of *p*-phenylenediamine with various alkylating agents such as methyl iodide,[3] methanol in the presence of hydrochloric acid at 170–200°,[4] or formaldehyde and formic acid.[5] In addition it has been prepared by methylating *p*-dimethyl-aminoaniline using methanol in the presence of hydrochloric acid at 170–200°,[6,7] followed by treatment of the resulting salts with aqueous ammonia at 180–190°.[7] In the most recent procedure, *p*-phenylenediamine was alkylated with sodium chloroacetate. Decarboxylation of the *p*-phenylenediaminetetraacetic acid at 180° gave 28% of tetramethyl-*p*-phenylenediamine based on the starting diamine.[8]

The present procedure[9] combines two general methods described earlier.[10,11] It is conveniently carried out and gives a substantially higher yield than previous methods. Dimethyl sulfate in the presence of aqueous sodium bicarbonate selectively methylates aromatic amines under mild conditions to give quaternary salts without affecting phenolic hydroxy groups present in the molecule. If the quaternization step is sterically hindered, the reaction stops at the tertiary amine stage.[10] Heterocyclic compounds may also be converted to quaternary salts in high yield, two or more methyl groups being introduced in one step.[12] The rate of reaction may conveniently be followed by observing the carbon dioxide evolution.

Dealkylation of quaternary ammonium salts using ethanolamine is more convenient than the use of aqueous ammonia in sealed tubes at high temperatures.[10] Ethanolamine may be replaced by other ethanolamines.[13] The reaction leads to preferential removal of methyl groups.

1. Chemisches Institut der Universität Würzburg, Germany.
2. Private communication from H. T. Clarke.
3. A. W. Hofmann, *Compt. Rend.*, 56, 994 (1863); *Jahresber. Fortschr. Chem.*, 422 (1863).
4. R. Meyer, *Ber.*, 36, 2979 (1903).
5. J. N. Ashley and W. G. Leeds, *J. Chem. Soc.*, 2706 (1957).
6. C. Wurster, *Ber.*, 12, 522 (1879).
7. J. Pinnow, *Ber.*, 32, 1401 (1899).
8. J. R. Cox, Jr., and B. D. Smith, *J. Org. Chem.*, 29, 488 (1964).
9. S. Hünig, H. Quast, W. Brenninger, and E. Schmitt, *Ber.*, 102, 2874 (1969).
10. S. Hünig, *Ber.*, 85, 1056 (1952).
11. S. Hünig and W. Baron, *Ber.*, 90, 395 (1957).
12. H. Quast and E. Schmitt, *Ber.*, 101, 4012 (1968).
13. K. Menzel, Ger. Patent 953170 (1953) [*Chem. Abstr.*, 53, 8071 (1959)].

2,3,4,5-TETRAMETHYLPYRROLE

(Pyrrole, 2,3,4,5-tetramethyl-)

$$\underset{\underset{CH_3}{|}}{\overset{CH_3CO}{\diagdown}}C{=}NOH \; + \; \underset{\underset{\underset{CH_3}{\diagdown}}{CO}}{\overset{CH_3}{\diagup}}CO{-}\underset{}{\overset{CH_3}{\diagup}}CH \quad \xrightarrow[\downarrow Zn]{CH_3CO_2H}$$

Submitted by A. W. JOHNSON and R. PRICE.[1]
Checked by VIRGIL BOEKELHEIDE and M. KUNSTMANN.

1. Procedure

In a 2-l. three-necked flask fitted with a stirrer, thermometer, and reflux condenser are placed 250 ml. of glacial acetic acid, 54.5 g. (0.84 g. atom) of zinc dust, and 52.5 g. (0.46 mole) of 3-methylpentane-2,4-dione (Note 1). The contents of the flask are stirred vigorously (Note 2), and a solution of 42 g. (0.415 mole) of diacetyl monoxime [2] in 150 ml. of glacial acetic acid is added from a separatory funnel at a rate to maintain the temperature of the mixture at 65–70°. The addition takes 1 hour. When the addition is complete, the mixture is refluxed with stirring for an additional 30 minutes. The flask is then fitted for distillation with steam under nitrogen; 500 ml. of water is added and steam is introduced. Steam distillation (Note 3) is continued until no more tetramethylpyrrole comes over. This takes 1–2 hours and the distillate amounts to 1–2 l. The tetramethylpyrrole crystallizes from the steam distillate and is collected by filtration, washed with water, and dried over phosphorus pentoxide in a vacuum desiccator. There is obtained 15–18 g. of white plates, m.p. 110–111° (lit.,[3] m.p. 112°).

By neutralizing the filtrate with sodium hydroxide solution, a second crop of 4–5 g. of tetramethylpyrrole, m.p. 109–110°, is obtained. The total yield is 20.5–22.5 g. (40–44%) (Note 4).

2. Notes

1. 3-Methylpentane-2,4-dione is prepared by the methylation of acetylacetone.[4, 5]

2. It is essential that the zinc dust be stirred effectively or the reaction may become violent.

3. Tetramethylpyrrole must be prevented from blocking the condenser. From time to time the condenser is cleared by turning off the coolant water.

4. 2,3,4,5-Tetramethylpyrrole is very readily oxidized in the air to a green resinous substance. If it is not used immediately, it should be stored under nitrogen or sealed in a glass vial under vacuum.

3. Methods of Preparation

2,3,4,5-Tetramethylpyrrole has been prepared by the action of sodium methoxide on 2,3,5-trimethylpyrrole,[6] by the reduction of 2,3,5-trimethylpyrrole-4-aldehyde semicarbazone with sodium ethoxide,[7] by the reduction of 2,3,4-trimethylpyrrole-5-aldehyde with sodium ethoxide and hydrazine hydrate,[8] and by the reduction of 2,4-dimethylpyrrole-3,5-dicarboxylic acid with lithium aluminum hydride.[9] Direct ring synthesis by the condensation of 3-aminobutan-2-one and butan-2-one in alkaline solution gave very poor yields, the principal product being 2,3,5,6-tetramethylpyrazine.[3] The above modification of direct ring synthesis avoids this side reaction.[10]

4. Merits of Preparation

The present method possesses these advantages over those reported earlier: [5-8] it is less laborious, in that it is a single-stage preparation, and it gives a better over-all yield.

1. Department of Chemistry, The University of Nottingham, Nottingham, England.
2. W. L. Semon and V. R. Damerell, *Org. Syntheses,* Coll. Vol. 2, 204 (1943).
3. H. Fischer and B. Walach, *Ann.,* 447, 38 (1926).
4. A. W. Johnson, E. Markham, and R. Price, this volume, p. 785.
5. K. von Auwers and H. Jacobsen, *Ann.,* 426, 161 (1921).
6. H. Fischer and E. Bartholomäus, *Z. Physiol. Chem.,* 80, 10 (1912).
7. H. Fischer and W. Zerweck, *Ber.,* 56, 519 (1923).
8. M. Dennstedt, *Ber.,* 22, 1924 (1889).
9. A. Treibs and H. Derra-Scherer, *Ann.,* 589, 188 (1954).
10. A. W. Johnson, E. Markham, R. Price, and K. B. Shaw, *J. Chem. Soc.,* 4254 (1958).

2,2,5,5-TETRAMETHYLTETRAHYDRO-3-KETOFURAN

(3(2)-Furanone, 4,5-dihydro-2,2,5,5-tetramethyl-)

Submitted by MELVIN S. NEWMAN and WALTER R. REICHLE.[1]
Checked by VIRGIL BOEKELHEIDE and GRAHAM SOLOMONS.

1. Procedure

To the solution formed by dissolving 3 g. of mercuric oxide and 10 ml. of concentrated sulfuric acid (Note 1) in 1 l. of water is added 250 g. (1.76 moles) of 2,5-dimethyl-3-hexyne-2,5-diol (Note 2). The mixture is warmed with gentle swirling to dissolve the diol. At 80–90° the clear solution suddenly turns cloudy and the flask is immersed in a bath of water at about 20°. A colorless oil rises to the surface within a few minutes (Note 3).

The flask is then fitted for steam distillation, and 700–800 ml. of distillate is collected. After addition of 55 g. of sodium chloride to the distillate, the phases are separated in a 1-l. separatory funnel. The organic layer is dried by intermittent stirring with 25 g. of anhydrous magnesium sulfate for 6–10 hours.

The material is filtered and the residual solid washed twice with 50-ml. portions of low-boiling petroleum ether. The combined filtrates are concentrated and distilled through a short packed column to yield 190–205 g. (76–82%) of 2,2,5,5-tetramethyltetrahydro-3-ketofuran; b.p. 149–151°, n_D^{25} 1.4180.

2. Notes

1. Any other water-soluble mercury salt may be used.

2. Supplied by Air Reduction Chemical Company, 150 East 42nd Street, New York, N. Y.

3. The reaction is fairly exothermic. Cooling is advisable. An increase in acid and mercuric ion concentrations results in a faster reaction starting at a lower temperature.

3. Methods of Preparation

The procedure described is essentially that of Richet [2] which has been repeated.[3,4] The reaction is of interest since it provides a facile method of preparing tetrahydro-3-furanones which are useful reagents for alkylation in the Friedel-Crafts reaction.[4]

1. Department of Chemistry, Ohio State University, Columbus, Ohio.
2. M. Richet, *Ann. Chim.*, 3 (12), 317 (1948).
3. B. L. Murr, G. B. Hoey, and C. T. Lester, *J. Am. Chem. Soc.*, 77, 4430 (1955).
4. H. A. Bruson, F. W. Grant, and E. Bobko, *J. Am. Chem. Soc.*, 80, 3633 (1958).

$\alpha,\alpha,\alpha',\alpha'$-TETRAMETHYLTETRAMETHYLENE GLYCOL

(2,5-Hexanediol, 2,5-dimethyl-)

$$Fe^{2+} + H_2O_2 \xrightarrow{H_2SO_4} Fe^{3+} + OH^- + \cdot OH$$

$$(CH_3)_3COH + \cdot OH \rightarrow \cdot CH_2 - \overset{\overset{\displaystyle CH_3}{|}}{\underset{\underset{\displaystyle CH_3}{|}}{C}} - OH + H_2O$$

$$2 \cdot CH_2 - \overset{\overset{\displaystyle CH_3}{|}}{\underset{\underset{\displaystyle CH_3}{|}}{C}} - OH \rightarrow HO - \overset{\overset{\displaystyle CH_3}{|}}{\underset{\underset{\displaystyle CH_3}{|}}{C}} - CH_2CH_2 - \overset{\overset{\displaystyle CH_3}{|}}{\underset{\underset{\displaystyle CH_3}{|}}{C}} - OH$$

Submitted by E. L. JENNER.[1]
Checked by JOHN D. ROBERTS and M. PANAR.

1. Procedure

Tertiary butyl alcohol (900 ml., 702 g., 9.47 moles) is dissolved in a solution prepared by mixing 28 ml. (0.50 mole) of concentrated sulfuric acid with 1.5 l. of water in a 5-l. round-bottomed flask (Note 1) equipped with a thermometer, stirrer, gas inlet tube, and two addition burets. One buret is charged with 86 ml. (1 mole) of 11.6M hydrogen peroxide (Note 2), and the other with a solution of 278 g. (1 mole) of ferrous sulfate pentahydrate and 55.5 ml. (1 mole) of concentrated sulfuric acid in 570 ml. of water (Note 3). The reaction flask is swept out with nitrogen and cooled to 10° by means of an ice bath. Stirring is commenced and the two solutions are added simultaneously and equivalently over a period of 20 minutes. The temperature is held below 20°.

When the addition is completed, 50 ml. (1 mole) of 52% sodium hydroxide is added with stirring and cooling, and then 450 g. of anhydrous sodium sulfate (not all of the salt dissolves). The cold solution is transferred to a separatory funnel and the phases are separated. The organic layer is neutralized with 52% sodium hydroxide; approximately 20 ml. is required to bring the pH to 7. The aqueous layer, including the precipitated ferric hydroxide, is added to the aqueous portion of the reaction mixture and the whole is extracted with 400 ml. of t-butyl alcohol. This extract is similarly treated with 52% sodium hydroxide (about 5 ml. is required). The resulting aqueous layer is combined with the main aqueous fraction, which is again extracted with 400 ml. of t-butyl alcohol. This whole process is again repeated so that the organic phases comprise the three extracts and the phase which separated initially from the reaction mixture.

The four organic fractions are combined and distilled under reduced pressure. The distillation is continued until the temperature of the flask is about 70°/5 mm. in order to remove most of the t-butyl alcohol. The still residue is then extracted with 2 l. of ether and the extract is treated with decolorizing carbon and diatomaceous earth. Distillation of the ether at slightly reduced pressure from a water bath yields α,α,α′,α′-tetramethyltetramethylene glycol as a pale yellow crystalline residue weighing 30–45 g. (41–62% yield based on hydrogen peroxide employed). The crude product is digested at room temperature in a mixture of 30 ml. of ether and 70 ml. of cyclohexane. The resulting slurry is filtered to yield 29–34 g. (40–46%; Note 4) of the glycol as a white crystalline solid, m.p. 87–88°. The product, which is pure enough for most purposes, can be further purified by recrystallization from ethyl acetate (1 g. in 4 ml.), cyclohexane (1 g. in 20 ml.), or water (1 g. in 2 ml.).

2. Notes

1. The flask should have creased sides and a conical indentation in the bottom and should be equipped with a high-speed, propeller-type stirrer rotated to force the liquid downwards. The stirrer should be constructed of glass because metals may interfere with the generation and utilization of the hydroxyl free radicals.

2. Commercial 35% hydrogen peroxide was employed. Any concentration from 5% to 50% may be used.

3. It is convenient to calibrate the burets so that the liquid is divided into 20 equal portions. Then, in the addition of the reagents, these calibrations aid in synchronizing the rates.

4. The submitter reports yields of 48–55%, which are slightly higher than those given here.

3. Methods of Preparation

$\alpha,\alpha,\alpha',\alpha'$-Tetramethyltetramethylene glycol has been prepared by the action of methylmagnesium bromide on acetonylacetone,[2-4] on ethyl levulinate,[5] and on ethyl succinate.[6,7] It has also been made by the hydrogenation of 2,5-dimethyl-3-hexyne-2,5-diol over nickel [8,9] and over platinum [10,11] and by the hydrogenation of 2,5-dihydroperoxy-2,5-dimethyl-3-hexyne.[12] Other methods of preparation include the autoxidation of 2,5-dimethylhexane [13] and the alkaline hydrolysis of 2,5-dibromo-2,5-dimethylhexane.[6] The present method, the hydroxyl- radical coupling of t-butyl alcohol,[14] is a one-step synthesis using readily available starting materials. A similar technique may be used to synthesize $\alpha,\alpha,\alpha',\alpha'$-tetramethyladipic acid from pivalic acid, $\alpha,\alpha,\alpha',\alpha'$-tetramethyltetramethylenediamine from t-butylamine, and $\alpha,\alpha,\alpha',\alpha'$-tetramethyladiponitrile from pivalonitrile.[14]

1. Contribution No. 485 from Central Research Department, Experimental Station, E. I. du Pont de Nemours and Co., Wilmington, Delaware.
2. E. Pace, *Atti Accad. Lincei,* 7 (6), 760 (1928) [*C.A.,* 22, 3890 (1928)].
3. R. Locquin and R. Heilmann, *Bull. Soc. Chim. France,* 45 (4), 1128 (1929).
4. N. Zelinsky, *Ber.,* 35, 2139 (1902).
5. L. Henry, *Compt. Rend.,* 143, 496 (1906).
6. Z. Pogorjelsky, *Chem. Zentr.,* 1904 I, 578.
7. C. Harries and R. Weil, *Ann.,* 343, 364 (1905).
8. H. Adkins and D. C. England, *J. Am. Chem. Soc.,* 71, 2958 (1949).
9. I. G. Farbenindustrie A.-G., Brit. pat. 508,944 [*C.A.,* 34, 2866 (1940)].
10. G. Dupont, *Ann. Chim. (Paris),* 30, (8), 526 (1913).
11. Yu. S. Zal'kind and S. V. Bukhovets, *J. Gen. Chem. (U.S.S.R.),* 7, 2417 (1937) [*C.A.,* 32, 2086 (1938)].
12. N. A. Milas and O. L. Mageli, *J. Am. Chem. Soc.,* 74, 1471 (1952).
13. J. P. Wibaut and A. Strang, *Proc. Koninkl. Ned. Akad. Wetenschap.,* 54B, 229 (1951) [*C.A.,* 46, 4477 (1952)].
14. D. D. Coffman, E. L. Jenner and R. D. Lipscomb, *J. Am. Chem. Soc.,* 80, 2864 (1958).

2,4,5,7-TETRANITROFLUORENONE*

(9-Fluorenone, 2,4,5,7-tetranitro-)

Submitted by MELVIN S. NEWMAN and H. BODEN.[1]
Checked by WILLIAM E. PARHAM, PETER DELVIGS, and E. LEETE.

1. Procedure

A 5-l. three-necked flask fitted with an all-glass addition funnel and two condensers is charged with 770 ml. of concentrated sulfuric acid and 1.3 l. of 90% fuming nitric acid (Note 1). The solution is heated under gentle reflux, and a solution of 73 g. (0.4 mole) of 9-fluorenone (Note 2) in 840 ml. of concentrated sulfuric acid (Note 3) is added from the dropping funnel over a 1-hour period. After the fluorenone addition is complete, a solution of 950 ml. of fuming nitric acid in 1120 ml. of concentrated sulfuric acid is added dropwise during 8.5 hours to the gently refluxing reaction mixture. The heating jacket is turned off and the solution is allowed to stand for 10 hours. The reaction mixture is poured into 5 gallons of water in two 5-gal. crocks (Note 4). The light yellow precipitate is washed with water, twice by decantation, filtered, washed several times with water and sucked dry, and finally is dried in a vacuum oven at 80° for 10 hours (Note 5). The yield of crude 2,4,5,7-tetranitrofluorenone, m.p. 249–253°, is 105–117 g. (72–80%). This solid is recrystallized from 1.6 l. of acetic acid containing 100 ml. of acetic anhydride. The hot solution is filtered through a fluted filter and cooled rapidly to yield 80–86 g. (51–54%) of 2,4,5,7-tetranitrofluorenone, m.p. 253.0–254.5° cor. (Notes 6 and 7).

2. Notes

1. Baker Analyzed reagent grade fuming nitric acid may be added to the sulfuric acid without special precautions, since the heat effect is not large.

2. Eastman white label 9-fluorenone, m.p. 82–84°, was used. The checkers used material, m.p. 83.5–84.5°, prepared from fluorene.[2]

3. The deep purple-brown solution may have to be warmed in order to dissolve all the fluorenone.

4. This operation must be carried out in the hood.

5. The product may be dried under reduced pressure over calcium chloride for several days.

6. Additional product amounting to 15–17% may be obtained by recrystallization of further crops from the mother liquor.

7. Tetranitrofluorenone crystallizes with 0.5 mole of acetic acid which is readily lost on heating under reduced pressure.

3. Methods of Preparation

The procedure described here is essentially that of Newman and Lutz.[3] 2,4,5,7-Tetranitrofluorenone has been prepared by nitration of fluorenone,[4] 2,4,7-trinitrofluorenone,[5,6] and 4,5-dinitrofluorenone.[6] The preparation by Schmidt et al.,[4,5] which supposedly yielded the 2,3,6,7-isomer, has been shown [6] to yield the 2,4,5,7-isomer.

4. Merits of Preparation

The complexes which 2,4,5,7-tetranitrofluorenone forms with aromatic compounds are in general higher melting and less soluble than are the corresponding complexes of 2,4,7-trinitrofluorenone.[3,7]

1. Department of Chemistry, Ohio State University, Columbus, Ohio.
2. E. B. Hershberg and I. S. Cliff, *J. Am. Chem. Soc.*, 53, 2720 (1931).
3. M. S. Newman and W. B. Lutz, *J. Am. Chem. Soc.*, 78, 2469 (1956).
4. J. Schmidt, F. Retzlaff and A. Haid, *Ann.*, 390, 210 (1912).
5. J. Schmidt and K. Bauer, *Ber.*, 38, 3758 (1905).
6. F. E. Ray and W. C. Francis, *J. Org. Chem.*, 8, 52 (1943).
7. M. Orchin, L. Reggel, and E. O. Woolfolk, *J. Am. Chem. Soc.*, 69, 1225 (1947).

(+)- AND (−)-α-(2,4,5,7-TETRANITRO-9-FLUORENYLIDENEAMINOOXY)PROPIONIC ACID

(Propionic acid, 2-(2,4,5,7-tetranitrofluoren-9-ylideneaminooxy)-, (+)- and (−)-)

$$(CH_3)_2C{=}NOH \ + \ CH_3CHBrCOOC_2H_5 \xrightarrow{\ C_2H_5ONa\ }$$

$$\underset{\overset{|}{\text{CH}_3}}{(CH_3)_2C{=}NOCHCOOC_2H_5} \xrightarrow[\text{2. HCl}]{\text{1. NaOH}} \underset{\overset{|}{\text{CH}_3}}{(CH_3)_2C{=}NOCHCOOH}$$

Submitted by PAUL BLOCK, JR.,[1] and MELVIN S. NEWMAN[2]
Checked by WILLIAM G. DAUBEN, MILTON E. LORBER,
CARROLL S. MONTGOMERY, and GARY W. SHAFFER

1. Procedure

A. *Ethyl α-(isopropylideneaminooxy)propionate.* A 1-l. three-necked flask is equipped with a mechanical stirrer, a dropping funnel, and a thermometer that can be immersed in the contents of the flask. There is added to the flask 500 ml. of commercial absolute ethanol (Note 1) followed by 17.5 g. (0.76 g. atom) of sodium, which is added carefully in small portions. When a clear solution has been obtained, 55.0 g. (0.75 mole) of acetone oxime is added (Note 2). The flask is cooled in a water bath held at 5–10°, the stirrer is started, and 136 g. (0.75 mole) of ethyl α-bromopropionate (Note 3) is added during 20–30 minutes at a rate such that the temperature does not rise above 20°. The cooling bath is removed, and the stirring is continued until the contents of the flask reach room temperature.

After standing for 12 hours, the reaction mixture is filtered by gravity into a 1-l. round-bottomed flask, and the solid sodium bromide is washed with 50 ml. of ethanol. The combined filtrate and washings are concentrated to a volume of about 400 ml., and 250 ml. of water is added to the cooled concentrate. The mixture is extracted with 50 ml. of a 1:1 mixture of ether and benzene, and the aqueous layer is reextracted with 100 ml. of the same solvent mixture. The organic extracts are combined, washed with 100 ml. of water and 50 ml. of saturated aqueous sodium chloride, and filtered through a few grams of anhydrous magnesium sulfate. The solvent is removed on a rotary evaporator, and the residue is distilled to yield 71–77 g. (55–59%) of ethyl α-(isopropylideneaminooxy)propionate, b.p. 62–64° (4 mm.).

B. *d,l-α-(Isopropylideneaminooxy)propionic acid.* In a 1-l. three-necked flask fitted with a stirrer and a thermometer that can be immersed in the contents of the flask is placed 300 ml. of 5% aqueous sodium hydroxide (0.37 mole). The flask is heated on a water bath until the temperature of the solution reaches 70°, and 52 g. (0.30 mole) of ethyl α-(isopropylideneaminooxy)propionate is added. The mixture is stirred rapidly while the temperature is held at 70°; the stirring is continued for 20 minutes beyond the time necessary for the contents of the flask to become homogeneous (Note 4). The solution is cooled and acidified to Congo red paper with 5N hydrochloric acid, and 175 g. of ammonium sulfate is added. The mixture is extracted three times with a total of 300 ml. of a 1:1 mixture of ether and benzene. The combined extracts are dried rapidly over 5 g. of anhydrous magnesium sulfate and filtered (Note 5). The solvent is removed by distillation, and 160 ml. of petroleum ether (b.p. 30–60°) is added to the cooled residue. The resulting solution is placed in a refrigerator for several hours. The crystals that separate are removed by suction filtration and washed with a small volume of cold petroleum ether. The yield of colorless product is 35–37 g. (80–85%); m.p. 59–60.5° (Note 6).

C. (+)- and (–)-α-(*Isopropylideneaminooxy*)*propionic acid*-(–)-*ephedrine salts*. A solution of 36.6 g. (0.200 mole) of (–)-ephedrine monohydrate (Note 7) in 800 ml. of ethyl acetate containing 6% of ethanol (Note 8) is placed in a 2-l. beaker. *d,l-α*-(Isopropylideneaminooxy)propionic acid (29.0 g., 0.200 mole) is dissolved in this solution by stirring (Note 9). The beaker is covered securely with a rubber dam, cooled for a short period in an ice bath, placed in a refrigerator at 0–5°, and allowed to remain undisturbed for 8–16 hours after crystallization has begun (Note 10). The solid mass of crystals is filtered by suction, and the funnel is covered with a rubber dam to remove most of the solvent. The solid product is placed in a 500-ml. beaker, 250 ml. of ethyl acetate is added (Note 11), and the mixture is heated until all the solid has dissolved. The solution is cooled, placed in a refrigerator for several hours, and filtered; the crystalline precipitate is dried in air. The yield of the (–)-ephedrine-(+)-α-(isopropylideneaminooxy)propionic acid salt is 22–25 g. (71–81%); m.p. 115–119° (Notes 12 and 13); $[\alpha]^{20}_D$ −4.2° (c 1.5, chloroform).

The combined filtrates are diluted with an equal volume of petroleum ether (b.p. 30–60°), placed in a refrigerator for 8–16 hours, and filtered. The solid product is recrystallized from ethyl acetate (10 ml. per gram of the salt). The yield of the monohydrate of the (–)-ephedrine-(–)-α-(isopropylideneaminooxy)propionic acid salt is 19–26 g. (58–79%); m.p. 88–90°; $[\alpha]^{20}_D$ −57° (c 1.5, chloroform) (Notes 13 and 14).

D. (+)- *and* (–)-α-(*Isopropylideneaminooxy*)*propionic acid*. To a solution of 20 g. (0.064 mole) of the (–)-base-(+)-acid salt in 60 ml. of water is added 14 ml. (0.070 mole) of 5N hydrochloric acid. The solution is filtered to remove a slight insoluble residue and extracted with four 25-ml. portions of a 1:1 mixture of ether and benzene. The combined extracts are dried rapidly over 1–2 g. of anhydrous magnesium sulfate and filtered. The organic solvents are removed by distillation from a steam bath, the residue is dissolved in 75 ml. of petroleum ether (b.p. 30–60°), and the solution is allowed to stand in a refrigerator for 12 hours.

The crystalline product (7.0–7.5 g.; m.p. 75–81°) is collected and dissolved in hot acetone (0.5 ml. per gram), and the solution is diluted with hexane (5 ml. per gram). The solution is placed in a refrigerator for 8–16 hours, and the crystalline (+)-α-(isopropylideneaminooxy)propionic acid (5.5–6.5 g.; 59–70%) that separates is collected; m.p. 83–85°; $[\alpha]^{20}$D +32° (c. 1.6, water).

In a similar manner, from 20 g. (0.061 mole) of the monohydrate of the (−)-base-(−)-acid salt, there is obtained 6.4–6.7 g. (73–76%) of the (−)-acid, m.p. 83–85°, $[\alpha]^{20}$D −29° (c 1.44, water), directly from the crystallization from petroleum ether. Subsequent recrystallization from acetone-hexane is normally not required.

E. (+)- and (−)-α-(2,4,5,7-Tetranitro-9-fluorenylideneaminooxy)propionic acid (TAPA). To a solution of 5.5 g. (0.038 mole) of either optical antipode of α-(isopropylideneaminooxy)-propionic acid in 85 ml. of 96% acetic acid in a 250-ml. round-bottomed flask are added 9.0 g. (0.025 mole) of 2,4,5,7-tetranitrofluorenone,[3] 0.30–0.35 ml. of concentrated sulfuric acid, and a few boiling chips. The flask is fitted with an air condenser (Note 15), and the contents are heated under reflux so that the condensing liquid nearly reaches the top of the condenser (Note 16). After 2 hours, 18 ml. of water is added to the hot solution, and crystallization is allowed to take place slowly, first at room temperature and finally for 12 hours in a refrigerator. The yellow crystalline acid is filtered and dissolved in 70 ml. of hot acetic acid. The solution is diluted while hot with 60 ml. of water, cooled rapidly with stirring, and kept at 0° for several hours. The optically active TAPA is filtered and air-dried away from direct sunlight until the odor of acetic acid is negligible. The crystals are then dried in an oven at 110° (Note 17) and protected from light by storage in a suitable container; yield 7.8–10.0 g. (70–90%). The TAPA from the (−)-acid has $[\alpha]^{25}$D +97° and that from the (+)-acid $[\alpha]^{25}$D −97° (Note 18).

2. Notes

1. Pure anhydrous ethanol[4] offers no advantage over commercial absolute ethanol.

2. "Eastman grade" acetone oxime was used as obtained from Eastman Organic Chemicals.

3. "Eastman grade" ethyl α-bromopropionate was used as obtained from Eastman Organic Chemicals.

4. Usually 10–20 minutes are required to obtain complete reaction.

5. If the solution is not entirely colorless, it should be shaken with a small amount of activated carbon and filtered before distillation.

6. The checkers found 53–56°; m.p. 57–61° has been reported.[5]

7. "Ephedrine alkaloid hydrous," Merck, was used. If anhydrous ephedrine is employed, only 33 g. should be used, and 3.6 g. (0.20 mole) of water should be added. Anhydrous conditions lead to incomplete resolution.

8. Commercial absolute ethanol (48 ml.) is pipetted into a 1-l. graduated cylinder and diluted with 800 ml. of ethyl acetate ("Eastman grade").

9. Both components are soluble in ethyl acetate at room temperature; the resulting salt is not. By dissolving the components sequentially, precipitation of the salt is generally avoided. Should the salt form, however, it must be dissolved by gentle heating.

10. Prolonged standing must be avoided as the deposition of the (–)-ephedrine-(–)-acid salt can occur.

11. Ethanol is not added to the ethyl acetate at this point.

12. Highly purified samples have m.p. 124.0–124.5°.

13. The two diastereoisomeric salts can be readily distinguished from each other. The (–)-ephedrine-(+)-acid salt is formed as cottony crystals that grow in the solution and eventually become a solid, white opaque mass. The monohydrate of the (–)-ephedrine-(–)-acid salt consists of clear, chunky crystals that grow from, and adhere to, the bottom and sides of the flask.

14. The water of hydration is lost on standing in a desiccator over phosphorus pentoxide; the melting point eventually reached is 109–110°.[6]

15. A 250 × 15-mm. glass tube is satisfactory.

16. The suspended tetranitrofluorenone dissolves completely in about 25 minutes; the vigorous heating is required to bring about the solution and reaction.

17. One mole of acetic acid of solvation is lost only slowly at room temperature; the solvated product has m.p. *ca.* 123°.[5]

The submitters found that the air-dried material, on being dried at 110°, yielded essentially solvent-free compound, m.p. 201–203° (dec. with prior darkening). The checkers found that at 110° the air-dried material melted, turned brown, and then resolidified. They also found that the material, on being dried at 70–80° (1 mm.) over potassium hydroxide pellets for several days, remained yellow but melted over a range 110–125°, resolidified, and remelted at 190–195°.

18. The checkers used material dried at 70–80° for their determination of the rotation and obtained values in agreement with those reported by the submitters.

3. Methods of Preparation

TAPA has been prepared only as described in this procedure.[5] α-(Isopropylideneaminooxy)propionic acid has been prepared and resolved by the present procedure[6] and has been prepared directly from α-bromopropionic acid and resolved as the (−)-ephedrine salt by crystallization from hydrocarbon mixtures.[5]

4. Merits of the Preparation

The use of ethyl α-bromopropionate simplifies the preparation of α-(isopropylideneaminooxy)propionic acid. Resolution in ethyl acetate solution has proved less erratic than in the hydrocarbon solvents previously recommended,[5] and the isolation of both diastereoisomeric salts formed is facilitated. TAPA has found use in the resolution of polycyclic aromatic compounds that do not possess functional groups that would permit resolution by other methods.[5, 7, 8]

1. Department of Chemistry, University of Toledo, Toledo, Ohio.
2. Department of Chemistry, The Ohio State University, Columbus, Ohio 43210.
3. M. S. Newman and H. Boden, this volume, p. 1029.
4. R. H. Manske, *Org. Syntheses*, Coll. Vol. 2, 154 (1943).
5. M. S. Newman and W. B. Lutz, *J. Am. Chem. Soc.*, 78, 2469 (1956).
6. P. Block, Jr., *J. Org. Chem.*, 30, 1307 (1965).
7. M. S. Newman, R. G. Mentzer, and G. Slomp, *J. Am. Chem. Soc.*, 85, 4018 (1963).
8. M. S. Newman, R. W. Wotring, Jr., A. Pandit, and P. M. Chakrabarti, *J. Org. Chem.*, 31, 4293 (1966).

1,2,3,4-TETRAPHENYLNAPHTHALENE*

(Naphthalene, 1,2,3,4-tetraphenyl-)

Submitted by Louis F. Fieser and Makhluf J. Haddadin [1]
Checked by Joyce M. Dunston and Peter Yates

1. Procedure

A. *Diphenyliodonium-2-carboxylate.* An Erlenmeyer flask containing 80 ml. of concentrated sulfuric acid (Note 1) is placed in an ice bath to cool. To a 2-l. three-necked flask are added (Note 2) 20 g. (0.081 mole) of lump-free *o*-iodobenzoic acid (Note 3) and 26 g. (0.096 mole) of potassium persulfate (Note 4.) The flask is cooled in an ice bath, the chilled sulfuric acid is added, and the flask is swirled in an ice bath for 4–5 minutes to produce an even suspension and to control the initial exothermal reaction. The flask is then removed from the ice bath and the time is noted.

The reaction mixture foams somewhat and acquires a succession of colors. The flask is mounted in a pan of acetone, a mechanical stirrer with a curved Teflon® blade is placed in the center neck of the flask and operated slowly, and a 250-ml. separatory funnel is mounted into a side neck. Three flasks are put in an ice bath to cool: one containing 190 ml. of distilled water, another 230 ml. of 29% ammonium hydroxide, and another 400 ml. of methylene chloride.

After the oxidation has proceeded for 20 minutes, the acetone bath is brought to 10° by addition of crushed dry ice, and the solution is stirred for 2–3 minutes. There is added 20 ml. of thiophene-free benzene (17.6 g., 0.226 mole), and stirring at 10° is continued for 1 hour (Note 5). The temperature of the acetone bath is lowered to −15° by addition of crushed dry ice and the bath kept at this temperature while the chilled 190 ml. of distilled water is added with efficient stirring to precipitate the potassium bisulfate salt of diphenyliodonium-2-carboxylic acid. The 400 ml. of chilled methylene chloride (Note 6) is added, and a 100° thermometer is mounted in a side neck in such a way that the 15–25° section is visible. With efficient cooling (bath at −15°) and stirring, the 230 ml. of chilled 29% ammonium hydroxide (Note 7) is added at such a rate that the temperature of the reaction mixture remains between 15° and 25°. Approximately 10 minutes is required for the addition, and after the addition the aqueous layer should be alkaline to indicator paper (pH 9). The stirrer and the thermometer are removed and rinsed with water and methylene chloride. The bulk of the aqueous layer is decanted into a 500-ml. Erlenmeyer flask for temporary storage, and the reaction flask is emptied and rinsed into a 500-ml. separatory funnel. The pale tan lower methylene chloride layer is drained into a 1-l. Erlenmeyer flask, the decanted aqueous layer is added to the separatory funnel, and the combined aqueous layer is extracted with two 100-ml. portions of methylene chloride. The combined extract is dried over anhydrous sodium sulfate, filtered into a 1-l. Erlenmeyer flask, and the filtrate is evaporated on the steam bath until the product is left as a dry grayish cake. The cake is dislodged and broken up with a stainless steel spatula, and the bulk of the product is transferred to a paper. Material adhering to the flask and spatula is dissolved in boiling methylene chloride and the solution transferred to a tared 500-ml. flask and evaporated to dryness. The solid product is added to the flask and the combined solid is brought to constant weight at steam-bath temperature and water-aspirator pressure. The yield of crude product is 23.3–24.4 g. (Note 8).

Boiling water (275 ml.) is poured into the flask, the product brought into solution at the boiling point, and 0.4 g. of Norit® is added carefully to the slightly cooled solution. The solution is again heated to boiling, filtered, and allowed to stand for crystallization overnight, eventually at 0°. The colorless prisms of diphenyliodonium-2-carboxylate monohydrate (Note 9) are collected and air-dried to constant weight at room temperature. The yield of product, m.p. 220–222° (dec.), is 20–22 g. (72–79%, Note 10).

B. *1,2,3,4-Tetraphenylnaphthalene.*   To a   100-ml. round-bottomed flask equipped with an 11-cm. water-cooled condenser (Note 11) there is added 60 ml. of diethylbenzene (*meta* and *para* mixture) (Note 12), and the flask is heated in a fume hood with the free flame of a microburner until refluxing liquid rises well into the condenser. If a cloudy zone of condensate appears at the top of the condenser, the moisture is removed with an applicator stick wrapped with absorbent cotton (Note 13). The same technique is used later for removal of water of hydration which appears in the early stages of the reaction and causes hissing and eruption if allowed to drop back into the flask (Note 14). The flame is removed and 10 g. (0.026 mole) of tetraphenylcyclopentadienone [2] (Note 2) and 11.8 g. (0.035 mole) of diphenyliodonium carboxylate monohydrate are added to the flask. The mixture is heated over a microburner at a rate such as to maintain vigorous gas evolution and gentle refluxing. The water of hydration is eliminated in 8–10 minutes. The flask is then fitted with a normal reflux condenser and the heating is continued. After 30 minutes considerable undissolved diphenyliodonium carboxylate can still be seen, under illumination, at the bottom of the flask. In another 5 minutes the color changes to transparent red, and in a minute or two longer the solution becomes pale amber. Refluxing is continued until no solid remains (10–'5 minutes) (Note 15). The flask is then fitted for distillation, and 55 ml. of liquid (diethylbenzene and iodobenzene, b.p. 188°) is removed by distillation. The residue is cooled and dissolved in 25 ml. of dioxane. The solution is rinsed into a 125-ml. Erlenmeyer flask and diluted with 25 ml. of 95% ethanol. The solution is heated to boiling, and water (6–7 ml.) is added gradually until a few shiny prisms remain undissolved on boiling. Crystallization is allowed to proceed at room temperature and then for several hours at 0°. The precipitate is removed by filtration, and the mother liquor upon further standing deposits a small second crop of crystals (0.3 g., m.p. 1° low). The main product melts initially in the range 196–199°, solidifies on cooling, and remelts sharply at 203–204° (Note 16). The total yield is 9.2–10.2 g. (82–90%).

2. Notes

1. The amount of acid is half that called for in previous procedures.[3, 4]

2. The solids are added using a powder funnel or a rolled-up piece of glazed paper to prevent material from lodging on the neck or walls.

3. Obtained from Eastman Organic Chemicals.

4. The fine granular material supplied by Fisher Scientific Co. is satisfactory; any lumps present should be crushed. Persulfate in the form of large prisms should be ground prior to use.

5. If the reactants are mixed at room temperature, a rapid temperature rise of about 7° is noted. The reaction can then be brought to completion by swirling at 50° for 5 minutes or at room temperature for 20 minutes, but the product contains considerable brown pigment.

6. The methylene chloride is added for efficient extraction of the product as it is liberated on neutralization. The product is more soluble in this solvent than in chloroform.[3, 4]

7. Neutralization with sodium hydroxide [3, 4] leads to troublesome separation of sodium salts.

8. This crude product contains a little solid which will not re-dissolve in an organic solvent and it is unsuitable for procedure B. The checkers used a 500-ml. filter flask with sealed side arm for the evaporation and drying operation.

9. *Anal.* Calcd. for $C_{13}H_{11}O_3I$ (342.13): C, 45.63; H, 3.24, I, 37.10. Found: C, 45.48; H, 3.19; I, 37.16.

10. Anhydrous material, m.p. 215–216° (dec.), can be obtained in quantitative yield by extracting 12 g. of the monohydrate in a Soxhlet extractor with 80 ml. of methylene chloride and evaporating to constant weight.

11. A convenient condenser is a water-cooled Ace Glass Bearing, No. 8244.

12. Obtained from Eastman Organic Chemicals, b.p. 175–181°. This solvent, in which the benzyne precursor is very sparingly soluble, seemed slightly superior to trimethylene glycol dimethyl ether (b.p. 222°) in which the solubility is considerably higher. o-Dichlorobenzene (b.p. 179°), a still better solvent for the dipolar salt, is less satisfactory than the ether. Diethyl oxalate (b.p. 184°) and N,N-dimethylacetamide (b.p. 195°) are unsatisfactory.

13. The checkers found it impossible to prevent water from dropping into the hot mixture and leading to its eruption through the top of the condenser. They found it convenient to eliminate the small condenser and use only a normal condenser where the prolonged reflux period is needed.

14. The yield was not improved by use of anhydrous material (Note 10).

15. Solid adhering to the walls of the flask can be dislodged by loosening the clamp supporting the flask and using the condenser as a lever to swirl the flask.

16. The double melting point has been observed in only one [5] of the previous studies. The initial melting point varies with the state of subdivision and is not a reliable index of purity. Several recrystallizations did not change the melting behavior or the remelt temperature. The checkers did not observe the double melting point with the product initially obtained, but did so with material recrystallized once.

3. Methods of Preparation

1,2,3,4-Tetraphenylnaphthalene has been isolated by Wittig and co-workers by generation of benzyne in the presence of excess tetraphenylcyclopentadienone as trapping agent. Yields of hydrocarbon isolated by chromatography and based upon the precursor are as follows: from o-fluorobromobenzene, 17%;[6] from either o-iodophenylmercuric iodide or bis-(o-iodophenyl)-mercury, 25%.[7] The hydrocarbon has also been obtained in low yield as one of two products resulting from the reaction of diphenylacetylene with triphenylchromium.[5] The present method is due to Le Goff,[4] who reports a 68% yield of hydrocarbon from diphenyliodonium-2-carboxylate. However, this investigator states in a private communication that he refluxed the benzyne precursor with a large excess of tetraphenylcyclopentadienone in diethylene glycol dimethyl ether and isolated the hydrocarbon by tedious hexane extraction and chromatography. In the experience of the submitters the solvent selected has too low a boiling point (161°) for efficient conversion.

4. Merits of the Preparation

The procedure demonstrates a safe and simple method for the generation of benzyne from a stable and easily prepared precursor and its use in the synthesis of a hitherto difficultly accessible hydrocarbon.

1. Department of Chemistry, Harvard University, Cambridge 38, Massachusetts.
2. J. R. Johnson and O. Grummitt, *Org. Syntheses*, Coll. Vol. **3**, 806 (1955).
3. F. M. Beringer and I. Lillien, *J. Am. Chem. Soc.*, **82**, 725 (1960).
4. E. Le Goff, *J. Am. Chem. Soc.*, **84**, 3786 (1962).
5. H. Herwig, W. Metlesics, and H. H. Zeiss, *J. Am. Chem. Soc.*, **81**, 6203 (1959).
6. G. Wittig and E. Knauss, *Ber.*, **91**, 895 (1958).
7. G. Wittig and H. F. Ebel, *Ann.*, **650**, 20 (1961).

TETROLIC ACID *

(2-Butynoic Acid)

$$CH_3C{\equiv}CH + NaNH_2 \rightarrow CH_3C{\equiv}CNa + NH_3$$

$$CH_3C{\equiv}CNa + CO_2 \rightarrow CH_3C{\equiv}CCO_2Na$$

$$CH_3C{\equiv}CCO_2Na + HCl \rightarrow CH_3C{\equiv}CCO_2H + NaCl$$

Submitted by J. C. Kauer and M. Brown.[1]
Checked by W. E. Parham, Wayland E. Noland, and Richard J. Sundberg.

1. Procedure

A 3-l., three-necked, round-bottomed flask is equipped with a glass paddle stirrer, a condenser containing a mixture of acetone and solid carbon dioxide, and a gas inlet tube. The outlet of the condenser is protected from the atmosphere by a T-tube through which a slow stream of nitrogen is passed. The flask is purged with nitrogen, and about 1.5 l. of anhydrous liquid ammonia is either poured or distilled into the flask. A small crushed crystal of ferric nitrate nonahydrate is added, followed by 23 g. (1 g. atom) of freshly cut sodium in small pieces (Note 1).

Methylacetylene (44–48 g., 1.1–1.2 mole) (Note 2) is bubbled in through the gas inlet tube with rapid stirring. Sodium methylacetylide precipitates as a flocculent gray solid. The solid carbon dioxide is removed from the condenser, and the ammonia is evaporated overnight under a slow stream of nitrogen. A hot water bath may be used to drive off residual ammonia. One liter of dry tetrahydrofuran (Note 3) and 500 ml. of anhydrous ether are added, and with rapid stirring a slow stream of anhydrous carbon dioxide from a cylinder is passed into the mixture (Note 4). After 8 hours the rate of absorption of carbon dioxide is very slow. Any solid caked on the inside walls of the flask should be scraped off with the glass paddle stirrer. A very slow flow of carbon dioxide is continued overnight (Note 5).

The solvent is removed as completely as possible by distillation on a steam bath under water-pump vacuum. Two hundred milliliters of water is added, and the solid is dissolved by swirling the flask (Note 6). The solution is filtered if suspended solid is present. The aqueous solution is extracted twice with 100-ml. portions of ether. The aqueous layer in a 1-l. Erlenmeyer flask is then cooled in ice, and a mixture of 70 ml. of concentrated hydrochloric acid and 200 g. of ice is added slowly with swirling. The acidified solution is continuously extracted with 200 ml. (or more) of ether for 24–36 hours. The extract is evaporated in a stream of air or nitrogen to give tetrolic acid in the form of a mushy tan solid that is further dried in a vacuum desiccator over concentrated sulfuric acid for 2 days (Note 7). The product is a tan crystalline solid weighing 58–60 g. (69–71% based on sodium) and melting at 71–75°. It is purified further by addition to 700 ml. of boiling hexane. As soon as the tetrolic acid has dissolved, about 1 g. of activated carbon is added, and the solution is filtered through a heated funnel (Note 8). The filtrate is refrigerated (5°) overnight and 42–50 g. (50–59%) (Note 9) of tetrolic acid is collected in the form of white needles, m.p. 76–77°. A second recrystallization from hexane gives tetrolic acid melting at 76.5–77° (Note 10).

2. Notes

1. The first few pieces of sodium should be converted to sodium amide as evidenced by a color change from blue to gray. The rest of the sodium is then added over a period of 30 minutes.

2. An excess may be used if the purity of the methylacetylene is in doubt; however, a large excess will result in foaming when the liquid ammonia is later evaporated. Methylacetylene of satisfactory purity is available from the Matheson Company.

3. The tetrahydrofuran is distilled from sodium and stored under nitrogen.

4. When a flow rate of 70–100 ml. per minute is used, the internal temperature does not rise above 30° and most of the carbon dioxide is absorbed. A lower yield (50%) of product is obtained when carbon dioxide gas is generated by the slow evaporation of commercial solid carbon dioxide. [*Caution! See p. 976.*]

5. The reaction is complete when the addition of a small amount of the solid to a few drops of water yields a solution with a pH below 10.

6. Residual tetrahydrofuran may separate as a second (upper) phase. It is removed by the ether extraction.

7. To avoid spattering of the solid the desiccator is evacuated slowly. If drying is incomplete, an aqueous layer will be left in the hexane solution when the tetrolic acid is recrystallized.

8. Prolonged boiling should be avoided since some tetrolic acid is lost by volatilization.

9. The submitters obtained yields of tetrolic acid as high as 67.2 g. (80%).

10. In one run the submitters passed excess methylacetylene (1.6 moles) into a solution of sodium in liquid ammonia until the color turned from blue to white. No ferric nitrate was used. This somewhat shorter procedure yielded pure white sodium methylacetylide and did not diminish the yield of tetrolic acid. Excess methylacetylene is necessary because 0.5 mole is converted to propylene.

3. Methods of Preparation

Tetrolic acid has been prepared by treatment of acetoacetic ester with phosphorus pentachloride followed by dehydrochlorination of the reaction products;[2] by the base-catalyzed isomerization of 3-butynoic acid;[3] and by the treatment of 4,4-dibromo-3-methyl-2-pyrazolin-5-one with alkali followed by acidification.[4]

It has also been prepared by the carbonation of sodium methylacetylide under pressure,[5,6] in ether suspension,[7] and in the dry state.[8]

4. Merits of Preparation

The virtue of the present method is its convenience, especially when pressure equipment is not available. This method is probably generally applicable to the synthesis of acetylenecarboxylic acids from terminal acetylenes. Thus phenylpropiolic acid was prepared from phenylacetylene in 51% yield by the present procedure.

1. Contribution No. 555 from the Central Research Department, Experimental Station, E. I. du Pont de Nemours and Company.
2. F. Feist, *Ann.*, 345, 104 (1906).
3. G. Eglinton, E. R. H. Jones, G. H. Mansfield, and M. C. Whiting, *J. Chem. Soc.*, 3199 (1954).
4. L. A. Carpino, *J. Am. Chem. Soc.*, 80, 600 (1958).
5. H. B. Henbest, E. R. H. Jones, and I. M. S. Walls, *J. Chem. Soc.*, 3650 (1950).
6. A. D. Macallum, U. S. pat. 2,194,363 (March 19, 1940).
7. M. Bourguel and M. J. Yvon, *Bull. Soc. Chim. France*, [4] 45, 1075 (1929).
8. G. Lagermark, *Ber.*, 12, 853 (1879).

THIOBENZOYLTHIOGLYCOLIC ACID*

(Benzoic acid, dithio-, carboxymethyl ester)

$$C_6H_5CCl_3 + 2KHS + 2KOH \rightarrow C_6H_5\overset{\overset{S}{\|}}{C}SK + 3KCl + 2H_2O$$

$$C_6H_5\overset{\overset{S}{\|}}{C}SK + ClCH_2CO_2Na \rightarrow C_6H_5\overset{\overset{S}{\|}}{C}SCH_2CO_2Na + KCl$$

$$C_6H_5\overset{\overset{S}{\|}}{C}SCH_2CO_2Na + HCl \rightarrow C_6H_5\overset{\overset{S}{\|}}{C}SCH_2CO_2H + NaCl$$

Submitted by FREDERICK KURZER and ALEXANDER LAWSON.[1]
Checked by MAX TISHLER, G. A. STEIN, W. F. JANKOWSKI.

1. Procedure

A solution of alcoholic caustic potash, prepared by dissolving 59.3 g. of 85% potassium hydroxide (0.90 mole) in 400 ml. of absolute ethanol with warming, is divided into two equal portions, one of which is saturated with hydrogen sulfide at room temperature (Note 1). The recombined solutions are placed in a 1-l. three-necked flask fitted with a Hershberg stirrer (Note 2), a gas delivery tube (Note 3), and a reflux condenser carrying a dropping funnel (Note 4). The air is displaced from the apparatus by passing a stream of nitrogen through the stirred liquid (Note 5). The solution is warmed to approximately 45–50°, and 49 g. (35 ml., 0.25 mole) of benzotrichloride is added dropwise through the condenser from a dropping funnel at a rate to maintain the temperature of the reaction mixture at approximately 60°; this requires 1–1.5 hours (Note 6). The reaction mixture turns deep red soon after the addition is started. When all the benzotrichloride has been added, the stirred deep-red suspension is refluxed gently for 30 minutes. A solution of 33.1 g. (0.35 mole) of chloroacetic acid in 200 ml. of water, neutralized with 29.4 g. (0.35 mole) of solid sodium bicarbonate, is next rapidly added through the condenser, the stirred mixture heated to boiling as rapidly as possible and refluxed (Note 5) for 5 minutes.

The resulting brownish red suspension is added to 750–1000 g. of ice contained in a 2-l. beaker (Note 7) and the turbid orange solution slowly acidified (to Congo red) with good stirring (Note 8) by the addition of approximately 50 ml. (Note 9) of concentrated hydrochloric acid. The deep-scarlet crystalline precipitate is collected at the pump after 30 minutes at 0° and rinsed with small quantities of water.

The air-dried product is crystallized by dissolving it in chloroform (approximately 120 ml.), followed by dilution of the filtered boiling liquid (Note 10) with hot petroleum ether (boiling range 60–80°, 60–80 ml.). The crystalline product, which separates rapidly, is collected at 0°, rinsed on the filter with a mixture of chloroform and petroleum ether (1:3), and dried. The yield of magnificent deep-scarlet lustrous prisms, m.p. 127–128°, varies between 28.9 and 30.4 g. (54–57% of the theoretical). Concentration of the combined filtrates and wash liquids under reduced pressure to a small volume (50–80 ml.) yields an additional small quantity (1.5–3.0 g., 3–6%) of material of satisfactory purity, m.p. 121–124°.

2. Notes

1. Saturation is complete when a slow stream of gas is passed through the solution during 2.5–3 hours. The initially turbid liquid generally clears and remains nearly colorless during this process.

The checkers on several occasions obtained a small amount of a flocculent precipitate that most probably was potassium carbonate.

2. The checkers used a Trubore stirrer with a Teflon paddle.

3. The delivery tube is fitted to allow the stream of nitrogen to enter as far under the surface of the liquid as possible without obstructing the operation of the stirring device.

4. The checkers used a pressure-equalizing dropping funnel.

5. The passage of nitrogen is continued throughout the experiment.

6. The exothermic nature of the reaction maintains the temperature of the mixture between 50° and 60°, depending upon the rate of the addition of the benzotrichloride.

7. The checkers used a 3-l., wide-necked, round-bottomed flask equipped with a mechanical stirrer.

8. Some unmelted ice should remain during the acidification, which is carried out slowly at 0°, to prevent the separation of the crude material in the form of an oil.

9. The checkers found that 25–30 ml. of concentrated hydrochloric acid was sufficient.

10. The submitters filtered the solution rapidly with suction through a preheated Büchner funnel. The checkers found that the product often crystallized too rapidly and plugged the filter. As a result, the crude product was dissolved in an excess of chloroform (160–175 ml.), then filtered, and the excess solvent evaporated before dilution with light petroleum ether. In some cases, no filtration was necessary because the chloroform solution was clear.

3. Methods of Preparation

Thiobenzoylthioglycolic acid has been prepared by the interaction of potassium dithiobenzoate and alkali chloroacetate.[2-4] The required intermediate, dithiobenzoic acid, has been obtained from phenylmagnesium bromide and carbon disulfide,[2,3,5] or by the condensation of benzaldehyde and hydrogen polysulfides,[2,6] or most conveniently by treatment of benzotrichloride with potassium hydrogen sulfide.[2,4,7] The last procedure has been adapted here to afford improved yields.

4. Merits of Preparation

Thiobenzoylthioglycolic acid is a useful thiobenzoylating agent,[2-4,8-12] and the resulting products find application for the synthesis of various heterocycles. These applications of thiobenzoylthioglycolic acid have recently been reviewed.[13]

1. Royal Free Hospital School of Medicine, University of London, 8 Hunter Street, London W. C. 1, England.
2. B. Holmberg, *Arkiv Kemi, Mineral., Geol.,* 17A, No. 23 (1944).
3. A. Kjaer, *Acta Chem. Scand.,* 4, 1347 (1950).
4. J. C. Crawhall and D. F. Elliott, *J. Chem. Soc.,* 2071 (1951).
5. J. Houben, *Ber.,* 39, 3224 (1906).
6. I. Bloch and F. Höhn, *J. Prakt. Chem.,* 82, 486 (1910).
7. A. Engelhardt and P. Latschinoff, *Z. Chem.,* 11, 455 (1868).
8. B. Holmberg, "Svedberg Memorial Volume," Uppsala, 1944, p. 299.
9. A. Kjaer, *Acta Chem. Scand.,* 6, 1374 (1952).
10. B. Holmberg, *Arkiv Kemi,* 9, 47 (1955).
11. J. B. Jepson, A. Lawson, and V. D. Lawton, *J. Chem. Soc.,* 1791 (1955).
12. A. Lawson and C. E. Searle, *J. Chem. Soc.,* 1556 (1957).
13. F. Kurzer, *Chem. & Ind. (London),* 1333 (1961).

m-THIOCRESOL [1]

WARNING

Diazonium xanthates (ArN=NSCSOC$_2$H$_5$) can detonate, and this procedure should be followed carefully to ensure decomposition of the xanthate as it is formed. Under no circumstances should the diazonium solution and the potassium ethyl xanthate be mixed cold and the mixture subsequently heated. A severe detonation has been reported when such a procedure was employed during the preparation of thiocresol.

It has been observed [2] that the dropwise addition of an aqueous solution of potassium ethyl xanthate to a cold (0°) aqueous solution of diazotized orthanilic acid results in the immediate loss of nitrogen when a trace of nickel ion is present in the stirred diazonium solution.[3] The catalyst can be added as nickelous chloride or simply by using a nichrome wire stirrer. When no nickel ion is added and a glass stirrer is employed, the diazonium xanthate precipitates and requires heat (32°) to effect decomposition.

The use of a nichrome stirrer or a catalytic amount of nickel ion is recommended [1] for such reactions to minimize the accumulation of diazonium xanthate; however, the catalytic role of nickel ion has not been explored with other diazonium salts.

1. *Org. Syntheses*, Coll. Vol. **3**, 809 (1955).
2. William E. Parham and William R. Hasek, unpublished work.
3. William R. Hasek, Ph.D. Thesis, The University of Minnesota, 1953, p. 121.

2a-THIOHOMOPHTHALIMIDE

[3(2H)-Isoquinolone, 1,4–dihydro-1-thioxo-]

Submitted by P. A. S. SMITH and R. O. KAN [1]
Checked by MELVIN S. NEWMAN and R. L. CHILDERS

1. Procedure

A. *Phenylacetyl isothiocyanate.* Twenty-five grams (0.16 mole) of phenylacetyl chloride (Note 1), 100 ml. of benzene, and 53 g. (0.16 mole) of lead thiocyanate (Note 2) are placed in a 1-l., three-necked, round-bottomed flask equipped with a mechanical stirrer and a reflux condenser. The stirrer is started and the mixture is refluxed for 5 hours. A small amount of activated charcoal is added, and refluxing is continued for 5 minutes. The warm mixture is filtered through a Büchner funnel under suction (Note 3), and the solid on the filter is washed with two 50-ml. portions of benzene. The solvent is removed from the filtrate under reduced pressure, and the residue is distilled at once to yield 17.5–22.7 g. (61–79%) of phenylacetyl isothiocyanate, b.p. 83–91° at about 0.3 mm. It is a colorless liquid that rapidly darkens on standing (Notes 4 and 5).

B. *2a-Thiohomophthalimide.* In a 500-ml., three-necked, round-bottomed flask equipped with a mechanical stirrer, a reflux condenser, and a dropping funnel are placed 150 ml. of carbon disulfide (Note 6) and 29.3 g. (0.22 mole) of anhydrous powdered aluminum chloride. The stirrer is started, and 17.7 g. (0.10 mole) of phenylacetyl isothiocyanate is added dropwise at such a rate that the solvent refluxes gently. The total addition time is about 5 minutes. The mixture is refluxed gently for 2 hours (Note 7) and is cooled in an ice bath and treated with a solution of 10 ml. of 12N hydrochloric acid in 90 ml. of water; the addition is dropwise at first, more rapid later. Stirring is continued at room temperature for another hour. Crude 2a-thiohomophthalimide is collected by filtration on a 10-cm. Büchner funnel and is pressed dry and subsequently dried thoroughly, either in a vacuum desiccator or in an oven at 40–45° (*Caution! Note 8*). A solution of the imide in 300 ml. of boiling glacial acetic acid is boiled a few minutes with a small amount of activated charcoal, and the hot solution is filtered through a large fluted filter as rapidly as possible to prevent premature crystallization on the filter. Orange-yellow crystals of 2a-homophthalimide precipitate when the filtrate is cooled. They are separated by filtration and dried in an oven or a vacuum desiccator; weight 9.2–13.3 g. (52–75%); m.p. 221–222°.

2. Notes

1. Eastman Kodak Company white label grade of phenylacetyl chloride was used, but equally good results are obtained with the crude acid chloride obtained by treating phenylacetic acid with an excess of thionyl chloride and removing the latter under reduced pressure.

2. Lead thiocyanate was made by stirring together a solution of 45 g. (1.37 moles) of lead nitrate in 360 ml. of boiling water with a solution of 266 g. (2.74 moles) of potassium thiocyanate in 140 ml. of boiling water. The mixture was cooled to room temperature, and 437 g. (99%) of lead thiocyanate was separated by filtration and air-dried.

3. If the filtrate is not clear, filtration should be repeated through the same filter.

4. When large quantities are used, the distillation should be performed in parts, for on prolonged heating phenylacetyl isothiocyanate decomposes with a heavy loss in yield.

5. The distillation should be carried out just before commencing Part B.

6. *sym*-Tetrachloroethane may be substituted for carbon disulfide. In this case 5 minutes of heating on a steam bath, or even no heating at all, gives satisfactory results, although the product is of slightly lower purity. The solvent may be removed quickly by steam distillation of the reaction mixture after addition of dilute acid, and the product is isolated by filtration of the slurry remaining in the flask.

7. The best heating device has been found to be an infrared lamp placed about 20 cm. from the vessel.

8. Drying at higher temperatures can be dangerous because of the low flash-point of carbon disulfide.

3. Methods of Preparation

The only reported method of preparation of 2*a*-thiohomophthalimide is by the reaction described here.[2]

4. Merits of the Preparation

This is a general method of converting arylcarbonyl and arylacetyl isothiocyanates to the corresponding thioimides as the following examples show (percent yield and duration of the reaction follow each example): 6-methyl-1*a*-thiophthalimide [2] (45%, 4 days); 4,6-dimethyl-1*a*-thiophthalimide [2] (64%, 24 hours); 5-methyl-2*a*-thiohomophthalimide (42%, 4 hours); 4-methyl-2*a*-thiohomophthalimide (48%, 4 hours); 5-methoxy-2*a*-thiohomophthalimide (41%, 4 hours); 4-chloro-2-thiohomophthalimide (40%, 4 hours); 1*a*-phenyl-2*a*-thiohomophthalimide (40%, 30 minutes); 1*a*-thio-1,2-naphthalimide [2] (25%, 4 days); 2*a*-thio-1-homo-1,2-naphthalimide [2] (41%, 16 hours); thiophene-2*a*-thio-2,3-dicarboximide (12%, 24 hours).

The thioimides can be hydrolyzed to the corresponding dicarboxylic acids.[3] The thioimides can be converted to the corresponding imides, and thiohomophthalimides can be converted to phthalimides; both conversions are one-step processes.[4] Thus a variety of substituted phthalic and homophthalic acids and their derivatives are available from these thioimides.

Thiohomophthalimides can be reduced to tetrahydroiso-
quinolines.[2]

1. Department of Chemistry, University of Michigan, Ann Arbor, Michigan.
2. P. A. S. Smith and R. O. Kan, *J. Am. Chem. Soc.*, **82**, 4753 (1960).
3. P. A. S. Smith and R. O. Kan, this volume, p. 612.
4. P. A. S. Smith and R. O. Kan, *J. Am. Chem. Soc.*, **83**, 2580 (1961).

o-TOLUAMIDE

WARNING

It has been reported that a very violent explosion occurred
during the preparation of *o*-toluamide as described in this
series.[1] The temperature of the reaction mixture was allowed
to exceed 50°, and the rapidly evolved mixture of oxygen
and alcohol vapor was ignited accidentally.

It is recommended that this preparation be performed only
at a safe distance from open flames, and behind a substantial
safety shield. Since approximately 20 l. of oxygen may be
evolved within a relatively short period, it is important to
have efficient stirring and cooling (40–50°), and to provide
adequate venting for the flask.

1. *Org. Syntheses*, Coll. Vol. 2, 586 (1943).

p-TOLUENESULFONYLHYDRAZIDE*

(*p*-Toluenesulfonic acid, hydrazide)

$$SO_2Cl \qquad\qquad SO_2NHNH_2$$

$$+ 2N_2H_4 \rightarrow \qquad\qquad + N_2H_4 \cdot HCl$$

$$CH_3 \qquad\qquad CH_3$$

Submitted by Lester Friedmán, Robert L. Litle,[1]
and Walter R. Reichle.
Checked by Alan Black and Henry E. Baumgarten

1. Procedure

Into a 1-l. round-bottomed three-necked flask fitted with a thermometer, a mechanical stirrer, and a dropping funnel are placed 200 g. (1.05 moles) of *p*-toluenesulfonyl chloride and 350 ml. of tetrahydrofuran (Note 1). The stirred mixture is cooled in an ice bath to 10–15°; then a solution of hydrazine in water (135 ml. of 85% hydrazine hydrate, 2.22 moles; Note 2) is added at such a rate that the temperature is maintained between 10° and 20° (Note 3). Stirring is continued for 15 minutes after the addition is complete. The reaction mixture is transferred to a separatory funnel. The lower layer is drawn off, and discarded. The upper tetrahydrofuran layer is filtered with suction through a bed of Celite to remove suspended particles and foreign matter (if any). The Celite is washed with a little tetrahydrofuran to remove any absorbed tosylhydrazide. The clear, colorless filtrates are stirred vigorously during the slow addition of two volumes of distilled water. *p*-Toluenesulfonylhydrazide separates as fluffy white crystalline needles (Note 4). The product is filtered through a Büchner funnel, washed several times with distilled water, and air-dried. A yield of 175–185 g. (91–94%) is obtained; m.p. 109–110° (Note 5).

2. Notes

1. According to the submitters, commercial tetrahydro-furan (DuPont) is washed several times with 40% aqueous sodium hydroxide to remove peroxides and organic stabilizers and then dried over solid sodium hydroxide. The clear supernatant liquid is used without further purification. This procedure is no longer regarded as being free from hazard as several serious accidents have occurred during attempts to purify tetrahydrofuran in this fashion. The checkers used a good grade of commercially available tetrahydrofuran taken from a freshly opened bottle (without washing or drying) as recommended in this volume (see p. 976).

2. The submitters diluted 74 ml. of 95% hydrazine with 74 ml. of water. The dilution of hydrazine with water is exothermic. Hydrazine hydrate (50–100%) may be substituted if the volume of water is adjusted so that the resulting solution contains 50% hydrazine.

3. The addition is complete in 20–25 minutes.

4. Celite analytical filter aid, a product of the Johns-Manville Company, was used.

5. In some runs the checkers found it necessary to chill the mixture in the refrigerator for several hours or to stir the mixture in an ice bath until the product crystalized.

6. The product may be contaminated with trace amounts of N,N'-di-p-toluenesulfonylhydrazide. A more nearly pure product may be obtained by dissolving the crude product in hot methanol (4 ml. per gram of hydrazide), filtering through a bed of Celite, and reprecipitating the purified material by addition of 2 to 2.5 volumes of distilled water. This purification step is not necessary for most uses.

The submitters obtained the same yields with quantities as much as ten times those specified here.

3. Discussion

p-Toluenesulfonylhydrazide has been prepared by shaking 50% hydrazine hydrate and p-toluenesulfonyl chloride in benzene for several hours.[2,3] Ammonia has been used as an agent for removing the hydrogen chloride evolved.[4] The present procedure is a modification of one previously published in *Organic Syntheses*.[5]

A number of hydrazides have been prepared in comparable yields from their respective sulfonyl chlorides by the procedure cited.[5] These include *p*-bromobenzenesulfonylhydrazide, *p*-chlorobenzenesulfonylhydrazide, *p*-methoxybenzenesulfonylhydrazide, *m*-nitrobenzenesulfonylhydrazide, *p*-nitrobenzensulfonylhydrazide, *o*-nitrobenzenesulfonylhydrazide, benzenesulfonylhydrazide, and methanesulfonylhydrazide.

p-Toluenesulfonylhydrazide has been found to be an exceptionally useful reagent in the synthesis of diazo compounds, olefins, and acetylenes and in the generation of diimide, carbenes, and carbenoid intermediates. Examples are too numerous to cite here; however, leading references may be found in *Reagents for Organic Synthesis*.[6]

1. Department of Chemistry, Case Institute of Technology, Cleveland, Ohio.
2. A. Albert and R. Royer, *J. Chem. Soc.*, 1148 (1949).
3. K. Freudenberg and F. Bluemmel, *Ann.*, 440, 51 (1924).
4. N. K. Sundholm, U.S. pat. 2,640,853 (1953) [*C.A.*, 48, 6464 (1954)].
5. L. Friedman, R. L. Litle, and W. R. Reichle, *Org. Syntheses*, 40, 93 (1960).
6. L. F. and M. Fieser, *Reagents for Organic Synthesis*, 1, 257, 1185 (1967); 2, 417 (1969); 3, 293 (1972).

1-p-TOLYLCYCLOPROPANOL

$$ClCH_2COCH_2Cl + p\text{-}CH_3C_6H_4MgBr \longrightarrow ClCH_2C(OMgBr)CH_2Cl$$
$$p\text{-}CH_3C_6H_4$$

$$\xrightarrow[\text{FeCl}_3]{\text{C}_2\text{H}_5\text{MgBr}} \quad CH_2 \overset{CH_2}{\diagup\diagdown} COMgBr \quad \xrightarrow[\text{HCl}]{\text{NH}_4\text{Cl}} \quad CH_2 \overset{CH_2}{\diagup\diagdown} COH$$
$$p\text{-}CH_3C_6H_4 \qquad\qquad\qquad p\text{-}CH_3C_6H_4$$

Submitted by C. H. DePuy and R. A. Klein [1]
Checked by O. Aniline and K. B. Wiberg

1. Procedure

In a dry 3-l. three-necked, round-bottomed flask fitted with an efficient reflux condenser, a stirrer, a Y-tube holding a 1-l. and a 250-ml. addition funnel, and protected from moisture by calcium chloride tubes is placed 5.76 g. (0.237 mole) of magnesium turnings barely covered by anhydrous ether. p-Bromotoluene (40 drops) and ethyl bromide (20 drops) are added, and the reaction starts immediately. p-Bromotoluene (35.0 g., 0.205 mole) in 200 ml. of anhydrous ether is added at such a rate that reflux is maintained. To the resultant solution of p-methylphenylmagnesium bromide is added, over a 1-hour period, a solution of 25.4 g. (0.200 mole) of dichloroacetone in 200 ml. of anhydrous ether.

At the same time, in a separate 2-l. three-necked, round-bottomed flask equipped with reflux condenser, stirrer, and addition funnel, ethylmagnesium bromide is prepared from 128.6 g. (1.18 moles) of ethyl bromide and 30 g. (1.23 moles) of magnesium in 800 ml. of anhydrous ether. When the reaction is complete, the addition funnel is replaced by a rubber stopper containing a short glass tube, and the reflux condenser is replaced by an exit tube lightly plugged with a small amount of glass wool. The Grignard reagent solution is forced, under mild nitrogen pressure, through the glass wool plug into the 1-l. addition funnel (Note 1). In the 250-ml. addition funnel is placed a filtered solution of 2.5 g. (0.0154 mole) of anhydrous ferric chloride in 200 ml. of anhydrous ether. Stirring is resumed, and the two solutions are simultaneously added to the dichloroacetone-*p*-methylphenylmagnesium bromide solution over a 2-hour period (Note 2). Stirring is continued for an additional 14 hours under dry nitrogen.

The reaction mixture is added to a slurry of 1500 g. of ice and 600 ml. of 2*N* hydrochloric acid saturated with ammonium chloride. The ether layer is separated, and the aqueous layer is extracted three times with 200-ml. portions of ether. The combined organic layers are washed with three 200-ml. portions of water until a neutral reaction is obtained with litmus and the wash water is free of chloride. The solution is dried over anhydrous magnesium sulfate and stored in a refrigerator. After evaporation of the ether, the residue is distilled at a low pressure through a short Vigreux column (Note 3). The fraction, b.p. 70–78° (0.4 mm.) (oil bath 100–135°), is collected to give 15–17 g. (51–57%) of the crude carbinol which crystallizes upon standing in an ice box. The product is recrystallized from pentane (4 g. per g. of alcohol) in an ice-salt mixture to give the pure alcohol, m.p. 38–39°.

2. Notes

1. Since the ethyl Grignard reagent is used in large excess, no special precautions need to be taken in the transfer to prevent the loss of small amounts.

2. Large volumes of gas are generated, primarily ethane and ethylene, from the disproportionation of the ethyl radicals produced in the reaction of ethylmagnesium bromide with ferric chloride. The reaction should be carried out in an efficient hood, or else a tube should be run from the top of the reflux condenser to a hood.

3. 1-Arylcyclopropanols readily rearrange to propiophenones under the influence of acids and bases. In carrying out the distillation, care must be taken that the apparatus is clean and neutral.

3. Methods of Preparation

The method is that of DePuy and co-workers.[2]   No other syntheses of 1-arylcyclopropanols have been reported.

4. Merits of the Preparation

The procedure can be adapted to the preparation, in comparable yield, of a variety of 1-substituted cyclopropanols, alkyl as well as aryl.

1. Department of Chemistry, University of Colorado, Boulder, Colorado.
2. C. H. DePuy, G. M. Dappen, K. L. Eilers, and R. A. Klein, *J. Org. Chem.*, **29**, 2813 (1964).

o-TOLYL ISOCYANIDE

$$2 \text{ } \underset{\text{}}{\text{(o-CH}_3\text{C}_6\text{H}_4\text{)}}\text{NHCHO} + POCl_3 + 4(CH_3)_3COK \rightarrow$$

$$2 \text{ } \underset{\text{}}{\text{(o-CH}_3\text{C}_6\text{H}_4\text{)}}\text{N}{\equiv}\text{C} + 3KCl + KPO_3 + 4(CH_3)_3COH$$

Submitted by Ivar Ugi and Rudolf Meyr.[1]
Checked by B. C. McKusick and O. W. Webster.

1. Procedure

Caution! Isocyanides should be prepared in a hood since they have pungent odors and some are known to be toxic.

The reaction is conducted in a 2-l. round-bottomed flask equipped with a dropping funnel, Hershberg stirrer, thermometer, and reflux condenser. A T-tube attached to a cylinder of dry nitrogen is inserted in the top of the condenser in order to keep the reaction mixture blanketed with nitrogen.

A suspension of potassium *tert*-butoxide is prepared by a slight modification of the procedure of Johnson and Schneider,[2] *particular attention being paid to the precautions they recommend for safe handling of potassium.* Dry *tert*-butyl alcohol (1250 ml.) is distilled directly into the reaction flask under nitrogen. One hundred grams (2.6 g. atoms) of potassium cut into about ten pieces is added. The stirred mixture spontaneously warms to the melting point of potassium (62°) in the course of 15–60 minutes, whereupon the metal disperses into droplets. As the potassium gradually dissolves, the temperature of the mixture rises to the boiling point of *tert*-butyl alcohol. The rate of solution of the potassium should be such that the *tert*-butyl alcohol refluxes gently, and this rate is regulated by the speed of stirring. If the boiling becomes too vigorous, the stirring is stopped completely, and if necessary the reaction vessel is cooled by immersion in a bath of cold oil kept in readiness for this purpose. Potassium *tert*-butoxide gradually precipitates, and the mixture is a thick suspension when all the potassium has reacted (Note 1).

N-*o*-Tolylformamide (135 g., 1.00 mole) (Note 2) is added to the hot stirred suspension, which becomes a clear solution within a few minutes. The solution is cooled to 10–20° by means of an ice bath and maintained at this temperature while 92 g. (0.60 mole) of phosphorus oxychloride is added to it with stirring over the course of 30–40 minutes. The reaction mixture is stirred at 30–35° for 1 hour and poured into an ice-cold stirred solution of 50 g. of sodium bicarbonate in 5 l. of water (Note 3). *o*-Tolyl isocyanide precipitates as an oil. It is taken up in 300 ml. of petroleum ether (b.p. 40–60°), and the organic phase is separated in a separatory funnel. The aqueous phase is extracted with three 200-ml. portions of petroleum ether. The combined extracts are washed with 50 ml. of 5% sodium bicarbonate solution, dried over 50 g. of powdered potassium hydroxide, and distilled through a 30-cm. vacuum-jacketed Vigreux column. *o*-Tolyl isocyanide is collected as a colorless, vile-smelling liquid at 61–63°/10 mm.; n_D^{25} 1.5212–1.5222; weight 74–85 g. (63–73%) (Note 4).

2. Notes

1. In order to keep down the volume of the reaction mixture, less *tert*-butyl alcohol is used than is necessary to dissolve the potassium *tert*-butoxide.

2. The checkers prepared N-*o*-tolylformamide [3] as follows. A solution of 100 g. (0.94 mole) of *o*-toluidine and 82 ml. (100 g., 2.13 moles) of 98% formic acid in 300 ml. of toluene is refluxed under a condenser attached to a water separator.[4] After water stops collecting in the separator (about 3 hours), toluene and excess formic acid are removed by distillation under reduced pressure. The crude N-*o*-tolylformamide that remains is recrystallized from toluene to give 95–101 g. (75–80%) of N-*o*-tolylformamide, m.p. 60/61°. If a formamide that melts above the boiling point of *tert*-butyl alcohol is to be converted to an isocyanide by the present procedure, it should be finely pulverized.

3. *o*-Tolyl isocyanide is rather unstable, and in order to get a good yield one should work up the reaction mixture as quickly as possible and avoid unnecessary heating of the crude isocyanide. If the isocyanide is to be stored for a long time, it should be kept at the temperature of Dry Ice.

4. The equipment used in this preparation can be freed of the disagreeable odor of *o*-tolyl isocyanide by being washed with 5% methanolic sulfuric acid.

3. Discussion

o-Tolyl isocyanide has been prepared in 20% yield by the action of chloroform and potassium hydroxide on *o*-toluidine.[5] It has also been prepared by the dehydration of N-*o*-tolylformamide using the phosgene/tertiary amine system in good yield.[6] The present procedure is better than the carbylamine reaction[5] in terms of yield.

Although the use of pohsgene/tertiary amine systems is superior to the present procedure for production of isonitriles, the toxicity of phosgene and the difficulty with which it is handled by the inexperienced worker make it less convenient than the present procedure. For most purposes, this procedure illustrates the best way to prepare aryl isocyanides. It is quite general, having been used by Ugi and Meyr[7] to make the following isocyanides from the corresponding form-amides: phenyl (56%), *p*-tolyl (66%), 2,6-dimethylphenyl (88%), mesityl (80%), *o*-chlorophenyl (43%), *p*-chlorophenyl (54%), 2-chloro-6-methylphenyl (87%), *p*-methoxyphenyl (64%), *p*-di-ethylaminophenyl (75%), *p*-nitrophenyl (41%), and 2-naphthyl (50%). Aliphatic isonitriles are generally best prepared by a simpler procedure involving the action of phosphorus oxychloride on an N-alkylformamide in the presence of pyridine.[8]

1. Institute of Organic Chemistry, University of Munich, Munich, Germany.
2. W. S. Johnson and W. P. Schneider, *Org. Syntheses,* Coll. Vol. 4, 134 (1963).
3. A. Ladenburg, *Ber.,* 10, 1123 (1877).
4. S. Natelson and S. Gottfried, *Org. Syntheses,* Coll. Vol. 3, 381 (1955).
5. J. U. Nef, *Ann.,* 270, 309 (1892).
6. P. Hoffmann, G. Gokel, D. Marquarding, and I. Ugi, in I. Ugi, "Isonitrile Chemistry," Academic Press, New York, 1971, p. 9.
7. I. Ugi and R. Meyr, *Ber.,* 93, 247 (1960).
8. I. Ugi, R. Meyr, M. Lipinski, F. Bodesheim, and F. Rosendahl, this volume, p. 300; see also R. E. Schuster, J. E. Scott, and J. Casanova, Jr., this volume, p. 772.

2-(p-TOLYLSULFONYL)DIHYDROISOINDOLE

(Isoindoline, 2-p-tolylsulfonyl-)

Submitted by J. Bornstein and J. E. Shields [1]
Checked by Rosetta McKinley and R. E. Benson

1. Procedure

Caution! This reaction should be carried out in a good hood because hydrogen is evolved and o-xylylene dibromide is a powerful lachrymator (Note 6).

A 1-l. three-necked flask is fitted with an efficient stirrer (Note 1), thermometer, condenser, and a pressure-equalizing dropping funnel that carries an inlet for admission of dry nitrogen. The entire apparatus is dried by warming with a soft flame as a brisk stream of nitrogen is passed through the system. The flow of nitrogen is reduced to a slow stream, and in the cooled flask are placed 18.9 g. (0.42 mole) of 53% sodium hydride dispersed in mineral oil (Note 2) and 60 ml. of purified dimethylformamide (Note 3). The mixture is stirred at room temperature and a solution of 34.2 g. (0.20 mole) of p-toluenesulfonamide (Note 4) in 100 ml. of purified dimethylformamide is added dropwise over a period of 1 hour. The resulting suspension is stirred at room temperature for 1 hour and then at 60° for an additional hour (Note 5).

A solution of 52.8 g. (0.20 mole) of o-xylylene dibromide (Note 6) in 300 ml. of purified dimethylformamide is added dropwise with stirring at such a rate as to maintain a temperature of 60–70° (Note 7). Subsequently the reaction mixture is stirred at room temperature for 3 hours and then poured into 600 ml. of ice water in a 2-l. Erlenmeyer flask. After standing at room temperature overnight the product is collected by suction filtration, pressed on the funnel, and washed twice with 100-ml. portions of water. The crude product is air-dried on filter paper for 2–3 hours and is then dissolved in 1.2 l. of boiling 95% ethanol. The solution is filtered through a heated funnel, and the filtrate is refrigerated overnight. The crystals are collected on a Buchner funnel and washed on the funnel with 100 ml. of cold 95% ethanol. The product is dried over phosphorus pentoxide in a vacuum desiccator. The yield of white crystals of 2-(*p*-tolylsulfonyl)-dihydroisoindole is 41–46 g. (75–84%), m.p. 174–175° (dec.).

2. Notes

1. Either a paddle-type sealed stirrer or a heavy-duty magnetic stirrer is suitable.

2. Sodium hydride was obtained from Metal Hydrides Division of Ventron Corporation, Beverly, Massachusetts.

3. Dimethylformamide, b.p. 152–154°, purchased from Matheson, Coleman and Bell, was stirred for 5 minutes with solid potassium hydroxide, decanted, shaken briefly with lime, filtered, and distilled.

4. Commercial *p*-toluenesulfonamide of high purity was recrystallized from water and dried over phosphorus pentoxide in a vacuum desiccator, m.p. 134–135°.

5. It is necessary to maintain vigorous stirring at this stage to prevent excessive foaming due to the evolution of hydrogen.

6. Precautions to be observed in handling *o*-xylylene dibromide are described in *Org. Syntheses*, Coll. Vol. **4**, 984 (1963). The dibromide was purchased from Eastman Organic Chemicals, recrystallized from 95% ethanol (3 ml./g.), and dried over potassium hydroxide in a vacuum desiccator, m.p. 89–91°.

7. Control of the temperature at this point is critical; a deeply colored product is obtained if the temperature is allowed to exceed 70°. The addition of the dibromide requires about 1 hour.

3. Methods of Preparation

2-(*p*-Tolylsulfonyl)dihydroisoindole has been prepared by alkylation of *p*-toluenesulfonamide with *o*-xylylene dibromide in the presence of sodium methoxide in ethanol.[2, 3]

4. Merits of the Preparation

This is the most practical procedure for the preparation of 2-(*p*-tolylsulfonyl)dihydroisoindole. It is superior to earlier ones [2, 3] because it is more convenient and affords considerably higher yields (*ca.* 80% versus *ca.* 45%).

The method illustrates the ability of the sodium hydride-dimethylformamide system to effect the alkylation of aromatic sulfonamides under mild conditions and in good yield. The method appears to be fairly general. The submitters have prepared N,N-diethyl- and N,N-di-*n*-butyl-*p*-toluenesulfonamide as well as 2-(*p*-tolylsulfonyl)benz[*f*]isoindoline from 2,3-bis-(bromomethyl)naphthalene, and 1-(*p*-tolylsulfonyl)pyrrolidine from 1,4-dichlorobutane; the yield of purified product exceeded 75% in each case.

The reductive cleavage of 2-(*p*-tolylsulfonyl)dihydroisoindole to 1,3-dihydroisoindole constitutes the most convenient synthesis of this heterocyclic compound.[3] The sulfonamide is also useful in the synthesis of isoindole.[4]

1. Department of Chemistry, Boston College, Chestnut Hill, Massachusetts 02167.
2. G. W. Fenton and C. K. Ingold, *J. Chem. Soc.*, 3295 (1928).
3. J. Bornstein, S. C. Lashua, and A. P. Boisselle, *J. Org. Chem.*, **22**, 1255 (1957).
4. R. Kreher and J. Seubert, *Z. Naturforsch.*, **20b**, 75 (1965).

2,4,5-TRIAMINONITROBENZENE

(1,2,4-Benzenetriamine, 5-nitro-)

$$+ 2KNO_3 + H_2SO_4 \rightarrow$$

$$+ K_2SO_4 + 2HOH$$

$$+ 4NH_3 \xrightarrow[140°]{(CH_2OH)_2}$$

$$+ 2NH_4Cl$$

$$+ Na_2S_0 + HOH \rightarrow$$

$$+ Na_2S_2O_3$$

Submitted by J. H. Boyer and R. S. Buriks.[1]
Checked by James Cason and Taysir M. Jaouni.

1. Procedure

A. *1,5-Dichloro-2,4-dinitrobenzene (Caution!* Note 1). To a well-stirred solution of 140 g. (1.386 moles) of potassium nitrate in 500 ml. of concentrated sulfuric acid is added 100.0 g. (0.680 mole) of *m*-dichlorobenzene in one portion. The temperature of the reaction mixture rises during a few minutes to 135–140°, then drops slowly to 125°. The stirred mixture is kept at 120–135° for an additional hour. After the reaction mixture has been cooled to about 90° it is poured over 1.5 kg. of crushed ice. The precipitated product is collected by suction filtration, drained well on the funnel, and dissolved in about 1 l. of boiling 95% ethanol. A small amount of insoluble impurity is removed by filtration of the hot solution by gravity through a fluted filter paper, and the product is allowed to crystallize in the refrigerator at about 0° (Note 2). The yield of yellow needles is 112–115 g. (70–71.5%), m.p. 103–104°.

B. *1,5-Diamino-2,4-dinitrobenzene.* Ammonia gas from a tank is bubbled into a well-stirred, clear yellow solution of 60.0 g. (0.253 mole) of 1,5-dichloro-2,4-dinitrobenzene in 400 ml. of technical grade ethylene glycol (heated to 140°), at such a rate that the gas is just absorbed. Within 30 minutes the color of the solution changes through orange to deep red. About 1 hour after the start of the reaction an orange, crystalline precipitate begins to separate. Heating is continued for an additional 2 hours as a slow stream of ammonia gas is bubbled through the reaction mixture. Finally, the reaction mixture is cooled to room temperature, the product is collected by suction filtration, and the finely divided orange-brown crystals are washed with boiling water and boiling ethanol. The yield of dried product, m.p. 300° (subl.), is 44–48 g. (88–95.5%) (Note 3).

C. *2,4,5-Triaminonitrobenzene.* A well-stirred slurry (Note 4) of 22.5 g. (0.114 mole) of 1,5-diamino-2,4-dinitrobenzene in 150 ml. of water is heated (Note 4) to the boiling point under reflux in a 500-ml. three-necked flask. To this vigorously stirred mixture, a clear orange-red solution of sodium polysulfide (prepared by heating a mixture of 30.0 g. of sodium sulfide nonahydrate, 7.25 g. of sulfur, and 125 g. of water) is added dropwise during a period of 1.5 hours. After completion of the addition, reflux of the well-stirred reaction mixture is maintained for an additional 1.5 hours. The resultant deep red mixture is cooled to 0° and the total insoluble material is collected by suction filtration. This residue of product, sulfur, and some starting material is thoroughly extracted with five 200-ml. portions of boiling water. The combined hot extracts are filtered by gravity and cooled to room temperature to yield 9.5–10.0 g. (49.5–52%) of red needles, m.p. 200–207°, of 2,4,5-triaminonitrobenzene.

2. Notes

1. Unnecessary contact with dichlorodinitrobenzene should be avoided. It is a skin irritant and may cause severe blisters.

2. When a first crop of crystals was collected at room temperature and a second at 0°, the two lots exhibited the same melting point. When the filtrate from crystallization at 0° was concentrated to about 500 ml., the small crop of additional crystals had a much lower melting point.

3. The product is very slightly soluble in most solvents. It was used satisfactorily in the next step without further purification. Melting with sublimation occurs between 285° and 300°, depending on the rate of heating.

4. The submitters report that the best yield in this heterogeneous reaction depends upon particle size of the diaminodinitrobenzene and efficient stirring, and that the diamine should be thoroughly ground in a mortar before use. The checkers found that grinding had no effect on the yield if heating was in an oil bath. Heating with a flame or heating mantle caused some caking and charring on the bottom of the flask, even with rather efficient stirring, and in one run the bottom of the flask dropped out during the reaction.

3. Methods of Preparation

1,5-Dichloro-2,4-dinitrobenzene has been prepared from *m*-dichlorobenzene and nitric acid[2] or potassium nitrate[3] in the presence of sulfuric acid. 1,5-Diamino-2,4-dinitrobenzene has been prepared by the nitration of *m-bis*-acetamidobenzene followed by hydrolysis;[4] or from 1,5-dichloro-2,4-dinitrobenzene and alcoholic ammonia in a pressure bottle at 150° for 8 hours.[2,5] The preparation and characterization of previously unknown 2,4,5-triaminonitrobenzene will be published elsewhere.

The present procedures represent simplified methods for obtaining the subject compounds and for accomplishing the illustrated conversions.

1. Department of Chemistry, Tulane University, New Orleans, Louisiana.
2. R. Nietzki and A. Schedler, *Ber.*, **30**, 1666 (1897).
3. Footnote No. 7 in Th. Zincke, *Ann.*, **370**, 302 (1909) refers to this method used by Fries.
4. R. Nietzki and E. Hagenbach, *Ber.*, **20**, 328, 2114 (1887).
5. P. Ruggli and R. Fischer, *Helv. Chim. Acta*, **28**, 1270 (1945).

1,2,4-TRIAZOLE *

(1H-1,2,4-Triazole)

(A) $NH_2NHCSNH_2$ + HCO_2H ⟶ $HCONHNHCSNH_2$ + H_2O

(B) $HCONHNHCSNH_2$ $\xrightarrow{\text{NaOH}}$

(C)

  $\xrightarrow{\text{HNO}_3}$

Submitted by C. Ainsworth.[1]
Checked by B. C. McKusick and B. C. Anderson.

1. Procedure (Note 1)

A. *1-Formyl-3-thiosemicarbazide.* Four hundred milliliters of 90% formic acid contained in a 2-l. round-bottomed flask is heated on a steam bath for 15 minutes, and then 182 g. (2 moles) of colorless thiosemicarbazide (Note 2) is added. The mixture is swirled until the thiosemicarbazide dissolves. The heating is continued for 30 minutes, during which time crystalline 1-formyl-3-thiosemicarbazide usually separates. Boiling water (600 ml.) is added, and the milky solution that results is filtered through a fluted filter paper. After standing for 1 hour, the filtrate is cooled in an ice bath for 2 hours, and the 1-formyl-3-thiosemicarbazide that separates is collected by suction filtration and air-dried overnight. It weighs 170–192 g. (71–81%) and melts at 177–178° with decomposition.

B. *1,2,4-Triazole-3(5)-thiol.* A solution of 178.5 g. (1.5 moles) of 1-formyl-3-thiosemicarbazide and 60 g. (1.5 moles) of sodium hydroxide in 300 ml. of water in a 2-l. round-bottomed flask is heated on a steam bath for 1 hour. The solution is cooled for 30 minutes in an ice bath and then is treated with 150 ml. of concentrated hydrochloric acid. The reaction mixture is cooled in an ice bath for 2 hours, and the 1,2,4-triazole-3(5)-thiol that precipitates is collected by suction filtration. The thiol is dissolved in 300 ml. of boiling water and the solution is filtered through a fluted filter paper. The filtrate is cooled in an ice bath for 1 hour, and the thiol is collected by suction filtration and air-dried overnight. The 1,2,4-triazole-3(5)-thiol weighs 108–123 g. (72–81%) and melts at 220–222°.

C. *1,2,4-Triazole. Caution! This preparation should be carried out in a ventilated hood to avoid exposure to noxious fumes.*

A mixture of 300 ml. of water, 150 ml. of concentrated nitric acid, and 0.2 g. of sodium nitrite (Note 3) is placed in a 2-l. three-necked flask equipped with a stirrer and a thermometer. The stirred mixture is warmed to 45°, and 2 g. of 1,2,4-triazole-3(5)-thiol is added. When oxidation starts, as indicated by the evolution of brown fumes of nitrogen dioxide and a rise in temperature, a bath of cold water is placed under the reaction flask to provide cooling and an additional 99 g. (total, 101 g.; 1 mole) of 1,2,4-triazole-3(5)-thiol is added in small portions over the course of 30–60 minutes. The rate of addition and the extent of cooling by the water bath are so regulated as to keep the temperature close to 45–47° all during the addition. The water bath is kept cold by the occasional addition of ice.

When the addition is completed, the bath is removed and stirring is continued for 1 hour while the reaction mixture gradually cools to room temperature. Sodium carbonate (100 g.) is added in portions, followed by the cautious addition of 60 g. of sodium bicarbonate (Note 4). The water is removed from the slightly basic solution by heating the solution in a 3-l. round-bottomed flask under reduced pressure on a steam bath. To aid in removing the last traces of water, 250 ml. of ethanol is added to the residue and the mixture is heated under reduced pressure on a steam bath until it appears dry (Note 5).

The residue is extracted twice with 600 ml. of boiling ethanol to separate the triazole from a large amount of inorganic salts. This extract is evaporated to dryness on a steam bath under reduced pressure, and the resulting residue is extracted with two 500-ml. portions of boiling ethyl acetate. The ethyl acetate extract is evaporated to dryness on a steam bath under reduced pressure. The crude 1,2,4-triazole remaining in the flask is dissolved by heating it with 50 ml. of absolute ethanol, and then 1 l. of benzene is added. The mixture is heated under reflux for 15 minutes, and the hot solution is filtered through a fluted filter paper. This extraction procedure is repeated. The two extracts are combined, cooled in an ice bath for 30 minutes, and filtered to remove colorless crystals of 1,2,4-triazole (m.p. 120–121°), weighing 28–30 g. after being dried in air. About 300 ml. of the filtrate is removed by slow distillation through a Claisen stillhead to remove the bulk of the ethanol. The residual solution is cooled in an ice bath for 30 minutes and filtered to separate an additional 8–10 g. of colorless 1,2,4-triazole, m.p. 119–120°. The total weight of 1,2,4-triazole is 36–40 g. (52–58% yield).

2. Notes

1. This procedure is no longer regarded as the best available for the preparation of 1,2,4-triazole. See Discussion section.

2. The thiosemicarbazide must be of good quality or the yield and quality of 1-formyl-3-thiosemicarbazide will suffer. The thiosemicarbazide supplied by Olin Mathieson Chemical Corporation, obtained as a colorless free-flowing powder, can be used without purification.

3. The use of sodium nitrite helps to avoid an induction period.

4. A large flask is used to contain the vigorous effervescence that occurs upon the addition of carbonate. The final pH should be near 7.5, and it is reached after the addition of bicarbonate no longer causes bubbling.

5. Prolonged heating under reduced pressure should be avoided, since 1,2,4-triazole tends to sublime.

3. Discussion

1-Formyl-3-thiosemicarbazide has been prepared by the reaction of thiosemicarbazide and formic acid.[2]

1,2,4-Triazole-3(5)-thiol has been prepared by heating thiosemicarbazide and formic acid,[3] by heating 1-formyl-3-thiosemicarbazide,[3] and by heating 1,3,5-triazine and thiosemicarbazide.[4] The ring closure of 1-formyl-3-thiosemicarbazide using aqueous base was suggested by L. F. Audrieth and F. Hersman.

1,2,4-Triazole has been prepared by the oxidation of substituted 1,2,4-triazoles,[5] by the treatment of urazole with phosphorus pentasulfide,[6] by heating equimolar quantities of formylhydrazine and formamide,[7] by removal of the amino function of 4-amino-1,2,4-triazole,[8] by oxidation of 1,2,4-triazole-3(5)-thiol with hydrogen peroxide,[3] by decarboxylation of 1,2,4-triazole-3(5)-carboxylic acid,[9] by heating hydrazine salts with formamide,[10] by rapidly distilling hydrazine hydrate mixed with two molar equivalents of formamide,[11] by heating N,N'-diformylhydrazine with excess ammonia in an autoclave at 200° for 24 hours,[11] by the reaction of 1,3,5-triazine and hydrazine monohydrochloride,[12] and by the deamination of 3-amino 1,2,4-triazole with hypophosphorous acid.[13] In view of the availability of 3-amino-1,2,4-triazole in several grades and from several commercial sources, the last-cited procedure, that of Henry and Finnegan,[13] is considered to be preferable to that described here for the preparation of 1,2,4-triazole itself.[14] The Henry and Finnegan procedure has been found to be useful for the deamination of a wide variety of heteroaromatic amines.[15]

Modifications of the present procedure for the preparation of 1,2,4-triazole have been used to prepared 3-aryl-1,2,4-triazoles[16] and 3-alkyl-1,2,4-triazoles.[17]

1. The Lilly Research Laboratories, Indianapolis, Indiana.
2. M. Freund and C. Meinecke, *Ber.*, 29, 2511 (1896).
3. M. Freund and C. Meinecke, *Ber.*, 29, 2483 (1896).
4. C. Grundmann and A. Kreutzberger, *J. Am. Chem. Soc.*, 79, 2839 (1957).
5. A. Andreocci, *Ber.*, 25, 225 (1892).
6. G. Pellizzari and G. Cuneo, *Ber.*, 27, 407 (1894).

7. G. Pellizzari, *Gazz. Chim. Ital.,* **24** II, 222 (1894); *Ber.,* **27**, 801 (1894).
8. T. Curtius, A. Darapsky, and E. Müller, *Ber.,* **40**, 815 (1907).
9. J. A. Bladin, *Ber.,* **25**, 741 (1892).
10. H. H. Strain, *J. Am. Chem. Soc.,* **49**, 1995 (1927).
11. C. Ainsworth and R. G. Jones, *J. Am. Chem. Soc.,* **77**, 621 (1955).
12. C. Grundmann and R. Ratz, *J. Org. Chem.,* **21**, 1037 (1956).
13. R. A. Henry and W. G. Finnegan, *J. Am. Chem. Soc.,* **76**, 290 (1954).
14. W. J. Chambers, private communication.
15. H. E. Baumgarten, private communication.
16. E. Hoggarth, *J. Chem. Soc.,* 1163 (1949).
17. C. Ainsworth and R. G. Jones, *J. Am. Chem. Soc.,* **76**, 5651 (1954).

α,α,α-TRICHLOROACETANILIDE

(Acetanilide, α,α,α-trichloro-)

$$\text{CCl}_3\text{COCCl}_3 + \underset{\text{NH}_2}{\bigcirc} \rightarrow \underset{\text{NHCOCCl}_3}{\bigcirc} + \text{CHCl}_3$$

Submitted by BERNARD SUKORNICK.[1]
Checked by JOHN D. ROBERTS and EUGENE I. SNYDER.

1. Procedure

Into a 1-l. three-necked flask, fitted with a mechanical stirrer, a reflux condenser, a thermometer (Note 1), and a dropping funnel, is placed a solution of 265 g. (1 mole) of hexachloroacetone (Note 2) in 400 ml. of hexane (Note 3). To the stirred solution is added, dropwise, 93 g. (1 mole) of aniline (Note 4) over a period of 35–40 minutes. During this time the temperature rises to about 55°. After the addition is complete, stirring is continued at 65–70° for 45 minutes.

The hot solution is poured into a 1-l. beaker and cooled to 0–5°. The solid is collected on a filter and air-dried; it weighs 208–218 g. (87–91%) and melts at 90–92°. One recrystallization from 400 ml. of 90% ethanol (Note 5) yields 160–165 g. (67–69%) (Note 6) of product melting at 92.5–93.0° (Note 7). A second crop of 9–16 g. (4–7%), m.p. 93.5–95.5°, can be obtained by concentrating the filtrate to 200 ml. and cooling (Note 8).

2. Notes

1. The thermometer and the reflux condenser are fitted to a two-necked adapter.

2. Commercial hexachloroacetone (Allied Chemical Corporation) was distilled and the fraction boiling at 93–97°/24 mm. was used.

3. Technical grade hexane suffices.

4. Technical grade aniline was purified by simple distillation and the light-yellow distillate was used directly.

5. The 90% ethanol was prepared by adding 22 ml. of water to 378 ml. of 95% ethanol.

6. The submitter reports a 76–81% yield of product melting at 94–96°.

7. The melting point is raised to 93.5–94.0° by carefully washing the product on a Büchner funnel with 50 ml. of iced 90% ethanol.

8. The submitter has applied this procedure successfully to several amines[2] (see Table I).

TABLE I

Amine	Yield of N-Trichloro-acetylamine, %	M.p., °C *
3-Chloroaniline	93	101
4-Chloroaniline	97	125–127
2-Toluidine	83	98
3-Toluidine	79	102–103
4-Toluidine	70	115
4-Fluoroaniline	83	96
Benzidine	65	301 (dec.)
Benzylamine	84	87–90
2-Phenylethylamine	97	117–120
Ammonia	96	141
Dimethylamine	89	b.p. 110–111°/16 mm.
2-Aminothiazole	80	196–198 (dec.)

* All melting points are uncorrected.

3. Methods of Preparation

Trichloroacetanilide has been prepared from hexachloroacetone and aniline,[3] from trichloroacetyl chloride and aniline,[4] by the action of aniline magnesium iodide on ethyl trichloroacetate,[5] by heating N-phenyltrichloroacetimidyl chloride with dilute methanol,[6] and from trichloroacetic acid and aniline in the presence of phosphorus oxychloride[7] or dicyclohexylcarbodiimide,[8] and from trichloroacetyl diethyl phosphonate and aniline.[9]

1. Allied Chemical Corp., P.O. Box 405, Morristown, New Jersey.
2. Unpublished results.
3. Ch. Cloez, *Ann. Chim. Phys.,* 9 (6), 204 (1886).
4. W. E. Judson, *Ber.,* 3, 783 (1870).
5. F. Bodroux, *Compt. Rend.,* 140, 1598 (1905).
6. J. von Braun, F. Jostes, and R. W. Munch, *Ann.,* 453. 133 (1927).
7. F. A. Berti and L. M. Ziti, *Arch. Pharm.,* 285, 372 (1952).
8. A. Benzas, C. Egnell, and P. Freon, *Compt. Rend.,* 252, 896 (1961).
9. A. N. Pudovik, T. K. Gazizov, and A. P. Pashinkin, *Ah. Obschch. Khim.,* 38, 12 (1968).

1,1,3-TRICHLORO-n-NONANE

(Nonane, 1,1,3-trichloro-)

$$C_6H_{13}CH{=}CH_2 + HCCl_3 \xrightarrow[\text{FeCl}_3]{C_6H_5COCHOHC_6H_5}$$

$$C_6H_{13}CHClCH_2CHCl_2$$

Submitted by D. Vofsi and M. Asscher [1]
Checked by S. N. Eğe and Peter Yates

1. Procedure

A solution of 0.54 g. (2 mmoles) of ferric chloride hexahydrate and 0.33 g. (3 mmoles) of diethylammonium chloride (Note 1) in 5 g. of methanol is added to a solution of 11.2 g. (0.1 mole) of 1-octene (Note 2) and 0.42 g. (2 mmoles) of benzoin (Note 3) in 36 g. (0.3 mole) of chloroform (Note 4). The resulting homogeneous mixture is introduced into a Carius tube of about 100-ml. capacity. Air is displaced by dropping a few pieces of dry ice into the tube (Note 5). The tube is sealed (Note 6), heated to 130°, kept at that temperature for 15 hours, cooled to room temperature (Note 7), and opened. The contents of the tube are transferred to a separatory funnel, and the tube is rinsed with about 10 ml. of chloroform. The reaction mixture is washed with 40 ml. of water. The aqueous solution is extracted with 10 ml. of chloroform, and the extract is added to the original chloroform layer. Solvent is distilled at atmospheric pressure (bath temperature up to 130°). The distillation flask is allowed to cool, and distillation is continued at 25 mm. (bath temperature up to 120°) (Note 8). The flask is cooled again, and distillation is continued to dryness at 0.1 mm. (bath temperature up to 150°), giving crude 1,1,3-trichloro-n-nonane (19.4 g.) as a yellow oil, b.p. 60–85° (0.1 mm.), n^{25}D 1.4650. The purity of this product is 95% (Note 9), and the actual yield is 80%. Fractionation of this material through a 13-in. Vigreux column gives 15 g. (64%) of pure, colorless 1,1,3-trichloro-n-nonane, b.p. 61–62° (0.1 mm.), n^{25}D 1.4640 (Notes 10 and 11).

2. Notes

1. Pure diethylammonium chloride can be obtained from Fluka A. G., Buchs, S. G., Switzerland. If this salt is omitted, somewhat lower yields (about 75%) of adduct are obtained.

2. Phillips 1-octene of 99% minimum purity was used; however, it was freed of peroxide by percolating through acid-washed alumina.

3. Benzoin, Eastman Organic Chemicals, practical grade, can be used directly.

4. Reagent grade chloroform is used.

5. If air is not displaced before sealing, there is an induction period of about 1 hour.

6. The Carius tube has a short piece (about 4 in.) of heavy-walled tube (8-mm. external diameter) sealed to it. This greatly facilitates subsequent sealing and re-use of the tube. The solution is introduced by means of a funnel with a drawn-out stem.

7. On cooling, the contents of the tube separate into two layers.

8. Occasionally a few drops, consisting mainly of unconverted 1-octene, are collected. The receiver then must be changed before the distillation at 0.1 mm. is continued.

9. The purity was determined by gas chromatography (1.5-m. column packed with 25% silicone oil on Chromosorb W, at 180°,

and a flow rate of 60 ml. of helium per minute). The yellow color, which is due to traces of benzil, may be removed by diluting the product with three times its volume of pentane, percolating the solution through a column of about 30 g. of acid-washed alumina, washing the alumina with 50 ml. of pentane, and distilling the pentane at atmospheric pressure. The residue, which is colorless, boils at 61–63° (0.1 mm.), n^{25}D 1.4643; the recovery is 95%.

10. The checkers distilled the reaction product directly through a 4-in. Vigreux column to obtain 15.4–15.8 g. (66–68%) of colorless product, b.p. 95–97° (2.5 mm.), n^{25}D 1.4632.

11. The submitters have found that the reaction may be carried out on a much larger scale in an autoclave. The reaction must be run in a glass liner. As the hot reaction mixture is homogeneous, the autoclave may be heated while standing upright. The liner may be filled to three-quarters of its capacity.

3. Methods of Preparation

The method described, which is the only one available for the direct preparation of 1,1,3-trichloroalkanes, is applicable to aliphatic olefins and gives good yields, especially with terminal olefins.[2] With styrene or butadiene, yields are much lower.

4. Merits of the Preparation

1,1,3-Trichloroalkanes are potential starting materials for the preparation of unsaturated aldehydes.[3]

A similar method[2] can be used for the addition of carbon tetrachloride to nonpolymerizable olefins (e.g., 1-octene, 2-octene, 1-butene, 2-butene); pure adducts are obtained in yields of over 90% if the components are allowed to react at 100° for 6 hours. Adducts of carbon tetrachloride with vinylic monomers (styrene, butadiene, acrylonitrile, methyl acrylate, etc.) can be prepared in good yields by substituting cupric chloride dihydrate in acetonitrile for ferric chloride hexahydrate and benzoin.

In ordinary homolytic reactions (as distinguished from the reaction described here), chloroform adds to the double bond in the sense H—CCl₃.[4] Bromodichloromethane adds in the sense

Br—CHCl$_2$,[3] similar to the orientation of chloroform additions in the present method (Cl—CHCl$_2$). The present method has the advantage of giving high yields while using cheap reagents, and it is thought to proceed as shown in the following equations.

Initiation

$$2FeCl_3 + C_6H_5COCHOHC_6H_5 \rightarrow$$
$$2FeCl_2 + C_6H_5COCOC_6H_5 + 2HCl$$

Propagation

$$FeCl_2 + HCCl_3 \rightleftharpoons FeCl_3 + \cdot CHCl_2$$

$$\cdot CHCl_2 + CH_2{=}CHC_6H_{13} \rightarrow Cl_2CHCH_2CH{-}C_6H_{13}$$

$$Cl_2CHCH_2CH{-}C_6H_{13} + FeCl_3 \rightarrow$$
$$Cl_2CH{-}CH_2CH{-}C_6H_{13} + FeCl_2$$
$$\underset{Cl}{|}$$

Carbon tetrachloride can be substituted for chloroform in this reaction when the cupric chloride modification described above is used.

1. Plastics Research Laboratory, Polymer Department, The Weizmann Institute of Science, Rehovoth, Israel.
2. M. Asscher and D. Vofsi, *Chem. Ind. (London)*, 209 (1962); *J. Chem. Soc.*, **1963**, 1887, 3921; **1964**, 4962.
3. M. S. Kharasch, B. M. Kuderna, and W. Urry, *J. Org. Chem.*, **13**, 895 (1948).
4. M. S. Kharasch, E. V. Jensen, and W. H. Urry, *J. Am. Chem. Soc.*, **69**, 1100 (1947).

TRIETHYLOXONIUM FLUOBORATE*

(Oxonium compounds, triethyloxonium tetrafluoroborate)

$$4(C_2H_5)_2OBF_3 \ + \ 2(C_2H_5)_2O \ + \ 3ClCH_2CH\!\!-\!\!CH_2 \longrightarrow$$
$$\underset{O}{\diagdown}$$

$$3(C_2H_5)_3O^+BF_4^- \ + \ B(OCHCH_2OC_2H_5)_3$$
$$\qquad\qquad\qquad\qquad\qquad\qquad | $$
$$\qquad\qquad\qquad\qquad\qquad CH_2Cl$$

Submitted by H. MEERWEIN [1]
Checked by B. C. ANDERSON, O. H. VOGL, and B. C. McKUSICK

1. Procedure

A 2-l. three-necked flask, a stirrer, a dropping funnel, and a condenser provided with a drying tube are dried in an oven at 110°, assembled while hot, and cooled in a stream of dry nitrogen. Sodium-dried ether (500 ml.) and 284 g. (252 ml., 2.00 moles) of freshly distilled boron fluoride etherate (Notes 1 and 2) are placed in the flask. Epichlorohydrin (140 g., 119 ml., 1.51 moles) is added dropwise to the stirred solution at a rate sufficient to maintain vigorous boiling (about 1 hour is needed). The mixture is refluxed an additional hour and allowed to stand at room temperature overnight. The stirrer is replaced by a filter stick, and the supernatant ether is withdrawn from the crystalline mass of triethyloxonium fluoborate; nitrogen is admitted through a bubbler during this operation to prevent atmospheric moisture from entering the flask. The crystals are washed with three 500-ml. portions of sodium-dried ether. The flask is transferred to a dry box, and triethyloxonium fluoborate is collected on a sintered-glass filter and bottled in a stream of dry nitrogen. The fluoborate is colorless; m.p. 91–92° (dec.), yield 244–272 g. (85–95%) (Note 3).

2. Notes

1. The checkers obtained boron fluoride etherate and epichloro-hydrin from Eastman Organic Chemicals and redistilled each through a 23-cm. Vigreux column immediately before use.

2. It is convenient to measure the liquids with syringes using the densities: epichlorohydrin d_4^{25} 1.179; boron fluoride etherate, d_4^{25} 1.125.

3. Triethyloxonium fluoborate is very hygroscopic. It should be stored in a tightly closed screw-cap bottle at 0–5° and should be used within a few days of the time it is made. It should be weighed and transferred in a dry box. It can be stored indefinitely under ether or at −80°.

3. Methods of Preparation

The procedure used is essentially that described by Meerwein and co-workers.[2, 3] The salt also has been prepared from ethyl fluoride and boron fluoride etherate, and from silver fluoborate, ethyl bromide, and ether.[4]

4. Merits of the Preparation

This simple procedure easily provides large amounts of tri-ethyloxonium fluoborate. Triethyloxonium fluoborate readily ethylates such compounds as ethers, sulfides, nitriles, ketones, esters, and amides on oxygen, nitrogen, or sulfur to give onium fluoborates (often isolable) that can react with nucleophilic reagents to give useful products.[5] For example, dimethylforma-mide gives the imino ether fluoborate $[(CH_3)_2NCH—OC_2H_5]^+$ BF_4^-, which is converted to $(CH_3)_2NCH(OC_2H_5)_2$ by sodium ethoxide.[5] Since an imino ether fluoborate is easily hydrolyzed to the corresponding amine and ester, triethyloxonium fluoborate is a useful reagent for converting amides to amines under mild conditions.[8] Curphey provides a fuller discussion of the alky-lating properties of trialkyloxonium fluoborates.[9]

If there is no advantage in ethylation over methylation, tri-methyloxonium fluoborate[6] or trimethyloxonium 2,4,6-trinitro-benzenesulfonate[7] may be preferable alkylating agents; their preparation is more laborious, but they may be stored for a longer period of time.

1. Deceased October 24, 1965; formerly at University of Marburg, Marburg, Germany.
2. H. Meerwein, E. Bettenberg, H. Gold, E. Pfeil, and G. Willfang, *J. Prakt. Chem.*, [2] **154**, 83 (1940).
3. H. Meerwein, G. Hinz, P. Hofmann, E. Kroning, and E. Pfeil, *J. Prakt. Chem.*, [2] **147**, 257 (1937).
4. H. Meerwein, V. Hederich, and K. Wunderlich, *Arch. Pharm.*, **291**, 552 (1958).
5. H. Meerwein, P. Borner, O. Fuchs, H. J. Sasse, H. Schrodt, and J. Spille, *Ber.*, **89**, 2060 (1956).
6. H. Meerwein, this volume, p. 1096.
7. G. K. Helmkamp and D. J. Pettitt, this volume, p. 1099.
8. H. Muxfeldt and W. Rogalski, *J. Am. Chem. Soc.*, 87, 933 (1965).
9. T. J. Curphey, *Org. Syntheses*, 51, 142 (1971).

1,1,1-TRIFLUOROHEPTANE

(Heptane, 1,1,1-trifluoro-)

$$CH_3(CH_2)_5CO_2H + 2SF_4 \rightarrow CH_3(CH_2)_5CF_3 + 2SOF_2 + HF$$

Submitted by W. R. Hasek.[1]
Checked by John E. Baldwin and John D. Roberts.

1. Procedure

Caution! Sulfur tetrafluoride is toxic. This procedure should be carried out in a good hood. The pressure vessel should be heated in a well-ventilated area.

Twenty-six grams (0.20 mole) of heptanoic acid is placed in a 145-ml. pressure vessel lined with Hastelloy-C (Note 1). The air in the vessel is displaced with nitrogen, and the head of the vessel is secured in place. The vessel is cooled in a bath of acetone and solid carbon dioxide, and the nitrogen in the vessel is evacuated with a vacuum pump to a pressure of 0.5–1.0 mm. Sixty-five grams (95% pure, 0.57 mole) of sulfur tetrafluoride (Note 2) is transferred to the cold vessel. This is conveniently done by connecting a cylinder containing 65 g. of sulfur tetrafluoride to the pressure vessel by a length of copper tubing having a $\frac{1}{16}$-in. bore and $\frac{1}{8}$-in. outside diameter (Note 3).

The pressure vessel is heated with agitation at 100° for 4 hours and at 130° for 6 hours. The vessel is allowed to cool to room temperature and the volatile by-products [*Caution! Toxic!* (Note 4)] are vented. The crude, fuming, liquid product (Note 5) is poured into a stirred suspension of 10 g. of finely divided sodium fluoride in 60 ml. of pentane (Note 6), the mixture is filtered, and the filtrate is fractionated through a 6-in. Vigreux column. 1,1,1-Trifluoroheptane is collected at 100–101°/760 mm., n_D^{25} 1.3449. The yield is 21.7–24.6 g. (70–80%).

2. Notes

1. The pressure vessel should be lined with Hastelloy-C, stainless steel, or other metal resistant to attack by hydrogen fluoride, because the latter substance is a by-product of the reaction. The pressure vessel employed should be safe for use at 500 atm. pressure and should be equipped with a rupture disk rated at 500 atm. If the equipment available is rated for use only at lower pressure, the size of the charge should be reduced appropriately.

2. Directions for the synthesis of sulfur tetrafluoride by the action of sodium fluoride on sulfur dichloride in acetonitrile have been published,[2] and a more detailed version of these directions appears in *Inorganic Syntheses*, 7, 119 (1963).

3. It is also possible to connect the supply cylinder of sulfur tetrafluoride to the pressure vessel by a short length of butyl rubber vacuum tubing.

If the supply cylinder of sulfur tetrafluoride contains more than 65 g., it may be placed on a balance in order to determine when the required amount has been transferred to the pressure vessel.

4. Since the volatile gases include sulfur tetrafluoride and thionyl fluoride, which possess toxicities comparable to that of phosgene, caution must be exercised in their disposal. A suitable procedure is to condense the volatile gases in a trap cooled in a mixture of acetone and solid carbon dioxide, and then to allow this material to pass slowly through an empty polyethylene bottle, which serves as a safety trap, and into a stirred aqueous potassium hydroxide solution.

5. If it is found necessary to retain the crude product for any period of time before working it up, it may be conveniently stored in a polyethylene bottle or other container resistant to attack by hydrogen fluoride.

6. As indicated above, the crude product contains hydrogen fluoride. The sodium fluoride disposes of this by-product by the reaction $NaF + HF \rightarrow NaHF_2$. An alternative procedure is to pour the crude product into water and to separate the product by extraction with pentane.

3. Methods of Preparation

1,1,1-Trifluoroheptane has been prepared only by the action of sulfur tetrafluoride on heptanoic acid.[3]

4. Merits of Preparation

The described procedure is useful for the preparation of a wide variety of compounds containing trifluoromethyl groups from the corresponding carboxylic acids.[3] The yields are generally 60–90%. Some representative examples are listed in Table I. In the cases of the difunctional acids, only 0.1 mole of the compound should be used in the procedure.

TABLE I

Product	B.P., °C.	n_D^{25}
1,1,1-Trifluorododecane	92 (12 mm.)	1.3896
1,1,1-Trifluorohexadecane	107 (0.3 mm.)	1.4148
1,1,1-Trifluoro-3,5,5-trimethylhexane	121–122	1.3657
(4,4,4-Trifluorobutyl)cyclohexane	172–173	1.3987
1,1,1,10,10,10-Hexafluorodecane	183–184	1.3519
1,1,1,6,6,6-Hexafluoro-3-hexene	90–91	1.3131
p-Bis(trifluoromethyl)benzene	113–115	1.3767
2,4-Bis(trifluoromethyl)chlorobenzene	147	1.4130
p-Trifluoromethylnitrobenzene	(m.p. 41–43°)	

Carboxylic anhydrides and esters react with sulfur tetrafluoride to give the same products as the acids only at elevated temperatures, i.e., 200° to 300°.

1. |Contribution No. 572 from the Central Research Department, Experimental Station, E. I. du Pont de Nemours and Co., Wilmington, Del.
2. C. W. Tullock, F. S. Fawcett, W. C. Smith, and D. D. Coffman, *J. Am. Chem. Soc.*, **82**, 539 (1960).
3. W. R. Hasek, W. C. Smith, and V. A. Engelhardt, *J. Am. Chem. Soc.*, **82**, 543 (1960); W. C. Smith, U.S. pat. 2,859,245 (1958).

m-TRIFLUOROMETHYL-N,N-DIMETHYLANILINE

(*m*-Toluidine, α,α,α-trifluoro-N,N-dimethyl-)

Submitted by WILLIAM A. SHEPPARD [1]
Checked by G. B. BENNETT and K. B. WIBERG

1. Procedure

A solution of 16.1 g. (0.100 mole) of *m*-trifluoromethylaniline (Note 1) and 14.3 g. (0.102 mole) of trimethyl phosphate (Note 2) is added to a 300-ml. round-bottomed flask with a side arm. The flask is equipped with a thermometer, magnetic stirrer, and air condenser topped by a water condenser under a nitrogen atmosphere. The stirred reaction mixture is gradually heated by an oil bath to approximately 150° over 30–60 minutes; at this point there is a mild exothermic reaction such that the temperature of the reaction reaches 160–170° and reflux starts (Note 3). After 2 hours at reflux (reaction temperature 145–150°) with oil-bath temperature maintained at 180–200°, the reaction mixture is cooled to room temperature.

A solution of 15 g. of sodium hydroxide in 100 ml. of water is added, and the mixture is stirred vigorously for 1.5 hours to hydrolyze the phosphate ester. The hydrolysis is initially mildly exothermic, and the reaction temperature increases to 50–70°. An additional 200 ml. of water is added. The product, which separates as an oil, is extracted with two 150-ml. portions of ether (Note 4). The combined ether extracts are dried for at least several hours over a mixture of anhydrous magnesium sulfate and sodium hydroxide pellets, filtered, and concentrated by distillation of the ether through a Vigreux column. The residue is distilled at reduced pressure. *m*-Trifluoromethyl-N,N-dimethylaniline is collected at 66–67° (4.5 mm.) and weighs 10.4–11.0 g. (55–58%); n^{24}D 1.4834-1.4828 (Notes 5, 6).

2. Notes

1. *m*-Trifluoromethylaniline (under the name *m*-aminobenzotrifluoride) obtained from Columbia Organic Chemicals Co., Inc., Columbia, South Carolina, was employed. The aniline is also available from Eastman Kodak under the name α,α,α-trifluoro-*m*-toluidine.

2. Trimethyl phosphate obtained from Columbia Organic Chemicals was employed. Although the phosphate ester is reported to be nontoxic under normal handling conditions,[3] use of a hood is recommended.

3. Separation of the reaction mixture into two phases can be observed if the stirrer is stopped for a short period at this point and is also noted on cooling after completion of reflux.

4. The phosphate salts sometimes precipitate before or during the extraction and should be removed by suction filtration to facilitate the extraction. Precipitation may be avoided by addition of larger volumes of water before extraction.

5. A very small forecut is discarded, and only a small amount of tarry residue remains in the pot after the distillation is complete. A spinning-band distillation column was employed by the submitter, but a simple Claisen head is considered adequate because of lack of by-products.

6. The product is free from secondary aniline product on the basis of infrared and n.m.r. proton analysis. If equimolar amounts of aniline and phosphate are employed, the product is obtained in a higher yield (12.3 g., 65%), but it contains a trace of *m*-trifluoromethyl-N-methylaniline as detected by infrared

analysis. This secondary aniline is readily removed by heating the product to reflux with 1 ml. of acetic anhydride followed by redistillation. Use of a larger molar excess of trimethyl phosphate does not affect the purity but does decrease the yield significantly.

3. Discussion

The described method of dialkylation of anilines is essentially that of Billman and co-workers.[2, 3] It has not previously been applied to *m*-trifluoromethylaniline.   *m*-Trifluoromethyl-N,N-dimethylaniline has been prepared in 29% yield by alkylating *m*-trifluoromethylaniline with methyl iodide.[4]

The use of trialkyl phosphates for dialkylation of anilines has been found applicable to naphthylamines [3] and to a large number of anilines substituted in the ortho, meta, or para position by groups such as chloro, methoxy, and methyl [2] and in the meta position by fluoroalkyl (author's laboratory). The reaction has been used to introduce ethyl and *n*-butyl as well as methyl groups by employing the appropriate phosphate esters. The reported yields range from 50% to 95%.

This method has two major advantages over other alkylation procedures: much less manipulation and higher yields; and no troublesome by-products, such as monoalkylated or quaternary products. The Eschweiler-Clarke procedure [5] for alkylation of amines (formaldehyde-formic acid) also has these synthetic advantages for the aliphatic series but gives high molecular weight condensation products with anilines (anilines highly substituted in the ortho-para position may be employed successfully, but *m*-trifluoromethylaniline gives only a resin).

The phosphate method has not been synthetically useful for alkylation of anilines of low basicity such as *p*-nitro-[3] or *p*-trifluoroaniline. Only monoalkylation occurs in introducing branched-chain alkyl groups such as isopropyl.[3] Use of this method for alkylation of aliphatic amines has not been reported.

1. Contribution No. 940 from the Central Research Department, Experimental Station, E. I. du Pont de Nemours and Company, Inc., Wilmington, Delaware.
2. D. G. Thomas, J. H. Billman, and C. E. Davis, *J. Am. Chem. Soc.*, **68**, 895 (1946).
3. J. H. Billman, A. Radike, and B. W. Mundy, *J. Am. Chem. Soc.*, **64**, 2977 (1942).
4. J. D. Roberts, R. L. Webb, and E. A. McElhill, *J. Am. Chem. Soc.*, **72**, 408 (1950).
5. M. L. Moore, *Org. Reactions*, **5**, 309 (1949).

4,6,8-TRIMETHYLAZULENE*

(Azulene, 4,6,8-trimethyl-)

Submitted by K. HAFNER and H. KAISER [1]
Checked by KARL BANGERT and VIRGIL BOEKELHEIDE

1. Procedure

A. *Cyclopentadienylsodium.* A 1-l. four-necked flask (or a three-necked flask with a Y-tube connection) is outfitted with a Trubore® stirrer, a pressure-equalizing dropping funnel, a thermometer reaching to the bottom of the flask, and a reflux condenser in whose outlet is placed a T-tube, one side of which leads to a bubble counter and the other is connected to a source of pure nitrogen. The system is flushed with nitrogen, and a suspension of 23 g. (1.0 mole) of sodium in 350 ml. of dry tetrahydrofuran (Notes 1 and 2) is prepared in the flask. There is then added dropwise with stirring 73.0 g. (1.1 moles) of freshly distilled cyclopentadiene (Note 3). As the exothermic reaction begins, evolution of hydrogen through the bubble counter can be observed immediately. The temperature of the reaction mixture should be kept below 35–40° by intermittent cooling of the flask with an ice bath. At the end of the reaction the color of the solution should be a pale rose; exposure to air causes a rapid change in color to dark brown (Note 4).

B. *4,6,8-Trimethylazulene. Caution! 2,4,6-Trimethylpyrylium perchlorate is explosive. Operations with it should be conducted behind a shield.* The arrangements of the reaction flask used in the preparation of cyclopentadienylsodium are now altered for the next step. While increasing the nitrogen flow rate strongly, the dropping funnel is removed and replaced by a wide-mouthed powder funnel. The strong flow of pure nitrogen coming out of the flask and around the powder funnel prevents the atmosphere from diffusing into the flask to any appreciable extent. Then, with strong stirring of the reaction mixture, 142 g. (0.64 mole) of 2,4,6-trimethylpyrylium perchlorate (*Caution! Moistened with dry tetrahydrofuran, Note* 5) is added in small portions through the powder funnel at such a rate that the immediate exothermic reaction which ensues maintains the temperature of the reaction mixture between 42° and, at most, 48°. The color of the reaction mixture turns purple immediately on addition of the 2,4,6-trimethylpyrylium perchlorate. Usually the addition requires about 1 hour; then the reaction mixture is stirred for an additional 20 minutes. The powder funnel is replaced with a stopper, the condenser is turned downward for distillation, and about 130 ml. of tetrahydrofuran is removed by distillation while stirring is continued. For the distillation the flask is heated on a steam bath, and the temperature of the reaction mixture at the end of the distillation is about 68–70°. The color of the distillate is a weak violet owing to the co-distillation of a small amount of 4,6,8-trimethylazulene. After the reaction mixture has cooled, it is transferred to a 3-l. separatory funnel and diluted, first with 75 ml. of methanol and then with 1 l. of water. This causes the separation of a dark violet oil which is taken up in 400 ml. of petroleum ether (b.p. 60–70°) and separated from the aqueous phase. The aqueous layer is extracted again with 200 ml. of fresh petroleum ether, and the combined petroleum ether extracts are washed five times with 175-ml. portions of water. Since a small quantity of a greasy by-product separates at the interface during the washing with water, the petroleum ether extract, after the final washing, is purified by passing it through a Büchner funnel lined with asbestos fibers as a filtering aid. After the filtrate has been dried over calcium chloride, the solution is concentrated under reduced pressure, and the residue is carefully freed of solvent by heating on a steam bath under reduced pressure for 4 hours.

The crude product is then transferred to an apparatus suitable for distillation of solids (Note 6), and this is joined to a high-vacuum system capable of a vacuum in the range of 10^{-5} mm. Distillation begins when the bath temperature reaches about 190°; a boiling point of around 120° is usually observed. When the distillate first begins to appear brown rather than violet, the distillation is stopped immediately (Note 7). The crystalline distillate (ca. 70 g.) is dissolved in 20 ml. of hot ethanol, filtered while hot, and allowed to cool. The solid (about 60 g. of crystals, m.p. 74–76°) is recrystallized from 20 ml. of ethanol to yield 47–53 g. (43–49%) of 4,6,8-trimethylazulene as dark-violet plates, m.p. 80–81° (Note 8).

2. Notes

1. The suspension of sodium is best prepared as follows. In a three-necked flask fitted with a ground-glass stopper, a reflux condenser, and a Vibromischer (available from A. G. für Chemie-Apparatebau, Zurich, Switzerland) are placed 150 ml. of toluene and 23 g. of sodium. When the toluene is boiling under reflux, the melted sodium is dispersed by the Vibromischer, and the flask is quickly cooled. Under nitrogen atmosphere the toluene is removed by decantation and is replaced by 350 ml. of dry tetrahydrofuran.

2. The dry tetrahydrofuran can be prepared by allowing tetrahydrofuran to stand over sodium, decanting, and distilling from lithium aluminum hydride. *(Caution! See p. 976.)*

3. For the preparation of cyclopentadiene from its dimer, see M. Korach, D. R. Nielsen, and W. H. Rideout, this volume, p. 414.

4. If desired, the cyclopentadienylsodium concentration in solution can be determined by withdrawing 1 ml. of solution, diluting this with 100 ml. of water, and titrating the resulting aqueous sodium hydroxide solution with $0.1N$ hydrochloric acid using methyl red as an indicator.

5. The 2,4,6-trimethylpyrylium perchlorate, obtained and stored as described by Balaban and Nenitzescu [2] or by Hafner and Kaiser,[3] is used directly.

6. A round-bottomed, standard-taper flask with a Claisen head carrying an ebullition capillary and a thermometer and attached to a two-necked flask with one neck for vacuum takeoff is satisfactory. It is important that the setup allow for heating by either flame or infrared lamp to melt the solid distillate and prevent its clogging the vapor passage.

7. It is helpful to empty the brown tarry residue from the distillation flask while it is still hot. The flask can then be cleaned by using a sulfuric acid-chromic acid solution.

8. For purification of small amounts of 4,6,8-trimethylazulene it is advantageous to dissolve it in a small amount of methanol and treat the solution with activated carbon.

3. Methods of Preparation

This procedure is adapted from that described earlier by Hafner and Kaiser,[4] and apparently it is the only method that has been used for synthesizing 4,6,8-trimethylazulene.

4. Merits of the Preparation

This procedure illustrates a simple and general method for preparing azulenes. It is far more convenient and proceeds in much better yield than previous syntheses of azulenes involving dehydrogenation.[5] Also, it is superior to the alternative methods utilizing the monoanil of glutacondialdehyde [6] or pyridinium salts.[7, 8] In fact, this procedure has made the azulenes a readily available class of compounds for study and use as starting materials. Illustrative of the latter are the recent syntheses of pentalene,[9] heptalene,[9] and *peri*-benzazulene derivatives.[10]

1. Institut für Organische Chemie, Munich, Germany.
2. A. T. Balaban and C. D. Nenitzescu, this volume, p. 1106.
3. K. Hafner and H. Kaiser, this volume, p. 1108.
4. K. Hafner and H. Kaiser, *Ann.*, **618**, 140 (1958).
5. V. Prelog and K. Schenker, *Helv. Chim. Acta*, **36**, 1181 (1953).
6. K. Ziegler and K. Hafner, *Angew. Chem.*, **67**, 301 (1955).
7. K. Hafner, *Angew. Chem.*, **67**, 301 (1955).
8. K. Ziegler and K. Hafner, U.S. Patent 2,805,266 (Sept. 3, 1957) [*C. A.*, **52**, 6409 (1958)].
9. K. Hafner and J. Schneider, *Ann.*, **624**, 37 (1959).
10. K. Hafner and H. Schaum, *Angew. Chem.*, **75**, 90 (1963).

2,6,6-TRIMETHYL-2,4-CYCLOHEXADIENONE
(2,4-Cyclohexadiene-1-one, 2,6,6-trimethyl-)

Submitted by David Y. Curtin and Allan R. Stein [1]
Checked by William G. Dauben and Joel W. Rosenthal

1. Procedure

A. *Lithium 2,6-dimethylphenoxide.* In a 300-ml. flask, equipped with a magnetic stirrer and a reflux condenser and flushed with nitrogen, are placed 150 ml. of toluene (freshly distilled from sodium), 1.40 g. (0.202 mole) of lithium metal (Note 1) and 25.0 g. (0.205 mole) of resublimed 2,6-dimethylphenol. The mixture is heated under reflux with stirring for 36 hours; a nitrogen atmosphere is maintained for the reflux period (Note 2). The condenser is replaced by a distillation head with a condenser set for distillation, and a distillation capillary is inserted in the thermometer joint. The bulk of the toluene is removed under nitrogen at reduced pressure (Note 3). The distillation head is rapidly removed, and the flask is flushed with nitrogen and closed by a stopper. The stoppered flask is transferred to a dry box flushed with nitrogen, the stopper removed, and the slurry in the flask filtered with suction through a 65-mm. sintered-glass funnel. The collected lithium salt is washed with three 75-ml. portions of hexane (freshly distilled from lithium aluminum hydride), and the white powder (Note 4) is dried at 100–150° and 0.5 mm. pressure to constant weight (3–10 hours is required). The yield of lithium 2,6-dimethylphenoxide as a fine white or light gray powder is 25.0–25.5 g. (98–100%) (Notes 5 and 6).

B. *2,6,6-Trimethyl-2,4-cyclohexadienone.* In a nitrogen-filled dry box, 25.0 g. of lithium 2,6-dimethylphenoxide (0.195 mole) is transferred to an oven-dried, thick-walled Pyrex® bomb tube (650 x 19 mm.). The bomb tube is stoppered with a rubber stopper fitted with a drying tube, removed from the dry box, and 75 ml. of methyl iodide (170 g., 1.20 moles) (freshly distilled from calcium hydride) is quickly pipetted into the bomb under a dry nitrogen stream. The bomb is cooled in a dry-ice bath and sealed with an oxygen torch. After warming to room temperature, the bomb is shaken to disperse the salt cake and placed in a bomb furnace which has been preheated to 135° (Note 7).

After 36 hours the furnace is allowed to cool, the bomb is removed, cooled to dry-ice temperatures, and opened carefully as there may be residual pressure. The golden brown liquid is poured into a 200-ml. flask, and the methyl iodide is removed on a rotary evaporator (*Caution! Hood*). The residue from the flask and the bomb is washed into a 500-ml. separatory funnel with 100 ml. each of ether and 1:1 solution of Claisen's alkali and water (Note 8). The funnel is shaken, and the alkali layer is removed. The ether layer is extracted four additional times with 100 ml. portions of the alkali (Note 9), washed twice with 75-ml. portions of water, once with saturated aqueous salt solution, and dried by filtration through anhydrous sodium sulfate into a 200-ml. round-bottomed flask. The ether is removed on a rotary evaporator, and the 23.0 g. of yellow oil remaining is allowed to stand at room temperature for 7–10 days to permit dimerization of the dienone (Note 10).

The 2,6-dimethylanisole is removed by vacuum distillation (nitrogen capillary bubbler) at 35–50° (0.75 mm. or less pressure). The solid dimerized product [2] in the residue is either recrystallized twice from the minimum amount of hot hexane or vacuum-distilled, b.p. 175° (25 mm.) (Note 11) followed by a single recrystallization from hexane.

The yield of purified 2,6,6-trimethyl-2,4-cyclohexadienone dimer (white needles, m.p. 119.5–121.0°) is 4.8–5.1 g. (18–19%) (Note 12), and the yield of 2,6-dimethyl anisole is 13.5–14.0 g. (51–53%).

The dienone monomer may be regenerated from its dimer as desired by heating the dimer above 170° for several minutes in a test tube and quenching, or by distillation at 25 mm. (Note 11) through a short condenser into a dry ice-acetone cooled receiver. The monomer may be stored for several days at dry-ice temperatures without appreciable dimerization.

2. Notes

1. The lithium should be added in the form of very small pieces. The pieces are most conveniently prepared as follows. Trim the oxide layer off a small block of lithium metal under mineral oil, grip it with tweezers and rinse the mineral oil off in a beaker of dry ether. Hold the block in the ether vapors momentarily to dry, and then plunge it into a tared beaker of mineral oil for weighing. Cut the block into strips with a sharp knife, remove the pieces one by one, and squeeze them into long flat ribbons with pliers which are frequently dipped into the mineral oil. Cut the ribbons into short sections over another beaker of dry ether, swirl and transfer the pieces to a third beaker of ether to wash off the last traces of mineral oil before adding the lithium to the reaction flask.

2. A positive pressure of nitrogen is maintained by attachment of a mercury bubbler on the top of the reflux condenser.

3. If heating is desired to speed the toluene removal, a steam or an oil bath is used to prevent charring of the salt.

4. If the salt is lumpy, it is best to grind it into a fine powder in an agate mortar before washing it with the hexane. The fine powder is more easily washed and reacts more readily in the alkylation reaction.

5. The lithium phenoxide may be prepared in larger quantities and may be stored for some time sealed under nitrogen and protected from light by wrapping the flask with aluminum foil. Traces of water lead to a gummy salt, while traces of oxygen cause a purple coloration. Quite badly discolored salt has been used successfully in the alkylation procedure, but yields tend to be reduced by tar formation.

6. Titration of a portion of the salt with $0.1N$ hydrochloric acid to methyl orange end-point shows the salt to be 97–103% lithium 2,6-dimethylphenoxide.

7. The temperature used is not crucial, but the best yield is obtained in the 120–180° range. At higher temperatures considerable tar forms, while at lower temperatures dienone yields are sacrificed.[3]

8. Claisen's alkali is a solution of 350 g. of potassium hydroxide in 250 ml. of water made up to 1 liter with methanol.

9. About 4.5 g. or 18% of the starting 2,6-dimethylphenol may be recovered from the combined alkaline extracts by acidification with concentrated hydrochloric acid and extraction of the liberated phenol with ether.

10. The Diels-Alder dimerization of 2,6,6-trimethyl-2,4-cyclo-hexadienone to 1,4,6,6,9,9-hexamethyl-$\Delta^{3,11}$-tricyclo-[6.2.2.0 ^{2,7}]-dodecane-5,10-dione [2] facilitates its separation from the major alkylation product, 2,6-dimethylanisole.

11. In the distillation of the dienone it is necessary to maintain a pot temperature of 175–200° to reverse the dimerization (Note 10).

12. An additional 0.5–1.0 g. of the dienone dimer may be obtained by allowing the 2,6-dimethylanisole fraction to stand for several days and then redistilling it.

3. Methods of Preparation

This preparation of 2,6,6-trimethyl-2,4-cyclohexadienone is based upon the published procedure of the submitters,[2] and it is the only preparation of the 2-substituted-2,4-cyclohexadienones. The simpler 6,6-dimethyl-2,4-cyclohexadienone is more conveniently prepared by the method of Alder.[4]

4. Merits of the Preparation

The use of lithium in toluene for the preparation of alkali metal phenoxides appears to be the most convenient and least expensive procedure. The procedure also has the merit of giving the salt as a finely divided powder.

The alkylation procedure can be used to prepare a wide variety of 2-substituted-2,4-cyclohexadienones,[2, 3, 5, 6] which are useful starting materials. The compounds can serve either as dienes or dienophiles[7] in the Diels-Alder reaction and can be opened photochemically to yield substituted $\Delta^{3, 5}$-hexadienoic acids.[8]

1. Department of Chemistry and Chemical Engineering, University of Illinois, Urbana, Illinois.
2. T. L. Brown, D. Y. Curtin, and R. R. Fraser, *J. Am. Chem. Soc.*, **80**, 4339 (1958)
3. R. R. Fraser, doctoral dissertation, University of Illinois, Urbana, Illinois, 1958, p. 76.
4. K. Alder, F. H. Flock, and H. L. Lessenich, *Ber.*, **90**, 1709 (1957).
5. A. R. Stein, unpublished results.
6. D. Y. Curtin and D. H. Dybvig, *J. Am. Chem. Soc.*, **84**, 225 (1962); D. Y. Curtin and R. R. Fraser, *J. Am. Chem. Soc.*, **80**, 6016 (1958); D. Y. Curtin, R. J. Crawford, and M. Wilhelm, *J. Am. Chem. Soc.*, **80**, 1391 (1958); N. Kornblum and R. Seltzer, *J. Am. Chem. Soc.*, **83**, 3668 (1961).
7. D. Y. Curtin and R. R. Fraser, *J. Am. Chem. Soc.*, **81**, 662 (1959).
8. D. H. R. Barton and G. Quinkert, *J. Chem. Soc.*, 1 (1960).

TRIMETHYLOXONIUM FLUOBORATE

(Oxonium compounds, trimethyloxonium tetrafluoroborate)

$$2(C_2H_5)_3O^+BF_4^- + 3(CH_3)_2O \rightarrow 2(CH_3)_3O^+BF_4^- + 3(C_2H_5)_2O$$

Submitted by H. Meerwein[1]
Checked by O. Vogl, B. C. Anderson, and B. C. McKusick

1. Procedure

Freshly prepared triethyloxonium fluoborate[2] (170 g., 0.90 mole) is dissolved in 500 ml. of anhydrous methylene chloride in a 1-l. three-necked flask equipped with a stirrer, gas-inlet tube, and drying tube (Note 1). The reaction flask is immersed in an ice bath, the stirrer is started, and 138 g. (3.00 moles) of dry dimethyl ether is passed into the solution from a tared cylinder over a period of about 2 hours. The reaction mixture is allowed to stand overnight at room temperature. An hour after the addition of dimethyl ether is complete, trimethyloxonium fluoborate begins to separate. The initially liquid product solidifies slowly.

The stirrer is replaced by a filter stick, and the supernatant methylene chloride is withdrawn from the crystalline mass of trimethyloxonium fluoborate; nitrogen is admitted through a bubbler during this operation to prevent atmospheric moisture from entering the flask. The crystals are washed with three 100-ml. portions of anhydrous methylene chloride. The flask is transferred to a dry box, and trimethyloxonium fluoborate is collected on a sintered-glass filter, dried for 2 hours in a vacuum desiccator at 25° (1 mm.), and bottled in a stream of dry nitrogen. The fluoborate is colorless; yield 114–124 g. (86–94%). Rapidly heated in an open capillary tube, it sinters and darkens, with decomposition, at 141–143° (Note 3).

2. Notes

1. In order to obtain maximum yields, all operations must be carried out under rigorously dry conditions. The apparatus should be dried in an oven at 110°, assembled while hot, and cooled in a stream of dry nitrogen. The checkers dried the methylene chloride over PA 100 silica gel (12–28 mesh) obtained from Davison Chemical Co., Baltimore, Maryland.

2. Trimethyloxonium fluoborate is less hygroscopic and keeps better than triethyloxonium fluoborate, but it should be stored at 0–5° in a tightly closed screw-cap bottle.[2] So stored, it can be kept at least a few weeks.

3. The decomposition point varies widely, depending on rate of heating and apparatus. Professor S. H. Pine, California State College at Los Angeles, informed the checkers that he observed decomposition at 210–220°, with the salt totally disappearing and $(CH_3)_2OBF_3$ forming on the wall of the

capillary tube above the bath. This prompted the checkers to study the decomposition by differential thermal analysis. At a heating rate of 30°/min., there was an endotherm peak at 142°, with sample all gone by 200°. At 15°/min., the peak was at 155°, with sample all gone by 180°.

3. Methods of Preparation

The procedure used is essentially that described by Meerwein and co-workers.[3] The salt has also been prepared from the same reagents in a sealed tube.[4]

Curphey has described a convenient synthesis from boron trifluoride diethyl etherate, dimethyl ether, and epichlorohydrin.[5]

4. Merits of the Preparation

This facile preparation is suitable for preparation of large amounts of salt. Like triethyloxonium fluoborate,[2] trimethyloxonium fluoborate is a potent alkylating agent. In comparison with trimethyloxonium 2,4,6-trinitrobenzenesulfonate,[6] trimethyloxonium fluoborate is easier to make but does not keep quite as well on storage.

1. Deceased October 24, 1965; formerly at University of Marburg, Marburg, Germany.
2. H. Meerwein, this volume, p. 1080.
3. H. Meerwein, P. Borner, O. Fuchs, H. J. Sasse, H. Schrodt, and J. Spille, *Ber.*, **89**, 2071 (1956).
4. H. Meerwein, E. Battenberg, H. Gold, E. Pfeil, and G. Willfang, *J. Prakt. Chem.*, [2] **154**, 143 (1939).
5. T. J. Curphey, *Org. Syntheses*, **51**, 142 (1971).
6. G. K. Helmakamp and D. J. Pettitt, this volume, p. 1099.

TRIMETHYLOXONIUM 2,4,6-TRINITROBENZENESULFONATE

(Oxonium compounds, trimethyloxonium 2,4,6-trinitrobenzenesulfonate)

Submitted by G. K. Helmkamp and D. J. Pettitt [1]
Checked by O. Vogl, B. C. Anderson, and B. C. McKusick

1. Procedure

Caution! Diazomethane is hazardous. Follow the directions for safe handling of diazomethane given in earlier volumes.[2, 3] All operations are carried out in a hood.

The apparatus is shown in Fig. 1. Two 500-ml. round-bottomed flasks *without standard-taper joints* (which could cause diazomethane to detonate) are used. The gas inlet C is connected to a cylinder of dimethyl ether. Gas inlet C is long enough to reach near the bottom of flask A, but tubing D extends only about halfway into flask B. A calcium chloride drying tube is attached to the gas outlet E. Flask B contains a Teflon®-covered stirring bar. The pieces of the apparatus are dried in an oven at 110°; well-dried apparatus is essential for a good yield.

A solution of diazomethane in 200 ml. of xylene is prepared from 15.0 g. (0.146 mole) of nitrosomethylurea [4] (Note 1). The diazomethane solution is decanted into flask A, and about 20 g. of

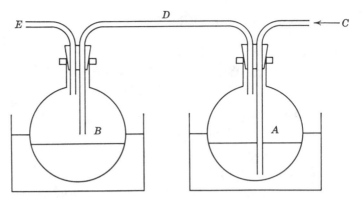

Fig. 1.

potassium hydroxide pellets is added to the solution. The mixture is swirled for a few seconds to ensure removal of most of the water. About 4.2 g. (0.10 mole) of diazomethane is present. Flask A is then immersed in a water bath at 20–25°.

2,4,6-Trinitrobenzenesulfonic acid (14.7 g., 0.050 mole) (Note 2), previously dried for at least 1 hour at 80–100° (1 mm.), is placed in flask B, which is then immersed to the level shown in Fig. 2 in a bath of acetone maintained at −35° to −40° by addition of small amounts of dry ice (Note 3). About 200 ml. of dimethyl ether is rapidly poured from an ampoule into flask B (Note 3). Flasks A and B are connected as shown in Fig. 2, and magnetic stirring is started in flask B. When most of the sulfonic acid has dissolved, gaseous dimethyl ether is introduced through C at such a rate that a rapid stream of individual bubbles passes through the diazomethane solution in flask A. In the course of the reaction all the acid goes into soution and is replaced by a fluffy precipitate of the oxonium salt. The introduction of dimethyl ether is discontinued as soon as the supernatant solution in flask B turns yellow (Notes 5 and 6).

Flask B is separated from the apparatus but kept in the cooling bath, and 200 ml. of anhydrous ethyl acetate is added; addition is slow so as to avoid excessive boiling of the dimethyl ether (Note 7). The flask, with a tube of calcium chloride attached, is gradually brought to room temperature; most of the dimethyl ether evaporates during this operation. Crystalline trimethyl-oxonium 2,4,6-trinitrobenzenesulfonate is separated on a coarse sintered-glass funnel, washed with two 25-ml. portions of ethyl acetate and with 50 ml. of high-boiling petroleum ether, and dried over phosphorus pentoxide at 25° (<1 mm.) (Note 8); yield 12–14 g. (68–79%) (Note 9), m.p. 181–183° (Note 10).

2. Notes

1. Xylene is used as the solvent instead of diethyl ether because of its considerably lower vapor pressure.

2. 2,4,6-Trinitrobenzenesulfonic acid from Nutritional Bio-chemical Corp., Cleveland, Ohio, can be used without any purification other than drying. The checkers observed m.p. 174–177° for the dried acid.

The acid can be prepared from picryl chloride according to the method described by Golumbic, Fruton, and Bergmann,[5] but the following modifications are recommended: sodium metabisulfite should be used in place of sodium bisulfite; the crude sodium salt is not recrystallized but is converted directly to the acid by the addition of hydrochloric acid to its acetone solution; the product is recrystallized by dissolving it in a minimum amount of hot acetone, adding chloroform until crystallization starts, and cooling to about 0°. Two recrystallizations yield a product with m.p. 194–196°.

3. If the level of the cooling bath is too high, or if the bath temperature is less than −40°, unnecessary condensation of dimethyl ether occurs. If the level of the bath is too low, a brownish ring of decomposition product forms in the flask. Since the brown material is soluble in dimethyl ether and ethyl acetate, it does not contaminate the trimethyloxonium salt.

4. The submitters first transferred the dimethyl ether from a cylinder to an ampoule in order to avoid the accumulation of excess water. The ampoule should have a moderately wide mouth in order to facilitate rapid transfer of dimethyl ether. The checkers made a mark on flask B corresponding to a volume of 220 ml., added the acid and stirrer, immersed the flask

in liquid nitrogen, and passed in gaseous dimethyl ether from a cylinder until the volume of condensate reached the mark.

5. Diazomethane reacts with 2,4,6-trinitrobenzenesulfonic acid with ring opening similar to that observed with 1,3,5-trinitrobenzene.[6] Hence an excess of the reagent is to be avoided. The yellow color is not due to the presence of diazomethane itself. The reaction time is highly sensitive to the temperature of the xylene solution and to the flow rate of gaseous dimethyl ether. The reaction time is usually 20–40 minutes.

6. The excess diazomethane in flask A should be destroyed by adding a few drops of glacial acetic acid.

7. The ethyl acetate acts only as a high-boiling material that makes the subsequent vacuum filtration easier to control.

8. If the product is air-dried for more than a few seconds on the filter, it may pick up a significant amount of water. Most of the solvent that remains with the crystals should be removed under vacuum.

9. The product at this stage of purification is sufficiently pure for synthetic applications. As measured by the amount of dimethyl ether evolved on heating, its purity is about 95%.

10. On very rapid heating, the compound effervesces at about 120–130°. It then resolidifies and melts again at 181–183°, which is the melting point of methyl 2,4,6-trinitrobenzenesulfonate. At low heating rates, the effervescence may not be noticed.

3. Methods of Preparation

This method for the preparation of trimethyloxonium 2,4,6-trinitrobenzenesulfonate is an adaptation of that described by the submitters.[7] The salt can also be prepared from trimethyloxonium fluoborate by anion exchange.[7] Trimethyloxonium fluoborate [8] and hexachloroantimonate [9-11] have been prepared by other methods.

4. Merits of the Preparation

Like triethyloxonium fluoborate,[12] trimethyloxonium 2,4,6-trinitrobenzenesulfonate is a potent alkylating agent. Trimethyloxonium 2,4,6-trinitrobenzenesulfonate is nonhygroscopic and hence keeps better than trimethyloxonium fluoborate,[8] but it is more laborious to make.

1 (a) Department of Chemistry, University of California, Riverside, California.
   (b) Pioneering Research Laboratory, Textile Fibers Department, Experimental Station, E. I. du Pont de Nemours & Co., Inc., Wilmington 98, Delaware. This work was supported in part by grant AM-08185 of the National Institutes of Health, U. S. Public Health Service.
2. Th. J. de Boer and H. J. Backer, *Org. Syntheses*, Coll. Vol. 4, 250 (1963).
3. J. A. Moore and D. R. Reed, this volume, p. 351.
4. F. Arndt, *Org. Syntheses*, Coll. Vol. 2, 165 (1943).
5. C. Golumbic, J. S. Fruton, and M. Bergmann, *J. Org. Chem.*, 11, 518 (1946).
6. Th. J. de Boer and J. C. van Velzen, *Rec. Trav. Chim.*, 78, 947 (1959).
7. D. J. Pettitt and G. K. Helmkamp, *J. Org. Chem.*, 28, 2932 (1963).
8. H. Meerwein, this volume, p. 1096.
9. F. Klages and H. Meuresch, *Ber.*, 85, 863 (1952).
10. F. Klages and H. Meuresch, *Ber.*, 86, 1322 (1953).
11. H. Teichmann and G. Hilgetag, *Naturwissenschaften*, 47, 39 (1960).
12. H. Meerwein, this volume, p. 1080; cf. T. J. Curphey, *Org. Syntheses*, 51, 142 (1971).

2,2,4-TRIMETHYL-3-OXOVALERYL CHLORIDE

(Valeryl chloride, 2,2,4-trimethyl-3-oxo-)

Submitted by EDWARD U. ELAM, P. GLENN GOTT,
and ROBERT H. HASEK [1]
Checked by V. BOEKELHEIDE and G. SINGER

1. Procedure

Caution! The starting material in this preparation, 3-hydroxy-2,2,4-trimethyl-3-pentenoic acid β-lactone, is a mild but deceptively persistent lachrymator.

A mixture of 5 g. of anhydrous zinc chloride (Note 1) and 280 g. (2.00 moles) of 3-hydroxy-2,2,4-trimethyl-3-pentenoic acid β-lactone [2] is placed in a 500-ml. three-necked flask equipped with a sealed stirrer (Note 2), a coarse fritted-glass gas-dispersion thimble, a thermometer immersed in the liquid, and an air-cooled reflux condenser (Note 3). The outlet of the condenser is connected to a bubble counter filled with concentrated sulfuric acid; this in turn is vented to the atmosphere through a water scrubber. The flask is immersed in an ice bath, and stirring is started. When the temperature of the mixture is about 10°, anhydrous hydrogen chloride is introduced through the gas-dispersion tube at such a rate that a slow stream of bubbles escapes through the bubble counter. Gas absorption is slow at first, but after a few minutes the zinc chloride dissolves and the rate of gas absorption increases sharply. The temperature rises rapidly to 50–70° at the same time (Note 4). After about 10 minutes the temperature falls and the rate of gas absorption decreases. The ice bath is revmoed, and the addition of hydrogen chloride is continued for 30 minutes.

The reaction mixture is distilled rapidly under reduced pressure through a short Vigreux column (Note 5). The yield of crude product, b.p. 50–80° (5–10 mm.), n^{20}D 1.4410–1.4416, is 310–330 g. (88–93%). This material is sufficiently pure for most purposes (Note 6). Fractionation through a 1 × 36-in. column packed with 8 × 8-mm. glass helices gives, after removal of a small amount of forerun, pure 2,2,4-trimethyl-3-oxovaleryl chloride, b.p. 86° (23 mm.), n^{20}D 1.4418.

2. Notes

1. Reagent grade anhydrous zinc chloride from a freshly opened bottle may be used without special drying.

2. A glass or Teflon stirrer should be used.

3. An air-cooled condenser long enough to trap escaping spray is sufficient since the mixture is never hot enough to reflux.

4. Overheating of the mixture at this point or during the subsequent distillation causes decomposition of the crude acid chloride with formation of tarry by-products. This decomposition in the presence of zinc chloride is fairly rapid at temperatures above 100°.

5. Since hydrogen chloride is evolved in the early stages of this distillation, no effort is made to control the pressure. The vacuum pump is protected from the hydrogen chloride by insertion of a 1.5 × 15-in. glass tube packed with sodium hydroxide pellets in the vacuum line. The purpose of this distillation is to remove dissolved hydrogen chloride and the zinc chloride catalyst; it should be completed as rapidly as possible, under the best vacuum attainable, with no attempt at fractionation. The distillate may be fractionally distilled for further purification if desired. 2,2,4-Trimethyl-3-oxovaleryl chloride is stable at its normal boiling point, 190° (730 mm.), after the zinc chloride has been removed. Contact of the acid chloride with metals should be avoided.

6. The chlorine content of the crude product varied from 19.9% to 20.1% in successive experiments; saponification analysis indicated a purity above 97%.

3. Methods of Preparation

The procedure described is related to that for preparing acetoacetyl chloride from diketene and hydrogen chloride.[3]

4. Merits of the Preparation

Unlike acetoacetyl chloride, which decomposes at temperatures above −20°, 2,2,4-trimethyl-3-oxovaleryl chloride is stable at elevated temperatures. It may find use as an intermediate; for example, it can be used in the preparation of acid chlorides by an exchange reaction which is forced to completion by decarboxylation of the by-product, β-keto acid:

$$RCO_2H + (CH_3)_2CHCOC(CH_3)_2COCl \rightleftharpoons$$
$$RCOCl + (CH_3)_2CHCOC(CH_3)_2CO_2H$$
$$\downarrow$$
$$(CH_3)_2CHCOCH(CH_3)_2 + CO_2$$

1. Research Laboratories, Tennessee Eastman Co., Kingsport, Tennessee 37662.
2. R. H. Hasek, R. D. Clark, and G. L. Mayberry, this volume, p. 456.
3. C. D. Hurd and C. D. Kelso, J. Am. Chem. Soc., 62, 1548 (1940).

2,4,6-TRIMETHYLPYRYLIUM PERCHLORATE

Method I

$$(CH_3)_3COH + 4(CH_3CO)_2O + HClO_4 \xrightarrow[-6CH_3CO_2H]{}$$

$$ClO_4^-$$

Submitted by A. T. BALABAN [1] and C. D. NENITZESCU [2]
Checked by KARL BANGERT and VIRGIL BOEKELHEIDE

1. Procedure

Caution! *2,4,6-Trimethylpyrylium perchlorate is explosive. Operations should be conducted behind a shield, and directions should be followed closely (see Note 1 of Method I and Section 4 before carrying out these preparations).*

In a 2-l. four-necked flask (or a three-necked flask with a Y-tube connection) outfitted with a stirrer, a short reflux condenser, a dropping funnel, and a thermometer reaching to the bottom of the flask, 148 g. (2.0 moles) of anhydrous *t*-butyl alcohol and 1020 g. (945 ml., 10.0 moles) of acetic anhydride are mixed with stirring and cooled to −10° by means of an ice-salt cooling bath. Then 250 g. (150 ml., 1.75 moles) of 70% perchloric acid is added rapidly from the dropping funnel to the stirred mixture over a period of 5–7 minutes (Note 2). With the first few drops a vigorous reaction begins which is manifested by evolution of fumes, coloration of the reaction mixture to orange and then reddish brown, and a rapid rise in temperature. When the temperature of the reaction mixture reaches 40–50°, crystals of 2,4,6-trimethylpyrylium perchlorate should begin to separate (Note 3); then the temperature is allowed to rise to 100°. The rate of perchloric acid introduction and the use of the cooling bath are then so controlled that the temperature of the reaction mixture is maintained between 100° and 105°. Toward the end of the addition the perchloric acid may be added quite rapidly and the desired temperature may still be maintained. After all the perchloric acid has been added, the cooling bath is removed and stirring of the mixture is continued. The temperature remains at about 90° for 10 or 15 minutes and then falls to about 75° after 30 minutes. The dark-brown stirred mixture is cooled once again until the temperature has fallen to 15°. The crystalline 2,4,6-trimethylpyrylium perchlorate, which has separated, is collected on a Büchner funnel and is washed on the funnel with a 1:1 mixture of acetic acid and ether and then washed twice with ether (Note 4). Suction is stopped before the crystals are dry. The product can be air-dried to give 195–210 g. (50–54%) of yellow crystals, m.p. 244° dec. (Notes 1, 5, and 6). For storing or for use in the preparation of 4,6,8-trimethylazulene, however, it is best to place the product in a cork-stoppered flask and moisten it with dry tetrahydrofuran.

2. Notes

1. The impact sensitivity of 2,4,6-trimethylpyrylium perchlorate was examined by Dr. T. E. Stevens at the Redstone Arsenal Division of Rohm and Haas Co., and the compound was found to be slightly more sensitive to detonation by impact than the commercial explosive RDX. This point should be kept constantly in mind. When the crystals are handled as a slurry or are wet with solvent, the hazard is considerably reduced. On the other hand, the dry perchlorate should be handled with great care and should never be crushed, rubbed, or pushed through a narrow opening.

2. The specified order of mixing the three reagents is critical. If the reagent added to the solution of the other two is t-butyl alcohol or acetic anhydride, large amounts of triisobutylenes are formed, separating as a colorless upper layer.

3. The rate of perchloric acid introduction should be slow at first so that in the range of 40–50° crystals of the 2,4,6-trimethylpyrylium perchlorate will begin to appear. Then the rate of addition should be increased to maintain the temperature in the optimum range of 100–105°. If the temperature rises too rapidly, no crystals will appear and the yield will be somewhat lower. Then seeding is helpful.

4. If crystallization is not complete, dilution of the filtrate by the ether washings will cause separation of additional crystals. These are collected separately because they are finer and less pure. Concentration of filtrates is to be avoided because severe explosions [3] have been reported when solutions of perchloric acid in acetic acid were concentrated.

5. The product is of satisfactory purity for use in the 4,6,8-trimethylazulene preparation without further purification. Recrystallization of a small sample of the 2,4,6-trimethylpyrylium perchlorate from a seven-fold amount of hot water, containing a few drops of perchloric acid and some carbon black, gives colorless crystals, m.p. 245–247° dec. However, the recrystallization of larger amounts in this way presents some hazard and is not recommended. Concentration of filtrates should be avoided (see Note 4).

6. The preparation of 2,4,6-trimethylpyrylium perchlorate may be carried out on a much smaller scale, such as one-tenth, with only a small lowering of the yield.

Method II

$$(CH_3)_2C{=}CHC\overset{\overset{\displaystyle O}{\|}}{-}CH_3 + 2(CH_3CO)_2O \xrightarrow[-3CH_3CO_2H]{HClO_4}$$

$$ClO_4{}^{\ominus}$$

Submitted by K. HAFNER and H. KAISER [4]
Checked by VIRGIL BOEKELHEIDE and H. FLEISCHER

1. Procedure

Caution!   2,4,6-Trimethylpyrylium perchlorate is explosive. Operations should be conducted behind a shield, and directions should be followed closely (see Note 1 of Method I and Section 4 before carrying out these preparations).

In a 2-l. four-necked flask (or a three-necked flask with a Y-tube connector) equipped with a stirrer, a reflux condenser, a dropping funnel, and a thermometer extending nearly to the bottom of the flask is placed 550 ml. (595 g., 5.83 moles) of acetic anhydride which is cooled to 0° with an ice-salt bath. Then 180 ml. (300 g., 2.09 moles) of a 70% solution of perchloric acid is added with stirring at a rate such that the temperature does not rise above 8° (Note 1). This step takes about 3 hours. The mixture is continually cooled and stirred, and 240 ml. (204 g., 2.09 moles) of mesityl oxide is then added slowly. The slow addition of 370 ml. (400 g., 3.92 moles) of acetic anhydride follows. The ice bath is then replaced by a water bath; the temperature of the reaction mixture will usually rise to 50–70° because of the heat liberated by the exothermic reaction, and the reaction mixture will turn dark. The reaction mixture is heated on a steam bath for 15 minutes to complete the reaction, and the mixture is then allowed to cool and stand at room temperature for 2 hours. The crystals, which have separated from the brown solution, are collected on a Büchner funnel and are washed on the funnel twice with 100-ml. portions of acetic acid, twice with 100-ml. portions of absolute ethanol, and twice with 100 ml. portions of absolute ether. This gives 250–260 g. (54–56%) of pale-yellow to light-brown crystals, m.p. 240° dec. (Note 2). For storage the crystals should be transferred, without drying, to an ordinary flask, moistened with dry tetrahydrofuran, and then kept in this state by stoppering the flask with an ordinary cork (Note 3).

2. Notes

1. Since the reaction is quite exothermic, the mixture must be well stirred to avoid developing any local hot spots which could lead to explosions. Although no difficulties were encountered in either the submitters' or checkers' laboratories, it is well to keep in mind that 2,4,6-trimethylpyrylium perchlorate is potentially hazardous; hence due precaution should be exercised at all times.

2. Although a small sample of 2,4,6-trimethylpyrylium perchlorate may with care be recrystallized from acetic acid to give

white crystals, m.p. 245–247° dec., it is recommended that this not be done with larger quantities. The 2,4,6-trimethylpyrylium perchlorate is of satisfactory purity for use in the 4,6,8-trimethyl-azulene preparation without further purification.

3. The hazard of handling 2,4,6-trimethylpyrylium perchlorate is greatly reduced if the crystals are kept moist with a solvent such as tetrahydrofuran. The flask used for storage should be stoppered with a cork rather than a ground-glass stopper to avoid the possibility of initiating an explosion by the grinding action of the stopper.

3. Methods of Preparation

2,4,6-Trimethylpyrylium perchlorate has been prepared from 2,6-dimethylpyrone and methylmagnesium halides;[5] from mesityl oxide and sulfoacetic acid;[6] from mesityl oxide (or less satisfactorily from acetone) and a mixture of acetic anhydride and perchloric acid;[7] from mesityl oxide, acetyl chloride, and aluminum chloride;[8] and from t-butyl chloride, acetyl chloride, and aluminum chloride.[8] The procedure given under Method I is adapted from that reported by Balaban and Nenitzescu[9] and is similar to that of Praill and Whitear.[10] The procedure given under II is adapted from that reported by Hafner and Kaiser.[11] The methods of preparation have been reviewed.[12]

4. Merits of the Preparation

2,4,6-Trimethylpyrylium perchlorate is a very versatile and useful starting material. Thus its reaction with cyclopentadienyl-sodium has made 4,6,8-trimethylazulene[13] easily available for general studies of the properties of azulenes[14] and for the synthesis of related compounds.[15] In addition, pyrylium salts are readily converted to a variety of pyridine derivatives,[9,16] 4H- and 2H-pyrans,[17] 2-acylfurans,[18,19] isoxazolines,[20] pyrazolines,[20] 1,2-diazepines,[20,21] 5-cyano-2,4-dien-1-ones,[22] and various benzene and naphthalene derivatives,[23] including derivatives of nitrobenzene[24] and phenol.[9,25,26] It is clear that its value as a starting material is such that it is receiving wide use.

In including this preparation in *Organic Syntheses*, it was felt that standard procedures which have been tested in more than one laboratory without difficulty and which attempt to point out as clearly as possible the potential hazards involved would serve a useful function for those who, despite the hazards, find this a necessary and important starting material.

1. Institute of Atomic Physics, R. P. R. Academy, Bucharest, Romania.
2. Chemical Institute, R. P. R. Academy, Bucharest, Romania.
3. E. Kahane, *Compt. Rend.*, **227**, 841 (1948).
4. Institut für Organische Chemie der Universität, Munich, Germany.
5. A. von Baeyer and J. Piccard, *Ann.*, **384**, 208 (1911); **407**, 332 (1915).
6. W. Schneider and A. Sack, *Ber.*, **56**, 1786 (1923).
7. O. Diels and K. Alder, *Ber.*, **60**, 716 (1927).
8. A. T. Balaban and C. D. Nenitzescu, *Ann.*, **625**, 74 (1959).
9. A. T. Balaban and C. D. Nenitzescu, *J. Chem. Soc.*, 3553 (1961).
10. P. F. G. Praill and A. L. Whitear, *J. Chem. Soc.*, 3573 (1961).
11. K. Hafner and H. Kaiser, *Ann.*, **618**, 140 (1958).
12. A. T. Balaban, W. Schroth, and G. Fischer, *Advan. Heterocycl. Chem.*, **10**, 241 (1969).
13. K. Hafner and H. Kaiser, this volume, p. 1088.
14. K. Hafner, H. Pelster, and J. Schneider, *Ann.*, **650**, 62 (1961).
15. K. Hafner and J. Schneider, *Ann.*, **624**, 37 (1959).
16. G. V. Boyd and L. M. Jackman, *J. Chem. Soc.*, 548 (1963).
17. A. T. Balaban, G. Mihai, and C. D. Nenitzescu, *Tetrahedron*, **18**, 257 (1962).
18. A. T. Balaban and C. D. Nenitzescu, *Ber.*, **93**, 599 (1960).
19. A. T. Balaban, *Tetrahedron*, **24**, 5059 (1968); **26**, 739 (1970).
20. O. Buchardt, C. P. Pedersen, V. Svanholm, A. M. Duffield, and A. T. Balaban, *Acta Chem. Scand.*, **23**, 3125 (1969).
21. A. T. Balaban and C. D. Nenitzescu, *J. Chem. Soc.*, 3566 (1961).
22. K. Dimroth and K. H. Wolf in W. Foerst, "Newer Methods of Preparative Organic Chemistry," Vol. 3, Academic Press, New York, 1964, p. 357.
23. K. Dimroth, G. Neubauer, H. Möllenkamp, and G. Oosterloo, *Ber.*, **90**, 1668 (1957).
24. K. Dimroth, *Angew. Chem.*, **72**, 331 (1960).
25. A. T. Balaban and C. D. Nenitzescu, *Studii Cercetari Chim. Acad. Rep. Populare Roumania*, **9**, 251 (1961); *Rev. Chim. Acad. Rep. Populaire Roumania*, **6**, 269 (1961).

2,4,6-TRIMETHYLPYRYLIUM TETRAFLUOROBORATE

$$(CH_3)_3COH + HBF_4 + 4(CH_3CO)_2O \xrightarrow[-6CH_3CO_2H]{}$$

Submitted by A. T. BALABAN and A. J. BOULTON[1]
Checked by DOROTHY G. McMAHAN and HENRY E. BAUMGARTEN

1. Procedure

In a 250-ml. Erlenmeyer flask provided with a magnetic stirrer bar and a thermometer containing 50 ml. (54 g., 530 mmol.) of acetic anhydride and 4.0 ml. (3.1 g., 41 mmol.) of dry *t*-butyl alcohol, 7 ml. (39 mmol.) of 40% fluoroboric acid (Note 1) are added, initially very cautiously, and then in *ca.* 0.2-ml. portions until all has been added, at such a rate that the final temperature reaches *ca.* 100°. A dark yellow-brown color develops. The solution is allowed to cool spontaneously to 80°, and then is chilled to 5° in an ice bath. Separation of the salt begins and is completed by the addition of 100 ml. of ether. After filtration on a Büchner funnel the salt is washed with 30 ml. of ether, yielding 3.9–4.1 g. (47–50%) of a colorless or pale yellow product, m.p. 218–220° (decomp.) (Note 2). The crude product may be recrystallized from ethanol-methanol 1:1 (*ca.* 70–80 ml.) containing a few drops of fluoroboric acid, affording 3.4–3.5 g. (41–43%) of colorless prisms, m.p. 224–226° (decomp.) (Note 3).

2. Notes

1. The checkers used 6 ml. of 48% fluoroboric acid obtained from the Ozark Mahoning Co.

2. The checkers collected two crops of crystals, the second precipitating during the filtration of the first. The crude product was dried between filter papers overnight.

3. The yields reported are those of the checkers obtained with 48% fluoroboric acid and are about 30% higher than those (27–32%) reported by the submitters. Larger scale preparations (five or more times the quantities described here) have been carried out by the submitters. Such preparations require external cooling.

3. Discussion

See the Discussion section of 2,4,6-trimethylpyrilium trifluoromethanesulfonate.[2]

1. Institute of Atomic Physics, Bucharest, Rumania.
2. A. T. Balaban and A. J. Boulton, this volume, p. 1114.

2,4,6-TRIMETHYLPYRYLIUM TRIFLUOROMETHANESULFONATE

$$(CH_3)_3COH + CF_3SO_3H + 4(CH_3CO)_2O \xrightarrow[-6CH_3CO_2H]{}$$

$$CF_3SO_3^{\ominus}$$

$CH_3 \quad \oplus \quad CH_3$

Submitted by A. T. BALABAN and A. J. BOULTON[1]
Checked by DOROTHY G. McMAHAN and HENRY E. BAUMGARTEN

1. Procedure

In a 150-ml. Erlenmeyer flask are mixed 18 ml. (19.4 g., 192 mmol.) of acetic anhydride and 3.2 ml. (2.5 g., 34 mmol.) of anhydrous *t*-butyl alcohol. Two to three drops of trifluoromethanesulfonic acid (Note 1) is added and the mixture is swirled rapidly until the initial esterification reaction is complete (temperature stops rising). Further trifluoromethanesulfonic acid (total acid: 3.0 ml., 5.0 g. (33 mmol.)) is added in 0.1–0.2-ml. portions, over 10 minutes, with swirling and pausing between each addition until the temperature begins to fall. At the end of the addition the temperature is 60–70° (Note 2), and the solution is brown in color. It is allowed to stand for a further 5 minutes and then is cooled in ice. Anhydrous ether (100 ml.) is added, whereupon the pyrylium salt separates. The mixture is filtered with suction using a sintered glass funnel and the crude product is washed with a further 50 ml. of dry ether (Note 3), giving 3.6–3.8 g. (40–42%) of mustard colored to pale brown plates, m.p. 116–118° (Note 4).

The crude product is recrystallized from dioxane-acetic acid (7:1) or chloroform-carbon tetrachloride (2:1). The crystal form is very variable; needles, plates, or prisms may be formed, depending on the rate of formation, and the temperature. After recrystallation the 2,4,6-trimethylpyrilium trifluoromethane sulfonate melts at 119–120° (Note 5).

2. Notes

1. Trifuloromethanesulfonic acid may be obtained from Aldrich Chemical Company or Pierce Chemical Company.

2. If the reaction is conducted at a lower temperature (*ca.* 20°), a lighter solution and a more nearly pure product result, but the yield is much reduced (1–2 g., 11–22%). Allowing the temperature to rise above 70° gives a darker brown mixture, but with no improvement in yield.

3. The washing ether should be added before the crystals are exposed to the air, and the ether and crystals should be stirred to ensure thorough mixing. Otherwise a troublesome tar may separate. The product should be isolated within 2–3 hours of its precipitation, or it will be accompanied by black impurities which are slowly deposited by the mother liquor, and are difficult to remove.

4. The checkers observed a variable range of melting points for the crude product, as broad as 108–113° and as sharp as 116–117°.

5. The checkers used chloroform-carbon tetrachloride for recrystallization and found that about 500 ml. of this mixture was required to dissolve the crude product and give a green colored solution. Their final yield after two recrystallations was 3.2–3.4 g. (36–38%), m.p. 118–119°.

3. Discussion

Although more expensive in materials, and so suitable mainly for small-scale work, the product does not have the explosive hazard of the corresponding perchlorate.[2] This advantage is shared with the tetrafluoroborate,[3] which, however, requires more acetic anhydride and may give poorer yields. The trifluoromethanesulfonate salt is also more soluble in organic solvents than the perchlorate or tetrafluoroborate (1 g. dissolves in 7 ml. of chloroform at 20°, and in 3 ml. at *ca.* 35°; it is also very soluble in alcohols and dichloromethane). For the usefulness of 2,4,6-trimethylpyrylium salts in general, see the notes pertaining to the perchlorate.[2]

1. Institute of Atomic Physics, Bucharest, Rumania.
2. A. T. Balaban and C. D. Nenitzescu, this volume, p. 1106; K. Hafner and H. Kaiser, this volume, p. 1108.
3. A. T. Balaban and A. J. Boulton, this volume, p. 112.

TRIPHENYLALUMINUM *

(Aluminum, triphenyl-)

$$3(C_6H_5)_2Hg + 2Al \rightarrow 2(C_6H_5)_3Al + 3Hg$$

Submitted by T. A. Neely, William W. Schwarz, and Herbert W. Vaughan, Jr.[1]
Checked by R. D. Lipscomb and B. C. McKusick

1. Procedure

Caution! Triphenylaluminum and its etherate undergo decomposition in the presence of air and moisture. Upon contact with water, vigorous heat evolution and sparking have been observed.

A 500-ml. one-necked flask with a side arm to admit nitrogen is fitted with a reflux condenser protected at the top by a T-tube through which nitrogen is slowly passed during the entire reaction (Note 1). In the flask there is placed 12 g. (0.44 g. atom) of aluminum wool (Note 2), and the system is thoroughly dried by flaming (Note 3) and then cooled to room temperature. A positive pressure of nitrogen is maintained in the system by admission of the gas through the side arm while the flask is detached from the condenser, stoppered, and transferred to a nitrogen-filled dry box. To the flask there is added 80 g. (0.23 mole) of diphenylmercury (Note 4) which is spread evenly on top of the aluminum wool, followed by 340 ml. of sodium-dried xylene.

The flask is stoppered, returned to the condenser, and immersed in a preheated oil bath, which is maintained at 140–150°. The reaction mixture is allowed to reflux for 24 hours, the water drained from the condenser, and the top of the condenser is connected to a vacuum system through a trap cooled in dry ice. The xylene is distilled by gradually reducing the pressure to 20–30 mm. The flask is cooled to room temperature and nitrogen readmitted.

The flask is returned to the dry box, and the nearly dry solid that remains in it is transferred to an extraction thimble (123 mm. x 43 mm.) previously dried in a vacuum oven. The product is extracted in a dried Soxhlet apparatus (250 mm. x 50 mm.) with 250 ml. of dry ether (Note 5) in a carefully dried 300-ml. flask. The extraction is continued for 15–20 hours (Note 6), during which time white crystals of triphenylaluminum etherate form in the flask. The flask and its contents are placed in a dry box, the ether is decanted, and the crystals are washed several times by decantation with small portions of dry ether. The triphenylaluminum etherate is dried at 25° under reduced pressure; m.p. 126–130°. The ether of crystallization is removed by heating the etherate at 150° (0.1 mm.) for about 13 hours (Note 7). Pure triphenylaluminum, m.p. 229–232°, is obtained; yield 23–27 g. (59–70%).

2. Notes

1. All operations are conducted in an atmosphere of prepurified nitrogen or in a nitrogen-filled dry box. A positive nitrogen pressure is maintained in the flask during all transfers and additions. Prepurified nitrogen is available from Matheson Co., East Rutherford, New Jersey.

2. Suitable aluminum wool is available from Custom Scientific Instrument Inc., Kearney, New Jersey. It is thoroughly cleaned with both methylene chloride and ether and dried before use.

3. All glassware used in this preparation must be dry. Flaming out while purging with prepurified nitrogen is sufficient.

4. Suitable diphenylmercury is available from Orgmet, Hampstead, New Hampshire, or it can be prepared by the procedure of Gilman and Brown.[2]

5. A new container of anhydrous ether from Mallinckrodt Chemical Works or Merck and Co. is satisfactory.

6. At the end of the extraction the residual aluminum wool should be disposed of very carefully because it is in a highly reactive condition. As the ether evaporates, the aluminum wool oxidizes rapidly and becomes quite hot. In one run, 84% of the theoretical amount of metallic mercury was liberated during this oxidation.

The checkers evaporated the ether in a stream of nitrogen, then allowed air to diffuse in gradually through a small opening during 2 days. The residue was then inert.

7. During this operation the etherate melts, effervesces, and then solidifies. The etherate sublimes to a small extent; this can be counteracted by immersing the flask in an oil bath only part way during the first half of the heating, and finally immersing it to the bottom of its neck.

3. Methods of Preparation

The general procedure for this reaction, which was first reported by Friedel and Crafts,[3] is essentially that of Hilpert and Grüttner [4] as modified by Gilman and Marple.[5] Upon laboratory examination of these methods, only water-reactive gums and tars were isolated. Nesmeyanov and Novikova [6] reported the preparation of triphenylaluminum by a similar method, but worked on a test-tube scale and did not report a yield. Wittig and Wittenberg [7] prepared crystalline triphenylaluminum in 43% yield by the action of phenyllithium on aluminum chloride. In most of the procedures found, the triphenylaluminum was used in solution; hence crystalline material was not isolated.

When aluminum wool was substituted for strips of aluminum foil, a 20% yield of triphenylaluminum was obtained. This yield was increased to 70% by the extractive isolation described above. An alternative method of isolation and purification, that of Krause and Polack,[8] was not attempted because of the lack of experimental detail and the complexity of the apparatus.

4. Merits of the Preparation

Triphenylaluminum is useful as a component of catalyst systems for ionic or coordination polymerization of vinyl compounds. This preparation of the material in solid form enables the purity of the compound to be easily determined. The availability of solid triphenylaluminum permits the user a choice of solvents for a reaction, and a variety of concentrations of the reagent. Storage and dispensation of the reagent are more convenient in the solid form.

1. Thiokol Chemical Corporation, Huntsville, Alabama. This work was supported by the United States Army Ordnance Corp.; Contract DA-01-021-ORD-5314, Mod. 3.
2. H. Gilman and R. E. Brown, *J. Am. Chem. Soc.*, **52**, 3314 (1930).
3. C. Friedel and J. M. Crafts, *Ann. Chim. et Phys.*, [6] **14**, 460 (1888).
4. S. Hilpert and G. Grüttner, *Ber.*, **45**, 2828 (1912).
5. H. Gilman and K. E. Marple, *Rec. Trav. Chim.*, **55**, 133 (1936).
6. A. N. Nesmeyanov and N. N. Novikova, *Izv. Akad. Nauk SSSR, Ser. Khim.*, 372 (1942).
7. G. Wittig and D. Wittenberg, *Ann.*, **606**, 13 (1957).
8. E. Krause and H. Polack, *Ber.*, **59**, 777, 1428 (1926).

TRIPHENYLENE*

Submitted by H. HEANEY and I. T. MILLAR.[1]
Checked by JOHN D. ROBERTS and L. K. MONTGOMERY.

1. Procedure (Note 1)

In a 600-ml. beaker fitted with a thermometer and mechanical stirrer are placed 150 ml. of concentrated hydrochloric acid and 55 g. (0.32 mole) of o-bromoaniline. After brief stirring, 100 g. of ice is added and the beaker is surrounded by an ice-salt bath. The solution is then diazotized by the dropwise addition with stirring of a solution of 24.3 g. (0.35 mole) of sodium nitrite in 100 ml. of water, the temperature being kept at 0–5°.

After stirring the diazotized solution for 15 minutes, it is slowly poured through a glass-wool filter into a solution of 180 g. (3.4 moles) of potassium iodide in 600 ml. of water. After standing overnight, the heavy dark oil is separated, washed successively with 10% aqueous sodium hydroxide, water, 5% aqueous sodium bisulfite and water, and then dried over magnesium sulfate.

Distillation under reduced pressure gives o-bromoiodobenzene as a nearly colorless liquid, b.p. 120–121° at 15 mm. Yield 65–75 g. (72–83%).

A 1-l. flask is fitted with a reflux condenser, dropping funnel, and a sealed mechanical stirrer. A nitrogen atmosphere is maintained in the flask during the entire reaction period.

In the flask is placed 150 ml. of anhydrous ether, and 5.7 g. (4.1 atomic equivalents) of lithium foil is then added (Note 2). A solution of 56.6 g. (0.2 mole) of o-bromoiodobenzene in 300 ml. of anhydrous ether is added dropwise (Note 3). When a vigorous reaction commences, the stirrer is started and the flask is cooled in ice water to maintain the temperature at about 10°. The reflux condenser is replaced by a thermometer, and the remainder of the o-bromoiodobenzene solution is added at a rate such that the temperature in the flask remains at about 10° (about 1.5 hours). When this addition is complete, 200 ml. of dry benzene is added; the mixture is stirred at 10° for 1 hour and finally at room temperature for 1 hour. The mixture is then poured through a glass-wool filter on 200 g. of ice.

The organic layer is separated, evaporated on a steam bath, and the dark semicrystalline residue is distilled with steam to remove biphenyl. The contents of the steam-distillation flask are then extracted with ether (Note 4), and the ethereal layer is separated, dried over magnesium sulfate, and percolated through a short column of chromatographic alumina (Notes 5 and 6). Evaporation of the ethereal solution gives crude triphenylene which is sublimed at 175–180° and 0.1 mm. pressure. After rejection of an initial sublimate of impure biphenyl, the sublimed material forms nearly colorless crystals, m.p. 186–194° (Note 7). Yield 8–9 g. (53–59%). It may be further purified by recrystallization from a mixture of methylene chloride and pentane yielding colorless crystals, m.p. 199° (Note 8).

2. Notes

1. Although the procedure described here is still a useful method of preparing triphenylene, it is in the opinion of the submitters no longer the best available. In their opinion a more nearly pure product and a higher yield (85%) may be obtained by the reaction of o-bromofluorobenzene and magnesium in tetrahydrofuran[12] (private communication from H. Heaney).

2. Slugs of lithium, coated with paraffin oil, are hammered into thin foil. They are washed free of oil with dry ether and cut by scissors into slips which are allowed to fall directly into the ether in the reaction flask.

3. By adding 10–20 ml. of the o-bromoiodobenzene solution to the metal before stirring is started, high local concentrations of

the dihalo compound which initiate reaction are built up. An induction period of about 10 minutes is usually observed before the vigorous reaction commences.

4. The checkers found that, in view of the limited solubility of triphenylene in ether (about 1 g. per 100 ml.), care must be exercised in the extraction to ensure that all the product is removed.

5. The volume of ether solution must be reduced to approach the solubility limit of the triphenylene in ether before the chromatographic procedure.

6. The checkers used a 30-mm. (I.D.) chromatographic column charged with approximately 250 g. of activated (400° for 12 hours), acid-washed, chromatographic aluminum oxide (Merck and Co., Inc.).

7. The contaminant in the product of m.p. 186–194° is present in low concentration and is very probably o-terphenyl. It is characterized by an absorption near 695 cm.$^{-1}$ in the infrared which is absent in the spectrum of pure triphenylene.

The submitters report that the purity of the product is rather dependent upon the purity of the lithium used and that results can vary from batch to batch. In addition, they state that side reactions may be catalyzed by traces of heavy metals, that the degree of vigor in the initial reaction may influence the purity of the product owing to local overheating, and that increase in scale of this reaction is deleterious.

8. The checkers carried out this recrystallization by dissolving triphenylene in a minimum of methylene chloride maintained at reflux. Pentane was slowly added to this solution; up to 90% recovery was achieved.

3. Methods of Preparation

Triphenylene has been prepared by self-condensation of cyclohexanone using sulfuric acid [2] or polyphosphoric acid [3] followed by dehydrogenation of the product, dodecahydrotriphenylene, using copper,[2] palladium-charcoal,[3] or selenium; [4] by electrolytic oxidation of cyclohexanone; [5] from chlorobenzene and sodium [6] or phenyllithium; [7] from 2-cyclohexyl-1-phenylcyclohexanol [8] or

2-(1-cyclohexene-1-yl)-1-phenylcyclohexanol [3,9] by dehydrogenation; from 9-phenanthryl magnesium bromide and succinic anhydride followed by reduction, cyclization, and dehydrogenation;[10] by the action of lithium on o-diiodobenzene[11] or magnesium on o-bromofluorobenzene,[12] and by the dehydrogenation of o-terphenyl.[13]

1. University College of North Straffordshire, Keele, Stratfordshire, England.
2. C. Mannich, *Ber.,* **40,** 153 (1907).
3. P. M. G. Bavin and M. J. S. Dewar, *J. Chem. Soc.,* 4479 (1955); see also J. Plěsek and P. Munk, *Chem. Listy,* **51,** 980 (1957), and C. C. Barker, R. G. Emmerson, and J. D. Periam, *J. Chem. Soc.,* 1077 (1958).
4. O. Diels and A. Karstens, *Ber.,* **60,** 2323 (1927).
5. F. Pirrone, *Gazz. Chim. Ital.,* **66,** 244 (1936).
6. W. E. Bachmann and H. T. Clarke, *J. Am. Chem. Soc.,* **49,** 2089 (1927).
7. G. Wittig and W. Merkle, *Ber.,* **75,** 1493 (1942).
8. C. D. Nenitzescu and D. Curcaneanu, *Ber.,* **70,** 346 (1937).
9. W. S. Rapson, *J. Chem. Soc.,* 15 (1941).
10. E. Bergmann and O. Blum-Bergmann, *J. Am. Chem. Soc.,* **59,** 1441 (1937).
11. H. Heaney, F. G. Mann, and I. T. Millar, *J. Chem. Soc.,* 1 (1956).
12. K. D. Bartle, H. Heaney, D. W. Jones, and P. Lees, *Tetrahedron,* **21,** 3289 (1965).
13. P. G. Copeland, R. E. Dean, and D. McNeil, *J. Chem. Soc.,* 1687 (1960).

2,3,5-TRIPHENYLISOXAZOLIDINE

(Isoxazolidine, 2,3,5-triphenyl-)

$$C_6H_5CHO \quad + \quad C_6H_5NHOH \quad \longrightarrow \quad \underset{C_6H_5}{\overset{H}{\diagdown}}C \overset{+}{=} \underset{O^-}{\overset{C_6H_5}{N}}$$

$$\underset{C_6H_5}{\overset{H}{\diagdown}}C \overset{+}{=} \underset{O^-}{\overset{C_6H_5}{N}} \quad + \quad H_2C = CHC_6H_5 \quad \longrightarrow$$

Submitted by Ingrid Brüning, Rudolf Grashey, Hans Hauck, Rolf Huisgen, and Helmut Seidl [1]
Checked by George E. Davis, Wayland E. Noland, and William E. Parham

1. Procedure

A. *N,α-Diphenylnitrone.* A solution of 27.3 g. (0.25 mole) of pure N-phenylhydroxylamine [2] (Note 1) in 50 ml. of ethanol is prepared in a 200-ml. Erlenmeyer flask by swirling a mixture of the two and warming it briefly to 40–60° (Note 2). To the clear, lightly colored solution is added 26.5 g. (25.3 ml., 0.25 mole) of freshly distilled benzaldehyde (exothermic reaction). The flask is stoppered and kept overnight at room temperature in the dark. The colorless needles of N,α-diphenylnitrone are collected on a Büchner funnel and washed once with 20 ml. of ethanol. There is obtained 42–43 g. (85–87%) of product (m.p. 111–113°), which can be further purified by dissolving the crude material in 80 ml. of ethanol and allowing the solution to cool for several hours in the ice box. In this manner there is produced 35–39 g. (71–79%) of pure crystalline nitrone, m.p. 113–114° (Note 3).

B. *2,3,5-Triphenylisoxazolidine.* In a 100-ml. two-necked flask provided with a reflux condenser and a gas-inlet tube are placed 20.0 g. (0.101 mole) of pure N,α-diphenylnitrone and 50 ml. (0.43 mole) of freshly distilled styrene (Note 4). The flask is heated at 60° for 40 hours under a slow nitrogen stream. The mixture is then cooled, and most of the excess styrene is removed (Note 5) from the clear orange solution by heating at a bath temperature of 55° (12 mm.). The warm residue is poured into 40 ml. of petroleum ether (40–60°), whereupon the isoxazolidine crystallizes immediately (Note 6). The flask is rinsed twice with 20-ml. portions of petroleum ether, and the washings are combined with the product. The resulting mixture is cooled for 1 hour in the ice box, and the lightly colored crystals are collected on a Büchner funnel and washed with two 20-ml. portions of petroleum ether. The yield of crude air-dried isoxazolidine (m.p. 96–98°) is 28–30 g. (92–99%).

For further purification the product is dissolved in 40 ml. of methylene chloride in a 250-ml. Erlenmeyer flask. The solution is heated to boiling, and 30 ml. of methanol is added (Note 7). When the solution has cooled to room temperature, 70 ml. of methanol is added to complete the crystallization, and the solution is kept in the ice box for 3 hours. The colorless needles are collected by vacuum filtration and washed with two 30-ml. portions of cold methanol. There is obtained 23–25 g. (76–82%) of product which melts at 99–100° (Notes 8–10).

2. Notes

1. The N-phenylhydroxylamine should be free of sodium chloride. This is easily attained by dissolution of the compound in benzene followed by filtration and then addition of petroleum ether to cause rapid crystallization.

2. On prolonged heating, N-phenylhydroxylamine begins to decompose.

3. The compound is light-sensitive and should be kept in a brown container.

4. The styrene should be redistilled and stabilized with 0.1% hydroquinone just prior to use; otherwise the final product will be contaminated with polystyrene. The checkers used approximately 60 ml. (0.52 mole) of styrene.

5. The checkers found that, if all the styrene is removed, the product may become too viscous to pour.

6. By this method the formation of a thick crustaceous material, which is difficult to pulverize or wash, is avoided.

7. In this way the boiling solution is kept clear.

8. From the mother liquor a second diastereoisomer can be isolated (m.p. 78.5–79.5°) in about 10% yield by fractional crystallization.

9. In an analogous manner several other isoxazolidines can be prepared. From the reaction of N,α-diphenylnitrone with 1,1-diphenylethylene, 2,3,5,5-tetraphenylisoxazolidine is obtained. As above, 10.0 g. (50.7 mmoles) of diphenylnitrone is heated under a nitrogen atmosphere for 24 hours at 85° with 15.3 g. (15.0 ml., 85.0 mmole) of 1,1-diphenylethylene.[3] The excess olefin is removed at 105–130° (bath temperature) under high vacuum (0.005 mm.). The yellow-gold viscous residue is dissolved by warming it in a mixture of 15 ml. of methylene chloride and 30 ml. of methanol; on cooling, crystallization commences. After 2 hours another 10 ml. of methanol is added, and the mixture is cooled overnight in the ice box. The colorless crystals are collected on a Büchner funnel and washed twice with 20-ml. portions of methanol. The yield of air-dried product (m.p. 113–115°) is 14–16 g. (73–84%). The compound can be further purified by adding methanol (30 ml.) to a boiling solution in methylene chloride (15 ml.). After the solution has cooled to room temperature, another 10 ml. of methanol is added; the mixture is kept in the ice box for several hours and then filtered. The pure compound melts at 115–116°, yield 13–15 g. (68–79%).

10. The preparation of the oily ethyl 2,3-diphenyl-5-methylisoxazolidine-4-carboxylate provides another example of this reaction. As in the procedure described with styrene, 10.0 g. (50.7 mmoles) of N,α-diphenylnitrone is heated under nitrogen for 24 hours at 90–100° with 35.0 g. (38.0 ml., 307 mmoles) of ethyl crotonate. The excess olefin, b.p. 45° (12 mm.) is removed on the water pump, and the red-orange residue, while still warm, is transferred to a 50-ml. Claisen flask using acetone as a rinse. After removal of the solvent, 13–14 g. (82–88%) of the isoxazolidine is obtained as an orange oil by high-vacuum distillation at 163–173° (0.003 mm.). Redistillation of this material yields 2–3 g. of fore-run and a purer product obtained as a yellow oil, b.p. 165–170° (0.003 mm.), n^{20}D 1.5602–1.5612.

3. Methods of Preparation

N,α-Diphenylnitrone was first obtained by Bamberger [4] from N-phenylhydroxylamine and benzaldehyde. The procedure described above is analogous to that of Wheeler and Gore.[5]

2,3,5-Triphenylisoxazolidine, 2,3,5,5-tetraphenylisoxazolidine, and ethyl 2,3-diphenyl-5-methylisoxazolidine-4-carboxylate have been prepared only by this method.[6]

4. Merits of the Preparation

The procedure described illustrates the use of 1,3-dipolar addition [7] of nitrones to olefins for the preparation of isoxazolidines. The preparations of 2,3,5,5-tetraphenylisoxazolidine and ethyl 2,3-diphenyl-5-methylisoxazolidine-4-carboxylate, as described in Notes 9 and 10, respectively, indicate the versatility of the method.

1. Institut für Organische Chemie der Universität München, München, Germany.
2. O. Kamm, *Org. Syntheses*, Coll. Vol. **1**, 445 (1941).
3. C. F. H. Allen and S. Converse, *Org. Syntheses*, Coll. Vol. **1**, 226 (1941).
4. E. Bamberger, *Ber.*, **27**, 1548 (1894).
5. O. H. Wheeler and P. H. Gore, *J. Am. Chem. Soc.*, **78**, 3363 (1956).
6. H. Hauck, Dissertation, Universität München, 1963.
7. R. Huisgen, *Angew. Chem.*, **75**, 604 (1963); *Angew. Chem. Intern. Ed.*, **2**, 565 (1963).

2,4,6-TRIPHENYLNITROBENZENE

(Benzene, 2-nitro-1,3,5-triphenyl-)

Submitted by K. Dimroth, A. Berndt, and C. Reichardt [1]
Checked by Saul Cherkofsky and Richard E. Benson

1. Procedure

In a 1-l. three-necked flask equipped with a mechanical stirrer, a reflux condenser, and a dropping funnel are placed 119 g. (0.30 mole) of 2,4,6-triphenylpyrylium tetrafluoroborate (Note 1), 21 ml. (24 g., 0.39 mole) of nitromethane (Note 2), and 350 ml. of absolute ethanol (Note 3). Triethylamine (70 ml., 51 g.) (Note 4) is added rapidly from the dropping funnel to the well-stirred suspension. The reaction mixture becomes reddish brown immediately, and the solid dissolves. After all the triethylamine has been added, the mixture is heated under reflux for 3 hours, cooled, and allowed to stand overnight in a refrigerator. The crystalline product that separates is collected on a Buchner funnel and washed with two 50-ml. portions of ice-cold methanol. The product (75–80 g.; m.p. 142–144°) is recrystallized from 200–250 ml. of glacial acetic acid to yield 70–75 g. (67–71%) of 2,4,6-triphenylnitrobenzene as slightly yellow crystals, m.p. 144–145° (Note 5).

2. Notes

1. The preparation of 2,4,6-triphenylpyrylium tetrafluoroborate is described on p. 1135.

2. Nitromethane is dried over anhydrous calcium sulfate (Drierite) or calcium chloride for 1 day and distilled; the fraction with b.p. 101.5–102.5° is used.

3. Commercial absolute ethanol is used without additional drying.

4. Triethylamine is dried over sodium hydroxide pellets and distilled; the fraction with b.p. 89.5–90° is used.

5. The n.m.r. spectrum (CDCl₃) shows singlets at 7.45 p.p.m. (15 H) and 7.65 p.p.m. (2 H) (downfield from internal tetramethylsilane reference).

3. Discussion

2,4,6-Triphenylnitrobenzene may be prepared by direct nitration of 1,3,5-triphenylbenzene[2-4] and by the reaction of 2,4,6-triphenylpyrylium tetrafluoroborate with nitromethane.[5] The present procedure is an adaptation of the latter method.

This procedure illustrates a general method for converting substituted pyrylium salts to nitrobenzene derivatives. The reaction has been the subject of several reviews.[6-8] The yields are generally high, and under these conditions only a single product is formed, in contrast to the nitration of 1,3,5-triphenylbenzene. The preparation of 2,4,6-triphenylnitrobenzene from the corresponding pyrylium salt eliminates isomer separation problems, which are encountered when the direct nitration procedure is used. Also, labeled compounds can readily be prepared by this method.[9]

1. Institut für Organische Chemie der Philipps-Universität Marburg (Lahn), Germany.
2. D. Vorländer, Z. Physik. Chem., **105**, 211 (1923); D. Vorländer, E. Fischer, and H. Wille, Ber., **62**, 2836 (1929).
3. K. Dimroth, G. Bräuniger, and G. Neubauer, Ber., **90**, 1634 (1957).
4. G. E. Lewis, J. Org. Chem., **30**, 2798 (1965).
5. K. Dimroth and G. Bräuniger, Angew. Chem., **68**, 519 (1956); K. Dimroth, G. Neubauer, H. Möllenkamp, and G. Oosterloo, Ber., **90**, 1668 (1957).
6. K. Dimroth, Angew. Chem., **72**, 331 (1960).
7. K. Dimroth and K. H. Wolf, in W. Foerst, "Newer Methods of Preparative Organic Chemistry," Vol. 3, Academic Press, Inc., New York, 1964, p. 357.
8. K. Dimroth, W. Krafft, and K. H. Wolf, in T. Urbánski, "Nitro Compounds," Pergamon Press, Oxford, 1964, p. 361 [Tetrahedron, **20**, Suppl. **1**, 361 (1964)].
9. K. Dimroth, A. Berndt, and R. Volland, Ber., **99**, 3040 (1966).

2,4,6-TRIPHENYLPHENOXYL

(Phenoxy, 2,4,6-triphenyl)

Submitted by K. DIMROTH, A. BERNDT, H. PERST
and C. REICHARDT [1]
Checked by E. K. W. WAT and R. E. BENSON

1. Procedure

A. *2,4,6-Triphenylaniline.* To a filtered solution of 70 g. (0.20 mole) of 2,4,6-triphenylnitrobenzene (Note 1) in 500 ml. of dioxane (total volume *ca.* 540 ml.) in a 1-l. pressure vessel equipped with a magnetic stirrer is added 10 g. of Raney nickel catalyst (Note 2) that has been previously rinsed with absolute ethanol. The head and fittings are attached, and the vessel is connected to a hydrogen cylinder. The system is alternately evacuated to 40–50 mm. and pressured with hydrogen to 30–40 p.s.i. three times (Note 3). After a final evacuation, hydrogen is introduced into the vessel until the pressure reaches 1000 p.s.i. (*ca.* 70 atm.). The reaction is allowed to proceed overnight (*ca.* 25 hours) (Note 4), during which time 0.6 mole (13.5 l.) of hydrogen is absorbed. The vessel is vented and the catalyst is removed by filtration from the reaction mixture and is washed with 30 ml. of dioxane. The filtrates are combined, and the solvent is removed by distillation under reduced pressure using a rotary evaporator at 40–50° (50 mm.) to leave an oil which solidifies on trituration with a small portion of methanol. The product is collected on a Buchner funnel and washed twice with 40-ml. portions of ice-cold methanol. The remaining light yellow 2,4,6-triphenylaniline (m.p. 135–136°) weighs 60–63 g. (94–98%) (Note 5).

B. *2,4,6-Triphenylphenol.* To a 2-l. three-necked flask equipped with a mechanical stirrer, a dropping funnel, and a thermometer are added 32 g. (0.10 mole) of 2,4,6-triphenylaniline and 300 ml. of glacial acetic acid. Stirring is begun and the contents of the flask are brought into solution by heating to 70°. Concentrated sulfuric acid (90 ml., *d* 1.84) is added dropwise while the temperature is lowered concurrently from 70° to 20° by cooling. After the addition is completed, the mixture is cooled to 0° with an ice-salt bath, and a solution of 9 g. (0.13 mole) of sodium nitrite in 50 ml. of water is added over a period of 20–30 minutes with stirring, the reaction temperature being kept at 0–5°. The stirring is continued for 20 minutes after the sodium nitrite solution has been added, then 300 ml. of ice-cold water and 3 g. of urea or amidosulphonic acid are added in small portions. The yellow diazonium salt solution is filtered with suction into an ice-cold flask and is kept cold (at 0°) while the next step is carried out.

To a 2-l. three-necked flask equipped with a mechanical stirrer, a reflux condenser, and a dropping funnel is added a mixture of 600 ml. of water and 150 ml. of concentrated sulfuric acid (*d* 1.84). The acid solution is vigorously stirred and heated to boiling, and the cold diazonium salt solution is added at such a rate that the boiling is not interrupted (Note 6). The time required for this addition should not exceed 30 minutes. After the addition is completed, boiling is continued for 10 minutes and then the mixture is allowed to cool to room temperature with stirring. The product is collected on a Buchner funnel, washed with water, and dried in a vacuum desiccator containing phosphorus pentoxide to give 28–32 g. of crude 2,4,6-triphenylphenol. After drying, the product is dissolved in *ca*. 200 ml. of benzene and filtered through a layer of 300 g. of alumina packed in a 30-mm. x 75-cm. chromatography column (Note 7). The product is eluted with benzene until about 500 ml. of eluate has been collected. The collected eluate is concentrated under reduced pressure using a rotary evaporator at 50 mm. pressure and 50° to yield light yellow crystals, which are recrystallized from glacial acetic acid (10 g. of 2,4,6-triphenylphenol requires *ca*. 30–35 ml. of acetic acid). The pure, nearly colorless product (13–15 g., 39–47%) melts at 149–150°.

C. *2,4,6-Triphenylphenoxyl.* In a 1-l. separatory funnel is placed a filtered solution of 10 g. (31 mmoles) of 2,4,6-triphenylphenol in 300 ml. of ether. To this solution is added 60 ml. of a filtered, saturated solution of potassium hexacyanoferrate(III) in 2N sodium hydroxide solution (Note 8), and the resulting mixture is vigorously shaken for about 10 minutes. After a few minutes the dimer of 2,4,6-triphenylphenoxyl begins to separate in the form of pink crystals. The crystals are isolated by filtration, washed with several portions of water (Note 9) and twice with ether. After drying in a vacuum desiccator over phosphorus pentoxide while protecting from light, the product weighs 8–9 g. (81–91%) and melts at 145–150° to a red liquid (Note 10). The purity of the 2,4,6-triphenylphenoxyl dimer (which in solution attains a rapid equilibrium with its red monomer) is established by titration with a solution of hydroquinone in acetone (Note 11). The radical titer of a freshly prepared solution of the dimer in benzene or acetone should be 98–99%.

2. Notes

1. The preparation of 2,4,6-triphenylnitrobenzene is described in *Organic Syntheses*, this volume, p. 1128.

2. The submitters used Raney nickel catalyst from the Badische Anilin- & Sodafabrik AG, Ludwigshafen (Rhein), Germany.

3. The checkers used a stainless steel pressure vessel that was cooled to −60° and then evacuated to 1 mm. The cold system was

purged with hydrogen three times, evacuated, and placed in the rocker assembly before pressuring it to 1000 p.s.i. with hydrogen.

4. The checkers used a rocking-motion autoclave and the reduction required 48 hours to complete.

5. The 2,4,6-triphenylaniline resulting from this procedure is sufficiently pure for use in the preparation of 2,4,6-triphenylphenol, but it may be recrystallized from 100 ml. of glacial acetic acid to give 50–55 g. (78–86%) of pure 2,4,6-triphenylaniline, m.p. 136–137°. On occasions a product with initial m.p. 121–122° is obtained which solidifies on further heating and then melts at 136–137°

6. Contact of the diazonium salt solution with the hot wall of the flask before decomposition in the solution should be avoided in order to prevent the formation of a brown resin.

7. The submitters used Aluminiumoxid WOELM neutral, Aktivitätsstufe I.

8. A saturated solution requires ca./35 g. (110 mmoles) of potassium hexacyanoferrate(III) per 100 ml. of $2N$ sodium hydroxide solution at room temperature.

9. The final wash water must be free of potassium hexacyanoferrate(III). The checkers washed the product until the filtrates were colorless.

10. The product obtained is analytically pure: Calcd. for $C_{24}H_{17}O$: C, 89.69; H, 5.33; O, 4.98. Found: C, 89.94; H, 5.27; O, 5.00. The product is stable for several months when stored in the dark. The 2,4,6-triphenylphenoxyl dimer is piezochromic; rubbing in a mortar produces a red color. Solutions of the colorless dimer in organic solvents are red owing to dissociation to the monomer radical.

11. The radical solution is titrated with $0.01M$ solution of analytically pure hydroquinone in pure acetone. The end point of the titration is marked by disappearance of the red color of the phenoxyl radical: 1 ml. of $0.01M$ hydroquinone solution is equivalent to 6.428 mg. of 2,4,6-triphenylphenoxyl dimer.

3. Discussion

The procedure for preparing 2,4,6-triphenylphenoxyl is based on the method described by Dimroth and co-workers.[2] This method represents the commonly used preparation of aroxyl radicals by oxidation of the corresponding phenol.[3] The chemistry of stable phenoxyl radicals has been reviewed.[4]

In solution the colorless 2,4,6-triphenylphenoxyl dimer attains a rapid equilibrium with its red monomer radical (dissociation constant in benzene 4×10^{-5} at 20°). The radical is surprisingly stable toward oxygen and can be stored in solution for a long time when it is protected from light. The stability of the 2,4,6-triphenylphenoxyl radical is ascribed to steric and mesomeric effects.[2, 5] The e.s.r. spectrum[5] and an ENDOR-spectrum[6] of the radical are described.

The dimer belongs to the rare group of compounds which are piezochromic. Rubbing in a mortar produces a red color due to mechanical bond-breaking and dissociation into the red-colored monomer. The p-quinol structure of the 2,4,6-triphenylphenoxyl dimer has been confirmed by infrared studies of ^{18}O-labeled material[7] and by X-ray analysis of 3-bromo derivative.[8]

Other aroxyl radicals, especially those with t-butyl groups at the phenyl ring, are described by Cook[9] and by Müller.[10]

1. Institüt für Organische Chemie der Philipps-Universität Marburg (Lahn), Germany.
2. K. Dimroth and G. Neubauer, *Angew. Chem.*, **69**, 95 (1957); K. Dimroth, F. Kalk, and G. Neubauer, *Ber.*, **90**, 2058 (1957); K. Dimroth, F. Kalk, R. Sell, and K. Schlömer, *Ann. Chem.*, **624**, 51 (1959); K. Dimroth, *Angew. Chem.*, **72**, 714 (1960); K. Dimroth and G. **Laubert**, *Angew. Chem.*, **81**, 392 (1969).
3. For a review see H. Musso, *Angew. Chem.*, **75**, 965 (1963).
4. E. R. Altwicker, *Chem. Rev.*, **67**, 475 (1967).
5. K. Dimroth, A. Berndt, F. Bär, R. Volland, and A. Schweig, *Angew. Chem.*, **79**, 69 (1967); *Angew. Chem. Intern. Ed. Engl.*, **6**, 34 (1967).
6. J. S. Hyde, *J. Phys. Chem.*, **71**, 68 (1967).
7. K. Dimroth and A. Berndt, *Angew. Chem.*, **76**, 434 (1964); *Angew. Chem., Intern. Ed. Engl.*, **3**, 385 (1964); K. Dimroth, A Berndt, and R. Volland, *Ber.*, **99**, 3040 (1966); K. Dimroth and A. Berndt, *Ber.*, **101**, 2519 (1968).
8. R. Allmann and E. Hellner, *Ber.*, **101**, 2522 (1968).
9. C. D. Cook, *J. Org. Chem.*, **18**, 261 (1953); C. D. Cook and R. C. Woodworth, *J. Am. Chem. Soc.*, **75**, 6242 (1953); C. D. Cook and N. D. Gilmour, *J. Org. Chem.*, **25**, 1429 (1960).
10. E. Müller and K. Ley, *Z. Naturforsch.*, **8b**, 694 (1953); *Ber.*, **87**, 922 (1954); *Chemiker-Ztg.*, **80**, 618 (1956); E. Müller, H. Eggensperger, A. Rieker, K. Scheffler, H.-D. Spanagel, H. B. Stegmann, and B. Teissier, *Tetrahedron*, **21**, 227 (1965).

2,4,6-TRIPHENYLPYRYLIUM TETRAFLUOROBORATE

(Pyrylium tetrafluoroborate, 2,4,6-triphenyl-)

$$2C_6H_5CH\!\!=\!\!CHCOC_6H_5 + C_6H_5COCH_3 + HBF_4, \longrightarrow$$

$$+ C_6H_5CH_2CH_2COC_6H_5 + H_2O$$

Submitted by K. DIMROTH, C. REICHARDT, and K. VOGEL [1]
Checked by SAUL CHERKOFSKY and RICHARD E. BENSON

1. Procedure

In a 1-l. four-necked flask (or a three-necked flask with a Y-tube connector) equipped with a mechanical stirrer, a reflux condenser, a dropping funnel, and a thermometer are placed 208 g. (1.00 mole) of benzalacetophenone (Note 1), 60 g. (58.5 ml., 0.50 mole) of acetophenone, and 350 ml. of 1,2-dichloroethane. The contents of the flask are warmed to 70–75°, and 160 ml. of a 52% ethereal solution of fluoboric acid (Note 2) is added from the funnel with stirring during 30 minutes. With the first addition the mixture becomes orange; subsequently the color changes to brownish yellow. After the addition is completed, the mixture is stirred and heated under reflux for 1 hour (Note 3). The fluorescent mixture is allowed to stand overnight in a re-frigerator. The crystalline product that separates is collected on a Buchner funnel and washed well with ether. By addition of 250 ml. of ether (Note 4) to the mother liquor an additional quantity of 2,4,6-triphenylpyrylium tetrafluoroborate is ob-tained. A total yield of 125–135 g. (63–68%) of yellow crystals results; m.p. 218–225° (Note 5). The product can be recrystal-lized from 650–700 ml. of 1,2-dichloroethane, when it separates in the form of yellow needles, m.p. 251–257° (Note 6). The yield of product dried at 80° (10 mm.) for 3 hours is 102.5–107 g. (52–54%) (Note 6).

2. Notes

1. The preparation of benzalacetophenone is described in *Org. Syntheses*, Coll. Vol. **1**, 78 (1941).

2. Ethereal fluoboric acid can be prepared as follows: 19 ml. (19 g., 0.95 mole) of anhydrous hydrofluoric acid, b.p. 19.4° (760 mm.) [*Caution! Hydrofluoric acid in contact with the skin produces extremely painful burns. It is therefore necessary to use every precaution to protect exposed parts of the body, especially the hands and eyes. Cf. Org. Syntheses, Coll. Vol. **2**, 295 (1943), Note 3; this volume, p. 136, Note 1*] is added in small portions with shaking or stirring to 126 ml. (142 g., 1.00 mole) of distilled boron trifluoride etherate, b.p. 126° (760 mm.), contained in a 500-ml. polyethylene flask that is cooled in an ice bath to 0°. The concentration of the resulting yellowish solution of fluoboric acid in ether is about 52% by weight (*ca.* 6.6 moles per l.).

3. Some boron trifluoride is evolved during the first part of the refluxing; it may be disposed of by absorption in water in a gas trap [cf. *Org. Syntheses*, Coll. Vol. **2**, 3 (1943)].

4. The ether used for washing the product may be added to the filtrate.

5. The 2,4,6-triphenylpyrylium tetrafluoroborate resulting from this procedure is sufficiently pure for use in the preparation of 2,4,6-triphenylnitrobenzene.[2]

6. It is necessary to dry under reduced pressure in order to remove that portion of the solvent that is tightly held. *Anal.* Calcd. for $C_{23}H_{17}BF_4O$: C, 69.73; H, 4.33; B, 2.73; F, 19.18. Found: C, 69.38; H, 4.47; B, 3.07; F, 19.51. The n.m.r. spectrum (acetone-d_6) shows a singlet at 9.1 p.p.m. (2 H) and multiplets at 8.6 p.p.m. and 7.9 p.p.m. (15 H) (downfield from internal tetramethylsilane reference).

3. Discussion

The present procedure is an improved modification of that described by Balaban[3] for the corresponding perchlorate. 2,4,6-Triphenylpyrylium tetrafluoroborate has also been prepared from the corresponding tetrachloroferrate[4,5] with fluoboric acid,[5] from acetophenone and boron trifluoride,[6] and from acetophenone, benzaldehyde, and boron trifluoride etherate.[7] Additional methods for the preparation of pyrylium salts have been reviewed.[5, 8-13, 21]

2,4,6-Triphenylpyrylium tetrafluoroborate is a versatile and useful stable starting material. Its reaction with nitromethane under basic conditions has made 2,4,6-triphenylnitrobenzene easily available.[2, 14] In addition, pyrylium salts are readily converted to a variety of pyridine derivatives [15, 16, 20] including alkyl- and arylpyridinium salts,[16, 20] to thiopyrylium salts,[17] and to substituted azulenes.[18]

The chemistry and transformation of pyrylium salts have been reviewed.[5, 8-11, 19, 21]

1. Institut für Organische Chemie der Philipps-Universität Marburg (Lahn), Germany.
2. K. Dimroth, A. Berndt, and C. Reichardt, *Org. Syntheses*, this volume, p. 114.
3. A. T. Balaban, *Compt. Rend.*, **256**, 4239 (1963).
4. W. Dilthey, *J. Prakt. Chem.*, [2] **94**, 53 (1916).
5. K. Dimroth, *Angew. Chem.*, **72**, 331 (1960).
6. W. C. Dovey and R. Robinson, *J. Chem. Soc.*, 1389 (1935); R. C. Elderfield and T. P. King, *J. Am. Chem. Soc.*, **76**, 5437 (1954).
7. R. Lombard and J.-P. Stephan, *Bull. Soc. Chim. France*, 1458 (1958).
8. A. T. Balaban and C. D. Nenitzescu, *Rev. Chim. Acad. Rep. Populaire Roumaine*, **6**, 269 (1961); *Acad. Rep. Populare Romine, Studii Cercetari Chim.*, **0**, 251 (1961) [*Chem. Abstr.*, **57**, 8534 (1962)].
9. K. Dimroth and K. H. Wolf, in W. Foerst, "Newer Methods of Preparative Organic Chemistry," Vol. 3, Academic Press, Inc., New York, 1964, p. 357.
10. W. Schroth and G. Fischer, *Z. Chem.*, **4**, 281 (1964).
11. H. Meerwein, in E. Müller, "Methoden der Organischen Chemie (Houben-Weyl)," Vol. 6, Part 3, Georg Thieme Verlag, Stuttgart, 1965, p. 325.
12. G. N. Dorofeenko, S. V. Krivun, V. I. Dulenko, and Yu. A. Zhdanov, *Usp. Khim.*, **34**, 219 (1965); *Russ. Chem. Rev. (English Transl.)*, 88 (1965) [*Chem. Abstr.*, **62**, 12993 (1965)].
13. A. T. Balaban and C. D. Nenitzescu, this volume, p. 1106.
14. K. Dimroth, G. Neubauer, H. Möllenkamp, and G. Oosterloo, *Ber.*, **90**, 1668 (1957).
15. A. Baeyer, *Ber.*, **43**, 2337 (1910); W. Dilthey, *J. Prakt. Chem.*, [2] **102**, 209 (1921).
16. W. Schneider, W. Döbling, and R. Cordua, *Ber.*, **70**, 1645 (1937); K. Dimroth, C. Reichardt, T. Siepmann, and F. Bohlmann, *Ann.*, **661**, 1 (1963).
17. R. Wizinger and P. Ulrich, *Helv. Chim. Acta*, **39**, 207 (1956).
18. K. Hafner and H. Kaiser, *Ann.*, **618**, 140 (1958); this volume, p. 1088.
19. A. T. Balaban and C. D. Nenitzescu, *Ann.*, **625**, 74 (1959).
20. A. T. Balaban and C. Toma, *Tetrahedron*, Suppl. **7**, 1 (1966); C. Toma and A. T. Balaban, *Tetrahedron*, Suppl. **7**, 9, 27 (1966).
21. A. T. Balaban, W. Schroth, and G. Fischer, in A. R. Katritzky and A. J. Boulton, *Advan. Heterocycl. Chem.*, **10**, 241 (1969).

TROPYLIUM FLUOBORATE*

(Cycloheptatrienocarbonium fluoborate)

$$2 \quad \bigcirc \quad + \quad 3PCl_5 \quad \longrightarrow \quad (C_7H_7{}^+ PCl_6{}^-)(C_7H_7{}^+ Cl^-) + 2PCl_3 \quad + \quad 2HCl$$

$$(C_7H_7{}^+ PCl_6{}^-)(C_7H_7{}^+ Cl^-) \quad + \quad 2HBF_4 \quad \longrightarrow \quad 2C_7H_7{}^+ BF_4{}^- + PCl_5 + 2HCl$$

Note: $C_7H_7{}^+$ is $\bigcirc{}^+$

Submitted by KENNETH CONROW.[1]
Checked by D. W. WILEY and B. C. McKUSICK.

1. Procedure

A suspension of 100 g. (0.48 mole, 33% excess) of phosphorus pentachloride in 800 ml. of carbon tetrachloride is prepared in a 1-l. flask equipped with an efficient stirrer and an exit valve for the hydrogen chloride that is evolved (Note 1). Tropilidene (cycloheptatriene; 24.2 g. of 91% material; 0.24 mole) (Note 2) is added all at once, and the mixture is stirred for 3 hours at room temperature (Note 3).

Absolute ethanol (400 ml.) is vigorously stirred in a 1-l. wide-necked Erlenmeyer flask immersed in an ice bath (Note 4). The tropylium hexachlorophosphate-tropylium chloride double salt [2] is separated from the reaction mixture by suction filtration, washed briefly with fresh carbon tetrachloride, and transferred as rapidly as possible into the cold, well-stirred ethanol (Note 5). The salt dissolves rapidly and exothermally to give a reddish solution. Fifty milliliters (0.39 mole) of 50% aqueous fluoboric acid is added rapidly to the cold stirred solution (Note 6). The dense white precipitate of tropylium fluoborate that forms is separated by suction filtration, washed with a little cold ethanol and with ether, and air-dried at room temperature (Note 7); weight 34–38 g. (80–89%); decomposition point about 200°; $\lambda_{max}^{0.1N\ HCl}$ 218 mμ (log ϵ 4.70), 274 mμ (log ϵ 3.61). The product is 98–100% pure (Notes 8 and 9).

2. Notes

1. The use of a flask just large enough to hold the reaction mixture obviates the necessity for an inert atmosphere, for the evolving hydrogen chloride soon displaces the small amount of air over the reaction mixture.

2. Cycloheptatriene containing 9% toluene is available from the Shell Chemical Company, New York. Less pure cycloheptatriene, obtained by pyrolysis of bicycloheptadiene followed by a crude distillation, has been used successfully in this preparation. The quantity of the tropilidene/toluene mixture is adjusted in accord with its purity as estimated by vapor-phase chromatography on didecyl phthalate.

3. The mixture thickens rapidly. After about an hour, even an efficient stirrer often fails to stir the whole mixture, but after a time the mixture thins again and the reaction is completed without incident.

4. Stirring is most conveniently accomplished with a magnetic stirrer. A large plastic bucket is used to contain the ice used for cooling.

5. Exposure of this salt to the atmosphere causes discoloration that may persist in the final product. A slight discoloration at this stage does not appear to affect the quality of the final product. A rubber dam is helpful on days of high humidity.

6. Use of perchloric acid gives the perchlorate. However, the perchlorate is so dangerously explosive that its use should be avoided.[3]

7. Additional salt is precipitated by the addition of ether to the ethanolic filtrate, but the quantity is so small that this treatment is not worth while.

8. The product may be crystallized from a large volume of ethyl acetate or from acetonitrile-ethyl acetate. However, there

is little reason to do this, for losses are heavy and the purity, as measured by ultraviolet spectroscopy, is hardly affected.

9. In a variation of this procedure that gives a nearly quantitative yield of good material, the intermediate salt is dissolved in 250 ml. of glacial acetic acid in a 2-l. beaker, and 100 g. of 50% fluoboric acid is added with stirring. When the evolution of gas has stopped, 1 l. of ethyl acetate is added to precipitate tropylium fluoborate. The fluoborate is separated by filtration, washed successively with ethyl acetate and ether, and dried in an oven at 40°.[4]

3. Methods of Preparation

This method is a modification of the method originally published by Kursanov and Vol'pin.[5] Tropylium salts have also been prepared by bromination-dehydrobromination of tropilidene,[6] and by the hydride-exchange reaction between tropilidene and triphenylmethyl carbonium ion.[7]

4. Merits of the Preparation

Tropylium salts are starting materials for the preparation of a wide range of substituted tropilidenes. The fluoborate is the salt of choice for work involving the tropylium ion because it is indefinitely stable, non-hygroscopic, and, unlike the perchlorate, non-explosive. Its preparation by this method avoids the use of triphenyl carbinol, which is an unnecessarily expensive reagent in the quantities required for tropylium ion preparation.

1. Department of Chemistry, Kansas State University, Manhattan, Kansas.
2. D. Bryce-Smith and N. A. Perkins, *J. Chem. Soc.*, 1339 (1962).
3. P. G. Ferrini and A. Marxer, *Angew. Chem., Intern. Ed. Engl.*, 1, 405 (1962).
4. C. V. Wilson, Eastman Kodak Co., private communication.
5. D. N. Kursanov and M. E. Vol'pin, *Doklady Akad. Nauk SSSR*, 113, 339 (1957) [*C.A.*, 51, 14572f (1957)].
6. W. von E. Doering and L. H. Knox, *J. Am. Chem. Soc.*, 79, 352 (1957).
7. H. J. Dauben, L. R. Honnen, and K. M. Harmon, *J. Org. Chem.*, 25, 1442 (1960).

UNSOLVATED *n*-BUTYLMAGNESIUM CHLORIDE

$$CH_3CH_2CH_2CH_2Cl + Mg \xrightarrow{\text{Methylcyclohexane}} CH_3CH_2CH_2CH_2MgCl$$
(Note 1)

Submitted by D. Bryce-Smith and E. T. Blues [1]
Checked by William E. Parham and Siemen Groen

1. Procedure

In a 250-ml. round-bottomed flask fitted with a glass-link or Teflon stirrer, thermometer, reflux condenser with outlet to an oil seal, dropping funnel, and inlet for nitrogen (Notes 2, 3) are placed 3.22 g. of magnesium powder (0.132 mole) (Note 4) having a particle size of 64–76 μ (Notes 5, 6), and 60 ml. of methylcyclohexane (Note 7). The apparatus is flushed with nitrogen. A slow stream of nitrogen is introduced, and the methylcyclohexane is heated to reflux temperature (Note 8).

About one-fifth of a solution of 9.26 g. of 1-chlorobutane (0.1 mole) (Notes 9, 10) in 20 ml. of methylcyclohexane is added to the vigorously stirred refluxing mixture. Reaction commences (gray turbidity) within 2–8 minutes (Notes 5, 11), and the remainder of the halide solution is then added steadily over about 12 minutes to the heated mixture, the rate being adjusted so that the inner temperature of the refluxing mixture does not fall appreciably below 99–100°. Stirring and heating under reflux are continued for an additional 15 minutes. The resulting product contains approximately 0.073 mole (73% yield) of *n*-butylmagnesium chloride as determined either by hydrolysis to *n*-butane (Notes 12, 13) or by titration (Note 14).

2. Notes

1. The empirical formula of n-butylmagnesium chloride prepared in methylcyclohexane cannot readily be determined because of the virtual insolubility of the reagent in this medium. The reagent is somewhat more soluble in aromatic media such as toluene or isopropylbenzene, and, although the empirical formula of the solute may initially approach C_4H_9MgCl, there is a tendency for precipitation of magnesium chloride from solution. This process appears to be catalyzed by traces of alkoxides, which are liable to be formed after contact of oxygen with the solution. In practice, products will tend to contain less halogen than is required by the simple formula C_4H_9MgCl. The reagents are associated (see reference 2 for a fuller discussion).

2. The nitrogen used (British Oxygen Co., White Spot) contained about 10 p.p.m. of oxygen and was dried by passage through a glass spiral cooled in acetone and solid carbon dioxide. For the most precise work, the submitters reduced the proportion of oxygen to about 0.1 p.p.m. by scrubbing the nitrogen with chromous chloride solution in a Nilox apparatus (Southern Analytical Ltd., Camberley, Surrey, England).

3. The apparatus should preferably be baked at 120° for several hours immediately before use. The uppermost region of condensing methylcyclohexane should not be cloudy; if it is, a few milliliters should be allowed to distil.

4. The yield of n-butylmagnesium chloride is increased to 80% (analyzed by evolution of n-butane) if twice the stated amount of magnesium is used.

5. Magnesium powder (grade 4, Magnesium Elektron Ltd., Manchester, England) was used within 6 months of its grinding by the manufacturer, and was sieved to the stated particle size. The use of unsieved material often gives results nearly as good, but exact reproducibility is more difficult because of variations of the particle size distribution from sample to sample. In general, the more freshly ground the magnesium, the shorter are the induction periods before reaction and, to a limited extent, the higher are the yields of organomagnesium product.

6. Fresh magnesium turnings for Grignard reaction can be used if suitable powder is unavailable, but initiation of reaction

is likely to be prolonged, and the subsequent addition of the halide solution should occupy at least 30 minutes, longer if possible.

7. Methylcyclohexane is purified by shaking with 3 portions of concentrated sulfuric acid, washing successively with water, sodium carbonate solution, and water, drying over calcium sulfate (Drierite), and distilling. The material boiling at 100–101° is used. Other nonsolvating media which can be used are toluene, xylenes, cumene, tetralin, light petroleum (b.p. 80°), decalin, and kerosene; aliphatic media are preferred, for reasons given in references 2, 3, and 4.

8. The rate of flow of nitrogen should be just sufficient to maintain a positive pressure in the apparatus. Too rapid a flow leads to loss of 1-chlorobutane.

9. 1-Chlorobutane is purified with sulfuric acid as for methyl-cyclohexane (Note 7), dried over calcium chloride, and fractionated. A middle fraction is collected.

10. The yield of product is increased to 81% (analyzed by evolution of n-butane) if 0.67 g. (0.0033 mole) of aluminum isopropoxide is added to the suspension of magnesium before addition of the halide solution. Alternatively, an equivalent amount of 2-propanol and iodine (giving 0.01 mole of C_3H_7OMgI) may be added. These modified procedures (particularly the second) also shorten the induction periods and render unnecessary any special drying of the reagents and apparatus and the use of fresh magnesium.

The products in such cases contain complexes between n-butyl-magnesium chloride and the particular alkoxide employed. With the stated low proportions of alkoxides, these complexes broadly resemble the alkoxide-free materials, but increased proportions of the alkoxide component give complexes having generally decreased chemical reactivity (see references 3 and 4).

11. The reaction generally starts without addition of iodine as an initiator, but the use of a crystal of iodine (no stirring) may occasionally be necessary with "old" magnesium or insufficiently dried materials or apparatus. A slower rate of addition of 1-chlorobutane gives slightly higher yields; for example, addition over a period of 60 minutes gave yields of 82–87%.

12. The reflux condenser was connected by an adaptor and

Teflon tube to a trap of known weight which was cooled by a mixture of acetone and solid carbon dioxide. The flow of nitrogen was stopped, and an excess of water (about 15 ml.) was added dropwise through the dropping funnel to the stirred reaction product. The resulting mixture was heated at the reflux temperature, and the butane was collected in the trap. The weight of butane, b.p. $-1°$ to $0°$, was 4.23–4.35 g. (73–76% yield).

13. The submitters have detected traces of trans-butene-2 and propylene among the gases (mainly n-butane and hydrogen) formed on hydrolysis.

14. Sufficient dry ether (approximately 100 ml.) is added to bring the organomagnesium products into solution. Aliquot portions of the solution are then added to a known volume of standard hydrochloric acid, and the excess acid is determined by titration with standard base. Yields determined in this way tend to be a few percent higher than those determined by collection of n-butane (Note 12).

3. Methods of Preparation

The method is an extension of the well-known Grignard synthesis in ethers to the use of nonsolvating media, and is a development of procedures previously reported.[2-6] A version of it has been employed with straight-chain primary alkyl chlorides, bromides, and iodides from C_2 to C_{14},[5-7] and in solvents (or an excess of the halide) which permit reaction temperatures above $120°$, with simple aryl halides such as chlorobenzene and 1-chloronaphthalene. Branched-chain primary, secondary, and tertiary alkyl halides, allyl, vinyl, and benzyl halides either fail to react or give extensive side reactions. Better results are reported to be obtained in such cases with the use of catalytic quantities of a mixture of an alkoxide and an ether such as diethyl ether or tetrahydrofuran in a hydrocarbon medium, but the products are not, of course, completely unsolvated.[4]

4. Merits of the Preparation

Unsolvated organomagnesium compounds have been recommended for the synthesis of organometallic derivatives of mercury, boron, aluminum, silicon, germanium, tin, phosphorus, arsenic, and antimony [3, 6, 8] and have been used in procedures for the alkylation of aromatic rings and for the production of various polymerization catalysts.[4, 6, 9]

1. Chemistry Department, The University, Reading, England.
2. D. Bryce-Smith and G. F. Cox, *J. Chem. Soc.*, 1175 (1961).
3. E. T. Blues and D. Bryce-Smith, *Chem. Ind.* (*London*), 1533 (1960); Brit. Patent 955,806 (1964).
4. D. Bryce-Smith, *Bull. Soc. Chim. France*, 1418 (1963).
5. L. I. Zakharkin, O. Yu. Okhlobystin, and B. N. Strunin, *Tetrahedron Lett.*, 631 (1962).
6. D. Bryce-Smith and G. F. Cox, *J. Chem. Soc.*, 1050 (1958).
7. F. J. Buescher, A. H. Frye, G. J. Goepfert, and V. G. Soukup, Abstracts of Papers presented at the Symposium on Current Trends in Organometallic Chemistry, University of Cincinnati, June, 1963, p. 10.
8. L. I. Zakharkin, O. Yu. Okhlobystin, and B. N. Strunin, *Izv. Akad. Nauk SSSR, Otd. Khim. Nauk*, 2002 (1962).
9. E. T. Blues and D. Bryce-Smith, *Proc. Chem. Soc.*, 245 (1961); Brit. Patent 955,807 (1964).

VINYL TRIPHENYLPHOSPHONIUM BROMIDE*

(Phosphonium bromide, triphenylvinyl-)

$$C_6H_5OCH_2CH_2Br + (C_6H_5)_3P \xrightarrow{C_6H_5OH}$$

$$C_6H_5OCH_2CH_2\overset{+}{P}(C_6H_5)_3 \ Br^- \xrightarrow{CH_3CO_2C_2H_5}$$

$$CH_2{=}CH\overset{+}{P}(C_6H_5)_3 \ Br^- + C_6H_5OH$$

Submitted by EDWARD E. SCHWEIZER and ROBERT D. BACH[1]
Checked by FRANÇOIS X. GARNEAU and PETER YATES

1. Procedure

Caution! Because phenol and its solutions are corrosive, rubber gloves should be used in the following operations. The product, vinyl triphenylphosphonium bromide, has been found to induce a sneezing, allergic reaction and contact with it should be avoided.

In a 1-l. three-necked flask is placed 1 lb. of reagent grade phenol (Note 1), 100 g. (0.50 mole) of β-bromophenetole, and 131 g. (0.50 mole) of triphenylphosphine (Note 2). The flask is equipped with a sealed stirrer, a thermometer, and a reflux condenser fitted with a calcium chloride drying tube. The mixture is stirred and heated to $90° \pm 3°$ (Note 3) and kept at this temperature for 48 hours.

The solution is cooled to room temperature and added slowly (during 45 minutes) from a dropping funnel to vigorously stirred anhydrous ether (3 l.) in a 4-l. beaker. Material that adheres to the sides of the beaker is scraped down with a long steel spatula, and the mixture is filtered by suction. The solid product is transferred into a second 3-l. portion of anhydrous ether, and the mixture is stirred vigorously for 15 minutes and filtered by suction. The product is washed with three 250-ml. portions of warm anhydrous ether.

The white crystalline residue of crude phenoxyethyltriphenylphosphonium bromide (Note 4) is placed in a 3-l. two-necked flask equipped with a sealed stirrer and a reflux condenser fitted with a calcium chloride drying tube. Reagent grade (Note 5) ethyl acetate (1.5 l.) is added, and the solution is stirred under reflux for 24 hours. The mixture is cooled to room temperature, and the ethyl acetate layer is decanted (or filtered if the salt is crystalline). This procedure is repeated until the filtered salt, vinyl triphenylphosphonium bromide, melts at 186° and higher (Note 6).

After the final filtration the product is washed with two 100-ml. portions of ethyl acetate and two 100-ml. portions of anhydrous ether and dried for 24 hours at 80°. The dried, analytically pure vinyl triphenylphosphonium bromide, m.p. 186–190°, weighs 122–158 g. (66–86%).

2. Notes

1. The checkers found that it was important to use phenol free of colored impurities; in a run in which phenol, m.p. 39.5–41°, with a slight rose tinge was used the yield of product was reduced to 56%.

2. The β-bromophenetole was obtained from Aldrich Chemical Co.; the triphenylphosphine was obtained from M and T Chemicals, Inc., or Carlisle Chemical Works, Reading, Ohio, and recrystallized once from anhydrous ether (with filtration).

3. It is important that the temperature does not rise above 95°. There is a slight exotherm on initial heating that may necessitate the removal of the heating mantle in order to maintain a temperature below 95°.

4. Pure samples of this material may be obtained by using acetic acid as solvent instead of phenol.[2]

5. Because the vinyl salt reacts with ethanol, and decomposition of the phenoxyethyl precursor is inhibited by acetic acid,[2] reagent grade ethyl acetate is recommended.

6. Four treatments have always been necessary. The residue has always crystallized on cooling after the third treatment.

3. Methods of Preparation

The present procedure is that described by the submitters.[2] Vinyl triphenylphosphonium bromide has also been prepared by dehydrobromination of 2-bromoethyltriphenylphosphonium bromide, but no preparative details or yields have been disclosed.[3]

4. Merits of the Preparation

This salt has been used as a general reagent for the preparation of a large number of heterocyclic and carbocyclic systems.[4-17] A variety of salts of type $XCH_2CH_2P^+(C_6H_5)_3Br^-$ has been prepared from the vinyl salt by treatment with alcohols, thiophenol, and diethylamine.[2]

1. Department of Chemistry, University of Delaware, Newark, Delaware 19711.
2. E. E. Schweizer and R. D. Bach, *J. Org. Chem.*, 29, 1746 (1964).
3. D. Seyferth, J. S. Fogel, and J. K. Heeren, *J. Am. Chem. Soc.*, 86, 307 (1964).
4. E. E. Schweizer, *J. Am. Chem. Soc.*, 86, 2744 (1964).
5. E. E. Schweizer and K. K. Light, *J. Am. Chem. Soc.*, 86, 2963 (1964).

6. E. E. Schweizer, L. D. Smucker, and R. J. Votral, *J. Org. Chem.*, **31**, 467 (1966).
7. E. E. Schweizer and G. J. O'Neill, *J. Org. Chem.*, **30**, 2082 (1965).
8. E. E. Schweizer and K. K. Light, *J. Org. Chem.*, **31**, 870 (1966).
9. E. E. Schweizer and L. D. Smucker, *J. Org. Chem.*, **31**, 3146 (1966).
10. E. E. Schweizer and J. G. Liehr, *J. Org. Chem.*, **33**, 583 (1968).
11. E. E. Schweizer, J. Liehr, and D. J. Monaco, *J. Org. Chem.*, **33**, 2416 (1968).
12. E. E. Schweizer, W. S. Creasy, J. G. Liehr, M. E. Jenkins, and D. L. Dalrymple, *J. Org. Chem.*, **35**, 601 (1970).
13. E. E. Schweizer and C. S. Kim, *J. Org. Chem.*, **36**, 4033 (1971).
14. E. E. Schweizer and C. S. Kim, *J. Org. Chem.*, **36**, 4041 (1971).
15. E. E. Schweizer and C. M. Kopay, *J. Org. Chem.*, **37**, 1561 (1972).
16. S. Brandange and C. Lundin, *Acta Chem. Scand.*, **25**, 2447 (1971).
17. D. Johnson and G. Jones, *J. Chem. Soc.*, **[PI]** 840 (1972).

TYPE OF REACTION INDEX

This index lists the preparations contained in this volume in accordance with general types of reactions. Only those preparations are included which can be classified under the selected headings with some definiteness. The arrangement of types and of preparations is alphabetical.

TYPE OF COMPOUND INDEX

Preparations are listed by functional groups or by ring systems. Phenyl, ethylenic, and acetylenic groups are not considered as substituents unless otherwise stated. Salts are included with the corresponding acids and bases.

1159

FORMULA INDEX

All preparations listed in the Contents are recorded in this index. The system of indexing is that used by *Chemical Abstracts*. The essential principles involved are as follows: (1) The arrangement of symbols in formulas is alphabetical except that in carbon compounds C always comes first, followed immediately by H if hydrogen is also present. (2) The arrangement of formulas is also alphabetical except that the number of atoms of any specific kind influences the order of compounds: e.g., all formulas with one carbon atom precede those with two carbon atoms, thus: CH_2I_2, CH_3NO_2, CH_5N, C_2H_2O. (3) The arrangement of entries under any heading is strictly alphabetical according to the names of the isomers. (4) Inorganic salts of organic acids and inorganic addition compounds of organic compounds are listed under the formulas of the compounds from which they are derived.

1169

PREPARATION OR PURIFICATION OF SOLVENTS AND REAGENTS INDEX

Organic Syntheses procedures frequently include notes describing the purification of solvents and reagents. These have been placed in a single index for convenience. The preparations of useful reagents and catalysts, as well as some techniques, determinations, and tests are included also.

APPARATUS INDEX

A number of the procedures in Organic Syntheses describe the use of special or less common pieces of apparatus and equipment. References to many of them are recorded in this index. Illustrations are indicated by numbers in boldface type.

Apparatus, for assay of *n*-butylmagnesium chloride, 1143
for bromination or chlorination using aluminum chloride catalyst, 117, 120
for chromous sulfate oxidations, 994
for decarboxylation of coumalic acid, 982, 983
for generation of nitrous gases, 651
for handling boron trifluoride, 600
for handling hydrogen fluoride, 66, 136, 332
for handling hydrogen fluoride and boron trifluoride, 480
for high speed stirred reactions, 1027
for introduction of chlorine gas, 371
for ketene generation from diketene, 679
for Kolbe electrolysis, 363, 446, 464
for monitoring carbon dioxide evolution, 127
for ozonolysis, 489
for photochemical reactions, 298, 529, 952
for preparation and use of degassed Raney nickel catalyst, 103
for preparation of diazomethane, 351
for preparation of hydrogen bromide, 546
for preparation of methyl magnesium carbonate, 439
for preparation of potassium salt of dimethyl sulfoxide, 938
for preparation of trimethyloxonium 2,4,6-trinitrobenzenesulfonate using diazomethane, 1100
for preparation of zinc-copper couple, 857
for pyrolysis of 2,3-diazabicyclo[2.2.1]-hept-2-ene, 99
for reactions in liquid ammonia, 12
for reactions of cyanogen chloride and sulfur trioxide, 227
for solvent stripping, 974

for thermal cycloaddition of ethylene and 5,5-dimethoxy-1,2,3,4-tetra-chlorocyclopentadiene, 425
for transfer of Grignard reagents, 1058
Apparatus, pyrolysis, for diazonium hexafluorophosphates, 134
for dimethyl 3-methylenecyclo-butane-1,2-dicarboxylate, 734
for Hofmann decomposition, 316
for thermal dehydroacetoxylation, 236
Autoclave, Hastellay C, 1083
rocking, for cycloaddition reaction, 235, 393
use of Hastelly-C shaking autoclave for alkaline rearrangement, 814
use of hydrogenation autoclave for aqueous oxidation, 811
Autoclave, stirred, for cycloaddition reaction, 459

Bath, cooling, large-scale, 312
Baths, heating, 619
Beaker, stainless steel, 617
Buret, for dispensing diazomethane, 247

Chromatography, apparatus for continuous, 730
Condenser, low temperature, 947

Distillation apparatus, for low-boiling liquids, 23, 682
for rapid steam distillation, 1020
use of vacuum desiccator for, 980
Dropping funnel, calibrated for phosgene addition, 201
dry-ice cooled, 675
Hershberg, 891

Electrolysis, two-compartment cell for, 446
Extractor, continuous, 630
stirred, in isolation of 3-hydroxyglutaro-nitrile, 615

AUTHOR INDEX

GENERAL INDEX

This index includes references to the reagents, chemicals, catalysts, and solvents used in the various preparations as well as to the intermediates and final products obtained from the preparations. It includes also selected references to reaction types and compound types involved in or referred to in the preparations, but the Reaction Index or Compound Index should also be referred to for possible additional entries in these categories. This index does not include references to apparatus, for which there is a separate index. Most entries in this index refer to the preparative parts of the procedures or to extensions thereof; thus the discussion sections of the procedures have not been extensively indexed.

The name of a compound in capital letters together with a number in boldface type indicates complete preparative directions for the substance named. The name of a compound in ordinary lightface type together with a number in boldface type indicates directions, usually adequate but not in full detail, for preparing the substance named. A name in lightface type together with a number in lightface type indicates a compound or an item mentioned in connection with a preparation.